COMPUTATIONAL METHODS IN ENGINEERING AND SCIENCE

PROCEEDINGS OF THE NINTH INTERNATIONAL CONFERENCE ON ENHANCEMENT AND PROMOTION OF COMPUTATIONAL METHODS IN ENGINEERING AND SCIENCE (EPMESC IX), 25–28 NOVEMBER 2003, MACAO, CHINA

Computational Methods in Engineering and Science

Edited by

V.P. Iu
University of Macau

L.N. Lamas
*National Civil Engineering Laboratory of Portugal
(Formerly at the Civil Engineering Laboratory of Macau)*

Y.-P. Li
University of Macau

K.M. Mok
University of Macau

A.A. BALKEMA PUBLISHERS   LISSE / ABINGDON / EXTON (PA) / TOKYO

Published by: A.A. Balkema, a member of Swets & Zeitlinger Publishers
        www.balkema.nl and www.szp.swets.nl

For the complete set of two volumes: ISBN 90 5809 567 3

Printed in the Netherlands

Table of Contents

Keynote lecture

Numerical methods

Solid mechanics

Structural mechanics

Geotechnics

Geographical information systems

Artificial intelligence and software engineering

Software technology

Computer-aided design, engineering and instruction

Fluid mechanics and applications

Physics and material science

Student paper competition

Editors' Foreword

The 9th International Conference on Enhancement and Promotion of Computational Methods in Engineering and Science (EPMESC IX) was organized by the University of Macau, and held in Macao from November 25 to 28, 2003.

The series of EPMESC international conferences devoted to computational methods in engineering and science was created by a group of prominent scholars in the field: Prof. E. Arantes e Oliveira, Prof. Y.K. Cheung, Prof. T. Kawai and the late Prof. Luo Songfa. The first was held in Macao in 1985 and, thereafter, held alternately in Macao and a city in China including Guangzhou, Dalian and Shanghai. The acronym for EPMESC has evolved from the conference title of "Education, Practice and Promotion of Computational Methods in Engineering using Small Computers" to "Enhancement and Promotion of Computational Methods in Engineering and Science". The conference has also grown from a Chinese/European happening to an event that attracts researchers from all continents. The conference aims to promote exchange of technical knowledge and experiences among engineers and scientists from all over the world, with a common interest in development and application of computational methods in engineering and science.

In response to the call for papers, over 200 abstracts from 30 countries and regions were submitted to the Secretariat. By a panel of referees, 125 papers are selected for inclusion in the Proceedings. It contains 16 invited keynote-lecture papers by distinguished scholars and researchers in their fields of expertise from the world, 91 regular full papers, and 18 students papers that have been submitted for a competition that is held for the first time in the EPMESC series. The themes covered by these Proceedings include: artificial intelligence and software engineering; computer aided design, engineering and instruction; fluid mechanics and applications; geographical information systems; geotechnics; numerical methods; physics and material science; software technology; solid mechanics and structural mechanics. All papers were reproduced directly from submitted camera-ready manuscripts; for some papers, minor corrections required for printing were made by the editors.

Together with the keynote lectures, the contributions reflect the robust computational methods research activity worldwide in recent years. They also provide a rich variety of examples of engineering and science problems to which computational methods can be usefully applied.

We would like to thank the valuable assistance of members of the EPMESC conference board and the organizing committee. The generous support of the financial sponsors is gratefully acknowledged. The advice and expert handling of the publication of the Proceedings by Swets & Zeitlinger Publishers/A.A. Balkema Publishers are well appreciated. Finally, we are grateful to all those who have shown interest in EPMESC IX.

V.P. Iu, L.N. Lamas, Y.-P. Li & K.M. Mok

Computational Methods in Engineering and Science, Iu et al. (eds)
© 2003 Swets & Zeitlinger, Lisse, ISBN 90 5809 567 3

Preface

Eighteen years ago, the series of EPMESC conferences on computational methods began her history in Macao. The pioneering promoters were Prof. E. Arantes e Oliveira, Prof. Y.K. Cheung, Prof. T. Kawai and the late Prof. Luo Songfa. The first EPMESC conference was also the very first international academic conference ever held in Macao. Since then Macao hosted the third, fifth and seventh EPMESC conference. As a tradition, the ninth conference is held in Macao this year, at the University of Macau.

The series of EPMESC conferences is devoted to the dissemination of development in computational methods for the solution of theoretical and practical problems in engineering and science. The conferences attract scientists and engineers from Europe, China, USA and other Asian countries and foster closer academic ties and cooperation within the international scientific community. Traditionally, the conferences are characterized by many keynote lectures and by large number of participants from Europe and China.

On behalf of the Organizing Committee and the University of Macau, I would like to thank Dr Chui Sai On, Secretary for Social Affairs and Culture of Macao SAR Government for his tremendous support to the conference. Sincere thanks are also owed to the Macau Foundation and to various sponsors in China, Portugal and Macao for their financial support. Last but not least, I would like to thank all members of the EPMESC Conference Board, the Organizing Committee and the Conference Secretariat for pivotal contribution to the conference and to the publication of the Proceedings.

V.P. Iu
Rector, University of Macau
Chairman, Organizing Committee of EPMESC IX

Computational Methods in Engineering and Science, Iu et al. (eds)
© 2003 Swets & Zeitlinger, Lisse, ISBN 90 5809 567 3

EPMESC IX Conference Board

Organizing Committee

V.P. Iu, Chairman (University of Macau, Macao)
Y.-P. Li, Scientific-Secretary (University of Macau, Macao)
K.M. Mok, Secretary-General (University of Macau, Macao)

Members
O.F. Botelho (Civil Engineering Laboratory of Macau, Macao)
K.P. Kou (University of Macau, Macao)
L.N. Lamas (National Civil Engineering Laboratory of Portugal and formerly at the
  Civil Engineering Laboratory of Macau, Macao)
M.H. Lok (University of Macau, Macao)
I.T. Ng (University of Macau, Macao)
E. Pereira (Instituto Superior Técnico, Portugal)
W.M. Quach (University of Macau, Macao)
U.W. Tang (University of Macau, Macao)
I.M. Wan (University of Macau, Macao)
M.W. Yuan (Peking University, China)

Conference Secretariat
M.H. Lok (University of Macau, Macao)
K.M. Mok (University of Macau, Macao)
U.W. Tang (University of Macau, Macao)

Keynote lecture

GBT-based computational approach to analyse the geometrically non-linear behaviour of steel and composite thin-walled members

D. Camotim & N. Silvestre

Department of Civil Engineering and Architecture, IST, Technical University of Lisbon, Portugal

ABSTRACT: The numerical implementation of a geometrically non-linear Generalised Beam Theory (GBT) is presented, discussed and illustrated. This GBT-based computational approach is intended to (i) perform buckling analyses in thin-walled members made of isotropic or orthotropic materials (e.g. cold-formed steel, composite or laminated plate profiles) and to (ii) analyse the post-buckling behaviour of initially imperfect isotropic members (mostly cold-formed steel profiles). A non-linear beam finite element formulation is specifically developed and numerically implemented to solve the GBT system of (linearised or non-linear) differential equilibrium equations. It employs Hermitean polynomials to approximate the mode amplitude functions and leads to a system of algebraic equations which either (i) defines a linear eigenvalue problem (buckling analysis) or (ii) is non-linear and must be solved by resorting to incremental-iterative techniques (post-buckling analysis). In order to validate and illustrate the application and capabilities of the proposed GBT-based computational approach, several numerical results are presented and thoroughly discussed. They concern (i) the buckling behaviour of composite lipped channel columns and beams and (ii) the post-buckling behaviour of cold-formed steel lipped channel columns. Some of these results are compared with values yielded by fully numerical (finite element or finite strip) analyses. A few words are also devoted to the computational efficiency of the GBT-based approach, mostly because it takes advantage of the unique mode decomposition nature of the theory.

1 INTRODUCTION

The Generalised Beam Theory (GBT) (i) extends Vlassov's classical thin-walled beam theory to incorporate cross-section deformations (plate flexure and distortion – see Fig. 1) and (ii) was originally developed by Schardt (1989, 1994), for members made of isotropic materials. Subsequently, Davies et al. (e.g. 1994) (i) employed this theory to study the local and global buckling behaviour of cold-formed steel profiles and (ii) showed that, in this context, GBT constitutes a rather powerful, elegant and clarifying method of analysis, mostly due to its unique mode decomposition feature. Quite recently, the range of problems that can be solved by means of GBT has been significantly widened by the authors. Indeed, GBT can now be used also (i) to perform first order or linear stability analysis in members made of arbitrary

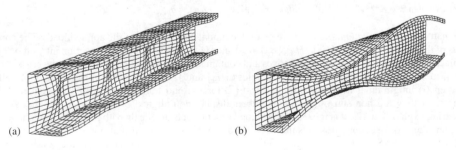

Figure 1. Thin-walled member cross-section deformations: (a) plate flexure and (b) distortion.

orthotropic materials, such as composite (FRP – fiber reinforced plastic) and laminated plate profiles (Silvestre & Camotim 2002b,c) and (ii) to analyse the post-buckling (geometrically non-linear) behaviour of initially imperfect cold-formed steel (isotropic) members (Silvestre & Camotim 2003a).

Besides accounting for the cross-section in-plane (local) deformations, which requires using genuine folded-plate theory, the most relevant GBT feature is the fact that the member buckling mode or deformed configuration are expressed as linear combinations of a set of pre-determined cross-section *deformation modes*. Due to this unique feature, GBT offers possibilities not available even through the use of powerful numerical techniques, such as the finite element (FEM) or finite strip (FSM) methods. Indeed, it is fair to say that GBT has now been established as a valid and often advantageous (mostly in terms of clarity) alternative to fully numerical analyses.

The steps and procedures involved in the application of GBT are associated to the performance of two main tasks, namely (i) a *cross-section analysis*, related to the identification of the deformation modes and determination of the corresponding modal mechanical properties, and (ii) a *member analysis*, which leads to the solution of the differential equilibrium equations and boundary conditions, for a specific set of member end support conditions and applied loads. As far as the first task is concerned, the use of symbolic computation has proven to be significantly more efficient than the standard numerical techniques. Concerning the second task, more commonly performed in the field of structural analysis, a general beam FE formulation has been developed by Silvestre & Camotim (2002d, 2003a), which (i) can be applied to perform buckling analyses (isotropic or orthotropic members) and post-buckling analyses (isotropic members) and (ii) is able to take into account any specified (limited) number of deformation modes.

The main objectives of this paper are the following: (i) to provide a brief overview of the GBT fundamentals, field of application and recent developments, (ii) to address the most relevant issues related to the computer implementation of GBT *buckling* (linear stability) and *post-buckling* (geometrically non linear) analyses, which is achieved through a combination of symbolic computation procedures and FEM incremental-iterative techniques, and (iii) to present and discuss a number of numerical results. These results (i) concern the buckling behaviour of composite lipped channel columns and beams and the post-buckling behaviour of cold-formed steel lipped channel columns and (ii) make it possible to illustrate and assess the elegance, efficiency and capabilities of the proposed GBT-based computational approach.

2 GBT CONCEPTS AND PROCEDURES

Since a detailed account of the GBT formulation is available in the literature (Silvestre & Camotim 2002b, c, 2003a), only a very brief overview is presented here. First, consider the coordinate system and displacement components shown in Figure 2, where x, s and z denote coordinates along the member length, cross-section mid-line and wall thickness, and u, v and w are the corresponding displacement components. In order to obtain a displacement representation compatible with the classical beam theory, $u(x,s)$, $v(x,s)$ and $w(x,s)$ must be expressed as

$$u(x,s) = u_k(s)\,\phi_{k,x}(x) \qquad v(x,s) = v_k(s)\,\phi_k(x) \qquad w(x,s) = w_k(s)\,\phi_k(x) \qquad , \qquad (1)$$

where $(\cdot)_{,x} \equiv d(\cdot)/dx$, the summation convention applies to subscript k and $\phi(x)$ is a "displacement amplitude function" defined along the member length. The application of the principle of virtual work leads, after considerable manipulation, to the system of equilibrium equations

$$C_{ik}\phi_{k,xxxx} + H_{ik}\phi_{k,xxx} - D_{ik}\phi_{k,xx} + F_{ik}\phi_{k,x} + B_{ik}\phi_k + X_{kij}W_k\phi_{j,xx} + h.o.t. = q_i \qquad , \qquad (2)$$

designated as *GBT equation(s)* and where qi are the modal components of the applied loads. The components of the stiffness matrices $[C_{ik}]$, $[H_{ik}]$, $[B_{ik}]$, $[F_{ik}]$ and $[D_{ik}]$ stem from the cross-section integration of the displacement components or their derivatives and contain all the information concerning the cross-section modal mechanical properties. Notice that matrices $[H_{ik}]$ and $[F_{ik}]$ account for material coupling effects between (i) longitudinal flexure and torsion and (ii) torsion and transverse flexure, respectively, which means that they are non-null only if the members display non-aligned orthotropy. As for the last term appearing explicitly in (2), it is related to the second order effects and deals with all interaction phenomena between the cross-section stress resultants W_k and the out-plane deformations – $[X_{kij}]$ are geometric stiffness matrices (Silvestre & Camotim 2002c). On the other hand, if one considers the non-linear terms of the kinematic (strain-displacement), several higher order terms (*h.o.t.*) appear in (2), thus generating

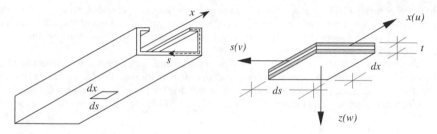

Figure 2. Coordinate system and displacement components.

a system of non-linear equilibrium equations. Finally, note that the developed non-linear GBT (Silvestre & Camotim 2003a) can also incorporate arbitrary initial geometrical imperfections.

The numerical implementation of the GBT system of equation (2) addressed in this paper makes it possible to perform:

(i) *Buckling* (linear stability) analyses of thin-walled members with arbitrary orthotropy (the *h.o.t.* are neglected and one has $q_i = 0$).

(ii) *Post-buckling* (geometrically non-linear) analyses of initially imperfect isotropic thin-walled members (one has $H_{ik} = F_{ik} = 0$). Notice that the performance of post-buckling analyses in perfect members requires the implementation of a path-following numerical technique to handle branch-switching at bifurcation points, which has not yet been done.

As mentioned earlier, the steps and procedures involved in the application of GBT analyses are related to the performance of two main tasks: (i) a *cross-section* analysis and (ii) a *member analysis* (Silvestre & Camotim 2002d, 2003a). The first one, carried out using mostly symbolic computation procedures, deals with identifying the relevant deformation modes and determining the associated modal mechanical properties (tensor quantities appearing in (2)). As for the second task, it involves establishing and solving the system of equilibrium equation (2) and their boundary conditions, on the basis of information concerning the (i) cross-section modal mechanical properties (known after completing the first task), (ii) member length and end support conditions and (iii) applied load nature and values. The corresponding results are the following:

(i) *Buckling analysis*: buckling mode shapes (defined by functions $\phi_k(x)$) and corresponding bifurcation loads ($\lambda_b W_{k0}$), obtained from the solution of a linear eigenvalue problem.

(ii) *Post-buckling analysis*: equilibrium paths describing the evolution of the member deformed configuration (functions $\phi_k(x)$) with the applied load parameter (λ), yielded by the incremental-iterative solution of a system of non-linear equations.

A description of the main procedures related to the (rather lengthy and relatively complex) set of operations required to perform the cross-section analysis can be found in works by Schardt (1989 – isotropic members) and Silvestre & Camotim (2002b – orthotropic members). Nevertheless, it is worth drawing the reader's attention to a few newly developed features concerning the cross-section modal analysis, which are not documented in the above publications:

(i) Two families of GBT deformation modes are employed, namely, (i_1) the *fundamental* modes (FM) and (i_2) the *shear* modes (SM). These designations stem from the fact that the former modes were originally considered by Schardt (1989), while the latter ones have only very recently been incorporated in GBT (Silvestre & Camotim 2003b,c).

(ii) The FM constitute the core of GBT and are based on Vlassov's hypothesis of neglecting the in-plane (membrane) shear deformations ($\gamma_{xs}^M = 0$) along the member walls, which means that they are *shear undeformable*. Notice also that they can still be divided into two sub-groups, depending on whether they stem from warping displacements u (FM$_u$) or flexural displacements w (FM$_w$).

(iii) Up until very recently, the idea of incorporating SM in GBT had never been raised, most likely because shear deformation does not affect either the 1st order or the linear stability behaviour of isotropic thin-walled members. The SM, which obviously violate Vlassov's hypothesis, can also be divided into two sub-groups, depending on whether they stem from warping displacements u (SM$_u$) or transverse membrane displacements v (SM$_v$).

5

(iv) Since the primary GBT system of non-linear equations is *highly coupled*, an appropriate coordinate transformation must be performed, in order to take full advantage of the GBT potential. It leads to the simultaneous diagonalisation of two among the three stiffness matrices (2nd order tensors) $[C_{kh}]$, $[B_{kh}]$ and $[D_{kh}]$.

(v) The above coordinate transformation requires solving an auxiliary eigenvalue problem and leads to the identification of sets of *orthogonal* cross-section deformation modes (eigenvectors). In order to illustrate the outcome of this procedure, Figures 3–6 display the most relevant features of the four types of GBT deformation modes (FM$_u$, FM$_w$, SM$_u$ and SM$_v$), for the particular case of an isotropic

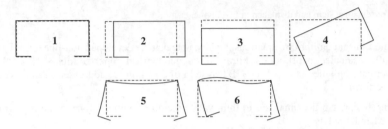

Figure 3.   In-plane (v_k, w_k) configurations of the FM$_u$ (rigid-body and distortional modes).

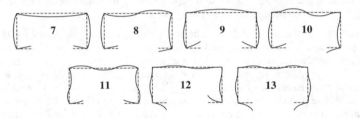

Figure 4.   In-plane (v_k, w_k) configurations of the FM$_w$ (plate-flexure modes).

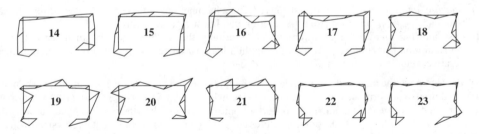

Figure 5.   Warping displacements (u_k) associated to the SM$_u$.

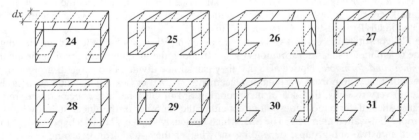

Figure 6.   Transverse (v_k) and flexural (w_k) displacements associated to the SM$_v$.

lipped channel cross-section (notice, however, that the GBT deformation modes of an orthotropic lipped channel cross-section are either "fully identical" or "almost identical" to these ones).

3 FINITE ELEMENT FORMULATION AND NUMERICAL IMPLEMENTATION

In order to develop a non-linear beam finite element (FE) formulation to solve the GBT system of equilibrium equations, one adopts Hermitean polynomials to approximate the deformation mode amplitude functions $\phi_k(x)$. After performing the usual integration procedures, the variational form of equation (2) becomes either (i) a system of algebraic equations defining a linear eigenvalue problem (buckling analysis) or (ii) a system of non-linear algebraic equations (post-buckling analysis), expressed in terms of a single load parameter λ.

Given the space limitations, it is just possible to present here a brief description of the FE formulation and numerical implementation of the two types of GBT analyses addressed in this paper. For a more detailed account, the interested reader is referred to previous work by the authors (Silvestre & Camotim 2002d, 2003a). In attempt to keep the presentation as clear as possible, it is convenient to begin by identify the most relevant tasks and procedures that must be performed prior to solving the GBT system of (linearised or non-linear) equilibrium equations:

(i) Selection of the GBT deformation modes to be included in the analysis, i.e. of the sub-system of equilibrium equations that must be solved. Indeed, accurate buckling/post-buckling results can be obtained by solving only a particular sub-system of (2), formed by the equilibrium equations involving the few (selected) relevant deformation modes. Since the member buckling mode or deformed configuration is then approximated exclusively on the basis of these few deformation modes, the computational effort can be significantly reduced without affecting the high level of accuracy of the analysis.
(ii) Longitudinal discretisation of the member in an adequate number of finite elements (n_e).
(iii) Identification of the member (nodal) degrees of freedom, which are grouped in a global displacement vector $\mathbf{d}$ and must take into account the member end support conditions.

Next, the tasks and procedures specifically related to the performance of a GBT *buckling analysis* are singled out:

(i) Assembly of the linear ($\mathbf{K_0}$) and geometric ($\mathbf{K_G}$) FE stiffness matrices to obtain the member global stiffness matrix ($\mathbf{K}_\lambda = \mathbf{K_0} + \lambda\mathbf{K_G}$).
(ii) Solution of the eigenvalue problem $\mathbf{K}_\lambda \mathbf{d} = 0$, which provides the bifurcation stress resultant values $W_{kb} = \lambda_b W_{k0}$ (λ_b are the eigenvalues – the lowest one, λ_{cr}, is designated as "critical") and the corresponding buckling modes $\mathbf{d}_b$ (eigenvectors).

Finally, the tasks and procedures involved in the performance of a GBT *post-buckling analysis* are the following:

(i) Assembly of the FE secant stiffness matrices (several per element) and external force vectors into global ones ($\mathbf{K(d)}$ and $\mathbf{f_e}$) to establish the system of non-linear equilibrium equations ($\mathbf{K(d)} \cdot \mathbf{d} = \lambda\mathbf{f_e}$).
(ii) Assembly of the FE tangent stiffness matrices into a global one ($\mathbf{T(d)}$) to establish the system of iterative equilibrium equations $\mathbf{T(d)} \cdot \Delta\mathbf{d} = \Delta\lambda\mathbf{f_e} - \mathbf{r}$, where $\mathbf{r}$ is the residual force vector.
(iii) Solution of the non-linear problem $\mathbf{K(d)} \cdot \mathbf{d} = \lambda\mathbf{f_e}$ by means of a predictor-corrector strategy. Since the Newton-Raphson method is adopted to obtain the equilibrium configuration (λ, $\mathbf{d}$) at the end of each increment, a sequence of linear problems $\mathbf{T(d)} \cdot \Delta\mathbf{d} = \Delta\lambda\mathbf{f_e} - \mathbf{r}$ must be solved, where matrix $\mathbf{T(d)}$ is updated after each iteration. An arc-length control scheme is employed, in which the current value ℓ_i (valid for increment i) is determined by an automatic incrementation procedure (the first load increment is a fixed percentage of the critical bifurcation value λ_{cr}).
(iv) Combining the solution of the non-linear problem (components of $\mathbf{d}$) with the FE shape functions, used to approximate each deformation mode amplitude $\phi_k(x)$, it is possible to determine the deformed configuration of a given cross-section, located at $x = x_0$. Notice that functions $u_k(s)$, $v_k(s)$ and $w_k(s)$ have already been obtained in the context of the cross-section analysis.

7

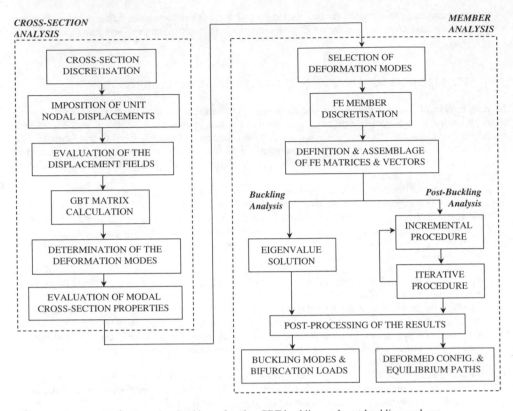

CROSS-SECTION
DISCRETISATION

IMPOSITION OF UNIT
NODAL DISPLACEMENTS

EVALUATION OF THE
DISPLACEMENT FIELDS

GBT MATRIX
CALCULATION

DETERMINATION OF THE
DEFORMATION MODES

EVALUATION OF MODAL
CROSS-SECTION PROPERTIES

SELECTION OF
DEFORMATION MODES

FE MEMBER
DISCRETISATION

DEFINITION & ASSEMBLAGE
OF FE MATRICES & VECTORS

Buckling Analysis

Post-Buckling Analysis

INCREMENTAL
PROCEDURE

EIGENVALUE
SOLUTION

ITERATIVE
PROCEDURE

POST-PROCESSING OF THE RESULTS

BUCKLING MODES &
BIFURCATION LOADS

DEFORMED CONFIG. &
EQUILIBRIUM PATHS

Figure 7.   Steps and procedures involved in performing GBT buckling and post-buckling analyses.

The sequential diagram shown in Figure 7 provides an overall view of the main steps and procedures involved in performing GBT (i) buckling (orthotropic members) and (ii) post-buckling (isotropic members) analyses, by means of the finite element approach just outlined.

4   ILLUSTRATION: BUCKLING BEHAVIOUR OF COMPOSITE COLUMNS & BEAMS

In order to illustrate the application and capabilities of the orthotropic (second order) GBT, a number of results concerning the stability behaviour of thin-walled lipped channel (i) columns (axial compression: $\lambda \equiv P$) and (ii) beams (major axis bending: $\lambda \equiv M$) are presented. All the members analysed have (i) a variable length L and (ii) pinned (locally and globally) and free-to-warp end sections. The cross-section mid-line dimensions are $b_{web} = 160\,mm$, $b_{flange} = 100\,mm$, $b_{lip} = 40\,mm$ and all the walls are formed by three orthotropic layers (i) displaying an asymmetric configuration, (ii) with equal thickness (0.1 cm for columns and 0.2 cm for beams) and (iii) exhibiting identical material properties ($E_x = 45\,GPa$, $E_s = 8\,GPa$, $G_{xs} = 4\,GPa$, $v_{xs} = 0.3$) and different fiber orientations ($0°-30°-60°$). The cross-section discretisation adopted involves 13 nodes (6 natural and 7 intermediate) and only the FM_u and FM_w deformation modes depicted in Figures 3 and 4 have been incorporated in the analysis (Silvestre & Camotim 2002a). It is worth noticing that the authors have recently become aware that, unlike in isotropic members, the shear modes (SM_u and SM_v) have a small-to-moderate influence in the buckling behaviour of orthotropic members. Therefore, the incorporation of these modes in the orthotropic (first and second order) GBT is presently under way (Silvestre & Camotim 2003b).

After (i) performing the cross-section analysis and (ii) adopting 8 finite elements to discretise the member longitudinally and (iii) solving the linear eigenvalue problem associated to the member buckling behaviour, one is led to the critical load value λ_{cr}. At this point, it is worth mentioning that the results (eigenvalues) yielded by a linear stability analysis may not be "true bifurcation loads", particularly in

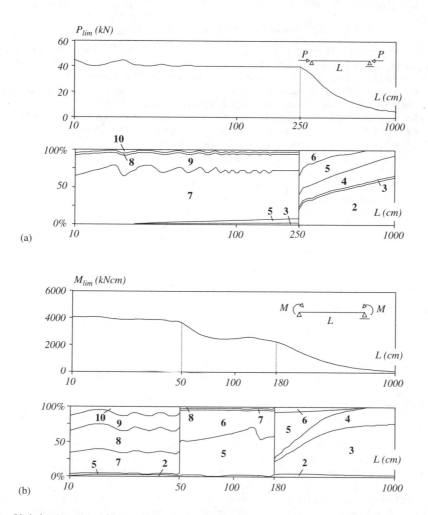

Figure 8.   Variation $\lambda_c \equiv \lambda_{lim}$ with L and buckling mode participation: (a) columns and (b) beams.

orthotropic members. Indeed, such members often do not exhibit a critical bifurcation and the lowest eigenvalue obtained simply stands for a "limit load", i.e. provides the load level to which the member primary path tends asymptotically (Silvestre & Camotim 2002c,d). In order to detect the real nature of the lowest eigenvalue of a given member, one must compare the deformation mode participation in the corresponding (i) primary path (first order analysis) and (ii) eigenmode (linear stability analysis). Although this issue won't be further addressed in this paper, one should point out that all the λ_{cr} values presented here, concerning the specific orthotropic columns and beams described above, are "limit loads". Indeed, due to membrane-flexural coupling effects (recall that the walls are asymmetrically layered plates), some deformation modes appear in both the member (i) primary path deformed configuration and (ii) critical "buckling mode" (eigenmode) shape. Therefore, the reader should keep in mind that (i) one has $\lambda_{cr} \equiv P_{lim}$ (columns) and $\lambda_{cr} \equiv M_{lim}$ (beams) and that (ii) the member follows a non-linear equilibrium path which leads to a deformed configuration akin to the corresponding "buckling mode" shape.

Figures 8(a) and 8(b) show, respectively for columns and beams, (i) the variation of P_{lim} and M_{lim} with the member length L (top curves) and (ii) the "degree of participation" (d.p.) of each individual GBT deformation mode in the member critical "buckling mode" (bottom graphs). Notice that the *exact* degree of participation of a GBT mode k in the cross-section deformed configuration must be evaluated on the basis of ϕ_k and, obviously, varies along the column length. The results reported here concern *average* degrees of participation, i.e. are based on the values of the quantities $\int_0^L |\phi_k(x)| dx$.

9

The observation of these figures leads to the following conclusions:

(i) Columns with $L < 250\,cm$ "buckle" in 1–17 half-wave "mixed" local-plate modes (LPM), comprising GBT modes **7** (large *d.p.*), **9** (fair *d.p.*) and **3 + 5 + 8 + 10** (small *d.p.*). Moreover, one has $P_{lim}^{LP} \approx 40\,kNcm$, a value exhibiting just a slight dependence on the length L.

(ii) Columns with $250 < L < 1000\,cm$ "buckle" in 1 half-wave flexural-torsional-distortional modes comprising GBT modes **2 + 4 + 5 + 6** (fair *d.p.*) and **3** (small *d.p.*).

(iii) Columns with $L > 1000\,cm$ "buckle" in 1 half-wave flexural-torsional modes (not shown).

(iv) Beams with $L < 50\,cm$ "buckle" in 1–3 half-wave "mixed" LPM, which comprise GBT modes **7 + 8 + 9 + 10** (fair *d.p.*) and **2 + 5** (small *d.p.*). Moreover, one has $M_{lim}^{LP} \approx 3990\,kNcm$, a value exhibiting again only a slight dependence on the beam length L.

(v) Beams with $50 < L < 180\,cm$ "buckle" 1–2 half-wave distortional modes (DM) comprising GBT modes **5** (large *d.p.*), **6** (fair participation) and **7 + 8** (small *d.p.*). A minimum moment value $M_{lim}^{D} \approx 2578\,kNcm$ is reached for $L \approx 85\,cm$.

(vi) Beams with $180 < L < 550\,cm$ "buckle" in 1 half-wave flexural-torsional-distortional modes comprising GBT modes **3 + 5** (fair-to-large *d.p.*), **4** (fair *d.p.*) and **2 + 6** (small *d.p.*).

(vii) Beams with $L > 550\,cm$ "buckle" in 1 half-wave flexural-torsional modes comprising GBT modes **3** (large *d.p.*), **4** (fair *d.p.*) and **2** (small *d.p.*).

5 ILLUSTRATION: LP & D POST-BUCKLING OF COLD-FORMED STEEL COLUMNS

This section addresses the application and capabilities of the isotropic non-linear GBT and the numerical results presented concern the post-buckling (geometrically non-linear) behaviour of cold-formed steel lipped channel columns with pinned (locally and globally) and free-to-warp end sections. In particular, stub-columns buckling in both the local-plate mode and the distortional mode are analysed. Their cross-section dimensions are $b_{web} = b_{flange} = 150\,mm$, $b_{lip} = 7.5\,mm$, $t = 1.5\,mm$ and the material (steel) behaviour is characterised by $E = 200\,MPa$ and $v = 0.3$. The cross-section and column (longitudinal) discretisations involve 23 nodes (6 natural and 17 intermediate) and 8 finite elements, respectively. Moreover, the post-buckling analyses (i) incorporate initial geometrical imperfections with the critical buckling mode shape and small amplitude values and (ii) include all the GBT deformation mode types addressed earlier (FM$_u$, FM$_w$, SM$_u$ and SM$_v$ – see Figs. 3–6) (Silvestre & Camotim 2003a).

Since one wishes to determine the local-plate and distortional post-buckling equilibrium paths associated to such cross-section geometry, a preliminary GBT linear stability analysis was performed, in order (i) to identify the stub-column lengths associated to critical single half-wave LPM and DM (L_{LP} and L_D) and (ii) to evaluate the corresponding bifurcation stresses σ_{cr}^{LP} and σ_{cr}^{D}. The results yielded by this analysis were: (i) $L_{LP} = 120\,mm$ e $\sigma_{cr}^{LP} = 70.6\,MPa$ and (ii) $L_D = 637\,mm$ and $\sigma_{cr}^{D} = 31.8\,MPa$. The (critical-mode-shape) initial geometrical imperfections exhibited by the stub-columns have amplitudes (i) $w_0 = 0.1t$ (LPM $-$ w_0 is the initial mid-web flexural displacement) and (ii) $v_0 = \pm 0.01\,t$ (DM $-$ v_0 is the initial lip transverse membrane displacement). The need to consider "positive" and "negative" initial DM imperfections stems from a non negligible dependence of the column distortional post-buckling behaviour on the v_0 sign, unveiled by Prola & Camotim (2002). This phenomenon, illustrated in Figure 9(b), will be addressed later in the paper. Since the degrees of non-linearity of the LPM and DM post-buckling equilibrium paths are known to be distinct (Prola 2001), different first load increments were adopted to start the incremental-iterative procedure: 15% (LPM) and 7.5% (DM) of the critical bifurcation load.

Figure 9(a) shows a comparison between the local-plate post-buckling paths (applied stress σ vs. normalised mid-web flexural displacement w/t) obtained by means of (i) the non-linear GBT formulation and (ii) the FEM, using ABAQUS (HKS 2003) and adopting a fine mesh of shear deformable shell FEs (S4 elements) to discretise the column. One observes that the two curves are always very close. For instance, for $\sigma = \sigma_{cr}^{LP} = 70.6\,MPa$ one has $(w/t)_{GBT} = 0.63$ and $(w/t)_{FEM} = 0.64$.

Figure 9(b), on the other hand, shows four distortional post-buckling equilibrium paths (applied stress σ vs. normalised lip transverse membrane displacement v/t), yielded by non-linear GBT and FEM (ABAQUS) analyses of two otherwise identical columns containing different initial imperfections: (i) $v_0 = +0.01\,t$ (the flange-lip assemblies "open") and (ii) $v_0 = -0.01\,t$ (the flange-lip assemblies "close"). After quickly glancing at the curves shown in Figures 9(a) and 9(b), one immediately realises that the distortional post-critical strength is much smaller than its local-plate counterpart (regardless of the v_0 sign),

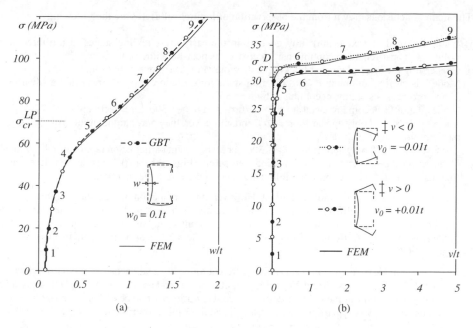

Figure 9. (a) Local-plate and (b) distortional post-buckling equilibrium paths of lipped channel columns.

as was recently reported by Prola (2001). Moreover, a close observation of the four curves depicted in Figure 9(b) leads to the following remarks:

(i) There is an almost perfect match between the post-buckling paths obtained by means of the non-linear GBT and FEM (ABAQUS) analyses. For instance, when $\sigma = \sigma_{cr}^D = 31.8\,MPa$, one has (i$_1$) $(v/t)_{GBT} = 0.320$ and $(v/t)_{FEM} = 0.326$, for $v_0/t = -0.01$, and (i$_2$) $(v/t)_{GBT} = 3.840$ and $(v/t)_{FEM} = 3.911$, for $v_0/t = +0.01$.

(ii) These results confirm the existence of a non negligible difference between the column distortional post-buckling behaviours associated to "positive" and "negative" v_0 values, as recently unveiled by Prola & Camotim (2002). These authors attributed this distortional post-buckling *asymmetry* (higher post-critical strength for $v_0/t < 0$) to marked differences in the warping behaviour of the flanges and (mostly) the lips.

(iii) The curves associated to $v_0/t = 0.01$ (dashed) and $v_0/t = -0.01$ (dotted) begin to "split" well below $\sigma = \sigma_{cr}^D$. Indeed, for $\sigma = \sigma_{cr}^D$, the difference is already very large: $v/t = 3.84$ ($v_0/t = 0.01$) against $v/t = 0.32$ ($v_0/t = -0.01$). This is due to the fact that the former post-buckling path becomes practically "horizontal".

Next, in order to underline the unique features of the non-linear GBT formulation, and also to illustrate how these features can be explored (i) to improve the understanding about the post-buckling behaviours under consideration or (ii) to develop alternative (computationally more efficient) methods of geometrically non-linear analysis, Figures 10(a)–(c) show "GBT modal analyses" of the equilibrium paths depicted in Figures 9(a) (LPM) and 9(b) (DM). In other words, these figures provide a detailed account of the evolution, along the post-buckling equilibrium path, of the "degree of participation" of the different GBT deformation modes on the column deformed configuration. This information significantly contributes to shed new light on several aspects concerning the column post-buckling behaviour and, in particular, supplies the means to provide an in-depth explanation of the distortional post-buckling asymmetry mentioned earlier.

The numbers 1–9 located along the top and bottom horizontal axes in Figures 10(a) to 10(c) correspond to the equilibrium points (configurations) identified by black dots on the post-buckling paths displayed in Figures 9(a) and 9(b). Therefore, by drawing a line segment uniting identical numbers, one is able to assess, quite easily, the degree of participation (in percentage terms) of each GBT deformation mode on the column deformed configuration.

11

The main conclusions drawn from the observation of Figures 10(a)–(c) are the following:

(i) Naturally, mode 1 (extension) fully governs in the early pre-buckling stages of the three stub-columns. Its participation continuously decreases as post-buckling progresses.

(ii) Concerning the local-plate post-buckling behaviour (Fig. 9(a)), one notices that:
Mode 7 is clearly dominant. Its participation starts for quite low applied stress values (near point 2) and progresses steadily, becoming as high as 86%.
Modes **9** and **11** become relevant for $\sigma > 0.70\sigma_{cr}^{LP}$ and $\sigma > 1.30\sigma_{cr}^{LP}$, respectively.
The SM$_u$ **15** and **17** only appear for $\sigma > \sigma_{cr}^{LP}$ and, moreover, they play a rather minor role (their joint participation never exceeds 4%).
The SM$_v$ **26** and **28** only appear for $\sigma > 1.30\sigma_{cr}^{LP}$, but their joint participation rapidly grows up to 5%.

(iii) Concerning the distortional post-buckling behaviour (Figs. 9(b)–(c)), considerably less known than its local-plate counterpart, the following aspects deserve to be mentioned:

– Naturally, mode **5** is the most important. However, its participation in the column deformed configuration is quite different for positive and negative initial imperfections. In the former case ($v_0/t = 0.01$), mode **5** appears at $\sigma \approx 0.73\sigma_{cr}^{D}$ (near point 4) and progressively grows up to about 85%. For $v_0/t = -0.01$, mode **5** only appears at $\sigma \approx 0.94\sigma_{cr}^{D}$ (near point 5), but grows (faster) to about 90%.

– The shear mode participation is also rather different for positive and negative initial imperfections. For $v_0/t = 0.01$, modes **15** and **23** emerge, together with mode **5**, at $\sigma \approx 0.73\sigma_{cr}^{D}$ and have a relevant participation in the column deformed configuration (it reaches 12%). For $v_0/t = -0.01$, modes **15** and **19** only participate for $\sigma > \sigma_{cr}^{D}$ and their joint contribution never exceeds 5%.

– The two previous items provide a clarifying and in-depth explanation for the distortional post-buckling asymmetry exhibited by the curves in Figure 9(b). Note that it is virtually impossible to extract such explanation from the results yielded by "conventional" numerical analyses (e.g. FSM or FEM analyses).

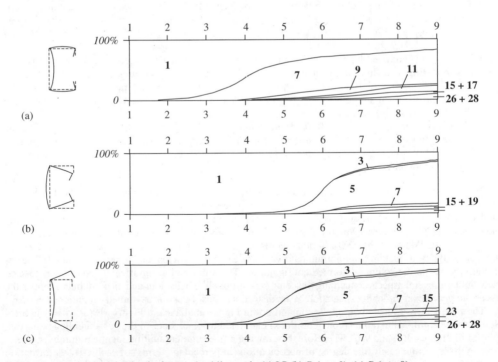

Figure 10.   Modal participation in the post-buckling path; (a) LP (b) D ($v < 0$); (c) D ($v > 0$).

– The roles played by modes **3** (FM$_u$), **7** (FM$_w$) and **26 + 28** (SM$_v$) are similar for positive and negative initial imperfections.

Taking into consideration the conclusions drawn from the above "GBT modal analyses", one is now in an ideal position to discuss the relevance, both in terms of accuracy and computational efficiency, of including the different deformation modes in the local-plate and distortional post-buckling analyses. In order to do this, it is convenient to compare the results obtained from various "approximate GBT analyses", in the sense that they include only a (selected) fraction of the relevant deformation modes identified in Figures 10(a)–(c). Therefore, several GBT post-buckling analyses were carried out and the corresponding results are presented in Figures 11 and 12 (to enable a better distinction between the different curves, their lower portions were omitted and a larger vertical scale was used). For comparison purposes, the "exact" curves (all deformation modes included), shown in Figures 9(a) and 9(b), are displayed once

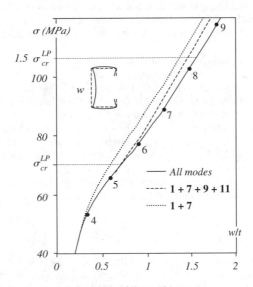

Figure 11.   Exact and approximate local-plate post-buckling paths.

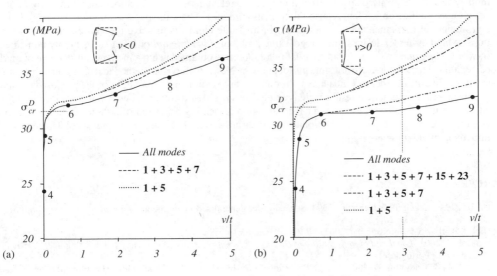

Figure 12.   Exact and approximate distortional post-buckling paths: (a) $v < 0$; (b) $v > 0$.

13

more. Observing the curves of Figure 11, concerning the column local-plate post-buckling behaviour, one is led to the following remarks:

(i) If only modes **1** and **7** are included in the analysis (dotted curve), the results are very accurate only up to $\sigma \approx 0.80\sigma_{cr}^{LP}$. For larger σ values, the accuracy progressively decreases, even if the w/t error never exceeds 15% (e.g. for $\sigma = \sigma_{cr}^{LP}$ and $\sigma = 1.50\sigma_{cr}^{LP}$, one has $(w/t)_{ap} = 0.54$ and $(w/t)_{ap} - 1.32$, instead of $(w/t)_{ex} = 0.63$ and $(w/t)_{ex} = 1.51$).

(ii) If the analysis includes modes **1, 7, 9** and **11** only (dashed curve), the results are practically exact up to $\sigma = \sigma_{cr}^{LP}$. After that, the w/t values remain fairly accurate, as the error never exceeds 5% (e.g. for $\sigma = 1.50\sigma_{cr}^{LP}$, one has $(w/t)_{ap} = 1.44$, instead of $(w/t)_{ex} = 1.51$.

Next, let us look at the curves shown in Figures 12(a)–(b), concerning the column distortional post-buckling behaviour ($v_0 = +0.01\,t$ and $v_0 = -0.01\,t$) and supporting the following conclusions:

(i) If only (i_1) modes **1** and **5** or (i_2) modes **1, 3, 5** and **7** are incorporated in the analysis, the corresponding post-buckling paths (dotted and dashed curves) related to $v_0 = +0.01\,t$ and $v_0 = -0.01\,t$ *coincide*, which indicates that the post-buckling *asymmetry* is exclusively due to the shear modes (SM$_u$ and SM$_v$).

(ii) If only modes **1** and **5** are included in the analysis, the results are practically exact up to $\sigma = 0.75\sigma_{cr}^{D}$ ($v_0 = 0.01t$) or $\sigma = \sigma_{cr}^{D}$ ($v_0 = -0.01t$). For larger σ values, v/t progressively looses accuracy, particularly for $v_0 = 0.01t$ (e.g. for $v_0 = 0.01t$ and $v/t = 3.0$, one has $\sigma_{ap} = 1.32\sigma_{cr}^{D}$, instead of $\sigma_{ex} = 0.98\sigma_{cr}^{D}$). By including the remaining FM (**3** and **7**), the solution improves only marginally (e.g. again for $v_0 = 0.01t$ and $v/t = 3.0$, one has now $\sigma_{ap} = 1.27\sigma_{cr}^{D}$).

(iii) For $v_0 = 0.01\,t$, a post-buckling analysis including modes **1, 3, 5, 7, 15** and **23** (i.e. all the participating fundamental and shear modes) greatly improves the accuracy (for $v/t = 3.0$, one has now $\sigma_{ap} = 1.05\sigma_{cr}^{D}$, only about 7% higher that the exact value). This confirms that, if $v_0/t > 0$, the incorporation of shear modes is essential to obtain accurate results.

(iv) For $v_0 = \pm 0.01\,t$, a post-buckling analysis which includes the 10 most relevant GBT modes (**1, 3, 5, 7, 15, 17, 23, 26, 28** and **30**) yields virtually exact results (errors below 1%).

6 CONCLUSION

A GBT-based computational approach to perform elastic (i) *buckling* analysis of orthotropic thin-walled members and (ii) *post-buckling* analyses of isotropic thin-walled members was presented and illustrated. First, the fundamental concepts and procedures involved in the application of GBT were briefly reviewed, namely the ones related to (i) performing the cross-section analysis (deformation modes and modal mechanical properties) and (ii) establishing the relevant system of member equilibrium equations, which may include only a few selected deformation modes. Then, the paper addressed the formulation and numerical implementation of non-linear beam finite elements to solve the above system of equations, which either (i) defines linear eigenvalue problem (*buckling analysis*) or (ii) is highly non-linear (*post-buckling analysis*). In the latter case, the solution was obtained using a incremental-iterative numerical technique (predictor-corrector strategy adopting the Newton-Raphson method and an arc-length control scheme). Finally, in order to validate and illustrate the application and capabilities of the proposed computational approach, numerical results were presented, discussed and compared with values yielded by FEM analyses. These results dealt with (i) the local and global buckling behaviour of orthotropic lipped channel columns and beams and (ii) the local-plate and distortional post-buckling behaviour of cold-formed steel lipped channel columns.

REFERENCES

Davies, J.M., Leach, P. & Heinz, D. 1994. Second-order generalised beam theory. *Journal of Constructional Steel Research* 31(2–3): 221–241.
Hibbit, Karlsson & Sorensen Inc. 2003. *ABAQUS Standard* (version 6.3).
Prola, L.C. 2001. *Local and global stability of cold-formed steel members* (Ph.D. Thesis). Civil Eng. Dept., IST, TU Lisbon (in Portuguese).
Prola, L.C. & Camotim, D. 2002. On the distortional post-buckling behavior of cold-formed lipped channel steel columns", *Proc. SSRC Annual Stability Conference, Seattle, 24–27 April*, 571–590.

Schardt, R. 1989. *Verallgemeinerte technishe biegetheorie*. Berlin: Springer-Verlag (in German).

Schardt, R. 1994. Generalized beam theory – an adequate method for coupled stability problems. *Thin Walled Structures* 19(2–4): 161–180.

Silvestre, N. & Camotim, D. 2002a. Stability behavior of composite thin-walled members displaying arbitrary orthotropy. *Proc. 15th ASCE Engineering Mechanics Conference (CD-ROM), New York, 2–5 June.*

Silvestre, N. & Camotim, D. 2002b. First order generalised beam theory for arbitrary orthotropic materials. *Thin Walled Structures* 40(9): 755–789.

Silvestre, N. & Camotim, D. 2002c. Second order generalised beam theory for arbitrary orthotropic materials. *Thin Walled Structures* 40(9): 791–820.

Silvestre, N. & Camotim, D. 2002d. GBT buckling analysis of pultruded FRP lipped channel members. *Computers & Structures* (accepted for publication).

Silvestre, N. & Camotim, D. 2003a. Non linear generalised beam theory for cold-formed steel members (submitted for publication).

Silvestre, N. & Camotim, D. 2003b. On the incorporation of shear deformation in first and second order generalised beam theory (in progress).

Computational Methods in Engineering and Science, Iu et al. (eds)
© 2003 Swets & Zeitlinger, Lisse, ISBN 90 5809 567 3

Modeling cementitious materials behavior

J.M. Catarino
LNEC – National Laboratory of Civil Engineering, Lisbon, Portugal

ABSTRACT: Cementitious materials are subject to loads related with the environment of civil engineering constructions. Such physical or chemical loads may lead to instantaneous or delayed changes in construction elements, including material degradation, as in the case of steel corrosion or chemical and abrasion attack to concrete. Modeling phenomena related with materials degradation has got significant improvement in the last decade, in the same time of a greater importance being given to the need of life time analysis of constructions, "designed" for periods of more than a century, in special cases, with proper care of maintenance and repair. Some models for concrete behavior are presented, taking into account the nature of degradation processes involved.

1 INTRODUCTION

Modeling materials behavior has always been a target for scientific research, as a tool for systematizing and explanation of knowledge, for analysis of experimental data and extrapolation of not yet known situations. Civil engineering has naturally pursued this objective. Unfortunately, even in traditional materials, are still partially unknown some basic mechanisms of degradation phenomena and their consequences in construction safety. It is evident the need of research in materials behavior in very small scales, so that materials structure can be related with the macroscopic properties significant for civil engineering constructions.

The number of papers presented in the last decade concerning "modeling materials behavior" is quite large, with a special mention to the work made by Acker, Bazant, Bentz, Garboczi, Jennings, Van Breugel and Wittmann. Modeling materials behavior means any system with the purpose of describing the response of a material to a set of loads, external or related with time evolution of the material itself, in different levels of precision, but always structured with the basis of the existing knowledge related with that material. The description of the response must be generic, including all aspects of physical and chemical behavior, based on the understanding of influent parameters in materials behavior.

The use of computational techniques for developing and exploring material models is quite new. In civil engineering these techniques have been used, in first place, in large scale models for structural analysis of bridges and dams, for example. It seems that the smaller is the scale, more complexity is in the analysis, due to the necessity of consideration of physical and chemical aspects, of a very different nature, being processed in different scales, in other words … *"it seems that the smaller models require larger computational facilities".*

Only in the last few years accessibility to large computational facilities has been generalized, making possible the use of simple formulations, easily programmed, with behavior easily understandable. The development of computer programs will become easier in near future and very useful for materials research.

Figure 1 symbolizes the type of models to develop for the behavior of materials, taking into consideration the complete characterization of constituents, production process, placing and curing, environment conditions and all physical and chemical loads during life time. The response of the model consists in time evolution of physical and chemical material properties.

Material models may be used isolate, in materials research, or coupled to structural analysis models, with the purpose of simulating the global behavior of a structure, taking into account all data related with its components, loads and construction history.

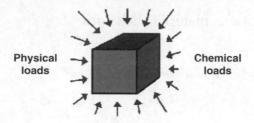

Figure 1.   Modeling materials behavior.

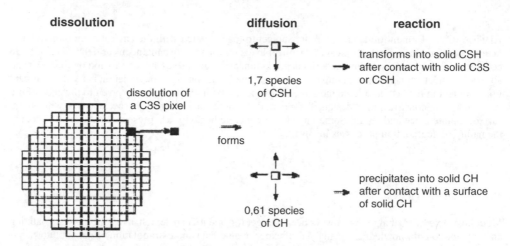

Figure 2.   Digital-image-based models for the hydration of cement pastes (Garboczi et al. 1998).

As well as in structural models, material models may be developed with the purpose of supporting designers, with friendly use and based on conservative criteria, or for research purpose, where first priority is given to interpretation and simulation of reality with the best available methods.

2   DIGITAL MODELS

As an example of cementitious material models with the purpose of representing directly microstructure and estimating transport properties or mechanical behavior, must be mentioned digital-image-based models, where all components of the system being analyzed (concrete, mortar or cement paste) are discretized in three-dimensional sets of "pixel".

The development of digital-image-based models is associated with the rapid increase of computational facilities, in particular the possibility of large availability of RAM memory.

For the development of such models it is necessary the establishment of initial conditions of "pixel" properties and the rules of change of those properties along time. Each "pixel" may randomly walk inside the system, reacting with other "pixel". In reality, this is "virtual walking" ... "pixels" remain in same positions – only changes of phases occur.

Figure 2 shows a scheme of a digital-image-based model. An important characteristic of such models is the possibility of representation of porous microstructure, opening large perspectives for global numeric modeling or transport problems related with chemical composition of materials and their physical and durability properties. The algorithm developed in these models includes four main routines: initiation, dissolution, diffusion and nucleation.

Initiation defines the starting distribution of C3S particles, in three dimensions. The dissolution routine includes the formation of C–S–H and CH, due to contact of C3S with water, and a random process of movement of particles, until nucleation conditions are fulfilled. Dissolution and diffusion routines

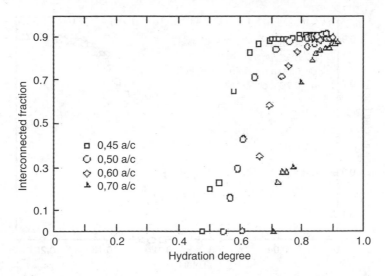

Figure 3. Influence of water/cement ratio and hydration degree in content of interconnected volume of CH (Bentz et al. 1991).

are performed until all CH and C–S–H have precipitated. Hydration process ends when no more C3S can be dissolved, or when a certain degree of hydration has been achieved.

The following short designation of chemical products is adopted in this paper:

C–S–H = gel of calcium hydrated silicates (C = CaO   S = SiO$_2$ H = H$_2$O)
CH   = calcium hydroxide
C3S   = tri-calcium silicate

3 LIXIVIATION OF CEMENTITIOUS MATERIALS

Concrete exposed to environment can have its microstructure attacked by loads such as pluvial or soil water, which may dissolve constituents as calcium hydroxide. Such effects can be simulated by computational methods, estimating changes in permeation of capillary porous media and in its ionic diffusion properties.

Hydration models that can simulate lixiviation processes, with random removal of CH, are already available. It can be observed that a proportion between 16% and 20% of the volume of capillary pores and calcium hydroxide phase is enough to form permeation paths after lixiviation.

Figure 3 shows that only in a late stage in the hydration process of cement paste enough CH exists to form a tri-dimensional microstructure interconnected, in superposition with tri-dimensional structure of C–S–H, formed in an early stage. The larger water/cement ratio, the larger is the necessary hydration degree to establish interconnection of CH, due to the lower amount of CH content.

Interconnected content of CH is never larger than 90%, due to the amount of CH involved by C–S–H that cannot form a path. When interconnected portion of CH is represented as a function of total CH, it can be observed that water/cement ratio is not a relevant factor (Fig. 4). Continuity of CH is possible after a volume of 12% of CH has been formed, with the mentioned inconvenient of being a permeation path in case a dissolution process of CH occurs.

In cement paste with low water/cement ratio, a discontinuous capillary structure may be transformed in a permeable media, in case a dissolution process occurs, with the increase, for example, of the diffusion coefficient of chloride ions. It is also possible the simulation of the increase of diffusion coefficients due to calcium hydroxide lixiviation of cement paste with completely hydrated C3S and different water/cement ratio. Removal of CH constitutes a multiplicative process, being possible an increase of the diffusion coefficient to 50 times the initial value, when lixiviation allows the formation of permeation paths originally blocked.

19

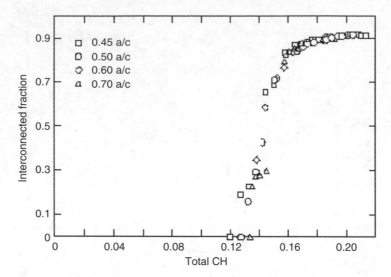

Figure 4. Content of interconnected portion in total volume of CH [2].

4 CORROSION OF STEEL BARS

Steel bars are naturally protected inside concrete, being corroded if a significant decrease of pH value occurs, due to carbonation, or if chloride content achieves a critical value at bar surface. If such conditions are fulfilled, corrosion occurs at a certain velocity, with a value function of the quality of concrete cover and humidity condition.

Tuutti studied deterioration index along time of concrete structures, with two distinct phases: the initiation period includes penetration of aggressive species in concrete and the propagation period includes beginning of corrosion of steel bars.

The majority of analytical models developed concern initiation period, establishing relationships between carbonation depth and penetration of chloride ions, along time, with relevant concrete properties.

In a simplified way, may be considered relationships of type:

$$X = CVt$$

where X = carbonation depth or chloride ion penetration; C = factor depending on concrete properties and boundary conditions; t = exposure time.

In case of carbonation, performance concrete properties being considered may be compressive strength, air permeability and diffusion coefficient of CO_2. Water/cement ratio has been also used as a parameter related with concrete composition (Broomfield 1997).

Another example is Parrot formula (CEN 1992)

$$d = \frac{ak^g t^n}{c^f}$$

where a = constant normally with value 64; k = concrete cover air permeability ($10–16\,m^2$); g, f = exponents, normally with value 0,4; t = age (years); n = exponent function of concrete cover relative humidity; c = calcium oxide content of hydrated cement paste of concrete cover (kg/m^3).

In this formula, besides air permeability, are also considered cement content, cement type, humidity of concrete cover (varying along time), which are very influencing parameters of the carbonation increase. Several research projects (Hilsdorf 1992) demonstrated a singular behavior of blast furnace slag cement (type III) where the carbonation is associated with an increase of pore diameter, facilitating penetration of CO_2, not observed in other cement types.

In CEB bulletin n. 238 (CEB 1997) are presented more elaborated carbonation models, based on diffusion coefficient of CO_2, taking into account local microclimate and the particular characteristics of exposed concrete surfaces. Another way to evaluate concrete performance for carbonation is based on accelerated tests, with CO_2 concentration much higher than in real conditions.

Concerning chlorides, the property parameter is the associated diffusion coefficient. Analytical models are generally based on the 2nd Fick's law, being the most important difficulty the variation along time of such coefficient (C), in particular with cements with additions (type II), and also the variable chloride concentration in concrete surface (CS).

Mangat and Molloy (Mangat 1994) proposed the following formula for this problem:

$$C = C_S \left\{ 1 - \mathrm{erf} \left[\frac{x}{2\sqrt{\dfrac{D_i}{1-m} t^{(1-m)}}} \right] \right\}$$

where m = coefficient function of water/cement ratio and cement type, with from 0,44 to 1,34; D_i = diffusion coefficient; x = depth.

Another relevant topic is the possibility of combination of chlorides with hydrated cement components, depending on cement type and chloride concentration (Tang 1996).

The complexity of formulations to consider other parameters, as is the case of temperature, requires numerical methods for its resolution. Chloride penetration in concrete cover is largely made through capillary absorption and other processes. Concrete quality in internal parts may be, in general, much different, what must be considered in numerical models. In case of chlorides, it is difficult to establish the end of initiation period, not being known with exactitude the associated chloride content.

Initiation of steel corrosion and its velocity depend on concrete properties, humidity, and the oxygen availability for cathodic reaction, although, in general, only a very small oxygen quantity is required.

Another relevant concrete property associated with steel corrosion is electrical resistivity, being accepted that values larger than $20\,K\Omega$ are associated with a small risk of corrosion. Electric resistivity depends a lot on humidity, increasing when humidity lowers its value.

Corrosion velocity has not been chosen for modeling, being calculated through "in situ" measurements, reduction velocity of steel bar section, using Faraday law, or the time need for first crack or concrete spalling.

5 EXPANSIVE REACTIONS

Alkali-aggregate reaction (AAR) is a chemical reaction between OH^- and Na_2O or K_2O, in gel of concrete pores, with a volume of reaction products larger than the sum of initial reagents, and so producing irreversible expansions in affected structures. Alkali may come from cement, admixtures, aggregates or other external sources.

Alkali-aggregate reaction may be classified in two types:

– alkali-silica reaction (ASR), that may occur with silica semi-crystalline minerals, volcanic or artificial glass, several quartz species and aggregates with slow expansion (ASSR)
– alkali-carbonate reaction (less frequent).

Alkali-silica reaction (ASR) may be considered with the following phases:

– formation of alkali-silica gel by chemical reaction
– expansion of gel by humidity abortion and expansion in the porous structure of cement paste in function of its viscosity
– internal micro cracking, due to pressure of confined gel in cement paste
– new migration of alkali-silica gel through cracks produced inside concrete, without loss of gel pressure.

Three simultaneous conditions must be fulfilled for developing alkali-silica reactions:

– a minimum amount of reactive aggregate
– a minimum amount of alkali
– a minimum value of humidity.

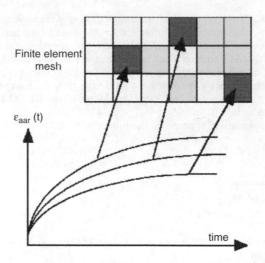

Figure 5.   Simulation of ε_{aar} along time (Capra 1994).

Time evolution of alkali-aggregate reaction is characterized by an initiation period, a development period and stabilization period (Nielsen 1993) (Hobbs 1990). It's much important the evaluation of strategies for intervention in structures damaged by this type of reaction. In certain cases the best solution is waiting, if possible, until degradation reaches the stabilization period, before making large investments in repair. An important difference between the behavior of a core and the behavior of real structures, in what concerns expansive reactions, may occur due to the influence of core extraction in humidity, strain distribution and curing conditions.

Alkali-aggregate reaction may finish after a certain period of time, in case one of the mentioned conditions for development of such reactions is no more fulfilled (lack of reactive material in aggregates, unavailability of alkalis or lowering of humidity). In accordance with Hobbs, when alkalis come from cement, eight to fifteen years are needed to complete alkali silica reaction (Hobbs 1990).

Time evolution of strains due to alkali-silica reaction has the following type:

$$\varepsilon_{aar} = \frac{a\,t}{1 + b\,t}$$

where ε_{aar} = expansion strain; t = time; a, b = coefficients depending of several parameters like relative humidity, stresses, temperature and alkali content.

For numerical simulation purpose Capra used this law in a finite element program (CASTEM 2000), considering concrete behavior with a damage factor, in accordance with Mazars (Capra 1994). Being a random problem, in each element the expansive strain ε_{aar} may change independently, in function of a random variable (a), with a statistical distribution based on experimental data, as mentioned in Figure 5.

Capra adopted a distribution of the random variable (a) adapted to a finite element mesh, for studying structural effects of alkali-silica reaction, in accordance with Sellier and Mebarki (Sellier 1993). Figure 6 shows an interesting experience of a cracking pattern that may be observed in structures affected by alkali-silica reaction. Cracks first appear in concrete surface and may extend to all volume.

The use of a probabilistic approach for simulating the load produces differential displacements, with curvature of the specimen that may be observed in the scheme of the deformed shape of a laminar element "numerically tested". In accordance with Capra several simulations should be made, with different distributions of the random variable, for a statistical analysis of results obtained and evaluation of structural effects of the expansive reaction.

Several research projects are being made by engineers, physicists and chemists about alkali silica reactions and numerical models for concrete durability. As mentioned by Capra, such studies must use probabilistic techniques dedicated to the analysis of situations comparable with observations made in real structures, for a better understanding of consequences in structural safety.

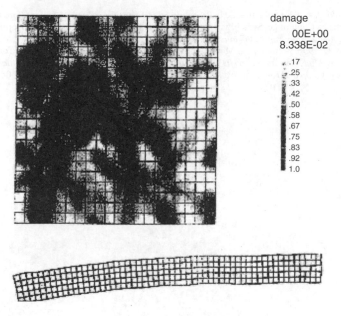

Figure 6.   Deterioration map of a concrete slab and its deformed shape (Capra 1994).

6   CONCLUSIONS

This paper presents only a very small part of what has been done recently about modeling cementitious materials behavior, particularly in what concerns the most well known types of degradation of reinforced concrete. However, the work already done shows it's important potentiality, not because of the possibility of obtaining theoretical results comparable with experimental data, but mainly because this tool helps a lot in improving knowledge of materials. That's the reason of the work being done at LNEC in this field, as well as in many universities and research laboratories all over the world, and the priority given by the European Commission in the 6th Framework Program to nanotechnology – "*the technology where dimensions of tolerances in the range of 0,1 nm to 100 nm play a critical role*".

REFERENCES

Bentz, D.P. & Garboczi, E.J. 1991. Percolation of Phases in a Three Dimensional Cement Paste Microstructure Model, *Cement and Concrete Research, vol. 21 n.° 2*, 325–344, *http://ciks.cbt.nist.gov/garbocz/paper22*.

Broomfield, J.P. 1997. Corrosion of Steel in Concrete – understanding, investigation and repair, *cap.3, E&FN Spon*, p. 16–29, London, UK.

Capra, B. 1994. Modeling of Alkali Aggregate Reaction Induced Mechanical Effects in Concrete, *Laboratoire de Mécanique et Technologie, E.N.S. Cachan/C.N.R.S.*

CEB 1997. Bulletin d' Information *n.° 238* – New approach to durability design. An example for carbonation induced corrosion.

CEN TC 104/TG1/WG1/Panel 1. 1992. Design for avoiding damage due to carbonation-induced corrosion, *Paper N62*.

Garboczi, E.J., Bentz, D.P. & Snyder, K.A. 1998. An Electronic Monograph: Modeling the Structure and Properties of Cement Based Materials, *ACBM/NIST Computer Modeling Workshop, http://ciks.cbt.nist.gov/garbocz/monograph*.

Hilsdorf, H.K., Schönlin, K. & Burieke, F. 1992. Dauerhafitgkeit von Betonen, *Institut für Massivbau und Baustoffechnologie*, Universität Karlsruhe.

Hobbs, D.W. 1990. Cracking and Expansion due to the Alkali-Silica Reaction: Its Effect on Concrete, *Structural Engineering Review, n.° 2*, 65–79.

Mangat, P.S. & Molloy, B.T. 1994. Prediction of long term chloride concentration in concrete, *Materials and Structures n° 27*, p. 338–346.

Nielsen, A.U., Gottfredsen, F. & Thogersen, F. 1990. Development of Stresses in Concrete Structures with Alkali-Reactions, *Materials and Structures, n.° 26*, 152–158.

Sellier, A. & Merbaki, A. 1993. Évaluation de la Probabilité de l'Ocurrence d'un Évènement Rare par Utilisation du Tirage d'Importance Conditioné. Sensibilisation aux Risques Liés aux Effects d'Échelle. Applications Actuelles et Perspectives, *Annales des Ponts et Chaussées, n.° 67.*

Tang, L. & Nilsson, O. 1996. A Numerical Method for Prediction of Chloride Penetration into Concrete Structures, *The Modeling of Microstructure and Its Potential for Studying Transport Properties and Durability, Ed. H. Jennings, Jörg Kroop and Karen Scrivener, NATO ASI Series E: Applied Sciences* – vol. 304 , p. 539–552.

Numerical modeling of metal forming – current practice and advanced applications

S. Cescotto, L. Duchêne, F. Pascon & A.M. Habraken
M&S Dep., University of Liège, Belgium

ABSTRACT: This paper first summarizes the current possibilities and the perspectives of effective modeling of metal forming processes. Practical considerations on manpower requirement and data availability are included in the discussion. Then, two examples of advanced applications are presented: the micro-macro approach for the modeling of sheet metal forming and the modeling of continuous casting.

1 CURRENT POSSIBILITIES AND PERSPECTIVES OF METALFORMING MODELING

1.1 *Introduction*

Nowadays, the specifications in terms of product quality and production time are such that modeling the metal forming of industrial parts has become compulsory. A large number of research and commercial computer codes are available and it is not always easy to make a choice among them. Hereafter, a few practical guidelines are proposed.

1.2 *General scheme of computer modeling*

Modeling must be considered as a set of operations including pre-processing, calculations, post-processing and optimization loop (Fig. 1).

Pre-processing consists in the definition of the data set. User friendliness is essential. One should merely specify the process parameters (tools used such as hydraulic press, drop hammer, etc; initial temperatures of dies and workpiece, reference numbers for the geometry of dies and workpiece which are independently defined in the CAD data base) and the material data including contact conditions (thermo-mechanical properties, friction coefficient, ...).

The finite element software is the core of the modeling: 2D and 3D analyses are possible. Today, automatic remeshing is fully available in 2D but there still remains to improve 3D remeshing in terms of robustness and computer time. In any case, 3D finite element simulation of the forming of complex industrial parts is time consuming. Parallel computation is normally required.

Post processing is, in most cases, the presentation of the finite element results in an attractive and easy-to-read way on which basis the user takes the decision to validate the process if the specifications are met or to modify it if necessary. Modifications are usually based on the experience of the user and are followed by a new finite element simulation. However, the automatic improvement the process parameters is sometimes available thanks to optimization techniques. So far, this is seldom proposed by general purpose commercial codes because optimization is closely linked with the particularities of the considered forming process, but intensive research work is currently performed in this direction.

1.3 *Material data*

Two aspects deserve careful attention: the type of constitutive model and the acquisition of material data: the more sophisticated the model, the more difficult the determination of the corresponding parameters. Table 1 lists some classical types of metal models.

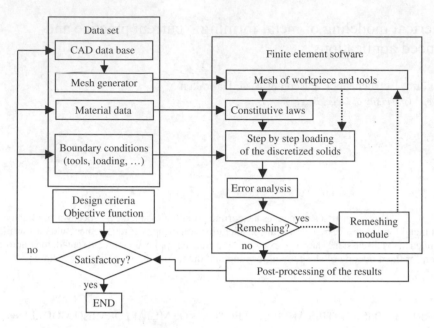

Figure 1. General scheme of metal forming modeling.

Table 1. Material models and corresponding potential results.

Model	Potential results
1. Rigid plastic	Cold bulk forming, shape of workpiece, stresses and strains distribution except at the end of closed die forging
2. Elasto-plastic	Same as 1plus stresses and strains distribution at the end of closed die forging and elastic springback
3. Rigid-visco-plastic	Same as 1 but applicable to hot forming, average temperature is considered
4. Elastic-visco-plastic	Same as 2 but applicable to hot forming, average temperature is considered
5. Thermo-visco-plastic (rigid or elastic)	Same as (rigid or elastic)-visco-plastic but the detailed distribution of temperature is considered
6. Damage	Can be combined with preceding models, damage distribution and fracture initiation can be predicted
7. Isotropic	Usually sufficient for bulk forming
8. Anisotropic	Necessary for sheet metal forming, models can be macro or micro; anisotropy can be considered for elastic, plastic and/or damage behaviors
9. Micro-macro	Macroscopic behavior (elastic, plastic, …) deduced from behavior of crystals by some averaging technique; important for sheet metal forming
10. Recrystallization	Can be combined with thermo-(visco)-plastic models, grain size and recrystallized zones can be predicted
11. Phase changes	Important for cooling processes (continuous casting), heat treatment, …
12. Contact	Coulomb's friction law or more advanced models

So, the constitutive model must be chosen according to the precision expected from the simulation. For models 1, 2, 3 and for usual metals, data tables are available as in Altan, Oh, Gegel (1983) or on the web (e.g. www.matweb.com), but for particular alloys, experiments are necessary. In addition, the validity of available data is sometimes questionable. For example, the scatter of literature data on Young's modulus at high temperature is important, as illustrated by Figure 2. For more complex models, data acquisition often requires expensive and time consuming laboratory tests. For example in the visco-plastic domain, Norton-Hoff law is widely used. Its identification requires compression or torsion tests at high temperature,

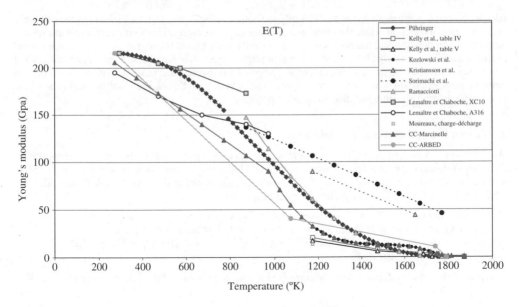

Figure 2.   Young's modulus of steel as a function of temperature from different data sources.

using non classical equipment at least at 3 temperatures and 3 strain rates. It takes about 1 week to get the results. An extreme case is encountered when metallurgical phase changes must be taken into account: getting all the necessary data for a new metal could take a few years in the worst cases!

For the micro-macro approach, which is well developing for the modeling of sheet metal forming, it is necessary to know on the one hand, the slip planes and the corresponding critical resolved shear stresses as well as the twinning modes of the crystal, and, on the other hand, the ODF (Orientation Distribution Function) of the metal sheet, which can be obtained by X-ray or neutron diffraction.

1.4   Conclusion

From the industrial point of view, when purchasing a computer code to model metal forming processes, it is vital to check the adequacy of the available material models and data with the types of metals used and the process conditions. The cost and time of data acquisition must be taken into account as an essential factor of the purchase decision.

2   ADVANCED MODELING OF SHEET METAL FORMING

2.1   Introduction

The goal is to integrate the influence of material texture in sheet metal forming simulations. The constitutive law describing the mechanical behavior is based on a microscopic approach at the crystallographic level. A large number of crystals must be used to represent correctly the global behavior. The transition from crystal level to macroscopic level is based on the full constraint Taylor's model. This model does not lead to a general law with a mathematical formulation of the yield locus. Only one point of the yield locus corresponding to a particular strain rate direction can be computed. To obtain such a stress point, it is assumed that the macroscopic stress results from the average of the microscopic stresses in each crystal belonging to a set of representative crystals, the orientations of which approximate the actual texture of the material. Repeating this for a large number of strain rate directions generates a set of stress points belonging to the yield surface in stress space. Fitting an equation on this set provides the analytical expression of the yield surface. For example, a least square fitting of a 6th order series (210 coefficients) on a large number of stress points (typically 70300) in the deviatoric stress space has been adopted by Munhoven et al. 1996. This is a global description of the yield locus that can be implemented in a FEM code. Unfortunately, taking account of texture evolution would imply the computation of the 210 coefficients of the 6th order series for each integration point, each time texture updating is necessary. This would require an impressive amount of computation and memory storage (210 coefficients for each integration point) which is only partially useful as, generally, for a given material point, the stress state remains in a local zone of the yield locus during the forming process. So, an original approach is investigated in which a direct interpolation between points in strain space and in stress space is achieved. With this method, only a local zone of the yield locus is used at a time for each integration point.

2.2 *Local description of a scaled yield locus*

We work in the 5 dimensional (5D) deviatoric strain and stress spaces. Let $\mathbf{s}^{*0}$ be one unit stress vector, direction of the central point of the local part of the yield locus that requires an approximation. $\mathbf{s}^{*(i)}$ are five unit stress vectors surrounding $\mathbf{s}^{*0}$ and determining the interpolation domain. They will be called the "domain limit vectors". They have the following properties:

– they are unit vectors: $\mathbf{s}^{*(i)} \cdot \mathbf{s}^{*(i)} = 1$ (no sum on i) and $\mathbf{s}^{*0} \cdot \mathbf{s}^{*0} = 1$
– there is a common angle between all $\mathbf{s}^{*(i)}$: $\mathbf{s}^{*(i)} \cdot \mathbf{s}^{*(j)} = 1 + \beta^2(\delta_{ij} - 1)$ $i, j = 1 \dots 5$
– there is a common angle between each $\mathbf{s}^{*(i)}$ and $\mathbf{s}^{*0}$: $\mathbf{s}^{*0} \cdot \mathbf{s}^{*(i)} = \cos\theta$ $i, j = 1 \dots 5$
– they determine a regular domain.

Hence, the central direction $\mathbf{s}^{*0}$ can be computed as a scaled average of the 5 domain limit vectors $\mathbf{s}^{*(i)}$:

$$\mathbf{s}^{*0} = \frac{1}{\cos\theta} \sum_{i=1}^{5} \mathbf{s}^{*(i)} \tag{1}$$

Both θ and β determine the size of the interpolation domain. They are linked by:

$$\beta^2 = \frac{6}{5}\sin^2\theta \tag{2}$$

As the 5 $\mathbf{s}^{*(i)}$ vectors are linearly independent, they constitute a covariant vector basis of the 5-dimensional space. The (non unit) contravariant base vectors are defined by:

$$\mathbf{ss}^{(i)} \cdot \mathbf{s}^{*(j)} = \delta_{ij} \tag{3}$$

and one can check that they depend linearly on vectors $\mathbf{s}^{*(i)}$ and $\mathbf{s}^{*0}$:

$$\mathbf{ss}^{(i)} = \frac{1}{\beta^2}\left(\mathbf{s}^{*(i)} - \frac{1-\beta^2}{\cos\theta}\mathbf{s}^{*0}\right) \tag{4}$$

The 5 η-coordinates of any vector $\mathbf{V} = \sum_{i=1}^{N} \eta_i \mathbf{s}^{*(i)}$ are determined from:

$$\mathbf{V} \cdot \mathbf{ss}^{(j)} = \sum_{i=1}^{N} \eta_i \mathbf{s}^{*(i)} \cdot \mathbf{ss}^{(j)} = \sum_{i=1}^{N} \eta_i \delta_{ij} = \eta_j \tag{5}$$

For a unit vector $\mathbf{V}$ coinciding with a domain limit vector $\mathbf{s}^{*(i)}$, the η-coordinates are:

$$\eta_j = \delta_{ij} \qquad \text{with} \quad j = 1..5 \tag{6}$$

The domain limit vectors represent the domain vertices. The 5 limit boundaries (or edges) of the interpolation domain have the equation: $\eta_i = 0$.

The above choices imply that any point belonging to the interpolation domain is associated to positive η-coordinates.

28

To determine the 5 domain limit vectors $\mathbf{s}^{*(i)}$, 5 unit vectors $\mathbf{s}'^{*(i)}$, are computed as a linear relation between $\mathbf{s}'^{*0} = <1, 1, 1, 1, 1>$ and successively, each vector of the Cartesian basis $\mathbf{e}^{(i)}$ (for example $\mathbf{e}^{(2)} = <0, 1, 0, 0, 0>$)

$$\mathbf{s}'^{*(i)} = \alpha\, \mathbf{s}'^{*0} + \beta\, \mathbf{e}^{(i)} \quad \text{with} \quad \alpha = \frac{\cos\theta}{5} - \frac{\sin\theta}{2} \quad \text{and} \quad \beta^2 = \frac{5}{4}\sin^2\theta \tag{7}$$

Then the rotation linking the real required central point $\mathbf{s}^{*0}$ and $\mathbf{s}'^{*0}$ is given by:

$$\mathbf{R} = \mathbf{I} + 2\,\mathbf{s}^{*0} \otimes \mathbf{s}'^{*0} - \frac{\left(\mathbf{s}^{*0} + \mathbf{s}'^{*0}\right) \otimes \left(\mathbf{s}^{*0} + \mathbf{s}'^{*0}\right)}{1 + \mathbf{s}^{*0} \cdot \mathbf{s}'^{*0}}, \tag{8}$$

where $\mathbf{I}$ is the second order unit tensor. It provides the domain limit vectors:

$$\mathbf{R}\, . \mathbf{s}'^{*(i)} = \mathbf{s}^{*(i)} \tag{9}$$

Now, let us consider both 5D stress and strain rate spaces. A regular domain is built in the strain rate space. It is defined by its 5 vertices $\mathbf{u}^{*(i)}$ (unit vectors). Thanks to 5 calls to Taylor's module, the associated stress vectors $\mathbf{s}^{(i)}$ can be defined. These 5 stress vectors define a non-regular domain in the stress space. The contravariant base vectors in each space are

$$\mathbf{uu}^{(i)} . \mathbf{u}^{*(j)} = \delta_{ij} \quad \text{and} \quad \mathbf{ss}'^{(i)} . \mathbf{s}^{(j)} = \delta_{ij}. \tag{10}$$

The contravariant vectors $\mathbf{ss}'^{(i)}$ computed above and $\mathbf{ss}^{(i)}$ given by (5) differ only because in (5), unit stress directions $\mathbf{s}^{*(i)}$ are used. Here the length of the stress vectors $\mathbf{s}^{(j)}$ is an important characteristic as it defines the yield locus anisotropy. These contravariant vectors $\mathbf{ss}'^{(i)}$ and $\mathbf{uu}^{(i)}$ give, in each space, the η-coordinates associated to any stress $\mathbf{s}$ or unit strain rate $\mathbf{u}^*$: $\eta_i = \mathbf{uu}^{(i)} \cdot \mathbf{u}^*$ and $\eta_i = \mathbf{ss}'^{(i)} \cdot \mathbf{s}$.

So any stress vector $\mathbf{s}$ or strain rate direction $\mathbf{u}^*$ can be represented with the help of the vector basis of their respective space and the η-coordinates:

$$\mathbf{u}^* = \sum_{i=1}^{5} \eta_i\, \mathbf{u}^{*(i)}. \quad \text{and} \quad \mathbf{s} = \sum_{i=1}^{5} \eta_i\, \mathbf{s}^{(i)} \tag{11}$$

Physically, a material state corresponds to one stress point $\mathbf{s}$ and one strain rate direction $\mathbf{u}^*$.

As a basic hypothesis, it is assumed that the η-coordinates computed by (11) are equal when the stress $\mathbf{s}$ and the strain rate direction $\mathbf{u}^*$ are physically associated. This property is exactly fulfilled on the domain limit vectors. The stress $\mathbf{s}^{(i)}$ corresponds to the strain rate direction $\mathbf{u}^{*(i)}$ and their η-coordinates are $\eta_i = 1$ and $\eta_j = 0$ ($i \neq j$) in both spaces. This property is extended inside the domains by convenience. The so-called stress–strain interpolation approach directly derives from this hypothesis. It provides the following interpolation relation:

$$\mathbf{s} = \sum_{i=1}^{5} (\, \mathbf{uu}^{(i)} . \mathbf{u}^*)\, \mathbf{s}^{(i)} = \mathbf{uu}^{(i)} \otimes \mathbf{s}^{(i)} : \mathbf{u}^* = \mathbf{C} : \mathbf{u}^* \tag{12}$$

For each domain, the $\mathbf{C}$ matrix is computed once from the stress domain limit vectors $\mathbf{s}^{(i)}$ and the contravariant vectors $\mathbf{uu}^{(i)}$ associated to the 5 strain rate vertices $\mathbf{u}^{*(i)}$. Inside one domain, (12) provides the stress state if the strain rate direction is given. The η-coordinates computed for a given $\mathbf{u}^*$ allow to check the domain validity. If any of them does not belong to the interval [0, 1], then the interpolation of (12) becomes an extrapolation and a new local domain is required.

29

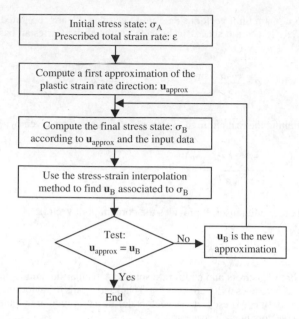

Figure 3. Stress integration scheme.

2.3 *Updating of the scaled yield locus*

When the current local description of the scaled yield locus does not any more cover the interesting zone, one has to find another local description enclosing the interesting part of the yield locus. Of course, the procedure described above could be repeated using one new strain rate direction $\mathbf{u}^*$ as central point. However, looking at the η-coordinate that does not any more belong to [0, 1], one can find the boundary which is not respected by the new explored direction. This boundary is identified by 4 domain limit vectors so that the definition of a new local domain only requires one additional domain limit vector. So, only one new vertex must be computed by Taylor's model to identify the neighbouring domain that probably contains the new explored strain rate direction.

2.4 *Stress integration scheme*

Since the stress–strain interpolation presented above does not use the concept of yield locus in a usual way, a specific integration scheme has been developed. The main ideas are summarized in the diagram of Figure 3 where obviously, no yield locus formulation is used.

At this level, the real stress and not the scaled one is necessary; so the size and the shape of the yield locus cannot any more be dissociated. As (12) only gives the shape corresponding to a reference level of hardening, an additional factor τ is introduced to represent the work hardening by a Swift law:

$$\mathbf{s} = \tau \, \mathbf{C} \cdot \mathbf{u}^* \quad \text{with} \quad \tau = K(\Gamma_0 + \Gamma)^n \tag{13}$$

where Γ is the total polycrystal slip. As in Winters 1996, this micro-macro hardening law is identified by a macroscopic uniaxial tensile test.

2.5 *Implementation of the texture updating*

In this model, not only is the texture used to predict the plastic behavior of the material, but the strain history of each integration point is taken into account in order to update the texture. This is summarized in Figure 4. It should be noticed that the constitutive law in the FEM code is based on the interpolation method described earlier and on Taylor's model applied on the actual set of crystallographic orientations through the yield locus. These crystallographic orientations are represented with the help of the Euler

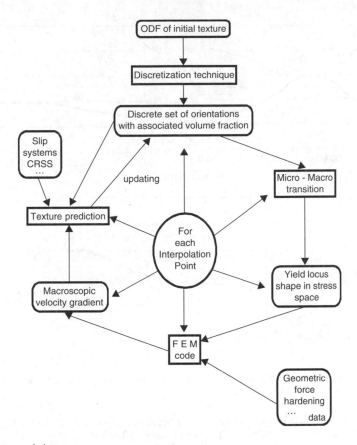

Figure 4. Texture updating.

Table 2. Maximum value of the ODF during deep-drawing.

Steel	Before deep-drawing	After deep-drawing
Mild steel	6	14.34
Dual phase steel	4.12	6.73
Complex phase steel	6.94	8.52

angles, ranging from $0°$ to $360°$ for φ_1 and from $0°$ to $90°$ for ϕ and φ_2 so as to take crystal cubic symmetry into account.

2.6 Deep-drawing simulations

In order to show up the influence of the texture evolution during a forming process, a deep-drawing simulation is examined. Three steels are compared. The first one is a mild steel, the second one is a dual phase steel and the third one is a complex phase steel. Their hardening exponents are respectively 0.2186, 0.2238 and 0.1397. Their tensile yield stresses are 136, 293 and 741 N/mm^2. Their ODF has been measured by X-ray diffraction. The maximum value of this function, i.e. the density of the most represented crystallographic orientation (see Table 2) is an indication of the anisotropy of the studied material.

The geometry of the deep-drawing process consists in a hemispherical punch with a diameter of 100 mm, a die with a curvature radius of 5 mm and a blankholder. The drawing ratio is 1.7, the blankholder force

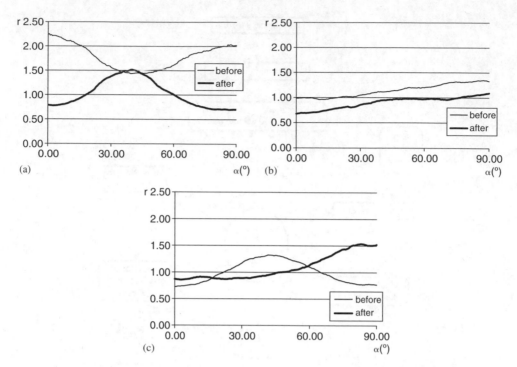

Figure 5. Evolution of the Lankford coefficient "r" for a particular element during the deep drawing process: (a) mild steel (b) dual phase steel (c) complex phase steel.

is 70 kN; the simulation is achieved up to a drawing depth of 50 mm. This geometry has already been used as a benchmark for the NUMISHEET'99 conference. A Coulomb law is used to model friction with a coefficient adapted to each steel. On the finite element mesh, a particular element chosen such that it completely undergoes the drawing process on the curvature of the die is examined. The texture evolution of that element is compared for the three steels. The values of the maximum of the ODF for each steel before and after the process are summarized in Table 2.

The initial anisotropy of the three steels is more or less the same but the behavior of these steels is quite different during deep-drawing, if we focus on the maximum of the ODF (a factor larger than 2 is found at the end of the forming process).

These differences can also be pointed out by the Lankford coefficient "r". This parameter is a good indicator for the drawability of a steel (a high "r" allows larger deep-drawing ratios). Figure 5 shows its evolution during the process. Here again, large differences between the three steels can be noticed. The mild steel is characterised by a high initial r coefficient (inducing a good formability) and a considerable evolution during the simulation. The two other steels have a lower value (around 1.0) and their evolution is also lower. These behaviors are in agreement with the conclusion drawn from the evolution of the maximum of the ODF (see Table 2).

3   ADVANCED MODELING OF CONTINUOUS CASTING

3.1   Context of the studies

Continuous casting is widely used nowadays for production of steel all around the world. Although the process has been developed for several decades, research is still needed to continue improving quality and yield. Two different ways are explored: the first one is the optimization of the caster performance, the second one is the management of existing problems in particular casting conditions.

In a first study, a thermo-mechanical F.E.M. model of the steel solidification in the area of the mould was developed, focusing on the effect of the mould taper on the primary cooling in order to optimise the mould for complex cross sections (beam blanks). Some papers have already been published on this part: see Pascon et al. (2000 & 2001).

In a second time, this model was enhanced to represent the behavior of the steel strand, after the exit of the mould, in the bending as well as in the straightening zone. This study aims at predicting transversal cracking sometimes observed in practice.

3.2 Macroscopic approach of thermo-mechanical behavior

3.2.1 Global approach

A thermo-mechanical macroscopic model has been worked out using a non-linear finite element code, called LAGAMINE, which has been developed since the early eighties at University of Liege for large strains/displacements problems, more particularly for metal forming modeling.

Since a complete 3D discretization of continuous casting seems difficult to manage (because of numerical stability and convergence reasons, and also because of computation time), a 2D½ model has been preferred.

This model belongs to the "slice models" family. With a 2D mesh, a set of material points representing a slice of the steel strand, perpendicular to the casting direction, is modeled. Initially the slice is at the meniscus level and its temperature is assumed to be uniform and equal to the casting temperature. Since this slice is moving down through the machine, heat transfer, stress and strain development as well as solidification growth are studied according to the boundary conditions imposed by the industrial process.

3.2.2 Mechanical model

3.2.2.1 Generalized plane strain state

From a mechanical point of view, the slice is in generalized plane strain state (2D½). That means that the thickness t of the slice is governed by the following equation:

$$t(x, y) = \alpha_0 + \alpha_1 x + \alpha_2 y \tag{14}$$

where $\alpha_0, \alpha_1, \alpha_2$ are degrees of freedom corresponding respectively to the thickness at the origin of the axes (α_0) and the thickness variations along axes x (α_1) and y (α_2), in the plane of the slice.

This formulation allows the development of stresses and strains in the out-of-plane direction, i.e. in the casting direction. It is thus more complete than classical 2D approaches (plane strain or plane stress states) and much less CPU expensive than a 3D approach.

Moreover, it allows the modeling of bending and straightening of the strand, enforcing a relation between the α_i degrees of freedom so that the correct radius of curvature of the machine is respected.

Last but not least, this 2D½ also permits to apply a force in the casting direction, which is of prime importance since one has to take into account the withdrawal force of the strand due friction in different places in the machine.

3.2.2.2 Constitutive law

The mechanical behavior of steel is modeled by an elastic-viscous-plastic law. The elastic part is governed by a classical Hooke's law. In the viscous-plastic domain, a Norton-Hoff type constitutive law is used, the expression of which (in terms of von Mises equivalent values) is:

$$\overline{\sigma} = \sqrt{3}.p_2.e^{-p_1\overline{\varepsilon}}.(\sqrt{3}.\overline{\dot{\varepsilon}})^{p3}.\overline{\varepsilon}^{p4} \tag{15}$$

where $p_{1,2,3,4}$ are temperature dependent parameters, which can be fit on experimental curves. With the usual assumptions of von Mises yield surface, associated plasticity and normality rule, expression (15) becomes, in tensorial form:

$$\dot{\varepsilon}_{ij}^{vp} = \frac{J_2^{p_5}.e^{\frac{p_1}{p_3}.\overline{\varepsilon}}.\overline{\dot{\varepsilon}}^{-\frac{p_4}{p_3}}}{2.\left(K_0.p_2\right)^{\frac{1}{p_3}}} \cdot \hat{\sigma}_{ij} \tag{16}$$

where $J_2 = \frac{1}{2} \cdot \hat{\sigma}_{ij} \cdot \hat{\sigma}_{ij} = \frac{1}{3} \cdot \overline{\sigma}^2$ and $p_5 = \frac{1 - p_3}{2.p_3}$.

The numerical integration of this constitutive law is based on an implicit scheme (see Habraken et al. 1998). All the parameters are thermally affected.

3.2.2.3 Ferrostatic pressure

The liquid pool in the center of the strand applies a pressure on the solidified shell. This pressure is called ferrostatic pressure p_f and it is equal to:

$$p_f = \gamma \cdot D \cdot (1 - f_s) \tag{17}$$

where γ is the volumetric weight of steel, D the depth under the meniscus level and f_s the solid fraction.

Since the studied steel is not a eutectic composition, solidification occurs over a range of temperature limited by the solidus temperature (T_{sol}) and the liquidus one (T_{liq}). A linear variation of the solid fraction according to temperature in this range is assumed:

$$0 \leq f_s(T) = \frac{T_{liq} - T}{T_{liq} - T_{sol}} \leq 1 \quad \forall T \in \left[T_{liq}, T_{sol}\right] \tag{18}$$

3.2.3 Thermal aspects

3.2.3.1 Internal heat conduction

The heat transfer in the material is governed by the classical Fourier's law, expressing the energy conservation and taking into account the release of energy during phase transformation (solidification is exothermic):

$$\frac{\Delta H}{\Delta T} \dot{T} = \nabla \cdot (\lambda \nabla T) \tag{19}$$

where H is the enthalpy, T the temperature and λ the thermal conductivity of the material. The enthalpy H is given by:

$$H = \int \rho \, c \, dT + (1 - f_s) . L_F \tag{20}$$

where ρ is the volumetric mass, c the specific heat and L_F the latent heat of fusion.

3.2.3.2 Thermal shrinkage

Thermal shrinkage ε^{therm} due to solidification is given by:

$$\dot{\varepsilon}_{ij}^{therm} = \alpha(T) . \dot{T} . \delta_{ij} \tag{21}$$

where α is the linear thermal expansion coefficient, which is thermally affected.

3.2.4 Boundary conditions

Different boundary conditions can occur, according to the position of the slice in the machine and to the contact conditions. Table 3 summarizes the different cases:

In case of mechanical contact, the normal pressure is calculated by allowing, but penalizing, the penetration of bodies into each other. The friction τ_c is then computed with Coulomb's friction law:

$$|\tau_c| = \mu . \sigma_n \tag{22}$$

Table 3. Boundary conditions.

Position of the slice in the caster	Contact conditions	Mechanical boundary conditions	Thermal boundary conditions
Primary cooling (slice in the mould)	Contact with the mould	Normal stress + tangential friction	Large heat transfer (direct contact)
	Loss of contact	Free surface (no stress)	Reduced heat transfer (through the slag)
Secondary cooling (under the mould) in the caster	Contact with the rolls	Normal stress + tangential friction	Heat transfer (direct contact with rolls)
	Between the rolls	Free surface (no stress)	Radiation + convection *or* water spray *or* flow

where μ is the friction coefficient and σ_n is the normal stress.

The heat transfer q from the strand to the ambient is given by the simple relation:

$$q = h\left(T_{strand} - T_{ambient}\right) \tag{23}$$

where h is the heat transfer coefficient according to the boundary conditions (see Table 3). In case of contact, h is the inverse of the contact resistance. In case of water spray cooling, h has been experimentally determined for different temperatures (700–1200°C), different shapes of spray, different rates of flow and at different distances from the nozzle.

3.3 Industrial application

A first industrial application was the study of the influence of the mould taper on the cooling rate during the primary cooling: it was reported in Pascon et al. (2000 & 2001). Here we want to study the influence of some local defects (such as nozzle perturbation, roll locking or roll misalignment) on the risk of transversal cracks initiation.

To achieve this, two "macroscopic" indicators of crack initiation are defined. Both combine the gap of ductility of steel in a given range of temperature (T_A–T_B for a steel composition) and the mechanical constraints in the direction of casting: the longitudinal stress for the first indicator I_1 and the longitudinal strain rate for the second one I_2.

$$I_1 = \begin{cases} \max\left(\sigma_{zz};0\right) & T \in \left[T_A;T_B\right] \\ 0 & T \notin \left[T_A;T_B\right] \end{cases}$$

$$I_2 = \begin{cases} \max\left(\dot{\varepsilon}_{zz};0\right) & T \in \left[T_A;T_B\right] \\ 0 & T \notin \left[T_A;T_B\right] \end{cases}$$

These indicators are different from zero only if the temperature corresponds to the gap of ductility and if the constraint tends to open a crack (tensile stress or elongation). In such a case, the higher the constraint, the higher the indicators. In any other case, the indicators are equal to zero, what means that the risk of transverse crack initiation vanishes.

The model provides many results, among others: temperature evolution (surface temperature, evolution of solidification, …), stress and strain states, crack initiation indicators, bulging between rolls, extracting force. Figure 6 represents a part (straightening zone) of the casting of a micro allied steel with standard conditions (without any local defect). This figure shows the value of the 2 indicators and the localization of the maximum values (maximum risk). These results are in agreement with observations on industrial products.

Other computations including local defects (water spray cooling perturbation, roll locking, roll misalignment) have been performed. Comparing the value of the indicators in the reference case (standard conditions) to those of each defect case, we studied the effect of each defect. This comparison allowed us to classify the defects from the less to the most critical.

35

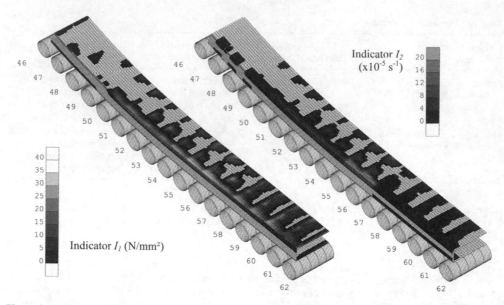

Figure 6.   Indicators of risk of transverse crack initiation with standard casting conditions (reconstituted 3D view of the surface of the cast product – ½ structure because of symmetry).

3.4   *Conclusion*

This study shows how powerful is the 2D½ approach: for the complete thermo-mechanical modeling of the strand, the computation time is of the order of 3 to 4 hours on a PC (Pentium) and accurate results are obtained which are confirmed by industrial observations.

ACKNOWLEDGEMENT

Dr. A.-M. Habraken, senior research associate, acknowledges the Belgian National Research Fund for its financial support.

REFERENCES

Altan, Oh, Gegel, 1983. Metal Forming, Fundamentals and Applications
Habraken, A.M., Charles, J.F., Wegria, J. and Cescotto, S. Dynamic Recrystallisation During Zinc Rolling, In: *Int.J. of Forming Processes*, Vol. 1 – n°1 (1998), 53–73
Munhoven, S., Habraken, A. M., Van Bael, A. and Winters, J. 1996. Anisotropic finite element analysis based on texture. *Proc. 3rd Int. Conf.: NUMISHEET'96*: 112–119
Pascon, F., Habraken, A.-M., Bourdouxhe, M. and Labory, F. Modélisation des phénomènes thermo-mécaniques dans une lingotière de coulée continue, In: *Mec. Ind.* Vol. 1 – n°1 (2000), 61–70 (in French)
Pascon, F., Habraken, A.M., Bourdouxhe, M. and Labory, F. Finite element modeling of thermo-mechanical behavior of a steel strand in continuous casting, In: *Proc. 4th ESAFORM Conference*, April 23–25 2001, Liège (Belgium), 867–870
Winters, J. 1996. *Implementation of Texture-Based Yield Locus into an Elastoplastic Finite Element Code*. Ph.D. Thesis, KUL, Leuven

Computational Methods in Engineering and Science, Iu et al. (eds)
© 2003 Swets & Zeitlinger, Lisse, ISBN 90 5809 567 3

Recent advances in plate vibration research at the University of Hong Kong

Y.K. Cheung
Department of Civil Engineering, The University of Hong Kong, Hong Kong

D. Zhou
Department of Mechanics and Engineering Science, Nanjing University of Science and Technology, China

ABSTRACT: In this paper, the recent advances of study on plate vibrations in The University of Hong Kong are introduced. The research work follows the well-known Rayleigh-Ritz method, which includes three parts. The first part is the study on two-dimensional vibrations of thin plates and shear deformable plates using the static beam functions as the basic functions. The second part is the study on three-dimensional vibrations of thick plates, based on exact elasticity theory and using the Chebyshev polynomials as the basic functions. And the third part is the study on plate-liquid interaction by using a semi-analytical method.

PART I TWO–DIMENSIONAL VIBRATION ANALYSIS OF RECTANGULAR PLATES

1 INTRODUCTION

Using beam eigenfunctions as the admissible functions to study the vibration of rectangular plates is one of the most successful applications of the Ritz method in dynamic analysis of structural elements (Young 1950). In addition, orthogonal polynomials have also shown wide applications in two-dimensional vibration analysis of various structural elements (Bhat 1985).

In recent years, sets of static beam functions have been developed as the admissible functions to analyze the free and forced vibrations of uniform/non-uniform beams and plates with/without internal point-supports or line-supports in the Rayleigh-Ritz method. The studies include uniform rectangular thin plates with line-supports (Zhou & Cheung 1999); uniform rectangular thin plates with elastic intermediate line supports and edge constraints (Cheung & Zhou 2000c); tapered Euler-Bernoulli beams (Zhou & Cheung 2000a); tapered Euler-Bernoulli beams with point-supports (Zhou & Cheung 1998); tapered rectangular thin plates (Cheung & Zhou 1999b); tapered rectangular thin plates with line-supports (Cheung & Zhou 1999a); tapered Timoshenko beams (Zhou & Cheung 2001b); uniform Mindlin rectangular plates (Cheung & Zhou 2000b); uniform Mindlin rectangular plates with intermediate line supports (Zhou et al. 2002c); Tapered Mindlin rectangular plates (Cheung & Zhou 2003). Moreover, the studies have been extended to symmetrically and unsymmetrically laminated composite rectangular plates with line-supports (Cheung & Zhou 2001a, b), point-supports (Cheung & Zhou 1999c) and thick composite rectangular plates with point-supports by using the finite layer method (Zhou et al. 2000).

2 SETS OF STATIC BEAM FUNCTIONS

2.1 *Uniform Euler-Bernoulli beam*

Consider a uniform Euler-Bernoulli beam with J intermediate elastic point-supports at x_j ($j = 1, 2, \ldots, J$) and acted upon by an arbitrarily distributed static load $q(x)$ as shown in Figure 1. The length of the beam is l and its flexural rigidity is EI. The stiffnesses of the intermediate point-supports are defined by k_{Tj}

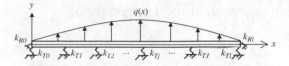

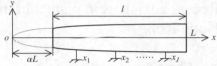

Figure 1. Elastically point-supported beam with elastic end constraints under an arbitrarily distributed load $q(x)$.

Figure 2. Tapered beam with internal point-supports.

($j = 1, 2,…, J$) and the reactions of these point-supports are defined by p_j ($j = 1, 2, …, J$). The rotational and translational stiffnesses of end constraints are defined, respectively, by k_{R0} and k_{T0} at the left end and k_{RI} and k_{TI} at the right end. Defining the following non-dimensional coordinate and parameters

$$\xi = x/l, \quad \xi_j = x_j/l, \quad Q(\xi) = l^4 q(l\xi)/EI; \quad K_{Tj} = l^3 k_{Tj}/EI, \quad K_{R0} = lk_{R0}/EI,$$

$$K_{RI} = lk_{RI}/EI, \quad K_{T0} = l^3 k_{T0}/EI, \quad K_{TI} = l^3 k_{TI}/EI, \quad P_j = l^3 p_j/EI \tag{1}$$

The deflection $y(\xi)$ of the beam should satisfy the following differential equation

$$d^4 y/d\xi^4 = \sum_{j=1}^{J} P_j \delta(\xi - \xi_j) + Q(\xi) \tag{2}$$

where $\delta(\xi - \xi_j)$ is the Dirac delta function.

Assume that the load $Q(\xi)$ can be expanded into a Fourier sinusoidal series as follows

$$Q(\xi) = \sum_{i=1}^{\infty} Q_i (i\pi)^4 \sin(i\pi\xi) \tag{3}$$

where Q_i are the unknown constants. Substituting the above equation into equation (2), the general solution can be easily given by

$$y(\xi) = \sum_{i=1}^{\infty} Q_i y_i(\xi), \quad y_i(\xi) = \sum_{k=0}^{3} C_k^i \xi^k + \sum_{j=1}^{J} P_j^i (\xi - \xi_j)^3 U(\xi - \xi_j)/6 + \sin(i\pi\xi) \tag{4}$$

where $y_i(\xi)$ ($i = 1, 2, 3, …$) are referred to as static beam functions of uniform Euler-Bernoulli beams, $U(\xi - \xi_j)$ is the step function. $P_j^i = P_j/Q_i$ ($j = 1, 2, 3, …, J$) and C_k^i ($k = 0, 1, 2, 3$) are the unknown constants, which can be uniquely determined by the following J intermediate support conditions

$$P_j^i = K_{Tj} y_i(\xi_j), \quad j=1,2,3,…,J, \tag{5}$$

and four boundary conditions

$$K_{R0} dy_i/d\xi = d^2 y_i/d\xi^2, \quad -K_{T0} y_i = d^3 y_i/d\xi^3, \qquad \text{at } \xi = 0,$$

$$-K_{RI} dy_i/d\xi = d^2 y_i/d\xi^2, \quad K_{TI} y_i = d^3 y_i/d\xi^3, \qquad \text{at } \xi = 1 \tag{6}$$

For the rigid intermediate point-supports and/or the classical boundary conditions, one only needs to take the corresponding stiffness coefficients as infinite or zero.

2.2 Non-uniform Euler-Bernoulli beam

Consider a tapered beam with J internal point-supports, respectively, at x_j ($j = 1, 2,…, J$) and acted by an arbitrarily distributed static load $q(x)$ as shown in Figure 2. The length of the sharply ended beam is L and the length of the truncated beam is $l = (1 - \alpha)L$ where α ($0 \leq \alpha < 1$) is referred to as truncation factor. The thickness variation of the beam is described by a power function

$$h(\xi) = h_1 \xi^s \tag{7}$$

where s is referred to as the taper factor, $\xi = x/L$. It is obvious that the flexural rigidity of the beam is $EI = EI_1 \xi^{3s}$ where EI_1 is that at the right end of the beam.

The deflection $y(\xi)$ of the beam should satisfy the following differential equation

$$(d^2/d\xi^2)(\xi^{3s}d^2y/d\xi^2) = \sum_{j=1}^{J} P_j \delta(\xi - \xi_j) + Q(\xi) \tag{8}$$

where the non-dimensional coordinates and parameters are identically defined as those in equation (1). The only difference is that EI is replaced by EI_1 and l by L.

Assume that the load $Q(\xi)$ can be expanded into a Taylor series as follows

$$Q(\xi) = \sum_{i=1}^{\infty} Q_i (\xi - \xi_c)^{i-1} = \sum_{i=1}^{\infty} Q_i \sum_{m=0}^{i-1} (-1)^{i-1-m} a_m^{i-1} \xi_c^{i-1-m} \xi^m \tag{9}$$

where Q_i are unknown constants, ξ_c is the center of the Taylor series expansion and $a_m^{i-1} = (i-1)!/[m!(i-1-m)!]$. Substituting the above equation into equation (8), the general solution can be derived as the follows

$$y(\xi) = \sum_{i=1}^{\infty} Q_i y_i(\xi), \quad y_i(\xi) = \sum_{j=0}^{3} C_i^j \bar{y}_j(\xi) + \sum_{j=1}^{J} P_j^i y_j^p(\xi) + y_i^q(\xi) \tag{10}$$

where $y_i(\xi)$ ($i = 1, 2, 3, \ldots$) are referred to as static beam functions of tapered Euler-Bernoulli beams, $\bar{y}_j(\xi)$ ($j = 0, 1, 2, 3$) are the homogeneous solutions of equation (8), $y_j^p(\xi)$ ($j = 1, 2, \ldots, J$) are the special solutions corresponding to the jth point-support reactions, while $y_i^q(\xi)$ are the special solutions corresponding to the Taylor series loads given in equation (9). As the uniform Euler-Bernoulli beam functions, $J + 4$ unknown constants C_k^i ($k = 0, 1, 2, 3$) and P_j^i ($j = 1, 2, 3, \ldots, J$) (for each i) in equation (10) can be uniquely decided by the J intermediate support conditions and four boundary conditions.

2.3 Uniform Timoshenko beam

Consider a uniform Timoshenko beam with J elastic intermediate point-supports and acted by an arbitrarily distributed static load $q(x)$ as shown in Figure 1. The transverse deflection $y(\xi)$ and the rotation-angle $\psi(\xi)$ satisfy the following differential equation

$$(\psi - dy/dx) = R_x d^2\psi/d\xi^2 \,,$$

$$(d\psi/d\xi - d^2y/d\xi^2) = R_x[Q(\xi) + \sum_{j=1}^{J} P_j \delta(\xi - \xi_j)] \tag{11}$$

In the above two equations, $R_x = EI/(\kappa GAl^2)$ where G is the shear modulus of rigidity, A is the cross-sectional area of the beam and κ is the shear correction factor. In this section, non-dimensional coordinates and parameters defined in equation (1) are still utilized. Under the Fourier sinusoidal loads given in equation (3), the general solution of equation (11) can be given as

$$y(\xi) = \sum_{i=1}^{\infty} Q_i y_i(\xi), \quad \psi(\xi) = \sum_{i=1}^{\infty} Q_i \psi_i(\xi)/l \tag{12}$$

where $y_i(\xi)$ and $\psi_i(\xi)$ ($i = 1, 2, 3, \ldots$) are referred to as static beam functions of uniform Timoshenko beams and

$$y_i(\xi) = C_0^i + C_1^i \xi + C_2^i \xi^2/2 + C_3^i(\xi^3/6 - R_x\xi)$$

$$+ \sum_{j=1}^{J} P_j^i[(\xi - \xi_j)^3/6 - R_x(\xi - \xi_j)]U(\xi - \xi_j) + [R_x(i\pi)^2 + 1]\sin(i\pi\xi) \,,$$

$$\psi_i(\xi) = C_1^i + C_2^i \xi + C_3^i \xi^2/2 + \sum_{j=1}^{J} P_j^i(\xi - \xi_j)^2 U(\xi - \xi_j)/2 + i\pi\cos(i\pi\xi) \tag{13}$$

39

In the above equations, $J + 4$ unknown coefficients C_k^i ($k = 0, 1, 2, 3$) and P_j^i ($j = 1, 2, 3, ..., J$) (for each i) can be uniquely decided by the following J intermediate support conditions

$$P_j^i = K_{Tj} y_i(\xi_j), \qquad j=1,2,3,...,J \tag{14}$$

and four boundary conditions

$$K_{R0}\psi_i = d\psi_i / d\xi, \quad -K_{T0} y_i = \psi_i(\xi) - dy_i / d\xi, \qquad \text{at} \ \ \xi = 0,$$

$$-K_{Rl}\psi_i = d\psi_i / d\xi, K_{Tl} y_i = \psi_i(\xi) - dy_i / d\xi, \text{at} \ \ \xi = 1 \tag{15}$$

2.4 Non-uniform Timoshenko beam

Consider a tapered Timoshenko beam without internal point-supports acted upon by an arbitrarily distributed static load $q(x)$ as shown in Figure 2. When the beam has a thickness variation as defined in equation (7), the governing differential equation is given by

$$R_x d[\xi^s (dy / d\xi - \psi)] / d\xi = Q(\xi) + \sum_{j=1}^{J} P_j \delta(\xi - \xi_j),$$

$$d(\xi^{3s} d\psi / d\xi) / d\xi + R_x \xi^s (dy / d\xi - \psi) = 0 \tag{16}$$

where $R_x = EI_1 / (\kappa GA_1 l^2)$. Bending moment $M(\xi)$ and transverse shear force $V(\xi)$ of the beam are, respectively, given as

$$M(\xi) = (EI_1 / l^2)\xi^{3s} d\psi / d\xi, V(\xi) = (\kappa GA_1 / l)\xi^s (dy / d\xi - \psi) \tag{17}$$

Under the Taylor series loads as defined in equation (9), the general solution of equation (16) can be given as

$$y(\xi) = \sum_{i=1}^{\infty} Q_i y_i(\xi), \psi(\xi) = \sum_{i=1}^{\infty} Q_i \psi_i(\xi) / l \tag{18}$$

where $y_i(\xi)$ and $\psi_i(\xi)$ ($i = 1, 2, 3, ...$) are referred to as static beam functions of non-uniform Timoshenko beams and

$$y_i(\xi) = \sum_{j=0}^{3} C_i^j \bar{y}_j(\xi) + \sum_{j=1}^{J} P_j^i y_j^p(\xi) + y_i^q(\xi),$$

$$\psi_i(\xi) = \sum_{j=0}^{2} C_i^j \bar{\psi}_j(\xi) + \sum_{j=1}^{J} P_j^i \psi_j^P(\xi) + \psi_i^q(\xi) \tag{19}$$

in which, $\bar{y}_j(\xi)$ ($j = 0, 1, 2, 3$) and $\bar{\Psi}_j(\xi)$ ($j = 0, 1, 2$) are the homogeneous solutions of equation (16), $y_j^p(\xi)$ and $\Psi_j^p(\xi)$ ($j = 1, 2, ..., J$) are the special solutions corresponding to the jth point-support reactions, while $y_i^q(\xi)$ and $\Psi_i^q(\xi)$ are the special solutions corresponding to the Taylor series loads given in equation (9). As the uniform Timoshenko beam functions, $J + 4$ unknown constants C_k^i ($k = 0, 1, 2, 3$) and P_j^i ($j = 1, 2, 3, ..., J$) (for each i) in equation (19) can be uniquely decided by J intermediate support conditions and four boundary conditions.

40

Table 1. The comparison of the first three eigenfrequencies.

r, s	Method	Terms	Ω_1	Ω_2	Ω_3
1, 0	STBF	5×5	8.4812	17.583	18.058
	VTBF	5×5	8.6657	18.083	18.769
		8×8	8.5417	17.771	18.287
1, 1	STBF	5×5	4.6896	8.6528	10.556
	VTBF	5×5	4.9265	9.5896	11.355
		8×8	4.7612	8.8453	10.795
2, 0	STBF	5×5	5.3347	8.1223	9.7717
	VTBF	5×5	6.1730	10.820	13.015
		8×8	5.7325	9.3740	11.194

3 AN EXAMPLE

Consider a simply-supported square Mindlin plates tapered in one or two directions. The plate has the truncation factors $\alpha = \beta = 0.1$ and the thickness ratio $h_0/b = 0.4$. The vibrating Timoshenko beam functions (VTBF) and the static Timoshenko beam functions (STBF) are, respectively, taken as the admissible functions to calculate the first three eigenfrequencies where r and s mean the tapered factors of the plate in the x and y directions, respectively. Shear correction factor is taken as $\kappa = 5/6$. A comparison study of the results is given in Table 1. It is shown that the static Timoshenko beam functions have more rapid convergence and higher accuracy than the vibrating Timoshenko beam functions.

PART II   THREE-DIMENSIONAL VIBRATION ANALYSIS OF PLATES

1 INTRODUCTION

In the past three decades, some attempts have been made for 3D vibration analysis of structural elements. The Ritz method is commonly used to study the 3D vibration of structural elements and shows advantages in simplicity, computational cost and accuracy. Leissa and his coworkers used simple algebraic polynomials (Leissa and So 1995) while Liew and his coworkers used generalized orthogonal polynomials (Liew et al. 1998) as admissible functions to analyze the three-dimensional vibration characteristics of structural elements. They have performed a lot of comprehensive studies.

In the recent two years, the authors presented Chebyshev-Ritz method to analyze the 3D free vibration of structural elements. High accuracy, rapid convergence and numerical robustness have been demonstrated. The studies include rectangular plates (Zhou et al. 2002b); solid and annular circular plates (Zhou et al. 2003a); solid and hollow cylinders (Zhou et al. 2003b) and torus (Zhou et al. 2002a) etc. The research work has been extended to triangular plates (Cheung & Zhou 2002b, c) and skew plates.

2 BASIC FORMULATION

According to the linear, small strain 3D elasticity theory, in a general orthogonal coordinate system (α, β, γ) the strain energy P and kinetic energy T of a homogeneous elastic body can be given as

$$P = \frac{1}{2} \iiint_{vol} \sigma^T \varepsilon dv , \qquad T = \frac{1}{2} \iiint_{vol} (\dot{u}^2 + \dot{v}^2 + \dot{w}^2) dv \tag{20}$$

in which, u, v and w are the displacements in the α, β and γ directions, respectively. σ and ε are the stresses and strains, respectively.

For free vibration, the displacements u, v, w can be written as

$$u = U(\alpha, \beta, \gamma)e^{i\omega t} , \quad v = V(\alpha, \beta, \gamma)e^{i\omega t} , \quad w = W(\alpha, \beta, \gamma)e^{i\omega t} \tag{21}$$

41

In such a case, the energy functional L can be written as

$$L = P_{max} - T_{max} \tag{22}$$

Assuming that the solutions can be approximately written in form of finite series as follows

$$U(\alpha,\beta,\gamma) = \sum_{i=1}^{I}\sum_{j=1}^{J}\sum_{k=1}^{K} A_{ijk} U_{ijk}(\alpha,\beta,\gamma), \quad V(\alpha,\beta,\gamma) = \sum_{l=1}^{L}\sum_{m=1}^{M}\sum_{n=1}^{N} B_{lmn} V_{lmn}(\alpha,\beta,\gamma),$$

$$W(\alpha,\beta,\gamma) = \sum_{p=1}^{P}\sum_{q=1}^{Q}\sum_{r=1}^{R} C_{pqr} W_{pqr}(\alpha,\beta,\gamma) \tag{23}$$

Substituting the above equations into equation (20) and minimizing the energy functional L in equation (22) with respect to the unknown coefficients

$$\frac{\partial L}{\partial A_{ijk}} = 0, \qquad \frac{\partial L}{\partial B_{lmn}} = 0, \qquad \frac{\partial L}{\partial C_{pqr}} = 0 \tag{24}$$

the following eigenfrequency equation is obtained

$$\{[K] - \omega^2[M]\}\{X\} = \{0\} \tag{25}$$

3   RITZ SOLUTIONS OF SOME COMMON STRUCTURAL ELEMENTS

3.1   *Rectangular plates*

The displacement functions of rectangular plates in the rectangular coordinate system (x, y, z), as shown in Figure 3, are assumed to be

$$U_{ijk}(\xi,\eta,\zeta) = f_u^1(\xi) f_u^2(\eta) P_i(\xi) P_j(\eta) P_k(\zeta),$$

$$V_{lmn}(\xi,\eta,\zeta) = f_v^1(\xi) f_v^2(\eta) P_l(\xi) P_m(\eta) P_n(\zeta), \tag{26}$$

$$W_{pqr}(\xi,\eta,\zeta) = f_w^1(\xi) f_w^2(\eta) P_p(\xi) P_q(\eta) P_r(\zeta),$$

where $\xi = 2x/a$, $\eta = 2y/b$, $\zeta = 2z/t$ and $P_s(\chi)$ $(s = 1, 2, 3....; \chi = \xi, \eta, \zeta)$ is the one-dimensional sth Chebyshev polynomial which can be written in terms of cosine functions as follows

$$P_s(\chi) = \cos[(s-1)\arccos(\chi)]; \quad (s = 1,2,3,...) \tag{27}$$

The boundary function components $f_\delta^i(\chi)$ $(\delta = u; v; w, i = 1; 2, \chi = \xi; \eta)$ corresponding to different boundary conditions are given in Table 2.

3.2   *Circular plates*

The displacement functions of an annular circular plate in the cylindrical coordinate system (r, θ, z) are assumed to be

$$U_{ijk}(\bar{r},\theta,\bar{z}) = F_u^0(\bar{r}) F_u^1(\bar{r}) P_i(\bar{r}) P_j(\bar{z}) \cos(k\theta),$$

$$V_{lmn}(\bar{r},\theta,\bar{z}) = F_v^0(\bar{r}) F_v^1(\bar{r}) P_l(\bar{r}) P_m(\bar{z}) \sin(n\theta), \tag{28}$$

$$W_{pqr}(\bar{r},\theta,\bar{z}) = F_w^0(\bar{r}) F_w^1(\bar{r}) P_p(\bar{r}) P_q(\bar{z}) \cos(r\theta),$$

Table 2. Boundary functions for different boundary conditions of rectangular plates.

B. C.	$f_u^1(\xi)$	$f_v^1(\xi)$	$f_w^1(\xi)$	$f_u^2(\eta)$	$f_v^2(\eta)$	$f_w^2(\eta)$
F-F	1	1	1	1	1	1
F-S	1	$1-\xi$	$1-\xi$	$1-\eta$	1	$1-\eta$
S-F	1	$1+\xi$	$1+\xi$	$1+\eta$	1	$1+\eta$
S-S	1	$1-\xi^2$	$1-\xi^2$	$1-\eta^2$	1	$1-\eta^2$
F-C	$1-\xi$	$1-\xi$	$1-\xi$	$1-\eta$	$1-\eta$	$1-\eta$
C-F	$1+\xi$	$1+\xi$	$1+\xi$	$1+\eta$	$1+\eta$	$1+\eta$
S-C	$1-\xi$	$1-\xi^2$	$1-\xi^2$	$1-\eta^2$	$1-\eta$	$1-\eta^2$
C-S	$1+\xi$	$1-\xi^2$	$1-\xi^2$	$1-\eta^2$	$1+\eta$	$1-\eta^2$
C-C	$1-\xi^2$	$1-\xi^2$	$1-\xi^2$	$1-\eta^2$	$1-\eta^2$	$1-\eta^2$

Table 3. Boundary functions for different boundary conditions of annular plates.

Boundary conditions	$F_u^0(\bar r)$	$F_v^0(\bar r)$	$F_w^0(\bar r)$	$F_u^1(\bar r)$	$F_v^1(\bar r)$	$F_w^1(\bar r)$
Clamped	$1+\bar r$	$1+\bar r$	$1+\bar r$	$1-\bar r$	$1-\bar r$	$1-\bar r$
Completely free	1	1	1	1	1	1
Hard simply-supported	1	$1+\bar r$	$1+\bar r$	1	$1-\bar r$	$1-\bar r$
Sliding	$1+\bar r$	1	1	$1-\bar r$	1	1
Soft simply-supported	1	1	$1+\bar r$	1	1	$1-\bar r$

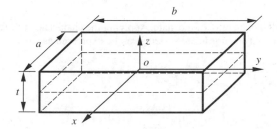

Figure 3. Geometry, dimensions and coordinates of a rectangular plate with uniform thickness.

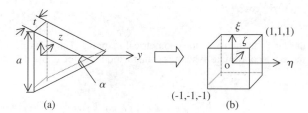

Figure 4. Domain transformation: (a) isosceles triangular plate (b) basic cubic domain.

where $\bar r = 2r/\bar R - \delta$, $\bar z = 2z/t$, $R = R_1 - R_0$ and $\delta = (R_1 + R_0)/(R_1 - R_0)$. The boundary function components $F_\delta^i(\bar r)$ ($\delta = u; v; w, i = 0; 1$) corresponding to different boundary conditions are given in Table 3.

3.3 Isosceles triangular plates

For simplicity, the actual isosceles triangular prism domain is mapped onto a basic cubic domain, as shown in Figure 4, using the following co-ordinate transformation

$$x = a\xi(1-\eta)/4, \quad y = a(1+\cos\alpha)(1+\eta)/(4\sin\alpha), \quad z = t\zeta/2 \tag{29}$$

Applying the chain rule of differentiation, the relation of the first derivative in the two coordinate systems can be expressed as

$$
\left\{\begin{array}{c} \dfrac{\partial()}{\partial x} \\ \dfrac{\partial()}{\partial y} \end{array}\right\} = J^{-1}\left\{\begin{array}{c} \dfrac{\partial()}{\partial \xi} \\ \dfrac{\partial()}{\partial \eta} \end{array}\right\}, \quad \dfrac{\partial()}{\partial z} = \dfrac{t}{2}\dfrac{\partial()}{\partial \zeta}
\tag{30}
$$

where

$$
J = \begin{bmatrix} \dfrac{\partial x}{\partial \xi} & \dfrac{\partial y}{\partial \xi} \\ \dfrac{\partial x}{\partial \eta} & \dfrac{\partial y}{\partial \eta} \end{bmatrix} = \dfrac{a}{4}\begin{bmatrix} 1-\eta & 0 \\ -\xi & (1+\cos\alpha)/\sin\alpha \end{bmatrix}
\tag{31}
$$

in which, J denotes the Jacobian matrix of the geometrical mapping. Through this mapping, the triangular prism domain in the x-y-z coordinate system is transformed into the cubic domain in the $\xi - \eta - \zeta$ coordinate system.

3.4 Skew plates

The actual skew plate domain (side lengths a and b, and skew angle α) can be mapped onto a basic cubic domain using the following coordinate transformation

$$
x = a(\xi + \eta\sin\alpha/\beta + 1 + \sin\alpha/\beta)/2,
$$
$$
y = b(\eta+1)\cos\alpha/2, \quad z = t\zeta/2
\tag{32}
$$

where $\beta = a/b$ is the side-length ratio of the plate. Applying the chain rule of differentiation, the relation of the first derivative in the two coordinate systems can be expressed as

$$
\left\{\begin{array}{c} \dfrac{\partial()}{\partial x} \\ \dfrac{\partial()}{\partial y} \end{array}\right\} = \bar{J}^{-1}\left\{\begin{array}{c} \dfrac{\partial()}{\partial \xi} \\ \dfrac{\partial()}{\partial \eta} \end{array}\right\}, \quad \dfrac{\partial()}{\partial z} = \dfrac{t}{2}\dfrac{\partial()}{\partial \zeta}
\tag{33}
$$

where

$$
\bar{J} = \begin{bmatrix} \dfrac{\partial x}{\partial \xi} & \dfrac{\partial y}{\partial \xi} \\ \dfrac{\partial x}{\partial \eta} & \dfrac{\partial y}{\partial \eta} \end{bmatrix} = \dfrac{a}{2}\begin{bmatrix} 1 & 0 \\ \sin\alpha/\beta & \cos\alpha/\beta \end{bmatrix}
\tag{34}
$$

in which, $\bar{J}$ denotes the Jacobian matrix of the geometrical mapping. Through this mapping, the triangular prism domain in the x-y-z coordinate system is transformed into the cubic domain in the $\xi - \eta - \zeta$ coordinate system.

3.5 Tori

The displacement functions of a torus in the toroidal coordinate system (r, θ, φ), as shown in Figure 5, are assumed to be

$$
U_{ijk}(\bar{r},\theta,\varphi) = P_i(\bar{r})\cos(j\varphi)\left\{\begin{array}{c} \sin(k\theta) \\ \cos[(k-1)\theta] \end{array}\right.,
$$
$$
V_{lmn}(\bar{r},\theta,\varphi) = P_l(\bar{r})\cos(m\varphi)\left\{\begin{array}{c} \cos[(n-1)\theta] \\ \sin(n\theta) \end{array}\right.,
\tag{35}
$$
$$
W_{pqr}(\bar{r},\theta,\varphi) = P_p(\bar{r})\sin(q\varphi)\left\{\begin{array}{c} \sin(r\theta) \\ \cos[(r-1)\theta] \end{array}\right.
$$

where $\bar{r} = 2r/a - 1$.

44

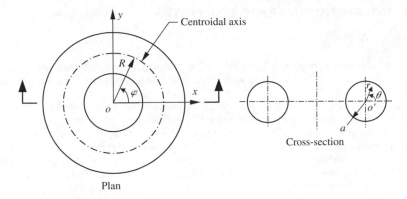

Plan   Cross-section

Figure 5.   A torus with circular cross-section.

Table 4.   Convergence of the higher-order eigenfrequency.

n	$I \times J$	Ω_{10}	Ω_{30}	Ω_{50}	Ω_{70}	Ω_{100}	Ω_{150}	Ω_{200}
Antisymmetric modes in the length direction								
0^a	10×10	9.3580	17.1484	23.6402	29.7285	43.3310	85.2013	–
	20×20	9.3580	17.1416	22.6639	27.4861	33.4351	41.5164	49.4094
	30×30	9.3580	17.1416	22.6639	27.4861	33.4350	41.5121	48.2049
	40×40	9.3580	17.1416	22.6639	27.4861	33.4350	41.5121	48.2049
0^t	10×10	12.0542	21.7238	30.9997	50.1407	163.8190	–	–
	20×20	12.0542	21.3506	27.4215	32.9873	39.6108	50.6521	62.5409
	30×30	12.0542	21.3506	27.4215	32.9867	39.4594	48.9844	56.6826
	40×40	12.0542	21.3506	27.4215	32.9867	39.4594	48.9844	56.6826
1	10×10	7.3911	13.0667	17.3740	21.3332	27.3258	38.1907	61.3575
	20×20	7.3911	13.0665	17.3401	21.0990	25.2579	31.1748	36.2781
	30×30	7.3911	13.0665	17.3401	21.0990	25.2579	31.1747	36.1932
	40×40	7.3911	13.0665	17.3401	21.0990	25.2579	31.1747	36.1932
Symmetric modes in the length direction								
0^a	10×10	9.0126	17.2103	23.9428	30.6621	42.0861	85.3880	280.6543
	20×20	9.0126	17.1088	22.8896	27.1655	33.4954	41.9581	49.8832
	30×30	9.0126	17.1088	22.8896	27.1655	33.4917	41.3004	48.1779
	40×40	9.0126	17.1088	22.8896	27.1655	33.4917	41.3004	48.1779
0^t	10×10	11.0949	21.3033	30.3155	49.5127	–	–	–
	20×20	11.0949	20.9505	26.7446	32.1330	38.6413	49.2256	61.4365
	30×30	11.0949	20.9505	26.7446	32.1330	38.6412	47.9158	55.7145
	40×40	11.0949	20.9505	26.7446	32.1330	38.6412	47.9158	55.7145
1	10×10	7.0682	13.1962	17.2163	21.1384	26.9759	38.6282	61.7339
	20×20	7.0682	13.1962	17.1711	20.7446	25.1702	30.8392	35.8783
	30×30	7.0682	13.1962	17.1711	20.7446	25.1702	30.8391	35.8639
	40×40	7.0682	13.1962	17.1711	20.7446	25.1702	30.8391	35.8639

4   AN EXAMPLE

Consider a free-free hollow circular cylinder with inner-outer radius ratio $R_0/R_1 = 0.25$ and length-radius ratio $H/R_1 = 2.0$. Table 4 gives the convergence of the higher-order eigenfrequencies $\Omega = \omega(\rho/E)^{1/2}$, respectively, for torsional vibration (0^t), axisymmetric vibration (0^a) and circumferential vibration $n = 1$. It is shown the present method has excellent robustness in numerical computation.

45

PART III   INTERACTION OF LIQUIDS AND PLATES

1   INTRODUCTION

Plate-liquid interaction is permanently interested in a lot of researchers. Exact solutions can be obtained only for some special cases and in general, approximate, numerical methods should be used (Zienkiewicz & Bettess 1978).

Recently, the authors developed a semi-analytical method to analyze the interaction of liquids with plates. The analytical method is applied to solve the liquid movement, while the approximate method is applied to solve the plate vibration. The governing eigenvalue equation is obtained by applying the Ritz method to the energy function of the liquid-plate system and the Galerkin method to the surface wave equation. Hydroelastic vibration of vertical rectangular plates in contact with liquid on one side (Zhou & Cheung 2000b), rectangular container bottom plates (Cheung & Zhou 2000a, Zhou & Cheung 2001a) and circular container bottom plate (Cheung & Zhou 2002a) have been studied, respectively. High accuracy and fast convergent rate have been observed.

2   BASIC FORMULATION

Consider a circular cylindrical container or a rectangular container, partially filled with liquid (domain V_f) having a free surface S_f. The bottom of the container is horizontal, which of a part is elastic (domain Ω_1) and the other is rigid (domain Ω_2). Assume that the liquid is inviscid, incompressible and irrotational, and the elastic part of the bottom can be modeled as a thin plate. For the small amplitude oscillation of the liquid, the velocity potential φ of liquid movement satisfies the Laplace equation

$$\nabla^2 \varphi = 0 \tag{36}$$

where ∇^2 is the Laplace operator. In the polar cylindrical coordinates, ∇^2 can be written as

$$\nabla^2 = \frac{\partial^2}{\partial r^2} + \frac{1}{r}\frac{\partial}{\partial r} + \frac{1}{r^2}\frac{\partial^2}{\partial \theta^2} + \frac{\partial^2}{\partial z^2} \tag{37}$$

and in the Cartesian coordinates, ∇^2 can be written as

$$\nabla^2 = \frac{\partial^2}{\partial x^2} + \frac{\partial^2}{\partial y^2} + \frac{\partial^2}{\partial z^2} \tag{38}$$

Taking the effect of the surface waves into account, the sloshing equation of the liquid is

$$\frac{\partial \varphi}{\partial z} + \frac{1}{g}\frac{\partial^2 \varphi}{\partial t^2} = 0, \qquad \text{at } z = H \tag{39}$$

where g is the gravity acceleration.

The consistency condition between the liquid movement and the bottom plate vibration can be given by

$$-\frac{\partial \varphi}{\partial z}\Big|_{z=0} = \begin{cases} 0, & \text{on } \Omega_2 \\ \dfrac{\partial w}{\partial t}, & \text{on } \Omega_1 \end{cases} \tag{40}$$

where w is the dynamic deflection of the plate.

For a container with rigid walls, one has

$$\frac{\partial \varphi}{\partial r}\Big|_{r=R} = 0, \quad \text{for a circular cylindrical container} \tag{41}$$

$$\frac{\partial \varphi}{\partial x}\Big|_{x=0,a} = 0, \quad \frac{\partial \varphi}{\partial y}\Big|_{y=0,b} = 0, \quad \text{for a rectangular container} \tag{42}$$

46

Assuming φ and w can be written as, respectively

$$\varphi = i\omega e^{-i\omega t}\Phi; \qquad w = e^{-i\omega t}W \tag{43}$$

where ω denotes the radian eigenfrequency of the liquid-plate system.

Applying the superposition principle, the velocity potential Φ can be decomposed into two parts

$$\Phi = \overline{\Phi} + \hat{\Phi} \tag{44}$$

where $\overline{\Phi}$ should satisfy the condition of no surface waves while $\hat{\Phi}$ should satisfy the condition of entirely rigid bottom, i.e.

$$\overline{\Phi} = 0, \qquad \text{at } z=H \tag{45}$$

$$\frac{\partial \hat{\Phi}}{\partial z} = 0, \qquad \text{at } z=0 \tag{46}$$

It is noted that $\overline{\varphi}$ and $\hat{\Phi}$ should, individually, satisfy equation (36) and equations (41) or (42). Inserting equation (44) into equations (39) and (40) and taking equations (45) and (46) into account, one has

$$\frac{\partial \Phi}{\partial z} - \frac{1}{g}\omega^2\Phi = \frac{\partial \overline{\Phi}}{\partial z} + \frac{\partial \hat{\Phi}}{\partial z} - \frac{1}{g}\omega^2\hat{\Phi} = 0, \qquad z=H \tag{47}$$

$$\frac{\partial \overline{\Phi}}{\partial z}\Big|_{z=0} = \begin{cases} 0, & \text{on } \Omega_2 \\ W, & \text{on } \Omega_1 \end{cases} \tag{48}$$

The method of separation of variables is applied to solve $\overline{\Phi}$ and $\hat{\Phi}$. For the circular cylindrical container, one has

$$\overline{\Phi} = \sum_{n=0}^{\infty} \overline{\varphi}_n(r,z)\cos(n\theta), \qquad \hat{\Phi} = \sum_{\hat{n}=0}^{\infty} \hat{\varphi}_{\hat{n}}(r,z)\cos(\hat{n}\theta) \tag{49}$$

in which, $\overline{\varphi}_n$ and $\hat{\varphi}_{\hat{n}}$ can be, respectively, given as

$$\overline{\varphi}_n(r,z) = \sum_{m=1}^{\infty} A_{mn}J_n(\varepsilon_{mn}r/R)[\cosh(\varepsilon_{mn}z/R) - \frac{\sinh(\varepsilon_{mn}z/R)}{\tanh(\varepsilon_{mn}H/R)}], \quad \text{for } n \geq 1 \tag{50}$$

$$\overline{\varphi}_0(r,z) = A_{00}(1-z/H) + \sum_{m=1}^{\infty} A_{m0}J_0(\varepsilon_{m0}r/R)[\cosh(\varepsilon_{m0}z/R) - \frac{\sinh(\varepsilon_{mn}z/R)}{\tanh(\varepsilon_{mn}H/R)}] \tag{51}$$

$$\hat{\varphi}_{\hat{n}}(r,z) = \sum_{\hat{m}=1}^{\infty} D_{\hat{m}\hat{n}}J_{\hat{n}}(\varepsilon_{\hat{m}\hat{n}}r/R)\cosh(\varepsilon_{\hat{m}\hat{n}}z/R), \qquad \text{for } \hat{n} \geq 1 \tag{52}$$

$$\hat{\varphi}_0(r,z) = D_{00} + \sum_{\hat{m}=1}^{\infty} D_{\hat{m}0}J_0(\varepsilon_{\hat{m}0}r/R)\cosh(\varepsilon_{\hat{m}0}z/R) \tag{53}$$

In the above equations, A_{mn} and $D_{\hat{m}\hat{n}}$ ($m, n, \hat{m}, \hat{n} = 0, 1, 2, \ldots$) are the unknown constants. ε_{mn} ($m = 1, 2, 3, \ldots, n = 0, 1, 2, \ldots$) are the roots of the first-order derivative of Bessel function of the first kind of order n, i.e.

$$J_n'(\varepsilon_{mn}) = 0, m=1,2,3,\ldots; n=0,1,2,\ldots \tag{54}$$

where ( ) denotes the derivative to the argument.

For the rectangular container, one has

$$\overline{\Phi} = \sum_{m=0}^{\infty} \sum_{n=0}^{\infty} B_{mn} \cos(m\pi x / a) \cos(n\pi y / b) F_{mn}(z) \tag{55}$$

$$\hat{\Phi} = \sum_{\hat{m}=0}^{\infty} \sum_{\hat{n}=0}^{\infty} C_{\hat{m}\hat{n}} \cos(\hat{m}\pi x / a) \cos(\hat{n}\pi y / b) \cosh(q_{\hat{m}\hat{n}} z / H) \tag{56}$$

where B_{mn} and $C_{m\hat{n}}$ $(m, n, \hat{m}, \hat{n} = 0, 1, 2, \ldots)$ are the unknown coefficients and

$$F_{mn}(z) = 1 - z / H, \quad \text{for } m=n=0 \tag{57}$$

$$F_{mn}(z) = \cosh(q_{mn} z / H) - \frac{\sinh(q_{mn} z / H)}{\tanh(q_{mn})}, \quad \text{for } m \neq n \neq 0 \tag{58}$$

in which

$$q_{mn} = \pi \sqrt{(mH / a)^2 + (nH / b)^2} \tag{59}$$

A set of local coordinates ξ, η is developed to describe the deflection of the elastic plate. It is assumed that the dynamic deflection of a plate with arbitrary shape can be written as

$$W = \sum_{i=1}^{\infty} \sum_{j=1}^{\infty} G_{ij} W_{ij}(\xi, \eta) \tag{60}$$

where G_{ij} $(i, j = 1, 2, 3, \ldots)$ are the unknown coefficients and $W_{ij}(\xi, \eta)$ are the vibrating shape functions of the plate in air, which can be obtained through various approaches such as analytical method, Ritz method or finite element method.

Substituting equation (44) into equation (48), and utilizing the Bessel-Fourier or trigonometric-Fourier expansions to the two sides of equation (48), one has

$$A_{mn} = \sum_{i=1}^{\infty} \sum_{j=1}^{\infty} G_{ij} Q_{ijmn}, \qquad B_{mn} = \sum_{i=1}^{\infty} \sum_{j=1}^{\infty} G_{ij} \tilde{Q}_{ijmn} \tag{61}$$

where Q_{ijmn} and $\tilde{Q}_{ijmn}$ are the integrals concerned with $W_{ij}(\xi, \eta)$.

3 EIGENVALUE EQUATION

The kinetic energy T_f of the liquid in container is defined by the volume integral

$$T_f = \frac{1}{2} \rho_f \iiint_{V_f} (\nabla \varphi)^2 \, dv \tag{62}$$

where ρ_f is the mass density of the liquid and ∇ is the gradient operator.

Considering equations (41) or (42) and applying the Green's theorem, the maximum kinetic energy T_f^* of the liquid can be transformed into a surface integral surrounding the liquid domain

$$T_f^* = \frac{1}{2} \omega^2 \rho_f \oiint_{S_f} \Phi \nabla \Phi \bullet \bar{n} ds \tag{63}$$

Considering equation (47), the maximum potential energy $\overline{U}_f$ of the liquid can be given by

$$\overline{U}_f = \frac{1}{2} \rho_f g \iint_{S_t} (\frac{\partial \Phi}{\partial z})^2 \, ds = \frac{1}{2} \rho_f \omega^2 \iint_{S_t} \frac{\partial \Phi}{\partial z} \hat{\Phi} ds \tag{64}$$

Considering equations (47) and (48), equation (64) can be written as

$$T_f^* = \frac{1}{2} \rho_f \omega^2 (\iint_{\Omega_1} \Phi W ds + \iint_{S_f} \frac{\partial \Phi}{\partial z} \hat{\Phi} ds) = \overline{T}_f + \overline{U}_f \tag{65}$$

48

where

$$\overline{T}_f = \frac{1}{2}\rho_f\omega^2\left(\iint_{\Omega_1}\overline{\Phi}Wds + \iint_{\Omega_1}\hat{\Phi}Wds\right) \tag{66}$$

Substituting equation (60) and the solutions $\overline{\Phi}$ and $\hat{\Phi}$ given in equations (49)–(59) into equation (66), one has

$$\overline{T}_f = \frac{1}{2}\omega^2\left(\sum_{i=1}^{\infty}\sum_{j=1}^{\infty}G_{ij}\sum_{\hat{i}}^{\infty}\sum_{\hat{j}}^{\infty}G_{\hat{i}\hat{j}}\widetilde{M}_{ij\hat{i}\hat{j}} + \rho_f\sum_{\hat{m}=0}^{\infty}\sum_{\hat{n}=0}^{\infty}D_{\hat{m}\hat{n}}\sum_{i=1}^{\infty}\sum_{j=1}^{\infty}G_{ij}\widetilde{Q}_{ij\hat{m}\hat{n}}\right) \tag{67}$$

where $\hat{M}_{ij\hat{i}\hat{j}}$ are the coefficients concerned with Q_{ijmn}.

It is well known that the maximum kinetic energy $\overline{T}_p$ of the elastic plate can be written as

$$\overline{T}_p = \frac{1}{2}\rho_p\omega^2\sum_{i=1}^{\infty}\sum_{j=1}^{\infty}G_{ij}\sum_{\hat{i}}^{\infty}\sum_{\hat{j}}^{\infty}G_{\hat{i}\hat{j}}M_{ij\hat{i}\hat{j}} \tag{68}$$

where ρ_p is the mass density of the plate.

The maximum strain energy $\overline{U}_p$ of the plate can be written as

$$\overline{U}_p = \frac{1}{2}\sum_{i=1}^{\infty}\sum_{j=1}^{\infty}G_{ij}\sum_{\hat{i}}^{\infty}\sum_{\hat{j}}^{\infty}G_{\hat{i}\hat{j}}K_{ij\hat{i}\hat{j}} \tag{69}$$

Define the energy function Π as follows

$$\Pi = \overline{U}_f + \overline{U}_p - (T_f^* + \overline{T}_p) = \overline{U}_p - (\overline{T}_f + \overline{T}_p) \tag{70}$$

Minimizing the energy function Π with respect to the unknown coefficients $G_{ij}(i, j = 1, 2, 3, \ldots)$ and taking i; $\hat{i}$ from 1 to I, j; $\hat{j}$ from 1 to J, m, $\hat{m}$ from 1 to M and n; $\hat{n}$ from 1 to N, one has

$$\sum_{i=1}^{I}\sum_{j=1}^{J}[K_{ij\hat{i}\hat{j}} - \omega^2(M_{ij\hat{i}\hat{j}} + \widetilde{M}_{ij\hat{i}\hat{j}})]G_{ij} +$$

$$\rho\omega^2\sum_{\hat{m}=0}^{M}\sum_{\hat{n}=0}^{N}D_{\hat{m}\hat{n}}\widetilde{Q}_{ij\hat{m}\hat{n}} = 0, \quad \hat{i} = 1,2,\ldots,I; \hat{J} = 1,2,\ldots,J \tag{71}$$

It is obvious that the above equation cannot be solved until $D_{\hat{m}\hat{n}}$ have been given. However, equation (47) would provide a set of additional Galerkin equations as follows

$$\sum_{i=1}^{I}\sum_{j=1}^{J}V_{ij\hat{m}\hat{n}}G_{ij} + \overline{V}_{\hat{m}\hat{n}}D_{\hat{m}\hat{n}} - \omega^2\widetilde{V}_{\hat{m}\hat{n}} = 0, \quad \hat{m} = 0,1,2,\ldots,M; \hat{n} = 0,1,2,\ldots,N \tag{72}$$

Combining equations (72) and (73), the eigenfrequencies and the corresponding mode shapes can be obtained by using the standard eigenvalue programs. It is clear that if the effect of the liquid surface waves is ignored, the eigenvalue equation degenerates into

$$\sum_{i=1}^{I}\sum_{j=1}^{J}[K_{ij\hat{i}\hat{j}} - \omega^2(M_{ij\hat{i}\hat{j}} + \widetilde{M}_{ij\hat{i}\hat{j}})]G_{ij} = 0 \tag{73}$$

4 AN EXAMPLE

Consider a circular plate clamped at the bottom of the container and with radius R1, thickness h and flexural stiffness D. The liquid depth-radius ratio $H/R = 1$, the liquid-plate density ratio ρ_f/ρ_p 5 0.28, the plate thickness-radius ratio $h/R_1 = 0.05$ and the Poisson's ratio n 5 0.3 are used in the analysis. Tables 5 and 6 give the first three dimensionless eigenfrequencies $\lambda_i = \omega_i R^2(\rho_p h/D)^{1/2}$ (i = 1, 2, 3) of bulging modes and sloshing modes for the axisymmetric vibration, respectively. Four different flexural

Table 5. The first three dimensionless eigenfrequencies λ of bulging modes for the axisymmetric vibration.

σ	t	λ_1	λ_2	λ_3
0.1	0.1	4.8059	26.191	66.817
	1	4.7460	26.189	66.816
	10	4.7397	26.188	66.816
	100	4.7391	26.188	66.816
1	0.1	5.3578	25.454	65.603
	1	4.7891	25.434	65.601
	10	4.7178	25.432	65.600
	100	4.7217	25.432	65.600

Table 6. The first three dimensionless eigenfrequencies λ of sloshing modes for the axisymmetric vibration.

σ	t	λ_1	λ_2	λ_3
0.1	0.1	6.1887	8.3759	10.086
	1	1.9565	2.6487	3.1896
	10	0.61872	0.83759	1.0086
	100	0.19566	0.26487	0.31896
1	0.1	6.1904	8.3759	10.086
	1	1.9565	2.6487	3.1896
	10	0.61872	0.83759	1.0086
	100	0.19566	0.26487	0.31896

stiffness-gravity ratio $t = D/(g\rho_p R_3^1 h)$ 5 0.1, 1, 10, 100 and two different plate-container radius ratio $\sigma = R_1/R = 0.1$, 1 are considered. From these tables, it is shown that the effect of surface waves decreases with the increase of the flexural stiffness-gravity ratio. Generally, the interaction of the bulging modes and the sloshing modes can be neglected unless the plate is extremely flexible (t , 10).

REFERENCES

Bhat, R.B. 1985. Natural frequencies of rectangular plates using characteristic orthogonal polynomials in the Rayleigh-Ritz method. *Journal of Sound and Vibration* 102: 493–499.
Cheung, Y.K. & Zhou, D. 1999a. Eigenfrequencies of tapered rectangular plates with line supports. International Journal of Solids and Structures 36(1): 143–166.
Cheung, Y.K. & Zhou, D. 1999b. Free vibrations of tapered rectangular plates using a new set of beam functions in Rayleigh-Ritz method. *Journal of Sound and Vibration* 223(5): 703–722.
Cheung, Y.K. & Zhou, D. 1999c. The free vibrations of rectangular composite plates with point-supports using static beam functions. *Composite Structures* 44(1–2): 145–154.
Cheung, Y.K. & Zhou, D. 2000a. Coupled vibratory characteristics of a rectangular container bottom plate. *Journal of Fluids and Structures* 14(3): 339–357.
Cheung, Y.K. & Zhou, D. 2000b. Vibrations of moderately thickness rectangular plates in terms of a set of static Timoshenko beam functions, *Computers and Structures* 78(6): 757–768.
Cheung, Y.K. & Zhou, D. 2000c. Vibrations of rectangular plates with elastic intermediate line-supports and edge constraints. *Thin-walled Structures* 37(4): 305–331.
Cheung, Y.K. & Zhou, D. 2001a. Vibration analysis of symmetrically laminated rectangular plates with intermediate line-supports. *Computers and Structures* 79(1): 33–41.
Cheung, Y.K. & Zhou, D. 2001b. Free vibrations of rectangular unsymmetrically laminated composite plates with internal line-supports. *Computers and Structures* 79(20–21): 1923–1932.
Cheung, Y.K. & Zhou, D. 2002a. Hydroelastic vibration of a circular container bottom plate using the Galerkin method. *Journal of Fluids and Structures* 16(4): 561–580.
Cheung, Y.K. & Zhou, D. 2002b. Three-dimensional vibration analysis of cantilevered and completely free isosceles triangular plates. *International Journal of Solids and Structures* 39(3): 673–687.

Cheung, Y.K. & Zhou, D. 2002c. Three-dimensional vibration analysis of right triangular thick plates using Chebyshev polynomial in Ritz method. *ACMSM*17: 24–26, June.

Cheung, Y.K. & Zhou, D. 2003. Vibrations of tapered Mindlin plates in terms of static Timoshenko beam functions. *Journal of Sound and Vibration* 260(4): 693–709.

Leissa, A.W. & So, J. 1995. Comparisons of vibration frequencies for rods and beam from 1-D and 3-D analysis. *Journal of the Acoustical Society of America* 98(4): 2122–2135.

Liew, K.M., Hung, K.C. & Lim, M.K. 1998. Vibration of thick prismatic structures with three-dimensional flexibility. *ASME Journal of Applied Mechanics* 65(3): 619–625.

Young, D. 1950. Vibration of rectangular plates by the Ritz method. *ASME Journal of Applied Mechanics* 17(1): 448–453.

Zhou, D., Au, F.T.K., Cheung, Y.K. & Lo, S.H. 2003a. Three-dimensional vibration analysis of circular and annular plates via the Chebyshev-Ritz method. *International Journal of Solids and Structures*, in press.

Zhou, D., Au, F.T.K., Lo, S.H. & Cheung, Y.K. 2002a. Three-dimensional vibration analysis of a torus with circular cross-section. *Journal of the Acoustical Society of America* 112(6): 2831–2840.

Zhou, D. & Cheung, Y.K. 1998. Eigenfrequencies of tapered beams with intermediate point supports. *International Journal of Space Structures* 13(2): 87–95.

Zhou, D. & Cheung, Y.K. 1999. Free vibration of line supported rectangular plates using a set of static beam functions. *Journal of Sound and Vibration*. 223(2): 231–245.

Zhou, D. & Cheung, Y.K. 2000a. The free vibration of a type of tapered beams. *Computer Methods in Applied Mechanics and Engineering* 188(1–3): 203–216.

Zhou, D. & Cheung, Y.K. 2000b. Vibration of vertical rectangular plates in contact with water on one side. *Earthquake Engineering and Structural Dynamics* 29(5): 693–710.

Zhou, D. & Cheung, Y.K. 2001a. Vibrations of elastic bottom plate of a sealed rectangular container fully filled with liquid. *International Journal of Structural Stability and Dynamics* 1(1): 145–162.

Zhou, D. & Cheung, Y.K. 2001b. Vibrations of tapered Timoshenko beams in terms of static Timoshenko beam functions. *ASME Journal of Applied Mechanics* 68(4): 596–602.

Zhou, D., Cheung, Y.K., Au, F.T.K. & Lo, S.H. 2002b. Three-dimensional vibration analysis of thick rectangular plates using Chebyshev polynomial and Ritz method. *International Journal of Solids and Structures* 39(26): 6339–6353.

Zhou, D., Cheung, Y.K. & Kong, J. 2000. Free vibration of thick, layered rectangular plates with point supports by finite layer method. *International Journal of Solids and Structures* 37(10): 1483–1499.

Zhou, D., Cheung, Y.K., Lo, S.H. & Au, F.T.K. 2003b. 3-D vibration analysis of solid and hollow circular cylinders via Chebyshev-Ritz method. *Computer Methods in Applied Mechanics and Engineering*, in press.

Zhou, D., Lo, S.H., Au, F.T.K. & Cheung, Y.K. 2002c. Vibration analysis of rectangular Mindlin plates with internal line supports using static Timoshenko beam functions. *International Journal of Mechanical Sciences* 44(12): 2503–2522.

Zienkiewicz, O.C. & Bettess, P. 1978. Fluid-structure dynamic interaction and wave forces, an introduction to numerical treatment. *International Journal of Numerical Methods in Engineering* 13(1): 1–16.

Instabilities in deformation processes of solids

Ioannis Doltsinis

Faculty of Aerospace Engineering, University of Stuttgart, Germany

ABSTRACT: Concepts related to the loss of stability in quasistatic deformation of solids are summarized from the point of view of the discretized representation by finite elements and the numerical following up of the process. Two classes of material models are considered: incrementally nonlinear materials as elastic–plastic and quasi-brittle damaging ones and nonlinear viscous solids. In either case the issues of interest regard the uniqueness of the solution, the stability of the state and of the deformation path, but the respective approaches depend on the material model. The various situations that can arise at a state marked as critical for the case of continuing loading in the incrementally nonlinear solid are pointed out and their significance for the deformation path is discussed. Stability of viscous deformation is examined by means of the temporal evolution of perturbations in the actual reference process. Reflection on the numerical stability of the time integration of viscous deformation is included.

1   INTRODUCTION

The stability of deformation processes of solids is examined for various types of material including elastic–plastic (Hill 1959, Rice 1976, Neilsen & Schreyer 1993) viscous (Biot 1965, Hart 1969, Stüwe 1998) and quasi-brittle resp. brittle damaging (Krajcinovic 1996). The analytical background is exposed from the point of view of computer process simulation and numerical analysis.

The issue of stability is investigated with respect to perturbations of the deformation path of the process (Petryk 2000). Uniqueness of solution and global stability (Bigoni 2000) as discussed for the discretized representation of the deforming solid base on the momentary tangent response operator of the system. The transition to local conditions for bifurcation and instability focuses on the constitutive level, intrinsic to the material (Bigoni 2000, Petryk 2000). The incremental nonlinearity arising from the essential difference between momentary inelastic loading and elastic unloading, inherent to elastic–plastic as well as brittle-damaging material behaviour, requires careful interpretation towards an assessment of states marked as critical during the course of the evolving deformation process because of a singular tangent operator associated with continuing loading. Depending on the direction in deformation space of the appertaining eigenvector the state can be unstable, or it allows for a multitude of solutions, resp. it is not critical in conjunction with the applied loading sequence.

If the material is viscous, stability is studied by the amplification of deviations from the reference deformation path in the passage of time. The evolution of disturbances depends on both the momentary viscosity matrix of the system and the geometrical stiffness. The former describes the sensitivity of the stress resultants with respect to variations in velocity, the latter that with respect to variations in geometry. In addition, a dependence of the applied loads on the deforming geometry of the solid is of relevance. The quantities entering an investigation of stability are as for the inviscid solid, but the essential difference of the two material models implies distinct concepts. The interference between physical and numerical instability in the computer simulation of the deformation process is outlined. It is seen that uniqueness of the velocity solution is prerequisite for stability against deviations from the reference path, and stability of the physics is a necessary condition for stability in numerics.

The subsequent considerations refer to quasistatic deformation processes of discretized, finite element systems (Doltsinis 2003) governed by the condition of equilibrium

$$S(\sigma, X) = P(t, X),$$ (1)

between the vector $\mathbf{P}$ of applied loads and the stress resultants $\mathbf{S}$ at the mesh nodal points at each instant t during the course of the deformation process. The vector $\mathbf{X} = {}^{\circ}\mathbf{X} + \mathbf{U}$ defines the geometry evolving from the initial ${}^{\circ}\mathbf{X}$ with the displacement $\mathbf{U}$, the vector σ defines the actual stress state in the system. The material constitutive law that relates stress with deformation is important for the process and for the investigation of stability.

2 ELASTIC–PLASTIC SOLID

The nature of the elastic–plastic constitutive approach suggests consideration of the rate equilibrium

$$\dot{\mathbf{S}} = \mathbf{K}(\dot{\mathbf{U}})\dot{\mathbf{U}} = \dot{\mathbf{P}}. \tag{2}$$

Here, the momentary tangent response matrix $\mathbf{K} = (\mathbf{K}_M + \mathbf{K}_G)$ can be detailed as follows

$$\dot{\mathbf{S}}(\sigma,\mathbf{X}) = \frac{\partial \mathbf{S}}{\partial \sigma}\dot{\sigma} + \frac{\partial \mathbf{S}}{\partial \mathbf{X}}\dot{\mathbf{U}} = (\mathbf{K}_M + \mathbf{K}_G)\dot{\mathbf{U}}, \tag{3}$$

the constituents distinguishing between changes experienced by the material reflected in $\mathbf{K}_M$ and those of purely geometrical origin (even in $\dot{\sigma}$) accounted for by the geometric stiffness matrix $\mathbf{K}_G$. Equally the applied loads

$$\dot{\mathbf{P}}(t,\mathbf{X}) = \frac{\partial \mathbf{P}}{\partial t} + \frac{\partial \mathbf{P}}{\partial \mathbf{X}}\dot{\mathbf{U}} = \frac{\partial \mathbf{P}}{\partial t} + \mathbf{K}_L\dot{\mathbf{U}}, \tag{4}$$

where $\mathbf{K}_L = \varpi\mathbf{P}/\varpi\mathbf{X}$ is known as the load correction matrix. It adds to the momentary tangent response matrix in the case of displacement-sensitive loading.

2.1 *Uniqueness of solution*

At given state and loading, two different solutions $\dot{\mathbf{U}}_1$ and $\dot{\mathbf{U}}_2$ of the rate equilibrium problem, eqn (2), can exist if

$$\mathbf{K}(\dot{\mathbf{U}}_1)\dot{\mathbf{U}}_1 - \mathbf{K}(\dot{\mathbf{U}}_2)\dot{\mathbf{U}}_2 = \mathbf{0}, \tag{5}$$

a displacement sensitivity of the applied loads included in the response matrix $\mathbf{K}$. A sufficient condition for the exclusion of multiple solutions can thus be stated as a work expression in the form

$$(\dot{\mathbf{U}}_1 - \dot{\mathbf{U}}_2)^t[\mathbf{K}(\dot{\mathbf{U}}_1)\dot{\mathbf{U}}_1 - \mathbf{K}(\dot{\mathbf{U}}_2)\dot{\mathbf{U}}_2] \neq 0 \qquad \text{for all } \dot{\mathbf{U}}_2 \neq \dot{\mathbf{U}}_1. \tag{6}$$

In elasticity, where the material behaviour is incrementally linear, the tangent response matrix is $\mathbf{K} \neq \mathbf{K}(\dot{\mathbf{U}})$ and uniqueness requires $\mathbf{K}$ to be a non-singular matrix. In elastic–plastic deformation the option of elastic unloading from a plastic state implies incremental nonlinearity. The fact that $\mathbf{K} = \mathbf{K}(\dot{\mathbf{U}})$ limits the utility of the uniqueness condition, but for the exclusion of an alternative solution $\dot{\mathbf{U}}_2 = \mathbf{0}$. Therefore, $\mathbf{K}(\dot{\mathbf{U}})$ must be a non-singular matrix for any $\dot{\mathbf{U}} \neq \mathbf{0}$. For further reaching conclusions Hill introduced a linear comparison solid adhering to the plastic loading branch at the state under investigation (Hill 1959). The concept has been extended by Raniecki to the use of non-associated plastic flow (Raniecki 1979) that deviates from the normality rule. Introduction of the linear, in loading comparison solid replaces the actual response matrix by $\mathbf{K}(\dot{\mathbf{U}}) \Leftarrow \tilde{\mathbf{K}}$ and the uniqueness condition becomes

$$(\dot{\mathbf{U}}_1 - \dot{\mathbf{U}}_2)^t[\mathbf{K}(\dot{\mathbf{U}}_1)\dot{\mathbf{U}}_1 - \mathbf{K}(\dot{\mathbf{U}}_2)\dot{\mathbf{U}}_2] \geq (\dot{\mathbf{U}}_1 - \dot{\mathbf{U}}_2)^t\tilde{\mathbf{K}}(\dot{\mathbf{U}}_1 - \dot{\mathbf{U}}_2) \neq 0 \qquad \text{for all } \dot{\mathbf{U}}_2 \neq \dot{\mathbf{U}}_1, \tag{7}$$

which requires the momentary tangent response matrix $\tilde{\mathbf{K}}$ of the system in loading to be non-singular.

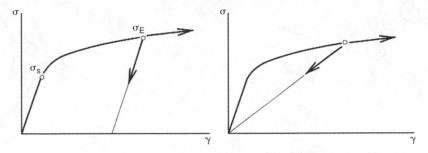

Figure 1. Loading and unloading branches in elastoplasticity (left) and quasi-brittle damage (right).

2.2 *Stability of equilibrium*

Departures from the equilibrium state under examination by a perturbation $\tau\dot{\mathbf{U}}$ of the displacement field in the discretized solid implies an increment of work in association with the stress resultants $\mathbf{S}$,

$$\Delta W_S = \tau\dot{\mathbf{U}}^t\mathbf{S} + \frac{1}{2}\tau^2\dot{\mathbf{U}}^t\dot{\mathbf{S}} + \cdots,\tag{8}$$

and analogously for the applied loads $\mathbf{P}$,

$$\Delta W_P = \tau\dot{\mathbf{U}}^t\mathbf{P} + \frac{1}{2}\tau^2\dot{\mathbf{U}}^t\dot{\mathbf{P}} + \cdots,\tag{9}$$

both approximated to the second order. The time increment τ measures the magnitude of the disturbance. Because of the assumed equilibrium prior to the perturbation the first terms in the work expressions are equal. Thus, to the second order

$$\Delta W_S - \Delta W_P = \frac{1}{2}\tau^2\dot{\mathbf{U}}^t(\dot{\mathbf{S}} - \dot{\mathbf{P}}),\tag{10}$$

where $\dot{\mathbf{S}} = \mathbf{K}(\dot{\mathbf{U}})\dot{\mathbf{U}} \neq \mathbf{P}$ since the imposed perturbation does not constitute a solution of the rate equilibrium. A sufficient condition for static stability in the energetic sense is obtained by the statement

$$\Delta W_S - \Delta W_P > 0 \Rightarrow \dot{\mathbf{U}}^t\big[\mathbf{K}(\dot{\mathbf{U}})\dot{\mathbf{U}} - \dot{\mathbf{P}}\big] > 0 \quad \text{for any } \dot{\mathbf{U}} \neq \mathbf{0}.\tag{11}$$

For stationary loads the above condition reduces to

$$\dot{\mathbf{U}}^t\mathbf{K}(\dot{\mathbf{U}})\dot{\mathbf{U}} > 0 \qquad (\partial\mathbf{P}/\partial t = \mathbf{0}),\tag{12}$$

which requires that the response matrix $\mathbf{K}(\dot{\mathbf{U}})$ is positive definite, a possible displacement sensitivity of the applied loads included. It is seen that this stability condition is contained in the uniqueness statement of eqn (6), if $\dot{\mathbf{U}}_2 = \mathbf{0}$.

2.3 *Analysis of the deformation path*

Observing along the deformation path as it evolutes the tangent response matrix $\widetilde{\mathbf{K}}$ appertaining to the solid in loading, let at a certain state the matrix become positive semi-definite with eigenvalues $\widetilde{\lambda}$ and associated eigenvalues $\widetilde{\mathbf{q}}$ ordered as

$$0 = \widetilde{\lambda}_1 < \widetilde{\lambda}_2 \leq \cdots \leq \widetilde{\lambda}_k, \qquad \widetilde{\mathbf{q}}_1, \widetilde{\mathbf{q}}_2, \cdots, \widetilde{\mathbf{q}}_k.\tag{13}$$

If the response matrix is unique as in elasticity, the state is unstable in that the solution of the rate problem is indeterminate. In solids responding differently to loading and unloading like elastic–plastic and quasi-brittle ones there are several possibilities to distinguish as a consequence of the incremental

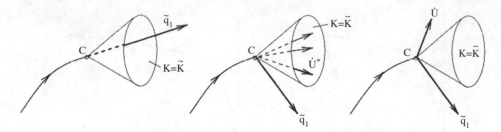

Figure 2.   Different situations at critical state.

nonlinearity. For deformation rates that induce continuing loading of the material it is $\mathbf{K} = \tilde{\mathbf{K}}$, whereas $\mathbf{K} \neq \tilde{\mathbf{K}}$ along paths that imply unloading from an inelastic state.

Let the eigenvector $\tilde{\mathbf{q}}_1$ associated with the eigenvalue $\tilde{\lambda}_1 = 0$ of the singular matrix $\tilde{\mathbf{K}}$ point to a direction of continuing loading. Then deformation $\dot{\mathbf{U}} = \beta \tilde{\mathbf{q}}_1$ along $\tilde{\mathbf{q}}$ is possible at constant applied load, and the state under examination is unstable. If the displacement rate $\dot{\mathbf{U}}^*$ along a different path of continuing inelastic loading of the material is associated with the rate $\dot{\mathbf{P}}^*$ of external loads, then a multitude of paths $\dot{\mathbf{U}}^* + \beta \tilde{\mathbf{q}}_1$ can be associated to the same $\dot{\mathbf{P}}^*$ as long as they maintain the loading condition for the inelastic material. Instability and bifurcation are relevant until the properties of $\tilde{\mathbf{K}}$ are altered by the ensuing deformation of the solid. At the same time deformation paths for which $\mathbf{K} \neq \tilde{\mathbf{K}}$ are not affected by the circumstances; they imply partial local unloading from inelastic states in the solid.

If the eigenvector $\tilde{\mathbf{q}}_1$ points into a deformation path that implies unloading from inelastic states, the appertaining response matrix is $\mathbf{K} \neq \tilde{\mathbf{K}}$ Then, deformation at constant applied load is not feasible and the state is stable. But, with a single path $\dot{\mathbf{U}}^*$ of continuing loading there are several other solutions $\dot{\mathbf{U}} = \beta \tilde{\mathbf{q}}_1$ for the solid in loading at the same rate of applied loads, and the state is one of stable bifurcation. Deformation along one of the paths in loading is stable and feasible, however, the solid bifurcates towards the region of uniqueness by partial unloading, thus overcoming the critical state. For deformation along directions governed by $\mathbf{K} \neq \tilde{\mathbf{K}}$, on the other hand, the state is not at all critical. In the numerical treatment of the deformation process critical states may be hidden if the solution method bases on updating the positive-definiteness of the tangent operator as does the Broyden–Fletscher–Goldfarb–Shanno (BFGS) algorithm (Press et al. 2002). In this case the deformation evolutes automatically along a non-critical path.

3   VISCOUS SOLID

Viscosity relates the stress to the deformation rate in the material. The stress resultants $\mathbf{S}$ thus depend on the velocity $\mathbf{V} = \dot{\mathbf{U}}$ which is governed by the condition of equilibrium, eqn (1), during the course of quasistatic deformation:

$$\mathbf{S}(\sigma, \mathbf{X}) = \mathbf{S}(\mathbf{V}, \mathbf{X}) = \mathbf{P}(t, \mathbf{X}). \tag{14}$$

The functional dependence of the stress resultants on the velocity can be presented in the form

$$\mathbf{S}(\mathbf{V}, \mathbf{X}) = \mathbf{D}(\mathbf{V}, \mathbf{X})\mathbf{V}, \tag{15}$$

where $\mathbf{D}(\mathbf{V}, \mathbf{X})$ is known as the viscosity matrix of the finite element system. The appearance of the velocity $\mathbf{V}$ reflects nonlinearity of the viscous material behaviour.

The positions in space that the mesh nodal points assume during the course of the deformation process follow integration in time of the velocity

$$\mathbf{X} = {}^{\circ}\mathbf{X} + \int_{{}^{\circ}t}^{t} \mathbf{V} \, \mathrm{d}t, \tag{16}$$

which is effected numerically by incrementation.

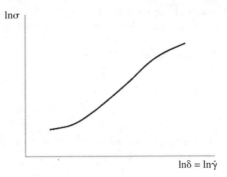

$\ln\delta = \ln\dot\gamma$

Figure 3. Viscous flow stress as a function of the deformation rate.

3.1 Uniqueness of velocity

Co-existence of two different velocity solutions V_1 and V_2 at a certain instant t during the course of deformation, requires that the equation

$$\mathbf{D}(\mathbf{V}_1)\mathbf{V}_1 - \mathbf{D}(\mathbf{V}_2)\mathbf{V}_2 = \mathbf{0}, \tag{17}$$

is fulfilled, geometry $\mathbf{X}$ and load $\mathbf{P}$ fixed. A sufficient condition for uniqueness stated in the work form reads

$$(\mathbf{V}_1 - \mathbf{V}_2)^t[\mathbf{D}(\mathbf{V}_2)\mathbf{V}_2 - \mathbf{D}(\mathbf{V}_1)\mathbf{V}_1] \neq 0 \qquad \text{for all } \mathbf{V}_2 \neq \mathbf{V}_1. \tag{18}$$

For a linear viscous solid $\mathbf{D} \neq \mathbf{D}(\mathbf{V})$, uniqueness requires a non-singular viscosity matrix. In the non-linear case exclusion of the solution $\mathbf{V}_2 = \mathbf{0}$ requires that the matrix $\mathbf{D}(\mathbf{V})$ is nonsingular. Further reaching conclusions as via the in loading comparison material for elastic–plastic and quasi-brittle solids are not possible since it is the value of the velocity and not the direction that here enters the nonlinearity. Small variations in the velocity solution can be excluded if

$$\delta\mathbf{V}^t\frac{\partial\mathbf{S}}{\partial\mathbf{V}}\delta\mathbf{V} \neq 0 \qquad \text{for all } \delta\mathbf{V} \neq \mathbf{0}, \tag{19}$$

which is associated with a non-singular derivative matrix $\partial\mathbf{S}/\partial\mathbf{V}$.

3.2 Stability of deformation

Displacement of a viscous solid from a state of quasistatic equilibrium will induce deformation deviating from the actual path as a function of time. The deviation may decay or amplify depending on the stability of the state. Let at a certain instant during the course of the motion a perturbation $\eta = \delta\mathbf{X}$ appear in the geometry of the solid, which implies a perturbation $\dot\eta = \delta\mathbf{V}$ in the velocity. Both, the actual state $\mathbf{X}$, $\mathbf{V}$ and the perturbated one $\mathbf{X} + \eta$, $\mathbf{V} + \dot\eta$, comply with the same boundary conditions so that for small perturbations the condition of equilibrium gives

$$\frac{\partial\mathbf{S}}{\partial\mathbf{V}}\dot\eta + \frac{\partial\mathbf{S}}{\partial\mathbf{X}}\eta = \frac{\partial\mathbf{P}}{\partial\mathbf{X}}\eta. \tag{20}$$

This relates the perturbations in velocity and geometry as

$$\dot\eta + \mathbf{N}\eta = \mathbf{0} \quad \text{where} \quad \mathbf{N} = \left(\frac{\partial\mathbf{S}}{\partial\mathbf{V}}\right)^{-1}\left(\frac{\partial\mathbf{S}}{\partial\mathbf{X}} - \frac{\partial\mathbf{P}}{\partial\mathbf{X}}\right). \tag{21}$$

The perturbation after a small interval in time τ is obtained by linearization

$$\eta(t+\tau) = (\mathbf{I} - \tau\mathbf{N})\eta(t). \tag{22}$$

57

The deformation process is stable at the considered instant if the perturbation decays. Taking the Euclidean norm resp. magnitude $\|\eta\|$ as a scalar measure for it, the requirement becomes

$$\eta(t+\tau)\| = \|(\mathbf{I} - \tau\mathbf{N})\eta(t)\| \leq \|(\mathbf{I} - \tau\mathbf{N})\| \|\eta(t)\| < \|\eta(t)\|. \tag{23}$$

Hence, for stability

$$\|(\mathbf{I} - \tau\mathbf{N})\| = \sigma(\mathbf{I} - \tau\mathbf{N}) < 1, \tag{24}$$

where σ is the spectral norm of the amplification matrix $\mathbf{M} = (\mathbf{I} - \tau\mathbf{N})$ for the perturbation

$$\sigma(\mathbf{M}) = \sqrt{\max \lambda_i(\mathbf{M}^t\mathbf{M})}, \tag{25}$$

and λ_i denotes the eigenvalues of the symmetric, positive semi-definite form $\mathbf{M}^t\,\mathbf{M}$.

Integration of the evolution equation (21) with $\mathbf{N}$ = const. furnishes

$$\eta = \exp(-\tau\mathbf{N})\eta_0 = \sum_{i=1}^{m} e^{-\tau\lambda_i} c_i \mathbf{q}_i. \tag{26}$$

Here, η_0 denotes the perturbation at the instant of appearance, τ the time elapsed since then, $\lambda_i = \lambda_i(\mathbf{N})$ $(i = 1, \ldots, m)$ the m eigenvalues of the matrix $\mathbf{N}$, and $\mathbf{q}_i$ the associated eigenvectors. The last expression relies on the representation of η_0 by the eigenvectors $\mathbf{q}_i(\mathbf{N})$ as

$$\eta_0 = \sum_{i}^{m} c_i\,\mathbf{q}_i. \tag{27}$$

Decay of the perturbation requires that the eigenvalues of the matrix $\mathbf{N}$ are all positive. Thus,

$$\lambda_i(\mathbf{N}) > 0 \quad \text{for stability}. \tag{28}$$

The matrix $\mathbf{N}$ is real but not necessarily symmetric, and can exhibit complex eigenvalues. In that case stability demands the real part of the eigenvalues to be positive.

Expanding in eqn (26) the exponential into a power series and retaining the linear terms gives

$$\eta = (\mathbf{I} - \tau\mathbf{N})\eta_0, \tag{29}$$

which corresponds to the previous local expression for the development of the perturbation by eqn (22). At this point, it is worth noticing that the amplification matrix and its properties depend on the time interval τ as well, a fact underlining the local nature of this stability statement. Furthermore, the existence of the matrix $\mathbf{N}$ is warranted by the uniqueness in the small of the velocity problem, see eqns (21) and (19).

3.3 Integration of motion

An incremental scheme that is frequently employed for the temporal integration of the motion along the deformation path reads

$$^b\mathbf{X} = {}^a\mathbf{X} + (1 - \varsigma)\tau \,{}^a\mathbf{V} + \varsigma\tau \,{}^b\mathbf{V}. \tag{30}$$

The time increment is $\tau = {}^bt - {}^at$, the value of the parameter $0 \leq \zeta \leq 1$ is chosen as for accuracy and numerical stability.

The issue of numerical stability refers to the integration of the equation of quasistatic motion as governed by the momentary equilibrium condition for the viscous solid. It is solved for the velocity at the beginning $t = {}^at$ and the end $t = {}^bt$ of the incremental step. Stability examines the sensitivity of the approximate integration to numerical disturbances of the solution. A numerical disturbance $^a\eta$ at the beginning of the increment is propagated by the integration scheme to $^b\eta$ at the end, as

$$^b\eta = {}^a\eta + (1-\varsigma)\tau\,{}^a\dot{\eta} + \varsigma\tau\,{}^b\dot{\eta}\,. \tag{31}$$

At the same time, the equilibrium condition equally holding for the actual and the disturbed solution relates η and $\dot{\eta}$ via the matrix $\mathbf{N}$ as by eqn (21). Substitution in eqn (31) gives

$$^b\eta = \left(\mathbf{I} + \varsigma\tau\,{}^b\mathbf{N}\right)^{-1}\left[\mathbf{I} - (1-\varsigma)\tau\,{}^a\mathbf{N}\right]{}^a\eta = \mathbf{M}\,{}^a\eta\,, \tag{32}$$

and the incremental amplification matrix of the disturbance is

$$\mathbf{M} = \left(\mathbf{I} + \varsigma\tau\,{}^b\mathbf{N}\right)^{-1}\left[\mathbf{I} - (1-\varsigma)\tau\,{}^a\mathbf{N}\right]. \tag{33}$$

Stability demands the magnitude of η to decay, which restricts the spectral norm of the amplification matrix to $\sigma(\mathbf{M}) < 1$. This requirement establishes a relationship between the length of the time increment τ and the parameter ζ of the integration scheme. More specific, assume the matrix $\mathbf{M} = \text{const.}$ and compose the initial disturbance η_0 as a linear combination of the eigenvalues of the matrix. Then, after n incremental steps

$$\eta_n = \mathbf{M}^n\eta_0 = \sum_{i=1}^{m}\lambda_i^n c_i\mathbf{q}_i\,. \tag{34}$$

The disturbance will decay with progressing n if the magnitude of all eigenvalues $\lambda_i\,(\mathbf{M})$ is less than unity. This leads to a condition that involves the eigenvalues of the matrix $\mathbf{N}$:

$$0 < \tau\lambda_i(\mathbf{N}) < \frac{2}{1 - 2\varsigma}\,. \tag{35}$$

Accordingly, stability of the numerical integration requires that the eigenvalues $\lambda_i(\mathbf{N})$ are all positive, as for physical stability of the system, and restricts in addition the time increment in dependence of the selected value for the parameter ζ.

The velocity solution has not been questioned that far. Implicit integration involves solution of eqn (14) for the velocity at the end of the incremental step, where the geometry follows by the integration scheme of eqn (30). The possibility of small variations in velocity can be excluded if

$$\left(\delta\,{}^b\mathbf{V}\right)^{\mathrm{t}}{}^b\!\left[\frac{\partial\mathbf{S}}{\partial\mathbf{V}} + \varsigma\tau\left(\frac{\partial\mathbf{S}}{\partial\mathbf{X}} - \frac{\partial\mathbf{P}}{\partial\mathbf{X}}\right)\right]\left(\delta\,{}^b\mathbf{V}\right) \neq 0 \quad \text{for all} \quad \delta\,{}^b\mathbf{V} \neq 0\,, \tag{36}$$

which requires the matrix expression in the brackets to be non-singular. For explicit integration ($\zeta = 0$) the requirement is as for uniqueness, eqn (19).

4 CONCLUSIONS

In elastic–plastic deformation, uniqueness of the solution and stability of a state along the path depend on the properties of the momentary response matrix of the rate problem. The response matrix accounts for material characteristics, and for geometrical effects on stress state and applied loads. The possibility of local elastic unloading from a plastic state is significant for the investigation of situations marked as critical on the assumption of continuous loading of the material. As a consequence of this, the situation may be indicative of instability, of a stable bifurcation, or be not at all critical in conjunction with the actually applied load. The considerations are of interest to quasi-brittle damaging materials as well, for which the elastic unloading branch distinctly differs from that of continuing damage.

In viscous deformation, uniqueness of the velocity solution depends on the viscosity matrix of the system. Nonlinearity of the material model restricts the utility of the statement to small variations of the solution. Stability examines the temporal evolution of perturbations and involves the relationship between perturbation and its time rate as deriving from the condition of equilibrium that governs the quasistatic motion. Physical stability is seen to be a necessary condition for the numerical stability of the time integration of viscous deformation processes.

REFERENCES

Bigoni, D. 2000. Bifurcation and instability of non-associative elastoplastic solids. In H. Petryk (ed.), *Material instabilities in elastic and plastic solids, CISM courses and lectures – No. 414*. Wien New York: Springer.

Biot, M.A. 1965. *Mechanics of incremental deformations*. New York: Wiley.

Doltsinis, I. 2003. *Large deformation processes of solids – From fundamentals to computer simulation and engineering applications*. Southampton: WIT Press.

Hart, E.W. 1967. Theory of the tensile test. *Acta Met.* 15: 351–355.

Hill, R. 1959. Some basic principles in the mechanics of solids without a natural time. *J. Mech. Phys. Solids* 7: 209–225.

Krajcinovic, D. 1996. *Damage mechanics*. Amsterdam: North-Holland, Elsevier.

Neilsen, M.K. and Schreyer, H.L. 1993. Bifurcations in elastic–plastic materials. *Int. J. Solids Structures* 30: 521–544.

Petryk, H. 2000. Theory of material instability in incrementally nonlinear plasticity. In H. Petryk (ed.), *Material instabilities in elastic and plastic solids, CISM Courses and Lectures-No. 414*. Wien New York: Springer.

Press, W.H., Teukolsky, S.A., Vetterling, W.T. and Flannery, B.P. 2002. *Numerical recipes in C++*, second edition, Cambridge: Cambridge University Press.

Raniecki, B. 1979. Uniqueness criteria in solids with non-associated plastic flow laws at finite deformations. *Bull. Acad. Polon. Sci. Sèr. Sci. Techn.* 27: 391–399.

Rice, J.R. 1976. The localization of plastic deformation. In W.T. Koiter (ed.), *Theoretical and applied mechanics*. Amsterdam: North-Holland.

Stüwe, H.P. 1998. Examples of strain localization. In P. Perzyna (ed.), *Inelastic solids, CISM Courses and Lectures – No. 386*. Wien New York: Springer.

On various numerical simulation techniques for perforation of concrete panel by hard projectile

S.C. Fan, X.Q. Zhou & Y.Y. Jiao
School of Civil & Environmental Engineering, Protective Technology Research Centre,
Nanyang Technological University, Singapore

ABSTRACT: This paper presents different numerical simulations of perforation process of a concrete slab impacted by a high-speed steel projectile. Perforation is a highly dynamic event. Results are obtained by different numerical methods, including the Finite Different Method (FDM), Discrete Element Method (DEM) and one of the meshless method called "Smooth Particle Hydrodynamics" (SPH). Their merits and demerits will be illustrated and discussed. In addition, a sound constitutive model for concrete material is presented.

1 INTRODUCTION

For protective concrete structures, the impact by a missile draws more attention to the phenomena of penetration and perforation. Other than experimental field tests, various methods have been developed to predict the phenomena. Till now, simple analytical methods or empirical formulae derived from experimental results remain popular. Those methods and formulae have been extensively used to predict the penetration depth, the perforation velocity, and the scabbing effect at the exit face due to the impact of rigid projectiles (Williams 1994). On the other hand, the advent of digital computer technology has encouraged the development of sophisticated numerical simulation techniques. It enables those highly dynamic events to be investigated through computerized simulations, at least partially. Results can be obtained by different numerical methods, including the Finite Element Method (FEM), Finite Different Method (FDM), Discrete Element Method (DEM) and one of the meshless methods called "Smooth Particle Hydrodynamics" (SPH). Nevertheless, a sound material model is essential to obtain good predictions.

In this work, a multi-surface stress-state model is employed to model the concrete target. Firstly, it is used to simulate perforation of a concrete panel by a steel rod using FDM continuum model. However, it is well known that the capability of a continuum model is limited. Its intrinsic assumptions prevent it from reproducing some phenomena often observed during the field tests, such as concrete spalling at the impact face, scabbing at the rear face, and shear plug formation ahead of the projectile. Against this background, the SPH method is used to model the perforation of concrete panel by hard projectile. The SPH technique is one of more mature meshless methods. Although it can reflect the phenomena of spalling and scabbing, nevertheless, results are not very satisfactory. Because of the circular shape of the particles, the flying parts somehow behave as sand particles (Wingate & Stellingwerf 1993, Johson 1996). Finally, the capabilities of DEM are studied. Illustrative examples are included.

2 PHYSICAL PHENOMENA

The physical phenomena associated with a projectile penetrating/perforating through a concrete target are very complicated. The nature of the deformations taking place in both the projectile and the target during the penetration process needs to be understood. It involves the material properties of both the projectile and the target. It also depends on the impact velocities and the nose shape of the projectile. In the penetration process, the stress waves initiated by the impact play an important role. The subsequent propagation of the stress waves creates a series of events including spalling, scabbing and fracture of the target.

When a projectile impacts upon a target, three types of projectile-response can happen. In the first type, the projectile ricochets, i.e. the projectile rebounds from the impacted surface or it skins through the impacted surface with a reduced velocity. In the second type, the projectile penetrates into the target without completing its passage through the body. In the third type, perforation occurs when the projectile has sufficient energy to penetrate through the target. Very often, perforation of the target (especially for relatively thin targets) can occur even at impact velocity well below that required to achieve complete penetration. It is probably due to the scabbing effect, which reduces the effective thickness of the target.

On the other hand, plain concrete is generally strong in compression and relatively weak in tension. Therefore the compressive wave generated in the target does not cause much damage to the target. The failure due to impact is governed by the dynamic tensile strength of concrete. At the impacting face, spalling occurs at the periphery of the impact area where maximum tensile stresses exist. When the compression pulse reaches the rear face of the target, it reflects as a tensile wave. The tensile wave travels back into the target and away from the rear face. Since concrete material is weak in withstanding such large tension, scabbing occurs close to the rear face (Laible 1980).

3 MAJOR DIFFERENCES AMONGST DIFFERENT METHODS

Riding on the advent of computer technology, various numerical methods have been extending their limits of problem-solving capability. Amongst them, the robustness of the Finite Element Method (FEM) is well recognized. Essentially, it obtains solution through discretization of the integral equation defined over a mesh of non-overlapping sub-domains. On the other hand, the FDM solves the problem by discretization of the governing differential equations directly over a set of points. These two methods are much more mature than the others. What follows will focus on the SPH and DEM.

3.1 *SPH method*

The SPH method is a node-based method. Be it in Lagrange or Euler formulation, it establishes some kinds of connectivity between nodes/particles and subsequently constructs the spatial derivatives. Within the problem domain, each node/particle "I" interacts with all other nodes/particles "J" which are within a given distance (usually assumed to be $2h$) from it. The distance h is called the smoothing length. The interaction is weighted by a function $W(x - x', h)$ which is called the kernel function. Using this principle, the value of a continuous function, or its derivatives, can be estimated at any node "I" based on known values at the surrounding nodes "J" through the following kernel approximants:

$$f(x) = \int f(x')W(x-x',h)dx' \tag{1}$$

$$\nabla \bullet f(x) = \int \nabla \bullet f(x')W(x-x',h)dx' \tag{2}$$

where f is a function of the three dimensional position vector x, and dx' is incremental volume.

By discretization of the integral equations, the continuous volume integrals can be converted to sums over discrete interpolation points. Eqs.(1) and (2) can be expressed in several forms. The commonly used symmetric formulation yields:

$$f(x^I) \approx -\rho^I \sum_{J=1}^{N} m^J \left(\frac{f(x^I)}{(\rho^I)^2} + \frac{f(x^J)}{(\rho^J)^2} \right) \bullet W(x^I - x^J, h) \tag{3}$$

$$\nabla \bullet f(x^I) \approx -\rho^I \sum_{J=1}^{N} m^J \left(\frac{f(x^I)}{(\rho^I)^2} + \frac{f(x^J)}{(\rho^J)^2} \right) \bullet \nabla W(x^I - x^J, h) \tag{4}$$

where the gradient ∇W is with respect to x^J; m denotes the mass; and ρ denotes the density.

It is worth noting that in the above derivation, no connectivity or spatial relation of the interpolation points is assumed. In addition, numerous viable choices of the kernel function are available. Very often, particularly in many hydrocodes, cubic B-spline kernel is employed.

3.2 The DEM

The DEM is an element-based method. Similar to the FEM, the problem domain is cut into sub-domains, namely blocks. The major characteristic is that the formulation is built over block-centers and inter-block surfaces. Some of the inter-block surfaces coincide with pre-existed or evolved joint faces within the domain mass. Discretization is applied to the governing differential equation over all block-centers as follows:

$$\dot{u}_I^{(t+\Delta t/2)} = \dot{u}_I^{(t-\Delta t/2)} + \left\{ \sum F_I^{(t)} + \alpha \left| \sum F_I^{(t)} \right| sign \, (\dot{u}_I^{(t-\Delta t/2)}) \right\} \frac{\Delta t}{m_I} \qquad (5)$$

where $\dot{u}_I^{(t+\Delta t)}$ and $\dot{u}_I^{(t-\Delta t)}$ are the velocities of block-center I at time $t + \Delta t$ and $t - \Delta t$ respectively. F_I^t is the resultant force at block-center I at time t. α is the damping coefficient, and m is the block mass. The resultant force F_I^t is derived from the contact forces at the inter-block surfaces. This formulation facilitates simulation of inter-block movements. Block separation occurs when the threshold stresses are exceeded. Usually, no failure is considered within a block. Instead of putting in place an algorithm for evolution of intra-block cracking surface, it is easier to have a more refined element mesh such that cracks evolve along the inter-block surfaces.

4   CONSTITUTIVE MODEL FOR CONCRETE MATERIAL

In this work, construction of the model is based on phenomenological behaviors at the macroscopic level. The constitutive model includes some or all of following constitutive relations: an equation of state (EOS) which relates the pressure to the mass density and internal energy, a deviatoric elastic constitutive relationship, a yield criterion and flow rule governing the plastic deformations, and a failure or damage model.

The hydrostatic and deviatoric stresses are de-coupled and represented by tensor invariants. The hydrostatic part is described either by the "Hugoniot" shock pressure/specific volume relationship (EOS), or piecewise-linear EOS. The deviatoric stresses are described using the so-called loading surface, failure envelope and residual strength surface. The surfaces or envelope are defined in the 3D principle stress space. Very often, the thermal effect due to the rises of internal temperature is so insignificant that it can be ignored (in contrast to metal targets).

In this section, the multi-surface stress-state model employed for concrete material is described. It governs the evolution of deviatoric stress and strain. The distinct feature lies in the construction of the failure criterion, which is based on the twin-shear strength theory. In the 3D stress space, stress states are demarcated by a few surfaces. The first is the elastic limit surface that is the envelope to all admissible elastic stress states. Beyond that, the stress state becomes plastic. The second is the maximum strength surface, namely the failure surface, which is the boundary limit to all admissible plastic stress states. Upon reaching the maximum stress state, damage begins to creep in and the strength starts weakening. Though it may survive on further damages, it suffers gradual degradation of strength. Ultimate fracture occurs when the stiffness is degraded to the unbearable states, defined by a residual strength surface. What follows will describe how to construct these critical surfaces.

4.1   Construction of the failure surface in stress space

The failure surface is defined first. A strategy of semi-empirical or semi-theoretical way of construction is adopted. A pair of empirical curves is chosen (Kosotvos 1995) and put in place as the skeleton meridians, and then the whole surface is completed via the twin-shear (TS) strength theory (Yu 1998). It is known and proved theoretically that the strength surface is 6-fold (6-sectorial) symmetric about the hydrostatic axis. In other words, the whole surface is a continuous repeated pattern of a typical 60° sectorial surface. The pair of Kotsovos meridian curves represents the failure stress states at 0° (tensile) and 60° (compressive) respectively. Then the 60°-sectorial surface is constructed by strength theory. The 3D TS strength criterion is established using two shear parameters and a unified weighted parameter "b", thus called unified Twin Shear strength theory. The construction details can be found in reference (Fan and Wang 2002). The two-step formulation is shown briefly below.

Step 1: Construction of the dynamic tensile and compressive meridian: r_t and r_c
Basically, the two meridians in the present model are obtained by applying the Dynamic Increase Factor (DIF) to the Kotsovos empirical static meridians with some modifications. We note that at high

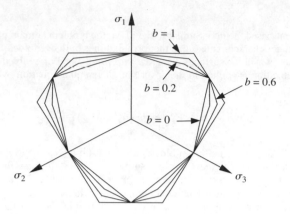

Figure 1.   Trajectories of failure surface by UTSS theory in a deviatoric plane.

pressure, the Kotsovos compressive meridian underestimates the experimental results and thus it is revised by multiplying a factor of 1.25 as follows:

$$r_t = 0.633\sqrt{3}\left(0.05 - \frac{\xi}{\sqrt{3}}\right)^{0.857} \times DIF \qquad\qquad \theta = 0° \tag{6}$$

$$r_c = 1.25 \times 0.944\sqrt{3}\left(0.05 - \frac{\xi}{\sqrt{3}}\right)^{0.724} \times DIF \qquad \theta = 60° \tag{7}$$

The DIF reflects the effect of strain rate. Experimental data are available and regression expression for the relationship can be employed (Williams, 1994). Values of DIF in Eqs. (6) and (7) are linear functions of the tensile DIF ($TDIF$) and compressive DIF ($CDIF$), depending on whether the stress state falls in the T-T zone, the T-C zone or the C-C zone (detail shown later).

Step 2: Construction of the 60°-sectorial surface between r_t and r_c meridians
The 60°-sectorial envelope is completed by piece-wise linear formulation between the two skeleton meridian curves in terms of a unified weighted parameter "b" (Fan and Wang 2002) (see Fig. 1). The cases of $b = 0$ and $b = 1$ correspond to the lower and upper limits of the convex shaped envelope admissible in the twin-shear strength theory. Actually, the parameter "b" is a material parameter, which has different values for different materials. For a specific material (with known "b" value), the envelope is defined as follows:

$$r_f = \frac{r_t r_c \sin 60°}{r_t \sin\theta + r_c \sin(60° - \theta)}(1 - b) + b\frac{r_t}{\cos\theta} \qquad \text{when} \qquad 0° \leq \theta \leq \theta_b \tag{8}$$

$$r_f = \frac{r_t r_c \sin 60°}{r_t \sin\theta + r_c \sin(60° - \theta)}(1 - b) + b\frac{r_c}{\cos(60° - \theta)} \qquad \text{when} \qquad \theta_b \leq \theta \leq 60° \tag{9}$$

where $\theta_b = \arctan\left[\frac{1}{\sqrt{3}}\left(\frac{2r_c}{r_t} - 1\right)\right]$, $\cos 3\theta = \frac{3\sqrt{3}}{2}\frac{J_3}{\sqrt{J_2^3}}$. J_2 and J_3 are the stress invariants.

4.2   Construction of the elastic-limit surface and loading surfaces

The elastic-limit surface is also called initial yield surface. A loading surface is an expanded yield surface taken into account the strain-hardening in association with the development of the effective plastic strain ε_p. It coincides with the initial yield surface when $\varepsilon_p = 0$. It expands non-uniformly with increasing ε_p until it reaches and coincides with the failure surface. In the present model, both the initial yield surface

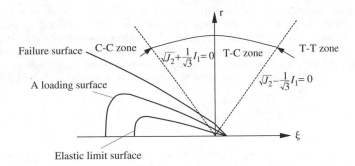

Figure 2. Construction of elastic-limit surface, loading surface and failure surface.

and the loading surface are capped at certain high hydrostatic pressures. The initial yield surface and the subsequent loading surfaces can be expressed in the same general form as follows:

$$r = \kappa r_f F_{cap} \tag{10}$$

where r is a function of the deviatoric stresses ($r = \sqrt{2J_2}/f_c$) normalized with respect to the uni-axial compressive strength f_c. The value of κ varies between κ_0 and 1 corresponding to the initial yield surface and the failure surface, respectively. F_{cap} is a dimensionless function, which defines the cap at high hydrostatic pressures as shown in Figure 2.

According to the stress state, each loading surface is sub-divided into three zones (see Fig. 2), namely the tension-tension (T-T) zone, the tension–compression (T–C) zone and the compression–compression (C–C) zone.

In the T-T zone, $\sqrt{J_2} - 1/\sqrt{3}\, I_1 < 0$. I_1 is the first stress invariant. The strain-hardening parameter κ is derived from uni-axial tensile test data as follows:

$$\kappa = \kappa_1 = \frac{\sigma_t}{f_t} = \kappa_{t0} + H_t' \varepsilon_p / f_t \tag{11}$$

where f_t is the static uni-axial tensile strength of concrete; H_t' is a hardening parameter determined experimentally; and κ_{t0} is a constant corresponding to the elastic limit under uni-axial tension (κ_{t0} is assumed to be 0.7 ~ 0.9). For stress states fallen in this zone, the *DIF* in Eqs.(6)–(7) is set equal to *TDIF*.

In the C-C zone, $\sqrt{J_2} + 1/\sqrt{3}\, I_1 < 0$. The strain-hardening parameter κ is derived from uni-axial compressive test data as follows:

$$\kappa = \kappa_2 = \frac{\sigma_c}{f_c} = \kappa_{c0} + H_c' \varepsilon_p / f_c \tag{12}$$

where f_c is the static uni-axial tensile strength of concrete; H_c' is a hardening parameter determined experimentally; and κ_{c0} is a constant corresponding to the elastic limit under uni-axial compression (κ_{c0} is assumed to be 0.4 ~ 0.6). For stress states fallen in this zone, the *DIF* in Eqs.(6)–(7) is set equal to *CDIF*.

In the T-C zone, $\sqrt{J_2} - 1/\sqrt{3}\, I_1 \geq 0$ and $\sqrt{J_2} + 1/\sqrt{3}\, I_1 \geq 0$. The strain-hardening parameter κ is treated as a weighted sum of κ_1 and κ_2 as follows:

$$\kappa = \alpha_1 \kappa_1 + \alpha_2 \kappa_2 \tag{13}$$

in which $\alpha_1 = \sigma_c + I_1/\sigma_c + \sigma_t$, $\alpha_2 = \sigma_t - I_1/\sigma_c + \sigma_t$, $\sigma_t = \kappa_{t0} f_t + H_t' \varepsilon_p$, $\sigma_c = \kappa_{c0} f_c + H_c' \varepsilon_p$. Note that in uni-axial tensile state, $I_1 = \sigma_t$ which leads to $\alpha_1 = 1$ and $\alpha_2 = 0$; while in uni-axial compressive state, $I_1 = -\sigma_c$ which leads to $\alpha_1 = 0$ and $\alpha_2 = 1$. For stress states fallen in this zone, the *DIF* in Eqs.(6)–(7) is thus set equal to (α_1 *TDIF* + α_2 *CDIF*).

4.3 Defining the residual strength surfaces

Upon reaching the failure stress state, damage begins to creep in and the strength starts weakening. It may survive on further damages, but it suffers gradual degradation of strength. The damaged material is considered as a new material having a reduced maximum strength, namely "residual strength". Upon further loading, the progressive damage states can be represented by a series of reducing residual strengths, each of which associates with a damage scalar D. When $D = 0$, the stress state is somewhere on the failure envelope (in the π-space) having a maximum admissible strength σ^*_{max}. At ultimate fracture, $D = 1$ and the fractured concrete material is assumed completely fractured at a minimum residual strength σ^*_{min}.

The damage variable D is defined as accumulated equivalent plastic strain (Holmquist & Johnson 1993), i.e.

$$D = \sum \frac{\Delta \varepsilon_p}{\varepsilon_p^f} \tag{14}$$

where $\Delta \varepsilon_p$ denotes the equivalent plastic strain during a cycle of integration; ε_p^f denotes the plastic strain to fracture under a constant pressure P, and is expressed in terms of 2 parameters.

5 FDM SIMULATION FOR PERFORATION OF CONCRETE SLAB BY PROJECTILE

The present constitutive model is incorporated to the hydrocode AUTODYN. Numerical results are checked against the experimental results obtained from perforation tests of a square concrete panel by a steel rod (Hanchak 1992). The concrete panel is lightly reinforced, having the dimensions of $680 \times 680 \times 178$ mm. The projectile is of ogival-nose shape, 143.7 mm long steel rods having a diameter of 25.4 mm and a 3.0 caliber-radius-head (CRH). Figure 3 shows the nominal geometric configurations of the steel projectile. Figure 4 shows the dimensions of the concrete target and reinforcements in it.

Two-dimensional axi-symmetric FDM analysis was carried out to simulate the dynamic response of the concrete panel, and thus a circular panel of 688 mm diameter is used to replace the square target. Since the target panel was only lightly reinforced, the effect of the reinforcement is negligible provided that the projectile does not hit at or near the reinforced bars. For this reason, the steel reinforcement bars were not included in the numerical simulation.

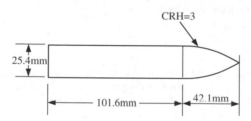

Figure 3.   Geometry of the steel projectile.

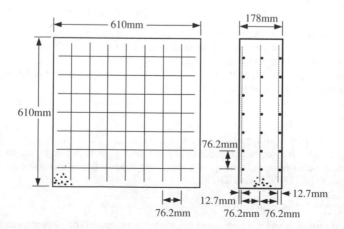

Figure 4.   Configuration of the concrete target.

Since experiments showed that little or no damage was inflicted on the projectile. It means that a first-order approximation for the projectile is accurate enough. Against this background, the steel projectile is assumed elastic-perfectly-plastic, having the following properties: mass density $\rho_s = 8020 \, \text{kg/m}^3$, bulk modulus $K_s = 175 \, \text{GPa}$, shear modulus $G_s = 80.8 \, \text{GPa}$, and yield stress $\sigma_{YS} = 1.72 \, \text{GPa}$. The concrete has an unconfined compressive strength of 48.0 MPa.

Four targets are hit by a projectile at different velocities: 1058 m/s, 750 m/s, 434 m/s, and 300 m/s. Figure 5 plots the impact velocities versus residual velocities. It can be seen that the numerical results agree very well with the test data. For easy visualization, development of damages (as contours of D, $0 < D < 1.0$) in the concrete target at different time cycle for one of the impact cases are shown in Figure 6.

Apparently, the present model with FDM analysis can yield good estimations of the projectile-target interactions. However, the whole process is based on continuum mechanics. In the process of perforation, some of the Lagrangian elements in the target can become grossly distorted. To alleviate this problem, a numerical "erosion" mechanism is put in place in the hydrocode such that a pre-defined strain value is set as the limit strain, beyond which, the highly distorted elements are removed as it progresses. Consequently,

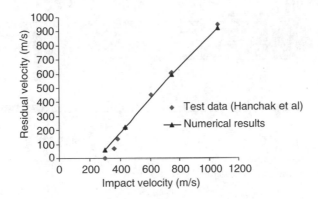

Figure 5.   Impact velocities versus residual velocities.

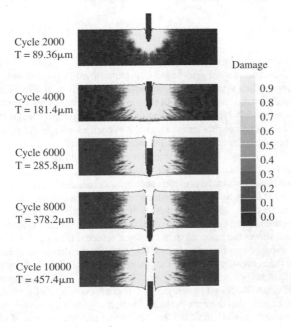

Figure 6.   Damage contour at selected time cycles.

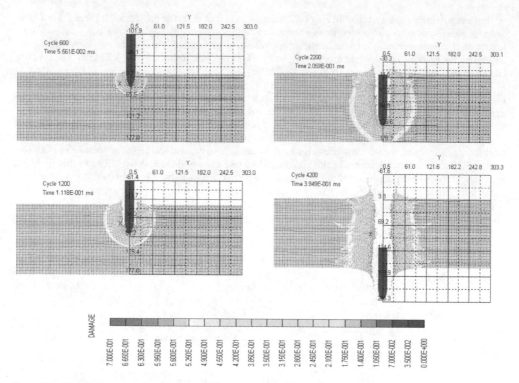

Figure 7.   Contour plots at selected time steps.

it is not a surprise that some phenomena, such as concrete spalling, scabbing and shear plug formation cannot be reproduced. In the next section, the SPH method is used to explore these phenomena.

6   SPH MODELING

The SPH has been recognized as one of the so-called "meshless" or "gridless" methods. It has significant advantages over the conventional grid-based Lagrange or Euler procedures. Particularly, it is free from grid/mesh tangling in the event of large deformation. In addition, it allows relatively easier incorporation of sophisticated constitutive models. In the simulation process for penetration/perforation, no erosion algorithms are needed. Nevertheless, it remains an evolving numerical technique, and currently it still suffers from some technical problems. The most notorious one is the stability of the algorithm caused by insufficient number of neighboring particles at the boundary or when its neighbors are unevenly distributed. However, it is fortunate that the instability is usually a local phenomenon, which would not emanate and thus would not dominate the overall solution.

The same impact configuration in the previous section is considered here. Both the projectile and the target regions are modeled using the SPH procedures. A 2D axi-symmetric analysis is carried out. The concrete target is discretized into 13528 particles while the projectile is represented by 1678 particles. For the steel projectile, the material constitutive relation adopts the linear EOS and the Johnson & Cook strength model. Figure 7 shows the material status at some selected time cycles. The contours of the progressive compressive damage can be seen clearly. The sizes of the damaged areas (craters) at both the impact and the exit surfaces are measured and compared with those shown in the post-test photographs in Reference (Hanchak et al. 1992). Good agreement is observed. Table 1 compares the exit velocities obtained from the SPH procedures and experiments. Good agreement is observed.

From Figure 7, it can be seen that there are particles flying away from the impact surface and the rear surface as well. To some extent, these flying particles can reflect the phenomena of spalling and scabbing. However, they behave like sand particles, not as blocks of fragment in various sizes.

Table 1. Comparison of exit velocities.

	Exit velocity (m/s)	
Impact velocity (m/s)	Experimental (Hanchak 1992)	Simulation using linear EOS
1058	947	936
750	615	635

Figure 8. Configuration of projectile and concrete target.

Figure 9. Target status at step 4425.

7 DEM MODELING

Since DEM is known to be good at simulating flying blocks, it is employed to investigate the perforation process of concrete slab by hard projectile. For simplicity, only a 2D simulation is carried out using a readily available code, namely UDEC. The concrete slab is 2 m wide and 0.5 m deep. The properties of concrete are: density $\rho = 2450\,\text{kg/m}^3$, bulk modulus $K = 24.1\,\text{GPa}$, and shear modulus $G = 14.5\,\text{GPa}$. Two sets of joints having the equivalent properties are specified. One dips at 45°, while another at −45°. The mesh in the near-impact region (0.4 m × 0.5 m) crossing the perforation path is more refined, having a joint-spacing of 0.02 m. In the far regions, the meshes are relatively coarser, having a joint-spacing of 0.1 m. The properties of joints are: normal stiffness $k_n = 100\,\text{GPa/m}$, shear stiffness $k_s = 100\,\text{GPa/m}$, cohesion $C = 12\,\text{MPa}$, friction angle $\phi = 45°$, and tensile strength $\sigma_t = 12\,\text{MPa}$. The projectile is a conical-nose shaped, 1.0 m long steel rod having a diameter of 0.1 m (c.f. ogival nose CRH ≈ 4). It impacts the concrete slab at a velocity of 100 m/s. Figure 8 shows the geometric configuration of the concrete target and the projectile.

Figure 9 illustrates the damage status of the concrete slab in an early stage of penetration at time step 4425. Two sets of shear cracks can be seen just in front of the projectile nose-tip, and spalling fragments are about to form near the impact point. Figure 10 shows the damage status when the projectile nose-tip is about half-depth into the concrete slab at time step 30975. It can be seen that some though not many spalling fragments are flying away from the impact face. In addition, scabbing fragments are seen flying away from the rear face, and shear plug is about to form in front of the projectile nose-tip. Figure 11 shows the damage status when the projectile penetrates through the whole thickness of the concrete slab at time step 70800. The shear plug can be seen separating from the target and it is about to leave behind a trumpet-shaped crater in the rear face of the target.

The purpose of this simulation is just to illustrate the potential of the DEM. We should note that the DEM is originally developed to simulate the mechanical responses of jointed rock masses under quasi-static circumstances or for low-velocity events (Cundall 1971). It is no surprise that it will encounter mesh/grid-tangling as in FEM/FDM when using for simulation of high-velocity impact events. Though a built-in algorithm in the UDEC is available to alleviate the inter-element overlapping (between the projectile and the target), it is not effective and the computation becomes unstable when the impact velocity is higher. To overcome this problem, a new algorithm needs to be put in place.

Figure 10.   Target status at step 30975.   Figure 11.   Target status at step 70800.

8   CONCLUSIONS

This paper discusses numerical simulations of perforation process of a concrete slab impacted by a high-speed steel projectile using different numerical methods, including the Finite Different Method, Smooth Particle Hydrodynamics and Discrete Element Method. In the FDM and SPH simulations, a sound constitutive model for concrete material is used. The results obtained by different methods are bound to inherit their own merits and demerits. The FDM can yield some good results such as residual velocity and damaged area. As it is based on continuum mechanics, it leads to poor representation of the spalling and scabbing effects. In this regard, the SPH method performs better. Nevertheless, because of its circular particle shape, all flying fragments behave somehow like sand particles. Finally, the DEM appears superior but further development is necessary.

REFERENCES

Bischoff, P. H. & Perry, S. H. 1991. Compressive Behavior of Concrete at High Strain Rate, *Materials and Structures* 24:425–450.
Cundall, P. A. 1971. A computer model for simulating progressive large scale movements in blocky rock systems. *Proc. Symp. Int. Soc. Rock Mech., Nancy, France* Vol. 1: paper II-8.
Eibl, J. & Schmidt-Hurtienne, B. 1999. Strain-Rate_Sensitive Constitutive Law for Concrete, *Journal of Engineering Mechanics* 125(12):1411–1420.
Hanchak, S. J., Forrestal, M. J., Young, E. R. & Ehrgott, J. Q. 1992. Perforation of Concrete Slabs with 48 MPa and 140 MPa Unconfined Compressive Strengths. *International Journal of Impact Engineering* 12(1):1–7.
Herrmann, W. 1969. Constitutive Equation for the Dynamic Compaction of Ductile Porous Material, *Journal of Applied Physics* Vol. 40, No. 6, pp. 2490–2499.
Holmquist, T. J. & Johnson, G. R. 1993. A Computational Constitutive Model for Concrete Subjected to Large Strains, High Strain Rates and High Pressures. *14th International Symposium on Ballistics, Quebec, Canada, September 1993*.
Hsieh, S. S., Ting E. C. & Chen, W. F. 1982. A Plasticity-fracture Model for Concrete. *International Journal of Solids and Structures* 18(3):181–197.
Johson, G. R. 1996. Artificial viscosity effects for SPH impact computations. *Int J Impact Eng* 18(5):477–488.
Kotsovos, M. D. & Pavlovic, M. N. 1995. *Structural Concrete – Finite Element Analysis for Limit State Design*. Thomas Telford Publications, London.
Laible, R. C. 1980. *Ballistic Materials and Penetration Mechanics, Methods and Phenomena 5*, Elsevier Scientific Publishing Company. 297pp.
Ottesen, N. S. 1977. A Failure Criteria for concrete, *Journal of the engineering Mechanics Division, ASCE* Vol. 103, No. EM4, August, pp. 527–535.
Riedel, W., Thomas, K., Hiermaier, S. & Schmolinske, E. 1999. Penetrating of Reinforced Concrete by BETA-B-500 Numerical Analysis Using a New Macroscopic Concrete Model for Hydrocodes. *Proceedings of 9th International Symposium IEMS, Berlin 1999*:315–322.
Williams, Martin S. 1994. Modeling of Local Impact Effects on Plain and Reinforced Concrete. *ACI Structural Journal* 91(S19):178–187.
Wingate, C.A. & Stellingwerf, R.F. 1993. Smooth particle hydrodynamics: The sphinx and SPHC codes. Advances in Numerical Simulation Techniques for penetration and Perforation of Solids, *ASME AMD*-Vol. 171:75–82.
Yu, M. H. 1998. *Twin Shear Theory and Its Application* (in Chinese). Chinese Science Publishing House, Beijing.

Computational Methods in Engineering and Science, Iu et al. (eds)
© 2003 Swets & Zeitlinger, Lisse, ISBN 90 5809 567 3

Applications of Toeplitz iterative solvers in science and engineering

Xiao-Qing Jin
Department of Mathematics, University of Macau, Macao

ABSTRACT: In this paper, we briefly survey some of the latest developments and applications by using Krylov subspace methods for solving Toeplitz-like systems. One of the main results is that the complexity of solving a large class of n-by-n Toeplitz systems can be reduced to $O(n \log n)$ operations. Some applications are studied.

1 INTRODUCTION

An n-by-n *Toeplitz matrix* is of the following form

$$
T_n = \begin{pmatrix}
t_0 & t_{-1} & \cdots & t_{2-n} & t_{1-n} \\
t_1 & t_0 & t_{-1} & \ddots & t_{2-n} \\
\vdots & t_1 & t_0 & \ddots & \vdots \\
t_{n-2} & \ddots & \ddots & \ddots & t_{-1} \\
t_{n-1} & t_{n-2} & \cdots & t_1 & t_0
\end{pmatrix},
\tag{1}
$$

i.e. T_n is constant along its diagonals. An mn-by-mn *block Toeplitz matrix* is of the following form

$$
T_{mn} = \begin{pmatrix}
T_{(0)} & T_{(-1)} & \cdots & T_{(2-m)} & T_{(1-m)} \\
T_{(1)} & T_{(0)} & \ddots & \ddots & T_{(2-m)} \\
\vdots & \ddots & \ddots & \ddots & \vdots \\
T_{(m-2)} & \ddots & \ddots & T_{(0)} & T_{(-1)} \\
T_{(m-1)} & T_{(m-2)} & \cdots & T_{(1)} & T_{(0)}
\end{pmatrix},
\tag{2}
$$

where $T_{(i)}$, $i = \pm 1, \cdots, \pm(m-1)$, are arbitrary n-by-n matrices. Particularly, if T_{mn} is a block Toeplitz matrix with Toeplitz blocks $T_{(i)}$, for $i = \pm 1, \cdots, \pm(m-1)$, then T_{mn} is said to be a BTTB matrix.

Toeplitz-like systems arise in a variety of applications in mathematics, scientific computing and engineering, (Chan & Ng 1996, Jin 2002). These applications have motivated mathematicians, scientists and engineers to develop fast and specific algorithms for solving Toeplitz systems. Such kind of algorithms are called *Toeplitz solvers*.

Most of current research works on Toeplitz solvers are focused on iterative method, especially on Krylov subspace methods such as the conjugate gradient (CG) method and the generalized minimal residual (GMRES) method. One of the main important results of this methodology is that the complexity of solving a large class of n-by-n Toeplitz systems can be reduced to $O(n \log n)$ operations provided that suitable preconditioners are chosen under certain conditions defined on Toeplitz systems. In this paper, we will survey some recent results of these iterative Toeplitz solvers. Applications to some practical problems will also be reviewed.

1.1 Background

Let us begin by introducing the background knowledge that will be used throughout the article. We assume that the diagonals $\{t_k\}_{k=-n+1}^{n-1}$ of T_n in (1) are Fourier coefficients of a function f, i.e.

$$t_k \equiv t_k(f) = \frac{1}{2\pi} \int_{-\pi}^{\pi} f(x) e^{-ikx} dx, \ i \equiv \sqrt{-1}.$$

The f is called the *generating function* of T_n and assumed in certain class of functions such that all the T_n are invertible. In practical problems from industry and engineering, we are usually given f first, not the Toeplitz matrix T_n (Chan & Ng 1996, Jin 2002).

In order to analyse the convergence rate of the CG method, we need to introduce the following definition of clustered spectrum (Chan & Ng 1996).

Definition 1.1 *The eigenvalues of a sequence of matrices $\{H_n\}_{n=1}^{\infty}$ are said to be clustered around a point $\gamma \in R$ if for any $\varepsilon > 0$, there exist positive integers M and N such that for all $n > N$, at most M eigenvalues of $H_n - \gamma I_n$ have absolute values greater than ε, where I_n is the identity matrix.*

If the eigenvalues of $\{H_n\}_{n=1}^{\infty}$ are clustered around a point γ, then the CG method has a fast convergence rate by the following theorem (Jin 2002).

Theorem 1.1 *Let u^k be the k-th iterant of the CG method applied to a symmetric positive definite system $H_n u = b$ and let u_* be the true solution of the system. If the eigenvalues λ_j of H_n are ordered such that*

$$0 < \lambda_1 \leq \cdots \leq \lambda_p \leq b_1 \leq \lambda_{p+1} \leq \cdots \leq \lambda_{n-q} \leq b_2 \leq \lambda_{n-q+1} \leq \cdots \leq \lambda_n,$$

then

$$\frac{\|u_* - u^k\|}{\|u_* - u^0\|} \leq 2 \left(\frac{\alpha - 1}{\alpha + 1} \right)^{k-p-q} \cdot \max_{\lambda \in [b_1, b_2]} \prod_{j=1}^{p} \left(\frac{\lambda - \lambda_j}{\lambda_j} \right).$$

Here $\| \cdot \|$ is the energy norm given by $\|v\|^2 = v^* H_n v$ and $\alpha \equiv \left(\frac{b_2}{b_1} \right)^{\frac{1}{2}} \geq 1$.

From Theorem 1.1, we know that the more clustered the eigenvalues are, the faster the convergence rate will be. Unfortunately, the spectra of matrices are not clustered around a certain point in general. Thus, the convergence rate of the CG method is slow usually. In order to accelerate the convergence rate, we need to precondition the system, i.e. instead of solving the original system $H_n u = b$, we solve the following preconditioned system

$$M_n^{-1} H_n u = M_n^{-1} b. \tag{3}$$

The preconditioner M_n is chosen with two criteria in mind (Golub & Van 1989, Jin 2002):

I. $M_n r = d$ is easy to solve;
II. The spectrum of $M_n^{-1} H_n$ is clustered and (or) $M_n^{-1} H_n$ is well-conditioned compared to H_n.

The main work involved in implementing the CG method to the preconditioned system (3) is the matrix-vector product $M_n^{-1} H_n v$ for some vector v. We will show that for Toeplitz matrices with circulant preconditioners, the cost of this matrix-vector product can be reduced dramatically.

1.2 Circulant preconditioners

In 1986, Strang (1986) and Olkin (1986) proposed independently the use of the preconditioned conjugate gradient (PCG) method with circulant matrices as preconditioners for solving Toeplitz systems. The circulant matrix is defined as follows:

$$C_n = \begin{pmatrix} c_0 & c_{n-1} & \cdots & c_2 & c_1 \\ c_1 & c_0 & c_{n-1} & & c_2 \\ \vdots & c_1 & c_0 & \ddots & \vdots \\ c_{n-2} & \ddots & \ddots & \ddots & c_{n-1} \\ c_{n-1} & c_{n-2} & \cdots & c_1 & c_0 \end{pmatrix}$$

It is well-known that circulant matrices can be diagonalized by the Fourier matrix F_n, (Davis 1994), i.e.

$$C_n = F_n^* \Lambda_n F_n . \qquad (4)$$

Here the entries of F_n are given by

$$(F_n)_{j,k} = \frac{1}{\sqrt{n}} e^{2\pi i jk/n} , \quad i \equiv \sqrt{-1} ,$$

with $0 \le j, k \le n - 1$ and Λ_n is a diagonal matrix holding the eigenvalues of C_n. Strang (1986) and Olkin (1986) noted that for any Toeplitz matrix T_n with a circulant preconditioner C_n, the product $C_n^{-1} T_n v$ can be computed in $O(n \log n)$ operations for any vector v as circulant systems can be solved efficiently by the Fast Fourier Transform (FFT) and the multiplication $T_n v$ can also be computed by FFTs by first embedding T_n into a 2n-by-2n circulant matrix. More precisely, we have a 2n-by-2n circulant matrix with T_n embedded inside as follows,

$$\begin{pmatrix} T_n & \times \\ \times & T_n \end{pmatrix} \begin{pmatrix} v \\ 0 \end{pmatrix} = \begin{pmatrix} T_n v \\ * \end{pmatrix},$$

and then the multiplication can be carried out by using the decomposition as in (4). The operation cost is, therefore, $O(2n \log(2n))$. Thus, the cost per iteration of the PCG method is still $O(n \log n)$.

A lot of circulant preconditioners have been proposed for solving Toeplitz systems since 1986. We only introduce two of them which have been proved to be good preconditioners.

1.2.1 *Strang's circulant preconditioner*

For Toeplitz matrix (1), Strang's preconditioner $s(T_n)$ is defined to be the circulant matrix obtained by copying the central diagonals of T_n and bringing them around to complete the circulant. More precisely, the diagonals of $s(T_n)$ are given by

$$s_k = \begin{cases} t_k, & 0 \le k \le [n/2] \\ t_{k-n}, & [n/2] < k < n \\ s_{n+k}, & 0 < -k < n. \end{cases}$$

1.2.2 *T. Chan's circulant preconditioner*
Let

$$M_F = \{ F^* \Lambda_n F | \Lambda_n \text{ is any n-by-n diagonal matrix } \}$$

where F is the n-by-n Fourier matrix. Note that M_F is the set of all circulant matrices (Davis 1994). For any arbitrary n-by-n matrix A_n, T. Chan's circulant preconditioner $c_F(A_n)$ (Chan 1988) is defined to be the minimizer of the Frobenius norm

$$\|A_n - W_n\|_F \qquad (5)$$

over all $W_n \in M_F$. The matrix $c_F(A_n)$ is called the optimal circulant preconditioner in Chan (1988). The diagonals of $c_F(A_n)$ are just the average of diagonals of A_n, with the diagonals being extended to length n by a wrap-around.

2   BLOCK PRECONDITIONERS

Let us consider a general block system $H_{mn} u = b$ first, where H_{mn} is mn-by-mn matrix partitioned as

$$H_{mn} = \begin{pmatrix} H_{1,1} & H_{1,2} & \cdots & H_{1,m} \\ H_{2,1} & H_{2,2} & \cdots & H_{2,m} \\ \vdots & \ddots & \ddots & \vdots \\ H_{m,1} & H_{m,2} & \cdots & H_{m,m} \end{pmatrix}$$

and blocks $H_{i,j}$ with $1 \le i, j \le m$ are square matrices of order n

2.1 Block preconditioners for block system

Several preconditioners that preserve the block structure of H_{mn} are constructed in Chan & Jin (1992). In view of the point case in Section 1.2, it is natural to define the *circulant-block preconditioner* of H_{mn} as follows,

$$c_F^{(1)}(H_{mn}) \equiv \begin{pmatrix} c_F(H_{1,1}) & c_F(H_{1,2}) & \cdots & c_F(H_{1,m}) \\ c_F(H_{2,1}) & c_F(H_{2,2}) & \cdots & c_F(H_{2,m}) \\ \vdots & \ddots & \ddots & \vdots \\ c_F(H_{m,1}) & c_F(H_{m,2}) & \cdots & c_F(H_{m,m}) \end{pmatrix},$$

where the blocks $c_F(H_{i,j})$ defined by (5) are just T. Chan's circulant preconditioners of $H_{i,j}$, $1 \leqslant i, j \leqslant m$. Actually, the matrix $c_F^{(1)}(H_{mn})$ is the minimizer of

$$\|H_{mn} - W_{mn}\|_F$$

over all matrices W_{mn} that are m-by-m block matrices with n-by-n circulant blocks.

Now, let $(H_{mn})_{i,j;k,l}$ denote the (i, j)-th entry of the (k, l)-th block of H_{mn} and let P be the permutation matrix that satisfies

$$\left(P^* H_{mn} P\right)_{k,l;i,j} = \left(H_{mn}\right)_{i,l;k,l},$$

for $1 \leqslant i,j \leqslant n, 1 \leqslant k,l \leqslant m$. The preconditioner $\tilde{c}_F^{(1)}(H_{mn})$ is a *block-circulant matrix* which is defined as

$$\tilde{c}_F^{(1)}(H_{mn}) \equiv P^* c_F^{(1)}\left(P H_{mn} P^*\right) P.$$

With the composite of operators $c_F^{(1)}$ and $\tilde{c}_F^{(1)}$, one can obtain a preconditioner

$$c_{F,F}^{(2)}(H_{mn}) \equiv \tilde{c}_F^{(1)} \circ c_F^{(1)}(H_{mn})$$

which is based on circulant approximations within each block and also on block level. The preconditioner $c_{FF}^{(2)}(H_{mn})$ is said to be a BCCB matrix. Actually, it is the minimizer of

$$\|H_{mn} - W_{mn}\|_F \text{ over all } W_{mn} \in M_{F \otimes F}. \text{ Here}$$

$$M_{F \otimes F} \equiv \left\{ \left(F_m \otimes F_n\right)^* \Lambda_{mn} \left(F_m \otimes F_n\right) \mid \Lambda_{mn} \text{ is any diagonal matrix } \right\}$$

where $\otimes$ denotes the tensor product and F_m, F_n are Fourier matrices. We note that $M_{F \otimes F}$ is the set of all BCCB matrices (Davis 1994). The BCCB preconditioners for solving BTTB matrices have been investigated in Chan & Jin (1992), Chan & Ng (1996) and Jin (2002).

Since any BCCB matrix C_{mn} can be diagonalized by the 2-dimensional Fourier matrix,

$$C_{mn} = \left(F_m^* \otimes F_n^*\right) \Lambda_{mn} \left(F_m \otimes F_n\right)$$

where Λ_{mn} is a diagonal matrix holding the eigenvalues of C_{mn}, therefore, the matrix-vector multiplication $C_{mn}^{-1} v$ can be computed in $O(mn \log mn)$ operations by using the 2-dimensional FFT. Moreover, any $mn \times mn$ BTTB matrix (2) can be enlarged into a $4mn \times 4mn$ BCCB matrix, so it can also be multiplied by a vector in $O(mn \log mn)$ operations (Chan & Jin 1992).

2.2 Convergence rate and operation cost

We analyse the convergence rate and operation cost of the PCG method when applied to solving BTTB systems $T_{mn}u = b$ where T_{mn} is given by (2). Let the entries of T_{mn} be denoted by

$$\left(T_{mn}\right)_{p,q;r,s} = t_{p-q}^{(r-s)},$$

for $1 \leq p, q \leq n, 1 \leq r, s \leq m$. The T_{mn} is associated with a generating function $f(x,y)$ as follows,

$$t_k^{(j)}(f) = \frac{1}{4\pi^2} \int_{-\pi}^{\pi} \int_{-\pi}^{\pi} f(x,y) e^{-i(jx+ky)} dx dy, \; i \equiv \sqrt{-1}.$$

We note that for any m and n, T_{mn}'s have the following important properties:

1. When f is real-valued, then $T_{mn}(f)$ are Hermitian, i.e. $t_k^{(j)}(f) = \bar{t}_{-k}^{(-j)}(f)$.
2. When f is real-valued with $f(x,y) = f(-x, -y)$, then $T_{mn}(f)$ are real symmetric, i.e. $t_k^{(j)}(f) = t_{-k}^{(-j)}(f)$.
3. When f is real-valued and even, i.e. $f(x,y) = f(|x|,|y|)$, then $T_{mn}(f)$ are level-2 symmetric, i.e. $t_k^{(j)}(f) = t_{|k|}^{(|j|)}(f)$.

Let $C_{2\pi \times 2\pi}$ denote the Banach space of all 2π-periodic (in each direction) continuous real-valued functions equipped with the supremum norm $\|\cdot\|\infty$. The following theorem gives the relation between the values of $f(x,y)$ and the eigenvalues of $T_{mn}(f)$ (Jin 1995a & Serra 1994).

Theorem 2.1 *If $f \in C_{2\pi \times 2\pi}$ with $f_{\min} < f_{\max}$ where $f_{\min}$ and $f_{\max}$ denote the minimum and maximum values of f respectively, then for all positive integers m and n, we have*

$$f_{\min} < \lambda_i(T_{mn}) < f_{\max}, \; for \; i = 1, \cdots, mn,$$

where $\lambda_i(T_{mn})$ is the i-th eigenvalues of T_{mn}. Moreover,

$$\lim_{m,n \to \infty} \lambda_{\max}(T_{mn}) = f_{\max} \; and \; \lim_{m,n \to \infty} \lambda_{\min}(T_{mn}) = f_{\min}.$$

From Theorem 2.1, we know that if $f \geq 0$, then $T_{mn}(f)$ is always positive definite. When f vanishes at some points $(x_0, y_0) \in [-\pi, \pi] \times [-\pi, \pi]$, then the condition number $\kappa(T_{mn})$ of T_{mn} is unbounded as m or n tend to infinity, i.e. T_{mn} is ill-conditioned.

Now, let us consider the class of level-2 symmetric BTTB systems

$$T_{mn}u = b \tag{6}$$

where

$$T_{mn} = \begin{pmatrix} T_{(0)} & T_{(-1)} & \cdots & T_{(2-m)} & T_{(1-m)} \\ T_{(1)} & T_{(0)} & \ddots & \ddots & T_{(2-m)} \\ \vdots & \ddots & \ddots & \ddots & \vdots \\ T_{(m-2)} & \ddots & \ddots & T_{(0)} & T_{(-1)} \\ T_{(m-1)} & T_{(m-2)} & \cdots & T_{(1)} & T_{(0)} \end{pmatrix}$$

and the blocks $T_{(i)}$, for $i = 0, \ldots, m - 1$, are themselves symmetric Toeplitz matrices of order n. For solving (6), by using the PCG method with the preconditioners $c_F^{(1)}(T_{mn}), \tilde{c}_F^{(1)}(T_{mn})$, and $c_{FF}^{(2)}(T_{mn})$ respectively, the cost per iteration requires $O(mn \log^2 m + mn \log n)$, $O(nm \log^2 n + nm \log m)$ and $O(mn \log mn)$ operations accordingly (Chan & Jin 1992).

The convergence rate of the PCG method was also analysed by R. Chan and Jin (Chan & Jin 1992, Jin 2002) for solving (6). It was shown that if the generating function $f > 0$ is in $C_{2\pi \times 2\pi}$, then the spectra of preconditioned matrices $(c_F^{(1)}(T_{mn}))^{-1}T_{mn}, (\tilde{c}_F^{(1)}(T_{mn}))^{-1}T_{mn}$ and $(c_{FF}^{(2)}(T_{mn}))^{-1}T_{mn}$ are clustered and the method converges linearly. Thus, the total operation cost for solving (6) only requires

$O(mn \log^2 m + mn \log n)$, $O(nm \log^2 n + nm \log m)$ and $O(mn \log mn)$ operations accordingly (Chan & Jin 1992). This is one of main important results of the PCG algorithm. However, Serra and Tyrtyshnikov proved theoretically in Serra & Tyrtyshnikov (1999) that any multilevel circulant preconditioners for multilevel Toeplitz matrices cannot produce a superlinearly convergence rate by using the PCG algorithms.

3 APPLICATIONS IN PDES

We study linear systems arising from discretizations of second-order partial differential equations (PDEs). Let us first consider elliptic equations of the form

$$\frac{\partial}{\partial x}\left(a\frac{\partial u}{\partial x}\right) + \frac{\partial}{\partial y}\left(b\frac{\partial u}{\partial y}\right) = f \tag{7}$$

defined on the unit square with given Dirichlet boundary condition, where $a = a(x,y)$, $b = b(x,y)$, and $f = f(x,y)$ are given functions with

$$0 < c_{\min} \leq a(x,y),\ b(x,y) \leq c_{\max} \tag{8}$$

for some constants $c_{\min}$ and $c_{\max}$. Usually, the numerical solution of 2-dimensional elliptic equation on a uniform grid, by using the five-point centered difference scheme, involves the solution of block tridiagonal system

$$Az = b .$$

The most important properties of the matrix A are its sparsity and bandwidth. It is desirable to retain the sparsity in the solving process and thus more attention has been devoted to iterative methods. We will use the CG method in solving such linear system.

Typically, the convergence rate of the CG method depends on the condition number $\kappa(A)$ of the coefficient matrix A. For elliptic problems of second order we usually have $\kappa(A) = O(m^2)$ and thus the growth is rapid with m where m is the size of matrix A. To alleviate this problem, we use a preconditioner. R. Chan and T. Chan in Chan & Chan (1992) proposed BCCB preconditioners which are based on averaging the coefficients of A to form a circulant approximation to A. The BCCB preconditioner of A is defined by

$$C = c_{F,F}^{(2)}(A) + \frac{\rho}{m^2} I ,$$

where $c_{FF}^{(2)}(A)$ is defined as in Section 2. Thus, the diagonals of C are obtained as the simple averages of the coefficients along the lines of the grid. The constant ρm^{-2} added to the main diagonal is to reduce the condition number of the resulting preconditioned systems.

Theorem 3.1 *Assume that (8) holds for elliptic equations (7). Then we have*

$$O(1) \leq \lambda\left(C^{-1}A\right) \leq O(m).$$

As a consequence, we have:

$$\kappa\left(C^{-1}A\right) \leq O(m).$$

The numerical results in Chan & Chan (1992) show that the preconditioned systems often have clustering of eigenvalues which is favorable to the convergence rate.

The idea of using circulant preconditioner for elliptic problems was extended to hyperbolic equations, parabolic equations and the first order PDEs (Jin & Chan 1992, Jin & Lei).

4 APPLICATIONS IN ODES

Consider the linear initial value problem (IVP)

$$\begin{cases} y'(t) = f(t, y(t)) \equiv J_m y(t) + g(t), & t \in (t_0, T] \\ y(t_0) = \eta, \end{cases} \tag{9}$$

where $y(t)$, $g(t) : R \rightarrow R^m$, $\eta \in R^m$ and J_m is an m-by-m matrix. Brugnano and Trigiante in Brugnano & Trigiante (1998) introduced a class of boundary value methods (BVMs) for (9). Such methods are based on the linear multistep formula (LMF). The matrix of the discretized system is nonsymmetric, large and sparse. In Chan et al. (2001), the Strang-type block circulant preconditioner with the GMRES method was proposed to solve this discretized linear system.

A BVM approximates the solutions of (9) by means of a discrete boundary value problem (BVP). By using a k-step LMF over a uniform mesh points $t_j = t_0 + jh, j = 0, \cdots, s$, with

$h = \dfrac{T - t_0}{s}$, we have the discrete BVP as follows,

$$\sum_{i=-v}^{k-v} \alpha_{i+v} y_{n+i} = h \sum_{i=-v}^{k-v} \beta_{i+v} f_{n+i}, \quad n = v, \cdots, s - k + v. \tag{10}$$

Here, the y_n is the discrete approximation to $y(t_n)$, $f_n = f(t_n, y_n) \equiv J y_n + g_n$ and $g_n = g(t_n)$. With the additional conditions which provide the following set of equations,

$$\sum_{i=0}^{k} \alpha_i^{(j)} y_i = h \sum_{i=0}^{k} \beta_i^{(j)} f_i, \quad j = 1, \cdots, v-1, \tag{11}$$

and

$$\sum_{i=0}^{k} \alpha_{k-i}^{(j)} y_{s-i} = h \sum_{i=0}^{k} \beta_{k-i}^{(j)} f_{s-i}, \quad j = s - k + v + 1, \cdots, s, \tag{12}$$

we then have a well-posed linear system with the equations (10), (11) and (12). This linear system defines the use of a BVM on the problem (9). The advantage in using BVMs is that they have much better stability properties (Brugnano & Trigiante 1998).

The BVMs in matrix form are given by introducing two $(s + 1)$-by-$(s + 1)$ matrices A and B. The matrix A is defined by

$$A = \begin{pmatrix} 1 & \cdots & 0 & & & & & \\ \alpha_0^{(1)} & \cdots & \alpha_k^{(1)} & & & 0 & & \\ \vdots & \vdots & \vdots & & & & & \\ \alpha_0^{(v-1)} & \cdots & \alpha_k^{(v-1)} & & & & & \\ \alpha_0 & \cdots & \alpha_k & & & & & \\ & \alpha_0 & \cdots & \alpha_k & & & & \\ & & \ddots & \ddots & \ddots & & & \\ & & & \ddots & \ddots & \ddots & & \\ & & & \alpha_0 & \cdots & \alpha_k & & \\ 0 & & & \alpha_0^{(s-k+v+1)} & \cdots & \alpha_k^{(s-k+v+1)} & \\ & & & \vdots & \vdots & \vdots & \\ & & & \alpha_0^{(s)} & \cdots & \alpha_k^{(s)} \end{pmatrix},$$

and B is obtained by replacing α's by β's in A with the first row being zero.

Therefore, the discretized problem generated by the application of the BVM (10)–(12) to problem (9) is given by

$$My - e_1 \otimes \eta + h(B \otimes I_m)g,\qquad(13)$$

where

$$M = A \otimes I_m - hB \otimes J_m, \quad e_1 = (1,0,\cdots,0)^T \in R^{(s+1)},$$

$$y = (y_0,\cdots,y_s)^T \in R^{(s+1)m}, \quad g = (g_0,\cdots,g_s)^T \in R^{(s+1)m}.$$

The matrix M in (13) turns out to be large and sparse when either $s \gg k$ or J_m is large and sparse. In those cases, the solution of the linear system (13) by using a direct method is expensive. We therefore use the GMRES method with a suitable preconditioner.

The Strang-type preconditioner S is defined by

$$S \equiv s(A) \otimes I_m - hs(B) \otimes J_m,$$

where

$$
s(A) = \begin{pmatrix}
\alpha_v & \cdots & \alpha_k & & & & \alpha_0 & \cdots & \alpha_{v-1} \\
\vdots & \ddots & & \ddots & & & & \ddots & \vdots \\
\alpha_0 & & \ddots & & \ddots & & 0 & & \alpha_0 \\
& \ddots & & \ddots & & \ddots & & & \\
& & \ddots & & \ddots & & \ddots & & \\
\alpha_k & & 0 & & \ddots & & \ddots & & \alpha_k \\
\vdots & \ddots & & \ddots & & & \ddots & & \vdots \\
\alpha_{v+1} & \cdots & \alpha_k & & & & \alpha_0 & \cdots & \alpha_v
\end{pmatrix}
$$

and $s(B)$ obtained by replacing α's by β's in $s(A)$. It was shown in Chan et al. (2001) that the eigenvalues of the preconditioned system are clustered. It follows that when the GMRES method is applied for solving the preconditioned system, the convergence rate is fast. Our method was extended to delay differential equations in (Lin et al. 2003).

5   APPLICATIONS IN IMAGE PROCESSING

The problem of elimination or minimizing degradations in a blurred noisy image is referred to as image restoration. Degradations include the blurring that can be introduced by optical systems or image motion, and the noise from electronic and photometric sources. When the quality of the recorded images is degraded by blurring and noise, important information remains hiding and cannot be directly interpreted without numerical processing.

The mathematical model of the linear image restoration problem is given as follows (Chan et al. 1993, Chan & Ng 1996),

$$g(\xi,\delta) = \int_{-\infty}^{\infty} \int_{-\infty}^{\infty} t(\xi,\delta;\alpha,\beta)f(\alpha,\beta)d\alpha d\beta + \eta(\xi,\delta)\qquad(14)$$

where $g(\xi,\delta)$ is the recorded (or degraded) image, $f(\alpha,\beta)$ is the ideal (or orginal) image, the vector $\eta(\xi,\delta)$ represents additive noise. The function t is called the point spread function and represents the degradation of the image. When the point spread function is a function t of $\xi - \alpha$ and $\delta - \beta$, the t is said to be spatially invariant.

In the digital implementation of (14) with $t(\xi - \alpha, \delta - \beta)$, the integral is discretized by using some quadrature rule to obtain the discretized scalar model

$$g(i,j) = \sum_{k=1}^{m} \sum_{l=1}^{n} t(i-k, j-l) f(k,l) + \eta(i,j).$$

In matrix form, we have the following BTTB system

$$g = T_{mn} f + \eta \tag{15}$$

where g, η and f are mn-vectors and T_{mn} is an mn-by-mn BTTB matrix. The PCG least-square method is proposed as main tool to solve (15) (Chan & Ng 1996). For these Toeplitz least squares problems, some different preconditioners based on the circulant approximations were studied in (Chan et al. 1993, Chan & Ng 1996, Kou et al. 2002).

REFERENCES

Bertaccini, D. 2000. A Circulant Preconditioner for the Systems of LMF-Based ODE Codes, *SIAM J. Sci. Comput.*, Vol. 22: 767–786.

Brugnano, L. & Trigiante, D. 1998, *Solving Differential Problems by Multistep Initial and Boundary Value Methods*, Amsterdam: Gordon and Berach Science Publishers.

Chan, R. 1989. Circulant Preconditioners for Hermitian Toeplitz Systems, *SIAM J. Matrix Anal. Appl.*, Vol. 10: 542–550.

Chan, R. 1991. Toeplitz Preconditioners for Toeplitz Systems with nonnegative generating functions *IMA J. Numer. Anal.*, Vol. 11: 333–345.

Chan, R. & Chan, T. 1992. Circulant Preconditioners for Elliptic Problems, *J. Numer. Linear Algebra Appl.*, Vol. 1: 77–101.

Chan, R. & Jin, X. 1992. A Family of Block Preconditioners for Block Systems, *SIAM J. Sci. Statist. Comput.*, Vol. 13: 1218–1235.

Chan, R., Jin, X. & Yeung, M. 1991. The Circulant Operator in the Banach Algebra of Matrices, *Linear Algebra Appls.*, Vol. 149: 41–53.

Chan, R., Nagy, J. & Plemmons, R. 1993. FFT-Based Preconditioners for Toeplitz-Block Least Squares Problems, *SIAM J. Sci. Numer. Anal.*, Vol. 30: 1740–1768.

Chan, R. & Ng, M. 1996. Conjugate Gradient Methods for Toeplitz Systems, *SIAM Review*, Vol. 38: 427–482.

Chan, R., Ng, M. & Jin, X. 2001. Strang-Type Preconditioners for Systems of LMF-Based ODE Codes, *IMA J. Numer. Anal.*, Vol. 21: 451–462.

Chan, R. & Yeung, M. 1992. Circulant Preconditioners for Toeplitz Matrices with Positive Continuous Generating Functions, *Math. Comput.*, Vol. 58: 233–240.

Chan, T. 1988. An Optimal Circulant Preconditioner for Toeplitz Systems, *SIAM J. Sci. Statist. Comput.*, Vol. 9: 766–771.

Davis, P. 1994. *Circulant Matrices, 2nd edition*, New York: John Wiley & Sons, Inc.

Golub, G. & Van Loan, C. 1989, *Matrix Computations, 2nd edition*, Baltmore: Johns Hopkins Univ. Press.

Gonalez, R. & Woods, R. 1992. *Digital Image Processing*, New York: Addison-Wesley Publishing Company, Inc.

Hemmingsson, L. & Otto, K. 1996. Analysis of Semi-Toeplitz Preconditioners for First Order PDE, *SIAM J. Sci. Comput.* Vol. 17: 47–64.

Huckle, T. 1992. Circulant and Skew Circulant Matrices for Solving Toeplitz Matrix Problems, *SIAM J. Matrix Anal. Appl.*, Vol. 13: 767–777.

Jin, X. 1994. Hartley Preconditioners for Toeplitz Systems Generated by Positive Continuous Functions, *BIT*, Vol. 34: 367–371.

Jin, X. 1995a. A Note on Preconditioned Block Toeplitz Matrices, *SIAM J. Sci. Comput.*, Vol. 16: 951–955.

Jin, X. 1995b. A Fast Algorithm for Block Toeplitz Systems with Tensor Structure, *Appl. Math. Comput.*, Vol. 73: 115–124.

Jin, X. 1996a. Fast Iterative Solvers for Symmetric Toeplitz Systems-A Survey and on Extension, *J. Comput. Appl. Math.*, Vol. 66: 315–321.

Jin, X. 1996b. A Preconditioner for Constrained and Weighted Least Squares Problem with Toeplitz Structure, *BIT*, Vol. 36: 101–109.

Jin, X. 1996c. Band-Toeplitz Preconditioners for Block Toeplitz Systems, *J. Comput. Appl. Math.*, Vol. 70: 225–230.

Jin, X. 1997. A Note on Construction of Circulant Preconditioners from Kernels, *Appl. Math. Comput.*, Vol. 83: 3–12.

Jin, X. 2002. *Developments and Applications of Block Toeplitz Iterative Solvers*, Beijing: Kluwer Academic Publishers & Science Press.

Jin, X. & Chan, R. 1992. Circulant Preconditioners for Second Order Hyperbolic Equations, *BIT, Vol. 32*: 650–664.

Jin, X., Sin, V. & Song, L. 2003. Circulant-Block Preconditioners for Solving Ordinary Differential Equations, *Appl. Math. Comput., Vol. 140*: 409–418.

Jin, X. & Lei, S. Sine Transform Based Preconditioners for Solving Constant-Coefficient First-Order PDEs, *Linear Algebra Appl.,* to appear.

Kou, K., Sin, V. & Jin, X. 2002. A Note on Fast Algorithm for Block Toeplitz Systems with Tensor Structure, *Appl. Math. Comput., Vol. 126*: 187–197.

Lapidus, L. & Pinder, G. 1982. *Numerical Solution of Partial Differential Equations in Science and Engineering*, New York: Wiley.

Lin, F., Jin, X. & Lei, S. 2003. Strang-Type Preconditioners for Solving Linear Systems from Delay Differential Equations, *BIT, Vol. 43*: 136–149.

Ng, M. 1997. Band Preconditioners for Block-Toeplitz–Toeplitz-Block-Systems, Linear *Algebra Appl., Vol. 259*: 307–327.

Olkin, J. 1986. *Linear and Nonlinear Deconvolution Problems*, Ph.D. thesis, Texas: Rice University, Houston.

Potts, D. & Steidl, G. 1999. Preconditioners for Ill-Conditioned Toeplitz Matrices, *BIT, Vol. 39*: 579–594.

Saad, Y. 1996, *Iterative Methods for Sparse Linear Systems*, Boston: PWS Publishing Company,.

Serra, S. 1994, Preconditioning Strategies for Asymptoically Ill-conditioned Block Toeplitz Systems, *BIT, Vol. 34*: 579–594.

Serra, S. & Tyrtyshnikov, E. 1999. Any Circulant-like Preconditioner for Multilevel Matrices Is Not Superlinear, *SIAM J. Matrix Anal. Appl., Vol. 21*: 431–439.

Strang, G. 1986. A Proposal for Toeplitz Matrix Calculations, *Stud. Appl. Math., Vol. 74*: 171–176.

Computational Methods in Engineering and Science, Iu et al. (eds)
© *2003 Swets & Zeitlinger, Lisse, ISBN 90 5809 567 3*

Multiscale methods and its applications to Nano-mechanics and materials

W.K. Liu
Professor and Associate Chair, Department of Mechanical Engineering, Northwestern University, Evanston, Illinois, USA

S. Hao
Senior Research Associate, Department of Mechanical Engineering, Northwestern University, Illinois, USA

F.J. Vernerey, H. Kadowaki, L. Zhang, Y. Liu & H. Park
Graduate students, Department of Mechanical Engineering Northwestern University, Evanston, Illinois

D. Qian
Assistant Professor Department of Mechanical, Industrial and Nuclear Engineering, University of Cincinnati, Cincinnati, OH, USA

ABSTRACT: Hierarchical and concurrent multiple scale modeling of heterogeneous systems and their governing equations are first derived followed by the development of hierarchical and concurrent multiscale simulation methods. These are fundamentally necessary to account for the multiple scale behavior observed in heterogeneous materials. One of the most interesting questions in computational mechanics is: how do you make a connection between what is going on at the atomic scale and the material behavior at the macroscale? In many cases it is enough to write down an equation of motion for the large scales, letting all of the *averaged* atomic behavior be captured in parameters like the elastic moduli. However in some situations, like material fracture, the dynamics is dominated by atomic-scale behavior that cannot be captured with the standard material laws. In our proposed research, we are taking a multiple-scale method for these problems in which we seek to capture both the atomic-scale and macro-scale dynamics simultaneously. The proper coupling of these scales requires innovations based physics, statistical mechanics and novel numerical methods. The above described heterogeneous materials systems all involve the concurrent coupling of dynamics at the atomic scale and the macroscale. The methods that we are developing to treat these problems must blend quantum and statistical physics with more conventional engineering concepts about continuum mechanics. We believe that it is only through this sort of multiscale, multi-disciplinary approach that these problems in engineering today can be solved. As an added benefit, we aim to use our models and methods to assist in the design and analysis of nano-materials and nanostructures.

1 CYBERSTEEL 2020

1.1 *Proposed approach*

Both strength and fracture toughness are the key property-indices for steels. Looking ahead to the projected requirements of new steels in the next twenty years, the primary design objectives will be the achievement of high dynamic fracture toughness at high strength levels in weldable, formable plate steels with high resistance to hydrogen stress corrosion cracking. This motivates the initial research of cybersteel 2020 that focuses on the multiscale-quantum design of new trip-phase ultrahigh-strength steel, in conjunction

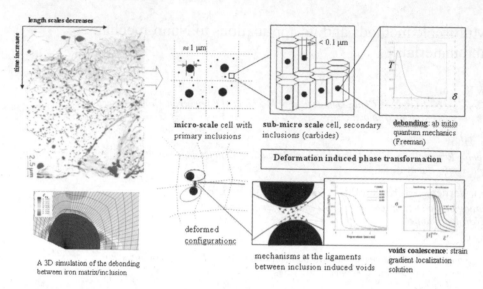

length scales decreases

time increases

≈ 1 μm

< 0.1 μm

micro-scale cell with primary inclusions

sub-micro scale cell, secondary inclusions (carbides)

debonding: ab initio quantum mechanics (Freeman)

Deformation induced phase transformation

deformed configurations

A 3D simulation of the debonding between iron matrix/inclusion

mechanisms at the ligaments between inclusion induced voids

voids coalescence: strain gradient localization solution

Figure 1. A multi-scale, multi-physics model for UHS steel design (Olson 1999), the TEM micrograph is from Krauss 1984.

with the developments of advanced numerical mechods (Cybersteel 2020, Liu et al. 1995, Liu et al. 1999, Belytschko et al. 1994, Olson 1997, Freeman 1991, Hao et al. 2002, Hao 2002).

Although modern technology provides many ways to achieve either high strength or high toughness respectively in one steel through manufacture process, it remains a challenge to gain both of them simultaneously for modern ultrahigh-strength steel design. This is because the toughness characterizes the capability of a material against fracture at a crack tip local. The average of such a kind of local property is presented as the strength of a material. The difference between the local and the global properties reflects the natural heterogeneousness of the micro-structures of steels.

At a microscopic level, steel is composed of crystalline grains mixed with small impurities called inclusions. It is well known that conventional steels contain "soft" iron matrix and "hard" inclusions (particles). For the ultra-high strength steel that we are interested in, the iron matrix comprises of ferrite (α), bainite (α'_B), and austenite (γ) that is designed to be transformable into BCC martensite (α') – a desirable hard substance (Olson 1999). The inclusions are divided into two groups: the primary particles with the diameter around microns and the secondary particles in the range of 10^0 to 10^2 nanometers. By adding alloys during metallurgical processes, the desirable dispersed secondary particles, such as M_2C carbides, are formed through precipitation or dissolved from primary particles, which blocks the dislocation path; thus, the strength of steel rise. Whereas the primary particles, such as nitrides, usually have lower toughness as compared with the iron matrix. The formation of primary particles is inevitable during manufacture process. The impurity elements, such as sulfur, may precipitate on the interface between the primary particle and iron matrix. This grain boundary precipitation weakens interfacial traction. The debonding between particles and the iron matrix, which is followed by the subsequent void nucleation, growth and coalescence, is the major cause of low fracture toughness.

A key-issue in steel design is to establish the relationships between the micro/nano-structures of steel and its macro-scale mechanical properties. To this end, a proposed multi-scale, multi-physics computational material design approach is outlined in Figure 1 (Olson et al. 2002). On the left side in this figure is a TEM micrograph of a carbon extraction replica from a fracture surface of a ultrahigh-strength steel (Krauss 1984), from which one can find that the particles can be divided into two groups according to their sizes. The rest of this figure introduces the notion of a hierarchical multi-scale, multi-physics model that is proposed to apply on the fracture process presented in the TEM micrograph. This notion can be interpreted as a bottom-up approach: starting from the right, the diagram with a curve at the upper corner is a traction-separation law through decohesion between secondary particles and iron matrix, which is computed at quantum scale (Krauss 1984). This first principle-based traction-separation law is embedded into the sub-micro cell that contains a secondary particle. The computational strategy introduced in (Liu et al. 1999,

82

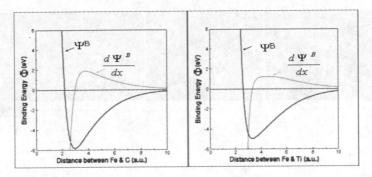

Figure 2. The decohesion (binding) energies and traction-separation laws for Fe/C and Fe/Ti systems.

Hao et al. 2000) in conjunction with a proposed RMMD scheme, which will be introduced in the next section, is applied to obtain a plastic potential and a corresponding constitutive law, which is used as the matrix properties in the micro-scale cell. This procedure is repeated hierarchically at the scale of micro-scale cell to obtain a corresponding macro-scale constitutive law that is used to simulate crack growth and to calculate the corresponding fracture toughness. Figure 2 illustrates the flow chart for the entire process.

The debonding causes void nucleation around both primary and secondary particles. Fracture in such a material is actually the accumulation of the damages in the forms of void nucleation, growth, and coalescence. This damage evolution process is also illustrated in Figure 1 where the vertical axis represents the time-scale. The coalescence between voids is usually accompanied with two competing mechanisms: brittle cleavage and ductile slipping (Tvargaard 1982). The dominant mechanism determines the quantity of the final output for the fracture toughness. Figure 1 proposes the coalescence on the matrix ligament between voids to be phenomenologically reproduced using a strain gradient theory (Fleck 1997, Gao 1999)-based localization solution (Hao et al. 2000); the underlying dislocation mechanism is included. Associated with the localized deformation during void coalescence, the plastic deformation drives the metastable austenite in iron matrix into martensite. As a competing mechanism against fracture, this plastic deformation induced phase transformation increases the strength of the iron matrix and disperses the strain concentration along shear bands, resulting in improvements for both strength and toughness. The Socarate-Parks model (Socarate 1995) is proposed to simulate this process.

In summary, the notions, approaches, and the underlying ideas illustrated in Figure 1 can be outlined as follows:

1) select alloys (chemical ingredient) to form carbon-compounds with the maximum interfacial energy to iron matrix;
2) adjust the sizes, orientations and distributions for both primary and secondary particles to obtain the optimized mechanical behavior; and
3) design the optimum trip-phase composition that can maximize the advantages of deformation induced phase transformation.

For these purposes, the proposed multi-scale, multi-physics steel design approach comprises of the following fundamental elements:

1) first principle principle-based analysis (start at quantum scale);
2) design numerical methods that can capture and bridge the mechanisms from quantum analysis, molecular dynamic simulation, and conventional continuum mechanics;
3) establish the relationship among micro/nano structures, bulk/bonding properties, and fracture toughness/strength;
4) verification by lab-scale test.

2 MULTI-SCALE, MULTI-PHYSICS ANALYSES

2.1 *Interfacial short range force – first principle calculation*

The titanium carbide (TiC), which provides considerable high interfacial traction (Argon et al. 1975, Argon & Im 1975), is applied in the cell models illustrated in Figure 1. The debonding at the grain

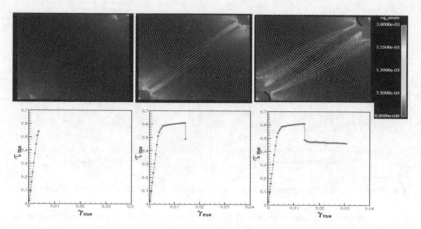

Figure 3. Snap-shots of the localization induced debonding process.

boundary between harder particles and the iron matrix is the controlling factor limiting the ductility and strength for metallic alloys. The first principle-based analysis indicates that the metallic decohesion energy is determined as the eigenvalue of the Hamiltonian operator that includes the quantum momentums and the interactions between nuclei (Z) and nuclei, nuclei and electron (e), electron and electron and the corresponding magnetic interaction. By neglecting the magnetic interaction, this Hamiltonian becomes (Born 1957):

$$H = -\sum_{i=1}^{ZN} \frac{\hbar^2}{2m} \nabla_i^2 - \sum_{i=1}^{ZN}\sum_{a=1}^{N} \frac{Ze^2}{|\mathbf{R}_a - \mathbf{r}_i|} + \frac{1}{2}\sum_{i \neq j}^{ZN}\sum_{i \neq j}^{ZN} \frac{e^2}{|\mathbf{R}_a - \mathbf{r}_i|} + \frac{1}{2}\sum_{a \neq b}^{N}\sum_{a \neq b}^{N} \frac{Z^2 e^2}{|\mathbf{R}_a - \mathbf{R}_b|} \tag{1}$$

where $\mathbf{r}_i$ is the position vector of election I, and R_a is that of nucleus a. Assuming the separation normal to an interface, the interfacial decohesion at the carbon site, i.e., the carbon ions are the closest neighbors to the iron ions at the grain boundary between TiC (titanium carbide) and the FCC iron matrix, has been calculated using FLAPW (Freeman 1991) – a density functional theory-based numerical procedure. This numerical analysis indicates that the C-site decohesion energy is 3.82 J/m². The detail of this simulation will be given in Krauss et al. 2002. Based on a similar analysis, the inter-atomic interaction relations, i.e. the decohesion energies and corresponding traction-separation relations between Fe (atom) and Ti (atom) as well as between Fe and C atoms, are also obtained, as shown in Figure 2; where Ψ is the decohesion energy and x is the distance between two nuclei.

2.2 Interfacial metallic decohesion/debonding

The C-site decohesion energy (3.82 J/M²) provides an upper bound estimate of the debonding process at TiC/FCC-Fe boundary (Krauss et al. 2002). This is because in steels the morphology at this kind of grain boundaries can be very complicated. By alternating crystal orientation, e.g. to the Ti-site (Figure 3), the dechesion energy will be reduced significantly. Also the tangent traction and corresponding dislocations motion play important roles during the grain boundary decohesion.

In order to consider these factors, a numerical approach, which we call RMMD (RKPM-MPFEM-Molecule Dynamic) (Liu et al. 1995 & 1999, Jao & Liu 2002), is proposed for studying the grain boundary debonding process. This approach can be briefly summarized as:

1) the Fe-FCC matrix and TiC carbide, respectively, are considered as continuum solids which are simulated using MPFEM to keep its crystal properties;
2) between the Fe atoms in the iron matrix and Ti or C atoms in particles, the molecule dynamic relations are imposed using the RKPM approach based on the inter-atomic interaction relations introduced in Figure 2;
3) when this approach is applied to the cell model in Figure 1, a lumped RKPM-MD scheme is applied to improve the computation efficiency.

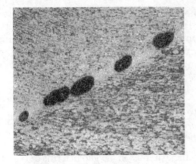

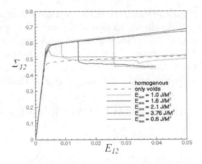

Figure 4.   An experimental observation [J.F. Mescall].

Figure 5.   The effect of decohesion energy on the global stress–strain curves.

The scheme is preformed by lumping a certain number of atoms into a particle and only the interaction between Fe and TiC particles are taken into account. Under this situation, the interaction force between the particle I with N_i lumped atoms in the iron matrix and the particle J with N_j lumped atoms in TiC carbide is defined by

$$F_{IJ} = \sum_{i=1}^{N_I} \sum_{j=1}^{N_J} \frac{\partial \Psi_{ij}}{\partial r_{ij}}$$ (2)

where r_{ij} is the distance between the atom i that is lumped into the particle I and the atom j that is lumped into particle J; Ψ_{ij} is the corresponding potential defined in the previous steps (1) and (2). When $N_i = N_j = 1$, 2 degenerates to the non-bond force based-molecular dynamic, such as L-J force. Based on the steps (1) and (2) of the RMMD approach, the interfacial traction-separation relation of the Ti-site has been obtained. The FDT calculation (Krauss et al. 2002) and the corresponding steps (1) and (2) of RMMD simulation give the short range separation force at Fe-FCC/TiC grain boundary. However, in steels a debonding will occur under the stress that is much lower than the first principle-based theoretical prediction due to the point defects such as dislocations and other kinds of impurities. According to the slip-line analysis in (Argon et al. 1975, Argon & Im 1975), the maximum separation stress is around $(1 + \sqrt{3})$ σ_{flow} where σ_{flow} is the flow stress of the iron matrix. This stress threshold, in conjunction with the first principle-based theoretical decohesion, provides the estimate of long range normal traction-separation laws.

The interfacial tangent traction is also crucial for such a grain boundary decohesion process, which is not yet taken into account either in the FDT simulation (Krauss et al. 2002) or in the previously introduced RMMD approach. By assuming that both tension and shear debonding at Fe-FCC/TiC boundary yield the same decohesion energy, then, according to dislocation analysis (Argon et al. 1995, Eshelby 1957, Rice & Wang 1989) the shear traction-separation law can be expressed as

$$\tau = \frac{\pi \gamma_b}{b} \sin\left(\frac{2\pi \Delta_t}{b}\right)$$ (3)

where γ_b denotes the decohesion energy that is calibrated when the normal separation is b (Burger's vector) per unit length from Figure 3b; Δ_t is the tangent separation. Both (3) and the normal long range traction-separation law can be expressed as the interfacial debonding potential defined by Needleman (Eshelby 1957).

Plotted in Figure 3 are a set of snapshots of process of debonding, voids nucleation and coalescence through shear banding between two particles, computed using RMMD on a volumetric representative element (cell model) for secondary particles. In these snapshots the equivalent plastic strain contours are plotted with the corresponding average stress–strain output. A suddenly stress drop occurs when debonding takes place. After the parameter studies in the cell modeling for both primary and secondary particles, we conclude that the volume fraction, orientations and distribution of the particle, also the stress state, and decohesion energy – these are the factors which determine the failure process of the steel model. Among those parameters the decohesion energy is especially important for steel design as it is directly associated with the strength of the cell. Plotted in Figure 5 is a set of computations where the decohesion

85

energy varies from zero (this corresponds to the case of voids) to its upper bound (C-site) while the spacing and volume fraction of the secondary particles are fixed. If we use the strain E_{12}^{debound} to represent the average (global) shear strain when debonding occurs, from Figure 5 one can find that under the imposed shear boundary condition E_{12}^{debound} increases substantially when the decohesion energy rises. When the decohesion reaches its upper bound ($3.8\,\text{J/M}^2$), no debonding takes place at all. At this end the debonding occurs when a certain amount of triaxial tension is imposed on the cell.

3. BRIDGING-SCALE CONCURRENT SIMULATION METHODS

3.1 *Overview of bridging scale concurrent simulation method*

The development of a bridging scale method is motivated by the extremes in the gap in terms of both length and time scales that are involved in the applications of nanoscale materials. Examples of such include nano-devices used in conjunction with other components that are larger, and have different response times. We propose a bridging-scale concurrent approach (Qian et al. 2002) for the correct simulation and, thus, design of such hybrid devices/structures. A unique feature of the method is that it combines the advantages of both continuum and atomistic simulation approaches, and enables tackling problems with larger length scales yet not losing any accuracy in describing the localized phenomena. Both efficiency and accuracy can be achieved in an optimum sense. The method is implemented with the formulation of the so-called "bridging scale" component. The bridging-scale component can be obtained by defining a multiscale projection, which effectively removes the overlaps between the two distinctive representations. In the proposed method, the mesh-free discretization co-exists with the molecular structure in these enriched regions. The methodology of the multiscale method can be traced back to the original paper by Liu et al. (1997), and has been successfully applied in the multiple-scale problems involving strain localization (Hao et al. 2000) boundary layers (Wagner et al. 2001) and coupling of finite elements with meshfree shape functions (Wanger & Liu 2001).

In the current problem setting, this is done by writing a multiple-scale decomposition of the atomic displacements u_α in terms of the coarse scale displacement d_I (approximated by either finite element or meshfree method (Li et al. 2002) and MD displacements d_α. To avoid any overlap between these two solutions, we should take projection of the MD displacements onto the shape function basis. Designating this projection operator as P:

$$u_\alpha = \sum_I N_I(x_\alpha)d_I + d_\alpha - Pd_\alpha \qquad (4)$$

where $N_I(x_\alpha)$ is the finite element/meshfree shape function for node I evaluated at atom α. The subtraction of the projection of d_α, is called the bridging scale and which allows for a unique decomposition into coarse and fine scales. With this decomposition, similar expressions in the form of a coarse scale equation and the standard MD equations of motion are derived. With a particular projection, the two sets of equation are only coupled through the forcing terms and boundary conditions, while the stiffness matrix is completely decoupled. This allows the use of existing FE/Meshfree code and MD codes along with suitable methods for exchanging information about internal forces and boundary conditions.

3.2 *Application to nanomaterials*

Figure 6(a) Qian et al 2002 shows a benchmark example of implementing the multiscale method for a particular nanostructure called a carbon nanotube (Qian et al. 2002). The molecular structure is subjected to pure bending as shown in Figure 6(b). It can be seen that the meshfree discretization and MD coexist and are coupled in the fine scale region. For the coarse scale region, a meshfree approximation is used. The multiwalled carbon nanotube system contains approximately three million atoms. This is replaced by approximately 30,000 nodes and another 80,000 atoms in the enriched region. The buckling pattern obtained from the simulation compares well with the TEM observation (Qian et al. 2002). A systematic description of this method is described in Qian, Wagner & Liu 2002, Qian et al. 2002.

3.3 *Application to nonlinear problem*

The bridging scale has recently been tested in a 1D model nonlinear problem. In this case, the Lennard-Jones 6–12 potential was used to describe the atomic interactions. The form of the LJ 6–12 potential energy function is well known, and is defined to be

(a) 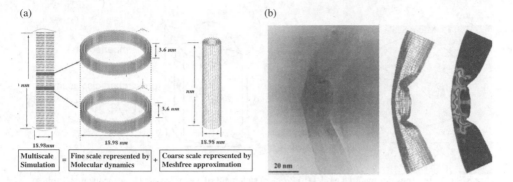 (b)

Figure 6.   (a) Multiscale analysis of nanostructures and (b) comparison of simulation results (right) with experimental observation for a carbon nanotube.

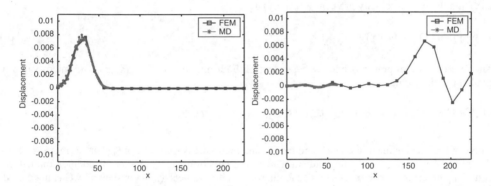

Figure 7.   Left: initial MD displacement with a fine scale component. Right: final MD displacement demonstrating transfer of fine scale wavelengths into the continuum.

$$\phi(r) = 4\varepsilon\left(\left(\frac{\sigma}{r}\right)^{12} - \left(\frac{\sigma}{r}\right)^{6}\right) \tag{5}$$

We show a simple 1D example in which the initial MD displacement follows the form of a Gaussian wave, but also has a significant fine scale component, which tests the ability of the bridging scale to transfer small wavelength information from the atomistic to the continuum. In this case, the finite element node to atom ratio was 8:1. As can be seen, the initial energy present in the MD system transfers into the surrounding continuum for both cases. In the first case, when no small wavelength information is present, the coarse scale is able to capture the long wavelengths of the system. When the fine scale perturbations are present in the MD displacement, the usage of a time history kernel allows the dissipation of the fine scale wavelengths into the continuum. Details on this coupled boundary condition on the MD displacement can be found in the work by Wagner et al. 2003.

4   MULTI-PHYSICS MATERIAL MODEL WITH BRIDGING-SCALE CONCURRENT SIMULATION METHODS

4.1   *Multi-physics material model*

The derivation of the multi-physics material model has its base on the so-called micromorphic material (Germain 1973, Eringen & Suhubi 1964) . To consider the micro-scale phenomena, a material point is treated as a continuum with finite volume called a micro cell (Fig. 8(a)). The total velocity at a point in

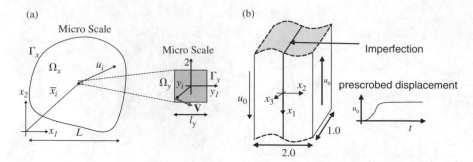

(a) Micro Scale

(b)

Figure 8. (a) Multi-physics material model (b) Problem statement of 1-D dynamic shear localization problem.

a micro-cell $\dot{\mathbf{u}}(\overline{\mathbf{x}} + \mathbf{y})$ is expressed as a superposition of the velocity at the mass center of a micro cell $\dot{\mathbf{u}}(\overline{\mathbf{x}})$ and the micro-velocity gradient $\chi_{ij}(\overline{\mathbf{x}} + \mathbf{y})$ as:

$$\dot{u}_i(\overline{\mathbf{x}} + \mathbf{y}) = \dot{u}_i(\overline{\mathbf{x}}) + \chi_{ij}(\overline{\mathbf{x}} + \mathbf{y})y_j \tag{6}$$

The internal virtual power is linearly dependent on both macro-velocity gradient and micro-velocity gradient in the form:

$$\delta\mathrm{P}^{\mathrm{int}} = \int_{\Omega_x} \sigma_{ji}(\overline{\mathbf{x}})\delta D_{ij}(\overline{\mathbf{x}}) + \sigma'_{ji}(\overline{\mathbf{x}})(\delta\chi_{ij}(\overline{\mathbf{x}}) - \delta L_{ij}(\overline{\mathbf{x}})) + \tau_{kji}(\overline{\mathbf{x}})\delta\chi_{ij,k}(\overline{\mathbf{x}})\mathrm{d}\Omega_x \tag{7}$$

where σ'_{ji}, τ_{kji} denote the micro-stress and the second micro-stress, and D_{ij} and L_{ij} denotes macro-rate-of-deformation and macro-velocity gradient. Governing equations become a set of coupled multi-field equations in terms of macro velocity $\dot{u}_i$ and micro velocity gradient χ_{ij}. As is shown by Chambon et al. 2001, these equations reduce to the Cosserat theory when the micro velocity gradient is restricted to be anti-symmetric, and reduce to the strain gradient theory of Fleck et al. (1994) when it is equal to the macro-rate-of-deformation.

To establish constitutive relations for these micro-stresses, a meso-cell (Kadowaki & Liu 2003) is added to this theory. Define β_{ji} as the gradient of the internal power density in terms of micro-velocity gradient. The internal power density can be expanded as follows:

$$\dot{e}^{\mathrm{int}}(\overline{\mathbf{x}} + \mathbf{y}) = \dot{e}^{\mathrm{int}}(\overline{\mathbf{x}} + \mathbf{y})\big|_{\chi_{ij}=L_{ij}} + \beta_{ji}(\overline{\mathbf{x}} + \mathbf{y})(\chi_{ij}(\overline{\mathbf{x}} + \mathbf{y}) - L_{ij}(\mathbf{x} + \mathbf{y})) \tag{8}$$

Micro stresses are defined by the zeroth and the first moment of β_{ji} over the meso-cell.

$$\sigma'_{ji}(\overline{\mathbf{x}}) \equiv \frac{1}{V_{\mathrm{meso}}} \int_{\Omega_{\mathrm{meso}}} \beta_{ji}(\overline{\mathbf{x}} + \mathbf{y})\mathrm{d}\Omega_{\mathrm{meso}}$$

$$\tau_{kji}(\overline{\mathbf{x}}) \equiv \frac{1}{V_{\mathrm{meso}}} \int_{\Omega_{\mathrm{meso}}} y_k \beta_{ji}(\overline{\mathbf{x}} + \mathbf{y})\mathrm{d}\Omega_{\mathrm{meso}} \tag{9}$$

To illustrate the response of this model, a one-dimensional dynamic shear-localization problem of granular material is analyzed with a strain softening constitutive law (Fig. 8(b)). In this example, we neglect the symmetric part of the micro velocity gradient. Proposed material model provided the finite width of a shear-localized zone (Fig. 9(a)).

4.2 Application with bridging scale concurrent calculation method

Although proposed multi-physics material model can provide a finite width of a shear band, very fine resolution is required to capture the phenomena correctly. A coupling of fine/coarse calculations is a common

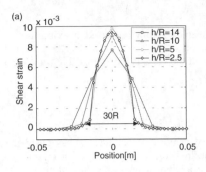

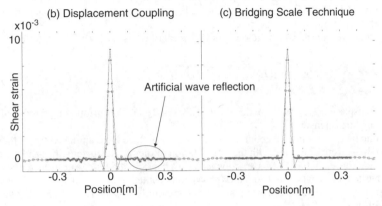

Figure 9.   Results of dynamic shear-localization calculation (a) finite width of a shear band (b) with direct coupling and (c) with bridging scale technique.

way to reduce the computation. However, if it is done by enforcing the continuity of the displacement, spurious wave reflections are often observed at the transition point of resolution level. Bridging scale concurrent simulation method (Wagner & Liu 2003, Wagner, Karpov & Liu 2002, Karpov, Wagner & Liu 2003) can deal with this difficulty. Figure 9(a) and (b) shows the shear strain distribution just after the waves induced by the failure went out of the fine-scale zone. While the displacement coupling method showed artificial reflection of high frequency waves (Fig. 9 (b)), bridging scale technique correctly suppress it and provide a converged solution within affordable computational cost (Fig. 9 (c)).

5   FLUID-STRUCTURE INTERACTIONS

The proposed Immersed Finite Element Method (IFEM) presented here is developed for modeling fluid-structure interactions (FSI) encountered in typical bioengineering problems as well as other physical models. In the Immersed Boundary Method (IB) that was originally developed by Peskin, the submerged structure is represented by a fiber network that are used to represent fibers and muscles. The main disadvantage of this method is the lack of an accurate solid representation that is capable of producing stress distributions for membrane-like or full-fledged three-dimensional solids. Problems still remain in treating complex bio-engineering problems that require arbitrary shapes and boundary conditions for the fluid domain. In the IFEM, the fluid dynamics is governed by the Navier Stokes equations of the viscous incompressible fluid with the Eulerian description of motion. The structure dynamics is governed by the Mooney-Rivlin rubber-like material model with the Lagrangian description of motion. In here, the discretized delta function needs to be able to handle non-uniform spaced fluid grids. The RKPM delta function reproduces nth order polynomial terms. The use of such kernel function may eventually open doors to the multi-scale and multi-resolution modeling of complex biological fluid involving molecular, cellular, and flexible vessel-flow interactions.

5.1 Governing equations

Consider a three-dimensional flexible structure immersed in a viscous incompressible fluid with arbitrary geometries and boundary conditions.

The Navier-Stokes equations of the viscous incompressible fluid are used for the fluid domain Ω:

$$\rho^f \left(\frac{\partial \mathbf{v}}{\partial t} + \mathbf{v} \cdot \nabla \mathbf{v} \right) = \nabla \cdot \boldsymbol{\sigma} + \mathbf{f} \quad \text{on } \Omega \tag{10}$$

$$\nabla \cdot \mathbf{v} = 0$$

where ρ^f, $\mathbf{v}$, $\boldsymbol{\sigma}$ and $\mathbf{f}$ stand for the density, velocity, Cauchy stress tensor and body force, respectively in the fluid domain. Moreover, the Cauchy stress tensor is defined as $\sigma_{ij} = -p\delta_{ij} + \tau_{ij}$ with $\tau_{ij} = \mu(v_{i,j} + v_{j,i})$, μ is the fluid dynamic viscosity. In addition to these equations of motion, we stipulate the following boundary conditions $v_i = g_i$ for $x \in \Gamma_{gi}$ and $\sigma_{ij} n_j = h_i$ for $x \in \Gamma_{hi}$ where g_i and h_i are given functions of prescribed velocities and surface tractions, respectively.

The equation of motion for the solid domain Ω_s is described as follows:

$$\rho^s \left(\frac{\partial^2 \mathbf{u}^s}{\partial t^2} \right) - \nabla \cdot \tau^s = \mathbf{f}^{sur} + \rho^s \mathbf{g} \quad \text{on } \Omega_s \tag{11}$$

where $\mathbf{u}^s$ is the displacement, ρ^s is the material density, $\mathbf{f}^{sur}$ and $\mathbf{g}$ are the surrounding forces and the gravity, τ^s is the Cauchy stress tensor.

The key concept of the immersed boundary method is to replace the coupling between the submerged structure and the surrounding fluid with the interpolation of the fluid grid velocities and the distribution of the solid nodal forces. In the IFEM, we no longer have a background fluid grid, instead we employ finite element meshes to represent the fluid domain with general geometries and boundary conditions. Therefore, the interpolation and the distribution take place among the fluid and solid finite element nodes, which are depicted as follows:

$$\mathbf{f} = \int_{\Omega_s} \mathbf{f}^{sur}(\mathbf{X}) \delta(\mathbf{x} - \mathbf{X}) dV \tag{12}$$

$$d\mathbf{X}/dt = \int_{\Omega} \mathbf{v}(\mathbf{x}) \delta(\mathbf{x} - \mathbf{X}) dV \tag{13}$$

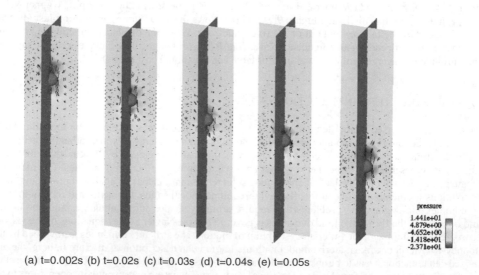

(a) t=0.002s (b) t=0.02s (c) t=0.03s (d) t=0.04s (e) t=0.05s

Figure 10.   A sphere with a diameter of 1.0 cm dropping in a fluid domain of $16 \times 4 \times 4 \, cm^3$. The fluid pressure is shown in the contour plots and the velocities are represented by the vectors.

where $\mathbf{f}^{sur}$ stands for the resultant nodal force due to the surround fluid at the Lagrangian node and $\mathbf{X}$ is the corresponding Lagrangian nodal position. The Reproducing Kernel Particle Method (RKPM) delta function is used here instead for non-uniform spacing where an arbitrary fluid domain can be represented.

5.2 *A sphere dropping in a channel*

One of the examples is a sphere dropping in a fluid channel. The fluid domain has a size of $16 \times 4 \times 4\,\text{cm}^3$. The channel contains two planes. The diameter of the sphere is 1.0 cm. The gravity is $980\,\text{cm/s}^2$. Shown in Figure 10 are the snapshots of the movements of the sphere and the pressure contours and the velocity vectors of the fluid domain. As seen in the figures, the gravity creates a force that drives the sphere in the channel, this force influences its surroundings in the fluid (shown in the pressure contour); and at the same time, the velocity of the fluid calculated from the fluid solver with the external and surrounding force applied drives sphere to move downward.

6 NANO-FLUIDICS

Advanced nanofabrication techniques will make it possible to build complex nano-scale devices. Some of these small devices are likely to contain fluids inside nanotubes or fluids around nanotubes, such as nanotubes immersed in water. The nanotube (or generalized nanochannel) can be both conduit for transport and sensor of flow. The nano-scale behavior of fluids may be fundamentally different from fluid behavior in a bulk system. Many current or envisioned applications such as biosensors, fuel storage devices and devices for which tribology plays a central role call for a deeper understanding of nano scale fluid behavior. We have studied the problem of flow past nanotube. We hope that our work could help in design and applications of the nanotube-based fluidic sensors.

6.1 *Water flow past nanotube*

We generated uniform helium and water fluid flow past a two-end clamped nanotube, for nanotubes of various sizes, at different flow velocities. The uniform fluid flow velocities we used were 1, 10, and 100 m/s. The example shown in Figure 11 is a (10,10) SWCNT with a diameter of 13.1 Å and a length of 60.3 Å. The fluid atoms fill the region, surrounding the SWCNT. Periodic boundary conditions are applied for the three orthogonal directions.

In this problem, three different interactions are considered: the potential energy of carbon–carbon interactions within the carbon nanotube (modelled with the Tersoff-Brenner potential), the carbon–water molecule interactions (modeled with the C–O Lennard-Jones potential) and the water molecule–water molecule non-bonded interactions (TIP4P rigid water model is used). Based on this model, we compared fluid flow for the three different fluid velocities, at a temperature of 298 K. Using different fluids or different fluid flow velocities lead to different dynamics behavior of the nanotube, such as drag force, boundary condition and deformation. For example, our simulation results show that the drag force on the nanotube will increase as flow velocity and nanotube diameter increase. An analytical solution based on Oseen's

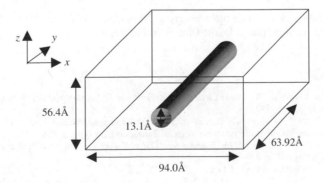

Figure 11.   Water flow past a (10,10) nanotube.

91

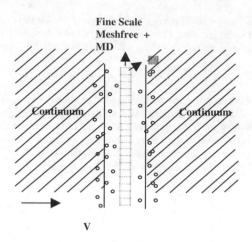

Figure 12.   Multiscale approach for the water flow past nanotube.

approximation for flow due to a moving body at small Reynolds number was also obtained. Comparison of simulation results with analytical solution shows that the analytical solution does not agree with the MD calculations, which demonstrates that the continuum assumption breaks down for the small Reynolds number case. But with increasing flow velocities or sizes of nanotubes, these two results show tendency of getting closer. By comparing MD results with continuum solutions, it should be possible to map out when the continuum equations are valid and when they are not.

6.2   *Multiscale approach*

MD simulations can only be employed for a short time and are useful for treating only small length scales due both computational cost and the time of the calculation. The nanotube used in experiments are multi-wall carbon nanotubes with lengths up to 2 μm, which involves billions of atoms. In order to increase the size of simulation and compare to experimental results, a multiscale approach for the flow past nanotube problem should be used (Fig. 12). This multiscale model consists of a continuum region far away from nanotube and a MD region close to nanotube. In continuum region, coarse scale nodes will be used to represent the fluid particles and continuum fluid solvers such as FEM or RKPM will be applied. In MD region, molecular dynamics will be applied. One challenge here is how to treat the flux of atoms between the MD region and the continuum region. One possible approach is to use some Monte-Carlo trial scheme, where an insertion or removal is accepted or rejected according to its influence with the energy. Techniques such as Schwarz iteration can be applied in the coupling region to achieve continuity.

ACKNOWLEDGEMENT

The authors gratefully acknowledge the supports of NSF, ONR, ARO, and NSF-IGERT. Dong Qian acknowledges the general support from the Ohio Board of Regents.

REFERENCES

Argon, A.S., Im, J. and Safoglu, R., 1975. *Cavity Formation from Inclusions in Ductile Fracture*. Metallurgical Transactions, A 6(4): p. 825–837.

Argon, A.S. and Im, J., 1975. *Separation of Second-Phase Particles in Spheroidized 1045 Steel, Cu-0.6pct Cr Alloy, and Maraging-Steel in Plastic Straining*. Metallurgical Transactions, A 6(4): p. 839–851.

Belytschko, T., Lu, Y.Y. and Gu, L., 1994. *Element-free Galerkin methods*. International Journal for Numerical Methods in Engineering, 37: p. 229–256.

Born, M., 1957. *Atomic Physics*. 6th ed.: Blackie.

Chambon, R., Caillerie, D. and Matsuchima, T., 2001. *Plastic continuum with microstructure, local second gradient theories for geomaterials: localization studies*. International Journal of Solids and Structures, 38(46–47): p. 8503–8527.

Cybersteel 2020, *ONR contract grant number: N00014-01-1-0953*, Northwestern University.

Eringen, A.C. and Suhubi, E.S., 1964. *Nonlinear Theory of Simple Micro-Elastic Solids.* International Journal of Engineering Science, 2: p. 189–203.

Eshelby, J.D., 1957. *The determination of the elastic filed of an ellipsoidal inclusion and related problems.* Proc. Roy. Soc. A., 241(1226): p. 376–396.

Fleck, N.A., et al., 1994. *Strain Gradient Plasticity – Theory and Experiment.* Acta Metallurgica Et Materialia, 42(2): p. 475–487.

Fleck, N.A. and Hutchinson, J.W., 1997. *Strain gradient plasticity.* Advances in Applied Mechanics, 33: p. 295–361.

Freeman, 1991. A.J., *FLAPW.*

Gao, H.J., Huang, Y., Nix, W.D. and Hutchinson, J.W., 1999. *Mechanism-Based Strain Gradient Plasticity – I. Theory.* Journal of Mechanics and Physics of Solids, 47: p. 1239–1263.

Germain, P., 1973. *Method of Virtual Power in Continuum Mechanics .2. Microstructure.* Siam Journal on Applied Mathematics, 25(3): p. 556–575.

Hao, S., Liu, W.K. and Qian, D., 2000. *Localization-induced band and cohesive model.* Journal of Applied Mechanics-Transactions of the Asme, 67(4): p. 803–812.

Hao, S., Liu, W.K. and Chang, C.T., 2000. *Computer implementation of damage models by finite element and meshfree methods.* Computer Methods in Applied Mechanics and Engineering, 187: p. 401–440.

Hao, S., Park, H.S., Liu, W.K., 2002. *Moving particle finite element method.* International Journal for Numerical Methods in Engineering, 53(8): p. 1937–1958.

Hao, S. and Liu, W.K., 2002. *Moving particle finite element method with global superconvergence.* submitted for publication.

Krauss, G. 1984. *Shear fracture of ultrahigh strength low alloy steels.* In *Innovations in Ultrahigh-strength steel technology.* Lake George, NY.

Krauss, G., Freeman, A.J., et. al. 2002, Northwestern University.

Kadowaki, H. and Liu, W.K., 2003. *Multiscale Analysis of Heterogeneous Materials.* Manuscript in preparation.

Karpov, E.G., Wagner, G.J., and Liu, W.K., 2003. *A Green's function approach to deriving wave-transmitting boundary conditions in moluecular dynamics simulations.* Submitted to Physical review letters.

Li, S.F. and Liu, W.K., 2002. *Meshfree and Particle Methods and Their Applications.* Applied Mechanics Reviews, 55: p. 1–34.

Liu, W.K., Jun, S. and Zhang, Y.F., 1995. *Reproducing Kernel Particle Methods.* International Journal for Numerical Methods in Fluids, 20: p. 1081–1106.

Liu, W.K., Hao, S., Belytschko, T., Li, S.F., Chang, C.T., 1999. *Multiple scale meshfree methods for damage fracture and localization.* Computational Materials Science, 16(1–4): p. 197–205.

Liu, W.K., Uras, R.A. and Chen, Y., 1997. *Enrichment of the finite element method with the reproducing kernel particle method.* Journal of Applied Mechanics-Transactions of the Asme, 64(4): p. 861–870.

Needleman, A., 1987. *A continuum model for void nucleation by inclusion debonding.* J. Appl. Mech., ASME, Trans, 54(3): p. 525–531.

Olson, G.B., 1997. *Computational design of hierarchically structured materials.* Science, 277(5330): p. 1237–1242.

Olson, G.B., 1999. *New directions in martensite theory.* Materials Science and Engineering a-Structural Materials Properties Microstructure and Processing, 275: p. 11–20.

Olson, G.B., Liu, W.K., Moran, B., Hao, S. 2002. *A multi-scale, multi-physics model in steel design.* In *SRG Meeting 2002*, Northwestern University, Evanston, IL, U.S.A.

Qian, D., et al., 2002. *Mechanics of carbon nanotubes.* Applied Mechanics Reviews, 55(6): p. 495–533.

Qian, D., Liu, W.K. and Ruoff, R.S., 2002. *Effect of interlayer interaction on the buckling of multi-walled carbon nanotube (under review).* Journal of nanoscience and nanotechnology.

Qian, D., 2002. *Multiscale simulation of the mechanics of carbon nanotubes*, in *Mechanical Engineering Department.*, Northwestern University: Evanston.

Qian, D., Wagner, G.J. and Liu, W.K., 2002. *A multiscale projection method for the analysis of carbon nanotubes (submitted).* Computer method in applied mechanics and engineering.

Rice, J.R. and Wang, J.S., 1989. *Embrittlement of Interfaces by Solute Segregation.* Materials Science and Engineering a-Structural Materials Properties Microstructure and Processing, 107: p. 23–40.

Socarate, S., Parks, D.M. 1995, MIT: Cambridge, MA.

Tvargaard, V. and Hutchinson, J.W., 1982. *On Localization in Ductile Materials Containing Spherical Voids.* International Journal of Fracture, 18: p. 237–252.

Wagner, G.J. and Liu, W.K., 2003. *Coupling of Atomistic and Continuum Simulations using a Bridging Scale Decomposition.* Accepted by Journal of Computational Physics.

Wagner, G.J., et al., 2001. *The extended finite element method for rigid particles in Stokes flow.* International Journal for Numerical Methods in Engineering, 51(3): p. 293–313.

Wagner, G.J. and Liu, W.K., 2001. *Hierarchical enrichment for bridging scales and mesh-free boundary conditions.* International Journal for Numerical Methods in Engineering, 50(3): p. 507–524.

Wagner, G.J., Karpov, E.G. and Liu, W.K., 2002. *Molecular Dynamics Boundary Conditions for Periodically Repeating Atomic Lattices.* Manuscript in preparation.

93

Computational Methods in Engineering and Science, Iu et al. (eds)
© 2003 Swets & Zeitlinger, Lisse, ISBN 90 5809 567 3

On the duality between hybrid and block elements

E.R. de Arantes e Oliveira & E.M.B.R. Pereira
Instituto Superior Técnico, Universidade Técnica de Lisboa, Lisboa, Portugal

ABSTRACT: In hybrid elements, the fields allowed within each element are equilibrated stress fields, the topology is defined by providing for each element the set of its nodes and the interaction between adjacent elements by setting compatibility conditions in terms of the nodal displacements. Hybrid elements are deformable, although rigidly connected along their common boundaries. The approximate solution is obtained by equalizing the displacements at the coinciding nodes of adjacent elements. In block elements, the fields allowed within each element are continuous displacement fields, the topology is defined by providing for each element the set of adjacent elements and the interaction between adjacent elements by equilibrium conditions in terms of the contact forces. Blocks are themselves rigid, although not rigidly connected to the adjacent elements along their common boundaries. The approximate solution is obtained by equilibrating the generalized contact forces associated to the interfaces of adjacent elements. The conclusion is drawn that the decomposition into hybrid elements provides convergence by approximate solutions stiffer than the exact solution; the decomposition into block elements provides convergence by approximate solutions less stiff than the exact solution.

1 INTRODUCTION

Let us consider a body decomposed into finite elements and suppose that, within each element, displacements, stresses and strains are continuous, and no body forces are assumed to be applied.

Let Ω denote the domain occupied by the body. Let Ω_e denote the subdomain corresponding to a current finite element e.

Let $\mathcal{H}$ be a Hilbert space whose points are structural fields on Ω, let the scalar product, the energy product, and the energy norm, in $\mathcal{H}$, be denoted respectively by (,), [,] and $\| \, \|$, the square of the energy norm being the strain energy, and the distance between two points in $\mathcal{H}$ representing the energy norm of their difference (Arantes e Oliveira 1968).

A general structural model involves three kinds of magnitudes – stresses, strains and displacements – the vectors of which are denoted by s, e and u, interrelated by the following equations:

equilibrium equations:

$$E\,s + f = 0; \tag{1}$$

strain-displacement equations:

$$e = D\,u; \tag{2}$$

stress–strain equations:

$$s = H\,e; \tag{3}$$

in which E and D are matrices of linear differential operators, and, for sake of material linearity and material stability, H is a symmetric, positive definite matrix of magnitudes depending on the constants which characterize the mechanical properties of the body. Vectors s and u are assumed such that the derivatives involved in E and D exist everywhere within each subdomain Ω_e.

On the boundaries $\partial\Omega$ and $\partial\Omega_e$ of domains Ω and subdomains Ω_e, the equilibrium equations become

$$N\,s = p ,$$
(4)

where N is a matrix, the elements of which are odd linear functions of the components of the unit normal vector at each point of the domain boundary, and p is the vector of the tractions applied to the boundary.

E, D, and N are such that the following identity,

$$\int_{\Omega_e} s^T D u \, d\Omega_e = \int_{\Omega_e} (E\,s)^T u \, d\Omega_e + \int_{\partial\Omega_e} (N\,s)^T u \, d\partial\Omega_e ,$$
(5)

i.e., the *work principle*, holds, for any subdomain Ω_e.

The strain energy associated with Ω_e is

$$U_e = \int_{\Omega_e} W \, d\Omega_e ,$$
(6)

where W is the strain energy density.

The complementary energy is

$$U_e^* = \int_{\Omega_e} W^* \, d\Omega_e ,$$
(7)

where

$$W^* = (s,e) - W$$
(8)

is the complementary energy density.

For sake of material linearity, W is a quadratic function of the strains and W^* a quadratic function of the stresses.

The general structural model that has been described admits two dual minimum theorems, the *total complementary energy theorem*, and the *total potential energy theorem*.

The total complementary energy theorem states that the solution to the model equation minimizes the *total complementary energy*:

$$T_e^* = U_e^* - \int_{\partial\Omega_e} p^T \underline{u} \, d\partial\Omega_e ,$$
(9)

on the *equilibrated fields* subspace. The total potential energy theorem states that the solution to the model equation minimizes the *total potential energy*:

$$T_e = U_e - \int_{\partial\Omega_e} \underline{p}^T u \, d\partial\Omega_e ,$$
(10)

on the *compatible fields* subspace. The underlined symbols denote prescribed magnitudes.

Conversely, any field that minimizes the potential (complementary) energy on the space of the compatible (equilibrated) fields is an exact solution to the equations and boundary conditions of the model.

2   DUALITY BETWEEN HYBRID AND BLOCK ELEMENTS

2.1   *Hybrid elements*

Hybrid elements (Pian 1964) are such that:

(i) the stress fields

$$s = \chi_e \, s_e \tag{11}$$

allowed within the elements are defined through polynomial stress fields assumed to equilibrate vanishing body forces, and depending on a set of generalized stresses s_e;

(ii)   the stresses s allowed within the elements are equilibrated but not necessarily compatible;

(iii) the nodes are located on the element boundaries, so that each node belongs to several elements; the topology of a system of hybrid elements can be defined by providing for each element the set of its nodes;

(iv)   the interaction between contiguous elements is defined by compatibility conditions expressed in terms of the nodal displacements q_e, such that, on $\partial \Omega_e$,

$$u = \Phi_e \, q_e. \tag{12}$$

Let us select a set of nodes on the hybrid element boundaries, and suppose that a set of nodal displacements is assigned to each node, such that the displacements along the segment between two contiguous nodes on the element boundary can be defined in terms of the nodal displacements at the ends of such segment. Then:

(i)   displacements along the boundary can be expressed in terms of the element nodal displacements;

(ii) the approximate solution associated to the decomposition of the body into hybrid elements is obtained by equalizing the generalized displacements at the coinciding nodes of adjacent elements;

(iii) hybrid elements are deformable, although rigidly connected along their boundaries;

(iv) deformability in a system of hybrid elements is essentially associated to the elements themselves, not to the element interfaces.

In hybrid elements, *the field that minimizes the element total complementary energy on a linear space of allowed fields* (11) *is the one belonging to that space that is nearest to the exact solution – equilibrated and compatible – associated to given boundary displacements*. In other words, introducing (11) and (12) in (9), differentiating and equating to zero, we obtain the *generalized stress-displacement equations* (Arantes e Oliveira 1975):

$$s_e = G_e D_e \, q_e, \tag{13}$$

where

$$G_e = \left[\int_{\Omega_e} \chi_e^T \, H^{-1} \, \chi_e \; d\Omega_e \right]^{-1}, \tag{14}$$

$$D_e = \int_{\partial \Omega_e} \chi_e^T \, N^T \, \Phi_e \; d\partial \Omega_e. \tag{15}$$

The nodal forces Q_e, associated with the nodal displacements q_e, can be defined as:

$$Q_e = \int_{\partial \Omega_e} \Phi_e^T \, p \; d\partial \Omega_e, \tag{16}$$

and, by virtue of (4) and (15),

$$Q_e = D_e^T \, s_e. \tag{17}$$

The decomposition of the body into hybrid elements leads to a system of elements satisfying (11), which behaves more and more like a system of compatible finite elements, as the degree of the polynomial shape functions in (11) becomes larger and larger. Conformity between elements is ensured by conditions (12).

97

Compatibility between elements is enforced by introducing the system displacement vector, U, and expressing vectors q_e as a linear function of U, i.e., by writing

$$q_e = A_e U .$$ (18)

Therefore,

$$K U = F$$ (19)

where

$$K = \sum_e (D_e A_e)^T G_e D_e A_e ,$$ (20)

and

$$F = \sum_e A_e^T Q_e .$$ (21)

2.2 Block elements

Block elements (Cundall 1971; Kawai 1977; Lemos 1987) are such that:

(i) the displacement fields

$$u = \Psi_e u_e,$$ (22)

allowed within the elements are defined through polynomial displacement fields continuous within each element, and depending on a set of generalized displacements u_e assigned to a nodal point located inside the element;
(ii) the fields allowed within the elements are compatible but not necessarily equilibrated;
(iii) the topology of a system of block elements can thus be defined by providing for each element, the set of adjacent elements;
(iv) the interaction between adjacent elements is defined by equilibrium conditions expressed in terms of the generalized tractions r_e, such that, on $\partial\Omega_e$,

$$p = \Gamma_e r_e,$$ (23)

Let us consider the block element boundaries decomposed into segments corresponding to the interfaces with the adjacent elements, and suppose that the tractions are distributed on each segment in such a way that they can be expressed in terms of generalized tractions r_e associated to such segment. Then:

(i) tractions acting on each segment can be expressed in terms of the generalized tractions;
(ii) the approximate solution associated to the decomposition of the body into block elements is obtained by equilibrating the generalized tractions associated to coinciding segments of the adjacent element boundaries;
(iii) block elements can themselves be rigid, although not rigidly connected to the adjacent elements along their common boundaries;
(iv) deformability in a system of block elements is, therefore, essentially associated to the element interfaces, not necessarily to the elements.

In block elements, *the field that minimizes the element total potential energy on a linear space of allowed fields* (22) *is the one belonging to that space that is nearest to the exact solution – compatible and equilibrated – associated to given boundary tractions*. In other words, introducing (22) and (23) in (10), differentiating and equating to zero, we obtain the *generalized displacement-traction equations*:

$$u_e = H_e^{-1} M_e r_e,$$ (24)

where

$$H_e = \int\limits_{\Omega_e} (D\,\Psi_e)^T \, H \, (D\,\Psi_e)\, d\Omega_e \,, \tag{25}$$

$$M_e = \int\limits_{\partial\Omega_e} \Psi_e^T \, \Gamma_e \, d\partial\Omega_e \,. \tag{26}$$

The decomposition of the body into block elements leads to a system of elements satisfying (22), which behaves more and more like a system of equilibrated finite elements, as the degree of the polynomial shape functions in (22) becomes larger and larger. Equilibrium between elements is ensured by conditions (23).

However, the remarkable difference between hybrid and block elements consists in that, in a system of block elements, deformability is essentially associated with the element interfaces, not necessarily with the elements. In case the elements are stiff, H^{-1} vanishes, as well as H_e^{-1}.

On the other hand, the term ΣU_e^* of the total complementary energy cancels, and, admitting that the prescribed displacements on the boundary vanish, the total complementary energy is reduced to

$$T^* = \frac{1}{2} \sum_{ef} r_{ef}^T \, H_{ef}^{-1} \, r_{ef} \,, \tag{27}$$

where the summation is extended over the whole set of interfaces between pairs e and f of adjacent elements, r_{ef} and r_{ef} are generalized contact force vectors, and H_{ef} a symmetric, positive definite matrix of magnitudes depending on the constants which characterize the mechanical properties of such interfaces.

For sake of equilibrium,

$$r_{ef} = -r_{fe} \,, \tag{28}$$

Let u_{ef} denote the corresponding generalized displacement vector at the interface of blocks e and f. As conformity is not supposed to be achieved,

$$u_{ef} \neq u_{fe} \,. \tag{29}$$

As the blocks are assumed to be rigid, the displacement field within a given block e can be completely defined by ascribing to an arbitrary nodal point N_e, selected within the block, the vector $U_e = u_e$ of the generalized displacements, (translations and rotations around N_e) of the block, so that the linear relation between u_{ef} and U_e may be expressed by

$$u_{ef} = A_{ef} \, U_e \,, \tag{30}$$

In order that each block e be in equilibrium under forces F_e and r_{ef}, the equilibrium equation for the block e assumes the linear form

$$\sum_f A_{ef}^T \, r_{ef} = F_e \,, \tag{31}$$

in which the summation in f extends over the whole set of blocks contacting with e, but the repeated indices summation convention is not used.

Combining (28), (30) and (31), the following equation results (Arantes e Oliveira et al. 1992)

$$\sum_f A_{ef}^T H_{ef} \left(A_{ef} \, U_e - A_{fe} \, U_f \right) = F_e \,, \tag{32}$$

in which the repeated indices summation convention is not used either. The system of equations obtained by putting together the equation (32) for the whole set of blocks provides magnitudes U_e. As it may easily be checked, the matrix of such system is symmetric.

3 ERROR ANALYSIS FOR HYBRID ELEMENTS

Let C_e denote the space of compatible structural fields on Ω_e equilibrating vanishing body forces.

Let $x_e \in C_e$ denote the compatible field on Ω_e that satisfies $\boldsymbol{u} = \boldsymbol{\Phi}_e \, \boldsymbol{q}_e$ on $\partial\Omega_e$.

Let C_e' denote the p-dimensional space of equilibrated fields on Ω_e satisfying $s_e - \boldsymbol{H}_e \boldsymbol{D}_e \boldsymbol{q}_e$ on Ω_e.

Let $x_e^* \in C_e'$ denote the compatible field on Ω_e that minimises T_e^* on C_e'.

Let $\boldsymbol{I}_e$ denote the operator with domain C_e and range C_e' associating x_q to its $\boldsymbol{I}_e$-image x_q', i.e.,

$$x_q' = \boldsymbol{I}_e \, x_q. \tag{33}$$

The stress vector s_q can be expressed within Ω_e by

$$s_q = s_t(\boldsymbol{O}) + \boldsymbol{O}\left(l_e^{p+1}\right), \tag{34}$$

where $s_t(\boldsymbol{O})$ may be called the tangent stress field (Arantes e Oliveira 1968) to the stress field s_q at point $\boldsymbol{O}$, a point within Ω_e. The stress field $s_t(\boldsymbol{O})$ is admitted to be a complete polynomial of the pth degree, all the terms of which are affected by independent arbitrary coefficients.

The completeness criterion consists in stating that $s(\boldsymbol{O})$ belongs to C_e', i.e., that $s_t(\boldsymbol{O})$ is allowed within Ω_e.

l_e denotes the diameter of element e, so that the second term in the right-hand side of equation (22) is of the order of l^p, where l is the maximum value of l_e on the whole set of elements.

The distance between an arbitrary element x_q of C_e and its $\boldsymbol{I}_e$-image x_q' in C_e' can be estimated by considering the distance of both fields to the tangent field x_t, and resorting to the triangular inequality

$$d_e\left(x_q, x_q'\right) \le d_e\left(x_q, x_t\right) + d_e\left(x_t, x_q'\right). \tag{35}$$

By virtue of (33),

$$d_e\left(x_q, x_t\right) = \boldsymbol{O}\left(l_e^{p+1}\right)\sqrt{\Omega_e}. \tag{36}$$

By virtue of (26),

$$d_e\left(x_t, x_q'\right) = \left\|x_t - x_q'\right\|_e = \left\|\boldsymbol{I}_e\left(x_t - x_q\right)\right\|_e \sqrt{\Omega_e} \le \left\|\boldsymbol{I}_e\right\|\left\|x_t - x_q\right\|_e \sqrt{\Omega_e}. \tag{37}$$

Therefore, by virtue of (15),

$$d_e\left(x_t, x_q'\right) \le \left\|\boldsymbol{I}_e\right\| d_e\left(x_q, x_t\right)\sqrt{\Omega_e} = \boldsymbol{O}\left(l^{p+1}\right)\sqrt{\Omega_e}, \tag{38}$$

and, by virtue of (34),

$$d_e\left(x_q, x_q'\right) = \boldsymbol{O}\left(l^{p+1}\right)\sqrt{\Omega_e}. \tag{39}$$

As, on the other hand,

$$d_n\left(x_q, x_q'\right) = \sqrt{\sum_e \left[\boldsymbol{O}\left(l^{p+1}\right)\right]\Omega_e}. \tag{40}$$

there results

$$d_n\left(x_q, x_q'\right) = \boldsymbol{O}\left(l^{p+1}\right)\Omega_n = \boldsymbol{O}\left(l^{p+1}\right). \tag{41}$$

Index n in equations (40) and (41) refers to a given degree of subdivision of Ω into subdomains Ω_e. The size of the elements, and therefore their maximum diameter, l, becomes smaller and smaller as n becomes larger and larger.

x_q satisfies equilibrium and compatibility within each element e. On the other hand, a set of nodal displacements having been assigned to each node, such that the displacements, which satisfy $u = \Phi_e q_e$ along the segments between each two contiguous nodes on element boundaries, can be defined in terms of the nodal displacements at the ends of such segments, compatibility is ensured along the element interfaces, so that a whole set of compatible fields x_q associated to the different subdomains Ω_e build up a compatible field on the whole domain Ω. In other words, conformity is not violated.

However, as the satisfaction of the conditions $u = \Phi_e q_e$ generally imply violation of equilibrium on the element interfaces, a set of fields x_q assigned to the different subdomains Ω_e is not susceptible of building up the exact solution x_0.

Let x_0 denote the approximate solution built up by compatible fields equilibrating vanishing body forces within each element, and satisfying the conditions $u = \Phi_e q_e$ on the element interfaces, i.e., minimizing the total potential energy $T(x)$ on the set C of the compatible fields on Ω, and satisfying $u = \Phi_e q_e$ on the element boundaries.

The completeness criterion consists in that the tangent field s_t corresponding to $u_t(O)$ may be allowed along each subdomain boundary $\varpi \Omega_e$. Admitting that the shape functions Φ_e contain a complete polynomial of the qth degree, the well-known reasoning used for conforming elements, leads to

$$d_n\left(x_0, x_{0t}\right) = O\left(l^{q+1}\right). \tag{42}$$

Then, as

$$d_n\left(x_0, x'_a\right) \le d_n\left(x_0, x_{0t}\right) + d_n\left(x_{0t}, x_a\right) + d_n\left(x_a, x'_a\right), \tag{43}$$

it comes out, by virtue of (40) and (42), that

$$d_n\left(x_0, x'_a\right) = O\left(l^{p+1}\right) + O\left(l^{q+1}\right). \tag{44}$$

Therefore, the least computational effort is achieved, for a given order of the error, making p equal to q which could not be done if the strain were derivatives of the displacements.

4   CONVERGENCE FOR HYBRID ELEMENTS

To test the convergence for hybrid elements consider the square plate subject to a traction load, referred to a cartesian referential, as represented in Figure 1.

Let us consider a general square element defined by the position of four corner nodes. For each element the following approximation can be defined for the stress field:

$$s = \begin{Bmatrix} \sigma_x \\ \sigma_y \\ \sigma_{xy} \end{Bmatrix} = \chi_e\, s_e = \begin{bmatrix} 1 & . & . & x & y & . & . \\ . & 1 & . & . & . & x & y \\ . & . & 1 & -y & . & . & -x \end{bmatrix} s_e. \tag{45}$$

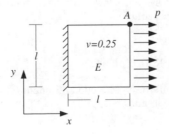

Figure 1.   Plate model.

Table 1. Results obtained for different meshes of hybrid elements.

Mesh	Nº Elements	$(E/pl)\,u_{x_A}$	$(E/p^2\,l^2)\,U$
1 × 1	1	0,982533	0,491266
2 × 2	4	0,986525	0,493480
4 × 4	16	0,992820	0,494774
8 × 8	64	0,994642	0,495326
16 × 16	256	0,995251	0,495553
"Exact"		0,995643	0,495706

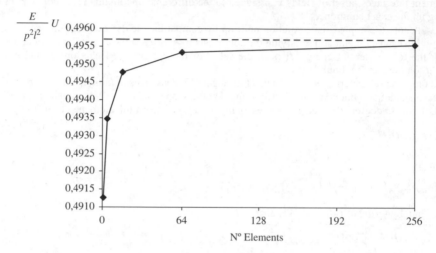

Figure 2. Strain energy convergence pattern obtained for the hybrid model.

Based on this approximation and considering equation (13), the generalized stresses can be defined in function of the nodal displacements for each element, and, therefore, the generalized nodal forces in terms of the nodal displacements.

The structure assemblage is performed considering, for adjacent elements, the compatibility conditions expressed in terms of the nodal displacements, and considering that the associated nodal force results from the sum of the contribution of each adjacent element and from the external load.

Based on those principles, the structure model governing system (19) is obtained.

For the square plate analysis four different uniform meshes, have been considered. For each mesh the x-displacement of point A and the value obtained for the strain energy have been evaluated. The results obtained are presented in Table 1 and compared with an "exact solution" obtained with a very refined hybrid-mixed mesh (Pereira & Freitas 1995, 1996).

In Figure 2 is presented the strain energy convergence pattern obtained.

The results presented in Table 1 and Figure 2 show that, for the present example, the approximate solution tends to the exact solution by values lower than the exact ones, which means that the approximate solution is stiffer than the exact solution.

5 CONVERGENCE FOR BLOCK ELEMENTS

Let us consider a beam with a uniform cross-section of length l. Submitting the beam to a constant moment M, the rotation Θ between its ends is

$$\Theta = \frac{M\,l}{EI}.\tag{46}$$

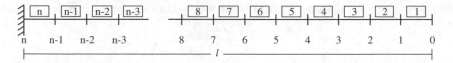

Figure 3.   Beam modeled with n block elements and n interfaces.

Let us now suppose that the beam is subdivided into n rigid beam elements. The deformation of the beam is concentrated in n cross-sections: the interface between the beam and its support, and the $n-1$ interfaces between the elements (see Fig. 3).

If the beam is submitted to a constant moment M, each two adjacent elements rotate with respect to each other, at joint p, by an angle Θ_p, which is related to M by

$$\Theta_p = \frac{M}{n\,EI}, \tag{47}$$

in such a way that, once the bending moment is constant, the exact solution (35) is provided.

Let us suppose now that the beam is buit-in at its left end, and acted by a concentrated force F at its right end. The bending moment at the left end of element p, i.e., at joint p, is

$$M_p = \frac{F\,p\,l}{n}. \tag{48}$$

The total approximate rotation Θ' becomes

$$\Theta' = \frac{F\,l}{n^2} \sum_{p=1}^{n} p. \tag{49}$$

But

$$\sum_{p=1}^{n} p = \frac{n(n+1)}{2}. \tag{50}$$

Therefore,

$$\Theta' = \frac{F\,l}{2\,n\,EI}(n+1). \tag{51}$$

As the exact solution is

$$\Theta = \frac{F\,l}{2\,EI}, \tag{52}$$

it turns out that

$$\Theta' = \frac{n+1}{n}\,\Theta. \tag{53}$$

The conclusion can thus be drawn that the approximate solution tends to the exact solution by values higher than the exact ones, which means that the exact solution is stiffer than the approximate solution.

6 CONCLUSIONS

It has been pointed out the duality between hybrid and block elements. Due to such duality:

A – the subdivision of the body into hybrid elements of decreasing size provides a sequence of approximate solutions stiffer than the exact solution;

B – the subdivision of the body into block elements of decreasing size provides a sequence of approximate solutions less stiff than the exact solution.

Using hybrid or block elements, lower or upper bounds can therefore be provided to kinematic magnitudes.

ACKNOWLEDGEMENTS

This work is part of the research activity of the *Instituto de Engenharia de Estruturas, Território e Construção (ICIST)*, and has been partially sponsored by the *Fundação para a Ciência e a Tecnologia (FCT)* and by the *Plurianual Funding* of the *FCT*. The support of Eng. Vieira de Lemos from the *NEE*, of the *Dams Department* of the *Laboratório Nacional de Engenharia Civil* is also acknowledged.

REFERENCES

Arantes e Oliveira, E.R. 1968. Theoretical Foundations of the Finite Element Method, *Int. J. Num. Meth. Engrng.* 4, 929–952.

Arantes e Oliveira, E.R. 1975. Foundations of the Mathematical Theory of Structures, International Center for Mechanical Sciences, *CISM Courses and Lectures*, N° 121, 7, 167–184, Springer Verlag, Wien.

Arantes e Oliveira, E.R., Pedro, J.O., Pina, C. & Lemos, J.V. 1992. Non-linear and Time-dependent Analysis in Three-dimensional Structures, In Zhang Yajun (ed), *EPMESC IV, Education, Practice and Promotion of Computational Methods in Engineering Using Small Computers*, II (Late Papers – III), 1325–1333, Dalian University Technology Press.

Cundall, P.A. 1971. A Computer Model for Simulating Progressive Large Scale Movements in Blocky Rock Systems, In *Symposium on Rock Fracture*, Nancy.

Kawai, T. 1977. New Element Models in Discrete Strucutural Analysis, *Japan Soc. Naval Arch.* 141, 174–180.

Lemos, J.V. 1987. A Distinct Element Model for Dynamic Analysis of Jointed Rock with Application to Dam Foundations and Fault Motion, PhD Thesis, Minnesota University.

Pereira, E.M.B.R. & Freitas, J.A.T. 1995. Implementation of a Mixed-Hybrid Finite Element Model Based on Legendre Polynomials, In Arantes e Oliveira, E., Bento, J., Pereira, E. (eds), *EPMESC V, Education, Practice and Promotion of Computational Methods in Engineering Using Small Computers*, 2, 987–992, Techno-Press.

Pereira, E.M.B.R. & Freitas, J.A.T. 1996. A Mixed-Hybrid Finite Element Model Based on Legendre Polynomials, *Int. J. Num. Meth. Engrng.* 39, 1295–1312.

Pian, T.H. 1964. Derivation of Element Stiffness Matrices by Assumed Stress Distributions, *AIAA Journal* 7.

Computational Methods in Engineering and Science, Iu et al. (eds)
© 2003 Swets & Zeitlinger, Lisse, ISBN 90 5809 567 3

Constitutive modelling of woven textile reinforcements and its application to high-pressure hydraulic hoses

H. Rattensperger, J. Eberhardsteiner & H.A. Mang
Institute for Strength of Materials, Vienna University of Technology, Austria

ABSTRACT: In this paper a numerical simulation model for hydraulic hoses reinforced by textile braids is presented. The components of the hose structure consisting of rubber as the matrix material and textile reinforcements are considered separately in the finite element simulation. For the description of the hyperelastic behaviour of rubber, well-known material models based on a potential function have been used. The textile reinforcement is represented by rebar elements. Their mechanical behaviour is accounted for by means of a so-called fabric lattice model. The developed analysis model is calibrated and verified by means of collapse load analyses of regular sections of hydraulic hoses and of numerical simulations of the crimping of a hose fitting.

1 INTRODUCTION

In many industrial areas, high-pressure hydraulic hoses are used for power transmission in steering, brake and drive systems. Another important field of application is fluid transport, e.g., in off-shore engineering. Such hoses must be tight, flexible, and resistant to high internal pressure. For the connection of flexible high-pressure hoses to the required equipment, threaded fittings are crimped to the hose. Such hydraulic hoses are composite structures. The fabric reinforcement, consisting of textile materials like polyester, results in a highly flexible behaviour as well as in high resistance to internal pressure. Rubber as base material guarantees the tightness of such hydraulic hoses. The textile reinforcement is arranged in braided form with plaited fibre bundles. The interaction between the individual fibre bundles influences the global stiffness of the reinforcement significantly.

The purpose of this work is to develop a realistic numerical simulation model for high-pressure hydraulic hoses with woven textile reinforcement. The aims of research are realistic predictions of the deformation and stress response under service loads, determination of the bursting pressure, and optimisation of the crimping of fittings. For that purpose, well-suited experimental tests for identification of the material parameters of rubber are essential. Concerning rubber, the main objective is to provide reliable material data for realistic loading conditions. For that reason, biaxial membrane tests in addition to the standard uniaxial tests are recommended for identification of material parameters.

The main topic of this paper is the mathematical description of the mechanical behaviour of the textile braid. Its stiffness and strength has a significant influence on global hose behaviour. The research concept is based on a trapezoidally-shaped fabric lattice model where the single fibres are represented by a spatial structure of rods. In addition to material and geometric nonlinearities, this constitutive model also takes into account the increase of the fabric stiffness because of the mutual obstruction of crossing textile fibre bundles. For consideration of the latter, a concept based on the Hertzian theory was used. The stress state in the reinforcement layer, including lateral compression of single fibres or fibre bundles, and the tangent stiffness according to this stress state, are determined by means of an incremental-iterative procedure within the framework of the finite element (FE) method.

Finally, the developed analysis model is calibrated and verified by means of collapse load analyses of regular sections of hydraulic hoses and numerical simulations of the crimping of a hose fitting.

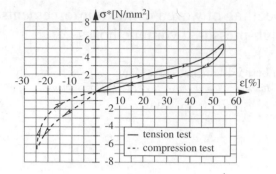

Figure 1. Stress–strain diagram of uniaxial tests.

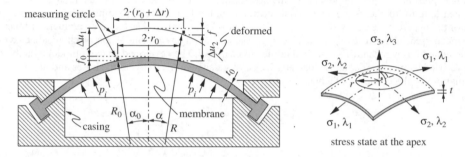

Figure 2. Axisymmetric biaxial membrane specimen and experimental set up.

2 MATERIAL BEHAVIOUR OF RUBBER

Rubber is used for the inside lining and the covering coat of hydraulic hoses. The mechanical properties of rubber are characterised by high elasticity and a nearly incompressible behaviour. Figure 1 shows a typical stress-strain diagram for rubber under uniaxial loading.

For the numerical representation of hyperelastic material behaviour, various constitutive models are available in the literature, e.g., constitutive models by Mooney-Rivlin, Ogden, van der Waals, or the tube model, and several polynomial laws. The parameters for the underlying strain energy potentials within the framework of hyperelasticity are determined through a least-square-fit procedure, which minimises the relative error in stress. In general, these parameters are identified on the basis of uniaxial tension and/or uniaxial compression tests. Material models with such parameters often fail when attempting to describe stress states, which are out of the range of experiments, as occur in engineering applications. Therefore, it is important to provide test data derived from more complex stress states [4]. For that reason, a biaxial membrane test was developed.

The experimental set-up of this biaxial membrane test is shown in Figure 2. The configuration of the test was optimised by numerical simulations with the objective of adequate accuracy of measurement and a pure membrane state of stress in the vicinity of the apex.

The membrane specimen, which is subjected to increasing internal pressure p_i, is clamped in a steel casing. During the test, the displacements Δu_1 and Δr of a marked circle with a diameter of $2r_0$ in its undeformed configuration, and the change of the rise, Δu_2, are recorded. With these quantities, the states of membrane strain and stress at the apex (Fig. 2) in terms of stretches λ_i, $i = 1, 2, 3$, and the stress components σ_i, $i = 1, 2, 3$, can be determined.

On the basis of uniaxial and biaxial test data, material parameters for the constitutive model by Ogden [6] were identified. Figure 3 shows the result from the re-analysis of the stress state in the membrane specimen. The validation of these parameters was done by comparison of experimentally and numerically obtained states of deformation of rubber under realistic stress conditions. For that purpose,

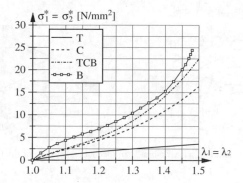

Figure 3.   Re-analysis of biaxial membrane test with material parameters based on uniaxial tension test (T), on uni-axial compression test (C), and on a combination of uniaxial tension, uniaxial compression and biaxial membrane test (TCB); B … biaxial test data.

a wire braid reinforced hydraulic hose was partially peeled and subjected to different levels of internal pressure. This loading resulted in significant rubber deformations between the woven wire bundles, determined from surface profile measurements by means of a laser sensor [7].

3   CONSTITUTIVE MODELLING OF TEXTILE BRAID REINFORCEMENT

Textile reinforcements of hydraulic hoses usually consist of a braid. Internal pressure leads to a complex load-carrying mechanism in the textile reinforcement. Several concepts are available for consideration of the mechanical behaviour of a braid in a numerical simulation, including the interaction between textile fibre bundles. In this paper a macro-mechanical concept for the description of the composite structure is used.

The assumption of a homogenised composite material is one mode of treatment of woven fabric reinforcements in FE analysis. This concept is based on the choice of a suitable anisotropic constitutive model. The equivalent material parameters are identified by means of experiments with fabric composites or by a so-called unit-cell method. In the latter case, a representative volume element is discretised using the mechanical properties of the different materials. Numerical analyses of characteristic deformation states yield equivalent material parameters, which are then introduced into a constitutive law for homogeneous anisotropic material used in a global FE model. In general, this concept is restricted either to linear constitutive behaviour, or to an orthotropic composite material. A unit-cell concept considering nonlinear material behaviour was suggested by Reese [7].

The fundamental idea of the concept used in this work is separate consideration of the matrix material (rubber) and reinforcement (textile or wire braid). Thereby, rebar elements represent the braided reinforcement which is embedded in the matrix material. The stiffness of the rebar layers is determined by the so-called fabric lattice model. In a paper by Kato et al. [4] the stiffness of the textile braid was integrated analytically at the integration point level. This resulted in restrictions regarding consideration of nonlinearities. In this work, a fabric lattice model based on a numerical formulation allows the taking of many forms of nonlinear material behaviour into account. It also permits numerical treatment of complex geometric situations, e.g., the simulation of the crimping of fittings.

3.1   *Mechanical concept of fabric lattice model*

In [8] the proposed fabric lattice model is used to represent wire reinforcements. In this paper textile braids have been considered. Two rebar layers, each with stiffness in only one of the two plaiting directions of the braid, represent the properties of the fabric lattice in the FE analysis. The respective rebar layers are interacting. For the calculation of stiffness and stress in the rebar layers by means of the proposed fabric lattice model, the following topics have to be considered:

- structural details of the textile braid and its mapping to the fabric lattice model,
- implementation of the fabric lattice model into a global FE analysis procedure,

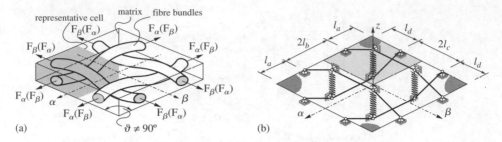

Figure 4.   (a) Details of the structure of a textile braid, (b) mechanical system of the fabric lattice model.

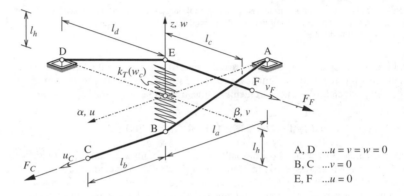

Figure 5.   Considered quarter of fabric lattice model.

- transfer of global quantities from the rebar layer to local quantities of the fabric lattice model,
- analysis of the fabric lattice model by means of a local nonlinear FE procedure, and
- back-transfer of local quantities from the fabric lattice model to global quantities of the rebar layer.

3.1.1   Mapping of the textile braid to the fabric lattice model

Details of the structure of a textile braid, shown in Figure 4a, illustrate the interaction of the fibre bundles. The forces acting in the fibres of one plaiting direction are influenced by the fibre forces in the other direction. This is reflected by notations such as $F_\alpha (F_\beta)$ in Figure 4a.

The shaded quarter of the detail in Figure 4a is treated as a representative cell. One half of a fibre bundle of this representative cell is treated as a bar in the fabric lattice model, as shown in Figure 4b. The mutual obstruction of the fibre bundles is represented by nonlinear springs between the respective bars. The fabric lattice cannot carry shear forces. These forces are carried entirely by the surrounding rubber material. The proposed fabric lattice model is not restricted to an orthogonal arrangement of fabrics. This is important for realistic numerical simulations of high-pressure hydraulic hoses, reinforced by textile braids.

3.1.2   Implementation of the fabric lattice model

Regarding the implementation of the fabric lattice model in a global FE analysis procedure, the model shown in Figure 4b can be reduced to the system illustrated in Figure 5. Thereby, the symmetry of the lattice model was utilised. Using this system, the tangent stiffness and the stresses in the textile braid are determined by means of a local analysis. Consideration of plastic deformations in the textile reinforcement and of a nonlinear spring behaviour requires an incremental-iterative solution algorithm for the local fabric lattice model. This local nonlinear analysis is performed at the integration point level of the global nonlinear FE analysis.

Table 1. Fabric lattice model – global and local variables.

Global	Local		Local	Global
$\epsilon^{k,i}$	$\mathbf{u}^n$	nonlinear	$\mathbf{F}^{n+1}$	$\sigma^{k,i+1}$
$\sigma^{k,i}$ $\Rightarrow$	$\mathbf{F}^n$	local analysis	$\Delta\mathbf{F}^{n+1}$ $\Rightarrow$	$\Delta\sigma^{k,i+1}$
$\Delta\epsilon^{k,i}$	$\Delta\mathbf{u}^{n+1}$	(increm./iter)	$\mathbf{K}^{n+1}$	$\mathbf{C}^{k,i+1}$

3.1.3 Transfer from global to local quantities

The state variables of a geometric nonlinear global FE analysis on the integration point level depends on the formulation of the FE program. This paper is based on a formulation using the Cauchy stress tensor σ and the logarithmic strain tensor ϵ. The first step of the numerical simulation of the textile braid is the transfer of the state variables from the rebar layers of the global FE analysis (ϵ, $\Delta\epsilon$, σ) to the corresponding quantities of the local fabric lattice model (vector of nodal displacements u, Δu, force vector F). In Table 1 the situation for the iteration steps i and $i + 1$ of increment k of the global nonlinear FE analysis is shown. The superscript of the local quantities n and $n + 1$, respectively, refers to the increment of the fabric lattice analysis.

The uniaxial stress state in a rebar layer is characterised by $\sigma \equiv \sigma_{11} = \sigma_1$. The corresponding strain quantities are $\epsilon \equiv \epsilon_{11} = \epsilon_1$, $\Delta\epsilon \equiv \Delta\epsilon_{11} = \Delta\epsilon_1$, and λ_1, denoting the longitudinal stretch. Hence, standard transfer relations between tensor components and the respective components of nodal displacements and forces of the fabric lattice model can be used. For the rebar layer connected with direction α of the fabric lattice model, the displacements u_C and Δu_C of node C (Fig. 5) are obtained as

$$u_C^n = (\lambda_1^{k,i} - 1)(l_a + l_b), \qquad \Delta u_C^n = (\lambda_1^{k,i+1} - \lambda_1^{k,i})(l_a + l_b), \qquad (1)$$

with the stretches of the rebar layer given as

$$\lambda_1^{k,i} = \exp(\epsilon_1^{k,i}), \qquad \lambda_1^{k,i+1} = \exp(\Delta\epsilon_1^{k,i+1})\,\lambda_1^{k,i}.$$

The nodal force acting at node C follows from

$$F_C^n = \frac{A^0\, J^{k,i}}{\lambda_1^{k,i}}\, \sigma_1^{k,i}, \qquad (3)$$

where A^0 denotes the initial section area, $J^{k,i}$ stands for the Jacobian determinant and $\sigma_1^{k,i}$ is the axial Cauchy stress of the rebar layer. The quantities associated with the β-direction of the lattice model are transformed analogous to Eqs. (1)–(3).

3.1.4 Components of the local fabric lattice model

3.1.4.1 Material behaviour of fibres

The components of the local fabric lattice model are bars and a nonlinear spring (Fig. 5). The bars represent the fibres of the woven braid made of textile materials like polyester or Kuralon®. The nonlinear elastic spring is used to consider the mutual obstruction of the fibre bundles. A characteristic stress–strain diagram for the textile material Kuralon® is shown in Figure 6. The constitutive relation for the textile materials is based on a hyperelastic formulation. The energy function has the form

$$W = \frac{1}{2}\left(K - \frac{2}{3}G\right)(\ln J)^2 + G\left[(\ln\lambda_1)^2 + 2(\ln\lambda_2)^2\right]. \qquad (4)$$

It was proposed in [5] and is used for determination of the constitutive law for Kuralon®. In Eq. (4) K and G correspond to the bulk and the shear modulus, respectively. J denotes the Jacobian determinant, λ_1 the axial stretch and λ_2 the transverse stretch of the textile fibre. This model is suitable for moderately large strains and provides a very good approximation of the true situation.

109

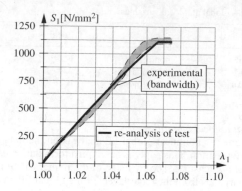

Figure 6.   Bandwidth of S_1-λ_1 – diagrams of Kuralon® fibres.

The partial derivative of W with respect to the Green strain component in the axial direction, E_1, results in the corresponding component of the Second Piola-Kirchhoff stress tensor,

$$S_1 = \frac{\partial W}{\partial E_1} = \frac{\partial W}{\partial \lambda_1} \frac{\partial \lambda_1}{\partial E_1} = \frac{9GK}{3K+G} \frac{\ln \lambda_1}{\lambda_1^2} = E \frac{\ln \lambda_1}{\lambda_1^2}. \tag{5}$$

S_1 is acting in the textile fibre. E denotes the initial elastic modulus. Equation (5) represents a linear relation between the Cauchy stress and the logarithmic strain. The second partial derivative of W with respect to E_1 yields the tangential modulus

$$E_T = \frac{\partial^2 W}{\partial E_1^2} == \frac{\partial S_1}{\partial \lambda_1} \frac{\partial \lambda_1}{\partial E_1} = E \frac{(1 - \ln \lambda_1)}{\lambda_1^4}. \tag{6}$$

With the Eqs. (5) and (6) the material behaviour of Kuralon® can be represented in the local FE analysis of the fabric lattice model.

3.1.4.2   Characterisation of the nonlinear spring

For consideration of the mutual obstruction of the fibre bundles a nonlinear spring is used to connect points B and E of the fabric lattice model (Fig. 5). Based on the Hertzian theory [2], which describes the contact of two elastic bodies with curved surfaces, the stiffness of the nonlinear spring is evaluated. The mechanical behaviour of the nonlinear spring is defined by

$$F_c = k_T(w_c)\, w_c \qquad \text{and} \qquad \frac{\partial F_c}{\partial w_c} = k_T\,, \tag{7}$$

where F_c denotes the spring force, w_c stands for the displacement and k_T is the tangent stiffness of the spring. Evaluating the equations for the interaction of two cylindrical elastic bodies [1] yields the relative displacement of the cylinders

$$w_c = \sqrt[3]{C\,(1-\nu^2)^2 \frac{\bar{F}_c^2}{E^2} \frac{r_1 + r_2}{r_1\, r_2}}\,, \qquad C = 1.03421\,, \tag{8}$$

with r_1 and r_2 denoting the radii of the cylinders, and $\bar{F}_c$ standing for the force acting at a single intersection point of the fibre bundles. This force is defined by $\bar{F}_c = 4\,F_c\,/(n_{fb} \cdot n_{fb})$, where n_{fb} denotes the number of fibres in a bundle [7]. Equation (8) is specified for a plaiting angle of 55°. In general, this value determines the orientation of the reinforcement of hydraulic hoses. Specification Eq. (8) for two fibres with equal diameter d_f yields

$$F_c = \frac{C}{8} n_{fb}^2 \sqrt{d_f} \frac{E}{1-\nu^2} \sqrt{w_c^3}\,. \tag{9}$$

110

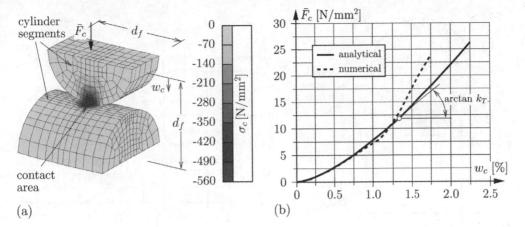

Figure 7. (a) contact area of two cylindrical bodies – contour plot of transverse stresses, (b) force-displacement diagram of the nonlinear spring.

The partial derivative of F_c with respect to the displacement w_c results in the tangent stiffness of the nonlinear spring

$$k_T = \frac{\partial F_c}{\partial w_c} = \frac{3\,C}{16}\,n_{fb}^2\sqrt{d_f}\,\frac{E}{1-\nu^2}\sqrt{w_c}. \tag{10}$$

3.1.4.3 Determination of ultimate load

The maximum loading of a textile braid occurs at the contact area of two crossing fibres (Fig. 7a). The stress state in the textile fibres of these parts of a braid is characterised by a combination of the longitudinal tensile stress and of the transverse compressive stress. For the evaluation of this stress state as well as for the determination of the ultimate load state, a semi-analytical method is used within the local fabric lattice model [7].

The failure mechanism of the fibres is illustrated in Figure 8a. The cross section of the fibre is divided into two parts, A_t and A_c. The former carries longitudinal tension and the latter transverse compression. Based on this assumption, the ultimate fibre force $\bar{F}_u$ is defined as

$$\bar{F}_u = \left(\frac{d_f^2\pi}{4} - d_c^2\zeta\right)f_{tu}, \tag{11}$$

where f_{tu} denotes the uniaxial tensile strength, d_c is the diameter of the contact area (Fig. 8a), and ζ stands for a parameter representing influence of the transverse compression. The latter depends on the individual plastic behaviour of the respective fibre material. It is determined by numerical analysis at the level of the intersection point (Fig. 7a). For vanishing transverse compression ($d_c = 0$), the maximum ultimate fibre force yields

$$\max \bar{F}_u = \left(\frac{d_f^2\pi}{4}\right)f_{tu}. \tag{12}$$

When reaching the limit load $\bar{F}_t = \max \bar{F}_u$, the pressure in the contact area is assumed to be equal to f_{tu}. From this, the diameter d_c is obtained as

$$d_c = \sqrt{\frac{4\bar{F}_c}{\pi f_{tu}}}. \tag{13}$$

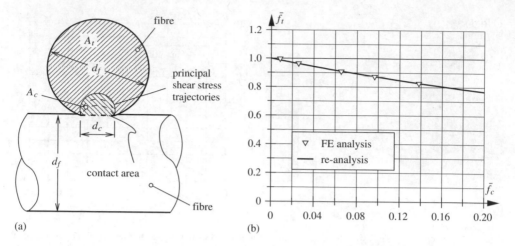

Figure 8. (a) characteristics of the principal shear stress trajectories, (b) $\bar{f}_t$-$\bar{f}_c$ – interaction.

Inserting Eq. (13) into Eq. (11) and introducing the dimensionless quantity $\bar{f}_t$, yields

$$\bar{f}_t = \frac{\bar{F}_u}{\max \bar{F}_u} = \left(1 - \frac{4\zeta\bar{f}_c}{\pi + 4\zeta\bar{f}_q}\right) ,$$ (14)

where $\bar{f}_c = \bar{F}_c/\bar{F}_t$ represents the ratio of the axial force and the transverse force. Figure 8b shows a comparison of results from FE analysis and results obtained by Eq. (14). Thereby, ζ was identified by one of the FE results. Parallel to the local analysis of the fabric lattice model, the failure criterion for the fibres,

$$\bar{F}_t \leq \bar{f}_t \max \bar{F}_u ,$$ (15)

was checked.

3.1.5 *Transfer from local to global quantities*
The results of the local analysis on the level of the fabric lattice model are the tangent stiffness K of the fabric lattice, the force F_C and its increment ΔF_C, acting at the node at which displacement increments are applied. According to Table 1, these quantities must be transformed to the following corresponding quantities on the level of the rebar layers of the global FE model: the tangent rebar stiffness C_1, the Cauchy stress σ_1, and its increment $\Delta\sigma_1$. The tangent rebar stiffness C_1 is obtained as

$$C_1^{k,i+1} = \frac{(l_a + l_b)(\lambda_1^{k,i+1})^2}{J^{k,i+1} A^0} K_1^{n+1} ,$$ (16)

where $(l_a + l_b)$ is the length of the lattice model in the α-direction (Fig. 5). The stress σ_1 and its increment $\Delta\sigma_1$ are obtained as

$$\sigma_1^{k,i+1} = \frac{F_C^{n+1} \lambda_1^{k,i+1}}{J^{k,i+1} A^0} ,$$ (17)

$$\Delta\sigma_1^{k,i+1} = \frac{F_C^{n+1} \lambda_1^{k,i+1}}{J^{k,i+1} A^0} - \frac{F_C^n \lambda_1^{k,i}}{J^{k,i} A^0} .$$ (18)

The quantities associated with the β-direction of the lattice model are transformed in an analogous manner.

112

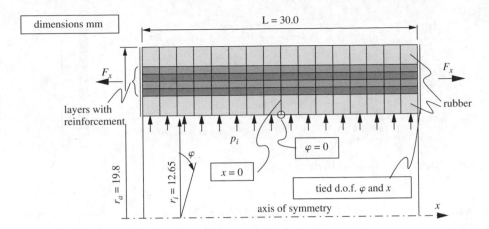

Figure 9. FE model of a representative part of a hydraulic hose.

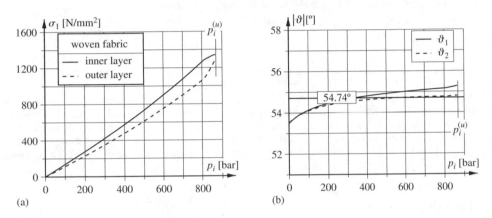

(a)

(b)

Figure 10. Representative part of a hose: (a) Cauchy stress in the fibres vs. internal pressure for two layers of textile braid, (b) change of the plaiting angle ϑ vs. internal pressure p_i.

4 NUMERICAL EXAMPLES

The fabric lattice model is used for the numerical simulation of (i) a representative part of a high-pressure hydraulic hose under internal pressure, and (ii) the crimping of a hose fitting.

Figure 9 illustrates the axisymmetric model with the boundary conditions of the considered part of the hose. Four rebar layers represent two layers of textile braid reinforcement. The axial tension force F_x depends on the internal pressure p_i, which is increased until the bursting pressure is reached. During the simulation the change of the hose volume, the stress state and the change of the plaiting angle of the reinforcement are observed.

Figure 10a shows the fibre stresses in the two layers of textile braid, as a function of the internal pressure. The inner textile braid has higher stress levels. Therefore, the bursting pressure of the hose is determined by the strength of the inner textile braid. The change of the braid angle in consequence of growing internal pressure is shown in Figure 10b. The theoretical optimum of the plaiting angle is $\vartheta = 54.74°$ [7]. The value of ϑ is a little higher when the bursting pressure is reached.

In a more complex numerical simulation the crimping of a hose fitting and the interaction between the fitting, rubber, and the reinforcement are investigated. An axisymmetric deformed model of a textile hose with a crimped fitting is shown in Figure 11. For the crimping process, the hose is partially peeled at its end. Then, a cylindrical metallic ferrule is pushed over this end and a nipple is inserted in

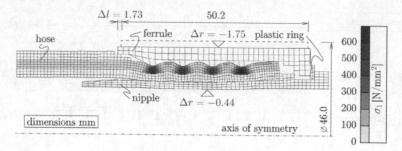

Figure 11.   Deformed FE model of the crimping of a hose fitting.

the hose. Thereafter, the ferrule is pressed onto the hydraulic hose by means of a press die. The resulting axial stresses of the fibres are shown in Figure 11 in the form of a contour plot. The peaks of the stresses are under the ribs of the ferrule. In case of too strong pressing in this area rupture of fibres will occur. A comparison of the deformation of the structure, as obtained from the numerical simulation, with the respective experimental observation shows very good agreement.

5   CONCLUSION

The objective of the presented work was to develop a realistic numerical simulation model for high-pressure hydraulic hoses with textile reinforcement in order to predict the deformation and stress response under service loads and the bursting pressure of such hoses. For consideration of textile woven reinforcements a local nonlinear fabric lattice model was developed. This model was applied to a special type of rebar layers. For verification of the model the results from numerical analysis of a representative part of a hose under internal pressure and from simulation of the crimping of a hose fitting were used. Good agreement with experimental results has confirmed the usefulness of the proposed model for numerical investigation of textile braids.

REFERENCES

[1] Geckeler, J. (1928). *Handbuch der Physik* Band 6, Kapitel 3: *Elastostatik*. Springer-Verlag, Berlin.
[2] Hertz, H. (1883). *Über die Verteilung der Druckkräfte in einem elastischen Kreiszylinder*. Zeitschrift für Mathematik und Physik, XXVIII. Jahrgang, Verlag B. G. Teubner, Leipzig.
[3] Kato, S., Yoshino, T. and Minami, H. (1999). *Formulation of constitutive equations for fabric membranes based on the concept of fabric lattice model*. Engineering Structures, **21**, pp. 691–708.
[4] Johannknecht, R. and Jerrams, S.J. (1999). *The need for equi-biaxial testing to determine elastomeric material properties*. Proceedings of the First European Conference on Constitutive Models for Rubber, 1999, Vienna, Austria.
[5] Meschke, G. and Helnwein, P. (1994). *Large-strain 3D-analysis of fibre-reinforced composites using rebar elements: hyperelastic formulations for cord*. Computational Mechanics, **13**, pp. 241–254.
[6] Ogden, R.W. (1984). *Non-linear elastic deformations*. John Wiley & Sons, New York.
[7] Rattensperger, H. (2003). *Berechnungsmodelle für verstärkte Elastomere und deren Anwendung für Hochdruckhydraulikschläuche*. PhD thesis, Vienna University of Technology, Vienna, Austria.
[8] Rattensperger, H., Eberhardsteiner, J. and Mang, H.A. (2001). Numerical Investigation of High-Pressure Hydraulic Hoses with Steel Wire Braid. Proceedings of the IUTAM Symposium on Computational Mechanics of Solid Materials at Large Strains 2001, Stuttgart, Germany, pp. 407–416.
[9] Reese, S. (2001). *Meso-macro modelling of fibre-reinforced composites exhibiting elastoplastic material behaviour*, CD-ROM Proceedings of the European Conference on Computational Mechanics 2001 (ECCM 2001), Cracow, Poland.

Computational Methods in Engineering and Science, Iu et al. (eds)
© 2003 Swets & Zeitlinger, Lisse, ISBN 90 5809 567 3

High-accuracy difference and meshless methods for fluid and solid mechanics

A.I. Tolstykh
Computing Center of Russian Academy of Sciences, Moscow, Russia

ABSTRACT: Two types of high-accuracy methods for PDE's are outlined. The first one is the Compact Differencing (CD) methods of a fixed high-order and their extension to multioperators-based both non-centered and centered schemes. The techniques show their peak performance in the cases of relatively simple geometries. As an illustration, a shear layer instability problem with vorticity rolling-up is considered. For complicated geometries, large deformations etc., a meshless method based on using Radial Basis Functions in a "finite difference mode" has been developed. Numerical examples are presented, the main emphasize being placed on solid mechanics problems.

1 INTRODUCTION

At present, developing high-order schemes is one of the avenue of investigation in the field of numerical simulation methods. The benefits of using the methods can be classified into two parts. First, they are needed in the context of some fundamental researches, for example, in the DNS area. Second, they can be very fruitful when solving real-life problems with engineering accuracy. In that case, considerably reduced number of nodes may lead to reduction of computational expenses measured by many orders of magnitude.

In (Tolstykh 2001), two approaches to constructing high-accuracy methods were outlined. The first one is based on Compact Differencing (CD) operators allowing to obtain fixed high-order discretizations of fluid or solid mechanics governing equations with sufficiently small numerical coefficients in their truncation errors. For the fluid dynamics applications, families of the so called Compact Upwind Differencing (CUD) schemes (Tolstykh 1994) were shown to be an efficient numerical tool. The schemes are appropriate for the skew-symmetric operators of the convection terms since they contain a built-in filter of high-frequency numerical noise generated by steep velocity gradients. They are beneficial for problems requiring good spatial resolution combined with sufficient robustness. Vorticity evolution and vortex wakes calculations as well as direct numerical simulations of turbulence may serve as examples.

A useful property of the CUD approximations is their dependence on upwinding parameter. It was found (Tolstykh 1998) that fixing distinct values of the parameter, it is possible to construct linear combinations of basis CUD operators corresponding to the chosen values and annihilate the lowest terms in the Taylor expansion series corresponding to the combinations. In this way, uniquely defined "multioperators" provide formally arbitrary-order discretizations. Moreover, multioperators calculations do not require extra time as compared with those for the basis operators if parallel computers are used. In (Tolstykh 2001), an example is presented showing that speed-up when using a seventh-order multioperators-based scheme can be amounted to many orders-of-magnitude . Theoretical investigations into multioperators can be found in (Tolstykh 2000a).

In the solid mechanics area, though the Finite Element Method (FEM) is an universally accepted numerical tool, the trend has been toward developing alternative technique in the context of problem-oriented methodologies (for example, for solving problems with large deformations and moving discontinuities). One of the approaches appropriate for limited classes of problems with simple geometries (for example, "simple" plates and shells) is, again, using CD methods which have the potential for being far more accurate than FEM. Since usually one encounters with self-adjoint operators due to the elliptical nature of the problems, centered CD formulas are the most suitable here.

In (Tolstykh 2001), an example of application of the fourth-order CD scheme based on the well-known Numerov approximations to second derivatives is presented. In the present paper, we show other examples concerning Kirchhoff plates with stiffeners. Moreover, we describe here a way of constructing centered multioperators by artificially introducing parameters in centered CD operators thus generating basis operators. It will be shown that the multioperators approximations to the Poisson and biharmonic equations can provide extremely high accuracy of numerical solutions.

Using high- and arbitrary-order difference methods for solid mechanics problems can be efficient only if very good quality smooth meshes are exploited. However, constructing such meshes in many cases is an independent and difficult task. Following problem adaptation principle, the second type of high accuracy numerical techniques outlined in (Tolstykh 2001) in the context of domain decomposition is a meshless method which can deal with complicated forms of boundaries and other cases when using scattered data points can be advantageous. The method is based on harnessing Radial Basis Functions in a "finite difference mode"(Tolstykh 2000).

Radial Basis Functions (RBF) are known to be an efficient tool for constructing interpolants with scattered nodes. Skipping an overview of numerous publications on the subject (the relevant references can be found, for example, in (Kansa & Carlson 1995), papers concerning their application to PDE's seem to be comparatively few in number, the main approaches being the boundary elements and collocation methods (Zerroukat 1998), (Kansa 1990), (Zhang et al. 2000). One of the problem typical for RBF methods is ill-conditioning of the resulting algebraic systems. As an alternative approach, the least square type of meshless methods should be mentioned (see, for example, survey (Belytchko et al. 1996)).

Below , the present RBF approach will be considered in the context of elasticity problems.

2 FIXED AND PRESCRIBED-ORDER COMPACT SCHEMES

2.1 *Centered and non-centered approximations*

Compact approximations can be viewed as rational functions of three-point (that is, three-diagonal in matrix formulations) operators containing parameters and/or numerical constants. In contrast, traditional approximations following from interpolation formulas are in fact polynomial functions of "elementary" three-point operators. The last, in turn, can be presented as linear combinations of the unity operator I and the central differences defined by

$$\Delta_0 f_j = f_{j+1} - f_{j-1}; \quad \Delta_2 f_j = f_{j+1} - 2f_j + f_{j-1}$$

where a uniform mesh $x_j = jh, j = 0, \pm 1, \pm 2, \ldots$, is assumed.

The simplest centered compact differencing (CD) formulas dated back to Collatz and Numerov read

$$\frac{\partial}{\partial x} = (I + \frac{1}{6} \Delta_2)^{-1} \Delta_0 / 2h + O(h^4), \quad \frac{\partial^2}{\partial x^2} = (I + \frac{1}{12} \Delta_2)^{-1} \Delta_2 / h^2 + O(h^4) \tag{1}$$

Approximations (1) are very accurate but unfortunately they are not appropriate for CFD problems with steep gradients solutions due to introducing spurious oscillations. To deal with such type of solutions, the author proposed in 70-ties non-centered compact upwind differencing (CUD) which gives

$$\frac{\partial}{\partial x} = (I + \frac{1}{6} \Delta_2 - \frac{1}{4} s\Delta_0)^{-1} (\Delta_0 - s\Delta_2) / 2h + O(h^3) \tag{2}$$

where s is the upwinding parameter. For $s > 0$ and $s < 0$ one has the left-sided or right-sided differencing respectively. To increase approximation orders of (1) and (2), more complicated formulas should be introduced. The procedure for (1) was described in (Lele 1992) while that for (2) was presented in (Tolstykh 1994). Skipping the details, the CUD discretizations of first derivatives can be cast in the form

$$\frac{\partial}{\partial x} = L_m(s) + O(h^m), \quad m = 3, 5$$

Denoting by $L_m^{(0)}$ and $L_m^{(1)}$ the operators of self-adjoint and the skew-symmetric parts of L_m, it was shown that

$$L_m(s) = L_m^{(1)}(s) + s\,L_m^{(0)}(s), \quad L_m^{(0)}(s) = L_m^{(0)}(-s), \tag{3}$$

where $L_m^{(0)}$ is a positive operator in an appropriate Hilbert space. It means that using L_m for discretization of convection terms can lead to positive operators (in the frozen coefficients sense) if the sign of s is properly chosen. Their self-adjoint components serve as a built-in filter of high-frequency numerical noise appearing due to steep gradients of exact solutions. Thus CUD-based schemes combine high accuracy and robustness. The details of the CUD algorithms and their realizations can be found in (Tolstykh 1994).

As an illustration, we consider unsteady double periodic problem investigated in the framework of the incompressible Navier-Stocks equations in (Brown & Minion 1995). Initially two shear layers were given by

$$u = \tanh(\rho(y - 0.25)) \quad \text{for} \quad y \le 0.5, \quad u = \text{tahh}(\rho(0.75 - y)) \quad \text{for} \quad y > 0.5$$

where δ is a perturbation coefficient while ρ is a "thickness" parameter.

The shear layers are unstable and at later time moments they begin to roll-up with forming vortices. An interesting feature of the problem is that even in the case of very fine mesh (256×256) the solution obtained is not "true". This fact manifested in a spurious vortex between two true vortices. It disappears only when refined (512×512) mesh is used in (Brown & Minion 1995). Trying several methods, the authors concluded that it is impossible to obtain "true" solution with coarser meshes then (512×512) one.

The calculations (Tolstykh & Chigirev 2000) were carried out using CUD-5 scheme with the fifth-order Runge-Kutta time-stepping for the same values of ρ, Re, δ ($\rho = 100$, Re $= 20000$, $\delta = 0.05$). The same flow fields for the same time moments were obtained, the difference being only due coarser mesh in our calculations. Namely, our 128×128 and 256×256 meshes gave the same vorticity fields as those (256×256 and 512×512) in (Brown & Minion 1995).

Figure 1 shows the vorticity isolines at time $t = 0.8$ in one half computational domain for the above mentioned meshes (the corresponding pictures from (Brown & Minion 1995) are not presented here because they are visually the same). The same type of calculations performed for the corresponding inviscid case showed that the CUD scheme can operate well without stabilization effect of the fluid viscosity under the conditions of steep gradients.

2.2 Non-centered arbitrary-order multioperators

Both centered and non-centered compact differencing formulas have fixed truncation error orders which prevent spectral convergence to exact solutions. Of course, the orders can be increased by using

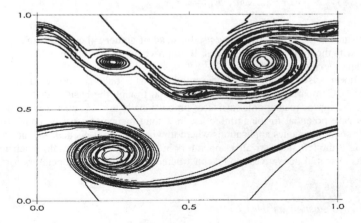

Figure 1. Shear layers rolling-up. The spurious vortex is seen in the upper part (128×128 mesh), below is the true solution (256×256 mesh).

more complicated rational functions of Δ_0 and Δ_2. Unfortunately, it can add considerable complexity to the methods and increase the numbers of required boundary conditions.

In (Tolstykh 1998), a way of constructing arbitrary-order CUD-based approximations was suggested and investigated. It looks as follows.

Suppose one has a set $S_M = (s_1, s_2, \ldots, s_M)$ of distinct values of the parameter s. Then it is possible to construct linear combinations of the basis operators $L_m(s_j)$ ("multioperators") which give $O(h^{m+M-1})$ truncation errors:

$$\frac{\partial}{\partial x} = L_M + O(h^{m+M-1}), \quad L_M = \gamma_1 L_m(s_1) + \gamma_2 L_m(s_2) + \ldots \gamma_M L_m(s_M) \tag{4}$$

The γ_j coefficients were shown to be uniquely defined in an analytical form as solutions of linear systems with the Vandermonde matrix. For decreasing the condition numbers of the systems, it was proposed to choose S_M as sets of zeros of Chebyshev polynomials for an interval $[s_{min}, s_{max}]$. The interval should be defined to meet the conditions similar to (3) thus preserving upwinding properties of the basis operators. Using parallel machines, the calculations of the action of a multioperator on a known grid function can be carried out in a parallel manner with the computational costs which do not exceed those for simple basis operators.

Considering multioperators as a way of discretization of derivatives in governing equations, various schemes can be constructed by introducing either time-stepping or iteration procedures. In all cases, to obtain numerical solutions for current time level or iteration, one should calculate actions of multioperators on known grid functions corresponding to the previous time level or iteration by using tri-diagonal Gauss eliminations.

Calculations for the benchmark problem of a vortex convected by a uniform flow using a seventh-order multioperators-based scheme ($M = 3$, $m = 3$) show that the numerical solutions preserve the vortex shape and position during large time intervals. Moreover, tremendous decrease of execution times has been seen when comparing with results of calculations with a standard second-order scheme. The multioperators scheme was used also in the case of the shear layer instability problem considered in the previous Subsection. It turned out that the numerical solutions give visually pictures displayed in Figure 1.

2.3 Centered multioperators

In the previous Subsection, an explicit dependence of the CUD formulas on the parameter s was used to define the basis operators. It is not the case if centered operators in (1) are considered. Nevertheless, it is possible to create basis operators changing numerical constants in the formulas by artificially introduced parameters. To outline the idea, consider some finite-difference operator E_h which is a second-order approximation to an exact one E. We define basis operators and the corresponding multioperator as

$$E_j = (I + c_j \Delta_2)^{-1} E_h, \quad j = 1, 2, \ldots M, \quad E_M = \sum_{j=1}^{M} E_j \gamma_j$$

where c_j are distinct values of the parameter used instead of numerical constants in (1) while γ_j satisfy linear systems with the Vandermonde matrices. To improve their condition numbers, c_j are supposed to be zeros of Chebyshev polynomials.

Solving a linear system with the Vandermonde matrix, one can prove (Tolstykh 2002) that $E = E_j + O(h^{2M})$. Comparing with (4), one may see that though the basis operators are only second-order accurate, the approximating order increases proportionally to $2M$ rather than to M. Thus centered multioperators have potential for being more accurate than their non-centered counterparts. They are appropriate for solid mechanics applications where upwinding is not needed. As particular cases, one may consider E as the left side derivatives operators in (1). In the case of multidimensional problems, approximations to spatial derivatives can be constructed using Δ_0 and Δ_2 operators for corresponding coordinates.

2.4 Application to elasticity problems

The above centered compact differencing and multioperators approximation to derivatives can be used for direct discretizations of any elasticity model without considering its week formulation. Again, numerical algorithms consist of calculations of multioperators actions and some iteration procedures. For engineering accuracy, they are expected to be very fast due to reduced numbers of grid points and self-adjoint operators typical for the governing equations.

To fix the idea of the approach, consider a Kirchhoff plate occupying in the (x, y) plane the domain Ω: $-1 \leqslant x \leqslant 1$, $0 \leqslant y \leqslant m$. The flexural rigidity of the plate $D(x, y)$ is supposed to be a sufficiently smooth function of its arguments, the only exception being its possible discontinuity at $x = 0$. It is supposed also that the plate may be fortified by a stiffener with the bending and torsional rigidities B and C respectively. Then the z-displacement w of the plate satisfies the biharmonic equation in both sub-domains $-1 \leqslant x \leqslant 0$, $0 \leqslant y \leqslant m$, $0 \leqslant x \leqslant 1$, $0 \leqslant y \leqslant m$ with proper boundary conditions at $\partial\Omega$. The variational principle gives the following "jump" conditions

$$[w] = 0, \quad [w_x] = 0, \quad [D(w_{xx} + \nu w_{yy})] = -C(w_{xy})_y,$$
$$[D(w_{xx} + \nu w_{yy})]_x + 2\{D(1-\nu)w_{xy}\}_y = -(Bw_{yy})_{yy} \tag{5}$$

where for a function $f(x, y)$, $[f]$ means $f(+0, y) - f(-0, y)$. In the particular case $B = C = 0$, $[D] = 0$, one has the interface conditions for the domain decomposition approach applied to plates with smoothly varying thickness.

Considering as an example a simply supported plate, we discretize the biharmonic equation using the above fourth-order compact differencing operators. To satisfy (5), a fifth-order formulas which relate $w_x(-0, y)$ and $w_x(+0, y)$ to w and w_{xx} at the "left" and "right" nodes respectively were constructed. Using them, a complete set of algebraic equations can be derived. In general, they can be solved by either direct or iterative solvers. The below presented results are obtained by alternately solving the "left" and "right" biharmonic equations.

The results of calculations with the standard-second order and the present fourth-order schemes for the square simply supported plate with a stiffener shown in Figure 2 are presented in Table 1. The rigidity $D = 1$ was assumed for both sides of the plate while the bending and torsional rigidity of the stiffener B and C were chosen as $B = 2$ and $C = 2$ respectively. In the Table, the L_2 norms of the solution errors δ and the corresponding mesh convergence order k are displayed for several $N \times N$ meshes, the reference "exact" solution being obtained using a very fine mesh. Similar results for the case of the different plate thickness at each side of the stiffener ($D = 1$, $x < 0$; $D = 2$, $x > 0$), Figure 2b, are shown in the Table.

As seen from the Table, using the fourth-order scheme instead of the second-order one leads to dramatic increase of accuracy.

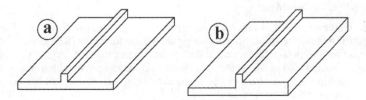

Figure 2.   Plates with stiffeners.

Table 1.   Plates with stiffeners: solution errors and mesh-convergence rates.

	Plate (a)				Plate (b)			
	Second-order		Fourth-order		Second-order		Fourth-order	
$N \times N$	δ	k	δ	k	δ	k	δ	k
8×8	6.574e-3		3.532e-5		5.883e-3		4.941e-5	
16×16	1.582e-3	2.06	4.715e-6	2.91	1.395e-3	2.08	5.045e-6	3.29
32×32	3.870e-4	2.03	3.100e-7	3.93	3.386e-4	2.04	3.212e-7	3.97
64×64	9.566e-5	2.02	1.818e-8	4.09	8.339e-5	2.02	1.874e-8	4.10

Table 2.   Solutions errors for the multioperators sixth-order scheme.

$N \times N$	Laplace equation		Biharmonic equation	
	δ	k	δ	k
4×4	2.31e-6		4.62e-6	
6×6	3.28e-7	4.81	6.57e-6	4.81
12×12	6.23e-9	5.72	1.24e-8	5.73
24×24	1.01e-10	5.95	2.01e-10	5.95

To illustrate possible peak performance of the multioperators method, we consider the BVP for the Poisson and biharmonic equations in the domain Ω: $0 \leqslant x \leqslant 1, 0 \leqslant y \leqslant 1$ with zero boundary conditions at $\partial\Omega$ and the right-hand side proportional to $\sin\pi x \sin\pi y$, the product being the exact solution. To approximate the Laplace operator, the sixth-order multioperators ($M = 3$) corresponding to x and y coordinates were used, the biharmonic operator being considered as the square of the Laplace one. The results for both problems for several meshes are shown in Table 2.

Extremely high accuracy displayed in the Table 2 allows one to use very coarse meshes which in turn leads to reducing of computational costs by orders-of-magnitude. For example, the solution for 4×4 mesh can be obtained almost instantaneously even when using the simplest explicit iteration procedure.

Using high-order compact approximations is advantageous in the cases when it is easy to construct smooth meshes without rapid variations of metric coefficients, examples being plates and shells of relatively simple forms. In some instances, it may be possible also to use the techniques in the framework of the domain decomposition strategy partitioning a computational domain into subdomains where "good" meshes can be constructed. To discretize governing equations in the "bad" subdomains or in single domains with complicated geometry it was proposed (Tolstykh 2000) to use differencing formulas based on radial basis functions. The resulting schemes, on the one hand, are compatible with finite difference schemes when using in the framework of the domain decomposition strategy and on the other hand can be viewed as quite accurate approximations offering advantages of meshless methods. Below, outlines of the idea and its application to benchmark elasticity problems are presented.

3   USING RADIAL BASIS FUNCTIONS IN A FINITE DIFFERENCE MODE

3.1   *Approximations to derivatives*

The main idea of the approach is constructing approximating formulas following finite difference methodology, the difference being only due to using radial basis functions (RBF) instead of polynomial ones for constructing interpolants for scattered data points. Denoting by, $x_1, x_2, \ldots x_N, x_j \in \Re^3$ a set of nodes, we consider each node x_j as the center of a cloud ω_j containing N_j neighbor nodes. Following the finite difference terminology, the cloud will be referred to as stencil. The stencil can be used as a local support for an interpolant for a function $u(\mathbf{x})$

$$s^{(j)}(\mathbf{x}) = \sum_{k=1}^{N_j} c_k \varphi(\|\mathbf{x} - \mathbf{x}_k\|), \quad s^{(j)}(\mathbf{x}_i^{(j)}) = u(\mathbf{x}_i^{(j)}), \quad \mathbf{x}_i^{(j)} \in \omega_j$$

where φ is a standard RBF while $\|\cdot\|$ is the Euclidean norm, that is the distance between two points. For any linear functional $Lu(\mathbf{x})$ its RBF approximation can be cast in the form $Lu(\mathbf{x}) = Ls(\mathbf{x}) + R$ where R is a residual. In particular, $Lu(\mathbf{x})$ may be viewed as the nth-order derivative in respect to the Cartesian coordinates x_i, $i = 1, 2, 3$ at the center $\mathbf{x}_j$. Denoting the derivative by $D_i^{(n)}u$, its approximation for jth stencil can be written as

$$D_i^{(n)}u = \sum_{l=1}^{N_j} c_l^{n,i} u(\mathbf{x}_l^{(j)}) + R, \quad c_l^{n,i} = \sum_{k=1}^{N_j} b_{kl} D_i^{(n)} \varphi(\|\mathbf{x} - \mathbf{x}_k\|) \tag{6}$$

where b_{kl} are the entries of the matrix which is the inverse of the matrix $\mathbf{A} = \{\varphi(\mathbf{x}_i - \mathbf{x}_j)\}$.

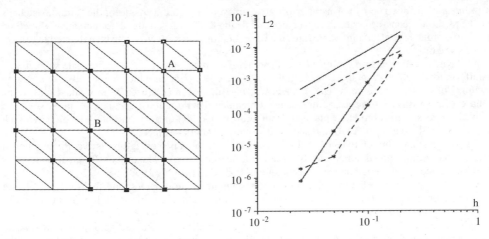

Figure 3. "Simple" and "enlarged" stencils (left) and solution errors (right); dashed lines correspond to the RBF "simple" and "enlarged" stencils (markers), solid lines correspond to FEM and compact scheme of fourth-order respectively (markers).

The examples presented in (Tolstykh 2001) show that in the case of a uniform mesh with the mesh size h and the multiquadrics (MQ) $\varphi(r) = \sqrt{r^2 + c}$ approximations errors R may behave as $O(h^m)$ where $m = 2, 4, 6$ depending on the stencils used in the calculations. Having in mind also that one may expect exponential convergence when increasing N_j, formula (6) has the potential for being quite accurate. It can be used for discretizations of PDE's in the same manner as in the case of the finite difference strategy. So, in contrast to the collocation method, PDE's are approximated rather than satisfied at each node.

The main features of the above approach are:

(i) the problem of ill-conditioning of the matrix $\mathbf{A}$ is greatly relaxed due to local supports;
(ii) there is considerable flexibility when choosing stencils; they can be problem-oriented and may vary with nodes and spatial direction. In particular, special types of stencils can be specified for boundary and near-boundary centers;
(iii) differencing formulas can be readily generalized by specifying functionals at nodes (for example, derivatives at certain points) thus allowing easy treatment of the Neumann boundary conditions.

Summing up, to solve a PDE using the present methodology, one should:

(i) Specify a nodes distribution in the considered computational domain;
(ii) For each node $\mathbf{x}_j$ considered as a center, specify a stencil with N_j nodes surrounding $\mathbf{x}_j$;
(iii) For each stencil, obtain "differencing" coefficients (for example, $c_l^{n,i}$ in (6)) by solving linear systems;
(iv) Substitute the approximations to derivatives at each node in the PDE and form the resulting "global" system by assembling the nodal approximations;
(v) Solve the global system.

3.2 Application to elasticity problems

The solutions presented below (Tolstykh et al. 2003) concerns mainly with benchmark problems allowing to estimate their accuracy. They have been obtained using the MQ RBF, the constant $c = 1$ being assumed since there is no theory describing its optimal choice. Judging from (Kansa & Hon 2000), considerable increase of accuracy can be expected when properly defining c. It provides additional possibility of improving the present results.

Kirchhoff plates. We consider below two cases of the Kirchhoff plates for which exact solutions are available. Their bending is described by the biharmonic equation. In the particular case of simply supported edges, a solution procedure can be reduced to successive solutions of two Poisson equations.

The first case is the square plate problem considered in Subsection 2.3 in the context of multioperators. The calculations were carried out using seven-points "simple" (or RBF-1) and nineteen points "enlarged" (or RBF-2) stencils shown in Figure 3, the above described fourth order technique and the FEM method with linear elements. The L_2 norms of errors are displayed in Figure 3. As seen, the "simple" stencils and FEM show second-order mesh convergence while the RBF with "enlarged" stencil and CD method are fourth-order accurate. At the same time, the RBF solutions are more accurate than their counterparts of the same order.

As the next testing example, we consider bending of a simply supported rhombic plate subjected to a uniform load. In that case, there is a singularity of the exact solution which has an adverse effect on accuracy of numerical methods. The problem was investigated in (Babusca & Scapolla 1989) in the context of performance of several finite element methods. Their the most accurate solutions obtained with 21 degrees of freedom elements using both uniform and non-uniform meshes are compared with the present RBF, fourth-order CD and sixth-order multioperators results.

Figure 4 displays the relative center displacement errors (on a percentage basis) vs. the number of nodes in the computational domain for two values of the rhomb angle. The exact solution considered as a reference one was obtained using the technique described in (Morley 1963).

While the most accurate are the solutions obtained with the fourth-order CD and sixth-order CD-based multioperators schemes, the mesh-convergence orders in all cases are not as high as in the previous example. Moreover, the performance of the fourth and sixth-order methods is approximately identical though the later is slightly more accurate. The results illustrate the influence of exact solution smoothness on accuracy of high-order scheme. To improve it in the present particular case, a special treatment of the singularity should be included in the algorithms. The examples show also that the present RBF technique fall into the category of high-accuracy methods.

Cantilever beam. Consider now the application of the described approach to one of the 2D elasto-statics problems, namely to the cantilever beam problem which is popular when verifying meshless methods (see for example (Belytchko et al. 1996), (Zhang et al. 2000)). The governing equations are

$$\sigma_{xx,x} + \sigma_{xy,y} = 0, \ \ \sigma_{xy,x} + \sigma_{yy,y} = 0$$

where, assuming plane-stress case

$$\sigma_{xx} = (u_{,x} + \nu w_{,y})E/(1-\nu^2), \ \ \sigma_{xy} = (u_{,y} + v_{,x})E/(2(1-\nu^2)),$$
$$\sigma_{yy} = (v_{,y} + \nu u_{,x})E/(1-\nu^2)$$

and u,v are displacements in the x- and y-directions while E is the elastic modulus. In the calculations, values $E = 1000$, $\nu = 0.3$ were assumed. As boundary conditions, the displacements defined by the exact solutions (Timoshenko & Goodier, 1970] were used. The equations were approximated at nodal points which were distributed in the same manner as those in the above cited publications. Though an

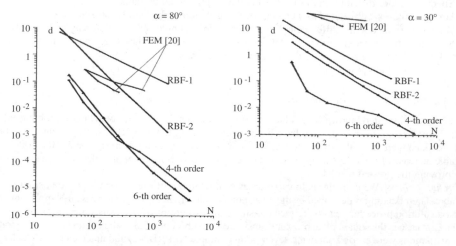

Figure 4.   Relative center displacement error for rhombic plate vs. number of nodes (on percentage basis).

optimal choice of stencils is beyond the scope of the present paper, different strategies of their forming were tried. One of them was as follows. For each center $\mathbf{x}_j$, the stencil was defined as a set of nodes which fall on a domain S_j: $\mathbf{x}_j \in S_j$ with a prescribed shape of its boundary and a prescribed characteristic length R (the latter was, for example, a circle radius, an edge of a rectangle etc.) or its characteristic area. In the present case, grid points of regular $M \times N$ meshes were used to define nodes. The beam length and width are $L = 12$ and $D = 2$ respectively. Figure 5 presents L_{uv} solution errors as functions of the mesh size h in the x-direction for several stencils which nodes fall on circles, squares, ellipses with the axis length ratio 2:1 and rectangles with the aspect ratio 2:1, the area of the supports being $20\,h^2$ and $50\,h^2$. The L_{uv} errors are defined as

$$L_{uv} = (\sum_i (u_i - u_{ei})^2 + (v_i - v_{ei})^2 / \sum u_{ei}^2 + v_{ei}^2)^{1/2}$$

where u_{ei} and v_{ei} are the exact values at a i-th node and the summation is carried out over the nodes of the computational domain.

As seen in Figure 5, the influence of the supports type is not very significant in the present case, the best choice being ellipses. As may be expected, enlarging stencils improves the accuracy and the convergence rate. However, it does not necessary mean that the improving will continue when including more and more nodes in stencils and keeping constant the mesh size.

Once numerical solutions for displacements are obtained, the corresponding stresses calculations may be viewed as a postprocessing procedure. A rich variety of RBF formulas for derivatives using different stencils can be used. In the present particular case, finite difference formulas were found to work well. Figure 6 presents the relative errors in the stresses σ_{xx} and σ_{xy} for the calculations with the elliptical supports.

Non-linear shell deformations. As another example of the present RBF technique application, we consider a non-linear shell problem described by the Karman-Fopple equations (Grigoluk & Mamai 1997). Based on the Kirchhoff assumptions, the equations for a plate having thickness $h = const$ read

$$u_{1,xx} + w_{,x}w_{,xx} + 0.5(1+v)(u_{2,xy} + w_{,y}w_{,xy}) + 0.5(1-v)(u_{1,yy} + w_{,x}w_{,yy}) = 0$$

$$u_{2,yy} + w_{,y}w_{,yy} + 0.5(1+v)(u_{1,xy} + w_{,x}w_{,xy}) + 0.5(1-v)(u_{2,xx} + w_{,y}w_{,xx})$$

$$D\Delta\Delta w = q + (Eh/(1-v^2))\{[u_{1,x} + vu_{2,y} + 0.5w_{,x}^2 + 0.5vw_{,y}^2]w_{,xx} +$$

$$[u_{2,y} + vu_{1,x} + 0.5w_{,y}^2 + 0.5vw_{,x}^2]w_{,yy} + (1-v)[u_{1,y} + u_{2,x} + w_{,x}w_{,y}]w_{,xy}\}$$

In the above equations, u_1, u_2, w are the displacements of a plate middle surface corresponding to the Cartesian coordinates x, y, z respectively. It is supposed that the coordinates origin is at the surface, the axis z being normal to it.

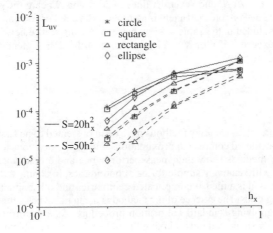

Figure 5. Relative displacements error.

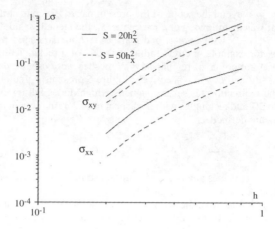

Figure 6. Relative errors in stresses.

Table 3. Non-linear round plate deformation.

Q	RBF-1			RBF-2			
	N = 6	N = 11	N = 21	N = 6	N = 11	N = 21	Ref
1	0.15725	0.16459	0.16704	0.16536	0.16744	0.16789	0.16785
2	0.30479	0.31693	0.32094	0.31906	0.32173	0.32234	0.32250
4	0.55499	0.56900	0.57538	0.57537	0.57501	0.57524	0.57625
6	0.75126	0.76252	0.76619	0.77366	0.76824	0.76773	0.76956

As an example of calculations using the above rather complicated non-linear equations, we consider bending of a round plate with clamped edges under uniform loading. Due to the central symmetry, the highly accurate solution can be obtained by solving ordinary differential equations. The solution was used as a reference one.

In the calculations, grid points of an unstructured triangulated grid were assumed as RBF nodes with the above described RBF-1 and RBF-2 stencils. To discretize the fourth derivatives, the RBF formulas for second derivatives were sequentially applied, special types of RBF operators being used near boundaries.

In Table 3, the center displacements of the plate $W = w_{center}/h$ are presented for various values of the dimensionless load $Q = q(R/h)^4/E$ and several values of N, h and R being the plate thickness and the plate radius respectively. The Poisson coefficient is assumed to be 0.3. For comparison, the reference solution (column "ref") is included in the Table. As can be seen, the difference between the RBF and reference solution does not exceed 0.5% for $N = 21$ in the case of the RBF-1 stencil and 0.2% for $N = 11$ in the case of the RBF-2 stencil.

4   CONCLUSIONS

Two types of non-standard high-accuracy methods have been presented . The first one is based on high-order non-centered and centered compact approximations and their extensions to arbitrary-order multi-operators schemes. The methods show their peak performance and are advantageous in the cases of computational domains with relatively simple forms of boundaries. In contrast, the second type follows a meshless methodology and is aimed at complicated geometries and other situations for which there is little point in constructing grids. The essence of the method is using radial basis functions in a finite difference mode rather than applying standard collocation procedure. As a result, the corresponding RBF interpolants are compactly supported thus relaxing the ill-conditioning problem. Both types of method can be used in concert in the framework of the domain decomposition strategy.

Numerical examples illustrating the high performance of the techniques are presented, the main emphasize being placed on elasticity problems.

ACKNOWLEDGMENTS

This work was supported by Russian Fund of Basic Researches (grant 01-01-00436) and INTAS (project 1150).

REFERENCES

Babuska, I. & Scapolla, T. 1989. Benchmark computation and performance evaluation for a rhombic plate bending problem. *International Journal for Numerical Methods in Engineering*. 28: 155–179.
Belytchko, T., Krongaus, Y., Organ, D., Fleming, M. & Krysl, P. 1996. Meshless methods: An overview and recent developments. *Computer Methods in Applied mechanics and Engineering*. 139: 3–47.
Brown, D.L. & Minion, M. 1995. Performance of underresolved two-dimensional incompressible flow simulations. *Journal of Computational Physics*. 121: 165.
Grigoluk & Mamai 1997. *Nonlinear Deformation of Thin-wall Construction*. (in Russian). Moscow: Nauka.
Kansa, E.J. 1990. Multiquadrics – a scattered data approximation scheme with applications to fluid dynamics – II: Solutions to parabolic, hyperbolic and elliptic partial differential equations, *Computers and Mathematics with Applications*. 19: 147–161.
Kansa, E.J. & Carlson, R.E. 1995.Radial basis function: a class of grid-free scattered data approximations *Computational Fluid Dynamics Journal*. 3: 479–496.
Kansa, E.J., & Hon, Y.C. 2000. Circumventing the ill-conditioning problem with multiquadric radial basis functions: Applications to elliptic partial differential equations, *Computer & Mathematics with Applications* 39: 123–137.
Lele, S.K. 1992. Compact finite-difference schemes with spectral-like resolution. *Journal of Computational Physics*. 103: 16–42.
Morley, L.S.D. 1963. Skew plates and Structures. *International Series of Monographs in Aeronautics*. New York: Macmillan.
Timoshenko, S.P. & Goodier, S.N. 1970 *Theory of Elasticity*. New York: McGraw-Hill.
Tolstykh, A.I. 1994. *High-accuracy non-centered Difference Schemes for Fluid Dynamics Applications*. Singapore: World Scientific.
Tolstykh, A.I. 1998. Multioperator high-order compact upwind methods for CFD parallel calculations In: *Parallel Computational Fluid Dynamics*. Emerson, D.R., Ecer, A., Periaux, J & Satofuka, N. (Eds.). 383–390. Amsterdam: Elsevier.
Tolstykh, A.I. 2000. On using RBF-based differencing formulas for unstructured and mixed structured-unstructured grid calculations. In: *Proceedings of 16th IMACS World Congress*. Deville M. & Owens (Eds.) Lausanne.
Tolstykh, A.I. 2000a. On constructing prescribed order schemes via linear combinations of compact differencing operators. *Journal of Computational Mathematics and Mathematical Physics*. 140: 1159–1172.
Tolstykh, A.I. 2001. High-accuracy and arbitrary-order methods for fluid and solid mechanics. In: *Proceedings EPMESC'VIII* San Lian Publisher: Shanghai.
Tolstykh, A.I. & Chigirev, E.N. 2000. On thin layers numerical simulation. *Journal of Computational Physics*. 166: 152–158.
Tolstykh, A.I. 2002. On integro-interpolation schemes of prescribed order of accuracy and other applications of the multioperator principle. *Journal of Computational Mathematics and Mathematical Physics*. 42: 1647–1660.
Tolstykh, A.I., Lipavskii, M.V. & Shirobokov, D.A. 2003. High-accuracy Discretization Methods for Solid Mechanics. In press.
Zerroukat, M.A. 1998. Fast boundary element algorithm for time-dependent potential problems, *Applied Mathematical Modeling*., 22: 183–196.
Zhang , X., Song, K.Z., Lu, M.W. Liu, X. 2000. Meshless methods based on collocation with radial basis functions. *Computational Mechanics*, 26: 333–343.

Computational Methods in Engineering and Science, Iu et al. (eds)
© *2003 Swets & Zeitlinger, Lisse, ISBN 90 5809 567 3*

Fuzzy finite element analysis of smart structures

S. Valliappan & K. Qi
School of Civil & Environmental Engineering, The University of New South Wales, Australia

ABSTRACT: This paper presents a unified method combining the finite element technique, the fuzzy set theory and the concepts of optimization for the analysis of "smart structures" which can be used to control the structural vibration due to earthquakes.

1 INTRODUCTION

Smart materials, such as Piezoelectric materials, have attracted growing attention in the past decades for shape, vibration and noise control because of interaction property between mechanical and electrical systems. Extensive investigations have been carried out to use smart materials as potential sensors and actuators for a wide range of application in aerospace and civil structures to actively control vibration and improve performance (Sirohi and Chopa, 1998, Valliappan and Qi, 2001).

Finite element method is widely used in the analysis of structures integrated with smart materials for shape, noise and vibration control. Ha et al. (1992) derived a finite element formulation for modelling static and dynamic responses of laminated composites containing distributed piezoelectric ceramics, from the variational principle. Hwang and Park (1993) developed a piezoelectric Kirchhoff type plate element with one electrical degree of freedom for a plate element together with active control system. Gosh and Batra (1994) formulated the problem for a fibre-reinforced laminated composite plate with piezoelectric ceramic element bonded symmetrically to its top and bottom surfaces. A new class of active controlled constrained layer damper has been proposed by Baz and Ro (1995) to vibration control of beams and plates using beam and plate elements, respectively. Samanta et al. (1996) developed a finite element model for active control of composite plates with piezoelectric sensor and actuator layers using a high order of shear deformable displacement plate theory. Varadan et al. (1996), Lim et al. (1996) and Kim et al. (1997) used three dimensional element to model the piezoelectric devices and flat shell elements to model the plate structure and transition elements to connect the three dimensional solid elements to flat shell elements. A coupled thermo-piezoelectric-mechanical theory was developed by Zhou et al. (2000) for the finite element analysis of composite plates with surface-bonded piezoelectric actuators. A higher-order laminate theory is used to describe the displacement fields of both composite laminate and piezoelectric actuator layers to accurately model the transverse shear deformation. Based on first-order shear theory and consistent methodology, Wang et al. (2001) developed a smart eight-noded isoparametric element to study the effect of the stretching-bending coupling of the piezoelectric sensor/actuator pairs on the system stability of smart composite plates. Recently, Zeng et al. (2003) proposed a fully coupled thermal electro-mechanical finite element approach which includes the nonlinearity in the response of piezoelectric materials. A hybrid meshless-differential order reduction method has been developed by Ng et al. (2003) and applied to shape control of smart structures with distributed piezoelectric sensors and actuators.

In the finite element analysis of "smart structures", most of the existing investigations assume the input parameters as deterministic variables and hence the responses of the system are also deterministic. However, it is well known that there are many uncertainties associated with the "smart structures" such as the piezoelectric, mechanical and physical properties. Besides these, there are also uncertainties associated with the properties of the materials used for structural components, foundations, soil/rock as well as the nature of earthquake loading. Some of these uncertainties are vague and imprecise and hence the use of fuzzy set theory is most appropriate.

Fuzzy set theory initiated by Zadeh (1965) has been used for a variety of engineering problems since then. Valliappan's team has been one of the pioneers in developing fuzzy finite element analysis for the solution of engineering problems (Valliappan et al. 1991, Valliappan & Pham 1992, 1994, 1995a, b). Chen and Rao (1997) proposed a fuzzy finite element approach for the vibration analysis of imprecisely-defined systems and suggested that their approach need to be extended for the finite element analysis of fuzzy engineering systems involving applied dynamic forces. Later Rao et al. (1998) developed a unified finite element method for engineering systems in the presence of hybrid uncertainties that are characterized by randomness and fuzziness. Pham et al. (1995) applied the fuzzy set theory to model damping in the dynamic finite element analysis. A computation for predicting the possibility distributions of the dynamic response and sensitivity coefficients of flexible multibody systems which include fuzzy parameters has been presented by Wasfy and Noor (1998). The uncertainty of strength and load values was considered when quantifying the safety and reliability of mechanical and structural systems by Sawyer and Rao (1999).

Dhingra et al. (1992) used a nonlinear membership function for multi-objective fuzzy optimization of mechanical and structural systems. Also, fuzzy sets have been adopted to analyse the fuzzy earthquake intensity, fuzzy response spectrum and structural fuzzy response in the optimum design of aseismic structures by Yuan and Quan (1985). Akpan et al. (2001) has presented a fuzzy element approach for static and dynamic analyses which include a sensitivity analysis to streamline the number of input fuzzy variables.

2 FINITE ELEMENT MODELLING FOR SMART STRUCTURES

Electro-mechanical coupling effects occur in certain materials when a strain is generated by an applied electric field or conversely an electric field is generated by an externally induced strain. The linear constitutive relations coupling the elastic and electric fields can be written in the following direct form (Tiersten, 1969)

$$\{\epsilon\} = \left\lfloor \mathbf{S}^E \right\rfloor \{\sigma\} + [\mathbf{d}]\{\mathbf{E}\} \tag{1a}$$

$$\{\mathbf{D}\} = [\mathbf{d}]^T \{\sigma\} + \left\lfloor \mathbf{b}^\sigma \right\rfloor \{\mathbf{E}\} \tag{1b}$$

where [S] is compliance matrices; [d] and [b] are the matrices of piezoelectric strain constant and electric permittivities, respectively; $\{\sigma\}$ and $\{\boldsymbol{\varepsilon}\}$ are stress matrix and strain matrix, respectively; $\{\mathbf{D}\}$ is the vector of electric displacement, and $\{\mathbf{E}\}$ is the vector of electric field intensity components (volt/length). The superscripts E, σ and ε represent constant electric field, stress and strain conditions, respectively. Superscript T represents the transpose of the matrix.

An eight-node brick element is used to discretize piezoelectric layers. This element has four degrees of freedom at each node (three translations and one electric potential). The overall system equations are obtained in terms of the global coordinates representing the global generalized mechanical displacement $\{\mathbf{U}_u\}$, the electric potentials, for the sensors $\{\mathbf{U}_s\}$, and for the actuators $\{\mathbf{U}_a\}$, as follows (Valliappan & Qi, 2003)

$$[\mathbf{M}_{uu}]\{\ddot{\mathbf{U}}_u\} + [\mathbf{K}_{uu}]\{\mathbf{U}_u\} + [\mathbf{K}_{ua}]\{\mathbf{U}_a\} + [\mathbf{K}_{us}]\{\mathbf{U}_s\} = \{\mathbf{F}_u\} \tag{2a}$$

$$[\mathbf{K}_{au}]\{\mathbf{U}_u\} + [\mathbf{K}_{aa}]\{\mathbf{U}_a\} = \{\mathbf{F}_a\} \tag{2b}$$

$$[\mathbf{K}_{su}]\{\mathbf{U}_u\} + [\mathbf{K}_{ss}]\{\mathbf{U}_s\} = \{\mathbf{0}\} \tag{2c}$$

in which $\{\mathbf{F}_u\}$ is the mechanical force vector as a result of the surface traction and $\{\mathbf{F}_a\}$ is the electric force vector as a result of the applied electric potential (voltage) distribution on the actuators. In control process, the input voltage, $\{\mathbf{U}_a\}$, to the actuators is regulated by the output of the sensors according to the specified control algorithm. Thus, Eq. (2a) can be expressed as

$$[\mathbf{M}_{uu}]\{\ddot{\mathbf{U}}_u\} + \left([\mathbf{K}_{uu}] - [\mathbf{K}_{us}][\mathbf{K}_{ss}]^{-1}[\mathbf{K}_{su}]\right)\{\mathbf{U}_u\} = \{\mathbf{F}_u\} - [\mathbf{K}_{ua}]\{\mathbf{U}_a\} \tag{3}$$

The output of sensor can be derived from Equation 2c as

$$\{U_s\} = -[K_{ss}]^{-1}[K_{su}]\{U_u\}$$ (4)

The system equation including damping can be given as

$$[M_{uu}]\{\ddot{U}_u\} + [C_{uu}]\{\dot{U}_u\} + ([K_{uu}] - [K_{us}][K_{ss}]^{-1}[K_{su}])\{U_u\} = \{F_u\} - [K_{ua}]\{U_a\}$$ (5)

3 FUZZY FINITE ELEMENT APPROACH

The theory of fuzzy sets theory was introduced by Zadeh (1965) when he found that real complex systems are difficult to model with conventional set theory. A fuzzy set is defined as a class of objects with a continuum of grades of membership between the values from zero to one. A fuzzy set allows a gradual change from one class to another instead of an abrupt boundary as in an ordinary set. Fuzzy-set theory provides a mathematical framework to deal with uncertainty that is caused by imprecise information rather than by randomness.

Fuzzy analysis involves two key issues: fuzzification of fuzzy variables and numerical solution of fuzzy equation.

3.1 Membership function

The membership function of a fuzzy set plays a key role in the theory of fuzzy sets. The membership function is considered as a possibility distribution function, providing information on the values that the described quantity can adopt. The membership function assigns each element in the universe of discourse a grade of belonging to the set in the closed interval between zero and one.

In engineering, parameters related to material, structure and geometry, as well as boundary conditions and loading, are fuzzy due to the imprecise information. Therefore, it is suitable to model such value as a fuzzy set with its membership function ranging from zero to one. A fuzzy variable is defined as a member of a fuzzy subset of a domain. For a fuzzy variable x, the membership function is defined as $\mu(x)$ for all x that belong to the domain. By constructing the membership, the imprecisely-defined variable is then fuzzified. Figure 1 shows the triangular membership function.

3.2 Fuzzy arithmetic

Based on the extension principle, calculation of fuzzy responses of a system, which are the functions of fuzzy variables, can be achieved by interval operation at certain α-cut levels (as shown in Fig. 1). The vertex method is used as the numerical procedure because it is cheaper than the discretization method in computing efforts. The membership function of a fuzzy variable can be "cut" by a certain number of α levels. At each α levels, the α-cut representation of a fuzzy variable can be obtained as $[x_l^\alpha, x_r^\alpha]$, where

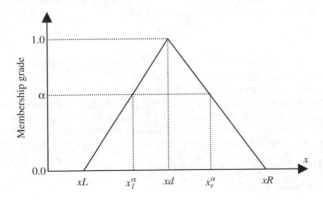

Figure 1. Triangular membership function.

129

x_l^α and x_r^α are the extreme left and right values (bound values) of membership grade for the α level. The interval operation is done for all the fuzzy variables to get the bound values at the specific α level. Then the fuzzy variable is discretized into two crisp values at the α level. Getting all the binary combinations for the bound values, finite element analysis then can be carried out using each binary combinations as the system input to get the responses of the system. Combining all the system responses calculated from all the binary combinations of fuzzy variables, the maximum and minimum values of system responses y^α can be obtained as the membership grade for the system responses,

$$y^\alpha = \begin{bmatrix} y_l^\alpha & y_r^\alpha \end{bmatrix} = \begin{bmatrix} \min\left|y^\alpha\left(B^\alpha\right)\right| & \max\left|y^\alpha\left(B^\alpha\right)\right| \end{bmatrix}$$

(6)

where B^α represents all binary combination of bound values of fuzzy variables at the α level. Performing above procedures for each α-cut levels results in the possibility distribution of system responses as the functions of fuzzy variable.

The procedure of fuzzy finite element analysis is summarized as following steps: (1) constructing membership function for each imprecisely-defined system variables; (2) performing interval operation to get the bound values for each fuzzy variables; (3) calculating system responses for all the binary combination of bound values of fuzzy variables, and getting the maximum and minimum values of system responses; (4) repeating steps (2) and (3) for all the α-cut levels from zero to one to get the fuzzy responses of the system.

3.3 Optimization of fuzzy parameters

As mentioned in the previous section, at specific α-cut level, running of finite element analysis has to be repeated many times to get the system responses for all the binary combinations of bound values of fuzzy variables, and this procedure must be done at all selected levels from zero to one. The computational cost of fuzzy finite element analysis depends on the number of fuzzy variables and the number of α-cut levels. The total number of finite element runs is, $N * 2^n$, where N is the number of α-cuts and n is the number of fuzzy variables. This indicates that the number of finite element analyses increases exponentially with the number of fuzzy parameters. To guarantee the accuracy of the possibility distribution of system responses, the number of α-cut levels has to be kept at certain level. Therefore, in order to reduce the cost of computing by reducing the required number of the finite element analyses, the number of fuzzy parameters has to be reduced.

The systematic way of doing this is by selecting the parameters which significantly affect the response of the systems as fuzzy variables. For this purpose, it is proposed to use sensitivity analysis which is often employed in optimization.

The design sensitivity analysis for static nonlinear problems in engineering has been proposed by Valliappan et al. (1997). The merits of simple finite difference method and a new semi-analytical sensitivity method have been discussed. For the dynamic analysis, it was found (Valliappan & Hakam, 2001a, b, Valliappan et al. 2001) that the finite difference method is computationally less expensive than the semi-analytical sensitivity method because of the number of iterations involved in the calculation. The forward difference approximation is favour to use as sensitivity index. The sensitivity value of a constraint function $g(x)$ with respect to the design variable x, is

$$\frac{\partial g(x)}{\partial x} = \frac{g(x + \Delta x) - g(x)}{\Delta x}$$

(7)

where Δx is the perturbation of the design variable x.

In the analysis of smart structures, the amplitudes of material properties, such as modulus of elasticity, Poisson's ratio, piezoelectric strain constants and electrical permittivity of piezoelectric materials, etc. vary significantly. Sensitivity analysis results obtained from Eq. 7 cannot reflect properly the relative importance of each parameter effect on system responses for the purpose of optimization of fuzzy parameters by reducing the number of fuzzy variables. Thus, the new sensitivity index is proposed in this paper as

$$abs\left(\frac{\Delta g(x)}{g(x)}\right) = \frac{abs(g(x + \Delta x) - g(x))}{abs(g(x))}$$

(8)

where Δx is taken as $0.05x$.

This type of sensitivity analysis provides a rational approach for reducing the number of fuzzy variables.

4 NUMERICAL SIMULATION

This section shows some numerical examples for fuzzy finite element analysis of structures, finite element analysis of "smart structures" and fuzzy finite element analysis of "smart structures".

4.1 *Fuzzy finite element analysis of a rigid frame*

Akpan et al. (2001a) proposed a method for practical fuzzy finite element analysis of structures. The methodology involves integrated finite element modeling, response surface analysis and implicit fuzzy analysis procedures. The uncertainties in the materials, loading and structural properties are represented by convex normal fuzzy sets. The response surface methodology is used to calculate the approximate values of the fuzzy finite element responses, then combinatorial optimization is carried out to determine the binary combinations of fuzzy variables that results in extreme responses at an α-level. The proposed methodology was used to solve a rigid frame shown in Figure 2 subjected to dynamic loading. Figure 3 shows the fuzzy frequencies for the first three eigenvalues. It is shown that the higher the number of fuzzy input variables, the wider the bounds of the possibility distribution of fuzzy responses.

4.2 *Finite element analysis of a shell with piezoelectric layers*

A generic finite element formulation was developed by He et al. (2002) for the static and dynamic control of functionally graded material shells with piezoelectric sensor and actuator layers. A constant velocity feedback control algorithm is used for closed-loop control of the integrated FGM shell. Figure 4 shows a simply supported graphite/epoxy cylindrical shell with a continuous PZT-4 actuator embedded. Figure 5 shows the time history of displacement response, sensor output and actuator input voltage for the cases without control and with velocity feedback control gain 0.001 and 0.01. It can be seen clearly that the amplitudes of vibration and actuator input reduce considerably faster with the increase of control gains. It can also be noted that the largest actuator voltage increases considerably (nearly 8 times) as the control gain is increased from 0.001 to 0.01.

4.3 *Coupled fuzzy finite element analysis of a steel damper integrated with piezoelectric materials*

The fuzzy finite element coupled with optimization method as proposed in this paper has been used to analyse a mild steel damper integrated with PZT material as shown in Figure 6 (Valliappan & Qi, in press).

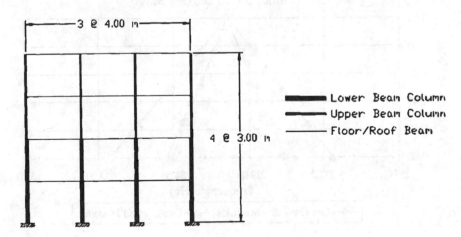

Figure 2. Elevation view of rigid frame

131

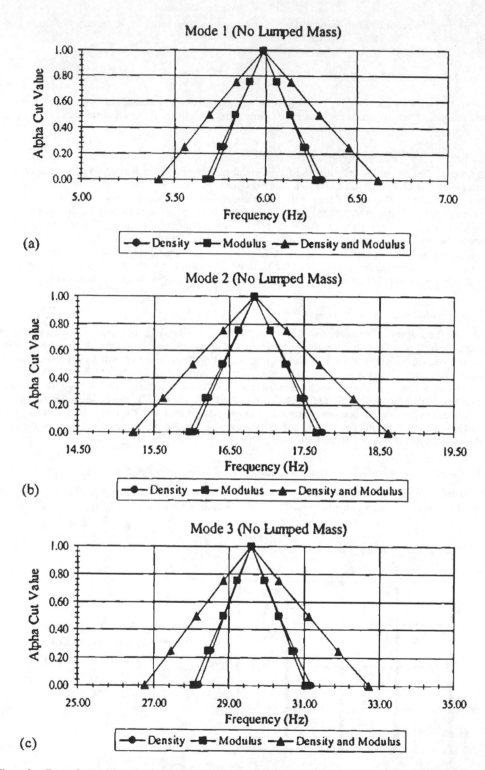

Figure 3.   Fuzzy frequencies of rigid frame when fuzzifying: E [0.90, 1.00, 1.10] and ρ [0.90, 1.00, 1.10].

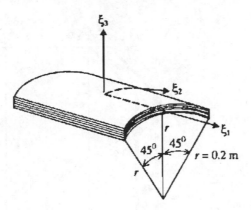

Figure 4. A cylindrical FGM shell with surface bonded piezoelectric sensor/actuator layers.

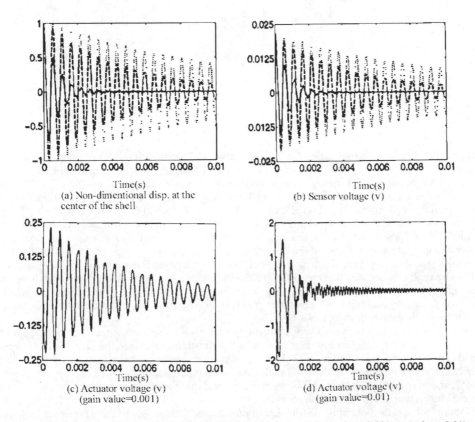

(a) Non-dimentional disp. at the
center of the shell

(b) Sensor voltage (v)

(c) Actuator voltage (v)
(gain value=0.001)

(d) Actuator voltage (v)
(gain value=0.01)

Figure 5. Response of the simply supported FGM shell (... without control, —gain = 0.001, ___gain = 0.01).

The PZT material is attached to both sides of the damper. The smart material can be used as sensors to detect structural deformation, or/and as actuators to increase the damping effect of the damper.

The steel damper is 1000 mm high, 50 mm wide and 4 mm thick. The thickness of the PZT patches is 0.4 mm. Both patches are used as sensors. The materials are: modulus of elasticity of mild steel

133

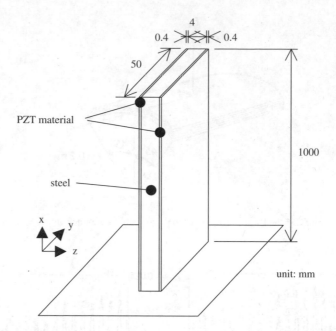

Figure 6. Smart steel damper.

(p1) $= 20 * 10^4$ mpa; Poisson's ratio of mild steel (p2) $= 0.25$; density of mild steel (p3) $= 7850$ kg/m^3; modulus of elasticity of PZT material (p4) $= 6.3 * 10^4$ mpa; Poisson's ratio of PZT material (p5) $= 0.30$; density of PZT material (p6) $= 7600$ kg/m^3; piezoelectric strain constant of PZT material d_{13}, d_{23} (p7 and p8) $= -9.6$ c/m^2 d_{33} (p9) $= 15.1$ c/m^2; d_{52} and d_{61} (p10 and p11) $= 12.0$ c/m^2; electric permittivity of PZT material ξ_{11}, ξ_{22} (p12, p13) $= 1.71 * 10^{-8}$ f/m and ξ_{33} (p14) $= 1.87 * 10^{-8}$ f/m.

The "smart" damper is modelled by 120 brick elements. Fuzzy finite element analysis was carried out to obtain the displacements and sensor output as the fuzzy responses of the system. A static load of 5N is applied in z-direction at the tip of the damper. Only the material properties were considered as fuzzy variables. The deviation from the mean values of the fuzzy variables is assumed to be 5%. Five α level cuts were used between zero and one. After sensitivity analysis, the fuzzy finite element analyses were carried for the case: (i) all of 14 material parameters were fuzzified; (ii) five parameters were fuzzified after the sensitivity analysis for displacement, and (iii) six parameters were fuzzified based on the sensitivity analysis for sensor output.

Figures 7 and 8 show the results of the sensitivity analysis for displacement and sensor output of the damper. Considering different system responses as the constraint functions, the fuzzy variables can be divided into two groups according to the results of sensitivity analysis: the primary group in which the constraint function is more sensitive to the fluctuation of input amplitude (such as p1 to p5 shown in Fig. 7, and p1, p2, p4, p5, p7 and p14 shown in Fig. 8); and the secondary group in which the input variables is less important for the system responses.

Figure 9 shows the possibility distribution of displacement responses of the damper for both cases where all 14 input variables were fuzzified and only 5 input variables (p1 to p5) which belongs to the primary group were fuzzified. There is only a slight difference between the two results. However, for the case of 14 fuzzy variables, it needs $5 * 2^{14} = 81920$ finite element runs and only $5 * 2^5 = 160$ is required for the case of 5 fuzzy variables. Figure 10 shows the possibility distribution of sensor output when 14 input variables were fuzzified and only 6 input variables (p1, p2, p4, p5, p7 and p14) of the primary group were fuzzfied. Very similar results are obtained for only 320 finite element runs. Therefore, it can be seen clearly that the sensitivity index is very effective in fuzzy variable optimization which can reduce the computing efforts dramatically.

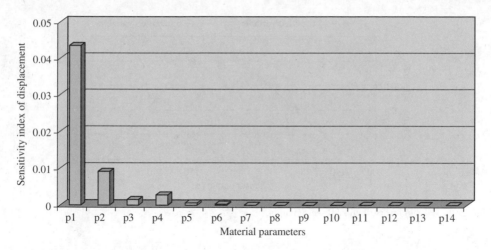

Figure 7.   Results of sensitivity analysis of displacement.

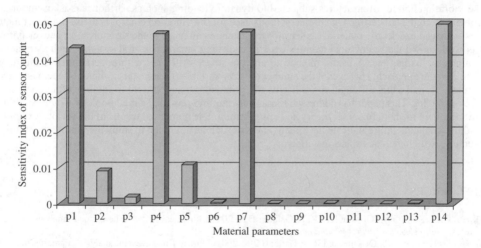

Figure 8.   Results of sensitivity analysis of sensor output.

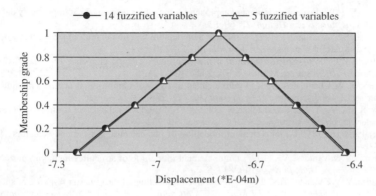

Figure 9.   Possibility distribution of displacement.

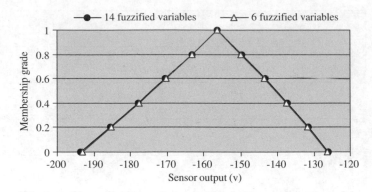

Figure 10. Possibility distribution of sensor output.

5 CONCLUDING REMARKS

The vibration control of structures subjected to dynamic loading such as earthquakes is an important problem in civil engineering. One of the methods used for the vibration control is to use the smart materials such as piezoelectric materials as sensors and actuators attached to the structures, for active control. It is well known that many uncertainties exist in analysing smart structural systems such as material parameters, loading etc. A useful method to analyse smart structures with uncertainties is the fuzzy finite element approach. However, if the number of fuzzy variables is large, this technique becomes uneconomical. To make this efficient, the concepts of optimization can be used to minimize the number of fuzzy variables. This paper has addressed these problems and provided a methodology which combines finite element method, fuzzy set theory and optimization. The numerical results of the various examples available in the literature indicate the effectiveness of such an approach to study the vibration control of structures which includes smart materials.

REFERENCES

Akpan, U.O., Koko, T.S., Orisamolu, I.R. & Gallant, B.K. 2001a. Practical fuzzy finite element analysis of structures. *Finite elements in analysis and design*. 38: 93–111.

Akpan, U.O., Koko, T.S., Orisamolu, I.R. & Gallant, B.K. 2001b. Fuzzy finite element analysis of smart structures. *Smart Mater. Struc.* 10: 273–284.

Baz, A. & Ro, J. 1995. Vibration control of plates with active constrained layer damping. *Proc. SPIE Conf. on Smart Structures and Materials 1995 – Passive Damping* (edited by C.D. Johnson), San Diego, California. 2445: 393–409.

Chen, L. & Rao, S.S. 1997. Fuzzy finite element approach for the vibration analysis of imprecisely defined systems. *Finite Elements in Analysis and Design*. 27: 69–83.

Dhingra, A.K., Rao, S.S. & Kumar, V. 1992. Nonlinear membership function in multiobjective fuzzy optimization of mechanical and structural systems. *AIAA Journal*. 30: 251–260.

Ghosh, K. & Batra, R.C. 1994. Shape control of plates using piezoelectric elements. *Proc. SPIE Conf. on Active Mat. and Smart Struct.*, Texas A&M Univ., USA. 2427: 107–121.

Ha, S.K., Keilers, C. & Chang, F.K. 1992. Finite element analysis of composite structures containing distributed piezoceramic sensors and actuators. *AIAA Journal*. 30: 772–780.

He, X.Q., Liew, K.M., Ng, T.Y. & Sivashanker, S. 2002. A FEM model for the active control of curved FGM shells using piezoelectric sensor/actuator layers. *AIAA Journal*. 30: 772–780.

Hwang, W.S. & Park, H.C. 1993. Finite element modelling of piezoelectric sensors and actuators, *AIAA Journal*. 31: 930–937.

Kim, J., Varadan, V.V. & Varadan. V.K. 1997. Finite element modelling of structures including piezoelectric active devices. *Int. J. for Num. Meth. in Engrg.* 40: 817–832.

Lim, Y.H., Varadan, V.V. & Varadan, V.K. 1996. Finite element modelling of the transient response of smart Structures. *Proc. SPIE Conf. on Smart Struct. and Materials, Mathematics and Control in Smart Structures*, San Diego, California. 2715: 233–242.

Ng, T.Y., Li, H., Cheng, J.Q. & Lam, K.Y. 2003. A new hybrid meshless-differential order reduction (hM-DOR) method with applications to shape control of smart structures via distributed sensors/actuators. *Engineering Structures*. 25: 141–154.

Pham, T.D., Valliappan ,S. & Yazdchi, M. 1995. Modelling of Fuzzy Damping in Dynamic Finite Element Analysis. *Proc. of Int. Conf. Fuzzy-IEEE/IFES'95*, Japan. 1971–1978.

Rao, S.S., Chen, L. & Mulkay, E. 1998. Unified finite element method for engineering systems with hybrid uncertainties. *AIAA Journal*. 36: 1291–1299.

Samanta, B., Ray, M.C. & Bhattacharyya, R. 1996. Finite element model for active control of intelligent structures. *AIAA Journal*. 34: 1885–1893.

Sawyer, J.P. & Rao, S.S. 1999. Strength-based reliability and fracture assessment of fuzzy mechanical and structural systems. *AIAA Journal*. 37: 84–92.

Sirohi, J. & Chopra, I. 1998. Fundamental behaviour of piezoceramic sheet actuators. *Proc., SPIE Conf. on Smart Struct. and Integrated Sys.*, San Diego, CA. 3329: 626–646.

Tiersten, H.F. 1969. Linear Piezoelectric plate vibration. New York: Plenum Press.

Valliappan, S. & Hakam, A. 2001a. Optimum design of structual foundations, Keynote Paper, *Proc. EPMESC'VIII*. 96–108.

Valliappan, S. & Hakam, A. 2001b. Finite element analysis of optimal design of foundations due to dynamic loading. *Int. J. Num. Meth. Eng.* 52: 605–614.

Valliappan, S., Hakam, A. & Khalili, N. 2001. Coupled optimization and finite element dynamic analysis of geotechnical problems. *Proc. Int. Conf. Comp. Meth. and Advances in Geomechnics*. Arizona. 2: 1051–1057.

Valliappan, S. & Pham, T.D. 1992. Constructing a membership function with objective and subjective information. *Microcomputers in Civil Engg.* 8: 75–82.

Valliappan, S. & Pham, T.D. 1994. Fuzzy finite element analysis of a foundation on an elastic soil medium. *Int. J. Num. Analyt. Meth. in Geomechanics*. 17: 771–789.

Valliappan, S. & Pham, T.D. 1995a. Elasto-plastic finite element analysis with fuzzy parameters. *Int. J. Num. Meth. in Engg.* 38: 531–548.

Valliappan, S. & Pham, T.D. 1995b. Fuzzy logic applied to numerical modelling of engineering problems. *Computational Mechanics Advances*. 2: 213–281.

Valliappan, S., Pham, T. D. & Murti, V. 1991. Application of fuzzy sets in the finite element analysis of geotechnical problems. *Computer Applications in Civil and Building Engineering*. Kozo Publishers, Tokyo: 111–118.

Valliappan, S. & Qi, K. 2001. Review of seismic vibration control using "smart materials". *Struc. Engg. & Mech.* 11: 617–636.

Valliappan, S. & Qi, K. 2003. Finite element analysis of a "smart" damper for seismic structural control. *Computers & Structures*. In Press.

Valliappan, S., Tandjiria, V. & Khalili, N. 1997. Design sensitivity and constraint approximation methods for optimum in nonlinear analysis, *Comm. Num. Meth. Engg.* 13: 999–1008.

Varadan, V.V., Lim, Y.H. & Varadan, V.K. 1996. Closed loop finite element modelling of active/passive damping in structural vibration control. *Smart Materials and Structures*. 5: 685–694.

Wang, S.Y., Quek, S.T. & Ang, K.K. 2001. Vibration control of smart piezoelectric composite plates. *Smart Mater. Struc.* 10: 637–644.

Wasfy, T.M. & Noor, A.K. 1998. Finite element analysis of flexible multibody system with fuzzy parameters. *Comput. Meth. Appl. Mech. Engrg.* 160: 223–243.

Yuan, W.G. & Quan, W.W. 1985. Fuzzy optimum design of aseismic structures. *Earthquake Engg. Struc. Dynamics*. 13: 827–837.

Zadeh, L.A. 1965. Fuzzy sets. *Inform. Control*. 8: 338–353.

Zeng, W., Manzari, M.T., Lee, J.D. & Shen, Y.L. 2003. Fully coupled non-linear analysis of piezoelectric solids involving domain switching. *Int. J. Numer. Meth. Engrg.* 56: 13–34.

Zhou, X., Chattopadhyay, A. & Gu, H. 2000. Dynamic responses of smart composites using a coupled thermo-piezoelectric-mechanical model. *AIAA Journal*. 38 (10): 1939–1948.

Computational Methods in Engineering and Science, Iu et al. (eds)
© 2003 Swets & Zeitlinger, Lisse, ISBN 90 5809 567 3

Recent discoveries related to the use of the Wittrick-Williams algorithm to solve transcendental eigenproblems

F.W. Williams & A. Watson
Department of Building and Construction, City University of Hong Kong, Hong Kong

W.P. Howson & D. Kennedy
Cardiff School of Engineering, Cardiff University, Cardiff, Wales, UK

ABSTRACT: Traditional finite element methods yield linear eigenvalue problems, e.g. when the eigenvalues are undamped natural frequencies. In contrast, when exact elements are available the overall stiffness matrix is a transcendental function of the eigenparameter. The Wittrick-Williams (W-W) algorithm must then be used to solve the resulting transcendental eigenproblem. This paper reports how analogies with linear eigenproblems have recently been used for transcendental eigenproblems to: discover a new "member stiffness determinant" property; develop an inverse iteration based recursive second order transcendental eigensolver; understand why "frequency squared" should be used and; to explore exact inverse iteration substructuring methods. The W-W algorithm is also applied to solve the topical homogeneous tree problem of mathematics. The resulting transcendental eigenproblem involves over 10^{12} linked Sturm-Liouville equations, is solved with negligible computational effort and is well conditioned. Additionally, a deeper understanding is obtained by using the engineer's physical understanding of the analogous structural problem.

1 INTRODUCTION

In 1970, the first author co-invented the Wittrick-Williams (W-W) algorithm (Williams & Wittrick 1970, Wittrick & Williams 1971), which remains the only way of solving the transcendental eigenproblems of structural analysis with the certainty that no eigenvalues are missed, these eigenvalues usually being the natural frequencies of undamped free vibration problems or the critical loads of buckling problems. Then the eigenvectors are, respectively, the vibration and buckling modes associated with the eigenvalues. Using the vibration case as an example, the transcendental eigenproblem differs from the familiar linear eigenproblem as follows.

Discretisation, e.g. by using the finite element method (FEM), gives a mass matrix $\mathbf{M}$ and static stiffness matrix $\mathbf{K}$ which are both symmetric and of the same finite order N, with $\mathbf{K}$ positive definite or semi-definite. Hence the eigenvalues and eigenvectors are found by solving the associated generalised linear eigenproblem

$$(\mathbf{K} - f\mathbf{M})\mathbf{D} = 0 \tag{1}$$

where $\mathbf{D}$ is the modal vector and contains amplitudes of quantities which must be multiplied by $\sin \omega t$, where ω = circular frequency and t = time, so that $f = \omega^2$.

In contrast, some methods are exact, i.e. the differential equations of the associated continuous problem are solved. They retain the infinite number of degrees of freedoms (DOF) of the problem and yield a dynamic stiffness matrix $\mathbf{K}(f)$ which includes both the stiffness and mass contributions and is a transcendental function of the eigenparameter, i.e. of f, so that the eigenproblem becomes

$$\mathbf{K}(f)\mathbf{D} = 0 \tag{2}$$

The W-W algorithm enables the eigenvalues of this transcendental eigenproblem to be extracted to any required accuracy.

The authors and others have extended and applied the algorithm in numerous ways for 32 years, e.g. those listed by Williams et al. (2003). However such work has dealt to only a small and/or unsatisfactory extent with analogies between transcendental eigensolvers and the numerous and widely used linear eigensolvers and has also been almost exclusively applied to structural mechanics problems. This paper reports recent and ongoing work on both such analogies and a major new application of the W-W algorithm in mathematics.

2   DERIVATION OF THE W-W ALGORITHM

The W-W algorithm was originally derived in four different ways, of which only one (Wittrick & Williams 1973) started from the Sturm sequence property of an exactly equivalent finite element model of (hypothetically) infinite order, as follows.

Because $\mathbf{K}$ and $\mathbf{M}$ are symmetric Equation 1 has a Sturm sequence property if $\mathbf{K}$ is positive definite and has the consequence that J, the number of eigenvalues lying between zero and any trial value f_t of the eigenparameter, equals the number of negative elements on the leading diagonal of $(\mathbf{K} - f_t\mathbf{M})^\Delta$, the upper triangular matrix obtained from $(\mathbf{K} - f_t\mathbf{M})$ by performing Gauss elimination in its usual computational form, i.e. pivotal rows are not scaled.

The first step of the proof considers the hypothetical *infinite order Sturm sequence problem* obtained by letting $N{\rightarrow}\infty$ in Equation 1, while assuming that the FEM model used to obtain $\mathbf{K}$ and $\mathbf{M}$ is such that accurate solution of this hypothetical problem would give the correct solution for the real, continuous, problem.

The second step is to apply the above form of Gauss elimination to this matrix but to arrest it after $N_i = N - N_c$ rows have been pivotal. For a framework, N_c can be envisaged to be the number of DOF at the joints, so that N_i is the infinite number of DOF which real members have between their ends. This is equivalent to starting from

$$\begin{bmatrix} \mathbf{K}_{ii} - f\,\mathbf{M}_{ii} & \mathbf{K}_{ic} - f\,\mathbf{M}_{ic} \\ \mathbf{K}_{ic}^T - f\,\mathbf{M}_{ic}^T & \mathbf{K}_{cc} - f\,\mathbf{M}_{cc} \end{bmatrix} \begin{bmatrix} \mathbf{D}_i \\ \mathbf{D}_c \end{bmatrix} = \begin{bmatrix} \mathbf{0} \\ \mathbf{0} \end{bmatrix} \tag{3}$$

where T denotes the transpose of a matrix, and hence obtaining

$$\begin{bmatrix} \left(\mathbf{K}_{ii} - f\,\mathbf{M}_{ii}\right)^\Delta & \mathbf{C} \\ \mathbf{0} & \mathbf{K}(f) \end{bmatrix} \begin{bmatrix} \mathbf{D}_i \\ \mathbf{D}_c \end{bmatrix} = \begin{bmatrix} \mathbf{0} \\ \mathbf{0} \end{bmatrix} \tag{4}$$

Here $\mathbf{C}$ is irrelevant because the problem reduces to

$$\mathbf{K}(f)\mathbf{D}_c = \mathbf{0} \tag{5}$$

Of course, if N had been finite this process for obtaining Equation 5 from Equation 3 is identical to that used with FEM to reduce the order of the problem by exact means, e.g. by exact substructuring, although $\mathbf{K}(f)$ is then often expressed as

$$\mathbf{K}(f) = \left(\mathbf{K}_{cc} - f\,\mathbf{M}_{cc}\right) - \left(\mathbf{K}_{ic}^T - f\,\mathbf{M}_{ic}^T\right)\left(\mathbf{K}_{ii} - f\,\mathbf{M}_{ii}\right)^{-1}\left(\mathbf{K}_{ic} - f\,\mathbf{M}_{ic}\right) \tag{6}$$

The third step is to recognise that the $\mathbf{K}(f)$ can alternatively be derived directly and exactly by solving the differential equations of the members of the structure, e.g. for vibration of rigidly jointed frames, to allow *exactly* for distributed mass, thus avoiding the discretisation errors of the traditional FEM. This removes the need for Equations 3–6 and hence also avoids the impossible task of assembling the eigenproblem of Equation 1 with $N{\rightarrow}\infty$, which is why it was called *hypothetical* earlier.

The fourth step uses s{ } to denote the Sturm sequence property, so that inspection of Equations 1, 3 and 4 and completion of the arrested Gauss elimination used to obtain Equation 4 gives

$$J = s\{\mathbf{K} - f_t\,\mathbf{M}\} = s\{\mathbf{K}_{ii} - f_t\,\mathbf{M}_{ii}\} + s\{\mathbf{K}(f_t)\} \tag{7}$$

The fifth and final step is to calculate $s\{\mathbf{K}(f_t)\}$ as the number of negative leading diagonal elements of $\mathbf{K}(f_t)^\Delta$ while finding $s\{\mathbf{K}_{ii} - f_t \mathbf{M}_{ii}\}$ via its physical interpretation, which is that it is the sum J_0 over all members of the structure of the contribution J_m that each would make to J if constraints clamped its ends, so that $\mathbf{D}_c = \mathbf{0}$. Hence Equation 7 becomes the W-W algorithm

$$J = J_0 + s\{\mathbf{K}(f_t)\} \tag{8}$$

where, with the summation being over all members,

$$J_0 = s\{\mathbf{K}_{ii} - f_t \mathbf{M}_{ii}\} = \Sigma J_m \tag{9}$$

Equation 8 yields many logical procedures for choosing successive values of f_t, such that convergence on any required eigenvalue(s) of the transcendental eigenproblem of Equation 2 is ensured because J is known for every f_t used, e.g. see Williams & Kennedy (1988), Kennedy & Williams (1991), Ye & Williams (1995).

To conclude, the transcendental eigenproblem to which the W-W algorithm, i.e. Equation 8, is applied involves assembling $\mathbf{K}(f_t)$ from member equations obtained by solving the differential equations governing the members and by obtaining J_m for each member from the same differential equations, enabling J_0 to be calculated as ΣJ_m.

3 APPLICATION TO RIGIDLY JOINTED PLANE FRAMES

The W-W algorithm applies to all structures for which transcendental member equations are available, but for illustrative purposes a vibrating Bernoulli-Euler beam with uncoupled axial and flexural behaviour is used. Figure 1 shows the amplitudes of the displacements d, rotations θ, forces p and moments m at the ends of the beam. The subscripts denote the ends of the beam and longitudinal (x) and transverse (y) displacements and forces. The member has length L, extensional rigidity EA, flexural rigidity EI and mass per unit length μ.

The transcendental member equations have the form (Lightfoot 1980)

$$
\begin{bmatrix}
\tau h & 0 & 0 & -\kappa h & 0 & 0 \\
0 & \gamma k/L^2 & \xi k/L & 0 & -\varepsilon k/L^2 & \delta k/L \\
0 & \xi k/L & \alpha k & 0 & -\delta k/L & \beta k \\
-\kappa h & 0 & 0 & \tau h & 0 & 0 \\
0 & -\varepsilon k/L^2 & -\delta k/L & 0 & \gamma k/L^2 & -\xi k/L \\
0 & \delta k/L & \beta k & 0 & -\xi k/L & \alpha k
\end{bmatrix}
\begin{bmatrix}
d_{x1} \\
d_{y1} \\
\theta_1 \\
d_{x2} \\
d_{y2} \\
\theta_2
\end{bmatrix}
=
\begin{bmatrix}
p_{x1} \\
p_{y1} \\
m_1 \\
p_{x2} \\
p_{y2} \\
m_2
\end{bmatrix}
\tag{10}
$$

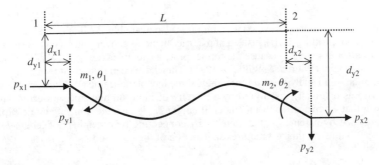

Figure 1. Deflected and undeflected positions of a beam member, showing the amplitudes of the displacements, rotations, forces and moments at its ends.

141

where

$$\left.\begin{array}{lll} \tau = v \cot v & \kappa = v \operatorname{cosec} v & h = EA/L & v^2 = f L^2 \mu/EA \\ \alpha = (SC' - CS')\lambda/\sigma & \beta = (S' - S)\lambda/\sigma & \gamma = (SC' + CS')\lambda^3/\sigma \\ \delta = (C' - C)\lambda^2/\sigma & \varepsilon = (S' + S)\lambda^3/\sigma & \xi = S'S \lambda^2/\sigma \\ k = EI/L & \lambda^4 = f L^4 \mu/EI & S = \sin\lambda & C = \cos\lambda \\ \quad S' = \sinh\lambda & C' = \cosh\lambda & \sigma = (1 - CC') \end{array}\right\} \tag{11}$$

As $N \to \infty$, $|\mathbf{K} - f\mathbf{M}|$ clearly becomes infinite, but it remains finite when normalised using its value at $f = 0$. After such normalisation has occurred but leaving the notation unaltered for convenience and simplicity, Equation 4 gives

$$\left|(\mathbf{K} - f\mathbf{M})\right| = \left|(\mathbf{K}_{ii} - f\mathbf{M}_{ii})\right| \times \left|\mathbf{K}(f)\right| \tag{12}$$

But $(\mathbf{K}_{ii} - f\mathbf{M}_{ii})$ is block diagonal because the internal DOF of a member are not coupled with those of other members. Each block has the form $(\mathbf{k}_{ii} - f\mathbf{m}_{ii})$, so that, when ordered with its (hypothetically) infinite number of internal DOF numbered first, this dynamic stiffness matrix of each member partitions into

$$\begin{bmatrix} \mathbf{k}_{ii} - f\,\mathbf{m}_{ii} & \mathbf{k}_{ic} - f\,\mathbf{m}_{ic} \\ \mathbf{k}_{ic}^T - f\,\mathbf{m}_{ic}^T & \mathbf{k}_{cc} - f\,\mathbf{m}_{cc} \end{bmatrix} \tag{13}$$

where $\mathbf{k}$ and $\mathbf{m}$ are the static stiffness and mass matrices.

Hence eliminating the internal DOF by the exact substructuring procedure used above (see Equation 6) gives

$$\mathbf{k}_c = \mathbf{k}_{cc} - f\,\mathbf{m}_{cc} - \left(\mathbf{k}_{ic}^T - f\,\mathbf{m}_{ic}^T\right)\left(\mathbf{k}_{ii} - f\,\mathbf{m}_{ii}\right)^{-1}\left(\mathbf{k}_{ic} - f\,\mathbf{m}_{ic}\right) \tag{14}$$

Hence Equation 12 becomes

$$\left|(\mathbf{K} - f\mathbf{M})\right| = \left(\Pi_m \Delta_m\right) \times \left|\mathbf{K}(f)\right| \qquad \left[= \left|\mathbf{K}_\infty(f)\right|, \, say \right] \tag{15}$$

where Π_m denotes multiplication together of the recently defined (Williams et al. 2003) member stiffness determinant Δ_m of each member. Clearly $\Delta_m = |(\mathbf{k}_{ii} - f\mathbf{m}_{ii})|$, but equally clearly it is a physical property of the member, for which an algebraic expression can be found when using the exact dynamic stiffnesses of Equations 10 and 11 (Williams et al. 2003). For flexural vibration, this expression is

$$\Delta_m = 6(1 - CC')/\lambda^4 \tag{16}$$

The expression for Δ_m for uncoupled axial vibration is also available (Williams et al. 2003).

The advantages of plotting $|\mathbf{K}_\infty(f)|$ instead of $|\mathbf{K}(f)|$ as an aid to converging on eigenvalues when using the W-W algorithm are obvious, because the former plot has no poles (i.e. vertical asymptotes) whereas the latter often has many (Wittrick & Williams 1971). They have been illustrated by a numerical example involving vibration of a rigidly jointed four storey, three bay plane frame (Williams et al. 2003).

Equation 16 was originally discovered by a mixture of intuition and trial-and-error. The test of its correctness was that when applied to a member of length $2L$, it gave the same algebraic expression, Δ_{2L} say, as was obtained by assembling the member by connecting two members of length L end-to-end. From Equations 15 and 10, this test requires satisfaction of the equation

$$\Delta_{2L} = \Delta_L^2 \times \begin{vmatrix} 2\gamma k/L^2 & 0 \\ 0 & 2\alpha k \end{vmatrix} \tag{17}$$

142

This "trial-and-error plus validation" approach has also been used to obtain Δ_m for flexure of the Timoshenko beam (Williams et al. submitted) and of some plates (Zare et al. submitted). However, more recently (Williams & Kennedy 2002) a relatively simple method was found for *deriving* the Bernoulli-Euler result, which should be more generally applicable. Its key points follow.

Imagine that Cramer's rule is used for the inversion of Equation 14. Then all elements (i, j) of $\mathbf{k}_c$ have the same denominator Δ_m, and it and the numerators k^n_{cij} are given by

$$\Delta_m = \left| \left(\mathbf{k}_{ii} - f\,\mathbf{m}_{ii} \right) \right| = \mathrm{fn}\,(f) \quad , \quad k^n_{cij} = \mathrm{fn}\,(f) \tag{18}$$

Clearly when values of k^n_{cij} which are zero for all values of f (e.g. due to lack of flexural and axial coupling) are excluded, Δ_m and the k^n_{cij} cannot be zero at $f = 0$. Therefore, normalising them using their values there and using the definition of λ given in Equation 11 gives polynomials of the form

$$\Delta_m = 1 + a_4 \lambda^4 + a_8 \lambda^8 + \dots \qquad k^n_{cij} = 1 + b_4 \lambda^4 + b_8 \lambda^8 + \dots \tag{19}$$

in which, crucially, the first term is always unity.

However, substituting the expansions of the trigonometric and hyperbolic functions of Equation 11 into its σ and into the six numerators of its second and third lines gives, in every case, polynomials with the form of Equation 19, except that their first term is in λ^4 instead of being unity. Hence dividing σ and all these six numerators by the first term $(= \lambda^4/6)$ in the expansion for σ gives

$$\Delta_m = 6(1 - CC')/\lambda^4$$
$$\alpha = \left\{ 6(SC' - CS')/\lambda^3 \right\}/\Delta_m \quad \beta = \left\{ 6(S' - S)/\lambda^3 \right\}/\Delta_m \quad \gamma = \left\{ 6(SC' + CS')/\lambda \right\}/\Delta_m$$
$$\delta = \left\{ 6(C' - C)/\lambda^2 \right\}/\Delta_m \quad \varepsilon = \left\{ 6(S' + S)/\lambda \right\}/\Delta_m \quad \xi = \left\{ 6(S'S)/\lambda^2 \right\}/\Delta_m \tag{20}$$

which forms the required derivation of Equation 16.

It is now apparent that f, i.e. frequency squared, rather than frequency, is fundamental to the transcendental eigensolution, both because the above FEM analogy uses f and also because the expansions of the expressions of Equation 11 contain only powers of f. This will seem obvious to those with an FEM background, but transcendental eigenproblem workers have usually thought of frequency (or even λ, implying (frequency)) as being the eigenparameter. Switching to f has practical advantages. For example both theory and experimental results have shown that using f accelerates a recent (Yuan et al, 2003) recursive adaptation of standard inverse iteration which enables, for the first time, (almost) machine accuracy modes to be found for the transcendental eigenproblem as follows.

First, the W-W algorithm is used to establish sufficiently close lower and upper bounds, ω_l and ω_u, on the required natural frequency. Then the steps of Yuan et al. (2003) essential to understanding the present paper are

$$f_a = \left\{ \tfrac{1}{2} (\omega_l + \omega_u) \right\}^2 \tag{21}$$

$$\mathbf{M}(f) = -\frac{d\mathbf{K}(f)}{df} = -\mathbf{K}'(f) \tag{22}$$

$$\mathbf{K}(f_a)\mathbf{D} = (f_a - f)\mathbf{K}'(f_a)\mathbf{D} \tag{23}$$

$$\mathbf{K}(f_a)\mathbf{D} = (f - f_a)\mathbf{M}(f_a)\mathbf{D} \tag{24}$$

Here, Equation 21 gives f_a for the first cycle of the recursive method and subsequent cycles use the best estimates from the previous cycle. Equation 21 is chosen in deference to the fact that users usually use ω, not ω^2, to specify their required accuracy. $\mathbf{M}(f)$ is the usual frequency dependent mass matrix of Leung's theorem (Leung 1993) and Equation 23 could be solved by inverse iteration, but instead the alternative form of Equation 24, obtained by substituting from Equation 22, is solved.

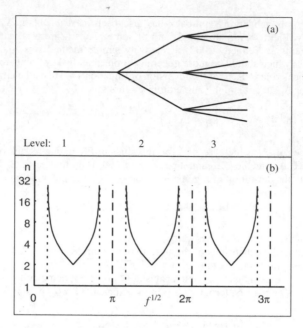

Figure 2. The tree problem: (a) a tree with $b = 3$ and $n = 3$ and; (b) its spectrum for $b = 3$ and $2 \leq n \leq 27$ (short dashed lines are limits of each band as $n \rightarrow \infty$).

The original proofs of the W-W algorithm included multi-level substructuring (Williams 1971), which retains the exact solution while greatly accelerating the eigenvalue calculations, particularly when identical substructures are present. The inverse iteration method of Equations 21–24 requires $\mathbf{D} \neq \mathbf{0}$, so that the stiffness coefficients of $\mathbf{K}(f)$ remain finite. Therefore additional, interior, nodes must be inserted within any members with a fixed-end natural frequency close to the sought natural frequency of the structure. It is tempting to make each such member and its interior node into a simple substructure and to eliminate the interior node prior to assembly of $\mathbf{K}(f)$. However, this does not work because eliminating such interior nodes results in the same $\mathbf{K}(f)$ as that obtained without using interior nodes. Therefore to date (Yuan et al. 2003) the interior nodes have been inserted without substructuring and numbered sequentially after the original nodes of the structure. However the authors are currently implementing an alternative inverse iteration approach, simulating a new substructuring approach which still inserts the interior nodes before the nodes of the structure and for which preliminary results show encouraging accuracy and computational efficiency.

4 THE MATHEMATICS TREE PROBLEM

Many disciplines involve eigenproblems and these often give spectra with alternating populated and void bands, e.g. pass- and stop-bands. The topical mathematics application reported here involves the tree topology shown in Figure 2(a). It illustrates the truly exceptional power of the W-W algorithm when its multi-level substructuring extension is used.

Specifically, a single component at level 1 branches into b identical components at level 2 which themselves branch similarly, etc. to level n. The components have their end conditions linked where branching occurs. The band spectrum shown in Figure 2(b) is for increasing values of n when all components obey the same simple second order Sturm-Liouville differential equation $-y'' = fy$. The identical bullet-shaped curves bound the populated bands and repeat to infinity with the intermediate gaps void except for extremely high multiplicity eigenvalues at $i\pi(i = 1, 2, \ldots)$. Note that the above equation is exactly that of an axially vibrating bar, if $f = \mu\omega^2/EA$.

A very recent mathematical study of this problem for n infinite (Sobolev & Solomyak 2002) (n finite is more probable in non-mathematical disciplines) could not obtain confirmatory numerical results for

144

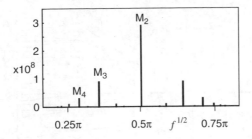

Figure 3.   Modal density and multiplicities for $b = 3$ and $n = 20$.

n large. This deficiency is currently being very successfully rectified by transfer from structural mechanics of the powerful multi-level substructuring (Williams 1971) extension of the W-W algorithm for solving such transcendental eigenproblems. Briefly, because the W-W algorithm calculates J_l, the number of eigenvalues below f_l, it can be used to converge on all required eigenvalues. A more relevant application for the tree problem is that the eigenvalue count for the interval from f_l to f_u equals $J_u - J_l$. Hence: using successive intervals reveals the eigenvalue distribution, or *modal density*, plot; this incidentally reveals the band structure shown and; using almost infinitesimal intervals identifies multiple eigenvalues and their multiplicity.

The multi-level W-W substructuring approach is most powerful when groups of components repeat frequently. Hence for the present problem a subsystem (i.e. substructure of the analogous tree of collinear bars) consisting of a set of b identical members at level n ($=3$ on Figure 2(a))and their linked member at level ($n - 1$) was analysed first to obtain equations relating solely to the left-hand end of the subsystem. b such subsystems (because they are identical only one was analysed) and a member at level ($n - 2$) were then similarly analysed, etc., until the complete tree was formed. For $n = 43$ and $b = 2$, a SUN Ultra 10 333 MHz computer using double precision arithmetic ($\cong 16$ significant figures) gave J in 0.000071 seconds without ill-conditioning, despite there being $\cong 10^{12}$ linked equations and no refinements such as pivoting. Eigenvalue multiplicities which limit as $n \to \infty$ to N_T divided by 2, 6, 14, 30 etc. (i.e. with multiplicities of order $\cong 10^{11}$) were found extremely straightforwardly, where N_T = total number of eigenvalues at or below $f^{1/2} = \pi$. These multiplicities agreed with the formulas for any b or n derived theoretically from an analogous structural mechanics problem, as briefly described in the next section.

Current studies include: fragmentation of all of the above multiple eigenvalues due to the introduction of a potential, which modifies the Sturm-Liouville equation to $- y'' + qy = fy$ and; trees of fourth and higher order Sturm-Liouville equations. For future work note that the most general form of the second order Sturm-Liouville equation, namely $- (p(x)y')' + q(x)y = fw(x)y$, can be solved by structural analogy with axial vibration of a bar. This is possible because $p(x)$, $q(x)$ and $w(x)$ can be represented by using non-uniform values for, respectively, EA, elastic (axial) foundation and mass per unit length. Similarly, Timoshenko beam theory is expected to relate to the general fourth order Sturm-Liouville equation case when elastic supports are included in the beam theory.

Figure 3 shows the modal density of the natural frequencies in any one of the populated bands of Figure 2(b), which is seen to be almost entirely dominated by the high multiplicity eigenvalues represented by the vertical bars with multiplicities M_2, M_3, M_4, etc. In addition, M_1 is an even higher multiplicity eigenvalue at $f^{1/2} = i\pi$ ($i = 1, 2, \ldots$), i.e. it bisects the gap between the populated bands.

5   MULTIPLICITIES DEDUCED BY STRUCTURAL ANALOGY

Now, as in Williams et al. (2002), consider Figure 4, remembering that all bars are collinear although clarity prevents this from being shown. The structure is always clamped at A and at all 27 of the nodes denoted by O–W. Each of O–W denotes three nodes, which are differentiated by using subscripts t, m and b for, respectively, the top, middle and bottom ones. Whenever numerical results are given in this paper they are for the left-hand and right-hand boundaries of the tree being clamped, which requires ten clamps for Figure 2(a) or 28 for Figure 4.

Consider first the multiplicities of the natural frequencies when all bars in the structure are identical. The modes of vibration at $\omega = i\pi$ clearly coincide with the clamped/clamped (C/C) natural frequencies

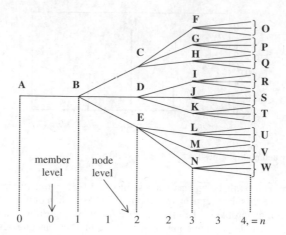

Figure 4. Four level tree with $b = 3$.

of each bar. Therefore any one of the 27 *paths* from any of nodes O–W to node A contains four bars which can clearly vibrate in-phase with each other and with equal amplitudes, while none of the other bars vibrate. This is because $y = 0$ at their three connection nodes, since the frequency is a C/C natural frequency, and clearly force equilibrium exists at the three nodes due to the amplitudes being equal. These 27 modes are obviously mutually independent, because each involves a different one of the 27 bars at level 3. Other modes are easily visualized, but are combinations of two or more of these 27, e.g. if the two bars of Q_tHQ_b (or Q_tHQ_m) vibrate in anti-phase they give $y = 0$ and equilibrium at H, but can be obtained by subtracting the mode for Q_tHCBA (or Q_mHCBA) from that of equal amplitude for Q_tHCBA. Similarly, the mode for $O_mFCBDJS_t$ is obtained by subtracting those for O_mFCBA and S_tJDBA, etc. Hence there are only 27 independent modes for this example, which equals the number of bars at level $n - 1$. Generalisation to all values of n and b obviously gives

$$M_1 = b^{(n-1)} \tag{25}$$

Similarly the eigenvalues at $\omega = (i - \frac{1}{2})\pi$ can be seen to correspond to clamped/free vibrations of a single bar as follows. Any path originating from all three nodes at any one of O–W, going to the node two levels to its left and returning to all three nodes at a different one of O–W defines a possible mode, a typical such path being P^*GCHQ^* where the asterisk denotes that P_tG, P_mG and P_bG are all present, etc. Hence, so long as all bars vibrate with equal amplitudes, compatibility (i.e. y is shared) is guaranteed at G and H if each of the six bars represented by P^*G and Q^*H vibrate in-phase with their right-hand ends clamped and their left-hand ends free, while CG and CH vibrate in anti-phase but with their right-hand ends free and C clamped. So long as CGP^* and CHQ^* vibrate with equal amplitudes in anti-phase to each other there is force equilibrium at C. Hence this mode satisfies all necessary compatibility and equilibrium conditions and so is one possible mode of the entire structure of Figure 4. Clearly $b - 1$ independent modes of this type exist for each of C, D and E, i.e. in general there are $(b - 1)b^{n-3}$ such modes. However in addition the four bars AB, BC, BD and BE form a similar pattern to the one just discussed and so contribute one further mode only, since bar AB is at level zero. In general for higher n, i.e. $n > 5$, there are $(b - 1)b^{n-5}$ of these, noting that the equilibrating force at C comes from one (or more) of CFO^*, CGP^* and CHQ^* vibrating in-phase with ABC and that the equilibrating forces at D and E arise similarly. So, if n is even $M_2 = (b - 1)\{b^{n-3} + b^{n-5} + ... + b\} + 1$, whereas for n odd the member at level 0 will not participate, so that summing the series for n odd or even gives

$$M_2 = \left\{b^{n-1} + (-1)^n\right\}/(b+1) \tag{26}$$

The above arguments can be extended to obtain the mode multiplicities M_k for any k, where k is the number of levels between nodes which act as clamped, i.e. in effect such modes are clamped at node levels

Table 1. The square roots of the eigenvalues and their multiplicities for a tree with $b = 3$ and for the values of n shown.

$f^{1/2}$	n						
	8	7	6	5	4	3	2
0.6433	1						
0.6757	2	1					
0.7227	6	2	1				
0.7945	18	6	2	1			
0.9117	55	18	6	2	1		
1.0004	2	1					
1.1230	168	56	19	6	2	1	
1.2330	1						
1.2999	18	6	2	1			
1.3769	2	1					
1.5708	547	182	61	20	7	2	1

Table 2. The square roots of the eigenvalues and their multiplicities for a tree with $b = 4$ and for the values of n shown.

$f^{1/2}$	n						
	8	7	6	5	4	3	2
0.7391	1						
0.7659	3	1					
0.8054	12	3	1				
0.8669	48	12	3	1			
0.9695	193	48	12	3	1		
1.0486	3	1					
1.1593	780	195	49	12	3	1	
1.2597	1						
1.3210	48	12	3	1			
1.3918	3	1					
1.5708	3277	819	205	51	13	3	1

$n - ik$ $(i = 1, 2, \ldots, \text{int}(n/k))$. Hence

$$\left.\begin{aligned} M_k &= (b-1)(b^{n-1} - b^{t-1})/(b^k - 1) + \text{int}(t/k) \qquad (k = 1,2\ldots) \\ t &= n - k\,\text{int}\{(n-1)/k\} \end{aligned}\right\} \tag{27}$$

6   ONGOING WORK ON THE MULTIPLE EIGENVALUES OF TREES

The physical arguments of the previous section are being extended to obtain many additional useful results including all those which may be observed from the computed results of Tables 1 and 2, which are respectively for $b = 3$ and $b = 4$ and which should be understood as follows.

The left-hand column gives all the distinct eigenvalues $f^{1/2}$ which occur in the left-hand half of a modal density plot which is identical to that of Figure 3 except that now n can have any value between 2 and 8 inclusive. Each of the remaining columns of the tables is for a particular value of n and indicates which of the values of $f^{1/2}$ given in the left-hand column is an eigenvalue for that value of n and what its multiplicity is. For example the $n = 4$ results column of Table 1 shows that a tree with $b = 3$ and $n = 4$ has three eigenvalues for which $0 < f^{1/2} \leqslant 1/2\pi$, their values (multiplicities) being 0.9117(1), 1.1230(2) and 1.5708(7).

Observed phenomena from these tables, all of which have been independently deduced from the physical analogy, include:

(1) every eigenvalue for any chosen value of n also occurs, but with higher multiplicity, for all higher values of n;

147

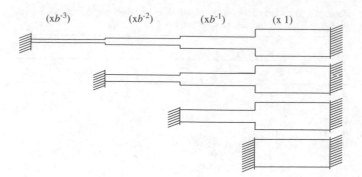

(xb^{-3}) (xb^{-2}) (xb^{-1}) $(x\,1)$

Figure 5.   The n clamped-ended stepped bars needed to calculate the eigenvalues for $n = 4$. The values in brackets show the factors by which EA and μ for the right hand uniform portion needs to be multiplied to obtain any of the uniform portions beneath the factor.

(2) reading from right to left every row in the tables starts with unity followed by $b - 1$ and;
(3) each succeeding number in the row is always equal to, or one different from, b times the number to its right.

The tables can be considered to consist of the left-hand column of eigenvalues and a table of multiplicities. The eigenvalues must be found by computation, whereas the table of multiplicities can be constructed directly by logic, based on physical interpretation of vibration of the tree of collinear bars. Moreover, although Tables 1 and 2 are for trees of identical uniform bars their tables of mutliplicities also apply without modification (for the relevant value of b) when the bars are non-uniform but are still all identical to one another.

A very helpful result which can be also be deduced from the physical interpretation of the tree of bars is that all of the eigenvalues (but not their multiplicities) can be found by collating all of the eigenvalues given by n stepped bars, as shown in Figure 5. Because it has been indicated above that tables of multiplicities can always be constructed, they can be used with these stepped bar eigenvalues to construct complete tables such as Tables 1 and 2.

7   CONCLUSIONS

The traditional FEM yields linear eigenproblems, e.g. with the eigenvalues being undamped natural frequencies. When exact elements are available, i.e. their governing differential equations have been solved analytically or numerically, the overall stiffness matrix is a transcendental function of the eigenparameter, e.g. it includes trigonometric/hyperbolic functions of the eigenparameter. The Wittrick-Williams algorithm must then be used to extract the eigenvalues reliably.

The algorithm can be derived from the Sturm sequence property of an exactly equivalent FEM model of (hypothetically) infinite order. Thus, analogies exist between the transcendental eigenproblem and the more usual linear eigenproblems which arise in many disciplines. Thus solution methods developed in either context can: be transferred to the other; give better understanding of existing methods and; be imported into other disciplines. All of these benefits are illustrated by this paper, which has outlined the authors' recent extensive work, including:

(1) the discovery of a new "member stiffness determinant" property for all transcendental formulations;
(2) an understanding of why "frequency squared", rather than "frequency", should be used as the transcendental eigenparameter;
(3) the development of a recursive second order inverse iteration method for transcendental eigenproblems;
(4) the development of exact inverse iteration substructuring methods, to aid both transcendental and linear eigensolvers;
(5) the mathematics problem of finding the eigenvalues of Sturm-Liouville differential equations linked to give a tree topology has been solved;

(6) the results were obtained for 10^{12} or more such linked equations in negligible computer time and without ill-conditioning and;

(7) many of these eigenvalues have extraordinarily high multiplicities, e.g. 10^{11}, which can be predicted exactly from theoretical consideration of a similar tree composed of axially vibrating bars.

ACKNOWLEDGMENTS

The authors are grateful to the U.K. Engineering and Physical Sciences Research Council for support under grants GR/R05406/01 and GR/R05437/01. They are also extremely grateful to Professor W.D. Evans and Dr B.M. Brown of Cardiff University for crucial help with the mathematical tree problem. The first and second authors hold posts at Cardiff University to which they will return upon completion of their appointments at City University of Hong Kong.

REFERENCES

Kennedy, D. & Williams, F.W. 1991. More efficient use of determinants to solve transcendental structural eigenvalue problems reliably. *Computers and Structures* 41(5): 973–979.

Leung, A.Y.T. 1993. *Dynamic stiffness and substructures*. London: Springer-Verlag.

Lightfoot, E. 1980. Exact straight-line elements. *Journal of Strain Analysis and Engineering Design* 15(2): 89–96.

Sobolev, A.V. & Solomyak, M. 2002. Schrödinger operators on homogeneous metric trees: spectrum in gaps. *Reviews in Mathematical Physics* 14(5): 421–467.

Williams, F.W. 1971. Natural frequencies of repetitive structures. *Quarterly Journal of Mechanics and Applied Mathematics* 24(3): 285–310.

Williams, F.W. & Kennedy, D. 1988. Reliable use of determinants to solve non-linear structural eigenvalue problems efficiently. *International Journal for Numerical Methods in Engineering* 26(8): 1825–1841.

Williams, F.W. & Kennedy, D. 2002. An overview of application of the recently discovered member stiffness determinant when solving the transcendental eigenproblems of structural engineering. *Proceedings 2nd international conference on structural stability and dynamics, Singapore, 16–18 December 2002*: 243–248.

Williams, F.W., Kennedy, D. & Djoudi, M.S. 2003. The member stiffness determinant and its uses for the transcendental eigenproblems of structural engineering and other disciplines. *Proceedings of the Royal Society: Mathematical, Physical and Engineering Sciences* (Accepted).

Williams, F.W., Kennedy, D. & Djoudi, M.S. (Submitted). Exact determinant for infinite order FEM representation of a Timoshenko beam-column via improved transcendental member stiffness matrices.

Williams, F.W., Watson, A. & Howson, W.P. 2002. Aspects of structural vibration behaviour related to solving by structural analogy the mathematical eigenvalue problem for homogeneous trees of differential equations. *Proceedings 2nd international conference on structural stability and dynamics, Singapore, 16–18 December 2002*: 249–253.

Williams, F.W. & Wittrick, W.H. 1970. An automatic computational procedure for calculating natural frequencies of skeletal structures. *International Journal of Mechanical Sciences* 12(9): 781–791.

Wittrick, W.H. & Williams, F.W. 1971. A general algorithm for computing natural frequencies of elastic structures. *Quarterly Journal of Mechanics and Applied Mathematics* 24(3): 263–284.

Wittrick, W.H. & Williams, F.W. 1973. New procedures for structural eigenvalue calculations. *Proceedings 4th Australasian conference on the mechanics of structures and materials, Brisbane, Australia, August 1973*: 299–308.

Ye, J. & Williams, F.W. 1995. A successive bounding method to find the exact eigenvalues of transcendental stiffness matrix formulations. *International Journal for Numerical Methods in Engineering* 38(6): 1057–1067.

Yuan, S., Ye, K., Williams, F.W. & Kennedy, D. 2003. Recursive second order convergence method for natural frequencies and modes when using dynamic stiffness matrices. *International Journal for Numerical Methods in Engineering* 56(12): 1795–1814.

Zare, A., Howson, W.P. & Kennedy, D. (Submitted). Member stiffness determinants for use with axially loaded isotropic plate assemblies.

Computational Methods in Engineering and Science, Iu et al. (eds)
© 2003 Swets & Zeitlinger, Lisse, ISBN 90 5809 567 3

Wave barriers for reduction of train-induced vibrations

Y.B. Yang
Department of Civil Engineering, National Taiwan University, Taipei, Taiwan

H.H. Hung & D.W. Chang
China Engineering Consultants, Inc., Taipei, Taiwan

ABSTRACT: The purpose of this study is to investigate the effectiveness of different vibration isolation countermeasures in reducing the ground vibrations induced by trains moving at speeds that may surpass the Rayleigh wave speed of the surrounding soils. The vibration isolation countermeasures considered herein include the installation of open trenches, in-filled trenches and wave impeding blocks (WIB). The finite/infinite element approach developed by Yang & Hung (2001) is employed in this study. This approach allows us to consider the load-moving effect of the train in the direction normal to the two-dimensional profile considered, and therefore to obtain three-dimensional results using only plane elements. The moving train is simulated by a sequence of moving wheel loads that may vibrate with certain frequency. Concerning the effect of isolation, the performance of the three types of barriers at different train speeds and excitation frequencies is investigated and discussed.

1 INTRODUCTION

The problem of vibration isolation has been the focus of a great deal of research since the 1950s. Common isolation countermeasures that have been adopted include the installation of open trenches, in-filled trenches and wave impeding blocks (WIB). Previous works related to trenches that may be cited include Woods (1968), Beskos et al. (1986), Ahmad et al. (1996), and Yang & Hung (1997), among others. As for the works on WIB, the following may be cited: Schmid et al. (1991), Antes & von Estorff (1994), and Takemiya & Fujiwara (1994).

As revealed by the results of the works cited above, the most important requirement for a trench to achieve a good effect of isolation is that the trench should have a depth on the order of the surface wave length. Consequently, the isolation of ground-born vibrations by trenches is effective only for moderate to high frequency vibrations. On the other hand, the wave impeding block is not good at reducing vibrations with very high frequencies, since the basic idea of using the wave impeding blocks comes from the vibration transmission behavior of the soil layer over a bedrock. According to Wolf (1985), no vibration eigenmodes can be induced below the cut-off frequency of the soil stratum, which equals $C/(4H)$, with H denoting the depth of the soil stratum and C the compressional or shear wave speed of the soil stratum. Thus, it is expected that an artificial solid block constructed underneath the soil can reveal the same effect to some extent and impede the spreading of vibrations with longer wavelengths, i.e., the low-frequency waves.

From the review of all the aforementioned works, we note that the isolation performance of these wave barriers depends mainly on the frequency range. Thus, for vibrations induced by machine foundations with a single vibration frequency, good performance of isolation can be easily achieved by simply adjusting the dimension of wave barriers to an optimized size. However, the vibrations induced by a moving train contain a wide range of frequencies, which varies with the train speed. Thus, as far as the performance of wave barriers in reducing the train-induced vibration is concerned, further investigations in time domain with due consideration of the train speed effect should be conducted.

In this paper, we will employ the 2.5-D finite/infinite element scheme developed by Yang & Hung (2001) to investigate the isolation efficiency of open trenches, in filled trenches and WIBs in reducing the

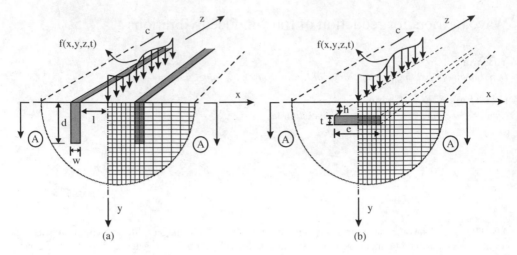

Figure 1. Typical models of the problem (a) open or in-filled trenches, (b) wave impeding block (WIB).

train-induced soil vibrations. Using such an approach, a problem with geometric and material irregularities can be dealt with in a rather easy way, compared with that of the homogeneous case. In this study, due to the limit of paper length, focus will be placed on the investigation of the isolation efficiency of the three wave barriers with respect to the moving speeds and excitation frequencies of the train, as well as for different fill materials. No attempt will be made to conduct a detailed parametric study for each of the dimensions of the three wave barriers, part of which was available in the study by Hung (2000).

2 PROBLEM FORMULATION AND BASIC ASSUMPTIONS

The problem to be considered was schematically shown in Figure 1, where a moving train with constant speed c is traveling along the z-axis on a soil stratum surface. For the purpose of reducing train-induced vibrations, in part (a) of the figure, two open or in-filled trenches with depth d and width w are placed at a distance l away from the railway center on each side. In part (b), however, a WIB with width e and thickness t is installed at depth h underneath the railway. Assume the material and geometry properties of the system, including the soils and trenches or WIB, to be uniform along the direction z, as shown in Figure 1. One may consider the whole system by only descretizing a profile perpendicular to the z-axis, i.e., the profile A-A as shown in Figure 1, within the framework of finite element formulation. This profile A-A contains a near field of finite irregular region and a semi-infinite far field. The near field containing the loads, soils and trenches is simulated by finite elements, while the far field covering the soils extending to infinity by infinite elements. Both the finite and infinite elements are plane-type elements but with 3-degrees of freedom per node to account for two horizontal and one vertical responses. The moving train is simulated by a sequence of wheel loads traveling on the ground surface along the z-axis. The following is the expression for the moving load:

$$f(x, y, z, t) = \delta(x)\delta(y)\phi(z - ct)\exp(i2\pi f_0 t) \tag{1}$$

where the exponential term is to account for the dynamic oscillation effect that may arise from the rail irregularity or mechanical system of the vehicles, the function $\phi(z)$ which represents the distribution pattern of the axle loads caused by N carriages can be obtained as the superposition of the load distribution function $q_0(z)$ of each axle as shown in Figure 2. Here, the axle load distribution function $q_0(z)$ is determined from the deflection curve of an infinite elastically supported beam with a point load T applied on the surface. Assume the bending stiffness of the beam to be EI, the foundation coefficient to be s, $q_0(z)$ can be given as

$$q_0(z) = \frac{T}{2\alpha}\exp(\frac{-|z|}{\alpha})\left[\cos(\frac{|z|}{\alpha}) + \sin(\frac{|z|}{\alpha})\right] \tag{2}$$

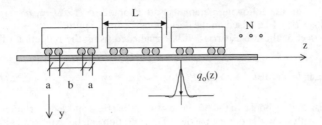

Figure 2.   Schematic of train induced loadings.

where α is the so called characteristic length determined by

$$\alpha = \sqrt{\frac{4EI}{s}} \tag{3}$$

It can be verified that the integration of $q_0(z)$ from $z = -\infty$ to ∞ gives one axle load T.

By performing the Fourier transformation to equation (1), one can express the external loading in frequency domain, i.e., $\tilde{f}(x, y, z, \omega)$, as

$$\tilde{f}(x = 0, y = 0, z, \omega) = \frac{1}{c}\exp(-ikz)\tilde{\phi}(-k) \tag{4}$$

in which $k = (\omega - 2\pi f_0)/c$ and $\tilde{\phi}(k)$ is the Fourier transform of $\phi(z)$. On the other hand, by the inverse Fourier transformation, the external loading in time domain can be recovered as

$$f(x, y, z, t) = \int_{-\infty}^{\infty} \frac{1}{c}\tilde{\phi}(-k)\exp(-ikz)\exp(i\omega t)d\omega \tag{5}$$

Equation (5) shows that the external loading can be expressed as the sum of a series of harmonic motions. For a linear system, the final time-domain steady-state response can be obtained by superposing the responses generated by each of the harmonic motions. Let the response generated by a harmonic term of the external load, $\exp(-ikz)\exp(i\omega t)$, be given as $H(i\omega)$. The final time-domain response of the system is

$$d(x, y, z, t) = \int_{-\infty}^{\infty} \frac{1}{c}\tilde{\phi}(-k)H(i\omega)\exp(-ikz)\exp(i\omega t)d\omega \tag{6}$$

The complex response function $H(i\omega)$ will be computed by the finite/infinite element approach in the frequency and wave number domain. Detailed formulation of such an approach is available in Yang & Hung (2001), which will not be recapitulated herein.

3   VIBRATION REDUCTION BY WAVE BARRIERS

3.1   *Major considerations*

To provide an equal base for comparison, the wave barriers with the same dimension will be adopted throughout the analysis, i.e., the trenches are assumed to have a depth $d = 4$ m, width $w = 1$ m and placed at a distance $l = 5$ m away from the railway center, and the WIB has a width $e = 4$ m, thickness $t = 1$ m, and installed at a depth $h = 1$ m underneath the railway. The underlying soil considered is a homogeneous half-space with the material properties listed in Table 1, together with the material properties typically used for the in-filled trench and WIB, assuming that the same material is used for the two barriers.

The shear wave, compressional wave and Rayleigh wave speeds computed of the underlying soil are $c_S = 100$ m/s, $c_P = 198.5$ m/s and $c_R = 93.2$ m/s. For the present case, the critical speed is 93.2 m/s.

As for the moving train, the following parameters are adopted: $a = 2.56$ m, $b = 16.44$ m, $L = 25$ m, $N = 4$ and wheel load $T = 10$ t, based on the definitions given in Figure 2.

The screening effect of the wave barrier will be evaluated using the reduction of vibration level in terms of dB, i.e.,

$$\text{Reduction of vibration level [dB]} = -20 \log (P_1/P_2) \tag{7}$$

where P_1 is the measured response in the presence of wave barriers and P_2 the reference value obtained from an associated analysis with no wave barriers. The vibration can be expressed either in terms of the displacement, velocity, or acceleration. In this paper, only the results in velocity will be presented, with the reference value of $v_0 = 10^{-8}$ m/sec suggested by Esveld (1989). As was mentioned previously, the effect of isolation depends generally on the content of frequencies. For this reason, the transfer function $H(i\omega)$ in frequency domain will also be presented in the following section so as to highlight the frequency-dependent characteristics of the problem considered.

3.2 Effect of train speed

First of all, the performance of wave barriers in reducing the vibrations induced by the moving static wheel loads, i.e., with $f_0 = 0$ Hz, at different speeds will be investigated. Under this assumption, the exponential term given in equation (1) can be dropped out. To investigate the effectiveness of screening of the wave barriers under different train speeds, the reduction of vertical velocity level for six different train speeds ranging from $M_2 = 0.7$ to 1.2 has been computed and plotted in Figure 3, in which the Mach number M_2 is defined as $M_2 = c/c_s$. As the shear wave speed of the underlying soils is $c_s = 100$ m/s, the train speeds considered in the figure range from $c = 70$ m/s (252 km/h) to 120 m/s (432 km/h). The average reduction of velocity level shown in the figure was obtained as the average of the vibration reductions computed over the distance from $x = 6$ to 20 m.

Figure 3 indicates that all the three wave barriers appear to be more effective in isolating the vibrations for a train speed greater than the critical speed (i.e., for $M_2 > 0.932$) than that caused by a sub-critical

Table 1. Material properties.

Material	Elastic constant E (MPa)	Poisson's ratio ν	Density ρ (kg/m³)	Damping ratio β
Soil	53.2	0.33	2,000	0.05
In-filled trench	11,760	0.25	2,400	0.05
WIB	11,760	0.25	2,400	0.05

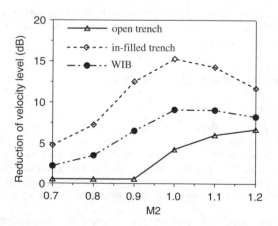

Figure 3. Effect of train speed on vibration reduction.

speed ($M_2 < 0.932$). The explanation for this phenomenon can be separated into two parts relating to the trenches and WIB.

First, it is well known that the trenches are effective only for isolating high-frequency vibrations, as has been indicated in previous studies, e.g., Yang & Hung (1997). However, for a static moving load with speeds lower than the Rayleigh wave speed, the vibrations of high frequencies decay rather fast with respect to the distance. For instance, let us examine the transfer function of velocity at $x = 10$ m for the train speed $c = 70$ m/s ($M_2 = 0.7$) plotted in Figure 4(a), the frequency content of the transfer function is localized only for a small range of frequencies, i.e., 0–10 Hz, for which the effect of trenches can be neglected. In contrast, the frequency content of the transfer function for $M_2 = 1.0$, as shown in Figure 4(b), is much wider than that for $M_2 = 0.7$. Thus, the strong influence of trenches in the high frequency range can be clearly explained.

Secondly, for the WIB, we observe from Figure 4(a) that the WIB is also ineffective for reducing the amplitude of the transfer function in the very low frequency range. Thus, similar to the trenches, the isolation efficiency of WIB for sub-critical speeds with frequency content concentrated mainly on the lower range frequencies will be poor. The reason that a WIB is not as good as a real bedrock in reducing the vibrations of very low frequencies is that it can not behave like a bedrock if its dimensions relative to the vibration wavelength is small. On the other hand, for the super-critical case with $c = 100$ m, as given in Figure 4(b), we note that the frequency content of the original system is concentrated mainly at the range below $f = 50$ Hz, and that the WIB is quite effective for isolating the vibrations in the range from $f = 10$ to 50 Hz, since the cut-off frequency for the soil stratum with an artificial bedrock installed at the depth $H = 1$ m is around 50 Hz. As a result, the performance of the WIB for the super-critical speed case is better than that of the sub-critical speed case.

In order to get an overview of the isolation effect of these wave barriers in reducing the vibrations caused by a static moving load with sub-critical or super-critical speed, the displacement over the distance from $x = -20$ to 20 m on the ground surface at the instant when the first wheel moves to the point $(x, z) = (0, 0$ m) were plotted in Figures 5 and 6 for the train speed $c = 70$ m/sec (sub-critical case) and $c = 100$ m/sec (super-critical case), respectively, where part (a) shows the displacement field without any isolation countermeasure, parts (b), (c) and (d) respectively depict the displacement field after the installation of open trenches, in-filled trenches and WIB. First, for the sub-critical case, as can be seen from Figure 5, due to the fact that for a static moving load with speeds lower than the Rayleigh wave speed, the vibration decays rather fast with respect to the distance, i.e., the displacements are trapped around the source even for the case with no wave barriers, the effect of both the open trench and WIB in reducing the vibrations induced by a train with a sub-critical speed ($c = 70$ m/s) is marginally small. As for the in-filled trenches, although the displacement behind the trenches shows a trend much more smooth than the case with no trenches, the construction of the in-filled trenches doesn't affect the displacement amplitude significantly.

On the other hand, for the super-critical case (Figure 6), i.e., with $c = 100$ m/s ($M_2 = 1.0$), we observe that there exists a Mach cone in the displacement field, and that the displacements are no longer

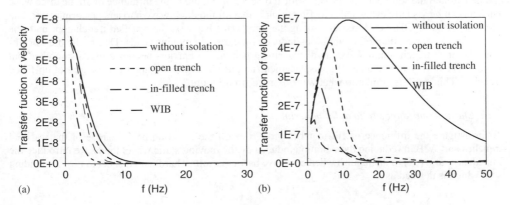

Figure 4. Influence of wave barriers on the transfer function of velocity for $x = 10$ m (a) $c = 70$ m/s ($M_2 = 0.7$), (b) $c = 100$ m/s ($M_2 = 1.0$).

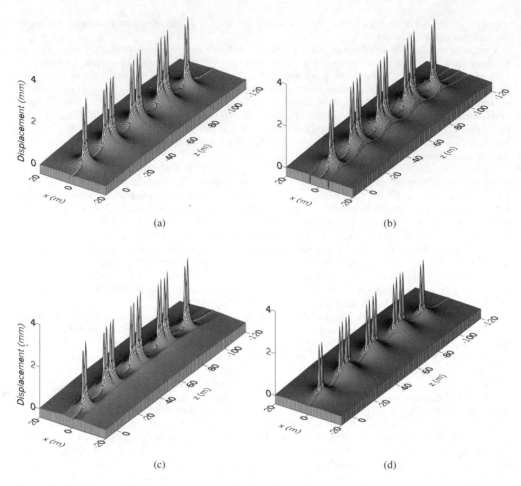

Figure 5. Displacement field under a moving train of speed $c = 70$ m/s ($M_2 = 0.7$) for the ground with (a) no isolation, (b) open trench, (c) in-filled trench, (d) WIB.

trapped around the source as for the sub-critical case. From Figure 6, the influence of all the three wave barriers can be visually evaluated. The first thing is that all the three wave barriers perform better than the case with $c = 70$ m/s. Moreover, among the three wave barriers, the in-filled trench appears to be most effective for reducing the vibrations induced by the static moving loads. Due to the presence of both the open trenches and WIB, the displacement behind the barriers becomes much smoother, but their effectiveness in reducing the displacement amplitude is not as apparent as that of in-filled trenches. This result is similar to what was observed from Figure 3.

3.3 Effect of impedance ratio of fill materials

To investigate the influence of fill material properties on the isolation performance of the in-filled trenches and WIB in reducing the vibration induced by the moving static wheel loads, we introduce the impedance ratio (IR) to distinguish whether a wave barrier is soft or hard with respect to the surrounding soil, which is defined as

$$\text{IR} = \frac{\rho' c_s'}{\rho c_s} \tag{8}$$

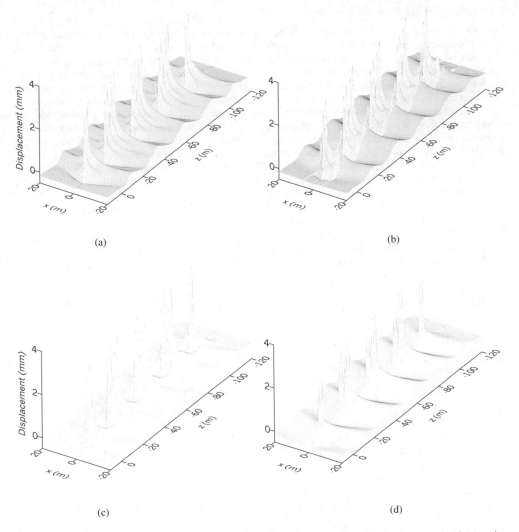

Figure 6. Displacement field under a moving train of speed $c = 100$ m/s ($M_2 = 1.0$) for the ground with (a) no isolation, (b) open trench, (c) in-filled trench, (d) WIB.

where ρ' and ρ denote the mass density of the barrier and the soil, respectively, and c_s' and c_s the shear wave speed of the two. By averaging the reduction of velocity level from distance $x = 6$ to 20 m, the influence of barrier impedance ratio, for both IR > 1 and IR < 1, was plotted in Figure 7, where parts (a) and (b) represents the results for $M_2 = 0.7$ and $M_2 = 1.0$, respectively. Evidently, for both the sub-critical and super-critical cases, both the in-filled trench and WIB appear to be much more effective when using stiffer fill materials, i.e., with IR > 1. Moreover, the stiffer the in-filled material, the higher the efficiency of isolation is. As for softer fill materials, i.e., with IR < 1, a minor reduction of velocity level can be observed for the in-filled trench, and the performance of isolation for the case with $M_2 = 1.0$ is better than that with $M_2 = 0.7$, a trend similar to that of an open trench. This is because in reality, the effect of an in-filled trench with IR < 1 can be regarded as the approximation of an open trench. On the other hand, for a soft WIB with IR < 1, the softer the fill material, the less the efficiency of isolation is. As a matter of fact, the presence of a softer WIB can inversely increase the level of vibrations. Thus, the inclusion of a soft WIB is not good at all for the purpose of reducing the vibrations, which therefore should always be avoided,

3.4 Effect of excitation frequency of moving loads

In order to investigate the effect of different wheel excitation frequencies f_0 on the screening efficiency of the three wave barriers, four different frequencies f_0 of excitation, i.e., f_0 = 5 Hz, 10 Hz, 20 Hz, 30 Hz, plus the case of zero frequency, are considered for a train speed of c = 70 m/s. By averaging the reduction of velocity level over the distance from x = 6 to 20 m, the average reduction of velocity level for the vertical response against the frequency f_0 has been plotted in Figure 8. From this figure, we find that the isolation efficiency of both the open and in-filled trenches improve as the excitation frequency f_0 increases. This result is in consistent with the previous finding that the trenches are good at reducing vibrations of short wavelengths. Also indicated by Figure 8 is that the trend of isolation efficiency improves generally with the increase in f_0, especially for the open trench. Thus, except for the case with f_0 = 0 Hz, the isolation efficiency of open trenches is better than that of in-filled trenches.

As shown in Figure 8, the performance of the WIB reaches its best when f_0 = 20 Hz, below which a decrease in f_0 can result in a poorer isolation efficiency. The reason is that the amplitude of the transfer function in the low-frequency range can hardly be reduced by the WIB, as was explained before. As for the poor performance of the WIB at f_0 = 30 Hz, it is due to the fact that the main frequency content induced by f_0 = 30 Hz and c = 70 m/s is concentrated in the range from 17 to 120 Hz, mostly beyond the cut-off frequency (around 50 Hz) of the bedrock with a depth of H = 1 m. Finally, by comparing the performance of these three wave barriers, one observes that the isolation efficiency of the open trench is the best among the three, especially for higher excitation frequencies f_0.

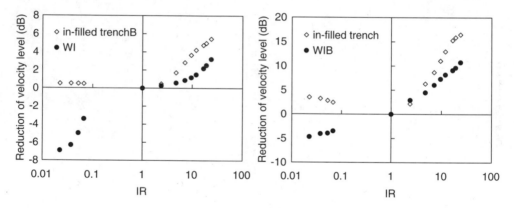

Figure 7.   Effect of impedance ratio of in-filled materials (a) c = 70 m/s (M_2 = 0.7), (b) c = 100 m/s (M_2 = 1.0).

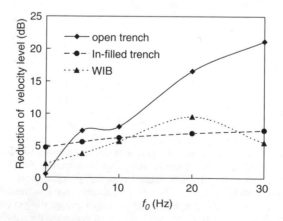

Figure 8.   Effect of excitation frequency f_0 of moving loads.

To further explore the influence of wave barriers in frequency domain, the velocity transfer function for the case studied in Figure 8, i.e., with $f_0 = 20$ Hz and $c = 70$ m/s, has been plotted in Figure 9, which indicates a drastic decrease in the amplitude for frequencies higher than 25 Hz due to the presence of both the open and in-filled trenches. Meanwhile, the existence of the WIB can significantly reduce the response for frequencies lower than $f = 50$ Hz, but as the frequency f decreases, the screening effect declines. Such a phenomenon can be easily conceived, if one realizes that as the frequency becomes smaller, i.e., as the wavelength becomes longer, the dimensions of the WIB become too small to trap the waves.

Another phenomenon revealed by Figure 9 is that the frequency content at $x = 10$ m for the case with no isolation is located mainly in the range braced by the two critical frequencies $f_{cr} = f_0(1 \pm c/c_R)$ due to the Doppler effect, which is proportional to the excitation frequency f_0. Thus, we can expect that with the increase in f_0, the frequency content will shift to a higher frequency range, and therefore the isolation efficiency of the trenches increases, while the effect of the WIB becomes rather irreverent.

Again, to have a clear picture on the influence of wave barriers in reducing the vibration caused by moving loads with self oscillation, the real-part displacement of the ground over the distance from $x = -20$ to 20 m, corresponding to the case shown in Figure 9, has been plotted in Figure 10, in which, parts (a)–(d) are computed for the cases before and after the installation of open trenches, in-filled trenches and WIB, respectively. A comparison between these figures indicates that the open trench is the most effective vibration countermeasure in reducing the ground displacement induced by the dynamic moving load. The performance of WIB appears to be the second best. As for the in-filled trenches, the existence of the fill material has changed the surface shape of vibration, but the change in vibration amplitudes can hardly be seen. This result is consistent with that observed in Figure 8 for the reduction of velocity level.

3.5 *Comparison and discussion*

In the preceding sections, the effectiveness of the open trench, in-filled trench and WIB in reducing the vibrations induced by the moving static and dynamic wheel loads have been investigated. In this section, a more realistic representation of the contact forces acting on the railway that includes both the static and dynamic terms will be adopted.

The following is the expression of the moving load exerted on the soil surface:

$$P(x = 0, y = 0, z, t) = \phi(z - ct) f(t) \tag{9}$$

where the function $\phi(z)$ representing the wheels distribution is the same as the one used in equation (1); the train speed c is taken as 70 m/s (252 km/hr); and the function $f(t)$ represents the contact forces

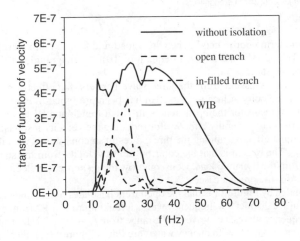

Figure 9. Influence of wave barriers on the transfer function of velocity for $r = 10$ m, $c = 70$ m/s and $f_0 = 20$ Hz.

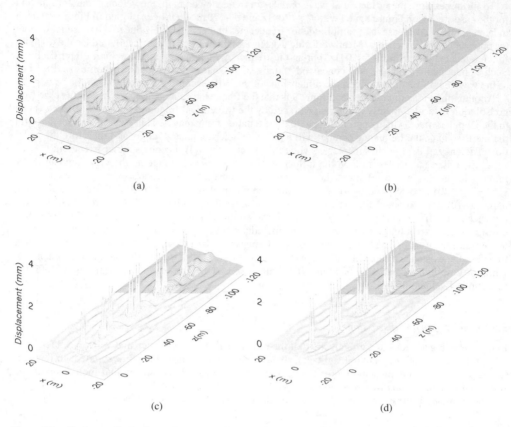

Figure 10.   Real-part displacement field under a moving train of speed $c = 70$ m/s and $f_0 = 20$ Hz for the ground with (a) no isolation, (b) open trench, (c) in-filled trench, (d) WIB.

between the wheels and rails, which moves with the wheels. Instead of treating $f(t)$ as a static or dynamic term, as was done previously, we assume

$$f(t) = 1 + \frac{1}{10}\left[\begin{array}{l} \cos(10\pi t) + \cos(20\pi t) + \cos(40\pi t) \\ + \cos(60\pi t) \end{array}\right] \qquad (10)$$

which includes a static term contributed by the wheel load and four dynamic terms with the vibration frequencies: 5 Hz, 10 Hz, 20 Hz and 30 Hz. In equation (10), the magnitude of the dynamic terms is assumed to be 1/10 of the static term.

Using the data previously assumed for the three wave barriers, the effectiveness of the three wave barriers in reducing the velocity at location $x = 10$ m has been presented in Figure 11, where parts (a), (b) and (c) represent the result for the open trench, in-filled trench and WIB, respectively. As can be seen, all the three barriers are generally effective for reducing the velocity response. However, the open trench appears to be most effective among the three, a result in consistence with that observed for the cases with non-zero f_0. The reason is that the contribution to velocity from the static load is too small, compared with that from the dynamic load. Consequently, the response of velocity is mainly controlled by the dynamic terms for the present case.

Another trend revealed by Figure 11 is that the open trench isolates mainly the high-frequency waves. Because of this, we can see distinct low-frequency waves left in Figure 11(a). Meanwhile, the WIB isolates mainly the middle-frequency waves, i.e., waves in the range from $f_0 = 10$ to 20 Hz. Thus, it can be observed that both the high-frequency and low-frequency waves are left in Figure 11(c). Finally, from the reduction of velocity level plotted for all the three barriers with respect to the distance x in Figure 12, one observes

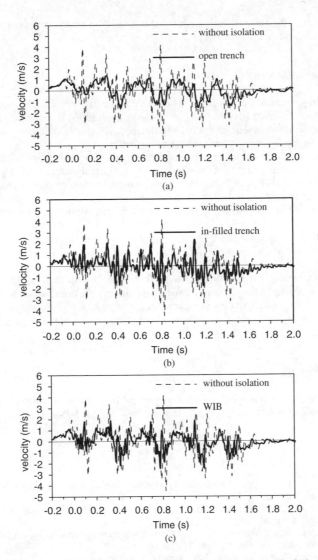

Figure 11.   Comparison of wave barriers in reducing velocity at $x = 10\,\mathrm{m}$ (a) open trench, (b) in-filled trench, (c) WIB.

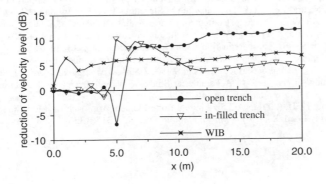

Figure 12.   Comparison between the isolation efficiency of three wave barriers.

161

that the open trench is the most effective among the three, and that this phenomenon is more pronounced with the increase in distance.

4 CONCLUDING REMARKS

In this paper, a comparison of the effectiveness of three wave barriers, i.e., the open trench, in-filled trench, and WIB with different fill materials, in isolating the train-induced vibrations under different train speeds and excitation frequencies has been performed. For the wave barriers with the parameters adopted in this study, the following conclusions can be drawn:

For isolating the vibrations induced by moving static loads ($f_0 = 0$ Hz), the in-filled trench appears to be the best choice. For isolating vibrations induced by moving dynamic loads ($f_0 \neq 0$ Hz), the open trench performs the best. When both the static and dynamic terms of moving loads are taken into account, the open trench remains the most effective.

For the case with moving static loads, all the three wave barriers perform better in reducing the vibrations induced by a train moving at super-critical speeds than at sub-critical speeds.

For the case with moving static loads, a stiffer in-filled trench or WIB with respect to the surrounding soil performs better than a softer one. The installation of a softer WIB can adversely increase the vibration.

For the case with moving dynamic loads, the performance of both the in-filled and open trenches improves with the increase of the exciting frequency f_0. Such a phenomenon is more pronounced for open trenches. The WIB is effective only in isolating the vibrations with wavelengths comparable to the dimensions of the WIB itself and the vibrations with frequencies below the cut-off frequency of its depth.

ACKNOWLEDGEMENT

The research reported herein has been sponsored in part by the National Science Council of the R.O.C. through Grant Numbers: NSC 89-2211-E-002-113 and NSC 90-2211-E-002-057.

REFERENCES

Ahmad, S., Al-Hussaini, T. M., and Fishman, K. L. (1996), "Investigation on Active Isolation of Machine Foundations by Open Trenches", *J. Geot. Eng.*, ASCE, **122**(6), 454–461.

Antes, H., and von Estorff, O. (1994), "Dynamic Response of 2D and 3D Block Foundations on a Halfspace with Inclusions", *Soil Dyn. & Earthq. Eng.*, **13**, 305–311.

Beskos, D. E., Dasgupta, B., and Vardoulakis, I. G. (1986), "Vibration Isolation Using Open or Filled Trenches, Part 1: 2-D Homogeneous Soil." *Comp. Mech.*, **1**(1), 43–63.

Esveld, C. (1989), *Modern Railway Track*. MRT-Productions, West Germany.

Hung, H. H. (2000), *Ground Vibration induced by High-Speed Trains and Vibration Isolation Countermeasures*, Ph.D. dissertation, Department of Civil Engineering, National Taiwan University, Taipei, Taiwan, R.O.C.

Schmid, G., Chouw, N., and Le, R. (1991), "Shielding of Structures from Soil Vibrations", *Soil Dyn. Earthq. Eng. V, Int. Conf. Soil Dyn. & Earthq. Eng.*, Computational Mechanics Publications, Southampton, UK, 651–662.

Takemiya, H., and Fujiwara, A. (1994), "Wave Propagation/Impediment in a Stratum and Wave Impeding Block (WIB) Measured for SSI Response Reduction", *Soil Dyn. & Earthq. Eng.*, **13**, 49–61.

Wolf, J. P. (1985), *Dynamic Soil-structure Interaction*, Prentice-Hall, Englewood Cliffs, NJ.

Woods, R. D. (1968), "Screening of Surface Waves in Soils", *J. Soil Mech. & Found. Div.*, **94**, SM4, 951–979.

Yang, Y. B., and Hung, H. H. (1997), "A Parametric Study of Wave Barriers for Reduction of Train-Induced Vibrations," *Int. J. for Numer. Meth. in Eng.*, **40**, 3729–3747.

Yang, Y. B., and Hung, H. H. (2001), "A 2.5D Finite/infinite Element Approach for Modelling Visco-elastic Body Subjected to Moving Loads," *Int. J. for Numer. Meth. in Eng.*, **50**, 1317–1336.

Computational Methods in Engineering and Science, lu et al. (eds)
© *2003 Swets & Zeitlinger, Lisse, ISBN 90 5809 567 3*

Advances in sparse solver for finite element analysis

Mingwu Yuan, Pu Chen & Shuli Sun
Department of Mechanics and Engineering Science, Peking University, Beijing, China

ABSTRACT: Symmetric positive definite equation solvers play a very important role in structural analysis. For a larger and larger structural system promoting the efficiency of the solvers is the key of application of finite element analyses (FEA). The focus of this paper is to compare the cell sparse solver that has been developed by authors with several top sparse solvers available in literatures. The performance in terms of elapsed time and memory requirement of different solvers are demonstrated by finding static displacement vectors for practical engineering problems. Numerical tests indicate that the cell-sparse solver has overall the best performance on PC machine.

1 INTRODUCTION

Research in solving symmetric positive definite systems of linear equations has always been motivated by the demand of practical engineering problems, especially problems arising in structural analysis, stress analysis, structural optimization, heat transfer, electromagnetic field etc. using FEA. The key to the reduction of the computational time in FEA is to reduce the time of solving the resultant linear system:

$$Ax = f \tag{1}$$

Here the matrix $A = (A_{ij}) \in R^{neq \times neq}$ is the symmetric positive definite global stiffness matrix. The vectors $x \in R^{neq}$ and $f \in R^{neq}$ stand for the unknown displacement and known load vector, respectively. We consider the numerical factorization of the matrix A, i.e.

$$A = LU = LDL^{T}, \tag{2}$$

where the factors $L = (L_{ij}) \in R^{neq \times neq}$ and $D = \text{diag}(D_{ii}) \in R^{neq \times neq}$ are a lower triangular matrix with unit diagonal and a diagonal matrix, respectively. In the sense of memory management, U shares the same memory locations as L^{T} and D together, since the unit diagonal of L^{T} does not necessarily occupy any storage space.

The techniques of the cell-sparse storage scheme, unrolling and out-of-core strategies have been implemented together to improve the numerical performance of sparse solvers. In this paper with 5 sections we will briefly give a description of the cell sparse solver and recent numerical results on PC.

2 SUPER-EQUATION AND CORRESPONDING STORAGE SCHEME

The successful implementation of any sparse equation solver depends considerably on the reordering method. Reversed Cuthill-Mckee (RCM), Minimum Front or Gipspoole-Stockmyer reordering algorithms can be used effectively in conjunction with skyline or variable bandwidth equation solution algorithms (Duff IS, 1989). Throughout this paper, it is assumed that the available sparse reordering algorithms, such as Modified Minimum Degree, Nested Di-section or Graph Partitioning (Schloegel K, 2000), have already been applied to the original coefficient matrix A.

Table 1. Conventional sparse storage scheme and super-equation concept.

Equation	1				2		3			4		5		6
ICA(*neq*)				4		6			9		11		13	14
JCA(*nzr*)	1	3	4	6	2	5	3	4	5	4	5	5	6	6
PCA(*nzr*)	11	a	b	c	22	d	33	e	f	44	g	55	H	66

2.1 Conventional sparse storage scheme

To facilitate the discussions in this section, assume the resultant 6×6 symmetric global stiffness matrix A as follows.

$$A = \begin{bmatrix} 11 & & a & b & & c \\ & 22 & & & & d \\ & & 33 & e & f & \\ & & & 44 & g & \\ & SYM & & & 55 & h \\ & & & & & 66 \end{bmatrix} = \begin{bmatrix} (11) & & \begin{pmatrix} a & b \end{pmatrix} & & (c) \\ & (22) & & & (d) \\ & & \begin{pmatrix} 33 & e \\ & 44 \end{pmatrix} & \begin{pmatrix} f \\ g \end{pmatrix} \\ & SYM & & (55) & (h) \\ & & & & (66) \end{bmatrix} \tag{3}$$

Here $a, b, \ldots, h$ are non-zeros (we use the terms "real" or "non-zero" in a generic sense so that it should be read as double precision or 8-byte real). The algorithms for a symmetric matrix only reserve memory for its upper triangular part. Conventionally, the sparse storage scheme, in which the upper triangle is treated as a sequential collection of compact row vectors, employs three arrays ICA(*neq*), JCA(*nzr*) and PCA(*nzr*) to represent the matrix A. Here *nzr* denotes the number of non-zeros in the upper triangle of A. The integer ICA(i) points towards the last entry of the i-th row in the column index array JCA(*nzr*) and the associated numerical array PCA(*nzr*), i.e. $\{JCA(j) \mid ICA(i-1) < j \leqslant ICA(i)\}$ and $\{PCA(j) \mid ICA(i-1) < j \leqslant ICA(i)\}$ build up the i-th compact row vector. Details of the scheme are illustrated in Table 1.

The sparse pattern of the matrix, which is represented by the arrays ICA and JCA, is called the *symbolic* matrix, and the numerical part represented by PCA is called the *numerical* matrix. A similar data structure with arrays ICU, JCU and PCU can be established for the factor U or D and L^T (shortened as U/DL^T) of A. The sizes of the numerical matrices PCA and PCU (at least for a given pivot sequence) are constant, but the total memory requirement can be reduced by reducing the symbolic matrices ICA, JCA and ICU, JCU. The pioneering work of this kind of reduction is Sherman's compressed storage scheme (Sherman AH, 1975).

2.2 Cell-sparse storage scheme

In order to reduce the memory requirement in engineering FEA, authors proposed a *cell-sparse* storage scheme (Chen P, 2000). Denoting submatrices in the second representation of the matrix A in Eqn.(3) as cells, the total number of non-zero cells (including 1×1 trivial cells) *nz* is smaller than the total number of non-zeros *nzr*. Replacing the sparse pattern of reals by the sparse pattern of cells, we obtain the cell-sparse storage scheme. If there are *mneq* super-equations, we employ five arrays, SUPER(*mneq*), IA(*mneq*), JA(*nz*), LA(*mneq*) and PA(*nzr*) to store the upper triangle of the matrix A. The integer SUPER(s) points towards the last equation label of the s-th super-equation, so that the difference SUPER(s) – SUPER($s-1$) is the size of the s-th super-equation. As illustrated in Table 2, IA(s) and LA(s) indicate the last super-column index in the cell index array JA and the last real of the s-th super-row in the numerical array PA, respectively. However, the reals in PA are still stored row by row; that means, PA(*nzr*) in Table 2 and PCA(*nzr*) in Table 1 have the same contents.

Considering the FEA of three-dimensional solids, there are a dominant number of 3×3 cells; that means $mneq \cong neq/3$. A simple calculation leads to an estimation $nz/nzr \cong (mneq/neq)^2 = 1/9$. In addition, the total length of IA, LA and SUPER roughly equals to the length of ICA. Thus, the cell-sparse storage scheme reduces the memory requirement by about 30% in comparison with the conventional sparse storage scheme, if 4-byte integers and 8-byte reals are used. Even for two-dimensional problems with a large number of 2×2 cells, the reduction ratio is still as high as 25%.

Table 2.   Cell-sparse storage scheme.

Super Eqn.		1		2		4		5	6
SUPER(*mneq*)		1		2		4		5	6
IA(*mneq*)			3		5		7	9	10
JA(*nz*)	1	3	5	2	4	3	4	4	5	5
LA(*mneq*)			4		6		11	13	14
PA(*nzr*)	11	(*a b*)	*e*	22	*d*	$\begin{pmatrix} 33 & e \\ & 44 \end{pmatrix}$	$\begin{pmatrix} f \\ g \end{pmatrix}$	55	*h*	66

3   SPARSE FACTORIZATIONS WITH SUPER-ROW UNROLLING TREATMENT

Applying unrolling strategy to *j*- and *k*- loop, a double-way loop unrolling implementation, the two-way unrolling JKI LDL^T can be outlined as follows:

Algorithm 1: "JKI" form of LDL^T factorization with two-way loop unrolling
  FOR mj = all master equations from 1 to neq
    FOR mk = all appropriate master equations from 1 to $mj-1$
      $S(0:m, 0:n) = -\text{diag}[A(mk:mk + m, mk:mk + m)] \cdot A(mk:mk + m, mj:mj + n)$
      $A(mj:mj + m, mj:neq) + = S^T (0:n, 0:m) \cdot A(mk:mk + n, mj:neq)$
    END
    FOR $l = 0:n$! treatment of super-row mj
      $S(0:l - 1, 0:0) = - \text{diag}[A(mj:mj + l, mj:mj + l)] \cdot A(mj:mj + l - 1, mj + l)$
      $A(mj:mj + l - 1, mj + l:neq) + = S^T (0:l - 1, 0:0) \cdot A(mj:mj + l - 1, mj + 1:neq)$
    $A(mj + l, mj + l + 1:neq)/ = A(mj + l, mj + 1)$
    END
  END

In case of full matrices, two-way unrolling strategies in JKI forms lead to a maximum speed-up of 140% for large problems on PC machines. Furthermore, the sparse factorization involves a great number of index manipulations; the sparse unrolling strategies should result in higher speed-up that can be made sure by numerical tests in section 4.

Assuming the prescribed maximum number of equations in a super-equation is 3, and then there are 9 cases of super-row elimination. Each case corresponds to a subroutine. Including self-reduction, forward reduction and back substitution, there are totally 30 subroutines to perform numerical operations in solution of linear system of equations. In general, these subroutines accelerate the numerical operation 3 to 5 times in solution process. For powerful machines with powerful compiler, such as HP workstation, in-line rather than subroutine code is preferred, since the subroutine call is sometimes costly compared to the amount of operation inside the subroutine.

4   NUMERICAL TESTS

4.1   *Test problems and computational platform*

Finite element models with the number of equations between 10,800 and 151,926 are illustrated in Table 3. Except for HSCT16KB, the problems are selected from Harwell-Boeing (BCSSTK*) (Duff IS, 1992), CYSHELL (S3DK*) (Kouhia R, 1998) and our home collection (PKUSTK*). All test problems arise in real engineering and industrial applications and have been extensively used in earlier research works.

In Table 3, *neq, nrhs* and *nzr* denote the number of equations, number of right hand sides and number of non-zeros in the upper parts of the matrices, respectively. The home test problems (PKUSTK*) were generated by SAP84. The non-zero off-diagonal values of the problems from Harwell-Boeing and the CYSHELL collection were generated using the pseudo-random number generator DRAND in Compaq Visual Fortran. Without losing generality, the randomly generated entries were placed into the interval $[-1, 1]$. The generated matrices were made diagonally dominant by replacing each diagonal by the absolute sum of all off-diagonals in its row and the real number 10.

Table 3. Test problems and their descriptions.

Problem	Description	*neq*	*nrhs*	*nzr*
PKUSTK01	Beijing botanical exhibition hall	22,044	5	500,712
PKUSTK02	Feiyue twin tower building	10,800	2	410,400
PKUSTK03	Dalian group silo	63,336	4	1,596,876
PKUSTK04	Yunsan Plaza	55,590	4	2,137,125
PKUSTK05	Cofferdam	37,164	5	1,121,154
PKUSTK06	Cofferdam	43,164	5	1,307,466
PKUSTK07	21 nodes solid element, $10 \times 10 \times 10$ mesh	16,860	1	1,217,832
PKUSTK08	21 nodes solid element, $11 \times 11 \times 11$ mesh	22,209	1	1,624,440
PKUSTK09	Group silo	33,960	4	808,800
PKUSTK10	4 tower silo	80,676	6	2,194,830
PKUSTK11	Cofferdam	87,804	5	2,652,858
PKUSTK12	Jijian Plaza, tall building	94,653	4	3,803,485
PKUSTK13	Machine element, 10 nodes tetrahedral solid	94,893	1	3,355,860
PKUSTK14	Tall building	151,926	4	7,494,215
HSCT16KB	A high speed civil transport model	16,146	1	515,651
BCSSTK17	Elevated pressure vessel	10,974	3	219,812
BCSSTK18	R.E. Ginna Nuclear Power Station	11,948	3	80,519
BCSSTK25	Columbia Center76-story skyscraper	15,439	3	133,840
BCSSTK29	Boeing 767 rear pressure bulkhead	13,992	3	316,740
BCSSTK30	Off-shore generator platform	28,924	3	1,036,208
BCSSTK31	Automobile component	35,588	3	608,502
BCSSTK32	Automobile chassis	44,609	3	1,029,655
S3DKQ4M2	Cylindrical shell, R/t = 1000	90,449	3	2,455,670
S3DKT3M2	Cylindrical shell, R/t = 1000	90,449	3	1,921,955

Table 4. In-core solvers in the numerical tests.

Solver	Storage scheme	Major operations	Unrolling
MA27	Coordinates	Outer-Product	
MA47	Coordinates	Outer-Product	BLAS-3
C	Row sparse with Sherman's scheme	DAXPY*	
D	Row sparse with Sherman's scheme	DAXPY	One-level
E	Row cell-sparse	DAXPY	Two-level

*DAXPY = Double Precision $\mathbf{AX} + \mathbf{Y}$.

The performance of the developed static solver is tested on a platform of Pentium III 850 PC using Compaq Visual Fortran 6.5, with compiler options/architecture:p6p/tune:p6p, 1 GB RAM.

Except for HSCT16KB, all problems were reordered by the METIS_NODEND algorithm for the sparse solvers.

Among the solvers in Table 4, MA27 and MA47 are selected from the well-known HSL Archive (Duff IS, 1983). Solver C is a simple JKI sparse solver. Solver D is a JKI one-level unrolling sparse solver that follows the concept of vector sparse solver proposed by Nguyen, (1997). Solver E represents the concept proposed in (Chen P, 2000).

The multi-frontal codes MA27 and MA47 were designated to solve indefinite sparse symmetric linear equations. During the factorization, the given pivot sequence may be modified to maintain numerical stability. However, if the matrix is known to be positive definite, the user can set a parameter CNTL(1) = 0 in the calling so that a logically simpler path is followed in the code. In all of our tests using MA27 and MA47, this option was used. Before entering MA27 and MA47, the matrix in all of our tests was permuted to the pivot sequence given by the METIS_NODEND algorithm. The switch of the minimum degree criterion embedded in MA27 and MA47 was turned off. Our experience showed that the minimum degree criterion produces generally more fill-ins than METIS_NODEND.

Table 5. Factorization time and solution time.

| Problem | Factorization time/solution time (sec) | | | | |
	MA27	MA47	C	D	E
PKUSTK01	4.78/0.44	2.58/0.39	4.40/0.22	2.91/0.12	0.98/0.12
PKUSTK02	4.88/0.11	2.58/0.05	4.01/0.06	2.80/0.11	0.83/0.06
PKUSTK03	31.25/1.21	17.13/0.99	26.43/0.46	18.51/0.38	5.05/0.22
PKUSTK04	73.33/1.20	44.00/1.15	55.03/0.66	41.08/0.55	12.19/0.49
PKUSTK05	89.69/1.60	51.80/1.26	68.11/0.82	60.81/0.66	10.66/0.66
PKUSTK06	119.79/2.14	70.97/1.42	91.89/0.94	84.64/0.82	14.83/0.50
PKUSTK07	98.04/0.28	63.73/0.27	75.91/0.27	71.30/0.33	20.08/0.27
PKUSTK08	164.99/0.39	106.39/0.44	130.29/0.44	121.77/0.44	38.89/0.39
HSCT16KB	24.68/0.17	14.12/0.11	11.43/0.16	7.52/0.17	5.54/0.11
BCSSTK17	2.86/0.16	2.03/0.16	2.84/0.11	2.25/0.05	1.15/0.11
BCSSTK18	2.14/0.11	0.71/0.11	1.04/0.06	0.77/0.05	0.99/0.11
BCSSTK25	9.89/0.22	3.13/0.22	4.34/0.11	3.57/0.11	3.02/0.11
BCSSTK29	5.44/0.22	2.47/0.25	4.51/0.11	3.74/0.11	2.09/0.11
BCSSTK30	20.65/0.55	10.99/0.55	17.24/0.33	15.10/0.28	4.84/0.27
BCSSTK31	22.74/0.60	11.37/0.60	18.68/0.33	16.31/0.33	8.19/0.27
BCSSTK32	38.19/0.77	22.19/0.77	31.20/0.44	28.12/0.44	8.79/0.33
S3DKQ4M2	158.43/2.10	99.30/2.09	92.93/2.09	89.31/1.15	16.36/0.83
S3DKT3M2	147.47/1.83	95.99/1.83	89.31/1.05	79.75/1.05	15.27/0.72

Table 6. Size of arrays and MFLOPS for solver E.

| Problem | | Array size (MB) | | | | MFLOPS |
	neq/mneq	JA	PA	JU	PU	PC
PKUSTK01	5.89	0.06	3.82	0.23	15.73	409.20
PKUSTK02	6.00	0.05	3.13	0.16	11.42	431.78
PKUSTK03	6.00	0.19	12.18	0.92	64.92	446.83
PKUSTK04	4.78	0.37	16.30	2.00	67.28	342.88
PKUSTK05	6.00	0.13	8.55	0.99	70.53	484.50
PKUSTK06	6.00	0.15	9.98	1.21	86.44	465.36
PKUSTK07	3.21	0.47	9.29	2.90	57.94	283.59
PKUSTK08	3.19	0.63	12.39	4.30	84.62	246.01
HSCT16KB	1.35	1.42	3.93	9.56	28.35	196.33
BCSSTK17	2.10	0.17	1.68	0.92	9.55	225.96
BCSSTK18	1.09	0.27	0.61	2.12	4.71	99.85
BCSSTK25	1.17	0.36	1.02	4.35	12.32	125.10
BCSSTK29	1.37	0.64	2.42	2.55	14.00	197.99
BCSSTK30	2.95	0.50	7.91	1.92	37.44	302.33
BCSSTK31	2.05	0.62	4.64	4.55	39.65	189.00
BCSSTK32	3.01	0.49	7.86	2.78	54.81	289.64
S3DKQ4M2	5.93	0.29	18.74	1.98	139.03	446.67
S3DKT3M2	5.93	0.23	14.66	1.81	126.82	456.14

Table 5 illustrates the elapsed time for the LDL^T factorization as well as time for solution, i.e., forward reduction and back substitution, on the computational platform. The times for the analysis phase of MA27 and MA47, as well as the times for the symbolic factorization of C, D and E are not taken into account. Generally, the times for the analysis phase of MA27 and MA47 are much longer than the times for the symbolic factorization of C, D and E.

On the platform, the proposed solver E needs on the average much less factorization time and less solution time than the other solvers. As an exception, the solver E needs slightly more time to complete the factorization and solution for BCSSTK18 than MA47 and solver D on the platform.

Table 6 summarizes the sizes of the major arrays and the MFLOPS (Million Floating Point Operations per Second) of the LDL^T factorizations related to solver E.

Here JA and PA are cell-sparse symbolic and numerical matrices of the original matrix A, respectively. The arrays JU and PU are these of the factor matrix U/DL^T. Table 6 indicates that the amount of savings in the symbolic matrix can be substantial when compared with Sherman's compressed storage scheme (Sherman AH, 1975). We find that the performance of solver E in terms of MFLOPS depends significantly on the ratio of the number of equations and the number of super-equations, i.e., $neq/mneq$. The MFLOPS for solver E reaches its lowest value at BCSSTK18 on the platform with $neq/mneq = 1.09$. Generally speaking, the larger the ratio $neq/mneq$ is, the faster the solver E is. The solver E takes advantage of the cell-sparse storage scheme. If there is a big percentage of 1×1 trivial cells in the matrices, the performance of solver E should not be better than solver C or solver D.

4.2 *Speedup analysis*

In modern computer architectures, many features, such as memory caching and instruction-level parallelism, are widely used to improve the computer's performance (Dowd K, 1998). The algorithms proposed in this study are well suited to take advantage of such features. Unrolling enables compilers to reduce the overhead of variable indexing to improve the performance of a program. However, the unrolling of the j- and k-loops in the two-level unrolling sparse LDL^T factorization have different additional benefits.

The unrolling of the k-loop in both one-level and two-level sparse LDL^T improves the algorithm's ability to use instruction interleaving. In modern computers, the CPU can issue a new instruction before the previous instruction is finished if the result of the previous instruction is not needed and there is no hardware conflict. Instruction interleaving is a technique to use such instruction-level parallelism. With instruction interleaving, if one instruction has to wait for the result of the previous instruction, other instructions are inserted between these two instructions so that the waiting time is not wasted. To use instruction interleaving, the loop body has to be large enough so that there are enough instructions independent of each other. With the unrolling of the k-loop, the size of the loop body is increased.

The unrolling of the j-loop on the other hand, improves memory cache hits. With ever increasing computer memory size, the time needed to load variables from the RAM (memory) into CPU registers become more and more critical. To solve this problem, modern computers have a so-called cache memory that holds duplications of the latest used variables. If the variables are used again when they are still in the cache memory, the variables are loaded directly from the cache memory into the CPU registers with a much faster speed. In LDL^T factorization, all appropriate L_{ik}'s are used only once to modify each target row, the row indicated by the j-loop. Since there are many L_{ik}'s associated with one target row and the size of the cache memory is relatively small, they cannot remain in the cache memory for the next target row. In the conventional sparse LDL^T factorization, only one target row is modified in each j-loop. As a result, L_{ik}'s are paged from RAM once for each target row. With the unrolling of the j-loop in the two-level unrolling sparse LDL^T, L_{ik}'s are used to modify several target rows in the same loop and they are paged from RAM only once for each super-equation.

5 CONCLUSION

In this paper, the sparse JKI LDL^T factorization with unrolling strategies is discussed. A definition of super-equations based on the non-zero pattern of the global stiffness matrix A is introduced and it leads to a cell-sparse storage scheme that can greatly reduce the size of the symbolic matrix. Besides, the factor U/DL^T can share the same set of super-equations and be stored in the cell-sparse storage scheme. The size ratio of the symbolic matrix to the numerical one that cannot be reduced for a given pivot sequence is around 1/18 for 3-dimensional solid models and 1/72 for spatial shell or frame models (4-byte integer and 8-byte real). In the conventional row sparse storage scheme the ratio takes the value of 1/2.

Corresponding to this cell-sparse storage scheme, the JKI LDL^T factorization has been conveniently ported to the form of super-rows for both the symbolic and numerical phases. Comparisons of the cell-sparse solver with the HSL Archive solution procedure MA27, MA47, and two another sparse solvers are given on the computational platform.

The numerical tests showed that the cell-sparse solver is very efficient in terms of elapsed time and memory requirement. It is expected that the proposed solution procedure can be used as the default direct solver in FEA.

ACKNOWLEDGMENTS

The authors would like to thank the support of the National Natural Science Foundation of China under grant No. 10172005.

REFERENCES

Chen B, Chen P, Sun SL, Yuan MW, Lu CK, Sheng P. 2000. Static and dynamic analysis of public plant conservatory at Beijing Botanical Garden. *Building Structure* 30(1): 53–56 (in Chinese).
Chen P, Runesha H, Nguyen DT, Tong P, Chang TYP. 2000. Sparse algorithms for indefinite systems of linear equations. *Computational Mechanics Journal* 25(1): 33–42.
Chen P, Sun SL, Yuan MW. 2000. A fast sparse static solver in finite element analysis. In Atluri SN & Brust FW (ed.), *Proceeding of the ICES'2000 Conference, Vol I*: 428–433.
Chen P, Zheng D, Sun SL, Yuan MW. 2003. High performance sparse static solver in finite element analyses with loop-unrolling. To appear in *Advances in Engineering Software*.
Damhaug AC, Reid J, Bergseth A. 1999. The impact of an efficient linear solver on finite element analysis. *Computers & Structures* 72: 594–604.
Davis T. *University of Florida Sparse Matrix Collection*, www.cise.ufl.edu/research/sparse/matrices.
Dowd K, Severance CR. 1998. *High Performance Computing* (2nd ed), Cambridge; Sebastopol, CA O'Reilly & Associates.
Duff IS, Erisman AM, Reid JK. 1989. *Direct Methods for Sparse Matrices*. Oxford: Clarendin Press.
Duff IS, Grimes R, Lewis J. 1992. *The User's Guide for the Harwell-Boeing Sparse Matrix Collection* (*Release I*), Technical Report TR/PA/92/86, CERFACS, Lyon, France.
Duff IS, Reid JK. 1983. The multifrontal solution of indefinite sparse symmetric linear equations. *ACM Trans Math Software* 9: 302–325.
Duff IS, Reid, JK. *A Fortran Code for Direct Solution of indefinite sparse symmetric linear systems*, Report RAL-95-001, Rutherford Appleton Laboratory, Oxfordshire.
Kouhia R. 1998. *Description of CYSHELL set*. Laboratory of Structural Engineering, Helsinki University of Technology.
Nguyen DT, Qin J, Chang TYP, Tong P. 1997. Efficient sparse equation solver with unrolling strategies for computational mechanics. In: *Proceeding of the ICES'97 Conference*: 676–681.
Ng EG, Peyton BW. 1993. Block sparse Cholesky algorithm on advanced uniprocessor computers. *SIAM J. Sci. Comput.* 14(5): 1034–1055.
Schloegel K, Karypis G, and Kumar V. 2000. Graph partitioning for high performance scientific simulations. In Dongarra J, Foster I, Fox G, Kennedy K, and White A (ed.), *CRPC Parallel Computing Handbook*. Morgan Kaufmann, New York.
Sherman AH. 1975. *On the efficient Solution of Sparse Systems of Linear and Nonlinear Equations*. Rept. No. 46 (Ph.D. Dissertation), Dept of Computer Science, Yale University, 1975.

Computational Methods in Engineering and Science, Iu et al. (eds)
© *2003 Swets & Zeitlinger, Lisse, ISBN 90 5809 567 3*

Energy band analysis for periodical wave-guides

W.X. Zhong
Dalian University of Technology, Dalian, Liaoning, China
(Visiting City University of Hong Kong)

A.Y.T. Leung & F.W. Williams
City University of Hong Kong, Hong Kong

ABSTRACT: The Hamiltonian system method is applied to electro-magnetic wave-guides to permit symplectic analysis to be applied to arbitrary anisotropic materials and to permit the interface conditions between segments of a periodical wave-guide to be treated. The transverse electric and magnetic fields compose the dual vectors. An electro-magnetic stiffness matrix is introduced which relates to the electric field vectors at both its ends. The pass- and stop-band solutions are given for a constant cross-section segment of a plane wave-guide and then the equations needed to combine two segments with different cross-sections are derived. The energy variational principle is applied to electro-magnetic stiffness matrix of a fundamental period of the wave-guide, which is computed by combining segments recursively until the shortest repeating length of the wave-guide has been assembled. Finally, the pass-band eigenvalues of the energy band are found by using the Wittrick-Williams algorithm throughout the calculations, starting with the combination of segments step.

1 INTRODUCTION

The eigenvalues for wave propagation along a periodical structure exhibits energy band behaviour similar to that in solid state physics (Kittel 1986). Thus waves with frequencies ω in the pass-band can propagate along the structure, whereas otherwise ω is in a stop-band and so the wave will decay to zero for long distance propagation. Therefore energy analysis of the pass- and stop-bands has very important practical applications. Because periodical electro-magnetic wave-guides exhibit such behaviour they are useful for filtering optical signals and so their analysis is important.

The basic unit in analysis of a periodical wave-guide is its fundamental period, i.e. its shortest repeating length. For a given frequency ω, its longitudinal wave constant analysis determines that when an eigenvalue is located on the unit circle, in which case the wave can propagate to infinity without decay. Hence this ω is said to be in the pass-band, whereas when ω is in the stop-band the wave constant is not on the unit circle and so the wave decays for long distance propagation. Therefore pass- or stop-band analysis is very important, because it can be used to filter optical signals. Note that the energy band analysis of periodical structures and of electro-magnetic wave-guides are analogous and so a method used in either field can be transformed to the other. The key analysis step for a periodical structure is to compute the dynamic stiffness matrix of its typical periodical sub-structure, whereas for a periodical wave-guide it is to generate the electro-magnetic stiffness matrix related to the two ends of a fundamental period.

A frequency ω must either lie in a pass-band or a stop-band. The pass-band solution can be derived from a Rayleigh quotient analysis. All the eigensolutions of the Rayleigh quotient laying in a given frequency range can be found by using the W-W (Wittrick-Williams) algorithm and its extension, which were developed in the context of the vibration and wave propagation problems of structural mechanics (Wittrick & Williams 1971, Zhong et al. 1997, Zhong & Williams 1992, Zhong & Williams 1993, Zhong & Williams 1995). The W-W algorithm is applied here to electro-magnetic wave-guide problems but is first summarised in the relevant structural mechanics context. During this summary, it is helpful to note that a sub-structure will subsequently be replaced by a segment of a wave-guide.

2 DYNAMIC STIFFNESS MATRIX OF THE FUNDAMENTAL SUB-STRUCTURE

The formulation is given in the frequency domain ω. A periodical structure is really a sub-structural chain (Zhong & Williams 1991), which is composed of fundamental periodical sub-structures linked together end-to-end. The left-hand and right-hand ends of a sub-structure are indicated by subscripts a and b respectively.

Let the dynamic stiffness matrix corresponding to the two end displacement vectors of a fundamental segment for a given frequency ω be

$$\mathbf{K}(\omega) = \mathbf{K} = \begin{bmatrix} \mathbf{K}_{aa} & \mathbf{K}_{ab} \\ \mathbf{K}_{ba} & \mathbf{K}_{bb} \end{bmatrix}, \quad \mathbf{K}_{aa} = \mathbf{K}_{aa}^T, \quad \mathbf{K}_{bb} = \mathbf{K}_{bb}^T, \quad \mathbf{K}_{ba} = \mathbf{K}_{ab}^T \tag{1}$$

here $\mathbf{K}$ is a symmetric matrix, the positive definiteness of which is not assured; $\mathbf{K}_{aa}$, $\mathbf{K}_{bb}$ and $\mathbf{K}_{ab}$ are $(n \times n)$ sub-matrices and; the fundamental period is the shortest repeating length of the structure. According to dynamic stiffness matrix theory (Leung 1993), knowing only the dynamic stiffness corresponding to the external displacements of the sub-structure is insufficient on its own to express the eigenvalue behavior, because the internal eigenvalue count $J(\omega_\#)$ is also needed (Wittrick & Williams 1971). The dynamic stiffness matrix is not the only available representation of the dynamic behavior of a sub-structure because the mixed energy representation is also available (Zhong et al. 1997, Zhong & Williams 1992, Zhong & Williams 1993, Zhong & Williams 1995) and is very important for the precise integration (Zhong & Williams 1994) of energy matrices.

3 ENERGY BAND OF A PERIODICAL STRUCTURE AND EIGENSOLUTIONS OF A SYMPLECTIC MATRIX

After the $(2n \times 2n)$ dynamic stiffness matrix $\mathbf{K}(\omega)$ of the fundamental sub-structure has been generated, the energy band analysis for the periodical structure or sub-structural chain (Zhong & Williams 1991) can be computed. The dynamic deformation energy $U_e(\mathbf{q}_a, \mathbf{q}_b; \omega)$ of a fundamental sub-structure can be expressed in terms of the dynamic stiffness matrix $\mathbf{K}(\omega)$ as

$$U_e(\mathbf{q}_a, \mathbf{q}_b; \omega) = \begin{Bmatrix} \mathbf{q}_a \\ \mathbf{q}_b \end{Bmatrix}^T \mathbf{K}(\omega) \begin{Bmatrix} \mathbf{q}_a \\ \mathbf{q}_b \end{Bmatrix} / 2 \tag{2}$$

where $\mathbf{q}_a$ and $\mathbf{q}_b$ are its two end displacement vectors and ω is only a parameter, which is sometimes not ever stated explicitly, as in equation (2) below which is obtained by applying the variational principle to the chain of m sub-structures

$$U(\mathbf{q}_0, \mathbf{q}_n) = \sum_{i=1}^{i=m} U_i(\mathbf{q}_{i-1}, \mathbf{q}_i), \quad \delta\Pi\big|_{\mathbf{q}_i, i=1,\cdots,m-1} = 0 \tag{3}$$

Using the method in (Zhong et al. 1997, Zhong & Williams 1991) to introduce the dual vector gives

$$\mathbf{p}_a = -\partial U_e / \partial \mathbf{q}_a = -(\mathbf{K}_{aa}\mathbf{q}_a + \mathbf{K}_{ab}\mathbf{q}_b) \tag{4a}$$

$$\mathbf{p}_b = \partial U_e / \partial \mathbf{q}_b = (\mathbf{K}_{ba}\mathbf{q}_a + \mathbf{K}_{bb}\mathbf{q}_b) \tag{4b}$$

The equilibrium equation of the i-th station is derived from equation (3) as

$$\partial U_i(\mathbf{q}_{i-1}, \mathbf{q}_i)/\partial \mathbf{q}_i + \partial U_{i+1}(\mathbf{q}_i, \mathbf{q}_{i+1})/\partial \mathbf{q}_i = \mathbf{p}_{b,i} - \mathbf{p}_{a,i+1} = 0$$

Note that equation (4) expresses the dual vectors $\mathbf{p}_a, \mathbf{p}_b$ by means of the original displacement vectors $\mathbf{q}_a$ and $\mathbf{q}_b$ of the displacement method. However, by introducing the *state vector*

$$\mathbf{v}_j \underset{\text{def}}{=} \{\mathbf{q}_j^T \quad \mathbf{p}_j^T\}^T$$

the equilibrium equation can be transformed into the following state vector transfer form, in which the right-hand end vectors $\mathbf{q}_b$ and $\mathbf{p}_b$ are expressed in terms of the left-hand end vectors $\mathbf{q}_a$ and $\mathbf{p}_a$. Hence from equation (4)

$$\mathbf{v}_b = \mathbf{S}\mathbf{v}_a, \quad i.e. \quad \mathbf{v}_{j+1} = \mathbf{S}\mathbf{v}_j \tag{5}$$

$$\mathbf{S} = \begin{bmatrix} \mathbf{S}_{11} & \mathbf{S}_{12} \\ \mathbf{S}_{21} & \mathbf{S}_{22} \end{bmatrix}, \quad \begin{aligned} \mathbf{S}_{11} &= -\mathbf{K}_{ab}^{-1}\mathbf{K}_{aa}, & \mathbf{S}_{12} &= -\mathbf{K}_{ab}^{-1} \\ \mathbf{S}_{21} &= \mathbf{K}_{ba} - \mathbf{K}_{bb}\mathbf{K}_{ab}^{-1}\mathbf{K}_{aa}, & \mathbf{S}_{22} &= -\mathbf{K}_{bb}\mathbf{K}_{ab}^{-1} \end{aligned} \right\} \tag{6}$$

where $\mathbf{S}$ is the transfer matrix, which is a symplectic matrix, i.e. it satisfies

$$\mathbf{S}^T \mathbf{J} \mathbf{S} = \mathbf{J} \tag{7}$$

where

$$\mathbf{J} = \begin{bmatrix} \mathbf{0} & \mathbf{I}_n \\ -\mathbf{I}_n & \mathbf{0} \end{bmatrix}, \quad \mathbf{J}^T = \mathbf{J}^{-1} = -\mathbf{J}, \quad \mathbf{J}^2 = -\mathbf{I}_{2n} \tag{8}$$

Using the method of separation of variables to solve the transfer matrix equation (5) yields the eigenvalue problem (Zhong & Williams 1992, Zhong & Williams 1993, Zhong & Williams 1995)

$$\mathbf{S}\psi = \mu\psi \tag{9}$$

after the eigen-pair (μ,ψ) has been found, the solution of the original transfer equation (5) is $\mathbf{V}_i = \psi\mu^i$, where i denotes the i-th junction of the sub-structural chain. Now the symplectic eigenvalue problem has the following characteristics.

If μ is an eigen-value then so also is μ^{-1}, as follows. Left multiplying equation (9) by $\mathbf{S}^T\mathbf{J}$ and using equation (7) gives $\mathbf{S}^T(\mathbf{J}\psi) = \mu^{-1}(\mathbf{J}\psi)$, which shows that μ^{-1} is an eigenvalue of $\mathbf{S}^T$ for which the eigenvector is $\mathbf{J}\psi$. However, any eigenvalue of $\mathbf{S}^T$ is also an eigenvalue of $\mathbf{S}$. Hence the $2n$ eigenvalues of $\mathbf{S}$ can be classified into the two groups

$$\alpha) \quad \mu_j, \quad \text{abs}(\mu_j)<1 \text{ or } \text{abs}(\mu_j)=1 \wedge \text{Im}(\mu_j)>0, \quad j=1,\cdots,n \tag{10a}$$

$$\beta) \quad \mu_{n+j}, \quad \mu_j^{-1}, \qquad j=1,\cdots,n \tag{10b}$$

where μ_j and μ_{n+j} are called mutually symplectic adjoint eigenvalues.

The following derivation applies for any two eigensolutions, denoted by j and k, of the symplectic matrix $\mathbf{S}$:

$$\mathbf{S}\psi_j = \mu_j\psi_j, \qquad\qquad \mathbf{S}\psi_k = \mu_k\psi_k$$

$$\mathbf{S}^T(\mathbf{J}\psi_j) = \mu_j^{-1}(\mathbf{J}\psi_j), \qquad \psi_k^T \cdot \mathbf{S}^T = \mu_k\psi_k^T$$

$$\psi_k^T \cdot (\mathbf{S}^T\mathbf{J})\cdot\psi_j = \mu_j^{-1}\psi_k^T \cdot \mathbf{J}\cdot\psi_j, \qquad \psi_k^T \cdot (\mathbf{S}^T\mathbf{J})\cdot\psi_j = \mu_k\psi_k^T \cdot \mathbf{J}\cdot\psi_j$$

Hence subtraction of the equations in the final line gives

$$(\mu_k - \mu_j^{-1})\psi_k^T \cdot \mathbf{J}\cdot\psi_j = 0$$

Therefore either the two eigensolutions j and $k = n + j$ are mutually *symplectic adjoint* to each other, so that two constant factors can be selected to achieve *adjoint symplectic normalization*, or the eigensolutions j and k are *symplectic orthogonal*. In other words

$$\psi_j^T \cdot \mathbf{J}\cdot\psi_k = \delta_{j,k-n} \tag{11}$$

which is called *adjoint symplectic ortho-normality*. Composing a $(2n \times 2n)$ matrix $\mathbf{\Psi}$ by using all the eigenvectors gives

$$\mathbf{\Psi} = \begin{bmatrix} \psi_1 & \cdots & \psi_n; & \psi_{n+1} & \cdots & \psi_{2n} \end{bmatrix} \tag{12}$$

173

and then using the adjoint symplectic ortho-normality relationship yields the matrix identity $\mathbf{\Psi}^T\mathbf{J}\mathbf{\Psi} = \mathbf{J}$ so that $\mathbf{\Psi}$ is called a *symplectic* matrix.

When the eigenvalue is located on the unit circle $\mu = e^{i\theta}$, the corresponding solution of the original equation is $\mathbf{v}_j = \psi\mu^j$. Therefore the vector $\mathbf{v}_j$ does not decay because $\text{abs}(\mu^j) = \text{abs}(e^{i(j\theta)}) = 1$ and so the solution gives a transmission wave.

For the simplest case of $n = 1$, the eigen-equation is $\mu^2 - (S_{11} + S_{22})\mu + 1 = 0$. Then for

$$(S_{11} + S_{22})^2 < 4 \quad \text{or} \quad \text{abs}[(K_{11} + K_{22})/K_{12}] < 2 \tag{13}$$

a transmission wave appears, i.e. ω is in the pass-band. Changing the above inequality to equality and solving with respect to ω gives the boundary between the pass- and stop-bands.

The general case is multi-dimensional (n-dimensional) and so solving the symplectic eigenvalue problem (9), requires prior solution of a skew-symmetric symplectic eigenproblem, as follows. Left multiplying equation (9) by $\mathbf{S}^T\mathbf{J}$ to gives $\mathbf{S}^T(\mathbf{J}\psi) = \mu^{-1}(\mathbf{J}\psi)$ and then using equation (7) gives $\mathbf{S}^{-1}\psi = \mu^{-1}\psi$. Combining this with equation (9) gives

$$\mathbf{A}\psi = (\mu + \mu^{-1})\mathbf{J}\psi, \quad \mathbf{A} = \mathbf{J}(\mathbf{S} + \mathbf{S}^{-1}), \quad \mathbf{A}^T = -\mathbf{A} \tag{14}$$

where the skew-symmetric nature of $\mathbf{A}$ is readily verified from equation (7). Solving to find the pairs of eigensolutions of equation (14) makes solving for the eigensolutions of the original equation (9) easy. Moreover an algorithm is available (Zhong & Williams 1993, Zhong et al. 1994, Zhong et al. 1993) for solving this symplectic eigenproblem for a skew-symmetric matrix.

4   PASS-BAND ANALYSIS FOR PERIODICAL WAVE-GUIDES

Because the electro-magnetic wave-guide analysis also generates a stiffness matrix, the above analysis can be used in wave-guide problems. The pass-band wave number of a periodical wave-guide is $\mu = \exp(i\theta)$. Here the phase angle θ is the only parameter and is real, so that the eigenproblem can be solved inversely, by finding the frequency ω for a given wave number μ. For a class α eigenvalue μ (see equation (10a)) this gives $0 \leqslant \theta \leqslant \pi$. In structural vibration the frequency eigenvalue problem is usually to find ω with the boundary conditions given. However, for wave propagation problems, the boundary condition becomes a given wave number and the requirement is to find the frequency ω. The boundaries between the pass- and stop-bands are of great concern and are given by solving to find ω for $\mu = 1$ and $\mu = -1$.

In practical applications, all the eigen-frequencies ω in a selected frequency range $0 < \omega^2 < \omega^2_\#$ are required and the W-W algorithm for counting the number of eigenvalues for a dynamic stiffness matrix is indeed for this purpose.

Substituting $\mu = 1$ into the eigen-equation $\mathbf{S}\psi = \mu\psi$ gives the eigen-equation for ω as

$$\mathbf{K}_{qq}(\omega)\mathbf{q} = 0, \quad \mathbf{K}_{qq}(\omega) = \mathbf{K}_{aa} + \mathbf{K}_{ab} + \mathbf{K}_{ba} + \mathbf{K}_{bb} \tag{15}$$

where all the sub-matrices are functions of ω. The other boundary between pass- and stop-bands is obtained by substituting $\mu = -1$, which gives

$$(\mathbf{K}_{aa} + \mathbf{K}_{bb} - \mathbf{K}_{ab} - \mathbf{K}_{ba})\mathbf{q} = 0 \tag{16}$$

Note that both of equations (15) and (16) are eigenvalue problems of ($n \times n$) symmetric matrices, and so the W-W eigenvalue counting algorithm is applicable. Note also that $\mathbf{K}_{aa}(\omega)$, $\mathbf{K}_{bb}(\omega)$ and $\mathbf{K}_{ba}(\omega)$ are all sub-matrices of the dynamic stiffness matrix $\mathbf{K}(\omega)$ of the segment (z_a, z_b).

Equations (15) and (16) apply only for the *boundaries* $\theta = 0$ and π *of the pass-band*, respectively. However for an arbitrary phase angle $0 \leqslant \theta \leqslant \pi$, the equation $\mathbf{S}\psi = \mu\psi$ yields the eigen-equation for ω as

$$\mathbf{K}_{qq}(\omega)\mathbf{q} = 0, \quad \mathbf{K}_{qq}(\omega) = \mathbf{K}_{ab}\mu + (\mathbf{K}_{aa} + \mathbf{K}_{bb}) + \mu^{-1}\mathbf{K}_{ba} \tag{17}$$

174

where μ is a given parameter and all the sub-matrices are functions of ω. Because the parameter $\mu = \exp(i\theta)$ is a complex number the dynamic stiffness matrix $\mathbf{K}_{qq}(\omega)$ is no longer real and symmetric but is instead Hermitian, as follows

$$\mathbf{K}_{aa}^{H} = \mathbf{K}_{aa}, \ \mathbf{K}_{bb}^{H} = \mathbf{K}_{bb}, \ \mathbf{K}_{ab}^{H} = \mathbf{K}_{ba} \tag{18}$$

$$\begin{aligned}\mathbf{K}_{qq}^{H} &= \mathbf{K}_{ab}^{H}\exp(-i\theta) + (\mathbf{K}_{aa}^{H} + \mathbf{K}_{bb}^{H}) + \mathbf{K}_{ba}^{H}\exp(i\theta)\\ &= \mathbf{K}_{ba}\exp(-i\theta) + (\mathbf{K}_{aa} + \mathbf{K}_{bb}) + \mathbf{K}_{ab}\exp(i\theta) = \mathbf{K}_{qq}\end{aligned} \tag{19}$$

Therefore equation (17) is an eigenproblem for a Hermitian matrix, which is a problem for which the W-W algorithm still applies and has often been used. The equation looked simple, however, its precise computation and the execution of the W-W algorithm need further explanation.

The description for the electric field vectors $\mathbf{q}_a$ and $\mathbf{q}_b$ at the two ends a and b of the fundamental period must coincide with each other, but unless $\theta = 0$ or π the values of $\mathbf{q}_a$ and $\mathbf{q}_b$ are different due to phase changes. The origin of the fundamental period should be chosen such that $\mathbf{q}_a$ (and also $\mathbf{q}_b$) has the least dimension possible. In applying W-W algorithm, the count of the number of eigenvalues internal to the fundamental period, $J(\omega_\#)$, is necessary, where $\omega_\#$ is a given upper bound of a frequency range for which the lower bound is zero. Hence there are $J_i(\omega_\#)$ eigenvalues in the range $0 \leqslant \omega^2 \leqslant \omega^2_\#$ when the two ends of the fundamental period are regarded as *clamped*, i.e. $\mathbf{q}_a = \mathbf{q}_b = \mathbf{0}$.

5 DYNAMIC STIFFNESS MATRIX FOR A FUNDAMENTAL PERIOD AND ITS EIGENVALUE COUNT

The eigen-equation of the fundamental segment (z_a, z_b) described above is based on the external dynamic stiffness matrix $\mathbf{K}(\omega)$ of the fundamental segment of the wave-guide. However, $\mathbf{K}(\omega_\#)$ represents only the external behaviour so that, as for the dynamic problem there is internal behaviour of the the fundamental period and so the internal eigenvalue count $J(\omega_\#)$ is required. Hence the boundaries between pass- and stop-bands in the frequency (ω)domain are still the eigenproblems of a symmetric matrix given by equations (15) and (16). Therefore the matrix $\mathbf{K}(\omega)$ should be computed precisely, by methods which depend on the type of wave-guide, etc. Below, a fundamental period of plane electro-magnetic wave-guide is considered with its frequency (ω) given.

Let the fundamental period be composed of several constant cross-section segments. Then combining two adjacent segments 1 and 2 gives a longer segment c, see Figure 1, with the equations for the combination being

$$\mathbf{K}_{aa}^{(c)} = \mathbf{K}_{aa}^{(2)} - \mathbf{K}_{ab}^{(2)}(\mathbf{K}_{bb}^{(1)} + \mathbf{K}_{aa}^{(2)})^{-1}\mathbf{K}_{ba}^{(2)} \tag{20a}$$

$$\mathbf{K}_{bb}^{(c)} = \mathbf{K}_{bb}^{(2)} - \mathbf{K}_{ba}^{(2)}(\mathbf{K}_{bb}^{(1)} + \mathbf{K}_{aa}^{(2)})^{-1}\mathbf{K}_{ab}^{(2)} \tag{20b}$$

$$\mathbf{K}_{ab}^{(c)} = -\mathbf{K}_{ab}^{(1)}(\mathbf{K}_{bb}^{(1)} + \mathbf{K}_{aa}^{(2)})^{-1}\mathbf{K}_{ab}^{(2)} \ , \ \mathbf{K}_{ba}^{(c)} = \mathbf{K}_{ab}^{(c)T} \tag{20c}$$

where the superscripts denote segments 1, 2 and c. Note that although the two ends of segment 1 are marked as a and b, it is not necessarily a fundamental segment. The above equation applies only for the combination of external ($n \times n$) stiffness matrices. However, for dynamic stiffness matrix analysis and hence also for the present analogous electro-magnetic stiffness matrix analysis, the internal eigenvalue count should also be combined, and the recurrence equation for this eigenvalue count is the W-W algorithm (Wittrick & Williams 1971)

$$J_c(\omega_\#) = J_1(\omega_\#) + J_2(\omega_\#) + s\{\mathbf{R}\} \tag{21}$$

where ($\omega_\#$) is a given frequency bound and $\mathbf{R}(\omega_\#) = \mathbf{R} = (\mathbf{K}_{bb}^{(1)} + \mathbf{K}_{aa}^{(2)})$ is the internal stiffness matrix when combining the two segments. $J_1(\omega_\#)$ and $J_2(\omega_\#)$ are the internal eigenvalue counts of segments 1

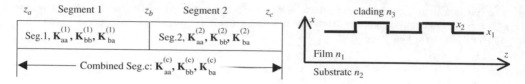

Figure 1. Segment combination.

Figure 2. Plane wave-guide.

and 2 respectively, and $s\{\cdots\}$ represents the eigenvalue count operator such that $s\{\mathbf{R}\}$ is the number of negative entries of the diagonal matrix $\mathbf{D}$ obtained by the factorization $\mathbf{R} = \mathbf{L}^T\mathbf{DL}$, see (Zhong et al. 1997). Equation (21) enables all the eigenvalues in the given range $0 \leqslant \omega^2 \leqslant \omega^2_\#$ to be found with certainty by many alternative methods of which the simplest is the bisection method.

When the fundamental period has more than two segments it can be assembled by applying the above equations recursively to add one segment out a time.

The problem has now been reduced to performing computations for each constant cross section segment. The electro-magnetic segment is a continuum and so has an infinite number of degrees of freedom in its cross-section. However, in practical computation these are usually approximated by a finite number of degrees of freedom, n, via a discretisation procedure, i.e. the semi-analytical approach is used. This discretisation has two alternative forms. In the first, which is called the spectral method and is the one adopted in this paper, the field is assumed to be the linear combination of n independent basis functions which represents the transverse cross-section distribution. Thus the n multipliers of these basis functions are the unknown functions of z which need to be found. A set of n simultaneous differential equations derived from the variational principle are then solved by a numerical method, e.g. the precise integration method (Zhong et al. 1997, Leung 1993, Zhong 1995). The second discretisation procedure is identical to the first except that at the first step FEM is used for the cross-section, to give the precise form (Zhong et al. 1998) of the finite strip method (Zienkiewicz & Taylor 1989).

For ease of presentation, a simple problem is now introduced as an example. Consider the plane electro-magnetic wave-guide shown in figure 2, where z is the longitudinal coordinate along which the wave propagates. Suppose that the core is vacuum surrounded by perfect conductors with $\varepsilon\varepsilon_1 = \varepsilon_2 = \varepsilon_0$, $\mu_0\varepsilon_0 = 1/c^2 c = 2.998 \times 10^8$ m/sec ($=$ the velocity of light in a vacuum), $x_2 = 1.25 \times 10^{-6}$ m, $x_1 = 0.8x_2$ and $l_1 = l_2 = 1 \times 10^{-7}$ m. The problem is to find the eigenvalue ω/c for any given wave phase angle for which $0 \leqslant \theta \leqslant \pi$.

For this problem, the fundamental sub-structure is composed of two uniform segments between which the cross-section changes abruptly. Let the fundamental period of the wave-guide be $-l_1 < z < l_2$, with segments $- l_1 < z < 0$ and $0 < z < l_2$ which are of thickness x_1 and x_2, respectively, and are represented by m and n basis functions, respectively. Now the abrupt change of cross-section $x_1 < x_2$ at $z = 0$, results the following conditions there

$$e_y(x,+0) = \begin{cases} e_y(x,-0), & x < x_1 \\ 0, & x_1 \leq x < x_2 \end{cases}, \quad h_x(x,+0) = \begin{cases} h_x(x,-0), & x < x_1 \\ 0, & x_1 \leq x < x_2 \end{cases}, \text{ at } z = 0$$

The variational method can be used to process these junction conditions by transforming the segment $0 < z < l_2$ to $-0 < z < l_2 + 0$, i.e. the thickness of the two external cross-sections changes to x_1 and the number of terms in the expansion changes to m. For brevity the derivation of this transformation matrix is omitted, but the resulting matrix $\mathbf{T}$ at the stations $z = 0$ and $z = l_2$ needed for treating the cross-section change gives

$$q^{(2)}_{ai} = \sum_{j=1}^{m} T_{ij} \cdot q^{(1)}_{aj}, \quad i = 1, \cdots, n, \qquad \text{or} \qquad \mathbf{q}^{(2)}_a = \mathbf{T}\mathbf{q}^{(1)}_a$$

$$T_{ij} = \sin\left[(j - ix_1/x_2)\pi\right]/\left[\pi(jx_2/x_1 - i)\right] - \sin\left[(j + ix_1/x_2)\pi\right]/\left[\pi(jx_2/x_1 + i)\right] \tag{22}$$

where the vectors $\mathbf{q}^{(2)}_a$ and $\mathbf{q}^{(1)}_a$ are of orders n and m, respectively, so that $\mathbf{T}$ is $n \times m$. A similar derivation for the right-hand end gives $\mathbf{q}^{(2)}_b = \mathbf{T}\mathbf{q}^{(1)}_b$. Here subscripts a and b represent the left- and right-hand ends, and the superscripts (1) and (2) denote series expansion for thickness x or x_2, respectively.

176

Therefore the electro-magnetic segment stiffness matrix $\mathbf{K}_{(2)}^{(2)}$, which is for the end vectors $\mathbf{q}_a^{(2)}$ and $\mathbf{q}_b^{(2)}$ of a wave-guide thickness x_2, should be transformed by using the equations

$$\begin{Bmatrix} \mathbf{q}_a^{(2)} \\ \mathbf{q}_b^{(2)} \end{Bmatrix} = \begin{bmatrix} \mathbf{T} & \mathbf{0} \\ \mathbf{0} & \mathbf{T} \end{bmatrix} \begin{Bmatrix} \mathbf{q}_a^{(1)} \\ \mathbf{q}_b^{(1)} \end{Bmatrix}, \quad \mathbf{K}_2^{(1)} = \begin{bmatrix} \mathbf{T}^T & \mathbf{0} \\ \mathbf{0} & \mathbf{T}^T \end{bmatrix} \mathbf{K}_2^{(2)} \begin{bmatrix} \mathbf{T} & \mathbf{0} \\ \mathbf{0} & \mathbf{T} \end{bmatrix} \tag{23}$$

for the two end electric field vectors $\mathbf{q}_a^{(1)}$ and $\mathbf{q}_b^{(1)}$ with thickness x_1. The matrix $\mathbf{K}_2^{(2)}$ is a $(2n \times 2n)$ matrix composed of four diagonal $(n \times n)$ matrices, whereas the transformation makes $\mathbf{K}_2^{(1)}$ into a fully populated $2m \times 2m$ matrix.

(Note, that for a general cross-section wave-guide the semi-analytical method discretizing the cross-section results in a $\mathbf{K}_2^{(2)}$ which is fully populated, but consideration of such problems, which still use $\mathbf{T}$, is beyond the scope of this paper.)

After transformation, the stiffness matrix $\mathbf{K}_2^{(1)}$ represents the behavior of segment 2 but relates to the contracted cross-section. However, the fundamental period segment is composed of both segments 1 and 2, so that the electro-magnetic stiffness matrix $\mathbf{K}_1^{(1)}$ is also needed by the segment combination algorithm (see equations (20) and (21) as well as see figure1) which is applied to give (by a recursive procedure when there are more than two segments in the fundamental period) the combined electro-magnetic stiffness matrix $\mathbf{K}_c$ for the fundamental period of wave-guide and its associated J_c. Thus the computation of $\mathbf{K}_c$ at given ω is a fundamental step in the analysis of a periodical wave-guide and corresponds to the dynamic stiffness matrix in structural analysis of the fundamental sub-structure (Zhong et al. 1993), which is also computed for given ω.

Because the stiffness matrix is fundamental to FEM, after introducing potential energy and its stiffness matrix the analysis of electro-magnetic wave-guides can use the same methodology as in structural mechanics. Thus computation of the eigenvalue count for a fundamental period of a wave-guide has three principal steps. The first two are to compute the eigenvalues counts internal to each uniform segment and for the combination of segments within the fundamental period. Then the final step gives equation (24) as follows. For a pair θ, $\omega_\#$ of given phase angle and given frequency, the eigenvalue count of the pass-band wave-guide J_p can be computed from the equation

$$J_p = J_f + s\{\mathbf{K}_{qq}\} \tag{24}$$

where $s\{\mathbf{K}_{qq}\}$ is the eigenvalue count of the Hermitian matrix given by equation (19) and J_f is the value of the eigenvalue count for the fundamental period yielded by the first two steps.

6 COMPUTATION OF THE PASS-BAND EIGENVALUES

As shown in section 4, the boundaries of the pass-band are given by equations (15) and (16), for which the phase angles are $\theta = 0, \pi$ respectively. For other values of θ and a given frequency ω, the electro-magnetic stiffness matrix $\mathbf{K}_c(\omega)$ of the fundamental period of a wave-guide can be computed as described in the previous sections. Note that $\mathbf{K}_c(\omega)$ is a Hermitian matrix, i.e. $\mathbf{K}_c(\omega) = \mathbf{K}_c^H(\omega)$, where superscript H denotes Hermitian transposition. Because the frequency domain is multiplied by $\exp(-i\omega t)$, equation (2) for the energy should be slightly modified, by replacing the superscript T as H to give

$$U_e(\mathbf{q}_a, \mathbf{q}_b; \omega) = \begin{Bmatrix} \mathbf{q}_a \\ \mathbf{q}_b \end{Bmatrix}^H \mathbf{K}(\omega) \begin{Bmatrix} \mathbf{q}_a \\ \mathbf{q}_b \end{Bmatrix} / 2 \tag{2'}$$

To find all the pass-band eigenvalues of a periodical wave-guide, the electro-magnetic stiffness matrix $\mathbf{K}_c(\omega)$ of the fundamental period should be used as the matrix $\mathbf{K}(\omega)$ in equations (15–17). Hence the matrix $\mathbf{K}_{qq}(\omega)$ in equation (17) is Hermitian, so that the W-W algorithm still applies. For a given $\omega_\#$ the eigenvalue count of the Hermitian matrix $\mathbf{K}_{qq}(\omega_\#)$ can be computed by the triangular factorization

$$\mathbf{K}_{qq}(\omega_\#) = \mathbf{L}\mathbf{D}\mathbf{L}^H, \quad s\{\mathbf{K}_{qq}\} = \text{number of negative entries in } \mathbf{D} \tag{25}$$

where $\mathbf{D}$ is a real diagonal matrix. Therefore all the eigenvalues can be found, e.g. by the bisection method. Continuing the computation of the previous numerical example gives the results in Table 1.

Table 1. The value of $10^{-6} \cdot \omega/c$ solved with $n = 24$, $m = 24$ are selected for the example.

$\theta = 0$	15°	30°	45°	60°	75°	90°	105°	120°	135°	150°	165°	180°
3.12	4.85	4.88	4.99	6.06	7.21	8.39	9.60	10.83	12.05	13.24	14.34	14.93
6.23	6.37	6.76	7.35	8.12	9.01	9.98	11.02	12.10	13.20	14.30	15.33	15.89
9.35	9.44	9.70	10.13	10.70	11.39	12.17	13.04	13.96	14.93	15.91	16.85	17.10
12.46	12.53	12.73	13.06	13.50	14.06	14.70	15.42	16.21	17.06	17.93	17.70	17.37
15.57	15.63	15.79	16.05	16.42	16.87	17.41	18.03	18.71	19.44	18.81	18.49	17.91
18.68	18.72	18.86	19.08	19.39	19.77	20.23	20.76	21.27	20.02	19.55	18.77	19.19
21.77	21.81	21.93	22.12	22.38	22.71	23.11	22.55	21.36	20.73	20.21	19.74	19.25
24.85	24.88	24.98	25.15	25.37	25.11	23.83	23.17	21.95	21.83	20.75	20.97	20.84

7 CONCLUDING REMARKS

Periodical electro-magnetic wave-guides have important applications and determining their characteristics involves an eigenproblem which this paper has shown can beneficially be solved by using symplectic mathematics and the methodology applied in structural mechanics, especially the eigen-value counting W-W algorithm. Thus the electro-magnetic stiffness matrix (or impedence) of a fundamental period of a periodical wave-guide has been introduced and then symplectic mathematics has been applied as in the analysis of sub-structural chains. The energy band analysis for the periodical wave-guide is then carried out on the same basis as in structural analysis, with the W-W algorithm used for the resulting Hermitian matrix.

ACKNOWDGEMENTS

Thanks are for support from NKBRSF (#G1999032805) of China, and also from British Research Council.

REFERENCES

Kittel, C. 1986. *Introduction to solid-state physics*. New York: J. Wiley & Sons.
Leung, A.Y.T. 1993. *Dynamic stiffness & sub-structures*. London: Springer.
Wittrick, W.H. & Williams, F.W. 1971. A general algorithm for computing natural frequencies of elastic structures. *Quart. J. Mech. Appl. Math.* 24:263–284.
Zhong, W.X. & Williams, F.W. 1991. On the localization of the vibration mode of a sub-structural chain-type structure. *Proc. of the Inst. of Mech. Engineering, Part-C* 205:281–288.
Zhong, W.X. & Williams, F.W. 1992. Wave propagation for repetitive structures and symplectic mathematics. *Proc. Inst. Mech. Engrs. part C* 206:371–379.
Zhong, W.X. & Williams, F.W. 1993. The eigensolutions of wave propagation for repetitive structures. *Structural engineering and mechanics* 1(1):47–60.
Zhong, W.X. & Williams, F.W. 1994. A precise time step integration method. *Proc. Inst. Mech. Engr,* 208:427–430.
Zhong, W.X. & Williams, F.W. 1995. On the direct solution of wave propagation for repetitive structures. *J sound & vib.* 181(3):485–501.
Zhong, W.X., Williams, F.W. & Bennett, P.N. 1997. Extension of the Wittrick-Williams algorithm to mixed variable systems. *J. Vib. & Acoust., Trans ASME* 119:334–340.
Zhong, W.X., Cheung, Y.K. & Li, Y. 1998. The precise finite strip method. *Computers & structure,* 69:779–783.
Zhong, W.X., Howson, W.P. & Williams, F.W. 2001. Precise solutions for surface wave propagation in stratified material. *Trans. ASME, J. of Vibr. & Acous.* 123:198–204.
Zhong, W.X., Ouyang, H.J. & Deng, Z.C. 1993. *Computational structural mechanics and optimal control.* Dalian: Dalian Univ. of Tech. Press.
Zhong, W.X., Lin, J.H. & Zhu, J.P. 1994. Computation of gyroscopic systems and symplectic eigensolutions of skew-symmetric matrices. *Computers & structures* 52(5):999–1009.
Zhong, W.X. 1995. *A new systematic methodology for theory of elasticity.* Dalian: Dalian University of Technology press.
Zienkiewicz, O.C. & Taylor, R.L. 1989. *The finite element method.* New York: McGraw-Hill.

Numerical methods

Computational Methods in Engineering and Science, Iu et al. (eds)
© *2003 Swets & Zeitlinger, Lisse, ISBN 90 5809 567 3*

Procedure for non-stationary PDF solution of nonlinear stochastic oscillators

G.K. Er & S. Frimpong
University of Alberta, Edmonton, Alberta, Canada

V.P. Iu
University of Macau, Macao

ABSTRACT: In this paper the recently proposed exponential polynomial closure method is extended for obtaining the non-stationary probability density function (PDF) of nonlinear stochastic oscillators. Numerical results are presented to show the effectiveness of this method. The extension of this procedure to multi-degree-of-freedom system is straightforward for obtaining the non-stationary PDF solution.

1 INTRODUCTION

The investigation on probability density function (PDF) solutions of nonlinear stochastic dynamic systems (NSDS) attracted many researchers in the past half century since the problems arising from more and more areas need a method with which better solution can be obtained. One of the challenging problems is how to improve the PDF solution in its tails because the tails of the PDF of the system responses govern the precision of the reliability or other statistical analysis of the system. Another challenging problem is how to obtain the non-stationary PDF solution of the nonlinear stochastic dynamic systems. Some methods were proposed for the stationary PDF solution of NSDS, such as equivalent linearization method (Booton 1954) and their modified versions (Caughey 1959, Lin 1967, Iwan 1973), stochastic average method (Stratonovich 1963), non-Gaussian closure method (Assaf & Zirkie 1976, Sobczyk & Trebicki 1990), multi-Gaussian closure method (Er 1997) and recently proposed exponential polynomial closure (EPC) method (Er 1998a, b, 1999, Er & Iu 1999, 2000a, b, Er 2000). As known, only the linearization method can be used to obtain the non-stationary PDF solutions of weakly nonlinear systems. For highly nonlinear system, new method is needed for obtaining non-stationary PDF solution. It is also known that EPC is not limited to weakly nonlinear system in obtaining stationary PDF solution of NSDS. Hence in this paper, we attempt to extend EPC method for obtaining the non-stationary PDF solution of NSDS.

2 EXPONENTIAL-POLYNOMIAL CLOSURE PROCEDURE

Exponential-polynomial closure (EPC) method was proposed recently, it was initially used for the PDF solution of nonlinear stochastic single-degree-of-freedom (SDOF) systems (Er & Iu 1999, 2000a). After that, it was extended to the nonlinear stochastic multi-degree-of-freedom (MDOF) systems with external excitations (Er & Iu 2000b), and further extended to the nonlinear stochastic MDOF systems with both external and parametric excitations (Er 2000). The effectiveness of the method has been verified in the publications mentioned above. For some systems, even exact stationary solution can be obtained with EPC. In the following discussion, EPC method is further extended to obtain the non-stationary PDF solution of nonlinear stochastic oscillators.

Consider the following nonlinear stochastic oscillator:

$$\ddot{Y} + h_0(Y,\dot{Y},t) = h_j(Y,\dot{Y},t)W_j(t) \tag{1}$$

where $W_j(t)$ is white noise.

Setting, $X_1 = Y, X_2 = \dot{Y}$, then Eq. (1) can be expressed as

$$\ddot{X}_1 + h_0(\mathbf{X},t) = h_j(\mathbf{X},t)W_j(t) \tag{2}$$

where $\mathbf{X} = \{X_1, X_2\} \in R^2$. Eq. (2) can be expressed in Stratonovich's form as

$$\frac{d}{dt}X_1 = X_2$$
$$\frac{d}{dt}X_2 = -h_0(\mathbf{X},t) + h_j(\mathbf{X},t)W_j(t) \qquad j = 1,2,...,n \tag{3}$$

where functions h_0 and h_j are generally nonlinear, and their functional forms are assumed to be deterministic. When the excitations $W_j(t)$ are Gaussian white noises with zero mean and cross correlation $E[W_j(t)W_k(t + \tau)] = S_{jk}\delta(\tau)$, in which $\delta(\tau)$ is the Dirac function and S_{jk} are constants, representing the cross-spectral density W_j and W_k. Eq. (3) may also be expressed in Ito's form as

$$\frac{d}{dt}X_1 = X_2$$
$$\frac{d}{dt}X_2 = f(\mathbf{X},t) + h_j(\mathbf{X},t)W_j(t) \qquad j = 1,2,...,n \tag{4}$$

where

$$f_i(\mathbf{X},t) = -h_0(\mathbf{X},t) + \frac{1}{2}\frac{\partial h_j(\mathbf{X},t)}{\partial X_2}h_j(\mathbf{X},t) \tag{5}$$

The system response $\mathbf{X}$ of (4) is a Markov vector and the PDF of the Markov vector is governed by the following FPK equation (Soong 1973):

$$\frac{\partial p}{\partial t} + x_2\frac{\partial p}{\partial x_1} + \frac{\partial}{\partial x_2}[f(\mathbf{x},t)p] - \frac{1}{2}\frac{\partial^2}{\partial x_2^2}[G(\mathbf{x},t)p] = 0 \tag{6}$$

where $\mathbf{x}$ is the state vector, $p = p(\mathbf{x}, t)$ and

$$G(\mathbf{x},t) = S_{ls}h_l(\mathbf{x},t)h_s(\mathbf{x},t) \tag{7}$$

It is assumed that $p(\mathbf{x}, t)$ fulfills the following conditions:

$$\begin{cases} p(\mathbf{x},t) \geq 0 & \mathbf{x} \in R^2 \\ \lim_{x_i \to \pm\infty} p(\mathbf{x},t) = 0 & i = 1,2 \\ \int_{R^2} p(\mathbf{x},t)d\mathbf{x} = 1 \end{cases} \tag{8}$$

Usually the exact solution of Eq. (6) is not obtainable. Hence the approximate PDF solution $\tilde{p}[\mathbf{x}, \mathbf{a}(t)]$ for Eq. (6) is assumed to be

$$\tilde{p}(\mathbf{x},\mathbf{a}|t) = \begin{cases} \exp^{Q_n(\mathbf{x},\mathbf{a})} & \mathbf{x} \times \mathbf{a} \in \Omega \times R^{N_p+1} \\ 0 & \mathbf{x} \notin \Omega \end{cases} \tag{9}$$

where $\mathbf{a} = \{a_0(t), a_1(t), a_2(t), \dots, a_{N_p}(t)\}$, being $a_i(t)$ the parameters to be determined; $N_p + 1$ equals the total number of unknown parameters and $\Omega = [m_1 - \alpha_1\sigma_1, m_1 + \beta_1\sigma_1] \times [m_2 - \alpha_2\sigma_2, m_2 + \beta_2\sigma_2] \subset R^2$, in which m_i and σ_i denotes the mean value and standard deviation of X_i, respectively; $\alpha_i > 0$ and $\beta_i > 0$ are defined such that $m_i - \alpha_i\sigma_i$ and $m_i + \beta_i\sigma_i$ are located in the tails of the PDF of $X_i(t)$; and $Q_n[\mathbf{x}, \mathbf{a}(t)]$ is an n-degree polynomial in the state variables x_1 and x_2, which can be expressed as

$$Q_n(x, a \mid t) = a_0(t) + a_1(t)x_1 + a_2(t)x_2 + a_3(t)x_1^2 + a_4(t)x_1x_2 + a_5(t)x_2^2 + \dots + a_{N_p}(t)x_2^n \tag{10}$$

Hence this method is named exponential-polynomial closure method. If the approximate PDF expression replaces the exact PDF in Eq. (6), the following residual error is yielded:

$$\Delta(\mathbf{x},\mathbf{a},t) = \frac{\partial \tilde{p}}{\partial t} + x_2\frac{\partial \tilde{p}}{\partial x_1} + \frac{\partial f}{\partial x_2}\tilde{p} + f\frac{\partial \tilde{p}}{\partial x_2} - \frac{1}{2}\left(\frac{\partial^2 G}{\partial^2 x_2}\tilde{p} + 2\frac{\partial G}{\partial x_2}\frac{\partial \tilde{p}}{\partial x_2} + G\frac{\partial^2 \tilde{p}}{\partial^2 x_2}\right) \tag{11}$$

Substituting Eq. (9) into Eq. (11) leads to

$$\Delta(\mathbf{x},\mathbf{a},t) = \delta(\mathbf{x},\mathbf{a},t)\tilde{p}(\mathbf{x},\mathbf{a} \mid t) \tag{12}$$

where

$$\delta(\mathbf{x},\mathbf{a},t) = \frac{\partial Q_n}{\partial t} + x_2\frac{\partial Q_n}{\partial x_1} + f\frac{\partial Q_n}{\partial x_2} - \frac{1}{2}[G\frac{\partial^2 Q_n}{\partial^2 x_2} + 2\frac{\partial G}{\partial x_2}\frac{\partial Q_n}{\partial x_2} + G(\frac{\partial Q_n}{\partial x_2})^2] + \frac{\partial f}{\partial x_2} - \frac{1}{2}\frac{\partial^2 G}{\partial^2 x_2} \tag{13}$$

If $\delta(\mathbf{x}, \mathbf{a}, t) \neq 0$, then Eq. (6) can be fulfilled. However, usually. $\delta(\mathbf{x}, \mathbf{a}, t) \neq 0$. A set of mutually independent functions $H_k(\mathbf{x}|t)$, $(i = 0, 1, 2, \dots, N_p)$, which span the space $R^{N_p + 1}$ can be introduced to make the projection of $\delta(\mathbf{x}, \mathbf{a}, t)$ on $R^{N_p + 1}$ vanish, which yields

$$\int_{R^2} \delta(\mathbf{x},\mathbf{a},t)H_k(\mathbf{x}\mid t)d\mathbf{x} = 0 \qquad k = 0,1,2,\dots,N_p \tag{14}$$

or

$$\int_{R^2} \{\frac{\partial Q_n}{\partial t} + x_2\frac{\partial Q_n}{\partial x_1} + f\frac{\partial Q_n}{\partial x_2} - \frac{1}{2}[G\frac{\partial^2 Q_n}{\partial^2 x_2} + 2\frac{\partial G}{\partial x_2}\frac{\partial Q_n}{\partial x_2} + G(\frac{\partial Q_n}{\partial x_2})^2] + \frac{\partial f}{\partial x_2} - \frac{1}{2}\frac{\partial^2 G}{\partial^2 x_2}\}$$
$$H_k(\mathbf{x}\mid t)d\mathbf{x} = 0 \qquad k = 0,1,2,\dots,N_p \tag{15}$$

Eq. (15) means that the reduced FPK equation is fulfilled with $\tilde{p}(\mathbf{x}, \mathbf{a}|t)$ in the average sense of integration if $\delta(\mathbf{x}, \mathbf{a}, t)H_k(\mathbf{x}|t)$ are integrable in R^2. The function $H_k(\mathbf{x}|t)$ can be selected as $x_1^{k_1}x_2^{k_2}G(\mathbf{x}|t)$, being $k = k_1 + k_2 = 0, 1, 2, \dots, N_p$ and $k_1, k_2 = 0, 1, 2, \dots, N_p$. Numerical experience shows that an effective choice for $G(\mathbf{x}|t)$ is the PDF solution from stochastic equivalent linearization procedure or the PDF solution from exponential closure method when $n = 2$.

3 NUMERICAL EXAMPLE

The following Duffing oscillator is used to verify the effectiveness of EPC method for non-stationary PDF solution:

$$\ddot{X} + c\dot{X} + kX + \varepsilon X^3 = W(t) \tag{16}$$

For $c = 0.2$, $k = 1$, $S = 1$ and $\varepsilon = 1$, the system is strongly nonlinear. The initial PDF of $X(t)$ and $\dot{X}(t)$ is Gaussian with mean values being zero and standard deviations of both $X(t)$ and $\dot{X}(t)$ being 0.5. The time step used in analysis is 0.02 seconds.

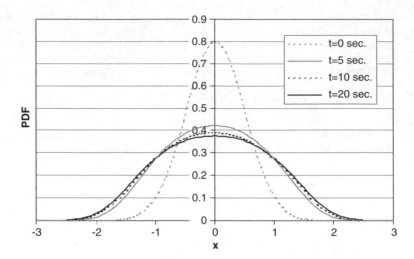

Figure 1. Probability density functions of X at different time instants.

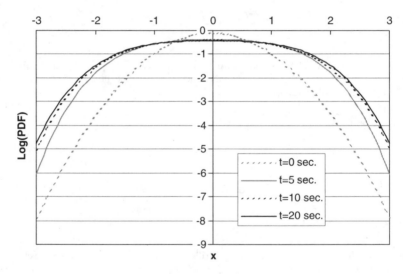

Figure 2. Logarithmic probability density functions of X at different time instants.

The obtained PDFs of $X(t)$ at $t = 0$ sec., $t = 5$ sec., $t = 10$ sec. and $t = 20$ sec. are shown in Figure 1. The logarithmic values of the PDFs are shown in Figure 2. From the two figures it is seen that the PDF of $X(t)$, obtained with EPC, converges to stationary state as expected. The exact stationary PDF solution of this system is obtainable. The PDF of $X(t)$ for $t = 20$ sec. is almost same as the exact stationary PDF of $X(t)$, which means that at $t = 20$ sec. the PDF of $X(t)$ is very close to the stationary PDF solution.

The obtained PDFs of $\dot{X}(t)$ at $t = 0$ sec., $t = 5$ sec., $t = 10$ sec. and $t = 20$ sec. are shown in Figure 3. The logarithmic values of the PDFs are shown in Figure 4. Numerical results showed that the PDFs of $\dot{X}(t)$ at different instants are Gaussian and converge to stationary solution, too. The PDF of $\dot{X}(t)$ at $t = 20$ sec. is also very close to the obtainable exact PDF solution.

From Figures 1–4, it is seen that the non-stationary PDF solutions of both $X(t)$ and $\dot{X}(t)$, obtained with EPC, converge to exact stationary solutions as expected.

184

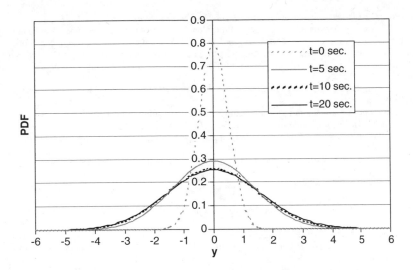

Figure 3. Probability density functions of Y at different time instants.

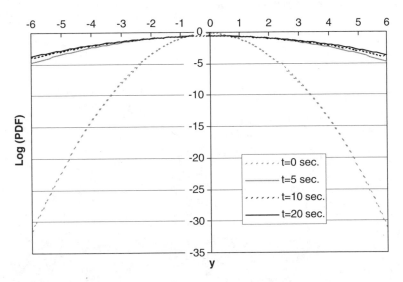

Figure 4. Logarithmic probability density functions of Y at different time instants.

4 CONCLUSIONS

From the above discussion it is seen that the PDFs of state variables, obtained with EPC, converge to the stationary solution. Numerical analysis shows that it is suitable for highly nonlinear system besides weakly nonlinear system. The availability of this procedure for obtaining the non-stationary PDF solutions of other more complicated oscillators need being further investigated with detailed problems. The extension of this procedure to MDOF system is straightforward.

REFERENCES

Booton, R. C. 1954. Nonlinear Control Systems with Random Inputs. IRE Trans. *Circuit Theory* CT-1, 1: 9–19.
Caughey, T. K. 1959. Response of a Nonlinear String to Random Loading, ASME *J Applied Mechanics* 26(3): 341–344.

Lin, Y. K., 1967. *Probabilistic Theory of Structure Dynamics*. McGraw-Hill, New York.

Iwan, W. D. 1973. A Generalization of the Concept of Equivalent Linearization. *Int. J. NonLinear Mechanics* 5: 279–287.

Stratonovich, R. L. 1963. *Topics in the Theory of Random Noise*. Vol. 1, Gordon and Breach, New York.

Assaf, S. A. & Zirkie, L. D. 1976. Approximate Analysis of Nonlinear Stochastic Systems. *Int. J. Control* 23(4): 477–492.

Sobczyk, K. & Trebicki, J. 1990. Maximum Entropy Principle in Stochastic Dynamics. *Probabilistic Engneering Mechanics* 5: 102–110.

Er, G. K. 1997. Multi-Gaussian Closure Method for Randomly Excited Nonlinear Systems. *Int. J. NonLinear Mechanics* 33: 201–214.

Er, G. K. 1998a. A new non-Gaussian closure method for the PDF solution of nonlinear random vibrations. Proc. *12th Engineering Mechanics Conference*, ASCE, May, 1998, San Diego, USA.

Er, G. K. 1998b. An improved non-Gaussian closure method for nonlinear stochastic systems. *Nonlinear Dynamics* 17(3): 283–297

Er, G. K. 1999. A consistent method for the PDF solutions of random oscillators. ASCE *Journal of Engineering Mechanics* 125(4): 443–447.

Er, G. K. & Iu, V. P. 1999. Probabilistic solutions to nonlinear random ship roll motion. ASCE *Journal of Engineering Mechanics* 125(5): 570–574.

Er, G. K. & Iu, V. P. 2000a. Stochastic response of base-excited Coulomb oscillator. *Journal of Sound and Vibration* 233(1): 81–92.

Er, G. K. & Iu, V. P. 2000b. A consistent and effective method for nonlinear random oscillations of MDOF systems. Proc. of IUTAM *Symposium on Recent Developments in Nonlinear Oscillations of Mechanical Systems*, Kluwer Academic Publishers.

Er, G. K. 2000. The probabilistic solutions to nonlinear random vibrations of multi-degree-of-freedom systems. ASME *Journal of Applied Mechanics* 67(2): 355–359.

Soong, T. T. 1973. *Random Differential Equations in Science and Engineering*, Academic Press, New York.

Computational Methods in Engineering and Science, Iu et al. (eds)
© *2003 Swets & Zeitlinger, Lisse, ISBN 90 5809 567 3*

Improvement for weighting functions in element free Galerkin method (EFGM)

X.Z. Lu & J.J. Jiang
Department of Civil Engineering, Tsinghua University, Beijing, China

ABSTRACT: A new method to set up the weighting functions used in element free Galerkin method (EFGM) is introduced in this paper. The traditional weighting functions in EFGM only have relationships with the distance between the nodes and the integration points without considering different node distributions along various axes and may bring more errors. Hence, two parameters are introduced to modify the weighting functions from round vesicular functions to elliptical vesicular functions. With this approach, the influence of node distributions on the accuracy is reduced and the precision is greatly improved for some commonly used weighting functions.

1 INTRODUCTION TO EFGM

A new numerical method which is called element free Galerkin method (EFGM) was developed by Belytschko (1994). Consider a field function $u^*(x)$ in a domain of Ω. We can establish its approximate function as follows

$$u^*(x) \approx u^h(x) = \sum_{j=1}^{m} p_j(x)a_j(x) \equiv \mathbf{p}^T(x)\mathbf{a}(x) \tag{1}$$

where $\mathbf{a}(x)$ is the factor vector, which is a function of x, and $\mathbf{p}(x)$ is an m dimension polynomial.
 The approximation around point x^* is

$$u^h(x,x^*) = \sum_{j=1}^{m} p_j(x^*)a_j(x) \equiv \mathbf{p}^T(x^*)\mathbf{a}(x) \tag{2}$$

$\mathbf{a}(x)$ will be determined by weighted least square fitting for the local approximation, which minimizes the difference between the local approximation and the original function. This yields the quadratic form

$$J = \sum_{j=1}^{n} w(\|x - x_j\|)[u^h(x,x_j) - u^*(x_j)]^2 = \sum_{j=1}^{n} w(\|x - x_j\|)\left[\sum_{i=1}^{m} p_i(x_j)a_i(x) - u^*(x_j)\right]^2 \tag{3}$$

Here $w(\|x - x_j\|)$ is the weight function, and $u^*(x_j)$ is the value of $u^*(x)$ on point x_j.
 Minimizing J to the factor $\mathbf{a}(x)$, that is, $\varpi J/\varpi a_i = 0$ yields

$$\mathbf{a}(x) = \mathbf{A}^{-1}(x)\mathbf{B}(x)\mathbf{u}^* \tag{4}$$

where

$$\mathbf{A}(x) = \sum_{i=1}^{n} w_i(\|x - x_j\|)\mathbf{p}(x_i)\mathbf{p}^T(x_i) \tag{4a}$$

$$\mathbf{B}(x) = [w_i(\|x-x_j\|)\mathbf{p}(x_1),\ldots\ldots,\ \ w_n(\|x-x_j\|)\mathbf{p}(x_n)] \tag{4b}$$

$$\mathbf{u}^* = [u^*(x_1), u^*(x_2),\ldots, u^*(x_n)]^T \tag{4c}$$

So, Equation (1) can be re-written as

$$\mathbf{u}^h(x) = \sum_{i=1}^{n} n_i(x) u_i^* \tag{5}$$

$$n_i(x) = \sum_{j=1}^{m} p_j(x)[\mathbf{A}^{-1}(x)\mathbf{B}(x)]_{ji} \tag{5a}$$

The partial difference of function $n_i(x)$ is

$$n_{i,k}(x) = \sum_{j=1}^{m} \{p_{j,k}(x)[\mathbf{A}^{-1}(x)\mathbf{B}(x)]_{ji} + p_j(x)[\mathbf{A}_{,k}^{-1}(x)\mathbf{B}(x) + \mathbf{A}^{-1}(x)\mathbf{B}_{,k}(x)]_{ji}\} \tag{6}$$

$$\mathbf{A}_{,k}^{-1}(x) = -\mathbf{A}^{-1}\mathbf{A}_{,k}\mathbf{A}^{-1} \tag{7}$$

And the global stiffness matrix will be obtained by

$$\mathbf{K} = \int_{\Omega} \mathbf{B}^T \mathbf{D} \mathbf{B} d\Omega \tag{8}$$

where, for planar problems,

$$\mathbf{B} = \begin{bmatrix} n_{1,x} & 0 & n_{2,x} & 0 & \ldots & n_{n,x} & 0 \\ 0 & n_{1,y} & 0 & n_{2,y} & \ldots & 0 & n_{n,y} \\ n_{1,y} & n_{1,x} & n_{2,y} & n_{2,x} & \ldots & n_{n,y} & n_{n,x} \end{bmatrix} \tag{9}$$

2   MODIFICATION ON WEIGHTING FUNCTIONS

Traditional weight functions in EFGM only have relationship with the distance r between integration points and nodes.. For example, the respective weighting functions adopted by Belytscho (1994), Kou (1998) and Li (2000) are as follows,

$$w_i(r_i) = 1 - 6\left(\frac{r_i}{r_{mi}}\right)^2 + 8\left(\frac{r_i}{r_{mi}}\right)^3 - 3\left(\frac{r_i}{r_{mi}}\right)^4 \tag{10}$$

$$w_i(r_i) = \frac{r_{mi}^2}{r_i^2 + \varepsilon^2 r_{mi}^2}\left(1 - \frac{r_i^2}{r_{mi}^2}\right) \tag{11}$$

$$w_i(r_i) = \frac{e^{-(r_i/c)^k} - e^{-(r_{mi}/c)^k}}{1 - e^{-(r_{mi}/c)^k}} \tag{12}$$

All these functions are related to r_i and do not consider the influence of node distribution along different axes. For example, for such problems as shown in Figure 1, the node distribution is quite irregular, and the maximal nodal spacing along the y-axis is much smaller than that along the x-axis. So using a round vesicular function (shown in Fig. 2 (a)) as the weighting function may bring relatively larger errors.

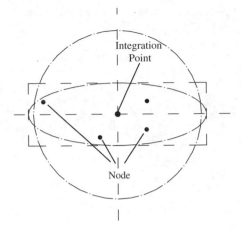

Figure 1.   Distribution of integration point and node.

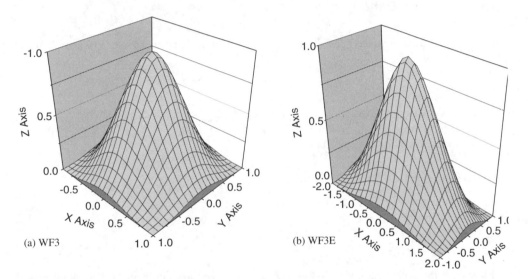

(a) WF3

(b) WF3E

Figure 2.   WF3 and WF3E.

However, if we modify the weighting function with an elliptic function to measure the distance between nodes and integration points,

$$\hat{r}_i = \sqrt{\frac{(x-x_i)^2}{a^2} + \frac{(y-y_i)^2}{b^2}} \qquad (13)$$

the result could be improved.

The two new parameters a and b are determined based on the following procedure,

(1) Find the largest differences D_x and D_y in x-direction and y-direction between the nodes and the integration points
(2) Find the minimum value η_{min} based on the following inequality for all nodes, all

$$\eta^2 \left(\frac{x_i^2}{D_x^2} + \frac{y_i^2}{D_y^2} \right) \le 1.0 \qquad (14)$$

(3) Calculate a and b as follows obtain the value for parameter a and b.

$$a = \frac{D_x}{\eta_{\min}} \tag{15}$$

$$b = \frac{D_y}{\eta_{\min}} \tag{16}$$

The weighting functions of WF1, WF2, and WF3 can be re-written as

WF1E: $\quad w_i(r_i) = \dfrac{1}{\hat{r}_i^2 t^2 + \varepsilon^2}\left(1 - \hat{r}_i^2 t^2\right)$ \hfill (17)

WF2E: $\quad w_i(r_i) = \dfrac{e^{-(\hat{r}_i r_{mi}/c)^k} - e^{-(r_{mi}/c)^k}}{1 - e^{-(r_{mi}/c)^k}}$ \hfill (18)

WF3E: $\quad w_i(r_i) = 1 - 6\left(\hat{r}_i t\right)^2 + 8\left(\hat{r}_i t\right)^3 - 3\left(\hat{r}_i t\right)^4$ \hfill (19)

where $t, \varepsilon, r_{mi}, c, k$ are parameters determined before computation. For comparison, WF3 and WF3E are plotted and shown in Figure 2.

3 EXAMPLES

A cantilever beam shown in Figure 3 is analyzed. The parameters for the beam are: Young's Modulus $E = 3.0 \times 10^7\,Pa$, Poisson Ratio $\mu = 0.25$, $P = 1000\,N/m^2$, $D = 1\,m$, $L = 5.8\,m$.
The analytical result for this problem is

$$\left\{ \begin{array}{ll} \sigma_x = -\dfrac{P}{I}(L-x)(y-\tfrac{1}{2}D) & \text{(20a)} \\[2ex] \sigma_y = 0 & \text{(20b)} \\[2ex] \tau_{xy} = -\dfrac{Py}{2I}(y-D) & \text{(20c)} \end{array} \right.$$

Error τ_{xy} at the target node versus parameters r_m, c and t are shown in Figure 4. For uniform nodal distribution, that is $a = b$, the elliptical vesicular functions are regressived to the traditional round vesicular

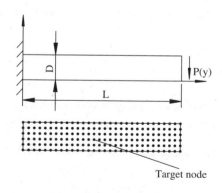

Figure 3. Cantilever beam.

functions. For WF1E, $\varepsilon = 0.01$, $t = 1.2$. For WF2E, $r_m = 1.2$, $r_m/c = 4$. For WF3E, $t = 1.2$. And these results are consistent with those of Kou (Kou, 1998), Li (Li, 2000) and Gavete (Gavete, 2000).

For a non-uniform and a rather random nodal distribution shown in Figure 5 and Figure 6, the errors in τ_{xy} at the target node with different weighting functions are shown in Table 1.

The table clearly shows that the modification of the weight functions can improve calculation precision, especially when there is large difference on node distribution along various axes.

The second numerical example is a semi-infinite plate under a point load. Because of symmetry, only the positive half along the Y axis of the plate is shown in Figure 7. The analytical results are:

$$\sigma_x = -\frac{2Px^3}{\pi(x^2 + y^2)^2} \tag{21a}$$

$$\sigma_y = -\frac{2Pxy^2}{\pi(x^2 + y^2)^2} \tag{21b}$$

$$\tau_{xy} = -\frac{2Px^2y}{\pi(x^2 + y^2)^2} \tag{21c}$$

The value of τ_{xy} at the point $(-3, 0)$ is shown in Table 2 with different weighting functions. It should be mentioned that the result of the finite element method with same DOF (degree of freedom) is 17–19 Pa. So we can see that the EFGM is much more accurate than FEA in this problem. And the precision is again greatly improved with modified weight functions for non-uniform node distributions.

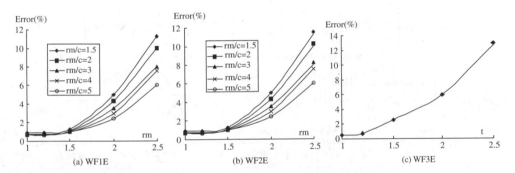

Figure 4.   Errors in τ_{xy} and parameters values.

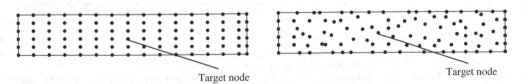

Figure 5.   Non-uniform node distribution.

Figure 6.   Random node distribution.

Table 1.   Errors for various weight functions.

	WF1 (%)	WF2 (%)	WF3 (%)	WF1E (%)	WF2E (%)	WF3E (%)
Non-uniform node distribution	1.6	1.9	2.1	0.65	0.79	0.90
Random node distribution	2.7	3.2	4.0	1.61	1.83	2.38

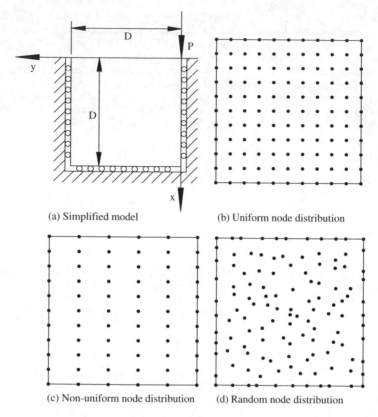

(a) Simplified model   (b) Uniform node distribution

(c) Non-uniform node distribution   (d) Random node distribution

Figure 7.   Semi-infinite planar problem.

Table 2.   Values of τ_{xy} (in Pa) with various weighting functions.

	Eq. (21c) Result	WF1	WF2	WF3	WF1E	WF2E	WF3E
Uniform node distribution	0	3.547	4.714	5.333	3.260	4.986	4.957
Non-uniform node distribution	0	8.013	9.115	10.748	4.248	5.723	5.640
Random node distribution	0	9.740	11.030	10.901	4.746	6.585	6.457

4   CONCLUSIONS

By modifing the weighting functions in EFGM from round vesicular functions to elliptical vesicular functions, the influence of node distribution along various axes can be considered. Numerical examples shows that the influence of node distributions on the accuracy is reduced and the precise is greatly improved for some commonly used weighting functions. And this method also can be applied for the spatial problems.

ACKNOWLEDGEMENT

This research is financial supported by the China National Natural Science Fund (project number: 59938180).

REFERENCE

Belytschko T, 1996, Meshless method: An overview and recent development. *Comput. Methods. Appl. Mech. Engrg*, 139: 3–47.

Belytschko T, 1994, Lu YY, Element Free Galerkin method. *Int. J. Numer. Methods. Engrg*, 37: 229–256.

Liu WK, 2001, Multi-scale meshfree particle method. In LIN SP (eds), *Proceeding of EMPESC VIII*, Shanghai: Sanlian Press, 61–71.

Li Wodong, 2000, Theoretical Study and Engineering Application of Meshless Method, *Wuhan: Doctor Thesis of Huazhong University of Science and Technology*.

Zhou Weiyuan, Kou Xiaodong, 1998, Meshless Method and Engineering Application, *Acta Mechanica Sinia*, 30(2): 193–202.

Kou Xiaodong, Mesh Free Method Tracing Structure Fracture and Arch Dam Stability Analysis, 1998, *Beijing: Doctor Thesis of Tsinghua University*.

Luis Gavete, Juan J. Benito, Santiago Falcon, etc. 2000, Penalty functions in constrained variational principles for element free Galerkin method. *Eur. J. Mech. A/Solids,* 19: 699–720.

Numerical solution of a singular Volterra integral equation by piecewise polynomial collocation

T. Diogo, S. Valtchev & P. Lima
Centro de Matemática e Aplicações, Instituto Superior Técnico, Lisboa, Portugal

ABSTRACT: This work is concerned with the numerical study of a singular Volterra integral equation by collocation type methods. Under certain conditions this equation has a family of solutions of which only one is differentiable and all the others have infinite gradient at the origin. Previously we have proved convergence of numerical methods to the smooth solution. In this work we investigate a way of choosing a particular trajectory which consists of fixing an initial interval where we specify the values of a chosen solution. A new equation is obtained to which standard numerical methods can be applied and, under certain conditions, there is convergence to the fixed exact solution. Some numerical examples are given which illustrate the performance of the methods employed.

1 INTRODUCTION

Consider the following second kind Volterra integral equation

$$u(t) = \int_0^t \frac{s^{\mu-1}}{t^\mu} u(s)ds + g(t), \quad t \in (0,T] \tag{1.1}$$

where $\mu > 0$ and g is a given function. Equations of the above type arise from certain heat conduction problems with time dependent boundary conditions (Bartoshevich 1975). The behavior of the exact solution has been analyzed in Diogo et al (1991) and Han (1994). If $\mu > 1$ and $g \in C^m[0, T]$ then Equation 1.1 has a unique solution in $C^m[0, T]$. In this case, certain classes of product integration methods were studied in Diogo et al (1991), Diogo et al (2000) and Tang et al (1992) for its numerical treatment. We note that the kernel of Equation 1.1, that is, the function $p(t,s) = s^{\mu-1}/t^\mu$, is integrable with respect to s in $[0,T]$. However, certain properties (such as the lack of compactness of the integral operator and all the iterated kernels associated with p being unbounded) do not allow us to use the classical arguments of stability proofs of numerical methods for weakly singular equations (see Brunner and van der Houwen (1986)). An interesting feature of the above equation is that it loses uniqueness of solution in the case $0 < \mu \leq 1$; it was shown in Han (1994) that Equation 1.1 has a family of solutions in $C[0,T]$ (given by Equation 1.2), one of which has C^1 continuity and all other solutions have infinite gradient at $t = 0$. In Lima & Diogo (2002) we have been able to prove that the Euler's method converges to the C^1 solution and we obtained asymptotic error expansions in fractional powers of the stepsize allowing the application of general extrapolation algorithms. In this work we consider a technique which enables us to approximate any particular solution u. After fixing an initial interval where we specify the values of u, a collocation method is then applied and convergence to the chosen solution is obtained, under certain conditions. Some numerical examples illustrate our theoretical results.

The following explicit representation for the solutions of Equation 1.1 is given in Han (1994).

Lemma 1.1. (a) If $0 < \mu \leq 1$ and $g \in C^1[0,T]$ (with $g(0) = 0$ if $\mu = 1$) then Equation 1.1 has a family of solutions $u \in C[0,T]$ given by the formula

$$u(t) = c_0 t^{1-\mu} + g(t) + \gamma + t^{1-\mu} \int_0^t s^{\mu-2}(g(s) - g(0))ds \tag{1.2}$$

where c_0 is an arbitrary constant and $\gamma = g(0)/(1-\mu)$ (if $\mu < 1$) or $\gamma = 0$ (if $\mu = 1$). Out of the family of solutions there is one particular solution. $u \in C^1[0,T]$. Such a solution is unique and can be obtained from Equation 1.2 by taking $c_0 = 0$.

(b) If $\mu > 1$ and $g \in C^m[0,T]$, $m \geqslant 0$, then the unique solution $u \in C^m[0,T]$ of Equation 1.1 is given by the formula

$$u(t) = g(t) + t^{1-\mu} \int_0^t s^{\mu-2} g(s) ds \tag{1.3}$$

We note that Equation 1.3 can be obtained from Equation 1.2 with $c_0 = 0$. Indeed, it follows from Equation 1.2 that $c_0 = \lim_{t \to 0^+} t^{\mu-1} u(t)$, this limit being zero when $\mu > 1$.

2 NUMERICAL METHODS

Let $\varepsilon > 0$ be a fixed real number. Substituting t by $t + \varepsilon$ in Equation 1.1 and defining

$$I_\varepsilon := \int_0^\varepsilon s^{\mu-1} u(s) ds, \quad \text{we get} \tag{2.1}$$

$$u(t+\varepsilon) = \frac{I_\varepsilon}{(t+\varepsilon)^\mu} + \int_\varepsilon^{t+\varepsilon} \frac{s^{\mu-1}}{(t+\varepsilon)^\mu} u(s) ds + g(t+\varepsilon) \tag{2.2}$$

The idea is to evaluate I_ε exactly for a chosen solution and then apply a discretization method to Equation 2.2. In this work we shall be concerned with collocation and iterated collocation methods in certain piecewise polynomial spaces, which have been used in Brunner (1980), Brunner & Norsett (1981) and Brunner & van der Houwen (1986) for second kind Volterra integral equations with regular kernels. We note that the kernel of the transformed Equation 2.2 is regular in $\{(t, s): 0 \leqslant s \leqslant t \leqslant T - \varepsilon\}$ and some ideas of the above references can be extended to this particular equation.

Let $\varepsilon = t_0 < t_1 < \cdots < t_N = T$ be a uniform mesh of the interval $[\varepsilon, T]$ into the subintervals $\sigma_0 = [t_0, t_1]$ and $\sigma_n = [t_n, t_{n+1}]$ with $n = 1, \ldots, N-1$. Set $Z_N := \{t_i = ih + \varepsilon, 1 \leqslant i \leqslant N - 1\}$, $h = (T - \varepsilon)/N$, and define the piecewise polynomial space $S_{m-1}^{(-1)}(Z_N) := \{v: v|_{\sigma_n} := v_n \in \pi_{m-1}, 0 \leqslant n \leqslant N - 1\}$, whose elements reduce to polynomials of degree not exceeding $m - 1$ on each subinterval σ_n and may have jump discontinuities at the knots t_n. Rewriting Equation 2.2 as

$$u(t) = \frac{I_\varepsilon}{t^\mu} + \int_\varepsilon^{t_n} \frac{s^{\mu-1}}{t^\mu} u(s) ds + \int_{t_n}^t \frac{s^{\mu-1}}{t^\mu} u(s) ds + g(t), \quad t \in \sigma_n \tag{2.3}$$

the solution of the above equation will be approximated, using collocation techniques, by an element u^h of the space $S_{m-1}^{(-1)}(Z_N)$. On each σ_n, $n = 0, \ldots, N - 1$, we consider a set of m collocation points, defined through the m collocation parameters $\{\lambda_j\}_{j=1,\ldots,m}$ by $X_n = (t_{nj} = t_n + \lambda_j h, 0 \leqslant \lambda_1 < \cdots < \lambda_m \leqslant 1\}$.

The approximation u^h will then be required to satisfy Equation 2.3 on each set X_n. Letting u_n^h denote the restriction of u^h to the subinterval σ_i ($i = 0, \ldots, N - 1$), we have the following collocation equation

$$u_n^h(t_{nj}) = \frac{I_\varepsilon}{t_{nj}^\mu} + g(t_{nj}) + \sum_{i=0}^{n-1} \int_0^1 \frac{(i+\tau+\varepsilon/h)^{\mu-1}}{(n+\lambda_j+\varepsilon/h)^\mu} u_i^h(t_i + \tau h) d\tau$$

$$+ \int_0^{\lambda_j} \frac{(n+\tau+\varepsilon/h)^{\mu-1}}{(n+\lambda_j+\varepsilon/h)^\mu} u_n^h(t_n + \tau h) d\tau \quad j = 1, \ldots, m \quad n = 0, \ldots, N-1 \tag{2.4}$$

If on each subinterval u^h is given by its Lagrange formula

$$u_n^h(t_n + \tau h) = \sum_{l=1}^m L_l(\tau) U_{nl} \text{ with } L_l(\tau) = \prod_{\substack{i=1 \\ i \neq l}}^m \frac{\tau - \lambda_i}{\lambda_l - \lambda_i} \text{ for } t_n + \tau h \in \sigma_n, \tag{2.5}$$

196

where $U_{nl} := u_n^h(t_{nl})$, then Equation 2.4 represents a sequence of N linear systems in the unknowns $\{U_{i1},...,U_{im}\}$ for n = 0,...,$N-1$, which are obtained in a recursive way. Thus, at every step (on every subinterval σ_n) one has to solve a system of m linear equations. We note that the first integral in Equation 2.4 uses the known values of the approximated solution (from previous steps) and the second integral contains the unknowns (the coefficients of u_n^h).

3 CONVERGENCE ANALYSIS

We have the following global convergence result.

Theorem 3.1. Consider Equation 2.2 with $0 < \mu \leq 1$ and $g \in C^m[0,T]$, ε being a fixed positive real number. Let us assume that the integral I_ε has been evaluated exactly for a chosen particular solution (corresponding to a certain value of the parameter c_0). Then there exists $\bar{h} > 0$, such that the collocation Equation 2.5 defines for each $h \in (0, \bar{h})$ a unique element $u^h \in S_{m-1}^{(-1)}(Z_N)$. The approximation thus obtained converges with order m ($h \to 0$) to that particular solution. Moreover, for any choice of the collocation parameters $\{\lambda_i\}_{i=1,...,m}$, the error $e(t) = u(t) - u^h(t)$ satisfies

$$\| e \|_\infty := \sup\{| e(t) |: t \in \sigma_n, \ n = 0,...,N-1\} \leq Ch^m$$

Local superconvergence and iterated collocation: The global order m is optimal for the space $S_{m-1}^{(-1)}(Z_N)$. However, for some particular choices of the collocation parameters $\{\lambda_j\}_{j=1,...,m}$ a higher order of convergence can be achieved either at the mesh knots $\bar{Z}_N = Z_N \cup \{\varepsilon,T\}$ or at the collocation points. For example, if the collocation parameters are the Radau II points or the Lobatto points (see Brunner & van der Houwen 1986), the orders of local superconvergence on $\bar{Z}_N$ are $2m - 1$ and $2m - 2$, respectively. On the other hand, the choice of the m Gauss points on [0,1] (that is, the m zeros of $P_m(2s - 1)$, where P_m is the Legendre polynomial of degree m) as collocation parameters does not lead to local superconvergence at the knots $\bar{Z}_N$. However, if $g \in C^{2m}[0,T]$ and we associate with $u^h \in S_{m-1}^{(-1)}(Z_N)$ the so-called iterated approximation u^I, defined by

$$u^I(t) = \frac{I_\varepsilon}{t^\mu} + \int_\varepsilon^t \frac{s^{\mu-1}}{t^\mu} u^h(s)ds + g(t) \tag{3.1}$$

then the local order $2m$ can be attained on $\bar{Z}_N$, that is, the error satisfies $\max_{t_n \in \bar{Z}_N} |e^I(t_n)| = O(h^{2m})$ where $e^I(t_n) := u(t_n) - u^I(t_n)$.

4 NUMERICAL EXAMPLES

In this section we consider two equations which we have solved in the interval [0,2] by the collocation and the iterated collocation methods, given by Equations 2.4 & 3.1, respectively. As an estimate of the convergence order we have used the number

$$k := \log_2\left(\| u - u^{2h} \|_\infty / \| u - u^h \|_\infty\right)$$

Example 1: If we set $g(t) = 1 + t + t^2 + t^3$ and $\mu = 0.5$ in Equation 1.1, then, using Equation 1.2, we obtain the general form of its family of solutions

$$u(t) = c_0 t^{0.5} - 1 + 3t + (5/3)t^2 + (7/5)t^3 \quad c_0 \in \Re \tag{4.1}$$

Example 2: We set $\mu = 0.75$ and choose g (with the help of *Mathematica*) such that the general solution of Equation 1.1 is given by

$$u(t) = c_0 t^{0.25} + t^{0.5} \mathrm{Sin}(2\pi t) \quad c_0 \in \Re \tag{4.2}$$

Example 3: If we set $g(t) = 1 + t^3$ then $u(t) = c_0 t^{1-\mu} + (\mu)/(\mu - 1) + [(\mu + 3)/(\mu + 2)]t^2$

Table 1.　Ex. 1, $c_0 = -2$ and $m = 2$.

N	$\varepsilon = 0.01$		$\varepsilon = 0.02$		$\varepsilon = 0.05$	
	$\|e\|_\infty$	K	$\|e\|_\infty$	K	$\|e\|_\infty$	K
100	4.40E−02	**	1.70E−02	**	5.74E−03	**
200	1.45E−02	1.60	4.98E−03	1.77	1.50E−03	1.94
400	4.35E−03	1.74	1.35E−03	1.88	3.84E−04	1.97
800	1.20E−03	1.86	3.53E−04	1.94	9.70E−05	1.98

Table 2.　Ex., $\varepsilon = 0.02$, $c_0 = -2$ and $m = 2$ (Radau II).

	$|e(t)|$				Order
h	t = 0.5	t = 1.0	t = 1.5	t = 2.0	K
0.02	3.69E−03	5.21E−03	6.39E−03	7.38E−03	**
0.01	4.98E−04	7.05E−04	8.64E−04	9.97E−04	2.89
0.005	6.30E−05	8.91E−05	1.09E−04	1.26E−04	2.99
0.0025	7.82E−06	1.11E−05	1.36E−05	1.57E−05	3.01

Table 3.　Ex. 1, $\varepsilon = 0.02$, $c_0 = -2$ and $m = 2$. Iterated Collocation on $\bar{Z}_N$.

	$|e^I(t)|$				Order
h	t = 0.5	t = 1.0	t = 1.5	t = 2.0	K
0.02	6.76E−04	9.56E−04	1.17E−03	1.35E−03	**
0.01	5.37E−05	7.60E−05	9.31E−05	1.07E−04	3.65
0.005	3.66E−06	5.17E−06	6.33E−06	7.31E−06	3.88
0.0025	2.34E−07	3.31E−07	4.06E−07	4.68E−07	3.97

We start with example 1 and consider collocation in $S_{m-1}^{(-1)}(Z_N)$, with $m = 2$, using $\lambda_1 = 0.2$ and $\lambda_2 = 0.4$ as the collocation parameters. In Table 1 the error norms are shown, for several fixed values of the parameter (ε, confirming the second order convergence of Theorem 3.1. Local superconvergence of order $k = 3$ on $\bar{Z}_N$ is illustrated in Table 2, with the Radau II points ($\lambda_1 = 1/3$ and $\lambda_2 = 1$) as collocation parameters. In Table 3 the results show the improvement obtained with the iterated collocation method (3.1) (employing the Gauss points $\lambda_{1,2} = (3 \pm \sqrt{3})/6$ as collocation parameters in $S_1^{(-1)}(Z_N)$, confirming the (local) order $k = 4$ at the mesh points. These results are even better than the ones obtained with collocation in $S_{m-1}^{(-1)}(Z_N)$, with $m = 3$. Here we have used $\lambda_1 = 0.2$, $\lambda_2 = 0.4$ and $\lambda_3 = 0.6$ and Table 4 confirms the predicted third order of convergence.

Finally, for example 2, we have applied the iterated collocation method with $m = 2$, yielding the expected local superconvergence order $K = 4$ (see Table 5). This was superior to the (global) order $K = 3$ obtained with collocation in $S_2^{(-1)}(Z_N)$, as we can see from Table 6.

We finish this section with a comparison of three second order methods (based on the technique of fixing an initial interval), applied to example 3. For further details on these methods, we refer to Valtchev (2002).

In the Trapezium method (a multistep method denoted MS), the repeated trapezium rule is used to approximate the integral in Equation 2.2:

$$u_n^h = \frac{I_\varepsilon}{t_n^\mu} + \frac{h}{2} \sum_{j=0}^{n-1} \left[t_{j+1}^{\mu-1} u_{j+1}^h + t_j^{\mu-1} u_j^h \right] / t_n^\mu + g(t_n)$$

Table 4. Ex. 1, $\varepsilon = 0.02$, $c_0 = -2$ and $m = 3$.

		$|e(t)|$			Order
h	$t = 0.5$	$t = 1.0$	$t = 1.5$	$t = 2.0$	K
0.02	1.74E−04	2.46E−04	3.02E−04	3.48E−04	**
0.01	4.70E−05	6.64E−05	8.14E−05	9.40E−05	1.89
0.005	7.93E−06	1.12E−05	1.37E−05	1.65E−05	2.51
0.0025	1.13E−06	1.59E−06	1.96E−06	2.37E−06	2.80

Table 5. Ex. 2, $\varepsilon = 0.02$, $c_0 = -2$ and $m = 2$. Iterated Collocation on $\bar{Z}_N$.

		$|e^l(t)|$			Order
h	$t = 0.5$	$t = 1.0$	$t = 1.5$	$t = 2.0$	K
0.02	5.69E−04	6.77E−04	7.49E−04	8.05E−04	**
0.01	4.54E−05	5.39E−05	5.97E−05	6.41E−05	3.65
0.005	3.10E−06	3.68E−06	4.08E−06	4.38E−06	3.87
0.0025	2.08E−07	2.48E−07	2.74E−07	2.95E−07	3.89

Table 6. Ex. 2, $c_0 = 5$ and $m = 3$.

	$\varepsilon = 0.01$		$\varepsilon = 0.02$		$\varepsilon = 0.05$	
N	$\|e\|_\infty$	K	$\|e\|_\infty$	K	$\|e\|_\infty$	K
100	1.61E−02	**	5.59E−03	**	7.90E−04	**
200	4.82E−03	1.74	1.20E−03	2.21	1.30E−04	2.61
400	1.04E−03	2.22	2.07E−04	2.54	1.87E−05	2.79
800	1.78E−04	2.54	3.07E−05	2.75	2.54E−06	2.88

In the product Trapezium method (a product integration method denoted PI), a product integration rule is used to approximate the integral. The following algorithm is obtained:

$$u_n^h = \frac{I_\varepsilon}{t_n^\mu} + \frac{1}{t_n^\mu} \sum_{j=0}^{n-1} \left[D1_j^\varepsilon u_{j+1}^h - D2_j^\varepsilon u_j^h \right] / h + g(t_n)$$

where

$$D1_j^\varepsilon = \int_{z_j}^{z_{j+1}} (s+\varepsilon)^{\mu-1}(s-z_j)ds$$

$$D2_j^\varepsilon = \int_{z_j}^{z_{j+1}} (s+\varepsilon)^{\mu-1}(s-z_{j+1})ds$$

can be calculated analytically.

We have also considered the linear collocation method (C2), with the collocation parameters $\{0.2, 0.5\}$ and have applied the three second order methods to example 3 with $\varepsilon = 0.02$, $\mu = 0.8$, $c_0 = -10$ and $T = 2$. The results are displayed in Table 7.

The errors are of the same magnitude and the CPU time seems of order $1/h^2$. On the other hand we see that the collocation C2 is slower than the product integration, which is slower than the multistep Trapezium MS.

Table 7.   Ex. 3. Error norms and CPU time in a PIII with 400 MHz and 256 Mb RAM.

N	h	Trapezium MS		Trapezium PI		C2	
		$\|e\|_\infty$	CPU	$\|e\|_\infty$	CPU	$\|e\|_\infty$	CPU
99	0.02	7.80E−02	0.48	6.80E−02	2.09	6.76E−02	8.30
198	0.01	2.17E−02	1.81	1.86E−02	8.26	2.36E−02	32.90
396	0.005	5.60E−03	7.02	4.78E−03	32.74	7.22E−03	131.49
792	0.0025	1.41E−03	27.90	1.20E−03	130.72	2.02E−03	523.24
1584	0.00125	3.54E−04	112.07	3.01E−04	533.95	5.37E−04	2093.52

5   CONCLUDING REMARKS

This work was concerned with the solution of the singular Volterra integral Equation 1.1, which has a family of solutions in the case $0 < \mu \leq 1$. We have introduced a technique based on splitting up the integral $\int_0^t (\cdot) ds = \int_0^\varepsilon (\cdot) ds + \int_\varepsilon^t (\cdot) ds$ and evaluating $I_\varepsilon = \int_0^\varepsilon (\cdot) ds$ exactly for a chosen particular exact solution. The transformed equation was then solved by collocation in $S_{m-1}^{(-1)}(Z_N)$ and global convergence of order m, to the chosen solution, was obtained for any choice of the collocation parameters. However, using the Gauss points as collocation parameters, local superconvergence of order $2m$ was achieved at the mesh points with the iterated collocation approximation. This impressive improvement may be considered as an advantage of the collocation methods.

Here we have only considered a fixed ε. Under investigation is the case when epsilon varies with h and the question of knowing how large must epsilon be so that convergence is guaranteed. Finally we need to consider the possibility of knowing I_ε only approximately and to estimate how this approximation will affect the error in the solution.

REFERENCES

Bartoshevich, M. A. 1975. *On a heat conduction problem*, Inz. – Fiz. Z. 28, No 2, 340–346.

Brunner, H. 1980. *Superconvergence in collocation and implicit Runge-Kutta methods for Volterra-type integral equations of second kind*, in: *Numerical treatment of integral equations* (Albrecht and Collatz, eds.), Internat. Ser. Numer. Math., 53, Birkhauser Verlag, Bassel–Boston–Stuttgart, 54–72.

Brunner, H. & Nørsett, S. P. 1981. *Superconvergence of collocation methods for Volterra and Abel equations of the second kind*, Numer. Math., 36, 347–358.

Brunner, H. & van der Houwen, P. J. 1986. *The numerical solution of Volterra equations*, North Holland.

Diogo, T., McKee, S. & Tao Tang 1991. *A Hermite-type collocation method for the solution of an integral equation with a certain weakly singular kernel*, IMA Journal of Num. Analysis 11, 595–605.

Diogo, T., Lima, P. & Franco, N. B. 2000. *High order product integration methods for a Volterra integral equation with logarithmic singular kernel*, Preprint 24, Departamento de Matemática – Instituto Superior Técnico.

Han, W. 1994. *Existence, uniqueness and smoothness results for second-kind Volterra equations with weakly singular kernels*, J. Int. Eq. Appls. 6, 365–384.

Lima, P. & Diogo, T. 2002. *Numerical solution of a nonuniquely solvable Volterra integral equation using extrapolation methods*, J. Comp. Appl. Maths 140, 537–557.

Tang, T., McKee, S. & Diogo, T. 1992. *Product integration methods for an integral equation with logarithmic singular kernel*, Appl. Numer. Math. 9. 259–266.

Valtchev, S. 2002. Métodos numéricos para uma equação integral de Volterra com infinitas soluções, Diploma Thesis, Departamento de Matemática, Instituto Superior Técnico, Lisboa.

A variational approach to boundary element solution of Helmholtz problems

Y. Kagawa & N. Wakatsuki
Department of Electronics and Information Systems, Akita Prefectural University, Honjo, Japan

Y. Sun
Delco Electronics Systems, Kokomo, IN, USA

L. Chai
Department of Electrical and Electronic Engineering, Okayama University, Okayama, Japan

ABSTRACT: The boundary element method is a discretized version of the boundary integral equation method. A variational formulation is presented for the boundary element approach to Helmholtz problems. The numerical calculation of the eigenvalues in association with hollow waveguides demonstrates that the variational approach provides the upper and lower bounds of the eigenvalues. The drawback of solving the eigenvalues by trial and error is removed by the introduction of the dual reciprocity method.

1 INTRODUCTION

The finite element method was historically formulated based on the variational principle, while the boundary element method was a synonym of the boundary integral equation in association with Green's function. We presented the variational boundary element formulation for two-dimensional Laplace problems in which the convergence of the capacitance calculation was demonstrated for its upper and lower bounds (Sun & Kagawa 1994), in which a hybrid-type formulation was incorporated (Tong & Rossettos 1977). Penman et al. made the formulation for the finite elements based on the dual and complimentary energy principle (Hammond & Penman 1976) (Penman & Fraser 1982).

The singular integral difficulty associated with the boundary elements was removed by the introduction of the charge simulation method (Sun & Kagawa 1995). The regular integral approach was then discussed for Helmholtz problems (Kagawa et al. 1996).

In the present paper, the dual and complimentary energy approach is applied to a two-dimensinal Helmholtz problems in which the bound and convergence of the eigenvalues are demonstrated for some rectangular waveguides.

2 VARIATIONAL BOUNDARY ELEMENT FOMULATION

We consider a Helmholtz equation in two-dimension

$$\nabla^2 \phi + \kappa^2 \phi = 0 \qquad \text{in } \Omega \tag{1}$$

where ϕ is potential and κ is the wave number, under the boundary conditions

$$\begin{aligned} \phi &= \tilde{\phi} & \text{on } \Gamma \\ \phi &= \hat{\phi} & \text{on } \Gamma_{\mathrm{I}} \end{aligned} \tag{2}$$

$$\begin{aligned} q &= \tilde{q} & \text{on } \Gamma \\ \tilde{q} &= \hat{q} & \text{on } \Gamma_{\text{II}} \end{aligned} \tag{3}$$

where $\Gamma(= \Gamma_{\text{I}} + \Gamma_{\text{II}})$ is the boundary of a closed domain Ω.

$\tilde{\phi}$ and $\tilde{q}(= \partial q/\partial n)$ are the potential and flux defined only on the boundary Γ and n is the unit vector outward normal to the boundary.

$\hat{\phi}$ and $\hat{q}$ are the known values prescribed on the boundary Γ_{I} and Γ_{II} respectively.

We consider the following functional corresponding to equations (1),(2) and (3).

$$\Pi\left(\phi, \tilde{\phi}, \tilde{q}\right) = \frac{1}{2} \int_{\Omega} \{(\nabla \phi)^2 - \kappa^2 \phi^2\} \, d\Omega - \int_{\Gamma} \left(\phi - \tilde{\phi}\right) \tilde{q} \, d\Gamma - \int_{\Gamma_{\text{II}}} \tilde{\phi} \hat{q} \, d\Gamma \tag{4}$$

It is easy to show that the equations (1) to (3) are obtained from the stationary condition of the functional. Integrating by parts gives, as the domain integral vanishes, in the first term by the help of equation (1),

$$\Pi\left(\phi, q, \tilde{\phi}, \tilde{q}\right) = \frac{1}{2} \int_{\Gamma} \phi q - \int_{\Gamma} \left(\phi - \tilde{\phi}\right) \tilde{q} \, d\Gamma - \int_{\Gamma_{\text{II}}} \tilde{\phi} \hat{q} \, d\Gamma \tag{5}$$

The boundary Γ is divided into M elements Γ_j ($i = j, 2, \ldots, M$), and potential ϕ_i at arbitrary point i in the domain is obtained as the superposition of the fundamental solutions ϕ_{ij}^* in association with the elements

$$\phi_i = \sum_{j=1}^{M} \phi_{ij}^* \beta_j = \{\phi^*\}_i^T \{\beta\} \tag{6}$$

where

$$\{\phi\}_i^T = [\phi_{i1}^* \ \phi_{i2}^* \ \cdots \ \phi_{iM}^*] \tag{7}$$

$$\{\beta\}^T = [\beta_1 \ \beta_2 \ \cdots \ \beta_M] \tag{8}$$

ϕ_{ij}^* is Green's function of the Helmholtz equation in two-dimensional space, which is the fundamental solution when the source is placed at point j. β is the coefficient corresponding to the magnitude of the source.

The fundamental solution is

$$\phi_{ij}^* = \frac{1}{4j} H_0^{(2)}\left(\kappa r_{ij}\right) \tag{9}$$

and its normal derivative outward to the boundary is

$$\frac{\partial \phi_{ij}^*}{\partial n} = \frac{n_x X + n_y Y}{4j r_{ij}} \kappa H_1^{(2)}(\kappa r_{ij}) \tag{10}$$

where $r_{ij} = \sqrt{X^2 + Y^2}$ and $X = x_j - x_i$, $Y = y_j - y_i$. n_x and n_y are the x and y directional components of n.

For a boundary element i, potential $\hat{\phi}_i$ and flux $\hat{q}_i$ are interpolated so that

$$\tilde{\phi}_i = \{N_{\tilde{\phi}}\}^T \{\tilde{\phi}\}_i \tag{11}$$

$$\tilde{q}_i = \{N_{\tilde{q}}\}^T \{\tilde{q}\}_i \tag{12}$$

where, for the two nodes constant element,

$$\{\tilde{\phi}_i\} = \{\tilde{\phi}_{i1} \ \tilde{\phi}_{i2}\}, \quad \{\tilde{q}_i\} = \{\tilde{q}_{i1} \ \tilde{q}_{i2}\} \tag{13}$$

and $\{N_{\tilde{\phi}}\} = \{1/2 \ 1/2\}, \quad \{N_{\tilde{q}}\} = \{1/2 \ 1/2\}$ (14)

The one half is as the result of the average of the nodal potential or flex values. Substituting equation (6) into equation (5), one has

$$\Pi(\beta, \tilde{\phi}, \tilde{q}) = \sum_{i=1}^{M} \left(\frac{1}{2} \int_{\Gamma_i} \phi_i \frac{\partial \phi_i}{\partial n} \, d\Gamma_i - \int_{\Gamma_i} \phi_i \tilde{q}_i \, d\Gamma_i + \int_{\Gamma_i} \tilde{\phi}_i \tilde{q}_i \, d\Gamma_i - \int_{\Gamma_{\mathrm{II}i}} \tilde{\phi}_i \hat{q}_i \, d\Gamma_{\mathrm{II}i} \right)$$

$$= \sum_{i=1}^{M} \left(\frac{1}{2} \int_{\Gamma_i} \sum_{j=1}^{M} \phi_{ij}^* \beta_j \sum_{j=1}^{M} \frac{\partial \phi_{ij}^*}{\partial n} \beta_j \, d\Gamma_i - \int_{\Gamma_i} \sum_{j=1}^{M} \phi_{ij}^* \beta_j \tilde{q}_i \, d\Gamma_i \right.$$

$$\left. + \int_{\Gamma_i} \tilde{\phi}_i \tilde{q}_i \, d\Gamma_i - \int_{\Gamma_{\mathrm{II}i}} \tilde{\phi}_i \hat{q}_i \, d\Gamma_{\mathrm{II}i} \right) \tag{15}$$

Integration is performed for each element and the connection of the elements are made in such a way that the potential is equal at the connecting nodes, while the net flux is zero at the connecting nodes.

The functional discretized is

$$\Pi = \frac{1}{2} \{\beta\}^T [H] \{\beta\} - \{\beta\}^T [G] \{\tilde{q}\} + \{\phi\}^T [L] \{\tilde{q}\} - \{\phi\}^T [N] \{\hat{q}\} \tag{16}$$

where

$$\{\tilde{\phi}\} = \{\tilde{\phi}_1 \ \tilde{\phi}_2 \cdots \tilde{\phi}_M\} \qquad \{\tilde{q}\} = \{\tilde{q}_1 \ \tilde{q}_2 \cdots \tilde{q}_M\} \qquad \{\hat{q}\} = \{\hat{q}_1 \ \hat{q}_2 \cdots \hat{q}_N\} \ (N \leq M)$$

The components of matrices $[H]$, $[G]$, $[L]$ and $[N]$ are respectively evaluated as

$$H_{jk} = \sum_{i=1}^{M} \int_{\Gamma_i} \phi_{ij}^* \frac{\partial \phi_{ik}^*}{\partial n} \, d\Gamma_i \tag{17}$$

$$G_{jk} = \int_{\Gamma_k} \phi_{jk}^* \, d\Gamma_k \quad (j, k = 1, 2 \cdots, M) \tag{18}$$

$L_{jk} = $ element length of element $j \qquad (j = k)$
$L_{jk} = 0 \qquad\qquad\qquad\qquad\qquad (j \neq k)$ (19)

$N_{jk} = $ element length of element j on $\Gamma_{\mathrm{II}} \quad (j = k)$
$N_{jk} = 0 \qquad\qquad\qquad\qquad\qquad\qquad\quad (j \neq k)$ (20)

Taking variation for functional (16) with respect to $\{\beta\}$ and $\tilde{q}$ respectively and making stationary, one has

$$\partial \Pi_\beta = [F] \{\beta\} - [G] \{\tilde{q}\} = \{0\} \tag{21}$$

$$\delta\Pi_{\tilde{q}} = [L]\{\tilde{\phi}\} - [G]^T\{\beta\} = \{0\} \tag{22}$$

Eliminating $\{\beta\}$ from them, one has the system equation

$$[K]\{\tilde{\phi}\} - [G]\{\tilde{q}\} = \{0\} \tag{23}$$

where

$$[K] = [F][R] \qquad [R] = ([G]^{-1})^T[L] \tag{24}$$

The component of $[F]$ is

$$F_{jk} = \frac{1}{2}\sum_{i=1}^{M}\int_{\Gamma_i}\phi_{ij}^*\frac{\partial\phi_{ik}^*}{\partial n}\,d\Gamma_i + \frac{1}{2}\sum_{i=1}^{M}\int_{\Gamma_i}\phi_{ik}^*\frac{\partial\phi_{ij}^*}{\partial n}\,d\Gamma_i \tag{25}$$

3   DUAL AND COMPLEMENTARY FORMULATION

The dual and complementary energy functional to the expression (4) is

$$\Pi(\phi,\tilde{\phi},q,\tilde{q}) = -\frac{1}{2}\int_{\Omega}\{(\nabla\phi)^2 - \kappa^2\phi^2\}\,d\Omega + \int_{\Gamma}(q-\tilde{q})\tilde{\phi}\,d\Gamma + \int_{\Gamma_I}\hat{\phi}\tilde{q}\Gamma \tag{26}$$

The integration by parts leads to the boundary integral expression

$$\Pi(\phi,\tilde{\phi},q,\tilde{q}) = -\frac{1}{2}\int_{\Gamma}\phi q\,d\Gamma + \int_{\Gamma}(q-\tilde{q})\tilde{\phi}\,d\Gamma + \int_{\Gamma_I}\hat{\phi}\tilde{q}\,d\Gamma \tag{27}$$

This expression has the form in which ϕ and q are interchanged in equation (5). Now, $\hat{\phi}$ is the forcing term, while $\hat{q}$ is the forcing term in equation (5). In the similar manner, the discretized expression to this is

$$\Pi = -\frac{1}{2}\{\beta\}^T[H]\{\beta\} + \{\beta\}^T[U]\{\tilde{\phi}\} - \{\tilde{q}\}[L]\{\tilde{\phi}\} + \{\tilde{q}\}^T[M]\{\tilde{\phi}\} \tag{28}$$

where

$$U_{jk} = \int_{\Gamma_k}\frac{\partial\phi_{jk}^*}{\partial n}\,d\Gamma_k \qquad (j,k = 1,2,\cdots,M) \tag{29}$$

$$\{\hat{\phi}\} = [\hat{\phi}_1\,\hat{\phi}_2\,\cdots\,\hat{\phi}_N] \qquad (N \leq M) \tag{30}$$

$$M_{jk} = \text{length of the element on } \Gamma_I \;\; (j = k)$$
$$M_{jk} = 0 \;\; (j = k) \tag{31}$$

Varying the functional (28) with respect to $\{\beta\}$ or $\{\tilde{\phi}\}$, and making them stationary, one has

$$\delta\Pi_{\beta} = -[F]\{\beta\} + [U]\{\tilde{\phi}\} = \{0\} \tag{32}$$

$$\delta\Pi_{\tilde{\phi}} = -[L]\{\tilde{q}\} + [U]^T\{\beta\} = \{0\} \tag{33}$$

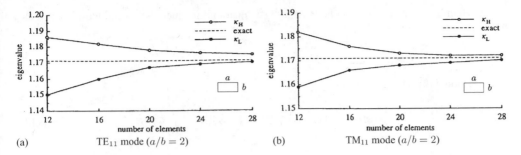

Figure 1. Convergence of eigenvalues against the number of elements.

Eliminating $\{\beta\}$, one arrives at the system equation

$$[A]\{\tilde{q}\} - [U]\{\tilde{\phi}\} = \{0\} \tag{34}$$

where

$$[A] = [F][B] \qquad [B] = ([U]^{-1})^T[L] \tag{35}$$

This is a dual and complementary expression to equation (23).

For the numerical demonstration, the eigenvalues of rectangular waveguides are considered. For the TE modes, equation (23) becomes

$$[K]\{\tilde{\phi}\} = \{0\} \tag{36}$$

The dual and complementary equation (34) becomes

$$[U]\{\tilde{\phi}\} = \{0\} \tag{37}$$

For TM modes, equation (23) becomes, as $\{\tilde{\phi}\} = \{0\}$

$$[G]\{\tilde{q}\} = \{0\} \tag{38}$$

The dual and complementary energy counterpart is

$$[A]\{\tilde{q}\} = \{0\} \tag{39}$$

As the wavenumber is implicitly contained in the system matrix, eigenvalues must be solved by the determinant search technique. The convergence of the eigenvalues against the number of elements is shown in Figure 1. The roots to the determinants in equations (36) and (38) give the lower bound κ_L, while the roots to the determinants in equations (37) and (39) give the upper bound κ_H.

4 DUAL RECIPROCITY FORMULATION

As seen in the previous section, the wavenumber is implicitly contained in the system matrix. The eigenvalues must therefore be found by the determinant search technique in complex domain.

This is very laborious and time-consuming. With the dual reciprocity method, the boundary element formulation results in the typical system matrix equation, to which standard eigenvalue solvers could be used. The dual reciprocity boundary element approach is the method with which the domain integral associated with the nonhomogeneous term is transformed into the boundary integral by way of approximate functions.

The Helmholtz equation (1) can be Poisson-type equation, in which the second term is taken as a nonhomogenous term

$$b = -\kappa^2 \phi \tag{40}$$

The functional (4) is then rewritten as

$$\Pi\left(\phi, \tilde{\phi}, \tilde{q}\right) = \frac{1}{2}\int_\Omega (\nabla\phi)^2 \, d\Omega - \int_\Gamma \left(\phi - \tilde{\phi}\right)\tilde{q}\,d\Gamma - \int_{\Gamma_{\text{II}}} \tilde{\phi}\hat{q}\,d\Gamma + \int_\Omega b\phi\,d\Omega \tag{41}$$

Integrating by parts, so that the first domain integral term in the right hand side vanishes, one has

$$\Pi(\phi, \tilde{\phi}, q, \tilde{q}) = \frac{1}{2}\int_\Gamma \phi q \, d\Gamma - \int_\Gamma \left(\phi - \tilde{\phi}\right)\tilde{q}\,d\Gamma - \int_{\Gamma_{\text{II}}} \tilde{\phi}\hat{q}\,d\Gamma + \int_\Omega b\phi\,d\Omega \tag{42}$$

The boundary is again divided into M elements and in the domain N points of interest are also allocated. The potential ϕ_i at an arbitrary point i can be expressed by the linear combination of the general solution ϕ_{ij} of the homogeneous equation, and the particular solution φ_{il}^* of the nonhomogeneous equation in such a way that

$$\phi_1 = \sum_{j=1}^M \phi_{ij}^*\beta_j + \sum_{l=1}^{M+N} \varphi_{il}\alpha_l = \{\phi^*\}_i^T\{\beta\} + \{\varphi^*\}_i^T\{\alpha\} \tag{43}$$

where

$$\{\phi^*\}_i = \{\phi_{i1}^* \, \phi_{i2}^* \, \cdots \, \phi_{iM}^*\} \qquad \{\beta\} = \{\beta_1 \, \beta_2 \, \cdots \, \beta_M\} \tag{44}$$

$$\{\varphi^*\}_i = \{\varphi_{i1}^* \, \varphi_{i2}^* \, \cdots \, \varphi_{i(M+N)}^*\} \qquad \{\alpha\} = \{\alpha_1 \, \alpha_2 \, \cdots \, \alpha_{M+N}\} \tag{45}$$

ϕ_{ij}^* is the fundamental solution of Laplace equation. When the source is placed at point j on the boundary, β_j is a corresponding source intensity which is unknown. φ_{il}^* is the particular solution of

$$\nabla^2\varphi_{il}^* = f_{il} \tag{46}$$

where f_{il} is an approximate function, which will be discussed later. When the source is placed at point l, α_l is a corresponding source intensity, which is also unknown.

The potential and flux at i in an element on the boundary interpolated as

$$\tilde{\phi}_i = \{N_{\tilde{\phi}}\}^T\{\tilde{\phi}\}_i \qquad \tilde{q}_i = \{N_{\tilde{q}}\}^T\{\tilde{q}\}_i \tag{47}$$

The interpolation functions are the same as in the previous case for constant elements used. Equations (44) to (47) are substituted into equation (43), which is then substituted into equation (42).

One has the functional in the discretized form

$$\Pi = \frac{1}{2}\{\beta\}^T[S]\{\beta\} + \{\beta\}^T[W]\{\alpha\} - \{\beta\}^T[G]\{\tilde{q}\} + \frac{1}{2}(\{\beta\}^T[H][\Phi] - [G][\Theta])\{\alpha\}$$
$$+ \{\tilde{q}\}[L]\{\phi\} + \{\tilde{q}\}[E]^T\{\alpha\} + \text{last term of the domain integral} \tag{48}$$

Taking variation to the functional with respect to $\{\beta\}$ or $\{\tilde{q}\}$ and making them stationary, one has

$$\delta\Pi_\beta = [S]\{\beta\} + [W]\{\alpha\} - [G]\{\tilde{q}\} + \frac{1}{2}([H][\Phi] - [G][\Theta])\{\alpha\} = \{0\} \tag{49}$$

206

$$\delta\Pi_{\tilde{q}} = [L]\{\tilde{\phi}\} + [G]^T\{\beta\} - [E]^T\{\alpha\} = \{0\} \tag{50}$$

Eliminating $\{\beta\}$ from equations (49) and (50), one has the expression

$$[Z]\{\tilde{\phi}\} - [G]\{\tilde{q}\} - [Y]\{\alpha\} = \{0\} \tag{51}$$

where

$$[Z] = [S]([G]^{-1})^T[L] \tag{52}$$

$$[Y] = [S]([G]^{-1})^T[E]^T - [W] - \frac{1}{2}([H][\Phi] - [G][\Theta]) \tag{53}$$

where, the cmponents of the matrices and vectors are

$$S_{jk} = \frac{1}{2}\sum_{i=1}^{M}\int_{\Gamma_i}\phi_{ij}^*\frac{\partial\phi_{ik}^*}{\partial n}\,d\Gamma_i + \frac{1}{2}\sum_{i=1}^{M}\int_{\Gamma_i}\phi_{ik}^*\frac{\partial\phi_{ij}^*}{\partial n}\,d\Gamma_i \tag{54}$$

$$H_{jk} = \int_{\Gamma_k}\frac{\partial\phi_{jk}^*}{\partial n}\,d\Gamma_k \qquad G_{jk} = \int_{\Gamma_k}\phi_{jk}^*\,d\Gamma_k \qquad E_{lk} = \int_{\Gamma_k}\varphi_{lk}^*\,d\Gamma_k \tag{55}$$

$$W_{jl} = \frac{1}{2}\sum_{i=1}^{M}\int_{\Gamma_i}\phi_{ij}^*\frac{\partial\varphi_{il}^*}{\partial n}\,d\Gamma_i + \frac{1}{2}\sum_{i=1}^{M}\int_{\Gamma_i}\varphi_{il}^*\frac{\partial\phi_{ij}^*}{\partial n}\,d\Gamma_i \tag{56}$$

$$\Phi_{jl} = \varphi_{jl}^* \qquad \Theta_{jl} = \frac{\partial\varphi_{jl}^*}{\partial n} \tag{57}$$

This is the discretized system equation in which $\{\alpha\}$ is to be determined in consideration of the approximate function properly chosen.

5   DISCRETIZED SYSTEM EQUATION OF STANDARD FORM

The nonhomogeneous term is now expanded in terms of the approximate functions as

$$b_i = \sum_{l=1}^{M+N}f_{il}\alpha_l = \sum_{l=1}^{M+N}(\nabla^2\varphi_{il}^*)\alpha_l \tag{58}$$

Two kinds of the approximate functions are chosen to be considered (Partridge & Brebbia 1992),

$$f_{il} = 1 + r_{il}^2 \tag{59}$$

$$f_{il} = 1 + r_{il}^2(\ln r_{il} - 1) \tag{60}$$

Equation (58) in the vectorial form is

$$\{b\} = [F]\{\alpha\} \quad \text{or} \quad \{\alpha\} = [F]^{-1}\{b\} \tag{61}$$

207

where

$$\{b\} = [b_1 \ b_2, \ \cdots \ b_{M+N}]$$ (62)

The component of $[F]$ is

$$F_{il} = f_{il}$$ (63)

Substituting equation (61) into equation (51), one has the system equation

$$[Z]\{\tilde{\phi}\} - [G]\{\tilde{q}\} - [Y][F]^{-1}\{\phi\} = \{0\}$$ (64)

For the TE modes, the boundary condition is $\{\tilde{q}\} = \{0\}$ so that equation (64) becomes

$$([Z] - \kappa^2[Y][F]^{-1})\{\phi\} = \{0\}$$ (65)

This is the expression of standard form for eigenvalue problems. For the TM modes, the boundary condition is $\{\tilde{\phi}\} = \{0\}$. Knowing $\nabla^2\phi = -\kappa^2\phi$, and from equations (58), one has the relation

$$-\kappa^2\phi_i = \sum_{l=1}^{M+N} f_{il}\alpha_l$$

Taking derivative with respect to n,

$$-\kappa^2 q_i = \sum_{l=1}^{M+N} \frac{\partial f_{il}}{\partial n}\alpha_l \ \text{ or } \ -\kappa^2\{q\} = [Q]\{\alpha\}$$ (66)

where the component of $[Q]$ is

$$Q_{il} = \frac{\partial f_{il}}{\partial n}$$ (67)

that is $\quad \{\alpha\} = -\kappa^2[Q]^{-1}\{q\}$ (68)

With this, the system equation (64) becomes

$$[G]\{\tilde{q}\} - \kappa^2[Y][Q]^{-1}\{\tilde{q}\} = \{0\}$$ (69)

This is another system equation of standard form.

6 DUAL AND COMPLEMENTARY FORMULATION

The dual and complementary expression to functional (42) is

$$\Pi(\phi, \tilde{\phi}, q, \tilde{q}) = -\frac{1}{2}\int_\Gamma \phi q \, d\Gamma + \int_\Gamma (q - \hat{q})\phi \, d\Gamma + \int_{\Gamma_I} \hat{\phi}\tilde{q} \, d\Gamma + \frac{1}{2}\int_\Omega b\phi \, d\Omega$$ (70)

The boundary is divided into M elements and equations (43) to (47) are substituted, and one has the similar discretized expression

$$\Pi = \{\beta\}^T[S]\{\beta\} + \{\beta\}^T[W]\{\alpha\} - \{\beta\}^T[H]\{\tilde{\phi}\} - \frac{1}{2}\{\beta\}^T([H][\Phi]$$
$$- [G][\Theta])\{\alpha\} + \{\tilde{\phi}\}^T[L]\{\tilde{q}\} - \{\tilde{\phi}\}^T[U]\{\alpha\} + \text{last term of the domain integral}$$ (71)

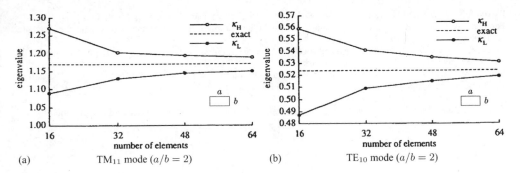

| (a) | TM$_{11}$ mode ($a/b = 2$) | (b) | TE$_{10}$ mode ($a/b = 2$) |

Figure 2.   Convergence of eigenvalues against the number of elements.

The functional is varied with respect to $\{\beta\}$ or $\{\tilde{\phi}\}$, which are then made stationary,

$$\delta\Pi_\beta = [S]\{\beta\} + [W]\{\alpha\} - [H]\{\tilde{\phi}\} - \frac{1}{2}([H][\Phi] - [G][\Theta])\{\alpha\} = \{0\} \tag{72}$$

$$\partial\Pi_{\tilde{\phi}} = [L]\{\tilde{q}\} - [H]^T\{\beta\} - [U]\{\alpha\} = \{0\} \tag{73}$$

where

$$U_{jk} = \int_{\Gamma_k} \frac{\partial\phi_{jk}^*}{\partial n} \, d\Gamma_k \quad (j, k = 1, 2, \cdots, M) \tag{74}$$

Eliminating $\{\beta\}$ from equations (72) and (73), one has the equation

$$[A]\{\tilde{q}\} - [H]\{\tilde{\phi}\} - [P]\{\alpha\} = \{0\} \tag{75}$$

where

$$[A] = [S]([H]^{-1})^T[L] \tag{76}$$

$$[P] = [S]([H]^{-1})^T[U]^T - [W] + \frac{1}{2}([H][\Phi] - [G][\Theta]) \tag{77}$$

The same rectangular waveguides as in the previous examples are considered. For the eigenvalue and eigenvector calculation, the standard solver commercially available is used. The convergence of the eigenvalues with the number of elements is shown in Figure 2.

7   CONCLUDING REMARKS

The variational boundary element modelling was examined for the eigenvalues of the Helmholtz problems. The dual and complimentary energy approach was demonstrated to show the upper and lower bound of the eigenvalues. This enables the efficient calculation ensuring the solutions' convergence.

REFERENCES

Hammond P. & Penman J. 1976. Calculation of Inductance and Capacitance by means of Dual Energy Principles, Proc. IEE, 123(6): 554–559.
Kagawa Y., Sun Y. & Mohmood Z. 1996. Regular Boundary Integral Formulation for the Analysis of Open Dielectric/Optical Waveguides, IEEE Trans. MTT, 44(8): 1441–1450.

Partridge D. W. & Brebbia C. A. 1992. *The Dual Reciprocity Boundary Method*, Comp. Mech. Pub.

Penman J. & Fraser J. R. 1982. Complimentary and Dual Energy Finite Element Principles in Magnetostatics, IEEE Trans. MAG-18(2): 319–324.

Sun Y. & Kagawa Y. 1994. A Variational Approach to Boundary Elements – Two dimensional Laplace Problems, Int. J. Numerical Modelling, Electronic Networks, Devices and Fields, 7(1): 1–14.

Sun Y. & Kagawa Y. 1995. Regular Boundary Integral Solution with Dual and Complementary Variational Formulations Applied to Two-dimensional Laplace Problems, Int. J. Numerical Modelling, Electronic Networks, Devices and Fields, 8(1): 127–137.

Tong P. & Rossettos J. N. 1977. *Finite elements Method*, MIT Press, Cambridge, Massachusetts and London.

Computational Methods in Engineering and Science, Iu et al. (eds)
© 2003 Swets & Zeitlinger, Lisse, ISBN 90 5809 567 3

Computational methods for a singular boundary-value problem

P.M. Lima
Centro de Matemática e Aplicações, Instituto Superior Técnico, Lisboa, Portugal

M.P. Pato
Instituto Superior de Engenharia de Lisboa, Departamento de Eng. Química, Lisboa, Portugal

ABSTRACT: In this paper we apply the finite differences and the shooting methods to the numerical solution of singular boundary value problems which describe the deformation of a membrane cap.
The considered differential equation may be written in the form

$$r^2 S_r'' + 3r S_r' = \frac{\lambda^2}{2} r^{2\gamma - 2} + \frac{\beta v r^2}{S_r} - \frac{r^2}{8 S_r^2}, \qquad r \in (0,1] \tag{1}$$

where S_r, r and v denote a the radial stress, the membrane radius and the Poisson coefficient, respectively, γ, λ and β are known positive constants.
We shall look for a positive solution S_r which satisfies the following boundary conditions

$$S_r \text{ is bounded at } 0 \quad \text{and} \quad S_r'(1) + (1-v)S_r(1) = \Gamma \quad \text{or} \quad S_r(1) = S, \tag{2}$$

where Γ and S are real numbers.
By using well-known iterative methods, like the Picard and Newton's method, the nonlinear problem is reduced to a sequence of linear ones, which are then discretized by the finite differences method.
The numerical results obtained by the finites difference method are compared with the ones presented in other papers and the efficiency of different numerical methods is discussed.

1 INTRODUCTION

In this work we apply efficient numerical methods to the solution of nonlinear boundary value problems for shallow membrane caps. Consider a membrane cap in the undeformed state.

Assuming that the strains and the vertical pressure are small, a rotationally symmetric deformation will be described by the exact equation

$$\frac{d}{dr}\left(r\sqrt{2\varepsilon_\theta + 1}\right) = \frac{mS\sqrt{2\varepsilon_r + 1}}{\Sigma_r} \tag{3}$$

where ε_θ and ε_r are, respectively, the radial and circumferential strains (Dickey 1987). Moreover, we have $m^2 = 1 + (z')^2$, $\Sigma_r = \sigma_r \sqrt{2\varepsilon_r + 1}$ and $\Sigma_\theta = \sigma_\theta \sqrt{2\varepsilon_\theta + 1}$, where σ_r and σ_θ denote, respectively, the radial and circumferential stresses of the membrane.

Assuming that the strains are small, the membrane is shallow and the vertical pressure is low, Dickey has shown, using the Hook's law, that an approximated equation may be derived from (3) (Dickey 1987).

More precisely, Dickey has proved that, under the mentioned conditions, the values of $(v, \varepsilon_\theta, \varepsilon_r, z',$ are small. By expanding both sides of Equation 3 in Taylor series and ignoring powers of v, with exponent equal or higher than 3 , and powers of ε_θ, ε_r, z', with exponent equal or higher than 2, he obtained the following equation:

$$r^2\frac{d^2}{dr^2}\sigma_r+3r\frac{d}{dr}\sigma_r=\frac{E}{2}\left(\frac{d}{dr}z\right)^2+vE^2\frac{d}{dr}(rG)\frac{G}{\sigma_r}-\frac{E^3}{2}\left(\frac{G}{\sigma_r}\right)^2 \tag{4}$$

where E is the elasticity modulus and $G = F/E$, with F defined by $rF = (1/h)\int_0^r \rho m P(\rho)d\rho$, where P is the pressure.

Two different kinds of boundary value problems may be considered: 1) the stress problem, when the radial stress at the boundary is known ($\sigma_r(a) = \sigma$) and 2) the displacement problem, when the radial displacement is given at the boundary ($u(a) \equiv (a/E)(a\sigma'_r(a) + (1 - v)\sigma_r(a)) - \mu$).

For the sake of simplicity, we may assume without loss of generality that P is constant ($P(r) \equiv (P)$) and the undeformed shape of the surface is given by $z(r) = C[1 - [(r/a)]^\gamma]$, where a and C are, respectively, the membrane radius and the height from the center to the surface, with $\gamma > 1$. Suppose that $a = 1$ and let us introduce the following notations:

$$S_r(a)=\frac{\sigma_r(r)}{K};\quad K=\left(\frac{EP^2}{h^2}\right)^{1/3};\quad \lambda=C\gamma\left(\frac{P}{Eh}\right)^{-1/3};\quad \beta=\frac{1}{2}\left(\frac{P}{Eh}\right)^{2/3}.$$

Then (4) takes the form

$$r^2S_r''+3rS_r'=\frac{\lambda^2}{2}r^{2\gamma-2}+\frac{\beta vr^2}{S_r}-\frac{r^2}{8S_r^2},\qquad r\in(0,1]. \tag{5}$$

If S_r is a solution of Equation 5 then the final form of the membrane may be obtained from the equations:

$$u(r)=\frac{Kr}{E}\left[rS_r'(r)+(1-v)S_r(r)\right]\quad w(r)=\frac{P}{2EhK}\int_r^1\frac{t}{S_r(t)}dt,$$

where u and w are the radial and the vertical displacement of the membrane, v is the Poisson coefficient (Ratio between the tangential and the radial deformation in an elastic domain.).

Since the boundary of the membrane corresponds to $r = 1$, the boundary condition for the stress problem is $S_r(1) = S \equiv (\sigma/K)$ ($S > 0$) and the boundary condition for the displacement problem is $S_r'(1) + (1 - v)S_r(1) = \Gamma \equiv (E\mu/K)$, $\Gamma \in R$.

In this paper, we shall consider three different cases:

Problem A: Analysis of large deformations of a spherical membrane (this is a particular case of problem B, which arises when $\gamma = 2$ and $\beta = 0$ in Equation 5).
Problem B: Analysis of all kinds of deformations of a spherical membrane, with $\gamma = 2$ and $\beta > 0$ (this is a particular case of problem C, which arises when $\gamma = 2$ in Equation 5).
Problem C: Analysis of deformations of a membrane with the shape given by any function of the form $z(r) = C[1 - [(r/a)]^\gamma]$, with $\gamma > 1$.

Since λ depends on P as $P^{-1/3}$ and β depends on P as $P^{2/3}$, we may conclude that when the pressure is low λ is large and β is small; therefore, the second term of the right-hand side of (5) may be ignored and we obtain problem A (see Baxley 1988).

For $\gamma = 2$ (problem B), Baxley & Gu (1999) have proved that the limit $\lim_{r\to 0^+} S_r(r)$ is positive, as well as $\lim_{r\to 0^+} r^{-1}S'_r(r)$.

In the case of the stress problem, with $\gamma = 2$, there exists an unique positive solution, if $S \le (1/4\beta v)$; and there is at least a positive solution, if $S > (1/4\beta v)$. For the displacement problem, if $(\Gamma/1 - v) \le (1/4\beta v)$ the problem has an unique positive bounded solution; this condition is equivalent to $4\Gamma\beta v \le 1 - v$, which is satisfied for sufficiently small v. If $(\Gamma/1 - v) > (1/4\beta v)$, there may exist multiple solutions.

When $\gamma \uparrow 2$ (problem C), Baxley & Robinson (1998) have shown that for certain values of Γ and v the solution is not monotone. The same authors have obtained sufficient conditions for existence and unicity of solution, as well as qualitative information on the solution. The problem has been divided into three different cases, depending on the value of $\gamma : \gamma > 2, 4/3 < \gamma < 2, 1 < \gamma < 4/3$.

In order to obtain a numerical solution, the variable substitutions $x = r^2$, $u(x) = S_r(r)x$ are introduced. Then the differential Equation 5 reduces to

$$u''(x) = \frac{\lambda^2 x^{\gamma-2}}{8} - \frac{x^2}{32u^2} + \frac{\beta vx}{4u} \tag{6}$$

and the boundary conditions may be written in the form

$$u(0)=0, \tag{7}$$

$$a_0 u(1) - a_1 u'(1) = A. \tag{8}$$

In the case of the stress problem, we have $a_0 = 1$, $a_1 = 0$ and $A > 0$; for the displacement problem, $a_0 = 1 - v$, $a_1 = 2$ and A may be any real number.

2 COMPUTATIONAL METHODS

Numerical results for the considered problems were obtained by the shooting method in Baxley (1988), Baxley & Gu (1999) and Baxley & Robinson (1998).

In the present work, numerical approximations for the same problems will be obtained not only by the shooting method, but also using iterative methods and finite differences schemes. The shooting method is often used to reduce a boundary value problem to an initial value one. In the case of the problem (6)–(8), this cannot be done directly, since the equation is singular at $x = 0$. Hence the initial conditions are given at a certain point x_1, close to 0. In order to do this, we use the asymptotic behavior of the solution near the origin. For example, in the case $\gamma = 2$, it has been shown in Baxley (1988) that if u is a solution of (6)–(7), then $u(x_1) \approx Lx_1 + (M/2) x_1^2, (x \to 0)$, where $M = -(1/32L^2) + (\lambda^2/8)$, L being a certain real constant. Therefore, the initial conditions for u may be written as $u(x_1) = Lx_1 + (1/2)[-(1/32L^2) + (\lambda^2/8)]x_1^2$ and $u'(x_1) = L + [-(1/32L^2) + (\lambda^2/8)]x_1$, where L is the shooting parameter. Then the boundary value problem reduces to finding the value of L, for which u satisfies the boundary condition (8).

The finite differences method offers alternative algorithms for the numerical approximation of problem (6)–(7)–(8). In this case, it is convenient to reduce the nonlinear problem to a sequence of linear ones and to apply a finite difference scheme to each linear problem.

Let us consider a second order nonlinear differential equation in the general form $Lu = f(x,u)$, where L is a linear differential operator. Suppose that we are looking for a solution of this equation, which satisfies the boundary conditions

$$\alpha_1 u(a) + \beta_1 u'(a) = \Gamma_1, \alpha_2 u(b) + \beta_2 u'(b) = \Gamma_2, \tag{9}$$

where $\alpha_i, \beta_i, \Gamma_i$ are given real numbers, $i - 1, 2$. It is well-known that a nonlinear problem of this form may be reduced to a sequence of linear ones by the Picard or the Newton method. In the case of the Picard method, a sequence of iterates $\{u^{(n)}(x)\}$, $n \geqslant 0$ is constructed by solving the linear boundary value problems

$$Lu^{(n+1)}(x) = f(x, u^{(n)}(x)), x \in (a, b) \tag{10}$$

$$\alpha_1 u^{(n+1)}(a) + \beta_1 u^{(n+1)\prime}(a) = \Gamma_1, \alpha_2 u^{(n+1)}(b) + \beta_2 u^{(n+1)\prime}(b) = \Gamma_2, \tag{11}$$

$n = 0, 1, \ldots$, where $u^{(0)}(x)$ is a certain function, defined on $[a, b]$. If we use the Newton method, the sequence of linear boundary value problems to be solved has the form

$$Lu^{(n+1)}(x) = f(x, u^{(n)}(x)) + \frac{\partial f}{\partial u}(x, u^{(n)})\Big|_{u=u^{(n)}} (u^{(n+1)} - u^{(n)}), x \in (a, b) \tag{12}$$

$$\alpha_1 u^{(n+1)}(a) + \beta_1 u'^{(n+1)}(a) = \Gamma_1, \alpha_2 u^{(n+1)}(b) + \beta_2 u^{(n+1)'}(b) = \Gamma_2. \tag{13}$$

In general, the Newton method converges faster than the Picard method, since its convergence is quadratic. These iterative methods were analysed in Mooney (1979), where sufficient conditions were impose on f, such that the iterative scheme converges monotonically to the solution of the nonlinear problem. In order to apply the iterative methods, proposed in the cited works, it is essential to know a suitable initial approximation, which is usually a subsolution or a supersolution of the considered problem.

A function $u(x)$ is said to be a *subsolution* of the considered nonlinear problem if it satisfies the inequalities $Lu(x) - f(x, u(x)) \leq 0, x \in (a, b), \alpha_1 u(a) + \beta_1 u'(a) \leq \Gamma_1, \alpha_2 u(b) + \beta_2 u'(b) \leq \Gamma_2$.

If we reverse the signs of these inequalities, we shall obtain the definition of a supersolution. According to the results of Mooney (1979), if we start the Picard method or the Newton method with a subsolution, the method will converge monotonically upwards to the exact solution. If we start the Picard method with a supersolution, it will converge downwards to the true solution. In the case of problem (6)–(7)–(8), these results may be applied if and $\beta = 0$, $\gamma = 2$ (problem A). For other values of β and γ the convergence of the iterative methods was not proved theoretically, but the numerical experiments have shown that the methods are applicable if β is not too large (see section 4).

For the particular case of the problem (6)–(7)–(8), we have used subsolutions of the form, suggested by the authors in Baxley (1988), Baxley & Gu (1999), and Baxley & Robinson (1998). The form of these subsolutions is shown in Table 1, where a and C are adjustable parameters, and the constants, satisfy $1 \leq \delta \leq 1 + 2/(1 - \nu)$, $r = 3/2\gamma - 9/8$, $q = 3/2\gamma - 15/8$, respectively.

Once a subsolution of the considered problem is found, the iterative schemes of the Picard and Newton methods may be written for the boundary value problem (6)–(7)–(8). In the case of the stress problem, the equations of the Newton method (see (12)–(13)) will have the following form:

$$\begin{cases} u^{(n+1)''}(x) + \left(\dfrac{\beta \nu x}{4u^{(n)}(x)^2} - \dfrac{x^2}{16u^{(n)}(x)^3} \right) u^{(n+1)}(x) = \dfrac{\lambda^2 x^{\gamma-2}}{8} \\[2mm] -\dfrac{x^2}{32u^{(n)}(x)^2} + \dfrac{\beta \nu x}{4u^{(n)}(x)} + \left(\dfrac{\beta \nu x}{4u^{(n)}(x)^2} - \dfrac{x^2}{16u^{(n)}(x)^3} \right) u^{(n)}(x), \\[2mm] u^{(n+1)}(0) = 0, \quad u^{(n+1)}(1) = S \end{cases} \tag{14}$$

If we choose the Picard method to solve the problem (6)–(8), then, in order to guarantee the convergence, we must first transform the nonlinear equation, by adding to each side of Equation 6 the term

$$\left(\frac{\beta \nu x}{4u^{(0)}(x)^2} - \frac{x^2}{16u^{(0)}(x)^3} \right) u(x),$$

Table 1. Subsolutions of the boundary value problems.

Problems A and B $\gamma = 2$	Problem C	
	$4/3 < \gamma < 2$	$\gamma > 2$
$u^{(0)}(x) = Cx(\delta - x)$	$u^{(0)}(x) = (1 - a)^{r-q} x^{1-\gamma/2}$	$u^{(0)}(x) = Cx(\delta - x)$

214

where $u^{(0)}(x)$ is a subsolution. Then the equations of the iterative scheme (10)–(11) take the form:

$$\begin{cases} u^{(n+1)''}(x) + \left(\dfrac{\beta v x}{4u^{(0)}(x)^2} - \dfrac{x^2}{16u^{(0)}(x)^3} \right) u^{(n+1)}(x) = \dfrac{\lambda^2 x^{\gamma-2}}{8} \\[2mm] - \dfrac{x^2}{32u^{(n)}(x)^2} + \dfrac{\beta v x}{4u^{(n)}(x)} + \left(\dfrac{\beta v x}{4u^{(0)}(x)^2} - \dfrac{x^2}{16u^{(0)}(x)^3} \right) u^{(n)}(x), \\[2mm] u^{(n+1)}(0) = 0, \quad u^{(n+1)}(1) = S \end{cases} \tag{15}$$

The iterative schemes for the case of the displacement problem are obtained just by changing the boundary conditions.

Given a linear boundary value problem of the form (14) or (15), a finite difference method is used to discretize the equation. By applying a three-point scheme with central differences in an uniform grid, we reduce the problem to a system of linear equations of the form AU = R, where

$$A = \begin{bmatrix} b_{1,1} & c_{1,2} & & & & \\ a_{2,1} & b_{2,2} & c_{2,3} & & \text{\Large 0} & \\ & a_{3,2} & b_{3,3} & c_{3,4} & & \\ & & \ddots & \ddots & \ddots & \\ \text{\Large 0} & & & \ddots & \ddots & \\ & & & & \ddots & \ddots & c_{N-2,N-1} \\ & & & & & a_{N-1,N-2} & b_{N-1,N-1} \end{bmatrix}.$$

In the case of the Newton method, we have:

$$a_{i+1,i} = 1/h^2, \qquad\qquad i=1, ..., N-2$$

$$b_{i,i} = \frac{\beta v x_i}{4u_i^{(n)2}} - \frac{x_i^2}{16u_i^{(n)3}} - \frac{2}{h^2}, \qquad i=1, ..., N-1$$

$$c_{i,i+1} = 1/h^2, \qquad\qquad i=1, ..., N-2$$

$$R = [r_1, r_2, ..., r_{N-1}, r_N - S/h^2]^T,$$

where $r_i = \dfrac{\lambda^2 x_i^{\gamma-2}}{8} - \dfrac{3x_i^2}{32u_i^{(n)2}} - \dfrac{\beta v x_i}{2u_i^{(n)}}, \qquad i=1, ..., N.$

According to our notations, the approximate solution of the $(n + 1)$-th linear boundary value problem is given by the vector

$$U^{(n+1)} = [u_1^{(n+1)}, u_2^{(n+1)}, ..., u_{N-1}^{(n+1)}, u_N^{(n+1)}]^T,$$

where $u_i^{(n+1)}$ is the approximate value of $u^{(n+1)}(x_i)$, x_i being the knots of a uniform grid of stepsize h ($x_i = ih$).

The schemes (14) and (15) are iterated until the approximate solution satisfies the condition

$$\left\| U^{(n+1)} - U^{(n)} \right\| \le \varepsilon, \tag{16}$$

where the Euclidean norm is considered and ε is the needed precision (in our computations we have used $\varepsilon = 0.5 \cdot 10^{-10}$).

Since the discretization error allows an expansion in powers of the stepsize, the accuracy of the numerical results may be further improved by means of the Richardson extrapolation (see, for example, Brezinski & Zaglia 1991).

3 NUMERICAL RESULTS

In this section, we shall present numerical results for some examples of problems B and C. We have considered some different sets of values of the equation parameters. Since the numerical results obtained by the Picard method do not differ significantly from those obtained by the Newton method, we present only the last ones.

In Table 2, we display some numerical results for problem B with displacement boundary conditions, in the cases of $\lambda = 0.233$ and $\lambda = 0.991$, $\beta = 1.0$, $\nu = 0.3$, $\Gamma = 0.2$. The results obtained by using Richardson extrapolation with two grids are also shown.

The graphic of the numerical solution corresponding to this case is displayed in Figure 1, where the corresponding initial approximation and the first Newton iterate are also shown.

Table 2. Approximate values of $u(1)$ with different stepsizes.

Stepsize	$\lambda = 0.233$	Richardson extrapolation	$\lambda = 0.9991$	Richardson extrapolation
1/20	0.3397907604		0.2847918027	
1/40	0.3397839105	0.3397816272	0.2847919571	0.2847920085
1/80	0.3397821975	0.3397816266	0.2847919957	0.2847920086
1/160	0.3397817693	0.3397816265	0.2847920053	0.2847920086
1/320	0.3397816622	0.3397816265	0.2847920078	0.2847920086

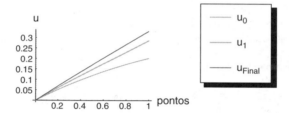

Figure 1. Graphics of the Newton iterates in the case of problem B with $\lambda = 0.233$, $\beta = 1.0$, $\nu = 0.3$, $\Gamma = 0.2$.

Table 3. Numerical approximation of $u(0.5)$ in the case of $\lambda = 0.233$, $\beta = 1.0$, $\nu = 0.3$, $\Gamma = 0.2$.

Stepsize	Approximation by the finite difference method	Richardson extrapolation with 2 grids	Richardson extrapolation with 3 grids
1/20	0.1743120026		
1/40	0.1743092897	0.1743083851	
1/80	0.1743086113	0.1743083852	0.1743083851
1/160	0.1743084417	0.1743083851	0.1743083851
1/320	0.1743083993	0.1743083851	0.1743083851

Table 4. Number of iterations and computing time for different methods and different values of γ ($\lambda = 1.0$, $\beta = 1.0$, $\nu = 0.3$, $S = 0.38$, $h = 1/80$).

γ	Number of iterations			Computing time		
	Shooting	Picard	Newton	Shooting	Picard	Newton
1.6	7	33	5	0.03	0.972	0.3
4	6	27	5	0.03	0.761	0.28

Table 3 illustrates the improvement of the accuracy obtained by the Richardson extrapolation, when 2 or 3 different grids are used.

The convergence speed of the iterative methods is compared in Table 4. This table refers to the numerical solution of problem C with stress boundary conditions and different values of γ. The computations were performed in a PC with a Pentium III processor (665 MHz). The algorithms were implemented in the language *Mathematica*.

4 CONCLUSIONS

The finite differences method, combined with the Picard and Newton iterative methods, enabled us to reduce the original problem to the solution of a sequence of linear systems of N equations, where $N + 1$ is the number of gridpoints of the considered mesh.

The analysis of the numerical results has shown that the finite difference method has second order convergence, as it should be expected. Moreover, the Richardson extrapolation, using up to 3 different grids enabled us to improve the accuracy and obtain results with 10 significant digits.

Concerning the iterative methods, as it may be seen in Table 3, the fastest convergence is obtained with the Newton method, for all the considered cases.

The shooting method is also very efficient, but its application requires that the initial value of the shooting parameter is sufficiently close to the exact one, and this condition may be not easy to satisfy.

REFERENCES

Baxley, J.V. 1988. A singular nonlinear boundary value problem: membrane response of a spherical cap, Siam J. Appl. Math., 48, 497–505.
Baxley, J.V. & Gu, Y. 1999. Nonlinear boundary value problems for shallow membrane caps", Comm. Appl. Anal., 3, 327–344.
Baxley, J.V. & Robinson 1998. Stephen B. Nonlinear boundary value problems for shallow membrane caps, II, Siam J. Appl. Math., 88, 203–224.
Brezinski, C. & Zaglia M. 1991. Extrapolation methods, theory and practice, Elsevier Science Publishers B.V.
Dickey, R.W.1987. Membrane caps, Quart. of Appl. Math., 45, 697–712.
Mooney, J.W.1978. Monotone methods for the Thomas-Fermi equation, Quart. of Appl. Math., 36, 305–314.
Mooney, J.W. 1979. A unified approach to the solution of certain classes of nonlinear boundary value problems using monotone iterations, nonlinear analysis, theory, methods & applications, Pergamon Press Ltd., Vol. 3, No. 4 , 449–465.

Computational Methods in Engineering and Science, Iu et al. (eds)
© 2003 Swets & Zeitlinger, Lisse, ISBN 90 5809 567 3

A linear complexity algorithm to solve large sparse linear systems

J.L. Almará
Universidad Nacional de Entre Ríos, Argentina

ABSTRACT: Many physical systems can be modeled using linear relations, and some direct methods to solve linear algebraic equation systems have an $O(n^3)$ complexity. Optimized versions of faster solutions for sparse symmetric banded systems handle the data through the skyline schema. This paper comments a way to organize the data and the algorithm structure to solve sparse diagonal dominant linear systems in linear time. The advantage of the present proposal is that the values distinct from zero do not have to be in a band close to the main diagonal. The solution is fast enough to solve a system with 25 non-zero entries per row, 64,000 equations, in 21.3 seconds, with a precision of 10^{-12}.

1 INTRODUCTION

The numerical solution of partial differential equations and integral equations in modeling various physical phenomena often requires the solution of large sparse or structured systems of linear algebraic equations. The direct solution of these linear systems using Gaussian elimination (Nakamura 1993) is prohibitively expensive in terms of both time and storage. Iterative methods are so-called Krylov projection type methods and they include popular methods as Conjugate Gradients, Bi-Conjugate Gradients, LSQR and GMRES (Saad & Schultz 1986). They have better performance, though not linear complexity.

This paper comments an algorithm and the supporting data structure to obtain solutions in linear time proportional to the size of the sparse system, regardless of its symmetry.

2 DATA STRUCTURES

The algorithms detailed in 3 are designed to operate on a specific data structure, based on dynamic memory allocation and trees.

2.1 *Sparse matrix*

A matrix with a large number of null elements is called a sparse matrix (Buzzi-Ferraris 1994). That is, the information is in a small percentage of non-zero elements. Some applications generate banded sparse matrices, having the non-zero elements grouped near the main diagonal. If the elements of the sparse matrix are arranged as a rectangular block, the storage of null elements is, usually, an inefficient use of resources.

The proposed data structure uses a column vector to store the information of each row. Each element of the column vector has several attributes: number of the row, an active indicator, several past values of the unknown (assuming a square matrix), an instance of the AVL tree class, and some other.

There is a row vector of doubles for the unknowns and a column vector of doubles for the independent terms.

2.2 *AVL trees*

Binary trees grown from random data may be pretty well balanced. But matrix data is frequently generated with some kind of order, and sorted insertions in regular binary trees result in degenerate structures, affecting access speed.

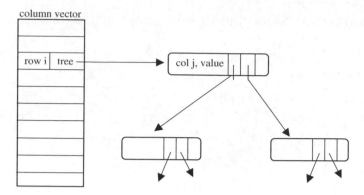

column vector

row i | tree

col j, value

Figure 1.   Data structure schema.

An AVL tree (Adel'son-Vel'Skiï & Landis 1962) is a binary tree in which the heights of the left and right subtrees differ by at most 1, and the right and left subtrees are also AVL trees.

It can be shown (Parsons 1994) that the average search time for an AVL tree with n nodes is proportional to 1.4 log n. The height of an AVL tree, at worst, is only about 44% higher than a perfect binary tree, and its worst search time is comparable to the average search time of a random binary tree (Kruse 1994).

However, the logic to keep the tree balanced after insertions and deletions is not straightforward, and the algorithm requires that the nodes of the tree include an additional field. But once it is programmed and tested (Wirth 1986), works as well as any other routine.

In this project, the sparse matrix of the coefficients of the linear system to solve is stored as a column vector of AVL tree rows, as seen in Figure 1.

3   NUMERICAL ALGORITHMS

To solve linear systems, there are two classical approaches.

The direct solution processes each element, row by row, column by column, applying linear transformations, to convert the system in an equivalent, more tractable form. Examples of direct methods are Gauss, LU, Cholesky, and others (Atkinson & Harley 1983).

Iterative solvers start with an initial approximation, and progress to the final answer, enhancing the values in each step.

For this project, the Gauss-Seidel algorithm (Barroso, L. et al. 1987) was selected, because it can properly manage sparseness and the underlying data structure.

The idea is to evaluate each unknown as a function of the others (and coefficients and constants), using the latest result for each. It can be shown (Barroso, L. et al. 1987) that, to assure convergence, there must exist diagonal dominance: the absolute value of the main diagonal element must be greater than the sum of the absolute values of the other row elements.

4   IMPLEMENTATION

This project was developed using Objet Oriented Programming (Cantù, 2001), having the advantages of isolation, inheritance and polymorphism. The classes and routines are written in Delphi(TM) (Inpris 1999) (about 2000 lines), a Pascal dialect, because the AVL tree procedures and others were readily available and tested. As it is a modern language, translation to C++ or Java should present no problems.

The linear system is implemented as a compound class: a class that contains instances of classes. The column vector of tree instances is allocated in the constructor, and the information of each row of the sparse matrix is stored in an AVL tree, that is created and initialized. Conversely, in the destructor, each row tree is emptied, and the vector of tree instances is disposed. There are some other private and public attributes, to get the linear system solved.

Each row of the sparse matrix is stored in an instance of an AVL tree, with its own constructor, destructor, and access methods.

4.1 Element access

The elements of the sparse matrix are read and written using methods of the main class, which then calls methods of the tree class.

To read the content of element at position (i, j), the position is checked to be inside the matrix and, using the i AVL tree instance from the column vector of trees, the j element of the tree, same as the j column of the matrix, is searched. If the element is found to exist, its content is retrieved; if the element is not found, 0.0 is assumed for the returned value.

There is a method to write or store a value at position (i, j), for completness. The position is verified to be inside the matrix, and using the i AVL tree instance from the column vector of trees, the j element of the tree is searched. If it is present, its value is modified if the value is greater than epsilon, or deleted otherwise. If it is not present, and the value is significative, is added.

From the external interface point of view, there are two methods to access the elements: GetValue(i, j) and PutValue(i, j, v). A C++ implementation could use operator overloading to simplify element access.

4.2 Element processing

The rows of the matrix are processed with a simple for loop. The columns are accessed using an iterative implementation of the inorder traversal algorithm for a binary tree (Flaming 1993), to avoid recursion. This allows this approach in sparse, non-symmetrical matrices, not necessarily banded.

4.3 Gauss-Seidel algorithm

The classical Gauss-Seidel algorithm (Melosh 1990) assumes control. The rows of the matrix are processed in a simple for loop, with a few enhancements.

4.4 Aitken acceleration

Aitken's Δ^2 process (Atkinson & Harley 1983) is a well-known acceleration technique. This is an enhancement of the successive over relaxation method (SOR), based on the idea of producing a new approximation using several previous values of the same unknown, instead of one, as this:

$$x_{k+3} = x_k - \frac{(x_{k+1} - x_k)^2}{x_{k+2} - 2x_{k+1} + x_k} \tag{1}$$

To apply this formula, some previous values of each unknown must be stored. In this implementation, they are part of the data structure held as an element of the column vector, and start working after the 5th iteration.

4.5 Freezing logic

At the end of each iteration step, the previous value of the unknown is compared with the actual one. If they differ in less than a prefixed epsilon, the row is marked as disabled. This means that, in future iterations, the disabled rows will not be used in the arithmetic calculations, improving performance.

After a certain initial number of approximations, the count of active equations diminishes, and the algorithm quickly ends up approximating a small number of unknowns. In the tests, only about 10 iterations are required in most of the cases.

4.6 The final approximation

After all the unknowns have been approximated, a final step is entered, in which a standard Gauss-Seidel method is applied, starting with very close initial values to the end ones, with no relaxation. Usually, this final step does not improve precision, but is there as a security resort.

5 RESULTS

The proposed methodology was tested with random generated sparse linear systems, with known unitary solution. The data was randomly distributed on a limited number of columns. The only constraint,

221

that of diagonal dominance, was achieved with the criteria of a diagonal element being only 0.01% greater than the sum of the absolute values of its row neighbors.

Tests were performed with symmetric and non-symmetric systems, banded and not banded, with positive and negative coefficients, in a personal computer, based on a Pentium III, 600 MHz, 128 Mb RAM. To keep all the system in main memory, the following relation (valid for this particular environment), which depends on the compiler, operating system, and processor, should hold:

$$\frac{number\ of\ main\ memory\ bytes}{number\ of\ matrix\ values} \approx 100 \tag{2}$$

If the number of matrix values is high compared with main memory, virtual memory and disk swapping starts working, resulting in performance degradation.

Table 1. Time to solve sparse linear systems [seconds].

Number of rows	Number of not null column elements								Lin.Corr
	5	10	15	20	25	30	35	40	
1,000	0.10	0.51	0.21	0.26	0.32	0.34	0.41	0.44	0.518
2,000	0.51	0.32	0.43	0.85	0.66	0.84	0.75	0.79	0.749
4,000	0.54	0.64	1.94	1.10	1.26	1.48	1.53	1.86	0.706
8,000	1.09	1.29	2.51	2.23	2.51	3.10	3.38	3.40	0.948
16,000	2.16	2.59	3.74	4.35	5.23	5.31	6.71	7.59	0.991
32,000	4.42	5.24	6.95	8.90	10.73	12.31	10.48	15.64	0.953
64,000	7.74	10.39	12.01	20.40	21.28	–	–	–	0.959
Lin.Corr.	0.997	0.999	0.993	0.998	1.000	0.997	0.990	1.000	

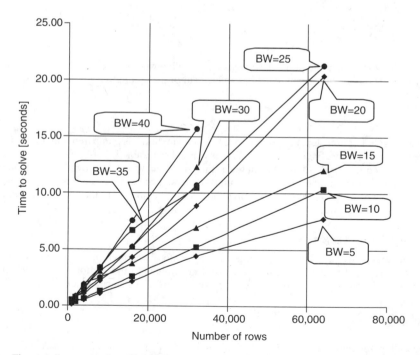

Figure 2. Time to solve vs. number of rows.

Table 1 shows measured time, in seconds, to solve sparse linear systems, between 1,000 and 64,000 rows, with non-zero columns in each row between 5 and 40. These time values are the average of 5 runs in similar conditions, and the time period is provided by the operating system, up to the millisecond.

The variations in measured time can be consequence of operating system memory managing, background data processing, random data generator and error propagation, and others. In systems with small number of non-zero columns, the data managing routines tend to be more important, relative to the arithmetic processing.

For each count of not null elements, the linear correlation coefficient between number of rows and time is presented, showing a strong linear relation.

The correlation between count of not null values and time is not so clear; though it is pretty high. The linear tendency gets greater as the number of rows increases.

The graph in Figure 2 shows an almost linear dependence between the number of rows of the sparse linear systems, and the time to obtain a solution, in seconds, with precision of 10^{-12}. The graph was parameterized with the number of not null values per row (loosely named BW "bandwidth"), from 5 to 40.

6 EXTENSIONS

The Gauss-Seidel method is one of the iterative possibilities to solve linear systems. The techniques commented here could be tested using other algorithms, to verify their applicability, and to gather more information to select the best one.

Other science fields could use these techniques. Training back propagation neural networks is, usually, a lengthy process. The author has made some tests, using the Aitken formulae, as well as freezing weights after some training cycles without changes.

7 CONCLUSIONS

Choosing a solution strategy to solve large sparse linear systems in realistic applications is still an issue. This paper shows that a solution for such systems can be obtained in linear time with high precision, combining problem analysis, data structures and algorithms, resulting in high speed and precision answers.

Object Oriented Programming, with graphic user interfaces, can be excellent tools to produce stable and reliable software solutions in the science area. The design of the involved classes increases the insight, and their interrelation helps to manage complexity. The implementation is faster and simpler, as tests and error corrections are easily performed.

REFERENCES

Adel'son-Vel'Skiï, G.M. & Landis, E.M. 1962. *Doklady Akademia Nauk SSSR*, 146, 263–66. English translation in Soviet Math, Doklady 3, (1962), 1259–1263.
Atkinson, L.V. & Harley, P.J. 1983. *An introduction to Numerical Methods with Pascal*. Finland, Addison-Wesley.
Barroso, L. et al. 1987. *Cálculo numérico (com aplicações)*. Brasil, Harbra.
Buzzi-Ferraris, G. 1994. *Scientific C++*. Cambridge, Addison-Wesley.
Cantù, M. 2001. *Mastering Delphi 6*. Alameda, Sybex.
Fleming, B. 1993. *Practical Data Structures in C++*. New York, Wiley.
Inprise Corp. 1999. *Delphi 5. Developer's Guide*, Scotts Valley, Inprise Corporation.
Kruse, R. 1994. *Data Structures and Program Design*. Englewood Cliffs, Prentice-Hall.
Melosh, R. 1990. *Structural engineering analysis by finite elements*. Englewood Cliffs , Prentice-Hall.
Nakamura, S. 1993. *Applied Numerical Methods in C*. Englewood Cliffs, Prentice-Hall.
Parsons, T.W. 1995. *Introduction to algorithms in Pascal*. New York, Wiley.
Saad, Y. & Schultz, M.H. 1986. *GMRES: A Generalized Minimal Residual Algorithm for Solving Nonsymmetrical Linear Systems*. SIAM J. Scientific and Statistical Computing, Vol. 7, No. 3, 105–126.
Wirth, N. 1986. *Algoritmos + Estructuras de Datos = Programas*. Spain, Ediciones del Castillo.

Transformed-space non-uniform pseudospectral time-domain method

W.K. Leung
Department of Electronic and Information Engineering, The Hong Kong Polytechnic University, Hong Kong

ABSTRACT: In this paper we describe a new non-uniform pseudospectral time-domain (NU-PSTD) method for electromagnetic simulations. In this method we transform a non-uniform grid $\{x_i\}$ into a uniform grid $\{u_i\}$ before applying the fast Fourier transform (FFT) to obtain the spatial derivatives. These transformed-space derivatives are then converted back to the real space using simple interpolation formulae. The resultant scheme differs from the standard uniform PSTD algorithm by a single factor of du/dx only, and is therefore equally efficient with a computational complexity of $O(N \log N)$. We demonstrate the new method with calculations of the reflection coefficients of a single and a coated dielectric slab. In both cases, the computed results are in excellent agreement with the analytic solution for frequencies that correspond to as few as 3 cells per wavelength.

1 INTRODUCTION

The central equations in electromagnetics are the Maxwell's curl equations:

$$\varepsilon \frac{\partial \mathbf{E}}{\partial t} = \nabla \times \mathbf{H} - \sigma \mathbf{E} - \mathbf{J} \tag{1}$$

$$\mu \frac{\partial \mathbf{H}}{\partial t} = -\nabla \times \mathbf{E} - \mathbf{M} \tag{2}$$

where $\mathbf{E}$ and $\mathbf{H}$ are the electric and magnetic field intensities, ϵ, μ and σ are the permittivity, permeability and conductivity of the medium, and $\mathbf{J}$ and $\mathbf{M}$ are the externally imposed electric and magnetic current densities, respectively. One of the most popular and powerful algorithms for solving the Maxwell's equations is the finite-difference time-domain (FDTD) method (Yee 1966, Shlager & Schneider 1995), which approximates the spatial and time derivatives by finite differences and updates the electromagnetic fields iteratively. More recently, a pseudospectral time domain (PSTD) method was proposed (Liu 1997), which approximates the spatial derivatives using one-dimensional (1D) fast Fourier transforms (FFT). The PSTD method is highly efficient, as the Nyquist sampling theorem dictates that two cells per wavelength are required at the highest frequency of interest. A major limitation of the PSTD method in its original form is that the FFT requires a uniform spatial grid, which means that fine-scale structures such as a thin dielectric coating cannot be handled accurately without using a very dense grid (and hence an enormous amount of computer memory). A similar problem exists when the structures are not commensurate with the spatial discretization, i.e. they are not simple integral multiples of the cell sizes. In these cases, the cell size is governed by the dimensions of the fine structures, and the main advantage of PSTD, namely requiring only 2 cells per wavelength, is lost. A non-uniform formulation of PSTD is therefore highly desirable.

In this paper we will describe our new formulation of NU-PSTD based on spatial transformation (Leung & Chen 2001).

2 NON-UNIFORM SPATIAL GRID

Consider an N-point non-uniform grid $\{x_j\}$ spanning the interval $[x_1, x_N]$. We can transform this grid into a uniform grid $\{u_j\}$ via a spatial transformation or mapping $M(x_j) = u_j = j$, where we have chosen to set $u_j = j$ for convenience. A few examples of such mappings M are shown in Figure 1.

In this paper, we will demonstrate our method using the following grid:

$$x_{j+1} - x_j = \begin{cases} 1 & \text{for } j = 1,2,\cdots,25 \\ s^{j-25} & \text{for } j = 26,27,\cdots,75 \\ s^{125-j} & \text{for } j = 76,77,\cdots,125 \\ 1 & \text{for } j = 126,127,\cdots,150 \end{cases} \tag{3}$$

where s is a parameter which can be used to control the degree of non-uniformity of the grid. In essence, the grid is composed of four sections and is symmetric about the center of the grid. In the first and fourth sections, the cell sizes are uniform while in the second and third sections, the cell sizes decrease gradually toward the center of the grid (controlled by the scaling parameter s). In general, the more s deviates from 1, the more non-uniform the grid becomes. The grid is illustrated in Figure 2. In all of our simulations, we set up a 10-layer anisotropic perfectly matched layer absorbing boundary at both ends of the grid in order to suppress the artificial "wrap-around" effect due to the use of Fourier transform in PSTD (Berenger 1994, Gedney 1996).

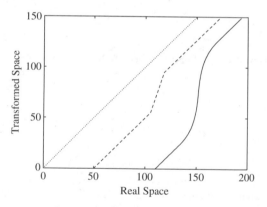

Figure 1.   Typical examples of transformation mapping M. (dotted line: uniform grid; dashed line: subgrid with a denser central region; solid line: the non-uniform grid in Figure 2 with $s = 0.95$).

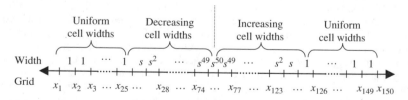

Figure 2.   The non-uniform grid (Eq. 3) as used in this paper.

226

3   TRANSFORMED-SPACE NON-UNIFORM PSTD METHOD

With the spatial transformation $M(x_j) = u_j = j$ in place, we can calculate the spatial derivatives of the simulated electromagnetic fields in the transformed space using standard FFTs:

$$\frac{\partial E}{\partial u} = FFT^{-1}(ik \times FFT(E)) \tag{4}$$

where E represents an electromagnetic field component, k's are spatial frequencies in the transformed space, and FFT (FFT^{-1}) denotes the forward (backward) FFT operation. The transformed-space derivatives are then converted back to the real space P as follows:

$$\left.\frac{\partial E}{\partial x}\right|_{x=x_j} = \left.\frac{du}{dx}\right|_{x=x_j} \cdot \left.\frac{\partial E}{\partial u}\right|_{u=u_j} = \left.\frac{du}{dx}\right|_{x=x_j} \cdot FFT^{-1}(ik \times FFT(E))\Big|_{u=u_j}. \tag{5}$$

These are the real-space derivatives that should be used in the PSTD update equations when the grid is non-uniform. In essence, the new method differs from the standard PSTD algorithm by a single factor of du/dx only and is therefore equally efficient with a computational complexity of $((N\log N)$. We remark that the values of du/dx depend entirely on the non-uniform grid $\{x_j\}$ but not the field component values. Therefore they need to be computed only *once* for the entire simulation. In practical calculations we do not expect the spatial transformation $M(x_j) = u_j$ is known analytically. However, we can approximate M at each x_j with a quadratic expression determined by the grid position x_j and its two nearest neighbors x_{j-1} and x_{j+1}. More explicitly, we use the following interpolation formula:

$$M(x) = a_j x^2 + b_j x + c_j \qquad \text{at} \quad x \approx x_j \tag{6}$$

where the parameters a_j, b_j and c_j are determined using the three conditions:

$$\begin{cases} M(x_{j-1}) = u_{j-1} \\ M(x_j) = u_j \\ M(x_{j+1}) = u_{j+1} \end{cases} \qquad \text{for} \quad j = 1, 2, 3, \ldots, N. \tag{7}$$

After some straightforward manipulation, the solution for du/dx at $x = x_j$ is given by:

$$\left.\frac{du}{dx}\right|_{x=x_j} = \frac{(x_{j+1}-x_j)^2 + (x_j-x_{j-1})^2}{(x_{j+1}-x_j)(x_j-x_{j-1})(x_{j+1}-x_{j-1})}, \tag{8}$$

which is simple to calculate. An even simpler expression for du/dx can be obtained by modeling the inverse function $x = M^{-1}(u)$ instead, as follows:

$$M^{-1}(u) = a_j u^2 + b_j u + c_j. \tag{9}$$

This interpolation approach results in the following alternative formula for du/dx:

$$\left.\frac{du}{dx}\right|_{x=x_j} = \frac{2}{x_{j+1}-x_{j-1}}. \tag{10}$$

When tested with different non-uniform grids, we find that both Equations (9) and (10) produce practically identical results. This corroborates the accuracy of our quadratic interpolation, and in the following we will report numerical results computed with Equation (9).

It is worth noting that our new NU-PSTD method is fundamentally different from previous NU-FFT-based algorithms (Dutt & Rokhlin 1993, Liu 1998), which interpolate the electromagnetic fields in real space directly. In our method, we model the grid transformation M instead. Our approach is advantageous because the non-uniform grid $\{x_j\}$ is chosen by the users themselves and is thus expected to be reasonably smooth and well-behaved. In contrast, the simulated electromagnetic fields may contain highly irregular variation, and their values may be unstable and inaccurate for direct interpolation. We therefore believe that our new algorithm should yield a better performance than the direct application of NU-FFT. Note also that our new NU-PSTD algorithm requires only the standard uniform FFT, which can be found in most numerical libraries. This makes the new method much easier to implement than the more specialized NU-FFT algorithms.

4 APPLICATIONS OF TRANSFORMED-SPACE NON-UNIFORM PSTD METHOD

We have tested our new NU-PSTD algorithm in simulating the propagation of a Gaussian pulse in a 1D homogeneous medium (taken to be free space). The computation domain is discretized non-uniformly using the grid in Figure 2 with $s = 0.95$, giving a largest-to-smallest cell size ratio of $s^{-50} = 13$. The pulse propagates in the positive x-direction (with electric and magnetic fields pointing along the positive y- and z-directions, respectively) and has a narrow half-width of 2 m, or 2 cell widths in the uniform region of the grid. The time step size is expected to obey a similar stability criterion as in the uniform PSTD algorithm (Liu 1997), given by:

$$\Delta t \le \frac{2\min|\Delta x|}{c\sqrt{D}\,\pi} \tag{11}$$

where $\min|\Delta x|$ is the minimum cell size, c is the speed of light in the medium and D is the dimensionality of the problem. We have used a very small time step of $\Delta t = 10^{-11}$ s in our simulations to minimize the time step errors so that we can focus on the possible effects of the non-uniform discretization. The pulse is observed at three grid points, x_{30}, x_{75} and x_{130}, located in the beginning, middle and ending section of the non-uniform grid, respectively. The results are shown in Figure 3 below, where it can be seen that the three pulses have a practically identical shape, differing from one another by no more than 0.0007 (cf. peak height of the pulse = 1). This observation suggests that only very small dispersion errors are present in our simulation even though a highly non-uniform grid has been used.

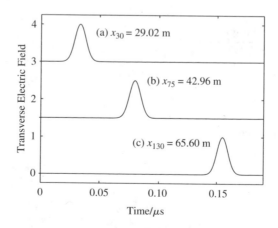

Figure 3. NU-PSTD simulation of a Gaussian TEM pulse (half-width = 2 m) propagating in vacuum, discretized using the grid in Figure 2 with $s = 0.95$ and largest-to-smallest cell size ratio of 13. The transverse electric fields are measured at (a) $x_{30} = 29.02$ m, (b) $x_{75} = 42.96$ m, and (c) $x_{130} = 65.50$ m, shifted upward by 3.0, 1.5 and 0.0 units, respectively, for clarity.

We have also applied the new NU-PSTD method to calculate the reflection coefficient of an infinite dielectric slab embedded in vacuum. The 1D computation domain is discretized using the same non-uniform grid in Figure 2 with $s = 0.96$ (and hence a largest-to-smallest cell size ratio of 7.70). The dielectric slab spans the grid points x_{76} through x_{92} (i.e. 3.19 m thick) and has a dielectric constant of $\epsilon_r = 2$. With a time step size of 2.5×10^{-10} s, we simulate the reflection of a TEM Gaussian pulse off the dielectric slab and measure both the incident and reflected pulses. The reflection coefficients at different frequencies are then calculated as the ratios between their respective Fourier transforms. Figure 4 shows the computed results along with the analytic solution. It can be seen that the NU-PSTD simulation produces very accurate results for frequencies up to 100 MHz, which corresponds to only 3 cells per wavelength. This is substantially better than the conventional Yee's FDTD method, which requires about 8–16 cells per wavelength.

As a final example, we have calculated the reflection coefficient of the same dielectric slab as in the previous example but now coated on the front side with a thin layer of dielectric material. The coating has a thickness of 0.13 m and a dielectric constant of $\epsilon_r = 5$. The computed results are shown in Figure 5, compared with the analytic solution. Again, the NU-PSTD results are in excellent agreement with the analytic solution for frequencies up to 100 MHz, which corresponds to only 3 cells per wavelength. We remark that the uniform PSTD or FDTD methods would require a much finer grid and about 5 times

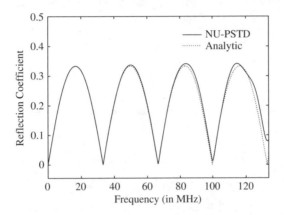

Figure 4. Reflection coefficient of an infinite dielectric slab of thickness 3.19 m and dielectric constant 2. solid line: NU-PSTD simulation; dotted line: analytic solution.

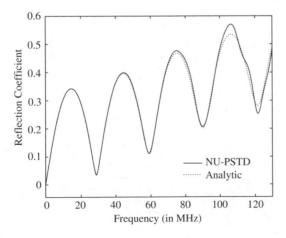

Figure 5. Same as Figure 4, except that the dielectric slab is coated with a thin dielectric layer (thickness = 0.13 m, dielectric constant $\epsilon_r = 5$).

more cells (and hence computer memory) for simulating the same system. The saving of NU-PSTD will be even more significant in 2D and 3D simulations. Note also that the dielectric slab and coating are generally difficult to represent in a uniform grid because their thicknesses may not be simple multiples of a common cell size. This demonstrates the advantage of using non-uniform grids in PSTD simulations.

5 CONCLUSIONS

In this paper, a new NU-PSTD algorithm based on spatial transformation has been presented. The new algorithm differs from the uniform PSTD method by a single factor of du/dx, which can be approximated accurately via quadratic interpolation. Since the values of du/dx do not depend on the electromagnetic fields, we only need to compute them once for the entire simulation. The new algorithm has a computational complexity of $O(N \log N)$ and is as practically efficient as the uniform PSTD algorithm. We have applied the new method to calculate the reflection coefficients of two test cases involving dielectric slabs. In both cases, the computed results have been found to be in excellent agreement with the analytic solution for frequencies that correspond to as few as 3 cells per wavelength. This suggests that a much larger cell size can be used in transformed-space NU-PSTD than in conventional FDTD simulations. In view of these features, the new NU-PSTD algorithm is expected to be useful for a wide range of electromagnetic simulations, particularly for systems containing fine-scale structures.

REFERENCES

Berenger, J.-P. 1994. A perfectly matched layer for the absorption of electromagnetic waves. *J. Comput. Phys.* 114: 185–200.
Dutt, A. & Rokhlin, V. 1993. Fast Fourier transforms for nonequispaced data. *SIAM J. Sci. Comp.* 14(6): 1368–1393.
Gedney, S.D. 1996. An anisotropic perfectly matched layer-absorbing medium for the truncation of FDTD lattices. *IEEE Trans. Antennas Propagat.* 44: 1630–1639.
Leung, W.K. & Chen, Y. 2001. Transformed-space non-uniform PSTD algorithm. *Microwave Opt. Technol. Lett.* 28: 391–396.
Liu, Q.H. 1997. The PSTD algorithm: a time-domain method requiring only two cells per wavelength. *Microwave Opt. Technol. Lett.* 15: 158–165.
Liu, Q.H. 1998. An accurate algorithm for nonuniform fast Fourier transforms (NU-FFT's). *IEEE Microwave Guided Wave Lett.* 8(1): 18–20.
Shlager, K.L. & Schneider, J.B. 1995. A selective survey of the finite-difference time-domain literature. *IEEE Antennas and Propagation Magazine* 37(4): 39–56. (see also the online FDTD database at http://www.fdtd.org/).
Yee, K.S. 1966. Numerical solution of initial boundary value problems involving Maxwell's equations in isotropic media. *IEEE Trans. Antennas Propagat.* AP-14(4): 302–307.

Computational Methods in Engineering and Science, Iu et al. (eds)
© 2003 Swets & Zeitlinger, Lisse, ISBN 90 5809 567 3

A particle-based mechanistic model for drape simulation

K.Y. Sze
Department of Mechanical Engineering, The University of Hong Kong, Hong Kong

X.H. Liu & C. Wang
Department of Mechanics, Huazhong University of Science & Technology, Wuhan, China

ABSTRACT: This paper presents a particle-based model for cloth drape simulation. Compared to its precursors, the present model employs the corotational framework. Using the framework and some commonly employed assumptions for large-displacement but small-strain problems, the tensile, trellising and bending energy of the fabric cloth are derived. By taking derivative of the total internal energy with respect to the displacement variables, a system of nonlinear equations can be obtained. It can be noted that these equations are considerably simpler than that of the conventional particle-based models. A few examples are examined and the predicted results conform well with our daily perception of fabric drapes.

1 INTRODUCTION

It is generally accepted that one of the key elements in the CAD/CAM and E-marketing of garments is an efficient and robust computational model which can predict the mechanistic response of fabrics. Intensive research efforts have been drawn towards the development of these models in the last two decades. Most of the previous work can broadly be categorized into geometry methods (Hinds et al. 1991, Rodel et al. 1998, Dai et al. 2001), particle- or grid-based methods (Breen et al. 1992, House et al. 1996, Stylios et al. 1996, Meyer et al. 2001, Chen et al. 2001) and the shell finite element methods (Collier et al. 1991, Kang & Yu 1995, Eischen et al. 1996, Tan et al. 1999).

In this paper, a particle-based fabric deformation model is presented. Compared to other particle-based methods, the model employs the corotational framework (Crisfield 1991, Belytschko et al. 2000) to formulate the nonlinear system of equilibrium equations. By virtue of the framework and some commonly employed assumptions for large-displacement small-strain problems, the internal energy does not involve any inverse trigonometrical functions which are the origins of the tedious equilibrium equations in other particle-based methods. The full Newton-Raphson method will be employed to solve the equilibrium equations. To improve the convergence, scaled rather than full displacement refinements are employed. Three numerical examples are presented to examine the capability of the model. These include a hanged square fabric sheet, a square fabric sheet draped over a circular table and a curtain under sideward force. It can be seen that the predictions conform well with our daily perception of cloth objects.

2 THE CLASSICAL METHOD OF SOLUTION FOR NONLINEAR ELASTIC SYSTEMS

For an elastic deformable body whose displacement field is characterized by the kinematic d.o.f.s of non-physical particles distributed over the body, its elastic energy can be written as:

$$U = U(\mathrm{U}) \tag{1}$$

where $\mathrm{U} = \{U_i\}$ is the system displacement vector that contains all the active kinematic d.o.f.s of the particles. In general, the total potential energy of the system can be expressed as:

$$\Pi = U(\mathrm{U}) - \mathrm{P}^T \mathrm{U} \tag{2}$$

in which $P = \{P_i\}$ is the system external force vector that accounts for the prescribed body and surface forces. The equilibrium condition can be obtained by the minimum potential energy principle as:

$$\frac{\partial \Pi}{\partial U} = \{\frac{\partial \Pi}{\partial U}\}_i = \mathbf{0} \quad \text{or, equivalently,} \quad F - P = \mathbf{0} \tag{3}$$

where

$F = \{F_i\} = \{\partial U / \partial U_i\}$ is system internal force vector arising from the internal forces.

To solve the above equation, the full Newton-Raphson iterative scheme can be used, i.e.

$$U^{(0)} = \mathbf{0} \ , \quad \Delta U^{(i)} = (K\big|_{U=U^{(i)}})^{-1}(P - F\big|_{U=U^{(i)}}) \quad \text{and} \quad U^{(i+1)} = U^{(i)} + \Delta U^{(i)} \tag{4}$$

in which
$U^{(i)}$ is the i-th iterative solution of U, $\Delta U^{(i)}$ is the iterative refinement of $U^{(i)}$,
$K = [K_{ij}] = [\partial F_j / \partial U_i] = \partial F^T / \partial U = \partial / \partial U (\partial U / \partial U)^T$ is the system tangential stiffness matrix.
The iteration can be stopped when the iterative displacement refinement or the unbalanced force $F - P$ is sufficiently small.

3 INTEGRATION ZONE AND ELASTIC ENERGIES

Figure 1a shows an undeformed fabric sheet. Similar to most of the particle-based methods for fabric drapes (Breen et al. 1992, House et al. 1996, Stylios et al. 1996, Meyer et al. 2001, Chen et al. 2001), the gridlines are orthogonal and aligned with the wraps and wefts. Particles are located at the intersection points of the gridlines.

To evaluate the elastic energy, particle-based integration zones are defined. For a particle away from the boundary, the edges defining an integration zone are parallel and equidistant from their adjacent gridlines, see Figure 1b. For a particle right next to the boundary, its integration zone is extended to the boundary, see Figure 1a. The elastic energy in a particle-based integration zone (see Figure 1b) can be expressed as:

$$U^N = U^N_{AB} + U^N_{CD} + U^N_{BD} + U^N_{DA} + U^N_{AC} + U^N_{CB} \tag{5}$$

where U^N_{AB} is the sum of the tensile and bending energies along $A^O B^O$, and U^N_{CD} is the sum of the tensile and bending energies along and $C^O D^O$. Moreover, U^N_{BD}, U^N_{DA}, U^N_{AC} and U^N_{CB} are the trellising elastic energies in the first, second, third and fourth quadrants of the integration zone, respectively. Once the expressions for the energy components are determined, the nonlinear equilibrium condition in (3) and the system tangential stiffness matrix in (4) can be derived by differentiation.

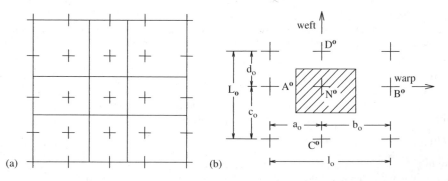

(a)   (b)

Figure 1.   (a) Particle-based integration zones □ in a fabric sheet. (b) Particle-based integration zone for particle N. — — denotes gridline and particles are located at grid points +.

4 TENSILE AND BENDING ENERGIES

To calculate the tensile strain and curvature along the warp direction inside the integration zone in Figure 1b, the translations of particles A, N and B are considered. Restricting ourselves to zero translation along the weft direction, the curvature and the tensile Green strain (Crisfield 1991) are:

$$\kappa_{AB} = -\frac{1}{[1+(d\bar{w}/d\bar{x})^2]^{3/2}}\frac{d^2\bar{w}}{d\bar{x}^2} \quad , \quad \varepsilon_{AB} = \frac{d\bar{u}}{d\bar{x}} + \frac{1}{2}(\frac{d\bar{u}}{d\bar{x}})^2 + \frac{1}{2}(\frac{d\bar{w}}{d\bar{x}})^2 \tag{6}$$

where $\bar{x}$ is the warp coordinate, $\bar{u}$ is the translation along $\bar{x}$ and $\bar{w}$ is the translation normal to fabric sheet. Assuming the fabric is linear, the elastic energy arising from ε and κ per unit fabric area is:

$$\frac{1}{2}(D_t\varepsilon_{AB}^2 + D_b\kappa_{AB}^2) \tag{7}$$

where D_t and D_b are respectively the tensile and bending rigidities along the A-B-direction. Owing to presence of the term:

$$1/[1+(d\bar{w}/d\bar{x})^2]^{3/2}$$

in the curvature, the derivatives of U_{AB}^N with respect to the nodal displacements are tedious. On the other hand, the above term cannot be neglected as the magnitude of $d\bar{w}/d\bar{x}$ can be much larger than unity. The worst is that the curvature will become much more complicated when the translation along the weft direction is not taken to be zero.

Owing to the above reasons, the corotational formulation is adopted (Crisfield 1991, Belytschko et al. 2000). Figure 2 shows the positions of three consecutive particles A, N and B along the warp. In the figure, (X,Y,Z) are the global Cartesian coordinates. The initial configuration is $A^O\text{-}N^O\text{-}B^O$ and is assumed to be straight. The current or deformed configuration is $A\text{-}N\text{-}B$. A corotating configuration $A^C\text{-}N^C\text{-}B^C$ is formed by translating and rotating $A^O\text{-}N^O\text{-}B^O$ as a rigid body such that A^C coincides with A and $A^C\text{-}N^C\text{-}B^C$ is parallel to $A\text{-}B$. Properties of the corotating frame $x\text{-}y\text{-}z$ are: (i) its origin is A, (ii) the $x\text{-}z$-plane is defined by $A\text{-}N\text{-}B$, (iii) the x-axis is defined by $A\text{-}B$ and (iv) the local transverse displacement w_N of particle N points

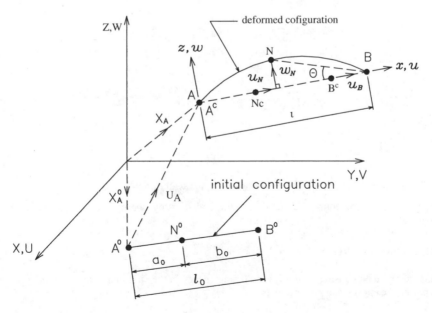

Figure 2. $A^O\text{-}N^O\text{-}B^O$, $A^C\text{-}N^C\text{-}B^C$ and $A\text{-}N\text{-}B$ are the initial, corotating and deformed configurations, respectively. A^C and A are coincident.

along the positive direction of z. The initial coordinates, current coordinates and the global displacement vector of the i-th particle are denoted respectively as:

$$\mathbf{X}_i^o = \{X_i^o, Y_i^o, Z_i^o\}^T \, , \quad \mathbf{X}_i = \{X_i, Y_i, Z_i\}^T \quad \text{and} \quad \mathbf{U}_i = \mathbf{X}_i - \mathbf{X}_i^o = \{U_i, V_i, W_i\}^T \tag{8}$$

The local displacement components along x and z obtained by subtracting the corotating configuration from the current configuration can be interpolated quadratically as:

$$u = N_A u_A + N_N u_N + N_B u_B = N_N u_N + N_B u_B \tag{9}$$

and

$$w = N_A w_A + N_N w_N + N_B w_B = N_N w_N \tag{10}$$

where

$$N_A = (x - a_o)(x - l_o)/(a_o l_o) \, , N_N = (l_o - x)x/(a_o b_o) \quad \text{and} \quad N_B = x(x - a_o)/(b_o l_o)$$

are the interpolation functions. In the functions, a_o, b_o and l_o are respectively the lengths of A^O-N^O, N^O-B^O and A^O-B^O as shown in Figure 1b and Figure 2. For the chosen corotating frame, there is no local displacement component transverse to the x-z-plane. With $\bar{x}$ and $\bar{w}$ replaced respectively by x and w, the expressions in (6) can thus be used to evaluate the tensile strain and curvature. Provided l_o is sufficiently small, the local displacements measured with respect to the corotating frame are small. This feature leads to the following commonly adopted approximate axial Green strain and curvature in the shallow arch element (Crisfield 1991):

$$\varepsilon_{AB} = \frac{du}{dx} + \frac{1}{2}(\frac{dw}{dx})^2 \, , \quad \kappa_{AB} = -\frac{d^2 w}{dx^2} \tag{11}$$

By incorporating (10) into (11),

$$\varepsilon_{AB} = N_N' u_N + N_B' u_B + \frac{1}{2}(N_N' w_N)^2 \, , \quad \kappa_{AB} = -N_N'' w_N \tag{12}$$

Thus, the integral of (7) over the integration zone of particle N is:

$$U_{AB}^N = \frac{L_o}{4} \int_{a_o/2}^{a_o + b_o/2} (D_t \varepsilon_{AB}^2 + D_b \kappa_{AB}^2) dx = \frac{1}{2} \begin{Bmatrix} u_N \\ w_N^2 \\ u_B \end{Bmatrix}^T \mathbf{C} \begin{Bmatrix} u_N \\ w_N^2 \\ u_B \end{Bmatrix} + \frac{1}{2} c_4 w_N^2 \tag{13}$$

where

$$\mathbf{C} = \frac{D_t L_o}{2} \int_{a_o/2}^{a_o + b_o/2} [N_N' , (N_N')^2/2 , N_B']^T [N_N' , (N_N')^2/2 , N_B'] dx ,$$

$$c_4 = \frac{D_b L_o}{2} \int_{a_o/2}^{a_o + b_o/2} (N_N'')^2 dx .$$

$L_o/2$ is the dimension of the integration zone along the weft direction (see Figure 1b).

Following the same procedure, U_{CD}^N or the tensile and bending energy in the weft direction can be formulated.

5   TRELLISING ENERGY

The next energy to be considered in (5) is trellising energy in the first quadrant of the integration zone of Figure 1b. The energy can be written as:

$$U_{BD}^N = \frac{D_s}{2} \int_{Area} \gamma_{BD}^2 dA \tag{14}$$

where D_s is the shear rigidity, *Area* denotes the area of the first quadrant and γ_{BD} is the shear strain. With the small strain assumption, the previous particle-based models calculate shear strain using the shear angle (Breen et al. 1992, House et al. 1996, Stylios et al. 1996, Meyer et al. 2001, Chen et al. 2001), i.e.

$$\gamma_{BD} = \frac{\pi}{2} - \cos^{-1}\frac{\mathbf{X}_{BN}^T\mathbf{X}_{DN}}{bd} \quad , \quad U_{BD}^N = \frac{D_s}{2}\frac{b_o d_o}{4}(\frac{\pi}{2} - \cos^{-1}\frac{\mathbf{X}_{BN}^T\mathbf{X}_{DN}}{bd})^2 \tag{15}$$

in which

$$b = (\mathbf{X}_{BN}^T\mathbf{X}_{BN})^{1/2} \quad , \quad d = (\mathbf{X}_{DN}^T\mathbf{X}_{DN})^{1/2} \; .$$

For higher computational efficiency and in compliance with the small strain assumption, the following approximation is adopted:

$$\gamma_{BD} \, \square \, \sin\gamma_{BD} = \frac{\mathbf{X}_{BN}^T\mathbf{X}_{DN}}{bd} \quad , \quad U_{BD}^N = \frac{D_s}{2}\frac{b_o d_o}{4}\gamma_{BD}^2 = \frac{D_s}{2}\frac{b_o d_o}{4}(\frac{\mathbf{X}_{BN}^T\mathbf{X}_{DN}}{bd})^2 \tag{16}$$

Following the same procedure, the trellising energy in other quadrants can be determined.

6 EXAMPLES

Having deriving the energy components, the nonlinear equilibrium condition in (3) and the system tangential stiffness matrix in (4) can be derived by differentiation. In this section, three numerical examples are considered to explore the capability of the model. These include a hanged square fabric sheet, a circular fabric sheet draped over a circular table, a square fabric sheet draped over a circular table, a curtain under sideward force and two tensile fabric structures. The following force convergence criterion is employed:

$$|F - P|/|P| \leq 10^{-4} \tag{17}$$

which is considered to be strict from engineering point of view. In all examples, the properties of the anisotropic wool fabric reported by Kang & Yu (1995) and listed in Table 1 are employed.

Although the full Newton-Raphson method works well for finite element analyses of conventional shells, it often fails in three-dimensional drapes. The underlining reason is that fabrics have very low resistance to bending. When a transverse load is applied to a fabric, its bending stiffness cannot resist the load and a very large iterative displacement refinement will be obtained. The iterative solution is too far away from the equilibrium state(s). The full Newton-Raphson method either diverges away from or requires a huge number of iterations in bringing the iterative solution close to the equilibrium state(s). In this light, the iterative refinement is taken to be only, typically, 1% of the one solved in (4).

6.1 *Hanging postures of a square fabric sheet*

In this example, a 150×150 cm square fabric sheet made of wool is hanged by fully restrained 25×25 cm square area in one corner, as shown in Figure 3. Before deformation, a diagonal of the sheet is vertical. A uniform 30×30 grid is employed. The nominal free-hanging length L_f is 20 cm. Small unidirectional perturbing forces transverse to the undeformed plane of the fabric sheet are applied to all particles. The forces are removed after the first iteration. Without the forces, the iteration stops at the equilibrium state in which there is only in-plane but not any out-of-plane deformation. Of course, such a state is only a mathematical artifact which never occurs in reality. Figures 5a and 5b show the front

Table 1. Fabric properties of wool from Kang & Yu [13].

Tensile rigidity, D_t (gf/cm)		Bending rigidity, D_b (gf cm²/cm)		Shear rigidity, D_s (gf/cm)	Weight (gf/cm²)
warp	weft	warp	weft		
1118.2	759.5	0.083	0.063	41.8	0.019

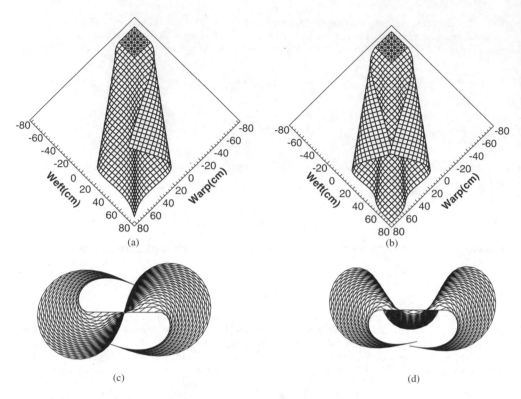

Figure 3. (a) and (c) show the front and top views of a hanged square fabric sheet when the initial perturbing force is $10^{-8}\,gf$ per particle. (b) and (d) show the front and top views of the same fabric sheet when the initial perturbing force is $5 \times 10^{-8}\,gf$ per particle. The hanged corner is hatched.

view of the fabric sheet when the perturbation forces are taken to be 10^{-8} and $5 \times 10^{-8}\,gf$ per particle, respectively. Figures 3c and 3d show the top views. It is interesting to note that different equilibrium states are obtained by using different magnitudes of the perturbation force. Both states conform to our real life perception.

6.2 Square fabric sheet draped over a circular table

In this example, a 400×400 cm square fabric sheet is draped over a $\phi256$ cm table. The entire fabric sheet is modeled by a uniform 50×50 grid. The example is particularly challenging due to its large nominal free-hanging length. Examples of similar degree of drapeability can rarely be seen in the literature. Figures 4a and 4b show respectively the top and an isometric view of the deformed fabric sheet. The draped pattern appears to be elegant and conforms with our daily perception. It can also be seen that the pattern is symmetric.

6.3 Deformation of a curtain under concentrated force

In this example, a curtain in the form of a 250×70 cm fabric sheet is considered. The curtain is hanged at the eleven points as shown in Figure 5a. Among them, point A is fully restrained and the remaining ten points can only move horizontally along A-K. K is pushed towards A by a $40.375\,gf$ concentrated force. Any higher forces will introduce self-contact which has not been considered in the present model. Gravity is acting along the weft direction. Initial perturbation forces $10^{-8}\,gf$ per particle transverse to the undeformed fabric sheet are applied to all particles within a-c-C-A, d-e-E-D, f-g-F-G and h-i-H-I. The remaining particles are loaded with perturbation forces of the same magnitude but opposite direction. All the perturbation forces are removed after the first iteration. Without the forces, the iteration stops at the equilibrium state in which there is only in-plane but not any out-of-plane deformation. Again, such a state is only a mathematical artifact which never occurs in reality. An isometric view of the deformed curtain is shown in Figure 5b.

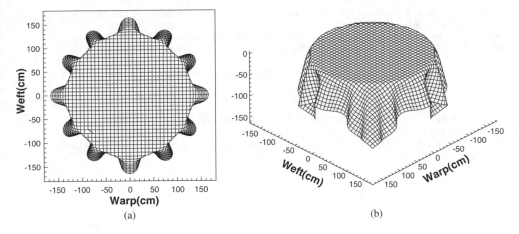

Figure 4. (a) Top view (b) and an isometric view of a square fabric draped over circular table.

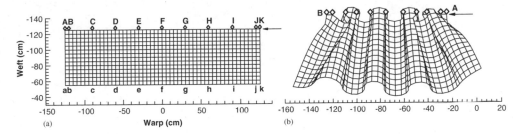

Figure 5. (a) The undeformed grid and (b) the deformed grid for a 250 × 70 cm curtain under gravity and a sideward pushing force at K.

7 CONCLUSION

Particle-based methods for fabric drape simulation often resort to simple particle mechanics and formulate the elastic energy in terms of the inter-particle distances and the inverse trigonometrical functions of the angles between the straight lines that join adjacent particles. In this paper, the corotational approach and some commonly employed assumptions for small strain problems are used to formulate the elastic energy. It can be seen that inverse trigonometrical functions are totally exempted in the elastic energy. As a result, the pertinent internal force vector and the tangential stiffness matrix are considerably simpler than other particle-based models. The draped and stretched fabric geometries appear to be natural and agree with what is being observed in real life. The full derivation of the equilibrium condition and the tangential stiffness matrix as well as addition numerical examples can be found in (Sze & Liu 2003).

REFERENCES

Belytschko, T., Liu, W.K. & Moran, B. 2000. *Nonlinear Finite Elements for Continua and Structures*. John Wiley & Sons.
Breen, D.E., House, D.H. & Getto, P.H. 1992. A particle-based particle model of woven cloth. *The Visual Computer*, **8**: 264–277.
Collier, J.R., Collier, B.J., O'Toole, G. & Sargand, S.M. 1991. Drape prediction by means of finite element analysis. *J. Textile Institute*, **82**: 96–107.
Crisfield, M.A. 1991. *Nonlinear Finite Element Analysis of Solids and Structures – Volume 1: Essentials*. John Wiley & Sons.
Dai, X.Q., Furukawa, T., Mitsui, S., Takatera, M. & Shimizu, Y. 2001. Drape formation based on geometric constraints and its application to skirt modeling. *Inter. J. Cloth. Sci. & Tech.*, **13**: 23–37.

Eischen, J.W., Deng, S. & Clapp, T.G. 1996. Finite element modeling and control of flexible fabric parts. *IEEE Computer Graphics & Application*, **16**: 71–80.

Hinds, B.K., McCartney, J. & Woods, G. 1991. 3D CAD for garment design. *Inter. J. Cloth. Sci. & Tech.*, **4**: 6–14.

House, D.H., De, Vaul, R.W. & Breen, D.E. 1996. Towards simulating cloth dynamics using interacting particles. *Inter. J. Cloth. Sci. & Tech.*, **8**: 75–94.

Kang, T.J. & Yu, W.R. 1995. Drape simulation of woven fabric by using the finite-element method. *J. Textile Institute*, **86**: 635–648.

Meyer, M., Debunne, G., Desbrun, M. & Barr, A.H. 2001. Interactive animation of cloth-like objects in virtual reality. *J. Visualization & Computer Animation*, **12**: 1–12.

Rodel, H., Ulbricht, V., Krzywinski, S., Schenk, A. & Fischer, P. 1998. Simulation of drape behavior of fabrics. *Int. J. Cloth. Sci. & Tech.*, **10**: 201–208.

Stylios, G., Wan, T.R. & Powell, N.J. 1996. Modeling the dynamic drape of garments on synthetic humans in a virtual fashion show. *Inter. J. Cloth. Sci. & Tech.*, **8**: 95–112.

Sze, K.Y., Liu, X.H. 2003. A corotation grid-based model for fabric drapes. *Inter. J. Numer. Methods Engrg.*, Accepted for publication.

Tan, S.T., Wong, T.N., Zhao, Y.F. & Chen, W.J. 1999. A constrained finite element method for modeling cloth deformation. *The Visual Computer*, **15**: 90–99.

Computational Methods in Engineering and Science, Iu et al. (eds)
© 2003 Swets & Zeitlinger, Lisse, ISBN 90 5809 567 3

Numerical implementation of a Hybrid/Mixed stress finite element model based on Legendre polynomials for the elastic analysis of 3-D structures

M.J.V. Silva
Faculdade de Ciências e Tecnologia, Universidade Nova de Lisboa, Monte de Caparica, Portugal

E.M.B.R. Pereira
Instituto Superior Técnico, Universidade Técnica de Lisboa, Lisboa, Portugal

ABSTRACT: This paper presents the implementation of a Hybrid/Mixed Stress finite element model for 3-D problems, using Legendre Polynomials as approximation functions. The present formulation is characterized by the use of independent approximations for three different fields. To implement this model, symbolic programming codes were used, thus reducing considerably computation time. The performance of the current model is demonstrated through the analysis of an arch dam structure. In order to validate the results, they are compared with the solutions obtained by other authors.

1 INTRODUCTION

The Structural Analysis research group of ICIST, has been involved in the development of a set of non-conventional finite element formulations. The work presented here is part of the research produced in this framework.

The present article focuses on the implementation of a Hybrid/Mixed Stress (HMS) finite element model, for the elastic analysis of 3-D Structures, using Legendre Polynomials as approximation functions.

The HMS finite element model defines three different and independent field approximations. Two approximations are used for the displacement field, one in the element domain and another on the static boundaries. Thus the element, according to Pian (1978) terminology, is regarded to be Hybrid. The third approximation is defined for the stress field in the domain. Since two different fields are approximated in the domain (stress and displacement) this element is also considered to be a Mixed element. Finally, it is termed as stress model because the finite element assemblage is based on equilibrium conditions.

This formulation is based on a weighted residual method. The HMS finite element governing system (Pereira & Freitas 1996) is established imposing, in a non-locally form, the fundamental principals of solid mechanics (equilibrium, compatibility and constitutive relations). Using a condensation technique for the stress resultant degrees of freedom, it is possible to obtain a compact form for the HMS governing system.

The HMS formulation has the advantage of allowing for a wide range of functions to be chosen for the approximation of the different fields. In this paper, Legendre polynomials are adopted. Due to the properties of the approximation functions closed form solutions, developed using symbolic programming codes, are used to compute the coefficients of the structural matrices. Time consuming numerical integration procedures are thus avoided. Highly sparse matrices are obtained, which are manipulated using efficient procedures to store, operate and solve that kind of systems. The implementation of the HMS finite elements is based on the use of eight and sixteen-node serendipian cubic master elements.

In order to illustrate the performance of the finite element formulation, numerical applications are presented based on the analysis of a 3-D arch dam structure.

2 FINITE ELEMENT FORMULATION

Consider a 3D body, V, delimited by a boundary, Γ, where the stress and displacement fields, referred to a cartesian coordinate system (x_1, x_2, x_3), are denoted by:

$$\sigma = \{\sigma_{11} \quad \sigma_{22} \quad \sigma_{33} \quad \sigma_{23} \quad \sigma_{13} \quad \sigma_{12}\}^T ; \qquad u = \{u_1 \quad u_2 \quad u_3\}^T .$$

In the Hybrid/Mixed Stress (HMS) finite element, two different fields are simultaneous and independently approximated in the element domain, the stress and the displacement fields:

$$\sigma = S_V X_V \tag{1}$$

$$u = U_V q_V \tag{2}$$

A third approximation, also independent from the previous ones, is used for the displacement field on the element static boundary,

$$u = U_\Gamma q_\Gamma \tag{3}$$

wherein matrices S_V, U_V and U_Γ gather the approximation functions, and vectors X_V, q_V and q_Γ are the associated weights. In a physical approach, these vectors can be interpreted as the generalized stresses and generalized displacements in the domain and on the static boundary (Freitas et al. 1999), respectively.

The HMS finite element governing system is established imposing non-locally the fundamental equations based on the Galerkin method (Freitas 1989, Pereira & Freitas 1996).

Therefore in the element domain the equilibrium conditions and the compatibility conditions are enforced in a weighted residual form, using as weight functions the displacement and stress field approximation functions, respectively. Applying this same method, the equilibrium boundary conditions are established, using as weight functions the displacement approximation functions, thus guarantying the link between adjacent elements.

Since every fundamental equation is enforced in a weighted residual form, except for the boundary compatibility conditions, which are verified locally, there is no need to establish any kind of restriction as to the nature of the approximation functions used. So a wide range of functions can be chosen for this purpose (Pereira 1993, Castro 1996).

The described method leads to the following governing system for the HMS element:

$$\begin{bmatrix} F & A_V & -A_\Gamma \\ A_V^T & \cdot & \cdot \\ -A_\Gamma^T & \cdot & \cdot \end{bmatrix} \begin{Bmatrix} X_V \\ q_V \\ q_\Gamma \end{Bmatrix} = \begin{Bmatrix} \cdot \\ -Q_V \\ -Q_\Gamma \end{Bmatrix} \tag{4}$$

A more detailed explanation on how to establish the HMS governing system can be found in (Silva & Pereira 2002).

The structural operators presented in (4) can be computed by the following expressions:

$$F = \int_V S_V^T f S_V \, dV \tag{5}$$

$$A_V = \int_V (DS_V)^T U_V dV \tag{6}$$

$$A_\Gamma = \int_{\Gamma_\sigma} (NS_V)^T U_\Gamma \, d\Gamma \tag{7}$$

$$Q_V = \int_V U_V^T \, b \, dV \tag{8}$$

$$Q_\Gamma = \int_{\Gamma_\sigma} U_\Gamma^T \, t \, d\Gamma \tag{9}$$

where f represents the flexibility matrix, t the stress prescribed on the boundaries, D the equilibrium differential operator. Matrix N contains the components of the unit outward normally associated with the differential operator D. For 3-D elastic and isotropic problems, these matrices take the form:

$$f = \frac{1}{E}
\begin{bmatrix}
1 & -v & -v & \cdot & \cdot & \cdot \\
-v & 1 & -v & \cdot & \cdot & \cdot \\
-v & -v & 1 & \cdot & \cdot & \cdot \\
\cdot & \cdot & \cdot & 2(1+v) & \cdot & \cdot \\
\cdot & \cdot & \cdot & \cdot & 2(1+v) & \cdot \\
\cdot & \cdot & \cdot & \cdot & \cdot & 2(1+v)
\end{bmatrix}; \tag{10}$$

$$D =
\begin{bmatrix}
\dfrac{\partial}{\partial x_1} & \cdot & \cdot & \cdot & \dfrac{\partial}{\partial x_3} & \dfrac{\partial}{\partial x_2} \\
\cdot & \dfrac{\partial}{\partial x_2} & \cdot & \dfrac{\partial}{\partial x_3} & \cdot & \dfrac{\partial}{\partial x_1} \\
\cdot & \cdot & \dfrac{\partial}{\partial x_3} & \dfrac{\partial}{\partial x_2} & \dfrac{\partial}{\partial x_1} & \cdot
\end{bmatrix}; \tag{11}$$

$$N =
\begin{bmatrix}
n_{x_1} & \cdot & \cdot & \cdot & n_{x_3} & n_{x_2} \\
\cdot & n_{x_2} & \cdot & n_{x_3} & \cdot & n_{x_1} \\
\cdot & \cdot & n_{x_3} & n_{x_2} & n_{x_1} & \cdot
\end{bmatrix}; \tag{12}$$

where E and v represent the Young modulus and the Poison ratio, respectively.

2.1 Condensed governing system

Assuming that the stress field approximation functions form a linearly independent base, the flexibility matrix, F, becomes a non-singular symmetric matrix. Thus the generalized stresses can be expressed in terms of the generalized displacements,

$$X_V = F^{-1}(-A_V q_V + A_\Gamma q_\Gamma) \tag{13}$$

Expression (13) allows for the elimination, from the original HMS format, of the degrees of freedom related to the generalized stresses, leading to a condensed form for the governing system:

$$\begin{bmatrix} K_{VV} & K_{V\Gamma} \\ K_{V\Gamma}^T & K_{\Gamma\Gamma} \end{bmatrix}
\begin{Bmatrix} q_V \\ q_\Gamma \end{Bmatrix} =
\begin{Bmatrix} -Q_V \\ -Q_\Gamma \end{Bmatrix} \tag{14}$$

where,

$$K_{VV} = -A_V^T F^{-1} A_V, \quad K_{V\Gamma} = A_V^T F^{-1} A_\Gamma \quad \text{and} \quad K_{\Gamma\Gamma} = A_\Gamma^T F^{-1} A_\Gamma.$$

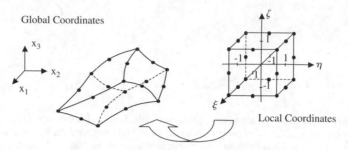

Global Coordinates

Local Coordinates

Figure 1.   Master element mapping.

Note that the condensed form (14) requires more time to compute the equation system as it involves matrix/matrix products and the inversion of the flexibility operator, and leads to a less sparse system. However, when applied to 3-D problems, these disadvantages can be counterbalanced during the solution phase, since the condensed equation system has less degrees of freedom and is more stable, as it can be shown in the following examples.

3   NUMERICAL IMPLEMENTATION

To implement the HMS model, eight and sixteen-node serendipian cubic master elements were used to represent the finite elements shape.

The mapping between the local coordinate system (ξ, η, ζ) and the global coordinate system (x_1, x_2, x_3), represented in Figure 1, can be obtained using the serendipian shape functions (Zienkiewicz & Taylor 1997),

$$x_k = \sum_{i=1}^{n} \varphi_i X_{i,k} \tag{15}$$

where k subscript denotes the global coordinate, φ_i stands for the ith nodal serendipian shape function, and $X_{i,k}$ represents the global coordinate for node i. The parameter n, can assume the value 8 for the eight-node master element or 16 for the sixteen-node master element.

Note that the HMS finite element is not isoparametric, *i.e.*, the shape and field approximation functions are not the same.

3.1   *Approximation functions*

As previously referred, the HMS model enables the use of a wide range of approximation functions. In the present work Legendre polynomials are used for this purpose. Legendre polynomials can be generated recursively (Spiegle 1968):

$$P_0(\xi) = 1 ; \qquad P_1(\xi) = \xi ; \qquad P_{n+1}(\xi) = 2\xi P_n(\xi) - P_{n-1}(\xi) - \frac{\xi P_n(\xi) - P_{n-1}(\xi)}{n+1} \tag{16}$$

where $P_n(\xi)$ represents the Legendre polynomial of degree n. Legendre polynomials present the advantage of forming an orthogonal base in $[-1, 1]$,

$$\int_{-1}^{1} P_i(\xi)P_j(\xi)d\xi = \begin{vmatrix} 0 & i \neq j \\ \dfrac{2}{2i+1} & i = j \end{vmatrix} \tag{17}$$

thus contributing to obtain highly sparse governing system matrices.

The approximation functions are defined in the local coordinate system (ξ, η, ζ), in the form:

$$S_V = \left[\overline{P}_\xi^0 \overline{P}_\eta^0 \overline{P}_\zeta^0 \langle 1 \rangle_{6\times6} \middle| \overline{P}_\xi^0 \overline{P}_\eta^0 \overline{P}_\zeta^1 \langle 1 \rangle_{6\times6} \middle| \overline{P}_\xi^0 \overline{P}_\eta^1 \overline{P}_\zeta^0 \langle 1 \rangle_{6\times6} \middle| \cdots \middle| \overline{P}_\xi^l \overline{P}_\eta^l \overline{P}_\zeta^l \langle 1 \rangle_{6\times6} \right]$$

$$U_V = \left[\overline{P}_\xi^0 \overline{P}_\eta^0 \overline{P}_\zeta^0 \langle 1 \rangle_{3\times3} \middle| \overline{P}_\xi^0 \overline{P}_\eta^0 \overline{P}_\zeta^1 \langle 1 \rangle_{3\times3} \middle| \overline{P}_\xi^0 \overline{P}_\eta^1 \overline{P}_\zeta^0 \langle 1 \rangle_{3\times3} \middle| \cdots \middle| \overline{P}_\xi^m \overline{P}_\eta^m \overline{P}_\zeta^m \langle 1 \rangle_{3\times3} \right]$$

$$U_\Gamma = \left[\overline{P}_{s_1}^0 \overline{P}_{s_2}^0 \langle 1 \rangle_{3\times3} \middle| \overline{P}_{s_1}^0 \overline{P}_{s_2}^1 \langle 1 \rangle_{3\times3} \middle| \overline{P}_{s_1}^1 \overline{P}_{s_2}^0 \langle 1 \rangle_{3\times3} \middle| \overline{P}_{s_1}^1 \overline{P}_{s_2}^1 \langle 1 \rangle_{3\times3} \middle| \cdots \middle| \overline{P}_{s_1}^n \overline{P}_{s_2}^n \langle 1 \rangle_{3\times3} \right]$$

where $\langle 1 \rangle$ stands for a $j \times j$ identity matrix, (s_1, s_2) represents the boundary local coordinates, and l, m, n are the maximum degrees adopted for the approximation functions.

3.2 Structural arrays

To implement the HMS model, it is necessary to express the structural arrays (5–9) referred to the local coordinate system,

$$F = \int_{-1}^{1} \int_{-1}^{1} \int_{-1}^{1} S_V^T f S_V |J| \, d\xi \, d\eta \, d\zeta \tag{18}$$

$$A_V = \int_{-1}^{1} \int_{-1}^{1} \int_{-1}^{1} (D'S_V)^T U_V \, d\xi \, d\eta \, d\zeta \tag{19}$$

$$A_\Gamma = \int_{-1}^{1} \int_{-1}^{1} (N'S_V)^T U_\Gamma \, ds_1 \, ds_2 \tag{20}$$

$$Q_\Gamma = \int_{-1}^{1} \int_{-1}^{1} U_\Gamma^T \hat{J} t' \, ds_1 \, ds_2 \tag{21}$$

$$Q_V = \int_{-1}^{1} \int_{-1}^{1} \int_{-1}^{1} U_V^T b |J| \, d\xi \, d\eta \, d\zeta \tag{22}$$

where J is the Jacobian matrix of the transformation (15), t' the external force vector written in the local coordinate system. D' and N' are support matrices, resulting from the transformation of the differential equilibrium operator D, and from the unit outward normal matrix, N, into the local coordinate system,

$$D = \frac{D'}{|J|} \tag{23}$$

$$N = \frac{N'}{|n'|} \tag{24}$$

where,

$$D' = \begin{bmatrix} \hat{J}_{11} & \cdot & \cdot & \cdot & \hat{J}_{31} & \hat{J}_{21} \\ \cdot & \hat{J}_{21} & \cdot & \hat{J}_{31} & \cdot & \hat{J}_{11} \\ \cdot & \cdot & \hat{J}_{31} & \hat{J}_{21} & \hat{J}_{11} & \cdot \end{bmatrix} \frac{\partial}{\partial \xi} +$$

$$+ \begin{bmatrix} \hat{J}_{12} & \cdot & \cdot & \cdot & \hat{J}_{32} & \hat{J}_{22} \\ \cdot & \hat{J}_{22} & \cdot & \hat{J}_{32} & \cdot & \hat{J}_{12} \\ \cdot & \cdot & \hat{J}_{32} & \hat{J}_{22} & \hat{J}_{12} & \cdot \end{bmatrix} \frac{\partial}{\partial \eta} + \begin{bmatrix} \hat{J}_{13} & \cdot & \cdot & \cdot & \hat{J}_{33} & \hat{J}_{23} \\ \cdot & \hat{J}_{23} & \cdot & \hat{J}_{33} & \cdot & \hat{J}_{13} \\ \cdot & \cdot & \hat{J}_{33} & \hat{J}_{23} & \hat{J}_{13} & \cdot \end{bmatrix} \frac{\partial}{\partial \zeta}$$

and

$$N' = \begin{bmatrix} n'_{x_1} & \cdot & \cdot & \cdot & n'_{x_3} & n'_{x_2} \\ \cdot & n'_{x_2} & \cdot & n'_{x_3} & \cdot & n'_{x_1} \\ \cdot & \cdot & n'_{x_3} & n'_{x_2} & n'_{x_1} & \cdot \end{bmatrix}$$

with,

$$\hat{J} = |J|\left(J^T\right)^{-1} \quad \text{and} \quad \left\{n'_{x_1} \quad n'_{x_2} \quad n'_{x_3}\right\}^T = \hat{J}\left\{n_\xi \quad n_\eta \quad n_\zeta\right\}^T$$

being n_ξ, n_η and n_ζ the components of the unit outward normal in the local coordinate system.

Taking into account the mapping defined in (15), it is possible to express the Jacobian, $|J|$, and the coefficients of matrix $\hat{J}$, in a polynomial form given by:

$$|J| = \sum \beta^{lmn} \xi^l \eta^m \zeta^n \tag{25}$$

$$\hat{J}_{ij} = \sum B_{ij}^{lmn} \xi^l \eta^m \zeta^n \tag{26}$$

where β and B represent polynomial coefficients.

Therefore all the arrays (18–22) present in the HMS finite element governing system can be determined as a linear combination of the following integrals,

$$I_{ij}^k = \int_{-1}^1 \xi^k P(\xi)_i P_j(\xi) \, d\xi \tag{27}$$

$$H_{ij}^k = \int_{-1}^1 \xi^k P_i(\xi) P_j'(\xi) \, d\xi \tag{28}$$

Based on (16) and (17) it is possible to establish analytical expressions for the former integrals (27) and (28) (Silva 2002).

Defining finite elements geometry through serendipian master elements, expressing the approximation functions and structural arrays on the element local coordinate system, and writing transformation operators (25) and (26) in a polynomial form, it is possible to implement the HMS model based on symbolic programming codes. These procedures avoid the use of numerical integration procedures and consequently the computation time in the structural governing system assemblage is substantially reduced.

4 TEST CASE

In order to test the performance of the HMS finite element model, the 3-D structure CIRIA5, a symmetric arch dam, used as a test structure by Ergatoudis et al. (1968), has been selected. The considered load is the hydrostatic pressure resulting from water at the dam maximum level.

To represent the structure shape, a mesh composed by 9 sixteen-node elements is adopted, as illustrated in Figure 2a. The structure analysis is performed using the three different levels of p-refinement, summarized in Table 1. The relation between the degrees of approximation used for the different fields is maintained.

Two different kind of static boundaries were considered: the inter-element element boundaries, and the external back and frontal boundaries of the dam, symbolized by i and e, respectively.

The entire tests were performed using a computer with the following characteristics: Alpha EV56 processor, with 600 MHz clock-rate and 128 Mb of RAM.

Using S2, S3 and S4 approximations, the performance between the standard (4), and the condensed (14), governing system format was tested. The results are summarized in Table 2.

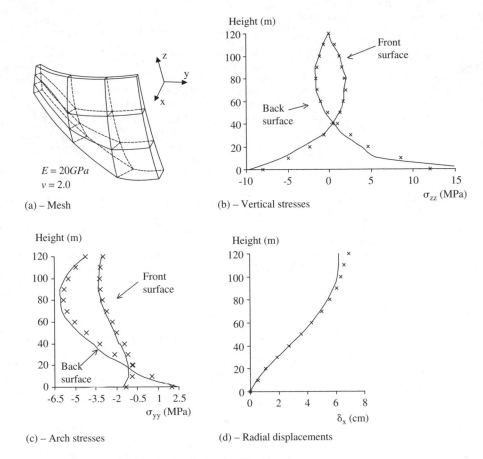

$E = 20GPa$
$v = 2.0$

(a) – Mesh

(b) – Vertical stresses

(c) – Arch stresses

(d) – Radial displacements

Figure 2.   Results obtained for CIRIA5 analysis using S3 refinement.

Table 1.   Approximations used in the problem analysis.

| | Maximum Polynomial Degree | | | | Structural Arrays Dimension | | | |
Approx.	S_V	U_V	U_Γ^i	U_Γ^e	F	A_V	A_Γ	NDOF
S2	2	1	1	2	1458	216	675	2349
S3	3	2	2	3	3456	729	1302	5487
S4	4	3	3	4	6750	1728	2139	10617

The presented results show that for 3-D problems the HMS condensed form proved clearly to be more efficient in terms of CPU time.

Figures 2b-d show the results obtained with approximation S3 for the vertical stresses (σ_{zz}), for the arch stresses (σ_{yy}) and for the radial displacements in the central cantilever, respectively. As a term of reference the solution obtained by Ergatoudis et al. (1968) is also presented, marked as "x".

5   CONCLUSIONS

The numerical implementation of a Hybrid/Mixed Stress model using Legendre polynomials for the analysis of 3D-problems was presented. For this purpose, eight and sixteen node serendipian master

Table 2. Performance of the condensed governing system format.

Approx.	Condensed form	NDOF	n° Iterations	CPU Time Mounting	Solving	Total	Deformation Energy (MN.m)
S2	Yes	891	939	0'05,5"	0'06,5"	0'12,0"	70.581
	No	2349	2928	0'02,6"	0'29,3"	0'32,0"	
S3	Yes	2031	1470	1'04,9"	1'04,2"	2'09,1"	70.150
	No	5487	4511	0'15,6"	3'28,3"	3'43,9"	
S4	Yes	3867	1762	9'33,1"	4'14,0"	13'47,1"	69.911
	No	10617	8923	2'47,3"	32'14,1"	35'01,4"	

elements were developed. The use of a symbolic programming code, to compute the governing system arrays, avoids numerical integration procedures for the structural matrices.

To test the model performance an arch dam structure was used. The numerical results show a good performance of the model when compared with other formulations.

The HMS governing system model was implemented using two different forms: the standard form, and a condensed form. Results proved that the last one is more efficient in terms of computation time.

ACKNOWLEDGEMENTS

This work is part of the research activity of the Instituto de Engenharia de Estruturas, Território e Construção (ICIST), and has been partially sponsored by Fundação para a Ciência e a Tecnologia (FCT) as part of the project POCTI/33066/ECM/2000 and by the Plurianual Funding of FCT. The support of Eng. Carlos Pina from the NEE – Dams Department of the Laboratório Nacional de Engenharia Civil is also acknowledged.

REFERENCES

Castro, L.S.C. 1996. Wavelets e Séries de Walsh em Elementos Finitos, *Doctoral Thesis, Universidade Técnica de Lisboa*, Lisboa.
Ergatoudis, J., Irons, B.M. & Zienkiewicz, O.C. 1968. Three dimensional analysis of arch dams and their foundations. *In Symposium on Arch Dams – A review of British Research and Development*, CIRIA, London.
Freitas, J.A.T. 1989. Duality and Symmetry in Mixed Integral Methods of Elastostatics, *International Journal For Numerical Methods in Engineering* 28, 1161–1179.
Freitas, J.A.T., Moitinho de Almeida, J.P. & Pereira, E.M.B.R. 1999. Non-conventional formulations for the finite element method, *Computational Mechanics* 23, 488–501.
Pereira, E.M.B.R. 1993. Elementos Finitos de Tensão – Aplicação à Análise Elástica de Estruturas, *Doctoral Thesis, Universidade Técnica de Lisboa*, Lisboa.
Pereira, E.M.B.R. & Freitas, J.A.T. 1996. A Mixed-Hybrid Finite Element Model Based on Legendre Polynomials, *Int. J. Num. Meth Engrng.* 39, 1295–1312.
Pian, T.H.H. 1978. A Historical Note About 'Hybrid Elements', *Int. J. Num. Meth Engrng.* 12, 891–892.
Silva, M.J.V. 2002. Elementos Finitos Híbridos-Mistos de Tensão – Aplicação à Análise de Barragens Abóbada, *MSc Dissertation, Universidade Técnica de Lisboa*, Lisboa.
Silva, M.J.V. & Pereira, E.M.B.R. 2002. Hybrid-Mixed Finite Element Model Based on Legendre Polynomials for 3-D Structures, *WCCM V, Fifth Word Congress on Computational Mechanics*. Vienna.
Spiegel, M.R. 1968. *Mathematical Handbook of Formulas and Tables*, Shaum's Outline Series in Mathematics, McGraw-Hill.
Zienkiewicz O.C. & Taylor R.L. 1997. (4ª Edition). *The Finite Element Method: Basic formulation and linear problems*, Vol. 1, McGraw-Hill.

A stiffness equation transfer method in frequency domain for transient structural response

H. Xue
Department of Physics, Suzhou University, Suzhou, Jiangsu, China

ABSTRACT: This paper deals with the transient structural response problems. An alternative to the direct integration techniques of transforming the equations of motion into the frequency domain is used. In this paper, the application of the stiffness equation transfer method is extended from the eigenvalue analysis of structures to transient structural response analysis in the frequency domain. In combination with fast Fourier transform, the stiffness equation transfer method in the frequency domain can be used to obtain the transient response of structures. Some numerical examples are proposed and their results are compared with those obtained by direct integration method.

1 INTRODUCTION

Numerical analysis of transient structural response problems continues to be the subject of ongoing research. General structural dynamics problems are often discretized in space with powerful techniques such as the finite element (FE) method. However, the finite element method has the disadvantage that in the case of a complex structure, a large number of nodes must be used which results in very large matrices that require large computers and more computation time. For this reason, the combined use of the finite element and transfer matrix (FE-TM) was proposed by Dokanish for the free vibration problems (Dokainish 1972). Since the publication of Dokanish's paper, several authors have proposed refinements and extensions of this method (Ohga et al. 1987, Bhutani & Loewy 1999). This method has the advantage of reducing stiffness matrix size to much smaller than that obtained with the FE method. However, it is pointed out that, in the standard FE-TM method, recursive multiplication of the transfer and point matrices are main sources of round-off errors. Particularly, in calculating high resonant frequencies or the response of a long structure, the numerical instability would be appeared and it leads to an unwanted solution. In addition, most of the previous formulations of the combined FE-TM method are only applicable to the models which have the same number of nodes on all the substructure boundaries.

In order to overcome simultaneously both these two disadvantages in the ordinary FE-TM method, the author proposed a stiffness equation transfer (SET) method for the steady state vibration response and eigenvalue analysis of structures (Xue, in press). This study is an extension of this method to the transient analysis of structures subjected various excitations. In the present method, the transfer of state vectors from left to right in the FE-TM method is transformed into a transfer of general stiffness equations in every section from left to right. The propagation of round-off errors occurred in recursive multiplication of the transfer and point matrices is avoided. At the same time, it allows different number of nodes on the right and the left boundary of each strip.

The transient response is usually determined by numerically integrating the discretized equations in time. Direct integration methods are commonly used and they are usually categorized as explicit ones and implicit ones. Explicit techniques, such as the central difference method, are most computationally efficient. However, these methods are conditionally stable. The time step must be less than some critical time step. In contrast, implicit methods are unconditionally stable. The Newmark-Beta and Wilson-Theta methods are the most popular implicit techniques. For these methods, the accuracy of the response depends on selecting an appropriate time step size.

A common feature of all direct integration schemes is that the transient response at each time step is based on the solutions computed for previous time steps. For long response times, solution accuracy can degrade if too large a step size is chosen, i.e. small time steps must be used to obtain response of the structure accurately.

To overcome these drawbacks, an alternative to the direct integration techniques of transforming the equations of motion into the frequency domain can be used. In this paper, the application of the stiffness equation transfer (SET) method is extended from eigenvalue analysis of structures to the transient structural response analysis in the frequency domain. In combination with fast Fourier transform, the SET method in the frequency domain can be used to obtain the transient response of structures. Some numerical examples are proposed and their results are compared with those obtained by direct integration method.

2 A STIFFNESS EQUATION TRANSFER METHOD IN FREQUENCY DOMAIN

2.1 *Motion equation in the frequency domain*

Without losing generality, we consider the plate shown in Figure 1. It is divided into n strips and the each strip is further subdivided into finite elements. The vertical sides dividing or bordering the strips are called sections. It is apparent that the right of section i is also the left of strip i.

The discretized finite element equations for substructure i at time t take the form

$$[M]_i\{\ddot{U}\}_i + [C]_i\{\dot{U}\}_i + [K]_i\{U\}_i = \{N\}_i + \{Q\}_i \tag{1}$$

where $[M]_i$, $[C]_i$ and $[K]_i$ represent the mass, damping and stiffness properties of substructure i, respectively. And $\{U\}_i$, $\{\dot{U}\}_i$, $\{\ddot{U}\}_i$, $\{N\}_i$ and $\{Q\}_i$ are the displacement, velocity, acceleration, internal force and external force vectors at time t, respectively. If the structure is subjected to a ground acceleration $\{\ddot{U}\}_0$, not a force, then $\{Q\}_i = -[M]_i\{\ddot{U}\}_0$, $\{U\}_i$, $\{\dot{U}\}_i$ and $\{\ddot{U}\}_i$ are the relative displacement, velocity and acceleration vectors at time t, respectively. In our study, only the nodal displacements caused by forces $\{Q(t)\}$ are concerned, so let $\{U\}_0 = \{\dot{U}\}_0 = \{0\}$, in which $\{U\}_0$ and $\{\dot{U}\}_0$, refer to the vectors of initial displacement and velocity. The time dependence in this equation may be removed by applying the Laplace transformation to obtain

$$(s^2[M]_i + s[C]_i + [K]_i)\{U(s)\}_i = \{N(s)\}_i + \{Q(s)\}_i \tag{2}$$

where s is the Laplace transform parameter, $\{U(s)\}_i$, $\{N(s)\}_i$ and $\{Q(s)\}_i$ are the Laplace transform of $\{U(t)\}_i$, $\{N(t)\}_i$ and $\{Q(t)\}_i$, respectively.

Now let $s = j\omega$, this equation may be rewritten in frequency form as follows:

$$([K]_i + j\omega[C]_i - \omega^2[M]_i)\{U(\omega)\}_i = \{N(\omega)\}_i + \{Q(\omega)\}_i \tag{3}$$

in which ω is the Fourier transform frequency parameter, $\{U(\omega)\}_i$, $\{N(\omega)\}_i$ and $\{Q(\omega)\}_i$ are the Fourier transform of $\{U(t)\}_i$, $\{N(t)\}_i$ and $\{Q(t)\}_i$, respectively. The discrete Fourier transform $\{Q(\omega)\}_i$ can be efficiently computed from $\{Q(t)\}_i$ using the Cooley-Tukey algorithm known as the FFT.

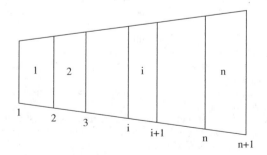
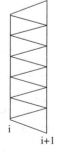

Figure 1.   Subdivision of structure into strips and finite elements.

Equation 3 can be written as

$$[G]_i\{U(\omega)\}_i = \{N(\omega)\}_i + \{Q(\omega)\}_i \tag{4}$$

where

$$[G]_i = ([K]_i + j\omega[C]_i - \omega^2[M]_i) \tag{5}$$

For each frequency parameter ω, Equation 4 can be efficiently solved with the stiffness equation transfer (SET) method described in the following. Subsequently the Fourier transform is numerically inverted to recover the time domain response.

2.2   An ordinary finite element-transfer matrix (FE-TM) method

Let $\{U\}_i^R$, $\{N\}_i^R$ and $\{Q\}_i^R$ be the right displacement, internal force and external force vectors of section i, and $\{U\}_{i+1}^L$, $\{N\}_{i+1}^L$ and $\{Q\}_{i+1}^L$ be the left corresponding vectors of section $i + 1$, so that we have

$$\begin{aligned}
\{U\}_i &= [\{U\}_i^R, \{U\}_{i+1}^L]^T \\
\{N\}_i &= [\{N\}_i^R, \{N\}_{i+1}^L]^T \\
\{Q\}_i &= [\{Q\}_i^R, \{Q\}_{i+1}^L]^T
\end{aligned} \tag{6}$$

Substituting Equation 6 into Equation 4, the latter can be written as

$$[G]_i \begin{bmatrix} \{U\}_i^R \\ \{U\}_{i+1}^L \end{bmatrix} = \begin{Bmatrix} \{N\}_i^R \\ \{N\}_{i+1}^L \end{Bmatrix} + \begin{Bmatrix} \{Q\}_i^R \\ \{Q\}_{i+1}^L \end{Bmatrix} \tag{7}$$

Matrix $[G]_i$ is the dynamic stiffness matrix for the strip i and it may be partitioned into four sub-matrices and Equation 7 may be rewritten as

$$\begin{bmatrix} [G_{11}] & [G_{12}] \\ [G_{21}] & [G_{22}] \end{bmatrix}_i \begin{Bmatrix} \{U\}_i^R \\ \{U\}_{i+1}^L \end{Bmatrix} = \begin{Bmatrix} \{N\}_i^R \\ \{N\}_{i+1}^L \end{Bmatrix} + \begin{Bmatrix} \{Q\}_i^R \\ \{Q\}_{i+1}^L \end{Bmatrix} \tag{8}$$

The displacements are continuous across section i, so that we obtain

$$\{U\}_i^R = \{U\}_i^L \tag{9}$$

Without losing generality, we suppose that there is no concentrated external load acting on section i (concentrated external load acting on section i may be treated as generalized external force on the left of strip i), due to the continuity of force at section i, we obtain

$$\{N\}_i^R = -\{N\}_i^L \tag{10}$$

Substituting Equation 9 & 10 into Equation 8 and with a little algebraic manipulation, Equation 8 can be rearranged to the form

$$\begin{Bmatrix} \{U\}_{i+1}^L \\ \{N\}_{i+1}^L \\ 1 \end{Bmatrix} = \begin{bmatrix} [T_{11}] & [T_{12}] & \{Q_1\} \\ [T_{21}] & [T_{22}] & \{Q_2\} \\ 0 & 0 & 1 \end{bmatrix}_i \begin{Bmatrix} \{U\}_i^L \\ \{N\}_i^L \\ 1 \end{Bmatrix} = [T]_i \begin{Bmatrix} \{U\}_i^L \\ \{N\}_i^L \\ 1 \end{Bmatrix} \tag{11}$$

where

$$\begin{aligned}
[T_{11}]_i &= -[G_{12}]_i^{-1}[G_{11}]_i \\
[T_{12}]_i &= -[G_{12}]_i^{-1} \\
[T_{21}]_i &= [G_{21}]_i - [G_{22}]_i[G_{12}]_i^{-1}[G_{11}]_i \\
[T_{22}]_i &= -[G_{22}]_i[G_{12}]_i^{-1} \\
\{Q_1\}_i &= [G_{12}]_i^{-1}\{Q\}_i^R, \quad \{Q_2\}_i = [G_{22}]_i[G_{12}]_i^{-1}\{Q\}_i^R - \{Q\}_{i+1}^L
\end{aligned} \tag{12}$$

Proceeding as in Dokainish's paper (Dokainish 1972), we obtain the transfer matrix of the state vectors for the total structure.

$$\begin{Bmatrix} \{U\}_{n+1}^L \\ \{N\}_{n+1}^L \\ 1 \end{Bmatrix} = [P] \begin{Bmatrix} \{U\}_1^L \\ \{N\}_1^L \\ 1 \end{Bmatrix} \tag{13}$$

in which

$$[P] = [T]_n [T]_{n-1} \cdots [T]_1 \tag{14}$$

Equation 13 relates the section variables of the left boundary of the structure to those of its right boundary. The boundary conditions of the left edge of the structure would require some components of the state vectors to be zeros. Similarly, the boundary conditions of the right edge of the structure would also require certain components of the state vectors to be zeros. The known state variables at right boundary are substituted into the above relationship to determine the unknown state variables at left boundary. After the initial state vector at left boundary is known, the state vector at the section can be obtained by recursively applying Equation 11 until all the state vectors are known. In this method, it is obvious that the sub-matrix $[G_{12}]_i$ must be a square matrix in order to obtain $[T]_i$. In addition to this, propagation of round-off errors occurs due to recursive multiplications of transfer matrix $[T]_i$ in Equation 14.

2.3 Transfer matrix for stiffness equations

In order to overcome the drawback in the ordinary FE-TM method, the present method makes a change of the transfer of state vectors from left to right in the ordinary FE-TM method to the transfer of stiffness equations of every section from left to right. At the same time, the recursive multiplications of the transfer matrix $[T]_i$ are then avoided.

Similarly as in generalized Riccati transformation of state vectors (Horner & Pilkey 1978), we assume that the generalized stiffness equations which relate the force vectors to the displacement vectors on the left of section i are given by

$$\{N\}_i^L = [S]_i \{U\}_i^L + \{E\}_i \qquad (i \geq 2) \tag{15}$$

where $[S]_i$, the coefficient matrix of the stiffness equation for section i, and $\{E\}_i$, the equivalent external force vectors on the section i.

Substituting Equation 9 & 10 into Equation 15, we obtain

$$\{N\}_i^R = -[S]_i \{U\}_i^R - \{E\}_i \tag{16}$$

Equation 16 describes the relation between the internal force vectors and the displacement vectors on the right of section i.

By expanding Equation 8, we obtain

$$[G_{11}]_i \{U\}_i^R + [G_{12}]_i \{U\}_{i+1}^L = \{N\}_i^R + \{Q\}_i^R \tag{17}$$

$$[G_{21}]_i \{U\}_i^R + [G_{22}]_i \{U\}_{i+1}^L = \{N\}_{i+1}^L + \{Q\}_{i+1}^L \tag{18}$$

Substituting Equation 16 into Equation 17, we obtain

$$\{U\}_i^R = -([G_{11}] + [S])_i^{-1} [G_{12}]_i \{U\}_{i+1}^L + ([G_{11}] + [S])_i^{-1} (-\{E\}_i + \{Q\}_i^R) \tag{19}$$

Substituting Equation 19 into Equation 18, we have

$$\{N\}_{i+1}^L = [S]_{i+1} \{U\}_{i+1}^L + \{E\}_{i+1} \tag{20}$$

where

$$[S]_{i+1} = [G_{22}]_i - [G_{21}]_i ([G_{11}] + [S])_i^{-1} [G_{12}]_i \tag{21}$$

$$\{E\}_{i+1} = [G_{21}]_i ([G_{11}] + [S])_i^{-1} (\{Q\}_i^R - \{E\}_i) - \{Q\}_{i+1}^L \tag{22}$$

Equation 20 represents the relationships between the internal force vectors and the displacement vectors on the left of section $i + 1$.

2.4   Transfer of entire structure

Supposing $[S]_2$ and $\{E\}_2$ are known, using Equation 21 & 22, $[S]$ and $\{E\}$ are transferred from the left of the second section to the right of the total structure. Hence we have

$$\{N\}_{n+1}^L = [S]_{n+1} \{U\}_{n+1}^L + \{E\}_{n+1} \tag{23}$$

By considering boundary conditions, the known force or displacement variables on right boundary of the total structure are substituted into Equation 23 to determine the unknown force or displacement variables. After the force and displacement vectors on right boundary of the total structure are solved, the force and displacement vectors at any section i are calculated from right to left by Equation 19 & 16.

It is worth notice that the transfer matrix $[P]$ for the ordinary FE-TM method is replaced by the transfer matrix $[S]_{n+1}$ in Equation 23 for the SET method. The dimension of the matrix $[S]_{n+1}$ is only half that of the matrix $[P]$. In the SET method, the storage requirements would only about half of the FE-TM method. In addition, the transfer matrix $[S]_{n+1}$ is obtained by recursively using Equation 21 & 22, and not by recursive multiplication of transfer and point matrices, so the propagation of round-off errors occurred in recursive multiplication of transfer and point matrices is thus avoided.

2.5   The method of determining $[S]_2$ and $\{E\}_2$

For strip 1, by expanding Equation 8, we have

$$[G_{11}]_1 \{U\}_1^R + [G_{12}]_1 \{U\}_2^L = \{N\}_1^R + \{Q\}_1^R \tag{24}$$

$$[G_{21}]_1 \{U\}_1^R + [G_{22}]_1 \{U\}_2^L = \{N\}_2^L + \{Q\}_2^L \tag{25}$$

It is obvious that $\{U\}_1^R$ and $\{N\}_1^R$ may be determined by using left boundary conditions of the total structure.

$\{U\}_1^R$ is known in a displacement boundary condition, Expanding Equation 25 and solving the relations between $\{N\}_2^L$ and $\{U\}_2^L$, we obtain

$$[S]_2 = [G_{22}]_1 \tag{26}$$

$$\{E\}_2 = [G_{21}]_1 \{U\}_1^R - \{Q\}_2^L \tag{27}$$

It is obvious that $\{N\}_1^R$ is known in a force boundary condition, similarly we have

$$[S]_2 = [G_{22}]_1 - [G_{21}]_1 [G_{11}]_1^{-1} [G_{12}]_1 \tag{28}$$

$$\{E\}_2 = [G_{21}]_1 [G_{11}]_1^{-1} (\{N\}_1^R + \{Q\}_1^R) - \{Q\}_2^L \tag{29}$$

In mixture boundary condition, we suppose $\{U\}_1^R = [\{U'\}_1^R, \{U''\}_1^R]^T$ and the corresponding $\{N\}_1^R = [\{N'\}_1^R, \{N''\}_1^R]^T$. If $\{U'\}_1^R$ is unknown and $\{U''\}_1^R$ is known, the corresponding $\{N'\}_1^R$ is known and $\{N''\}_1^R$ is unknown. For strip 1, Equation 8 is rearranged and repartitioned, so we have

$$\begin{bmatrix} [H_{11}] & [H_{12}] & [H_{13}] \\ [H_{21}] & [H_{22}] & [H_{23}] \\ [H_{31}] & [H_{32}] & [H_{33}] \end{bmatrix} \begin{Bmatrix} \{U'\}_1^R \\ \{U''\}_1^R \\ \{U\}_2^L \end{Bmatrix} = \begin{Bmatrix} \{N'\}_1^R \\ \{N''\}_1^R \\ \{N\}_2^L \end{Bmatrix} + \begin{Bmatrix} \{Q'\}_1^R \\ \{Q''\}_1^R \\ \{Q\}_2^L \end{Bmatrix} \tag{30}$$

Expanding Equation 30 and solving the relations between $\{N\}^L_2$ and $\{U\}^L_2$, we obtain

$$[T]_2 = [H_{33}] - [H_{31}][H_{11}]^{-1}[H_{13}] \tag{31}$$

$$\{E\}_2 = [H_{31}][H_{11}]^{-1}\left(\{N'\}^R_1 + \{Q'\}^R_1\right) + [H_{32}]\{U''\}^R_1$$
$$\quad - [H_{31}][H_{11}]^{-1}[H_{12}]\{U''\}^R_1 - \{Q\}^L_2 \tag{32}$$

3 NUMERICAL EXAMPLE

In order to investigate the accuracy and the computational efficiency of our method, we develop a program SETTDR2 based on this method for use on an IBMPC586 microcomputer and give two numerical examples for illustration.

As a first example, a simply supported beam was subjected to a step load P of 100 kg at its middle point. For this sample problem, the physical parameters of this beam are as follows: length $L = 20$ m, flexure strength $EJ = 6.4 \times 10^7$ kgm^2 and $\rho F = 987$ kg $\cdot$ s$^2 \cdot$ m^{-2}, here ρ is the mass density and F the area. The damping is neglected. The beam is divided into 20 elements and the number of nodes is 21. The total response time is 10 sec.

Figure 2 shows the displacement response of middle point of the beam for the first 10 cycles. The exact solution is shown as a solid line. Superimposed on the exact solution is the response obtained using Newmark-Beta (+) method and the SET (*) method. For this example, a time step of 0.005 sec was chosen for the Newmark-Beta method. As can be seen from Figure 2, the accuracy of the Newmark-Beta method progressively worsens for long response times. The SET method does not suffer from this progressive lessening of accuracy.

In comparing these methods, in addition to the accuracy, the computer time required to obtain each solution should also be noticed. To compute the response points at the same instants in time over the

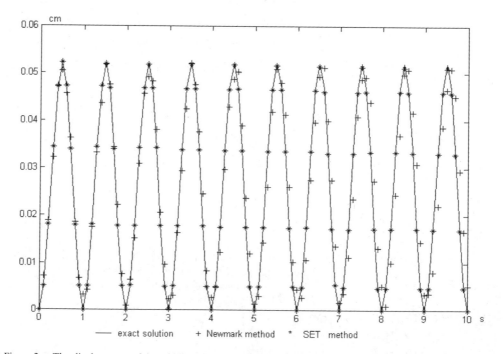

Figure 2.   The displacement of the middle point for a simply supported beam.

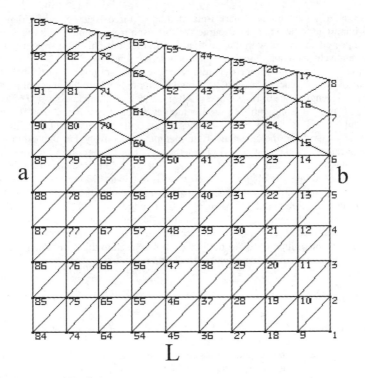

Figure 3.   A cantilever trapeziform plate.

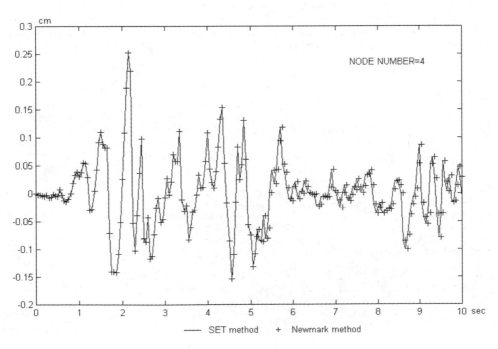

Figure 4   Vertical earthquake displacement for the node 4 of the trapeziform plate.

total response time of 10 sec, the Newmark-Beta method required 10.2 sec and SET method required 12.4 sec. It is important to note that the accuracy of the response computed with the SET method was significantly improved compared with the Newmark-Beta method. This improvement in accuracy was obtained for a relatively small additional cost in computation time.

The second example is to obtain the dynamic response of a cantilever trapeziform plate under EL Centro earthquake wave (vertical direction) as shown in Figure 3, where the physical parameters of the plate are as follows: length $l = 90$ cm, width $a = 90$ cm, $b = 60$ cm, thickness $t = 0.635$ cm, a specific weight $\gamma = 78$ kN/m^3, Poisson's ratio $\nu = 0.3$, modules of elasticity $E = 2.0 \times 10^5$ MPa, Rayleigh damping constant $\alpha = 0.1, \beta = 0.05$. Figure 4 shows the dynamic response displacement of node 4 computed by the SET method and the Newmark-Beta method, respectively. For this example, a time step of 0.005 sec was chosen for the Newmark-Beta method. As can be seen from Figure 4, very little difference exists between the results for the first 5 sec. But for a long response time, the discrepancy between the results become more obvious. It seems this is due to a shift in phase in the Newmark-Beta method. To compute response points at the same instants in time over the total response time of 10 sec, the Newmark-Beta method required 102.5 sec, the SET method required 121 sec. In this example, the number of nodes on the left boundary is 10, and that on the right boundary is 8. Most of the ordinary FE-TM method can only be applied to the chain-like structure with equal number of degrees of freedom on the boundaries, so the ordinary FE-TM method cannot be used in this case. The present method has potentially wider application than the ordinary FE-TM method.

4 CONCLUSION

A combination of the SET method and the FFT method for solving transient structural response problems was proposed and illustrated by two examples.

From the numerical results some conclusions can be drawn. Our method can offer many advantages for computing the transient response of structures. In particular, long response calculations can be determined with far greater accuracy than a step by step direct integration method. Another important use is when the response is sought for at a particular point in time. Because our method does not depend on previous time solutions, an accurate response can be computed at any required point in time.

Another topic that was not addressed in the paper is the possibility of implementing our method on a parallel processing computer. Because the equation for any point in time is independent of all other solution times, the equation for all instants in time can be solved simultaneously. This makes the computation ideally suited for parallel processing computer, our method will become more attractive from a computational cost standpoint.

REFERENCES

Bhutani, N. & Loewy, R.G. 1999. Combined finite element-transfer matrix method. *Journal of Sound and Vibration* 226 (5): 1048–1052.

Dokanish, M.A. 1972. A new approach for plate vibration: combination of transfer matrix and finite element technique. Transactions of the American Society of Mechanical Engineers, *Journal of Engineering for Industry* 94: 526–530.

Horner, G.C. & Pilkey, W.D. 1978. The Riccati transfer matrix method. Transactions of the American Society of Mechanical Engineers, *Journal of Mechanical Design* 100: 297–302.

Ohga, M., Shigematsu, T. & Hara, T. 1993. A finite element-transfer matrix method for dynamic analysis of frame structures. *Journal of Sound and Vibration* 167(3): 401–411.

Xue, H.Y. 2003. A stiffness equation transfer method for natural frequencies of structures. To be published in *Journal of Sound and Vibration* (in press).

Computational Methods in Engineering and Science, Iu et al. (eds)
© 2003 Swets & Zeitlinger, Lisse, ISBN 90 5809 567 3

Steady state of deformation of plane elliptical particles: a discrete element model simulation

A. Eliadorani
Research Associate, University of British Columbia, Vancouver, Canada

ABSTRACT: The concept of steady-state deformation in granular media is investigated by numerical simulation of plane elliptical particles. The numerical simulation that is adopted is the discrete element model (DEM). The advantage of numerical simulation is that complete micro-feature information on the assembly can be obtained as well as the continuation of the tests up to 60 percent axial strain. Several biaxial tests were carried out on 1001 particles assembly to a constant vertical strain rate along with a constant confining pressure in horizontal direction. It was found that the mechanical behaviour of assemblies of elliptical particles is not significantly different from that of real granular materials. For real sands, compression line crosses the steady state line while for simulated materials the compression line likely does not cross the steady state. The likely reason is that for real sands the contact force-interparticle displacement relationship are non-linear while for simulated materials a linear contact force-displacement is considered.

1   INTRODUCTION

Micromechanics of granular materials is a discipline that approaches the study of materials such as sands from the point of view of particulate mechanics. Similar to continuum mechanics, micromechanics is concerned with the engineering properties of materials but approaches the problem description on a more-fundamental level.

A description of the behaviour of granular materials can be characterized by simulating the movement of individual particles in a large assembly of particles using discrete particle (element) models. The discrete element models (DEM) involve the solution of equations of motion for an assembly of particles. The advantage of the discrete element model over continuum models and physical tests in that the overall or net behaviour of the assembly due to the collective effect of all of the particle–particle interactions throughout the assemblage can be predicted. Thus, the physics of the various particle contacts, normal stresses, and tangential stresses dictate the ultimate behaviour of the assembly.

Rothenburg and Bathurst (1992) have shown that the co-ordination number for an assembly of elliptical particles is dependent on the ellipse eccentricity. The co-ordination number is defined as the average number of contacts per particles. They have shown that there is an optimum particle eccentricity that results in a packing of maximum density (i.e. eccentricity = 0.2).

This paper presents the results of numerical simulation on an assembly of elliptical particles. The simulations were made to investigate the behaviour of granular materials as a function of confining pressure and fabric and to interpret the numerical results in terms of the steady-state. The steady-state deformation tests, which requires biaxial tests at large strains, were carried out on 1001 particles assemblies. The obtained isotropic compression and steady state lines are compared with others reported in the literature. The differences of biaxial simulations and physical tests are discussed. Also, the contribution of the three induced anisotropy strength components (fabric & force components) to the overall shearing resistance at all strain levels under different confining pressures are examined. The effect of increase in co-ordination number on shear strength and their relationship are explained.

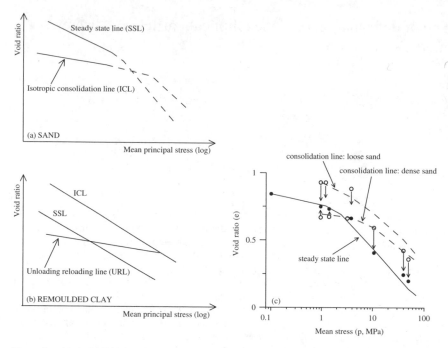

Figure 1.   (a) & (b) Initial and steady state lines in sand and remoulded clay (c) isotropic compression and steady state data for Chattahoochee River sand (modified from Vesic & Clough 1968).

2   STEADY-STATE DEFORMATION

The steady state of deformation is that state where the material deforms continuously at constant volume with no change in stress level. If the various stress conditions which correspond to steady state are plotted on a graph of void ratio versus log mean stress, they form a line in this co-ordinate space. This line is referred to as the critical state line (CSL) or the steady state line (SSL).

A central theme of steady state of deformation in soil mechanics is that it is not sufficient to consider the mechanical behaviour of soils in terms of stresses alone; some structural variables such as void ratio or specific volume are also required. A single steady state point on a void ratio – mean stress plot is defined by a triaxial test in which a sample deforms under conditions of constant effective stress, void ratio and velocity. A series of simulated biaxial tests were carried out on sample at different void ratios and confining stress levels to define a number of steady state points. When the stress conditions which correspond to steady state are plotted on a graph of void ratio versus mean stress, they form a line in this co-ordinate space which is referred to as the critical state line or the steady state line.

A steady (critical) state model has been developed by remolded clays (Roscoe & Burland 1968) and defines "parallel lines" in void ratio (e) – log mean stress (ln P) space for consolidation and shear strength of normally consolidated samples and incorporates much of the more complicated behaviour associated with overconsolidation (Fig. 1). In sands, each isotropic compression line is a virgin compression line (by contrast clays have only one virgin compression line) so that there are infinite number of virgin consolidation lines which are shown in Figure 1c (Vesic & Clough 1968) and Figure 2 (Jefferies & Been 1987).

The steady state approach can be very instrumental and useful in determining the anticipated response (i.e. dilative and work-softening or contractive and work-hardening) of granular materials. However, such an approach actually offers insight into the deformation responses of the assembly and the type and magnitude of volumetric distortion which can be anticipated. This approach provides a basis for much of the engineering design of hydraulic sand fill structures for artificial islands (Jefferies et al. 1988, Sladen et al. 1985). The widely used liquefaction evaluation procedure (Poulos et al. 1985) is based on the strength at steady state.

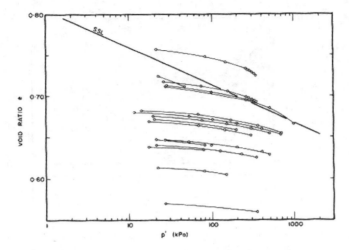

Figure 2.   Isotropic compression and steady state lines from triaxial tests on Kogyuk sand (Jefferies & Been 1987).

3   NUMERICAL SIMULATION

The discrete numerical simulation presented in this paper is based on Cundall and Strack (1979)'s discrete element model (DEM), in which discrete particles are modeled as particle in two dimensions, each obeying Newton's equation of motion. The discrete element method implements an explicit time-finite-difference solution in which each calculation cycle includes the application of two force-displacement laws at all particle contacts. Thus, in DEM, the calculation alternate between the application of Newton's second law of motion for particles and the force-displacement law for contacts. The movement of particles in a stressed assembly of particles are the results of the propagation of the disturbance due to the external application of forces and displacements. The contact forces at equilibrium and displacements are found through a series of calculations which track the movements of individual particles. Since the particles are not in perfect equilibrium with their neighbors, there are unbalanced forces between particles. Because of these unbalanced forces, the particles try to rearrange themselves in such a way that equilibrium is obtained. The task of the solution scheme is to determine a set of displacements that will bring all particles to equilibrium.

The existence of a contact is examined through the assembly, after application of the motion law to all particles. If a contact exists between two particles, based on defined normal and shear contact stiffness, the new contact forces and displacements are computed. Rothenburgh and Bathurst (1991) have shown that the contact stiffness, as well as the displacement response in contacts for ellipses, are similar for discs except that relative displacements used to determine contact forces are controlled by motions of centers of curvature at contact points. However, to determine the forces between particles, the linear contact stiffness is used.

3.1   Biaxial shear test simulations

The numerical simulation was made on assemblies of elliptical particles consisting of 1001 ellipse systems with an eccentricity equal to 0.2 and 20 different particle mean radii. For this study, a system with an eccentricity of 0.2 and an average co-ordination number (i.e. $\gamma = 5.8$) corresponding to the densest state of the system for the prescribed grain size distribution was chosen (Rothenburg & Bathurst 1991). The number of particles was chosen to approximate a log-normal size-distribution that is considered typical for a well-graded granular media. The number of particles (i.e. 1001) was a compromise between a desire to have as large an assembly as possible, to ensure a statistically representative system but, at the same time, to ensure that computation time was not excessive.

After assembly generation, a compacted dense isotropic assembly was created as the starting configuration for all tests reported in this study. This isotropically compacted dense assembly was subsequently

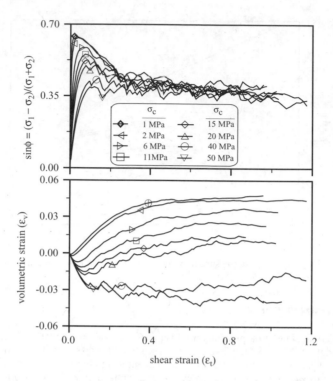

Figure 3.   Relationship between sin $\phi_{mobilized}$, shear strain, and volume change for elliptical particles.

subjected to biaxial compression under constant horizontal stress and with an interparticle friction set to 0.5 ($\mu = 0.5$). Several test samples were created by subjecting the initial sample to different hydro-static confining stresses. The steady-state deformation tests, which required biaxial tests at large strains, were carried out on all samples to determine the steady state line.

3.2   *Macroscopic results*

Following the isotropic consolidation tests, the assemblies were subjected to biaxial compression tests under constant horizontal stress (σ_2 = constant) and constant vertical strain rate ($\dot{\varepsilon}_1$ = constant). The macroscopic behavior measured at the boundaries of the tests is shown in Figure 3. In Figure 3, sin $\phi_{mobilized} = (\sigma_1 - \sigma_2)/(\sigma_1 + \sigma_2)$, defined in terms of principal stresses σ_1 and σ_2, is plotted against the shear strain ($\varepsilon_1 - \varepsilon_2$) for all simulation tests.

The qualitative features of Figure 3 indicate that the behavior of the biaxial tests is comparable to the behavior of dense granular systems. Changes in the stress–strain behavior of dense sand in triaxial tests due to changes in confining pressure are well known. Sands of the same initial relative density expand at low stresses and compress at high stresses. Tests with high confining stresses show a very weak attempt to strain harden before leveling off. On the other hand, other tests show definite strain harden-ing up to a plateau at which time failure occurs and steady state is reached. Tests with low confining pressures contract initially and subsequently dilate, but tests with high confining pressures contract continuously to steady-state.

The numerical simulations indicated that under monotonic loading, the peak friction angle decreased and the strain required to reach this peak increased with increasing confining stress level. The sample's initial state had a very large effect on this behaviour. With reference to Figure 3, an increase in consol-idation pressure (σ_c) reduced the dilative tendencies of the sample and increased the strain to failure. The peak dilatancy rate and volumetric strain values at steady state were decreased with increasing confining stress level.

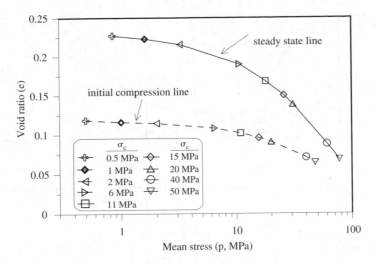

Figure 4. Isotropic compression and steady state lines from biaxial simulation tests.

3.3 Determined steady state line

Figure 4 shows the measured isotropic compression and steady state lines in the compression plane void ratio (e) – mean stress, $p = (\sigma_1 + 2\sigma_2)/3$, with the mean stress plotted on a logarithmic scale. The isotropic compression tests have a flatter slope than the steady state line at the same stress level. The general shape of isotropic compression line can be described as approximating the elastic response at low confining stress and smoothly increasing with confining stress to become parallel to the steady state line at large confining stresses resembling the compression behavior of clays. In clays, the virgin compression line and steady state line are parallel at any stress level (Roscoe et al. 1958, Fig. 1b) but in sands compression lines and SSL are parallel only at large stress levels because at this range, sand cannot remember any higher preconsolidation pressure.

Due to excess of overlapping of particles, the continuation of the tests at confining pressures of more than 50 MPa was not possible. Comparison of these two lines from numerical simulation in this study with real sands (e.g. Vesic & Clough, 1968, Jefferies & Been 1987, see Fig. 1c and Fig. 2) shows that for real sands, compression lines cross the steady state line while for simulated tests the steady state line likely does not cross the compression line. This indicates that simulated materials are too compressible compared to real sands. The likely reason is that for real sands the contact force-interparticle displacement relationship is non-linear while in the current simulation, a linear contact force-interparticle displacement was used.

3.4 Microscopic results

To present the microscopic results, it is useful to introduce basic concepts and terminology related to the characterization of the microstructure.

The term fabric is defined as the arrangement of particles and includes the description of particle contacts and their orientations. An "anisotropy system" is generally taken to refer to a system where material behavior is dependent upon direction. Anisotropy may be the result of the mode of deposition during the formation process (inherent anisotropy) or it may be induced during deformation. In this study, anisotropy is limited to that which is induced during deformation. Macroscopic behavior anisotropy must be the result of the existence of anisotropy of interparticle contacts (fabric anisotropy) and the magnitude of the force carried by those contacts (force anisotropy).

The isotropic distribution of contact forces for the initial compacted assembly clearly reflects the hydrostatic stress state under which it was created. The thickness of each contact force line is proportional to the magnitude of the force which each contact carries. At peak sand steady state, the orientation and intensity of contact force chains is clearly biased in the direction of the maximum principal stress.

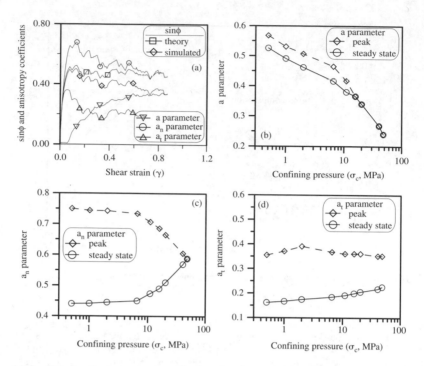

Figure 5. (a) Coefficients of anisotropy and verification of stress-force-fabric relationship (b) contact orientation anisotropy (parameter a) (c) average normal force anisotropy (parameter an) (d) average shear force anisotropy (parameter at) at peak and steady state versus confining pressure.

From plots in the form of a polar histogram, the distribution of contact and average inter particle force components with respect to orientation is visually apparent. This histogram can be approximated by second Fourier component of a truncated Fourier series (Rothenburg 1980). The degree of contact orientation and contact force anisotropies can be measured by the coefficients a (contact orientation anisotropy), a_n (normal force anisotropy), a_t (tangential force anisotropy) from the second Fourier component. The initial distributions of contact normals and average interparticle force components show that under a hydrostatic boundary stress condition, anisotropy coefficients, a, a_n, and a_t are essentially near zero (i.e. isotropic system). The test results showed that concurrent with increasing deviatoric stress and dilatancy there was an increase in contact anisotropy and decrease in force components anisotropy. On the other hand, the directions of anisotropy are essentially coincident with the direction of maximum principal stress; coincidence of force, fabric, and stress tensor was observed at all stages.

3.5 Parameters of anisotropy and deviatoric stress

Static equilibrium dictates that external loads applied to any system must be balanced by internal forces. Figure 5a shows the typical measured secant angle of friction together with coefficients of anisotropy a, a_n, a_t. Superimposed on the figure is the predicted curve for $\sin \phi$ using (Rothenburg & Bathurst 1989):

$$\sin \phi = \frac{\sigma_1 - \sigma_2}{\sigma_1 + \sigma_2} = \frac{1}{2} \frac{(a + a_n + a_t)}{1 + (\frac{a a_n}{2})} \qquad (1)$$

The term $1 + a\, a_n/2$ in Equation 1 which is small for disk-shaped particles is significant for elliptical particles. Equation 1 implies that the shear strength of the assembly is related to its aggregate ability to (i) develop anisotropy in contact orientation and (ii) withstand direction dependent average contact forces. The predicted curve appears to be a reasonable approximation to the measured $\sin \phi$. The discrepancies at large strains may relate to development of anisotropy in the orientation of ellipse axes. From the same figure, it is possible to trace the relative contributions of assembly anisotropies to the shear capacity of the system. It appears from Figure 5a that the major contribution to mobilized shear strength in these assemblies at all strain levels is due to anisotropy in normal contact forces plus stress-induced anisotropy in fabric (contact normals). Physically, this means that assembly microstructure evolves so that the capacity of the system is due to contact orientations and interparticle forces which attempts to align themselves with the direction of maximum load.

3.6 *Effect of confining stress on anisotropy coefficients*

Figure 5b through 5d summarizes peak and steady state values of coefficient terms $\mathbf{a}$, $\mathbf{a_n}$, $\mathbf{a_t}$ determined from the results of numerical simulation on elliptical particles having a range of confining stress. A study of the data shows that generally anisotropy in these system is sensitive to the magnitude of confining stress.

As confining pressure increased, the assembly was less able to sustain higher levels of anisotropy in fabrics; higher confining pressure prevents disruption of interparticle contacts (Fig. 5b). The development of the parameter $\mathbf{a}$ is a manifestation of damage owing to preferential loss of interparticle contacts which ultimately reduces $\mathbf{a_n}$.

By reference to Figure 5c, as confining stress increased, the ability of the assembly to sustain normal force anisotropy ($\mathbf{a_n}$) at peak shear was decreased; this is consistent with macroscopic results. Also, it is shown that at higher stress levels, the anisotropy at peak and steady state are matched and characterize loose behavior. At lower confining stresses, after the peak point when the assembly dilates, the conglomerates of particles that transfer high normal forces lose their lateral support and can no longer sustain high contact forces. This is shown in Figure 5c as lower values of $\mathbf{a_n}$ at steady state are obtained at low confining stress. The large values of $\mathbf{a_n}$ determined from the simulated test point out that assemblies tend to transfer load along preferential paths characterized by high normal forces. The parameter $\mathbf{a_t}$ shows a slight increase in steady state values with increasing confining stress (Fig. 5d). The peak values of $\mathbf{a_t}$ shows independency of $(\mathbf{a_t})_{max}$ on confining stress. The peak of $\mathbf{a_t}$ is the point when interparticle slip initiates.

Comparison of parameters $\mathbf{a}$ and $\mathbf{a_n}$ at steady state shows that while anisotropy in contact orientations (parameter $\mathbf{a}$) is the largest contributor to the overall shearing resistance at low confining stress (52% of $\sin\phi_{ss}$), at high confining stresses the parameter $\mathbf{a_n}$ becomes the major contributor (58% of $\sin\phi_{ss}$). Also the contribution of shear contact forces to assembly shear capacity at high confining stress is comparable to normal contact orientation. This confirms the pressure sensitivity of granular materials; the assembly at high confining stress can sustain significant interparticle forces since there is no dilatancy in the system.

4 CONCLUSIONS

Determination of the steady state line of sand is important to several aspects of the engineering of sand fills or natural sand deposits. For the purpose of this study, several biaxial simulation tests on elliptical particles were performed to investigate the steady state deformation. In general, it was found that the mechanical behavior of assemblies of elliptical particles at steady state is not significantly different from that of real granular materials. Although there is similarity in strength properties with real sands, deformation characteristics of planar systems are somewhat different due to higher density of contacts in assemblies of elliptical particles.

The presented study addressed the important topic of pressure-sensitivity of granular materials. This feature is very important to characterize the behavior of granular materials as it is well known that even a very dense sand can behave as a seemingly loose material when sheared under high confining pressure. Experimental studies by Been et al. (1991) convincingly demonstrated that major feature of sand behavior are controlled not so much by pressure and density but by a potential for volume change under given confining stress conditions. The present study has confirmed this result.

The simulated biaxial tests on elliptical particles showed a similar initial compression and steady state lines as real sands. For real sands, compression line crosses the steady state line while for simulated materials the compression line likely does not cross the steady state line. This indicates that simulated materials are too compressible compared to real sands. The likely reason is that for real sands the contact force-interparticle displacement relationship are non-linear while for simulated materials a linear contact force-displacement is considered.

The numerical results show that the three induced anisotropy strength components contribute to the overall shearing resistance at all strain levels. It appears, while anisotropy in contact orientation (parameter $\mathbf{a}$) is the largest contributor to the overall shearing resistance at low confining stress, at high confining stress the parameter $\mathbf{a_n}$ becomes the major contributor. It is also shown that the contribution of the parameter $\mathbf{a_t}$ at high confining stresses is comparable to normal orientation anisotropy. This confirms the pressure sensitivity of granular materials; the assembly at ultimate shear strength can sustain significant interparticle forces with increasing confining stress.

The steady decrease in the peak angle of friction with increasing confining stress is related to reduction of fabric anisotropy and contact normal force anisotropy which is consistent with microscopic results at higher confining stress can prevent disruption of interparticle contacts.

ACKNOWLEDGEMENTS

Represented herein is a portion of a thesis submitted in partial fulfillment of the requirements for the degree of Master of Applied Science prepared by the author in the Department of Civil Engineering of University of Waterloo. The author wishes to thank his supervisors, L. Rotheburg and E.L. Matyas, for their guidance during the course of this work.

REFERENCES

Been, K., Jefferies, M.G. & Hachey, J. 1991. The critical state of sands. *Géotechnique* 41(3): 365–381.
Jeffereis, M.G. & Been, K. 1987. Use of critical state representations of sand in the method of stress characteristics. *Canadian Geotechnical Journal* 24(3): 441–446.
Jefferies, M.G., Rogers, B.T., Stewart, H.R., Shinde, S.B., James, D.A. & Williams-Fitzpatric, S. 1988. Island construction in the Canadian Beaufort Sea. *ASCE Specialty Conference Hydraulic Fill Structures*: 816–883. Fort Collins.
Poulos, S.J. 1981. The steady state of deformation. *Journal of the Geotechnical Engineering Division, ASCE* 107(5): 553–562.
Poulos, S.J., Castro, G. & France, J.W. 1985. Liquefaction evaluation procedure. *Journal of the Geotechnical Engineering Division, ASCE* 111(6): 772–792.
Roscoe, K.H. & Burland, J.B. 1968. On the generalized stress–strain behavior of "wet" clays. In *Engineering Plasticity* 535–609. Cambridge: Cambridge University Press.
Roscoe, K.H., Schofield, M.A. & Wroth, C.P. 1958. On the yielding of soils. *Géotechnique* 8(1): 22–53.
Rothenburg, L. 1980. *Micromechanics of idealized granular systems*. PhD Thesis, Department of Civil Engineering, Ottawa: Carleton University.
Rothenburg, L. & Bathurst, R.J. 1989. Analytical study of induced anisotropy in idealized granular materials. *Géotechnique* 39(4): 601–614.
Rothenburg, L. & Bathurst, R.J. 1991. Numerical simulation of idealized granular assemblies with plane elliptical particles. *Computer and Geotechnics* 11: 315–329.
Rothenburg, L. & Bathurst, R.J. 1992. Micromechanical features of granular assemblies with plane elliptical particles. *Géotechnique* 42(1): 79–95.
Sladen, J.A., D'Hollander, R.D., Krahn, J. & Mitchell, D.E. 1985. Back analysis of the Nerlerk berm liquefaction slides. *Canadian Geotechnical Journal* 22(4): 462–466.
Strack, O.D.L. and Cundall, P.A. 1979. The distint element method as a tool for research in granular media, Part I, *Report to NSF concerning Grant Eng. 76-20711*, Dept. of Civil and Mining Eng., U of Minnesota, USA.
Vésic, A.S. & Clough, G.W. 1968. Behaviour of granular material under high pressures. *Journal of the Soil Mechanics and Foundation Engineering Division ASCE* 94(3): 661–688.

Computational Methods in Engineering and Science, Iu et al. (eds)
© 2003 Swets & Zeitlinger, Lisse, ISBN 90 5809 567 3

Modelling of shape and profile by using finite element method in cold rolling of thin strip

Z.Y. Jiang, A.K. Tieu & C. Lu
Faculty of Engineering, University of Wollongong, Wollongong NSW, Australia

ABSTRACT: In this paper, a new analysis for cold rolling of thin strip by a 3-D rigid plastic finite element method is developed. This model takes into account the friction variation in the roll bite. The computed rolling force, spread and forward slip are compared with the measured values. The effect of tension on rolling pressure, spread and forward slip is discussed. A characteristic of the distribution of the strip speed at the exit of the roll bite and the strip shape and profile along the width are also presented for the case of friction variation.

1   INTRODUCTION

A significant consumption of thin strip each year and the high quality requirement make it necessary to model accurately thin strip rolling. A robust and accurate mathematical model of the roll bite can predict key parameters such as load, torque and forward slip as a function of the rolling parameters (Le & Sutcliffe 2001). Figure 1 shows a schematic of the metal rolling process and edge drop of the rolled strip. In order to obtain a significant benefit and remain competitive in the strip rolling industry, many metal manufacturers have paid more attentions to improve its set-up and control algorithms on rolling mills. One of the important features of thin strip is its shape and profile (e.g. an edge drop in Fig. 1(b)), which affects significantly the operation of rolling mills and the quality of the strip for user.

Fleck & Johnson (1987) developed a new theory of cold rolling of thin foil and this theory is improved in the model of Fleck et al. (1992). Sutcliffe & Rayner (1998) verified a theoretical predicting model with experiments. A robust model for rolling of thin strip and foil was developed by Le & Sutcliffe (2001). Yuen and co-workers (1996, 1995) have extended a model for thin strip rolling to include strain hardening of the strip. The above analysis of the rolling of thin strip and foil is based on slab method and a constant friction was employed. Hartley et al. (1987, 1989) used an elastic-plastic finite element method to analyse the slab and flat rolling, and Hartley & Pillinger et al. (1997) also

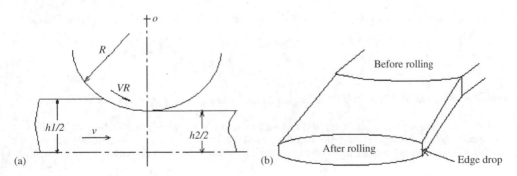

Figure 1.   (a) Schematic of metal rolling process and (b) edge drop of the rolled strip.

investigated the section rolling. In the analysis a friction-layer technique was used and a friction factor introduced. Osakada et al. (1982) developed a slightly compressible material formulation for the analysis of metal forming by the rigid plastic FEM. Mori et al. (1982) used the rigid plastic FEM with slightly compressible material to analyse the plane-strain rolling. Some researchers have also used the rigid plastic/visco-plastic FEM to solve the strip rolling, shape rolling and the edge rolling (Liu 1994, Kobayashi et al. 1989). Jiang et al. (2000a, b, 2001) has investigated the special strip rolling by rigid plastic/visco-plastic FEM. Jiang et al. (2003a, b) have also developed a finite element model which considers the friction variation in thin strip rolling.

Previous research work has shown that friction at the interface of strip and work rolls has a significant effect on the mechanics of rolling. The traditional approach is to assume that the frictional force in the roll bite is proportional to the normal force, with the friction coefficient remaining constant in the roll bite. However, this will incur a loss of accuracy in the roll gap model, and affect the shape and profile of the strip. In order to understand the friction mechanism in metal rolling processes, research has been carried out in both experiments and theoretical modelling. Lenard (1992) and Liu et al. (2001) have shown that the friction varies in the roll bite, and it affects the rolling force and the accuracy of on-line models. Therefore, in this study an analysis for cold rolling of thin strip by a 3-D rigid plastic finite element method is developed. This model takes into account the friction variation in the roll bite. The computed rolling force, spread and forward slip are compared with the measured values. The effect of tension on rolling pressure, spread and forward slip is discussed. A characteristic of the distribution of the strip speed at the exit of the roll bite and the strip shape and profile along the width for the case of friction variation in the roll bite are also presented.

2 THEORETICAL BACKGROUND

2.1 *Basic equations*

The analysis of thin strip rolling makes use of the finite element method based on the flow formulation model for slightly compressible materials (Mori & Osakada 1982). The main advantage of this approach over the conventional flow formulation based on the Levy-Mises rigid-plastic constitutive equations is the use of the stress–strain rate relationship directly by

$$\sigma_{ij} = \frac{\bar{\sigma}}{\dot{\bar{\varepsilon}}} \left(\frac{2}{3} \dot{\varepsilon}_{ij} + \delta_{ij} (\frac{1}{g} - \frac{2}{9}) \dot{\varepsilon}_v \right) \tag{1}$$

where δ_{ij} = Kronecker delta; g = a small positive constant varying between 0.01 to 0.0001, which is a factor conferring a small degree of compressibility; $\bar{\sigma}$ = effective stress and $\dot{\bar{\varepsilon}}$ = effective strain rate are defined, respectively, by

$$\bar{\sigma} = \sqrt{\frac{3}{2} \sigma'_{ij} \sigma'_{ij} + g \sigma_m^2} \tag{2}$$

where σ'_{ij} = deviatoric stress tensor; σ_m = hydrostatic stress. The equivalent strain rate is written as

$$\dot{\bar{\varepsilon}} = \sqrt{\frac{2}{3} \dot{\varepsilon}'_{ij} \dot{\varepsilon}'_{ij} + \frac{1}{g} \dot{\varepsilon}_v^2} \tag{3}$$

where $\dot{\varepsilon}'_{ij} = \dot{\varepsilon}_{ij} - \delta_{ij} \dot{\varepsilon}_v / 3$ is the deviatoric strain rate tensor and $\dot{\varepsilon}_v = \dot{\varepsilon}_x + \dot{\varepsilon}_y + \dot{\varepsilon}_z$ is the volumetric strain rate.

If tractions are applied to the entry and exit surface of the strip, the solution of the material flow in the plastically deforming region must be of a geometrically self-consistent pattern that ensures the slightly compressibility requirements on the velocity field and minimizes the following functional:

$$\Phi = \iiint_v \bar{\sigma} \dot{\bar{\varepsilon}} dv + \iint_{s_f} \tau_f \Delta V_f \, ds \pm \iint_{s_v} T_1 v ds = \phi^p + \phi^f + \phi^t \tag{4}$$

where the first term on the right hand side is the work rate of plastic deformation (ϕ^p), $\bar{\sigma}$ the equivalent stress and $\dot{\bar{\varepsilon}}$ the equivalent strain rate. The second term on the right hand side is the work rate of friction (ϕ^f); ΔV_f being the relative slipping velocity along the surface of contact between the rolled material and the rolls, where the frictional shear stress τ_f is applied. The third term on the right hand side is the work rate (ϕ^t) of tension. T_1 is the tension and v is the velocity of the cross section with tension. Here "$-$" indicates the front tension, and "$+$" the back tension.

2.2 Friction modelling

In the friction variation model the friction varies along the contact length of the deformation zone. The sense of the friction shear stress is opposed to the relative slipping velocity, Δv_f. In this study, the frictional shear stress model is as follows.

$$\tau_f = -K_i m \sigma_s \frac{\Delta v_f}{\sqrt{\Delta v_f^2 + e^2}} \tag{5}$$

where m = friction factor; σ_s = yield stress; K_i = a coefficient describing the changes of frictional shear stress in the deformation zone with K_1 used in the forward zone and K_2 in the backward zone; e is a small positive constant. The relative slip velocity, Δv_f, between the rolled material and roll is given by

$$\Delta v_f = \sqrt{\left(v_x \sec\alpha - V_R\right)^2 + v_y^2} \tag{6}$$

where v_x and v_y are the velocity components in the x and y directions respectively, α = angular position of the node (see Fig. 1a); V_R = the tangential velocity of the roll. If $e = 0.0$ and $K_1 = K_2 = 1.0$, Equation (5) shows a constant friction model.

2.3 Calculation of forward slip and spread

The forward slip can be calculated by Equation (7):

$$S = \frac{\sum\limits_{i=1}^{m_2} v_{xi}}{m_1 V_R} - 1 \tag{7}$$

where v_{xi} = node velocity at exit of rolled material; m_1 = number of nodes at the exit of strip. The spread can be formulated by Equation (8):

$$\Delta\bar{b} = \frac{1}{m_2} \sum\limits_{k=1}^{m_2} \Delta b_{zk} \tag{8}$$

where m_2 = number of elements in thickness direction; Δb_{zk} = spread in thickness direction at kth element.

3 SIMULATION EXPERIMENTAL CONDITIONS

3.1 FE mesh

As the deformation is symmetric about both the y–x and z–x planes, one quarter of the deforming workpiece as shown in Figure 2, was studied. Isoparametric hexahedral elements with eight Gauss points were used throughout the strip, including the corner at entry to the roll gap. The number of elements along the x, y and z directions are 10, 5 and 4 respectively, resulting in a total number of 200 elements and 330 nodes. There is a thin element layer (Jiang et al. 2001) at the entry of the roll bite.

3.2 Velocity boundary conditions

The velocity boundary conditions are shown in Table 1. v_x, v_y, and v_z are the velocity components in directions x, y, and z, respectively.

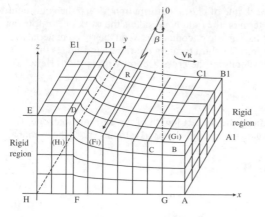

Figure 2.   FE mesh used for analysis of thin strip rolling.

Table 1.   Velocity boundary conditions (see Fig. 2).

Position	Velocity
AA_1B_1B	$v_y = v_z = 0$
EHH_1E_1	$v_y = v_z = 0$
CDD_1C_1	$v_z/v_x = -tg\beta$
BCC_1B_1	$v_z = 0$
HH_1A_1A	$v_z = 0$
$EHAB$	$v_y = 0$

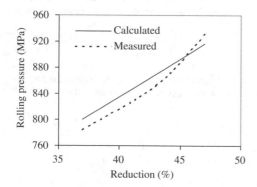

Figure 3.   Comparison of calculated rolling pressure and measured value.

Figure 4.   Comparison of calculated spread and measured value.

3.3   Flow model of material

Low carbon steel was used to simulate the strip rolling. The constitutive model (Liu 1994) used to describe the behaviour of this low carbon steel is

$$\sigma = 245.98 + 702.66\,\bar{\varepsilon}^{0.58} \quad (MPa) \tag{9}$$

where $\bar{\varepsilon}$ = plastic strain.

4   RESULTS

4.1   Comparison of calculated results with measured values

For the rolling of strip with thickness 1.9 mm, width 30.0 mm before rolling, $K_1 = 0.8$ and $K_2 = 1.2$, $e = 0.1$, $m = 0.25$ and $g = 0.01$ and rolling speed of experimental mill 0.316 m/s. Figure 3 shows a comparison of calculated rolling force and measured value. It can be seen that the rolling pressure increases with reduction, and the calculated rolling pressure is in good agreement with the measured value. Figure 4 shows a comparison of calculated spread with measured value. It can be seen that the spread of strip increases as reduction increases, and the calculated result is closer to the measured value. A comparison of the calculated forward slip for friction variation model with measured value is shown in Figure 5. It can be seen that the calculated forward slip is closer to the measured result for reductions over 45%.

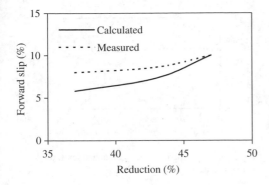

Figure 5. Comparison of calculated forward slip and measured value.

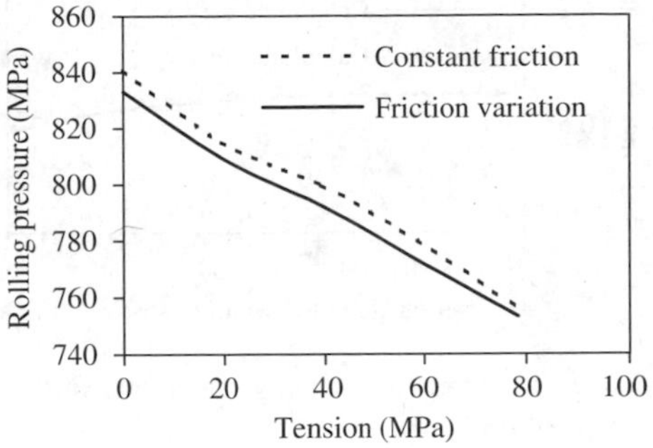

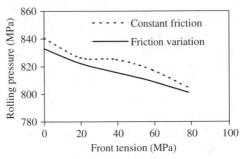

Figure 6. Effect of tension on rolling pressure for $T_f = T_b$.

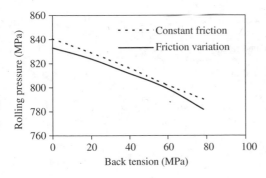

Figure 7. Effect of back tension on rolling pressure.

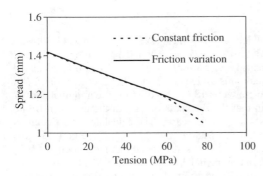

Figure 8. Effect of front tension on rolling pressure.

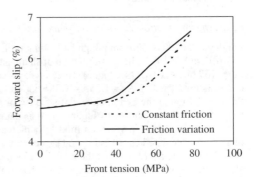

Figure 9. Effect of tension on spread for $T_f = T_b$.

Figure 10. Effect of front tension on forward slip.

4.2 Effect of tension on simulation results

For cold rolling of low carbon steel, when $m = 0.25$, thickness of strip before rolling is 1.9 mm, $K_1 = K_2 = 1.0$, $e = 1.0$, the effect of tension on rolling force is shown in Figures 6–8 (the range of front tension T_f and back tension T_b is 0–78.4 MPa which is about 1/3 yield stress). It can be seen that the rolling pressure decreases when the tension increases, and the decrease of rolling pressure becomes small when only front tension is applied (see Fig. 8). It can also be seen that when the friction variation model was used in the simulation, the rolling pressure is less than that with constant friction ($\tau_f = m\sigma_s$). The tension also has an influence on spread as shown in Figure 9. It can be seen that the spread decreases when the tension

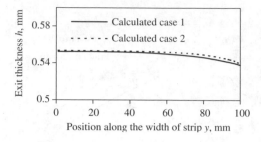

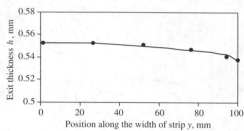

Figure 11. Effect of friction variation on exit thickness of strip.

Figure 12. Comparison of calculated exit thickness with measured value.

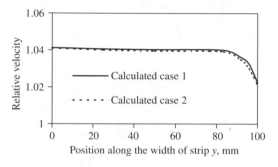

Figure 13. Effect of friction variation on relative velocity of the exit.

increases and the calculated value of spread with constant friction is less than that with the friction variation model. However, when the front tension increases, the forward slip increases significantly (see Fig. 10) and the calculated value of forward slip with constant friction is less than that with friction variation model.

4.3 *Effect of friction variation on shape and flatness of strip*

For $\phi 50/100/200 \times 300$ mm 6-high UC rolling mill, aluminum strip thickness 0.68 mm and width 100 mm was rolled (Liu 1994). In the simulation of shape and profile, the velocity boundary condition at the exit of the roll bite AA_1B_1B (see Fig. 2) is unknown. Calculated case 1 is for $K_1 = K_2 = 1.0, e = 0.1$; calculated case 2 is for $K_1 = 0.8, K_2 = 1.2, e = 0.1$. Friction factor m is 0.12. The effect of the friction variation on the distribution of the exit thickness of strip for ratio crown 0.018 (a ratio of strip crown to thickness of strip) (Liu 1994) is shown in Figure 11. It is found that the thickness of strip at the exit of roll bite decreases when friction variation model K_1 increases and K_2 decreases. The calculated exit thickness of strip for calculated case 1 is in good agreement with the measured value (see Fig. 12). Figure 13 gives the effect of the distribution of the relative velocity of the exit. It is also found that the velocity of strip at the exit of roll bite increases when the friction variation model K_1 increases and K_2 decreases. This means that the velocity and thickness along the width of strip are affected when the friction at the forward slip zone increases and that at backward slip zone decreases, as it affects the buckle of strip after rolling.

5 CONCULSIONS

A model for the shape and profile of thin strip in cold rolling considering the friction variation model is presented. The results show that the friction variation model has a significant effect on the simulation results of rolling pressure, spread and forward slip when the tension is applied. The friction variation model has an effect on the shape and profile of strip. The friction variation model constants K_1 and K_2 affect the strip velocity at exit, and the buckle of strip after rolling. This is useful to determine the rolling conditions which can result in a good shape and profile of strip, particularly for rolling thin strip. When a suitable friction variation model is employed, the calculated rolling pressure, spread and forward slip are in good agreement with the measured value. The calculated exit thickness along the width of strip is close to the measured value. The modelling of the shape and profile by FEM in cold rolling of thin strip is applicable.

ACKNOWLEDGEMENTS

This work is supported by an Australian Research Council (ARC) Discovery-Project grant including an Australian Research Fellowship, and a UoW Start Up Grant, Australia.

REFERENCES

Dixon, A.E. & Yuen, W.Y.D. 1995. A computationally fast method to model thin strip rolling, *Proceedings of Computational Techniques and Application Conference*: 239–246.

Fleck, N.A. & Johnson, K.L. 1987. Towards a new theory of cold rolling thin foil. *Int. J. Mech. Sci.* 29(7): 507–524.

Fleck, N.A., Johnson, K.L., Mear, M.E. & Zhang, L.C. 1992. Cold rolling of foil. *Proc. of Inst. Mech. Engrs Part B: J. Engng Manufact.* 206(B2): 119–131.

Hartley, P., Sturgess, C.E.N., Liu, C. & Rowe, G.W. 1989. Experimental and theoretical studies of workpiece deformation, stress, and strain during flat rolling. *Int. Mater. Reviews* 34: 19–34.

Jiang, Z.Y. & Tieu, A.K. 2000a. Modelling of rolling of strips with longitudinal ribs by 3-D rigid visco-plastic finite element method. *ISIJ Int.* 40(4): 373–379.

Jiang, Z.Y., Liu, X.L., Liu, X.H. & Wang, G.D. 2000b. Analysis of ribbed strip rolling by rigid-viscoplastic FEM. *Int. J. Mech. Sci.* 42(4): 693–703.

Jiang, Z.Y. & Tieu, A.K. 2001. A method to analyse the rolling of strip with ribs by 3-D rigid visco-plastic finite element method. *J. Mater. Proc. Technol.* 117: 146–152.

Jiang, Z.Y. & Tieu, A.K. 2003a. Modelling of thin strip rolling with friction variation by a 3-D finite element method. *JSME Int.* in press.

Jiang, Z.Y. & Tieu, A.K. 2003b. A 3-D finite element method analysis of cold rolling of thin strip with friction variation. *Tribology Int.* in press.

Kobayashi, S., Oh, S.I. & Altan, T. 1989. *Metal forming and the finite element method*. New York: Oxford University Press.

Le, H.R. & Sutcliffe, M.P.F. 2001. A robust model for rolling of thin strip and foil. *Int. J. Mech. Sci.* 43: 1405–1419.

Lenard, J.G. 1992. Friction and forward slip in cold strip rolling. *Tribol. Trans.* 35(3): 423–428.

Liu, C., Hartley, P., Sturgess, C.E.N. & Rowe, G.W. 1987. Finite element modelling of deformation and spread in slab rolling. *Int. J. Mech. Sci.* 29(4): 271–283.

Liu, X.H. 1994. *Rigid-plastic FEM and its application in steel rolling*. Beijing: Metallurgical Industry Press.

Liu, Y.J., Tieu, A.K., Wang, D.D. & Yuen, W.Y.D. 2001. Friction measurement in cold rolling. *J. Mater. Proc. Technol.* 111: 142–145.

Mori, K., Osakada, K. & Oda, T. 1982. Simulation of plane-strain rolling by the rigid-plastic finite element method. *Int. J. Mech. Sci.* 24(9): 519–527.

Osakda, K., Nakano, J. & Mori, K. 1982. Finite element method for rigid-plastic analysis of metal forming – formulation for finite deformation. *Int. J. Mech. Sci.* 24(8): 459–468.

Sutcliffe, M.P.F. & Rayner, P.J. 2001. Experimental measurements of load and strip profile in thin strip rolling. *Int. J. Mech. Sci.* 40(9): 887–899.

Wen, S.W., Hartley, P., Pillinger, I. & Sturgess, C.E.N. 1997. Roll pass evaluation for three-dimensional section rolling using a simplified finite element method. *Proc. of Inst. Mech. Engrs Part B: J. Engng Manufact.* 211(B2): 143–158.

Yuen, W.Y.D., Dixon, A.E. & Nguyen, D.N. 1996. The modelling of the mechanics of the deformation in flat rolling. *J. Mater. Proc. Technol.* 60: 87–94.

Computational Methods in Engineering and Science, Iu et al. (eds)
© *2003 Swets & Zeitlinger, Lisse, ISBN 90 5809 567 3*

Numerical simulation for cold rolling of thin strip during work roll edge contact

Z.Y. Jiang, H.T. Zhu & A.K. Tieu
Faculty of Engineering, University of Wollongong, Wollongong NSW, Australia

ABSTRACT: In this paper, an influence function method is developed to analyse the cold rolling of thin strip when the work roll edges contact, which forms a new deformation feature in rolling. The effect of reduction on specific force such as rolling force, intermediate force, edge contact force, the length of edge contact and the shape and profile of thin strip in cold rolling when the work roll edges contact is discussed. The research shows that the friction, particularly the transverse friction has a significant influence on the shape and profile of the thin strip. Numerical simulation tests have verified the validity of this developed method.

1 INTRODUCTION

A cold rolled thin strip is widely used in the electronic and instrument industries. The problem on how to improve its flatness and the dimensional accuracy has always been of major interest to the steel manufacturers. In some cold rolling mills, for example, it has often been found that the work rolls touch and deform (see Fig. 1) when the thin strip is rolled. Work roll contact at the edges should be considered in an analysis of the cold rolling of thin strip, which forms a new deformation feature. In this case, the models of deformation and rolling mechanics are different from the traditional cold rolling processes of strip. Not only will the distribution of the roll pressure change when the work rolls contact beyond the edges of the strip, but also the deformation model of work rolls (Kuhn & Weinstein 1970, Edwards & Spooner 1973), friction variation at the interface of the rolls and the rolled strip (Hwu & Lenard 1988) and work roll wear. The main features of this study concentrate on the distribution of rolling pressure, shape and profile of the strip, and to improve its quality when the work rolls contact beyond the edges

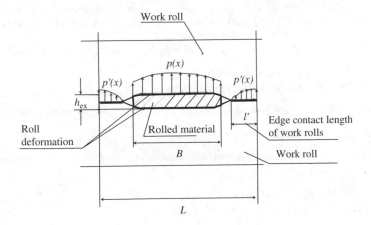

Figure 1. Edge contact of work rolls in cold rolling of thin strip

of the strip. The determination of the contact length between work rolls is a difficult problem and is a highlight of this paper.

Sutcliffe et al. (1998, 2001) developed a robust model for rolling of thin strip and foil and has carried out the experimental measurements of load and strip profile. In thin strip rolling, a comparison of methods to estimate the roll torque and a modified method for lateral spread has also been conducted (Shi et al. 2001, Abo-Elkhier 2002, Utsunomiya 1999). Jiang et al. (2002, 2003a) has calculated the elastic deformation of strip, and the resulting shape, profile and flatness in cold rolled thin strip. Elastic deformation of the rolls brings about problems of profile, shape and flatness (Wang & Wu 1990). The problem on how to improve its shape and flatness, and the dimensional accuracy has a major interest to the steel manufacturers. Researchers have found solutions to these problems by introducing new types of mills, such as Continuous Variable Crown (CVC) and Pair Cross (PC) mills equipped with roll shifting, roll crossing and work roll bending (Paolo et al. 1999, Anonymous 1997, Ginzburg & Azzam 1997). These are rolling processes where the work rolls do not contact each other when relatively thick strip is rolled.

Based on the assumption that the thickness is uniform at every point across the width of the strip, the compatibility in the rolling direction and lateral direction has been considered separately by Kuhn & Weinstein (1970). In Kuhn's work, a simplified 3-D analysis of roll deformation was used, and the roll shape was taken as conforming to some observed or desired strip shape, such as uniform thickness. In the real rolling process of thin strip, the shape and profile, including that of the roll thermal crown, wear, and the initial crown of the strip, affects the model. Edwards & Spooner (1973) also gave deformation compatibility relationship for the cold rolling of thin strip when the work rolls contact beyond the edges of the strip. But detailed results were not reported. In this study, the effects of reduction and friction on specific force such as rolling force, intermediate force, edge contact force and the shape of thin strip in cold rolling when the work roll edges contact are discussed. In particular the effect of transverse friction on the shape and profile of thin strip are presented. A formulation of the influence function for this special rolling was derived. The effect of friction on rolling mechanics and deformation of the cold rolled thin strip are also considered in the simulation. Numerical simulation tests have verified the validity of this developed method.

2 BASIC EQUATIONS

The calculation of the deformed rolls is based on the displacement compatibility relationships between the work roll and backup roll, work roll and thin strip, and the work rolls. Due to symmetry of the left and right sides of the rolls at the central line of the roll barrels, one half of the roll barrels is studied, and the equally divided zone Δx is shown in Figure 2. The rolling pressure and the pressure between the work roll and backup roll are uniform in zone Δx. The dimensions of the rolls and strip are also described in Figure 2.

The deformed work roll profile is obtained by calculating the roll deflections due to bending, shear, effect of Poisson's ratio, bending moment, interference between work and backup rolls, and work roll flattening.

2.1 Roll deflection due to bending

Beam theory for the bending and shear components has been widely employed to calculate the roll deflections (Yuen & Wechner 1990, O'connor & Weinstein 1972, Cresdee 1991). A typical roll deflection model under the effect of point loads is shown in Figure 3.

The roll deflection of the beam under the effect of bending at a position x can be described as follows:

$$y(x) = \begin{cases} [q(z) - p(z)](L-z)^2[3(L-x) - L+z]/(6EI) & x \leq z \\ [q(z) - p(z)][(x-z)^3 - (L-y)^2(3x-y-2L)]/(6EI) & x > z \end{cases} \tag{1}$$

where E = Young's modulus; I = second moment of area; and $p(z)$, $q(z)$ = point loads respectively.

2.2 Roll deflection due to shear

According to O'Connor (1972), the deflection of the neutral axis for short stubby beams due to shear is given by:

$$y(x) = \begin{cases} 4(L-z)[q(z) - p(z)]/(3AJ) & x \leq z \\ 4(L-x)[q(z) - p(z)]/(3AJ) & x > z \end{cases} \tag{2}$$

where A = cross-sectional area; and J = shear modulus of the beam.

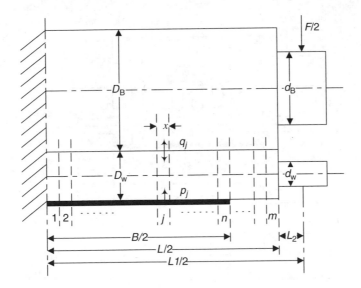

Figure 2.   Mechanics model of 4-high rolling mill.

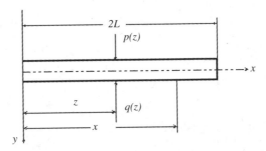

Figure 3.   Deflection of the neutral axis due to point loads.

Given in Figure 4, the left and right hand sides beyond the edges of the strip being rolled are named roll edge contact region. $p'(x)$ is contact pressure between work rolls at the left and right hand contact regions. The roll deflection due to bending moment, the effect of Poisson's ratio, interference between work and backup rolls and work roll flattening is referred to reference (Jiang et al. 2003b).

3   SIMULATION CONDITIONS

Given below are values of the important parameters used in the simulation for thin strip cold rolling on Hille 100 experimental rolling mill.

Work roll diameter: 62.88 mm
Work roll barrel: 249 mm
Work roll crown: 0 μm
Poisson's ratio of work roll: 0.3
Young's modulus of work roll: 22000 Kg/mm²
Distance between housing screw: 340 mm
Backup roll diameter: 228 mm
Backup roll barrel: 249 mm
Backup roll crown: 0 μm

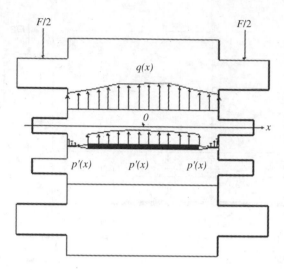

Figure 4. Work roll edges contact.

Poisson's ratio of backup roll: 0.3
Young's modulus of backup roll: 22000 kg/mm²
Central distance between bending cylinder: 340 mm
Slab thickness of strip: 0.5 mm
Entry thickness of strip: 0.30 mm
Exit thickness of strip: 0.10 and 0.08 mm
Width of strip: 140 mm
Back tension: 0 kN
Front tension: 0 kN
Rolling speed: 1 m/s
Friction coefficient: 0.15, 0.17, 0.2, parabolic increasing and decreasing
Initial crown of strip at entry: 0.0 mm
Defining point of strip crown from edge: 10 mm
Bending force of work roll: 0 kN/chock
Deformation resistance equation

$$k = 91.40 + 151.84 \cdot \varepsilon^{0.2168} \text{ MPa} \tag{3}$$

Rolling force is calculated by using Bland-Ford-Hill model (Jiang et al. 2003b). Deflections of the work roll and backup roll are calculated by using simple beam theory for bending and shear.

4   RESULTS

Based on the analysis of influence function method, a simulation program was developed and performed on a PC. For different exit thickness of strip and friction, the rolling force, intermediate force, edge contact force and the profile of strip are obtained.

4.1   *Effect of exit thickness of strip on specific force and profile of strip*

The slab thickness is 0.50 mm, entry thickness of strip is 0.30 mm, exit thickness h_{ex} 0.10 and 0.08 mm respectively, bending force is zero. The effect of exit thickness of strip on specific force is shown in Figure 5. It can be seen that the rolling force P increases when the exit thickness of strip decreases, which indicates the reduction increases when the exit thickness of strip decreases. It is also seen that the intermediate force Q increases when the exit thickness of strip h_{ex} decreases. When the exit thickness of

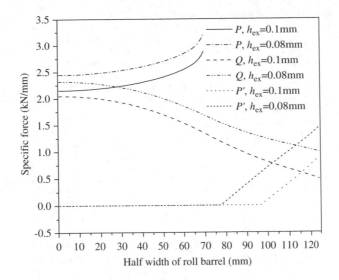

Figure 5.   Effect of exit thickness of strip on specific force.

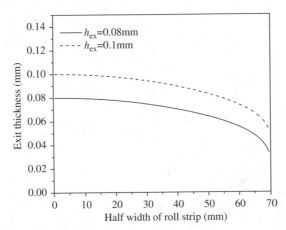

Figure 6.   Effect of exit thickness of strip on profile of strip.

strip decreases, the edge contact force P' and the length of edge contact increase. Therefore, the edge contact of work rolls becomes more significant when the exit thickness of strip decreases.

Given Figure 6 is the distribution of the exit thickness of strip. It can be seen that the exit crown of strip decreases when the exit thickness decreases (see Table 1). Therefore, when the bending force is zero, the shape and profile of the rolled strip becomes better with a higher reduction.

4.2   *Effect of friction on simulation results*

The slab thickness is 0.50 mm, entry thickness 0.30 mm, exit thickness h_{ex} 0.10 mm and bending force is zero. The effect of friction coefficient f on specific force is shown in Figure 7. It can be seen that the rolling force P increases significantly when the friction coefficient increases. It can also been seen that the intermediate force Q and edge contact force P' rise substantially when the friction coefficient increases, and affects the length of edge contact.

The effect of friction coefficient f on profile of strip is shown in Figure 8. It can be seen that the exit crown of strip increases when the friction coefficient f increases. Because the friction coefficient increases, the rolling force and edge contact force increase, so the edge drop of strip becomes smaller.

275

Table 1. Comparison of specific force (kN) and exit crown of strip.

Exit thickness (mm)	Rolling force (kN)	Edge contact force (kN)	Intermediate force (kN)	Exit crown of strip (μm)
0.2	325.73	24.90	350.63	26.00
0.18	367.39	69.51	436.89	24.63

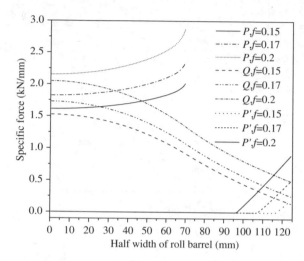

Figure 7.   Effect of friction coefficient on specific force.

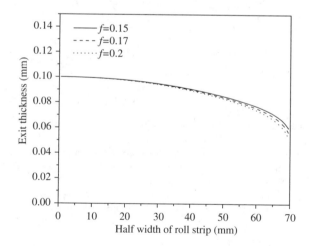

Figure 8.   Effect of friction coefficient on profile of strip.

Numerical testing has been conducted for different transverse friction along the width of strip. The distribution of friction coefficient is set as f, f_d and f_i (see Fig. 9). Figure 10 gives the effect of transverse friction on profile of strip. It can be seen that the transverse friction has a significant effect on profile of strip. The friction coefficient at the edge of strip increases, the exit crown of strip reduces, which indicates that the shape of strip becomes better. This is very useful to improve the shape and profile of strip for cold rolling of thin strip by increasing the edge friction along the width of strip.

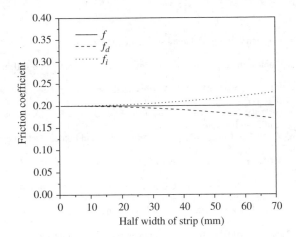

Figure 9.   Distribution of transverse friction variation.

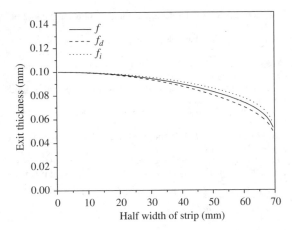

Figure 10.   Effect of transverse friction on profile of strip.

CONCLUSIONS

A new model for the shape and profile of thin strip in cold rolling when the work rolls touch and deform beyond the side of strip was developed successfully. The results show that the specific force such as rolling force, intermediate force and the strip profile for this special rolling is significantly different with the change of exit thickness of strip, and the edge contact length is affected by the reduction. Friction has a significant effect on the rolling force, edge contact force and the length of edge contact. It also affects the strip shape and profile, particularly the transverse friction has a significant influence on the strip shape and profile. When the transverse friction coefficient at the edge of strip increases, the exit crown of strip reduces, which indicates that the shape of strip becomes better. This would be useful to improve the strip shape and profile by increasing the edge friction along the width of strip in cold rolled thin strip.

ACKNOWLEDGEMENTS

This work is supported by an Australian Research Council (ARC) Discovery-Project grant, which includes an Australian Research Fellowship.

REFERENCES

Abo-Elkhier, M. 2002. A modified method for lateral spread in thin strip rolling. *J. Mater. Proc. Technol.* 124(1–2): 77–82.

Anonymous. 1997. Shape control in cold rolling by combining variable crown and roll bending. *Steel Times Int.* 21(2): 34–35, 40.

Cresdee, R.B., Edwards, W.J. & Thomas, P.J. 1991. An advanced model for flatness and profile prediction in hot rolling. *Iron and Steel Engineer*: 41–51.

Edwards, W.J., & Spooner, P.D. 1973. Analysis of strip shape. In G.F. Bryand (ed.), *Automation of Tandem Mills*: 177–212. London: Iron and Steel Institute.

Ginzburg, V.B. & Azzam, M. 1997. Selection of optimum strip profile and flatness technology for rolling mills. *Iron & Steel Engineer*: 74(7): 30–38.

Hwu, Y.J., & Lenard, J.G. 1988. Finite element study of flat rolling. *J. Eng. Mater. & Technol.-Transactions of the ASME*: 110(1): 22–27.

Jiang, Z.Y. & Tieu, A.K. 2002. Elastic-plastic finite element method simulation of thin strip with tension in cold rolling. *J. Mater. Proc. Technol.* 130–131: 511–515.

Jiang, Z.Y. & Tieu, A.K. 2003a. Modelling of thin strip rolling with friction variation by a 3-D finite element method. *JSME Int.* in press.

Jiang, Z.Y., Zhu, H.T. & Tieu, A.K. 2003b. Effect of rolling parameters on cold rolling of thin strip during work roll edge contact. *The Sixth Asia Pacific Conf. on Mater. Proc.*, Taipei, Taiwan, Sept. 22–25, 2003.

Kuhn, H.A. & Weinstein, A.S. 1970. Lateral distribution of pressure in thin strip rolling. *J. Eng. for Industry*: 453–460.

Le, H.R. & Sutcliffe, M.P.F. 2001. A robust model for rolling of thin strip and foil. *Int. J. Mech. Sci.* 43(6): 1405–1419.

O'Connor, H.W. & Weinstein, A.S. 1972. Shape and flatness in thin strip rolling. *ASME J. Eng. for Industry*: 1113–1123.

Paolo, B., Roberto, B. & Massimo, R. 1999. New mill stand with flexible crown control for rolling ultra-thin hot strip. *Metallurgical Plant & Technology Int.* 22(3): 7.

Shi, Jingyu, McElwain, D.L.S. & Langlands, T.A.M. 2001. A comparison of methods to estimate the roll torque in thin strip rolling. *Int. J. Mech. Sci.* 43(3): 611–630.

Sutcliffe, M.P.F. & Rayner, P.J. 1998. Experimental measurements of load and strip profile in thin strip rolling. *Int. J. Mech. Sci.* 40(9): 887–899.

Utsunomiya, H., Saito, Y. & Matsueda, S. 1999. Proposal of a new method for the spread rolling of thin strips. *J. Mater. Proc. Technol.* 87(1–3): 207–212.

Wang, G.D. & Wu, G.L. 1990. *Rolling Theory and Practice of Plate and Strip*. Beijing: China Railway Press.

Yuen, W.Y.D. & Wechner, B. 1990. A physical model for on-line roll bending set up in cold rolling. *Proc. of the 5th Int. Rolling Conf.*: 234–241.

Finite element analysis for bulging characteristics of radial slot safety diaphragms

G.F. Gao & X.W. Ding

Institute of Special Chemical Equipments, Dalian University of Technology, Dalian, China

ABSTRACT: With MITC4 degenerated shell element and based on T.L. formulations, finite element analysis was made for radial slot safety diaphragms. The simulation results and experimental observations demonstrated that the biggest stresses and strains focused on the small areas, which locate on the margin of pinholes in the direction of bridge. And there is remarkable linear relationship between ultimate pressure and the diameter of inner concentric circles of pinholes. In addition, the $pa/(s_0b)$ has significant meaning. For the different inner diameter, it's found that the maximum equivalent stresses show nearly coincident variety rule relative to $pa/(s_0b)$. And the values of $pa/(s_0b)$ are about 5.60–5.90 MPa·mm^{-1} just before the rupture.

1 INTRODUCTION

Metallic sheets under lateral pressure are widely used to protect equipments from overpressure or to meet some particular needs of technical processes. The plain safety diaphragm is the simplest of all, but for the applications of low design ultimate pressure (rupture pressure) together with small vent sizes, a plain diaphragm would often be impracticable owing to the extreme thinness of sheet required. Other types of safety diaphragms would be used; the type of slot diaphragms is the main one.

Slot safety diaphragms are mainly composed of slot sheets, seal membranes, and backpressure brackets. Several slots are cut on the sheet; pinholes are at each end of slots (Fig. 1). So the ultimate pressure could be decreased to the design value. In an attempt to airproof the working medium, non-metallic membrane is lined on the pressured side. In addition to support the backpressure, the brackets also has the function of fixing seal membranes.

Just the contrary of usual engineering design, safety diaphragms cannot but be ruptured in a specific pressure range. For example the margin tolerance is ±5% for rated rupture pressure from 0.3 to 100 MPa. From elastic deformation, plastic deformation up to rupture, the material would come through the entire course of deformation. Geometrically, the problem has the character of large strains. It is very difficult to acquire analytic resolutions. There are currently only some semi-empirical equations that could be used to predict the ultimate pressure, and the design and fabrication of slot safety diaphragms depends severely on the experiments.

With MITC4 (Mixed Interpolation of Tensorial Components) shell element and based on T.L. (Total Lagrange) formulations, finite element analysis was made for the bulging behavior and ultimate pressure of radial slot diaphragms in this study. The results of numerical simulation are discussed and compared with the experimental results.

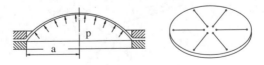

Figure 1. Loading device and configuration of the slot sheet.

2 THIN SHELL FEM

Ahmad et al. (1970) introduced the degeneration concept to the finite element modeling of curved shell structures, and the concept of degenerate solid shell element has established its supremacy over all other existing approaches. It was soon noticed that the degenerate shell element based on the conventional assumed displacement method has its deficiencies which suffers from the shear and membrane locking as the ratio of thickness to span becomes small, and the rate of convergence is slow. In an attempt to avoid the effect of locking, techniques of reduced integration (RI) and selective reduced integration (SRI) were developed successively (Zienkiewicz et al. 1971, Haughes et al. 1978). RI/SRI, however, are not always successful in overcoming locking behavior. The solutions may still be over-stiff for problems with highly constrained boundaries, especially with coarse meshes. Furthermore, the use of RI/SRI reduces the rank of the elemental matrices and thus possesses spurious zero energy modes for problems with lightly constrained boundaries.

Several alternative attempts to alleviate locking have been made by many scholars. Assumed natural strain method is one of them. Using Lagrange multipliers and with independent interpolation of the membrane strains and transverse shear strains, Dvorkin & Bathe (1984) developed a valuable MITC shell element. A 4-node general shell element, called MITC4 element, was developed at first. Subsequently, Bathe and co-workers developed higher-order MITC quadrilateral shell elements such as 8-node, 9-node and 16-node elements. Bathe et al. (2000) evaluated the performance of the MITC shell elements. The MITC4 element displays robust convergence, and stays virtually unaffected by the decrease of the ratio of thickness to span.

3 MITC4 SHELL ELEMENT

MITC4 shell element, a three-dimensional quadrilateral element, is based on degenerated concept and has 4 nodes (Fig. 2). Each node has five degrees of freedom, three displacements in the directions of global Cartesian coordinate axes and two rotations about nodal Cartesian coordinate axes. The displacements of any particle in the element are

$$u_i = h_k u_i^k + \frac{r_3}{2} s_k h_k (V_1^k \beta_k - V_2^k \alpha_k), \ i = 1, 2, 3; \ k = 1, \dots, 4 \tag{1}$$

where the r_i are the curvilinear coordinates of the element; h_k are shape functions based on two curvilinear coordinates (r_1, r_2) corresponding to node k; s_k is the shell thickness at node k; and V_i^k are nodal coordinates system at node k. The vector V_3^k is constructed from the nodal coordinates of the top and bottom surfaces at node k. The vector V_1^k is perpendicular to V_3^k and parallels to the global x_1-, x_3-plane, so that

$$V_1^k = e_2 \times V_3^k \big/ \big| e_2 \times V_3^k \big| \tag{2}$$

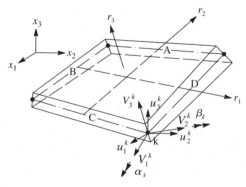

Figure 2. MITC4 shell element.

$$V_2^k = V_3^k \times V_1^k \tag{3}$$

here e_2 is the unit vector along the x_2 direction. α_k and β_k are rotations of V_3^k about the V_1^k and V_2^k axes.

The membrane strain components are calculated as usual from the displacement interpolations, while the transverse shear strain components are interpolated distinguishingly.

$$\varepsilon_{13} = \frac{1}{2}\left(1 + r_2\right)\varepsilon_{13}^A + \frac{1}{2}\left(1 - r_2\right)\varepsilon_{13}^C \tag{4}$$

$$\varepsilon_{23} = \frac{1}{2}\left(1 + r_1\right)\varepsilon_{23}^D + \frac{1}{2}\left(1 - r_1\right)\varepsilon_{23}^B \tag{5}$$

where the variables, ε_{13}^A, ε_{13}^C, ε_{23}^D and ε_{23}^B are the strain components at points A, C, D, and B, computed directly from assumed displacement fields.

4 FINITE ELEMENT FORMULATIONS

In the T.L. formulation, the virtual work principle expressing the equilibrium requirements of the element at time $t + \Delta t$ is

$$\int_{^0V} {}_{0}^{t+\Delta t}S_{ij}\delta\, {}_{0}^{t+\Delta t}\varepsilon_{ij}d^0V = {}^{t+\Delta t}R \tag{6}$$

here the ${}_{0}^{t+\Delta t}S_{ij}$ and ${}_{0}^{t+\Delta t}\varepsilon_{ij}$ are the components of the 2nd Piola-Kirchhoff stress tensor and Green-Lagrange strain tensor at time $t + \Delta t$ respectively, both referred to the initial configuration at time 0. ${}^{t+\Delta t}R$ is the external virtual work due to the applied surface tractions and body forces.

For the incremental solution of geometric and material nonlinear problems, the stresses and strains are decomposed into the known quantities, ${}_{0}^{t}S_{ij}$ and ${}_{0}^{t}\varepsilon_{ij}$, and unknown increments, ${}_{0}S_{ij}$, and ${}_{0}\varepsilon_{ij}$, that is

$${}_{0}^{t+\Delta t}S_{ij} = {}_{0}^{t}S_{ij} + {}_{0}S_{ij}$$

$${}_{0}^{t+\Delta t}\varepsilon_{ij} = {}_{0}^{t}\varepsilon_{ij} + {}_{0}\varepsilon_{ij}$$

In addition, the strain increments, ${}_{0}\varepsilon_{ij}$, can be decomposed into two parts, the linear part, ${}_{0}e_{ij}$, and the nonlinear part, ${}_{0}\eta_{ij}$, thus

$${}_{0}\varepsilon_{ij} = {}_{0}e_{ij} + {}_{0}\eta_{ij}$$

The basic equilibrium relation of element can be written as follows:

$$\int_{^0V} {}_{0}S_{ij}\delta\, {}_{0}\varepsilon_{ij}d^0V + \int_{^0V} {}_{0}^{t}S_{ij}\delta\, {}_{0}\eta_{ij}d^0V = {}^{t+\Delta t}R - \int_{^0V} {}_{0}^{t}S_{ij}\delta\, {}_{0}e_{ij}d^0V \tag{7}$$

In order to build up element stiffness matrices, appropriate constitutive relations are needed. The constitutive equations can be expressed as the relation between stress rate and strain rate.

$$\dot{\sigma}_{ij} = C_{ijkl}^{ep}\dot{\varepsilon}_{kl} \tag{8}$$

where C_{ijkl}^{ep} is elastic-plastic tangent modulus, $\dot{\sigma}_{ij}$ is the rate of Cauchy stress.

After the linearization of equilibrium equation and the discretization of finite element, the incremental equation of element stiffness matrix can be expressed as

$$\left({}_{0}^{t}K_L + {}_{0}^{t}K_{NL}\right)\Delta u = {}^{t+\Delta t}R - {}_{0}^{t}F \tag{9}$$

here $_0^t K_L$ and $_0^t K_{NL}$ are the linear and nonlinear strain incremental stiffness matrix respectively, and the $_0^t F$ is the nodal force vector.

The load was subdivided into a series of load increments, and Newton-Raphson equilibrium iterations were used to solve this nonlinear problem. Numerical integration was carried out with Gauss-Legendre full integration.

As discussed in former study (Gao & Ding 2002), the maximum pressure which can satisfy the equations of constitution, geometry and static equilibrium all together should be the ultimate pressure. The precondition is that the convergence could be ensured. It's similar to the lower limit theorem in plastic limit analysis. Which can also be verified with the changing of maximum stresses and strains.

5 RESULTS AND DISCUSSION

The material of the circular radial slot sheet is *1Cr18Ni9Ti*. The material constants, hardness factor, strain-hardening exponent, and Young's modulus are as follows: $A = 1557.35$ MPa, $n = 0.417$, $E = 1.95 \times 10^5$ MPa. In this study, the Mises associative plasticity and multi-linear isotropic hardening model was adopted.

The geometrical dimensions, sheet diameter, diameter of outer concentric circle of pinholes, diameter of pinholes, and initial thickness of sheet are as follows: $D = 220$ mm, $D = 210$ mm, $\lambda = 2$ mm, $s_0 = 0.25$ mm. There are six slots with the width of 0.20 mm. The diameter of inner concentric circles of pinholes, d, is 10, 15, 20, 25, and 30 mm respectively.

In view of the symmetry of configuration, and in an attempt to avoid the mesh distortions in local pole areas, 1/6 of configuration was taken for the geometry modeling, and two slots are selected as the borderlines of the model. Figure 3 shows the deformation of bulging (h is the deflection at the crown of sheets). Relative deflection at the crown of sheet against $pa/(s_0b)$ under different d is shown in Figure 4

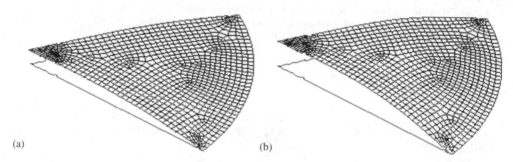

(a)   (b)

Figure 3. The bulging of 1/6 slot sheet. (a) $d = 30$ mm, $p = 0.05$ MPa, $h = 7.60$ mm; (b) $d = 30$ mm, $p = 0.17$ MPa, $h = 15.59$ mm.

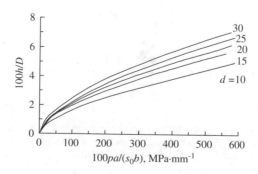

Figure 4. Relative deflection at the crown.

(here a is radius of sheets, p is hydrostatic pressure, and b is the length of bridge). It is obvious that the deflection would get larger with the increase of the diameter of inner concentric circle.

Simulation results demonstrated that the highest stresses and strains fasten on the small areas, which located on the margin of pinholes in the direction of bridge. Stresses and strains near the outer pinholes are much less than that near the inner pinholes. Figure 5 shows the contour of equivalent stresses just before the rupture of diaphragms. Observed from experiments, the start position of rupture is located at the local areas of the highest stresses as illustrated in Figure 5.

For $d = 15$–30 mm, the simulation results of ultimate pressure are 0.070, 0.105, 0.140, and 0.170 MPa respectively, the relative errors compared with the experimental values are -7.89%, $+4.69\%$, -0.36%, and $+2.41\%$ respectively.

There is remarkable linear relationship between ultimate pressure and the diameter of inner concentric circle (Fig. 6, Table 1). That is

$$p_b = 0.0066d - 0.027$$

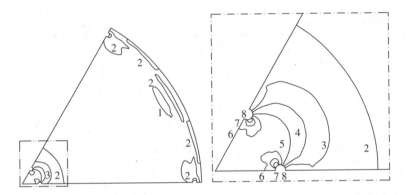

Figure 5. Contour of equivalent stresses (MPa). $d = 20$ mm, $p = 0.105$ MPa; 1: 110.49, 2: 322.53, 3: 428.55, 4: 534.58, 5: 640.60, 6: 746.62, 7: 852.64, 8: 958.66.

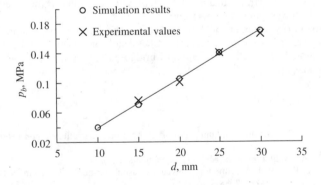

Figure 6. Simulation and experimental results of ultimate pressures.

Table 1. Evaluating of the linear relationship.

R^{2*}	$F_{0.99}(1,3)^*$	F	Verify results
0.9991	34.1	3330.33	Excellent

* Coefficient of determination. ** Critical value

283

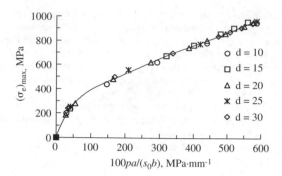

Figure 7. The maximum equivalent stresses.

The linear relation was also approved with experiments (Hu et al. 1993, Ma et al. 1996). When $d = 2\lambda = 4$, the bridge equals to zero, here the theoretical resolution of ultimate pressure should be zero. Ultimate pressure extrapolated from above linear relation was $d = 4.09$, which demonstrated that the linear relationship is existed also in the situation of small value of d.

If use m to denote the number of slots, thus

$$b = d \sin \frac{\pi}{m} - \lambda$$

For circular radial slot sheets, simulation results demonstrated that $pa/(s_0 b)$ has significant meaning. It is clear that the maximum equivalent stress shows nearly coincident variety rule for different d relative to $pa/(s_0 b)$ (Fig. 7). Furthermore, the values of $pa/(s_0 b)$ are about $5.60 \sim 5.90\,\mathrm{MPa\cdot mm^{-1}}$ near the rupture of diaphragms. It's useful for the practical engineering.

6 CONCLUSIONS

As a matter of fact, the strength of safety diaphragms is composed of the strength of slot sheets and seal membranes. The material of seal membranes is nonmetal such as Teflon, and the thickness is about 0.05 mm. Experiments demonstrated that the strength of seal membranes is much less than the strength of slot sheets. So which was ignored in this study.

The authors also did finite element analysis for other types of slot safety diaphragms such as safety diaphragms with square cross-slots or rectangle peripheral slots. It is found that the location of the highest stresses and strains, and the effect of length of bridge on the ultimate pressure are similar to the results depicted in this paper.

REFERENCES

Ahmad, S. et al. 1970. Analysis of thick and thin shell structures by curved finite elements. *Int. J. Numer. Methods Eng.* 2(3): 419–451.
Bathe, K.J. et al. 2000. An evaluation of the MITC shell elements. *Comput. Struct.* 75(1): 1–30.
Dvorkin, E.N. & Bathe, K.J. 1984. A continuum mechanics based four-node shell element for general nonlinear analysis. *Eng. Computations* 1(1): 77–88.
Gao, G.F. & Ding, X.W. 2002. Analysis method of the mechanism of the clamped metallic membrane with large strain. *Chem. Eng. Machinery*, 29(2): 72–77 (in Chinese).
Haughes, T.J.R. et al. 1978. Reduced and selective integration techniques in the finite element analysis of plates. *Nucl. Eng. Design* 46(1): 203–222.
Hu, Z.J. et al. 1993. A study on bursting characteristics of hole-slot flat rupture discs. *Pressure Vessel Technology* 10(5): 46–50, 63 (in Chinese).
Ma, Y. & Ding, X.W. 1996. Strength design of cracking-type bursting discs. *Petro-chemical Equipment* 15(1): 30–34 (in Chinese).
Zienkiewicz, O.C. et al. 1971. Reduced integration technique in general of plates and shells. *Int. J. Numer. Methods Eng.* 3(2): 275–290.

Computational Methods in Engineering and Science, Iu et al. (eds)
© *2003 Swets & Zeitlinger, Lisse, ISBN 90 5809 567 3*

A moving mesh mixed method for advection–diffusion equations

Y. Liu
School of Mathematics, Georgia Institute of Technology, Atlanta GA, USA

R.F. Santos
Department of Mathematics, University of Algarve, Faro, Portugal

ABSTRACT: A moving finite element method for solving an advection–diffusion problem is presented. The problem is assumed to be in a mixed weak formulation and stability for the numerical method is shown via symmetric error estimates, the first for mixed methods for parabolic equations. Results of numerical experiments, especially in the non-linear case, are also shown.

1 INTRODUCTION

In this talk, following Liu et al. 2003, we present a moving finite element method for solving an advection–diffusion problem. It is well known that solutions of time dependent parabolic equations may present large gradients or discontinuities locally in space that may sweep the space domain, as time goes on. To achieve and maintain high accuracy, it is advisable to adapt the mesh along the numerical integration. One way, is to use moving methods.

Moving finite element methods have been one of the numerical analysis tools for solving partial differential equations since the early eighties, Miller 1981, Miller & Miller 1981. From then on, several methods, as well as their error analysis, based on Galerkin formulations, have been provided. Stability for the method shown here is obtained by proving a symmetric error estimate. While more often encountered in the elliptic setting, this type of best approximation result is not widely seen for parabolic problems. In related work, see Dupont 1982, Bank & Santos 1993, Dupont & Liu 2003.

Time derivatives may develop singularities at the edges between elements, in mixed formulations. For this reason, directional time derivatives are used instead to obtain the estimates. It is also important to point out that since one may have diffusion and advection present at the same time, one should have a mesh movement that reflects this situation. Also, a fixed mesh or one that follow the characteristics may not be the best one. This is why we use a more general procedure. The technique is explained in the section dealing with the numerical results.

In Liu et al. 2003, one deals with a linear problem only. However, one can show, following the same argument as in Dupont 1982, that the theory is still true in a special non-linear case. We also show the results of the correspondent new numerical experiments.

In the next section we formulate our problem and state the main assumptions. In the third section the symmetric estimate is proved. Finally, in the last section we show the results of numerical experiments.

2 PROBLEM FORMULATION

Let Ω be an open bounded domain of R^n and $Q = \Omega \times (0,T)$. We want to find u such that

$$\begin{cases} \partial_t u - \nabla \cdot (a\nabla u + bu) = f, & \text{in} \quad \Omega, \\ u = 0, & \text{on} \quad \partial\Omega \times (0,T), \\ u = u_0, & \text{for} \quad t = 0, \end{cases} \tag{1}$$

with $a(x)$, $b(x)$ and $f(x,t)$ are smooth, bounded; and $a_1 \geqslant a(x) \geqslant a_0 > 0$.

Given a fixed polyhedron, $\bar{P} = \cup P_i$, where the P_i's are closed and have nonempty disjoint interiors, we will assume that each P_i is a simplex and that they form a tessellation of $\bar{P}$. For simplicity, we also assume that Ω is a polyhedron. Mesh evolution, in time, is then modeled by giving a continuous mapping $S: \bar{P} \times [0,T] \to \bar{\Omega}$ such that

i) $S(\cdot,t)$ is, for each t, a one-to-one piecewise linear mapping, with respect to $\{P_i\}$, from $\bar{P}$ onto $\bar{\Omega}$;
ii) S is continuously differentiable on each $P_i \times [0,T]$;
iii) $\partial\Omega = S(\partial P, t)$.

In this way we have constructed a mesh that is the image, in time, of a fixed reference mesh.

We emphasize that, with these assumptions, Ω_i is also a simplex and $\{\Omega_i(t)\}$ is a moving partition of Ω. Sometimes, instead of considering the mesh movement as the image of a fixed mesh, it is convenient to view it as generated by a mapping of Ω onto itself. In this way, we consider $S^{-1} = S^{-1}(\cdot, t)$ the inverse of S as a map of P onto Ω; it is, therefore, defined on $\bar{Q}$. By doing so, the finite element mesh is transported with velocity $\dot{x}(t) = S_t(S^{-1}(x, t), t)$, a continuous piecewise linear function on the partition $\{\Omega_i\}$. Let $\tilde{V}_h$ be a finite dimensional subspace of $L^2(P)$. Our finite element space on Ω is defined by

$$V_{h(t)} = \left\{ \phi(x,t) : \phi(S(\cdot,t),t) \in \tilde{V}_h \right\}$$

Letting $H(div, \Omega)$ be the space of those L^2 functions whose divergence is also in L^2, we take for $H_h(t)$ a finite dimensional subspace of $H(div, \Omega)$ such that $div\, H_h = V_h$, for any t. In fact, we will assume that V_h is the space of discontinuous polynomials of degree at most m and take for H_h the Raviart-Thomas flux space. If P_h is the L^2 projection onto V_h, we let Π_h be the linear operator form $H(div, \Omega)$ to H_h satisfying $(div(W - \Pi_h W), r) = 0$, $\forall r \in V_h$ and $div\, \Pi_h = P_h\, div$, Raviart & Thomas 1977.

We also define a particular directional derivative as

$$\frac{DF(x,t)}{Dt} \equiv \frac{\partial F(x,t)}{\partial t} + \dot{x} \cdot \nabla_x F(x,t).$$

Let us now consider the mixed formulation for problem (1): Find u and σ such that

$$\begin{cases} (\alpha\sigma + (\beta + \alpha\dot{x})u, \chi) = 0, & \forall\chi \in H(div,\Omega), \\ \left(\dfrac{Du}{Dt} + div\,\sigma + (\nabla \cdot \dot{x})u, r\right) = (f,r), & \forall r \in L^2(\Omega). \end{cases} \tag{2}$$

The mixed approximation problem is: Find functions $u_h:[0,T] \to V_h$ and $\sigma_h: [0,T] \to H_h$ such that, with $u_h(0) = P_h u(0)$,

$$\begin{cases} (\alpha\sigma_h + (\beta + \alpha\dot{x})u_h, \chi) = 0, & \forall\chi \in H_h, \\ \left(\dfrac{Du_h}{Dt} + div\,\sigma_h + (\nabla \cdot \dot{x})u_h, r\right) = (f,r), & \forall r \in V_h. \end{cases} \tag{3}$$

The usual Sobolev norms are used throughout. If $\underline{h}(x, t)$ a function whose restriction to $\Omega_i(t)$ is $h_i(t)$, we define, for any function w whose restriction to Ω_i is in $H^s(\Omega_i)$, the following norm:

$$\|w\|^2_{\underline{H}^s} = \sum \|w\|^2_{H^s(\Omega_i)}.$$

We will also assume:
Condition 1: There exists a positive constant C_1 such that for any $w \in H^{s_1}(\Omega)$, $s_1 \geq 0$, and any $t \in [0,T]$,

$$\|w - P_h w\| \leq C_1 \left\|\underline{h}^{\min\{m+1,s_1\}} w\right\|_{\underline{H}^n};$$

and for any $W \in (H^{s_2}(\Omega))^n$, $s_2 \geq 1$, and for any $t \in [0, T]$,

$$\|W - \Pi_h W\| \leq C_1 \left\|\underline{h}^{\min\{m_1+1,s_2\}} W\right\|_{\underline{H}^{s_2}};$$

$m_1 = m + 1$ in one dimension; and $m_1 = m$ in higher space dimension.

Condition 2: There exists a constant C_2 such that for any $W \in (H^1(\Omega))^n$ and any $t \in [0,T]$,

$$\left\| \Pi_h W \right\| \leq C_2 \left\| W \right\|_1.$$

Condition 3: the well known elliptic H^2-regularity for Ω.

Let A be the pseudo-inverse of div, the continuous operator $A:L^2(\Omega) \to H_h$, defined by $\varphi - div(A\varphi) \perp V_h$ and $\|A\varphi\|$ minimal. Note that if Conditions 1 and 3 hold then A behaves like a smoothing operator. Moreover, for $\rho \in H_h$,

$$\|A div \rho\| \leq \|\rho\|. \tag{4}$$

Finally, one can show that

$$\left(\frac{D\varphi}{Dt}, \varphi \right) = \frac{1}{2} \frac{d}{dt} \|\varphi\|^2 - \frac{1}{2} \left(\varphi, \varphi (\nabla_x \cdot \dot{x}) \right).$$

3 MAIN RESULTS

Let F_h the linear flux operator associated with the space V_h, that is, $F_h : V_h(t) \to H_h(t)$ such that for any $v_h \in V_h(t)$, $\alpha F_h(v_h) + (\beta + \alpha \dot{x}) v_h, \chi) - (v_h, div \chi) = 0, \forall \chi \in H_h$.

We also define the following norm, naturally associated with this problem:

$$\left\| (\eta, \psi) \right\|_*^2 = \|\eta\|_{L^\infty(0,T:L^2(\Omega))}^2 + \left\| A \frac{D\eta}{Dt} \right\|_{L^2(0,t:L^2(\Omega))}^2 + \left\| A(div \psi) \right\|_{L^2(0,t:L^2(\Omega))}^2$$

The next theorem gives the symmetric estimate. In a way it says that if we have a good control of the mesh movement and the equation's coefficients are well behaved, then we are able to obtain a best approximation solution.

Theorem: Assume Condition 2 and that there exist constants c_1, c_2 such that for all $(x, t) \in Q$, $|\nabla_x \cdot \dot{x}| \leq c_1$ and $|\beta + \alpha \dot{x}| \leq c_2$. Then there exists a constant $C > 0$, which depends only on C_2, c_1, c_2, T, the bounds of coefficient a, and Ω, such that for any piecewise smooth function v_h with $v_h(\cdot,t) \in V_h(t)$,

$$\left\| (u - u_h, \sigma - \sigma_h) \right\|_* \leq C \left\| (u - v_h, \sigma - F_h(v_h)) \right\|_*.$$

The main steps for the proof of this theorem are the following.

Let v_h be a piecewise C^1 function such that $v_h(\cdot,t) \in V_h(t)$. We put $S_h = F_h(v_h)$ and

$$v = u_h - v_h, \qquad \rho = \sigma_h - S_h,$$

$$\eta = u - v_h, \qquad \psi = \sigma - S_h.$$

Subtracting Equation 3 from Equation 2 obtain

$$\begin{cases} (\alpha \rho + (\beta + \alpha \dot{x})v, \chi) - (v, div \chi) = 0, & \forall \chi \in H_h, \\ \left(\frac{Dv}{Dt} + div \, \rho + (\nabla \cdot \dot{x})v, r \right) = \left(\frac{D\eta}{Dt} + div \, \psi + (\nabla \cdot \dot{x})\eta, r \right), & \forall r \in V_h. \end{cases}$$

Take, as test functions, $\chi = \rho$ and $r = v$ to get, via an energy-type argument, that

$$\frac{d}{dt} \|v\|^2 + \frac{1}{a_1} \|\rho\|^2 \leq C \left\{ \|v\|^2 + \left\| A \left(\frac{D\eta}{Dt} + div \psi \right) \right\|^2 + \|\eta\|^2 \right\};$$

using the well known Gronwall's inequality,

$$\|v\|^2_{L^\infty(0,T;L^2(\Omega))} + \|\rho\|^2_{L^2(0,T;L^2(\Omega))} \leq C\left\{\|v(0)\|^2 + \|(\eta,\psi)\|^2_*\right\}.$$

By choice, the norm of $v(0)$ is bounded by $\|(\eta,\psi)\|_*$. Therefore, from Equation 4,

$$\|v\|^2_{L^\infty(0,T;L^2(\Omega))} + \|\rho\|^2_{L^2(0,T;L^2(\Omega))} \leq C\left\{\|v(0)\|^2 + \|(\eta,\psi)\|^2_*\right\}.$$

With the pseudo-inverse properties one can show that:

$$\begin{aligned}
\left(A\frac{Dv}{Dt}, A\frac{Dv}{Dt}\right) &= \left(\frac{Dv}{Dt}, A*A\frac{Dv}{Dt}\right) \\
&= -\left(\operatorname{div}\rho + (\nabla\cdot\dot{x})v, A*A\frac{Dv}{Dt}\right) \\
&\quad + \left(\frac{D\eta}{Dt} + \operatorname{div}\psi + (\nabla\cdot\dot{x})\eta, A*A\frac{Dv}{Dt}\right) \\
&\leq C\left\|A\frac{Dv}{Dt}\right\|\left\{\|A\operatorname{div}\rho\| + \|v\| + \left\|A\frac{D\eta}{Dt}\right\| + \|A\operatorname{div}\psi\| + \|\eta\|\right\}.
\end{aligned}$$

Then

$$\left\|A\frac{Dv}{Dt}\right\|^2_{L^2(0,T;L^2(\Omega))} \leq C\|(\eta,\psi)\|^2_*$$

and

$$\|(v,\rho)\|_* \leq C\|(u-v_h, \sigma-S_h)\|_*.$$

The proof is finished with the triangle inequality.

Now, assuming that Condition 1 is satisfied, we can, adding more constrains to the mesh, show the following superconvergence result, where $h = \max_i h_i$ and $h_i = \operatorname{diam}\Omega_i$:

$$\begin{aligned}
\|P_h u - u_h\| &\leq C\left\{\left\|h\underline{h}^{\min\{m+1,s\}}u\right\|_{L^\infty[0,T;\underline{H}^s]} + \left\|h\underline{h}^{\min\{m+1,s-1\}}\frac{Du}{Dt}\right\|_{L^2[0,T;\underline{H}^{s-1}]} \right. \\
&\quad + \left\|h\underline{h}^{\min\{m+2,s-1\}}\sigma\right\|_{L^2[0,T;\underline{H}^{s-1}]} + \left\|h^2\underline{h}^{\min\{m+1,s-2\}}\right\|_{L^2[0,T;\underline{H}^{s-1}]} \\
&\quad \left. + \left\|h^2\underline{h}^{\min\{m+1,s-2\}}\frac{D\sigma}{Dt}\right\|_{L^2[0,T;\underline{H}^{s-1}]}\right\}.
\end{aligned}$$

4 NUMERICAL EXPERIMENTS

We now present the results of some numerical experiments. In all of them we compare the results of calculations done with both moving and a fixed, uniform mesh. Strategy for the moving case is simple: the mesh is specified at the initial and final times and both are connected afterwards.

Example 1. Consider the following linear problem:

$$\begin{cases}
u_t - (u_x - b_1 u)_x = 0, & (x,t) \in (0,10)\times(0,1), \\
u(0,t) = u(10,t), & t \in [0,1], \\
u(x,0) = u_0(x), & x \in [0,10].
\end{cases}$$

For the initial condition, u_0 is a smooth non-negative function with support in the interval [3, 5]; b_1 is a C^2 non-negative function such that $b_1 = 3.5$ on [2,7]; $b_1 = 0$ on $[0, 1] \cup [8, 10]$; and b_1 is a 5th order polynomial on (1, 2) and (7, 8). Initially, the density of mesh points in (0, 6) was about one third higher than the average density across the entire interval. Here we first use a coarse solution to calculate the grid node distribution, at the times $t = 0.5$ and $t = 1$. The local mesh density is proportional to $\varepsilon + |u_{xx}|$, where u is approximated on a coarse uniform grid and ε is between the maximum absolute value and the average value of the second derivative. For all the other time levels the grid position is the quadratic interpolation of the corresponding grid positions at $t = 0$, $t = 0, 5$ and $t = 1$.

The mesh evolution is shown in Figure 1. In Figure 2 the solution, at $t = 1$, is plotted for a number of 40 cells.

The exact solution is obtained through a very fine mesh calculation.

Table 1 compares the L^2 and L^∞ norms for this example. By $\tilde{u}_h$ we mean a piecewise linear approximation connecting the points $(x_i, u_h(x_i))$, with x_i the cell center.

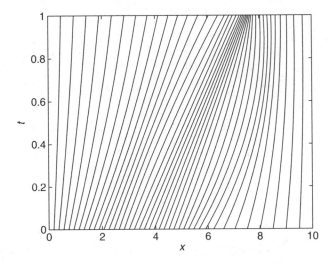

Figure 1. Moving mesh for example 1.

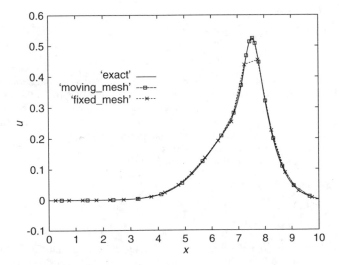

Figure 2. Solution for example 1.

Table 1. A comparison between L^2 and L^∞ norms for example 1.

	Moving mesh			Fixed mesh		
n	$\|u - u_h\|$	$\|u - \tilde{u}_h\|_\infty$	$\|P_h u - u_h\|$	$\|u - u_h\|$	$\|u - \tilde{u}_h\|_\infty$	$\|P_h u - u_h\|$
20	0.052	0.012	0.0035	0.083	0.075	0.013
40	0.026	0.0031	0.00082	0.040	0.025	0.0027
80	0.013	0.0010	0.00026	0.020	0.0066	0.00070
160	0.0066	0.00026	9.1e-05	0.010	0.0018	0.00020

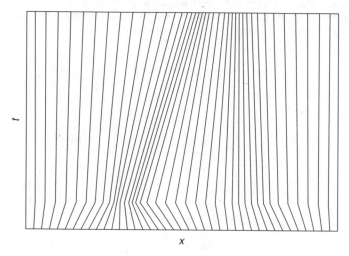

x

Figure 3. Moving mesh for example 2.

Example 2. Consider now the Burger's type problem: $u_t + uu_x - \varepsilon u_{xx} = 0$, with Dirichlet boundary conditions; and with the initial condition

$$u(x,0) = u_0(x) \equiv \begin{cases} 2*\sin(4\pi x), & 0.0 \le x \le 0.25, \\ \sin(4\pi x), & 0.25 \le x \le 0.5, \\ 0.5*\sin(4\pi x), & 0.5 \le x \le 0.75, \\ \sin(4\pi x), & 0.75 \le x \le 1.0. \end{cases}$$

We take $\varepsilon = 0.01$. Initially, we have two fronts that move quickly in opposite directions. After some time, one overcomes the other and goes rapidly to the boundary, decaying to zero. In this example, the mesh density function is $d(x) = (M - m)/3$ with $M = \max|u_{xx}|$ and $m = \min|u_{xx}|$.

In Figure 3 we have the moving mesh, with $n = 40$ cells. Figure 4 shows the solution and Figure 5 an enlargement of it. The fine solution was computed on a mesh with 1000 cells and the Crank-Nicolson scheme. Table 2 compares again the L^2 and L^∞ norms.

Example 3. Here we solve the following combustion problem:

$$u_t - u_{xx} - R\frac{e^\delta}{\alpha\delta}(1 + \alpha - u)e^{-\delta/u} = 0,$$

$$u_x(0,t) = 0, \ u(1,t) = 1,$$

$$u(x,0) = 1, \ 0 \le x \le 1.$$

This problem models evolution of a reacting mixture, R being the reaction rate. The unknown is the temperature u; α is heat release and δ the activation energy. Temperature increases slowly from its initial value, with a very large hot "point" forming at $x = 0$.

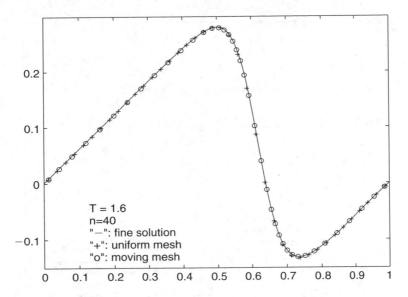

Figure 4.   Solution for example 2.

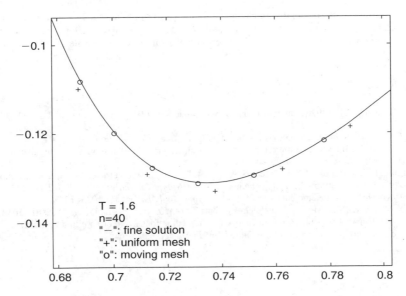

Figure 5.   Enlargement of figure 4, for example 2.

Table 2.   A comparison between L^2 and L^∞ norms, for example 2.

Moving mesh				Fixed mesh		
n	$\|u - u_h\|$	$\|u - \tilde{u}_h\|_\infty$	$\|P_h u - u_h\|$	$\|u - u_h\|$	$\|u - \tilde{u}_h\|_\infty$	$\|P_h u - u_h\|$
20	0.089	0.13	0.031	0.14	0.26	0.059
40	0.043	0.031	0.0057	0.060	0.11	0.016
80	0.021	0.012	0.0025	0.030	0.032	0.0044

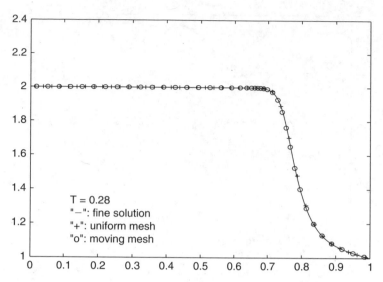

Figure 6.   Solution for example 3.

After a finite time ignition happens and temperature jumps from 1 to near $1 + \alpha$ and the front propagates to the boundary $x = 1$. For the experiments we took $R = 5$, $\alpha = 1$, and $\delta = 20$. In Figure 6 we show the solution at time $t = 0.28$. The mesh density function is the same as in Example 2.

REFERENCES

Bank, R.E. & Santos, R.F. 1993. Analysis of some moving space-time finite element methods. *SIAM J. Numerical Analysis* 30: 1–18.
Dupont, T.F. 1982. Mesh modification for evolution equations. *Math. Computation* 39: 85–107.
Dupont, T.F. & Liu, Y. 2003. Symmetric error estimates for moving mesh Galerkin methods for advection–diffusion equations. *SIAM J. Numerical Analysis* 40: 914–927.
Liu, Y. & Bank, R.E. & Dupont, T.F. & Garcia, S. & Santos, R.F. 2003. Symmetric error estimates for moving mesh mixed methods for advection–diffusion equations. *SIAM J. Numerical Analysis* 40: 2270–2291.
Miller, K. 1981. Moving finite elements, part II. *SIAM J. Numerical Analysis* 18: 1033–1057.
Miller, R. & Miller, K. 1981. Moving finite elements, part I. *SIAM J. Numerical Analysis* 18: 1019–1032.
Raviart, P.A. & Thomas, J.M. 1977. A mixed finite element method for second order elliptic problems. In Springer-Verlag (eds), *Mathematical Aspects of the Finite Element Method*: 292–315.

Computational Methods in Engineering and Science, Iu et al. (eds)
© *2003 Swets & Zeitlinger, Lisse, ISBN 90 5809 567 3*

Finite element simulation of a tri-axial piezoelectric sensor

N. Wakatsuki & Y. Kagawa
Department of Electronics and Information Systems, Akita Prefectural University, Honjo, Japan

M. Haba
Hokuto Denko Corporation, Atsugi, Kanagawa, Japan

ABSTRACT: Electromechanical transducers for sensing and actuating disturbances or vibrations have been used in many fields of applications. There have been transducers of different configuration developed for the uni-directional transduction. Present paper demonstrates a single element transducer for three axial components, made of a piezoelectric cylindrical shell. The separation of the three axial transductions is achieved by devising a proper electrode arrangement. The structure and the fundamental idea are first presented and then follows the numerical analysis by means of the finite element modeling, and their characteristics and behaviors are then experimentally verified.

1 INTRODUCTION

Piezoelectric devices are used in various fields. Sensors, actuators and electromechanical filters are some examples among others. The sensors and actuators are called transducers. Electromechanical transducers utilize a function of converting the energy from the electrical system to the mechanical one and vice versa. The mechanism of sensors and actuators is identical, in which the system is governed by the same constitutive equations.

Sensors for detecting mechanical motions or loadings are sometimes needed. Piezoelectric sensors can be applied to sensing the load in robot arms. In order to make for a robot to move smoothly, mechanical loads should be simultaneously measured in three-dimensional space. On the other hand, though piezoelectric actuators only provide microscopic motion, they could be used for driving the stage and the needle in scanning tunneling microscopes (Binning & Rohrer 1985).

In the present paper, the characteristics of a three axial piezoelectric sensor and actuator are analyzed using three-dimensional finite element modeling in which the piezoelectric effect is included.

2 PRINCIPLE OF THREE AXIAL SENSING AND ACTUATION

The sensor investigated here is a cylindrical shell made of piezoelectric ceramic, polarized in the radial direction with several pairs of electrodes placed on both surfaces of the shell as shown in Figure 1. Inner surface is entirely covered with an uniform electrode used for a common grounding (GND). In the figure, there are shown eight electrodes on the outer surface. Lower four are only used, named $x1$, $x2$, $y1$ and $y2$ as a matter of convenience. Upper four are left only for the future purpose, which are electrically open.

The independent sensing of the three axial components depends on the symmetrical arrangement of the electrodes, in which the lateral piezoelectric effect in the shell is utilized. Suppose that a force is applied at the tip of the sensor while the other end is fixed to the base. The stress in the z-direction causes the same electric voltage on all of the electrodes. When the force is applied to the x-direction, the voltage appeared on $x1$ is reversal to that on $x2$, while the voltage appeared on $y1$ is the same as that on $y2$. Now, applied force is expressed in terms of three axial components as $\{f\} = \{f_x, f_y, f_z\}$, and the voltages on electrodes $x1$ to $y2$ against GND are referred to V_{x1} to V_{y2}, which are the open-circuited

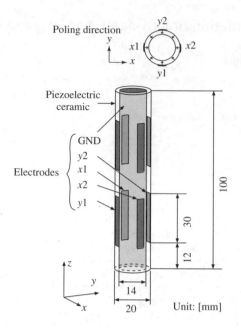

Figure 1. A piezoelectric sensor.

voltages. By virtue of the symmetrical arrangement of the electrodes, one has the following relations,

$$
\begin{bmatrix} V_{x1} \\ V_{x2} \\ V_{y1} \\ V_{y2} \end{bmatrix} = \begin{bmatrix} \alpha & 0 & \beta \\ -\alpha & 0 & \beta \\ 0 & \alpha & \beta \\ 0 & -\alpha & \beta \end{bmatrix} \begin{bmatrix} f_x \\ f_y \\ f_z \end{bmatrix}
\tag{1}
$$

α, β are the coefficients which depends on piezoelectric constants and electrode configuration. Solving the equation for the forces yields $f_x = (V_{x1} - V_{x2})/2\alpha$, $f_y = (V_{y1} - V_{y2})/2\alpha$, $f_z = (V_{x1} + V_{x2})/2\beta$, or $f_z = (V_{y1} + V_{y2})/2\beta$. Last two equations show that f_z is determined by using either a pair of V_{x1} and V_{x2} or a pair of V_{y1} and V_{y2}. However, averaging will help reducing the measurement errors, so that

$$
f_z = (V_{x1} + V_{x2} + V_{y1} + V_{y2})/4\beta .
\tag{2}
$$

We have, by measuring four voltage V_{x1} to V_{y2}, the three axial components, f_x, f_y and f_z. That is,

$$
\begin{bmatrix} f_x \\ f_y \\ f_z \end{bmatrix} = \begin{bmatrix} \frac{1}{2\alpha} & -\frac{1}{2\alpha} & 0 & 0 \\ 0 & 0 & \frac{1}{2\alpha} & -\frac{1}{2\alpha} \\ \frac{1}{4\beta} & \frac{1}{4\beta} & \frac{1}{4\beta} & \frac{1}{4\beta} \end{bmatrix} \begin{bmatrix} V_{x1} \\ V_{x2} \\ V_{y1} \\ V_{y2} \end{bmatrix}
\tag{3}
$$

This relation is also obtained by using the least squares procedure. Coefficients α and β are to be determined by experiment or numerical simulation.

On the other hand, when electric potentials are applied to the electrodes, the displacements caused at the tip of the actuator will be expressed as

$$
\begin{bmatrix} u_x \\ u_y \\ u_z \end{bmatrix} = \begin{bmatrix} -a & a & 0 & 0 \\ 0 & 0 & -a & a \\ -b & -b & -b & -b \end{bmatrix} \begin{bmatrix} V_{x1} \\ V_{x2} \\ V_{y1} \\ V_{y2} \end{bmatrix} .
\tag{4}
$$

Coefficients a and b in the equation here are again to be determined by the numerical analysis or experiment. If these equations are solved for V_{x1} to V_{y2}, the voltages to be applied to obtain arbitrary motion at the tip can be determined. As they can not uniquely be determined only from these equations, we take the condition,

$$V_{x1} + V_{x2} = V_{y1} + V_{y2} \tag{5}$$

The equation (4) and (5) are simultaneously solved to yield

$$\begin{bmatrix} V_{x1} \\ V_{x2} \\ V_{y1} \\ V_{y2} \end{bmatrix} = \begin{bmatrix} -\frac{1}{2a} & 0 & -\frac{1}{4b} \\ \frac{1}{2a} & 0 & -\frac{1}{4b} \\ 0 & -\frac{1}{2a} & -\frac{1}{4b} \\ 0 & \frac{1}{2a} & -\frac{1}{4b} \end{bmatrix} \begin{bmatrix} u_x \\ u_y \\ u_z \end{bmatrix}. \tag{6}$$

3 FINITE ELEMENT MODELING

Constitutive equations or the piezoelectric equations with e-type are expressed as follows

$$\begin{aligned} \{T\} &= [c^E]\{S\} - [e]^T\{E\} \\ \{D\} &= [e]\{S\} + [\varepsilon^S]\{E\} \end{aligned} \tag{7}$$

where $\{T\}$: stress vector, $[c^E]$: stiffness matrix when $\{E\} = 0$, $\{S\}$: strain vector, $[e]$: piezoelectric constant matrix, $\{E\}$: electric field vector, $\{D\}$: electric displacement vector and $[\varepsilon^S]$: permittivity matrix when $\{S\} = 0$.

The discretized equation of the piezoelectric system with finite element modeling is expressed as

$$\begin{bmatrix} [K] & [\Gamma] \\ [\Gamma]^T & -[G] \end{bmatrix} \begin{bmatrix} \{u\} \\ \{\phi\} \end{bmatrix} = \begin{bmatrix} \{F\} \\ \{Q\} \end{bmatrix} \tag{8}$$

where $[K]$: stiffness matrix, $[\Gamma]$: electro-mechanical coupling matrix, $[G]$: capacitance matrix, $\{u\}$: displacement vector, $\{\phi\}$: electric potential vector, $\{F\}$: force vector and $\{Q\}$: electric charge vector. The code PIEZO3D (Kagawa 1981; Yamabuchi & Kagawa 1989) has been developed for the three-dimensional finite element modeling including the piezoelectric effect. In the code, an isoparametric cubic element with 8 nodes (Kagawa et al. 1996; Kagawa et al. 2001) is used, and has four degrees of freedom are allocated to each node for the displacements in x, y and z directions, and for the electric potential. While electric potentials are linearly interpolated, displacements are interpolated with second order polynomials but without increasing the nodes in order to include the bending effect.

4 THREE AXIAL FORCE SENSOR

The forces must be resolved into three axial components. In this section, a numerical simulation using three-dimensional FEM is demonstrated. The results are verified by the experiment.

4.1 Finite element simulation

The prototype transducer is a cylindrical shell as illustrated in Figure 1 with its dimensions, which is divided for the finite element modeling into 16 elements along the circumference, and 17 elements for the length. Since the stress changes are greater near the fixed bottom end, such a part is divided into smaller elements. The present model consists of 272 elements, requiring 1912 degrees of freedom and the system matrix has the band width of 505.

The boundary and driving conditions must be prescribed both at mechanical and electrical terminals, depending on the way of the use. When it is used as a sensor, mechanical excitation is assumed to be applied evenly at the top surface, while the electric potentials on the open circuited electrodes are taken to be unknown against the common ground electrode. When it is used as an actuator, the electric potentials are properly prescribed at the electrodes.

Material properties used for numerical analysis of the prototype are shown in Table 1, which are supplied by the company.

The displacements are illustrated in Figure 2 when the force of 1 [N] is respectively applied to the top surface of the shell in x, y and z direction.

The electric potential distributions on the outer surface are shown in Figure 3(a)–(c) respectively, which are shown projected into a rectangular plane. The displacements in x and y directions are 7.7×10^{-7} [m] while the displacement in z direction is only 7.4×10^{-9} [m]. The electric potential distributions are identical when the shell is rotated by 90° along the circumference for the former case, while electric potential is almost constant over the outer surface for the latter case. Open-circuited voltages at the electrodes for the axial forces are shown in Table 2.

Table 1. Material properties of piezoelectric ceramic. (C-201, by Fuji ceramics corp.)

$c_{11}^E = 14.9$	$c_{12}^E = 8.7$	$c_{13}^E = 9.1$	$c_{33}^E = 13.2$	$c_{44}^E = 2.5$	$c_{66}^E = 3.1$
$e_{15} = 15.2$	$e_{31} = -4.5$	$e_{33} = 17.5$	$\varepsilon_{11}^S/\varepsilon_0 = 904$	$\varepsilon_{33}^S/\varepsilon_0 = 818$	

c^E: stiffness ($\times 10^{10}$ N/m^2) e: piezoelectric constants (C/m^2) $\varepsilon^S/\varepsilon_0$: relative permittivity.

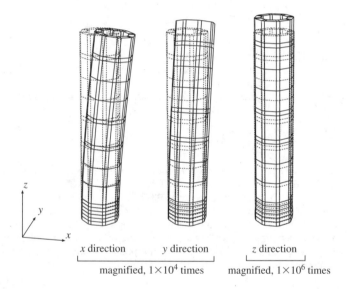

x direction y direction z direction

magnified, 1×10^4 times   magnified, 1×10^6 times

Figure 2. Displacements when 1 [N] is applied at the tip.

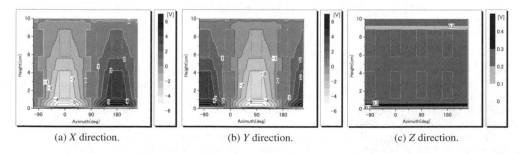

(a) X direction.   (b) Y direction.   (c) Z direction.

Figure 3. Potential distribution to the force [1 N] applied at the tip.

296

4.2 *Experimental verification*

The numerical results are verified by the experiments. The prototype sensor is fixed on a steel plate of the thickness of 10 mm using epoxy resin. Input force is applied to the tip of the sensor using electrodynamic exciter. The experimental setup is shown in Figure 4.

Figure 5(a) shows the frequency responses of the sensitivity for each horizontal and vertical excitation. The sensor has almost constant sensitivities over the frequency above 10 [Hz], which are 3.3 [V/N] for the horizontal excitation and 0.21 [V/N] for the vertical excitation.

Table 2.   Response voltages to the force of 1 [N] in each direction.

Direction	V_{x1}	V_{x2}	V_{y1}	V_{y2}	[V]
x	3.3	−3.3	0.00	0.00	
y	0.00	0.00	3.3	−3.3	
z	0.20	0.20	0.20	0.20	

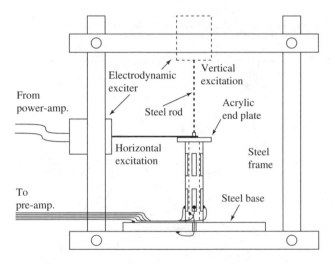

Figure 4.   The experimental setup.

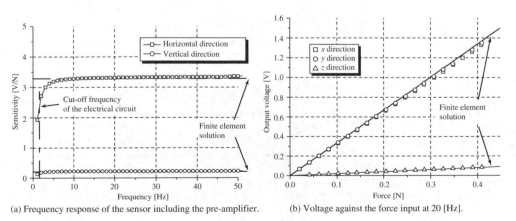

(a) Frequency response of the sensor including the pre-amplifier.   (b) Voltage against the force input at 20 [Hz].

Figure 5.   Response of the sensor.

As in the present analysis, zero output current was assumed, the pre-amplifiers with very high input impedance of about 100 [MΩ] is used in the experiment. Thus the sensitivity of the sensor depends on the frequency, showing high-pass filtering characteristic.

Figure 5(b) shows the output voltage to the input force at 20 [Hz]. The symbols and the lines respectively indicate the measured data and finite element solutions. The output voltage linearly depends on the input force.

Sensitivity coefficients α and β are obtained by the numerical analysis and the experiment, which are given as in Table 3. The agreement is good.

The resolution capability into the three axial components is then confirmed. Now, the case when the force is applied to an arbitrary direction is considered.

The azimuth and the elevation angle of the direction are defined as shown in Figure 6. The three axial components of the excitation force are estimated according to the equation(3) with the use of the values of α and β obtained in the previous experiment. The magnitude and the frequency of the excitation force applied to the sensor is respectively fixed to 0.4 [N] at 20 [Hz] during the experiments. Figure 7(a) and 7(b) shows the three axial force components obtained from the voltage measurement, in which the elevation and azimuth angles are varied respectively. Symbols and lines respectively show the experimental and the numerical results. The agreement is reasonable. In Figure 7(b), f_z, which is the z component

Table 3. Coefficients about sensitivity of the sensor.

	α	β	[V/N]
Numerical	3.3	0.20	
Experimental	3.3	0.21	

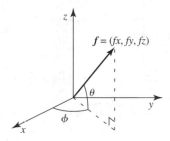

Figure 6. Definition of the azimuth and the elevation angles, ϕ and θ.

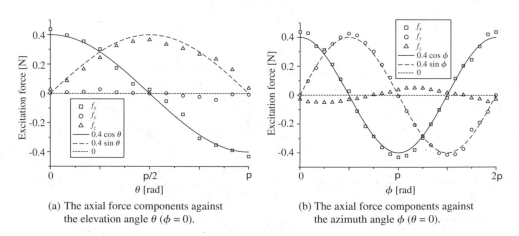

(a) The axial force components against the elevation angle θ ($\phi = 0$).

(b) The axial force components against the azimuth angle ϕ ($\theta = 0$).

Figure 7. The solution capability into the three axial components.

298

of the excitation force, should be zero, while in the experiment, the obtained value depends on the direction, which is the highest at the angle $\phi = (1/4)\pi$ and $(5/4)\pi$. This may be caused due to the fact that the sensitivity is not even for every electrode. The difference may be caused due to the fabrication, polarization process, electrode deposition and the experimental setting. Since the sensitivity to the vertical force is much smaller than to the horizontal one, the error is more pronounced for the vertical force measurement.

5 THREE AXIAL ACTUATOR

Piezoelectric actuators are often utilized to produce a small mechanical displacement. The actuator considered here provides a three-dimensional movement using a single piezoelectric element.

5.1 Finite element simulation

The shell is divided into 16 elements along the circumference, and into 14 elements along the height direction. The regions near the electrode edges are divided into smaller elements. It consists of 224 elements, requiring 1477 degrees of freedom. The system matrix has the band width of 397. For mechanical boundary condition, the shell is fixed at the bottom surface. The driving electrical input potentials are applied to the electrodes on the outer surface refering to the GND electrode.

The displacements obtained by the numerical analysis are illustrated in Figure 8. The input electric potentials are indicated in the captions.

The finite element simulation gives the displacements of 6.7×10^{-9} [m] both in x and y directions and 8.4×10^{-10} [m] in z direction at the tip of the shell, respectively.

5.2 Experimental verification

The numerical results are verified by the experiment. The actuator is fixed on a steel plate in the same manner as in the case of the sensor. Movement is measured at the tip with a laser Doppler vibrometer (LDV). A cubic aluminum post is fixed on the end disk for better light reflection. The experimental setup is illustrated in Figure 9.

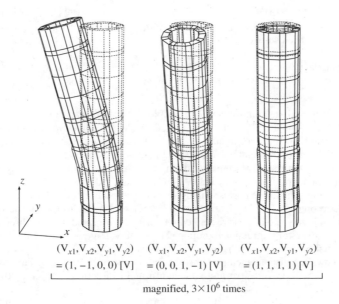

$(V_{x1}, V_{x2}, V_{y1}, V_{y2})$ $(V_{x1}, V_{x2}, V_{y1}, V_{y2})$ $(V_{x1}, V_{x2}, V_{y1}, V_{y2})$
$= (1, -1, 0, 0)$ [V] $= (0, 0, 1, -1)$ [V] $= (1, 1, 1, 1)$ [V]

magnified, 3×10^6 times

Figure 8. Displacements to the applied voltages, magnified 3×10^6 times.

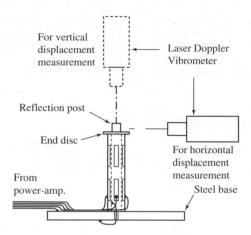

Figure 9. The experimental setup.

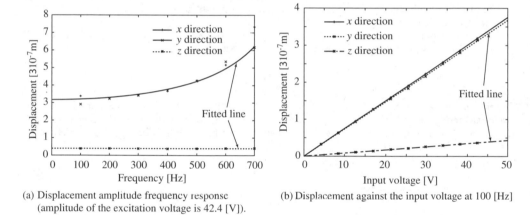

(a) Displacement amplitude frequency response (amplitude of the excitation voltage is 42.4 [V]).

(b) Displacement against the input voltage at 100 [Hz]

Figure 10. Experimental result.

The device is capable of static movement in principle, while LDV can not detect the static displacements. The amplitude of vibration velocity is measured for the sinusoidal wave excitation, and then the response is extrapolated down to zero frequency. The frequency responses measured are shown in Figure 10(a). Here, the amplitude of the input voltage is 42.4 [V] at peak value and 30 [V] in the effective value.

Figure 10(a) shows that the displacements in x and y direction depend on the frequency and they become almost independent of frequency below 200 [Hz], and independent of frequency from 100 to 700 [Hz] in z direction. The curve corresponding to x and y directional displacements is drawn as the function of single resonance, fitted by least squares. In the experiments to follow, the voltages with the same amplitudes but inverse polarities are applied into electrodes $x1$ and $x2$ for x directional motion, or into electrodes $y1$ and $y2$ for y directional motion. The same voltage is applied into all four electrode pairs for z directional motion.

Figure 10(b) shows the displacements against the input voltage at 100 [Hz]. The response displacements at the tip are 7.5×10^{-9} [m] par unit voltage in x and y direction, and 8.6×10^{-10} [m] par unit voltage in z direction.

These results give the coefficients a and b in equations (4) and (6), which are shown in Table 4. Good agreement is also achieved.

Table 4. Coefficients about movement of the actuator.

	a	b	[m/V]
Numerical	3.4×10^{-9}	2.1×10^{-10}	
Experimental	3.8×10^{-9}	2.2×10^{-10}	

6 CONCLUDING REMARKS

The sensitivity and the frequency characteristics of three axial piezoelectric sensors and actuators are considered. The numerical solution with the three-dimensional finite element modeling is presented together with the experimental verification.

The numerical simulation agrees reasonably well with the experimental result. The error is pronounced for the vertical component due to the smaller sensitivity, for which some compensation should be devised.

As the present prototype transducers are made of a solid ceramic and have high mechanical impedance, they are of advantage to sensing the force rather than the displacement, or to achieving a very small displacement.

ACKNOWLEDGMENT

H. Kuroda and K. Inoue should be acknowledged for their help of the experiments. Dr. Tsuchiya is also acknowledged for his advice during the course of this study.

REFERENCES

Binning, G. & Rohrer, H. 1985. The scanning tunneling microscope. *Scientific American*, 50.
Kagawa, Y. 1981. *Finite Elements for Acoustical Engineers/Fundamentals and Applications*. Baifukan Press, Tokyo. (in Japanese).
Kagawa, Y., Tsuchiya, T. & Kataoka, T. 1996. Finite element simulation of dynamic responses of piezoelectric actuators. *Journal of Sound and Vibration*, 191(4):519–538.
Kagawa, Y., Tsuchiya, T. & Sakai, T. 2001. Three-dimensional finite element simulation of a piezoelectric vibrator under gyration. *IEEE Transaction on Ultrasonics, Ferro-electrics and Frequency Control*, 48(1):180–188.
Yamabuchi, T. & Kagawa, Y. 1989. Numerical simulation of a piezoelectric ultrasonic motor and its characteristic. *Journal of the Japan Society for Simulation Technology*, 8(3):69–76. (in Japanese).

Computational Methods in Engineering and Science, Iu et al. (eds)
© *2003 Swets & Zeitlinger, Lisse, ISBN 90 5809 567 3*

Parametric analysis and optimization for Flip Chip package

Q.Y. Li, J.S. Liang & D.G. Yang
Department of Electronic machinery and Traffic Engineering, Guilin University of Electronic Technology, Guangxi, China

ABSTRACT: This paper focus on the methodology of parametric influences analysis and optimization for the VonMises equivalent residual stress induced in major packaging process of the Flip Chip microelectronic assembly. Based on the results of a 3D Finite Element Method (*FEM*) simulation loops for the packaging process, a Response Surface analysis by using the technique of Central Composite Design Sampling (*CCDS*) is introduced. An approximate function between the chief packaging parameters and the maximum value of the equivalent stress in the solder joints has been built. Probabilistic analysis as well as parametric sensitivities study is also implemented on the ground of a random Monte Carlo Simulation results. Finally, the better solution of the chief packaging parameters is obtained from a global parametric optimization strategy of the Genetic Algorithm.

1 INTRODUCTION

Flip chip technology refers to the interconnection technique in which the active side of the silicon die is faced downwards and mounted onto the substrate by solder joints. It is an attractive approach in the field of microelectronics packaging because of merits such as high-density assembly, small outside dimensions, rapid signal treatment and low assembly noise. A schematic diagram of Flip Chip assembly as well as the three major packaging parameters is shown in Figure 1.

However, because solder joints are directly connected to the substrate, when circumstance temperature is changed, mismatch in coefficient of thermal expansion (*CTE*) between different packaging materials will induce high residual stress in silicon die and solder joints. Meanwhile, during the Flip Chip major

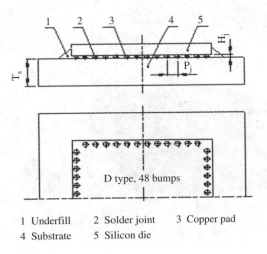

1 Underfill   2 Solder joint   3 Copper pad
4 Substrate   5 Silicon die

Figure 1. A schematic diagram of Flip Chip package.

packaging process, the curing of the underfill is accompanied by partly constrained shrinkage, which will increase stress level mentioned above. Previous researches have showed that the maximum VonMises equivalent stress in solder joints during the thermal loading history serves a very important role to evaluate the reliability of the Flip Chip package. These researches also indicated that the major packaging parameters are crucial to the level of VonMises stress in the solder joints. So it is feasible to investigate the causal relationship between the stress and the packaging parameters and gain a better solution of the Flip Chip packaging by minimize the VonMises stress in the solder joints.

2 FEM SIMULATION FOR FLIP CHIP PACKAGE

2.1 Geometry model

The 3-D parametric *FEM* model for a 48-bumps Flip Chip package is set up in a commercial *FEM* code, by taking the dimension of the stand-off height of solder joints (H_j), pitch between solder joints (P_j) and thickness of the *FR-4* substrate (T_s) as input variables, so that the *FEM* model are changeable during the simulation loop of the response surface analysis. At the same time, a symmetrical condition of the package is also a consideration to reduce the simulation time, so only 1/8 of the Flip Chip package is modeled and meshed in this study. The boundary conditions are used as follows: the bottom area of the FR-4 substrate is fixed in all Degrees of Freedom (*DOF*), and normal displacements on the axial and diagonal symmetry faces are set to zero. The *FEM* mesh and the constrained faces of the model are shown in Figure 2.

2.2 Material mode

Three kinds of material modes are involved in the Flip Chip model. An isotropic linear elastic mode is used for silicon die, copper pads and substrate, and only the Young's modulus, Poisson ratio and coefficient of thermal expansion are needed to describe their thermal–mechanical properties. Meanwhile, viscoplastic Anand mode for 63Sn37Pb eutectic solder joints and cure-dependent Maxwell mode for viscoelastic underfill resin are coded in the *FEM* model respectively.

When the surrounding temperature is higher than 40% of its melting point (about 183°C), the SnPb eutectic solder will present viscoplastic behavior that can be described by the following constitutive relations:

$$dp = Ae^{\frac{-Q}{R\theta}} \left[\sinh \left(\zeta \frac{\sigma}{s} \right) \right]^{\frac{1}{m}} \tag{1}$$

$$\dot{s} = \left\{ h_0 \left(|B| \right)^\alpha \frac{B}{|B|} \right\} dp \tag{2}$$

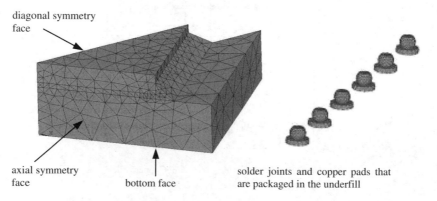

diagonal symmetry face

axial symmetry face

bottom face

solder joints and copper pads that are packaged in the underfill

Figure 2.   *FEM* mesh and the constrained faces of the Flip Chip model.

$$B = 1 - \frac{s}{s^*} \qquad (3)$$

$$s^* = \hat{s} \left[\frac{dp}{A} e^{\frac{Q}{R\theta}} \right]^n \qquad (4)$$

Here, dp is effective inelastic deformation rate. A is pre-exponential factor, Q is activation energy. R is universal gas constant. ζ is multiplier factor. σ is effective Cauchy stress. s is deformation resistance. $\dot{S}$ is time derivative of deformation resistance. s^* is saturation value of deformation resistance. $\hat{S}$ is coefficient for deformation resistance saturation value. h_0 is hardening/softening constant. α is strain rate sensitivity of hardening/softening. m is strain rate sensitivity of stress, n is strain rate sensitivity of deformation resistance saturation value.

A curing-dependent viscoelastic constructive relation that based on the Boltzman superposition principle is adopted to describe the stress–strain relation in underfill resin as follows:

$$S_i(t) = \int_{-\infty}^{t} C_{ij} \{ \alpha(\xi), (t - \xi) \} \cdot \left\{ \left(\frac{\partial E_j}{\partial \xi} \right)_\xi - \left(\frac{\partial E_j^*}{\partial \xi} \right)_\xi \right\} d\xi \qquad (5)$$

Here S_i, E_j and E_j^* represent the components of vectors of stress, strain and initial strain tensors, respectively. $\alpha(\xi)$ represents curing parameter versus curing time $\xi \cdot C_{ij}$ is the cure dependent relaxation modulus functions. Because of the isotropic assumption of the underfill resin, C_{ij} can be described by bulk and shear relaxation modulus functions as:

$$C_{ij} \{ \alpha(\xi), (t - \xi) \} = K \{ \alpha(\xi), (t - \xi) \} \cdot V_{ij} + G \{ \alpha(\xi), (t - \xi) \} \cdot D_{ij} \qquad (6)$$

wherein V_{ij} and D_{ij} are the volumetric and deviatoric component matrices. K and G are bulk and shear relaxation modulus, respectively. They can be described by a generalized Maxwell model with limited elements (Prony series) in an approximating form:

$$G \{ \alpha(\xi), (t - \xi) \} \approx \sum_{n=1}^{N} G^n \{ \alpha(\xi) \} \cdot e^{-(t-\xi)/\tau_n} \qquad (7)$$

$$K \{ \alpha(\xi), (t - \xi) \} \approx \sum_{n=1}^{N} K^n \{ \alpha(\xi) \} \cdot e^{-(t-\xi)/\tau_n} \qquad (8)$$

where $K^n\{\alpha(\xi)\}$ and $G^n\{\alpha(\xi)\}$ are cure dependent stiffness coefficients and are determined by Dynamic Mechanical Analysis (DMA) method, τ_n is the relaxation time.

2.3 Thermal loading history and simulation results

The Flip Chip package is subjected to a thermal loading history contains curing stage (point a to b), cooling stage (point b to d) and thermal cyclic stage (point d to h) as shown in Figure 3 (a). It should be noted that the influence of the temperature gradient is neglected here. This hypothesis is made because previous experiments have shown that the time for different parts of the Flip Chip package take to become uniformly heated is much less than that for temperature changing itself during the thermal loading history.

Figure 3 (b) shows the FEM simulation result of the VonMises equivalent stress that distributes in the solder joints at the end point of the thermal loading history. The simulation result was gained form a nominal-dimensioned Flip chip package, which three major package parameters, namely, H_j, P_j and T_s is set to 0.15 mm, 0.45 mm and 1.50 mm respectively. The result indicates that the maximum VonMises stress is always located in the solder joint nearest to the central point of the Flip Chip package. This phenomenon is mainly due to the CTE mismatch between the silicon die and the substrate.

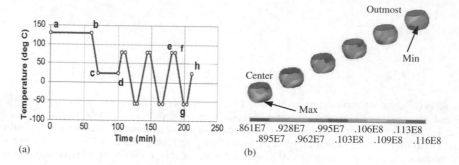

(a)

(b)

Figure 3. Thermal loading history and the *FEM* simulation result. (a) Thermal loading history (b) VonMises stress distribution.

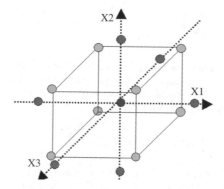

Figure 4. Central composite design for a problem has three input variables.

3 PARAMETRIC ANALYSIS

3.1 *Response surface approximation for Flip Chip package*

In order to construct the approximate function between the packaging parameters and the output stress, a response surface driven Design of Experiment (*DOE*) loop is implemented. Three major packaging parameters, H_j, P_j and T_s, are used as input variables, and their valid intervals are set to:

$$H_j \in [0.1, 0.2], \quad P_j \in [0.4, 0.5], \quad T_s \in [1.0, 2.0] \quad \text{(units in mm)}$$

The three design variables form a 3-D designing space of the Flip Chip package. For a purpose to acquire more information by a least *FEM* Simulation looping times, the Central Composite Design Sampling (*CCDS*) technology is adopted in this research. A central composite design of three input variables includes a central point (the original designing point), six axial points located on each axis symmetrical to the central point, plus eight factorial points located at the corners of a cube. Hence there are totally fifteen sampling points in a central composite design with three input variables. The locations of the sampling points for a problem with three random input variables is illustrated in Figure 4.

Fifteen sampling points in the designing space of the Flip Chip package created by using the *CCDS* strategy are listed in Table 1.

After the sampling points have been made, a FE analysis-processing file is applied to these designing points, thus simulation result for each point is produced. Response Surface Methods (*RSM*) locate the sample points in the space of random input variables so that an appropriate approximation function for the input/output parameters of the package system can be found most efficiently. Typically, the response surface approximate function takes on a form of fully quadratic polynomial with cross terms

Table 1. Sampling points created by *CCDS*.

Loop	H_j	P_j	T_s	Loop	H_j	P_j	T_s	Loop	H_j	Pj	T_s
1	0.1500	0.4500	1.500	6	0.1005	0.4500	1.500	11	0.1050	0.4950	1.950
2	0.1500	0.4005	1.500	7	0.1995	0.4500	1.500	12	0.1950	0.4050	1.050
3	0.1500	0.4995	1.500	8	0.1050	0.4050	1.050	13	0.1950	0.4950	1.050
4	0.1500	0.4500	1.005	9	0.1050	0.4950	1.050	14	0.1950	0.4050	1.950
5	0.1500	0.4500	1.995	10	0.1050	0.4050	1.950	15	0.1950	0.4950	1.950

Note: Units in mm.

Table 2. Coefficients in equation (9) given by the *RSM* analysis.

Constant	H_j	P_j	T_s	$H_j * H_j$	$P_j * P_j$	$T_s * T_s$	$H_j * P_j$	$H_j * T_s$	$P_j * T_s$
1.43E6	9.24E9	4.95E8	−1.15E9	−3.30E13	−2.29E12	3.32E11	2.37E12	−5.13E11	5.50E11

Note: (1) Unit of VonMises stress is in Pa; (2) $SSE = 2.016E10\,Pa^2$.

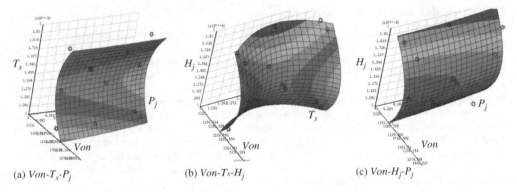

(a) *Von-T_s-P_j*   (b) *Von-Ts-H_j*   (c) *Von-H_j-P_j*

Figure 5. Contour plots of the response surfaces.

and the linear regression analysis methods are adopted. In this case the approximation function can be described by the following equation:

$$\hat{y} = C_0 + \sum_{i=1}^{m} C_i x_i + \sum_{i=1}^{m}\sum_{j=1}^{m} C_{ij} x_i x_j \tag{9}$$

Herein $\hat{y}$ represents the response value of the maximum VonMises stress in solder joints, m is the number of the design variables, $x_i(i = 1$–$m)$ are the design variables, C_0 is the constant term of the approximation function, $C_i(i = 1$–$m)$ represent the coefficients of the linear terms, $C_{ij}(i, j = 1$–$m)$ are the coefficients of the quadratic terms. The Goodness-of-Fit of equation (9) is usually measured by the Sum of Square Error (*SSE*) as follows:

$$SSE = \sum_{j=1}^{n}(y_i - \hat{y}_i)^2 = [\{y\} - \{\hat{y}\}]^T [\{y\} - \{\hat{y}\}] \tag{10}$$

where y_i and $\hat{y}_i$ are values of the output parameter and value of the regression model at i-th sampling point. When taking the maximum VonMises stress in solder joints as the output variables, the coefficients of equation (9) obtained in a response surface driven Design of Experiment (*DOE*) procedure are given in Table 2.

The contour plots of the response surface between the maximum VonMises stress in solder joints and the corresponding input variables are shown in Figure 5,

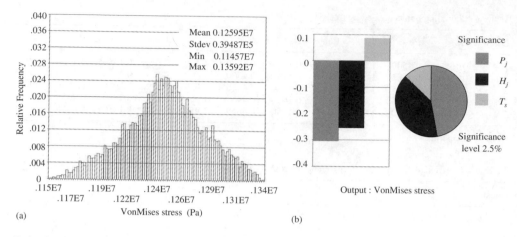

(a)

(b)

Figure 6.   Results of parametric analysis (a) Histogram of the output stress (b) Parametric sensitivity result.

3.2   *Probabilistic and parametric sensitivities analysis*

The random Monte Carlo simulation method based on response surface approximate function is introduced to carry out the probabilistic analysis of the Flip Chip package in this paper. A samples space contains ten thousand designing points is made by using the Latin Hypercube Sampling (*LHS*) method, which has a sample memory function that can reduce the random simulation loops. The input variables of H_j, P_j and T_s are assumed to obey the uniform distribution form and only characterized by their own lower and upper limits. No correlation effects are considered between each input variable. This assumption is based on the fact that each possible designing point located within the valid domain of the input variable has a uniform probability density. The response stress of the Flip Chip package at each design point is firstly calculated by the approximate function during the random simulation loop, then parametric statistics and trends analysis for input/output parameters are implemented. The histogram of the relative frequency of the output parameters and the sensitivity diagram for input/output value at a significance level of 2.50% studied here are showed in Figure 6.

Figure 6 (a) shows that the maximum value of the VonMises stress in the solder joints distributes in the region from 0.115E+07 to 0.134E+07 (Pa), and the highest relative frequency is dropped in the sphere of about 0.124E+07 to 0.126E+07 (Pa). In our research, a parametric sensitivity analysis is also performed to make an investigation to find how and on what degree that each input parameter will take to influence the output stress. Figure 6 (b) shows the sensitivity charts as well as the pie diagram for the maximum VonMises stress at a significance level of 2.50%. From Figure 6 (b), we notes that pitch between solder joints (P_j) is the first ranked negative factor that impacts the output VonMises stress, while the stand-off height of solder joints (H_j) is ranked the second. So it is foreseeable that the reduction of the value of P_j and H_j will significantly bring down the maximum VonMises stress in the solder joints. However, because the thickness of the FR-4 substrate (T_s) only has a small positive sensitivity value to the output stress, so the increase of the T_s will not lead to a significantly synchronous increase of the maximum VonMises stress in the solder joints.

4   GLOBAL OPTIMIZATION FOR FLIP CHIP PACKAGE

In order to obtain the approximate global optimum solution for Flip Chip package based on the minimum VonMises equivalent stress induced in the solder joints, Genetic Algorithm (*GA*) is adopted in our research. The approximate function derive from the response surface analysis is used as the objective function, and the valid interval of each input variable serve as the constrained conditions. After make $x_1 = H_j$, $x_2 = P_j$ and $x_3 = T_s$, the optimization model can be given as follows:

$$Min.\hat{y} = C_0 + \sum_{i=1}^{3} C_i x_i + \sum_{i=1}^{3}\sum_{j=1}^{3} C_{ij} x_i x_j \qquad (11)$$

308

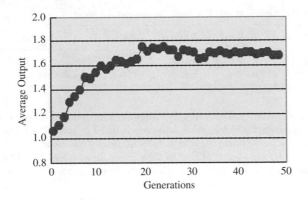

Figure 7.   GA iterations procedure for Flip Chip package.

$$s.t. \quad \begin{cases} 0.1 \le x_1 \le 0.2 \\ 0.4 \le x_2 \le 0.5 \\ 1.0 \le x_3 \le 2.0 \end{cases}$$

Herein the coefficients of $C_i(i = 0\text{--}3$) and $C_{ij}(i, j = 0\text{--}3)$ are listed in Table 2.

In most case, the fitness functions for GA are usually take on a maximum and no-constrained form, so here we need to convert equation (11) by means of penalty functions to meet these demands. Consider the general minimum problem with n inequality constrains as follows:

$Min.\, f(x)$

$s.t. \quad g_i(x) \ge 0(i = 1,2,\cdots n)$ \hfill (12)

It can be converted to a maximum and no-constrained problem:

$Max.F(x) = G(x) \cdot P(x)$ \hfill (13)

where $G(x)$ is an evaluation function of the solutions' quality, $P(x)$ is a penalty function.

$$G(x) = \begin{cases} \dfrac{K}{1 + (1.1)^{\beta f}} & \text{when } f \ge 0 \\[2mm] \dfrac{K}{1 + (0.9)^{\beta f}} & \text{when } f < 0 \end{cases}$$ \hfill (14)

$$P(x) = \dfrac{1}{(1.1)^{\omega(x)}}$$ \hfill (15)

$$\omega(x) = \sum_{i=1}^{m} |g_i(x)|$$ \hfill (16)

Here K and β are experiential constants and can be set to $K = 2, \beta = 0.1$. Now we can change f in equation (14) to $\hat{y}$ (≥ 0) in equation (11) and then get $G(x)$. Calculate $\omega(x)$ from the constrained function in equation (11) and get $P(x)$ in equation (15). Hence equation (13) can be obtained finally.

In our research, a trial and error process is applied to select the population size, crossover rate, mutation rate and the maximum generations of the Genetic Algorithm, and they are ultimately set to 30, 0.8, 0.05 and 50 respectively. The iterations curve of the GA for Flip Chip package is showed in Figure 7.

Table 3. Optimum solution versus the original designing.

Solutions	H_j (mm)	P_j (mm)	T_s (mm)	VonMises stress (Pa)
Original	0.1500	0.4500	1.500	1.4083E6
Optimum	0.1997	0.4949	1.0078	1.2269E6

From Figure 7, the algorithm starts to converge after the 35th generation. The optimum solution got by the above *GA* comparing with the original designing is listed in Table 3. Comparing with the initial design, optimum solution bring down the maximum VonMises stress by about 13%, and the optimum solution is gained when H_j and P_j are set near to their upper limits while T_s is close to its lower limit. This conclusion is consistent to parametric sensitivity results we have made on the above paragraphs.

5 CONCLUSIONS

In this paper, a combined methodology of *FEM*, *RSM*, Monte Carlo simulations and Genetic Algorithm are successfully implemented on the parametric analysis and optimization procedure for Flip Chip package. Mechanical correlations between the maximum VonMises stress in solder joints and three major packaging parameters have been revealed, and approximate global optimum solution for the package is obtained by *GA*. It is convinced that the above work will provide a good foundation for further research on a larger scale and multi-effect parameters group in the field of complex engineering systems like Flip Chip package.

ACKNOWLEDGEMENT

The research work in this paper is financially supported by the National Natural Science Foundation of China (*NSFC*) (No. 60166001).

REFERENCES

Ernst, L.J. & Hof, C. vant. 2000. Mechanical modeling and characterization of the curing process of underfill materials. The 50th International Conference of Electronic Component Technology (ECTC'50), *Proc. intern. symp.*, Las Vegas, USA. May 25–28, 2000.

Lau, J.H. & Lee, S.R. 2002. Modelling and analysis of 96.5Sn-3.5Ag lead-free solder joints of wafer level chip scale package on build-up microvia circuit board. *IEEE Transactions on Electronics Packaging Manufacturing*, Vol. **25**, No.1.

Liang, J.S & Li, Q.Y. 2002. The probabilistic design for parameters of Flip Chip microelectronic package. *Journal of Guilin Institute of Electronic Technology*, **22(6)**: 38–42.

Liang, J.S & Li, Q.Y. 2002. Optimal structural parameters for Flip Chip assembly based on minimum residual stress induced in major package process. The Second China-Japan-Korea Joint Symposium on Optimization of Structural and Mechanical Systems (CJK-OSM2), *Proc. intern. symp.*, Busan, Korea. November 4–8, 2002.

Yang, D.G & Zhang, G.Q. 2001. Combined experimental and numerical investigation on Flip Chip solder fatigue with cure dependent underfill properties, the 51st International Conference of Electronic Component Technology (ECTC'51), *Proc. intern. symp.*, Orlando, FL USA. May 29–June 1, 2001.

Yang, D.G & Zhang, G.Q. 2002. Parameter sensitivity study of cure-dependent underfill properties on Flip Chip failures. The 52nd International Conference of Electronic Component Technology (ECTC'52), *Proc. intern. symp.*, San Diego, USA. May 28–31, 2002.

Solid mechanics

Computational Methods in Engineering and Science, Iu et al. (eds)
© 2003 Swets & Zeitlinger, Lisse, ISBN 90 5809 567 3

Application of Green's function method to plane symmetrical contact mechanics problems

Menghua Zhang
Jiangmen Construction Bureau, Jiangmen, Guangdong, China

Mengshi Jin
Huazhong University of Science and Technology, Wuhan, Hubei, China

ABSTRACT: Based on the Flament's solution to the plane problems in elasticity under the action of uniformly distributed load q and of symmetrically uniformly distributed shear force τ on the symmetrical segments of the boundary, and after integration and superposition the equations for calculating the normal displacement and the shear displacement of the boundary point are derived. When two bodies are pressed together the continuous contact boundary may be separated into 2N symmetrical segments. On each of them meanwhile act the uniformly distributed load q_i and the uniformly distributed shear force τ_i. According to the 2N compatibility conditions between N contact point couples, 2N equations may be established. By solving the equations, N contact stress q_i and N shear resistance τ_i are obtained. Some results obtained by this method are in good agreement with those of analytical solutions.

1   INTRODUCTION

The analytical solution (Belajef's solution) to the plane contact mechanics problems can only be applied to the transverse contact problems between two cylinders, in which the gaps between the contact point couples form a quadratic curve. In this paper there is no such restriction. It can be used to solve any plane symmetrical contact problems. First, based on the Flament's solution to the plane problems in elasticity, under the action of uniformly distributed normal load q and of symmetrically, uniformly distributed shear force τ on the symmetrical segments of the boundary, after integration and superposition the equations for calculating the normal displacement and shear displacement of the boundary point are derived separately. Second, when two bodies are pressed together, the continuous contact boundary may be separated into 2N symmetrical segments. On each of them meanwhile act the uniformly distributed load (i.e. contact stress) q_i and the uniformly distributed shear force (i.e. shear resistance) τ_i. According to the 2N compatibility conditions between N contact point couples, 2N equations may be established. Among them N conditions show that there are equal relative normal displacements between each contact point couple, the other N conditions show that there is no relative shear displacement between each contact point couple. By solving the 2N equations, N contact stress q_i and N shear resistance τ_i are obtained. Finally, the total pressure p' may be easily obtained by superposition.

If the influence of shear resistance on the contact region is neglected, N contact stress q_i may be obtained by solving the N compatibility equations, which show that there is the equal relative normal displacement between N contact point couples. The total pressure p' may be calculated by superposition.

2   BASIC RELATIONS

Based on the Flament's solution to the half plane problem in elasticity, when a normal concentrated force p' acts on the boundary (as shown in Fig. 1), the normal displacement of the boundary point is

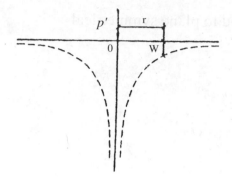

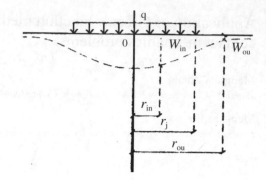

Figure 1. Normal displacement curve of a semi-infinite plane due to a concentrated force P' at the original point.

Figure 2. Normal displacement curve of a semi-infinite plane due to a uniformly distributed load q on a symmetrical segment.

$$W = \frac{p'}{\pi E}[(1 + \mu) - 2\ln\frac{d}{r}] \tag{1}$$

where E = Young's modulus, μ = Poisson's ratio, r = polar distance of the point under consideration, p' = concentrated force per unit thickness, d = a designated depth where the vertical displacement of each point is assumed to be zero.

In the case of symmetrically, uniformly distributed load q (as shown in Fig. 2) acting on the boundary, the normal displacements of the inner and the outer boundary points are:

$$W_{in} = \frac{2q}{\pi E}[(r_j + r_{in})\ln\frac{d}{(r_j + r_{in})} + (r_j - r_{in})\ln\frac{d}{(r_j - r_{in})} + (1 - \mu)r_j] \quad (r_{in} \le r_j) \tag{2}$$

$$W_{ou} = \frac{2q}{\pi E}[(r_{ou} + r_j)\ln\frac{d}{(r_{ou} + r_j)} - (r_{ou} - r_j)\ln\frac{d}{(r_{ou} - r_j)} + (1 - \mu)r_j] \quad (r_{ou} \ge r_j) \tag{3}$$

In the case of symmetrically, discontinuously and uniformly distributed load q (as shown in Fig. 3(a)) acting on the boundary, according to the principle of superposition, by subtracting 3(c) from 3(b), the normal displacements of the inner and the outer boundary points may be written as:

$$W_{in} = \frac{2q}{\pi E}[(r_j + r_{in})\ln\frac{d}{(r_j + r_{in})} + (r_j - r_{in})\ln\frac{d}{(r_j - r_{in})} - (r_{j-1} + r_{in})\ln\frac{d}{(r_{j-1} + r_{in})}$$

$$- (r_{j-1} - r_{in})\ln\frac{d}{(r_{j-1} - r_{in})} + (1 - \mu)(r_j - r_{j-1})] \quad (r_{in} \le r_{j-1}) \tag{4}$$

$$W_{ou} = \frac{2q}{\pi E}[(r_{ou} + r_j)\ln\frac{d}{(r_{ou} + r_j)} - (r_{ou} - r_j)\ln\frac{d}{(r_{ou} - r_j)} - (r_{ou} + r_{j-1})\ln\frac{d}{(r_{ou} + r_{j-1})}$$

$$+ (r_{ou} - r_{j-1})\ln\frac{d}{(r_{ou} - r_{j-1})} + (1 - \mu)(r_j - r_{j-1})] \quad (r_{ou} \ge r_j) \tag{5}$$

314

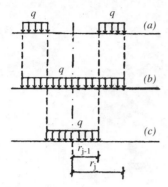
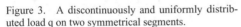

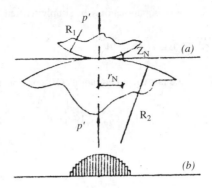

Figure 3. A discontinuously and uniformly distributed load q on two symmetrical segments.

Figure 4. Contact stress between two cylinders when pressed together.

All expressions above may be used only for plane stress problems. As far as plane strain problems are concerned, $1/E$ and μ in the above expressions should be replaced by $(1 - \mu^2)/E$ and $\mu/(1 - \mu)$ respectively; and the stress $\sigma_z = \mu(\sigma_x + \sigma_y)$.

When two cylinders are pressed together, the continuous contact boundary may be separated into 2N symmetrical segments. On each of them acts the uniformly distributed load (i.e. contact stress) $q_i (i = 1, 2, \cdots, n)$. Because the contact boundary is very small, the load q_i may be treated as acting on the half plane boundary directly (as shown in Fig. 4(b)).

On the maximum polar distance r_n of the contact boundary, the relative normal displacement due to deformation may be denoted as

$$W_n = W_{1n} + W_{2n} \tag{6}$$

where W_{1n} and W_{2n} are the unknown normal displacements of cylinder 1 and 2 at the point respectively. The gap between the contact point couple before compression is known and denoted as Z_n. The relative normal displacement of two cylinders at the point is the sum of W_n and Z_n, i.e.

$$W = W_n + Z_n \tag{7}$$

Considering the continuity of the contact region, the relative normal displacements of all the contact point couples are equal, i.e.

$$W = W_i + Z_i \qquad (i = 1, 2, \cdots n) \tag{8}$$

Substituting equations (8) into equation (7) gives

$$W_i = W_n + (Z_n - Z_i) \tag{9}$$

There are $(N+1)$ compatibility conditions between the contact point couples. Among them N conditions lie in the polar distance $r_i (i = 1, 2, \cdots, n)$; one condition lies in the original point. Finally, the following equations are obtained

$$\overset{(n+1)\times n}{\begin{bmatrix} G_{11} & G_{12} & \cdots & G_{1n} \\ G_{21} & G_{22} & \cdots & G_{2n} \\ \vdots & \vdots & & \vdots \\ G_{(n+1)1} & G_{(n+1)2} & \cdots & G_{(n+1)n} \end{bmatrix}} \overset{n\times 1}{\begin{Bmatrix} q_1 \\ q_2 \\ \vdots \\ q_n \end{Bmatrix}} = \overset{(n+1)\times 1}{\begin{Bmatrix} Z_n - 0 \\ Z_n - Z_1 \\ \vdots \\ Z_n - Z_n \end{Bmatrix}} + \overset{(n+1)\times 1}{\begin{Bmatrix} W_n \\ W_n \\ \vdots \\ W_n \end{Bmatrix}} \tag{10}$$

The above equations may be simply written as

$$[\ \overset{(n+1)\times n}{G}\ ]\overset{n\times 1}{\{Q\}}=\overset{(n+1)\times 1}{\{Z\}}+\overset{(n+1)\times 1}{\{W_n\}} \tag{11}$$

From the last row in equations (10), one obtains

$$[G_{(n+1)1}\ \ \overset{1\times n}{G_{(n+1)2}}\ \ \cdots\ \ G_{(n+1)n}]\overset{n\times 1}{\{Q\}}=W_n \tag{12}$$

so that,

$$\begin{bmatrix} G_{(n+1)1} & \overset{n\times n}{G_{(n+1)2}} & \cdots & G_{(n+1)n} \\ G_{(n+1)1} & G_{(n+1)2} & \cdots & G_{(n+1)n} \\ \vdots & \vdots & & \vdots \\ G_{(n+1)1} & G_{(n+1)2} & \cdots & G_{(n+1)n} \end{bmatrix}\overset{n\times 1}{\{Q\}}=\overset{n\times 1}{\{W_n\}} \tag{13}$$

Equations (13) can be simply written as

$$[\overset{n\times n}{G_{(n+1)j}}]\overset{n\times 1}{\{Q\}}=\overset{n\times 1}{\{W_n\}} \tag{14}$$

Substituting equations (14) into equations (11) gives

$$[\overset{n\times n}{G'-G_{(n+1)j}}]\overset{n\times 1}{\{Q\}}=\overset{n\times 1}{\{\overline{Z}\}} \tag{15}$$

where $[G']$ is a square matrix of N order obtained by eliminating the last row in $[G]$, $\{\overline{Z}\}$ is a column matrix of N order obtained by eliminating the last term "0" in $\{Z\}$.

If the shear resistance on the contact boundary is taken into account, it may also cause the relative normal displacements and the relative shear displacements of the inner and the outer contact point couples. Taking Figure 5 as an example, the displacements of the inner and the outer points may be written as:

$$W_{in}=-\frac{(1-\mu)}{E}\tau_j(r_j-r_{j-1})$$

$$U_{in}=\frac{2\tau_j}{\pi E}[(r_j-r_{in})\ln\frac{d}{(r_j-r_{in})}-(r_{j-1}-r_{in})\ln\frac{d}{(r_{j-1}-r_{in})}$$

$$-(r_j+r_{in})\ln\frac{d}{(r_j+r_{in})}+(r_{j-1}+r_{in})\ln\frac{d}{(r_{j-1}+r_{in})}]\qquad (r_{in}\le r_{j-1}) \tag{16}$$

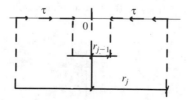

Figure 5.   Uniformly distributed shear stress on the contact boundary.

$$W_{ou} = 0$$

$$U_{ou} = \frac{2\tau_j}{\pi E}\left[-(r_{ou}-r_j)\ln\frac{d}{(r_{ou}-r_j)} + (r_{ou}-r_{j-1})\ln\frac{d}{(r_{ou}-r_{j-1})}\right.$$

$$\left. -(r_{ou}+r_j)\ln\frac{d}{(r_{ou}+r_j)} + (r_{ou}+r_{j-1})\ln\frac{d}{(r_{ou}+r_{j-1})}\right] \qquad (r_{ou}\geq r_j) \tag{17}$$

The normal contact stress q_j as shown in Figure 3(a) may also cause the relative shear displacements U of the contact point couples

$$U_{in} = 0 \qquad (r_{in}\leq r_{j-1}) \tag{18}$$

$$U_{ou} = -\frac{(1-\mu)}{E}q_j(r_j-r_{j-1}) \qquad (r_{ou}\geq r_j) \tag{19}$$

There are $(2N+1)$ compatibility conditions in total. Among them $(N+1)$ conditions show that there are equal relative normal displacements between each contact point couple at r_i $(i = 1, 2, \cdots, n)$ and at the original point "0". The other N conditions show that there is no relative shear displacement between each contact point couple at r_i $(i = 1, 2, \cdots, n)$. Finally, the following $(2N+1)$ equations can be obtained

$$\overset{(2n+1)\times 2n}{\begin{bmatrix} G_{11} & G_{12} & \cdots & G_{1n} & T_{1(n+1)}\cdots T_{1(n+2)}\cdots T_{1,2n} \\ G_{21} & G_{22} & \cdots & G_{2n} & T_{2(n+1)}\cdots T_{2(n+2)}\cdots T_{2,2n} \\ \vdots & \vdots & & \vdots & \vdots \quad \vdots \quad \vdots \\ G_{(2n+1)1} & G_{(2n+1)2} & \cdots & G_{(2n+1)n} & \cdots\cdots \quad \cdots\cdots \quad T_{(2n+1)2n} \end{bmatrix}} \overset{2n\times 1}{\left\{\begin{matrix} \{q_i\} \\ \{\tau_i\} \end{matrix}\right\}} = \overset{(2n+1)\times 1}{\left\{\begin{matrix} \{Z_i\} \\ \{0\} \end{matrix}\right\}} + \overset{(2n+1)\times 1}{\left\{\begin{matrix} \{W_n\} \\ \{0\} \end{matrix}\right\}} \tag{20}$$

Through similar treatment as before, a matrix equation with 2N order is obtained. By solving the 2N equations, all contact stress q_i and shear resistance τ_i $(i = 1, 2, \cdots, n)$ are obtained.

The total pressure p' between the contact bodies may be written as

$$p' = 2\sum_{j=1}^{n}(r_j-r_{j-1})q_j \tag{21}$$

where $r_0 = 0$.

For cylinders, the gap between the contact point couple before compression can be expressed as

$$Z_i = \frac{(R_1+R_2)}{2R_1R_2}r_i^2 \tag{22}$$

where R_1 and R_2 are the radii of the cylinders respectively, r_i is the polar distance of the ith contact point couple. For contact problems of a cylinder and a concave surface, the radius of the concave surface should be taken as negative.

According to Belajef's theory, for the transverse contact problems of two cylinders, there exist the following relations among the maximum contact stress q_{max}, the maximum polar distance r_n and the total pressure p'

$$q_{max} = \sqrt{\frac{p'(R_1+R_2)}{\pi^2 R_1R_2(K_1+K_2)}} \tag{23}$$

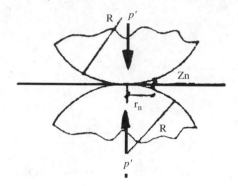

Figure 6.   Two cylinders are pressed together by a concentrated force p'.

Table 1.   Contact stress and shear resistance for two cylinders.

| r_i (mm) | Without considering the influence of shear resistance | With considering the influence of shear resistance | |
	q_i (kN/mm²)	q_i (kN/mm²)	τ_i (kN/mm²)
0.0180	0.4559	0.4453	0.1806
0.0360	0.4814 (5.08%)	0.5329	0.0456
0.0540	0.4799	0.5044	−0.0539
0.0720	0.4727	0.4959	0.0395
0.0900	0.4715	0.5167	−0.0421
0.1080	0.4447	0.4489	−0.0063
0.1260	0.4309	0.4688	−0.0153
0.1440	0.4252	0.4509	0.0024
0.1620	0.4215	0.4774	−0.0122
0.1800	0.3754	0.3889	−0.0789
0.1980	0.3669	0.3930	0.0586
0.2160	0.3678	0.4634	−0.0573
0.2340	0.2842	0.2764	−0.0855
0.2520	0.2537	0.2875	0.0450
0.2700	0.1677	0.3004	−0.2499

$$r_n = \sqrt{\frac{4p'R_1R_2(K_1 + K_2)}{(R_1 + R_2)}} \qquad (24)$$

where $K_1 = \sqrt{\dfrac{1 - \mu_1^2}{\pi E_1}}$ and $K_2 = \sqrt{\dfrac{1 - \mu_2^2}{\pi E_2}}$.

3   NUMERICAL EXAMPLES

Example 1. Figure 6 shows two cylinders with the same radius and Poisson's ratio. $R_1 = R_2 = R = 50$ mm, $\mu_1 = \mu_2 = \mu = 0.3$, $E_1 = 200$ kN/mm² and $E_2 = 150$ kN/mm². When the maximum polar distance r_n of the contact region is 0.27 mm, calculate the maximum contact stress q_{max} and total pressure p' in two cases: case 1 without considering the influence of shear resistance and case 2 with considering it. The results are compared with those of analytical solutions.

In calculation, the designated depth d equals 50 mm; the contact boundary is divided into 15 equal segments. The contact stress and shear resistance are listed in Table 1.

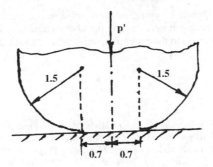

Figure 7. A die and a semi-infinite plane are pressed together by a concentrated force p'.

Table 2. Contact stress and shear resistance for two bodies.

r_i (mm)	z_i (mm)	Case 1 q_i (kN/mm²)	Case 2 q_i (kN/mm²)	τ_i (kN/mm²)
0.1	0	4.42	4.45	−0.73
0.2	0	7.34	7.34	1.18
0.3	0	8.55	8.86	−0.15
0.4	0	2.13	1.88	−1.18
0.5	0	8.97	9.15	1.58
0.6	0	7.45	7.66	−0.58
0.7	0	11.57	11.60	−0.78
0.8	0.003	9.22	9.28	0.14
0.9	0.013	9.66	9.72	−0.12
1.0	0.033	13.55(q_{max})	13.78(q_{max})	−0.79
1.1	0.053	6.35	6.55	−1.54
1.239	0.100	3.71	3.71	−1.11

The total pressure in case 1 is 0.209 kN/mm; and it is 0.232 kN/mm in case 2. The analytical results are $\bar{p}' = 0.2151$ kN/mm and $\bar{q}_{max} = 0.5072$ kN/mm².

In Table 1, the percentage in the parenthesis is the error of the maximum contact stress. It is seen that the maximum contact stress q_{max} in case 2 is larger than that in case 1. The reason is that the total pressure p' increases when the maximum polar distance r_n of the contact region remains the same.

Example 2. Two elastic bodies with the same Poisson's ratio are shown in Figure 7. $E_1 = 200$ kN/mm², $E_2 = 150$ kN/mm² and $\mu_1 = \mu_2 = 0.3$. When the maximum polar distance r_n of the contact region equals 1.239 mm, calculate the maximum contact stress q_{max} and the total pressure p'.

In calculation, two cases are considered. In case 1 the influence of shear resistance is not taken into account; in case 2 it is taken into account. The designated depth d is taken as 500 mm; the contact boundary is divided into 12 segments. The contact stress and shear resistance are listed in Table 2. The total pressures in case 1 and case 2 are 18.87 kN/mm² and 19.09 kN/mm² respectively. Comparison of results can not be made because no analytical solution to the problem is available at present.

4   CONCLUSION

Plane symmetrical contact mechanics problems may be met very commonly in engineering practice. In this paper, the normal displacement and shear displacement of the inner and outer points under the action of uniformly distributed normal load and of symmetrically uniformly distributed shear force on symmetrical segments are taken as their Green's functions. This greatly simplifies the problem. The present method can be used to solve any plane symmetrical contact problems with arbitrary forms. It has many features, such as less unknowns, less computation time, fine precision and simple programming.

All normal contact stress should be compressive to ensure the continuity of the contact region. Tensile stress is not allowed because it may destroy the above continuity hypothesis. For this purpose, proper modification should be made to the program to ensure the contact stress of the non-contact point couple to be zero.

REFERENCES

Bezoohoff N.E. 1956. *Theory of Elasticity and Plasticity*. High Education Press (Chinese translation).
Longfu Wang. 1984. *Theory of Elasticity*. 2nd ed., Science Press (in Chinese).
Mengshi Jin. 1999. Green's Function Method to Contact Mechanics Problems. *Proc. 8th National Conference on Structural Engineering* 1: 233–239. Kunming, China, 22–25 Oct. 1999 (in Chinese).
Mengshi Jin. 2000. Green's Function Method to Plane Symmetrical Contact Mechanics Problems. *Proc. 10th National Conference on Computer Application in Civil Eng.*: 275–280. Guangzhou, China, 22–24 Nov. 2000 (in Chinese).
Timoshenko S. & Goodier J. N. 1951. *Theory of Elasticity*. 2nd ed., Eng. Societies Monographs.

Computational Methods in Engineering and Science, Iu et al. (eds)
© *2003 Swets & Zeitlinger, Lisse, ISBN 90 5809 567 3*

Modeling and analysis of mechanical problems by hemi-variational inequalities

S. Migórski

Institute of Computer Science, Faculty of Mathematics, Physics and Computer Science Jagiellonian University, Cracow, Poland

ABSTRACT: The paper presents results on the mathematical modeling of non-convex and non-smooth problems arising in mechanics and engineering. The approach to such problems is based on the notion of hemi-variational inequality. First we present the methods for elliptic hemi-variational inequalities. Then a new result on well posedness of a dynamic viscoelastic problem is delivered.

1 INTRODUCTION

The notion of hemi-variational inequality was introduced and studied in the early eighties (Panagiotopoulos 1985, 1991, 1993). The hemi-variational inequalities are a generalization of classical variational inequalities and have their origin in non-smooth mechanics. They allow to give variational formulations and study engineering problems which involve non-monotone multi-valued relations. In the stationary case the hemi-variational inequalities can be formulated as the substationary point problems for the corresponding non-smooth, non-convex and non-differentiable energy functionals. The results obtained for the hemi-variational inequalities give answers to several unsolved or partially unsolved problems, e.g. the delamination problem and the adhesive contact in cracks. The hemi-variational inequalities are derived from non-convex super-potentials by using the generalized gradient of Clarke. For the description of origins of hemi-variational inequalities, the mathematical theory and the computational techniques, we refer to Panagiotopoulos (1985, 1993), Naniewicz & Panagiotopoulos (1995), Motreanu & Panagiotopoulos (1999), Miettinen et al. (1995) and Haslinger et al. (1999).

The approach based on the notion of hemi-variational inequalities allows to consider sub-differential boundary value problems and to study a wide class of problems which are interesting from the physical point of view. In this note we describe the methods for stationary hemi-variational inequalities and then we provide a new result on the existence and uniqueness of solutions to dynamic hemi-variational inequality of hyperbolic type.

2 ELLIPTIC HEMI-VARIATIONAL INEQUALITIES

We describe the destruction support problem as a model example of non-monotone contact problem leading to the elliptic hemi-variational inequality.

Let $\Omega \subset R^2$ be a bounded domain occupied by a linear elastic body in its undeformed state. Suppose that the boundary $\Gamma = \partial\Omega$ of Ω consists of three open disjoint parts Γ_U, Γ_F and Γ_s, $\Gamma = \bar{\Gamma}_U \cup \bar{\Gamma}_F \cup \bar{\Gamma}_S$. A point $x \in \Omega$ is refereed to a fixed Cartesian coordinate system. For $i, j = 1, 2, 3$ we use the standard notation: $n = \{n_i\}$ is the outward unit normal vector to Γ, $\sigma = \{\sigma_{ij}\}$ is the stress tensor, $\varepsilon = \{\varepsilon_{ij}\}$ is the strain tensor, $u = \{u_i\}$ denotes the displacement vector, $f = \{f_i\}$ is the volume force vector and $S = \{S_i = \sigma_{ij}n_j\}$ stands for the boundary force (the stress vector on Γ). We decompose the stress vector S

and the displacement u on Γ into the normal and tangential components: $S_N = \sigma_{ij}n_in_j$, $S_T = S - S_N$ and $u_N = u_in_i$, $u_T = u - u_N$. The small deformation theory implies the following pointwise relations in Ω:

$$\sigma_{ij,j}(u) = -f_i \quad \text{(the equilibrium equation)},$$

$$\varepsilon_{ij}(u) = \frac{1}{2}(u_{i,j} + u_{j,i}) \quad \text{(the strain-displacement law)}, \tag{1}$$

$$\sigma_{ij}(u) = c_{ijkl}\varepsilon_{kl}(u) \quad \text{(the constitutive equation, Hooke's law)},$$

where $\sigma_{ij,j}(u) = (\partial/\partial x_j)\sigma_{ij}(u)$ and the elasticity tensor c_{ijkl} is assumed to satisfy the usual elliptic and symmetric properties. On the boundary we consider the following conditions. On Γ_U: $u = 0$ (prescribed displacement, homogeneous for simplicity), on Γ_F: $S = F$ (i.e. $S_i = F_i(x)$ are prescribed tractions), on Γ_S: $S_T = C_T$ (the tangential forces $C_T = C_T(x)$ are known) and S_N satisfies the idealized law

$$\begin{cases} S_N = 0 & \text{if} \quad u_N < 0 \\ S_N + k_0 u_N = 0 & \text{if} \quad 0 \le u_N < \varepsilon \\ -k_0\varepsilon \le S_N \le 0 & \text{if} \quad u_N = \varepsilon \\ S_N = 0 & \text{if} \quad u_N > \varepsilon. \end{cases} \tag{2}$$

The Winkler coefficient constant $k_0 > 0$ is supposed to be positive and $\varepsilon > 0$. Conditions (2) describe unilateral contact between the body and the Winkler-type support: in the noncontact region $S_N = 0$, in the contact region the support generates a reaction force which is proportional to its deformation $-S_N \sim u_N$, the destruction of the support appears when the tractions reach the limited value (the maximal value of reactions that can be maintained by the support in $k_0\varepsilon$) and again $S_N = 0$ in the region where the support is destroyed. Since the diagram $(u_N, -S_N)$ on Γ_S is non-monotone, it is not possible to formulate the destruction support problem as a variational inequality. In order to give the variational formulation of the above problem, we introduce the function $\beta: R \to R$ such that

$$\beta(t) = \begin{cases} 0 & \text{if} \quad t < 0 \\ k_0 t & \text{if} \quad t \in [0,\varepsilon) \\ 0 & \text{if} \quad t \ge \varepsilon. \end{cases}$$

and define the multivalued map $\bar{\beta}: R \to 2^R$ which is obtained from β by filling in the jump at $t = \varepsilon$. Generally, given $\beta \in L^\infty_{loc}(R)$ we define $\bar{\beta}$ as follows

$$\bar{\beta}(t) = [\lim_{\delta \to 0^+} \operatorname*{ess\,inf}_{|t-s| \le \delta} \beta(t), \ \lim_{\delta \to 0^+} \operatorname*{ess\,sup}_{|t-s| \le \delta} \beta(t)] \subset R.$$

It is known, cf. (Chang 1981), that there exists a locally Lipschitz function $j: R \to R$ determined (up to an additive constant) by the relation $j(t) = \int_0^t \beta(s)ds$ and $\partial j(t) \subset \bar{\beta}(t)$. Additionaly, if $\lim_{t \to s^\pm} \beta(t)$ exist for every $s \in R$, then we have $\partial j(t) = \bar{\beta}(t)$ for $t \in R$. Here the subdifferential ∂j is understood in the sense of Clarke. We recall, cf. (Clarke 1983) that if $h: E \to R$ is a locally Lipschitz function defined on a Banach space E, then the generalized directional derivative of h at $x \in E$ in the direction $v \in E$ is defined by

$$h^0(x;v) = \limsup_{y \to x, t \downarrow 0} \frac{h(y + tv) - h(y)}{t}$$

and the generalized gradient of h at x is a subset of a dual space E^* given by

$$\partial h(x) = \{\zeta \in E^* : h^0(x;v) \ge \langle \zeta, v \rangle \text{ for all } v \in E\}.$$

Taking the above into consideration, the boundary condition (2) can be written as

$$-S_N \in \overline{\beta}(u_N) = \partial j(u_N) \text{ on } \Gamma_S, \tag{3}$$

where $j: R \to R$ is of the form

$$j(t) = \int_0^t \beta(s)\,ds = \begin{cases} 0 & \text{if} & t < 0 \\ \dfrac{1}{2}k_0 t^2 & \text{if} & 0 \le t \le \varepsilon \\ \dfrac{1}{2}k_0 \varepsilon^2 & \text{if} & t > \varepsilon. \end{cases}$$

Now let us introduce the space V of kinematically admissible displacements $V = \{v \in H^1(\Omega;R^3): v_i = 0 \text{ on } \Gamma_U\}$. Multiplying the equilibrium equation (1) by $v - u$, where $v \in V$, integrating over Ω, applying the Green formula cf. (Zeidler 1990) and using the definition of the Clarke subdifferential, we arrive to the following problem: find $u \in V$ such that

$$a(u, v-u) + \int_{\Gamma_S} j^0(u_N; v_N - u_N)\, d\sigma \ge \langle l, v-u \rangle \quad \text{for all} \ \ v \in V, \tag{4}$$

where

$$a(u,v) = \sum_{i,j=1}^{n} \int_\Omega \sigma_{ij}(u)\, \varepsilon_{ij}(v)\, dx,$$

$$\langle l, v \rangle = \sum_{i=1}^{n} \int_\Omega f_i v_i\, dx + \sum_{i=1}^{n} \int_{\Gamma_F} F_i v_i\, d\Gamma + \sum_{i=1}^{n} \int_{\Gamma_S} C_{T_i} v_{T_i}\, d\Gamma.$$

The problem (4) is called *hemi-variational inequality*. An analogous hemi-variational inequality can be obtained when on Γ_S a pointwise boundary condition for the tangential force $-S_T \in \partial j(u_T)$ is assumed and the normal forces $S_N = C_N$, $C_N = C_N(x)$ are prescribed. Such relation between u_T and $-S_T$ appears, for instance, in rock interface analysis in geomechanics and describes a non-monotone friction law (Panagiotopoulos 1985, 1993).

We remark that the hemi-variational inequalities arise in applied mechanics and engineering where the constitutive laws are expressed by means of subdifferential relations:

$$\sigma \in \partial w(\varepsilon) \tag{5}$$

where ∂w denotes the Clarke subdifferential of a function $w: R^6 \to R$. It is well known, see Chapters 3 and 6 in (Panagiotopoulos 1993), that with the appropriate choices of convex superpotential w, the relation (5) describes the elastic ideally locking materials, the elastic work-hardening materials, the elasto-perfectly plastic materials (Hencky's theory), the polygonal $\sigma - \varepsilon$ laws, rigid viscoplastic materials and the materials obeying the law of the deformation theory of plasticity. In the case w is assumed to be convex, the hemi-variational inequality reduces to the variational inequality. For a description of a large number of materials with non-monotone $\sigma - \varepsilon$ relations of the form (5), e.g. composite material structures (cracking, crushing, slip, delamination), the reinforced concrete in tension (tension-stiffening effects), beton reinforced by steel fibres, non-convex locking materials and fuzzy material laws, we refer to Chapter 3 of (Panagiotopoulos 1993) and the references therein.

The following three methods can be used to establish the existence of solutions to static hemi-variational inequalities of the form (4): the regularization method, the method based on the deformation lemma and the surjectivity method.

2.1 The method of regularization

It was first suggested by (Rauch 1977) who considered a discontinuous semi-linear elliptic equation. Then the method was adopted by (Panagiotopoulos 1985, 1991, 1993) for more general hemi-variational inequalities. First we associate with the function j the multifunction $\overline{\beta}: R \to 2^R$ such that $\partial j(t) = \overline{\beta}(t)$ for

$t \in R$. Then the function β is regularized and the finite dimensional approximation of (4) with this regularization is considered. We get the existence of solutions of (4) by passing to the limit in a regularized finite-dimensional problem.

2.2 The method based on the deformation lemma

We associate with the problem (4) the following substationarity problem: find $u \in V$ such that

$$0 \in \partial E(u) \tag{6}$$

where $E: V \to R$ is the energy functional of the form

$$E(u) = \frac{1}{2}a(u,u) - \langle l,u \rangle + J(u) \quad \text{and} \quad J(u) = \int_{\Gamma_s} j(u(x))\,d\Gamma \;.$$

Under suitable hypotheses every solution of (6) is a solution of (4). The existence of solutions to the problem (6) follows from the following two results of (Chang 1981). The second one is a consequence of the deformation lemma for a locally Lipschitz functions.

Theorem 1. If $j: V \to R$ is a locally Lipschitz function such that $|J(v)| \leqslant k_1\|v\|^2 + k_2$ with $k_1 > 0$, $k_2 \in R$, then the energy functional E satisfies the Palais-Smale, shortly (PS), condition and E is bounded from below.

Theorem 2. Let X be a reflexive Banach space and $F: X \to R$ be a locally Lipschitz function satisfying (PS) condition and bounded from below. Then $m = \inf_X F$ is a critical value of F, i.e. there is $x_0 \in X$ such that $F(x_0) = m$ and $0 \in \partial F(x_0)$.

2.3 The method based on the surjectivity results for pseudo-monotone operators

We consider the following type of hemi-variational inequalities: find $u \in V$ such that

$$f \in Au + \partial J(u) \tag{7}$$

where $A: V \to V^*$ is given by $\langle Au, v \rangle = a(u, v)$ for $u, v \in V$. This problem is equivalent to the problem (6), since (7) holds if and only if $0 \in Au - l(u) + \partial J(u) = \partial E(u)$. The existence of solutions to (7) is based on two facts: the sum of two pseudo-monotone operators is a pseudo-monotone operator and a pseudo-monotone and coercive operator is surjective cf. (Denkowski et al. 2003) and on the following result.

Theorem 3. Let V be a reflexive Banach space. Suppose that $A: V \to V^*$ and $\partial J: V \to 2^{V^*}$ are pseudo-monotone operators such that $A + \partial J$ is coercive. Then for every $f \in V^*$ there exists $u \in V$ such that $f \in (A + \partial J)(u)$, i.e. (7) admits a solution.

For results on the uniqueness of solutions to hemi-variational inequality we refer to Miettinen and Panagiotopoulos (1999) and Migórski (2001a, 2003a, b).

3 DYNAMIC HEMI-VARIATIONAL INEQUALITIES

In this section we present a hemi-variational inequality of hyperbolic type arising in the dynamic process of frictional contact between a deformable body and a foundation. We deliver a new result on the existence of solutions to such model. Our result answers an open question stated by Panagiotopoulos (1993).

We suppose the body is viscoelastic with a linear elasticity operator and a nonlinear viscosity operator. The contact is modeled with a general normal damped response condition. The dependence of the normal stress on the normal velocity is assumed to have non-monotone character of the subdifferential form. We model the friction assuming that the tangential shear on the contact surface is given as a non-monotone and possibly multivalued function of the tangential velocity.

Suppose that a deformable viscoelastic body occupies the reference configuration $\Omega \subset R^d(d = 2, 3)$. As in Section 2, the boundary of Ω is divided into three mutually disjoint Γ_D, Γ_N and Γ_C. The body is clamped on Γ_D, so the displacement field vanishes there. Volume forces of density f_1 act in Ω and surface tractions of density f_2 are applied on Γ_N. The body may come in contact with an obstacle, the so-called foundation, over the potential contact surface Γ_C. We denote by $u = (u_1, \cdots, u_d)$ the displacement vector, by σ the stress tensor and by ε the linearized (small) strain tensor. We suppose the Kelvin-Voigt viscoelastic constitutive relation $\sigma(u, u') = C(\varepsilon(u')) + G(\varepsilon(u))$, where C and G are given non-linear and linear constitutive functions, respectively. As concerns the conditions on the contact surface Γ_C, we consider subdifferential boundary conditions. We assume that the normal stress σ_N and the normal velocity u'_N satisfy the non-monotone normal damped response condition of the form

$$-\sigma_N \in \partial j_N(x,t,u'_N) \quad \text{on} \quad \Gamma_C \times (0,T) \tag{8}$$

where j_N: $\Gamma_C \times (0,T) \times R \to R$ is locally Lipschitz in its last variable and the Clarke sub-differential is taken with respect to the third variable. The friction relation is given by

$$-\sigma_T \in \partial j_T(x,t,u'_T) \quad \text{on} \quad \Gamma_C \times (0,T) \tag{9}$$

and describes the multi-valued law between the tangential force and the tangential velocity, where j_T is locally Lipschitz in the third variable.

The condition (8) models the motion of a deformable body on a support of granular material. It also generalizes the condition studied by (Rochdi et al. 1998), where the contact surface Γ_C was assumed to be covered with a lubricant that contains solid particles, such as one of the new smart lubricants, or with worn metallic particles. The condition (9) was considered in its simple one dimensional form (Panagiotopoulos 1993). Such condition appears in the tangential direction of the adhesive interface and describes the partial cracking and crushing of the adhesive bonding material. We refer to Panagiotopoulos (1993) for several examples of the zig-zag friction laws which can be put in the form (9). Multidimensional cases were considered via minimum type and maximum type functions, cf. Naniewicz and Panagiotopoulos (1995) for concrete examples.

Denoting by u_0 and u_1 the initial displacement and the initial velocity, the classical formulation of the contact problem may be stated as follows: find a displacement field u: $Q = \Omega \times (0,T) \to R^d$ and a stress field σ: $Q \to S_d$ (S_d is the space of second order symmetric tensors) such that

$$\begin{cases} u'' - div\,\sigma = f_1, \quad \sigma = C(\varepsilon(u')) + G(\varepsilon(u)) & in & Q \\ u = 0 & on & \Gamma_D \times (0,T) \\ \sigma n = f_2 & on & \Gamma_N \times (0,T) \\ -\sigma_N \in \partial j_N(x,t,u'_N), \; -\sigma_T \in \partial j_T(x,t,u'_T) & on & \Gamma_C \times (0,T) \\ u(0) = u_0, u'(0) = u_1 & in & \Omega. \end{cases} \tag{10}$$

Introducing the space $V = \{v \in H^1(\Omega;R^d): v = 0 \text{ on } \Gamma_d\}$, the variational formulation of the problem (10) reads as follows: find a displacement field u: $(0,T) \to V$ and a stress field σ:$(0,T) \to L^2(\Omega;S_d)$ such that

$$\begin{cases} \langle u''(t),v \rangle + (\sigma(t),\varepsilon(v)) + \int_{\Gamma_C}(j_N^0(x,t,u'_N,v_N) + j_T^0(x,t,u'_T,v_T))d\Gamma(x) \geq \langle f(t),v \rangle \\ \qquad\qquad\qquad\qquad\qquad \text{for all } v \in V \text{ and a.e.}\,t \in (0,T) \\ \sigma(t) = C(\varepsilon(u'(t))) + G(\varepsilon(u(t))) \text{ for a.e.}\,t \in (0,T) \\ u(0) = u_0, \; u'(0) = u_1, \end{cases}$$

where $\langle f(t),v \rangle = (f_1(t),v) + (f_2(t),v)$ for all $v \in V$ and a.e. $t \in (0,T)$.

In the study of the problem we need the following hypotheses.

The viscosity operator C: $Q \times S_d \to S_d$ satisfies the Caratheodory conditions (it is measurable in (x, t) and continuous in ε) and

$\| C(x,t,\varepsilon) \| \leq c_1(b(x,t) + \| \varepsilon \|)$ for $\varepsilon \in S_d$, a.e. $(x,t) \in Q$ with $b \in L^2(\Omega)$, $c_1 > 0$;

$(C(x,t,\varepsilon_1) - C(x,t,\varepsilon_2)) : (\varepsilon_1 - \varepsilon_2) \geq 0$ for all $\varepsilon_1,\varepsilon_1 \in S_d$ and a.e. $(x,t) \in Q$;

$C(x,t,\varepsilon) : \varepsilon \geq c_2 \| \varepsilon \|^2$ for all $\varepsilon \in S_d$ and a.e. $(x,t) \in Q$ with $c_2 > 0$.

The elasticity operator $G: Q \times S_d \to S_d$ is of the form $G(x,t,\varepsilon) = E(x)\varepsilon$ with a symmetric and positive elasticity tensor $E \in L^\infty(\Omega)$.

The functions j_N and j_T satisfy the properties: $j_N: \Gamma_C \times (0,T) \times R \to R$ is a function such that

j_N is measurable in (x,t) for all $\xi \in R$ and $j_N(\cdot,\cdot,0) \in L^1(\Gamma_C \times (0,T))$;

j_N is locally Lipschitz in ξ for all $(x,t) \in \Gamma_C \times (0,T)$;

$|\eta| \le c_N(1+|\xi|)$ for all $\eta \in \partial j_N(x,t,\xi)$, $(x,t) \in \Gamma_C \times (0,T)$ with $c_N > 0$;

$j^0_N(x,t,\xi;-\xi) \le d_N(1+|\xi|)$ for all $\xi \in R$, $(x,t) \in \Gamma_C \times (0,T)$ with $d_N \ge 0$;

and $j_T : \Gamma_C \times (0,T) \times R^d \to R$ is a function such that

j_T is measurable in (x,t) for all $\xi \in R^d$ and $j_T(\cdot,\cdot,0) \in L^1(\Gamma_C \times (0,T))$;

j_T is locally Lipschitz in ξ for all $(x,t) \in \Gamma_C \times (0,T)$;

$\|\eta\| \le c_T(1+\|\xi\|)$ for all $\eta \in \partial j_T(x,t,\xi)$, $(x,t) \in \Gamma_C \times (0,T)$ with $c_T > 0$;

$j^0_T(x,t,\xi;-\xi) \le d_T(1+\|\xi\|)$ for all $\xi \in R^d$, $(x,t) \in \Gamma_C \times (0,T)$ with $d_T \ge 0$.
Moreover, we suppose

$f_1 \in L^2(0,T;L^2(\Omega;R^d))$, $f_2 \in L^2(0,T;L^2(\Gamma_N;R^d))$, $u_0 \in V$ and $u_1 \in L^2(\Omega;R^d)$.

In the above hypotheses the symbol ∂j denotes the Clarke sub-differential of j with respect to the variable ξ.

The existence and uniqueness theorems for the problem (10) under the above hypotheses are a consequence of a more general result which is stated and proved (Migórski 2003b), Theorem 10.

Recent results on dynamic hemi-variational inequalities and their numerical treatment can be found in Denkowski & Migórski (1998), Denkowski et al. (2003), Goeleven et al. (1999), Miettinen & Panagiotopoulos (1999), Migórski (1998, 2001b, 2003c) and the references therein.

4   CONCLUSIONS

The results presented in this paper on hemi-variational inequalities are important in the computational methods in mechanics (see (Haslinger et al. 1999)), where the mathematical analysis is fundamental to find solutions numerically. The paper is merely the first step of our research in this direction. Computational aspects involving finite dimensional (Galerkin techniques) approximation for dynamical hemi-variational inequalities are at present under investigation. The ultimate goal of this research may be to develop methods so that one can more efficiently and promisingly solve mechanical and engineering problems for structures which are characterized by non-monotone, possibly multi-valued relations between stresses and strains or between reactions and displacements. Methodologies and programs will be further developed to achieve this goal.

REFERENCES

Chang, K.C. 1981. Variational methods for nondifferentiable functionals and applications to partial differential equations. *J. Math. Anal. Appl.* 80: 102–129.
Clarke, F.H. 1983. *Optimization and Nonsmooth Analysis*. New York: Wiley-Interscience.
Denkowski, Z. & Migórski, S. 1998. Optimal shape design for elliptic hemivariational inequalities in nonlinear elasticity. In: W.H. Schmidt et al. (eds.), *International Series of Numerical Mathematics*: Vol. 124, 31–40. Basel, Boston, Berlin: Birkhäuser Verlag.

Denkowski, Z., Migórski, S. & Papageorgiou, N.S. 2003. *An Introduction to Nonlinear Analysis: Applications*. Boston, Dordrecht, London: Kluwer/Plenum.

Goeleven, D., Miettinen, M. & Panagiotopoulos, P.D. 1999. Dynamic hemivariational inequalities and their applications. *J. Optimiz. Theory and Appl.* 103(3): 567–601.

Haslinger, J., Miettinen, M. & Panagiotopoulos, P.D. 1999. *Finite Element Method for Hemivariational Inequalities. Theory, Methods and Applications*. Boston, Dordrecht, London: Kluwer.

Miettinen, M., Mäkëla, M. & Haslinger, J. 1995. On numerical solution of hemivariational inequalities by non-smooth optimization methods. *J. Global Optimization* 6: 401–425.

Miettinen, M. & Panagiotopoulos, P.D. 1999. On parabolic hemivariational inequalities and applications. *Nonlinear Analysis* 35: 885–915.

Migórski, S. 1998. Identification of nonlinear heat transfer laws in problems modeled by hemivariational inequalities. In: M. Tanaka & G.S. Dulikravich (eds.), *Proceedings of International Symposium on Inverse Problems in Engineering Mechanics 1998 (ISIP'98)*, 27–37, Amsterdam: Elsevier Science B.V.

Migórski, S. 2001a. On existence of solutions for parabolic hemivariational inequalities. *Journal Comp. Appl. Math.* 129: 77–87.

Migórski, S. 2001b. Parameter identification for evolution hemivariational inequalities and applications. In: M. Tanaka & G.S. Dulikravich (eds.), *Proceedings of International Symposium on Inverse Problems in Engineering Mechanics 2001 (ISIP2001)*, 211–219, Amsterdam: Elsevier Science B.V.

Migórski, S. 2003a. Modeling, analysis and optimal control of systems governed by hemivariational inequalities. In: *Industrial Mathematics and Statistics* dedicated to commemorate the Golden Jubilee of Indian Institute of Technology, Kharagpur, India, 2003.

Migórski, S. 2003b. Dynamic hemivariational inequality modeling viscoelastic contact problem with normal damped response and friction, in press.

Migórski, S. 2003c. Modeling and identification in a dynamic viscoelastic contact problem with normal damped response and friction. In: M. Tanaka & G.S. Dulikravich (eds.), *Proceedings of International Symposium on Inverse Problems in Engineering Mechanics 2003 (ISIP2003), in press*, Amsterdam: Elsevier Science B.V.

Motreanu, D. & Panagiotopoulos, P.D. 1999. *Minimax Theorems and Qualitative Properties of the Solutions of Hemivariational Inequalities and Applications*. Boston, Dordrecht, London: Kluwer.

Naniewicz, Z. & Panagiopopoulos, P.D. 1995. *Mathematical Theory of Hemivariational Inequalities and Applications*. New York, Basel, Hong Kong: Marcel Dekker.

Panagiotopoulos, P.D. 1985. *Inequality Problems in Mechanics and Applications. Convex and Nonconvex Energy Functions*. Basel: Birkhäuser.

Panagiotopoulos, P.D. 1991. Coercive and semicoercive hemivariational inequalities. *Nonlinear Analysis* 16: 209–231.

Panagiotopoulos, P.D. 1993. *Hemivariational Inequalities, Applications in Mechanics and Engineering*. Berlin: Springer-Verlag.

Rauch, J. 1977. Discontinuous semilinear differential equations and multiple valued maps, *Proc. Amer. Math. Soc.* 64: 277–282.

Rochdi, M., Shillor, M. & Sofonea, M. 1998. A quasistatic contact problem with directional friction and damped response. *Applicable Analysis* 68: 409–422.

Zeidler, E. 1990. *Nonlinear Functional Analysis and Applications II A/B*. New York: Springer.

327

Computational Methods in Engineering and Science, Iu et al. (eds)
© 2003 Swets & Zeitlinger, Lisse, ISBN 90 5809 567 3

Computation and comparison of instability and bifurcation modes in frictional contact systems

A.P. da Costa & J.A.C. Martins
Instituto Superior Técnico, Depº de Engª Civil e Arquitectura and ICIST, Lisboa, Portugal

ABSTRACT: This paper has two objectives. The first objective is to relate the conditions for the occurrence of angular bifurcations in quasi-static trajectories of finite dimensional frictional contact systems with those for the occurrence of directional instabilities at equilibrium configurations of the same systems. The second objective of this paper is to check that the algorithm used earlier by the authors to compute bifurcation modes in these non-smooth systems and the one used to determine directional instability modes in the same systems do lead to the same result in situations at which such modes should coincide.

1 KINEMATICS, ADMISSIBLE SETS AND EQUILIBRIUM STATES

We consider finite element discretisations of linear elastic solids with plane motion whose configuration at each time $t \geqslant 0$ is described by the values $u_i(t)$, $1 \leqslant i \leqslant N$, of their nodal displacements. Rigid body modes are eliminated by prescribing some time independent nodal displacements. The corresponding displacement vector at time t, denoted by $\mathbf{u}(t) \in |^N$, groups the displacements of the nodes p ($p \in \boldsymbol{P}_F$) that are *Free* from kinematic constraints in the sub-vector $\mathbf{u}_F$, and groups the displacements of the nodes $p \in \boldsymbol{P}_C$ that may establish *Contact* with a fixed, flat and rigid obstacle in the sub-vector. The displacements that are normal (n) and tangential (t) to the obstacle are grouped in the sub-vectors $\mathbf{u}_n$ and $\mathbf{u}_t$ of $\mathbf{u}$, respectively. During the smooth portions of the evolution of a contact candidate node $p \in \boldsymbol{P}_C$, the components of the displacement vector ($\mathbf{u}^p(t) \in |^2$), of the velocity vector ($\dot{\mathbf{u}}^p(t)$) and of the reaction vector ($\mathbf{r}^p(t)$) of the node p satisfy the *unilateral contact conditions*

$$u_n^p(t) \leq 0, \qquad r_n^p(t) \leq 0, \qquad u_n^p(t)\, r_n^p(t) = 0, \tag{1}$$

and the *friction law of Coulomb*

$$r_t^p(t) \in \mu\, r_n^p(t)\, \sigma[\dot{u}_t^p], \tag{2}$$

where $\mu \geqslant 0$ is the coefficient of friction, (˙) denotes derivation with respect to time and $\sigma[.]$ denotes the multi-valued application such that, for each $x \in |$, $\sigma[x] = x/|x|$, if $x \neq 0$ and $\sigma[x] = [-1, +1]$, if $x = 0$. The *set of admissible displacements* is defined by $\boldsymbol{K}_\mathbf{u} = \{\mathbf{u} \in |^N : u_n^p - d_n^p \leq 0, p \in \boldsymbol{P}_C\}$. For each $\mathbf{u} \in |^N$ the *set of admissible reactions* $\boldsymbol{K}_\mathbf{r}(\mathbf{u}) = \{\mathbf{r} \in |^N : r_n^p = r_t^p = 0, p \in \boldsymbol{P}_F \cup \boldsymbol{P}_f(\mathbf{u}); r_n^p \leq 0$ and $|r_t^p| + \mu r_n^p \leq 0, p \in \boldsymbol{P}_C(\mathbf{u})$ such that $u_n^p - d_n^p \geq 0\}$; d_n^p is the distance in the reference configuration between the node p and the obstacle. For a given equilibrium state we need to characterise the various frictional contact states that each node currently in contact may have in its smooth near future evolutions. It is then useful to decompose the set of contact candidate nodes: for each $\mathbf{u} \in |^N$ and each $\mathbf{r} \in \boldsymbol{K}_\mathbf{r}(\mathbf{u})$ we consider the following partition of the set of contact candidate particles: $\boldsymbol{P}_C = \boldsymbol{P}_f(\mathbf{u}) \cup \boldsymbol{P}_z(\mathbf{u}, \mathbf{r}) \cup \boldsymbol{P}_d(\mathbf{u}, \mathbf{r}) \cup \boldsymbol{P}_s(\mathbf{u}, \mathbf{r})$, where $\boldsymbol{P}_f(\mathbf{u}) = \{p \in \boldsymbol{P}_C : u_n^p - d_n^p < 0\}$ [nodes currently out of contact ("free")], $\boldsymbol{P}_z(\mathbf{u}, \mathbf{r}) = \{p \in \boldsymbol{P}_C : u_n^p - d_n^p \geq 0, r_n^p = r_t^p = 0\}$ [nodes in contact with zero reaction], $\boldsymbol{P}_d(\mathbf{u}, \mathbf{r}) = \{p \in \boldsymbol{P}_C : u_n^p - d_n^p \geq 0, r_n^p < 0$ and $|r_t^p| < -\mu r_n^p\}$ [nodes in contact with reaction strictly inside the friction cone and consequent vanishing (right) displacement rate] and $P_s(\mathbf{u}, \mathbf{r}) = \{p \in P_C : u_n^p - d_n^p \geq 0, r_n^p < 0$ and $|r_t^p| = -\mu r_n^p\}$ [nodes in contact with non-vanishing reaction on the boundary of the friction cone and consequent possible slip in the near future].

A static equilibrium state $(\mathbf{u}, \mathbf{r})$ is characterised by a displacement vector $\mathbf{u} \in K_\mathbf{u}$ and a reaction force vector $\mathbf{r} \in K_\mathbf{r}(\mathbf{u})$ such that the equilibrium equations $\mathbf{Ku} = \Lambda\mathbf{P} + \mathbf{r}$ are satisfied. The matrix $\mathbf{K}$ is the (constant) symmetric positive definite stiffness matrix and the vector $\mathbf{P}$ is the partial derivative of the external forces with respect to the load parameter Λ.

For $\mathbf{u} \in |^N$ and $\mathbf{r} \in K_\mathbf{r}(\mathbf{u})$ we define the *(displacement and reaction dependent) cone of admissible right velocities* (Martins et al. 2002)

$$K_\mathbf{v}(\mathbf{u}, \mathbf{r}) = \{\mathbf{v} \in |^N: \quad v_n^p \leq 0, \qquad\qquad \text{in } P_z;$$
$$v_n^p = v_t^p = 0, \qquad\qquad \text{in } P_d;$$
$$v_n^p = 0 \text{ and } \sigma[r_t^p]v_t^p \leq 0, \quad \text{in } P_s\} \tag{3}$$

Having in mind the laws (1) and (2) the *(displacement and reaction dependent) cone of admissible right reaction rates* is defined by

$$K_\mathbf{w}(\mathbf{u}, \mathbf{r}) = \{\mathbf{w} \in |^N: \quad \mathbf{w}^p = \mathbf{0}, \qquad\qquad\qquad \text{in } P_F;$$
$$w_n^p = w_t^p = 0, \qquad\qquad\qquad \text{in } P_f;$$
$$w_n^p \leq 0 \text{ and } |w_t^p| + \mu w_n^p \leq 0, \quad \text{in } P_z;$$
$$\sigma[r_t^p]w_t^p + \mu w_n^p \leq 0, \qquad\qquad \text{in } P_s\}. \tag{4}$$

A visualisation of the decomposition of P_C into P_f, P_z, P_d and P_s is shown in Figure 1, together with the admissible right rates of change of u_n, r_n and r_t.

The definitions (3) and (4) take into account the conditions that involve solely the kinematic variables $\mathbf{v}$ and solely the static variables $\mathbf{w}$. However, these quantities have to satisfy also the complementarity conditions: $v_n w_n = 0$ and $w_t v_t - \mu w_n |v_t| = 0$ in P_z and $(\sigma[r_t]v_t)(\sigma[r_t]w_t + \mu w_n) = 0$ in P_s, which involve both velocities and reaction rates. These conditions may be written in the form $\mathbf{w} \cdot g(\mathbf{v}) = 0$, where the map $g: |^N \to |^N$ results from the maps $g^p: |^2 \to |^2$ given, for each contact candidate node, by

$$g^p(\mathbf{v}^p) = \mathbf{v}^p, \qquad\qquad \text{for } p \in P_F \cup P_f(\mathbf{u}) \cup P_d(\mathbf{u}, \mathbf{r}).$$

$$g^p(\mathbf{v}^p) = \begin{Bmatrix} v_n^p - \mu|v_t^p| \\ v_t^p \end{Bmatrix}, \qquad \text{for } p \in P_z(\mathbf{u}, \mathbf{r}), \tag{5}$$

$$g^p(\mathbf{v}^p) = \begin{Bmatrix} v_n^p + \mu\sigma[r_t]v_t^p \\ v_t^p \end{Bmatrix}, \qquad \text{for } p \in P_s(\mathbf{u}, \mathbf{r}),$$

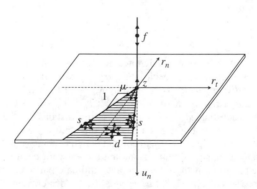

Figure 1. Admissible normal displacement (u_n), normal and tangential reactions $(r_n$ and $r_t)$, and their admissible first order (right) rates of change for a contact candidate node $p \in P_C$.

330

2 NON-LINEAR AND LINEAR COMPLEMENTARITY FORMULATIONS

2.1 The first order quasi-static rate problem

For given equilibrium state and load rate at some time, the quasi-static rate problem consists of determining the right rate of change of state at that time due to that rate of change of load. The study of the rate problem is important to understand the local structure of the equilibrium trajectories. In particular, when the solution of the rate problem for a given load rate is not unique, the equilibrium trajectory exhibits an *angular bifurcation*.

The first order quasi-static rate equilibrium equations are $\mathbf{K}\mathbf{v} = \dot{\Lambda}\mathbf{P} + \mathbf{w}$. The rate problem for an equilibrium state $(\mathbf{u}, \mathbf{r})$ may be formulated as the following *mixed explicit complementarity problem* (Hyers et al. 1997):

Find $(\mathbf{v}, \mathbf{w}) \in |^N \times |^N$, *such that*
$$\mathbf{K}\,\mathbf{v} = \dot{\Lambda}\,\mathbf{P} + \mathbf{w} \; and \tag{6}$$

$$K_w(\mathbf{u}, \mathbf{r}) \ni \mathbf{w} \perp g(\mathbf{v}) \in g(K_v(\mathbf{u}, \mathbf{r})) = (K_w(\mathbf{u}, \mathbf{r}))^*, \tag{7}$$

where ( )* denotes the dual cone of ( ) and $\perp$ denotes orthogonality.

From now on we assume that, at the equilibrium state, $\mathbf{P}_z = \varnothing$. Eliminating the degrees of freedom that are currently free (F and f) and performing the following linear transformation of variables, $\Xi_{st} = \mathbf{S}_s\mathbf{v}_{st}$ and $\Psi_{st} = \mathbf{S}_s\mathbf{w}_{st} - \mu_s\mathbf{w}_{sn}$ where $\mathbf{S}_s = -diag(\sigma[r_t^n], p \in \mathbf{P}_s)$, the first order quasi-static rate problem transforms into a linear complementarity problem (LCP):

Find $(\Xi_{st}, \Psi_{st}) \in |^{n_s} \times |^{n_s}$, *such that*
$$\bar{\mathbf{K}}^*_{st,st}\,\Xi_{st} = \Psi_{st} + \dot{\Lambda}\bar{\mathbf{P}}^*_{st} \; and \tag{8}$$
$$0 \leq \Xi_{st} \perp \Psi_{st} \geq 0,$$

where $\#\mathbf{P}_s = n_s$, $\bar{\mathbf{K}}^*_{st,st} = \mu_s\bar{\mathbf{K}}_1 + \bar{\mathbf{K}}_0$ is a linear pencil of matrices in μ_s. The bars over $\mathbf{K}$ represent Schur complements due to the elimination of the currently free degrees of freedom (Pinto da Costa et al. 2001, Cottle et al. 1992).

The above LCP has a unique solution for any loading rate vector $\dot{\Lambda}\bar{\mathbf{P}}^*_{st}$ if and only if all the principal minors of the coefficient matrix are strictly positive, i.e., if and only if $\bar{\mathbf{K}}^*_{st,st}$ is of class P (Cottle et al. 1992). Since for $\mu_s = 0$ the coefficient matrix is positive definite (hence of class P), the occurrence of an angular bifurcation (multiplicity of solution of problem (8)) may occur only for a sufficiently high coefficient of friction combined with appropriate stiffness couplings.

To find all the solutions of the quasi-static rate problem we used in (8) the algorithm of De Moor. This is a non-iterative algorithm that does not involve any matrix inversion; it is a geometrically inspired algorithm, which performs intersections between the geometric objects that represent the solution sets of a sequence of sub-problems of growing dimension and hyper-planes that represent equilibrium equations successively added to those problems (De Moor et al. 1992). This algorithm is capable of computing the solution set of the rate problem even in the degenerated cases when the number of solutions is infinite. In the case of a bifurcation that occurs at vanishing load rate the rate solution corresponds to an infinite number of solutions with the same shape (bifurcation mode).

2.2 The directional instability problem

During the smooth portions of a dynamic evolution, the system is governed by the Lagrange equations
$$\mathbf{M}\,\ddot{\mathbf{u}}(t) + \mathbf{K}\,\mathbf{u}(t) = \Lambda\mathbf{P} + \mathbf{r}(t) \tag{9}$$

together with laws (1) and (2); $\mathbf{M}$ is the $N \times N$ symmetric positive definite mass matrix. We discuss now the existence of non-oscillatory growing dynamic solutions $(\mathbf{u}(t), \mathbf{r}(t))$ to (9), (1), (2) with initial conditions arbitrarily near equilibrium states *for constant applied external forces* (divergence instability). The stability is studied by analysing the behaviour of the directionally linearized system in the neighbourhood of the static equilibrium state $(\mathbf{u}, \mathbf{r})$. For admissible directional increments $\delta\mathbf{u}$ and $\delta\mathbf{r}$, the equations of motion (9) take the linearized form

$$\mathbf{M}\,\delta\ddot{\mathbf{u}}(t) + \mathbf{K}\,\delta\mathbf{u}(t) = \delta\mathbf{r}(t), \tag{10}$$

where $\mathbf{K}$ is the stiffness matrix (the same as in (6)). We search perturbed solutions to (11) in the form

$$(\delta\mathbf{u}(t),\,\delta\mathbf{r}(t)) = (\alpha(t)\mathbf{v},\,\beta(t)\mathbf{w}) \tag{11}$$

where $\mathbf{v} \in K_v(\mathbf{u},\mathbf{r})$, $\mathbf{w} \in K_w(\mathbf{u},\mathbf{r})$, $\mathbf{w} \cdot g(\mathbf{v}) = 0$ and $\alpha(t)$ and $\beta(t)$ are sufficiently smooth non-decreasing non-negative functions. It is possible to show that the linearized system (9) admits a solution of the form (11) during some time interval if and only if

$$\exists \lambda \geq 0,\, \mathbf{v} \in \mathbb{P}^N,\, \mathbf{v} \neq \mathbf{0} \text{ and } \mathbf{w} \in \mathbb{P}^N,\, \text{such that}$$
$$(\lambda^2\mathbf{M} + \mathbf{K})\,\mathbf{v} = \mathbf{w} \text{ and} \tag{12}$$

$$K_w(\mathbf{u},\mathbf{r}) \ni \mathbf{w} \perp g(\mathbf{v}) \in g(K_v(\mathbf{u},\mathbf{r})) = (K_w(\mathbf{u},\mathbf{r}))^*. \tag{13}$$

In these circumstances the equilibrium state $(\mathbf{u},\mathbf{r})$ is unstable by divergence. The occurrence of a directional instability (existence of a non-trivial solution of problem (12), (13)) occurs for a sufficiently high coefficient of friction combined with appropriate stiffness and/or mass couplings.

It is possible to compute directly the values of μ_s and the modes that correspond to the $\lambda = 0$ transition to instability, which leads to the problem

$$Find\ \mu \geq 0,\, \mathbf{v} \in \mathbb{P}^N,\, \mathbf{v} \neq \mathbf{0} \text{ and } \mathbf{w} \in \mathbb{P}^N,\, \text{such that}$$
$$\mathbf{K}\,\mathbf{v} = \mathbf{w} \text{ and} \tag{14}$$

$$K_w(\mathbf{u},\mathbf{r}) \ni \mathbf{w} \perp g(\mathbf{v}) \in g(K_v(\mathbf{u},\mathbf{r})) = (K_w(\mathbf{u},\mathbf{r}))^*. \tag{15}$$

Eliminating the free degrees of freedom and using the same transformation of variables used to obtain (8), the problem (14), (15) is transformed into the following *complementarity eigenproblem in μ_s* (*CEIP* $-\mu$)

$$Find\ \mu_s \geq 0 \text{ and } (\Xi_{st},\Psi_{st}) \in \mathbb{P}^{n_s} \times \mathbb{P}^{n_s} \text{ with } \Xi_{st} \neq 0,\, \text{such that}$$
$$(\mu_s\bar{\mathbf{K}}_1 + \bar{\mathbf{K}}_0)\,\Xi_{st} = \Psi_{st} \text{ and} \tag{16}$$

$$0 \leq \Xi_{st} \perp \Psi_{st} \geq \mathbf{0}. \tag{17}$$

Note that the existence of a non-trivial solution to the problem (16, 17) occurs for a sufficiently high coefficient of friction combined with appropriate stiffness couplings (note that the mass is not relevant in this transition case because it is multiplied by $\lambda^2 = 0$).

In (Pinto da Costa et al. 2001) we computed the solutions to problem (16, 17) with the algorithm *PATH*, which is based on a Newton method adapted to solve non-smooth systems of equations. Problem *CEIP* $-\mu$ may be rewritten as a *mixed complementarity problem* by the introdution of a new non-negative auxiliary variable η that is complementary to the eigenvalue μ_s; the definition of η provides an additional equation that normalizes the solution. Dirkse and Ferris (1995) demonstrated a global convergence result for that algorithm, even in some problems with non-monotonous operators, which is our case.

2.3 *Some relations between the rate problem and the directional instability problem*

The coefficient matrix of the previous problem is the same as the one of the rate problem (8). Consequently, for vanishing load rate ($\dot{\Lambda} = 0$) the problems (8) and (17) are the same. This means that an eigenmode in the transition $\lambda = 0$ to instability is a solution to the rate problem *for vanishing load rate* ($\dot{\Lambda} = 0$), and vice-versa.

If the homogeneous *LCP* (16, 17) has only the trivial solution, then the $\lambda = 0$ onset of instability is not possible. By definition, a homogeneous *LCP* has the trivial solution as its unique solution if and only if its coefficient matrix is of class R_0 (Cottle et al. 1992). Consequently, when $\mu_s\bar{\mathbf{K}}_1 + \bar{\mathbf{K}}_0 = \bar{\mathbf{K}}^*_{st,st}$ is of class R_0, the $\lambda = 0$ transition to instability is not possible.

Since every P matrix is also of class R_0, we conclude from the above that, for equilibrium states with no nodes in contact with zero reaction ($\mathbf{P}_z = \varnothing$), *the onset of directional instability by vanishing of an eigenvalue λ cannot occur while the first order quasi-static rate problem has a unique solution for every load rate direction.*

Since the properties P and R_0 are sensitive to the non-symmetric part of a matrix, the above statement based on those properties provides sharper results than previous power rate statements of Chateau & Nguyen (1991) and Bigoni (2000).

3 TWO FINITE ELEMENT EXAMPLES

The purpose of the two following examples is to show numerically the coincidence between *(i)* the bifurcation modes that are computed with the algorithm of De Moor and that solve the rate problem (9) for vanishing load rate $\dot{\Lambda} = 0$, and *(ii)* the instability modes that are computed with the *PATH* algorithm and that solve the directional instability problem (16, 17) at the $\lambda = 0$ transition to instability.

The first example is a model of tribology experiments with rectangular polyurethane blocks sliding on an araldite base which may be considered a rigid obstacle, described in (Progri & Villechaise 1984) and (Zeghloul & Villechaise 1991). The elastic block is discretized with a non uniform mesh of 776 linear P1 finite elements that has 65 equally spaced contact candidate nodes (see Figure 2). The block is assumed to be in a state of plane stress. Its elastic properties are: modulus of elasticity = 5 MPa, Poisson's ratio = 0.48. The geometric parameters are: length 80 mm and height 40 mm. The density of the material is 1.2 kgdm^{-3}. The block is submitted to a quasi-static loading consisting first of prescribed displacements on the upper side, which is symmetrically pressed against the obstacle until the resultant of the normal reactions on the lower side is -55 N. Then the loading proceeds by prescribing an horizontal motion of the upper side towards the left. In this tangential loading phase, the successive equilibrium states have a growing region of nodes in impending slip spreading from right to left.

In the final steady sliding configuration the 52 nodes on the left are in impending slip to the left and the 13 nodes on the right are free. *Using the PATH algorithm,* a nontrivial instability mode solution of $CEIP-\mu$ (16, 17) could be found for a coefficient of friction $\mu_s = 1.707375$, which is represented in Figure 2. Substituting then this value of μ_s in (8) and solving for $\dot{\Lambda} = 0$ *with the algorithm of De Moor* we obtain a bifurcation mode that coincides with the instability mode, up to four digits. The numerical values of the eigenmode represented in Figure 2 at the contact nodes are shown in Table 1: node 1 is the first contact node on the left and the mode is normalized such that the sum of the components of Ξ_{st} is equal to 2500.

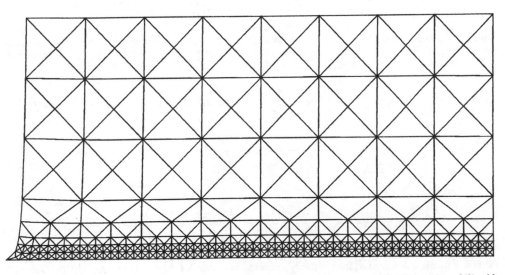

Figure 2. Graphical representation of an instability and bifurcation mode for $\mu_s = 1.7073747$ (solution of (8) with $\dot{\Lambda} = 0$ and of (16, 17)).

333

Table 1. Values at the contact nodes of an instability and bifurcation mode for $\mu_s = 1.7073747$ (solution of (8) with $\Lambda = 0$ and of (16, 17)).

Node	Ξ_{st}	Node	Ξ_{st}	Node	Ξ_{st}	Node	Ξ_{st}
1	191.5833	14	52.4597	27	36.7259	40	28.3344
2	142.6062	15	50.6190	28	35.9199	41	27.8090
3	114.3948	16	48.9899	29	35.1520	42	27.2853
4	100.5868	17	47.5262	30	34.4224	43	26.7656
5	90.8390	18	46.1379	31	33.7292	44	26.2552
6	83.1192	19	44.8023	32	33.0677	45	25.7575
7	76.8346	20	43.5344	33	32.4300	46	25.2694
8	71.6505	21	42.3564	34	31.8027	47	24.7914
9	67.2185	22	41.2641	35	31.1844	48	24.3228
10	63.2884	23	40.2509	36	30.5795	49	23.8662
11	59.8670	24	39.3070	37	29.9926	50	23.4378
12	56.9737	25	38.4190	38	29.4227	51	23.0897
13	54.5541	26	37.5607	39	28.8706	52	22.9728

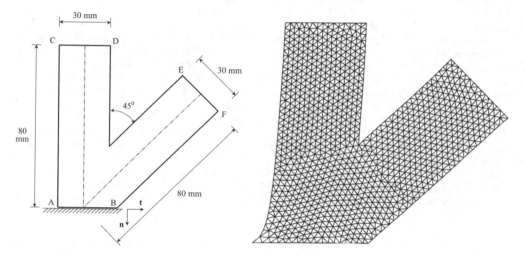

Figure 3. (a) Geometry of the domain of a plane finite element model exhibiting bifurcations and directional instabilities; (b) Graphical representation of an instability and bifurcation mode for $\mu_s = 1.4523123$ (solution of (8) with $\Lambda = 0$ and of (16, 17)).

Table 2. Values at the contact nodes of an instability and bifurcation mode for $\mu_s = 1.4523123$ (solution of (8) with $\Lambda = 0$ and of (16, 17)).

Node	Ξ_{st}	Node	Ξ_{st}	Node	Ξ_{st}	Node	Ξ_{st}
1	225.3752	6	140.9660	11	115.7307	16	102.2593
2	193.8377	7	134.4293	12	112.3109	17	100.3741
3	175.1744	8	128.6954	13	109.3462	18	98.7303
4	161.5462	9	123.6647	14	106.7348	19	97.3266
5	149.6701	10	119.3915	15	104.4365		

The second example consists of a finite element version of the single particle example of Klarbring (1990). This plane domain has the geometry indicated in Figure 3(a) and is discretised in 1997 linear $P1$ finite elements (19 equally spaced contact nodes). The boundary segments CD and EF are fixed. The nodes in segment AB may have contact with a horizontal rigid flat obstacle. This structure is assumed to be in a state of plane stress and have linear elastic behaviour (modulus of elasticity = 5 MPa, Poisson's ratio = 0.48). The self-weight of the structure is neglected. The equilibrium configuration is the unde-formed configuration represented in Figure 3(a) that corresponds to a state of impending slip to the left of all the contact nodes in segment AB. This state of impending slip is obtained by applying on AB a uni-formly distributed inclined load that makes an angle with the vertical equal to the friction angle $tg^{-1}\mu$. Figure 3(b) represents the *coinciding* instability *and* bifurcation mode solutions of (16), (15) and of (8) ($\mu_s = 1.4523123$). The numerical values at the contact nodes of the eigenmode represented in Figure 3(b) are shown in Table 2. Node 1 is point A and node 19 is point B; the mode is normalized such that the sum of the components of Ξ_{st} is equal to 2500.

4 CONCLUSIONS

The present study addressed the first order quasi-static rate problem (relevant for the study of angular bifur-cations) and the directional instability problem (relevant for the study of divergence instabilities), both formulated at an equilibrium state of an unilateral frictional contact system.

We showed that when there are no nodes in contact with zero reaction, the uniqueness of solution of the rate problem for every loading rate direction excludes the onset of directional instability through a null eigenvalue λ.

We also checked that the algorithm used earlier in (Pinto da Costa & Martins, 2003) to compute bifur-cation modes in these *non-smooth* systems and the one used in (Pinto da Costa et al. 2001) to determine modes of directional instability in the same systems do lead to the same result in situations at which such modes should coincide.

REFERENCES

Bigoni, D. 2000. Bifurcation and instability of non-associated elastoplastic solids. In H. Petryk (ed.), *Material Instabilities in Elastic and Plastic Solids*; *CISM courses and lectures nº 414*. Springer Verlag, 1–52.
Chateau, X. & Nguyen, Q.-S. 1991. Buckling of elastic structures in unilateral contact with or without friction. *Eur. J. Mech., A/Solids* 10(1): 71–89.
Cottle, R.W., Pang, J.-S. & Stone, R.E. 1992. *The Linear Complementarity Problem*. Boston: Academic Press.
De Moor, B., Vandenberghe, L. & Vandewalle, J. 1992. The generalized linear complementarity problem and an algorithm to find all its solutions. *Mathematical Programming, Series A* 57(3): 415–426.
Dirkse, S.P. & Ferris, M.C. 1995. The PATH solver: A non-monotone stabilization scheme for mixed complementarity problems. *Optimization Methods and Software* 5: 123–156.
Hyers, D.H., Isac, G. & Rassias, T.M. 1997. *Topics in Nonlinear Analysis & Applications*. Singapore, New Jersey, London: World Scientific.
Klarbring, A. 1990. Examples of non-uniqueness and non-existence of solutions to quasi-static contact problems with friction. *Ingenieur-Archiv.* 56: 529–541.
Martins, J.A.C., Pinto da Costa, A. & Simões, F. 2002. Some notes on friction and instabilities. In João A.C. Martins and Michel Raous (ed.), *Friction and Instabilities*; *CISM courses and lectures nº 457*. Springer Verlag, 65–136.
Pinto da Costa, A., Figueiredo, I.N., Júdice, J.J. & Martins, J.A.C. 2001. A complementarity eigenproblem in the sta-bility analysis of finite dimensional elastic systems with frictional contact. In Michael C. Ferris, Olvi L. Mangasarian and Jong-Shi Pang (ed.), *Complementarity: Applications, Algorithms and Extensions*. Kluwer Academic Publishers, 67–83.
Pinto da Costa, A. & Martins, J.A.C. 2003. The evolution and rate problems and the computation of all possible evolutions in quasi-static frictional contact. *Comput. Methods Appl. Mech. Engrg.* (accepted).
Progri, R. & Villechaise, B. 1984. Analyse des glissements dans un contact sec. *C.R. Acad. Sci. Paris,* t. 299, Série II, nº 12: 763–768.
Stewart, D.E. & Trinkle, J.C. 1996. An implicit time-stepping scheme for rigid body dynamics with inelastic collisions and Coulomb friction. *Int. J. Num. Meth. Engrg.* 39: 2673–2691.
Zeghloul, T. & Villechaise, B. 1991. Phénomènes de glissements partiels découlant de l'usage de la loi de Coulomb dans un contact non lubrifié. *Matériaux et Techniques*, Spécial Tribologie Décembre: 10–14.

Computational Methods in Engineering and Science, Iu et al. (eds)
© 2003 Swets & Zeitlinger, Lisse, ISBN 90 5809 567 3

Precise solution for the surface wave propagation in stratified anisotropic materials

J. Xie & Y. Sun

Department of Engineering Mechanics, Shanghai Jiaotong University, Shanghai, China

ABSTRACT: This paper presents the propagation of semi-space Rayleigh surface wave for multilayer anisotropic materials. The displacements and stresses are combined into a mixed state vector and discussed in system of Hamilton, while not that of Lagrange. Therefore, the orders of differential equations are reduced. The eigenvalues of wave equations are solved by the precise integral and extended Wittrick–Williams algorithm. Numerical examples demonstrate the efficiency and precision of this method in an- and isotropic materials.

1 INTRODUCTION

The analysis of wave propagation in layered material is important in many engineering applications, especially when the media have free surfaces parallel to the transmission of wave. There are many studies on the propagation of surface and other waves, yielding some typical solutions, Ewing et al. 1957, Timoshenko & Gooier 1951, Achenbach 1973, and Kennett 1983. Unfortunately, the complexity of described problems on waves limits the range of solutions. Till now, there are many numerical methods applied, such as Finite Element Method (FEM) and similar matrix approaches, Saitoh et al. 2000, Xue et al. 1997, but few suitable for multilayer anisotropic materials. Therefore, it is necessary to develop a precise and efficient method for surface wave propagation.

At presented paper, the propagation is discussed in the frequency domain. Hence, the dynamics equations of problems are converted into second-order ordinary differential ones. Both displacements and stresses are combined into a mixed state vector, different from the FEM for displacements. Therefore, system of Lagrange is changed into that of Hamilton and, correspondingly; the orders of differential equations are reduced from two down to one. The method can be used to find the eigen solutions exactly. It can simplify the problems and may gain increase of processed variables to a certain extent, yet it is not the key problem considering the high and fast computational ability of personal computer nowadays. Numerical examples are given to illustrate the application of this presented study.

2 DESCRIPTIONS OF EQUATIONS

Let the z-axis point downwards with $z = 0$ at the surface, see Figure 1. The l layers are separated by the horizontal planes $z = z_i < 0$, ($i = 0, 1, \dots, l$), where $z_i < z_j$, for $j > i$, $z = z_l$ is the lowest boundary. The

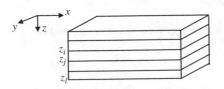

Figure 1. The computational model of multilayered materials.

material of each layer is assumed to be homogeneous anisotropic, but can differ from layer to layer. The surface wave problem is solved below for such layered materials. Suppose the wave propagate along the direction x, described in frequency domain, the dynamics equations are given as follows,

$$\frac{\partial \bar{\sigma}_x}{\partial x} + \frac{\partial \bar{\tau}_{xy}}{\partial y} + \frac{\partial \bar{\tau}_{xz}}{\partial z} + \rho \omega^2 \bar{u} = 0,$$

$$\frac{\partial \bar{\tau}_{xy}}{\partial x} + \frac{\partial \bar{\sigma}_y}{\partial y} + \frac{\partial \bar{\tau}_{yz}}{\partial z} + \rho \omega^2 \bar{v} = 0,$$

$$\frac{\partial \bar{\tau}_{xz}}{\partial x} + \frac{\partial \bar{\tau}_{yz}}{\partial y} + \frac{\partial \bar{\sigma}_z}{\partial z} + \rho \omega^2 \bar{w} = 0. \tag{1}$$

where ω = wave frequency; $\bar{\sigma}_x$ and $\bar{\sigma}_z$ = longitudinal stresses; $\bar{\tau}_{xy}$ = shear stresses; $\bar{u}$, $\bar{v}$, $\bar{w}$ = displacements parallel to x, y, and z axis respectively; ρ = density of mass; ω = frequency of wave; and "$\bar{}$" denotes all these kinds of variables are functions of x, y and z.

For anisotropic materials, the stress–displacement relationship is given by,

$$\begin{Bmatrix} \bar{\sigma}_x \\ \bar{\sigma}_y \\ \bar{\sigma}_z \\ \bar{\tau}_{yz} \\ \bar{\tau}_{xy} \\ \bar{\tau}_{xz} \end{Bmatrix} = \begin{bmatrix} C_{11} & C_{12} & C_{13} & C_{14} & C_{15} & C_{16} \\ & C_{22} & C_{23} & C_{24} & C_{25} & C_{26} \\ & & C_{33} & C_{34} & C_{35} & C_{36} \\ & & & C_{44} & C_{45} & C_{46} \\ & \Delta_{ij} & & & C_{55} & C_{56} \\ & & & & & C_{66} \end{bmatrix} \begin{Bmatrix} \bar{\varepsilon}_x \\ \bar{\varepsilon}_y \\ \bar{\varepsilon}_z \\ \bar{\gamma}_{yz} \\ \bar{\gamma}_{xy} \\ \bar{\gamma}_{xz} \end{Bmatrix} \tag{2}$$

where Δ_{ij} = symmetry section to the upper triangle; C_{11}, C_{12}, ..., C_{66} = constants of materials, a total number of 21.

Boundary conditions are supposed as,

when $z = 0$, $\bar{\sigma}_z = 0$, $\bar{\tau}_{xz} = 0$, $\bar{\tau}_{yz} = 0$ \hfill (3)

when $z = h$, $\bar{u} = 0$, $\bar{v} = 0$, $\bar{w} = 0$ \hfill (4)

Geometric equations are as follows,

$$\begin{cases} \bar{\varepsilon}_x = \dfrac{\partial \bar{u}}{\partial x}, \bar{\varepsilon}_y = \dfrac{\partial \bar{v}}{\partial y}, \bar{\varepsilon}_z = \dfrac{\partial \bar{w}}{\partial z} \\ \bar{\gamma}_{xy} = \dfrac{\partial \bar{u}}{\partial y} + \dfrac{\partial \bar{v}}{\partial x}, \bar{\gamma}_{yz} = \dfrac{\partial \bar{w}}{\partial y} + \dfrac{\partial \bar{v}}{\partial z} \\ \bar{\gamma}_{xz} = \dfrac{\partial \bar{w}}{\partial x} + \dfrac{\partial \bar{u}}{\partial z} \end{cases} \tag{5}$$

Continuity conditions between interfaces,

when $z = z_i$ $(i = 1, 2, \cdots, l)$, $\bar{\sigma}_x, \bar{\sigma}_z, \bar{\tau}_{xz}, \bar{u}, \cdots, \bar{w}$ are continuous \hfill (6)

Suppose wave number is k, then displacements are illustrated in frequency domain as,

$$\bar{u} = u(y,z)e^{ikx}, \quad \bar{v} = v(y,z)e^{ikx}, \quad \bar{w} = w(y,z)e^{ikx} \tag{7}$$

To simplify the problem, suppose stresses and displacements are homogeneous parallel to y axis, then Equations 1 and 5 are equal to zero under the differential of y. $\bar{u}$, $\bar{v}$ and $\bar{w}$ in Equation 7 are simplified as functions of x and z. The components in Equations 1, 2 and 5 are resumed except for $\varepsilon_y = 0$; therefore, the number of constants in Equation 2 is decreased to 15. We can define this solution as 2.5 dimensions.

Substitute Equations 2, 5 and 7 into Equation 1, we get,

$$\mathbf{K}_{22}\ddot{\mathbf{q}} + (\mathbf{K}_{21} - \mathbf{K}_{12})\dot{\mathbf{q}} - \mathbf{K}_{11}\mathbf{q} = 0 \tag{8}$$

where $\mathbf{q} = \{u\ v\ w\}^T$; "$\cdot\cdot$" = differential of z;

$$\mathbf{K}_{22} = \begin{bmatrix} C_{66} & C_{46} & C_{36} \\ C_{46} & C_{44} & C_{34} \\ C_{36} & C_{34} & C_{33} \end{bmatrix} ; \ \mathbf{K}_{21} = \begin{bmatrix} ikC_{16} & ikC_{56} & ikC_{66} \\ ikC_{14} & ikC_{45} & ikC_{46} \\ ikC_{13} & ikC_{35} & ikC_{36} \end{bmatrix} ; \ \mathbf{K}_{21}^H = \mathbf{K}_{12} ;$$

$$\mathbf{K}_{11} = \begin{bmatrix} k^2C_{11} - \rho\omega^2 & k^2C_{15} & k^2C_{16} \\ k^2C_{15} & k^2C_{55} - \rho\omega^2 & k^2C_{56} \\ k^2C_{16} & k^2C_{56} & k^2C_{66} - \rho\omega^2 \end{bmatrix} .$$

where upper script^H = Hermit transmission.

In order to reduce the order of Equation 8, the coupling vector $\mathbf{p}$ of the displacement vector $\mathbf{q}$ are constructed as,

$$\mathbf{p} = -(\mathbf{K}_{22}\dot{\mathbf{q}} + \mathbf{K}_{21}\mathbf{q}) = -\begin{Bmatrix} C_{66}\dot{u} + C_{46}\dot{v} + C_{36}\dot{w} + ikC_{16}u + ikC_{56}v + ikC_{66}w \\ C_{46}\dot{u} + C_{44}\dot{v} + C_{34}\dot{w} + ikC_{14}u + ikC_{45}v + ikC_{46}w \\ C_{36}\dot{u} + C_{34}\dot{v} + C_{33}\dot{w} + ikC_{13}u + ikC_{35}v + ikC_{36}w \end{Bmatrix} = -\begin{Bmatrix} \tau_{xz} \\ \tau_{yz} \\ \sigma_z \end{Bmatrix} \tag{9}$$

From Equations 8 and 9, the coupling equations of $\mathbf{q}$ and $\mathbf{p}$ are constructed as,

$$\begin{cases} \dot{\mathbf{q}} = \mathbf{A}\mathbf{q} - \mathbf{D}\mathbf{p} \\ \dot{\mathbf{p}} = -\mathbf{B}\mathbf{q} - \mathbf{A}^H\mathbf{p} \end{cases} \tag{10}$$

where $\mathbf{A} = -\mathbf{K}^{-1}_{22}\mathbf{K}_{21}$; $\mathbf{B} = \mathbf{K}_{11} - \mathbf{K}_{12}\mathbf{K}_{22}^{-1}\mathbf{K}_{21}$ and $\mathbf{D} = \mathbf{K}_{22}^{-1}$.

Equations 10 can be rewritten as follows,

$$\dot{\mathbf{v}} = \mathbf{H}\mathbf{v}\ , \mathbf{H} = \begin{bmatrix} \mathbf{A} & -\mathbf{D} \\ -\mathbf{B} & -\mathbf{A}^H \end{bmatrix}, \mathbf{v} = \begin{Bmatrix} \mathbf{q} \\ \mathbf{p} \end{Bmatrix} \tag{11}$$

where H = complex Hamilton matrix; A, B and D = matrices of system.

With the boundary equations of (3), (4) and (6), we can get the solutions after solving Equation 11. When $z_l = h$ is large enough, there exists surface wave on the free surface of $z = 0$.

3   SOLUTIONS OF EQUATIONS

It is a problem with double boundaries to Equation 11. The traditional method is to use difference to simplify differential equations and then integrate it. But, unfortunately, we always cannot get the satisfactory solutions when facing a large number of integrals. At presented paper, the precise integral, Zhong 2002, and extended Wittrick–Williams algorithms, Zhong et al. 1997, are exploited, which yields precise solutions with efficient numerical digital of the personal computer.

It makes the number of axes finite when substituting difference for differential. Let $z_a < z < z_b$ as an interval; correlate the original variables of $\mathbf{q}_a$ and $\mathbf{p}_a$ in z_a, and those of $\mathbf{q}_b$ and $\mathbf{p}_b$ in $\mathbf{z}_b$. Thence, it yields as follows,

$$\mathbf{q}_b = \mathbf{F}\mathbf{q}_a - \mathbf{\Gamma}\mathbf{p}_b$$
$$\mathbf{p}_a = \mathbf{Q}\mathbf{q}_a + \mathbf{F}^H\mathbf{p}_b \tag{12}$$

where $\mathbf{F}$, $\mathbf{\Gamma}$ and $\mathbf{Q}$ = pending matrices. When the matrices of $\mathbf{A}$, $\mathbf{B}$ and $\mathbf{D}$ are irrelative to z, $\mathbf{F}$, $\mathbf{\Gamma}$ and $\mathbf{Q}$ are the functions of interval $\Delta z = z_b - z_a$.

Considering Equation 10 and 12, it gets,

$$\begin{cases} d\mathbf{F}/d(\Delta z) = \mathbf{AF} - \mathbf{\Gamma BF} \\ d\mathbf{\Gamma}/d(\Delta z) = \mathbf{D} + \mathbf{A\Gamma} + \mathbf{\Gamma A}^H - \mathbf{\Gamma B\Gamma} \\ d\mathbf{Q}/d(\Delta z) = \mathbf{B} + \mathbf{QA} + \mathbf{A}^H\mathbf{Q} - \mathbf{QDQ} \end{cases} \tag{13}$$

Equation 13 belongs to an interval. Two contact intervals should be combined. Suppose the first interval as (z_a, z_b), and the second (z_b, z_c). Considering Equation 12, it yields,

$$\begin{cases} \mathbf{q}_b = \mathbf{F}_1\mathbf{q}_a - \mathbf{\Gamma}_1\mathbf{p}_b \\ \mathbf{p}_a = \mathbf{Q}_1\mathbf{q}_a + \mathbf{F}_1^H\mathbf{p}_b \end{cases}, \quad \begin{cases} \mathbf{q}_c = \mathbf{F}_2\mathbf{q}_b - \mathbf{\Gamma}_2\mathbf{p}_c \\ \mathbf{p}_b = \mathbf{Q}_2\mathbf{q}_b + \mathbf{F}_2^H\mathbf{p}_c \end{cases} \tag{14a, b}$$

After both combined, it gets,

$$\begin{cases} \mathbf{q}_c = \mathbf{F}_c\mathbf{q}_a - \mathbf{\Gamma}_c\mathbf{p}_c \\ \mathbf{p}_a = \mathbf{Q}_c\mathbf{q}_a + \mathbf{F}_c^H\mathbf{p}_c \end{cases} \tag{15}$$

Therefore, according to Equations 14 and 15, the interval matrices are as follows,

$$\begin{cases} \mathbf{F}_c = \mathbf{F}_2\left(\mathbf{I} + \mathbf{Q}_2\mathbf{\Gamma}_1\right)^{-H}\mathbf{F}_1 \\ \mathbf{\Gamma}_c = \mathbf{\Gamma}_2 + \mathbf{F}_2\left(\mathbf{\Gamma}_1^{-1} + \mathbf{Q}_2\right)^{-1}\mathbf{F}_2^H \\ \mathbf{Q}_c = \mathbf{Q}_1 + \mathbf{F}_1^H\left(\mathbf{Q}_2^{-1} + \mathbf{\Gamma}_1\right)^{-1}\mathbf{F}_1 \end{cases} \tag{16}$$

All matrices in above equations are unknown interval ones. In order to define it, system matrices of A, B and D should be used. So the algorithms of 2^N multilayer substructures in computational structure mechanics, Zhong 1995, can be applied. The sublayer or interval $\Delta z = z_b - z_a$ can be divided into 2^N sublayers, N takes value of 20 then $2^N = 1048576$. Therefore, the length of sublayer is $\tau = \Delta z/1048576$. And the combination of sublayers is similar to the element eliminating or condensation. Since all considered sublayers are the same, once condensing, the number of sublayers is halved, then after $N = 20$, it remains a sublayer of Δz. For τ is very small, the matrices of sublayers $\mathbf{F}$, $\mathbf{\Gamma}$ and $\mathbf{Q}$ can be solved by the expansion of power series as follows,

$$\begin{cases} \mathbf{F}(\tau) = \mathbf{I} + \mathbf{f}_1\tau + \mathbf{f}_2\tau^2 + \mathbf{f}_3\tau^3 + \mathbf{f}_4\tau^4 + \cdots \\ \mathbf{\Gamma}(\tau) = \mathbf{g}_1\tau + \mathbf{g}_2\tau^2 + \mathbf{g}_3\tau^3 + \mathbf{g}_4\tau^4 + \cdots \\ \mathbf{Q}(\tau) = \boldsymbol{\theta}_1\tau + \boldsymbol{\theta}_2\tau^2 + \boldsymbol{\theta}_3\tau^3 + \boldsymbol{\theta}_4\tau^4 + \cdots \end{cases} \tag{17}$$

Considering Equation 13, it yields,

$$\begin{cases} \mathbf{f}_1 = \mathbf{A}, \\ \mathbf{f}_2 = \dfrac{1}{2}(\mathbf{Af}_1 - \mathbf{g}_1\mathbf{B}), \\ \mathbf{f}_3 = \dfrac{1}{3}(\mathbf{Af}_2 - \mathbf{g}_2\mathbf{B} - \mathbf{g}_1\mathbf{Bf}_1), \\ \mathbf{f}_4 = \dfrac{1}{4}(\mathbf{Af}_3 - \mathbf{g}_3\mathbf{B} - \mathbf{g}_2\mathbf{Bf}_1 - \mathbf{g}_1\mathbf{Bf}_2), \end{cases} \qquad \begin{cases} \mathbf{g}_1 = \mathbf{D}, \\ \mathbf{g}_2 = \dfrac{1}{2}(\mathbf{Ag}_1 + \mathbf{g}_1\mathbf{A}^H), \\ \mathbf{g}_3 = \dfrac{1}{3}(\mathbf{Ag}_2 + \mathbf{g}_2\mathbf{A}^H - \mathbf{g}_1\mathbf{Bg}_1) \\ \mathbf{g}_4 = \dfrac{1}{4}(\mathbf{Ag}_3 + \mathbf{g}_3\mathbf{A}^H - \mathbf{g}_1\mathbf{Bg}_2 - \mathbf{g}_2\mathbf{Bg}_1) \end{cases}$$

$$\begin{cases} \boldsymbol{\theta}_1 = \mathbf{B}, \\ \boldsymbol{\theta}_2 = \dfrac{1}{2}(\mathbf{f}_1^H\mathbf{B} + \mathbf{Bf}_1), \\ \boldsymbol{\theta}_3 = \dfrac{1}{3}(\mathbf{f}_2^H\mathbf{B} + \mathbf{Bf}_2 + \mathbf{f}_1^H\mathbf{Bf}_1) \\ \boldsymbol{\theta}_3 = \dfrac{1}{4}(\mathbf{f}_3^H\mathbf{B} + \mathbf{Bf}_3 - \mathbf{f}_2^H\mathbf{Bf}_1 - \mathbf{f}_1^H\mathbf{Bf}_2) \end{cases} \tag{18a, b, c}$$

Substituting Equation 18 into Equation 17, we can get the definite values of $\mathbf{F}$, $\boldsymbol{\Gamma}$ and $\mathbf{Q}$. All deductions of equations are precise before the above expansions. Though, there are round-off errors after τ^4, $\tau = \Delta z/1048576$ is small enough. Hence, considering the digital efficiency of personal computer, the solutions are precise enough in engineering applications.

4 NUMERICAL EXAMPLES

In order to illustrate the application of the presented method, three numerical examples are given as follows.

4.1 Example 1

For the aim of comparison with regular solutions, firstly, we consider the isotropic materials. The computational model is shown in Figure 1. The parameters are given as: the Young's modulus of the material is $E = 1.5 \times 10^7$ Pa, Poisson ratio is $v = 0.25$, density of mass is $\rho = 2 \times 10^3$ kg/m^3, and wave number $k = 0.02$. From the former discussion, Zhong 1995, we can know the fundamental frequency of surface wave is $\omega^2 = 0.845299$ (rad/s)^2. As the method mentioned above, it is divided into 64 sublayers parallel to z-axis, $\Delta z = h/64$ and $\tau = \Delta z/1048576$. The power of the 1st–8th order frequencies, ω_i^2, is computed and listed in Table 1. From the Figure 2, with the increasing of depth from 1000 m to 400 m, the first order ω_1^2 remains almost unchanged; especially when $h = 1000$ m, correspondingly it is $\omega_1^2 = 0.845299$ (rad/s)^2, which is the same to the regular solution. Also, it is clear that, with the more decreasing of depth h, the bottom constraints affect every order frequency greatly. When in $h = 200$ m and 100 m, the first order frequency gains large change. It is explained that, when the depth h is less than the wavelength, a value of 314.159 m, the bottom affects the frequency evidently.

Table 1. The ω^2 of elastic wave in isotropic materials ((rad/s)^2).

Nth\h(m)	1000	800	600	400	200	100
1	0.845299	0.845307	0.845482	0.849605	0.973725	1.616850
2	1.006168	1.009638	1.017135	1.038553	1.154213	2.022005
3	1.037938	1.065684	1.135944	1.346978	2.387913	3.254228
4	1.055517	1.086745	1.154213	1.380045	2.487131	6.551652
5	1.144179	1.240507	1.428368	1.963829	4.070333	8.697175
6	1.154213	1.240957	1.462338	2.095566	4.098617	13.89726
7	1.302257	1.472276	1.839602	2.799091	4.855314	16.42125
8	1.308683	1.502023	1.920831	2.889104	7.620928	21.70306

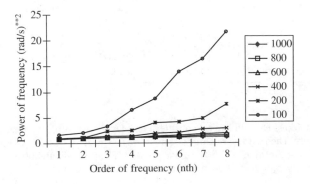

Figure 2. The curve of ω^2 elastic wave in isotropic materials

Table 2. The ω^2 of elastic wave in layered materials $((rad/s)^2)$.

Nth\h(m)	1000	800	600	400	300
1	0.972001	0.972001	0.972029	0.974833	0.994190
2	1.139857	1.139859	1.139877	1.140941	1.148183
3	1.540508	1.548811	1.575529	1.715654	1.948322
4	1.575285	1.595266	1.637323	1.758293	1.982638
5	1.661375	1.757329	1.954608	2.500718	3.136413
6	1.684606	1.767015	1.989432	2.522052	3.262786
7	1.874406	2.073467	2.499860	3.443281	3.881368
8	1.884710	2.104762	2.510298	3.700325	5.264383

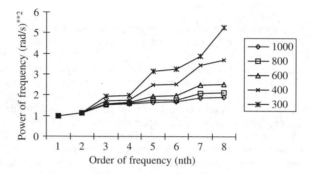

Figure 3. The curve of ω^2 elastic wave in layered materials.

Table 3. The constant of materials (10^7)

C_{11}	C_{13}	C_{14}	C_{15}	C_{16}	C_{33}	C_{34}	C_{35}	C_{36}	C_{44}	C_{45}	C_{46}	C_{55}	C_{56}	C_{66}
1.5	0.5	0.4	0.1	0.2	1.5	0.1	0.2	0.4	0.7	0.3	0.2	0.7	0.2	0.3
1.0	0.5	0.08	0.08	0.2	1.0	0.08	0.08	0.2	0.5	0.08	0.1	0.4	0.1	0.3
2.0	0.7	0.5	0.2	0.3	2.0	0.3	0.1	0.4	1.0	0.2	0.3	1.0	0.3	0.8

Table 4. The ω^2 elastic wave in multilayered anisotropic materials $((rad/s)^2)$.

Nth\h(m)	1000	800	600	400	300
1	0.233916	0.233916	0.233916	0.234051	0.236664
2	0.804202	0.804210	0.804424	0.809956	0.830433
3	1.156779	1.168117	1.190352	1.259337	1.396606
4	1.258179	1.328387	1.462587	1.620393	1.750641
5	1.447581	1.568163	1.704520	1.904560	2.110483
6	1.633871	1.757194	1.893735	2.285657	2.651133
7	1.791819	1.924970	2.173554	2.608714	2.994032
8	1.951627	2.151092	2.340269	2.835819	3.276546

4.2 Example 2

Compared with the above Example 1, we consider the 3 layers of isotropic materials, different elastic constants respectively. The first layer, depth 100 m, $z_1 = 100$ m, $E = 1.25 \times 10^7$ Pa, and $\nu = 0.25$. The second, depth 100, $z_2 = 200$ m, $E = 1.5 \times 10^7$ Pa, $\nu = 0.3$. The third, depth 800 m, 600 m, 400 m, 200 m and 100 m, respectively, $z_3 = h = 1000$ m, 800 m, 600 m, 400 m and 300 m, $E = 2.0 \times 10^7$ Pa, $\nu = 0.3$. In all three layers, the density of mass is $\rho = 2 \times 10^3$ kg/m^3, wave number $k = 0.02$. The solutions are listed in Table 2, where h is the sum of depth. From the Figure 3, with the increasing of depth from

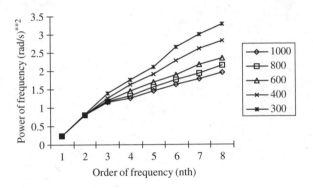

Figure 4. The curve of ω^2 of elastic wave in multilayered anisotropic materials.

1000 m to 400 m, the first order remains almost unchanged, while the second has the small change. When h is 300 m, there is a trend of increasing on the first two orders. It is still explained that, when the depth h is less than the wavelength, the bottom affects the frequency greatly.

4.3 Example 3

Considering 3 sublayers, every anisotropic, and the depth is supposed as example 2. But the density is the same $\rho = 2 \times 10^3\,\text{kg/m}^3$; wave number is $k = 0.02$. Then, the constants in Equation 2 are computed as Table 3, and corresponding values are listed in Table 4. From Figure 4, with the decrease of depth h, the first order changes little, and the second change more, but the bottom still affects frequency more largely with the decrease of depth h.

5 CONCLUSIONS

The presented paper introduces system of Hamilton on the wave propagation, and discusses the elastic wave propagation in semi-infinite body with free surfaces. The order of differential equations is reduced in system of Hamilton, then the solutions given easily simultaneously. The n orders of eigenvalues, from low to high, are computed with the extended Wittick–Williams algorithms, losing no root. And the precise integral makes the precise solutions in the usual computational methods. The numerical examples prove the accuracy of this study with great precise and tell that this method can be used in different materials.

REFERENCES

Ewing, W.M., Jardetzky, W.S. & Press, F. 1957. *Elastic waves in layered media.* New York: McGraw-Hill.
Timoshenko, S.P. & Gooier, J.N. 1951. *Theory of elasticity.* New York: McGraw-Hill.
Achenbach, J.D. 1973. *Wave propagation in elastic solids.* Amsterdam: North-Holland.
Kennett, B.L.N. 1983. *Seismic wave propagation in stratified media.* Cambridge, UK: Cambridge Univ. Press.
Saitoh Kunimasa, Koshiba Masanori. 2000. Beam propagation method for three-dimensional surface acoustic waveguides. *Japanese Journal of Applied Physics* 39:2999–3003.
Xue T., Lord, W. & Udpa, S.1997. Finite element simulation and visualization of leaky Rayleigh waves for ultrasonic NDE. *IEEE Transactions on Ultrasonic Ferroelectrics and Frequency Control* 44 (3):557–564.
Zhong, W.X. 2002. *The coupling system in applied mechanics.* Beijing, China: Science Press (in Chinese).
Zhong, W.X., Williams, F.W. & Bennett, P.N. 1997. Extension of the Wittick–Williams algorithm to mixed variable systems. *J. Vib. & Acous. Trans ASME* 119:334–340.
Zhong, W.X. 1995. *A new systematic methodology for theory of elasticity.* Dalian, China: Dalian Univ. of Technology Press (in Chinese).
Zhong, W.X. 1995. Precise integration of eigen-waves for layered media. *Proc. EPMESC-5* 2:1209–1220. Macao.
Williams, F.W., Zhong, W.X. & Bennett, P.N. 1993. Computation of the eigenvalues of wave propagation in periodic substructural systems. *J. Vib. & Acous. Trans ASME* 115.422–426.

Seismic behaviour of gas and oil buried pipelines subjected to ground liquefaction

M.J. Falcão Silva
LNEC, Lisbon, Portugal

R. Bento
IST, Lisbon, Portugal

ABSTRACT: The simulation of seismic performance of buried pipelines of gas and oil networks fall upon a fragment of the principal line of natural gas network – TRANSGÁS – and of oil's network – CLC – when subjected to large ground deformation due to liquefaction. For the analytical model of the buried pipelines it is proposed a discrete model, such as the Winkler, and two distinct analysis: in the first one it is imposed a displacement's history and in the second it is considered an acceleration's history. These analyses aim the definition of the vulnerability functions for buried pipelines and compare them with existent empirical relations. Various studies are performed using different parameters such as effective soil mass coefficient, pipeline diameter and width of PGD. Several conclusions are stated, mainly related to the influence of the different factors studied on the seismic behaviour of buried pipelines.

1 INTRODUCTION

The assessment of the performance of buried pipelines, when subjected to permanent ground deformations (PGD) due to the occurrence of an earthquake, may have great importance not only for the design phase of this kind of structures but also in a rehabilitation program. While the seismic behaviour of pipes when submitted to seismic waves propagation is significantly well known both on the basis of case analysis and of the results obtained from analytic models, the answer to this kind of elements when subject to permanent ground deformations is more complex, mainly due to:

– Considerable uncertainty concerning the possible amplitude and direction of PGD;
– Insufficient use of simple and realistic methods to allow extending sensitivity studies;
– The fact that existing empirical relations to characterize the seismic vulnerability of buried pipes were deduced on the basis of a small number of cases, in comparison to the relations obtained to evaluate the effect of seismic wave propagation.

The need to correctly evaluate the seismic behaviour of buried pipelines submitted to high values of ground deformations, with the purpose of clearly evaluate the seismic vulnerability of these structures, requires that analytic studies are carried out in order to characterize adequately the level of damage when phenomena such as ground liquefaction, lateral spread or geological faults occur. Of the mentioned phenomena permanent ground deformations induced by liquefaction are considered as the main cause of damage in buried structures (Lim et al. 2000), having liquefaction precisely been adopted in this work as cause of ground rupture and as the origin of the high values of ground deformations considered.

For the analytical modelling of pipelines it is proposed a discrete model (Winkler type) and two different analyses; in analysis A_1 it was imposed a displacement history obtained by corrected integration of accelerograms, while in analysis A_2 it was used a history of accelerations (accelerograms). It is also important to refer that in the analysis A_2 it was necessary to consider the mass of the soil involving the pipe in order to better simulate the movement of the ensemble soil/pipe as a whole. With this purpose the mass evaluated is concentrated at the level of the extremity nodes of the several elements of the model. The programme used in both analyses is SAP2000 (SAP2000 2000).

2 ANALYTICAL MODEL FOR THE BURIED PIPELINES SUBJECTED TO SOIL'S LIQUEFACTION

2.1 Seismic action modelling

When modelling the seismic action, a time analysis is carried out through a step-by-step integration, being the seismic action defined by histories of ground displacements and accelerations along time. As a way of simulating the seismic action through time it is possible to use deterministic and stochastic methods. For the artificially accelerograms' generation existing programmes in ICIST are used (Guerreiro 2002). The referred accelerograms are generated on the basis of spectral density functions of potency presented in RSA (1983) for type 1 and type 2 actions which represent respectively earthquakes of moderate magnitude to a small focal distance where high frequencies are predominant as well as earthquakes of a higher magnitude to a higher focal distance where smaller frequencies are observed. Also considered are soils type I and III which include rocks and stiff coherent soils (type I) and soft, very soft cohesive soils and also loose cohesionless soils (type III) (RSA 1983). The reason for this choice is that is intended to determine the maximum and minimum displacements observed at the ground level when a certain seismic action occurs.

For both seismic actions and both considered soils, 10 accelerograms are generated corresponding to the horizontal component of acceleration. When generating the accelerograms, potency spectra referring to the horizontal component of ground acceleration indicated in RSA are considered. Selection of artificially generated accelerograms is made in order that response spectra corresponding to those accelerograms would present pacing similar to those observed for code response spectra purpose by RSA.

After having defined the artificially accelerograms, the displacement histories were evaluated by means of corrected integration. In Figure 1 it is presented, as examples, three displacement histories used in the parametric studies presented in this work.

2.2 Buried pipelines modelling

As referred, for the behaviour modelling of buried pipelines it is used a discrete model, Winkler type. These models do not reproduce the continuous character of pipeline's involving environment neither takes into account existing three-dimensional effects. Furthermore, they are a function of a parameter (reaction coefficient), which depends on several factors (pipe diameter and length of the considered element). In spite of these constraints associated to the empirical parameter of the model, these calculation methods have had big practical applicability, taking into account that they are easy to use (Santos 1999). Opposite to this, they may allow to consider easily the introduction of the non-linearity of the system, by considering non-linear empirical curves "p–u" which relate reaction and ground displacement shown in Figure 2.

With the discrete model mentioned different characteristic analyses are made: firstly the seismic action simulated by means of displacements histories corresponding to the corrected integration of the selected accelerograms taking into account the effect of the ground damping; and secondly it is admitted that the

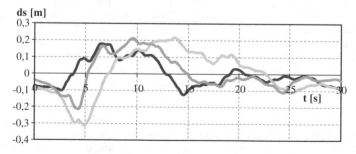

Figure 1.   Three different displacement histories correspondent to earthquake type 2 and soil type III.

seismic action is simulated by means of artificially generated or previously chosen accelerograms. In the latter further the damping, both the pipeline mass and the involving ground is considered. After having been conveniently determinate, the involving ground mass must be concentrated at the extremity nodes of each model elements.

In the analysis performed it is considered a 500 m length pipeline, having admitted that the zone where the liquefaction phenomenon occurs presents a 100 m length. On the basis of these conditions – reference situation – models are studied by varying the pipe diameter, the thickness of the pipe wall, the length of the liquefaction zone, the ground rigidity constants so that one could draw some conclusions related to value variation of damage of buried gas and oil pipelines resulting from the occurrence of an earthquake causing ground liquefaction. As shown in Figure 3, a three distinct zone piping is considered: a central zone (corresponding to the ground liquefaction), a transition zone and the external zones (zones without liquefaction).

2.3 *Description of the structural model*

Although this is a pipeline (hollow a cylinder section) it was noticed that based on some sensitivity studies carried out, formulation based on finite elements such as beams lead to much satisfactory results. The elements' length is presented in Figure 4. For that reason, in all studied models this kind of finite elements are considered. As referred, the ground is modelled by means of variable rigidity springs placed at the end of each finite element.

The ground rigidity value initially considered in the example for each zone is determined on the basis of the ground constant used in the calculation of the ground bordering strength in the pipeline's transversal direction (ERDCJ 2000). It is considered that this ground constant takes an 11000 kN/m³ value, being designated by k_2. As k_2 corresponds to a reaction coefficient it is necessary to transform it to the corresponding rigidity value. For that purpose, k_2 is multiplied by the corresponding external pipe diameter – ϕ (711.2 mm for gas network and 406.4 mm for oil network. The obtained value, k, is named reaction module (Equation 1).

$$k = k_2\, \phi \qquad\qquad (1)$$

Finally, in order to know the rigidity of each spring (K) placed in each one of the three possible displacement direction at the extremity of each finite element it is necessary to consider its influence

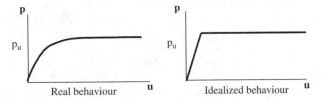

Figure 2.   "p–u" curve based on the elastic perfectly plastic behaviour model.

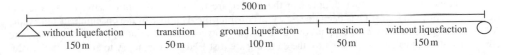

Figure 3.   Analysed pipeline zones.

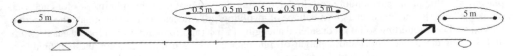

Figure 4.   Finite element network: sixty 50 m elements and four hundred 0.5 m elements.

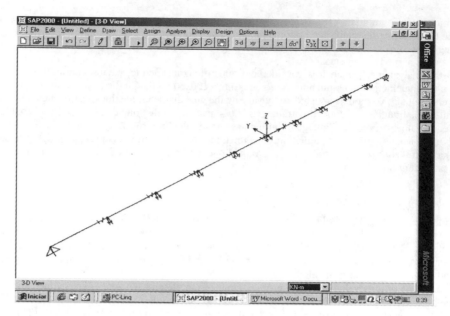

Figure 5.   Standard pipeline for the base model.

length (Equation 2). To make it easier, it is considered that the rigidity of each spring would take the same value for the three considered directions (x, y and z) in which L corresponds to the influence length of each spring.

$$K = k\,L \tag{2}$$

In the liquefaction zone one assume that the ground, even at the end of the liquefaction process, keeps a residual rigidity value corresponding to about 100 times smaller than the one obtained for elements having the same length in non liquefaction zones. Finally, in transition zones – (between different length elements and between non liquefied/liquefied zones) the spring rigidity is calculated according to Equation 3.

$$K = k'\,L' + k''\,L'' \tag{3}$$

In which k' and k'' correspond to values of the reaction module for each one of the referred zones, being the L' and L'' values of the respective influence lengths. Having determined the ground rigidity values it has been possible to build the base structural model (SAP2000 1997) used in both analyses and shown in the Figure 5.

The geometry characteristics adopted in the model are the same for both analyses, A_1 and A_2. Besides, in both situations it has been necessary to consider the damping of the ensemble ground/pipe. Although the pipelines are in steel, it is considered a damping coefficient value of 2% as well as 0% and 20%. The different values considered for the damping are justified as a way to try to evaluate their influence in each analysis.

For analysis A_2 it is still necessary to determine the mass of ground which is assumed to vibrate together with the pipeline, so that the displacement value of the buried structure is the same as the displacement value of the surrounding soil. In determining the value of the vibrating masses for each element it is important to consider a given volume of involving soil – necessary to determinate the effective mass. For the surrounding soil it is considered a length of 20 φ (approximately 15 m for the gas network and 8 m for the oil network) and a height of 3φ (respectively 2 m and 1.2 m for the two analysed networks). In Table 1 are indicated the masses determined for the elements belonging to each different zones and placed at the extremity of each of the mentioned elements.

Table 1. Effective mass for all pipelines zones.

Network	φ (mm)	Mass (ton)				
		ZWL (5 m)	ZWL (5/0.5)	ZWL (0.5 m)	ZWL/LZ	LZ (0.5 m)
TRANSGÁS	711.2	266.4	146.5	26.6	28.1	29.6
CLC	406.4	87.8	48.3	8.8	9.3	9.8

Table 2. Soil's rigidity constants for all pipeline's zones.

Network	φ (mm)	k_2 (kN/m³)	K (kN/m)				
			ZWL (5 m)	ZWL (5/0.5)	ZWL (0.5 m)	ZWL/LZ	LZ (0.5 m)
		5000	17780.0	9779.0	1778.0	897.9	17.8
TRANSGÁS	711.2	11000	39116.0	21513.8	3911.6	1975.3	39.1
		50000	177800.0	97790.0	17780.0	8978.9	177.8
CLC	406.4	11000	22352.0	12293.6	2235.2	1128.8	22.4

In the extremity nodes of the model it is assumed a mass of about half of the obtained value for the elements of 5 m width, corresponding to the zone without liquefaction because only half of the element contributes for the mass of the referred nodes.

3 PARAMETRIC STUDY

As reference, it is assumed the gas pipeline model (TRANSGÁS) of 500 m length and 12.7 mm wall thickness, in which the ground reaction coefficient is 11000 kN/m³ and the liquefaction zone spread along a 100 m width. The CLC network is also modelled because it is the only oil network existing in the zone under study, besides the fact that it presents a diameter of about half (406.4 mm) of the gas network modelled (711.2 mm). Also implemented is the model for pipelines having different wall thickness (8.7 mm, 11.1 mm and 15 mm) from the considered one (12.7 mm) either for the base model of the mentioned network and for the oil pipelines. It is also carried out a study intending to analyse the influence of soil's rigidity variation on results. For that it is used the reference model considering k_2 values of about 5000 kN/m³ and 50000 kN/m³. Table 2 shows the values obtained for the soil's rigidity constants.

It is also studied the influence of liquefaction width. As the width considered in the base model is 100 m, it is adopted two additional situations corresponding to 10 m and 50 m liquefaction width.

In the following it is presented the graphic representations of the two analyses carried out for the various parametric studies adopted. In the first case, the variation of the external diameter of the pipeline is studied, assuming as constants the wall thickness (12.7 mm), the soil reaction coefficient of 11000 kN/m³ and the 100 m length of the liquefied zone shown in Figure 6.

Comparing the values obtained in the two analyses it is possible to conclude that both for smaller external diameter values (406.4 mm) and higher values (711.2 mm) the analysis that leads to higher design stresses is analysis A_2 for a damping coefficient value of 0%. On the analyses under study, the one that always leads to smaller strength values, and therefore less probable to induce breaks in pipelines, is analysis A_2 in which a damping of 20% is assumed. Also to be mentioned is an almost coincidence between design strength obtained for a diameter of 406.4 mm in analysis $A_1\_2\%$, $A_1\_20\%$ and $A_2\_2\%$. In what concerns the largest studied diameter (711.2 mm) it is observed similar results in analysis $A_1\_2\%$ and $A_1\_20\%$. More reliable results could be obtained if it has been studied intermediate diameters, although the studied diameters of pipelines are the most significant in the area defined for the study of the gas and oil networks.

Similar to what is made for the external diameter variation, several analyses are compared when varying the pipeline wall thickness (Fig. 7). The external diameter corresponding to this parametric study is 711.2 mm (gas network), the soil's reaction coefficient is 11000 kN/m³ and the width of the liquefaction zone is 100 m.

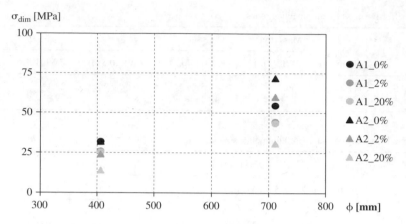

Figure 6.   Design stresses as function of the external diameter of the pipeline.

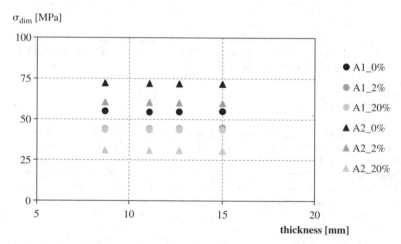

Figure 7.   Design stresses as function of pipeline wall thickness.

As it can be verified in Figure 7, analysis $A_2\_0\%$ is the one, which expresses higher, and therefore more serious, values of design stress, whatever is the thickness of the pipeline the analysed value. On the other hand, it is verified that the analysis that gives more reduced stress values is $A_2\_20\%$. This difference within the same analysis may be justified by the fact the mass considered for the ensemble soil/pipeline is responsible for the increase of inertia and consequently for the increase of stress in the pipeline in a non damped situation, whereas considering a higher value for the damping coefficient of the ensemble. An almost perfect coincidence between $A_1\_2\%$ and $A_1\_20\%$ is also noticed, therefore it is possible to conclude that the damping values doesn't influence in a significant way an analysis when the action is considered as a displacement history. In general design stress values obtained in analysis A_1 and A_2 are constant for any variation of the pipeline's wall thickness.

For the situation in which the soil's reaction coefficient and consequently the soil's rigidity varied, mantaining constant the external diameter of the pipeline (711.2 mm), the thickness of the pipeline wall (12.7 mm) and the length of the liquefaction zone (100 m), the following results are obtained and presented in Figure 8.

In general it is noticed that the design stress values decrease when the soil's reaction coefficient, k_2, increases. This way, for increasing k_2 values, the soil presents a higher rigidity, and so the correspondent internal forces tend to be to be smaller as the k_2 coefficient increases. A possible explanation for what

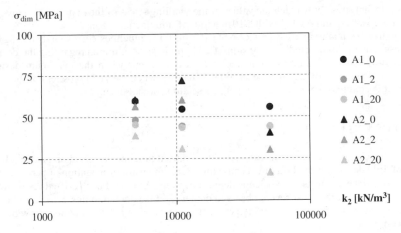

Figure 8.   Design strength as function of the soil's reaction coefficient.

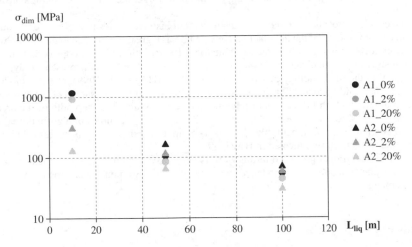

Figure 9.   Design stresses as function of ground liquefaction length.

happens is that relative deformation between the liquefaction zone and the remaining zones reduces, thus leading to stress values smaller. The analysis performed lead to higher design stress values, both for the more reduced reaction coefficient values ($5000 \, kN/m^3$) as well as for the highest one is $A_1\_0\%$, while the dominant analysis for intermediate values is $A_2\_0\%$. For low damping the values of the design strength obtained in the analysis A_2 can result from an excessive value of dynamic amplification resulting from resonance phenomena that might have appeared as a result of the alteration of the vibration frequencies induced by the variation of soil's reaction coefficient. For better results it would be necessary to make a more exhausting parametric study for intermediate values of damping and k_2. In what concerns results obtained in $A_2\_20\%$, it can be stated that in any circumstance they are the ones that raise more reduced design stress values and therefore generate lower probabilities for breaks to occur in buried pipelines.

Finally, making vary the width of the zone where liquefaction occurs (L_{liq}) and admitting the external diameter (711.2 mm), the thickness of the pipe wall (12.7 mm) and the reaction coefficient constant (11000 kN/mm^3) different analyses are considered. The results obtained are presented in Figure 9.

Stress values decrease in what concerns the analyses made, whatever the damping coefficient considered, weakens as the width of the liquefaction zone increases. A close proximity is noticed in what concerns the design stress obtained in all analyses where a 100 m width of liquefied zone is considered, although in analysis $A_2\_20\%$ a considerable deviation is observed. As it can be verified, the values obtained for design stress only reach the values of the yielding stress of the material in analysis A_1 (for all the damping values) and $A_2\_0\%$ with a 10 m width of liquefaction's zone.

Yielding stress of materials used in TRANSGÁS and CLC pipelines corresponds to 482 MPa and 448 MPa respectively. In this study a very simplified hypothesis is admitted regarding the definition of pipelines ruptures. It is adopted that pipeline' breaks only occur when the maximum design stress reaches the yielding stress of the pipeline material. This way, when the yielding stress of the material is not reached it is admitted that ruptures in pipeline's sections do not occur.

4 CONCLUSIONS

In this study the evaluation of seismic vulnerability of buried pipelines is based on analytical analyses. The vulnerability definition can be a direct cause of the seismic wave propagation or permanent ground deformations, and these effects must be considered separately. As it has been verified that PGD are systematically the cause of the most severe observed damages, it is adopted in this work only the effect of permanent ground deformation due to liquefaction phenomena. The results obtained allow reaching the following conclusions:

- The differences observed between A1 and A2 are mainly because of the input is displacement and acceleration, respectively. Besides that is important not to forget the value of damping considered in each analysis;
- The analyses A1, corresponding to impose displacement histories, leads to more coherent outcomes;
- The influence of variation of the liquefaction zone width has a great importance in the final results. The change from a 100 m width of liquefaction zone to an 10 m width leads to a considerable increase of the design stress in the pipelines, belonging to both analysed networks in the present study;
- Not identical to the methodology proposed by HAZUS99 (HAZUS99 1999), analytic modelling of buried pipelines could account for (as is the case in this study) factors such as the variation of the external diameter and wall thickness of pipeline, the soil's rigidity and the width of the area where liquefaction occurs. All these must not, in any case, be neglected in a model of a given buried pipeline in a given continuous environment, like the soil. The pipeline vulnerability is necessarily influenced by more factors than those indicated in HAZUS99;
- With the analytical model proposed in this work is very easy to include the non-linear behaviour of the soil. This phenomenon can be modelled by means of imposing a non-linear constitutive law for the spring elements.

REFERENCES

ERDCJ – Earthquake Resistant Design Codes in Japan 2000. Japan Society of Civil Engineers.
Falcão Silva, M.J. 2002. Vulnerabilidade sísmica de redes de gás e combustível, Tese de Mestrado, Instituto Superior Técnico, Universidade Técnica de Lisboa (in portuguese).
Guerreiro, L. 2002. Geração de séries de acelerações. DTC Report n°½ ICIST Instituto de Engenharia de Estruturas, Território e Construção. Lisbon (in portuguese).
HAZUS99 1999. Earthquake loss estimation methodology. Federal Emergency Management Agency, Washington, D.C.
Lim, Y.M., Kim, M.K., Kim, T.W., Park, A.S.W. 2000. Seismic behaviour of pipelines during transverse liquefaction considering effects of ground deformation. 12WCEE *Proceedings of the 12th World Conference of Earthquake Engineering.* Auckland, New Zeland.
RSA – Regulamento de Segurança e Acções, (1983), Decreto Lei n° 235/83, Lisboa.
Santos, J. 1999. Caracterização de solos através de ensaios dinâmicos e cíclicos de torção – Aplicação ao estudo de estacas sob acções horizontais estáticas e dinâmicas. *Tese de Doutoramento.* Instituto Superior Técnico, Universidade Técnica de Lisboa (in portuguese).
SAP2000 2000. Integrate finite element analysis and design of structures. *Computers and Structures Inc., Berkeley.* California, USA.

Structural mechanics

Computational Methods in Engineering and Science, Iu et al. (eds)
© *2003 Swets & Zeitlinger, Lisse, ISBN 90 5809 567 3*

The use of a non-linear numerical analysis to investigate the behaviour of reinforced concrete flat slabs in the vicinity of edge columns

K.A. Murray
Dundalk Institute of Technology, Dundalk, Co. Louth, Ireland

D.J. Cleland & S.G. Gilbert
Queen's University, Belfast, Northern Ireland

R.H. Scott
University of Durham, UK

ABSTRACT: The development of a non-linear numerical model of reinforced concrete in the vicinity of edge columns is presented. The model uses a "stepped" grillage analysis that idealises the slab as an interconnected assembly of beam elements. The layout of the grillage members, simulation of the boundary conditions, column support and loading are discussed. The determination and distribution of member properties associated with the different loading stages are considered. The distribution of deformations and moments in the slab and column obtained from the non-linear numerical model is compared with the experimental results obtained from a series of one-third scale experimental slabs constructed and tested at Queen's University, Belfast (QUB). The numerical model is also applied to the full-scale in-situ concrete framed building, the European Concrete Building Project, constructed by the Building Research Establishment at Cardington, England.

1   INTRODUCTION

A flat slab consists of a reinforced concrete slab that is directly supported by concrete columns without the use of intermediate beams. The design of these structures is governed by national codes of practice that have developed as a result of empirical research. The bulk of this research to date has tended to concentrate on interior slab-column connections with little investigation of the connection between the slab and the exterior column having been conducted. This exterior connection is more critical than at the interior column due to unbalanced loading effects and the fact that the edge column has a smaller peripheral area to resist the loading.

A computational model was developed (Murray 2001) to simulate the behaviour of the test slabs. The computational modelling of reinforced concrete has the advantage over experimental modelling that the effect of variables can be studied more economically.

2   TEST PROGRAMME

2.1   *Scaled test slabs at Queen's University Belfast*

Nine reinforced concrete slabs were tested to destruction in order to investigate the behaviour of the flat slab in the vicinity of edge columns, throughout the entire loading history. The main characteristics of the series are presented in Table 1. The test models were constructed at one-third scale to represent a portion of a flat slab structure. The dimensions of the test models are as shown in Figure 1. This model type was used previously by Gilbert & Glass (1987) and was found to behave in a manner that closely represents

Table 1. Characteristics of test series.

Series	Model	c_1 (mm)	c_2 (mm)	c_1/c_2	H_c (mm)	ρ_c (%)	ρ_s (%)	f_{cu} (MPa)	f'_c (MPa)	f_t (MPa)
1	S1-1	100	200	0.5	540	0.792	0.643	45.10	41.51	3.07
	S1-2	100	150	0.67	540	0.792	0.643	42.90	38.18	3.24
	S1-3	100	100	1	540	0.792	0.643	43.74	39.94	3.39
	S1-4	150	100	1.5	540	0.792	0.643	42.70	37.76	2.96
	S1-5	200	100	2	540	0.792	0.643	47.32	43.31	3.65
	S1-6	250	100	2.5	540	0.792	0.643	41.20	36.18	3.22
2	S2-1	100	200	0.5	540	1.715	0.643	47.98	43.84	3.59
	S2-2	100	150	0.67	540	1.715	0.643	52.80	50.69	4.12
3	S3-4	150	100	1.5	540	0.792	0.643	45.09	41.50	3.58

c_1 is the column side perpendicular to free edge; c_2 is the column side parallel to free edge; H_c is the column height; ρ_c is the reinforcement concentration in column strip; ρ_s is the reinforcement concentration in middle strip; f_{cu} is the characteristic cube compressive strength; f'_c is the equivalent cylinder compressive strength; f_t is cylinder tensile strength.

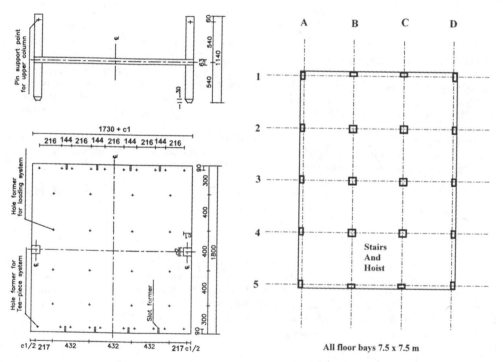

Figure 1. Dimensions of QUB test models (mm).   Figure 2. Cardington slab floor layout.

that of a real structure, particularly in relation to slab deformation at mid-span in each direction, and in relation to the shape of the failure cone when punching occurs. A unique feature of these test series was the incorporation of internally strain gauged reinforcing bars in all the test specimens. The technique for installing internal strain gauges in reinforcement was developed by Scott & Gill (1987) at the University of Durham, England. The main advantage of the system is that the strain gauges can be arranged at much closer intervals along the bar so that a more detailed and reliable picture of the behaviour of the reinforcement can be obtained. Fourteen of these reinforcing bars were fixed in each specimen; at the top and bottom of the slab at each of the locations. Therefore strain could be measured at top and bottom at the same location on plan and at very close intervals along the bar which permitted the calculation of slab curvature and the corresponding value of bending moment. Other measurements made on each specimen were slab deflection, slab rotation along the free edge, column deflection and column end reactions. To simulate a uniformly distributed load, load was applied to the slab at 16 points through hydraulic jacks and a system of spreader beams.

2.2 Full-scale test at European Concrete Building Project, Cardington, England

The Building Research Establishment constructed a seven storey full-scale in-situ concrete framed building, the European Concrete Building Project, at its test facility at Cardington, England. A programme of service load testing was undertaken on the sixth floor of the building.

Seven internally strain gauged reinforcing bars were cast into an internal and external bay of one floor during the construction process. Internal bar strain readings and slab deflections were measured during a series of six tests. Construction of the test facility at Cardington provided an ideal and unique opportunity to correlate data from a full scale building with the results generated from the stepped grillage model developed in this work and which was based on scaled laboratory test slabs. The Cardington floor layout is shown in Figure 2.

3   THE NUMERICAL MODEL

3.1   Grillage analysis

The computational method which uses beam elements only for the analysis of slabs is known as grillage analysis (GA). This model has been widely used in the analysis and design of slab and pseudo-slab bridge decks. Whittle (1994) published recommendations on the use of GA in flat slab design. It is convenient for the engineer to visualise and prepare the data required for the grillage analysis. This makes it easier than finite element analysis to vary member properties at any loading stage. This is particularly true for the members which frame into the side faces of the column where the progressive reduction in the flexural and torsional stiffnesses can be applied effectively in a stepped manner.

Grillage analysis software called "Enfram" produced by Encad Systems Ltd. was employed. The software allowed for a maximum of 500 joints and 60 section properties. The input data was the modulus of elasticity, unit weight of the concrete and the section properties of depth, area, flexural and torsional stiffnesses. Inputting the stiffnesses directly allowed a non-linear analysis to be conducted.

3.2   Member layout

The layout of the grillage members was based on the dimensions of the structural models tested at QUB. A fine division of the slab was adopted, especially at the slab–column connection because of the need to model the slab cracking at the column in detail. With regard to the QUB slabs, a half panel of the model was considered by taking advantage of symmetry. Figure 3 presents a typical layout for test model S1-4. By symmetry, the boundary conditions were set out as follows:

(i) All the joints along the continuous edge lines (nodes 1 to 106 and 15 to 120) have no rotation about the x-axis. They are free to rotate about the y-axis and are free to deflect.
(ii) All the joints along the transverse symmetry line (nodes 106 to 120) have no rotation about the y-axis. They are free to rotate about the x-axis and are free to deflect.
(iii) All the joints along the free edge line (nodes 1 to 15) are unrestrained.

Originating from the columns, there are three parallel grid lines along the longitudinal span. A similar pattern was applied to the arrangement of three grid lines through the column region in the transverse direction. The provision of the three members framing into the sides of the columns has two significant advantages over a single member connection. Firstly, the effect of the progressive torsional cracking of the slab at the side face of the column can be modelled more realistically. Secondly, the progressive application of pin joint connections between the front face of the column and the three connecting slab members permits the progressive yielding of the slab to column connection in the later stages of the loading history to be modelled effectively. Near the column to column centre line, a finer grid division is adopted than that close to the continuous edges in order to produce a higher accuracy of analysis within the column strip, which is believed to have a stronger influence than the middle strip on the behaviour of the whole structure.

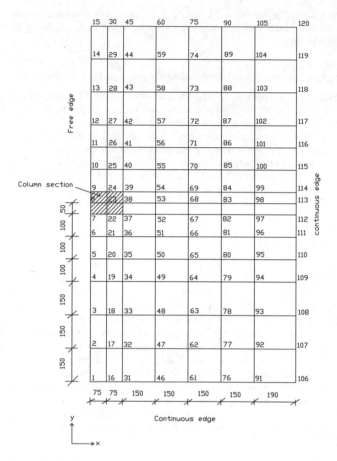

Figure 3. Grillage layout for QUB slab S1-4.

3.3 *Simulation of the columns*

The modelling of the restraint provided by the columns is of critical importance in attempting to simulate accurately the behaviour of the slab–column connection. A widely used representation of the column restraint is a single central point support which is either rigid or elastic. Other researchers proposed modelling the column with four springs, one spring at each joint coinciding with a corner.

The QUB laboratory test models in this project contained both upper and lower columns with a pin support at mid-storey height. An investigation of six different computer column models was conducted in order to ascertain the system which best simulated the behaviour of the slabs. Figure 4 illustrates the different column simulations which were investigated. Typical elastic uncracked grillage analysis results are presented in Figure 5. Further investigation of the simulation of moments in both the slab and the column revealed that the uniform thickness column model with a single central support was the most accurate. As simulation of moments was of importance in developing a prediction method for the failure load in the slab, this column model was adopted. Close examination of Figure 5 indicates a change in slope of the experimental data at a load of approximately 6 kN. This was later deemed to be the load at which the uncracked linear stage of the analysis stopped and the cracked stage commenced.

3.4 *Stepped grillage analysis of the slab*

In order to simulate the behaviour of the slab throughout its entire loading history, five stages or steps were adopted to reflect the changing pattern of cracking in the slab.

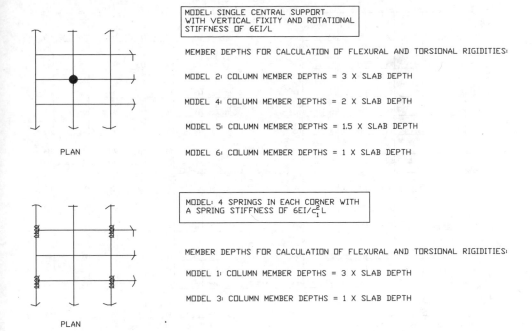

Figure 4. Column grillage models.

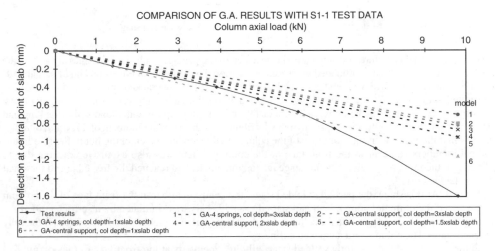

Figure 5. Effect of different column models on deformation.

Step 1: Figure 3 illustrates step 1, the elastic stage. Up to a load of 6 kN (between 12% and 17% of the final experimental loads), it was assumed that all grillage members in the model were uncracked and the full uncracked flexural and torsional stiffnesses of all the members were employed. The column was also assumed to be uncracked.

Step 2: Classical reinforced concrete analysis of the test slab section allowed calculation of the cracking moment. A study of the internal strain gauges in the steel reinforcing bars revealed which sections of the test slab had reached this cracking moment and at what load. This marked the end of the uncracked linear stage of the analysis and corresponded well with the load at which the slope of the deflection graphs started to change slope, as seen in Figure 5. The pattern of cracked grillage members could then be

359

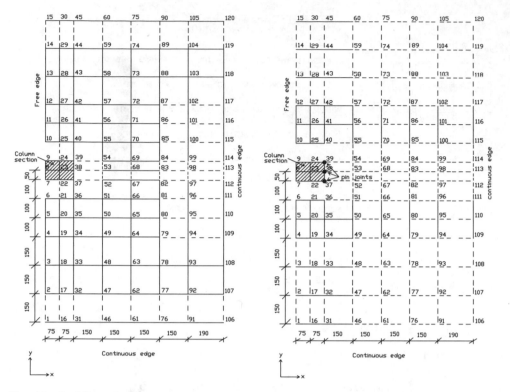

Figure 6. Partially cracked stage.　　Figure 7. Fully yielded stage.

determined from this picture and applied to the next step of the grillage analysis. This cracking affects the stiffness of the slab which is simulated in the numerical model by reducing the flexural and torsional stiff-nesses of the relevant grillage members. At a load of 6 kN, a "partially cracked" phase was introduced as illustrated in Figure 6. The members in Figure 6 drawn with a dotted line represent the cracked members.

Stage 3: At a load of 15 kN (between 30% and 43% of the final experimental loads), the cracking had developed much more fully. Investigation of the distribution of the cracking moment of 2 kNm/m or more, reveals that the bending at the front face of the column had extended the cracking out to the five longi-tudinal members adjacent to the front face of the column. There was also extensive cracking due to bending in the mid-span region. By this stage the torsion cracks had reached the free edge and torsional release had occurred.

Stage 4: At this stage, the partially yielded stage, the centre member in the front face of the column had reached the ultimate moment and had started to yield. This was simulated by inserting a pin at the joint connecting this member with the column. The distribution of the cracked members was as for the fully cracked stage. The column had a cracked stiffness.

Stage 5: At this stage, the fully yielded stage, all three members at the front face of the column had reached yield and were all modelled with pins where they connected with the column. The distribution of the cracked members was as for the fully cracked stage. The column had a cracked stiffness. The layout is shown in Figure 7.

4   RESULTS

4.1   Grillage modelling of QUB slabs

A comparison was made between the experimental and grillage deflections and rotations at critical points in the test slabs. Figure 8 illustrates a typical comparison. The different stages of the non-linear grillage analysis can be seen along with the effect that cracking has on the data. It can be seen that there is good

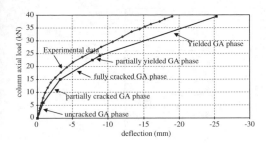

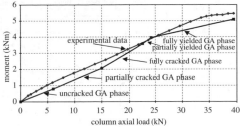

Figure 8. Comparison of experimental and 5 step GA data – deflection at panel centre of slab S1-4.

Figure 9. Comparison of experimental and 5 step GA data – moment in column of slab S1-4.

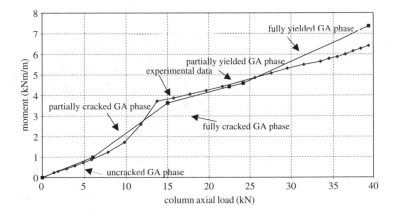

Figure 10. Comparison of experimental and 5 step GA data – moment in panel centre of slab S1-4.

correlation up to the end of the partially cracked phase and thereafter the grillage becomes more flexible than the test model. In general there was excellent correlation up to an average load of 64% of the final test load.

Figure 9 illustrates a typical comparison of the moment in the column as found in the test slab and as simulated by the grillage analysis. It can be seen that there is excellent comparison throughout the different phases of the analysis.

Figure 10 illustrates the excellent correlation between the moment as found in the test slab and that as simulated by the grillage analysis throughout the different phases of the analysis.

4.2 Grillage modelling of Cardington slab

The Cardington slab provided an ideal and unique opportunity to correlate the results from the stepped grillage model developed in this project and based on laboratory scaled test slabs with the experimental data obtained from the full-scale building. Figure 11 illustrates a typical comparison of experimental and computed deflections.

Figure 12 illustrates a typical comparison between the experimental and computed moments. It is interesting to note the large reduction in experimental moment on the side of column C3 opposite to the loading. This is due to absorption of the moment by the interior column. This phenomenon is also reflected in the grillage output albeit to a much greater extent.

5 CONCLUSIONS

1. A grillage layout is proposed with three longitudinal and transverse members framing into the column to allow simulation of progressive yielding and torsional cracking.
2. A column model is proposed with a single central support with vertical fixity and rotational flexibility.

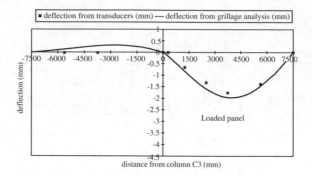

Figure 11. Comparison of experimental and GA data – test A deflection along grid line 3 of Cardington

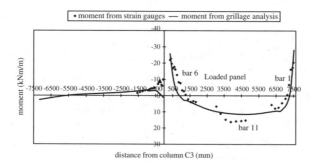

Figure 12. Comparison of experimental and GA data – test A moment along grid line 3 of Cardington.

3. The layout and stiffness of cracked grillage members is based on classical reinforced concrete analysis.
4. A 5 step grillage analysis is proposed that simulates initial uncracked linear and subsequent non-linear cracked behaviour.
5. In general there was excellent simulation of deformation and moment in both the scaled QUB test slabs and the full scale Cardington slab.
6. The proposed numerical model provides a useful design tool for engineers particularly for loading up to design ultimate.

REFERENCES

Gilbert, S.G. & Glass, C. 1987. Punching failure of reinforced concrete flat slabs at edge columns. *The Structural Engineer* Vol. 65B, No. 1, March 1987: 16–21.
Murray, K.A. 2001. An investigation of the behaviour of reinforced concrete slabs in the vicinity of edge columns. *Ph.D. thesis*, Queen's University Belfast.
Scott, R.H. & Gill, P.A.T. 1987. Short-term distributions of strain and bound stress along tension reinforcement. *The Structural Engineer* Vol. 65B, No. 2, June 1987: 39–43.
Whittle, R.T. 1994. Design of reinforced concrete slabs to BS 8110. *CIRIA report 110*.

Computational Methods in Engineering and Science, Iu et al. (eds)
© 2003 Swets & Zeitlinger, Lisse, ISBN 90 5809 567 3

Object-oriented finite element programming for reinforced concrete nonlinear analysis

F. Jiang, B. Li & L. Ding
School of Civil & Hydraulic Engineering, Dalian University of Technology, China

A. Deeks
School of Civil and Resource Engineering, University of Western Australia

ABSTRACT: Classes in object-oriented programming are made for the reinforced concrete nonlinear finite element structural analysis. The program developed in this paper is based on reinforced concrete nonlinear finite element structural analysis models and MFC (Microsoft Foundation Class Library). These classes include concrete element class, bar element class, bond element class and nonlinear element class. They construct a new reinforced concrete finite element analysis system, combining with other basic classes related to the concepts of finite element method. The descriptions and implement approach for the classes also are given in detail. All the work finished by this paper provides an approach to employ the more reasonable object-oriented programming for the reinforced concrete nonlinear finite element structural analysis.

1 INTRODUCTION

Finite element methods viewed as effective methods have been widely applied in engineering designs and research work. Traditional finite element analysis programs are used to be created by means of procedural-oriented programming and program structures have a module characteristic. Every section of structured programs has to handle a great number of various data according to the given algorithm. The data and the procedures to operate on the data should have close relationships but they are completely separated. Furthermore, if any data needs adding or changing in form, a serious of relevant procedural changes will happen. Therefore, the procedural-oriented programming has limited expandability and the ratio of reused codes is low. Besides it is not only very complicated to debug but also difficult to maintain. In recent years, object-oriented programming has been applied in finite element analysis. Ford proposed the conceptual model of finite element methods based on object-oriented programming and explained the object-oriented earlier methods in finite element analysis (Forde et al. 1990). Since then, many scholars have dedicated themselves to improving the object-oriented finite element methods following the Ford's idea (Dubois-Pelerin & Pegon 1990, Zimmermann, T. & Yves, D.-P., Bomme, P. 1992, Mackie 1999, Cross, Masters & Lewis 1999, Bittencourt 2000, Eyheramendy 2000, Tan, 1999). In reinforced concrete finite element analysis, reinforced concrete is a kind of composite material with such characteristics as elastic-plastic behavior, cracking and creep, and its stress–strain laws influenced by many parameters are complicated and various (Kang 1996). Moreover, computing needs nonlinear iterative procedure. Therefore, it is difficult that both data structures and computing methods are integrity handled. Improving computing methods and expanding programs is not easy, either. But, object-oriented programming has many advantages in these aspects. Recently, some scholars begin to study the data design in reinforced concrete analysis making use of object-oriented programming (Jeremic & Sture 1998).

In this paper, classes related to reinforced concrete finite element analysis are made based on the idea proposed before and programs for reinforced concrete structure nonlinear analysis are constructed with these classes in Visual C++ 6.0. Compared with programs constructed in traditional methods, the program expandability can be increasingly improved, the program can be maintained more conveniently and the period of programming and debugging can be reduced. In addition, pre- and post-processing functions can be easily added and interfaces for other pre- and post-processing programs can be built.

2 OBJECT-ORIENTED FINITE ELEMENT METHODS

2.1 *Main characteristics of object-oriented programming*

Object-oriented programming is a method that describes practical problem in an object-oriented view. Then the problem will be described in a computer language and handled. The description and handling arc realized by classes and objects, which are the high-level conclusion, classification and abstraction for

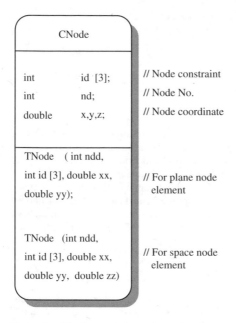

Figure 1. Member varieties and member functions of CNode class.

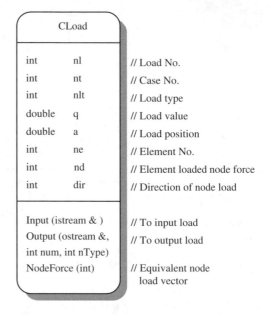

Figure 2. Member varieties and member functions of CLoad class.

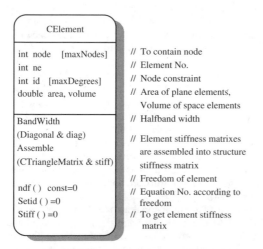

Figure 3. Member varieties and member functions of CElement class.

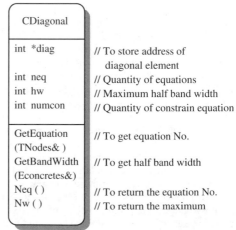

Figure 4. Member varieties and member functions of CDiagonal class.

364

practical problems. A class is an encapsulation including functions and data between which there are logical relationships between them. In fact, a class can be considered as a data type defined by you. The difference from other basic data type is that the class, a kind of special type, includes the functions operating on data. In object-oriented programming, a program module is composed of classes. Object-oriented programming has the main character as follows:

1. Abstraction: It means a process that specific problems are generalized and then common characteristics of the objects are abstracted and described.
2. Encapsulation: Object-oriented programming encapsulates specific kinds of data with the specific procedures to operate on that data.
3. Inheritance: C++ languages provide inheritance mechanism for classes. Classes can inherit both data and procedures from their ancestors. Developers can give objects more descriptions based on the inherent characteristics of the classes.
4. Polymorph: The functions having similar behaviors in the classes can occupy the same name.

2.2 *Definitions and implement methods of the main basic classes in finite element analysis*

The main basic classes in finite element analysis include CNode class (Fig. 1), CLoad class (Fig. 2), CElement class (Fig. 3), CDiagonal class (Fig. 4) abstracted from the coral conception of finite element methods. They also include the classes such as matrix class, vector class and equation class.

The basic classes usually can be combined with the tool classes in MFC and build data structures such as arrays, lists and so on, so it is easy to expand functions of the basic classes. The principal classes in object-oriented reinforced concrete finite element programming can be made by means of inheritance of the basic classes and be applied in reinforced concrete nonlinear finite element analysis.

3 REINFORCED CONCRETE FINITE ELEMENT ANALYSIS PROGRAMMING

3.1 *Ground plan of programming*

In reinforced concrete finite element analysis, element models, stress–strain laws and criteria of failure of materials, which have to be applied in a certain domain, are all complicated and various. Therefore, to develop a general reinforced concrete finite element analysis software, it is a good ideal that those often used element models, stress–strain laws and criteria of failure of materials are all put into the same software so that they can be chosen to use according to users' requirements. Figure 5 shows the ground

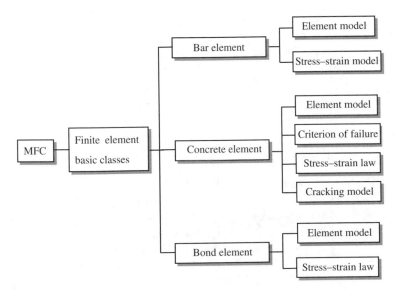

Figure 5. Relationship among all classes in program.

365

plan of the idea. It is difficult to implement the plan in procedural-oriented programming. Classes are abstracted from all kind of methods by means of abstraction, encapsulation, inheritance and polymorph, encapsulating data members and function members. Users can only care about interface data regardless of inner details.

Local expanding of programs will not affect other parts of primitive programs because of dependence of every class.

3.2 *Instance program*

A practical program is given in order to illustrate how to realize the plan.

- Computing models are separate models, concrete elements are triangle elements, bar elements are line elements. Double spring bond elements are set up between reinforced elements and concrete elements.

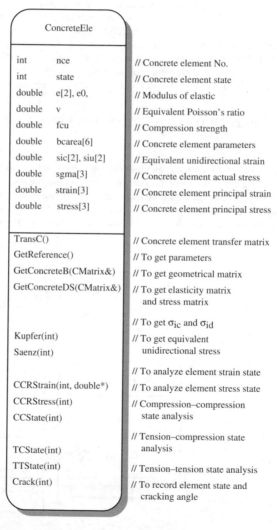

Figure 6.   Member varieties and member functions of ConcreteEle class.

- The Stress–strain law for concrete is the Darwin orthotropic increment model; the criterion of failure is the Kupfer criterion of failure; bars are assumed to be idealized elastic-plastic mass. The bond–slop relations between concrete and bars are viewed as the Houde-Mizra empirical equation involving the effect of concrete compressive strength.
- The Computing method is the variable matrix increment iteration method.
- Programming in Visual C++ 6.0 is based on message handling mechanism provided by Windows programming.

Figures 6, 7 and 8 show the main varieties and implement method of ConcreteEle class, BarEle class and BindEle class respectively.

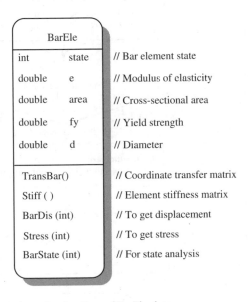

Figure 7.   Member varieties and member functions of BarEle class.

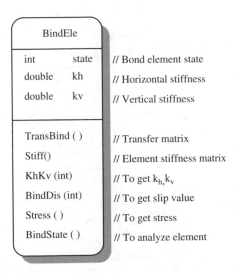

Figure 8.   Member varieties and member functions of BindEle class

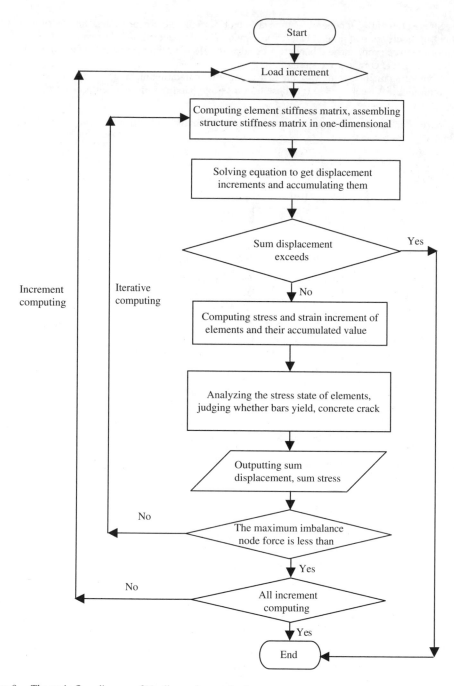

Figure 9.   The main flow diagram of Nonlinear class method.

3.3   *Nonlinear model*

Problems concerning nonlinear computing are usually complicated in reinforced concrete finite element analysis programming. In the course of nonlinear increment iteration, a great deal of information on finite element basic classes, ConcreteEle class, BarEle class and BindEle class is needed. It is necessary that a

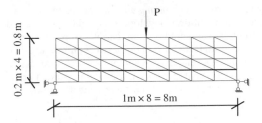

Figure 10. Map of the structural form and the element mesh.

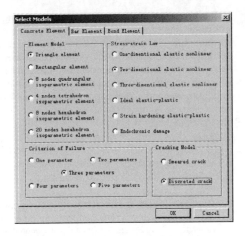

Figure 11. Interface for selecting the model of concrete element.

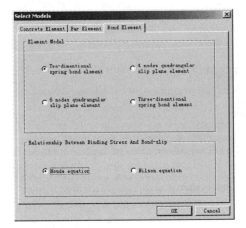

Figure 12. Interface for selecting the model of bond element.

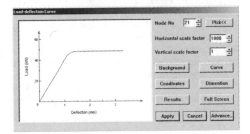

Figure 13. Relationship between loads and deflection in the center of span.

Nonlinear class for nonlinear computing is created. To input data and to output results are realized through other interfaces. The data interfaces are combined with CArchive class, CView class in MFC and with other software like AutoCAD. This kind of program like this has not only good transplantation but also good pre- and post-processing functions. The variable matrix increment iteration method is used in the instance program. The main flow chart is shown in Figure 9.

4 EXAMPLE

The diagram of structural form of beam with simple supported ends and element mesh is shown in Figure 10. The concrete is C30 and the bars are 2Φ20. Computing model interfaces are shown in Figure 11 and Figure 12. The center of span is subjected to 50 KN concentrated force. The relationships between loads and deflection in the center of span are shown in Figure 13.

5 CONCLUSION

In this paper, reinforced concrete nonlinear finite element analysis programs are constrcted by means of object-oriented programming in Visual C++ 6.0. Nonlinear analysis classes, concrete element classes and bond element classes are made on the basis of the work proposed by other scholars. These developed classes provide more convenience for the development of reinforced concrete finite element analysis software in object-oriented programming.

It is by means of object-oriented programming that the errors in debugging are constrained in local part of classes and programs can be maintained conveniently. In addition, developers can add new data and methods into classes but do not have to change inherent data and methods, so the ratio of reused codes is high and the results from various reinforced concrete models can be compared with one another. Functions that hide the inner detail are easy to be assembled. Developers can add pre-and post-processing functions making full use of MFC. All advantages of object-oriented programming demonstrated in the system developed in this paper make it possible to explore a better visual reinforced concrete nonlinear finite element structural analysis system.

REFERENCES

Bittencourt, M.L. 2000. Using C++ templates to implement finite element classes. *Engineering Computations.* 17(6–7): 775–788.

Cross, J.T., Masters, I., & Lewis, R.W. 1999. Why you should consider object-oriented programming techniques for finite element methods. *International Journal of Numerical Methods for Heat & Fluid Flow.* 9(3): 333–347.

Dubois-Pelerin, Y. & Pegon, P. 1990. Object-oriented programming in nonlinear finite element analysis. *Computers & Structures.* 67(1): 25–241.

Eyheramendy, D. 2000. An object-oriented hybrid symbolic/numerical approach for the development of finite element codes. *Finite Elements in Analysis and Design.* 36(3–4): 315–334.

Forde, B.W.R., Foschi, R.O. & Stiemer, S.F. 1990. Object-Oriented finite element analysis. *Computers & Structures.* 34(3): 355–374.

Jeremic, B. & Sture, Stein. 1998. Tensor objects in finite element programming. *International Journal for Numerical Methods in Engineering.* 41: 113–126.

Kang, Q. 1996. *Finite element analysis in reinforced concrete. Beijing: China.* Water Resources and Hydropower Press.

Mackie, R.I. 1999. Object-oriented finite element programming-the importance of data modelling. *Advances in Engineering Software.* 30(9–11): 775–782.

Tan, Y. 1999. *Computing in industry structure and OOP programming. Beijing: China.* Building Material Industry Press.

Zimmermann, T. Yves, D.-P. & Bomme, P. 1992. Object-oriented finite element programming: I. Governing principles. *Comput. Meth. Appl. Mech. Engin.* 98(2): 291–303.

Computational Methods in Engineering and Science, Iu et al. (eds)
© *2003 Swets & Zeitlinger, Lisse, ISBN 90 5809 567 3*

Optimal design of composite thin-walled beam structures with geometrically nonlinear behavior

J. Barradas Cardoso & L.G. Sousa
Instituto Superior Técnico, Lisboa, Portugal

A.J. Valido
Escola Superior de Tecnologia, Setúbal, Portugal

ABSTRACT: This paper presents a finite element model for optimal design of composite laminated thin-walled beam structures with geometrically nonlinear behavior, including post-critical behavior, where the beams are closed cross-section Euler-Bernoulli beams. A general continuum formulation based on the virtual work principle, considering the Updated Lagrangean procedure, is used to describe the deformation of the structure. The structural discretization is formulated throughout three-dimensional two-node Hermitean finite beam elements. The beams are made from an assembly of thin flat-layered panels. Design sensitivities of the structural performance cost and constraint functionals, namely displacement and stress (failure index) are imbedded into the finite element modeling and assembled in order to perform the structural design sensitivity analysis by the adjoint structure method. Design optimization is performed throughout nonlinear programming techniques. The lamina orientation and the laminate thickness are selected as the design variables.

1 INTRODUCTION

Today, the use of composite materials is generalized to a large variety of different industries due to their good ratios of strength/weight and stiffness/weight. From aerospace to automotive, from naval to construction industries, there are attempts to replace classical materials like steel and concrete by composite materials. With the modern design requirements for using the materials in extreme environments, the structural behavior may be highly nonlinear, allowing large displacements.

Several works have been presented during the last years related to the analysis and/or optimization of thin-walled composite beams (Bauld & Tzeng 1984, Woolley 1989, Bhaskar & Librescu 1995, Floros & Smith 1997, Davalos & Qiao 1999, Lee & Kim 2001, Valido et al. 2001, Valido 2001).

The aim of this work is to present a finite element model for analysis and optimization of thin-walled composite laminated beams, with geometrically nonlinear behavior, where the beams are closed cross-section Euler-Bernoulli beams. Since the beams are long enough and have closed cross-section, then they have no significant warping as well as not significant shear deformation.

An Updated Lagrangean description (Bathe & Bolourchi 1979) and a generalized displacement control method (Yang & Shieh 1990) are used to describe the nonlinear structural deformation.

The structural discretization is formulated throughout three-dimensional two-node Hermitean finite beam elements. The beam cross-sections are made from an assembly of thin flat-layered panels (Datoo 1991). The cross-section stiffness is dependent both on its geometry and on the individual panel stiffness, which depend by their side on the laminate configuration and lamina material distribution. In this case, effective cross-section stiffness properties have to be calculated.

The adjoint structure method is used to derive the design sensitivities. The elemental sensitivities are assembled for the entire structure in order to calculate the design sensitivities of the structural performance cost and constraint functionals, namely displacement and maximum failure index defined according with Tsai-Hill theory. Design optimization is performed throughout nonlinear programming techniques. Lamina orientation and laminate thickness are treated as design variables.

2 NONLINEAR STRUCTURAL ANALYSIS

Using the Updated Lagrangean formulation to describe the motion of the continuum, the governing equation of a body at the time (load level) $t + \Delta t$ is

$$\int_t^{t+\Delta t} \mathbf{S} \cdot \delta_t^{t+\Delta t} \boldsymbol{\varepsilon} \, d^t V = \int_t^{t+\Delta t} \mathbf{f} \cdot \delta^{\,t+\Delta t} \mathbf{u} \, d^t V + \int_t^{t+\Delta t} \mathbf{T}^0 \cdot \delta^{\,t+\Delta t} \mathbf{u} \, d^t \Gamma_T \tag{1}$$

where δ refers to arbitrary variation, '$\cdot$' refers to the standard tensor product, $_t^{t+\Delta t}\mathbf{S}$ = 2nd Piola-Kirchhoff stress tensor; $_t^{t+\Delta t}\boldsymbol{\varepsilon}$ = Green-Lagrange strain tensor, $_t^{t+\Delta t}\boldsymbol{\varepsilon}$ = Green-Lagrange strain tensor; $_t^{t+\Delta t}\mathbf{f}$ = body force, $_t^{t+\Delta t}\mathbf{T}^0$ = prescribed surface traction and $^{t+\Delta t}\mathbf{u}$ = displacement field. The left superscript and the left subscript stand for the configurations where the quantities are measured and referred, respectively.

Equation (1) is solved by using an incremental/iterative procedure assuming the equilibrium for all time steps, from the initial time to time t, has been obtained, and the equilibrium solution at time $t + \Delta t$ is required. Considering the incremental decomposition of the different quantities and linearization, we get the incremental virtual work equation

$$\int \left({}_t\mathbf{S} \cdot \delta_t \boldsymbol{\varepsilon}_L + {}^t\boldsymbol{\sigma} \cdot \delta_t \boldsymbol{\varepsilon}_N \right) d^t V = \int_t \mathbf{f} \cdot \delta \mathbf{u} \, d^t V + \int_t \mathbf{T}^0 \cdot \delta \mathbf{u} \, d^t \Gamma_T \tag{2}$$

where $^t\boldsymbol{\sigma} = {}^t\mathbf{S}$ is the Cauchy stress measure and the Green-Lagrange strain tensor increment is given as

$$_t\boldsymbol{\varepsilon} = {}_t\boldsymbol{\varepsilon}_L + {}_t\boldsymbol{\varepsilon}_N, \quad {}_t\boldsymbol{\varepsilon}_L = {}_t\beta(\mathbf{u}^T), \quad {}_t\boldsymbol{\varepsilon}_N = {}_t\eta(\mathbf{u}^T, \mathbf{u}^T)/2 \tag{3}$$

with the operators

$$_t\beta(\,) = \frac{1}{2}\left\{ \left[{}_t\nabla(\,) \right] + \left[{}_t\nabla(\,) \right]^T + \left[{}_t\nabla(\,) \right]\left[{}_t\nabla\left({}^{t+\Delta t}\mathbf{u}^T \right) \right]^T + \left[{}_t\nabla\left({}^{t+\Delta t}\mathbf{u}^T \right) \right]\left[{}_t\nabla(\,) \right]^T \right\}$$

$$_t\eta(\mathbf{a},\mathbf{b}) = \frac{1}{2}\left\{ \left[{}_t\nabla(\mathbf{a}) \right]\left[{}_t\nabla(\mathbf{b}) \right]^T + \left[{}_t\nabla(\mathbf{b}) \right]\left[{}_t\nabla(\mathbf{a}) \right]^T \right\} \tag{4}$$

and ∇ the space gradient operator.

3 DESIGN SENSITIVITY ANALYSIS

In the present work, design sensitivity analysis has been performed by the adjoint structure method, which may be generally defined with basis on a variational principle (Arora & Cardoso 1992). This way, consider the augmented functional

$$L(\mathbf{b}, \mathbf{z}, \mathbf{z}^a) = \Psi(\mathbf{z}, \mathbf{b}) + W^a(\mathbf{b}, \mathbf{z}, \mathbf{z}^a) \tag{5}$$

where $\Psi(\mathbf{z}, \mathbf{b})$ is a general performance functional, representing an objective function or a constraint, and $W^a(\mathbf{b}, \mathbf{z}, \mathbf{z}^a)$ is the equilibrium state equation after the compatible adjoint fields $\mathbf{z}^a$ have substituted the virtual fields. The total design variation of L is

$$\bar{\delta}L = \bar{\delta}\Psi + \bar{\delta}W^a \tag{6}$$

Satisfying the equilibrium after a design variation, $\bar{\delta}W^a = 0$, the Equation (6) is reduced to

$$\bar{\delta}\Psi = \bar{\delta}L = \bar{\bar{\delta}}L + \tilde{\delta}L = \bar{\bar{\delta}}L \tag{7}$$

after one imposes L to be stationary with respect to the primary state field $\mathbf{z}$, i.e., $\tilde{\delta}L = 0$, this constituting the equilibrium equation of the adjoint structure. The application of this approach to a nonlinear structure leads to an auxiliary problem which stiffness is the tangent stiffness of the primary structure at load level where sensitivity is required:

$$\int \left({}^t\boldsymbol{\sigma} \cdot {}_t\beta(\mathbf{u}^{aT}, \tilde{\delta}^t\mathbf{u}^T) + \boldsymbol{\sigma}^a \cdot \tilde{\delta}_t\boldsymbol{\epsilon} \right) d^t V = \int \mathbf{f}^a \cdot \tilde{\delta}^t\mathbf{u} \, d^t V + \int \mathbf{T}^{a0} \cdot \tilde{\delta}^t\mathbf{u} \, d^t \Gamma_T \tag{8}$$

The Equation (8) is the adjoint structure equilibrium equation, which is a linear equation on the adjoint fields but dependent on the primary state fields at load level t, corresponding to the performance measure Ψ. The explicit sensitivities of the equilibrium equation are calculated step-by-step along with the incremental analysis procedure (Valido 2001).

4 FINITE ELEMENT MODEL

The structural discretization is formulated throughout three-dimensional two-node Hermitean finite beam elements. The Bernoulli-Euler hypothesis is considered. The element may have large displacements and rotations, but strains are small. Using the finite element modeling, the Equation (1) of incremental virtual work becomes

$$^t\mathbf{K}\,\mathbf{U} = {}_t\mathbf{P} \tag{9}$$

where $_t\mathbf{P}$ is the incremental vector of external forces, $^t\mathbf{K} = {}^t\mathbf{K}_L + {}^t\mathbf{K}_{NL}$ is the tangent stiffness matrix (with its linear and nonlinear parts) and $\mathbf{U}$ is the incremental displacement vector.

If $\mathbf{U}^a$ is the adjoint vector corresponding to the measure Ψ requiring sensitivity analysis, the adjoint equilibrium equation, Equation (8), is discretized as

$$^t\mathbf{K}^T\mathbf{U}^a = \left(\Psi,_U\right)^T \tag{10}$$

where $^t\mathbf{K}$ is the primary tangent stiffness matrix at the time t.

The beams are made from an assembly of thin flat-laminate panels (Figure 1). A beam referential system (x, y, z) for the member geometry and cross-section properties, a local panel coordinate system (x, s, n), wherein the n axis is normal to the middle surface of a panel and the s axis is directed along the cross-section contour middle line, and a material axes system (1, 2) for each ply, are used. The following assumptions are considered:

- The composite ply is orthotropic, macroscopically homogeneous, linear elastic and a plane stress condition exists;
- Each laminate is symmetric about its mid-plane, thus all the coupling stiffness terms vanish;
- Each panel is orthotropic in the membrane mode, thus there are no shear coupling;
- The thickness can vary rom panel to panel in a cross-section, but it is constant in each beam element along x;
- Symmetry (or anti-symmetry) of the cross-section is kept, both in terms of geometry and material properties.

Since beams are assembled from flat-layered panels, effective cross-section stiffness, which account for this situation, are calculated as

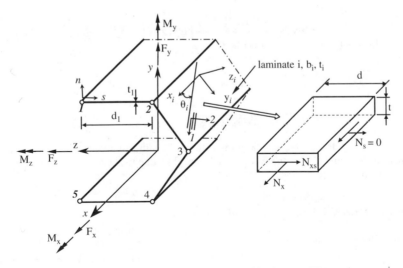

Figure 1. Thin-walled beam. Forces acting in the cross section, and laminate membrane stress resultants.

$$\overline{EA} = \sum_{i=1}^{NE} E_{x_i}^m A_i \ , \quad \overline{EI}_z = \sum_{i=1}^{NE} E_{x_i}^m I_{z_i} \ , \quad \overline{EI}_y = \sum_{i=1}^{NE} E_{x_i}^m I_{y_i} \ , \quad \overline{EI}_{yz} = \sum_{i=1}^{NE} E_{x_i}^m I_{yz_i}$$

$$\text{For open sections } \overline{GJ} = \frac{1}{3} \sum_{i=1}^{NE} G_{xs_i}^b b_i t_i^3; \quad \text{For single-cell sections } \overline{GJ} = \frac{4\Omega^2}{\displaystyle\oint_s \frac{ds}{G_{xs}^m t}} \tag{11}$$

where NE is the number cross-section panels, t_i is the thickness of the i-panel, Ω is the total area enclosed by the central line of the closed profile and E_{xi}^m, G_{xsi}^m, G_{xsi}^b are the equivalent elastic constants of the i-laminate, given as

$$E_{x_i}^m = 1/(t a_{11}), \quad G_{xs_i}^m = 1/(t a_{66}), \quad G_{xs_i}^b = 12/(t^3 d_{66}) \tag{12}$$

where the superscripts "m" and "b" indicate the membrane and bending modes, respectively. The compliance coefficients a_{ij} and d_{ij} are obtained by inverting the classical laminate constitutive equations (Datoo 1991).

The computational analysis of stress and correspondent ply failure index is calculated in the points that define the laminate position in the cross-section (points 1, 2, 3, 4 and 5 in Figure 1). The laminate forces are obtained by accumulating the incremental values calculated step-by-step during the integration process of the non-linear structural equilibrium equations. We assume the forces acting in the cross-section cause essentially membrane stress resultants in the laminates constituting it, as represented in Figure 1. The incremental stresses in the ply material axis are

$$S_1 = \left[Q_{11} \left(c^2 a_{11} + s^2 a_{12} \right) + Q_{12} \left(s^2 a_{11} + c^2 a_{12} \right) \right] N_x + \left[Q_{11} c s a_{66} - Q_{12} c s a_{66} \right] N_{xs}$$

$$S_2 = \left[Q_{12} \left(c^2 a_{11} + s^2 a_{12} \right) + Q_{22} \left(s^2 a_{11} + c^2 a_{12} \right) \right] N_x + \left[Q_{12} c s a_{66} - Q_{22} c s a_{66} \right] N_{xs} \tag{13}$$

$$S_6 = Q_{66} 2 c s \left(a_{12} - a_{11} \right) N_x + Q_{66} \left(c^2 - s^2 \right) a_{66} N_{xs}$$

where the incremental stress resultants at an arbitrary point (y, z) in the contour middle line of the cross-section are given as

$$N_x = E_x^m t \left(\frac{F_x}{\overline{EA}} - \frac{M_z \overline{EI}_y - M_y \overline{EI}_{yz}}{\overline{EI}_y \overline{EI}_z - \overline{EI}_{yz}^2} y + \frac{M_y \overline{EI}_z - M_z \overline{EI}_{yz}}{\overline{EI}_y \overline{EI}_z - \overline{EI}_{yz}^2} z \right) \tag{14}$$

$$N_{xs} = N_{xs}^{(F)} + N_{xs}^{(T)} \tag{15}$$

with

$$N_{xs}^{(F)} = -\left(\frac{F_y \overline{EI}_y + F_z \overline{EI}_{yz}}{\overline{EI}_y \overline{EI}_z - \overline{EI}_{yz}^2} \int_0^s E_x^m t \, y \, ds + \frac{F_z \overline{EI}_z + F_y \overline{EI}_{yz}}{\overline{EI}_y \overline{EI}_z - \overline{EI}_{yz}^2} \int_0^s E_x^m t \, z \, ds \right) + q_{s=0}^{(F)} \tag{16}$$

$$N_{xs}^{(T)} = \pm G_{xs}^b t^2 \frac{M_x}{\overline{GJ}} \qquad \text{for open cross-sections}$$

$$N_{xs}^{(T)} = \frac{M_x}{2\Omega} \qquad \text{for single cell cross-sections} \tag{17}$$

where F_x, M_y, M_z, F_y, F_z and M_x are the normal force, bending moments, shear forces and torsion moment, respectively, and $q_{s=0}^{(F)}$ is the incremental shear flow due to shear forces at the point chosen as the origin of the local coordinate s.

5 NUMERICAL EXAMPLES

5.1 Cantilever flat composite beam

A cantilever flat composite beam is optimized for minimum mass and complete 360-degree circular deflection (minimize $|\theta_{z11} - 2\pi|$) when subject to a tip bending moment (M = 1700 N.m) and to a maximum

Tsai-Hill failure index (FI ≤ 1). The cross section is formed by the laminate **1** with eight layers $[0/45/ - 45/90]_s$ and total thickness $t_1 = 1$ mm, and by the laminate **2** with four layers $[45/45]_s$ and total thickness $t_2 = 0.5$ mm. The material properties are $E_1 = 140$ Gpa, $E_2 = 10$ GPa, $\nu_{12} = 0.3$, $G_{12} = 5$ GPa, $\rho = 1900$ Kg/m³, $\sigma_{1t}^u = 1500$ MPa, $\sigma_{2t}^u = 50$ MPa, $\sigma_{1c}^u = 1200$ MPa, $\sigma_{2c}^u = 250$ MPa and $\sigma_6^u = 70$ MPa.

The optimization procedure is performed in a two-level approach: in a first level the optimization is performed with respect to the thickness of the laminates for the minimum mass design and in a second level it proceeds with respect to the lamina orientations for the complete circle deflection. Figure 2 shows the cantilever beam and its cross-section. Figures 3–5 show the Moment–Displacement curves and the deformed configurations for both the starting and optimum designs. Starting and optimum designs are presented in the Table 1, where U_{11} and V_{11} means node 11 displacement in x and y direction respectively, FI_{10} means maximum failure index in element 10, and lower and upper indices of lamina orientation stand respectively for lamina number and laminate number.

When the deformed configuration of the beam corresponds to a complete circumference, the radius of curvature is R = L/2π = 10/2π = 1.59 m. Since this is a pure bending problem, the theoretical flexural

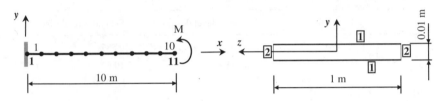

Figure 2. Cantilever flat composite beam.

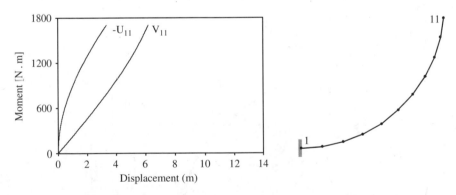

Figure 3. Moment–Displacement curves and deformed configuration for initial design.

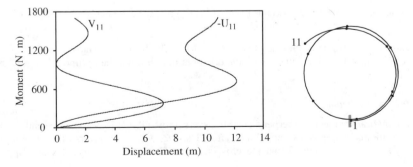

Figure 4. Moment–Displacement curves and deformed configuration after 1st level optimization.

375

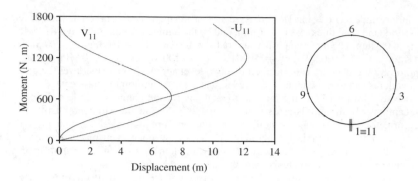

Figure 5. Moment–Displacement curves and deformed configuration after the 2nd level optimization.

Table 1. Initial and optimum designs for cantilever flat composite beam.

	Initial design	First level design	Second level design
t_1 [mm]	4	0.582	0.582
θ_2^1	45°	45°	19.6°
FI_{10}	0.0193	$1^{(*)}$	0.359
$\overline{EI}_{ZZ}$ [N.m^2]	11 390	1577	2702
Mass [Kg]	123.92	12.99	12.99
$|\theta_{Z11} - 2\pi|$	4.791	3.660	4.64E-5

(*) – Active constraint.

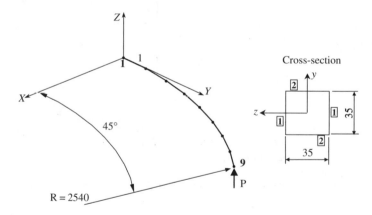

Figure 6. Cantilever arc tubular cross-section beam subject to a perpendicular transversal force.

stiffness value can be evaluated knowing that $1/R = M/EI_{zz}$. In this case we have $1/1.59 = 1700/\overline{EI}_{zz} \Rightarrow \overline{EI}_{zz} = 2702$ N . m^2. This value agrees with the one obtained through the optimization process, that is 2702 N . m^2.

5.2 Cantilever arc beam

A cantilever arc beam subject to a perpendicular transversal force P = 150 N at the free end, see Figure 6, is considered in this example. Two equal laminates, **1** and **2**, with eight layers $[45/-45/45/-45]_s$ and total thickness $t_1 = t_2 = 4$ mm, forms the cross-section. The material properties are $E_1 = 75$ GPa, $E_2 = 6$ GPa, $\nu_{12} = 0.34$, $G_{12} = 2$ GPa, $\rho = 1400$ Kg/m^3, $\sigma_{1t}^u = 1300$ MPa, $\sigma_{2t}^u = 30$ MPa, $\sigma_{1c}^u = 280$ MPa, $\sigma_{2c}^u = 140$ MPa and $\sigma_6^u = 60$ MPa. The beam is discretized by eight finite elements with the same length.

376

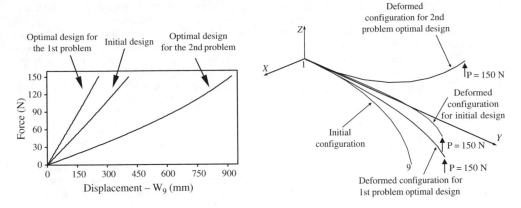

Figure 7. Cantilever arc beam Force–Displacement curves. Starting and optimal designs: initial and deformed configurations.

Table 2. Results of cantilever arc beam optimization.

	Initial Design	1st Problem		2nd Problem
		Design after 1st level optimization	Optimal Design	Optimal Design
t_1, t_2 [mm]	4, 4	4, 4	0.95, 1.3	1.43, 1.43
θ_1^1, θ_3^1	45^0	22.3^0, 22.3^0	22.3^0, 22.3^0	45^0, 45^0
θ_1^2, θ_3^2	45^0, 45^0	14.4^0, 14.4^0	14.4^0, 14.4^0	45^0, 45^0
$\overline{EA}$ [N]	4097469	28911050	8375106	1464845
$\overline{EI}_y$ [N.mm^2]	839298305	5362974832	1425415549	299197392
$\overline{EI}_z$ [N.mm^2]	839298305	6480918712	1995416353	299197392
$\overline{GJ}$ [N.mm^2]	3321661664	1316787399	377767289	1187494045
FI_1	0.16	0.10	1 (*)	1 (*)
U_9 [mm]	-32.6 / -27.5	-1.36 / -	-14.3 / -12.6	-167.0 / -148.7
V_9 [mm]	-40.6 / -43.1	-1.56 / -	-16.9 / -17.8	-226.5 / -238.2
W_9 [mm]	409.2 / 409.9	79.4 / -	259.0 / 258.7	921.3 / 929.2
θ_{x9}	0.303 / 0.304	0.031 / -	0.097 / 0.096	0.702 / 0.709
θ_{x9}	-0.107 / -0.107	0.066 / -	-0.222 / -0.223	-0.261 / -0.262
θ_{x9}	0.017 / 0.013	-0.001 / -	-0.011 / -0.013	0.086 / 0.071
Mass [Kg]	1.56	1.56	0.440	0.602

(*)Active constraint.Obs: The lamination sequences for laminates 1 and 2 in 1st problem optimal design are respectively, $[22.3 / -22.3 / 22.3 / 22.3 / -22.3]_s$ and $[14.4 / -14.4 / 14.4 / -14.4]_s$

Two optimization problems are considered. In the first one the beam is optimized for minimum mass and maximum stiffness, considering lamina orientations and thickness of both laminates as design variables. The optimization procedure is performed in a first level with respect to the lamina orientations for the minimum z displacement in the free end (W_9), and in a second level with respect to the laminates thicknesses for minimum mass. For both levels constraints on Tsai-Hill failure index (FI ≤ 1) are considered. In the second problem the beam is optimized for minimum mass, subject to constraints on Tsai-Hill failure index, and considering only the laminates thicknesses as design variables.

Figure 7 shows the Force–Displacement curves and initial and deformed configurations for initial and optimal designs. Table 2 presents the optimization results for both problems.

The failure index constraint for element 1, became active in both optimal designs. In the first problem one gets a smaller mass and torsion stifness and a larger axial and bending stiffness than in the second one. This means the lamina orientation is an important stiffness factor for the laminated structures. The shadow cells correspond to results obtained with the commercial code COSMOS/M using the 3D-beam element (BEAM3D) considering the composite section stiffness properties indicate in the Table 2.

6 CONCLUSIONS

A finite element model of analysis and sensitivity analysis has been applied sucessfully to several optimal design examples of composite beam structures. The applications consider 3-D closed cross-section thin-walled beams. When lamina orientations are design variables, the optimization problem is a non-convex problem. In this case, different starting designs were considered in order to obtain the optimum point.

As one expected, the stiffness of laminate composite structures is strongly dependent on the lamina orientation, hence this orientation is fundamental to these structures.

Generally, thin-walled composite structures are modeled as plate-shell type structures. The utilization of the beam model for these structures reduces significantly the problem dimension, what turns convenient for optimal design. However, some local information is loss. Indeed this beam model may be a good choice for preliminary design.

REFERENCES

Arora, J.S. & Cardoso, J.B. 1992. Variational Principle for Shape Design Sensitivity Analysis. *AIAA Journal* 30(2): 538–547.

Bathe, K.J. & Bolourchi, S. 1979. Large Displacement Analysis of Three-Dimensional Beam Structures. *I.J. Num. Meth. Engng.* 14: 961–986.

Bauld, N.R. & Tzeng, L.S. 1984. A Vlasov Theory for Fiber-Reinforced Beams with Thin-Walled Open Cross Sections. *I.J. of Solids and Structures* 20(3): 277–297.

Bhaskar, K. & Librescu, L. 1995. A Geometrically Non-Linear Theory for Laminated Anisotropic Thin-Walled Beams. *I.J. Engng. Science* 33(9): 1331–1344.

Datoo, M.H. 1991. *Mechanics of Fibrous Composites*. Elsevier Applied Science.

Davalos, J.F. & Qiao, P.A. 1999. Computational Approach for Analysis and Optimal Design of FRP Beams. *Computers and Structures* 70: 169–183.

Floros, M.W. & Smith, E.C. 1997. Finite Element Modeling of Open-Section Composite Beams with Warping Restraint Effects. *AIAA Journal* 35(8): 1341–1347.

Lee, J. & Kim, S.E. 2001. Flexural-Torsional Buckling of Thin-Walled I-Section Composites. *Computers and Structures* 79: 987–995.

Valido, A.J., Sousa, L.G. & Cardoso, J.B. 2001. Optimum Design of Geometrically Nonlinear Composite Beam Structures. Proc. *WCSMO-4 – Fourth World Congress of Structural and Multidisciplinary Optimization, Dalian, 4–8 June 2001*.

Valido, A.J. 2001. *Optimal Design of Composite Beam Structures with Geometrically Nonlinear Behavior* (Ph.D. Thesis, in Portuguese). Technical University of Lisbon, Portugal.

Woolley, G.J. 1989. *Design Analysis of Composite Beams* (Msc Thesis). Cranfield Institute of Tchnology, UK.

Yang, Y.B. & Shieh, M.S. 1990, Solution Method for Nonlinear Problems with Multiple Critical Points. *AIAA Journal* 28(12): 2110–2116.

Nonlinear dynamic calculation of a new type flexible suspension bridge

W.B. Yao
Faculty of Engineering, Zhejiang Forestry University, Zhejiang, China

ABSTRACT: The conception and characteristics of a new type flexible suspension bridge are presented. According to deformation and forced characteristics of the flexible suspension bridge, the finite element model of flexible cable is established and a static computing method of a flexible suspension bridge under applied load eccentrically is brought forward. In nonlinear dynamic computation way, the paper develops a new method that is called rational approximate method and nonlinear finite element equations of the system are solved with the help of the method. Finally, simulated calculation of the new type flexible suspension bridge is made. Numerical computations for a real flexible suspension bridge show that the method is correct and reliable.

1   INTRODUCTION

The new type flexible suspension bridge is a special bridge mainly made by steel wire rope which is used as loaded cable as well as stable and supported cable. Being its structure simple and bridge's appearance beautiful and stability well and cost low and building period short, it is widely used for mountainous area and tourism area in China.

As load transfer principle and pre-tension technology are adopted, traditional flexible suspension bridge is "solidified" (see Figure 1). In order to meet necessity with engineering, the state and structure of the new type suspension bridge have to be designed and computed precisely.

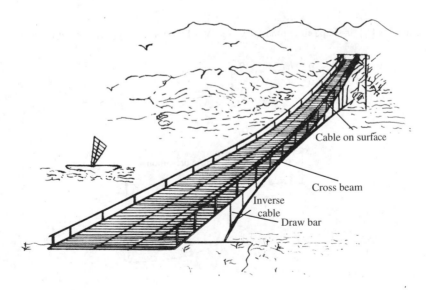

Figure 1.   The new flexible suspension bridge

So far the key solved the problem by the finite element is how to solve nonlinear finite element dynamic equations. Being original methods such as Houbolt method and Newmark method and Wilson-θ method are discovered to exist many problems, it is very important to seek ask for other more effective and reliable methods.

Recently a new method that is called rational approximate method was used to solve nonlinear finite element dynamic equations, but relative specific results are lack. The paper develops the new method. According to deformation and characteristics applied force of the flexible suspension bridge, nonlinear finite element dynamic equations are established and the dynamic response problem of the new type flexible suspension bridge is researched. The computational results and results obtained by experiments are basically coincident, thus the method is proved to be correct and reliable.

2 THE FINITE ELEMENT MODEL OF A NEW TYPE FLEXIBLE SUSPENSION BRIDGE

In order to use finite element method computing the cable bridge, let cable separate into a number of two orders curvilinear element which per element has three nodes (see Figure 2). Wirerope is seen as one dimension body. The place of nodes inner element before deformation can use local coordinate S determining. The local coordinate is called Lagrange coordinate. Whole coordinate is OXYZ.

2.1 Displacement model and shape function

In Figure 2 the element has three nodes, midpoint 2 locates in middle of node 1 and node 3. Using two orders inserting value functions:

$$
\left.
\begin{aligned}
x &= a_1 + a_2 s + a_3 s^2 \\
y &= b_1 + b_2 s + b_3 s^2 \\
z &= c_1 + c_2 s + c_3 s^2
\end{aligned}
\right\}
\tag{1}
$$

Let $(x_i\ y_i\ z_i)$ $(i = 1, 2, 3)$ represents coordinate of three nodes inner element after deflection. Thus using nodes coordinates determines coefficients a_i, b_i, c_i $(i = 1, 2, 3)$ of equation (1). Substituting them into equation (1) yields:

$$
\{X\}^e = [N]_s \{X\}_{Ne}
\tag{2}
$$

In equation (2), $[N]_s$ is shape function matrix and N is function of Lagrange coordinates.

$$
[N]_s =
\begin{bmatrix}
N_1 & 0 & 0 & N_2 & 0 & 0 & N_3 & 0 & 0 \\
0 & N_1 & 0 & 0 & N_2 & 0 & 0 & N_3 & 0 \\
0 & 0 & N_1 & 0 & 0 & N_2 & 0 & 0 & N_3
\end{bmatrix}
\tag{3}
$$

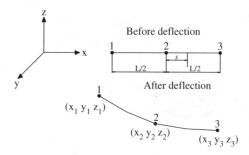

Figure 2.   The finite element of cable.

$$N_1 = -\frac{S}{L} + 2\frac{S^2}{L^2} , N_2 = 1 - 4\frac{S^2}{L^2} , N_3 = \frac{S}{L} + 2\frac{S^2}{L^2} \tag{4}$$

2.2 Cable tension

According to the theory of nonlinear elasticity, under the condition of one dimension, yields tension:

$$T = AE(1 + E_s)\varepsilon_s \tag{5}$$

Substituting $E_s = (\partial S^*/\partial S) - 1$ into equation (5) yields

$$T = AE(\frac{\partial S^*}{\partial S})\varepsilon_s \tag{6}$$

Since $\varepsilon_s = [(\partial S^*/\partial S)^2 - 1]/2$ results

$$T = \frac{1}{2}AE(\frac{\partial S^*}{\partial S})[(\frac{\partial S^*}{\partial S})^2 - 1] \tag{7}$$

where * represents value after deformation, S^* represents Lagrange coordinate after deformation. According to equation (2) yields

$$(\frac{\partial S^*}{\partial S})^2 = \{X\}^{*^T}_{Ne}[N']^T[N']\{X\}^*_{Ne} \tag{8}$$

where $[N'] = (\partial/\partial S)[N]_s$, therefore

$$T = \frac{AE}{2}\sqrt{\{X\}^{*^T}_{Ne}[N']^T[N']\{X\}^*_{Ne}}(\{X\}^{*^T}_{Ne}[N']^T[N']\{X\}^*_{Ne} - 1) \tag{9}$$

2.3 Static equation of the finite element

According to variational principle yields static equation of cable bridge

$$[K_E]\{X\}^*_N - \{R\} = 0 \tag{10}$$

where $\{R\} = \{g\} + \{p\} + \{q\} - \{r\}$, $\{g\}$ = equivalent node load vector of cable bridge' self weight; $\{p\}$ = alive load vector; $\{q\}$ = distributed load intensity vector; $\{r\}$ = node concentrated load vector which includes reaction force between cable and girder, weight of girder and load subjected to bridge.

In stiffness matrix $[K_E]$ existing coordinates $\{X\}^*_N$ to be solved, equation (9) is nonlinear equation. The equation can be solved by Newton-Raphson method.

2.4 Newton-Raphson method

Letting $U = \{X\}^*_N$, thus

$$R_{(U)} = R_{(\{X\}^*_N)} = \{R\} \tag{11}$$

$$F_{(U)} = F_{(\{X\}^*_N)} = [K_E]\{X\}^*_N \tag{12}$$

Thus $f_{(U)} = F_{(U)} - R_{(U)} = [K_E]\{X\}^*_N - \{R\}$ (13)

$$\frac{\partial f}{\partial \{X\}_N^*} = \frac{\partial F}{\partial \{X\}_N^*} = \frac{\partial}{\partial \{X\}_N^*} ([K_E]\{X\}_N^*) = [K_T] \tag{14}$$

$$\frac{\partial}{\partial \{X\}_N^*} ([K_E]\{X\}_N^*) = [K_E] + \{X\}_N^* \cdot \frac{d[K_E]}{\partial \{X\}_N^*} \tag{15}$$

By deducing yields

$$\{X\}_N^* \cdot \frac{\partial [K_E]}{\partial \{X\}_N^*} = [K_L] \tag{16}$$

Hence $[K_T] = [K_E] + [K_L]$ \hfill (17)

$[K_T]$ is called as tangential stiffness matrix.

Notes, $[K_T]\Delta\{X\}_N^i = \{e\}^i$ \hfill (18)

Finally according to above equations yields $\Delta\{X\}_N^i$ and $\{X\}_N^i = \{X\}_N^{i-1} + \Delta\{X\}_N^i$

2.5 Computational model of a new type flexible suspension bridge under symmetric load

As the bridge is forced on symmetric load, the states of any cables on surface and inverse cables are all same. Therefore, the bridge can be called as the structure made up of a cable on surface and an inverse cable. When the bridge is balanced, computational models of the cable on surface and the inverse cable are written in turn:

Cable on surface : $[K_D]_1\{X\}_{Ne}^* - \{g\}_1 - \{q\}_1 - \{P\} + \{r\} = 0$ \hfill (19)

Inverse cable : $[K_D]_2\{X\}_{Ne}^* - \{g\}_2 - \{r'\} = 0$ \hfill (20)

where $\{g\}$ = equivalent node load vector of cable bridge self weight; $\{P\}$ = work load vector; $\{q\}$ = distributed load intensity vector; $\{r\}$ = node concentrated load vector which includes action force between cable and girder; $\{r'\}$ = reaction force vector of $\{r\}$.

As self weight of cable and girder is ignored, $\{r\} = \{r'\}$.

Let equation (19) and (20) expand to whole structure, yields

$$[K_D]\{X\}_{Ne}^* - \{g\}_1 - \{g\} - \{q\}_1 - \{P\} = 0 \tag{21}$$

Finally nonlinear static equations are solved by Newton-Raphson method.

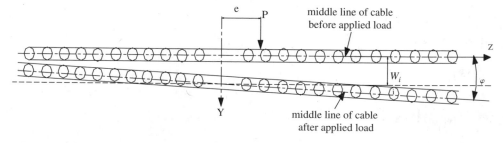

Figure 3.   Deformation of cable on surface.

2.6 *Static computation of a new type flexible suspension bridge under eccentric load*

When the bridge is acted on eccentric live load, the forces on cable will be contributed over again and the girder will be state slanted. If the slanting angle is too large, vehicles will exist danger. Therefore static computation will be necessary.

Let i represent number of section applied load, j represent number of cable, Y coordinate of j cable in i section is

$$Y_{ij} = W_i + Z_j tg\varphi \tag{22}$$

where W_i is droop in i section of cable. As coordinates of other sections change, droop in other sections of cable is an unknown quantity. As applied load conditions in i section is known and other sections are not known. Thus cable on surface can be computed according to below equations:

$$[K_D]\{X\}_{Ne}^{\bullet} - \{g\}_1 - \{q\}_1 + \{r\} = \{e\} \tag{23}$$

and $[K_T]_1\delta\{X\}_N^* = -\{e\}^i$ \hfill (24)

Similar inverse cable can be also computed with the help of above method.

3 A NEW SOLVING METHOD OF NONLINEAR DYNAMIC EQUATIONS

3.1 *Rational approximate method*

The method tries hard to adopt some approximate method solving analytic solution of differential equations. Assuming that one order constant coefficient differential equation:

$$[A]\frac{d\{U\}}{dt} + [B]\{U\} = \{P\} \qquad U|_{t=0} = U_0 \tag{25}$$

The solution of differential equations is

$$\{U\}_t = [\exp(-t[A]^{-1}[B])](\{U\}_0 - [B]^{-1}\{P\}) + [B]^{-1}\{P\} \tag{26}$$

Writing as differential format:

$$\{U\}_{t+\Delta t} = [\exp(-\Delta t[A]^{-1}[B])](\{U\}_t - [B]^{-1}\{P\}) + [B]^{-1}\{P\} \tag{27}$$

where $\exp(-\Delta t[A]^{-1}[B])$ is difficult to be calculated. Taking Norsett approximate formula:

$$\exp(-[s]) = \sum_{k=0}^{m-1} L_k(\frac{1}{\alpha})\frac{(\alpha[s])^k}{([I]+[s])^{k+1}} \tag{28}$$

where: $L_k(1/\alpha)$ is a polynomial. Generally taking $n = 2$, $\alpha = 1 - (1/\sqrt{2})$
Substituting (28) into (26) yields:

$$\{U\}_{t+\Delta t} = [\sum_{k=0}^{m-1} L_k(\frac{1}{\alpha})\frac{(\alpha\Delta t[A]^{-1}[B])^k}{([I]+\alpha\Delta t[A]^{-1}[B])^{k+1}}](\{U\}_t - [B]^{-1}\{P\}) + [B]^{-1}\{P\} \tag{29}$$

Equation (29) can be also expressed as below:

$$([A] + \alpha\Delta t[B])\{U\}_{t+\Delta t} = [A](\{U\}_t + \sum_{i=1}^{m-1} L_j(\frac{1}{\alpha})\{W\}_j) + \alpha\Delta t\{P\} \tag{30}$$

There $\{W\}_j (j = 1, 2, 3, \ldots)$ can be solved by below formula:

$$\begin{cases} ([A] + \alpha \Delta t [B])\{W\}_1 = \alpha \Delta t([B]\{U\}_t - \{P\}) \\ ([A] + \alpha \Delta t [B])\{W\}_{j+1} = \alpha \Delta t \{W\}_j \end{cases} \tag{31}$$

According to equation (30) $\{U\}_{t+\Delta t}$ can be solved. The method can be also generalized to two orders differential equation.

3.2 *Solving method of nonlinear dynamic equations*

The nonlinear dynamic equation is

$$[M]\{\ddot{X}\}_{n+1} + \{P_{n+1}\} = \{R_{n+1}\} \tag{32}$$

$$\{P_{n+1}\} = \int_v [B^{n+1}]^T \{\sigma_{n+1}\} dv \tag{33}$$

Because $\{P_{n+1}\}$ is function of $\{X\}$, equation (32) can not be solved directly.
Variating equation (33) yields

$$[M]\{\ddot{X}\}_{n+1} + \delta\{P_{n+1}\} = \delta\{R_{n+1}\} \tag{34}$$

Since $\delta\{P_{n+1}\} = [K_T]\delta\{X\}$ \hfill (35)

Substituting equation (35) into equation (34) yields

$$[M]\{\ddot{X}\}_{n+1} + [K_T]\{X\} = \{R\} \tag{36}$$

Equation (36) can be solved with the help of the rational approximate method and Newton-Raphson method.

4 CALCULATIONAL EXAMPLE

A old type cable bridge: bridge span $L = 200\,\text{m}$, modulus of elasticity $E = 1 \times 10^{11}\,\text{N/m}^2$, area of cable cross section $A = 358 \times 10^{-6}\,\text{m}^2$, distributing load $Q_1 = 105\,\text{N/m}^2$, alive load $P = 200\,\text{kN}$, self weight of cable $Q_0 = 33.9\,\text{N/m}^2$, eccentric distance $e = 20\,\text{cm}$.

In order to compare with their properties between new type bridge and old one. The paper adopts directly above data simulating new type bridge. The computational results are written as in Table 1.

By all appearance, when the bridge is acted on eccentric live load, slanting angle of new type cable bridge is smaller than old one's. It is shown that security and stability distinctly increase after new technology is adopted. It is approved by Lanping and Xuanwei new type bridge in Yunnan province.

Dynamic analysis: impact load is applied in center of bridge

$$P_D = \begin{cases} \dfrac{s}{t_1} \times t & 0 \le t \le 0.5\,\text{sec} \\ 5 & t > 0.5\,\text{sec} \end{cases}$$

Dynamic response of nodal 15 and nodal 66 (see Figure 4). According to the figure, the calculating results of new method are very close to Newmark's. Therefore the method developed is proved as being correct.

Table 1. Data analysis between new type cable bridge and old type cable bridge.

	Average tension of cable (kN)		Slant angle	
Condition applied load	New bridge (computational value)	Old bridge (computational value)	New bridge (computational value)	Old bridge (computational value)
No load	15.637	15.064	–	–
Fully loaded	18.749	18.325	–	–
Eccentric load	–	–	1°17″	4°24″

Note: *represents experimental results.

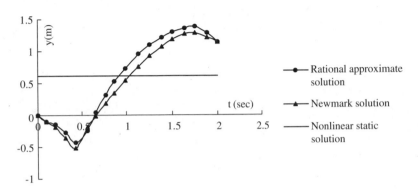

Figure 4. Simulated computational results.

5 CONCLUSION

1. The paper presents a new method by which cable bridge can be conveniently calculated. The nonlinear computational results can make engineer satisfied.
2. Numerical computation shows that the technological property parameters in the new type flexible suspension bridge are superior to ones of other flexible suspension bridges.
3. The rational approximate method has been proved as being correct by theoretical study and practice application. It can be also used for solving other nonlinear dynamic problems.
4. It should be obvious that the computational model built by the paper is simple, direct and practical, by which the geometrical nonlinear questions are easily handled.

REFERENCES

Liu B.C. & Zhang J. 1982. Large deflection computation of flexible cable. *Computer application in construction engineering*. China science publisher, 212–215.
Qu B.N. et al. 1995. Nonlinear finite element approach for stable type suspended bridge. *Proceedings of EPMESC*. 1995. 8, 223–228.
Yao W.B. & Liu B.C. 1998. The computational theory of single cable taking into account steel cable bending stiffness. *Journal of Kunming University of Science and Technology*, No. 1, 1998. 5–11.
Qu B.N. et al. 1994. The structure of stable suspended bridge and its calculating theory of nonlinear Finite element. *Mechanics and engineering*, Science and Technology publisher, 41–45.

Static and dynamic testing of a skewed overpass bridge

M. Xu, J. Rodrigues & L.O. Santos
National Laboratory for Civil Engineering, Lisbon, Portugal

ABSTRACT: This paper presents the static and dynamic testing of the skewed overpass PS3. The static load test was performed with truck loads up to a total of 1268 kN. The vertical displacements were measured. The dynamic tests were carried out in three set-ups. 16 modes of the natural vibration of the structure were identified, using the technique of output-only modal identification, and the dynamic characteristics for each mode were estimated. For interpretation of the test results a finite element model for the bridge was built. The experimental results were compared with the analytical values computed by the FE model.

1 INTRODUCTION

The overpass bridge PS3 is located at highway A2 between Lisbon and Algarve, Portugal (Figure 1). In June of 2002, before the opening to the traffic, the static and dynamic tests were carried out. The tests were conducted in order to perform an evaluation of the static behavior of the bridge and to identify its dynamic characteristics, namely, the vibration frequencies, mode shapes and damping ratios. The experimental results of both tests are compared with a finite element model of the bridge. This paper presents the results of the static and dynamic tests in association with the numerical model.

The PS3 is a skewed prestressed concrete overpass bridge, 102 m long, with two central spans of 34 m and two lateral spans of 17 m. The bridge is supported monolithically by tree concrete piers. The axes of the supports are skewed at an angle of 40.19° to the bridge's longitudinal axe (TRIEDE 2000). The deck has 9.40 m width (Figure 2). It is a slab with a central rib, which has a constant height of 0.90 m on the lateral spans and a varied height on the central spans. The maximum height is 1.75 m over the central pier. The piers are rectangular with 0.40 m width. The length is 1.60 m for the lateral piers and 2.00 m for the central one. The central pier has a maxim height of 9.5 m. Figure 3 and Figure 4 present the elevation and plan views of the bridge.

2 ANALYTICAL MODEL

A finite element model for the bridge PS3 was developed to evaluate its response to the static and dynamic tests. The FE model was built via SAP2000 (CSI 2000).

Figure 1. The panorama of the PS3

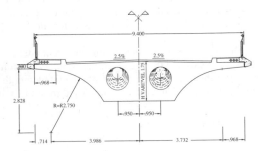

Figure 2. The deck transverse section.

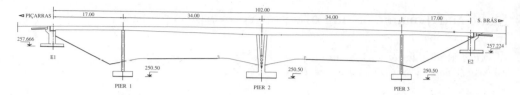

Figure 3.　Elevation view of the overpass bridge PS3.

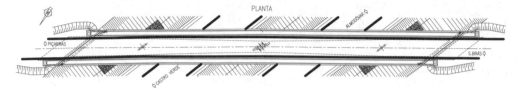

Figure 4.　Plan view of the overpass bridge PS3.

Figure 5.　The FE model.

Figure 6.　The FE model (plan).

Figure 7.　The truck loads on the position 5.

Figure 8.　The truck loads on the central pier.

The FE model consists of 399 shell elements and 147 frame elements. The shell elements are used for the lateral cantilevers and the piers and the frame elements are used for the central rib. The connection between the lateral cantilevers and the central beam or the piers was modelled with body constraints. Six link elements were used for modelling the bearings at the abutments. Figures 5 and 6 present the FE model of PS3.

Before the load tests the preliminary FE model was used to estimate the deformation of the structure on static loads and the shapes of the natural vibration modes.

After the tests the FE model was calibrated with the static load tests results at first, and then adjusted to the dynamic characteristics identified from the dynamic tests.

3   STATIC TEST

The static test was performed with 4 truck loads up to a total of 1268 kN. These loads were placed in accordance to the load plan that maximizes the most important effects in the structure (Figures 7 and 8),

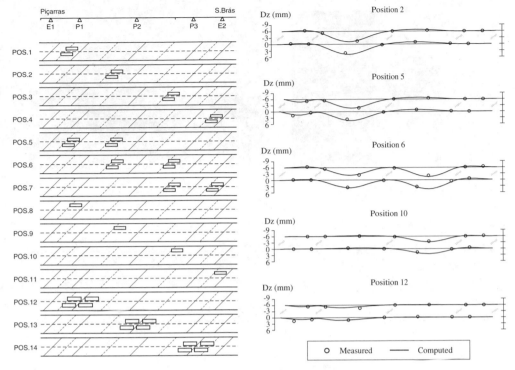

Figure 9.   The distribution of truck loads.

Figure 10.   The vertical displacements.

however without inducing unwanted situations of early cracking in the structure, the 14 load positions presented in Figure 9 were performed. During the test vertical displacements at mid spans and at the supports were measured.

The FE model was used for interpretation of the experimental results of the static test. The modulus of elasticity of the concrete was considered as 35 GPa.

Figure 10 shows the vertical displacements of the deck on some load positions and the computed structural deformation. It is evident that a good agreement between experimental values and analytical results has been obtained.

4   DYNAMIC TESTS

4.1   Testing procedure

Dynamic tests were performed to obtain experimentally the dynamic characteristics of the structure, namely, the vibration frequencies, mode shapes and damping ratios. During the tests, accelerations induced in the structure by the truck movement were measured using Kinemetrics Uniaxial Episensor (ES-U) accelerometers (Figure 11) and other equipments for signal conditioning and data acquisition (Figure 12) (Rodrigues 2002).

The dynamic tests of PS3 were carried out in three set-ups. During the tests 14 accelerometers were used. Three of them were used as reference sensors, always in the same points, 2 for vertical acceleration (points 13 and 14) and the other for transverse acceleration (point 13). In total, vertical acceleration was measured in 32 points, transverse in 3 points and longitudinal in 1 point. The localization of the points is illustrated in Figure 13. The points instrumented in each test set-up and the corresponding acceleration directions are described in Table 1.

The sampling rate for data acquisition was selected as 200 Hz. The records at each set-up had 163 840 lines corresponding to a time length of about 14 minutes.

389

Figure 11.   The accelerometers.   Figure 12.   The dynamic test.

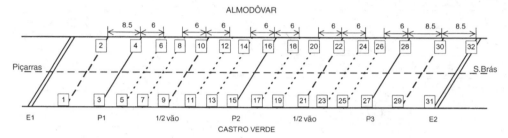

Figure 13.   Instrumented points in the dynamic tests.

Table 1.   Dynamic tests.

| | Box 1 | | | Box 2 | | | Box 3 | | | Box 4 | | | Box 5 | |
Set-up	Cn1	Cn2	Cn3	Cn1	Cn2	Cn3	Cn1	Cn2	Cn3	Cn1	Cn2	Cn3	Cn2	Cn3
1	1 V	7 V	13 V	2 V	8 V	14 V	15 V	19 V	27 V	16 V	20 V	28 V	9 T	13 T
2	3 V	9 V	(ref.)	4 V	10 V	(ref.)	17 V	23 V	29 V	18 V	24 V	30 V	21 T	(ref.)
3	5 V	11 V		6 V	12 V		21 V	25 V	31 V	22 V	26 V	32 V	21 L	

4.2   Modal identification

The software ARTeMIS – Output-only modal identification was employed for the modal identification of the bridge (SVS 2002). This program allows to accurately estimate natural frequencies of vibration and associated mode shapes and modal damping of a structure from measured response only. It is fast and simple to use.

Before the signal processing for modal identification, the test signals were pre-processed with the following operations (Rodrigues 2002): trend removal; low-pass filtering at 20 Hz with a 4 poles Butterworth filter; decimation of the signals from 200 Hz to 50 Hz. The advantage of decimating the signals was to reduce the size of the records, speeding up all the following computed processes without loosing information in the frequency range of interest.

Based on the test data, the spectral densities and correlation functions were estimated. The power spectral density (PSD) matrix was computed from independent samples with 2048 data points each one, with 66.67% overlap. For the sampling frequency of 50 Hz, the frequency resolution of the spectra is therefore 0.024 Hz.

The technique of Frequency Domain Decomposition (FDD) implemented in ARTeMIS was applied to estimate natural frequencies and mode shapes. In this technique the power spectral density (PSD) matrix

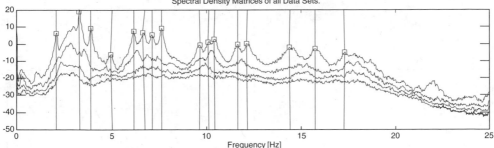

Figure 14. FDD: Spectral of singular values of the PSD matrix and identified natural frequencies.

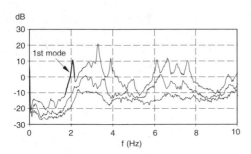

Figure 15. EFDD: PSD identified for 1st mode (MAC = 0.94).

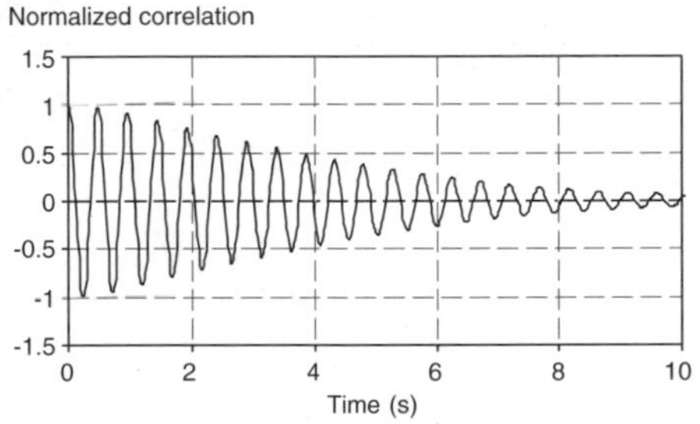

Figure 16. EFDD: Auto correlation function for 1st mode.

is decomposed at each frequency line via singular value decomposition (SVD). The singular values (SV) plots, as functions of frequency, estimated from SVD can be used to determine modal frequencies and mode shapes. The peaks of singular values plots indicate the existence of structural modes. The singular vector corresponding to the local maximum singular value is the respective unscalled mode shape (Tamura 2002).

Figure 14 shows the spectra of the first 4 singular values of the PSD matrix. In this figure the natural frequencies identified by FDD are also presented.

The technique Enhanced Frequency Domain Decomposition (EFDD) was mainly developed to estimate the damping of vibration modes (Brincker 2000). Actually the singular values in the vicinity of each natural frequency are equivalent to the PSD function of the corresponding modes. Based on a MAC criterion, each PSD function is identified around the peak of the singular values by comparing the respective mode shape estimate with the singular vectors for the frequency lines around the peak (Figure 15). The piece of the SDOF spectral density function obtained around the peak is taken back to time domain (with the IFFT algorithm), and the frequency and the modal damping ratio are simply estimated from the zero crossing times and the logarithmic decrement of the corresponding SDOF auto correlation function (Figure 16).

A total of 16 vibration modes of PS3 were identified from the analysis of the dynamic tests data, using the techniques FDD and EFDD. The 16 mode shapes are depicted in Figure 17. The dynamic characteristics for each identified mode are presented in Table 2.

4.3 Analysis with the finite element model

The FE model verified by the static load tests results was also used to interpret the experimental results of the dynamic tests

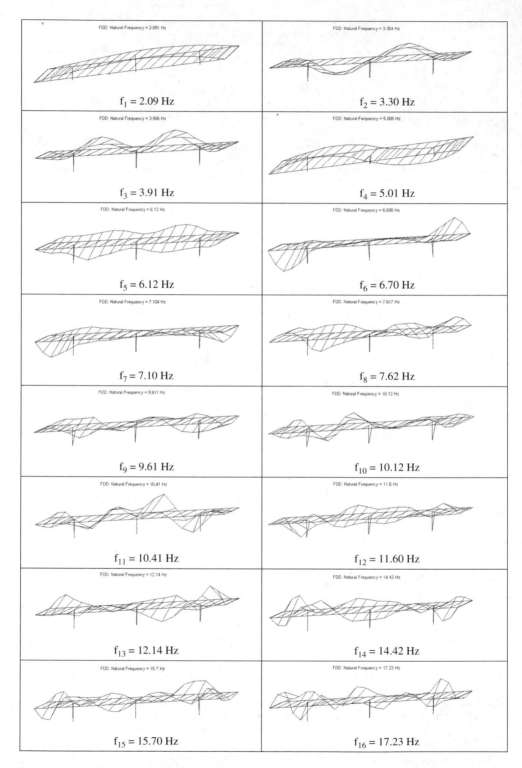

Figure 17.　Identified 16 mode shapes of PS3.

Table 2.   Dynamic characteristics of PS3.

| No | Frequency | | | Damping | |
	FDD	EFDD	Model	ξ (%)	Type of mode
1	2.09	2.11	2.07	1.74	Fundamental transverse mode
2	3.30	3.31	3.23	1.05	Fundamental vertical mode
3	3.91	3.90	3.87	1.38	Vertical mode
4	5.01	4.99	5.11	1.90	Transverse mode
5	6.12	6.12	6.06	1.25	Fundamental torsional mode
6	6.70	6.70	7.90	1.44	Torsional local mode
7	7.10	7.11	7.22	1.20	Torsional mode
8	7.62	7.60	9.15	0.92	Coupled (torsional/vertical)
9	9.61	9.62	–	0.98	Coupled (torsional/vertical)
10	10.12	10.12	9.92	1.15	Vertical mode
11	10.41	10.40	11.06	1.18	Vertical mode
12	11.60	11.62	11.93	1.03	Torsional mode
13	12.14	12.09	14.76	1.20	Torsional mode
14	14.42	14.42	17.69	1.49	Torsional mode
15	15.70	15.75	21.79	1.43	Torsional mode
16	17.23	17.31	24.09	0.92	Torsional mode

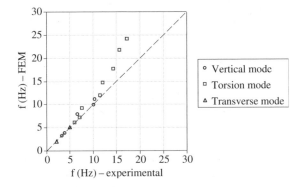

Figure 18.   FE model vs. experimental frequencies.

The FE model was tuned according to the modal identification from the dynamic tests. However the modulus of elasticity of the structural concrete was kept with the value previously referred (35 GPa). The mass density of the structural elements is 2291 kg/m^3 and the additional mass is 400 kg/m, corresponding to the mass of the lateral sidewalks and the guard rail. The stiffness of the link elements, for modelling the supports at the abutments, was adjusted to fit the computed modal characteristics to the ones identified from the tests.

The natural frequencies computed by the updated FE model were compared to the frequencies identified from the tests and are also presented in Table 2. A graphical comparison between the experimentally identified frequencies (EFDD method) and the frequencies computed with the FE model is also presented in Figure 18. In the case that a perfect agreement had been achieved, all the data points represented in Figure 18 would be in the 45 degrees line that is also represented in that figure. Figure 18 shows, therefore, that a good agreement exists between the experimental and the FE model frequencies, especially in what concerns the vertical and transverse modes.

5   CONCLUSIONS

The experimental results obtained in the static load test and the dynamic tests have a good correlation with the analytical values computed by FE model.

The information that was experimentally obtained about the structural behaviour of this overpass, is an important contribution to the characterization of its actual condition at the end of the construction and before its opening to the traffic. It is important to note that dynamic tests, similar to the ones that were presented here, can be performed during the lifetime of the structure, without the need to impose traffic restrictions.

REFERENCES

Brinker, R., Zhang, L.-M. & Andersen, P. 2000. Output-only modal analysis by frequency domain decomposition, *Pro. Of the ISMA25 conference*, Leuven.
CSI, 2000. *SAP2000 – Integrated Finite Element Analysis and Design of Structures.*
Rodrigues, J. 2002. Dynamic performance of a steel truss bridge under railway traffic, IMAC2002, Los Angeles.
Structural Vibration Solution (SVS), 2002. *ARTeMIS Extractor Handy, Release 3.1*, Denmark.
TRIEDE, SA 2000. Overpass Bridge PS3, Design, Lisbon, July 2000.
Tamura, Y., Zhang, L.-M., Yoshida, A., Cho, K., Nakata, S. & Naito, S. 2002. Ambient vibration testing & modal identification of an office building, IMAC 2002, Los Angeles.

Computational Methods in Engineering and Science, Iu et al. (eds)
© 2003 Swets & Zeitlinger, Lisse, ISBN 90 5809 567 3

Discrete optimum design of hollow section lattice girders

P.J.S. Cruz
Civil Engineering Department, University of Minho, Guimarães, Portugal

ABSTRACT: Despite the guidance and simplifications introduced by the new eurocodes the design of hollow section lattice girders is still time consuming. In order to make the design phase less time demanding, a linear optimisation model is proposed. Given the structural geometry and several constraints (displacement, size, stress, etc.) it is possible to calculate the optimum cross-section for each member. Although this research was restricted to the analysis of planar lattice girders, composed by rectangular hollow sections with welded joints, it is easy to generalize this formulation to three-dimensional structures with other sections and new kinds of joints. The example presented illustrates the practical applicability of the model.

1 INTRODUCTION

According to Eurocode 3 (EC 3) the distribution of forces in the elements of a hollow section lattice girder can be evaluated assuming that pinned joints connect the members. This code provides detailed rules to evaluate the design axial resistance of uniplanar joints. These rules comprise an appreciate number of joint configurations (K, N, T, KT and Y).

In this code the design axial resistance of welded joints is given in terms of the axial resistance of the members that meet at this node. As an example, it is herein presented a K type joint between square hollow section brace members and square hollow sections chords (Fig. 1):

$$N_{i,Rd} = \frac{8.9 f_{yo} t_0^2}{\sin\theta_i} \left[\frac{b_1 + b_2}{2b_0}\right] \gamma^{0.5} k_n \left[\frac{1.1}{\gamma_{Mj}}\right] \tag{1}$$

in which,

$$k_n = 1 \qquad \text{for } n \geq 0 \quad \text{(tension)} \tag{2.1}$$

$$k_n = 1.3 + \frac{0.4\,n}{\beta} \qquad \text{for } n \leq 0 \quad \text{(compression)} \tag{2.2}$$

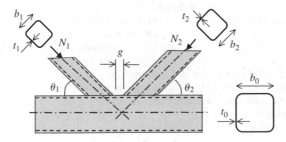

Figure 1. Gap type K joint.

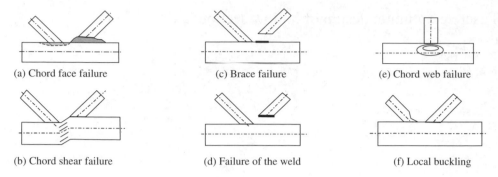

(a) Chord face failure   (c) Brace failure   (e) Chord web failure

(b) Chord shear failure   (d) Failure of the weld   (f) Local buckling

Figure 2.   Failure modes of square hollow sections.

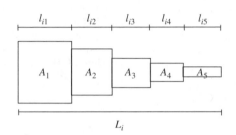

Figure 3.   Segmental analysis concept.

where $\gamma_{Mj} = 1.1$, $\gamma = b_0 / (2t_o)$, $n = s_0 / f_{y0}$ and $b = b_1 / b_0$ or $b = (b_1 + b_2) / (2b_0)$; being σ_0 the maximum compressive stress in the chord at the joint and f_{y0} the design value of the yield strength of this member.

This resistance verification format is valid if the joint geometry satisfies a set of conditions imposed in EC3. For joints outside this range of validity all the failure modes presented in Figure 2 should be considered.

If a joint does not have the necessary resistance and if it is not possible to change its geometry, or the dimension of the cross-sections, it is possible to increase the strength by means of appropriate stiffeners. However, these stiffeners significantly increase the labour costs and can jeopardize the aesthetic. The choice of the retrofitting configuration depends on the mechanism type that conditioned the strength.

2   OPTIMISATION ALGORITHM

According to the concept of segmental analysis, initially formulated by Templeman and Yates (Templeman et al. 1983) and used in this research as in previous works (Cruz 2000, Cruz & Santos 1990), all the bars are constituted by a set of segments with a known cross-section and with lengths to be calculated (Fig. 3).

For optimisation, it is required the minimization of the global volume of the structure (objective function) restrained to different constraints (displacements, strengths, etc.). The optimum solution tends to the one in which each bar is composed by only one section, i.e. in which all the bars have constant cross section.

In steel frames the solution obtained considering the global weight minimization criterion can be 20% more expensive than the solution obtained taking into account the real behaviour of the joints and the labour costs (Cruz 1999). In lattice girders the minimization of the global weight seems to be enough. In fact, in this kind of structures the influence of the labour costs is much smaller than in the case of steel frames, since the joints are significantly simpler and the production uses to be quite industrialized. On the other hand, for the kind of cross sections adopted in this research (square hollow sections), the weight minimization almost guarantees the minimization of the lateral surface and consequently the minimization of the coating and fire protection.

The "Revised Simplex Method" was adopted simultaneously with the concept of "Segmental Analysis". The detailed description of the linear-programming and of several versions of the Simplex method can be found in (Hillier & Lieberman 1990).

Besides all other constraints, which will be presented in this section, it is necessary to impose that in each bar the sum of the lengths of the segments is equal to the total length of the bar:

$$\Sigma\, l_{ij} = L_i \tag{3}$$

The global volume of the structure is given by the following equation:

$$V = l_{11}A_1 + l_{12}A_2 + l_{13}A_3 + l_{14}A_4 + l_{15}A_5 + \ldots + l_{ij}A_j + \ldots \tag{4}$$

where i = index of the bar; j = index of the segment; l_{ij} = length of the segment j of the bar i (which will be evaluated) and the A_j = area of the cross-section of this segment.

In the case of statically indeterminate structures, any change in a cross-section implies a modification of the forces installed in the bars. In this case, the solution is obtained through an iterative process, in which in each cycle the structure is recalculated, having into account the sections of the bars obtained in the previous cycle.

2.1 Displacement constraints

In each node with restrained displacement is imposed that the value of the displacement, calculated by the virtual work principle, is lower than the acceptable displacement (δ).

$$\sum \frac{N_i \bar{N_i}}{E_j A_j}\, l_{ij} \le \delta \tag{5}$$

where N_i is the force in bar i, for a given external action, and $\bar{N_i}$ is the force in the same bar, due to the virtual unit force associated to the restrained displacement.

2.2 Strength constraints

The strength constraints are introduced imposing a zero length to the segments that presents a strength lower than the installed stress.

Knowing the type of steel used the calculation of the strength in a compressed bar is immediate:

$$N_{b,Rd} = \chi\, A\, \frac{f_y}{\gamma_M} \tag{6}$$

in which A = area of the cross-section; f_y = nominal value of the yield stress and γ_M = partial safety factor. The bar segments that does not obey to this condition must be excluded.

According to the EC3, the design-buckling load of a compression bar is:

$$N_{b,Rd} = \chi\, A\, \frac{f_y}{\gamma_M} \tag{7}$$

in which χ = reduction factor for the relevant buckling method is given by the following equation:

$$\chi = \frac{1}{\phi + \sqrt{\phi^2 - \bar{\lambda}^2}} \tag{8}$$

where ϕ is given by the following equation:

$$\phi = 0.5\left[1 + \alpha\left(\bar{\lambda} - 0.2\right) + \bar{\lambda}^2\right] \tag{9}$$

in which α assumes the value 0.21 for hot rolled hollow sections, in the case of rolled tubular sections, and $\bar{\lambda}$ = non-dimensional slenderness shall be obtained from:

$$\bar{\lambda} = \frac{l_b / i}{93.9\,\sqrt{235 / f_y}} \tag{10}$$

in which i = radius of gyration of the section and l_b = buckling length of the bar.

2.3 *Imposition that some bars have the same section*

By aesthetical and constructively reasons it is frequent to impose that some groups of elements have the same section. The inclusion of this type of constraint increases the practical applicability of the model and is particularly useful in the case of statically indeterminate structures. These constraints are introduced imposing that the homologous segments of these bars have the same section.

3 IMPLEMENTATION OF THE ALGORITHM

The constraints presented above can be written in the following matrix notation:

$$\boldsymbol{M} . \boldsymbol{x} \leq \boldsymbol{b} \tag{11}$$

or under the form:

$$
\begin{array}{c}
\text{NELEM} \left\{ \\
\text{NRGRU} \left\{ \\
\text{NRDES} \left\{ \\
\text{NRTEN} \left\{ \\
\text{NRLIG} \left\{
\end{array}
\underbrace{
\begin{bmatrix}
\boldsymbol{M}_1 & \text{Length} \\
\boldsymbol{M}_2 & \text{Groups} \\
\boldsymbol{M}_3 & \text{Displacem.} \\
\boldsymbol{M}_4 & \text{Strength} \\
\boldsymbol{M}_5 & \text{Joints}
\end{bmatrix}
}_{\text{NELEM} \times \text{NSEGM}}
\times \{ \boldsymbol{X} \} =
\begin{Bmatrix}
\boldsymbol{b}_1 \\
\boldsymbol{b}_2 \\
\boldsymbol{b}_3 \\
\boldsymbol{b}_4 \\
\boldsymbol{b}_5
\end{Bmatrix}
\tag{12}
$$

in which NELEM = number of elements; NRGRU = number of constraints necessary to guaranty that a given number of bars have the same section; NRDES = number of displacement constraints; NRTEN = number of strength constraints; NRLIG = number of constraints imposed by the fact that some segments do not verify geometrical conditions or by the fact that the strength of those segments is lower than the value given by equation (1) and X = vector of the unknown quantities (length of all segments).

$$\boldsymbol{X} = \underbrace{\left\{ l_{11} \ \ l_{12} \ \ l_{13} \ \ l_{14} \ \ l_{15} \ \cdots \ l_{ij} \ \cdots \right\}}_{\text{NELEM} \times \text{NSEGM}}{}^{\text{T}} \tag{13}$$

The meaning of the sub-matrixes presented in equation (12) will be exposed in the following sections.

3.1 *Length constraints*

The number of rows of the matrix $\boldsymbol{M}_1$ is equal to the number of elements of the lattice girder. The number of columns is equal to the number of elements multiplied by the number of segments. This matrix is diagonal with the following aspect:

$$
\boldsymbol{M}_1 =
\left.
\underbrace{
\begin{bmatrix}
\boxed{1\ 1\ 1\ 1\ 1} & & & & \\
& \boxed{1\ 1\ 1\ 1\ 1} & & [0] & \\
& & \cdots & & \\
& [0] & & \boxed{1\ 1\ 1\ 1\ 1} & \\
& & & & \boxed{1\ 1\ 1\ 1\ 1}
\end{bmatrix}
}_{\text{NELEM} * \text{NSEGM}}
\right\} \text{NELEM}
\tag{14}
$$

The components of the vector $\boldsymbol{b}_1$ are the lengths of the elements.

$$\boldsymbol{b}_1 = \underbrace{\left\{ L_1 \ \ L_2 \ \cdots \ L_i \ \cdots \ L_n \right\}}_{\text{NELEM}}{}^{\text{T}} \tag{15}$$

398

3.2 Imposition that several bars have the same section

$$
M_2 = \begin{bmatrix}
+1 & 0 & 0 & 0 & 0 & -1 & 0 & 0 & 0 & 0 & \cdots & 0 \\
0 & +1 & 0 & 0 & 0 & 0 & -1 & 0 & 0 & 0 & \cdots & 0 \\
0 & 0 & +1 & 0 & 0 & 0 & 0 & -1 & 0 & 0 & \cdots & 0 \\
0 & 0 & 0 & +1 & 0 & 0 & 0 & 0 & -1 & 0 & \cdots & 0 \\
\cdots & \cdots & \cdots & \cdots & \cdots & \cdots & \cdots & \cdots & \cdots & \cdots & \cdots & \cdots
\end{bmatrix} \Big\} \text{ NRGRU} \tag{16}
$$

$$\underbrace{\hphantom{M_2 = \begin{bmatrix} +1 & 0 & 0 & 0 & 0 & -1 & 0 & 0 \end{bmatrix}}}_{\text{NELEM * NSEGM}}$$

For each pair of bars to which is imposed the same section, it is necessary to define a number of constraints equal to the number of materials minus one. In these constraints it is imposed the same lengths to the homologous segments. As an example, the length of the segment 1 of the bar i must be equal to the length of the segment 1 of the bar j, and thus successively.

All the components of the vector b_2 are equal to zero.

$$
b_2 = \underbrace{\{ 0 \quad 0 \quad \cdots \quad 0 \quad \cdots \quad 0 \}}_{\text{NRGRU}}{}^{\text{T}} \tag{17}
$$

3.3 Displacements constraints

The product of each row of the matrix M_3 by the vector of the unknown quantities (length of the segments of all the elements) must be lower or equal to the displacement δ of a given node. Therefore the components of the matrix M_3 are equal to $(N_i\, \bar{N}_i)/(E_j\, A_j)$. In this matrix, the bar (or the two bars) over a variable identifies the first (or the second) virtual unit force.

$$
M_3 = \begin{bmatrix}
\dfrac{N_1\bar{N}_1}{E_1 A_1} & \dfrac{N_1\bar{N}_1}{E_2 A_2} & \dfrac{N_1\bar{N}_1}{E_3 A_3} & \dfrac{N_1\bar{N}_1}{E_4 A_4} & \dfrac{N_1\bar{N}_1}{E_5 A_5} & \cdots & \dfrac{N_i\bar{N}_i}{E_j A_j} & \cdots \\[3mm]
\dfrac{N_1\bar{\bar{N}}_1}{E_1 A_1} & \dfrac{N_1\bar{\bar{N}}_1}{E_2 A_2} & \dfrac{N_1\bar{\bar{N}}_1}{E_3 A_3} & \dfrac{N_1\bar{\bar{N}}_1}{E_4 A_4} & \dfrac{N_1\bar{\bar{N}}_1}{E_5 A_5} & \cdots & \dfrac{N_i\bar{\bar{N}}_i}{E_j A_j} & \cdots \\[3mm]
\cdots & \cdots & \cdots & \cdots & \cdots & \cdots & \cdots & \cdots
\end{bmatrix} \Big\} \text{ NRDES} \tag{18}
$$

$$\underbrace{\hphantom{M_3 = \begin{bmatrix} \dfrac{N_1\bar{N}_1}{E_1 A_1} & \dfrac{N_1\bar{N}_1}{E_2 A_2} & \dfrac{N_1}{E_3} \end{bmatrix}}}_{\text{NELEM * NSEGM}}$$

The components of the vector b_3 are equal to the maximum values of the imposed displacements.

$$
b_3 = \underbrace{\{ \delta_1 \quad \delta_2 \quad \cdots \quad \delta_n \}}_{\text{NRDES}}{}^{\text{T}} \tag{19}
$$

3.4 Strength constraints

$$
M_4 = \begin{bmatrix}
1 & 0 & 0 & 0 & 0 & 0 & 0 & \cdots & 0 \\
0 & 0 & 1 & 0 & 0 & 0 & 0 & \cdots & 0 \\
\cdots & \cdots & \cdots & \cdots & \cdots & \cdots & \cdots & \cdots \\
0 & 0 & 0 & 0 & 1 & 0 & 0 & \cdots & 0 \\
0 & 0 & 0 & 0 & 0 & 1 & 0 & \cdots & 0
\end{bmatrix} \Big\} \text{ NRTEN} \tag{20}
$$

$$\underbrace{\hphantom{M_4 = \begin{bmatrix} 1 & 0 & 0 & 0 & 0 & 0 & 0 \end{bmatrix}}}_{\text{NELEM * NSEGM}}$$

The strength constraints are introduced imposing a zero length to the segments in which the installed stress is greater than the resistance stress.

The matrix M_4 has the same number of rows than the number of segments in which the resistance is not verified.

All components of the vector b_4 are equal to zero.

$$b_4 = \underbrace{\{\, 0 \quad 0 \quad \cdots \quad 0 \quad \cdots \quad 0 \,\}^{\mathrm{T}}}_{\text{NRTEN}} \tag{21}$$

3.5 Geometrical constraints (verification of the joints)

The matrix M_5 has the same number of rows than the number of segments of the braces and chords, in which the geometrical constraint conditions are not verified neither the value given by equation (1).

$$M_5 = \left.\underbrace{\begin{bmatrix} 1 & 0 & 0 & 0 & 0 & 0 & 0 & \cdots & 0 \\ 0 & 0 & 1 & 0 & 0 & 0 & 0 & \cdots & 0 \\ \cdots & \cdots & \cdots & \cdots & \cdots & \cdots & \cdots & \cdots & \cdots \\ 0 & 0 & 0 & 0 & 0 & 1 & 0 & \cdots & 0 \end{bmatrix}}_{\text{NELEM} * \text{NSEGM}} \right\} \text{NRLIG} \tag{22}$$

All components of the vector b_5 are equal to zero.

$$b_5 = \underbrace{\{\, 0 \quad 0 \quad \cdots \quad 0 \,\}^{\mathrm{T}}}_{\text{NRLIG}} \tag{23}$$

The consideration of those constraints increases the complexity of the model, as it is not trivial the decision of which segments cannot be used in each bar. In fact, as it can be observed in equation (1) the characteristics of each brace depend on the characteristics of the other bars converging in this joint. Thus, if any segment does not satisfy any of the mentioned conditions, the solution could be the elimination of this segment or the increase of the section of any of the contiguous bars.

4 NUMERICAL APPLICATION

A crane, composed by 38 bars and 21 nodes, with the geometry illustrated in Figure 4, was analysed, considering the use of four square hollow sections of steel S355: 60-4; 100-5; 140-8 and 180-12. In this notation the first number represents the width of the section and the second the thickness of these tubes. The areas of these sections are, respectively: 8.79; 18.73; 41.55 and 79.09 cm^2.

The nodes of the lower chord are submitted to the action of vertical descending forces of 10 KN, with the only exception of the node of the cantilever end where the force has a value of 100 KN.

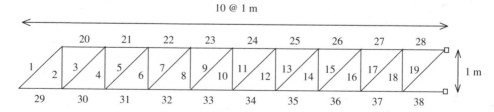

Figure 4.   Geometry of the crane.

Table 1.   Results of the different examples.

Bar	10 DC	15 DC	15 DC & SC	20 DC
Weight (kN)	926.3	608.4	616.0	501.6

This example was solved considering three values of the maximum vertical displacement of the cantilever end: 10, 15 and 20 cm. Additionally the strength constraints were considered.

The results of those analyses are summarised in Table 1. Starting by the analysis of the last column, referring to case in which a maximum displacement of 20 cm is allowed and the strength is not constrained, the consequence of the high value imposed to the displacement constraint is that almost all bars are constituted by the smaller cross-section. It must be stated out that the weight of the structure increases in 22.7% and 82.8% in the case of considering, respectively, the values of 15 or 10 cm to the displacement constraint (columns 3 and 2).

Taking as a reference the solution obtained imposing the maximum displacement of 15 cm, without any strength constraints, it can be concluded that the weight of the structure increases 8.3% if the strength constraints are also considered (fourth column of Table 1).

It must be mentioned that the consideration of the strength constraints does not change the results in the case in which the maximum displacement is equal to 10 cm. This means, that in this case the structure is conditioned by the displacement constraint.

In the last case analysed, the consideration of strength constraints does not change the solution with the maximum displacement of 20 cm. This means that to displacements equal or greater than 15 cm the solution is conditioned by the strength constraints. A more complete discussion of these results can be found in Cruz (2000).

5   CONCLUSIONS

A model of linear optimisation was applied to the design of hollow section lattice girders.

Imposing the geometry of the structure and some constraints to the displacements, the strengths and dimensions of the cross sections, the model finds the ideal section of each bar, making the design process quicker and easier. This includes the procedures proposed in the Eurocode 3 for the strength verification of the welded joints. The optimising methodology proposed is simple and elegant.

Although this research was restricted to the study of planar structures, with square hollow cross sections and welded joints, it is easy to extend it to three-dimensional structures with other sections and different types of joints.

The example presented is elucidative of the practical applicability of the model. The results obtained with this model were recently assessed with the use of genetic algorithms (Almeida 2001).

REFERENCES

Almeida, J.R. & Cardoso, J.M.B.B. 2001. Dimensionamento de estruturas metálicas utilizando algoritmos genéticos. In A. Lamas, P. Vila-Real & L. Simões da Silva (eds), *Construção Metálica e Mista 3*: 669–678, Aveiro, *6–7 December 2001*. Portugal.

Cruz, P.J.S. 2000. Optimização discreta de estruturas tubulares. *Revista Portuguesa de Engenharia de Estruturas* (48): 39–47.

Cruz, P.J.S. 1999. Economic studies of steel building frames, *EPMESC VII – Education, Practice and Promotion of Computational Methods in Engineering using Small Computers*: 615–623, Macau, *2–5 August 1999*, Portugal.

Cruz, P.J.S. & Santos, R. 1990, Análise segmental de estruturas articuladas planas. *Revista de Materiais de Construção*, 30: 42–44.

Eurocode 3 1994. ENV–1993-2, Design of Steel Structures – Bridges, CEN, European Committee for Standardization, Document CEN/TC 250/SC 3 – N 419 E, Brussels.

Hillier, F.S. & Lieberman, G.J. 1990. *Introduction to operations research*, McGraw-Hill International Editions, Industrial Engineering Series, 5th Edition.

Templeman, A.B. & Yates, D.F. 1983. The segmental method for the discrete optimum design of structures, *Engng. Optim.*, 6, pp. 145–155.

Flexural torsional analysis of thin-walled curved beams with open cross section accounting warping

H.J. Duan & Q.L. Zhang
Department of Building Engineering, Tongji University, China

ABSTRACT: A finite element formulation for the geometric nonlinear analysis of thin-walled curved beams with open sections is performed. Using the Update Lagrangian formulation, the displacement field is described, and the warping degree of freedom and curvature effects are taken into consideration to simulate the structural behaviors of thin-walled curved beams. The proposed model is based on the fundamentals of solid mechanics and follows the basic principles of the virtual work. All displacement parameters including the warping deformation are defined the centroid axis, so that the couple terms of bending and torsion are added to elastic strain energy. The improved arc-length method is adopted to trace the nonlinear load-deflection paths. Finally numerical examples are presented to illustrate the validity and accuracy of the proposed numerical approach, also found that warping deformation is not negligible in the analysis of thin-walled curved beams with open section.

1 INTRODUCTION

Thin-walled beams with open section made of high strength materials are used extensively in aerospace industry, civil engineering, ship construction and etc. Most thin-walled structures are slender and have open section shapes. Where the centroid and shear center don't coincide when a transverse load is applied away from the shear center it cause torque. Because of the open nature of the sections, this torque induces warping in the beam.

Compared with straight beam, deformation behaviors of curved beam are coupled. So exact analysis of thin-walled curved beams is usually very complex for calculations. Timoshenko & Gere (1961) derived the governing equations for buckling of curved beams neglecting the effect of warping. Attard (1986) developed two finite element formulations for the calculation of the lateral buckling load for elastic straight prismatic thin-walled open beams under conservative static loads. Attard & Somervaille (1987) presented of finite element formulation for elastic nonlinear static analysis of thin-walled open beams under combined bending and torsion. Chan & Kitipornchai (1987) investigated geometric nonlinear analysis of structures comprising members of asymmetric thin-walled open section. Many literatures dealt with the thin-walled straight beams, only a few works have done to analyze thin-walled curved beams accounting to warping.

In this paper, a numerical model for analyzing thin-walled curved spatial beams with open cross section is presented. The proposed model is based on the fundamentals of solid mechanics and the basic principles of virtual work. Large deformations with small strains are taken into formulation, while considering the effecting of warping and curvature effects to simulate the structure behaviors of thin-walled curved members more exactly. An improved arc-length method in conjunction with a combined initial-iterative method was adopted to cope with the numerical instability in the snap-through buckling analysis. Finally, the reliability of proposed model is verified through numerical examples.

2 BASIC THEORY

2.1 Basic assumptions

In this paper, the following assumptions are adopted:

1. The thin-walled curved beam is linearly elastic and prismatic.

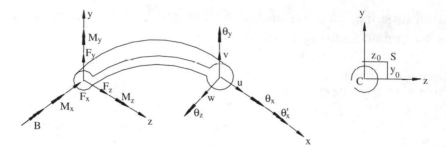

Figure 1.   Coordinate system of thin-walled curved beam.

2. The cross section is rigid with respect to in-plane deformation except for warping deformation.
3. The effects of local buckling deformations are negligible.
4. Shearing deformation of the middle surface of the member is negligible.
5. Strains are small but displacement and rotations can be large.

2.2   Kinematics and strains

Figure 1 shows an element of general thin-walled curved beam with open section. The right-hand orthogonal coordinate system x, y, z has been chosen such that y and z pass through the undeformed end cross-section centroid. There are seven actions (F_x, F_y, F_z, M_x, M_y, M_z and B) with corresponding displacement component ($u, v, w, \theta_x, \theta_y, \theta_z$ and θ_x') that can be applied at each end of thin-walled curved element.

U, V and W denote the displacement of any point p (x, y, z) of cross section in the x, y and z direction, respectively.

$$U = u - v'y - \left(w' - \frac{u}{R}\right)z - \left(\theta_x' + \frac{v'}{R}\right)\omega \tag{1a}$$

$$V = v - \theta_x(z - z_0) \tag{1b}$$

$$W = w + \theta_x(y - y_0) \tag{1c}$$

Where u, v and w are the displacement components of the centroid and θ_x is the rotation of the cross-section around the shear centers. y_0 and z_0 are the distance between centroid and shear center in the direction of y and z, respectively. ω is the sectorial area, R is radius of curved beam.

2.3   Strain–displacement relations

Linear strain–displacement and nonlinear strain–displacement relation are expressed as follows:

$$e_{xx} = \frac{\partial U}{\partial x} + \frac{W}{R}, \quad e_{xy} = \frac{\partial V}{\partial x} + \frac{\partial U}{\partial v}, \quad e_{xz} = \frac{\partial W}{\partial x} - \frac{U}{R} + \frac{\partial U}{\partial z} \tag{2a, b, c}$$

$$\eta_{xx} = \frac{1}{2}\left[\left(\frac{\partial V}{\partial x} + \frac{W}{R}\right)^2 + \left(\frac{\partial V}{\partial x}\right)^2 + \left(\frac{\partial W}{\partial x} - \frac{U}{R}\right)^2\right] \tag{2d}$$

$$\eta_{xy} = \frac{1}{2}\left[\frac{\partial U}{\partial y}\left(\frac{\partial U}{\partial x} + \frac{W}{R}\right) + \frac{\partial V}{\partial y}\frac{\partial V}{\partial x} + \frac{\partial W}{\partial y}\left(\frac{\partial W}{\partial x} - \frac{U}{R}\right)\right] \tag{2e}$$

$$\eta_{xz} = \frac{1}{2}\left[\frac{\partial U}{\partial z}\left(\frac{\partial U}{\partial x} + \frac{W}{R}\right) + \frac{\partial V}{\partial x}\frac{\partial V}{\partial z} + \frac{\partial W}{\partial z}\left(\frac{\partial W}{\partial x} - \frac{U}{R}\right)\right] \tag{2f}$$

Corresponding to assumption 2, we get for stress resultants of following expression:

$$F_x = \int_A \sigma_{xx}\,dA, \quad F_y = \int_A \tau_{yx}\,dA, \quad F_z = \int_A \tau_{zx}\,dA \tag{3a, b, c}$$

$$M_x = \int_A \left[\tau_{zx}\left(y - y_0 - \frac{\partial \omega}{\partial z}\right) - \tau_{yx}\left(z - z_0 - \frac{\partial \omega}{\partial y}\right) \right] dA \tag{3d}$$

$$M_y = \int_A \sigma_{xx} z\,dA, \quad M_z = \int_A \sigma_{xx} y\,dA, \quad B = \int_A \sigma_{xx}\omega\,dA \tag{3e, f, g}$$

2.4 Virtual work equation

For general continuum, the virtual work equation is expressed as:

$$\int_{t_V} C_{ijkl} e_{kl}\delta\, e_{ij}{}^t dV + \int_{t_V} {}^t\sigma_{ij}\delta\eta_{ij}{}^t dV = {}^{t+\Delta t}Q - \int_{t_V} {}^t\sigma_{ij}\delta\, e_{ij}{}^t dV \tag{4}$$

For beam element, the first term and second term of Eq. (4) are expressed as respectively:

$$\Pi_E = \int_{t_V} (E e_{xx}\delta\, e_{xx} + 2G e_{xy}\delta\, e_{xy} + 2G e_{xz}\delta\, e_{xz})^t dV \tag{5a}$$

$$\Pi_G = \int_{t_V} (\sigma_{xx}\delta\eta_{xx} + 2\sigma_{xy}\delta\eta_{xy} + 2\sigma_{xz}\delta\eta_{xz})^t dV \tag{5b}$$

Substituting Eqs. (2a, b, c) into Eq. (5a), the equations can be derived as:

$$\Pi_E = \frac{1}{2}\int_0^L \left[EA\delta\left(u' + \frac{w}{R}\right)^2 + EI_y\delta\left(w'' + \frac{w}{R}\right)^2 + EI_z\delta\left(v'' - \frac{\theta_x}{R}\right)^2 \right.$$
$$+ EI_\omega\delta\left(\theta_x'' + \frac{v''}{R^2}\right)^2 + GJ\delta\left(\theta_x' + \frac{v'}{R}\right)^2 + 2EI_{\omega y}\delta\left(w'' + \frac{w}{R^2}\right)\left(\theta_x'' + \frac{v''}{R}\right)$$
$$\left. + 2EI_{\omega z}\delta\left(v'' - \frac{\theta_x}{R}\right)\left(\theta_x'' + \frac{v''}{R}\right) + 2EI_{yz}\delta\left(v'' - \frac{\theta_x}{R}\right)\left(w'' + \frac{w}{R^2}\right) \right]dx \tag{6a}$$

Substituting Eqs. (2d, e, f) into Eq. (5b), the equations can be derived as:

$$\Pi_G = \frac{1}{2}\int_0^L \left\{ F_x\delta\left[v'^2 + \left(w' - \frac{u}{R}\right)^2 + (y_0^2 + z_0^2)\theta'^2\right] + F_y\delta\,\theta\left(w' - \frac{u}{R}\right) + F_z\delta(v'\theta) \right.$$
$$+ M_x\delta\left[\left(w' - \frac{u}{R}\right)\left(v'' - \frac{\theta_x}{R}\right) - v'\left(w'' - \frac{u'}{R}\right)\right] + M_y\delta\left[\frac{1}{R}\left(w' - \frac{u}{R}\right)^2 - 2v'\theta'\frac{v'^2}{R}\right]$$
$$\left. + M_z\delta\left[\left(\theta_x' + \frac{v'}{R}\right)\left(w' - \frac{u}{R}\right)\right] + 2\frac{B}{R}\delta\left(w' - \frac{u}{R}\right)\left(\theta_x - \frac{v'}{R}\right) + \bar{K}\delta\left(\theta_x + \frac{v'}{R}\right)^2 \right\}dx \tag{6b}$$

where $\bar{K}$ is the Wagner efficient. The geometrical properties of cross-section are defined by following quantities:

$$A = \int_A dA, \quad I_y = \int_A z^2\,dA, \quad I_z = \int_A y^2\,dA, \quad I_{yz} = \int_A yz\,dA$$

$$I_\omega = \int_A \omega^2\,dA, \quad I_{\omega y} = \int_A y\omega\,dA, \quad I_{\omega z} = \int_A z\omega\,dA$$

3 THE STIFFNESS MATRIX

Figure 1 shows the nodal displacement vector of thin-walled curved beam element including restrained warping effect. To accurately express element deformation, pertinent shape functions are necessary. In this study, a linear displacement field is adopted for u, and a cubic displacement field for other displacements. The incremental displacements can be expressed as:

$$u = [f_1][u_1 \ \ u_2]^T \tag{7a}$$

$$v = [f_2][v_1 \ \ \theta_{z1} \ \ v_2 \ \ \theta_{z2}]^T \tag{7b}$$

$$w = [f_2][w_1 \ \ \theta_{y1} \ \ w_2 \ \ \theta_{y2}]^T \tag{7c}$$

$$\theta_x = [f_2][\theta_{x1} \ \ \theta'_{x1} \ \ \theta_{x2} \ \ \theta'_{x2}]^T \tag{7d}$$

where:

$$[f_1] = [1 - \xi \ \ \xi], \ \xi = x/l$$

$$[f_2] = [2\xi^3 - 3\xi^3 + 1 \quad (\xi^3 - 2\xi^3 + 1)l \quad -2\xi^2 + 3\xi^3 \quad (\xi^3 - \xi^2)l]$$

Substituting Eqs. (7a, b, c, d) into Eqs. (6a, b) and integration along the element length, the equilibrium equation of a thin-walled curved beam element is obtained as:

$$(\mathbf{K}_E^e + \mathbf{K}_G^e)\Delta\mathbf{U}^e = \mathbf{P}^e - \mathbf{F}^e \tag{8}$$

where:

$$\Delta\mathbf{U}^e = [u_1 \ v_1 \ w_1 \ \theta_{x1} \ \theta_{y1} \ \theta_{z1} \ \theta'_{x1} \ u_2 \ v_2 \ w_2 \ \theta_{x2} \ \theta_{y2} \ \theta_{z2} \ \theta'_{x2}]$$

$$\mathbf{P}^e = [F_{x1} \ F_{y1} \ F_{z1} \ M_{x1} \ M_{y1} \ M_{z1} \ B_1 \ F_{x2} \ F_{y2} \ F_{z2} \ M_{x2} \ M_{y2} \ M_{z2} \ B_2]$$

In Eq. (8), $\mathbf{K}_E^e$ and $\mathbf{K}_G^e$ are 14 × 14 element elastic and geometric stiffness matrices for the thin-walled curved beam element in local coordinate respectively. Definition of the nodal displacement, evaluation of element stiffness matrices, and assembly of element stiffness matrices for entire structure using the coordinate transformation lead to global equilibrium matrix equation as follows:

$$(\mathbf{K}_E + \mathbf{K}_G)\Delta\mathbf{U} = \mathbf{P} - \mathbf{F} \tag{9}$$

4 ITERATION PROCEDURE

When the external loads P assumed to be a proportional load, P can be replaced by λP_0. Where P_0 is any reference load vector and λ is loading parameter. Kweon & Hong (1994) proposed an improved arc-length method while conducting the postbuckling analysis of composite laminate cylindrical panels.

For the $(n + 1)$-th iteration of m-th load step, Eq. (9) can be rewritten using an increment of load parameter $\Delta\lambda^{n+1}$ as:

$$\mathbf{K}_T \cdot \Delta\mathbf{U}^{n+1} = \Delta\mathbf{P}(\lambda^n) + \Delta\lambda^{n+1} \cdot \mathbf{P}_0 \tag{10}$$

where

$$\Delta\mathbf{P}(\lambda^n) = \lambda^n\mathbf{P}_0 - \mathbf{F}^{n+1} \tag{11}$$

The increment displacement $\Delta\mathbf{U}^{n+1}$ can be calculated by:

$$\Delta\mathbf{U}^{n+1} = \Delta\mathbf{U}^{n+1}(\lambda_n) + \Delta\lambda^{n+1}\mathbf{U}_m \tag{12}$$

where

$$\Delta\mathbf{U}^{n+1}(\lambda^n) = \mathbf{K}_T^{-1}\Delta\mathbf{P}(\lambda^n), \ \ \mathbf{U}_m = \mathbf{K}_T^{-1}\mathbf{P}_0$$

Based on following the relation:

$$\Delta_m\mathbf{U}^{n+1} = \Delta_m\mathbf{U}^n + \Delta\mathbf{U}^{n+1}, \ \Delta_m\mathbf{U}^T\Delta_m\mathbf{U} = (\Delta l)^2 \tag{13a, b}$$

The incremental load parameter can be determined from Eqs. (13a, b):

$$\Delta\lambda^{n+1} = \frac{-B \pm \sqrt{B^2 - 4AC}}{2A} \tag{14}$$

where

$$A = \mathbf{U}_m^T \cdot \mathbf{U}_m, \quad B = 2\Delta_m\mathbf{U}^n + (\Delta\mathbf{U}^{n+1}(\lambda^n))^T \cdot \mathbf{U}_m$$

$$C = (\Delta_m\mathbf{U}^n + \Delta\mathbf{U}^{n+1}(\lambda^n))^T \cdot (\Delta_m\mathbf{U}^n + \Delta\mathbf{U}^{n+1}(\lambda^n)) - \Delta l^2$$

Kweon & Hong (1994) recommended that both solutions are discarded and the corresponding load step is automatically restarted with the arc-length reduced to a half. To prevent the number of iterations from being too large, a maximum number of iteration was 15 in this study.

5 NUMERICAL EXAMPLES

5.1 *Large displacement 3D analysis of a 45° bend*

As shown in Figure 2, the large displacement response of cantilever 45° bend subject to a concentrated end load has been calculated. This example was first analyzed by Bathe (1979) using ADINA in 1979. This solution will be referred to ADINA1. The structural was modeled with eight equal straight beam ele-

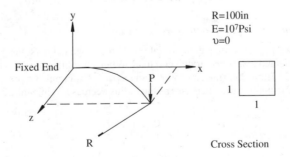

Figure 2. 3D details of 45° bend

407

Table 1. Bend displacement at the tip.

	Load P = 0 lbs			Load P = 300 lbs			Load P = 600 lbs		
	x	y	z	x	y	z	x	y	z
ADINA1	70.7	0.0	29.3	59.2	39.5	22.5	47.2	53.4	15.9
ADINA2	70.7	0.0	29.3	58.5	40.4	22.2	46.8	53.6	15.7
NACS1	70.7	0.0	29.3	59.2	39.5	22.6	47.2	53.4	15.9
NACS2	70.7	0.0	29.3	58.6	40.3	22.3	46.7	53.6	15.7
Spillers	70.7	0.0	29.3	60.0	43.8	22.9	46.2	59.2	19.1
This study	70.7	0.0	29.3	58.6	40.4	22.2	47.0	53.5	15.7

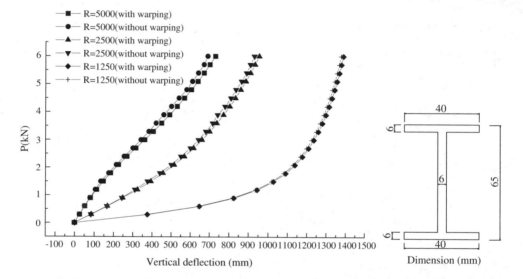

Figure 3. Variation of warping effect with radius R.

ments and loaded up to p = 600 with 60 equal load steps. Another improved solution was provided by Bathe (1990) and this will be referred to ADINA2. They calculated this problem by 3D solid elements. Since the two results agree closely with each other, their solutions are widely accepted as a precise solution. Chen (1993) also analyzed this example structure using NACS. The obtained results are referred to in Table 1 as NACS[1] and NACS[2] according to the consideration and/or no consideration of iteration at each load step. In this study, the structure is modeled with four equal curved beam elements and loaded up to p = 600 with only 10 equal load steps. All results are listed in Table 1. The result of this study compares well with Bathe's solution, using the fewer element and load steps.

5.2 Large displacement 3D analysis of thin-walled beam

To study the effect of warping according to the variation of an average radius R, the cantilever curved beam I – shaped section subjected to a concentrated end load. When chord length is constant (L = 1990 mm), the radius of beam is 1250 mm, 2500 mm, and 5000 mm respectively. Figure 3 representing obtained results shows that the geometric nonlinearity increases as the radius decrease. While the effect of warp increased as the radius increase.

5.3 Clamped–clamped arch

A clamped shallow arch, which has the snap-through buckling behavior, was analyzed. The arch is loaded by a point load at top center, and the geometry and dimensions are shown in Figure 4. Two kinds

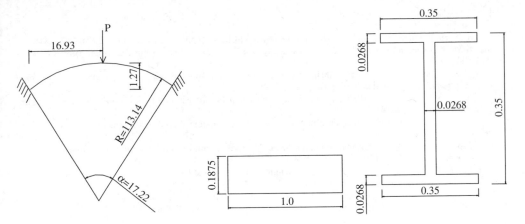

Figure 4.   Clamped–clamped shallow arch and cross section.

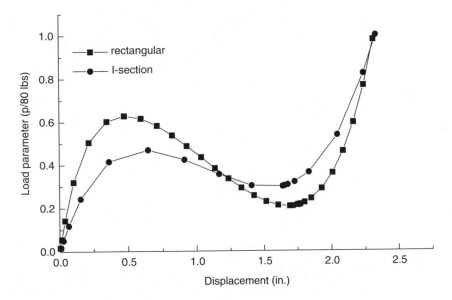

Figure 5.   Load–displacement relation of top center.

of cross section (rectangle and I-shape) with the same bending stiffness were selected. From Figure 5 can be seen, there is a notable discrepancy between solid and thin-walled section the structural behaviors can be different in accordance with the sectional type even if the bending stiffness is the same values.

6   CONCLUSIONS

Based on principal of virtual work, an Update Lagrangian formulation of thin-walled curved beam element for geometric nonlinear analysis has developed. Especially by including the effect of warping into formulation more exactly. The improved arc-length method is adopted to trace the nonlinear load-deflection paths. Finally numerical examples are presented to illustrate the validity and accuracy of the proposed numerical approach, also found that warping deformation isn't negligible in the analysis of thin-walled curved beams with open section.

REFERENCES

Attard, M.M. 1986. Lateral buckling analysis of beams by the FEM. *Computer and structures* 23(2): 217–231.
Attard, M.M. & Somervaille, I.J. 1987. Non-linear analysis of thin-walled open beams. *Computer and structures* 25(3): 437–443.
Bathe, K.J. & Boloourchi, S. 1979. Large displacement analysis of three-dimensional beam structures. *International journal of numerical method in engineering* 14: 961–986.
Chan, S.L. & Kitipornchai, S. 1987. Geometric nonlinear analysis of asymmetric thin-walled beam-columns. *Engineering Structures* 9(4): 243–254.
Chen, Z.Q. & Agar, T.J.A. 1993. Geometric nonlinear analysis of flexible spatial beam structures. *Computer and structures* 49(6): 1083–1094.
Kang, Y.J. & Yoo, C.H. 1994. Thin-walled curved beams. I: Formulation of nonlinear equations. *Journal of engineering mechanical* ASCE 120(10): 2072–2101.
Kweon, J.H. & Hong, C.S. 1994. An improved arc-length method for post buckling analysis of composite cylindrical panels. *Computer and structures* 53: 541–549.
Spillers, W.R. 1990. Geometric stiffness matrix for space frames. *Computer and structures* 36: 29–37.
Timoshenko, S.P. & Gere, J.M. 1961. *Theory of elastic stability*. New York: McGraw-Hill.

Computational Methods in Engineering and Science, Iu et al. (eds)
© 2003 Swets & Zeitlinger, Lisse, ISBN 90 5809 567 3

Behaviour factors of MR steel frames with semi-rigid connections – a parametric study

C. Rebelo
Faculty of Sciences and Technology, University of Coimbra, Coimbra, Portugal

L. Magalhães
Polytechnic Institute of Castelo Branco, Castelo Branco, Portugal

ABSTRACT: The paper presents a parametric study of the q-factor of steel moment resistant frames. The study considers semi-rigid beam-to-column joints with standard behaviour, characterized by their stiffness, resistance and deformation capacity. Three performance levels or limit states are considered, Serviceability Limit State (SLS), Damageability Limit State (DLS) and Ultimate Limit State (ULS).

1 INTRODUCTION

The structural safety check of building steel frames concerning seismic actions is usually carried out on an elastic-linear basis using design response spectra, which take into account the ductility (deformation capacity) and the energy dissipation capacity of the structure by means of the behaviour factor q. Adequate resistance of the structural elements (beams and columns) is required, including joints, whereas the necessary ductility to develop any post-elastic mechanism should be guaranteed.

Despite the inherent limitation of such a factor to take into account in a separate manner all the possible source and type of non-linearities that can arise from the extreme loading developed during an earthquake, it has been considered that it is very useful in design. In so far an effort has been made to clarify and identify every source of non-linearity which can have a decisive contribution to the behaviour factor, such as the type of structure, bracing, type of cross section, etc.

According to Eurocode 8 the characteristic semi-rigid behaviour of the joints in moment resisting (MR) steel frames is generally not included in their seismic behaviour. Connections in dissipative zones must have enough overstrength in order to allow for yielding of the beams, avoiding storey mechanisms and maximising the energy dissipation. However, connections exhibit, in general, semi-rigid behaviour and may contribute to the energy dissipation. Therefore they should be taken into account to the characterization of the q-factor.

2 BEHAVIOUR FACTOR EVALUATION

2.1 *Characterization of semi-rigid connections*

The moment-rotation behaviour of the connections are modelled as a bilinear relation characterized by three parameters: initial stiffness ($S_{j,ini}$), post-elastic stiffness ($S_{j,pl}$) and maximum elastic bending moment (M_j). The parametric study used the characteristics given in Table 1 for the connections, which were all combined to give 96 types of connections. Only those presented in bold were used in all the frames represented in Table 3. The beams were in every case of the type IPE330 ($M_{pl,Rd} = 221$ kNm).

2.2 *Seismic action*

Since a non-linear dynamic time-history analysis had to be performed, accelerograms were simulated according to the specification included in Eurocode 8 (EC8) and to an adapted procedure described in

Table 1. Characteristics of the connections.

	Elastic moment M_j [KNm]	Initial stiffness $S_{j,ini}$ [KNm/rad]	Post-elastic stiffness $S_{j,pl}$ [% of $S_{j,ini}$]
Pinned	50	$8EI_b/L = 39547$	0
	100	**12,5EI_b/L = 61792,5**	**0.1**
Paritial resistant	**150**	**25EI_b/L = 123585**	**1**
	200	37,5EI_b/L = 185377,5	**10**
Full resistant	250		
	300		

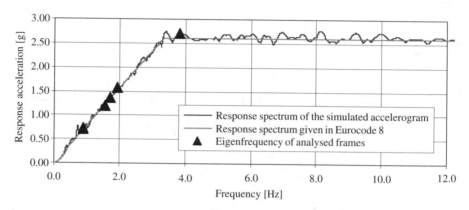

Figure 1. Comparison of response spectra for 5% damping, soil type A and 1 g peak ground acceleration.

Clough & Penzien (1993). Starting from a predefined time interval and a smooth power spectrum of the seismic acceleration, a time series is generated and modulated with a time function. Since the response spectrum of the resulting accelerogram will not have, in general, a form close to those given in EC8, the original power spectrum is repeatedly modified in order to generate a better approximation of the EC8 response spectrum considered.

Although an exact approximation is not possible, depending on the frequency interval chosen for the power spectrum, only a few iterations are usually needed to get a very close response spectrum, as the one presented in Figure 1 for soil condition type A as defined in EC8, which is within ±5% of the EC8 response spectrum.

2.3 Performance criteria

According to Grecea et al. (2002) structures designed against earthquakes have to comply several criteria related to strength, stiffness and ductility, which can be referred to conditions of interstorey drift, residual drift and rotation capacity of the connections.

Three limit states are considered in this study, in order to evaluate the corresponding behaviour factors. The Serviceability Limit State (SLS) concerns the case of low return period earthquakes (e.g. <20 years) and is quantified through a maximum allowed interstorey drift of 0.6% of the storey height. The Damageability Limit State (DLS) refers to rare earthquakes (475 years return period) and corresponds to serious structural and non structural damages, which can, however, be repaired without high costs or especially difficult repair techniques. Its quantification is based on an interstorey drift of 3% of the storey height. The Ultimate Limite State (ULS) is considered for very rare earthquakes (970 years return period). Although this limit state corresponds to very serious structural damages, it is expected that safety of people is guarantied avoiding collapse of the structure. Since large deformations are expected, the local ductility criteria are determinant for safety conditions to be verified. In this case maximum plastic rotations of 0.03 rad in the connections are considered, according to AISC (1997).

Table 2. Limit state criteria for global behaviour factor evaluation.

Safety level	Plastic rotation in connections [rad]	Plastic rotation in members [rad]	Interstorey drift [% of storey height]
SLS	–	–	0.6
DLS	–	–	3.0
ULS	0.03	(*)	–

(*) values depending on frame type and cross section geometry given in Table 3.

Table 3. Properties of the MR steel frames.

Type	P1	P3 × 1	P3 × 2	P6 × 2	P6 × 3	P3-2 × 2
Geometry						
Columns				HEB 260 $M_{pl,Rd}$ = 352.8 kNm		
Beams				IPE 330 $M_{pl,Rd}$ = 221.2 kNm		
Natural frequency [Hz]$^{(*)}$	3.81	1.71	1.58	0.9	0.88	1.71
θ_u [rad]	0.101	0.075	0.075	0.063	0.063	0.075
θ_y [rad]	0.012	0.011	0.011	0.010	0.010	0.011

(*) obtained for rigid connections.

The member ductility depends directly on the material ductility and on the cross section class, and it can be expressed in terms of rotational capacity. The rotation capacity can be evaluated as the ratio between the plastic rotation at the collapse state to the elastic one, according to the following formula:

$$R = \frac{\theta_u}{\theta_y} - 1$$

where θ_u is the ultimate plastic rotation and θ_y is the yielding rotation.

To calculate the ultimate and yielding rotation in order to compute the rotation capacity, different methods has been proposed. In this study we used the Mazzolani-Piluso semi-empirical method, recommended by ECCS (1994). This method depends on the slenderness of the member cross-section, to take into account the lateral-torsional buckling, and on the axial force. The values obtained for the six different frame types are presented in Table 3.

2.4 Methodology

The evaluation of the behaviour factors is based on the following methodology (Dubina et al. 2000):

1. The structure is submitted progressively to the accelerogram presented above multiplied by an amplification factor λ; the maximum amplification λ_y in elastic phase is registered;
2. The amplification is incremented to λ_u corresponding to reach the criterion established for the pertinent Limit State; λ_u is the value of the amplification corresponding to the inter-storey limit drift for SLS an DLS and the amplification corresponding to the limit rotation in the connection or members for ULS.

The behaviour factor is defined by the following relation. $q = \lambda_u/\lambda_y$.

413

Table 4. Comparison of q-factors considering the variation of the three connections parameters for ULS.

2.5 Geometric properties of the frames

In this parametric research several types of moment resisting steel frames were considered. Their characteristics are presented in Table 3. The natural frequency is plotted in Figure 1 together with the corresponding spectral value.

2.6 FE analysis

The FE analysis took into account the geometric non-linearities and the non-linear behaviour of the beam-column connections and of the columns' bases. The analysis was performed by means of the computer program LUSAS.

3 PARAMETRIC STUDY AND RESULTS

3.1 Influence of the connections using ULS criterion

In respect to the variation of the plastic moment, lower levels of M_j allow greater plastic to elastic rates of rotation capacity of the connections. Therefore, increasing values of q are expected and verified when M_j decreases (Table 4). The eventual exception is the frame P1, where the behaviour factor is controlled by the yielding of the columns' base and not by the non-linear behaviour of the connection as in all other cases.

When the initial stiffness ($S_{j,ini}$) is increased the q-factor increases, as expected, except for the 6-storey frames where the tendency is inverse of that one.

Increasing the post-elastic stiffness ($S_{j,pl}$) of the connections the q-factor increases in all situations.

Table 5.   Comparison of q-factors considering both limit states DLS and ULS and the connections stiffness.

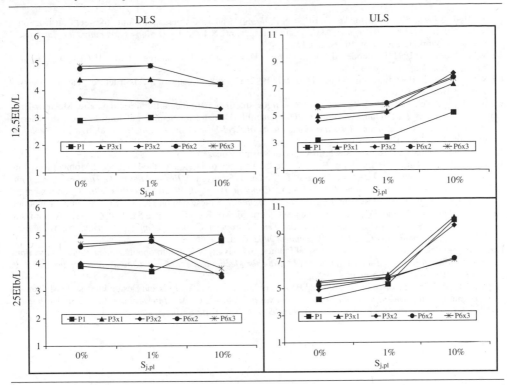

3.2   Influence of limit states and methodology in q-factor evaluation

When SLS is the used criterion, the q-factor is in general equal to 1, except for the case of connections with high $S_{j,ini}$ and low M_j in which the plastification occurs for low acceleration multipliers.

As expected, the q-factors for DLS are, in general, lower then those for ULS. Exceptions are the cases of very low M_j used together with frame type P1, since in this case the limit state was attained through plastic rotation at the column basis.

The hardening stiffness of the connections in the post-elastic phase has different effect when DLS and ULS are considered (Table 5), since an increase of $S_{j,pl}$ leads to an increase of the q-factor only when ULS is considered.

The methodology used here for the q-factor evaluation was compared with the Ballio-Setti method (ECCS 1994) and some differences were detected. This method leads, in general, to lower q-factors when low values of the connections' resistance and stiffness are considered.

4   CONCLUDING REMARKS

In this paper some preliminary results are presented concerning a parametric study of seismic behaviour factors for MR steel frames, obtained by non-linear dynamic time-history analysis and considering three limit states.

Conclusions about the influence of the connections' stiffness and resistance are according to those expected when the criterion used is the ULS. When DLS is considered, q-factors are in general lower and the influence of the post-elastic connection behaviour does not follow the one found when ULS is the criterion adopted.

REFERENCES

AISC. 1997. *Seismic Provisions for Structural Steel Buildings.* Chicago: American Institute of Steel Construction.

CEN/TC 250/SC 8/N306. 2001. Eurocode 8, Final Draft n.° 4 prEN 1998-1 – *Design of Structures for Earthquake resistance.* Working document. Brussels: CEN.

CEN/TC 250/SC 3. 1992. Eurocode 3 ENV 1993-1-1 – Design of Steel Structures – Part 1.1: *General Rules and Rules for Buildings.* Brussels:CEN.

CEN/TC 250/SC 3. 1992. Eurocode 3 ENV 1993-1-1: 1992/A2:1998 – Design of Steel Structure – Part 1.1: *Revised Annex J.* Brussels:CEN.

Clough, R.W. & Penzien, J. 1993. *Dynamics of Structures.* International Student Edition. Singapore: McGraw-Hill.

Dubina, D., Ciutina, A., Stratan, A., Dinu, F. 2000. Ductility Demand for Semi-Rigid Joint Frames. In Federico M. Mazozolani (ed.), *Moment Resistant Connections of Steel Frames in Seismic Areas, Design and Reability*: 371–408. New York: E & FN Spon.

ECCS. 1994. Manual on Design of Steel Structures in Seismic Zones, n.° 76.

Faggiano, B., De Matteis, G., Landolfo, R. 2002. On the Efficacy of Design Methods for Steel Moment Resisting Frames According to Eurocode. In A. Lamas & L. Simões da Silva (eds.), *Proceedings of the Third European Conference on Steel Structures, volume II: 1247–1258, Coimbra, Portugal, 19–20 September 2002.* Coimbra: cmm.

FEA Ltd, Lusas Finite Element System. *Lusas – Theory Manual 1.* Kingston upon Thames, Surrey, United Kingdom.

Grecea, D., Dinu, F., Dubina, D. 2002. Performance Criteria for MR Steel Frames in Seismic Zones. In A. Lamas & L. Simões da Silva (eds.), *Proceedings of the Third European Conference on Steel Structures, volume II: 1269–1278, Coimbra, Portugal, 19–20 September 2002.* Coimbra: cmm.

Ivany, M. 2000. Semi-Rigid Connections in Steel Frames. In M. Ivany & C. Baniotopoulos, *Semi-Rigid Connections in Structural Steelwork, CISM – International Centre for Mechanical Sciences – Courses and Lectures n.° 419, Part I:1–101.* New York: Springer.

Jaspart, J-P. 2000. Integration of the Joint Actual Behaviour Into the Frame Analysis and Design Process. In M. Ivany & C. Baniotopoulos, *Semi-Rigid Connections in Structural Steelwork, CISM – International Centre for Mechanical Sciences – Courses and Lectures n.° 419, Part I:103–166.* New York: Springer.

Computational Methods in Engineering and Science, Iu et al. (eds)
© 2003 Swets & Zeitlinger, Lisse, ISBN 90 5809 567 3

Analysis of complex parametric vibrations of plates and shells using the Bubnov-Galerkin approach

J. Awrejcewicz
Technical University of Lodz, Department of Automatics and Biomechanics, Poland

A.V. Krys'ko
Saratov State University, Department of Mathematics, Saratov, Russia

ABSTRACT: The Bubnov-Galerkin method is applied to reduce partial differential equations governing flexible plates and shells dynamics to a system with finite degrees-of-freedom. Chaotic behaviour of systems with various degrees-of-freedom is analyzed.

1 MULTIBODY DYNAMICAL SYSTEMS

Chaotic vibrations exhibited by lumped systems with many degrees of freedom are quite rarely investigated. However, recently remarkable progress in this field has been observed: hydrodynamic processes governed by ordinary differential equations have been investigated (Swinney & Gollub 1981), the finite dimensional discretized (with respect to spatial coordinates) models of Ginzburg-Landau equations (L'vov et al. 1981), multidimensional models of radiophysical systems governing the dynamics of coupled oscillators and generators (Waller & Kapral 1984), as well as chains of oscillators and generators have been analysed. In the majority of the cited works a problem of modeling a continuous system by a lumped (discrete) system governed by ordinary differential equations is addressed.

Nowadays various approximational methods are applied to construct lumped systems. In this work we use the Bubnov-Galerkin method, which has been successfully applied to different types of differential equations: elliptic, hyperbolic and parabolic ones. In the monographs (Krysko & Kutsemako 1999, Awrejcewicz & Krysko 2003) a review of the Bubnov-Galerkin method (MBG) is given including a discussion of its convergence for various classes of differential equations.

2 PROBLEM FORMULATION AND THE BUBNOV-GALERKIN METHOD

The equations governing dynamics of a rectangular shell including both transversal and longitudinal harmonic excitations have the following form (Krysko & Kutsemako 1999, Awrejcewicz & Krysko 2003):

$$\frac{\partial^2 w}{\partial t} + \varepsilon \frac{\partial w}{\partial t} + \frac{1}{12(1-\mu^2)}\left[\frac{1}{\lambda^2}\frac{\partial^4 w}{\partial x^4} + \lambda^2 \frac{\partial^4 w}{\partial y^4} + 2\frac{\partial^4 w}{\partial x^2 \partial y^2}\right] +$$

$$+ \left[P_x \frac{\partial^2 w}{\partial x^2} + P_y \frac{\partial^2 w}{\partial y^2}\right] - L(w,F) - \nabla_k^2 F + k_x P_x + k_y P_y - q = 0, \tag{1}$$

$$\frac{1}{\lambda^2}\frac{\partial^4 F}{\partial x^4} + \lambda^2 \frac{\partial^4 F}{\partial y^4} + 2\frac{\partial^4 F}{\partial x^2 \partial y^2} + \nabla_k^2 w + \frac{1}{2}L(w,w) = 0, \tag{2}$$

where:

$$\nabla_k^2 = k_x \frac{\partial^2}{\partial x^2} + k_y \frac{\partial^2}{\partial y^2}, \qquad L(w,w) = 2\left[\frac{\partial^2 w}{\partial x^2}\frac{\partial^2 w}{\partial y^2} - \left(\frac{\partial^2 w}{\partial x \partial y}\right)^2\right],$$

$$L(w,F) = \frac{\partial^2 w}{\partial x^2}\frac{\partial^2 F}{\partial y^2} + \frac{\partial^2 w}{\partial y^2}\frac{\partial^2 F}{\partial x^2} - 2\frac{\partial^2 w}{\partial x \partial y}\frac{\partial^2 F}{\partial x \partial y}. \tag{3}$$

The system of equations (1)–(2) is already in the non-dimensional form, whereas the relations between dimensional and non-dimensional parameters read:

$$w = 2h\overline{w}, \quad x = a\overline{x}, \quad y = b\overline{y}, \quad \lambda = a/b, \quad F = E(2h)^3 \overline{F},$$

$$k_x = \frac{2h}{a^2}\overline{k}_x, \quad k_y = \frac{2h}{b^2}\overline{k}_y, \quad q = \frac{E(2h)^4}{a^2 b^2}\overline{q},$$

$$t = \frac{ab}{2h}\sqrt{\frac{\rho}{gE}}\,\overline{t}, \quad \varepsilon = \frac{2h}{ab}\sqrt{\frac{gE}{\rho}}\,\overline{\varepsilon}, \quad P_x = \frac{E(2h)^3}{b^2}\overline{P}_x, \quad P_y = \frac{E(2h)^3}{a^2}\overline{P}_y. \tag{4}$$

Let us denote the left hand sides of equations (1)–(2) by ϕ_1 and ϕ_2, respectively. Hence, the equations have the form:

$$\Phi_1\!\left(\frac{\partial^2 w}{\partial^2 x}, \frac{\partial^2 F}{\partial^2 x}, k_x, k_y, P_x, P_y, q, t,...\right) = 0, \qquad \Phi_2\!\left(\frac{\partial^2 w}{\partial^2 x}, \frac{\partial^4 F}{\partial^4 x}, k_x, k_y...\right) = 0. \tag{5}$$

In addition, the corresponding boundary conditions should be attached. Since the exact solution of the formulated boundary value problem is not known, the MBG method with higher approximations is applied. We assume the following form of the unknown functions

$$w = \sum_{i,j} A_{ij}(t)\varphi_{ij}(x,y), \qquad F = \sum_{i,j} B_{ij}(t)\psi_{ij}(x,y), \tag{6}$$

$$i = 1,2,...,M_x; j = 1,2,...,M_y.$$

Applying the MBG procedure to equations (5) and taking into account (6), we get

$$\sum_{vz}\left[\sum_{ij}(\frac{\partial^2 A_{ij}}{\partial t^2} + \varepsilon\frac{\partial A_{ij}}{\partial t} - q + k_x P_x + k_y P_y)I_{3,vz\,ij} + \sum_{ij} A_{ij} I_{1,vz\,ij} - \right.$$

$$\left. - \sum_{ij} B_{ij} I_{2,vz\,ij} + \sum_{ij} A_{ij} I_{5,vz\,ij} - \sum_{ij} A_{ij}\sum_{kl} B_{kl} I_{4,vz\,ijkl}\right] = 0,$$

$$\sum_{vz}\left[\sum_{ij} A_{ij} I_{7,vz\,ij} + \sum_{ij} B_{ij} I_{8,vz\,ij} + \sum_{ij} A_{ij}\sum_{kl} A_{kl} I_{6,vz\,ij\,kl}\right] = 0, \tag{7}$$

$$v,i,k = 1,2,...,M_x; \quad z,j,l = 1,2,...,M_y$$

In the above symbol $\underset{vz}{\sum}[*]$, standing before each of the equations of the system (7), should be interpreted as the vz system with similar form, and the integrals of the MBG procedure are:

$$I_{1,vzij} = \int_0^1\int_0^1 \frac{1}{12(1-\mu^2)}\left[\frac{1}{\lambda^2}\frac{\partial^4\varphi_{ij}}{\partial x^4} + \lambda^2\frac{\partial^4\varphi_{ij}}{\partial y^4} + 2\frac{\partial^4\varphi_{ij}}{\partial x^2\partial y^2}\right]\varphi_{vz}\,dxdy,$$

418

$$I_{2,vzij} = \int_0^1 \int_0^1 \left[k_y \frac{\partial^2 \psi_{ij}}{\partial x^2} + k_x \frac{\partial^2 \psi_{ij}}{\partial y^2} \right] \varphi_{vz} \, dxdy, \qquad I_{3,vzij} = \int_0^1 \int_0^1 \varphi_{vz} \, dxdy,$$

$$I_{4,vzijkl} = \int_0^1 \int_0^1 \left[\frac{\partial^2 \varphi_{ij}}{\partial x^2} \frac{\partial^2 \psi_{kl}}{\partial y^2} + \frac{\partial^2 \varphi_{ij}}{\partial y^2} \frac{\partial^2 \psi_{kl}}{\partial x^2} - 2 \frac{\partial^2 \varphi_{ij}}{\partial x \partial y} \frac{\partial^2 \psi_{kl}}{\partial x \partial y} \right] \varphi_{vz} \, dxdy,$$

$$I_{5,vzij} = \int_0^1 \int_0^1 \left[\frac{\partial^2 \varphi_{ij}}{\partial x^2} P_x + \frac{\partial^2 \varphi_{ij}}{\partial y^2} P_y \right] \varphi_{vz} \, dxdy, \tag{8}$$

$$I_{6,vzijkl} = \int_0^1 \int_0^1 \left[\frac{\partial^2 \varphi_{ij}}{\partial x^2} \frac{\partial^2 \varphi_{kl}}{\partial y^2} - \frac{\partial^2 \varphi_{ij}}{\partial x \partial y} \frac{\partial^2 \varphi_{kl}}{\partial x \partial y} \right] \psi_{vz} \, dxdy,$$

$$I_{7,vzij} = \int_0^1 \int_0^1 \left[k_y \frac{\partial^2 \varphi_{ij}}{\partial x^2} + k_x \frac{\partial^2 \varphi_{ij}}{\partial y^2} \right] \psi_{vz} \, dxdy,$$

$$I_{8,vzij} = \int_0^1 \int_0^1 \left[\frac{1}{\lambda^2} \frac{\partial^4 \psi_{ij}}{\partial x^4} + \lambda^2 \frac{\partial^4 \psi_{ij}}{\partial y^4} + 2 \frac{\partial^4 \psi_{ij}}{\partial x^2 \partial y^2} \right] \psi_{vz} \, dxdy.$$

The integrals (8) except for (possibly) $I_{3,vzij}$, if the transversal load q is applied not to the whole shell surface, are computed along the whole middle shell surface.

To conclude, the derived system (7) consists of $M_x \times M_y$ second order differential equations with respect to time and of the linear algebraic equations with respect to B_{ij}.

The initial conditions have the following form

$$w\big|_{t=0} = w_0, \qquad \frac{\partial w}{\partial t}\bigg|_{t=0} = 0, \tag{9}$$

where w_0 is either taken from the corresponding statical problem or is defined using another approach.

Assuming the loading terms, the system of equations (7) is solved using the numerical method, and A_{ij} and B_{ij} are obtained. Next, the found values of A_{ij} and B_{ij} are substituted into (6), and the being sought functions w, F are finally found.

3   RESULTS

As an example the squared plate ($\lambda = 1$, $\varepsilon = 1$), supported by balls on its contour on flexible non-stretched (non-compressed) ribs and excited longitudinally by $P_x = P_{x_0}(1 - \sin \omega_2 t)$, is investigated. The computations are carried out for fixed ω_2 with variations of the control parameter $P_{x_0} = 1 \pm 18$. The numerical results are used to construct the dependencies $w_{max}(P_{x_0})$, $w_{ij}(\dot{w}_{ij})$, power spectrum, the Poincaré sections and the Lyapunov exponents.

Some of the mentioned characteristics are shown in Figures 1–4. In Figure 1 the dependence of maximal deflection in the center of squared plate, versus the longitudinal load P_{x_0}, is reported. The curve 1 is obtained using a first order approximation (i.e. in the relation (6) $\varphi_{ij} = \sin(i\pi x)\sin(j\pi y)$, $\Psi_{ij} = \sin(i\pi x)\sin(j\pi y)$, and $i = j = 1$); the curve 2 is obtained using the 9th order approximation ($i = j = 3$); the curve 3 is obtained using the 25th order approximation ($i = j = 5$); and finally, the curve 4 is obtained using 49th order approximation ($i = j = 7$). The derived results are divided into four intervals: $I - 1 \leqslant P_{x_0} \leqslant 4.5$, $II - 4.5 \leqslant P_{x_0} \leqslant 5.5$, $III - 5.5 \leqslant P_{x_0} \leqslant 7.5$, $IV - 7.5 \leqslant P_{x_0} \leqslant 18.5$.

The intervals $0 \leqslant P_{x_0} \leqslant 1$ and II correspond to intervals of stable equilibrium. In the interval $I–II$, for all approximations (practically) the same results are obtained for all earlier mentioned characteristics

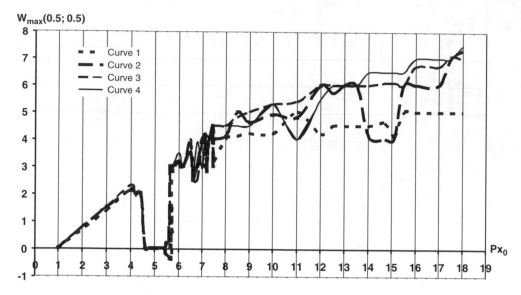

Figure 1. The maximal deflection w_{max} (0.5; 0.5) versus the longitudinal load P_{x_0}.

(the dependencies $(w_{ij}(t))$, $(\dot{w}_{ij}(t))$, power spectra, Poincaré sections and Lyapunov exponents). In the interval *III*, in practice a convergence is achieved when forty nine series terms are used. The dependencies $\Sigma_{ij}^{M_x = M_y} A_{ij}(t)$; $\Sigma_{ij}^{y} A_{ij}[\Sigma_{ij}\dot{A}_{ij}]$ and power spectra are reported in Figures 2–4 (the associated parameters are attached to the figures). Analysis of the data in Figure 2 for $P_{x_0} = 5.65$ shows that the results obtained using the first approximation are qualitatively different from the results obtained for $M_x = M_y = 3$; 5; 7.

4   CONCLUSIONS

All of the characteristics for $M_x = M_y = 3$; 5; 7 practically overlap; the vibrations are quasi-periodic and (for higher approximations) a strange chaotic attractor is detected, which consists of intervals of fast and slow time scales. The relaxation character of vibrations is typical for all modes and has high frequency inclusions on the top of impulses. In zones *I–IV* the same results are obtained for all approximations. In the first approximation (one-degree-of-freedom) the solutions overlap with higher approximations in zones *I–III*. Beginning with zone *IV*, the solutions in the first approximation are not useful to approximate the vibration process and they are qualitatively different from higher order approximations. The reported dependencies obtained for $M_x = M_y = 3$; 5; 7 in practice are the same for both $\Sigma_{ij}A_{ij}(t)$, as well as for each term of the series $A_{ij}(t)$, phase portraits and power spectra. Increasing the parameter P_{x_0} the system begins to lose regular vibrations in the vicinity of region *V*, and to "forget" its initial state, and it transits into a zone of chaotic vibrations. All of approximations, in spite of the first one, characterize the chaotic vibrations. In addition, zones where neutral curve of vibrations change appear in "stiff" manner, i. e. series of "stiff" stability loss is observed in the vibration process (see Fig. 3). Note that the results using approximation of 25 and 49 terms are similar. Recall that in the Lorenz model the most sensitive parameter is that associated with modes number. In our model governed by von Kármán equations, this property is not detected. On the contrary, the system behaves similarly for all modes for $P_{x_0} \in [0.5, 5]$. For $P_{x_0} \in [5.5; 7.5]$ higher approximations also converge to one solution, and for $P_{x_0} > 7.5$ higher approximations describe a chaotic plate dynamics. Therefore, here a coupling scheme is observed. The investigation of the characteristics reported shows that each term of the series A_{ij}, for fixed P_{x_0} values, fully describes the character of vibrations (a synchronization of subsystems is observed). One may also conclude, that beginning from $P_{x_0} > 7.5$ the so called "multimode" turbulence (or "true" turbulence) is observed.

420

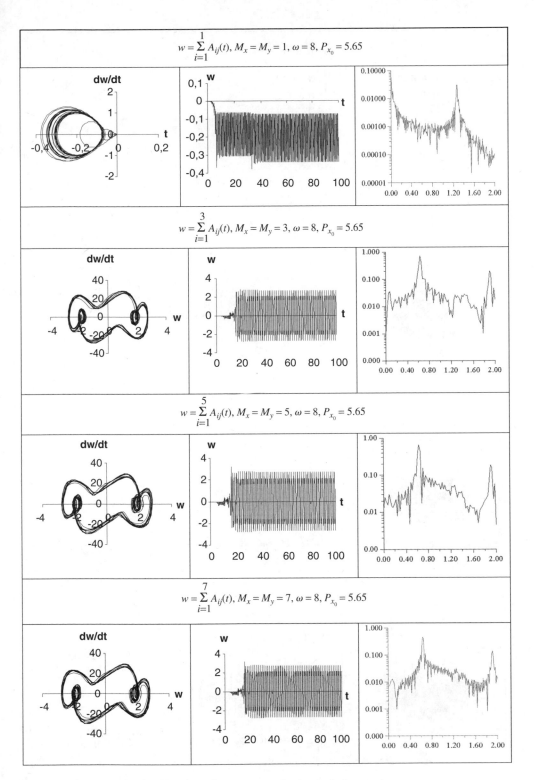

Figure 2. Phase portraits, time histories and power spectra for the attached parameters.

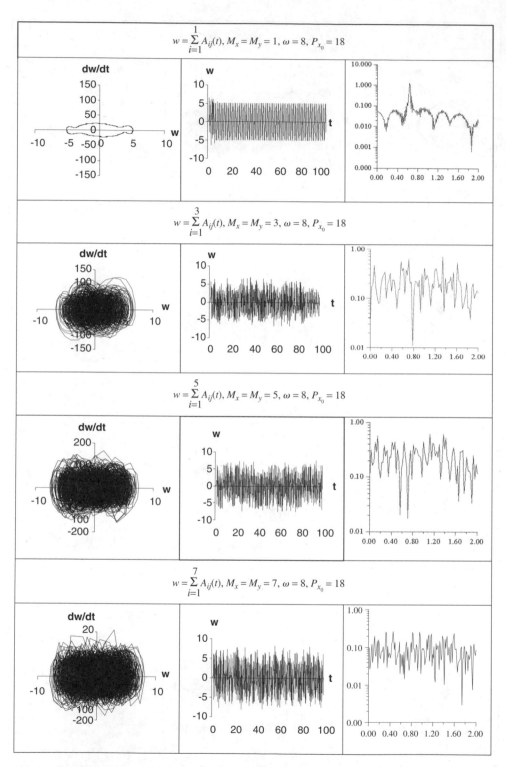

Figure 3.   Phase portraits, time histories and power spectra for the attached parameters.

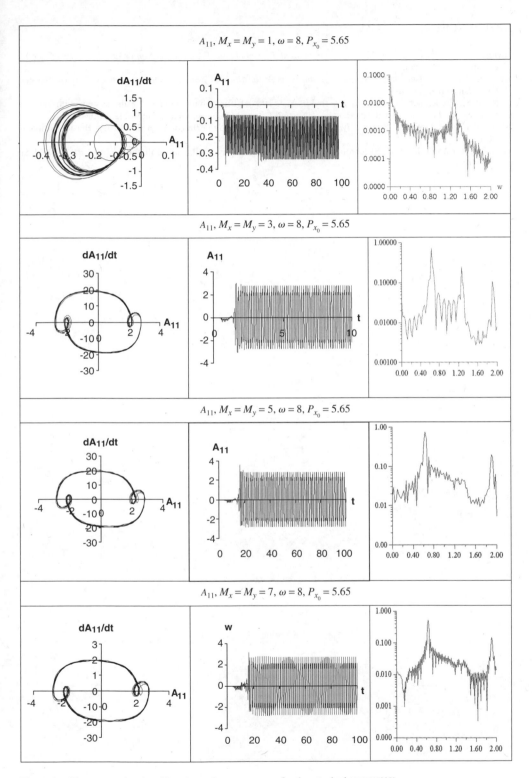

Figure 4. Phase portraits, time histories and power spectra for the attached parameters.

REFERENCES

Awrejcewicz, J. & Krysko, A. V. 2003. *Nonclassical Thermoelastic Problems in Nonlinear Dynamics of Shells.* Berlin, Springer-Verlag.

Krysko, V. A. & Kutsemako, A. N. 1999. *Stability and Vibrations of Non-Homogeneous Shells.* Saratov, Saratov University Press, in Russian.

L'vov, V. S., Predtechenskiy, A. A. & Chernykh, A. I. 1981. Bifurcations and chaos in the system with Taylor's vortices: Real and numerical experiments. *Izviestia VUZov, Radiophysics* 80(3): 1099–1121, in Russian.

Swinney, H. L. & Gollub, J. P. 1981. *Hydrodynamic Instabilities and the Transition to Turbulence.* Berlin, Springer-Verlag.

Waller, I. & Kapral, R. 1984. Synchronization and chaos in coupled nonlinear oscillators. *Phys. Lett. A* 105(4/5): 163–168.

Computational Methods in Engineering and Science, Iu et al. (eds)
© 2003 Swets & Zeitlinger, Lisse, ISBN 90 5809 567 3

Flexible multibody systems models using composite materials components

M.A. Neto

Faculdade de Ciência e Tecnologia, Universidade de Coimbra, Coimbra, Portugal

J.A.C. Ambrósio

Instituto de Engenharia Mecânica – Instituto Superior Técnico, Lisboa, Portugal

ABSTRACT: The use of a multibody methodology to describe the large motion of complex systems that experience structural deformations enables to represent the complete system motion, the relative kinematics between the components involved, the deformation of the structural members and the inertia coupling between the rigid body motion and the elastodynamics. Here, a flexible multibody formulation of complex models is extended to include elastic components made of composite laminated or anisotropic materials. The deformation of the structural member must be elastic and linear if described in a coordinate frame fixed to its domain, regardless of the complexity of its geometry. To achieve the proposed flexible multibody formulation a finite element model for each flexible body is obtained. For beam composite material elements, the sections properties are found using an asymptotic procedure that involves a two-dimensional finite element analysis of their cross-section. The equations of motion of the flexible multibody system are solved using an augmented Lagrangean formulation and the accelerations and velocities are integrated in time using a multi-step integration algorithm.

1 INTRODUCTION

The need for more accurate models to describe the complex behavior of flexible systems experiencing large motion and small elastic deformations led to the development of powerful analysis techniques. The most popular formulations use time variant mass matrices to describe the inertia coupling between the rigid body motion and the elastodynamics (Shabana 1982). Some coefficients of these inertia coupling matrices are dependent on the type of finite elements used. They do not appear in standard finite element developments and consequently need to be specially derived when these methodologies are used. A procedure proposed by Ambrósio and Gonçalves (2001) allows for these coupling matrices to be evaluated in a pre-processing stage and to eliminate its dependency in the type of finite element used in the model.

The finite element formulation for the flexible bodies use a body fixed coordinate frame to describe the deformation field of each body. The flexible bodies together with the rigid bodies of the multibody system are represented by a set of Cartesian coordinates and have their relative motion restrained by a set of kinematic constraints (Nikravesh 1988). It is necessary to define the kinematic constraints between the flexible and the rigid bodies. The description based on the use of virtual bodies is applied in this work (Gonçalves & Ambrósio 2002).

The formulations used for the description of large motion of flexible members have been used for systems made of standard materials. Recent efforts have been made to describe composite and laminated materials in the framework of multibody systems (Bauchau & Hodges 1999). The work now presented is a contribution to the representation of flexible multibody systems using composite materials. In particular, it is presented a methodology to describe composite beam elements that follows the work proposed by Cesnik and Hodges (1997). This methodology is applied to demonstrative examples where the flexibility of members of a multibody system play a decisive role in the behavior of the system.

2 MULTIBODY FORMULATION AND EQUATIONS OF MOTION

2.1 *Flexible body motion*

Let it be assumed that a flexible body is composed of a coordinate system rigidly attached to a point on the flexible body, as depicted by Figure 1. Let it also be assumed that the flexible body is represented using the finite element method, with lumped formulation for the mass matrix, and where the reference frame, described by vector $\mathbf{r}_i$ is located in the body center of mass (Ambrósio & Gonçalves 2001). The flexible body kinetic energy may be expressed as

$$T_i = \tfrac{1}{2}\dot{\mathbf{q}}_i{}^T\mathbf{M}_i\ \dot{\mathbf{q}}_i \tag{1}$$

The velocity vector for a flexible body, is $\dot{\mathbf{q}}_i = [\mathbf{r}_i^T \quad \omega_i^T \quad \dot{\mathbf{u}}_i^{\prime T}]^T$, containing the rigid body velocities and the local nodal velocities $\dot{\mathbf{u}}' = [\tilde{\boldsymbol{\delta}}^T \quad \dot{\theta}'^T]^T$. The flexible body mass matrix is

$$\mathbf{M}_i = \begin{bmatrix} \sum m_k\mathbf{I} & -\sum m_k\mathbf{A}\tilde{\mathbf{b}}_k' & \sum m_k\mathbf{A}\mathbf{I}_k^T & \mathbf{0} \\ \sum m_k\tilde{\mathbf{b}}_k'\mathbf{A}^T & -\sum m_k\tilde{\mathbf{b}}_k'^T\tilde{\mathbf{b}}_k' + \sum \mu_k\mathbf{I} & \sum m_k\tilde{\mathbf{b}}_k'\mathbf{I}_k^T & \sum \mu_k\mathbf{I}_{-k}^T \\ \sum m_k\mathbf{I}_k\mathbf{A}^T & -\sum m_k\mathbf{I}_{-k}\tilde{\mathbf{b}}_k' & \sum m_k\mathbf{I}_k\mathbf{I}_k^T & \mathbf{0} \\ \mathbf{0} & \sum \mu_k\mathbf{I}_k & \mathbf{0} & \sum \mu_k\mathbf{I}_k\mathbf{I}_{-k}^T \end{bmatrix} \tag{2}$$

When a consistent mass formulation is used in the finite element description of the flexible body, the inertia coupling terms depend on the particular finite element shape functions used.

The elastic energy for the flexible body, expressed by U, is written as

$$U = \tfrac{1}{2}\mathbf{q}^T\underline{\mathbf{K}}\ \mathbf{q} \tag{3}$$

where $\underline{\mathbf{K}}$ is an augmented stiffness matrix written as

$$\underline{\mathbf{K}} = \begin{bmatrix} \mathbf{0} & \mathbf{0} \\ \mathbf{0} & \mathbf{K}_{ff} \end{bmatrix} \tag{4}$$

and where $\mathbf{K}_{ff}$ is the standard finite element stiffness matrix or any other equivalent matrix, depending on the adopted description of the body flexibility. However, if matrix $\mathbf{K}_{ff}$ is not dependent on the deformation of the flexible body, the formulation implies that only linear elastic deformations are assumed. This is the case for all that follows.

As all coordinates defined in equation (1) are independent, the Lagrange equations of motion for the flexible body are given by

$$\frac{d}{dt}\left(\frac{\partial T_i}{\partial \dot{\mathbf{q}}_i}\right) - \left(\frac{\partial T_i}{\partial \mathbf{q}_i}\right) + \left(\frac{\partial U_i}{\partial \mathbf{q}_i}\right) - \mathbf{g}_i = \mathbf{0} \tag{5}$$

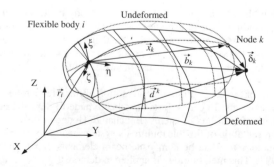

Figure 1. Global position of node k.

The definitions of the kinetic and elastic energy, given by equations (2) and (3) respectively, are now substituted in equation (5) leading to

$$\mathbf{M}_i\,\ddot{\mathbf{q}}_i = \mathbf{g}_i + \mathbf{s}_i - \underline{\mathbf{K}_i}\mathbf{q}_i \qquad (6)$$

2.2 System equations of motion

For a multibody system, it is necessary to define a set of kinematic constraints that restrict the relative motion between the bodies. The kinematic constraints are defined by algebraic equations that are added to the Lagrange equations by using a vector of Lagrange multipliers, λ, as

$$\frac{\mathrm{d}}{\mathrm{dt}}\left(\frac{\partial T}{\partial \dot{\mathbf{q}}}\right) - \left(\frac{\partial T}{\partial \mathbf{q}}\right) + \left(\frac{\partial U}{\partial \mathbf{q}}\right) + \mathbf{\Phi}_{\mathbf{q}}^{\mathrm{T}}\lambda - \mathbf{g} = \mathbf{0} \qquad (7)$$

which leads to

$$\mathbf{M}\ddot{\mathbf{q}} + \mathbf{\Phi}_{\mathbf{q}}^{\mathrm{T}}\lambda = \mathbf{g} + \mathbf{s} - \underline{\mathbf{K}}\mathbf{q} \qquad (8)$$

The constraints acceleration equations are given by

$$\ddot{\mathbf{\Phi}}\;\left(\mathbf{q}_r,\mathbf{q}_f,t\right) \equiv \mathbf{\Phi}_{\mathbf{q}}\ddot{\mathbf{q}} - \gamma = \mathbf{0} \qquad (9)$$

adding these equations to the system equations of motion, leads to

$$\begin{bmatrix} \mathbf{M}_r & \mathbf{M}_{rf} & \mathbf{\Phi}_{\mathbf{q}_r}^{T} \\ \mathbf{M}_{fr} & \mathbf{M}_{ff} & \mathbf{\Phi}_{\mathbf{q}_f}^{T} \\ \mathbf{\Phi}_{\mathbf{q}_r} & \mathbf{\Phi}_{\mathbf{q}_f} & 0 \end{bmatrix} \begin{Bmatrix} \ddot{\mathbf{q}}_r \\ \ddot{\mathbf{u}}' \\ \lambda \end{Bmatrix} = \begin{Bmatrix} \mathbf{g}_r \\ \mathbf{g}_f \\ \gamma \end{Bmatrix} - \begin{Bmatrix} \mathbf{s}_r \\ \mathbf{s}_f \\ 0 \end{Bmatrix} - \begin{Bmatrix} 0 \\ \mathbf{K}_{ff}\mathbf{u}' \\ 0 \end{Bmatrix} \qquad (10)$$

The derivation of kinematic joints involving one or more flexible bodies is generally a complex task that must be repeated for different sets of flexible coordinates. The virtual bodies are applied in the definition of the joints that involve flexible bodies, and require only for a rigid joint to be derived (Ambrósio & Gonçalves 2002).

2.3 Coordinates reduction by component mode synthesis

The equations of motion for the flexible multibody systems, in the form described by equation (10) lead to a inefficient numerical implementation due to the large number of generalized coordinates necessary to describe complex models. This problem is overcome by using a component mode synthesis methodology (Ambrósio & Gonçalves 2001). Though only the modes of vibration are used in this formulation, other modes could also be considered in order to improve numerical precision and efficiency (Yoo & Haug 1986, Pereira & Proença 1991, Cavin & Dusto 1977). Let the nodal displacements of the flexible part of the body be described by a weighted sum of the modes of vibration associated to the flexible bodies natural frequencies

$$\mathbf{u}' = \mathbf{X}\mathbf{w} \qquad (11)$$

The vector $\mathbf{w}$ represents the contributions of the vibration modes towards the nodal displacements and $\mathbf{X}$ is the modal matrix. Due to the reference conditions, the modes of vibration used in this formulation are constrained modes. Moreover, due to the assumption of linear elastic deformations the modal matrix is invariant. By applying the orthonormality of the modes of vibration with respect to the mass matrix, i.e. $\mathbf{X}^T\mathbf{M}_{ff}\mathbf{X} = \mathbf{I}$ and $\mathbf{X}^T\mathbf{K}_{ff}\mathbf{X} = \Lambda$, where Λ is a diagonal matrix containing the squares of the flexible body natural frequencies the equations of motion are

$$
\begin{bmatrix}
\mathbf{M}_r & \mathbf{M}_{rf}\mathbf{X} & \boldsymbol{\Phi}_{\mathbf{q}_r}^T \\
\mathbf{X}^T\mathbf{M}_{fr} & \mathbf{I} & \mathbf{X}^T\boldsymbol{\Phi}_{\mathbf{q}_f}^T \\
\boldsymbol{\Phi}_{\mathbf{q}_r} & \boldsymbol{\Phi}_{\mathbf{q}_f}\mathbf{X} & 0
\end{bmatrix}
\begin{Bmatrix}
\ddot{\mathbf{q}}_r \\
\ddot{\mathbf{w}} \\
\lambda
\end{Bmatrix}
=
\begin{Bmatrix}
\mathbf{g}_r \\
\mathbf{X}^T\mathbf{g}_f \\
\gamma
\end{Bmatrix}
-
\begin{Bmatrix}
\mathbf{s}_r \\
\mathbf{X}^T\mathbf{s}_f \\
0
\end{Bmatrix}
-
\begin{Bmatrix}
0 \\
\Lambda\mathbf{w} \\
0
\end{Bmatrix}
\tag{12}
$$

This is a smaller set of equations that instead of having close to 6 times the number of nodes as generalized coordinates only involves as many generalized coordinates as modes of vibration. Depending on the structure a few number of modes may be used, in comparison to the large number of nodal coordinates, to represent with accuracy the structural response. The finite element code supporting this methodology can be viewed as a pre-processor to obtain the coefficients used in equation (12). Alternatively, experimental modes of vibration can be used.

3   BEAM ELEMENTS WITH COMPOSITE MATERIALS

The flexibility is included in the multibody system assuming the hypothesis of small displacements and rotations with respect to a body attached reference frame. Under such conditions, linear finite elements can be easily extended to handle problems in a nonlinear range.

3.1   *Beam analysis*

Due to their geometries, rotor blades, wings, space frames, and many other structural components have one dimension that is much larger than the other dimensions. Such flexible structures can often be treated as beams. The kinetic and the strain energies of the beam are

$$
T = \frac{1}{2}\int_0^l \mathbf{V}^T \ \mathbf{M} \ \mathbf{V} \ dx
\tag{13}
$$

$$
K = \frac{1}{2}\int_0^l \boldsymbol{\epsilon}^T \ \mathbf{D} \ \boldsymbol{\epsilon} \ dx
\tag{14}
$$

where l is the length of the beam in the direction x along the beam reference line, $\mathbf{M}$ and $\mathbf{D}$ are the beam cross-section section inertia and stiffness tensors and $\mathbf{V}$ and ϵ are the components of the velocity and strain vectors respectively. In the computation of matrix $\mathbf{D}$ for composite materials it must be considered the elastic couplings that might arise.

3.2   *Constitutive law of the beam*

The idealization of the beam structure leads to a much simpler mathematical formulation than would be obtained if complete 3-D elasticity were used. Therefore, the challenge is to capture correctly the beam behavior associated with the two dimensions that are being eliminated. The Variational Asymptotic Beam Sectional Analysis (VABS), proposed by Bauchau and Hodges (1999) is a methodology that splits the geometrically nonlinear 3-D elasticity problem for composite beam-like structure into two problems. The first problem is the two-dimensional asymptotically correct analysis over the cross section, which is typically linear, providing elastic constants. The second problem is a one-dimensional geometrically-exact beam analysis. The 2-D standard FEM computer code, VABS, can produce the generalized 1-D constitutive law for various beam theories. A classical beam stiffness model is described by

$$
\begin{bmatrix}
N_1 \\
M_1 \\
M_2 \\
M_3
\end{bmatrix}
=
\begin{bmatrix}
d_{11} & d_{12} & d_{13} & d_{14} \\
d_{12} & d_{22} & d_{23} & d_{24} \\
d_{13} & d_{23} & d_{33} & d_{34} \\
d_{14} & d_{24} & d_{34} & d_{44}
\end{bmatrix}
\begin{bmatrix}
\varepsilon_{11} \\
k_1 \\
k_2 \\
k_3
\end{bmatrix}
\tag{15}
$$

where N_1 is the axial force, M_1 is the twisting moment and M_2 and M_3 are the bending moments in the y and z directions respectively. ε_{11} is the axial stretching measure of the beam, k_1 is the elastic twist per unit length, and k_2 and k_3 are the elastic components of the curvature.

A Timoshenko-like beam model is described by

$$
\begin{bmatrix} N_1 \\ N_2 \\ N_3 \\ M_1 \\ M_2 \\ M_3 \end{bmatrix} = \begin{bmatrix} d_{11} & d_{12} & d_{13} & d_{14} & d_{15} & d_{16} \\ d_{12} & d_{22} & d_{23} & d_{24} & d_{25} & d_{26} \\ d_{13} & d_{23} & d_{33} & d_{34} & d_{35} & d_{36} \\ d_{14} & d_{24} & d_{34} & d_{44} & d_{45} & d_{46} \\ d_{15} & d_{25} & d_{35} & d_{45} & d_{55} & d_{56} \\ d_{16} & d_{26} & d_{36} & d_{46} & d_{56} & d_{66} \end{bmatrix} \begin{bmatrix} \varepsilon_{11} \\ 2\varepsilon_{12} \\ 2\varepsilon_{13} \\ k_1 \\ k_2 \\ k_3 \end{bmatrix} \tag{16}
$$

where the new quantities included in the equation are the transverse shear forces N_2 and N_3 and the transverse shear measures ε_{12} and ε_{13}. VABS is a standard 2-D Finite Element code that can compute the matrix $\mathbf{D}$ for arbitrary materials and geometries.

3.3   Displacement field of the beam

In the FEM analysis of the 1-D beam problem the Euler-Bernoulli theory (EBT) and the Timoshenko first order shear deformation theory (FSDT) are used. The Euler-Bernoulli theory, ignores transverse shear deformation, which have significant effects on the behavior of the fiber reinforced lamined structures, due to the large difference in the elastic properties between fiber and matrix material. This leads to high ratios of in-plane Young`s modulus to transverse modulus for most applications. When these high ratios are coupled with depth effects this theory is inadequate for the analysis of highly anisotropic beams. To improve the situation the Timoshenko first order shear deformation theory (FSDT), is applied to multilaminated anisotropic beams. In this theory, the transverse strains are constants through the depth of the beam, hence, the transverse shears are also constant. This discrepancy is overcome by introducing shear correction factors.

   Consider a set of unit vectors $\mathbf{e}_1$, $\mathbf{e}_2$, and $\mathbf{e}_3$ attached to a point of the beam cross-section being $\mathbf{e}_1$ aligned with the beam axis while $\mathbf{e}_2$ and $\mathbf{e}_3$ define the plane of the cross-section. Let $u_1(x_1,x_2,x_3)$, $u_2(x_1,x_2,x_3)$ and $u_3(x_1,x_2,x_3)$ be the displacement of an arbitrary point of the beam in the $\mathbf{e}_1$, $\mathbf{e}_2$, and $\mathbf{e}_3$ directions respectively. The displacement field in the plane of the cross-section solely consists of two rigid body translations $U_2(x_1)$ and $U_3(x_1)$, hence we consider that the cross-section is infinitely rigid in its own plane.

$$
u_2(x_1,x_2,x_3) = U_2(x_1) \tag{17}
$$

$$
u_3(x_1,x_2,x_3) = U_3(x_1) \tag{18}
$$

   The second Euler-Bernoulli assumption states that the cross-section remains plane after deformation. This implies an axial displacement field consisting of a rigid body translation $U_1(x_1)$, and two rigid body rotations $\phi_2(x_1)$ and $\phi_3(x_1)$.

$$
u_1(x_1,x_2,x_3) = U_1(x_1) + x_3 \ \phi_2 \ - \ x_2 \ \phi_3 = U_1(x_1) + x_3 \ k_2 \ - \ x_2 \ k_3 \tag{19}
$$

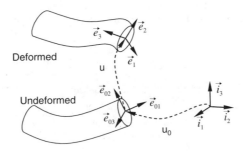

Figure 2.   Beam local coordinate frames and point displacements.

429

where the rigid body rotations of the cross sections $\phi_2(x_1)$ and $\phi_3(x_1)$ are positives about the axes e_2 and e_3, respectively. In the case of the Euler-Bernoulli theory the third assumption states that the cross-section remains normal to the deformed axis of the beam. This implies the equality of the slopes of the beam and the rotation of the section

$$\phi_2 = -\frac{\partial U_3}{\partial x_1}$$
(20)

$$\phi_3 = \frac{\partial U_2}{\partial x_1}$$
(21)

With the displacement field described by equations (17) through (19) the mass and stiffness matrices of the beam elements are obtained. When the classic constitutive law is used the element considered is cubic beam element and it has two nodes. In the case of the constitutive law of Timoshenko model, a quadratic element with three nodes is used.

4   APPLICATION CASES

4.1   *The actuated beam problem*

A cantilevered beam with a tip mass is actuated at its mid-point by a crank link mechanism represented in Figure 3. The model consists in one revolute joint in points A, B and R, and a spherical joint in point M. The crank is a rigid body with a constant angular velocity of 3.14 rad/sec. The beam and the link are modeled by finite elements. The cross-sections of the beam and link are thin-walled rectangular sections depicted in Figure 3, with the dimensions of 0.953×0.537 in. The cross-section walls consist six layers of graphite/epoxy material with the material properties $E_l = 20.59 \times 10^6$ psi, a transverse modulus of $E_t = 1.42 \times 10^6$ psi, a shearing modulus of $G_{lT} = 0.87 \times 10^6$ psi, and a Poisson's ratio of 0.42. In this application it is considered that a lay-up of zero degrees the fibers are aligned with the axis of the beam.

The beam cross-section is modeled using the finite element procedure VABS. The sectional properties for lay-up, calculated with VABS, are

$$D = 10^3 \begin{bmatrix} 1770 & 0 & 0 & 0 & 0 & 0 \\ 0 & 1770 & 0 & 0 & 0 & 0 \\ 0 & 0 & 1770 & 0 & 0 & 0 \\ 0 & 0 & 0 & 8.16 & 0 & 0 \\ 0 & 0 & 0 & 0 & 215 & 0 \\ 0 & 0 & 0 & 0 & 0 & 86.9 \end{bmatrix}$$
(22)

Due to the fact that all fibers are aligned in the same direction for this application case this lay-up presents no elastic coupling. Two multibody models of the system are simulated. In the first model, designated by full model, there is no reduction technique used for the finite element representation of the cantilever beam. In the second model the modal superposition technique is used to reduce the number of generalized flexible coordinates. The response of the system is simulated for a time period of 6 sec.

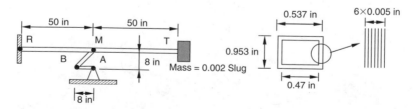

Figure 3.   The actuated beam problem.

The tip transverse deflection of the beam obtained with the full model is represented in Figure 4. No difference between the solution obtained with the classic and Timoshenko beam theories are observed. This is justified by the fact that the beam can be considered to have a thin walled cross section and, therefore, the influence of shear forces is not important.

The same case is simulated with the reduced model where the modal superposition formulation of the finite element model is used with four modes of vibration. For the case of the model using the mode component synthesis there is no difference in response of the system using either of the beam theories. The qualitative response of the system is similar to that obtained with the full model but the amplitude of the vibration is smaller, suggesting a stiffer cantilever beam. This observation suggests that more modes of vibration should be used in the reduced model.

4.2 The double pendulum

This example consists of a double pendulum made of two flexible beams, with tip point masses at their ends, connected by revolute joints. The initial positions of the system bodies are such that they are aligned in the horizontal position, as shown in Figure 5. Table 1 presents the geometrical characteristics of the model and also the material properties used. The simulated case is such that the arm beam rotates with a constant angular velocity of 20 rad/s.

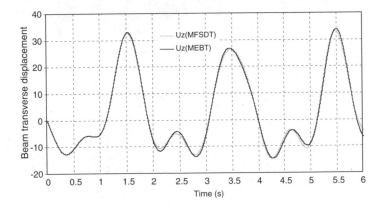

Figure 4. Tip transverse deflection of the beam considering the formulation with mode component synthesis using four modes of vibration.

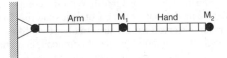

Figure 5. Initial position of the double pendulum.

Table 1. Characteristics of the double pendulum.

	Arm	Hand	Isotropic material	Composite material
				$E_2 = E_3 = E_1 = 129.207\,\text{GPa}$
Length (L)	0.545 m	0.675 m	$E = 73\,\text{GPa}$	$G_{12} = 5.15658\,\text{GPa}, G_{23} = 2.5414\,\text{GPa}$
b	0.06 m	0.04 m	$\rho = 2700\,\text{Kg/m}^3$	$G_{13} = 4.3053\,\text{GPa}, v_{12} = v_{13} = 0.3$
h	0.015 m	0.01 m		$v_{23} = 0.218837, \rho = 1550.066\,\text{Kg/m}^3$
Total mass Arm			1.324 Kg	0.760 Kg
Total mass Hand			0.729 Kg	0.418 Kg
Point mass M_1	1 Kg			
Point mass M_2	3 Kg			

431

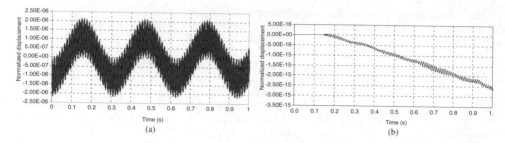

Figure 6. Double pendulum mid-node dimensionless deflection with isotropic material: (a) arm; (b) hand.

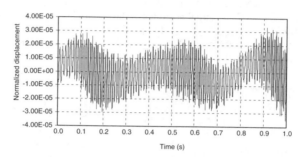

Figure 7. Double pendulum hand mid-node dimensionless deflection with composite material.

The dimensionless deflection of the mid node of the arm and hand obtained in the simulation of the double pendulum model where the beams are made of isotropic material are displayed in Figures 6 and 7. The output parameter is a dimensionless value obtained as the ratio of the beam mid node deflection with respect to the body fixed coordinate frame by the beam length. The multibody model of the double pendulum uses four vibration modes in order to reduce the number of generalized elastic coordinates.

The deformation of the arm is several orders of magnitudes higher than the deformations of the hand. The hand can be represented by a rigid body model when the isotropic material is used because the hand displacement field is almost null.

The same multibody model is used to simulate a double pendulum where the beams are made of a composite material, with the characteristics shown in Table 1. With the composite materials the weight of the arm and hand is lower than for the model that uses the isotropic material. In this case the arm has a lower deformation than the deformation observed for the arm made of isotropic material. However, the hand deformation for the case of the double pendulum made of composite material has a magnitude similar to that observed for the arm, as seen in Figure 7. When a composite material is used for the double pendulum it is not possible to represent the hand by a rigid body as it is suggested for the isotropic material.

5 CONCLUSIONS

A multibody based methodology for the analysis of flexible multibody systems made of composite materials was described in this work. The formulation used a finite element based description of the flexibility of each body referred to a body fixed coordinate system. Special emphasis was put in the representation of beam elements made of composite materials where a three step procedure, proposed by Bauchau and Hodges (1999) was used. The material models used in the foreseen applications of the methodology are limited to be linear elastic. With these assumptions the mode component synthesis can be used in the flexible multibody formulation. The application of the methodology was demonstrated through two numerical examples consisting in an actuated cantilever beam and a double pendulum. The results were analyzed with reference to the different multibody models to the use of isotropic and composite materials. These preliminary results show that the formulation proposed is effective offering a good potential to be applied to more complex cases of structures made of composite materials and experiencing large rigid body rotations.

REFERENCES

Ambrósio, J. & Gonçalves, J. 2001. Complex Flexible Multibody Systems with Application to Vehicle Dynamics, Multibody System Dynamics, **6**(2): 163–182.

Bauchau, O.A. & Hodges, D.H. 1999. Analysis of Nonlinear Multibody Systems with Elastic Couplings, Multibody System Dynamics, **3**: 163–188.

Cavin, R.K. & Dusto, A.R. 1977. Hamilton's Principle: Finite Element Method and Flexible Body Dynamics, AIAA Journal **15**(12): 1684–1690.

Cesnik, C.E.S. & Hodges, D.H. 1997. VABS: A New Concept for Composite Rotor Blade Cross-Sectional Modeling, Journal of the American Helicopter Society **42**(1): 27–38.

Gonçalves, J. & Ambrósio, J. 2002. Advanced Modeling of Flexible Multibody Dynamics Using Virtual Bodies, Computer Assisted Mechanics and Engineering Sciences, **9**(3): 373–390.

Hodges, D.H. 1990. A Review of Composite Rotor Blade Modeling, AIAA Journal, **28**(3): 561–565.

Nikravesh, P.E. 1988. Computer-Aided Analysis of Mechanical Systems, Prentice Hall, New Jersey.

Pereira, M.S. & Proença, P.L. 1991. Dynamic Analysis of Spatial Flexible Multibody Systems Using Joint Coordinates, *Int. J. Num. Meth. In Engng.* **32**: 1799–1812.

Shabana, A. 1982. Dynamic Analysis of Large-Scale Inertia Variant Flexible Systems, Ph.D. Thesis, University of Iowa.

Yoo, W.S. & Haug, E.J. 1986. Dynamics of Flexible Mechanical Systems Using Vibration and Static Correction Modes, Journal of Mechanisms, Transmissions and Automation in Design, **108**: 315–322.

Computational Methods in Engineering and Science, Iu et al. (eds)
© *2003 Swets & Zeitlinger, Lisse, ISBN 90 5809 567 3*

Damage detection in concrete dams using dynamic continuous monitoring and numerical models

S. Oliveira, J. Rodrigues & A. Campos Costa
Laboratório Nacional de Engenharia Civil, Lisboa, Portugal

P. Mendes
Instituto Superior de Engenharia de Lisboa, Lisboa, Portugal

ABSTRACT: Ever since various decades LNEC has been trying to identify the processes of evolutive deterioration in concrete structures, namely in concrete dams, by establishing a correlation between the modal parameters (natural frequencies and modal configurations), which refer to the dynamic structural behaviour, and the deterioration of the structural characteristics. The experience obtained by LNEC, in various forced vibration tests in concrete dams, makes it possible to conclude that the intended objective cannot be reached on the basis of measurements with 5-to-10 year intervals. In fact, the modal parameters to be identified are influenced not only by the deterioration of the structure over time, but also by the environmental thermal variations and by the variations in the reservoir water level. Therefore, it is necessary to measure the dynamic response using new systems for continuous dynamic monitoring, using only the vibrations in the dams under operational conditions.

1   INTRODUCTION

The present work intends to demonstrate that, currently, it is possible to characterise with good precision the dynamic response of arch dams under the action of environmental loads, even though the vibrations corresponding to that excitation are of very low amplitude in this type of structures. This has been made possible by the technological development achieved at the level of: (i) vibration measurement equipments (namely, accelerometers); (ii) signal conditioning, acquisition and storage systems; (iii) modal identification techniques under environmental excitation.

Therefore, the results presented refer to an ambient vibration test conducted at the Cabril dam, in February 2002, with the support of EDP (Portuguese Electricity Company). The results of dynamic measurements performed at the Cabril dam under the action of environmental excitation are compared with the results previously obtained with forced vibration tests. The observed results are also compared with those of a 3-D finite element model, based on the hypothesis of elastic-linear behaviour and assuming that the hydrodynamic effect of water is properly simulated through associated water masses, in accordance with Westergaard's formula (Oliveira 1991). In order to define future strategies aiming at the use of continuous dynamic monitoring results to identify changes associate with deterioration phenomena in arch dams, a discussion is presented about the influence of the main actions, hydrostatic pressure and temperature variations, on the fundamental parameters of the dynamic response (natural frequencies and modal configurations). The influence of possible changes associated to deterioration processes on the same parameters is also discussed with the same purpose.

2   MEASUREMENT AND ANALYSIS OF ENVIRONMENTAL VIBRATIONS IN LARGE DAMS

The measurement of vibrations in concrete dams has been carried out with the purpose of performing the modal identification of the dam-foundation-reservoir system: determination of natural frequencies, modal configurations and modal damping.

The general purpose is to use the modal identification results to: (i) assess the appropriateness of the hypothesis adopted at the level of mathematical models to simulate the dynamic behaviour of structures; (ii) calibrate the parameters of the models (modulus of elasticity, parameters referring to the boundary conditions, to the behaviour of retraction joints, to the inter-action with the reservoir, etc.); and (iii) detect possible changes in the overall dynamic characteristics that may be the sign of deterioration phenomena that are difficult to detect by the traditional monitoring methods.

Regarding the changes in the observed dynamic characteristics, mention must be made of the fact that these may correspond, on one hand, to gradual changes of the properties of materials, either due to usual phenomena, such as the maturation of concrete, or due to pathological phenomena, such as the development of cracking resulting from concrete swelling reactions.

In order to characterise properly the structural health of large dams, it is necessary to carry out the measurement and analysis of vibrations for different reservoir levels and in different yearly temperature conditions. This is to be preferably done under excitations of different amplitudes, whenever the aim is to study modal damping and not just natural frequencies and modal configurations. This type of very detailed characterisation, which is indispensable when the intent is to co-relate changes in the dynamic response with deterioration phenomena over time, is very difficult to achieve using only results from forced vibration tests. This is particularly due to the high cost associated with the installation and control of excitation equipment (input-output tests). Therefore, considering the low costs involved, the measurement and analysis of vibrations due only to the usual excitation sources under service conditions (associated to natural actions, such as wind or earthquakes and micro-earthquakes, or to actions due to operation activities, such as the operation of power units or to flow discharges) are currently of the highest interest.

3   AMBIENT VIBRATION TEST AT CABRIL DAM

The results presented in this paragraph refer to an ambient vibration measurement test performed at the Cabril dam (50-year old, large double curvature arch with 132 maximum height see Fig. 1). The test was conducted in February 20, 2002, with the reservoir at level 267 m. The main purpose of the test was to obtain elements to justify the submission of a project to the National Scientific Re-equipment Plan entitled "*Study of Evolutive Deterioration Processes in Concrete Dams. Safety Control Over Time*". Within the framework of that project it is intended to develop a continuous dynamic monitoring system for large concrete dams.

3.1   *Test equipment and acquisition parameters adopted*

Figure 2 shows a schematic drawing of the measurement system used. That system consists of the elements as follows:

- 12 uniaxial acceleration transducers of the type force-balance (Kinemetrics; Model: EpiSensor ES-U) with a 2.5 Volt/g sensitivity (installed in the radial sense, see Fig. 2);
- 4 power supply and signal conditioning units, developed in LNEC;
- data acquisition board DAQ Card AI16XE-50, National Instruments (16 bits);
- cables;
- 1 portable computer for acquisition and storage of measurements.

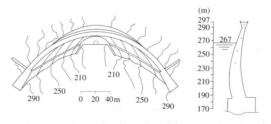

Figure 1.   Plan and cross-section by the central pier of Cabril dam.

This system was used with a maximum sensitivity of 2.5 Volt/mg; the saturation level of the acquisition system being around ±10 Volt. In the situation of maximum gain, the system makes it possible to measure accelerations until about ±4 mg (in fact, as will be demonstrated, maximum acceleration values of about 2 mg were measured in a situation where the power units were in operation), with a precision corresponding to $8 \, mg/2^{16}$ (16 bits).

The acceleration records were performed with the acquisition parameters as follows:

- Maximum gain in the conditioning: 1000
- Sampling frequency: 200 Hz
- Sample duration higher than 0.5 hour. Long duration samples were chosen with the purpose of obtaining good frequency resolutions.

A computer programme developed in LNEC was used in the pre-processing stage. That programme is in LabView and makes it possible to automate the removal of the average of signals recorded and to carry out filtering and decimation to obtain a re-sampling of the signal recorded from the original sampling frequency of 200 Hz to 50 Hz (application of a low-pass filter in the 20 Hz = 0.8 ×50 Hz/2 and decimation for the intended 50 Hz).

3.2 Analysis of the acceleration histories and modal identification

The acceleration histories at the 12 measurement points were initially recorded for about 45 minutes with the Hydroelectric power units, located at the upstream toe, in operation (25 minutes at 38 Mwatts and 20 minutes at full power – Fig. 3a) and, subsequently, after disconnection of the units, records were performed for about 35 minutes (Fig. 3b).

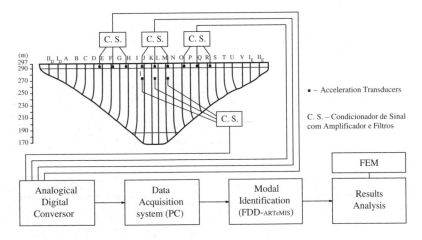

Figure 2.   Schematic drawing of the acquisition system.

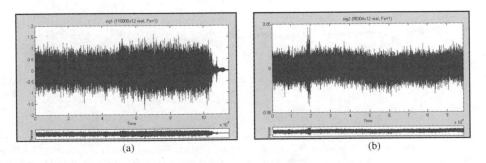

Figure 3.   Acceleration records obtained on transducer 1. (a) With power units in operation; (b) With units disconnected – with environment excitation only (wind).

437

In the first situation (units in operation), the maximum accelerations measured were about 2 mg (Fig. 3a) and with disconnected units (only with environment excitation, mainly due to the action of the wind), the maximum accelerations measured were of approximately 0.05 mg (Fig. 3b).

The computer program ARTeMIS Extractor, release 3.0 (Ambient Response Testing and Modal Identification Software) was used to carry out the modal identification of the system dam-foundation-reservoir. That software makes it possible to carry out the modal identification of structures based on the FDD technique. That technique consists of carrying out a decomposition of the response of the system represented by the matrix of spectral densities of signals measured (12×12 matrix, in this case), on a diagonal matrix corresponding to a system of independent degrees of freedom (SDOF). The frequencies are determined by selection of the resonance peaks in the spectra of singular values, whereas the modal configurations are estimated by the corresponding singular vectors.

In Figure 4 we can see the results of modal identification referring to the situation of power units both in operation and disconnected in terms of the average of normalised singular values of spectral density matrices of the 12 acceleration records. The results presented were obtained by using a decimation factor of 5 on the records corresponding to re-samplings to a 50 Hz frequency and by carrying out sub-sample averages of 1024 points ($T \approx 20\,s$) overlapped at 2/3.

The results obtained in these two distinct excitation situations show coherent results in terms of natural frequencies of peaks identified as corresponding to structural vibration modes (shown in the figure). In the case of Figure 4a, it is possible to observe a narrow and very well defined peak, in the 3.57 Hz frequency, which is exactly the rotation frequency of the power units, which demonstrates the good frequency resolution achieved.

Figure 5 shows the vibration modes corresponding to the 4 peaks indicated, as well as the value of the corresponding natural frequencies, special reference being made to the coherence of configurations obtained for the two situations mentioned.

3.3 *Finite element mathematical model for support to the interpretation of results from the modal identification*

In order to interpret the previous modal identification results, a 3-D finite element model (of the type cube, isoparametric of 20 nodal points) of the set dam-foundation-reservoir was used (Fig. 6), on which the hydrodynamic effect of water is considered through Westgaard's associated water masses. Also in that model, it is initially assumed that the concrete, of specific mass $\gamma = 24\,kNm^{-3}$, is a homogeneous and isotropic material of linear elastic behaviour with a modulus of elasticity $E_0 = 33\,GPa$ (about 30% higher than the value determined on basis of the usual deformability tests) and with a Poisson coefficient

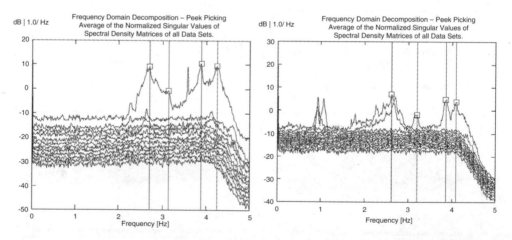

Figure 4. Modal identification based on the Frequency Domain Decomposition method. Representation of normalised singular values of spectral density matrices of the 12 acceleration records (results of the computer software *ARTeMIS*). Identification of peaks corresponding to the first vibration modes of the structure. Application to records obtained with power units either in operation (a) or disconnected (b).

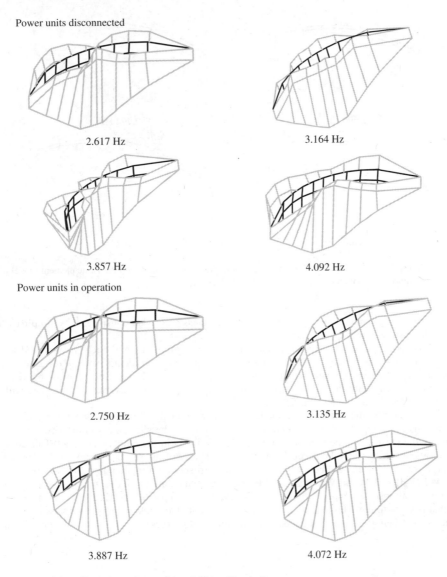

Power units disconnected

2.617 Hz   3.164 Hz

3.857 Hz   4.092 Hz

Power units in operation

2.750 Hz   3.135 Hz

3.887 Hz   4.072 Hz

Figure 5.   Modal configurations observed (modal identification).

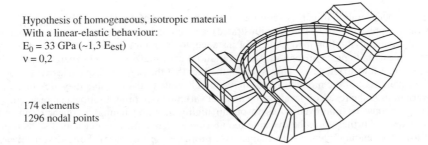

Hypothesis of homogeneous, isotropic material
With a linear-elastic behaviour:
$E_0 = 33$ GPa $(\sim 1{,}3\ E_{est})$
$\nu = 0{,}2$

174 elements
1296 nodal points

Figure 6.   Finite element model of the set dam foundation-reservoir.

439

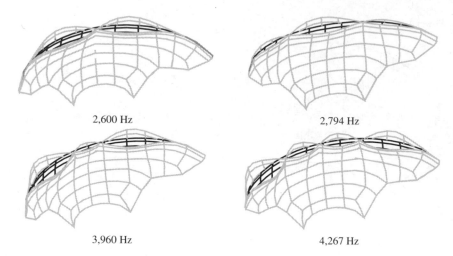

2,600 Hz   2,794 Hz

3,960 Hz   4,267 Hz

Figure 7. Configurations of the first vibration modes numerically calculated with the finite element model in the hypothesis of homogeneous material of linear elastic performance.

$v = 0.2$. Similar mechanical characteristics with a null mass were also considered, by simplification, for the rock mass, so that the latter may act as an elastic support.

With this model, the natural frequencies and modal configurations, numerically calculated for the three first vibration modes, are coherent with the results of the modal identification. Nevertheless, the mode clearly identified at the site with a frequency of about 4.1 Hz has a configuration that does not present any similarity with the configuration of the 4th mode that was numerically calculated with the elastic and homogeneous model (Fig. 7).

In order to interpret that difference at the level of the previously mentioned modal configuration, it has been considered that the horizontal cracking that occurs at the site, on a section between 7 m and 20 m below the crest, could be represented in a simplified way in the numerical model. This could be achieved by considering that the vertical modulus of elasticity of the finite elements at the zone of the section referred to above could be reduced to a value corresponding to a damaged = 0.9, being therefore $E_z = E(1-d) = 3.3$ GPa. With that alteration in the mathematical model, the three first vibration modes remain almost unchanged, which creates a new mode, for a 4,600 Hz frequency, with a configuration that is close to the 4th mode identified at the site. This fact makes it possible to assume that the hypothesis of it being a mode associated with the horizontal cracking could be considered in future studies.

4 MODELS FOR THE INTERPRETATION OF THE DYNAMIC RESPONSE PARAMETERS OVER TIME

The effects of deterioration of dams over time can be represented by changes in terms of modal configurations (as the present case has already illustrated), as well as by variations over time of natural frequencies of the main vibration modes (due, for instance, to variations in the modulus of elasticity over time, either of a pathologic nature or not). These variations are usually difficult to perceive, even in periods comprising a few dozen years. This is due to the fact that the natural frequencies vary proportionally to the square root of the modulus of elasticity and also due to the fact that they are affected by the natural variations in the reservoir level and by thermal variations. In order to solve that problem, the test methodologies must be improved so as to obtain highly accurate natural frequency values (long-duration records). Furthermore, the dynamic monitoring systems must be developed so as to make it possible to obtain enough elements to characterise properly the effects of the reservoir level and of yearly thermal variations on natural frequencies, so as to differentiate more easily the time effects on natural frequencies.

Configuration identified on the basis of measurements done (4th mode)

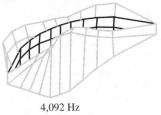

4,092 Hz

Configuration determined numerically in the hypothesis of homogeneous material

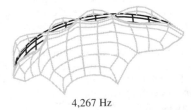

4,267 Hz

Configuration determined numerically considering cracking (in a simplified way)

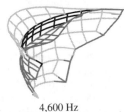

4,600 Hz

Figure 8. Analysis of the configuration of the 4th mode identified on the basis of measurements done: comparison with modal configurations numerically determined, either considering or not the effect of horizontal cracking on the upper zone of the work.

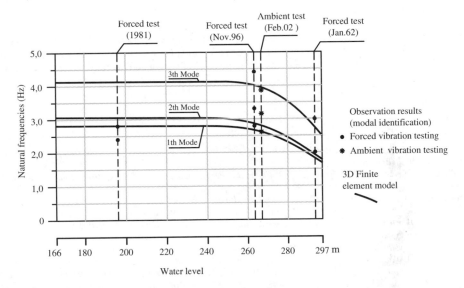

Figure 9. Effect of the reservoir level on natural frequencies. Comparison of numerical model results with results from the forced vibration test and with results from the present environmental vibration test.

In Figure 9 we can see the influence lines, which were numerically calculated with the finite element model previously mentioned. These lines represent the variation of natural frequencies of the 3 first vibration modes in accordance with the reservoir level. The results observed in three forced vibration tests that were conducted in 1962 are also presented (water level: 294 m), 1981 (empty reservoir), 1996 (water level: 264 m), as well as the results of the environmental vibration test that was referred to above (February 2002, water level 267 m).

441

5 FINAL CONSIDERATIONS

It has been demonstrated that, with the current equipment, the test methodologies for measuring environmental vibrations (output-only modal testing) make it possible to obtain results of characterisation of the dynamic performance of arch dams, which are coherent with the results obtained in forced vibration tests. The environmental vibration test results presented, which refer to the Cabril dam, have indicated that it would be appropriate to develop a research study, with a view to develop continuous dynamic monitoring systems for large dams. The main purpose of the latter would be to complement the current monitoring systems, of which the results have proved to be essential for the safety control of these structures. As is known, in order to initiate the research study previously mentioned, it is essential to carry out new tests at the Cabril dam and in other large dams. In those tests, a higher number of measurement points and new adjustments at the level of acquisition parameters are to be used.

The expected outcome is that the continuous dynamic monitoring, other than making it possible to observe and interpret the dynamic response of those structures during the occurrence of earthquakes, will also make it possible to identify more easily the possible changes in the structural performance over time. In fact, the latter may be co-related with deterioration effects induced by phenomena of different origins, ranging from changes in the concrete with a pathologic origin (swellings, for instance) to cracking resulting from exceptional actions, such as intensive phenomena or overtopping. Therefore, it has been considered of the highest interest to invest in the development of fixed dynamic monitoring systems, based on Smart-sensor networks, with serial connection achieved by low cost optical fibre cables. Furthermore, it is also essential to implement modal identification software based on FDD and/or SSI techniques, as well as to develop effect separation methodologies for interpreting the possible future changes, which might be detected on the basis of variations in the natural frequencies. Lastly, it is also important to improve mathematical models for support to the interpretation of the dynamic performance observed.

ACKNOWLEDGEMENTS

The authors wish to thank EDP for all the support provided in the preparation and execution of the test. Thanks are also due to the Principal Research Officer Almeida Garrett, from the Scientific Instrument Design Centre of LNEC, for his valuable support at the level of design and operation of instruments used in the test.

REFERENCES

Brincker, R., Ventura, C., Andersen, P. 2002. *Why Output-Only Modal Testing is a Desirable Tool for a Wide Range of Pratical Applications*.

Carvalhal, F.J., Oliveira, C., Schiappa, F. 1989. *Elementos de Sistemas e de Análise e Processamento de Sinais – Curso* LNEC.

Campos Costa, A., Rodrigues, J., 2001. *Structural Health Assessment of Bridges by Monitoring their Dynamic Characteristics*, Seminário Segurança e Reabilitação das Pontes em Portugal.

Oliveira, S. 1991 *Elementos finitos parabólicos para análise estática e dinâmica de equilíbrios tridimensionais*. LNEC. Trabalho de síntese, Lisboa.

Peeters, B. 2000 *System Identification and Damage Detection in Civil Engineering*. Tese de Doutoramento.

PNRC, LNEC-FEUP. *Estudo de processos de deterioração evolutiva em barragens de betão. Controlo da segurança ao longo do tempo,* Programa de Candidatura ao Programa Nacional de Re-equipamento Científico da FCT.

Rodrigues, J., Campos Costa, A. 2001. *Caracterização Dinâmica de Estruturas de Pontes com Base em Ensaios de Vibrações Ambiente,* LNEC.

Rodrigues, J. 2003. *Identificação Modal Estocástica. Métodos de Análise e Aplicações em Estruturas de Engenharia Civil* (em preparação*)*.

Computational Methods in Engineering and Science, Iu et al. (eds)
© *2003 Swets & Zeitlinger, Lisse, ISBN 90 5809 567 3*

Displacement-based design applied to a reinforced concrete building

R. Bento & S. Falcão
Instituto Superior Técnico, Lisbon, Portugal

ABSTRACT: A new technique to initiate the seismic design process by imposing an initial damage state, called Displacement-Based Design, is presented. This involves a step-by-step procedure, in which damage, expressing a limit, is used to evaluate internal force demand and finally the structures performance evaluation. This paper aims at applying the Displacement-Based Design methodology to a four-storey reinforced concrete building. Some results are compared with the non-linear dynamic procedure results, for which five accelerograms were used. The results obtained are presented in terms of global deformation demands like *maximum roof displacement* and *base shear*, and as well as local demands in the form of *inter-storey displacements, storey shear* and *storey displacement distribution*, and *rotation ductility* at critical sections. The results show that the described procedure provides adequate information on the seismic demand imposed by the ground motion on a regular structural system.

1 INTRODUCTION

It has been recognize that deformations and displacements can be better related to damage for a particular performance limit state, rather than force. Based on this idea, new approaches are being proposed (Calvi 1999, Cornell 2000, Kowalsky 1997, Priestley 2000, Loeding *et al.* 1998) in which the damage, in the form of strain limits for steel and concrete, are expressed as limit state. Using these strain limits, displacements for the structure are obtained and the structure finally designed, to resist the forces generated by these displacements. The process is exactly the opposite of a force based and displacement check approach i.e. the Modified FBD, wherein displacements, strains and hence damage are calculated, rather than imposed initially. The title of *Displacement-Based Design (DBD)* is thus chosen, as the displacements are decided right from the start of the design process. The DBD procedure for reinforced concrete structures can be considered as a step-by-step procedure, in which damage, expressing a limit, is used to evaluate internal force demand and finally the structures performance evaluation. The procedure is based on an effective structure (SDOF), which intends to represent the original MDOF structure.

The relationship between the MDOF and the SDOF system (Fig. 1) are derived using the equal work principle (Falcão 2002), which involves equating the work done by the MDOF to the SDOF system, equation (1), wherein F are the forces and Δ the displacements.

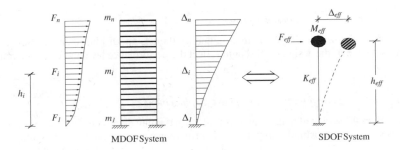

Figure 1. MDOF and SDOF systems.

$$(F\varDelta)_{MDOF} = (F\varDelta)_{SDOF} \tag{1}$$

In the following discussion, i represents a separate degree of freedom (DOF) or storey in the MDOF structure, and eff represents an effective characteristic of the SDOF system. The F_i is the force applied at each level/DOF to produce the displacement $\varDelta_i$. In addition, F_{eff} is the effective force of the effective SDOF system and should be equal to the sum of all the forces at each level/DOF – equation (2).

$$F_{eff} = \sum_{i=1}^{n} F_i \tag{2}$$

The forces at each level/DOF and the effective displacement ($\varDelta_{eff}$) are given by the following equations, where m_i is the mass at storey i.

$$F_i = F_{eff} \frac{m_i \varDelta_i}{\sum\limits_{i=1}^{n} m_i \varDelta_i} \tag{3}$$

$$\varDelta_{eff} = \frac{\sum\limits_{i=1}^{n} m_i \varDelta_i^2}{\sum\limits_{i=1}^{n} m_i \varDelta_i} \tag{4}$$

With the effective force and the effective displacement defined, all other characteristic of the SDOF system, like effective period (T_{eff}) of vibration and effective stiffness (K_{eff}) can be evaluated as:

$$T_{eff} = 2\,\pi \sqrt{\frac{M_{eff}}{K_{eff}}} \tag{5}$$

$$K_{eff} = \left(\frac{4\,\pi^2}{T_{eff}^2} \right) M_{eff} \tag{6}$$

2 STEPS IN DISPLACEMENT-BASED DESIGN

In the following, a step-by-step procedure for the DBD methodology is presented.

2.1 Define the mathematical model

The very fact that non-linear analysis will be used in the final stage requires that the mathematical model describing the structure should possess the capability of performing this type of analysis.

2.2 Select a displacement response spectrum

In this procedure, the seismic hazard is represented by the design Displacement Response Spectrum (DRS). The selection of the design DRS should reflect the performance level under consideration in the analysis.

2.3 Select a maximum drift level or a maximum displacement

This step involves the setting of maximum drift level or maximum displacement. The former is normally selected based on non-structural components deformation capacities or the strain limits in critical members like beams and column base hinges. The selection of the latter can be based on, for instance, the limits imposed by adjacent structures, aiming to prevent pounding effects.

444

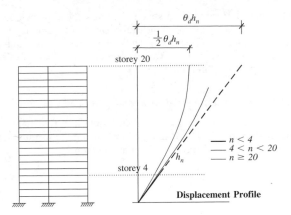

Figure 2.   Drift and displacement profile.

It should however be noted that maximum drift level and maximum displacement are inter-calculable as will be shown in the next step, and hence any of them can be defined here. The total drift rarely exceeds 3% to 4%, for all limit states.

2.4   Calculate the maximum displacement profile

By relating the maximum drift level and the maximum displacement, the displacement profile for the building can be estimated (Fig. 2). Since, for a frame structure, the maximum attained drift decreases as the number of stories increases, it is necessary to suitably adapt the probable displacement profile. Modifications suggested by Loeding *et al.* (1998), in the case of framed structures, are reproduced in the following equations where these are derived using the first three mode responses – equations (7), (8) and (9). In these equations Δ_i represents the storey displacement at level i, θ_d the selected design maximum drift, h_i the height of storey i, from base, h_n the total height of the frame and n the number of stories. The evolution of drift along the building height depends greatly on the pattern of plastic hinge distribution along the height.

$$\Delta_i = \theta_d\, h_i \qquad\qquad (\, n \le 4) \tag{7}$$

$$\Delta_i = \theta_d\, h_i \left(1 - \frac{(n-4)}{32}\frac{h_i}{h_n}\right) \qquad (4 < n < 20) \tag{8}$$

$$\Delta_i = \theta_d\, h_i \left(1 - \frac{h_i}{2\,h_n}\right) \qquad\quad (n \ge 20) \tag{9}$$

2.5   Calculate the system displacement

The system displacement (Δ_{sys}) can be calculated by defining an effective SDOF system (Δ_{eff}) to represent the MDOF system – equation (4).

2.6   Select an appropriate level of system damping

An appropriate level of damping for the effective SDOF system is selected, to represent the equivalent viscous damping of the MDOF system. In case of reinforced concrete structures, damping values of 20–25% are proposed (Loeding *et al.* 1998), for frames with beam-sway mechanisms (with hinges at beam-ends and base columns) and a drift of 2.5%. However, a value of 15% to 20% is considered sufficient for the level of expected damage, when a drift of 1.5% is chosen (ATC-40 1996).

With the system damping defined, it is possible to estimate the expected displacement ductility for the SDOF system, which should be a function of the equivalent viscous damping within the MDOF system. The hysteretic energy dissipation of all members can be converted into an equivalent viscous damping, as a function of the ductility of the members. Various proposals for the equivalent viscous damping *vs.* displacement ductility are presented in Figure 3.

2.7 Calculate the effective structural period, mass and stiffness

In order to calculate the period of the effective SDOF system it is first required to reduce the elastic response spectrum defined (step 2.2). This can be achieved using the formulation proposed by Boomer *et al.* (2000) – equation (11), or by any other appropriate method.

$$S_{D,\xi} = S_{D,5\%} \sqrt{\frac{10}{5 + \xi}}$$

(10)

Once the reduced spectral displacement is defined, the effective period (T_{eff}) can be obtained, as shown in Figure 5.

Other characteristic of the effective SDOF system, like the effective mass (M_{eff}) and stiffness (K_{eff}), can be easily established as shown in the following equations.

$$M_{eff} = \sum_{i=1}^{n} m_i \left(\frac{\Delta_i}{\Delta_{eff}} \right)$$

(11)

$$K_{eff} = \frac{4 \pi^2}{T_{eff}} M_{eff}$$

(12)

The effective force (F_{eff}) is calculated according to equation (13).

$$F_{eff} = K_{eff} \, \Delta_{eff}$$

(13)

2.8 Calculate the base shear

Assuming that the effective force of the SDOF is the base shear force (V_b) for which the MDOF system will be eventually designed at the selected limit state. Consequently, the analysis shifts back to the MDOF system by equating $V_b = F_{eff} = K_{eff} \Delta_{eff}$.

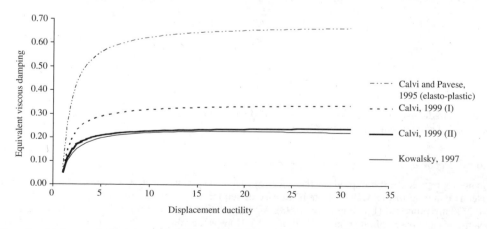

Figure 3. Various equivalent viscous damping *vs.* displacement ductility relationships.

2.9 Distribute the base shear over the building height

The base shear can now be distributed along the building height, proportional to the storey displacement according to the assumed displacement profile, as given in equation (15). Other similar formulations can also be used to define the displacement profile and, therefore, the force distribution.

$$F_i = V_b \frac{m_i \, \Delta_i}{\displaystyle\sum_{i=1}^{n} m_i \, \Delta_i} \tag{14}$$

2.10 Perform analysis and design the members

Finally, to obtain the internal member forces for design, a static analysis can be performed on the frame by applying the lateral forces along with the gravity loads. The analysis may be performed by suitability assuming a relative stiffness between the framing elements. The example under consideration in this work, adopts a beam to column relative stiffness (I_b/I_c) as shown in equation (15). This assumption is used as the beam-ends are expected to perform beyond their yield values, according to a weak beam-strong column structural behaviour, as desirable for greater hysteretic energy dissipation. More precisely, it is assumed that:

– Beam and column elastic or cracked stiffness are close;
– Beams will attain a ductility level of 5, while the columns behave elastically, for a seismic action.

$$\left(\frac{I_b}{I_c} = 0.2\right) \tag{15}$$

The design forces obtained from the static analysis are used to define the reinforcements for all sections. These reinforcements are obtained at the ultimate moment capacities.

3 CASE STUDY

To analyse the applicability of the DBD methodology a four-storey reinforced concrete building is used. Some results are compared with the non-linear dynamic procedure results.

3.1 Description of the structure

The structural geometry and the direction of analysis of the four-storey reinforced concrete structure considered are presented in Figure 4. The structure was designed according to Eurocode 8 – EC8 (CEN

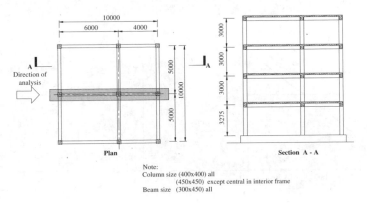

Note:
Column size (400x400) all
            (450x450) except central in interior frame
Beam size (300x450) all

Figure 4   Capacity General Layout (dimensions in mm) and direction of analysis.

1998), assuming a subsoil class B (medium), a peak ground acceleration equal to 0.3 g and considering a high ductility class (behaviour factor of 5). Dead and live loads were also considered in the design. The program IDARC version 5.0 (Valles *et al.* 1996) is adopted. Considering the direction of the analysis, the structure is split into two typical frames, the external and internal frame. The discretization was such that single elements were used for columns and beams. In IDARC version 5, beam and column elements are modelled similarly for flexural and shear deformations. Axial deformation is also considered in columns, but neglected in beams. The elements include a rigid zone length to simulate increase in stiffness at the joint, specified according to the connecting elements. In addition, the spread plasticity concept of the program aimed at capturing the variation of inelastic deformations, due to formation of cracks, which tend to spread from the joint interface along the element, is also used. The stiffness and mass distribution used to model the structure are as presented in Falcão (2002).

3.2 Design of the structure

The seismic hazard in the form of displacement response spectrum is defined for a peak ground displacement (PGA) of 0.3 g and 5% viscous damping (Fig. 5), from the EC8 acceleration response spectrum. For the non-linear dynamic analysis five artificial accelerograms were used.

Based on the selected performance objective, *i.e.* damage control, a drift range of 1% to 2% is expected to ensure that the strain limits in critical members do not attain their design values, as proposed in ATC-40 (1996). In this study a drift of 1.5% is considered.

Since the structure consists of only four stories, a linear distribution is considered to evaluate the displacement profile – equation (7). Thus, the storey displacements obtained are as follows: $\Delta_1 = \theta_d$ $h_i = 1.50\% \times 3.275 = 0.049$ m; $\Delta_2 = 0.0941$ m; $\Delta_3 = 0.1391$ m; and $\Delta_4 = 0.184$ m.

The system displacement is calculated as $\Delta_{sys} = \Delta_{eff} = 0.139\,m$ and the effective system displacement occurs at a height called as effective height, estimated as $h_{eff} = 9.243\,m$ (75.30%)

It is assumed that the system damping is 20%. The period evaluation for the effective SDOF system requires, firstly, the reduction of the elastic response spectrum according, for instance, the equation (10). The reduced displacement response spectrum is represented in Figure 5. The effective period for the damping level estimated is obtained from this figure and expressed as $T_{eff,20\%} = 1.92$ s.

The effective mass is calculated according equation (11) – $M_{eff} = 288$ ton (84.53%) – while the effective stiffness for the 20% damping is given as $K_{eff,20\%} = 3084.40\,kN$ – equation (12).

The maximum base shear force (V_b) of the SDOF system, *i.e.* the base shear for which the MDOF system will be eventually designed for the selected limit state, is given as: $V_{b,20\%} = 427.6$ kN. The base shear defined can now be distributed along the building height, proportional to the adopted displacement profile: $F_1 = 43.7$ kN; $F_2 = 86.6$ kN; $F_3 = 128.0$ kN; and $F_4 = 169.4$ kN.

Finally, to obtain the member forces for design, a static analysis is performed on the frame, applying the lateral forces. The gravity loads are also considered for the design purpose. An appropriate stiffness ratio between the beams and columns is assumed to reflect the probable state that the structure can achieve – equation (15). With the member forces obtained the member sections are design.

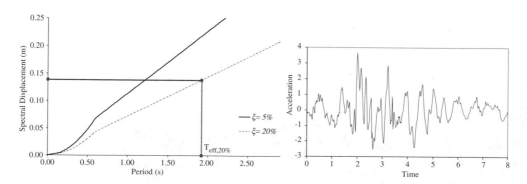

Figure 5.   Displacement response spectrum; Reference accelerogram.

3.3 Performance evaluation

In order to evaluate the performance of the designed structure various methodologies may be used; namely dynamic inelastic time history analysis (THA) or the pushover analysis (Modal Adaptive 1), as referred previously. For this study, the pushover analysis, *i.e.* nonlinear static analysis in the CSM format, is used for performance evaluation by means of the program IDARC – (Falcão 2002, Falcão & Bento 2002). Once the performance point is evaluated (Fig. 6) the demands in terms of storey shear, storey displacements, interstorey displacements (Fig. 7), beam ductility distribution and damage state of the frames (Fig. 8) can be obtained.

The total base shear from the initial analysis (427.6 kN) seems not to be very far to that obtained during the performance evaluation (517.44 kN).

Similarly, it can also be observed that a decrease of about 37% in the case of the top storey displacement. Initially a maximum top storey displacement of 184.0 mm is adopted (1.5% drift) while the performance evaluation resulted in 131.9 mm (Fig. 7b). In fact, the interstorey drift used in the initial analysis (1.5% drift) is not reached at the performance point, which provides different results as can be

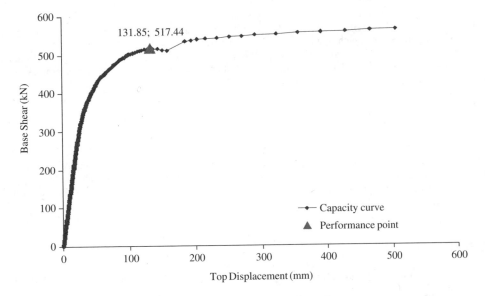

Figure 6. The performance point on the capacity curve for the 20% damped structure.

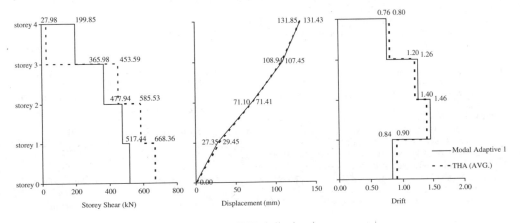

Figure 7. Storey shear, storey displacements and storey drifts at performance point.

449

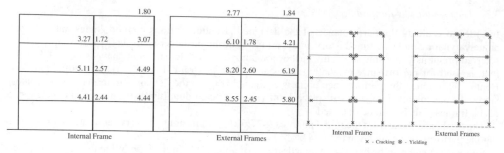

Figure 8.   Beam Ductility distribution at the performance point and Damaged state of the frames – IDARC output.

seen in Figure 7; the maximum value achieved is 1.45% between the 1st and 2nd storey. The results obtained with DBD are similar to the non-linear dynamic procedure results – Figure 7 – mainly in terms of storey displacements and interstorrey displacements (Drift).

From the damaged state of the structure, Figure 8, it is evident that the columns do not attain its yield strength, while the beams reach its plastic state, just as desired, while using the weak beam-strong column criteria for design. This also reflects on the ductility obtained at the beam-ends (Fig. 8). The maximum value of beam ductility demand attained is 8.55 at the first storey level. It should be noted that the column ductility are not presented, as they do not yield.

4   CONCLUSIONS

The DBD procedure incorporates the concept that a structure should be designed by defining its permissible damage state. Defining a suitable drift and distribute it over the structure height one can represent a strain limit to designate a damage state and deformation. The results obtained show that the described procedure provides adequate information on the seismic demand imposed by the ground motion on a regular structural system.

REFERENCES

ATC-40. 1996. *Seismic Evaluation and Retrofit of Concrete Buildings*, Vol. 1 & 2 ATC-40. Applied Technology Council, CA 94065.
Boomer, J.J., Elnashai, A.S. & Weir, A. 2000. Compatible Acceleration And Displacement Spectra For Seismic Design Codes, 12 WCEE, Auckland.
Calvi, G.M. 1999. *A Displacement-Based Approach for Vulnerability Evaluation of Classes of Buildings*. Journal of Earthquake Engineering, Vol. 3, No. 3.
CEN. 1998. *Eurocode 8*, Design provisions for earthquake resistance of structures, ENV 1998-1-1:1994, Commission of the European Communities, Brussels.
Cornell, C.A. & Krawinkler, H. *Progress and Challenges in Seismic Performance Assessment*, PEER Center News, Vol. 3, No. 2, Spring 2000.
Falcão, S. 2002. Performance Based Seismic Design, An application to a Reinforced Concrete Structure, MSc thesis, Technical University of Lisbon, IST. Lisbon.
Falcão, S. & Bento, R. 2002. Analysis Procedures for Performance-based Seismic Design, Proceedings of the 12th ECEE, European Association for Earthquake Engineering & Society for Earthquake and Civil Engineering Dynamics, Elsevier.
Kowalsky, M.J. 1997. *Direct Displacement-Based Design: A Seismic Design Methodology and Its Application to Concrete Bridges*, PhD dissertation, University Of California, San Diego.
Loeding, S., Kowalsky, M.J. & Priestley, M.J.N. 1998. *Displacement-based design methodology applied to R.C. Building frames*. Report SSRP 98/06 Structures Department, UCSD.
Priestley, M.J.N. 2000. Performance Based Seismic Design, 12th World Conference on Earthquake Engineering, Auckland, New Zealand.
Valles, R.E., Reinhorn, A.M., Kunnath, S.K., Li, C. & Madan, A. 1996. IDARC 2D Version 5.0: A Program for the Inelastic Damage Analysis of Buildings, Technical Report NCEER-96-0010, National Center for Earthquake Engineering Research, State University of New York at Buffalo, Buffalo, NY 14261.

Geotechnics

Computational Methods in Engineering and Science, Iu et al. (eds)
© 2003 Swets & Zeitlinger, Lisse, ISBN 90 5809 567 3

Solving geomechanical problems with UWay FEM package

Yu.G. Yanovsky, A.N. Vlasov, M.G. Mnushkin & A.A. Popov
Institute of Applied Mechanics, Russian Academy of Sciences, Moscow, Russia

ABSTRACT: The best known, effective and sometimes the single available methods for practical problems solution in many branches are the numerical ones. In given paper the functionality and structure of UWay software package based on Finite Element Method is described briefly.

Package is developed in accordance with object-oriented paradigm. Such approach provides an ability to utilize efficiently vector nature of forces and displacements and also tensor nature of stresses and strains. All data used in work is a set of class instances. This fact increases code stability, reduces its volume and enables developers to easily support and enhance their product. In paper the developed class hierarchy and brief description of essential data entities that form the core of FEM will be carried out.

As an example of successful solution done with UWay we perform planar consolidation problem of soil protection dam DUNG QUAT in Vietnam. Important solution results in graphical form will be plotted.

1 INTRODUCTION

Machine modeling is used widely in the world of science as a helpful tool for understanding of the nature. It is also very productive for the purposes of industrial design.

We can approximately subdivide all the software packages created for modeling of various processes into two groups. First group consists of products capable to solve complex industrial and scientific problems (commercial or industrial packages). Developers of these systems usually implement only well-studied material models, algorithms and techniques. By the way, it should be noticed that they attempt to satisfy requirements and needs of professionals form extremely different branches of science and engineering. There are situations when one method convenient, for example, in aircraft design would be absolutely useless for solving problems of soil mechanics. It is obvious, that any area possesses its unique specialties, which technically will not ever be implemented within single, even fantastically powerful, analysis system.

It occurs to us, that packages of first group possess also another very important disadvantage. They are commercial, and this fact eliminates potential ability of serious enhancements. Because the most important feature of such program is constant readiness to perform desired analysis. There is no time for experiment; there is no ability of dramatic changing, because each large step usually leads to necessity to proof users that action taken was reasonable and often shocks them.

There is also another valuable reason why these software packages are not easily modifiable. They are large, and often various parts of them are written in different programming languages. With such approach the Babylon tower problem grows – the problem of languages mixing (Revuzhenko, 1999). It does not surprise, but in accordance with results of Fifth World Congress on Computational Mechanics (Vienna, Austria, 2002) and others congresses and symposiums, there are many development projects in the field of computational mechanics initiated in research centers worldwide.

Definitely, these packages are usually less functionally powerful and posses worse user interface options etc. But in the same time their modification and tuning for the purposes of new specific problems is dramatically easier. It is very significant circumstance but possibility of such systems creation is within the scope of small group of professionals in the field of development, numerical methods and computational mechanics. All these specialized software packages, in our viewpoint, do form the second group. UWay is also supposed to be associated with it.

2 OBJECT-ORIENTED APPROACH

Product of second group development project is nearly impossible without preliminary investigations processed by various professionals. Start point of investigation is a hypothesis, but its main goal is hypothesis verification. Initial information used in investigation is a set of incorrectly formalized models (wide range of models) i.e. models, which properties had not been completely understood. In the beginning of investigation it is usually impossible to offer anything else then just to move forward without a plan and form it while amount of gathered facts is being grow. The most important thing is to try and observe the response. This means that it is required to establish and support constant feedback between developer and problem being investigated.

Software packages implementing FEM are unavoidably complex. It is practically impossible to cover all aspects of system by single developer. In other words, complexity of such systems is beyond the scope of human intelligence. Unfortunately, but complexity is obviously necessary property of entire large software systems. Here necessity supposed to be understood as the following: it is possible to create the complexity but it is unthinkable to avoid it (Booch, 1991).

An object-oriented approach in present is the single methodology that enables to handle complexity generated even by the largest systems. It brings the ability of relatively easy modification and development of software product.

3 PACKAGE APPLICATION AREA

Numerical analysis is the final step of any theoretical research. Software market for engineering and scientific analysis offers a lot of products that enable users to perform solution of practical problems in the field of traditional continous media mechanics and obtain quite good results. But unfortunately everything is not so well, for example, in geomechanics and geophysics. In this area the problems involving interaction between the materials in different phase states or strongly heterogeneous rocks and soils are usually considered. There are still not many of qualitative, intuitively understandable and user-friendly professional systems in this branch. Those ones that exist do not cover entire set of problems that needed to be solved.

Software package UWay is supposed mostly to perform Finite Element Analysis of stress-deformed state and stability of groundmasses considered in the problems of geotechnical engineering and geophysics.

Universality of package is determined by common methodology of solution wide range physical problems. It is seems that problems of continous media mechanics – static, dynamic, buckling in linear and nonlinear formulation should be within the scope of UWay. It should also be able to solve problems involving time factor (steady-state and transient heat transfer and seepage analyses), and of course coupled problems.

From the point of view of functionality UWay is supposed to be a complete system. In other words, users, after a while, would be able to perform all steps of complicated analysis: to create or import geometry model, mesh it automatically, define all additional model parameters, perform the analysis and obtain all required results in predefined form. In present not all of the steps mentioned above are available.

A size of problems analyzed does depend only on hardware resources (storage, CPU speed, bandwidth), because there are a lot of various algorithms that utilize efficiently properties of different hardware platforms and specialties of various problems.

4 CLASS HIERARCHY

Here we would like to describe in more detail a few principal aspects. The most significant UWay feature, which distinguishes it from other FEM systems, is its object-oriented nature. All its methods, algorithms and functions were developed in accordance with lows of mathematical abstractions. UWay is written totally in C++ object-oriented programming language. Hierarchies of classes and rules of interaction between objects of these classes were designed and implemented. This fact provided an ability to utilize efficiently vector nature of forces and displacements and also tensor nature of stresses and strains.

In Figures 1–3 the fragments of class hierarchies used in package development are displayed. The complete description of all developed classes is probably beyond the scope of this paper, so only brief description of essential ones, that form the core of FEM, is listed.

- Element – finite elements data structure and functionality.
- Node – the nodes of the mesh.

- Tensor – tensors class hierarchy (stiffness tensor, stress tensor, strain tensor etc.).
- ShapeFunction – Shape functions class hierarchy.
- DOperator – differential operators class hierarchy (differential operators do transform, for example, displacements vectors into strain tensors or temperatures into heat fluxes vectors etc.).

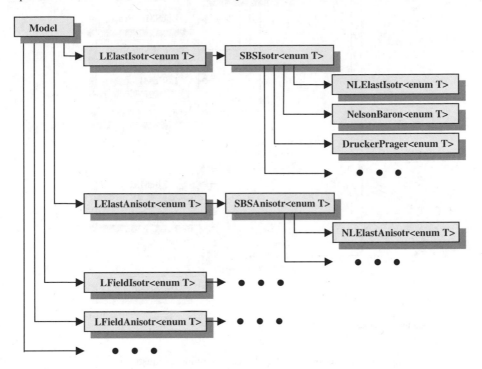

Figure 1. Constitutive lows class hierarchy.

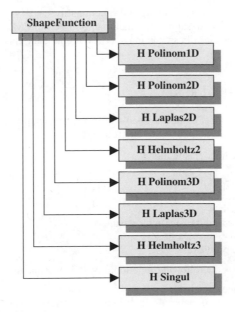

Figure 2. Shape functions class hierarchy.

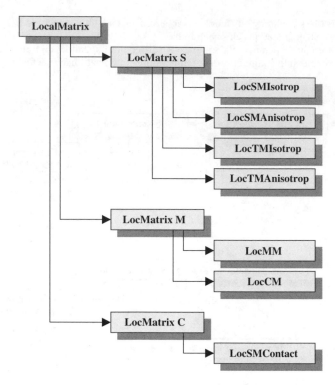

Figure 3. Local matrices class hierarchy.

- Model – templated class hierarchy of material behavior models (constitutive lows).
- LocalMatrix – local (elemental) matrices class hierarchy (stiffness, mass, damping, heat conductivity etc.).
- GlobalMatrix – sparse matrix data structure and functionality.
- Vector – vectors class hierarchy (supported functionality used in FEM).
- Load – class hierarchy that describes wide range types of boundary conditions.
- …

5 CAPABILITIES AND FEATURES

There are following mathematical models available in UWay that enable solution of complicated problems involving various types of nonlinearities and time factor:

- Isotropic and anisotropic liner elastic model.
- Nonlinear elastic model.
- Incremental models theory of plasticity.
- Model of elasto-plastic yielding.
- Visco-elasto-plastic models.
- Steady state and transient heat transfer and seepage.
- Phase change heat transfer.
- Solid-state phase change problem.

In general, package brings the following functionality. It is possible to:

- Perform linear and nonlinear analyses.
- Consider in model a number of materials described with different constitutive lows and posses various characteristics.

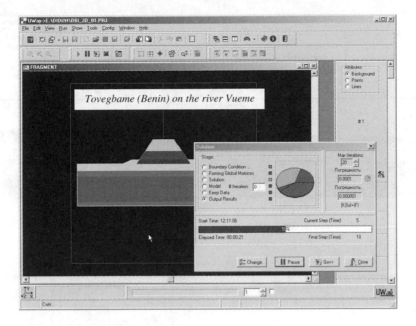

Figure 4. Main viewer of UWay user interface system.

- Utilize different finite element formulations within single model.
- Combine elements of different dimensions.

UWay is capable to consider problems involving:

- Analysis domain geometry modification (for example, stages of construction).
- Variation of material properties.
- Changing of boundary conditions.
- Strongly nonlinear soil materials.

There are following shape functions used for field approximation within elements:

Planar problems:
- Polynomial approximation (4–8 node elements).
- Approximation built on the exact solution of Laplas equation (arbitrary number of nodes).
- Approximation with control parameter built on the exact solution of Hemholtz equation.

Spatial problems:
- Polynomial approximation (8–20 node elements).
- Approximation with control parameter built on the exact solution of Hemholtz equation.

UWay is designed and built to be able to treat with large (millions DOF) real life problems. To maintain this feature we were supposed to implement fast and efficient linear equation solvers and, therefore, advanced matrix storage formats.

Here are linear solvers for single processor machines implemented in package:

- Dual threshold Incomplete LU factorization (ILUT) first implemented by Y. Saad coupled with Generalized Minimal Residual (GMRES) method for nonsymmetric case.
- Robust Incomplete Cholesky Second order Stabilized (RIC2S) first implemented by I.E. Kaporin coupled with usual Conjugate Gradient (CG) method for symmetric one.

For vector and multiprocessor computers (especially for SMP) some explicit approximate inverse preconditioning techniques are being implemented. This work is not completed yet.

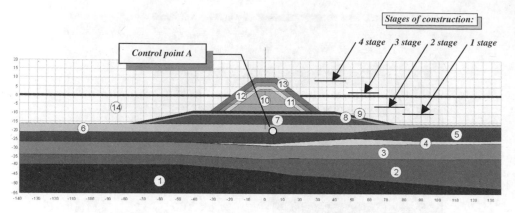

Figure 5.   Geometry of analysis model. Soil types are marked with consecutive numbers.

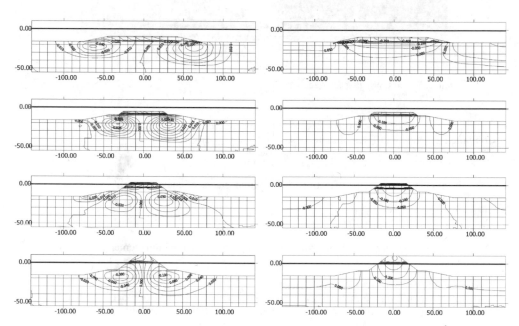

Figure 6.   Vertical (right) and horizontal (left) displacements isolines (m) at various stages.

Matrix storage formats used are:

- Modified Compressed Sparse Row (MSR) – for preconditioners.
- All global matrices are stored in single dimension array and indexed with instance of class Graph. Such schema saves storage and speeds up matrix to vector multiplications due to less bandwidth required to transfer index data.
- SkyLine storage format is still used, but it is about to be removed completely.

For solution of contact problems including various friction effects a direct constraint procedure is being implemented.

UWay provides its original pre- and post-processing services, but it is also possible to convert EDS/FEMAP Neutral files into UWay input files format and analyze models built there. Users are

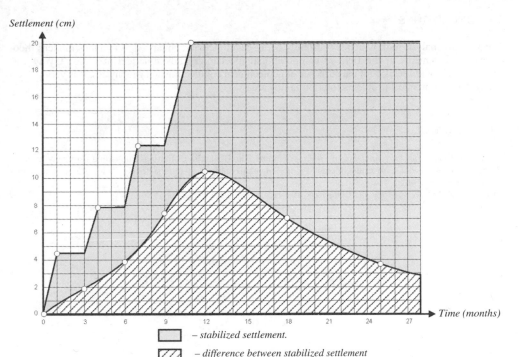

Settlement (cm)

Time (months)

▭ – *stabilized settlement.*

▨ – *difference between stabilized settlement and one computed for given moment of time.*

Figure 7.   Settlements of control point A via time.

enabled to view and post-process the results in FEMAP after/during the analysis. Results files could be converted into Neutral Format automatically and FEMAP started for post-processing.

List of available preprocessors is going to be filled soon with:

- MSC/MARC Mentat.
- ANSYS GUI.

Operating systems supported are:

- Microsoft Windows 9x, ME, 2000/XP.
- Linux Red Hat 8.0 (highest priority).

6   EXAMPLE OF USE

As an example of successful solution done with UWay we perform planar consolidation problem of soil protection dam (breakwater) DUNG QUAT in Vietnam. Geometry of analysis model is displayed in Figure 5. Some important solution results in graphical form are plotted in Figures 6 and 7.

Parameters of Marine clay (layer 5) for FEM analyses are: unit weight $\gamma = 7.2\,kN/m^3$, modulus of elasticity $E = 5\,MPa$, Poisson ratio $\nu = 0.4$, angle of internal friction $\varphi = 22°$ cohesion $c = 0.001\,MPa$, coefficient of consolidation $c_v = 0.00432\,m^2/day$, consolidation of permeability $k_f = 0.000056\,m/day$.

The stabilized values of horizontal and vertical displacement within the limits of settlement area are shown as lines of levels on Figure 6.

From the analysis of results of numerical calculation follows, that ~85% settlement are realized within 2nd years from the moment of the beginning of construction. Maximal stabilized a settlement on the bottom of a breakwater will make ~60 om (see Fig.7).

459

7 CONCLUSIONS

UWay is an analysis tool, which is constantly being developed and enhanced. It includes modern theoretical achievements in the field of computational mechanics and numerical methods. UWay was designed and built specific for the purposes of geotechnical engineering and geophysics. It is very powerful system for practical problems solution in this area.

REFERENCES

Booch, G. 1991. *Object oriented design with applications*. The Behjamin/Cummings Publishing Company Inc.
Revuzhenko, A.F. 1999. Concerning methods of non-standard analysis in solid state mechanics. *Physical mesomechanics*, 2(6): 51–62.
Yufin, S.A. 2000. *Geoecology and Computers*. Rotterdam: Balkema.

Computational Methods in Engineering and Science, Iu et al. (eds)
© 2003 Swets & Zeitlinger, Lisse, ISBN 90 5809 567 3

Three-dimensional finite element analysis of multi-element composite foundation

J.J. Zheng, J.H. Ou & X.Z. Wang
College of Civil Engineering and Mechanics, Huazhong University of Science and Technology
Wuhan, China

ABSTRACT: A numerical analysis is studied for examining a five-pile composite foundation (one CFG (Cement Flyash Gravel) pile and four cement–soil piles) based on a three-dimension model in this paper. By changing the values of the parameters, including the thickness and stiffness of cushion, the lengths and diameters of piles, their effects on the settlement and stress distribution of the composite foundation are revealed. The results show that the settlement is greatly affected by the length of the CFG pile, and thus it is safe to say that the CFG pile acts as the reducing-settlement pile, conforming to the design principle of a multi-element composite foundation. It is also demonstrated that the stress distribution is markedly affected by the variables of cushion.

1 INTRODUCTION

A multi-element composite foundation is the foundation reinforced by two or three types of piles (i.e. flexible piles, rigid piles, and granular material piles, all of which are often used in composite ground for vertical reinforcement). Since the bearing capacity and the deformation properties of each kind of piles are different, the settlement mechanism and the stress distribution of the foundation are complicated.

To date, many simplified designing methods have been proposed. Though these methods have provided satisfactory engineering practice results, their mechanisms are not clear, especially in terms of the settlement computation (Zheng 2002). To evaluate the contribution of each element (key piles, secondary piles and soils) on the multi-element composite foundation, many numerical analyses were conducted recently. But most were limited to two-dimensional (2D) models (Cheng et al. 2000, 2001). Although a three-dimensional (3D) problem can sometimes be simplified into a 2D model, the results will be inaccurate in a multi-element composite foundation because the effects of the relative locations of piles within the foundation cannot be modeled.

In the present study, the 3D finite element analyses are implemented for examining a five-pile composite foundation composed of one CFG pile and four cement–soil piles. According to the obtained achievement and engineering practice (Liu 2003), the major factors affecting the settlement and the stress distribution of the composite foundation include the thickness and the stiffness of cushion, and the length and diameters of piles. Thus they were taken as the focus of the study. To make the analyses clear, only one factor was varied each time while no change for the others.

2 3D FINITE ELEMENT MODEL

The five-pile composite foundation used in the numerical simulations is sketched in Figure 1. For convenience of meshing, quadrate section piles are adopted. The distance between a CFG pile and each cement–soil pile is 1.5 meters in both horizontal and vertical directions. The thickness of the 4 m × 4 m plate is 0.3 m, while that of the 4 m × 4 m cushion is 0.2 m. To minimize the influence of the boundary constraint, a 20B × 20B × 3L2 soil block is considered in analyses (where B is the side length of the plate, and L2 is the length of the CFG pile).

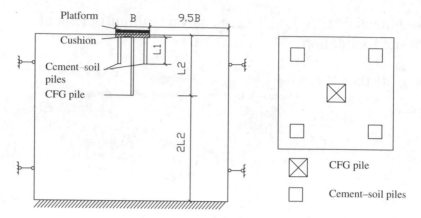

Figure 1. Sketch map of the foundation.

Table 1. Material properties used in the study.

Material	Plate	Cushion	CFG pile	cement–soil pile	Soil
E(Pa)	$E_c = 2 \times 10^{11}$	$E_m = 2 \times 10^7$	$E_{P1} = 2.5 \times 10^{10}$	$E_{P2} = 1 \times 10^8$	$E_s = 5 \times 10^6$
μ ·	$\mu_c = 0.26$	$\mu_m = 0.30$	$\mu_{P1} = 0.20$	$\mu_{P2} = 0.30$	$\mu_s = 0.30$

The finite element package ANSYS is used to simulate this five-pile composite foundation. Since the foundation is symmetric, only a quarter of it is considered (including a quarter of CFG pile and one cement–soil pile). Eight-node isoparametric elements are adopted. To improve the calculation, a relatively fine mesh is used near the piles, and it becomes sparser further from the piles.

Since plastic properties of soil are neglected in general in engineering design theory of multi-element composite foundation, an elastic model is used for convenience of the analyses. Table 1 summarizes the typical material properties used in the study. Unless otherwise stated, these properties have been used throughout.

3 DIAMETER OF PILE

The diameter of pile is a crucial factor in a composite foundation design. In the study, the diameter of CFG pile is gradually increased, while other factors of the five-pile composite foundation keep unchanged. Assuming a pressure of 150 kPa on the foundation, a simulation is carried out through ANSYS. Then a similar process is conducted with the graduated increase of the diameter of cement–soil piles this time. The results of the two kinds of changes are shown in Table 2 and Table 3 respectively.

In the above Tables, D1 and D2 are the diameters of CFG pile and cement–soil piles respectively; n_1 is the stress ratio of the CFG pile to the soil, while n_2 is the stress ratio of the cement–soil piles to the soil; S is the settlement of the foundation; σ_1 is the stress on top of the CFG pile, and σ_2 is the stress on top of the cement–soil pile; σ_s is the stress on the superficial soil; $f_{sp,k}$ is the bearing capacity of the composite foundation. Unless otherwise stated, these symbols have been used throughout.

From the data above, It is seen that with the diameters of either CFG pile or cement–soil piles growing, the bearing capacity ($f_{sp,k}$) and the settlement of foundation (S) increases and decreases respectively. The stresses on soil, CFG pile and cement–soil piles decrease in the same time. To compare the effects brought by the increase of the diameters of two kinds of piles, the S–D curves of two kinds of piles are drawn, as shown in Figure 2.

According to Figure 2, we can also see that the growth of the diameter of CFG pile is more effective in reducing the settlement of foundation than that of the cement–soil piles. The conclusion is agreement with the result obtained by the settlement calculation theory in practice. In practice, the settlement of

Table 2. Simulation results by ANSYS (changing the diameter of CFG pile).

D_2 (m)	n_1	n_2	S (m)	σ_1 (MPa)	σ_2 (MPa)	σ_s (kPa)	$f_{sp,k}$ (kPa)
0.30	233.189	15.602	0.043211	10.525	0.70420	45.135	140
0.35	209.160	16.174	0.041633	8.4281	0.65173	40.295	145
0.40	136.748	12.978	0.040655	6.7946	0.64486	39.968	150
0.45	147.758	15.730	0.039175	5.7432	0.61140	38.869	158
0.50	278.196	31.014	0.038609	5.2120	0.58104	18.735	162

Table 3. Simulating results by ANSYS (changing the diameter of cement–soil piles).

D_1 (m)	n_1	n_2	S (m)	σ_1 (MPa)	σ_2 (MPa)	σ_s (kPa)	$f_{sp,k}$ (kPa)
0.30	173.443	15.712	0.042797	7.8641	0.71238	45.341	140
0.40	238.725	24.050	0.041976	7.2040	0.72575	30.177	145
0.50	136.748	12.978	0.040655	6.7946	0.64486	24.968	150
0.60	319.333	23.525	0.039244	7.0793	0.52152	22.169	155

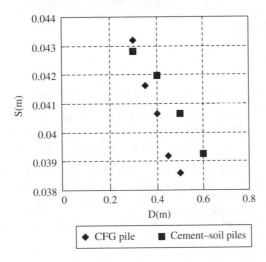

Figure 2. S–D curve.

soil in consolidation stratum S_1 is worked out by a layer-wise summation method (Zheng 2002), and the formula is shown as

$$S_1 = \sum \frac{\Delta P_i H_i}{E_{csi}} \tag{1}$$

where ΔP_i is the incremental value of additional stress on the i-th composite stratum, H_i is the thickness of the i-th composite stratum, and E_{csi} is the compound compression modular of the i-th composite stratum.

As for a three-element composite foundation, E_{cs} is determined by the weighted area method as follows

$$E_{cs} = m_1 E_{p1} + m_2 E_{p2} + (1 - m_1 - m_2) E_s \tag{2}$$

where $m_1 = A_1/A$; $m_2 = A_2/A$; $m_1 = A_s/A$; A_1, A_2, A_s are respectively the total section area of a key pile, of secondary piles and of soil; A is the total section area of the foundation.

According to the principle above, the growth of E_{cs} must result in a decrease of the settlement. Because both E_{p1} and E_{p2} are larger than E_{cs}, increasing each of them shall improve the value of E_{cs},

Table 4. Simulation results by ANSYS (changing the length of CFG pile).

L_2 (m)	n_1	n_2	S (m)	σ_1 (MPa)	σ_2 (MPa)	σ_s (kPa)
10	91.217	14.123	0.0823	7.996	1.238	87.659
15	123.562	14.018	0.0726	9.793	1.111	79.256
20	123.983	12.093	0.0691	11.380	1.110	91.787
25	134.330	12.797	0.0649	10.852	1.034	80.786

Table 5. Simulation results by ANSYS (changing the length of cement–soil piles).

L_1 (m)	n_1	n_2	S (m)	σ_1 (MPa)	σ_2 (MPa)	σ_s (kPa)
5	110.855	10.292	0.0691	11.03	1.024	99.499
10	123.983	12.093	0.0691	11.38	1.110	91.787
12	113.734	11.268	0.0687	10.82	1.072	95.134
15	118.541	11.132	0.0684	10.85	1.085	91.504

and accordingly reduces the settlement. In this study, since E_{p1} is significantly larger than E_{p2}, increasing the diameter of the CFG pile is more effective in reducing the settlement than increasing the diameter of cement–soil piles.

4 LENGTH OF PILES

Since the piles in the multi-element composite foundation have different bearing capacities and deformation properties, the lengths of piles are also different in most cases. So the lengths of piles are playing an important role in a foundation design. In the study, the CFG pile is gradually lengthened with other factors of the foundation unchanged. With an assumed pressure of 250 kPa on the foundation, a simulation is carried out through ANSYS. Then a similar process is conducted by lengthening the cement–soil piles this time. The obtained results of the two kinds of changes are displayed in Table 4 and Table 5 respectively.

As shown in Table 4 and Table 5, with the CFG pile lengthening, there is a significant reduction in the settlement of the foundation, and a great increase in the stress ratio of the CFG pile to the soil, while the stress ratio of the cement–soil piles to the soil changes little. As the cement–soil pile lengthening, there is no significant change in the settlement or the stress. The simulation results prove that the key pile with higher strength acts as the settlement-reducing pile, which conforms to the multi-element composite foundation theory.

5 STIFFNESS AND THICKNESS OF CUSHION

The cushion is a granular layer (e.g. sands or gravel) under the platform. The adoption of a cushion in a composite foundation enables the soil to share the loads, reduces the stress concentration of the foundation plate, and adjusts the load distribution on piles and soil (Liu Fenyong 2003). In the study, the role of the cushion in the bearing capacity of the composite foundation is specifically examined by varying its thickness and stiffness.

Firstly, the stiffness of the cushion is gradually increased. Assuming a pressure of 150 kPa on the foundation, the simulation is done through ANSYS results of which are displayed in Table 6.

As shown in Table 6, with the stiffness of the cushion increasing, the stress of the CFG pile increases greatly, while that of the cement–soil piles and that of the soil decrease. Besides, the settlement of the foundation decreased slightly.

Next the thickness of the cushion is gradually increased. The simulation results are displayed in Table 7 and Figure 3.

As shown in Table 7 and Figure 3, with the thickness of the cushion increasing, the stress ratio of the CFG pile to the soil decreases greatly, and that of cement–soil piles reduce slightly.

These results are of significance for engineering practice. For example, as for the stiff soil between piles with a larger bearing capacity, it is better to use a thicker and less stiff cushion. If the secondary piles are

464

Table 6. Simulation results by ANSYS (changing the stiffness of the cushion).

	Stiffness (MPa)							
	10	20	30	40	50	60	70	80
σ_1 (MPa)	5.1051	6.7946	7.7407	8.3404	8.7506	9.0463	9.2675	9.4337
σ_2 (MPa)	0.7091	0.6448	0.5957	0.5599	0.5327	0.5113	0.4940	0.4795
σ_s (kPa)	65.422	47.320	41.110	35.647	31.845	29.039	26.879	25.163
S (m)	0.0447	0.0407	0.0386	0.0374	0.0365	0.0358	0.0353	0.0349

Table 7. Simulation results by ANSYS (changing the thickness of the cushion).

	Thickness (m)						
	0	0.1	0.2	0.3	0.4	0.5	0.6
σ_1 (MPa)	9.8490	8.1448	6.7946	5.9733	5.745	5.2071	5.057
σ_2 (MPa)	0.3532	0.6123	0.6449	0.6698	0.5438	0.5697	0.5359
σ_s (kPa)	14.257	23.427	39.968	48.533	47.679	47.098	46.287
S (m)	0.0333	0.0386	0.0407	0.0414	0.0411	0.0413	0.0410

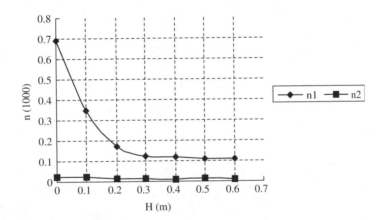

Figure 3. H–n curve.

expected to bear a large proportion of the load, the cushion should be much softer. It should be noted that in practice the stiffness of a cushion is adjusted by changing its constitution of aggregates and the diameters of the aggregates. In most cases, with the thickness and stiffness of the cushion taken into account, the corresponding engineering design will be more satisfactory.

6 CONCLUSION

By changing the variables of piles, soil and cushion, 3D numerical simulations are completed on a five-pile multi-element composite foundation. The results show that all the factors play important roles in the foundation design, but they differ in effects. Besides, some design theories of multi-element composite foundation are further confirmed. Nevertheless, all the results hold conducive significant to engineering practice. It is suggested that appropriate adjustments be made on the diameters and lengths of piles, thickness and stiffness of the cushion, so as to diotribute the load suitably on the key pile, secondary piles and soil, to make full use of the bearing capacities of the piles, and to control the settlement of the foundation effectively.

REFERENCES

Chen Qiang, Huang Zhiyi & Zuo Renyu. 2000. Analysis of deformation properties of ternary combined composite foundation. *Industrial Construction* 30(10): 52–55.
Chen Qiang, Huang Zhiyi & Zuo Renyu. 2001. Behavior of combined composite ground and simulation by study FEM. *China Civil Engineering Journal* 34(1): 50–55.
Liu Fenyong. 2003. Field test of a composite foundation including mixed pile. *Chinese Journal of Geotechnical Engineering* 25(1): 71–75.
Zheng Junjie. 2002. Theory and practice of multi-element composite ground. *Chinese Journal of Geotechnical Engineering* 24(2): 208–212.

Simplified methods for kinematic soil–pile interaction analysis

J.A. Santos

Departamento de Engenharia Civil, Instituto Superior Técnico, Lisboa, Portugal

ABSTRACT: During an earthquake pile foundations are subjected to inertia forces coming from the superstructure and to imposed curvatures by the soil deformations due to the passage of the seismic waves.

This paper focuses on the major aspects related to the kinematic soil–pile interaction during earthquake loading. Numerical results using the BDWF model (beam on dynamic Winkler foundation) will be presented to illustrate the key features of the problem.

1   INTRODUCTION

Nowadays there are an increased number of structures supported on piles with large diameter. During an earthquake the piles are subjected to inertia forces coming from the superstructure and to imposed curvatures by the soil deformations due to the passage of the seismic waves. The majority of research on the dynamic response of pile foundations is related with the study of inertia loading. The piles are usually designed to support only the inertial forces generated from the superstructure during earthquake loading.

However, recent observations on seismic foundation damage had indicated that in several cases the pile damage is caused by the lateral movement of the surrounding soil induced by the passage of seismic waves. This type of soil–pile interaction is called kinematic loading and is illustrated in Figure 1.

According to recent observations Mizuno (1986) and Mizuno et al. (1997) classified the pile damage patterns into four categories:

1. Damage with subsidence of pile head.
2. Ring-type crack due to bending moment.
3. Separation of pile from pile cap.
4. Buckling failure of welding joint.

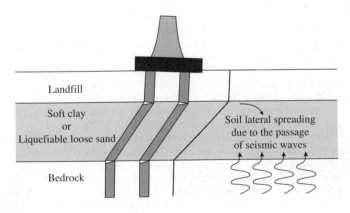

Figure 1.   Soil–pile seismic kinematic interaction.

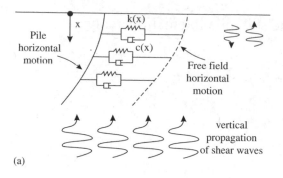 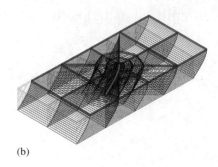

(a)   (b)

Figure 2.   (a) Dynamic discrete mean model.   (b) Dynamic 3-D continuum mean model.

In pattern 2 the damage zone can be located far from the pile cap. The damage is due to the kinematic loading and it can be significant in poor ground conditions and at interfaces of soil layers with sharply different stiffness (Fig. 1).

The kinematic soil–pile interaction can be studied using models that can be grouped into two classes: (i) discrete mean models and (ii) continuum mean models. In the first class, the soil is represented by a series of linear or non-linear springs and dashpots (Fig. 2a). Varying the parameters along pile length easily simulates the soil profile. Several authors developed significant research work on the BDWF model (Berrones & Whitman 1982, Kavvadas & Gazetas 1993, Makris & Gazetas 1992, Nikolaou & Gazetas 1997).

In the second class of models the soil is generally represented by an elastic half space continuum (Fig. 2b). Rigorous models based on boundary element or finite element method was applied to obtain the solution (Fan et al. 1991, Kaynia & Kausel 1991). These rigorous solutions were used to calibrate the parameters of the BDWF model.

This paper focuses on the major aspects related to the kinematic soil–pile interaction during earthquake loading and simplified methods will be presented.

2   BDWF MODEL

The importance of kinematic loading has been recognized in some recently published design codes: Eurocode 8 (1995), AASHTO (1983), JSCE (1988) and AFPS (1990). However only a few research work had done in this domain.

The BDWF model described before can be used very efficiently to model the kinematic soil–pile interaction. The dynamic equilibrium of the soil–pile system (Fig. 3) is governed by the following equation:

$$E_p I_p \frac{\partial^4 y}{\partial x^4} + m \frac{\partial^2 \overline{y}}{\partial x^2} + c \frac{\partial(\overline{y} - \overline{u})}{dt} + k(\overline{y} - \overline{u}) = 0 \tag{1}$$

where E_p = Young's modulus of pile; I_p = moment of inertia of pile section; y = relative pile displacement; x = depth; t = time; m = pile mass density per unit length; c = damping coefficient; $\overline{y}$ = absolute pile displacement; $\overline{u}$ = absolute free-field displacement; and k = subgrade reaction modulus.

A seismic wave propagation model can be used to obtain the free-field response. A finite element computer code named *CINEMAT* was developed by Santos (1999) and it combines the BDWF model with a 1-D seismic wave propagation model. The soil is discretized in horizontal multi-layers and the kinematic soil–pile response is obtained in frequency domain. To compute the response in time domain the FFT technique is used. This model was validated by comparison of results with rigorous 3-D finite element method and also boundary element method. The deviation between the models is usually less than 15%, which is acceptable from the practical point of view. The equivalent linear method is incorporated in the code to approximate the nonlinear behavior of the surrounding soil, during earthquake loading.

As described before, the kinematic loading can be very significant in two cases: (a) in poor ground conditions in which amplification can occur (b) at interface of soil layers with sharply different stiffness. These two cases can be easily visualized through the results of Figure 4 and Figure 5 obtained from the computer code *CINEMAT*.

468

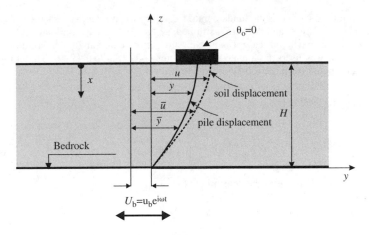

Figure 3. Kinematic soil–pile interaction: BDWF model.

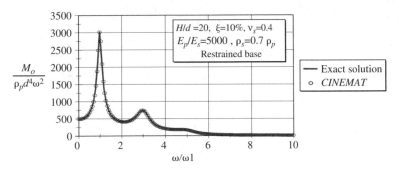

Figure 4. Transfer function for pile head bending moment in a homogeneous soil profile.

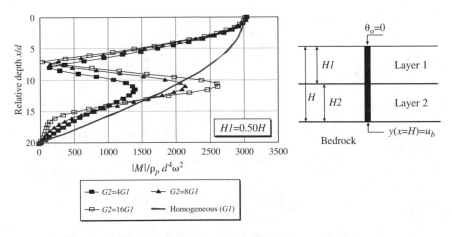

Figure 5. Bending moment distribution in a two-layers soil profile.

Figure 4 shows the transfer function for the bending moment of a single pile in a homogeneous elastic soil profile characterised by its parameters (E_s and ν_s). The other parameters are: ρ_s and ρ_p are the mass density of soil and pile, respectively; d is the pile diameter, ω is the frequency of a harmonic base

469

excitation with amplitude u_b; ω_1 is the fundamental frequency of soil deposit; ξ is the soil damping. The pile head is fixed (no rotation) and the base restrained, i.e., the pile and the soil have the same displacement at the tip. For this particular case it is possible to get the exact analytical solution for base harmonic excitation that was used to validate the computer code (Santos, 1999). The complex bending moment can be computed from the following equation:

$$\frac{M(x)}{E_p I_p} = 2\lambda_c^2\, m(x) + a^2 \Gamma u_b\, \frac{\cos(ax)}{\cos(aH)} \tag{2}$$

$$m(x) = e^{\lambda_c x}\left[C_1\, \sin(\lambda_c x) - C_2\, \cos(\lambda_c x)\right] - e^{-\lambda_c x}\left[C_3\, \sin(\lambda_c x) - C_4\, \cos(\lambda_c x)\right] \tag{3}$$

where C_1, C_2, C_3, C_4 = constants to be determined by:

$$\begin{bmatrix} 1 & 1 & -1 & 1 \\ -1 & 1 & -1 & -1 \\ A & B & C & D \\ -A & B & C & -D \end{bmatrix}\begin{Bmatrix} C_1 \\ C_2 \\ C_3 \\ C_4 \end{Bmatrix} = \begin{Bmatrix} 0 \\ 0 \\ 1-\Gamma \\ \dfrac{\Gamma a^2}{2\lambda_c^2} \end{Bmatrix} \tag{4}$$

$$A = e^{\lambda_c x}\, \cos(\lambda_c H) \tag{5}$$

$$B = e^{\lambda_c x}\, \sin(\lambda_c H) \tag{6}$$

$$C = e^{-\lambda_c x}\, \cos(\lambda_c H) \tag{7}$$

$$D = e^{-\lambda_c x}\, \sin(\lambda_c H) \tag{8}$$

$$\lambda_c = \sqrt[4]{\frac{k - m\omega^2 + ic\omega}{E_p I_p}} \tag{9}$$

$$a = \frac{\omega}{V_s\sqrt{1+i2\xi}} \qquad (V_s\ \text{is the shear wave velocity of soil}) \tag{10}$$

$$\Gamma = \frac{k + ic\omega}{E_p I_p\left(a^4 + 4\lambda_c^4\right)} \tag{11}$$

At pile head the bending moment can be easily obtained from equation (2): $M_o = |M(x = 0)|$.

A second case is analysed in which a layered soil profile is considered.

Figure 5 shows the distribution of the normalized bending moment of a single pile in a two-layers soil profile (G1 and G2 are the shear modulus of layers 1 and 2; H1 and H2 are the thickness of layers 1 and 2, respectively) subjected to a harmonic base excitation at resonant frequency. Three cases with different stiffness contrast are analysed: G2 = 4G1, G2 = 8G1 and G2 = 16G1. All the other parameters are equal to the previous case.

The results presented in Figure 4 and 5 are dimensionless from which can be pointed out the following two conclusions:

(a) The kinematic response is strongly affected by the frequency. The frequency contains of the excitation relatively to the fundamental frequency of soil deposit is one of the major aspects to be considered in the analysis (Fig. 4).
(b) In non-homogeneous soil profile the bending moment in layers transition can be as high as at the pile cap. These high bending moments or curvatures are concentrated in layers transition with sharply different stiffness (Fig. 5).

One of the simplify approach which is used widely in design neglects the mass and the damping effects on pile. It can be considered as a simplify version of equation (1), i.e.:

$$E_p I_p \frac{\partial^4 y}{dx^4} + k(\bar{y} - \bar{u}) = 0 \quad or \quad E_p I_p \frac{\partial^4 y}{dx^4} + k\,y = k\,u \tag{12}$$

The free-field response can be determined using a one-dimensional analysis as described before. From the displacements profiles $u(x)$, curvatures or bending moment can be obtained from an equivalent static analysis (equation 12).

In this approach the pile is constrained to displace in phase with the free-field response. This condition is valid only for flexible piles in relatively homogeneous soil profile.

3 CONCLUSIONS

This paper discusses the problems related with the kinematic soil–pile interaction during earthquake loading. The BDWF model is used to illustrate the key features of the problem. Numerical examples are used to demonstrate that high bending moments or curvatures can be generated in poor ground conditions or in layers transition with sharply different stiffness. In such regions the piles shall be designed to be ductile, using proper confining reinforcement.

ACKNOWLEDGEMENT

This work was developed under the research activities of ICIST (Institute for Structural Engineering, Territory and Construction of Instituto Superior Técnico) and was partially supported by pluriannual funding from FCT (Science and Technology Foundation), Portugal.

REFERENCES

AASHTO 1988. Guide *specifications for the seismic design of highway bridges*, Washington D.C.
AFPS 1990. Recommandations pour la redaction de regles relatives aux ouvrages et installations a realizer dans les regions sujettes aux seismes, Paris.
Berrones, R.F. & Whitman, R.V. 1982. Seismic response of end-bearing piles. *Journal of Geotecnical Engineering*, ASCE, 108(4): 554–569.
ENV 1998-5. *Eurocode 8*, Part 5: foundations, retaining structures and geotechnical aspects, Bruxelles.
Fan, K. & Gazetas, G. Kaynia, A. Kausel, E. Ahmad, S. 1991. Kinematic seismic response of single piles and pile groups. *JGED*, ASCE, 117, no. 12, 1860–79.
Fan, K. Gazetas, G. Kaynia, A. Kausel, E. Ahmad, S. 1991. Kinematic seismic response of single piles and pile groups. *JGED*, ASCE, 117(12): 1860–1879.
JSCE 1988. Earthquake engineering design for civil engineering structures in Japan. *Japanese Society of Civil Engineers*, Tokyo.
Kavvadas & Gazetas 1993. Kinematic seismic response and bending of free-head piles in layered soil. *Géotechnique* 43(2): 207–222.
Kaynia, A. & Kausel, E. 1991. Dynamics of piles and pile groups in layered soil media. *Soil Dynamics and Earthquake Engineering*, 10(8): 386–401.
Makris & Gazetas 1992. Dynamic pile–soil–pile interaction. Part II: lateral and seismic response. *Earthquake Engineering and Structural Dynamics*, 21(2): 145–162.
Mizuno, H. 1987. *Pile damage during earthquakes in Japan*. Dynamic response of pile foundations (ed. T. Nogami), ASCE: 53–78.
Mizuno, H., Iiba, M., & Hirade, T. 1996. Pile damage during 1995. Hyougoken-Nanbu earthquake in Japan. *11th World Conference on Earthquake Engineering*, Acapulco-Mexico, paper no. 977.
Nikolaou & Gazetas 1997. Seismic design procedure for kinematically stressed piles. Seismic Behaviour of Ground and Geotechnical Structures. *Proc. of Discussion Special Technical Session on Earthquake Geotechnical Engineering* during 14th International Conference on Soil Mechanics and Foundation Engineering, Hamburgo, Germany, P.S. Sêco e Pinto Editor, A.A. Balkema: 253–260.
Santos, J.A. 1999. Soil characterisation by dynamic and cyclic torsional shear tests. *Application to the study of piles under lateral static and dynamic loadings* PhD Thesis, Technical University of Lisbon, Portugal (in Portuguese).

Computational Methods in Engineering and Science, Iu et al. (eds)
© *2003 Swets & Zeitlinger, Lisse, ISBN 90 5809 567 3*

Numerical simulation of soils involving rotation of principal stress axes

Y.X. Liu
Department of Civil Engineering, Logistical Engineering University, Chongqing, China
Key Lab. for the Exploitation of Southwest Resources & the Environmental Disaster Control Engineering,
Ministry of Education, China

Y.R. Zheng
Department of Civil Engineering, Logistical Engineering University, Chongqing, China

X. Jiang
Key Lab. for the Exploitation of Southwest Resources & the Environmental Disaster Control Engineering,
Ministry of Education, China

ABSTRACT: Rotation of principal stress axes is a common aspect of many loading situations in geo-technical engineering. Based on the elastoplastic numerical simulation of plane strain problems of soils involving rotation of principal stress axes, the influences of rotation of principal stress axes on the distribution of stress and deformation of soils are provided, and the situation that rotation of principal stress axes should be considered is presented.

1 INTRODUCTION

Rotation of principal stress axes is a common aspect of many loading situations in geotechnical engineering. The issue of rotation of principal stress axes appears in numerous loading environments, such as earthquakes, vehicular traffic and sea waves. Numerous experimental studies have shown the important effects of rotation of principal stress axes on the response of soils. These studies have revealed that permanent strains steadily accumulate (Wong et al. 1986, Hicher et al. 1987, Tatsuoka et al. 1986) and in undrained condition, pore pressures build up continuously during rotation of principal stress axes even if the principal stress level is maintained constant (Shibuya et al. 1986, Ishihara et al. 1986). Thus, rotation of principal stress axes is a mechanical problem should be considered in the geotechnical field.

The influence of rotation of principal stress axes can not be computed reasonably by the ordinary constitutive models of soils. In the light of complete stress increment formulation of the elastoplastic theory of soils, a new method to consider the effects of rotation of principal stress axes strictly was provided by Liu et al. (1998). Based on this new model, elastoplastic numerical simulation of plane strain problems of soils involving rotation of principal stress axes is performed.

2 THE BASIC PRINCIPLE TO COMPUTE THE INFLUENCE OF ROTATION OF PRINCIPAL STRESS AXES

The influences of the variation of the value of principal stress are always considered in ordinary constitutive model of soils, but the influence of rotation of principal stress axes is ignored. Based on experimental results, a 2-D constitutive model formulated with the general stress increment and general strain increment was provided by Matsuoka at al. (1986). The influences of the rotation of principal stress axes is involved in the variation of the general stress increments The base for this model Is the experimental results of plane strain, and the extension of this model to 3-D stress state have not sufficient scientific foundation. A kinematic hardening model, in which rotation of principal stress axes is considered by implying a new kinematic hardening rule, was presented by Naki et al. (1991). The key factor of kinematic hardening model is the kinematic hardening rule. It is known to all that the kinematic hardening of yield surfaces is dependent on the stress path. It is difficult to give the kinematic hardening rule under complex stress path, and more difficult under 3-D case. Getierrez et al. (1993) put forward a bounding surface model to compute the influences of rotation of principal stress axes. The flow rule of this model was suspicious. Based on the decomposition of stress increment and the complete stress increment formulation of plastic stress–strain relation of geomaterials, a new model which can compute the influences of the rotation of principal stress axes was provided.

Based on matrix theory, the characteristic of the stress increment which triggers the rotation of principal stress axes is obtained, and any stress increment is decomposed into two parts: coaxial component whose principal axes are the same as that of stress, and the rotational component which cause the rotation of principal stress axes (Liu et al. 1998).

The 3-D general stress σ can be expressed as

$$\sigma = \begin{pmatrix} N_1 & N_2 & N_3 \end{pmatrix} \begin{pmatrix} \sigma_1 & 0 & 0 \\ 0 & \sigma_2 & 0 \\ 0 & 0 & \sigma_3 \end{pmatrix} \begin{pmatrix} N_1 \\ N_2 \\ N_3 \end{pmatrix} = T\Lambda_1 T^T \tag{1}$$

where $\sigma_1, \sigma_2, \sigma_3$ = values of principal stress; N_1, N_2, N_3 = the directions of principal stress axes.

The coaxial component of stress increment, $d\sigma_c$, and the rotational components, $d\sigma_{r1}, d\sigma_{r2}$ and $d\sigma_{r3}$ can be written respectively as

$$d\sigma_c = T \begin{pmatrix} M_1 & 0 & 0 \\ 0 & M_2 & 0 \\ 0 & 0 & M_3 \end{pmatrix} T^T \quad M_1 = d\sigma_1 \; M_2 = d\sigma_2 \; M_3 = d\sigma_3 \tag{2}$$

$$d\sigma_{r1} = T \begin{pmatrix} 0 & A_1 & 0 \\ A_1 & 0 & 0 \\ 0 & 0 & 0 \end{pmatrix} T^T \quad A_1 = d\theta_1(\sigma_1 - \sigma_2) \tag{3}$$

$$d\sigma_{r2} = T \begin{pmatrix} 0 & 0 & 0 \\ 0 & 0 & B_1 \\ 0 & B_1 & 0 \end{pmatrix} T^T \quad B_1 = d\theta_2(\sigma_2 - \sigma_3) \tag{4}$$

$$d\sigma_{r3} = T \begin{pmatrix} 0 & 0 & C_1 \\ 0 & 0 & 0 \\ C_1 & 0 & 0 \end{pmatrix} T^T \quad C_1 = d\theta_3(\sigma_1 - \sigma_3) \tag{5}$$

$$d\sigma = d\sigma_c + d\sigma_r = d\sigma_c + d\sigma_{r1} + d\sigma_{r2} + d\sigma_{r3} \tag{6}$$

where $d\theta_1, d\theta_2$, and $d\theta_3$ = the increments of rotational angle triggered by rotational stress increment $d\sigma_{r1}, d\sigma_{r2}$, and $d\sigma_{r3}$ respectively.

The decomposition of stress increment shows that rotational stress increment is a kind of deviatoric stress increment, and could not cause the variation of principal stress value. The essential reason for that the influence of the rotation of principal stress axes could not computed by ordinary constitutive model is that their yield surface is just the function of the principal stress values, certainly the yield of geomaterials generated by rotation of principal stress axes could not be reflected.

According to the decomposition of stress increment, Eq.(6), the total strain increment could be formulated by

$$d\epsilon = d\epsilon^e + d\epsilon^P = d\epsilon^e + d\epsilon^P_c + d\epsilon^P_{r1} + d\epsilon^P_{r2} + d\epsilon^P_{r3} \tag{7}$$

where $d\epsilon^e$, $d\epsilon^P$ = elastic strain increment and plastic strain increment respectively; $d\epsilon^P_c$ = the plastic strain increment caused by coaxial stress increment, $d\sigma_c$; $d\epsilon^P_{r1}$, $d\epsilon^P_{r2}$, and $d\epsilon^P_{r3}$ = the plastic strain increments caused by rotational stress increment $d\sigma_{r1}$, $d\sigma_{r2}$, and $d\sigma_{r3}$ respectively.

The elastic deformation $d\epsilon^e$ could be computed by the generalized Hook's Law.

The plastic deformation $d\epsilon^P_c$ triggered by the coaxial stress increment $d\sigma_c$ could be regarded as the plastic deformation ignoring the influence of rotation of principal stress axes, and could be computed by using the multiple yield surface theory (Liu 1997).

The constitutive model of geomaterials is always formulated in the plane of p (the mean stress)-q (deviatoric stress), and could be written as

$$\begin{cases} d\varepsilon_v^P = Adp + Bdq \\ d\varepsilon_s^P = Cdp + Ddq \end{cases} \tag{8}$$

where $d\varepsilon_v^P$ = volumetric strain increment; $d\varepsilon_s^P$ = deviatoric strain increment; dp = the increment of mean stress; dp = the increment of deviatoric stress; A, B, C, and D = plastic coefficients.

The influence of rotation of principal stress axes could not be computed by this formulation. A formulation of complete stress increment of elastoplastic stress–strain relation was provided by Liu (1997) & Liu et al. (1998), to consider the influence of rotation of principal stress axes.

$$\begin{cases} d\varepsilon_v^{P'} = A_0dp' + B_0dq' \\ d\varepsilon_s^{P'} = C_0dp' + D_0dq' \end{cases} \tag{9}$$

where A_0, B_0, C_0, and D_0 = plastic coefficients; dp' = the generalized mean components of stress increment; dq' = the generalized deviatoric components of stress increment; $d\varepsilon_v^{P'}$ = the generalized volumetric components of plastic strain increment; $d\varepsilon_s^{P'}$ = the generalized deviatoric components of plastic strain increment.

$$\begin{cases} dp' = (d\sigma_{11} + d\sigma_{22} + d\sigma_{33})/3 \\ dq' = \dfrac{1}{\sqrt{2}}\sqrt{(d\sigma_{11} - d\sigma_{22})^2 + (d\sigma_{11} - d\sigma_{33})^2 + (d\sigma_{33} - d\sigma_{22})^2 + 6(d\sigma_{12}^2 + d\sigma_{13}^2 + d\sigma_{23}^2)} \end{cases} \tag{10}$$

$$\begin{cases} d\varepsilon_v^{P'} = d\varepsilon_{11}^P + d\varepsilon_{22}^P + d\varepsilon_{33}^P \\ d\varepsilon_s^{P'} = \dfrac{\sqrt{2}}{3}\sqrt{(d\varepsilon_{11}^P - d\varepsilon_{22}^P)^2 + (d\varepsilon_{11}^P - d\varepsilon_{33}^P)^2 + (d\varepsilon_{33}^P - d\varepsilon_{22}^P)^2 + \dfrac{3}{2}(d\varepsilon_{12}^{P2} + d\varepsilon_{13}^{P2} + d\varepsilon_{23}^{P2})} \end{cases} \tag{11}$$

where the components of stress increment and plastic strain are that in the general stress space.

According to Eq.(9), the plastic deformation triggered by the rotation of principal stress axes can be computed strictly. For example the plastic deformation caused by rotational stress increment $d\sigma_{r1}$ is expressed as

$$\begin{cases} d\varepsilon_{vr1}^{P'} = A_0dp' + B_0dq' = B_0dq' = B_0\sqrt{3}|\sigma_1 - \sigma_2||d\theta_1| \\ d\varepsilon_{sr1}^{P'} = C_0dp' + D_0dq' = D_0dq' = D_0\sqrt{3}|\sigma_1 - \sigma_2||d\theta_1| \end{cases} \tag{12}$$

where the plastic coefficients could be determined by the model provided by Matsuoka et al. (1987). To take into account the Rowe's stress-dilatancy equation (Rowe 1962), the rotational plastic strain increment

$d\epsilon^P_{r1}$ triggered by $d\epsilon^P_{r1}$ can be obtained. According to the same principle, the plastic strain increments $d\epsilon^P_{r2}$, and $d\epsilon^P_{r3}$ caused by rotational stress increment $d\sigma_{r2}$, and $d\sigma_{r3}$ could be computed respectively.

3 THE DESIGN PRINCIPLE OF THE PLANE STRAIN PROGRAM OF NUMERICAL SIMULATION

In this paper, only the numerical simulation of plane strain was performed. The loads are classified as uniform load and local load. Every kind load was subdivided to a certain number parts. The deformation caused by every part load are computed by iterative algorithm with constant stiffness matrix, the computation procedures for every load component are provided in the following.

Given a certain load, the displacement increment of node, du, and the corresponding strain increment of element, $d\varepsilon(d\varepsilon_x, d\varepsilon_{xy}, d\varepsilon_y)$, are computed through elastic analysis. The strain increment is transformed into the former principal stress space before the next load component is applied, the transformed strain increment can be expressed as:

$$d\varepsilon = \begin{bmatrix} d\varepsilon_{11} & d\varepsilon_{12} \\ d\varepsilon_{21} & d\varepsilon_{22} \end{bmatrix}$$

Under $d\varepsilon_{12} = d\varepsilon_{21} = 0$, rotation of principal stress axes will not exist, otherwise rotation of principal stress axes is triggered by the applied load. According to Eq.(3), the rotation angle increment of principal stress axes is given as:

$$d\theta_1 = d\varepsilon_{12}/(\sigma_1 - \sigma_2)$$

The rotational components, and the coaxial components of elastoplastic deformation increment could be obtained. Then the increments of principal stress, $d\sigma_1$, $d\sigma_2$, and $d\sigma_3$, could be computed by the coaxial constitutive relation, and the true stress increment corresponding to the strain increment could be given by

$$\mathbf{d\sigma} = T\begin{pmatrix} d\sigma_1 & d\theta_1(\sigma_1 - \sigma_2) & 0 \\ d\theta_1(\sigma_1 - \sigma_2) & d\sigma_2 & 0 \\ 0 & 0 & d\sigma_3 \end{pmatrix}T^T$$

The load equivalent to the difference of the stress increment of elastic analysis and the true stress increment is applied into the system until the numerical simulation result is convergent.

4 NUMERICAL SIMULATION RESULTS OF THE PLANE STRAIN PROBLEMS INVOLVING ROTATION OF PRINCIPAL STRESS AXES

4.1 *Physical model*

There are many sort of steady problems in geotechnical engineering involving the rotation of principal stress axes. To consider the load triggered the rotation of principal axes rotation, these problems could be classified as two major categories: one is load-type caused by the local or non-uniform load, illustrated in Figure 1, for example the load of building; the other is the slope-type generated by unloading of some parts in the engineering, illustrated in Figure 4, for example the excavated slope or foundation ditch.

4.2 *Numerical simulation of the load-type problem involving the rotation of principal stress axes rotation*

The physical model for the load-type problem involving the rotation of principle stress axes is provided in Figure 1, where Q2 is the uniform load, and the rotation of principal stress axes is caused by the local load Q1.

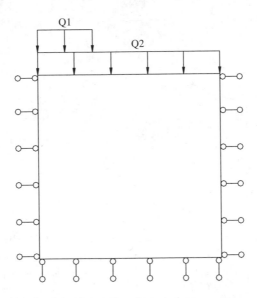

Figure 1. The load-type problem involving the rotation of principal stress axes.

For comparison, three cases were calculated: Only the uniform load Q2 was considered; All the loads include uniform load Q2, and local load Q1 were considered, but the influence of rotation of principal stress axes was ignored; All the loads and the influence of rotation of principal stress axes were considered. Through the comparison of the computed results of these three cases, the influences of rotation of principal stress axes could be observed clearly:

1) Under the action of uniform load, the deformation of soils is uniform, and the influence of rotation of principal stress axes can be ignored;
2) To consider the local load, the non-uniform displacement appears. The influences of the rotation of principal stress axes on the distribution law of deformation and stress are neglectable. The influences of the rotation of principal stress axes on the computed results of horizontal displacement, the second principal stress, and the third principal stress are relatively small. However the influences of rotation of principal stress axes on the computed values of vertical displacement, the first principal stress, and the rotational angle of principal axes are significant. For example, the value of the maximum vertical displacement would increase twenty percent if rotation of principal stress axes is considered. The computed results of the vertical displacement without the influence of rotation of principal stress axe is provided in Figure 2, and the computed results of the vertical displacement with the influence of rotation of principal stress is given in Figure 3.

4.3   Numerical simulation of the slope-type problem involving rotation of principal stress axes rotation

The physical model for the load-type problem involving rotation of principle stress axes is provided in Figure 4. Parameters of this model is the same as that of the load-type problem, just their boundary conditions are different: There are not uniform and non-uniform load on the top boundary of the slope-type model, and without the horizontal displacement restrain on the left boundary.

For comparison, two cases were calculated: under the action of self-weight of soils and the influence of the rotation of principal stress axes was ignored; under the action of self-weight of soils and the influence of rotation of principal stress axes was computed. Through the comparison of the computed results of these two cases, the influence of the rotation of principal stress axes could be obtained:

1) To take into account the computed results of displacement and stress, The influences of rotation of principal stress axes on the distribution law of deformation and stress are negligible.

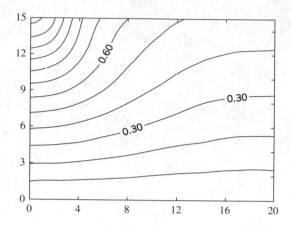

Figure 2.   The computed results of vertical displacement without the influence of the rotation of principal stress axes.

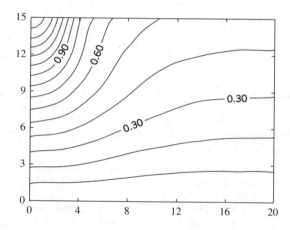

Figure 3.   The computed results of vertical displacement with the influence of rotation of principal stress axes.

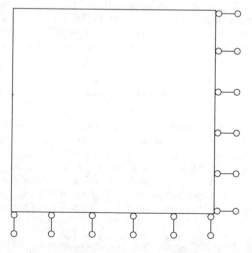

Figure 4.   The slope-type problem involving the rotation of principal stress axes.

478

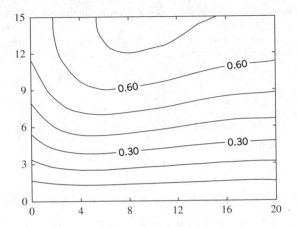

Figure 5. The computed results of horizontal displacement without the influence of rotation of principal stress axes.

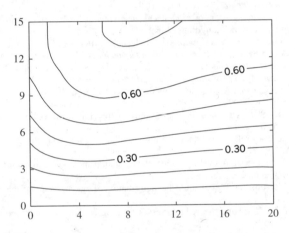

Figure 6. The computed results of horizontal displacement with the influence of rotation of principal stress axes.

2) The influence of rotation of principal stress axes on the computed results of horizontal displacement is relatively significant, and its influences on the computed results of vertical displacement and stress are small. The reason is the difference between the principal stress is small. The computed results of the horizontal displacement without the influence of rotation of principal stress axe is provided in Figure 5, and the computed results of the horizontal displacement with the influence of the rotation of principal stress axe is presented in Figure 6.

5 CONCLUSION

Based on the elastoplastic numerical simulation of plane strain problems of soils involving rotation of principal stress axes, the laws of the influences of principal stress axes rotation on the distribution of stress and deformation of soils are provided, and the situation that rotation of principal stress axes should be consider is presented:

1) Under the action of uniform load and the deformation of soils is uniform, the influence of rotation of principal stress axes could be ignored;
2) While the inhomogeous deformation of soils is obvious and the difference between principal stress is large, the influence of rotation of principal stress axes should be considered.

479

3) The influences of rotation of principal stress axes on the distribution laws of numerical results (stress, displacement) are relatively small.

4) Because the homogeneous deformation seldom exists in the actual engineering, rotation of principal stress axes would be a general problem in geotechnical engineering.

ACKNOWLEDGEMENTS

This work has been funded by Chongqing Committee of Science and Technology.

REFERENCES

Getierrez M., Ishihara K. & Towhata I. 1993. Model for the deformation of sand during rotation of principal stress direction. *Soils and Foundations* 33(3): 105–117.

Hicher P. & Lade P.V. 1987. Rotation of principal directions in K0-consolidation clay. *Journal of Geotechnical Engineering* 113(7): 774–788.

Ishihara K., Towhata I. & Yamazaki A. 1986. Sand liquefaction under rotation of principal stress axes. *2nd Int. Symp. on Numerical Models in Geotechnics*, Ghent: 1015–1018.

Liu Yuanxue. 1997. The general stress–strain relation of soils involving the rotation of principal stress axes. *PhD thesis*. Logistical Engineering University, Chongqing, China.

Liu Yuanxue & Zheng Yingren. 1998. A new method to analyze the influence of the rotation of principal stress axes on the stress–strain relation of soils. *Chinese Journal of Geotechnical Engineering* 20(2): 45–47.

Liu Yuanxue, Zheng Yingren & Chen Zhenghan. 1998. The general stress–strain relation of soils involving the rotation of principal stress axes. *Applied Mathematics and Mechanics* 19(5): 407–413.

Matsuoka H. & Sakakihara K. 1987. A constitutive model for sands and clays evaluating principal stress rotation. *Soils and Foundations* 27(4): 73–88.

Naki T., Fujii J. & Taki H. 1991. Kinematic hardening models for clay in three-dimensional stresses. *Computer Methods and Advances in Geomechanics* 36–45.

Rowe Y.W. 1962. Stress Dilatancy relation for static equilibrium of an assembly of particles. In *Contact. Proc. of Royal Soc.* London, Series A: 500–527.

Shibuya S. & Hight D.W. 1986. Paterns of cyclic principal stress rotation and liquefaction. *2nd Int. Symp. on Numerical Models in Geotechnics*, Ghent: 265–268.

Tatsuoka F., Sonoda S. & Matsuoka H. 1986. Failure and deformation of sand in torsional shear. *Soils and Foundations* 26(4): 79–97.

Wong R.K.S. & Arthor J.R.F. 1986. Sand sheared by stresses with cyclic variations in direction. *Geotechnique* 36(2): 215–226.

Computational Methods in Engineering and Science, Iu et al. (eds)
© *2003 Swets & Zeitlinger, Lisse, ISBN 90 5809 567 3*

Estimation of applied loads on a driven pile at different depths

H. Novais-Ferreira
Civil Engineering Laboratory of Macau, Macao

ABSTRACT: The definitions of driven diagram of driven piles, load-transfer curve, curve of loads on piles, and the importance of these curves are emphasized. Driven diagrams can be obtained for all piles. Some of them can be monitored and the curve of loads on the pile can be obtained directly. However, these monitoring is expensive and time consuming, and usually is not made for small or common works. Apart from this the extrapolation of results of one pile to all the piles of the same work does not take into account the eventual small local geological differences.

In this paper the idea to use the driven diagrams of all driven piles calibrated by some DLT-SM tests is presented. The use of N_i (accumulated value of blows) seems convenient and two relationships between F_{ei} and N_i can be used.

1   INTRODUCTION

During driving a pile, the number of blows per meter of penetration can be recorded and a curve of number of blows per meter (n_i) at the each depth (d_i) is obtained (*differential* or *density driving diagram*) – $n_i = f(d_i)$. The number n_i is correlated with the energy spent for the penetration from ($d_i - 1$) to d_i meters.

The sum of the number of blows per meter of penetration can be made and a curve of number of accumulated blows per meter (n_i) until the depth (d_i) is obtained (*cumulative driving diagram*) – $N_i = \Sigma n_i = f(d_i)$. The number N_i will be correlated with the total energy spent for the penetration from 0 to d_i meters.

When a pile is loaded, the total applied load (F_t) on the top is progressively transferred to the soil, section by section (at depths d_i), by friction (shear strength soil-pile, F_{si}) and finally to the toe (bottom strength, F_b). The load transferred is $F_i = F_{si}$ for $d_i < L$ (L = length of the pile) and, $F_i = F_{si} + F_b$ for $d_i = L$. The *load-transfer pile–soil curve* is $F_i = f(d_i)$. Normally the complement of this load-transfer curve is considered ($F_{ei} = F_t - F_i$) and named *curve of loads*, a curve of the load (F_{ei}) supported by a pile at each depth [d_i].

To obtain these curves it is necessary to make a static load test, applying load (F_t) on the pile monitored with strain gauges in order to measure strains (compression) of the pile by sections. Calibrating gauges to give stresses or loads, the stress or loads values can be recorded.

The dynamic load test with signal matching (DLT-SM) has a well known approach that permits to have an idea of: (a) the friction load by layer, f_{si} (usually around 5 layers); (b) the bottom load F_b. Considering all layers, the sum of f_{si} becomes F_s (total friction load on the pile). A *load-transfer pile–soil curve* is easy to obtain.

The curve could be estimated using parameters of the soil from laboratory, or results of static load tests. If the estimations are made, a check is possible of the DLT-SM results.

2   BEHAVIOR OF PILES

The behavior of the piles can be characterized by correlations between loads:

Friction ratio (f_r)	F_s/F_b
Relative friction strength (R_s)	F_s/F_t
Relative toe strength (R_b)	F_b/F_t

F_s = friction load; F_b = toe (bottom) load;
$F_t = F_s + F_b$ = total load

These definitions are similar to the definition of the CPT friction ratio (Bowles 1997: page 171): f_r, % = 100 * q_s/q_c (ratio friction over cone pressures).

These relations are not constants if F_t increases, depending from the differential movement between soil and pile. In a consolidated soil, the differential movement is higher at the top and lower at the bottom of the pile. Increasing the load F_t applied at the top of the pile, the percentage of load transferred to the soil layer by layer, will be constant during the time if the relation E is constant (linear relation between stress and strains) For a non linear plastic stress/strain relation, the load transfer at the low levels increases relating to the load transfer at the higher levels. Using the results of a DLT-SM test, only the ultimate load of the pile is measured, which does not mean that all the sections are at an ultimate situation but at a situation compatible with the deformation attained (remembering that the curve deformation/load for a soil presents a maximum and a residual strength later).

3 EQUATIONS FOR DIAGRAMS OF BLOWS – DRIVING DIAGRAMS

Four empirical equations are applied for the mathematical definition of the driving diagrams, and the accuracy of adjustment is estimated. Two equations are based in the individuals blow count per meter (n_i), and two equations are based in the accumulated blow count per meter (N_i).

Linear correlation with (n_i) – An assumption can be made *considering that $F_{si} = b * n_i + a$.* With these assumptions it is possible to formulate the following equation of the load (F_{ei}) in the pile at depth (d_i)

$$F_{ei} \text{ (from 0 to i)} = F_t - (n_i / n_f) * (F_t - F_b)$$

n_f – maximum number of blow, obtained at end of driving the pile.

Inverse correlation with (n_i) – The final set (set) of a pile is the mean penetration (in mm/blow) of the pile during the last 10 blows of the hammer. A simple correlation between n_i and Set_i can be established (see Bowles 1997: 992)

$$n_i/1000 = 1/ set_i$$

$$set_i = 1000 / n_i$$

The Hiley formula, the well known rational pile formula (Bowles 1997:973/978) can be written as:

$$P_u = \{e_h * E_h / [s+0.5*(k_1+k_2+k_3)]\} * \{(W+n^2 * W_p) /(W+W_p)\}$$

$$1/P_u = \{ [s+0.5*(k_1+k_2+k_3)]\} /\{e_h * E_h *(W+n^2 * W_p) /(W+W_p)\}$$

$$1/P_u = \{ [s+K)]\} *(1/EW) = s/EW + K/EW$$

The number of blows per meter (n_i) can be related with (s) and s = set_i/10

$$1/s = n_i/100$$

$$s = 100/ n_i$$

The Hiley formula results in:

$$1/P_u = (1/n_i) 100/EW + K/EW = M*(1/n_i) + B$$

Consequently the linear correlation between $(1/P_u)$ and $(1/n_i)$ can be suggested. The value of P_u includes

$$P_u = P_{us} + P_{ub}$$

where P_{us} = ultimate lateral load and P_{ub} = ultimate base load.

Let us consider that

$1/F_{ei} = A - 1/(B + n_i)$

$n_i = 0$ imply $1/F_{ei} = 1/F_t = A - (1/B)$

$n_i = n_f$ imply $1/F_{ei} = 1/F_b = A - 1/(B+n_f))$

$1/F_b - 1/F_t = 1/(B) - 1/(B+n_f)$

The following equation can be solved;

$B*(B+n_f) = (n_f * F_b * F_t)/ (F_b - F_t)$

The constant A is given by

$A = (1/F_t) - (1/B)$

Linear correlation with (N_i) – Let us consider the assumption that the friction load from deep 0 until i is related with the accumulated blows on layers from 0 until d_i

$N_i = \Sigma(n_i)$

It is possible to assume a linear correlation $F_{ei} = b + m + N_i$.
The equation passing for F_t and F_b must be

$F_{ei} = F_t - (N_i / N_t) * (F_t - F_b)$

The value of F_{ei} varies linearly with the difference $(F_t - F_b)$.
Exponential correlation with (N_i) – Following the common exponential correlations between forces and deformations, another correlation can be assumed:

$F_{ei} = a * exp(b*N_i)$

$n_i = 0$ imply $F_{ei} = F_t = a$

$n_i = n_f$ imply $F_{ei} = F_b = a * exp(b*N_t)$

$F_b / F_t = exp(b*N_t)$

$b = [LN(F_b / F_t)] / N_t$

$F_{ei} = F_t * exp\{[LN(F_b / F_t)] * (N_i / N_t)\}$

$F_{ei} / F_t = exp\{[LN(F_b / F_t)] * (N_i / N_t)\}$

$LN (F_{ei} / F_t) = [LN(F_b / F_t)] * (N_i / N_t)$

$F_{ei} / F_t = (F_b / F_t) ^ (N_i / N_t)$

The value of F_{ei} varies linearly with $(F_b/F_t) ^ (N_i/N_t)$, and almost linearly with (F_b/F_t). The friction ratio $f_r = P_{us}/P_{ub}$ and the ratio (F_b/F_t) are not constant. However it is possible *to accept that no big error is committed assuming a constant ratio for each pile if the geological conditions are similar.*
For piles with no DLT-SM, the estimation of loads F_t, F_s and F_b can be made using the tested piles with a similar driving diagram,

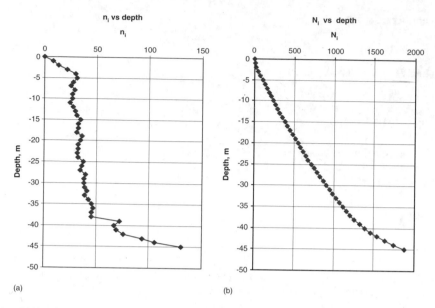

Figure 1.   (a) Density driving diagram (Pile No. a17), (b) Cumulative driving diagram (Pile No. a17).

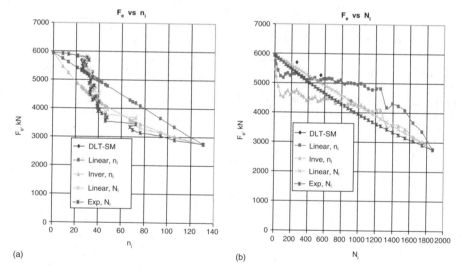

Figure 2.   (a) F_e vs n_i, (b) F_e vs N_i.

4   APPLICATION

The method was applied to several piles in Macau. One example of application includes:

– Characteristics of the pile, final set and driving diagrams, density (n_i) and accumulated (N_i).
– Table of driven diagram values and calculated values for the following values: (a) F_{ei} (DLT-SM); (b) F_{ei} (linear, n_i); F_{ei} (inverse, n_i); (c) F_{ei} (linear,N_i); F_{ei} (Exponential, N_i).
– Curves of load on pile (F_e) calculated (a) from individual values n_i, and (b) from accumulated values N_i. The values from DLT-SM are included for comparison.
  Values of F_{ei} result directly from the measurements of DLT-SM.
  Values of F_e include all the F_{ei} values and other estimated values for each depth.

484

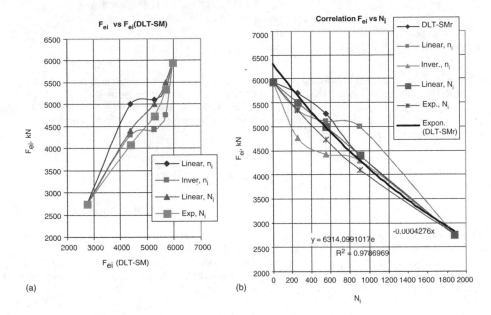

Figure 3.   (a) F_{ei} vs F_{ei}(DLT-SM), (b) F_{ei} vs N_i.

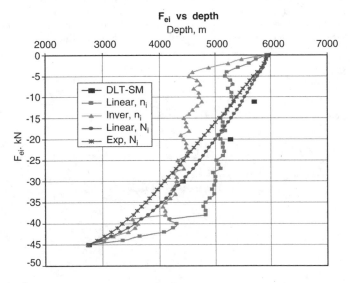

Figure 4. F_{ei} vs depth.

- Table of values of F_{ei}: Linear correlation of F_{ei} calculated versus F_{ei} (DLT- SM).
  Correlation curve between F_{ei} and F_{ei} (DLT-SM).
  Correlation curve between F_{ei} and N_i incuding F_{ei} (DLT-SM) and calculation of exponential relation F_{ei}(DLT-SM) versus N_i.
- Diagram of loads on the pile versus depth.

Only the drawings for one pile (pile a 43) are presented as example (Figure 1a – n_i vs depth; Figure 1b – N_i vs depth; Figure 2a – F_e vs n_i; Figure 2b – F_e vs N_i; Figure 3a – F_{ei} vs F_{ei} (DLT-SM); Figure 3b – F_{ei} vs N_i; Figure 4 – F_{ei} vs depth).

The values of the constants **a** and **b** are also calculated by the theoretical equations:

$$a = F_{ei} = F_t \text{ and } b = [LN \, (F_b \, / \, F_t)] \, / \, N_t$$

A synthesis of the constants of correlation (OLS – Ordinary Least Square method, using the obvious EXEL LINEST program) between values of F_{ei} calculated by the four methods used – (a) F_{ei} (linear, n_i); (b) F_{ei} (inverse, n_i); (c) F_{ei} (linear, N_i); (d) F_{ei} (Exponential, N_i) – and the measured values by DLT-SM is presented. The coefficient of determination R^2 is also presented. The highest values of R^2 are for the correlations against N_i.

5 CONCLUSIONS

1. The density driven diagram (n_i vs depth) defines very well the several layers of soil at the site of the pile.
2. The accumulated driven diagram (N_i vs depth) also defines the several layers of soil at the site of the pile but is not as clear.
3. Use of density diagrams (n_i) – The pile load versus depth estimated by linear procedure (F_{ei} vs n_i) adjusts roughly with the values obtained from the DLT-SM test. The inverse adjustment is more time consuming, and does not seems to present significant improvement.
4. Use of cumulative diagrams (N_i) – The values obtained from N_i adjust better. The coefficient of determination R^2 of the exponential adjustment is higher than 0.96.
5. The theoretical curves of the exponential relation (F_{ei} /F_t normalized value of F_{ei}) versus the normalized number of accumulated blows (N_i/N_t) are practically linear for (F_b/F_t) > 0.7 (toe piles).
6. The behavior of piles is quite different from one site to another, as expected.
7. It seems acceptable to estimate the curve of loads on a pile not tested by the method presented. The accuracy of the estimation depends on the accuracy of the adopted value of R_b.

REFERENCE

Bowles 1997. *Foundation Analysis and Designe*. New York: Mc Graw-Hill.

Computational Methods in Engineering and Science, Iu et al. (eds)
© *2003 Swets & Zeitlinger, Lisse, ISBN 90 5809 567 3*

Influence of some parameters on the behavior of diaphragm wall

H.S. Leong
Geotechnical Department, LECM, Macao

ABSTRACT: Due to the growth in population and commercial activities, there is an increasing trend to construct deeper basements in Macao. In deep excavation, diaphragm wall is extensively used worldwide to solve numerous foundations, sheeting, and dewatering problems. This investigation presents a study on the influence of various parameters on diaphragm wall response. A number of different cases were studied in which the following parameters were varied: soil friction angle, excavation depth versus total wall penetration depth ratio (β), ground water level (G.W.L.), surcharge and horizontal strut stiffness (H.S.S).

1 INTRODUCTION

In this investigation, we used simplified method to analyze the lateral deformation of diaphragm wall; the soil–structure interaction is analyzed by finite element method based on beam on elastic foundation. When the computed bending moment is greater than crack moment of the diaphragm wall, the wall stiffness will change to consider the crack effect. Using this method, different soil parameters and load conditions were input to study the influence of these factors to the lateral deformation of diaphragm wall.

2 ANALYTICAL MODELING

This investigation modeled the wall and the struts by linear elastic beam and spring respectively (Hetenyif 1946). Active soil pressure was used on the active side. While on the passive side, the soil reactions were represented by the at-rest soil pressure plus elasto-plastic soil springs. The stiffness of the soil spring was obtained from the coefficient of sub-grade reaction (Bowles 1988).

The method could be outlined in the following steps:

1. Draw the wall configuration and soil profile to a reasonable scale and showing the soil parameters γ, c, and $\varnothing$; compute and show the pressure coefficients K_a, K_0, and K_p. Tentatively locate nodes and brace points.
2. Input physical parameter of the structural member for the wall E_c, length, A and I. Wall friction from the horizontal component is neglected.
3. Estimate the modulus of subgrade reaction for the stages from the dredge line down and for the retained earth (behind the wall) for the several stages.
4. The stage analysis proceeds essentially as follows:
   – Stage 1. Treat wall as a cantilever pile for some depth of excavation. Use the active earth pressure – at-rest pressure) to the wall. Estimate k_s for the soil below dredge line for passive wall resistance. $k_s = 20 \, (cN_c + q'N_q + 0.5\gamma \, B \, N_\gamma)$.
   – Stage 2. Apply the strut. Estimate k_s for the steel strut. $k_s = E_s \, A/L$ (spacing).
   – Stage 3. Make the next excavation. From computer output we would have both node forces and wall displacement. Update the coordinates for next stage calculation; all springs would be replaced by (nodal forces + zero displacement springs).
   – Stage 4. Repeat stage 2 and stage 3 and following. This is done in turn for each subsequent stage to include stage N.

In this calculation model, when the soil spring reaction on the passive side reached the limit (passive earth pressure – at-rest earth pressure), the soil spring was removed and then the soil reaction acted as a constant value (passive earth pressure – at-rest earth pressure).

When the computed bending moment of wall was greater then the crack moment of the diaphragm wall, the wall stiffness would changed according to ACI Code, and the recalculation proceeded iteratively until convergence of wall stiffness was reached. It is known that the flexural rigidity $E_c I$ varies with the magnitude of bending moment in the general manner. Of course, the moment or inertia I_{cr} of the transformed cracked section increases roughly in proportion with an increase in the percentage of reinforcement. Sections with higher percentages of reinforcement exhibit less change in rigidity under increasing load than those with low percentages of reinforcement. For loads below the cracking load, defections may be based on the gross concrete section, with generally a small difference arising from whether or not the transformed area of reinforcement is also included. However, as the load increases above the cracking load, the moment of inertia approaches that of the cracked transformed section, although it may be greater between cracks. Generally, the use of gross section underestimates the deflection, and the use of transformed cracked section overestimates the deflection. However, the degree of accuracy is affected by the magnitude of the service load compared to the load which causes cracking. In order to provide a smooth transition between the moment of inertia I_{cr} of the transformed cracked section and the moment of inertia, I_g of the gross uncracked concrete section, since 1971 the ACI Code has used the expression developed by Branson,

$$I_e = \left(\frac{M_{cr}}{M_{max}}\right)^3 I_g + \left[1 - \left(\frac{M_{cr}}{M_{max}}\right)^3\right] I_{cr} \leq I_g \tag{1}$$

where I_e = Effective Moment of Inertia; $M_{cr} = f_r I_g / y_t$ = cracking moment; M_{max} = maximum service load moment acting at the condition under which deflection is computed; I_g = moment of inertia of gross uncracked concrete section about the centroidal axis, neglecting reinforcement; I_{cr} = moment of inertia of transformed cracked section; F_r = modulus of rupture of concrete, taken by ACI Code as for normal-weight concrete; y_t = distance from neutral axis to extreme fiber of concrete in tension.

As an approximation, a single value of effective moment of inertia is suggested for practical use when the variable I results from the variation in the extent of tension concrete cracking. In this investigation the following method have been used. Weighted average method, in this method the adjusted I is obtained by weighting the moments of inertia in accordance with the magnitudes of the end moments. The following weighted average expression has been recommended by ACI Committee 435 (1995) as giving a somewhat improved result over the use of the midspan value alone. For spans with both ends continuous, average $I_e = 0.70 I_m + 0.15[I_{e1} + I_{e2}]$; For spans with one end continuous, average $I_e = 0.85 I_m + 0.15 I_{e1}$, where I_{e1} and I_{e2} are the effective moments of inertia at the two ends of the span.

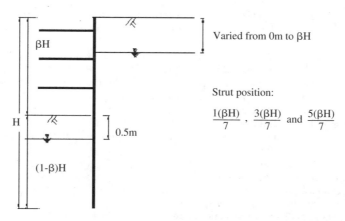

Figure 1. Three layers strutted wall.

Using the beam on nonlinear foundation method, different soil parameters and load conditions were input to study the influence of these factors to the lateral deformation and bending moment of diaphragm wall; a three layers strutted wall in granular soil shown in Figure 1 was analyzed.

Figure 2 shows the plot of depth versus k_s at the final stage of excavation for a range of friction angle values.

Estimate k_s for the soil below dredge line for passive wall resistance. Generally, we considered dewatering under the excavation surface about 0.5 m during construction.

3 RESULTS

Figure 3 and Figure 4 shows the variations of the maximum displacement and maximum moment of the strutted wall with different parameters, for a range of friction angle values.

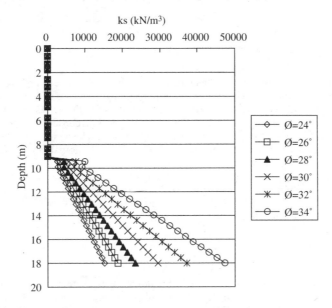

Figure 2. Depth versus k_s at the final stage of excavation for different friction angle.

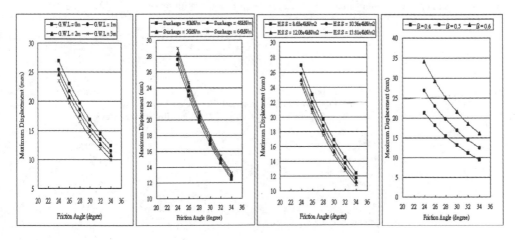

Figure 3. Variation of maximum displacement of strutted wall with different parameters.

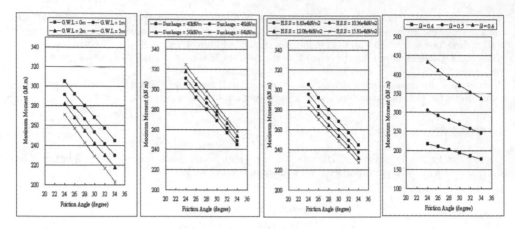

Figure 4. Variation of maximum moment of strutted wall with different parameters.

It was found that wall response (displacement and bending moment) decreases with increasing soil friction angle due to an increased resistance of soil; wall response decreases with lowering ground water, due to a decreased ground water pressure; wall response increases with increasing surcharge, due to an increased lateral pressure and larger lateral soil movements; wall response decreases with stiffer strut support because such support conditions result in smaller soil movements; wall response increases significantly with increasing "excavation depth/total wall penetration depth ratio (β)", due to a decreased lateral soil resistance pressure.

It can be seen that the displacement and bending moment decreases almost linearly with increment of friction angle, the slope is similar by changing the parameters of ground water level (G.W.L.), surcharge and horizontal strut stiffness (H.S.S.); the slope is decreased more rapidly with decreasing excavation depth/total wall penetration depth ratio (β).

4 CONCLUSIONS

The maximum displacement usually occurs near the bottom of the excavation. Wall displacement increases significantly with increasing "excavation depth/total wall penetration depth ratio (β)", due to a decreased lateral soil resistance pressure. This is the parameter that, in our analysis, appeared to be more influential on the completed maximum wall displacements and maximum moment, which, at the end of the excavation stage, displacement increase by 27–30% and moment increase by 37–42% when the value of β increases by 0.1.

REFERENCES

American Concrete Institute, 1995. ACI Committee 435, *Control of Deflection in Concrete Structures.*
Bowles, J.E., 1988. *Foundation Analysis and Design*, McGraw-Hill Book Company, pp. 662–668.
Hetenyi, M., 1946. *Beam on Elastic Foundations,* The University of Michigan Press, Ann Arbor, MI, p. 255.

Geographical information systems

Computational Methods in Engineering and Science, Iu et al. (eds)
© 2003 Swets & Zeitlinger, Lisse, ISBN 90 5809 567 3

Quick generation of geographic entities in an urban disaster prevention system

G. Pan & A. Ren
Department of Civil Engineering, Tsinghua University, Beijing, China

ABSTRACT: In order to generate road network data suitable for GIS application from digital maps rapidly, we need an effective method of storage and management in attribute data for geographic entity of road networks. Therefore, based on the analysis of the existing methods of storage and management of attribute data on AutoCAD, a new method of quick generation of geographic entities in an urban disaster prevention system was developed by the authors. The implement details employed by the authors such as custom entities, visualization algorithms and generation of topologic relations are presented in this paper.

1   INTRODUCTION

With the development of construction in urban information infrastructure, GIS (Geographic Information System) has been widely applied in various respects of our world, such as disaster prevention and mitigation, city planning, transportation, environmental protection, etc. Massive high quality fundamental geographic data are required to take full advantage of GIS. But at present most fundamental geographic data are mapping-oriented urban digital maps, which are ·dwg file format and acquired using AutoCAD as data-acquirement platform. Neither data quality nor data format of these digital maps where only vector graphic data are contained without corresponding attribute data could meet the demand of GIS. In order to set up GIS application system, we need to extract vector elements from these vector graphics to which attribute information is to be appended, to convert them to GIS geographic entities, and to build topological relationships among them. The task listed above plays a principal role in the construction of GIS application system. Therefore we need an effective method for the storage and management of attribute data of geographic entities in an urban road network to extract road network data available for GIS from digital urban maps rapidly.

2   GEOGRAPHIC ENTITIES AND TOPOLOGICAL RELATIONSHIPS

A geographic entity is the most fundamental element of GIS application systems, which could not be divided anymore. Compared with normal graphic data, a geographic entity has not only spatial features such as location, shape, magnitude and so on, but also non-spatial features. Furthermore, there are topological relationships among different geographic entities. Figure 1 describes geographic entities and the corresponding topological relationships among them.

We use the topological data structure shown in the following tables to describe the topological relationships among the geographic entities in Figure 1. The topological data structure can be used to describe the topological relationships among the geographic entities in Figure 1. The tables are presented as follows.

Because the road network involved in an urban fire-fighting commanding and decision system just contains two types of data: nodes and arcs, we only need to study the network topological relationships between nodes and arcs shown in Table 1 and Table 2. Furthermore, the construction of network topological relationships is the prerequisite for subsequent network analysis such as solution of the shortest path, resource allocation and positioning, etc.

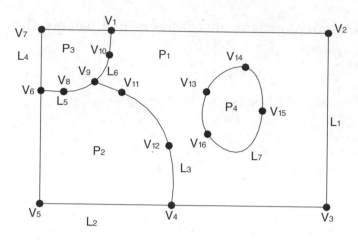

Figure 1.   Geographic entities and topological relationships.

Table 1.   The node table for topological coding.

Node ID	Coordinate	Arc ID
V_1	x_1, y_1	L_1, L_4, L_6
V_4	x_4, y_4	L_1, L_2, L_3
V_6	x_6, y_6	L_2, L_4, L_5
V_9	x_9, y_9	L_3, L_5, L_6
V_{13}	x_{13}, y_{13}	L_7

Table 2.   The arc table for topological coding.

Arc ID	Vertex coordinate	Start node	End node	Left polygon	Right polygon
L1	x1, y1;x2, y2;x3, y3;x4, y4	V1	V4	P0	P1
L2	x4, y4;x5, y5;x6, y6	V4	V6	P0	P2
L3	X4, y4;x12, y12;x11, y11;x9, y9	V4	V9	P2	P1
L4	x6, y6;x7, y7;x1, y1	V6	V1	P0	P3
L5	x6, y6;x8, y8;x9, y9	V6	V9	P3	P2
L6	x1, y1;x10, y10;x9, y9	V9	V1	P3	P1
L7	x13, y13;x14, y14;x15, y15;x16, y16	V13	V13	P1	P4

Table 3.   The polygon table for topological coding.

Polygon ID	Arc ID	Perimeter	Area	Center coordinate
P_1	L_1, L_3, L_6, L_7			
P_2	L_2, L_3, L_5			
P_3	L_4, L_5, L_6			
P_4	L_7			

A road is the essential component in a road network model. We extract the centerline of the road from digital maps to represent a road in the real world. A road is a sequence of connected segments, which consists of two distinguished points (the start and end points). All the road segments of which a road is comprised have the same name. From the perspective of geometry, a road corresponds to a multi-line entity in AutoCAD and the road segments of which the road is comprised correspond to the Line entities

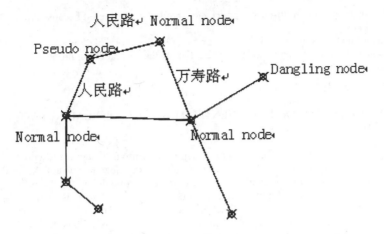

Figure 2.   Geographic entities and topological relationships.

of which the multi-line is comprised. In a road network model, nodes are used to describe intersections of roads in the real world. There are three types of nodes in the road network model: a pseudo node is such type of node at which only two unique road segments with certain similar attributes such as name intersect; a dangling node is another type of node which is the endpoint of just one road; the last type of node is a normal node, illustrated as Figure 2.

3   COMBINATION OF ATTRIBUTE DATA AND GEOMETRY DATA

The sources of data are the digital maps of ·dwg file format. The maps are comprised of various themes of data, such as buildings, roads, bridges, etc, which are represented as Line entities or Poly-line entities in AutoCAD to describe the geometry data of geographic entities. Generally, the attribute data of such geographic entities are represented as nearby Text entities, and there are no relationships between the attribute data and geometry data. Therefore an approach for combination of attribute data and geometry data is required. AutoCAD offers powerful functionality in graphic editing with poor functionality in management of attribute data, and so we need to conduct some development on the AutoCAD platform to accomplish the functionality in management of attribute data. In AutoCAD, there are two types of storage of attribute data: interior storage and exterior storage, which are listed below.

Interior storage means that attribute data are saved into AutoCAD ·dwg files. It attaches attribute data to a graphic entity directly in order to combine them seamlessly, although it perhaps results in data redundancy when attribute data are complicated.

Interior storage of attribute data can be implemented through several methods in AutoCAD. These methods include: block-attribute, extended-entity data, object-data and custom-entity. The block-attribute method can only associate string attributes with a block entity, which is the principal disadvantage. This disadvantage can be solved by the extended-entity data method, while the structure of attribute data is too complicated to deal with. The object-data method provides the significant functionality of query without the above limitations, but it simply can be used in AutoCAD Map. The method of custom entity offers many advantages compared with the other methods mentioned above, by which a seamless combination of the attribute data and spatial data can be obtained. AutoCAD offers the C++ programming interface called ObjectARX to customize the user-defined entity. In ObjectARX, it is allowed to define the data and operations of a class freely by object-oriented way. Thus, the graphic entity in AutoCAD and the geographic entity are uniform, that is, the geographic entity oriented problem, which arises when we are preparing the data of road network available for GIS can be solved.

The methodology of exterior storage is to save the attribute data of the geographic entity into the exterior database, by which the application can inherit all the advantages of database technology. Therefore the application can not only organize the structure of the complicated attribute data freely but also share data with other applications conveniently. It is important to note that a lot of issues, such as data integrity, data security, multi-user support, etc, should be considered because AutoCAD is a file-based application system.

In a word, no method is intrinsically superior among all the above methods; the sophistication level of attribute data will guide which method is the best. Finally the context of our problem shows the custom-entity method is the best choice.

4 IMPLEMENTATION OF OBJECT-ORIENTED MODELING OF ROAD NETWORK IN AUTOCAD

Although some GIS business software such as Arc/Info, MapInfo also support ·dwg file format, their support is too limited compared with AutoCAD. Therefore we decide to take AutoCAD as development platform to make full use of its powerful graphic editing and data modeling functions.

Compared with the other development tools provided by AutoCAD such as Visual LISP, VBA, etc, ObjectARX has better efficiency and more functionality yet with more complexity. It is a C++ program-ming environment that supports the object-oriented method. ObjectARX 5.0 is the latest release with a lot of enhancements including the complete support for COM and ActiveX, which means that the GIS ActiveX controls – MapObjects, MapX, etc – can be used to embed GIS functionality into applications to make the best of the advantages of component GIS.

The road network of a city is composed of a large number of roads. It is apparent that a road entity is the core component of the road network, which contains three types of data: geometry, attribute, and topology. When we design the framework of the road network classes, the independency and extensibility of the attribute data should be guaranteed in order to extend the road network application to other homologous applications, such as water utility network system, electrical utility network system, etc. For this reason the common attribute data of a road entity is separated as a base class of the attribute data, which encapsulates the essential attributes of a road entity, such as name, width, etc, and related operations. If a road entity has additional attributes, we only need to derive a new attribute data class from the base class and add the corresponding attribute data. The road entities can be classified by the structure of the attribute data, while they share the same structure of the geometry data and topology data. Hence we first build a base class of road entities that encapsulates all the data and related opera-tions which are irrelevant to the attribute data. If a sort of road entity which contains certain attribute data is required, we simply add a pointer to the object of the attribute data class and derive a new class of road entity from the base class, which inherit all the data and operations of the base class automatically without any redefinition. This is the benefit that object-orientation design brings.

The road network is comprised of three fundamental elements: *road entities*, *road segments* and *nodes*. It is lucky that there are three corresponding classes of geometry entities in the class library provided by ObjectARX, which can be used to describe the geometry data of above elements. Therefore, these types of geometry entities, *AcDbEntity*, *AcDbPolyline*, and *AcDbPoint*, are used as the base classes of above elements, which include all the functions, such as display, edit, transform, etc provided by these AutoCAD entities. In conclusion, the object-oriented modeling of road networks facilitates our develop-ment greatly.

According to above analysis and practical requirements of our project, the diagram which shows the inheritance relationships among the three fundamental elements is illustrated as Figure 3.

According to the idea behind components which is to divide software functionality into independent modules, the code that implements custom geographic elements is encapsulated into database (DB) component file with the extension ·dbx and the code that issues prompts, displays dialogs, modifies menus in AutoCAD, and so forth is encapsulated into user-interface (UI) component file with the exten-sion ·arx. The benefit of separating an application into UI and DB portions is the high level of software re-use.

5 IMPLEMENTATION OF VISUAL DISPLAY AND COMFORTABLE MODIFICATION OF ATTRIBUTE DATA USING OPM TECHNOLOGY

The theme of this section is to show how to provide a user-friendly interface convenient for management of the attribute data of geographic entities.

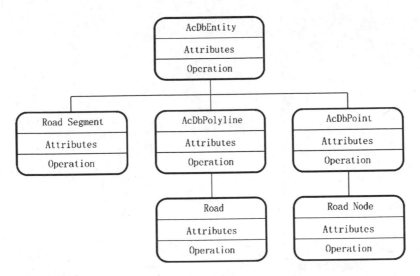

Figure 3.   Inheritance relationships among geographic entities.

The Object Property Manager (OPM) is a tool that allows users to view and modify properties of entities and objects easily. The OPM supports two kinds of properties: static properties, defined for an object statically at compile time; dynamic properties, can be added, configured and supported at runtime. For static properties, ObjectARX applications can provide the COM "wrapper" classes for their custom objects. The OPM will use these classes to determine which static properties are available for that object and how to "flavorize" those properties for display (such as, does the property require a custom combo box in the OPM, or need to bring up a dialog for property editing?). For dynamic properties, ObjectARX, VB, or VBA applications can create "property plug-ins" using the OPM Dynamic Properties API to configure and manage properties for any native or custom class at runtime.

The OPM is essentially a control that parses type information from COM objects to determine their properties. When objects in the drawing are selected, the selection set is converted into an array of IUnknown pointers representing the COM objects that wrap all native entities in AutoCAD. These COM object wrappers are the fundamental support for the ActiveX Automation interface and are the underlying objects that the OPM communicates with.

To implement COM object wrappers defining static properties for custom objects the easiest method is to use the ATL. The whole process consists of three major steps: create the base object COM wrapper for custom objects, add properties and methods, implement additional functional interfaces if necessary. See the *AutoCAD User's Guide* for more information on OPM. A typical interface of OPM is shown as Figure 4.

When a road entity is selected in the drawing area, the object name combo box will show the class name of the road entity illustrated as the above Figure. There are three additional categories of information, the attribute data of a road entity, the topology data of a road entity, and the display flag, except the general category in the OPM. The attribute data of a road entity contains five properties: Name, Material, Width, Length and ID, among which Length and ID are read-only while the other properties can be modified instantly in an edit box or from a combo box based on the property type; the topology data of a road entity shows the IDs of the start point and end point, which are also read-only; the display flag determines whether to show related information in the drawing area of AutoCAD. In addition, the detailed description of the selected property will be shown on the lowest part of OPM.

6   THE ALGORITHM OF GENERATING NETWORK TOPOLOGICAL RELATIONSHIPS AUTOMATICALLY

The essential of network topological relationships is the relationships between nodes and arcs shown in Table 1 and Table 2. Figure 5 is illustrated to present the algorithm of generating the network topological

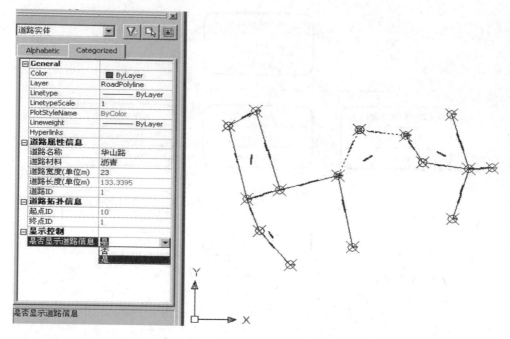

Figure 4. A typical interface of OPM.

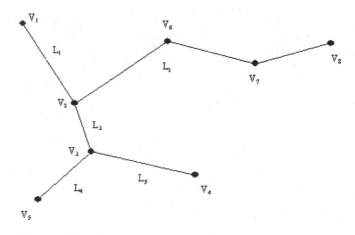

Figure 5. Automation of generating the network topological relationships.

relationships. First we calculate the number of the road segments connected to each node in Figure 5 to identify which type it belongs to, and save the result to an object of class CRoadNode. For example, there are a total of two normal nodes and four dangling nodes with the exception of two pseudo nodes, V6 and V7, illustrated as Figure 5. Additionally, the properties of a dangling node are the same as a normal node in the situation of such algorithm.

The definition of the CRoadNode class described as C++ language is shown as below:

```
class CRoadNode
{
public:
CRoadNode(){bUsed = FALSE; id = −1; flag = 1; numLines = 0;}
```

```
public:
int flag;// 0 represents the node to which the number of the road segments connected is
         // not equal to 2, ie, a normal node or a dangling node
         // 1 represents a pseudo node
         // 2 represents a normal node whose number of the connected road segments equals to 2
AcGePoint3d position;        // Location
AcDbObjectIdArray objIdArray; // ObjectId Array of the connected road segments
int numLines;  // Number of the connected road segments
BOOL bUsed; // Flag variable that show whether the pseudo node has been added to the vertices
         // array of a road entity, in order to avoid redudant search
long id;       // identifier
};
```

Then a good search approach must be required in order to incorporate every road segment into the corresponding road. The basic idea is to search the connected road segments from each node, of which the implementation is so complicated because of the variety of the node types. To solve above problem, much improvement has been made upon the approach to increase the efficiency. There are two kinds of road according to its composing: one consists of certain pseudo nodes, such as L_2 shown in Figure 5. To such type of road, we search from a pseudo node along the connected road segments until a normal node or dangling node is met, while appending a flag to each pseudo node to show whether it has been traversed; one does not contain any pseudo node, such as the other roads shown in Figure 5. To such type of road, we only need to search according to the road segments straightforward.

7   CONCLUSION

Compared with the traditional manual method of building an urban road network available for GIS, the new method of quick generation of urban road networks based on vector graphics in AutoCAD presented in this paper has brought some benefits such as visualization of attribute data for road geographic entities, easy modification and management of attribute data, and automation of building topological relationships for network, which can greatly increase the efficiency in the construction of GIS fundamental geographic information library with guarantee of high data quality. The application of this method in commanding and decision-making system of fire fighting for Shantou has proved to be successful. The approach of building network topological relationships proposed in this paper can also be applied in other homogeneous systems ranging from urban transportation management systems to electricity network management systems which share a similar network model.

REFERENCES

Autodesk, Ins. 1999. AutoCAD2000 ObjectARX developer's guide: 611–654. Autodesk, Ins.
Chao Zhang. 2000. Tutorials on application of GIS: 54–81. Beijing: Higher Education Press.
Jianghong Sun, LiWei Ding, et al. 1999. Application of AutoCAD ObjectARX development tools: 231–334. Beijing: Tsinghua University Press.
Scott McFarlane. 2001. Combination of AutoCAD and database: 2–69. Beijing: Mechanical Engineering Publishing House.
Yang Wen & Aizhu Ren. 2002. The 4D model of massive reinforced-concrete under construction. Computer Engineering and Design 23(1): 29–33.

Computational Methods in Engineering and Science, Iu et al. (eds)
© 2003 Swets & Zeitlinger, Lisse, ISBN 90 5809 567 3

Development of a database linked to a GIS for coupling with groundwater modeling tools

I. Ruthy & Ph. Orban
Hydrogeology, Department of GEOMAC, University of Liège, Belgium

R.C. Gogu
ETH Hoenggerberg, Institute of Cartography, Zurich, Switzerland

A. Dassargues
Hydrogeology, Department of GEOMAC, University of Liège also at Hydrogeology & Engineering Geology,
Department of Geology-Geography, Katholieke Universiteit Leuven, Belgium

ABSTRACT: Groundwater analysis strongly depends on the availability of large volumes of high-quality data. Putting all data in a coherent and logical structure supported by a computing environment helps ensure validity and availability and provides a powerful tool for hydrogeological studies. A hydrogeological Geographical Information System (GIS) database that offers facilities for hydrogeological modeling has been designed in Belgium for the Walloon Region. Interest is growing in the potential for integrating GIS technology and groundwater simulation models. A "loose-coupling" tool was created between the spatial database scheme and the groundwater numerical model interface GMS© (Groundwater Modeling System). Following time and spatial queries, the hydrogeological data stored in the database can be easily used within different groundwater numerical models.

1 INTRODUCTION

In recent years, the use of Geographical Information System (GIS) has grown rapidly in groundwater management and research. GIS is now widely used to create digital geographic databases, to manipulate and prepare data as input for various model parameters, and to display model output. These functions allow primarily overlay or index operations, but new GIS functions that are available or under development, could further support the requirements of process-based approaches.

A GIS-managed hydrogeological database has been developed by the HG team of the University of Liège (Belgium) in order to support data used in vulnerability-assessment techniques and numerical modeling for groundwater flow and contaminant transport studies. The database has been developed in agreement with the hydrogeological specificity of the environment of the Walloon Region, Belgium. The design of the coupling between the database and process-based numerical models was also performed. Subsequent projects have dealt also with the preparation of groundwater-quality maps and hydrogeological maps.

2 DATA, DATABASE, GIS AND GROUNDWATER MODELS

2.1 Data and databases

Data and information required by hydrogeological studies are complex. Information concerning geology, hydrology, geomorphology, soil, climate, land use, topography, and man-made (anthropogenic) features need to be analyzed and combined. Data are collected from existing databases and maps as well as through new field measurements.

Use of point automatic collecting systems for some of the physical and chemical parameters is more and more used. Remote-sensing techniques to assess parameters related to soil, unsaturated zone, geomorphology, and climate are increasingly being used. Some of the techniques for measurement of hydrogeological parameters (sampling, monitoring of hydraulic heads and flow rates, geophysical techniques) show a steady improvement. All these data need to be managed, and this can be done in databases and particularly in GIS databases.

Storing data implies data analysis, conceptual design of data models, and data representation. In hydrogeology, because of a limited number of sample locations, point-attribute data also need to be processed by applying adequate kinds of interpolation or modeling algorithms. The derived data also need to be managed.

2.2   Basic concepts of GIS as applied in hydrogeology

A GIS is defined as a system for input, storage, manipulation, and output of geographically referenced data (Goodchild 1996). GIS provides a means of representing the real world through integrated layers of constituent spatial information (Corwin 1996). Geographic information can be represented in GIS as objects or fields. The object approach represents the real world through simple objects such as point, lines, and areas. The objects, representing entities, are characterized by geometry, topology, and non-spatial attribute values (Heuvelink 1998).

In hydrogeology, some examples of spatial objects are wells, piezometers, boreholes, galleries, and zones of protection. Attribute values of objects could be the number of a well, the ownership, the diameter of a gallery or drain. The field approach represents the real world as fields of attribute data without defining objects, some examples are strata elevation, hydraulic head, and vulnerability zones. This approach provides attribute values in any location. In GIS, this distinction between objects and fields is often associated with vector data models and raster data models. The vector model represents spatial phenomena through differences in the distribution of properties of points, lines, and areas. In this system, each layer is an adapted combination of one or more classes of geometrical features. A raster model consists of a rectangular array of cells with values being assigned to each cell. In the raster model, each cell is usually restricted to a single value. Thus, representing the spatial distribution of a number of parameters or variables requires multiple layers.

2.3   Groundwater models

The process-based models used in hydrogeology include the simulation of steady or transient state groundwater flow, advection, hydrodynamic dispersion, adsorption, desorption, retardation, and multi-component chemical reaction. Very often, exchanges with the unsaturated zone and with rivers are also addressed. In these models, equations based on physical processes are solved.

Modeling groundwater flow and contaminant transport in aquifers represents a spatial and temporal problem that requires integration of process-based models. Each model parameter or variable can be represented on a three- or four-dimensional (x, y, z, and time) information layer. Due to the heterogeneity of the geology, managing these data can be done most effectively through GIS.

Data used in groundwater modeling consists of four categories: (1) the aquifer-system stress factors, (2) the aquifer-system and strata geometry, (3) the hydrogeological parameters of the simulated process, and (4) the main measured variables (historical data needed for calibration and validation). Stress factors for groundwater flow include: effective recharge, pumping volumes, water-surface flow exchanges, etc. In contaminant transport modeling, the input and output contaminant mass flows are stress factors. These stress factors are imposed on the model through "boundary" conditions or "source/sink" terms. An appropriate aquifer-system geometry can be determined using geological information (maps and cross sections), topographical maps, and contour maps of the upper and lower limits for the aquifer strata and aquitards. Initial estimates of the distributed values and spatial distributions of the hydro geological parameters (hydraulic conductivity, storage coefficient, dispersivity, etc.) need to be made using raw data and interpretations. Maps and vertical cross sections are used. For a groundwater flow problem, the main measured variable is usually the hydraulic head, and for contaminant transport problems, concentrations. These consist of point values measured at different time periods in the entire aquifer. Historical measurements are required for model calibration and validation.

2.4 Databases linked to a GIS

The geo-referenced database can be linked to a GIS project, by spatial queries and a Structured Query Language (SQL) connection (existing GIS function). The data stored in the database can be easily updated and represented on the map. All points of the map are linked to the database by their unique number.

Powerful spatial analysis is feasible once the database is established. Maps representing database attribute queries (time- and space-dependent parameter values) can be created. Simple statistics related to hydrogeological entities can be displayed on the screen or printed on paper support maps. Geo-statistical procedures (i.e. kriging) complete the analysis. Some of the tools needed to achieve the objectives are already implemented in the base software package, but most of them require knowledge of GIS techniques, database philosophy, and targeted programming using specific programming languages.

2.5 GIS linked to groundwater modelling

Links can be organized between models and GIS using three techniques: loose coupling, tight coupling, and embedded coupling. Loose coupling is when the GIS and the model represent distinct software packages and the data transfer is made through input/output model pre-defined files. GIS software is used to pre-process and post-process the spatial data. An advantage of this solution is that the coupled software packages are independent systems, facilitating potential future changes in an independent manner. In tight coupling, an export of data to the model from GIS is performed, but the GIS tools can interactively access input model subroutines. In this case, the data exchange is fully automatic. An example of this coupling is the Groundwater Modeler link (Steyaert & Goodchild 1994) between the ERMA Spatial database scheme (supported by Modular GIS Environment, Intergraph 1995) and MOD-FLOW, MODPATH, MT3D finite-difference software packages. When a model is created using the GIS programming language or when a simple GIS is assimilated by a complex modeling system, embedded coupling is used. Tight coupling as well as embedded coupling involves a significant investment in programming and data management that is not always justified. Also, this could be constraining when a high flexibility is required.

For simulating the groundwater flow and transport in saturated and unsaturated zones, our own developed finite element code SUFT3D (Carabin & Dassargues 1999) is used. The well known Groundwater Modeling System (GMS) is here used as a powerful pre-processor and post-processor (Engineering Computer Graphics Laboratory 1998) for our numerical modeling operations. The hydrogeological-attribute data can be directly introduced or they can be imported from a specific format file. The need for importing data in GMS exists in the three steps of groundwater flow modeling: conceptual model design, model construction, and calibration.

Starting from the defined conceptual model, the model can be built within GMS using "feature objects" defining the conceptual model to generate a 2D mesh. The 3D mesh is built using the 2D mesh and TINs.

3 ADVANCED APPROACH FOR MANAGING HYDROGEOLOGICAL DATA: THE HYGES DATABASE SCHEME

Recognizing that field hydrogeologists, modelers, and regulators all need to manage data, the purpose of developing the hydrogeological database concept (called HYGES) was to integrate the main data and information that the hydrogeologist uses. The objectives for the final database were: (1) to provide an organized scheme for capturing, storing, editing, and displaying geographically referenced hydrological data and information, (2) to process and analyze spatially distributed data, (3) to properly support aquifer vulnerability assessments, (4) to easily provide values for numerical models parameters and variables, and (5) to create hydrogeological maps.

Existing and required data types were examined in order to design the database scheme. Parameters and information were reclassified and regrouped several times. Many hydrogeological parameters and relationships were analyzed in order to be placed in the database. Maximum storage of data with a minimum data redundancy, reduction storage capacity, and optimum retrievability of data for analysis were the constraints that defined the final scheme. Data-integration limits were imposed because of different restrictions concerning the hardware and software storage capacity and limitations in current activities and in available information.

3.1 HYGES database construction

Data analysis is an important consideration in database construction. In order to identify the data needs and to provide the optimal data representation, accurate assessments of all types of data and data formats are extremely important before designing a database.

The data-collection operation showed that hydrological and hydrogeological data come from very different sources: water regulators, water companies, environmental agencies, geological services, research offices, and many others. In this case, the main data providers were Ministry of Walloon region, Walloon Society for Water Distribution (SWDE), Water Supply Company of Liege (CILE), Water Supply Company of Brussels (CIBE), Belgian Geological Survey, Laboratory of Engineering Geology, Hydrogeology and Geophysical Prospecting (LGIH), and others. These various sources have strong dissimilarities in data type, in quality and in quantity, as well as in storage media. All the data were analyzed for import to a single system. Data that appear to be redundant had to be specified in the database scheme to avoid loss of information. Such decisions were based on (1) pumping schedules, (2) data registration formats, (3) uncertainty of existing data (measures and registration), and (4) insufficiency in data registration system.

Depending on the accepted conceptual data-model (basic assumptions) and needs, additional data could appear. Also, data that were not explicit or sufficient needed to be flagged or supplied with fields of information or even entire tables. An example is the case where flow-rate registrations related to several wells were available, without distinguishing the pumping schedule of each well. There, a field containing wells sharing the same flow rate value had to be specified.

Data formats are also an important issue, because the pre-treatment of data consists of hours of encoding or of writing import/export codes. Data coming from paper sources, such as tables, maps, and singular data, as well as different spreadsheets and data existing in databases having distinct schemes, were analyzed in order to create a unified database system.

After structuring the spatial database scheme, hydrological and hydrogeological data coming from Ministry of Walloon region, SWDE, CILE, and elsewhere, were introduced into a GIS project using Arc/Info (ESRI) with Access (Microsoft). This solution was chosen after analyzing the software platforms used by different hydrological and hydrogeological research teams, Belgian regulators, water companies, and authorities, in order to ensure compatibility in future data-exchange operations.

In the first step, the information was collected for the following hydrogeological entities: wells and wells systems, piezometers, drains, water-supply galleries, and quarries and mines exploited for water. For these features, the following characteristics were incorporated: location (in Belgian Lambert coordinates), address, altitude, depth, local aquifer information, and owners. More than 50 years (1947–2003) of time-dependent data were encoded, including hydraulic heads and annual and monthly pumping rates. Quality data represented by 284 water-quality parameters determined on 4077 groundwater samples are now registered in the database. The information was supplemented with digital maps showing geology and strata elevation, land-use maps, zones of hydrogeological protection, and others.

3.2 HYGES database scheme

Data and information specific to geomorphologic, geologic, and hydrological conditions were divided in two parts, primary and secondary data. The primary-data section contains layers of general environmental information, such as topography, geological maps, soils maps, hydrological and hydrogeological raw data, or data undergoing an initial minor pre-treatment; and information related to hydrogeological investigations and development means, such as wells, piezometers, drains, mines, quarries; and land-use maps. Secondary data consist of data derived from processed primary data; examples are maps of hydraulic head, hydraulic-conductivity maps, vulnerability maps, etc. (Fig.1).

A spatial reference for the represented hydrogeological features was used based on the topographical map of Belgium with a scale of 1:25,000. This map is uses the Lambert Conformal Conic projection with the following parameters (Belgium Lambert): spheroid International 1909; 1st standard parallel 49°50'0.002" N; 2nd standard parallel 51°10'0.002" N; central meridian 4°22'2.952" E; latitude of projection's origin 90°0'0.000"; false easting (meters) 150000.01300; and false northing (meters) 5400088.43800. The current geological map of Belgium uses the same scale and the same projection as the topographical map.

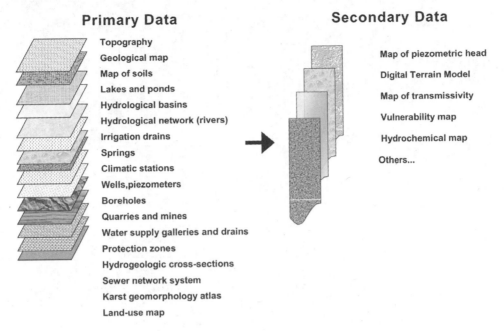

Primary Data

Topography
Geological map
Map of soils
Lakes and ponds
Hydrological basins
Hydrological network (rivers)
Irrigation drains
Springs
Climatic stations
Wells,piezometers
Boreholes
Quarries and mines
Water supply galleries and drains
Protection zones
Hydrogeologic cross-sections
Sewer network system
Karst geomorphology atlas
Land-use map

Secondary Data

Map of piezometric head
Digital Terrain Model
Map of transmissivity
Vulnerability map
Hydrochemical map
Others...

Figure 1.   Overview of the hydrogeological database structure.

The information contained in the Access database is divided in three main groups following the geo-metrical characteristic of the information (associated to a point, an arc or a polygon). An example of "point" tables contained in the primary database is shown in the Figure 2.

3.3   *Applications of the database "HYGES"*

One of the applications of this georelational database is the development of the hydrogeological maps. Coupling a Geographical Information System (GIS) and the hydrogeological database provides a powerful tool for groundwater management. Therefore these maps are realized with a GIS software (ArcView-ESRI). This connection allows an automatic update of the GIS project for punctual elements (wells, piezometers, springs, climatic stations, river-gauging stations, ...) every time a new record is added to the database (with its geographic coordinates). Polylines (galleries, isopiestic lines, ...) and polygons (protection zones, watersheds, ...) are digitized in the GIS project and their attributes are actively related to the database by a unique number. The hydrogeological database can be consulted starting from the map thanks to an ArcView extension called BDHydro, developed by the hydrogeolocical cell of the Faculty of Applied Sciences of Mons, (Belgium). It opens a window by clicking on one of the element on the map and allows the user to display in the GIS project for example a hydraulic head evolution, a hydrochemical analysis table, the lithological log diagram of a well, ...

4   COUPLING THE DATABASE TO THE GMS INTERFACE

Different programs were developed to automatically use the attributes of wells, rivers, and drains in GMS for the mesh of the finite-element models. They allow maintenance of the coupled software packages as independent systems, facilitating any future changes in the spatial database scheme or in any particular module of the software. The presented scheme can satisfy the hydrogeologist's immediate needs in term of research and various environmental studies. However, hydrogeologists are advised that a complete GIS structure is more that a database scheme.

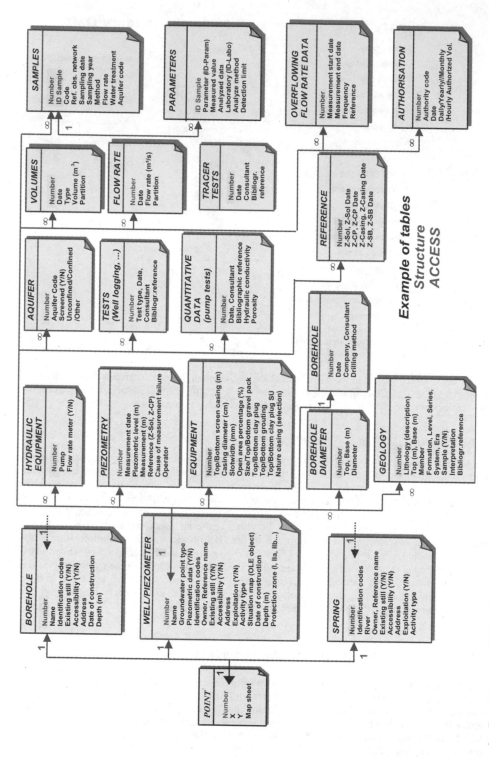

Figure 2. Example of a simplified version of the point attribute data scheme, particularly "Wells and Piezometers".

5 CONCLUSION

The hydrogeological GIS database described in this paper offers capabilities for hydrogeological modeling as well as other hydrogeological studies. The presented database schema is implemented in the Hydrogeology Group of the University of Liege in Belgium. The software support is Arc/Info (ESRI) in connection to Access (Microsoft). The schema could be applied to other GIS and Relational Database Management Systems (RDBMS) that can be connected. The spatial database was conceived as being modular. Users that are using only a RDBMS in the absence of a GIS tool can handle the attribute data. New developments are also underway.

ACKNOWLEDGEMENTS

We thank the Walloon Region of Belgium for the different projects supporting entirely the development of the exposed tools. A part of this work has also been funded through the EU-MANPORIVERS project (ICA4-CT-2001-10039).

REFERENCES

Gogu R.C., Carabin G., Hallet V., Peters V. & Dassargues A. 2001. GIS-based hydrogeological database and groundwater modelling. *Hydrogeology Journal* 9: 555–569

Carabin G. & Dassargues A. 1999. Modelling groundwater with ocean and river interaction. *Water Resources Research* 8: 2347–2358

Corwin L.D. 1996. GIS applications of deterministic solute transport models for regional-scale assessment of non-point source pollutants in the vadose zone. In: Corwin DL and Loague K (eds) *Applications of GIS to the modelling of non-point source of pollutants in the vadose zone, Soil Science Society of America*. Madison, USA. 48: 69–100

Engineering Computer Graphics Laboratory 1998. Groundwater modelling system reference manual (GMS v2.1). Brigham Young University, Provo, Utah, USA.

ESRI 1997. The Arc/Info version 7.1 & ArcView 3.2 software package documentation. Environmental Systems Research Institute, Inc., USA.

GMS version 2.1 1998. Groundwater Modeling System. The Department of Defense. Reference Manual. Engineering Computer Graphics Laboratory, Brigham Young University.

Goodchild M.F. 1996. The application of advanced information technology in assessing environmental impacts. In: Corwin DL and Loague K (eds) *Applications of GIS to the modelling of non-point source of pollutants in the vadose zone, Soil Science Society of America*. Madison, USA. 48: 1–17

Heuvelink G.B.M. 1998. Error propagation in environmental modelling with GIS. Taylor & Francis Ltd, UK

Intergraph 1995. Working with Environmental Resource Management Applications groundwater modeller for the Windows NT Operating system. Intergraph Corporation, Huntsville, USA.

Steyaert L.T. & Goodchild M.F. 1994. Integrating geographic information systems and environmental simulation models: A status review. In: Michener WK, Brunt JW and Stafford SG (eds) *Environmental Information Management and Analysis: Ecosystem to Global Scales, Taylor & Francis Ltd*, UK, 21: 333–355

Computational Methods in Engineering and Science, Iu et al. (eds)
© 2003 Swets & Zeitlinger, Lisse, ISBN 90 5809 567 3

Spatial geotechnical database for planning and design in the Lisbon area (Portugal)

I.M. Almeida & F.M.S.F. Marques
Department and Centre of Geology, Faculty of Science, Lisbon University, Lisbon, Portugal

G.B. Almeida
Câmara Municipal de Lisboa, Lisbon, Portugal

ABSTRACT: More than 50 years of geotechnical site investigation practice in the Lisbon area have created a large and complex set of information, dispersed among several institutions. The objective of this ongoing project is to combine the existing geotechnical information into a single uniform database. To achieve this objective, after an all-encompassing analysis of the existing data, it was necessary to set up a format for a comprehensive geotechnical database, combining the power of interactive computer technologies, including the use of a geographic information system, with geotechnical expertise and knowledge of how data is used in practical engineering decisions. Both the structure and content of the final, unified database have been prepared in order to be useful for a wide range of research and practice areas, including civil engineering planning, design and maintenance, and geological hazards assessment.

1 INTRODUCTION

Nowadays is commonly recognized that engineering and environmental decisions must be based on the detailed knowledge of the subsurface, emphasizing the major role of the knowledge acquired through decades of geotechnical exploration. The wealth of geotechnical data, most of which is in cumbersome paper format, can only be used if it is accessible, ready for application and easily and frequently updated. On the other hand, information technology in combination with the classical expert judgment and qualitative data is becoming increasingly important in geotechnical engineering.

A major advance in data management (Toll 2001) has been the development of a standard format for data exchange developed by the Association of Geotechnical and Geoenvironmental Specialists in the UK.

More than half a century of geological and geotechnical site investigation practice in the Lisbon area have created a large and complex set of data (Almeid 1991), dispersed over several different institutions which should urgently be stored in a single uniform database.

Due to the legal constraints for the free use of a large part of this data, mainly because it belongs to different private entities, the first steps must be made by public institutions. In this case the project was promoted by the Lisbon city council (CMLisboa).

2 GEOLOGICAL BACKGROUND

2.1 *Geological setting*

The geology of the Lisbon metropolitan area is characterized by the differences between the southwestern area, composed of Cretaceous limestones and Late-cretaceous basalt, and the remaining area, dominated by overlying Palaeogene and Miocene weak formations.

The main structure is a complex W–E anticline, in the southwestern area, that causes the Cretaceous limestone to outcrop. The northwards extension of this structure is expressed by a buried sequence of undulating anticlines and synclines. The eastern part of the area forms a quite regular monocline dipping gently ($<10°$) eastwards.

The geomorphological evolution of the area produced a very characteristic landscape. The strong Mesozoic formations gave place to a dominant relief while the Tertiary formations were shaped in irregular small hills. A superficial cover of recent sediments smoothens the relief.

2.1.1 *Superficial recent deposits*
Superficial recent deposits, mainly composed by normally consolidated soils, can be identified in almost all boreholes with very variable values of thickness:

- SF – Superficial fills. Present in reclaimed areas and filling natural or artificial depressions, such as valleys and old quarries.
- Ad – Alluvium from the small valleys crossing the town.
- As – River Tagus alluvium, composed by very heterogeneous lenticular sandy muds and muddy sands, usually is saturated and below water level. Their thickness, in the deepest zones near the margins, can be greater than 30 metres.

2.1.2 *Miocene formations*
The Miocene in the Lisbon area presents a quite complete estuarine sequence, with alternate marine and continental facies, mainly composed by hard soils and soft, weakly cemented carbonate rocks, superficially decompressed. The thickness of the complete sequence is up to approximately 300 metres being thinner in the West, and becoming thicker eastwards.

The Miocene series is usually divided into several lithostratigraphic units, with variable thickness:

- M_{VIIb} to M_I – Silty sands, silty clays, calcareous sandstones and limestones.

2.1.3 *Palaeogene*
The Palaeogene terranes are clastic continental sediments, with a long stress history, corresponding to hard soils and soft to medium strong rocks. The thickness of this complex can be locally greater than 300 meters. In general it is thicker north- and westwards, and is sometimes absent in the South:

- ϕ – Clayey sandy gravels and marly clays.

2.1.4 *Late-cretaceous*
The Late-cretaceous Lisbon Volcanic Complex corresponds to a set of basaltic lava flows and pyroclastic beds. The thickness of the basaltic sequences is higher in the northern and western areas, where it can reach 300 meters:

- β – Basaltic lavas and pyroclasts.

2.1.5 *Cretaceous*
The Cenomanian complex, composed by limestones and marls, up to 330 meters thick, outcrops in the southwest part and forms the mesozoic basement of the area:

- C_C – Compact and crystalline limestones and alternating marly limestones and marls.

2.2 *Geological-geotechnical sketch map*

Figure 1 presents a simplified geological map of the Lisbon town, adapted from Almeida (1986). In this map the outcropping formations are divided in superficial deposits (alluvium and superficial fills, I − As + SF; II − Ad) and bedrock formations. These last formations were divided according to their main lithological composition and geotechnical properties, and they are grouped in three main categories:

- Softer overconsolidated Tertiary formations (mainly sands and sandstones) – M_{VIIa}, M_{VIIb}, M_{VIb}, M_{Vb}, M_{Va}, MI_{Vb}, M_{II};
- Harder overconsolidated Tertiary formations (mainly clays, silts, marls and limestones) – M_{VIc}, M_{VIa}, M_{Vc}, M_{IVa}, M_{III}, M_I, ϕ;
- Mesozoic formations (mainly marls, limestones and basalts) – β, C_C.

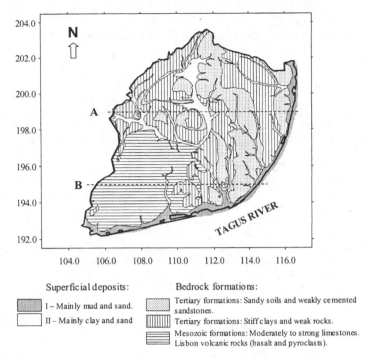

Figure 1. Outline of the geology of Lisbon, adapted from Teves-Costa et al. 2001. Line A and B – approximate locations for W–E geological profiles A-199 and B-195 (Rectangular Gauss Coordinates, International Ellipsoid, Lisbon Datum).

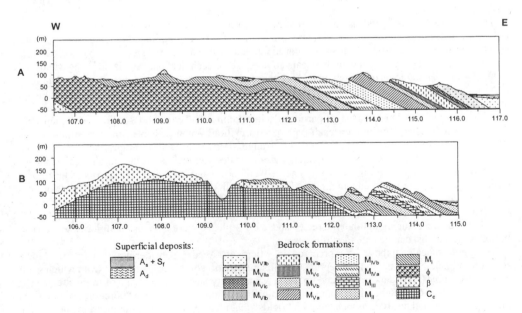

Figure 2. Geological profiles A-199 and B-195 (see Fig. 1) plotted with vertical exaggeration (5X) to illustrate geological structure (adapted from Teves-Costa et al. 2001).

2.3 Geological profiles

The outline of the geological structure is illustrated by the two east-west geological profiles included in Figure 2.

3 DATA CHARACTERISATION

The prime geotechnical information can be found in a very large number of geological and geotechnical site investigation reports, mainly containing borehole and pit log descriptions, "in situ" tests data and ground water level data. The main advantage of this data is the large amount of existing boreholes but it should always be taken in account that some of the information, often in qualitative form, is subjective and often incomplete, requiring a full interpretation before use.

A second source of information less extensive includes reports of laboratory and field "in situ" tests, and geophysical data. A third group of data can be derived from design parameters of excavations, retaining walls and foundations, and also critical observation and monitoring during and after construction.

Finally, it is also important to allow the inclusion of other data sources, such as the location and extent of old filled quarries and old landslides deposits.

The work developed by Almeida (1991) allowed a global perspective of the existing information and a first approach to the selection, hierarchy and design of the geotechnical database.

4 DATABASE DESIGN

The implementation of a geotechnical database in a GIS environment must consider the existing geotechnical information and fulfil some fundamental aspects: archival of existing records; addition of new records; retrieval/query of existing records; to handle spatiotemporal variability, not only visually but also with interactivity.

Two main solutions can be adopted, the use of commercial geotechnical software or the adaptation of current database software. In a review of the available software for geotechnical data input Spink (1996) quotes four systems in Windows (HoleBASE+ and SID) and two in DOS (TECHBASE and gINT).

Considering the need of a friendly format, adapted to the common practice in the geotechnical firms working in the area, it was decided to implement a database in Microsoft Access (IST 2000), named BDG-CML.

The GIS was implemented using an ArcView platform. The system consists on three range data layers, with different sub-layers. The first layer is the map of the region, including buildings and street centerline network to facilitate navigation and interactivity. The second layer is the geological map. The third is a set of point layers with the localization of the geotechnical data available in the database.

One of the requirements of the database design was the need to confine its contents to raw data from the original sources, keeping data interpretation to a minimum. On the other hand, due to the uncertainty in the quality and reliability of the information, it is necessary to minimize the range of variability of each parameter. This problem is particularly important in the case of descriptive and qualitative data that need special solutions for each type of information. When possible, the best solution is to adopt a list of answers (☑ in Fig. 3), limiting the range of errors, as in the stratigraphic interpretation of the log sequence, and the subjective interpretation of descriptions, as in the lithological sequence (soil type) and color. The stratigraphic sequence is well known and shouldn't be modified. The color is a very subjective parameter but sometimes very important in the interpretation of the structure or even the stratigraphic unit. A list of main soil colors was initially introduced but it can be extended if necessary. The soil types can be described using two "fields" in the corresponding form. The first "field" includes the name of the main lithotype (clay, sand, etc.) while the second one is conditioned to the first. In both cases the list can be extended if necessary.

The geotechnical database was prepared in Microsoft Access considering a hierarchical data structure (Fig. 3). The main entity "Report" contains information of the project report from which geological/geotechnical raw data was picked up. Each report entity has associated one or more site investigation methods, including boreholes, and also administrative information. Each "borehole" or other site investigation method entity is associated with "fields" that identify its geographic location and drilling method. The borehole entity is related to "stratigraphic" and "lithology" entities, which contain descriptions of the soil layers encountered during drilling. Layers from which samples were collected for

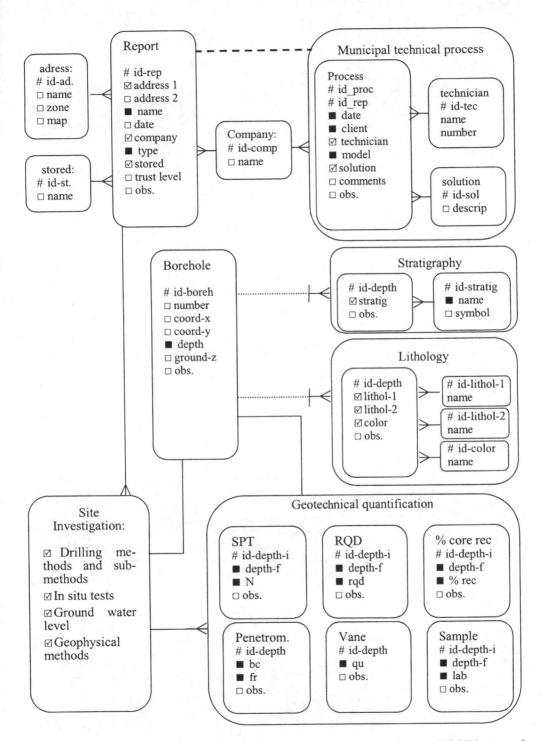

Figure 3.   General view of the geotechnical database design. The bullets represent the type of "field" in terms of filling: #– auto-number entity identification; ☑ limited choice list; ■ – obligatory; ☐ – optional.

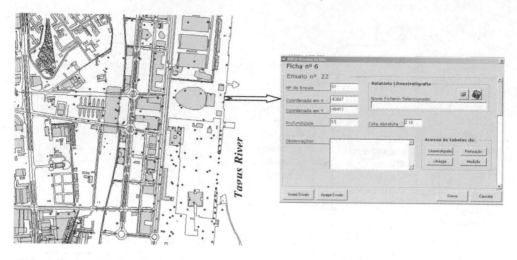

Figure 4. On the left, example of the SIG borehole localization in part of the northeast zone of Lisbon (EXPO '98). On the right, example of one of the database forms (borehole form).

laboratory testing were associated with "sample" entities, which in turn will be related to "test" entities containing the results of these tests.

The database is filled writing the respective content in each "field" of the different forms (Fig. 4), beginning by the main form "Report", which contain the identification of the geotechnical report and "buttons" to accede to next level complementary forms and next "Report".

In each form (Fig. 4) there are several "fields", including some of mandatory filling and others associated with a list of limited answers. The relationship between entities is processed through the identification auto-number.

5 CONCLUSIONS

The database presented was designed to store different types of data but maintaining input flexibility. All separate data is combined into a single electronic relational database, designed in Microsoft Access, in which all data fields can be transferred into appropriate fields and records in the unified database, linked to a digital map base. Available geotechnical data for any well or borehole can be viewed and processed using digital maps, and displayed and manipulated using ArcView.

The database contains actually more than 4,000 digitised geotechnical borehole logs gathered from the EXPO98, APL and Lisbon municipality, including evaluation of data integrity and quality. This work is still limited in scope, due to the difficulties associated with obtaining information from private companies, but efforts are being made to extend it to all existing information.

REFERENCES

Almeida, F.M. 1986. Carta Geológica do Concelho de Lisboa – 1:10.000, Serviços Geológicos de Portugal, Lisbon, Portugal.
Almeida, I.M. 1991. Características Geotécnicas dos Solos de Lisboa, *PhD Thesis*, University of Lisbon.
IST 2000. Manual de Utilização da Base de Dados Geotécnicos, Instituto Superior Técnico (internal report). Lisbon. Portugal.
Spink, T. 1996. Geotechnical Software Review. *The Electronic Journal of Geotechnical Engineering*, 1 (http://www.ejge.com/1996/Ppr9602/Abs9602.htm).
Teves-Costa, P., Almeida, I.M. and Silva, P.L. 2001. Microzonation of the Lisbon town: 1D theoretical approach. *Pure Applied Geophysics* 158: 2579–2596.
Toll, D.G. 2001. Computers and Geotechnical Engineering: A Review. In B.H.V. Topping (ed.), *Civil and Structural Engineering Computing* 2001: 433–458. Saxe-Coburg Publications.

Artificial intelligence and software engineering

Computational Methods in Engineering and Science, Iu et al. (eds)
© 2003 Swets & Zeitlinger, Lisse, ISBN 90 5809 567 3

Controlled solidification in continuous casting using neural networks

S. Bouhouche
Control Engineering Group, Applied Research Unit, SIDER Group, Annaba, Algeria

M.S. Boucherit
Dept.of Computer Science & Control Engineering, Algerian Institute of Technology, Algeries, Algeria

M. Lahreche
Iron and Steel Laboratory, Applied Research Unit, SIDER Group, Annaba, Algeria

J. Bast
Institute of Mechanical Engineering, Mining and Technology University, Freiberg, Saxony, Germany

ABSTRACT: In the present study, mould thermal monitoring (MTM) technology in continuous casting has been investigated in order to optimize casting control. Continuous measurements of the thermal profile are obtained by a matrix of thermocouples. In such way real-time monitoring and control of the process are considered. A breakout appears generally during metal sticking on the copper plate of the mould followed by perforation of the solid shell due to a solidification disturbance. The measurement and acquisition of temperature in different points at the mould surface constitute a tool for analysis and modeling of the phenomenon. A process database is used for training the neural network (NN) using back-propagation algorithm. After training the (NN) predict new breakouts. Using this approach the number of false alarms has been considerably reduced comparatively to the conventional system.

1   CONTROL AND SOLIDIFICATION IN CONTINUOUS CASTING PROCESS

The following developments have been realised on a continuous casting process. The principle of this process is given in Figure 1.

1.1   *Control and monitoring of solidification in the mould*

In the present study, (MTM) technology in continuous casting has been investigated in order to optimise casting control. The mould is the location of complex metallurgical reactions characterising the liquid – solid transformation forming the first crystals of solidification. The control algorithm predicts solidification defects using mathematical models. In this work, a new approach for a breakout detection system has been developed by means of NN. A process database is used for training the NN using a back-propagation algorithm. In continuous casting, the phenomenon of the breakout is generally caused by rupture of the solid crust due to an increase in temperature at various points of the mould. Both peak and temperature oscillations have a direct influence on the quality resulting from solidification (Blazek & Saucedo 1989, Chakraborti & Mukherjee 2000, Hemy & Smylie 2002). These phenomena appear at the time of slag incrustation, formation or propagation of cracks and in the case of poor friction and generally at the time of an imbalance of distributed thermal reactions in the mould. In this study, the monitoring and the detection of abnormal phenomena affecting thermal conditions in the mould have been developed using NN (Norgaard & Ravn 2000). The input of the time series network is formed by the measured temperature samples, while the output is formed by alarm defining the importance of defects. A new spatial network considers the combination of different time series models alarm. The training has been carried out by the exploitation of databases characterising the normal and deteriorated operating conditions of solidification process. Such databases contain information on the dynamics of process parameters

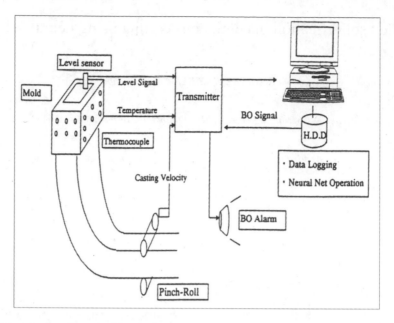

Figure 1. Principle of continuous casting process.

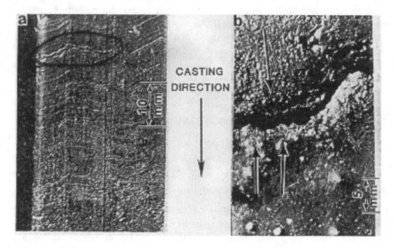

Figure 2. Example of breakout propagation.

and the operating state of the process (alarms, shutdown of production,). In the following training, the simulation tests based on cases of real defects are applied to estimate the model detection ability.

1.2 *Analysis of breakout phenomena*

1.2.1 *Breakout propagation process*
The mechanism for the original sticking can be explained by the existing conditions at the meniscus such as variations of casting speed, mould bath level of liquid steel, steel temperature and lubrification. Changes of casting speed have an important influence. A breakout appears generally during metal sticking on the copper plate of the mould followed by perforation of the solid shell due to a solidification disturbance. Sticking breakout is propagated with various speeds in various directions and particularly

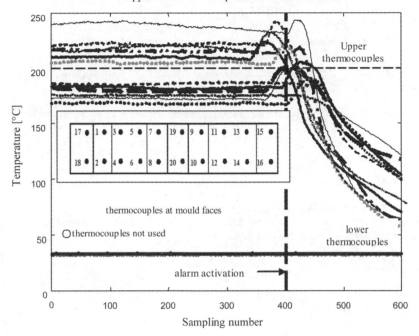

Figure 3. Breakout mould temperature variations.

in casting direction (Blazek & Saucedo 1989, Hemy & Smylie 2002, Kumar 1999, Pinheiro et al. 2000). Figure 2 shows an example of breakout propagation and Figure 2a a little crack which has been developed in a breakout affecting the slab quality (see Fig. 2b).

In this complex situation, it is practically impossible to describe the development of a breakout in the geometrical space of the mould using an analytical model based on heat transfer, solidification and the mechanical laws. The measurement and acquisition of temperature in different points at the mould surface constitute a tool for analysis and comprehension of the phenomenon. This experimental approach is also used for the development of a reliable system. The technique is the basis of the MTM system that considers the mould as a thermal reactor and the appearance of breakout is a result of an imbalance of the distributed thermal reactions. The dynamics of process data that have generated a breakout are affected by these random terms.

1.2.2 Breakout effect in the mould temperature field

Generally when a breakout is generated, the upper thermocouple records a higher temperature T^U due to the local breakout, followed by a reduction in temperature that is also due to a partial solidification. Under the effect of the casting speed, the crack propagates and the same phenomenon is observed at lower thermocouples T^L. Alarms and reductions of casting speed are activated. In the case of conventional techniques, when the difference between the measured temperatures and those calculated by a model reaches a fixed threshold, a series of alarms is activated. When the error reaches dangerous levels, the casting speed is automatically reduced to zero (Blazek & Saucedo 1989, Hemy & Smylie 2002).

Figure 3 gives an example of temperature field variation according to a breakout.

1.3 Breakout prediction and detection

Since the development and implementation of the breakout detection system on continuous casting processes, efforts have been focused on the simplification of instrumentation by reducing the number of thermocouples and the development of advanced models able to minimise the number of false alarms (Madill 1996). The principle of detection is based on the analysis of temperatures on the mould and their gradients.

2 MODELLING IN OF BREAKOUTS

2.1 *Conventional method*

In each point M of the copper mould, the variation of temperature $T(M,t)$ with time t, can be defined as:

$$T(M,t) = T(M,t_0) + \left[\frac{\partial T(M,t)}{\partial t}\right]_0 \Delta t + \left[\frac{\partial^2 T(M,t)}{\partial t^2}\right]_0 \Delta t^2 + \ldots \left[\frac{\partial^n T(M,t)}{\partial t^n}\right]_0 \Delta t^n + \Delta T(0) \tag{1}$$

The conventional approach approximates the temperature dynamics as a linear function of time:

$$T(M,t) = T(M,t_0) + \left[\frac{\partial T(M,t)}{\partial t}\right]_0 \Delta t \tag{2}$$

$$= T(M,t_0) + a.\Delta t \tag{3}$$

Figure 4 gives a geometrical interpretation of temperature dynamics.
   Three cases of temperature dynamics are taken into account by the conventional system: The gradients of upper and lower temperature are defined as:

$$a^U = \left[\frac{\partial T^U(t)}{\partial t}\right]_0 = \frac{T^U(t_0) - T^U(t_1)}{t_1 - t_0} \tag{4}$$

$$a^L = \left[\frac{\partial T^L(t)}{\partial t}\right]_0 = \frac{T^L(t_0) - T^L(t_1)}{t_1 - t_0} \tag{5}$$

The temperature difference between the upper and the lower thermocouples is expressed as:

$$\Delta T(t_1)^{U-L} = T^U(t_1) - T^L(t_1) \tag{6}$$

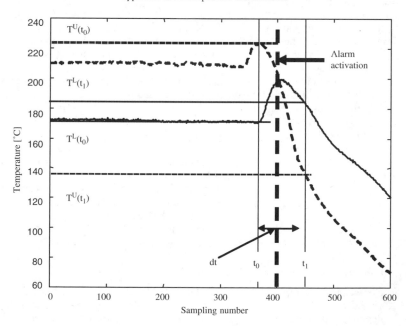

Figure 4.   Geometrical interpretation of a breakout.

The breakout detection algorithm is based on the analysis of the values of equations (4), (5) and (6). The limits of a^U, a^L and ΔT^{U-L} are predefined (Madill 1996). False alarms are generally due to thermal perturbations. Sometimes these variations cannot be well detected by the conventional system using a fixed error range or a predefined statistical characteristics of the error between measured and calculated temperatures in each point. This reduces the process reliability. The neural network permits to solve the problem by the learning process using the breakout data base related to the real and false alarm situation, respectively.

2.2 Advanced methods using neural networks modelling

The principle of breakout detection using (NN) is based on the analysis of a node of thermocouples regarding upper and lower processing units. Each unit considers the temperature variation in time (time series model) and the interaction between different thermocouple temperatures (spatial model). Figure 5 gives the principle of the neural network breakout detection system.

2.2.1 Time series model

The time series model takes into account the temperature variations that can be approximated by equation (1). The principle is to find the whole complex of relations between dynamic variations of

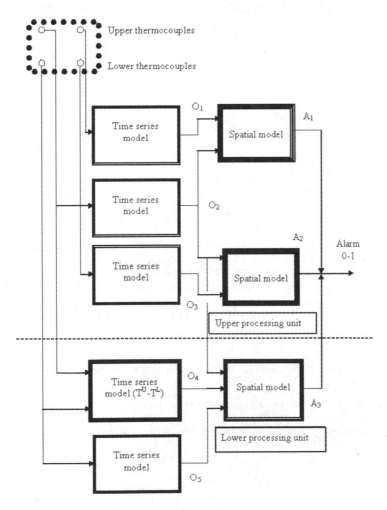

Figure 5.   Structure of breakout detection using neural networks.

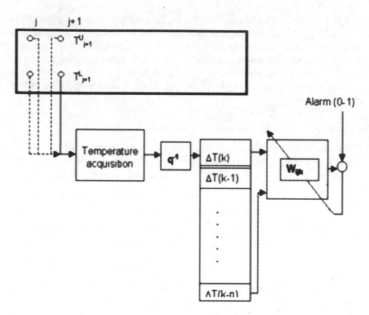

Figure 6.   Principle of learning process.

temperature and the appearance of defects (Kumar 1999, Pinheiro et al. 1998). This can be formulated by the following non-linear relationship:

$$Alarm = NN[\Delta T(k), \Delta T(k-1)....\Delta T(k-n)] \tag{7}$$

$\Delta T(k)$ is the temperature change which is defined as:

$$\Delta T(k) = T(k) - T(k-1) \tag{8}$$

The model is obtained by the learning process using the back-propagation algorithm and the characteristics of breakout temperature. Figure 6 gives the learning principle of the time series model.

After several trials an optimal (NN) with (n = 60) in the input layer, 15 nodes in the first layer and one node in the output layer that corresponds to the alarm output signal has been chosen. After the convergence, network weights W_{ijk} are collected in a file for further use with regard to other thermal profile breakout detection. Breakouts are detected by a conventional system at the sampling number 400. Alarm is released by passing from 0 to 1.

2.2.2   Spatial model

The Spatial Model assumes cross interaction between the thermocouples upper (j) − upper (j+1) and upper (j) − lower (j+1).

O_1 is the alarm output corresponding to the time series model of the upper thermocouple (j+1)
O_2 is the alarm output corresponding to the time series model of the upper thermocouple (j)
O_3 is the alarm output corresponding to the time series model of the lower thermocouple (j+1)

3   APPLICATION

Prediction of real breakouts (real alarms) is given in Figure 7. Application to more real and false breakouts are summarised in Table 1.

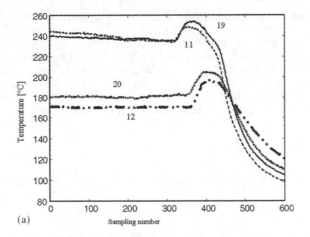

(a)

Figure 7a.   Real breakout Nr 2: thermocouple node [19-20-11-12].

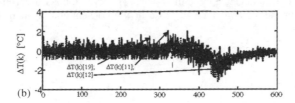

(b)

Figure 7b.   Differentiation of equation (8).

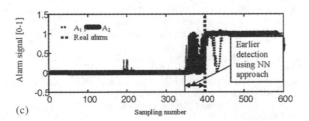

(c)

Figure 7c.   Calculated and real alarms.

Table 1.   Analysis of breakouts detection.

System*	Sticker	Misclassification	Breakout
Conventional system	10	4	0
Neural network system	10	0	0

Sticker: Alarm has been generated and was justified; Misclassification: Alarm has been generated but was not justified; Breakout: A breakout occurred but no alarm has been generated.

4   RESULTS ANALYSIS

As presented in Table 1, the neural network breakout system ability was tested through (10) real breakouts detected by the conventional system in continuous casting process. Neural network models have investigated all breakouts detected by the conventional system. The detection is achieved by the upper,

523

lower and upper-lower processing units. In this case, neural network and conventional systems are equivalent.

In the case of false breakout detection by the conventional system (misclassification), the neural network model has been tested through four (4) false breakouts detected by the conventional system. The NN model has not detected all false alarms. In this case the NN model has a greater ability not to take into account the field temperature variations that do not generated. As presented in Table 1, the objective of reducing the false alarm rate has been achieved. The neural network breakout detection system has been tested through real data. The obtained results confirm an improvement.

5 CONCLUSION

The NN breakout system based on the instrumentation of the mould by a matrix of thermocouples using a model of prediction by NN was developed, implemented and tested. The training of the model and tests of detection were carried out using real and false breakout data. In the case of real alarm, the results do not detect the presence of false alarms. The developed algorithm detected the dynamic behaviour of temperature profiles having generated real breakouts. In the case of false alarms, the developed model does not detect false breakouts.

All real alarms have been detected by both systems. However, the alarms from the neural network detection system occur earlier compared to those of the conventional system. Results confirm an improvement.

REFERENCES

Blazek, K.E. & Saucedo, I.G. 1989. Recovery of sticker type breakouts. *Proceedings of Steelmaking Conference*: 99–107.
Chakraborti, N. & Mukherjee, A. 2000. Optimisation of continuous casting mould parameters using genetic algorithms and other allied techniques. *Ironmaking and Steelmaking* 27(3): 243–247.
Hemy, P. & Smylie, R. 2002. Analysing casting problem by the on-line monitoring of continuous casting mold temperatures. *JOM*. www.umc.tms.org/pubs/journals/JOM/021/Hemy/Hemy-0201.html
Norgaard, M. & Ravn, O. 2000. Neural networks for modelling and control of dynmic systems. In Springer (eds). Netherlands.
Kumar, S. 1999. Development of intelligent mould for on-line detection defects in steel billets. *Ironmaking and Steelmaking*? 26(4):269–284.
Pinheiro, C.A.M. & Brimacombe, J.K. 2000. Mould heat transfer and continuously cast billet quality with mould flux lubrification Part2: Quality issues, *Ironmaking Steelmaking* 27(2):144–159.
Madill, J.D. 1996. Application of mould thermal monitoring to Avesta Sheffield's SMACC slab caster. *Ironmaking and Steelmaking* 23(3): 228–234.
Ngyena, H.T. & Sugeno, M. 1998. Fuzzy systems: Modelling and control. In Kluwer Academic Publishers.

Computational Methods in Engineering and Science, Iu et al. (eds)
© *2003 Swets & Zeitlinger, Lisse, ISBN 90 5809 567 3*

Semantic & morphological analysis and mouse tracking technology in translation tool

F. Wong, M.C. Dong & Y.P. Li
Department of Computer & Information Science, Faculty of Science and Technology,
University of Macau, Macao

Y.H. Mao
Speech and Language Processing Research Center, Tsinghua University, Beijing, China

ABSTRACT: This paper presents the research results of several technologies in Chinese/Portuguese machine translation system (*PCT translation system*). This includes the analysis at the morphology of Portuguese lexical items and the extraction of correlative semantic information. Comparing with English, Portuguese has a rich developed morphology and is more difficult in machine auto translation. The study of the mouse tracking technology to create a fast on-line reading environment in translation system is another topic that will be discussed in this paper. Finally, the resolution of conflicts caused by the different character coding systems of Portuguese and Chinese is studied.

1 INTRODUCTION

The study of using machine to automatically translate a language into another language has been recognized as an important area in the new era. As the fast development of Internet technology, people have a lot of opportunities to reach various information from the Internet. The barrier of languages has been the last obstacle that keeps different nation peoples in distance. This falls into the mission of machine translation system, how machine can efficiently convert a language into a target language that people can understand. The most benefit of it is for commercial purpose especially in the era of market globalization.

In the recent years, more people have recognized the importance of developing machine translation system for low-density languages. This allows people to learn more about these languages of different nations directly, instead of through an intermediate language like English. Macao, for example, is a city where the coexistence of Portuguese and Chinese cultures has been a unique characteristic in Macao. Considering that there is no any practical machine translation tool developed for Portuguese and Chinese, Tsinghua University, INESC-Macau and University of Macau had jointly started a research project in Portuguese-Chinese bi-directional machine auto translation (Li et al. 1999) (Wong et al. 2002).

In this paper, we will focus in discussing the problems, of translation system for Portuguese and Chinese, which are more complex when comparing with that for English and Chinese. These include the analysis of alternations of Portuguese word's morphemes and the platform for processing Portuguese and Chinese information, as the encoding systems of these two languages are incompatible. Finally, a fast reading environment to access the translation of text is introduced that can benefit user from instantly getting the translation of content from their working environments.

Morphology is concerned with the ways in which words are formed from basic sequences of phonemes. Basically, inflectional morphology and derivational morphology are concerning most in our study, while compound words is considered in syntactic analysis. Inflectional morphology is the system defining the possible variations on a root form to reflect the grammatical meaning, languages such as Portuguese is highly inflectional, for example, the verb *partir* (to leave) with present tense forms *parto*, *partes, parte, partimos, partis, partem.* While derivational morphology is concerning with the formation of root forms from other roots, often of different grammatical category. Thus, an adjective *verdade* (true) may be formed the noun *verdadeiro* (truth) with the ending *-eiro*. The objective of morphological analysis here is to conclude the conventions of various grammatical morphemes joined to a lexical item and hence to reduce the size of a dictionary from keeping their inflectional paradigms in full.

As the fast emerging of various MT systems, more and more non-professional peoples are using the MT system to do their translation work (Lehrberger et al. 1988). MT system is no more a stand-alone application; modern MT system has been treated as a tool and is usually embedded in user's working environment. When translation is needed, the system is triggered to do the translation and produces the result within user's application. That means, MT comes to the text and serves the user. Therefore, in our study, we have developed a fast reading environment by using the mouse tracking technology. Which allows user to easily get the translation of text that being pointed by the mouse cursor. By integrating this technique into a MT system, user can arbitrarily ask the system to translate the text of a language that he does not familiar with during document reading or Internet browsing. This translation manner can greatly improve the efficient in consulting the meaning of unknown words.

The platform for processing Portuguese and Chinese information in MT system is another problem that needed to be resolved. In English and Chinese translation system, the character-encoding scheme of English has no any conflict with that of Chinese. The range of encoding for English character remains in the range of 0~127, known as standard ASCII character set. While in Portuguese, part of its characters (with accent) in the encoding system across the extended character set, such as $\tilde{a}$, ς, $\hat{e}$, etc. The internal encoding of these symbols is overlapped with that of Chinese characters. Thus, as a result, causes the system to confuse when Portuguese and Chinese text are being displayed in the screen at the same time.

The rest of this paper is organized as followings. Next section, the component of morphological analysis for Portuguese is discussed, and the algorithm in recognizing a word from its variations is presented. The technology of mouse tracking is described in the third section, and the solution to resolve the platform for processing Portuguese and Chinese are presented in section four. Section five reviews the application of the discussed technologies in a word-based translation tool, while these technologies can be further applied to sentence-based translation system. Finally, a conclusion is drawn to end this paper.

2 MORPHOLOGICAL ANALYSIS OF PORTUGUESE WORD

Portuguese is one of the languages from the same cognate as Latin that has a highly developed morphology, and is relatively richer in inflectional variation. The lexical item is realized in various forms according to the grammatical category of number, gender, person, tense, etc. The richer the morphology it is, the higher the requirement it needs to the software analyzer. The analysis component is divided into two major modules (Hutchines et al. 1992): *inflectional module* and *derivational module*. The analyzing of morpheme inflection and the encoded grammatical information according to the ending morpheme is processed in *inflectional module*. While the *derivational module* analyses the prefix or suffix of a lexical item and hence extracts the semantic information (such as lexical category) from this lexical morpheme. Usually, lexical morpheme (prefix or suffix) is a semantic information marker.

Similar to earlier efforts, the morphological component needs a lexicon dictionary used by both *inflectional* and *derivational* modules. A *basic lexical dictionary* lists the lexical items occurring as much as possible, but all the entries are identified by a base form of the word (uninflected word). This prevents from the explosive growth of the item entries. In addition to the basic lexical dictionary, *inflectional module* enquires three more dictionaries. First, the component treats closed-class verbal items, such as irregular verbs, by listing their inflectional paradigms in full. For treatment of inflectable open-class lexical items, two relatively small-specialized lexicons are used in place of the usual large lexicon: lexicon of stems (root word) of words and lexicon of ending morphemes together with the grammatical information encoded. Much simpler, *derivational module* relies only on the basic lexical dictionary and a morphemes (prefixes and suffixes) lexicon where lexical category of base item and the derived item are encoded in additions.

The Morphotactis, the structure and content of the knowledge base in morphological analysis component have been designed: (1) to help solving analysis problems – inflection, derivation and homography paradigms, and (2) to minimize the knowledge acquisition efforts. Thus, as a result, the coverage of lexical item can be greatly improved, as well the computational performance.

2.1   The lexicons

The *lexicon of inflectional endings* (suffixes) consists of all the empirically derived inflections for regular verbs, nouns, adverbs, and adjectives. The entry in the lexicon of ending is of the form as:

$$
Suffix \rightarrow \begin{bmatrix} \{f_1, f_2, \ldots f_n\} \\ \ldots \\ \{f_1, f_2, \ldots f_n\} \end{bmatrix} \tag{1}
$$

where f_i are values of grammatical features associated with this ending, such as grammatical category of gender, number, person, tense, etc. This information is necessary for the analysis in *inflection module*. For example, the entry of a suffix of *-amos* is shown in Table 1.

The identifier "V1" indicates the verbal declension paradigm, "T1" demonstrates the tense of the current inflected verbal item and "P4" further identifies the person and number. These features provide partial semantic information, while others can be retrieved from the basic lexical dictionary. This lexicon is used to define the initial set of morphological features of a word for further testing. Table 2 shows another entries of *-issimo* and *-ibilissimo*, which can be used to analyse adjective. Some entries may contain strings of lexical item before the ending to be substituted.

The *lexicon of stems* consists a list of roots of the verbal items with the category inflection paradigms and their infinitive verb forms. The use of this lexicon is to identify the infinitive verb, and to verify if the stem, output from the ending analyzer, is belonging to the same paradigm. From another point of view, the software analyzer can prevent from misinterpreting an inflected verb caused by incorrect spelling. Table 3 shows partial entries.

The *lexicon of morphemes* consists a list of suffixes (or prefixes), which are empirically able to join to a lexical item to produce a new lexical item. This lexical morpheme, on the other hand, is a semantic information marker. Contrary to what happens in inflectional morphology, one of the features, part-of-speech is obligatory. Usually, the new lexical item resulting from this operation may have a different lexical category than the base item. Table 4 shows the entries of lexicon.

All of these lexicons are required in the morphological component for analysing the lexical morpheme, and extracting the semantic information from the morpheme paradigms.

Table 1.   Lexical entry for the suffix *-amos*.

Suffix	Grammatical features	Description
-amos	V1, T1, P4	Verb, 1st inflection rule, Present (indicative mode), Person (1st, plural)
	V3, T7, P4	Verb, 3rd inflection rule, Present (subjunctive mode), Person (1st, plural)
	V37, T11, P4	Verb, 37th inflection rule, Imperative, Person (2nd, plural)

Table 2.   Lexicon entries for suffixes *-issimos* and *-ibilissimo*.

Suffix	Replaced	Grammatical features	Description
-issimo	*-o, -e*	ASP, G1, N1	Superlative adjective, masculine, singular, substituted ending *-o* or *-e*
-ibilissimo	*-ivel*	ASP, G1, N1	Superlative adjective, masculine, singular, substituted ending *-ivel*

Table 3.   Lexicon entries for stems *ababalh* and *part*.

Stems	Word	Grammatical features	Description
ababalh	*ababalhar*	V1	Verb, 1st inflection rule
part	*partir*	V3	Verb, 3rd inflection rule

527

Table 4. Lexicon entries for suffixes *-dor* and *-amente*.

Suffix	Replaced	Grammatical features	Description
-dor	*-r*	N, V	Noun – new item, Verb – base item, substituted with ending *-r*
-amente	*-o*	A, D	Adjective – new item, Adverb – base item, substituted with ending *-o*

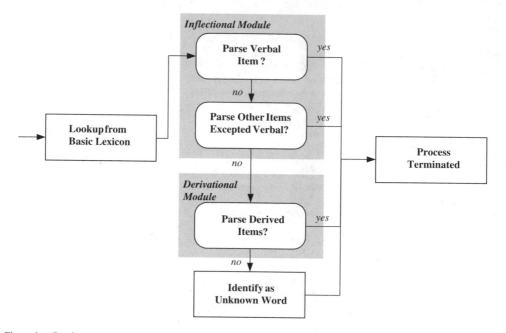

Figure 1. Parsing procedure.

2.2 *Analysis algorithm*

The morphological analyzer attempts to lookup the information from the basic lexicon to an input Portuguese word. Whether or not the attempt successes, the input word is passed to the analyzer for further testing. This compulsory process tries to explore the homography of a word in its inflected form, in order to maximize the coverage.

During the analysing process, when a word was not identified by a parser in any component, it is then fed into next component procedure. Figure 1 illustrates the procedure for analyzing a word. The order of the calls to component procedures in the algorithm in Figure 1 is established to minimize the processing time and effort. This is because the morphological conventions for verb is more completed than that of others, therefore, it is less effort to analyse the verbal items.

All of the parser components are running similar strategy in analysing. For example in verbal analyzer, the parser goes through the ending lexicon for a match, and then determines the stem of the word. The truncated stem will be identified from the stem lexicon to retrieve the semantic information encoded in the entry. If an item could not be parsed by any component, the item is identified as an unknown word at the output of the system and consequently will show no translated word to it.

3   FAST TRANSLATION ENVIRONMENT

As mentioned, present MT system is designed as a tool that is embedded inside the working environment of user, such as the word processing application, web browsing application (Hu et al. 1998), and etc. By using the computer technology, sophisticated translation interface can be developed that allows user to easily access the translation process. Compared with traditional keyboard input translating manner, mouse-tracking instant translation has obvious advantage. User can conveniently find the meaning of a word or sentence, during referencing electronic document or browsing the Internet with unfamiliar language, by solely using the mouse pointer. That means, when user wants to consult the translation of words, he just simply locates the mouse cursor over the text. After for a while, the system will capture the text that is being pointed, and presents the translation of it in the screen. Which, as a result, saves time from typing and, therefore, increase the user interests in reading or learning foreign language.

Since the interface technique is platform dependent, the principle of mouse tracking technology, here in our system, is to monitor and capture every *output text* that Windows is going to draw on the screen. Once the textual information has been obtained, it can be further analyzed and processed. In Windows, all the output application-programming interfaces (APIs) are integrated in a Graphics Device Interface dynamic link library, *GDI.dll*, including text-drawing functions (Richter 1996). Therefore, if the functions can be intercepted, then we can freely to monitor all the content of displayed text. That is the key technique in realizing the methodology of mouse tracking technology.

3.1 *Different ways in text monitoring*

Generally, there are three different ways to intercept a DLL function. We briefly describe the mechanism of each in following (Pietrek 1996):

- The first method is to use *Tool Help* library, which Windows provides as a means to debug other applications. This method can intercept all the functions of a DLL that the application calls. But it requires that the application must be under debugging. Apparently, this method cannot be applied to arbitrary application. Which means, this method is valid to a specified application's interface in the screen, but not to others.
- An alternative way is to find all the certain function callings in an application's executable codes. Since it causes the changes of coding in applications, we strongly do not recommend of using this method. Similarly, this mechanism validates to a specified application, and cannot applied to arbitrary application neither.
- The third method is to modify the coding of the DLL functions. Where, we can add some code to the function being intercepted. Then all the calls to this function can be traced and monitored by our code. Which can fulfill our requirements in: (1) be able to monitor the content of an output text, (2) can be applied to any application running in the same environment. But it must be very careful in operating this mechanism. What we are doing in our PCT system is to directly access the kernel functions of the Windows operating system, any fault may cause the system to halt. In our experiment, this technique has been proved stable, and has been widely applied to many of our research systems.

3.2 *The principle of word tracking*

The objective of mouse tracking is to capture the word under the mouse position after the mouse cursor has stayed for a while, and thus to provide a fast on-line and convenient input mannar to the translation system. The whole procedure of the operation can be divided into several steps, as illustrated in shaded region in Figure 2.

Since Windows works upon the message mechanism (Microsoft 1998), the technique of message hook is adopted to monitoring the mouse movement. The kernel receives messages from various kinds of input devices, such as keyboard, mouse, interrupt deveice, etc. It analyses the messages and forwards the operation to the corresponding message handlers. Upon receiving a message of mouse movement in our tracking engine, a timer is triggered and waits for the expiration of a time slice. Once the time runs off, the action of capturing text starts by: (1) cause the text underneath to be repainted, (2) monitor and capture content when Windows is trying to repaint the text with the *text-out* functions that is being intercepted, (3) identify the target text by calculating and comparing the coordinates with that of the mouse's position. Then a text is therefore to be captured by our mouse tracking engine, and used as input to the translation engine. After the translation, the meaning in target language will be presented to the user in a pop up window.

4 MULTI-LANGUAGE PLATFORM

When dealing with the translation system for Portuguese and Chinese, the first thing that needed to be considered is how these two languages can be processed as the encoding scheme of the languages are different. Some Portuguese characters (with accent) are sharing the same code with that of Chinese characters. Therefore, in order to properly display both languages in the same Windows environment requires special process. Following (Table 5) illustrates the conflicted situation, where the word "*Português (Portuguese)*" has an accent character that, without special processed, cannot be properly displayed together with Chinese text. Since the code of this accent character is being used in Chinese encoding system.

Before discussing the algorithm, Table 6 reviews the common codes that are shared by both Portuguese and Chinese. In the Chinese coding system, each character is represented by two bytes, high byte and low byte. In order to distinguish the characters of standard Latin symbol and Chinese character, the first byte of the character adopts the character set from extended one, that means, the code is larger than 0×95 (127 in decimal). While the second byte can be any number within the range of: $0 \times 40 \sim 0 \times 7e$ and $0 \times a1 \sim 0 \times fe$. Unfortunately, this coding system is only concerning with the coexistence of Chinese and English characters. When involving with Portuguese, the operating system does not take care of such combination, thus causes to the conflict in displaying.

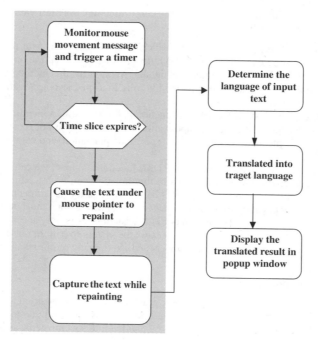

Figure 2. Mouse tracking instant translation mechanism.

Table 5. Display Portuguese and Chinese text at the same time.

Correct content	Display content
Português (葡萄牙語)	Portugu瘰 (葡萄牙語)

Table 6. Overlapped code area between Portuguese and Chinese.

Language	Code area of 1st byte	Code area of 2nd byte
Chinese	$0 \times a1 \sim 0 \times fe$	$0 \times 40 \sim 0 \times 7e$ && $0 \times a1 \sim 0 \times fe$
Portuguese		$0 \times 20 \sim 0 \times 7f$ && $0 \times c0 \sim 0 \times ff$

4.1 *Mechanism in properly displaying multi-language content*

The confusion in displaying Portuguese and Chinese text on a Windows environment is affected by two factors: *character set* and the associated *font face*. If an application uses a font with an unknown character set, it should not attempt to translate or interpret strings that are rendered with that font. Character set is important in the font mapping process. To ensure consistent results, specify a specific character set and make sure that the value matches the character set supported by the font face. An alternative method to fix this problem is to adopt the Unicode scheme; all the information should be converted and processed under the Unicode format. In our system, we use the first method to resolve the conflicts between different languages since all the data information is prepared and processed under the ASCII coding system.

5 APPLICATION TO PCT SYSTEM

In this section, the proposed solution applied to a word-based translation system is demonstrated. This solution can also be extended to the translation based on sentence or paragraph. More generally, this methodology can be transferred to other languages between Chinese and Romance, such as Chinese and French, etc. Figure 3 presents how a inflected verb, *partimos* (*leave* – PRESENT, INDICATIVE, 1ST PERSON, PLURAL), is analyzed and the meaning of it is consulted from the dictionary according to the lemma of the analyzed verb. In the explanation window of the system, only partial grammatical features, *present* and *indicative*, of the analysis is presented.

Figure 4 shows the scenario how the translation of a word is consulted and presented to the user in a small popup window once the user locate the mouse pointer over the target text in the screen. Observing the

Figure 3. Interface of the main program.

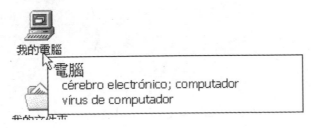

Figure 4. Snapshot of mouse tracking translation

content of the popup window, both Chinese and the Portuguese text with accent can be properly displayed without any conflict. This fast translation technique has been proved to be very efficient and convenient to the user.

6 CONCLUSION

In this paper, we have studied the morphological analysis of Portuguese. Compare with English, Portuguese has a rich morpheme alternations, and is relatively more complicated in analysis. Another problem that we have tackled in this study is to resolve the conflict occurred in processing Chinese and Portuguese information under Windows environment. Normally, Windows system does not handle the imcomptable character encoding system. By properly selecting the correct *character set* and associated *font face*, this conflict can be resolved. Another way to achieve the same result is through the Unicode encoding scheme. Finally, elaborated user interface of a translation system has been discussed and studied in our work. Which has created a new comprehensive interface that allows user to consult the meaning of text in a convenient way, without necessary to go to the MT application. These methodologies that we proposed here can be directly transferred to sentence-based translation system, and even can be applied to other translation systems for Chinese and Romance languages.

REFERENCES

Hu, H.W., He, Y.M., Wang, Q.X., Chen, J.J. & Yong, D.S. 1998. Realization of Janpanese to Chinese Net-translating Browser, *In Proceedings 1998 International Conference on Chinese Information Processing*, Tsinghua Univ. Beijing, China, Nov. 18–20, pp. 482–489.

Hutchines, W.J. & Somers, H.L. 1992. *An Introduction to Machine Translation*, Academic Press.

Lehrberger, J. & Bourbeau, L. 1988. *Machine Translation: Lingusitic characteristics of MT systems and general methodology of evaluation*, John Benjamins Publishing Company.

Li, Y.P., Pun, C.M. & Wu, F. 1999. Portuguese-Chinese Machine Translation in Macao, *Machine Translation Summit VII*, Singapore, pp. 236–243.

Microsoft. 1998. *Programmer's Guide to Microsoft Window*, Microsoft Corporation.

Pietrek, M. 1996. *Windows 95 System Programming Secrets*, IDG Books World Wide, Inc.

Richter, J. 1996. *Advanced Windows NT*, Microsoft Press.

Wong, F., Dong, M.C. & Li, Y.P. 2002. Portuguese to Chinese Machine Translation System, in *Proceedings of Symposium on Technological Innovation in Macau*, Macau, pp. 134–140.

Computational Methods in Engineering and Science, Iu et al. (eds)
© 2003 Swets & Zeitlinger, Lisse, ISBN 90 5809 567 3

Estimation of area population referring to other areas

M. Eto
Hachinohe University, Aomori-ken, Japan

ABSTRACT: To estimate the future population within an area, there are many ways from the application of a single regression to the consideration of the cohort-survival and the population flow to/from the area being analyzed. But it is difficult to estimate those numbers theoretically. Then we tried to use the multiple regression with other close areas, which have similar changing character or opposite character. When we choose a good combination of areas, multiple regression method is better fitting than single regression method. Though this method is useful for the areas, which have rather stable changing character in a long term, in other areas some improvement are needed. To consider the effects of more areas, the neural network is applied for the population forecast. The best result was obtained by the neural network method among those three methods.

1 INTRODUCTION

There are a number of models to estimate the future population within an area. They vary from simple one that is the application of a single regression line to complex one that analyzes the cohort-survival. In addition, the number of peoples who will move to/from the area being analyzed, must be taken into account.

Redferm (1999) discussed on emigrants and immigrants in a country and Heimsath (1991) discussed on the real estate absorption analysis.

But it is difficult to estimate that number theoretically. Then we tried to use the multiple regression referring to other close areas, which have similar changing character or opposite character. In the latter case, the coefficient of the multiple regression may be negative and it suggests there are population flows between those areas.

From this idea, the neural network is applied to estimate the population of each area using the population obtained from the census as the desired outputs.

The results show that the multiple regression method is better than the single regression method and the neural network method is the best among these methods.

2 FORECASTING BY SINGLE REGRESSION

A case study of population forecasting in a city was carried out with the simple regression line (1)

$P_{i(n)} = a_i + b_i n$

$P_{i(n)}$: Estimated population of nth year in the area i. $\qquad$ (1)

Constants a_i, b_i are obtained from censuses of last 1year ~7years in the area i applying the method of least squares.

The population of each area, changing characters and the map of Hachinohe City which is analyzed here are presented in Table 1 and Figure 1.

The comparison between $P_{i(n)}$ the calculated populations of each region and the populations obtained from the census of the target year shows that in some areas this method works well, but in some areas it does not work well.

In those areas, the flow of population must be taken into account.

Table 1. Populations in 17 areas in Hachinohe City.

Area	1992	1993	1994	1995	1996	1997	1998	1999	2000	2001	2002
A	12423	12304	12190	12087	11951	11915	11837	11783	11758	11644	11619
B	12487	12458	12594	12620	12499	12391	12268	12070	11966	11785	11601
C	16605	16813	16797	17047	17026	16964	16792	16993	16911	17007	17045
D	14750	14587	14547	14309	14364	14330	14175	13948	13713	13593	13419
E	18070	17861	17858	17792	17668	17526	17345	17220	17073	16839	16866
F	16712	16856	17070	17118	17268	17333	17364	17308	17289	17141	17198
G	30244	30265	30493	30514	30637	30639	30662	30789	30698	30740	30764
H	9927	9746	9577	9599	9550	9578	9562	9449	9464	9400	9310
I	21868	22116	22274	22617	22906	23337	23580	24020	24397	24787	25384
J	7200	7058	6926	6756	6645	6447	6285	6198	6118	6127	6070
K	10633	10638	10648	10621	10706	10816	10892	10781	10697	10615	10443
L	14123	14047	14081	13976	13951	14260	14137	14025	13911	13802	13662
M	4893	4875	4812	4765	4820	4895	4901	4858	4859	4882	4882
N	2175	2158	2112	2129	2135	2135	2131	2124	2098	2064	2072
O	20213	20237	20113	20202	20514	20750	20889	21323	21345	21595	21851
P	27834	28183	28315	28632	28829	28886	28979	29056	28940	28992	29195
Q	3976	4008	3946	3928	3897	3874	3892	3847	3836	3757	3715

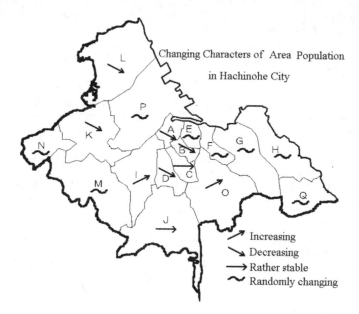

Figure 1. Map of Hachinohe City and rough illustration of population changing character.

3 FORECASTING BY MULTIPLE REGRESSION

Since it is very difficult to investigate the population flow from some area to other areas, multiple regression (2) was attempted between two regions which have similar changing characters with the area i.

$$Q_{i(n)} = \alpha_i + \beta_i U_{i(n-1)} + \gamma_i V_{i(n-1)}$$

$Q_{i(n)}$: Estimated population at nth year in area i. (2)

U_i, V_i: Populations in the area u and v which have similar (or opposite) changing characters with the area i.

Constants α_i, β_i, γ_i are obtained by the least squares method using the populations from the censuses of 1 year~7 years before.

The result shows that this method is better than single regression method but in some area it is not satisfactory. To improve this method, more areas must be referred to calculate the population in the area i.

Though it was considered to take into account more areas for calculation of area i, in some areas, it was difficult to select two or more areas that have similar changing character with the area i. So we tried to use the neural network method.

4 OUTLINE OF THE NEURAL NETWORK

The neural network is a simulation of leaning process of neurons in a brain. Figure 2 and Figure 3 are the schematic diagrams of neural network.

In Figure 2, x_1~x_2 are a set of input data, w_1~w_n are the weights of connection, θ is the threshold of the neuron and y is the output. It is calculated from (3)~(5).

$$s = w_1 x_1 + w_2 x_2 + w_3 x_3 + ----- + w_n x_n \qquad (3)$$

$$y = \text{sigmoid}(s-\theta) \qquad (4)$$

$$\text{sigmoid}(z) = 1/(1 + e^{-\alpha z}) \qquad (5)$$

Sigmoid function is introduced to shape the output into 0~1. The number α controls the effect of the function and usually it is taken as 0.5~1.

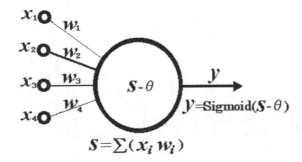

Figure 2.   Inputs and output of a neuron.

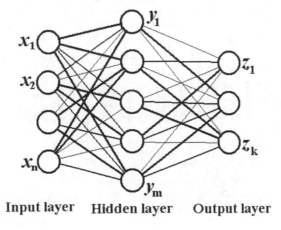

Input layer   Hidden layer   Output layer

Figure 3.   Three layered neural network.

535

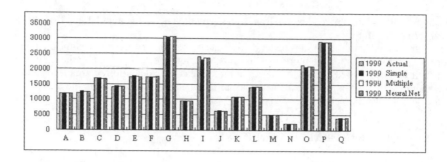

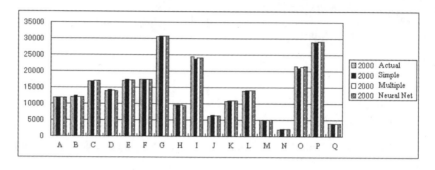

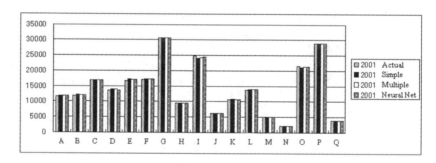

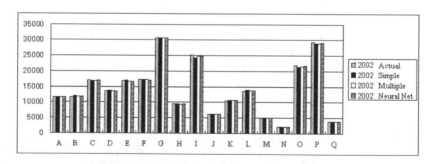

Figure 4.   Forecast of populations in 1999~2002.

In a Three layered structure (Fig. 2), the outputs of the 1st layer (Input layer) become the inputs of 2nd layer (Hidden layer), and the outputs of the hidden layer become the inputs of 3rd layer (Output layer). The similar process in Figure 1 is performed in each stage. Then the set of outputs in the 3rd layer is the output of this system.

If there are any errors when the outputs are compared with the desired outputs, the thresholds and the weights of connection are adjusted. In the first, the thresholds of hidden layer and the weights between the hidden and output layer are adjusted using the least squares method, then the thresholds of the input layer and the weights between the input and the hidden layer are adjusted. This method is called the back propagation method (Rumelhart et al. 1986). In the process, the sigmoid function provides simple formulae to calculate the correction of thresholds and weights.

5   TRAINING OF NEURAL NETWORK

Before performing the forecast of population, it is necessary to train the neural network with several sets of inputs and desired outputs, repeating enough times to get good results.

It is recommended to convert the inputs and the desired outputs into around 1/2 to use the sensitive part of the sigmoid function.

Populations by the census of 1989~2001 were used as the inputs and the desired outputs. To convert all data into around 1/2, each datum was divided by the corresponding datum of 1992 and subtracted 1/2. The results of outputs are reconverted with the reverse process.

For example the set of populations of 1999 is treated as the desired outputs for the set of populations of 1989 as the inputs. It means the forecasting populations of 1990 are calculated using the populations of 1989, and they are compared with the actual populations of 1990. If there are any errors, the weights of the neural network are adjusted. This process is the training of the network.

That is, the training is performed with the data of 1989 as the input, referring to the data of 1990 as the desired output. The same combinations of 1990 and 1991, 1991 and 1992, ..., 1999 and 2000 are provided. The process is repeated 10000 times to accomplish the training.

6   FORECASTING BY NEURAL NETWORK

The population forecasts for 1999~2002 are performed using the data of the last year, after the training using the data of previous 2 years~7 years before the corresponding year. The data of 1 year before are not actual data, but the estimated data by the simple regression method are used to avoid the noise.

7   RESULTS AND CONCLUSION

Populations from the censuses of 1999~2002 were compared with $P_{i(1999)}$~$P_{i(2002)}$, the population estimated from the single regression and $Q_{i(1999)}$~$Q_{i(2002)}$, the one estimated from the multiple regression. The results are presented in Figure 4 together with other results.

When we choose a good combination of u and v, usually $Q_{i(n)}$ estimated from the multiple regressions has better fitting than $P_{i(n)}$ estimated from the single regression comparing with the actual populations.

For example, the changing character of area A is similar to area B (regression coefficient = 0.985) and opposite to area I (regression coefficient = −0.959). In this case the variables mean i = a, u = B, v = I and the results $Q_{a(1999)}$~$Q_{a(2002)}$ are better than $P_{a(1999)}$~$P_{a(2002)}$ comparing with the actual populations.

The results of neural network are also presented in the Figure 4. To compare more precisely, the results of 2002 is partly magnified into Figure 5. Table 2 is the statistical table on the error percentages of these results in 2002.

Table 2.   Error percentage of forecasting.

	Single regression	Multiple regression	Neural network
Average	0.7387	0.2097	0.1491
σ	1.9059	1.2003	0.9891
σ^2	3.6323	1.4407	0.9784
Minimum	−4.0604	−2.3519	−2.1935
Maximum	3.1342	2.0896	1.6288
Range	7.1946	4.4415	3.8224
Number of data	17	17	17

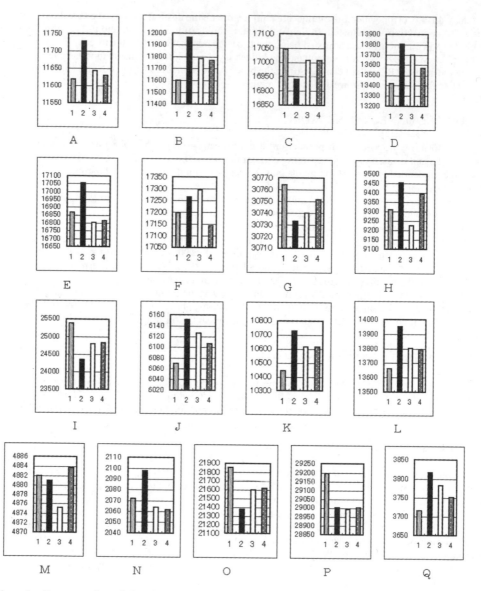

Figure 5.   Forecasts of populations in 2003. 1: Actual; 2: Single; 3: Multiple; 4: Neural Network.

It is generally said that the results of multiple regressions are better than the one of the single regression, and the one of neural network is the best among these results, at least about Hachinohe City.

REFERENCES

Heimsath, C. H. 1991. Small-Area Population Estimation in Absorption Analysis. *The Journal of Real Estate Research.* Vol.6, No.3: 315–326.
Redferm, P. 1999. A New Method of Estimating Undercount in the Census of Population, Using a Bayesian Approach. *The 52nd ISI Session: Invited Paper Meeting 22.*
Rumelhart, D. E., Hinton, G. E. & Williams, R. J. 1986. Learning Representations by Back Propagating Errors. *Nature,* Vol.323, No.9: 533–536.

Computational Methods in Engineering and Science, Iu et al. (eds)
© *2003 Swets & Zeitlinger, Lisse, ISBN 90 5809 567 3*

An application of expert system in real-estate industry

E. Wibisono
Department of Industrial Engineering, University of Surabaya, Surabaya, Indonesia

ABSTRACT: This paper presents an application of expert system in real-estate industry. The system is built using an active-server-page programming for the purpose of publication in the Internet. The key characteristic of the system is its ability to mimic the skills of marketing personnel in a real-estate company in guiding customers to find a house according to the customers' specifications. The system is also able to suggest the closest match of the desired house should the initial search fail. Finally, the system is also equipped with a mechanism to assess customers' financial capability in order to determine the amount of monthly credit. Using the power of graphics interface in modern browsers, customers are also able to view the front look of the house, its map, and its relative location within a real estate.

1 INTRODUCTION

Since its early development, expert systems have been developed mostly with orientation toward engineering/technical applications such as *Dendral* for chemical (Lindsay et al. 1980), *MYCIN* for medical (Buchanan & Shortliffe 1984, Shortliffe & Buchanan 1975, Shortliffe 1976), *Prospector* for geology (Duda et al. 1979, Hart et al. 1978), *ISIS* (Fox & Smith 1984) and *OPGEN* (Freedman & Frail 1987) for manufacturing, and R1 (or also known as XCON) for computer applications (McDermott 1982). The major drawback of expert system application in service industries is due to the fact that even with the current generation of artificial intelligence, it is still difficult to replace the critical function in service industries, which is the service itself. Service requires human touch in terms of behaviors, manners and other considerations and thinking process beyond logical context in decision-making. These elements will undoubtedly be an irreplaceable factor of human workforce. However, the application of expert systems in other important functions of service industries is still open, for example in information system.

Following the classification of Greenwell (1988), expert systems can be classified into several classes depending on the complexity level of the problems. Problems requiring the system to guide and direct the user in the process to search the solution need a system capable to support the decision-making process (decision-support systems). At the lower level, there exists a class of expert systems that may simply function as information provider with a characteristic of capability to access large databases (knowledge-based systems). It is in this area that this paper is based and developed.

This paper presents a simple model of an expert system application in a real-estate industry. This particular industry is chosen as the representation of service industries because it is commonly known that the industry has a relatively high turnover rate of employees, especially in marketing department. As such, the preservation of corporate knowledge is very essential to the sustainability of such industries. This knowledge and how it functions can be systematically documented in and later applied by a decision-making tool such as expert systems. Review of the literature also suggests that development of expert systems in this area is very limited. Most studies were carried out more at the macro level involving parties such as government that is expected to be able to make use of the systems in policy making for public benefits. These include studies on Geographical Information Systems (Barnett & Okoruwa 1993, Budic 1994, Dueker & DeLacy 1990, Rodriguez et al. 1995), Spatial Decision Support Systems (Peterson 1998), and forecasting and mass appraisal for real-estate valuation (Nawawi et al. 1997, Pacharavanich & Rossini 2001, Rossini 2000). At the business or micro level, application of expert systems gains very little

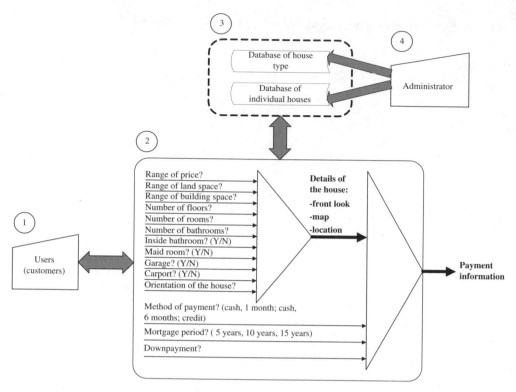

Figure 1. System diagram consisting of four main elements and their interrelationships: (1) users, (2) search interface, (3) databases, and (4) administrator.

exposure. This paper also extends the study of Wibisono (1998) where a similar application was also developed but with less friendly user interface. The contrast in that study from this paper is that in the previous study, the author used an expert system shell for the inference engine, whereas this paper uses an active-server-page programming for the engine. However, although more works are required in building the engine, the system takes advantage from the sophisticated interfaces that will be provided by the web browsers. The web browsers will be used as the medium to publish the system since it is expected that customers of the real-estate companies will use the system.

As in the nature of other similar expert-system designs, the database and inference engine in this model are separated in order to make easy further additions of information in the database, without needing to disturb the inference engine.

2 SYSTEM DESIGN

There are four major elements in the system: (1) users (customers) who will make inquiries, (2) an interface which will guide the users in the house search, (3) a database containing information of available houses, images of the front look of the houses, map of the houses, and map of the real-estate company, and (4) a system administrator who has authority to maintain and update the database. Details are shown in Figure 1.

The system runs in two stages. As shown in Figure 1, the first stage is depicted as the smaller triangle where the system will ask the users the characteristics of the house they are looking for. The characteristics are grouped into three types of input: range, entry, and yes/no questions. Range inputs include price, land space, and building space. Entry inputs are number of floors, number of rooms, and number of bathrooms. Inputs requesting yes/no answers cover the availability of a bathroom inside the main bedroom, maid room, garage, and carport. Further inputs at this stage are concerning the orientation of

Table 1. Tolerance values for enlarging search domain.

Level of priority	Input type	Variable	Tolerance
1	Range	Price	−25% to 25%
1	Range	Land area	−20% to 20%
1	Range	Building area	−20% to 20%
2	Entry	# of floors	−1 to 1
2	Entry	# of rooms	−1 to 1
2	Entry	# of bathrooms	−1 to 1
2	Yes/No	Inside bathroom	
2	Yes/No	Maid room	
2	Yes/No	Garage	
2	Yes/No	Carport	
2	Yes/No	Inside bathroom	
2	Yes/No	Main street	
2	Yes/No	Corner	
2	Yes/No	Dead end	

the house, i.e. facing north/south/east/west, and its relative position in the real estate (whether it is in the main street, corner, etc.). Bypassing the input(s) will imply that the particular variable(s) is of no interest to the users and therefore the system will disregard it in the search process.

The system will then search in the database a house that matches the criteria submitted by the users. Since to match all criteria has a small likelihood, the system is enhanced with a capability to enlarge the search domain using tolerance values in Table 1.

The level of priority in the leftmost column in Table 1 refers only to the sequence of the enlarged search within the system, and not necessarily indicates the degree of importance between levels of priority. The first level of priority will have to be tested first and if conditions are not met, the test will not continue to the second priority. Note also that in the Table, yes/no questions do not have tolerance values.

For example, if a user is looking for a house with price ranging from IDR 100 millions to IDR 150 millions, and there is no such a house in the database but there is one priced at IDR 175 millions (still within the range of +25% from the upper limit of the query of IDR 150 millions), then the system will respond that the search resulted in a close match of the desired house. The system can then go to the next stage.

If, after the search domain is enlarged with the tolerance values in Table 1, more than one variable from the range variables (variables having the first priority) do not meet the user's preferences, then the system will give up and respond that there is no match for the house being searched in the system database. If the first priority is passed, i.e. only one variable from the range variables do not meet the user's preferences, the system will check for the second priority variables. If more than two variables from this group do not meet the user's preferences, then the system will also and respond to the user that the search fails to find a house matching the user's preferences. At this point, however, provided that the users are targeted market segment from the company, the search may have covered 70–80% of success result in matching customers' criteria with the "products" that the company can offer. The intelligence of the system in deciding the expansion of the search domain, depending on the result of the first inquiry, is a critical attribute of the system in the interactive process with the customers. In a sense, the system can be considered capable to impersonate the skills of a real-estate marketing agent in servicing customers by offering a variety of choices.

In the second stage, given the selected house, questions and answers will be conducted to assess the financial capability of the users. The users will be asked the desired payment term (cash in 1 month, cash in 6 months, or credit). If the users decide to choose credit, they will then be asked the minimum amount of the initial payment (which should be higher than a certain percentage from the house price; this percentage is a parameter that can only be changed by the system administrator) and duration of the credit. This will determine the amount of monthly credit that the users will have to afford. Below is the equation for calculating the amount of monthly credit, given the amount of initial payment and duration of the credit.

Monthly credit = (House price − Initial payment) * Interest rate factors $\hspace{3em}$ (1)

The interest rate factors are determined by the company depending on the market interest rate. These factors have different value for different credit duration.

Table 2. Database of house type.

Type	Andora	Felicia	Kana	Leila	Monaco	Papirus
Land space (m^2)	105	200	135	152	202	525
Building space (m^2)	64	107	81	81	110	284
# of floors	1	1	1	1	1	2
# of rooms	2	3	2	2	3	3
# of bathrooms	1	1	1	1	2	3
Inside bathrooms	No	No	No	No	Yes	Yes
Maid room	Yes	Yes	Yes	Yes	Yes	Yes
Garage	No	No	No	No	No	Yes
Carport	Yes	Yes	Yes	Yes	Yes	Yes
Cash 1 month (IDR)	102,520,000	188,660,000	152,780,000	173,160,000	201,930,000	301,530,000
Cash 6 months (IDR)	105,440,000	194,460,000	155,700,000	176,190,000	205,810,000	307,880,000
Credit (IDR)	108,640,000	200,810,000	158,900,000	179,490,000	209,110,000	313,280,000

Again, this function of the system is an imitation of the skill that is possessed by a real-estate marketing agent in doing financial calculation. The inquiry process or the interactive session concludes after this point.

3 DATABASES

The system is connected with two databases consisting of the existing and available houses and also its other parameters. The first database consists of information on available house types. This covers the specifications of the houses such as the price, land and building spaces, number of floors, rooms, and bathrooms, and the availability of inside bathroom, maid room, garage, and carport. This database is linked with images showing the front look and map of each house type.

The second database consists of individual information of each particular house that the company has. This database has a one-to-many relationship with the first database and connected by the data field of house type. However, apart from having a field of house type, this database also covers more detailed information on each particular house, such as its orientation (facing north/south/east/west), and its location in the whole real estate (which block, what number, and also whether it is on the main street, corner, or dead end). This second database is also linked with images showing the location of an individual house in the real estate.

These two databases can only be maintained and updated by the system administrator. Maintenance of databases is an important aspect in ensuring the validity of the system. The system administrator can add information on a new house type and/or a new individual house, and also delete a house type and/or an individual house if the associated houses have been sold thus making them no longer relevant to be kept in the database.

Table 2 provides information of the fields contained in the database of house type and Table 3 provides information of the fields in the database of individual houses.

4 SAMPLE OF OUTPUT

This section depicts several screenshots of the system that will be published in the Internet. The publication in the Internet is necessary particularly for real-estate companies that have market segment coming from various geographical regions. As such, customers can browse the basic information of the real estate first before deciding to visit the company. More importantly, the expert system provided in the website serves as a powerful tool that enables the customers to have simulated interaction with the marketing agents of the company.

From the main page of the website, the expert system is linked through *Find Needed Home* menu. Selecting this option will bring the users to enter the system which its initial look is shown in Figure 2. Users can then input their preferences to begin the search. An example of a search result with closest matches is depicted in Figure 3, showing the front look of the houses matching the users' preferences.

Table 3. Database of individual houses.

Type	Block	Number	Facing	Main street	Corner	Dead end
Andora	H5	1	North	Yes	Yes	No
Andora	H5	2	North	No	Yes	No
...						
Felicia	F1	8	West	Yes	No	No
Felicia	F4	8	North	Yes	No	Yes
...						
Kana	D1	1	South	Yes	Yes	No
...						
Leila	C1	15	South	No	No	No
...						
Monaco	G1	7	West	Yes	No	No
...						
Papirus	A2	12	South	No	No	No
...						

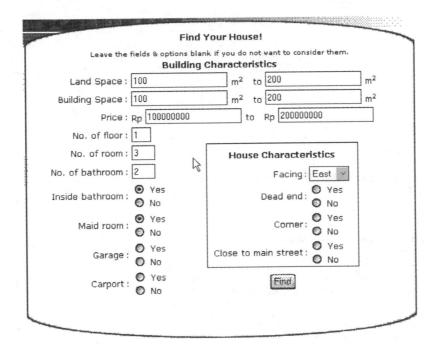

Figure 2. The real-estate expert system's initial window.

From here the users can choose to do credit calculation, look at the map of the house, or look at the map of the real estate showing the location of the house.

The credit assessment window is shown in Figure 4. Figure 5 shows the map of the real-estate company. If a particular section of the map is clicked, that particular section will be enlarged thus enable the users to get a detailed and closer look on that block.

5 CONCLUSION

The purpose of designing an expert system application for real-estate industry is to reduce dependency on human intellectual capital of the company. This is very important because real-estate industry is

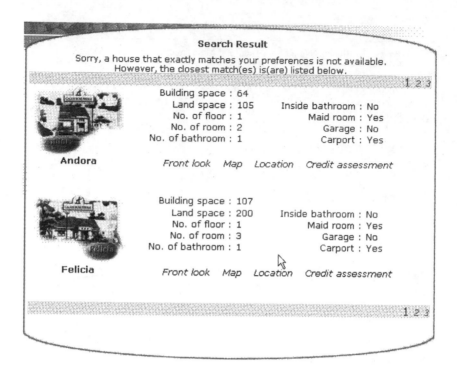

Figure 3.   An example of search result with closest matches.

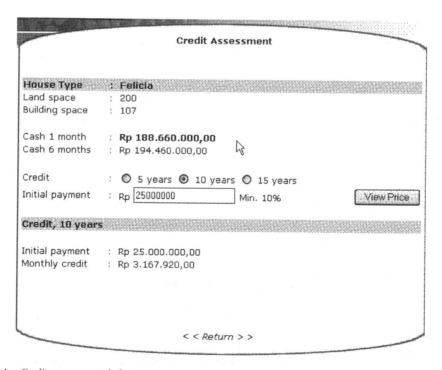

Figure 4.   Credit assessment window.

544

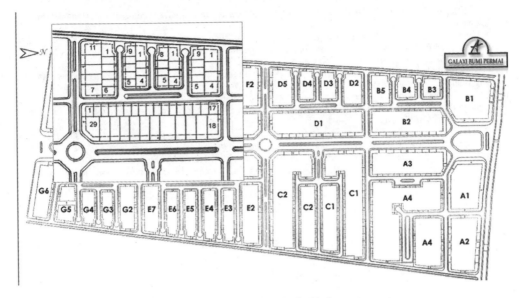

Figure 5.   Map of the real estate with enlarged view of a particular block.

commonly known as industry with high employee turnover, particularly in the marketing department. By creating a system that can mimic the skills of the marketing staff, the system does not only function as a substitute for the human workforce, but more importantly it also becomes a management tool that be used to preserve corporate knowledge.

The selection of a web-based platform for the publication of the system also provides distinct advantages for the companies. The companies can reach potential customers in various geographical regions. Inquiry process can be performed in a long-distance mode thus at the same time enlarging the companies' market.

By using an active-server-page programming in a web-based platform, the system design requires extra works in developing the search engine. However, this sacrifice is offset by the good support of graphics interface in today's web browsers, thus enhancing the user friendliness of the system. The system is also built with the capability to expand its search domain when initial inquiry fails. This capability is another personification of human work in a system that also adds a point to the system's user friendliness.

REFERENCES

Barnett, A.P. & Okoruwa, A.A. 1993. Application of Geographic Information Systems in site selection and location analysis. *Appraisal Journal* 61(2): 245–53.

Buchanan, B.G. & Shortliffe, E.H. 1984. *Rule-based expert systems: the MYCIN experiments of the Stanford Heuristic Programming Project*. Reading: Addison-Wesley.

Budic, Z.D. 1994. Effectiveness of Geographic Information Systems in local planning. *APA Journal Spring*: 244–63.

Duda, R.O. et al. 1979. A computer-based consultant for mineral exploration. *Technical report in SRI International*.

Dueker, K.J. & DeLacy, P.B. 1990. Geographic Information Systems in the land development process. *Journal of the American Planning Association* 56(4): 483–91.

Fox, M.S. & Smith, S.F. 1984. ISIS: a knowledge-based system for factory scheduling. *Expert Systems Journal* 1(1): 25–49.

Freedman, R.S. & Frail, R.P. 1987. OPGEN: the evolution of an expert system for process planning. *AI Magazine* 7(5): 58–70.

Gonzalez, A.J. & Dankel, D.D. 1993. *The engineering of knowledge-based systems: theory and practice*. New Jersey: Prentice Hall.

Greenwell, M. 1988. *Knowledge engineering for expert systems*. Chicester: Ellis Horwood.

Hart, P.E. et al. 1978. A computer-based consultation system for mineral exploration. *Technical report in SRI International*.

545

Lindsay, R.K. et al. 1980. *Applications of Artificial Intelligence for organic chemistry: the Dendral Project.* New York: McGraw-Hill.

McDermott, J. 1982. R1: A rule-based configurer of computer systems. *Artificial Intelligence* 19(1): 39–88.

Nawawi, A.H. et al. 1997. Expert system development for the mass appraisal of commercial property in Malaysia. *Journal of the Society of Surveying Technicians* 18(8): 66–72.

Pacharavanich, P. & Rossini, P. 2001. Examining the potential for the development of computerised mass appraisal in Thailand. In *7th Annual Pacific-Rim Real Estate Society Conference, Adelaide, 21–24 January 2001.*

Rich, E. & Knight, K. 1991. *Artificial Intelligence.* New York: McGraw-Hill.

Rodriguez, M. et al. 1995. Using Geographic Information Systems to improve real estate analysis. *Journal of Real Estate Research* 10(2): 163–72.

Rossini, P. 2000. Using expert systems and artificial intelligence for real estate forecasting. In *6th Annual Pacific-Rim Real Estate Society Conference, Sydney, 24–27 January 2000.*

Shortliffe, E.H. 1976. *Computer-based medical consultation: MYCIN.* New York: Elsevier.

Shortliffe, E.H. & Buchanan, B.G. 1975. A model of inexact reasoning in medicine. *Mathematical Biosciences* 23:351–79.

Wibisono, E. 1998. Sistem pakar pada industri real estate: sebuah aplikasi. Kristal 6(1): 66–80.

Computational Methods in Engineering and Science, Iu et al. (eds)
© *2003 Swets & Zeitlinger, Lisse, ISBN 90 5809 567 3*

Neural networks in automotive engine tuning

C.M. Vong & Y.P. Li
Department of Computer and Information Science, Faculty of Science and Technology,
University of Macau, Macao

P.K. Wong
Department of Electromechanical Engineering, Faculty of Science and Technology,
University of Macau, Macao

ABSTRACT: Automotive engine performance is significantly affected with effective tuning. Current engine tuning relies on the experience of the mechanics, and it is usually done by trial-and-error method. It may take days or even weeks, and may even fail to tune the engine optimally because a formal performance model of the engine has not been determined yet. With the aid of data mining methods, the engine tuning can be done automatically through the use of a computer. In addition, a nearly optimal engine setting can be obtained which dramatically increases the engine performance. An experiment has also been done to verify the usefulness of data mining methods in this application area.

1 INTRODUCTION

Automotive engine performance is significantly affected with effective tuning. In general, the original car manufacturers fix the control parameters on most car engines with electronic fuel-injection systems. It is impossible to adjust the setting for different engine conditions and driver practices. With the recent development of low-cost electronic components, programmable electronic control unit (ECU) is widely adopted by racing and super production car engines. The unit substitutes for the OEM ECU and allows the user to adjust the engine parameters via a PC.

Current technique of engine parameters tuning only relies on the experience of the mechanics, and it is usually done by trial-and-error method. It may take days or even weeks, and may even fail to tune the engine optimally because a formal performance model of the engine has not been determined yet. The difficulty of building such a performance model is due to the set of parameters, which incurred highly nonlinear correlations.

With the emerging data mining techniques, the engine parameters can be tuned to achieve a suboptimal performance even its model is not known.

2 AUTOMOTIVE ENGINE PARAMETERS

Currently, one of major goals in modern engine tuning is the careful selection of air/fuel ratio "∅" for different driving conditions (Crouse et al. 1995). Actually the performance of the electronic fuel-injection engine mainly depends on air/fuel ratio control.

For achieving maximum engine performance, various air/fuel ratios should be produced for different engine speeds and throttle positions. Some references (Crouse et al. 1995, Hartman et al. 1993) recommended the following air/fuel ratios for running four-stroke naturally aspirated engines optimally.

Equations (1) & (2) show the relationship of adjustable engine parameters in the modern programmable ECU and air/fuel ratio.

$$\phi = 0.147\lambda \tag{1}$$

Table 1. Parameters of air-fuel ratio and its corresponding condition.

∅	Running condition
6	Rich burn limit (cold start and fast idle speed)
11.5	Approximate rich best torque at wide-open throttle (for accceleration)
12.2	Safe best power at wide-open throttle
13.3	Approximate lean best torque
14.7	(Stoichiometric ratio) maximum conversion efficiency for catalytic converters
15.5	Lean cruise
16.5	Best fuel economy
22	Electronic fuel injection lean burn limit

$$\lambda = f(p, s, a, \frac{\partial p}{\partial t}, b, r, h, i, j, c) \quad (2)$$

where s, engine speed; p, throttle position/manifold air pressure; a, acceleration enrichment; b, temperature compensation; r, rate of exhaust gas recirculation; h, altitude compensation; i, ignition spark advance; j, fuel injection time; c, fast idle time; t, time; λ, % of stoichiometric.

Adjusting the parameters in Equation (2) can produce different air-fuel ratios. However, the multivariable function in Equation (2) is an unrecognized pattern – an unknown performance model. Currently, all tuning processes on Equation (2) are manually performed by trail-and-error method. The results totally depend on the mechanic's experience. In addition, different automotive engines have different performance models. That arises another obstacle for engine tuning.

3 DATA MINING

Data mining is an emerging technique that is also considered to be one of the ten most contributive scientific technologies in the future 20 years. Nowadays, more and more companies are employing data mining techniques (Han et al. 2000, Mitchell 1997) to digest the huge amount of data that they own, in order to find out valuable hidden knowledge. This knowledge is usually difficult or even impossible to be noticed.

Data mining is a multidisciplinary field, drawing work from areas including database technology, machine learning, neural networks, statistics, and data visualization. Because of this multidiscipline, data mining is a constitution of many methods, addressing different kinds of knowledge discovery problems. Moreover, upon different kinds of data, different data mining methods are employed.

In this paper, neural network (NN) – a method from the area of machine learning is employed to find out the sub-optimal settings of selected parameters defined in equation (2).

3.1 Neural networks

Neural network is a set of connected input/output units where each connection has a weight associated with it. Each of these units is called a neuron. Neural network is a kind of supervised learning methods. The input of a neural network is usually a large set of data. This set of data is called a training set. By passing this training set to the neural network, the network learns by adjusting the associated weights of each neuron so as to be able to predict the correct class of the input samples. As neural networks involve long training times, they are only suitable for batch-processing applications.

3.2 Multilayer feed-forward neural network

A multilayer feed-forward neural network (Johnson et al. 1992) contains three layers: input layer, hidden layer, and output layer. The inputs are fed simultaneously into a layer of units making up the input layer. The weighted outputs of these units are, in turn, fed simultaneously to a second layer of "neuron-like" units, known as a hidden layer. The weighted outputs of the hidden layer are input to units making up the output layer, which emits the network's prediction for given samples. The units in the output layer are referred to as output units. Figure 1 shows an example of a feed-forward neural network.

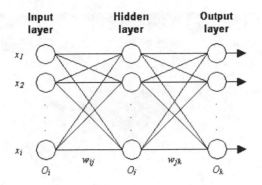

Input
layer
Hidden
layer
Output
layer

Figure 1.   A feed-forward neural network.

3.3   *Mathematical model and algorithm*

The algorithm used in neural networks is backpropagation (Russell et al. 1995). Backpropagation learns by iteratively processing a set of training samples, comparing the network's prediction for each sample with the actual known class label. For each training sample, the weights are modified so as to minimize the mean squared error between the network's prediction and the actual class. These modifications are made in the "backwards" direction, i.e., from the output layer back to the hidden layer. In general, the weights will eventually converge, and the learning process stops. To compute the net input to a certain unit in hidden layer and output layer, the formula is determined as:

$$I_j = \sum_j w_{ij} o_i + \theta_j$$

where w_{ij} is the weight of connection from unit i to unit j; O_i is the output of unit i; and θ_j is the bias of the unit. For input layer, $O_j = I_j$.

And the algorithm is summarized as below:

Input: The training samples, samples; the learning rate, l; a multiplayer feed-forward network, network.

Output: A neural network trained to classify the samples.

ALGORITHM

(1)   Initialize all weights and biases in network;
(2)   **While** terminating condition is not satisfied{
(3)   **for** each training sample X in samples{
(4)   **for** each hidden or output layer unit j{
(5) $I_j = \Sigma_i w_{ij} O_i + \theta_j;$
(6) $O_j = \dfrac{1}{1+e^{-1}}\}$
(7)   **for** each unit j in the output layer
(8) $Err_j = O_j (1 - O_j)(T_j - O_j);$
(9)   **for** each unit j in the hidden layer
(10) $Err_j = O_j (1 - O_j) \Sigma_k Err_k w_{kj};$
(11)   **for** each weight W_{ij} in the network{
(12) $\Delta w_{ij} = l * Err_j O_i;$
(13) $w_{ij} = w_{ij} + \Delta w_{ij};\}$
(14)   **for** each bias θ_j in network{
(15) $\Delta\theta_j = l * Err_j;$
(16) $\theta_j = \theta_j + \Delta\theta_j;\}$
(17) }}

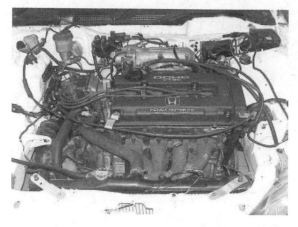

Figure 2.   HONDA B16A DOHC VTEC engine.

Figure 3.   Connection of notebook computer to MoTeC M4 ECU.

4   IMPLEMENTATION AND TESTING

As there are many existing data mining software in the market, it may even take longer time and cost for developing brand new software for the above NN algorithm. The software package chosen for the implementation is SPSS Clementine 4.0, which runs under Microsoft Windows XP. The engine used for testing is HONDA B16A DOHC VTEC (Fig. 2) and the chassis configuration is Honda Civic EG6 with 195/50-R15 tires. In order to set necessary data from the engine, a programmable ECU is used for substituting the OEM ECU. The programmable ECU chosen is MoTeC M4 Pro. A notebook computer is connected to the MoTeC ECU for data acquisition and tuning as shown in Figure 3. 50 sets of engine parameter tables have been taken as training data for the NN. Each table just takes the parameters related to air-fuel ratio. An instance of the table is shown in Figure 4.

Figure 4. An instance of engine set-up parameter.

Figure 5. A G-force meter for on-road measurement.

4.1 Discussion of results

After training by NN, a set of weights corresponding to the adjustment of air-fuel ratio parameters is returned. These new parameters are then input to the MoTeC ECU. The maximum road horsepower and wheel torque of the car before tuning is 128 hp and 1695 Nm respectively. After using the new setting extracted from the NN, the maximum road horsepower and wheel torque is increased to be 139 hp and 1789 Nm respectively. The horsepower shows an increase of 8.6%, while the wheel torque demonstrates an increase of 5.5%. The maximum horsepower and torque are measured on road using a state-of-the-art G-force meter as shown in Figure 5.

5 CONCLUSION

Automotive engine tuning is a very difficult task. For decades, nobody could tackle this problem effectively and only trial-and-error method can work. This method is also highly dependent on the mechanic's knowledge. However, this kind of knowledge is hard to be formulated as mathematical models because there are too many variables for human to handle. With the emerging data mining techniques, the problem of engine tuning can be augmented. Neural Network (NN) is one of the data mining techniques. NN is employed in this research, as it is an optimization tool that perfectly fits the current optimization problem. An experiment has been conducted to verify the improvement of engine performance after tuning with the parameters supplied by NN. Although no optimal performance can be achieved so far, a near optimal performance can be obtained at this stage.

REFERENCES

Han J. & Kamber M. 2000. *Data Mining: Concepts and techniques*. Morgan Kaufmann Publishers.

Mitchell T. 1997. *Machine Learning*. McGraw-hill companies, Inc.

Russell S. & Norvig P. 1995. *Artificial Intelligence: A modern approach*. Prentice Hall, Inc.

Johnson R. & Wichern D. 1992. Applied multivariate statistical analysis, 3rd edition. Prentice Hall, Inc.

Crouse W. H. & Anglin D. L. 1995. *Automotive mechanics*, 10th edition, McGraw-Hill.

Hartman J. 1993. *Fuel injection: Installation, performance tuning, modifications*, Motorbooks international powerpro series.

Computational Methods in Engineering and Science, Iu et al. (eds)
© 2003 Swets & Zeitlinger, Lisse, ISBN 90 5809 567 3

Case-based reasoning in airport terminal planning

C.M. Vong & Y.P. Li
Department of Computer & Information Science, Faculty of Science and Technology, University of Macau, Macao

P.K. Wong
Department of Electromechanical Engineering, Faculty of Science and Technology, University of Macau, Macao

ABSTRACT: Airport terminal planning is one of the crucial design stages of an airport because inappropriate design may lead to many undesirable situations. In order to avoid these situations, computer-based simulation is used. To do this process more efficiently, two reasoning paradigms, rule-based reasoning and case-based reasoning, were incorporated in the knowledge-based system supporting the simulation of the airport terminal planning.

1   INTRODUCTION

Case-based reasoning paradigm with assistance of rule-based reasoning, supporting modeling and consequent decision making processes, is studied in the paper. In some situations, the framework can also be partially assisted by rule-based support if appropriate. This combined technology has been adapted to airport terminal planning, predicting the peak hour passenger flow and the corresponding utilization of different counters, etc. Modeling airport terminal is an inevitable part of this effort and the supporting framework can contribute substantially to smooth the modeling. Case-based reasoning was adopted as the key paradigm to the prediction of the passenger flow.

2   CASE-BASED REASONING

Case-based reasoning has been adopted as a key paradigm of the framework supporting modeling in decision support system (Harmon 1993). Case-based reasoning is one of the most effective paradigms of knowledge-based systems (Allen et al. 1995). This promising concept addresses successfully main problems of the traditional knowledge-based systems. Despite great success of them in the last decades, these traditional knowledge-based systems face several serious problems. Watson and Marir (Watson 1994) outlined the main issues of the traditional paradigms as the following:

- Knowledge elicitation bottleneck (knowledge elicitation is a difficult and tedious process).
- Implementing a knowledge-based system, using the traditional paradigm is a lengthy process requiring cooperation of high-qualified experts.
- Traditional paradigms are mostly unable to manage effectively large volumes of knowledge.
- It is difficult to maintain a system based on a traditional paradigm during its life cycle.

The above problems can be partially solved by case-based reasoning. Differing from more traditional methods, case-based reasoning is one of the most effective paradigms of knowledge-based systems. Relying on case history is its principal feature. For a new problem, case-based reasoning strives for a similar old solution. This old solution is chosen based on a correspondence of the new problem to some old problem (which was successfully solved by this solution). From this perspective, case-based reasoning represents a strongly anthropomorphic approach. Solving a new problem, people usually center on some old solution that was successful for a similar problem in the past. So, previous significant cases are gathered and saved in a case library.

A case-based reasoning system can only be as good as its case library (Turban 1995). This implies two important issues. Only successful and sensibly selected old cases should be stored in the case library. Each candidate is chosen after an evaluation only. The case library serves as a knowledge base of the case-based reasoning system. The system can learn by acquiring knowledge from these old cases. Learning is basically achieved in two ways:

(i) accumulating new cases, and
(ii) through the assignment of indexes.

The first possibility basically means a quantitative improvement, while the second one can improve the case library rather qualitatively. When solving a new case, the most similar old case is retrieved from the case library. The suggested solution of the new case is generated in conformity with this retrieved old case. The search for a similar old case from the case library represents an important operation of the case-based reasoning paradigm. Retrieval relies basically on two methods: nearest neighbor, and induction. Complexity of the first method is linear, while the complexity of the search in a decision tree, developed by the latter method, is only logarithmic. This reasonable complexity enables the effective maintenance of large case libraries.

The case-based reasoning approach prefers rather verified ways, in accordance with the anthropomorphic and pragmatic approach. Why to start from scratch again if we can learn from the past? Once a suitable old solution, tied to the new context, can be retrieved by the case-based reasoning part of the framework, such a tested solution should be preferred.

Case-based reasoning relies on the idea that situations are often repeating during the life cycle of an applied system. This is usually a realistic assumption. Further, after a short period of time the most frequent situations should be identified (and documented in the case library). So, the case library can usually cover common situations after a short time. However, when relying on case-based reasoning exclusively, two problems can be encountered: (i) it is difficult to start from the very beginning with an empty case library, (ii) on the other hand, after some time the case library can become huge and contains much redundancy. Therefore it is better to combine the case-base reasoning with some other paradigm to compensate for these marginal insufficiencies. The rule-based reasoning component can cover these gaps with the help of implemented knowledge and heuristics. This means that the rule-based part is capable of suggesting its own solution. This suggestion will be requested especially under the following circumstances:

- No suitable old solution can be found for a current situation in the case library. For this event, the rule-based reasoning part can be activated. However, once generated by this part of the framework, such a solution is then evaluated and tested more carefully.
- Some situations may be almost the same but not identical. Such cases cause a high level of redundancy of the case library. To replace such a class of very similar cases by a set of rules can partially solve this problem.
- The rule-based reasoning part can also cooperate in solving some specific situations. So, in an effort to extend its capability, the rule-based paradigm is augmented (Babka 1992, Kolodner 1993). Along with knowledge in the form of rules, programs can also be integrated into the framework. As a built-in tool, a selective mechanism is implemented and encapsulated in the framework.

Figure 1 (Vong 1996) shows the structure of the framework integrating case-based reasoning and rule-based reasoning paradigms as a whole. By integrating these two paradigms, we intend to have a system whose reasoning process is closer to that of a human being with a large body of practical experience and fundamental knowledge about a certain problem domain that allows user to face and overcome new situations. The efficiency for finding out solution for a problem is then much increased comparing to that using traditional computer-based simulation. The large amount of computation done by simulation can be reduced by first consulting case-based reasoning system then rule-based reasoning system. The combined paradigm uses case-based reasoning as the central reasoning process. So the system will first try to find similar cases (solutions) in the knowledge base of the system. If similar past successful case can be found, the system will proceed with case-based reasoning process and finally return a solution. In the situation that no similar case was encountered in the past, rule-based reasoning can find a new acceptable solution by deriving rules from the knowledge base of the system. No matter which paradigm is used, the new generated result should be kept in the knowledge base as a past

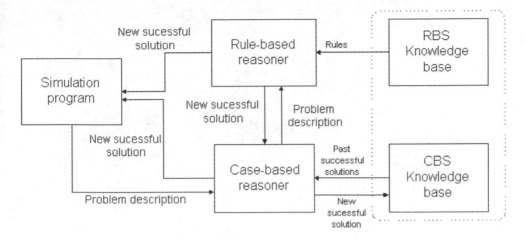

Figure 1. Case-based reasoner combining with Rule-based reasoner. RBS = Rule-Based System, CBS = Case-Based System.

successful case so that later the system may retrieve this case as a rough solution for another problem in the similar situation.

3 APPLICATION TO AIRPORT TERMINAL PLANNING

Airport terminal planning is an important topic in airport planning and management. Its primary objective is to achieve an acceptable balance between passenger convenience, operating efficiency, facility investment, etc. (Wells 1986, Ashford et al. 1995). There are several important factors affecting the air traveller. One of them is the speed of processing of check-in, security and baggage, etc. They are basically affected by the number of counters currently open. To determine the number of counters several factors must be taken into account, for example, number of airlines, the capacity of aircraft, interarrival time of passengers, size of the airport terminal, etc.

For airport terminal planning, this combined paradigm (Vong 1996, Galvão 1996) is applied to a system which helps the airport planner in his/her decision making process of how to allocate facilities on a day-to-day basis. During the daily operations of an airport terminal, many decisions have to be made on how to allocate facilities during a certain time in order to avoid passenger congestion. This job falls in the category of airport planning. Airport planners and managers are frequently confronted with two problems:

- How determine the number of various major functional elements of an airport terminal building that are needed to handle a given passenger throughput, and, conversely.
- How to determine the passenger throughput capability of existing airport terminal facilities.

If the problems were overestimated, then much human resources and facilities would be wasted. If the problems are underestimated, then passenger congestion would occur and the operating efficiency may be much lowered. So airport planners and managers must carefully consider the number of counters to open. In addition, the situation is changing over time. For instance, if there are a lot of passengers congesting at 9:00 p.m. but just a few after 11:00 p.m., then more counters are needed to open at 9:00 p.m. but fewer counters at 11:00 p.m.

The main terminal facilities profiling the processing of passenger flow in an airport terminal are as follows:

- Check-in Desks
- Departure Passport Control
- Security Check
- Arrival Passport Control

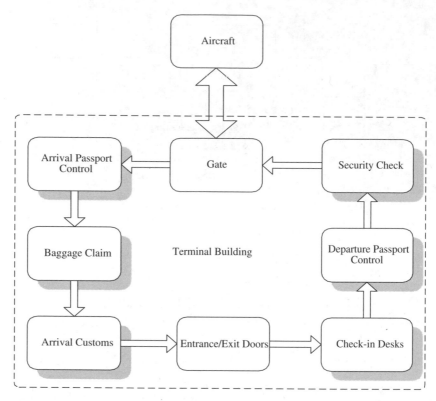

Figure 2. Passenger flow in an airport terminal.

- Baggage Claim Devices
- Arrival Customs

Then using the combined paradigm to solve the determination of counters to open for a certain time, a similar old case is first retrieved by Case-Based Reasoning from the case library. This retrieved case may not need exactly to match the current problem but at least the main features described in the case structure must be similar. For instance, the current date is Sunday 10:00 a.m., then the retrieved case should also be Sunday 10:00 a.m. because date of week and time are important points for consideration of counters to open. After that, the retrieved case is adapted to match the current situation based on the evaluation of this past case, where the evaluation is a part of the retrieved old case. If no old similar case can be found or the similarity of the retrieved case is low, then Rule-Based Reasoning can be used to derive a new solution for this problem. Finally input the result and the features of this problem and physical evaluation of this new generated solution into the case library so that the learning process of Case-Based Reasoning can refine this solution. Then a new experience is learnt by this decision-support system and can be applied to later situations.

4   CONCLUSION

A case-based reasoning system relies on its knowledge base (case library) to solve problems. Even if the knowledge base of such system is very comprehensive, containing a wide range of knowledge about a certain problem domain, there is a possibility that a certain problem is not similar to all the existing problems in the system's knowledge base.

A rule-based reasoning system contains generalized knowledge about a certain domain. The rule-based reasoner, using its knowledge base, will probably be able to derive a solution to this problem. Later on the problem and its solution can be put into the case-based reasoning system's knowledge base as a new case. Thus, a rule-based reasoning system provides a good support to a case-based reasoning system.

The integration of both paradigms proved to be of much interest. The combined technology can profit from advantages of both paradigms. This concept is also much closer to the way of human thinking and reasoning.

In this research, this combined paradigm was relatively successfully applied to airport terminal planning and several related tasks. In normal situation, modeling and simulation of airport terminal takes a long time (Wong et al. 1995). With this combined paradigm, the modeling and simulation phases are replaced by setting up a knowledge base (case library) and retrieving similar cases or deriving new solutions from the knowledge base the from that takes relatively short time compared to traditional simulation methods.

REFERENCES

Allen, Patterson, Mulvenna & Hughes. 1995. Case-Based Reasoning Research and Development. In *Proceedings of First International Conference, ICCBR-95*, Sesimbra, Portugal, October 1995. 1–10.

Ashford N. & Wright P. 1995. *Airport Engineering, Third Edition*. John Wiley & Sons, Inc.

Babka O. 1992. Knowledge-based System Supporting Design System. In *European Journal of Engineering Education*, Vol. 17: 181–187.

Galvão P. & Vong C.M. 1996. A step closer to modelling human reasoning: Case-based Reasoning. *Boletim de Engenharia de Macau*, No. 3: 54–56.

Harmon & Hall 1993. *Intelligent Software Systems Development*, John Wiley.

Kolodner J. 1993. *Case-based Reasoning*, Morgan Kaufman Publ., San Mateo, CA, U.S.A.

Turban 1995. *Decision Support and Expert Systems: Fourth Edition*, Prentice Hall International, Inc., London.

Vong C.M. & Galvão P. 1996. Case-based Reasoning. *Final project for fulfillment of B.Sc.*, FST, University of Macau.

Watson & Marir 1994. Case-based Reasoning: A Review. *The Knowledge Engineering Review*, Vol. 9(4).

Wells A. 1986. *Airport Planning and Management*. Tab Books.

Wong, Silva & Ng. 1995. Simulation of Airport Operations with Knowledge Based Support System. *Final project for fulfillment of B.Sc.*, FST, University of Macau.

Computational Methods in Engineering and Science, Iu et al. (eds)
© *2003 Swets & Zeitlinger, Lisse, ISBN 90 5809 567 3*

The model of the artificial immune response

X. Dai, C.K. Li & A.B. Rad
The Hong Kong Polytechnic University, Hong Kong

ABSTRACT: Above all, this paper gives the immune response of the immune system. After analysing the phenomena of bindings, stimulus and the replication of the B cells in the immune response, some new ideas about constructing the artificial immune response model (this model is actually an immune network) are presented. We discussed the model from the architecture, stimulus of the units, interaction of the units and the dynamic update of the immune network. Because of the good learning ability and memory, and also the characteristics of distributed and fault tolerant, the immune network has good potential for information processing of complex systems.

1   INTRODUCTION

The immune system of human being is a very complex natural defense system. It has the ability to analyze and learn the antigens, which invade the body, and kill the antigens by producing large antibodies. The main role of immune system is to recognize its cells or molecules in its body and, and distinguish these cells as self and non-self (Dasgupta & Attoh-Okine 1997).

Jerne (1973, 1974) proposed idiotypic network of immune system, and he described the concentration by differential equations. Based on the Jerne's work, a probability method describing the idiotypic network was given by Perelson (1989). Based on the principle of self and non-self discrimination, a negative selection algorithm was given by Forrest et al. (1994). Moreover, Hoffmann (1986) compared the immune system and the neural system, he gave some similarities on the system level between the immune system and neural system. Farmer (1986), Bersini and Varela et al. compared the immune system with learning classifier system. Gibert and Routen gave the immune memory. The various models of immune system, promote the application of artificial immune system (AIS). Therefore, more researchers pay attention to the models of immune system, which suit the engineering application.

The immune response is an important characteristic of immune system. This paper introduces the process of immune response first. From the bindings, stimulus, replication of B cells, we propose the idea of modeling of artificial immune response. This model (is an immune network essentially) has the good learning ability and memory, and has the characteristics have distributed and fault tolerant. So, this model has good potential to information processing of complex systems.

2   AN OVERVIEW OF IMMUNE SYSTEM

The immune system of human being protects its body from infection of bacteria and virus. There are two types of immunity, innate immunity and adaptive or acquired immunity. In contrast to the innate immune system (just like the skin), the adaptive immune system uses a specific immune response to antigens, and has the some memory after getting rid of a foreign antigen.

The immunity, which we discuss, means the adaptive immunity. According to the immunologists, there are lots of immune cells cycling around our bodies, and lymphocytes are the main immune cells, which participate the immune response. The lymphocytes have the features such as specialty, diversity, memory and adaptability, and the lymphocytes can divide into two types: T cells and B cells. B cells can produce antibody to kill the antigen, while the T cells have the function of promoting or suppressing the reaction of B cells, see Figure 1

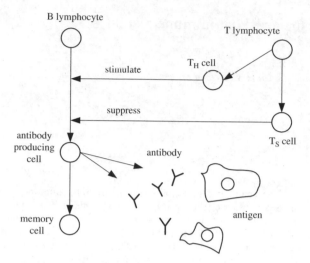

Figure 1.   Mechanism of producing antibody in immune system.

The immune system has the following characteristics (Hunt & Cooke 1996):

1. The immune system can produce more than millions of antibodies to avoid the infection of foreign antigens.
2. The immune system has the diversity.
3. The immune system is a distributed system which has no central controller.
4. The immune system can adapt to dynamic circumstance quickly.
5. The memory of immune system is content-addressable, so can recognize the similar antigen.

Now, we discuss the important phenomenon of immune system.

2.1   Process of the immune response

When the immune system encounters the antigen in the first time, only a small number of immune cells can recognize the epitope of antigens. The recognition can stimulate the replication of immune cells, and produce the antibody to match the antigen. This process is named clonal expansion. In the clonally expansion, there is some hypermutation, which means that some immune cells mutate in the replication process. And as a result of clonal expansion, large numbers of antibody-producing cells are produced aiming at the invading antigens, to kill the foreign antigen. At the same time, these cells are kept in the immune memory. When some similar antigens invade the body next time, more quick response and more antibodies are produced, this process is called secondary response. We should notice here, that not only the same, but also the similar antigen invades the body, will cause secondary response.

2.2   The stimulation of B cells

In the process of immune response, B cells reproduce themselves by their stimulation level. In the surface of the B cells, there are some antibodies, and when the antibodies bind the antigen, the B cells are stimulated. The level of stimulation is not only relative to the match between antigen and antibody, but also relative to the match between the antibody and the other interactional antibodies.

When the stimulation of B cells exceeds a special threshold, the clonal expansion occurs, which means the large replication of immune cells. And in the replication, hypermutation happens on some B cells for the sake of more adaptability of immune systems.

3   MECHANISM OF IMMUNE RESPONSE AND NETWORK MODEL

According to Rowe (1994), the function of immune system is just like the "second brain", because it has the ability of memory and response to the new pattern.

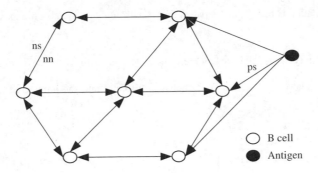

Figure 2. The concise architecture of the immune network.

3.1 Architecture of the immune network

Immune cells are not isolated, but communicate by the interval contact and dissolved molecules excreted by immune cells, which is similar to the communication of electrical signal between axons and dendrites in neural system (Chowdhury & Stauffer 1992, Dasgupta 1997).

So, the architecture of artificial immune system is a network essentially. And the immune network can also refer to the ideas of PDP (parallel distributed processing) network (Vertosick & Kelly 1989). First, the nodes (or units) of the immune network are not isolated, but connected by topological architecture. Second, there are some weights between the nodes. Finally, the output of one node is based on the interaction of the other nodes.

We think the nodes of immune network are the B cells, the input is the foreign antigen, and the output of the immune network is the stable state of the B cells. The architecture of the immune network illustrates as Figure 2.

In Figure 2, ns is defined as the stimulation of B cell, and nn is the suppression of B cell, and ps is the stimulus which antigens cause. The arrows mean the effects between one B cell and another B cell, or a B cell and an antigen.

We should mention here, every B cell could represent one, which used to learn, and every antigen represents one which used to train or test. And the immune network updates its unit according to the antigen, so the immune network can learn and produce some memory.

3.2 The representation of the process of immune response

To the foreign training data ("antigen"), the units of immune network ("B cells") match it, and calculate the stimulation level of "B cells" respectively by some special match algorithms. If the stimulation level exceeds the threshold, the "B cell" binds the "antigen" and replicates itself, the degree of replication based on the stimulation level of the "B cell". And in process of the replication, the children of the "B cells" is not identical to their father, some mutation occurs in the children of "B cells", so the diversity of immune network is got. Moreover, we introduce some new "B cells" into the immune network, in order to deal with some new "antigens".

When the similar "antigens" invade the immune network, because the immune network has been trained and the unit is distributed, the nodes of immune network can recognize the antigen quickly, this is just like the secondary response of the immune response.

3.3 Interaction between the units of immune network

In the natural immune system, every antigen has an epitope, and the main role of which is to represent the character of the antigen. And every antibody has a paratope to match with the epitope of antigen. The discrimination between antigen and antibody depends on the match between epitope and paratope, which is just like the relationship of "lock and key". In addition, the antibody has an idiotope, which can combine with other antibodies.

Now, we analyze the match algorithm between "antigen" and "antibody", "antibody" and "antibody" in our immune network.

The main stimulation is the affinity between the "B cell" and "antigen", which is defined as follows:

$$ps = (1 - pd) \tag{1}$$

where, ps is the affinity of "B cell" and "antigen", pd is defined as the distance of normalized data space ($0 \leq pd \leq 1$).

The second stimulation ns is the affinity between "B cell" and the connected "B cells", which is defined as follows:

$$ns = \sum_{x=0}^{n}(1 - dis_x) \tag{2}$$

where, n is the number of connected "B cells", and dis_x is the distance between xth "B cells" and this "B cell".

"B cell" cannot only be stimulated, but also be suppressed by the loosely connected "B cells", which is defined as follows:

$$nn = -\sum_{x=0}^{n} dis_x \tag{3}$$

3.4 Stimulation of units of immune network

From section 3.2, we know that the calculation of stimulation level is a very important aspect of the immune network. The stimulation level of each B cell affects the replication of this B cell.

The stimulation level of each B cell depends on interaction of the antigen and connected antibody, and consists of three main parts: (1) the stimulation of antigen to antibody, (2) the stimulation of connected antibodies in the network to this antibody, (3) the enmity of other antibodies to this antibody.

Moreover, the stimulation level also affected by the death rate. Considering the above four factors, the stimulation level of B cell (i.e. the unit of immune network) is calculated by the following equation.

$$s = c\left[\sum_{j=1}^{N} m(a, xe_j) - k_1 \sum_{j=1}^{N} m(a, xp_j) + k_2 \sum_{j=1}^{n} m(a, y)\right] - k_3 \tag{4}$$

where, s is the stimulation, the first term of the right hand of above equation is the affinity of the other B cells to this B cell in the immune network, the second term is the enmity of the other B cells to this B cell, the third term is the stimulus of antigen to this B cell, and the fourth term is the death of B cell. The notations used in Equation 4 are defined as follows:

N: the total number of B cells in the immune network
n: the number of antigens
c: the rate constant
a: the B cell which is calculated
xe_j: the idiotype of the jth B cell which causes stimulation of a
xp_j: the paratope of the jth B cell which causes suppression of a
y: the current antigen which causes stimulus of a
m: the function of affinity or enmity between B cell and B cell, B cell and antigen
k_i: the coefficient to avoid the unbalance of the four terms of right hand of Equation 4.

3.5 Dynamic update of the immune network

In the immune system of biology, the composition of the system is not fixed, but variable from time to time. Based on the dynamic environment, the immune system produces self-organization. This function is called metadynamics function (Varela et al. 1988, Bersini & Valera 1994), which is achieved by adding new cells and removing the useless cells (dead cells).

So, the architecture of immune network should be changeable according to the different "antigens". The dynamic update of the immune network mainly embody at: (1) new units added into the immune network (including the "B cells" reproduced by "B cells" and produced itself), (2) the dead units removed from the immune network. See Figure 3.

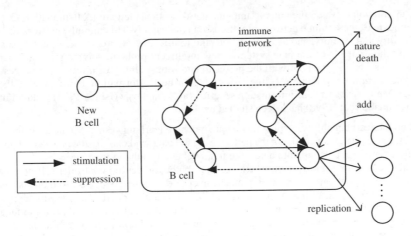

Figure 3.   Dynamic update of immune network.

When the stimulation level of B cells in the immune network exceeds a special threshold, B cells are reproduced to produce large number of the antibodies to kill the special antigen. In the process of replication, somatic hypermutation occurs (Farmer et al. 1986), this diversity can make the immune network more adaptive to the different antigens.

The replication rate of the B cells is defined as:

$$e_x = k(sl(x)) \tag{5}$$

where, sl is the stimulation of the B cells, and depends on the equation in section 3.4.

Based on the idea that immune system is also a microevolution system, we can use genetic algorithm (GA) to update the architecture of immune network. The target of the GA makes the B cells in the immune network match the foreign antigen better. We define the fitness of GA as the stimulation of the B cells. Considering the diversity of somatic hypermutation, we need high mutation rate in GA (about 10%).

So, the detailed algorithm to update the immune network is as follows:

1. Initial the parameters of all units (B cells) in the immune network.
2. Load the antigens
3. Repeat the following actions until the terminated condition is met
   Step 1: Select an antigen from the antigen randomly
   Step 2: Choose a B cell of immune network randomly
   Step 3: Select the percentage of the B cells around the selected B cell
   Step 4: For the above B cells, provided with the antigen and request immune response
   Step 5: Calculate the simulation level of each selected B cells
   Step 6: Reproduce the B cells, whose simulation level exceeds the threshold, and also make these
            B cells some mutation (using GA)
   Step 7: Remove the B cells whose simulation level is low
   Step 8: Add the new B cells to the immune network

We should mention here, in Step 3, we don't let all the B cells in the immune network to react with the antigen, because this is more similar to the natural immune response. When the immune response is repeated several times, nearly all the immune cells react with the antigen.

From the above algorithm, we can also see two ways which produce new B cells. The new B cells can make the immune network more adaptive to the antigens.

4   CONCLUSION AND FUTURE WORK

This paper gives some concepts of immunity, and analyzes the immune response first. Then, model of artificial immune response is given, which is an immune network. After that, the architecture of immune network, interaction between the units of immune network, the calculation of simulation level and the dynamic update of the immune network are discussed respectively. This immune network has the good learning ability and memory, and has a good potential to information processing of complex systems.

Based on the principle of immune system, a new intelligent problem-solving technique named artificial immune system is emerged with the powerful and robust ability of information processing. Just like artificial neural network (ANN), the AIS can learn new information and memorize the learned knowledge. Many fields such as information security, fault diagnosis (Ishida 1993), data mining, robot control (Watanabe et al. 1998) have been studied by AIS.

1. Based on the learning ability of immune system, machine learning system can be designed. Based on the existing knowledge, it can learn the pattern from the data, and classify the new data. This machine learning system can be applied to detect the fraud behavior in the mortagage.
2. Inspired from the immune system, which is a distributed, no central controller self-organizing system, we can design the decision-making mechanism of distributed behavior. And this has been used in the algorithm of controlling autonomous mobile robots and cooperation of several robots.
3. Based on the characteristics of diversity, self-modulating and immune memory of immune system, immune genetic algorithm can be designed. TSP problem and optimization of multimodal function have been solved by immune genetic algorithm.
4. Based on the advantage of quick response to the circumstance, intelligent controller can be designed.
5. Based on the ability of self and non-self discrimination of immune system, we can design algorithms in the application of virus detection and fault diagnosis.

But most of the current research on AIS only gives the heuristic algorithm. The mechanism of immune system is not very clear yet, and AIS hasn't developed its own mathematical bases. Based on the principle of immune system, we need to extract new ideas and algorithms for our engineering and computation further. We can consider more about the T cells' promotion and suppression to the B cells. And we also can consider the optimization of the immune network to improve the convergence speed of the learning procedure. The relevant work should be done more in the future.

REFERENCES

Bersini, H. & Valera, F.J. 1994. The Immune Learning Mechanisms: Reinforcement, Recruitment and their Applications. In R. Paton (ed.), *Computing with Biological Metaphors*: 166–192, Chapman & Hall.
Chowdhury, D. & Stauffer, D. 1992. Statistical physics of immune networks. *Physica A*, 186: 61–81.
Dasgupta, D. 1997. Artificial Neural Networks and Artificial Immune Systems: Similarities and Differences. In *Proceedings of the IEEE International Conference on Systems, Man, and Cybernetics*, 12–15 Oct. 1997, Orlando, Florida, vol.1, pages: 873–878.
Dasgupta, D. & Attoh-Okine, N. 1997. Immunity-based systems: a survey. In *Proceeding of the IEEE International Conference on Systems, Man, and Cybernetics*, 12–15 Oct. 1997, Orlando, Florida, vol.1, pages: 369–374.
Farmer, J.D., Packard, N.H. & Perelson, A.S. 1986. The immune system, adaptation, and machine learning. *Physica D*, 22: 187–204.
Forrest, S., Perelson, A.S., Allen, L. & Cherukuri, R. 1994. Self–Nonself Discrimination in a Computer. In *Proceedings of IEEE Symposium on Research in Security and Privacy*, pages: 202–212, Oakland, CA, 16–18 May 1994.
Hoffmann, G.W. 1986. A neural network model based on the analogy with the immune system. *Journal of Theoretical Biology*, 122: 33–67.
Hunt, J.E. & Cooke, D.E. 1996. Learning using an artificial immune system. *Journal of Network and Computer Applications*, 19: 189–212.
Ishida, Y. 1993. An immune network model and its applications to process diagnosis. *Systems and Computers in Japan*, 24(6): 38–45.
Jerne, N.K. 1973. The immune system. *Scientific American*, 229(1): 52–60.
Jerne, N.K. 1974. Towards a network theory of the immune system. *Ann. Immunol.* (Inst. Pasteur), 125C: 373–389.
Perelson, A.S. 1989. Immune network theory. *Immunological Review*, 110: 5–36.
Rowe, G.W. 1994. *The Theoretical Models in Biology*. Oxford University Press, first Edition.
Vertosick, F.T. & Kelly, R.H. 1989. Immune network theory: a role for parallel distributed processing? *Immunology*, 66: 1–7.
Varela, F., Coutinho, A., Dupire, B. & Vaz, N. 1988. Cognitive Networks: Immune, Neural, and Otherwise. In: A. Perelson (ed.), *Theoretical Immunology, Part II*: 359–375, New Jersey: Addison-Wesley.
Watanabe, Y., Ishiguro, A., Shirai, Y. & Uchikawa, Y. 1998. Emergent construction of a behavior arbitration mechanism based on the immune system. *Advanced Robotics*, 12(3): 227–242.

Software technology

Computational Methods in Engineering and Science, Iu et al. (eds)
© 2003 Swets & Zeitlinger, Lisse, ISBN 90 5809 567 3

VHDL modeling of MPEG audio layer 1 decoder

Mamun Bin Ibne Reaz & Mohd. S. Sulaiman
Multimedia University, Cyberjaya, Selangor, Malaysia

Mohd. Alauddin Mohd. Ali
Universiti Kebangsaan Malaysia, B.B. Bangi, Selangor, Malaysia

ABSTRACT: The decoding of the voice audio bit stream is an issue in terms of real-time transmission of high quality voice audio over the Internet. A stand-alone chip to perform decoding is a better solution over software approach. The MPEG audio compression provides high compression with minimal loss. This paper describes a VHDL model of MPEG audio layer 1 decoder that perform concurrent processing while receiving voice quality audio input bit stream at a constant bit rate and simultaneously producing a stream of 8-bit monopole PCM samples at a constant sampling frequency in real time to ease the description, verification and hardware realization. Simulation and compilation is done to verify the functionality and validity of the design outputs. To test the feasibility and correctness of the design the output is compared with the calculated output results from Perl implementation that proves the effectiveness of the design.

1 INTRODUCTION

The MPEG1 audio (ISO/IEC 1993) becomes a popular algorithm especially in its usage in the Internet for audio data compression due to its high quality with various compression rate, sampling rate and mode. The MPEG1 is classified into three different layers by its complexity. MPEG1 allows audio signal to be compressed up to 12 times so that it can be transmitting at low bitrate. It can compress 1.5 Mbit/sec CD quality audio data into 32 to 448 Kbit/sec for layer 1. It requires supporting sampling rates of 32, 44.1 and 48 Khz.

LAME developed their own MPEG encoder and decoder for both MPEG1 and MPEG2 that runs independently on its own codes. The developed algorithms are adaptive and automatically perform adjustments to optimize performance (Cheng 1998). MAD (MPEG Audio Decoder) is another organization that developed MPEG encoder and decoder using fixed-points. MAD provides a 24-bit PCM output to improve the quality of the audio signals. This allows 16-bit PCM applications to use the extra resolution to increase the audible dynamic range through the use of dithering (Leslie 1989). Joakim Enerstam and Jan Peman presented an MPEG Audio Encoder/Decoder using a DSP chip, the TMS320C6701 (Enerstam & Peman 1990). Both LAME and MAD developed their MPEG decoders for software implementation, which allows floating point implementation and the usage of pointers. Similarly, the hardware implementation of the encoder/decoder using a DSP chip by Enerstam and Peman also used floating-points in their calculation. In VHDL implementation, designs with floating points are not synthesizable. Hence, the implementation in VHDL had to be done using only integers. Furthermore, the range of the integers in VHDL is confined to $2^{31} - 1$ to -2^{31}. However, in VHDL implementation, the input is an input stream of bits at a fixed bit rate of 128 kbits/s. The input is computed in real-time and the output is 8-bit PCM samples at a frequency of 44.1 kHz. As a result, there is a need to perform concurrent processing whereby the input handling, processing and output handling processes run simultaneously.

This research is to develop a model of MPEG audio layer 1 decoder using VHDL so that new applications such as digital voice memo and communication devices can spur from this technology. The use of VHDL for modeling is especially appealing since it provides a formal description of the system and allows the use of specific description styles to cover the different abstraction levels (architectural, register transfer and logic level) employed in the design (Zamfirescu 1993).

2 DESIGN METHODOLOGY

2.1 Implementation choice

In modeling the voice audio decoder, MPEG audio layer 1 is chosen due to its less complexity. As human audible range is 20 kHz, therefore to avoid imaging according to nyquist theory a sampling frequency of 44.1 kHz is selected. The range of bit rates in MPEG audio layer as ISO standards are from 32 to 448 kbps. Since the model is for voice audio bit streams, therefore 128 kbps bit rate is chosen. 128 kbps require a storage device that can stream at 16 kbps. In calculation the integer point is used by upscale the values as the input and output of the decoder is bit streams and 8-bit PCM samples respectively which is a standard logic vector.

To avoid the trigonometric functions in calculations of decoding process a ROM is implemented to store the entire coefficient. This is possible since, the Han Window coefficient is only dependent on the sampling frequency, as the sampling frequency is predetermined.

The model is designed to handle the bit stream and not dependent on the streaming of the input. A separate process that runs concurrently with the decoder unit is designed to handle the bit streams. This allows the decoder to function even if the input of the bit streams are not in a constant rate. A separate output unit is designed to run concurrently with the decoder unit and a buffer is used to allow some tolerant in the output flow thus ensure the PCM samples would always be at the correct frequency.

2.2 MPEG audio decoding algorithm

The decoding process is the process of expanding the encoded MPEG audio layer 1 format into PCM samples. This decoding process is separated into three blocks, which are the frame unpacking, reconstruction and inverse mapping. The frame unpacking block of the decoding algorithm unpacks and recovers the various piece of information in the frame. The frame unpacking block also does error detection if error-check is applied (i.e. protection bit is cleared). The reconstruction block reconstructs the quantized version of the set of mapped samples using Equation 1 below:

$$S'' = \left(\frac{2^{nb}}{2^{nb} - 1} \right) \times \left(S''' + \frac{1}{2^{nb-1}} \right) \tag{1}$$

where S'' = requantized value; S''' = fractional number; nb = no. of bits allocated to the samples.

The inverse mapping block transforms these mapped samples back into uniform PCM. The ISO/IEC 11172-3:1993 is followed for MPEG audio layer decoder.

The MPEG audio layer 1 bit streams are separated into frames. Each frame consists of four parts the frame header, error checks, audio data and ancillary data. The frame header contains the basic decoding parameter of the stream includes sampling frequency and layer. The frame header used in this research is shown in Figure 1.

The error checks hold the optional 16-bit parity-check word used for error detection within the encoded bit stream. The cyclic redundancy check of the frame is calculated based on the length of the frame and

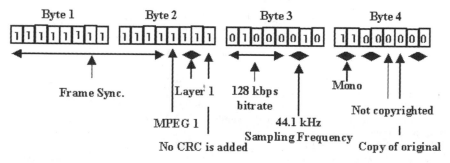

Figure 1. Expected frame header.

compared to verify the valid frame. The audio data consists of encoded information to be decoded. It is separated into three sections the bit allocation code, scalefactors code and subband samples.

The bit allocation determines the number of bits that represent the samples in a particular subband. The size of the bit allocation code is 4 bits each. Since there are 32 subbands, effectively it uses 128 bits. The number of bits allocated for each subband is the integer value of the bit allocation code +1, except for "0000" where no bits are allocated and "1111" which is invalid. The scalefactors scale the normalized sample value to its actual value for a particular subband is an index based on the ISO. Scalefactors code is only present if the bit allocated to that particular subband is not zero. The subband samples are the normalized value of each sample. Since there are 12 parts of 32 subband samples in one frame, the arrangements of the samples are in parts with the first 32 subband samples followed by the next subband samples. The number of bits used to represent each sample is based on the bit allocated to the subband of the sample (ISO/IEC 1993). The ancillary data is not defined by the ISO and it is user defined which is not significant in this research.

3 DESIGN AUTOMATION

Many programs are developed to automate the design by speeding up the required designing time. Most of the automation used in the development of the decoder involves formatting values as the decoding algorithm involves the usage of constants provided by the ISO standards. These constants are implemented as ROMs thus it is hard-coded into the VHDL model. A program is also designed to convert the PCM output into a wav format to evaluate as an audible output beside absolute values.

A program is developed to extracts the scalefactors from the ISO specifications into VHDL format to form the scalefactors constants up scaled by a factor of 10^4.

Another program is developed to calculate the N_{ik} coefficient since it involves trigonometric functions, which is not supported by VHDL. The N_{ik} coefficients are calculated based on the Equation 2 below:

$$N_{ik} = \cos\left[(16+i)\times\frac{(2k+1)\pi}{64}\right] \qquad \text{For i=0 to 63 and k=0 to31} \qquad (2)$$

The calculated values are typically in the range of -1 to 1; thus the value is up scaled by a factor of 10^4.

Another program is developed to extracts the Window Coefficient D_i from the ISO specifications into VHDL format to form the D_i constants upscale by a factor of 10^4.

Due to the limitation of file management functions in the VHDL, the test vectors are extracted from a text file. Since the MPEG audio file is typically a binary file, thus MPEG audio file is extracted bit by bit and formatting it as a text file, where each line contain either 1 or 0. By performing the formatting, the testbench read this text file and provide the test vectors to the decoder as a bit streams which is developed using another program.

4 VHDL MODELING

4.1 VHDL format consideration

VHDL language is capable of designing in hierarchies, reusing components, error management and verification allows describing huge complex circuitry efficiently (Sjoholm & Lindh 1997).

In designing VHDL model, three packages std_logic_1164, std_logic_arith and std_logic_unsigned is defined from the IEEE library. These packages include definition of standard logic data types, arithmetic operations involving standard logic data types and integer data types and integer to standard logic vector conversion and vice versa. The "shift" package performs bit-shifting operations on standard logic vector data type. The bit shifting operations is used for bit management and multiplication of power 2 purposes.

Both the entity and architecture of the MPEG audio decoder is "decoder". In the entity area, all the input and output ports of the decoder component is defined. There are 4 ports in the "decoder" component, which are Clock, Input, Output and Outputready. Both the Clock and Input are of std_logic, a single bit port. The Output is of type std_logic_vector (0 to 7), an 8 bit port. The Outputready is of type std_logic.

The architecture section follows the entity section and describes the behavioral relation between the inputs and the outputs.

In the architecture of the VHDL code, it is made up of both concurrent part and also sequential part (processes). The decoder has three processes, "Inputhandling", "Mainprocessing" and "Outputhandling". These three processes are running concurrently with each other (i.e. the second process does not need to wait for the first process to complete before execution). Within the process, the processing is done sequentially. The sensitivity list is a list of signal (concurrent) that the process monitors for. The process only start to process when there are some changes in the signals of the sensitivity list. For example, in the "Inputhandling" process, the sensitivity list has the signal "Clock", this means that whenever the clock signal changes, the process start and all the statements in that particular process execute sequentially. For the case of the "Mainprocessing" and "Outputhandlin" their sensitivity list consists of "Frameready" and "Outputclock" signal respectively.

Throughout the design of the decoder, many ROMs and buffer are used. These ROMs and buffer stores integer values, thus arrays of size required by the ROM or buffer are created for this purpose.

4.2 Top level design

The decoder has an input of type std_logic for the bit stream and an output of type 8-bit std_logic_vector for the PCM output. A clock signal of type std_logic is added to the decoder to set the bit stream data rate. In order to aid the testing of the MPEG Audio decoder, an "Outputclock" signal of type std_logic is added. This signal toggles whenever the new PCM output is outputted.

Two critical issues are considered to operate the decoder smoothly. First, the decoder is able to constantly accept the continuous bit streams of data even if there is some delay in the processing. Second, the decoder output the PCM samples at a constant rate of 44100 Hz, thus independent of the processing time required. In order to handle these two issues, the input portion, the processing portion and the output portion runs asynchronously and parallel to each other thus three processes "Inputhandling", "Mainprocessing" and "Outputhandling" of the decoder runs concurrently.

4.3 Behavioral description

The "Inputhandling" process is designed to handle the continuous input of bit stream. This process performs bit management and frame detection. Once the new frame is detected the whole frame passed to the frame buffer "buff2". This buffer is declared as a signal, thus it functions as an intermediate buffer between the "Inputhandling" and "Mainprocessing". The input is read from the "Input" port. The input rate is synchronized by the rising edge of an input clock, which runs at the desired data rate. The "Inputhandling" runs concurrently with the other processes in VHDL. An intermediate buffer is placed to pass the frame to the "Mainprocessing" process. The "Mainprocessing" process is the main processing unit of the decoder. At the top level, the "Mainprocessing" process is executed based on the "Frameready" signal, which is supplied by the "Inputhandling" process when a new frame is detected and pushed to the frame buffer.

In this model, the bits shifted left by the corresponding bits extracted using the shifting method. Since the function of power is not available in VHDL, therefore, by shifting the value of "0000000000000001" to the left by "nb" is effectively multiplying 1 by 2^{nb}. Thus, the values of 2^{nb} and $2^{(nb-1)}$ are substituted with two variables where their values are calculated via the shifting method.

In VHDL, integer data type is limited to the range of 2^{31} to -2^{31}. In order to make no overflow, all the values are up scaled by a factor of 10^4. Since, the calculations involve many multiplication operations; the convention is consistent by dividing the result of the multiplication by a factor of 10^4. Thus, in the VHDL codes, the scaling down of values is done in every multiplication operation.

In the VHDL model, the output timing is fixed at the sampling frequency of 44.1 kHz. In order to push the "Mainprocessing" process samples into the "Outputhandling" process a PCM samples buffer (Concurrent) is used, similar to the "Inputhandling" to "Mainprocessing" buffer. The PCM samples buffer provide the function of making the design more tolerant to timing deviation. The size of the buffer is set to 768 (2×384) PCM sample as each frame consists 384 PCM samples thus each time the "Mainprocessing" process output 384 samples. By having the buffer, the "Mainprocessing" process fills up the buffer alternatively from upper 384 to lower 384 PCM samples therefore the timing of the data rate input need not to be exactly matched to the output timing of 44.1 kHz. The data rate only needs an average of desired data rate. The "Outputhandling" process manage the outputting of the PCM samples at a rate of 44.1 kHz by inverting a "outputclock" signal at a rate of 44.1 kHz effectively the frequency of the "outputclock" at 22.05 kHz. The sensitivity list of the "Outputhandling" process is added with the "outputclock" signal therefore every time the "outputclock" is toggle (inverted), the "Outputhandling" process start execution. When executed, the process pumps out a PCM sample from the PCM buffer. An internal counter move the pointer of the buffer from 0 to 767 then back to 0 again. As a result, the PCM sample outputted at a constant rate of 44.1 kHz.

5 VHDL SIMULATION

5.1 Testbench development

The testbench is considered as a component with no ports and the decoder is encapsulated by the testbench. The test vectors used in the testing extracted from a text file, since the number of test vector bits are in terms of a few hundred thousand bits. The "std.textio" package in the standard library is used to develop testbench to read the test vectors from a file line by line (a bit per line) and also write the output of the VHDL model in to a file line by line (a PCM sample in integer per line).

For inputting of test vectors, the process reads a line from the file, then the value of that line is extracted as integer and converted into "std_logic" and fed to the "Input" port of the decoder at every rising edge of the clock which runs at the desired bit rate. After that, the next line is read and the process repeats. For outputting of the decoder to a file is done by first converting the "std_logic_vector" type value of the "Output" port into integer data type. It is then written in the file as a new line. This process contains the "Outputready" signal in its sensitivity list thus executes whenever the signal toggles. The process write the new PCM sample values to the file once notified by the decoder that the output is ready via the "Outputready" signal.

5.2 Simulation verification

There are two handshaking protocols, one at the "Inputhandling" and "Mainprocessing" processes interface and another at the "Outputhandling" process to the "Outputready" port of the decoder. For the first handshake, when the "Inputhandling" process detects the new frame, it is sent to the frame buffer (concurrent signal) and the "Frameready" signal is toggled. When the "Mainprocessing" process detects the toggling of the "Frameready" signal, it perform the decoding operations and output the 384 samples to the PCM samples buffer (concurrent signal) is illustrated in Figure 2.

For the second handshake, whenever the "Outputhandling" process outputs the new PCM sample, the "Outputready" signal is toggled thus indicate the new PCM samples are outputted is illustrated in Figure 3.

The outputting of the PCM samples is at a rate of 44.1 kHz, thus the "Outputready" signal is toggle at a rate of 44.1 kHz. As a result the "Outputready" signal is a square-wave of 50% duty cycle and a period of 45.3 µs. The input clock is also at a rate equal to the desired bit rate. The bit rate is 128 kbps, thus the input clock is effectively a square-wave of 50% duty cycle and a period of 8.8 µs. Figure 4 illustrates the timing verification on the simulation waveform.

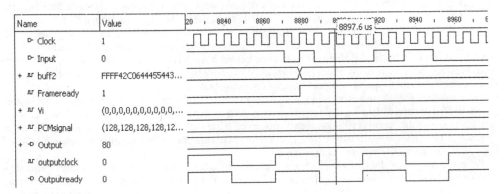

Figure 2. The detection of first sync sequence section of the simulation waveform.

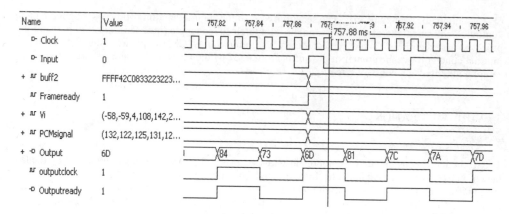

Figure 3. Random center section of the simulation waveform.

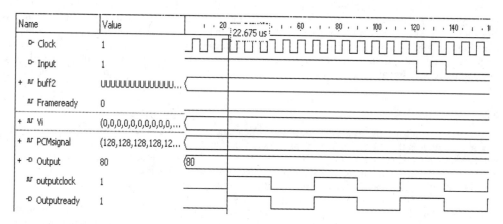

Figure 4. The simulation waveform.

6 RESULTS AND DISCUSSION

The analysis and discussion of each stage of the design flow is carried out in three different angles audio inspection, time domain waveform and frequency domain waveform. Comparison is made between the wav outputs of the stages. After obtaining the wav file from the conversion program, a first level audio inspection is performed. From the auditory inspection, it is concluded that there are many spikes in the sound, but there is no attenuation and frequency shift distortion heard. Since the Perl model's result is considered more accurate, the comparison is done between the VHDL simulation results and the Perl model results. In order to analyze the noise injected by the VHDL model, an analytical comparison is done on the time and frequency domain waveforms using Matlab.

From the time domain waveforms in Figure 5, it is verified that there are spikes being injected by the VHDL model as observed during auditory inspection. Although, there are spikes in the waveform, the general waveform of the sound is still dominantly observed. Thus, some filtering can be used to remove the spikes and satisfactory results can be obtained. These spikes are due to the lack of precision available in the VHDL model, as the integer range is limited.

From the frequency domain waveforms in Figure 6, spikes injected are observed as high frequency components. There are four major spikes in the frequency domain waveform of the VHDL simulation result, which occurs at around 2200 Hz, 5000 Hz, 8000 Hz and 12000 Hz. Since, the human voice's frequency response dominant mainly at 2000 Hz, thus a low-pass filter is able to remove the spikes with a

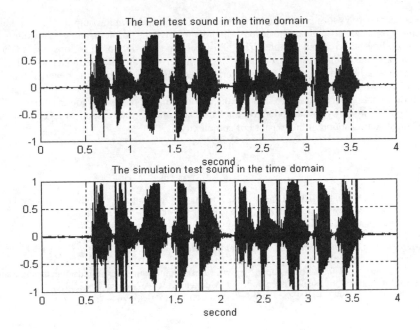

Figure 5. Time domain waveforms of the Perl output & simulation output.

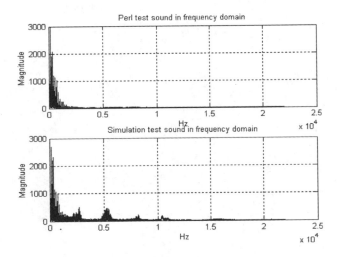

Figure 6. Frequency domain waveforms of the Perl and simulation output.

cut-off frequency at 2000 Hz. As conclusion, the results of the VHDL simulation was acceptable although distortion was suffered because these spikes.

7 CONCLUSION

In this research project, a model of MPEG Audio Layer 1 decoder using VHDL is successfully designed to get continuous MPEG audio layer 1 bit streams as input and output the decoded 8-bit PCM samples as a standard logic vectors. The decoder is able to function normally, even if there is some deviation of data rate in the input

The decoder portrays some errors and limitation such as the spikes at the output. These spikes are due to the limited precision possible of the decoder. In order to solve the problem, a low pass filter is required after the decoding of the output samples. Thus, this problem does not post a serious treat to the design. The decoder also has a limitation where it is unable to flag error if the bit stream is not an MPEG Audio layer 1 bit stream. On top of that, there is not error checking introduced in the decoder, thus if the bit streams suffer some error, the output of the PCM samples will also be distorted.

Currently we are experimenting to improve the design for a more precise and accuracy so that there is no injection of noise by the decoder. We also are experimenting to extend to layer 2 which would provide better quality sound at the expanse of complexity.

REFERENCES

ISO/IEC International Standard CD 11172-3 1993. Information Technology – Coding of Moving Pictures and Associated Audio for Digital Storage Media at up to about 1.5 Mbits/s – Part 3: Audio.

Cheng, M. 1998. The L.A.M.E. (Lane Ain't an MP3 Encoder) project, *http://www.mp3dev.org/mp3/*.

Leslie, R. 1989. MAD: MPEG Audio Decoder, *http://www.mars.org/home/rob/proj/mpeg/*.

Enerstam, J. & Peman, J. 1990. *Hardware Implementation of MPEG Audio Real-Time Encoder*. Unpublished Masters thesis. Luleo University of technology. Luleo. Sweden.

Zamfirescu, A. 1993. Logic and Arithmetic in Hardware Description Languages, Fundamentals and Standards in Hardware Description Languages. Edited by Jean P. Mermet. *NATO ASI Series*: 109–151.

Sjoholm, S. & Lindh, L. 1997. *VHDL for Designers*. Prentice Hall.

Computational Methods in Engineering and Science, Iu et al. (eds)
© 2003 Swets & Zeitlinger, Lisse, ISBN 90 5809 567 3

A measurement model for C++ program complexity analysis

S. Kanmani & P. Thambidurai
Department of Computer Science and Engineering, Pondicherry Engineering College, Pondicherry, India

V. Sankaranarayanan
Ramanujam Computing Centre, Anna University, Chennai, India

ABSTRACT: C++ programming language does not enforce the programmers to code with the Object-Oriented (OO) features. Consequently, the program characteristics may not be meeting the structural requirement even though it performs well for the given inputs. This paper presents a measurement model, which can be used to measure the OO features: encapsulation, inheritance, polymorphism and coupling in the C++ program using appropriate metrics. Measuring the static code gives the indices for each of the OO properties, which can be compared with the expected values. From these numerical values the behavior of the program can be analyzed.

1 INTRODUCTION

During 1990s Object-Oriented (OO) technology became the paradigm of choice for many product builders. As time passes, OO technology is replacing classical software development approaches. This technology offers support to provide software product with higher quality and lower maintenance costs. Since traditional software measures aim at procedure-oriented software development (Halstead 1977, Henry & Kafura 1981) and it cannot fulfill the requirement of the Object-Oriented software, a set of new software measures adopted for the characteristics of Object technology (Chidamber & Kemerer 1991, 1994, Li & Henry 1993, Lorenz & Kidd 1994, Abreau & Melo 1996, Brian Henderson 1996). Hence, OO metrics became the essential part of OO technology as well as good software engineering.

The six measures suggested by Chidamber and Kemerer (CK) are the most widely used design measures for OO systems focusing on class and class hierarchy (Chidamber & Kemerer 1991, 1994). Following CK measures many other metrics suite had been introduced. Metrics for Object-Oriented Design (MOOD) set of six measures, which can be used for measuring OO features were introduced (Abreau & Melo 1996). A group of measures identified for the maintenance of Object-Oriented systems (Li and Henry 1993). Lorenz and Kidd introduced measures in four categories namely size, inheritance, internals and externals (Lorenz & Kidd 1994).

As OO measures are many in number and most of them are not language specific, it is very difficult to find one single measure for a particular application. Hence, a two-dimensional metric model comprising of a group of measures is proposed, which could be applied on C++ code for the OO properties: encapsulation, inheritance, coupling and polymorphism. Also a tool is designed to automatically collect these measures from the source code.

The following section deals with the related work done. Section 3 introduces the proposed model. Section 4 discusses the design of the metric collection tool: Object-Oriented Metric Calculator (OOMC). Section 5 elaborates the results arrived at an experiment carried out. Section 6 concludes the conclusion along with possible future work.

2 RELATED WORK

The importance given to the measurement technologies lead to the introduction of numerous new software measures (Xeono et al 2000). To apply the OO metrics technology in practice, automation of

	Class	System
Size	2	2
Encapsulation	2	2
Inheritance	6	2
Coupling	10	1
Polymorphism	5	1

Figure 1.   Proposed model.

metric collection is necessary to study about the characteristics of the system. These automations can be applied on design document of the software being developed or on the source code. Many tools exist to aim at this objective. Some of them are commercially available (QMOOD++, MOODKIT-G1, MOOD-KIT-G2, Scanner, etc.) which support a specific framework or an arbitrary set of measures. Some of the other tools are derived from the static code analyzers. The results of such studies confirm that these measures can be used as quality indicators for fault proness, defect regions, rework, etc. Each of these studies is applied on large sized industrial/commercial projects (Briand et al 2000).

This work is confined to apply the OO measures to quantify the use of OO properties in the programs/projects developed in the laboratory courses. A metric collection tool is designed to automatically collect the measures given by the proposed model. On collecting these metric values the developer/evaluator can

- view the complexity of the program developed
- weigh the exploitation of the OO properties in the program (as C++ allows developers to solve the problem without Object-Orientation)
- understand the program characteristics and in turn its behavior

3   PROPOSED MODEL

Program quality can be measured by two important factors.

1. Correctness of the program and
2. Complexity of the program

For any type of programming environment, the correctness of the program is measured by test case analysis. This can be performed by executing the program for various test cases (with different input values) and matching the results (outputs) with the expected values.

But the complexity of the program requires the measurement of constructs (used in the program), which varies for different types of programming environment (structured programming, OO programming, etc). There are many standard tools and methods available to compute the metrics for structured programs but not for OO programs. Hence, a two-dimensional measurement model, which is suitable for the programs in C++ , is proposed for measuring the OO properties as shown in Figure 1. This model considers the OO properties in one dimension and components that are measured for the properties in another dimension. The number in each cell represents, the number of metrics identified for that category. Table 1 contains the name of each measure, its original definition and reference.

4   OBJECT-ORIENTED METRIC CALCULATOR (OOMC)

It is an automated tool, named OOMC, developed in C++ programming language to collect the metrics proposed in the above model for the inputted C++ program. It is a user-friendly, interactive OO metric data collection tool for C++ programming language. In addition to collecting metric data OOMC also provides options for querying and retrieving on metric data of the system under analysis.

TURBO C++ environment has been used for the development of OOMC. Visual Basic is used as a front-end tool to provide user-friendly interface. This tool consists of the following major components.

- Code parser
- Intermediate storage
- Metric computation
- Results presentation

Table 1. System level measures in the proposed model.

Metric definition	Reference
Size	
1. Number of classes (NC)	(Kanmani et al 2002)
Total number of classes in the system	
2. Number of class hierarchies (NCH)	(Kanmani et al 2002)
Total number of distinct class hierarchies in the system	
Encapsulation	
3. Attribute hiding factor (AHF)	(Abreau et al 1996)
Percentage of attributes that are not visible in the system (sum of the invisible attributes in all classes divided by total number of attributes defined in the system)	
4. Method hiding factor (MHF)	(Abreau et al 1996)
Percentage of methods that are not visible in the system (sum of the invisible methods in all classes divided by total number of methods defined in the system)	
Inheritance	
5. Attribute inheritance factor (AIF)	(Abreau et al 1996)
Percentage of attributes that are inherited in the system (sum of all inherited attributes in all classes divided by sum of inherited and non inherited attributes in all classes)	
6. Method inheritance factor (MIF)	(Abreau et al 1996)
Percentage of methods that are inherited in the system (sum of all inherited methods in all classes divided by sum of inherited and non inherited methods in all classes)	
Coupling	
7. Coupling factor (COF)	(Abreau et al 1996)
Percentage of coupling in the system (actual number of couplings not imputable to inheritance divided by maximum number of couplings possible in the system)	
Polymorphism	
8. Polymorphism factor (POF)	(Abreau et al 1996)
Percentage of polymorphic methods in the system (number of possible different polymorphic situations divided by number of possible distinct polymorphic situations)	

4.1 Code parser

The parser parses the essential statements of the given source code for the data used in the computation of each of the metric. It assumes that the program is syntactically correct. From the analysis of the measures listed in the Table 2, it is found that the data required for the computation of them lie in the following statements of the program only.

- The class derivation/definition statement
- The class's attribute declaration statement
- The class's method signature
- The class's method definition statement
- Friend class declaration statement

The generic statement patterns for each of the above statements are derived from the language syntax and are used in the design of the parser. After parsing, the collected information is stored appropriately in a temporary file.

4.2 Intermediate storage

It contains the information about each of the classes in the program and the details of inheritance relationship between the classes in the system. The following tables are produced for each of the classes at

Table 2. Class level measures in the proposed model.

Metric definition	Reference
Size	
1. Weighted class size (WCS) WCS = Number of attributes + Total method size	(Brian Henderson 1996)
2. Un weighted class size (UCS) UCS = Number of attributes + Number of methods	(Brian Henderson 1996)
Encapsulation	
3. Number of private attributes (Apriv) Apriv = Number of private attributes in the class	(Lorenz & Kidd 1994)
4. Number of private methods (Mpriv) Mpriv = Number of private methods in the class	(Lorenz & Kidd 1994)
Inheritance	
5. Number of children (NOC) NOC = Number of immediate children of a class	(Chidamber & Kemerer 1994)
6. Depth of inheritance (DOI) DOI = Depth of inheritance	(Chidamber & Kemerer1994)
7. Number of ancestor (NOA) NOA = Number of ancestor classes	(Lorenz & Kidd 1994)
8. Number of descendant (NOD) NOD = Number of descendant classes	(Kanmani et al 2002)
9. Total number of inherited attributes Ainh = Number of inherited attributes of the class	(Lorenz & Kidd 1994)
10. Total number of inherited methods Minh = Number of inherited methods of the class	(Lorenz & Kidd 1994)
*Coupling**	
11. Inverse friend *class-attribute* import coupling (IFCAIC)	(Briand et al 1999)
12. Ancestor *class-attribute* import coupling (ACAIC)	(Briand et al 1999)
13. Other *class-attribute* export coupling (OCAEC)	(Briand et al 1999)
14. Friend *class-attribute* export coupling (FCAEC)	(Briand et al 1999)
15. Descendant *class-attribute* export coupling (DCAEC)	(Briand et al 1999)
16. Inverse friend *class-method* import coupling (IFCMIC)	(Briand et al 1999)
17. Ancestor *class-method* import coupling (ACMIC)	(Briand et al 1999)
18. Other *class-method* import coupling (OCMIC)	(Briand et al 1999)
19. Friend *class-method* Export coupling (FCMEC)	(Briand et al 1999)
20. Other *class-method* export coupling (OCMEC)	(Briand et al 1999)
Polymorphism	
21. Overloading in stand alone system (OVO) Number of times each function member name is overloaded in the class	(Benlarbi and Melo 1999)
22. Static polymorphism in ancestor (SPA) Number of static polymorphic function members that appear in the class and its ancestors	(Benlarbi and Melo 1999)
23. Static polymorphism in descendants (SPD) Number of static polymorphic function members that appear in the class and its descendants	(Benlarbi and Melo 1999)
24. Dynamic polymorphism in descendants (DPD) Number of dynamic polymorphic function members that appear in the class and its descendants	(Benlarbi and Melo 1999)
25. Dynamic polymorphism in ancestors (DPA) Number of dynamic polymorphic function members that appear in the class and its ancestors	(Benlarbi and Melo 1999)

* Measures exclusive for C++-type of coupling given in italic.

the end of parsing and they are stored in a text file created (in the name of the class):

Class attribute table	contains the name of the attribute, data type, visibility, no. of items etc.
Class method table	contains name of the method, visibility, constructor/destructor, virtual, arguments, type etc.
Friend class table	contains name of the classes defined as friend to this class.

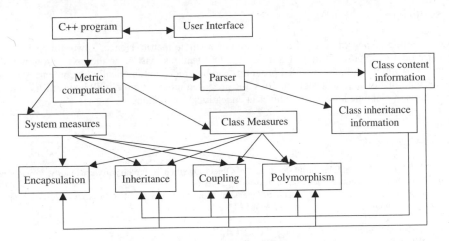

Figure 2.   Information flow between the components in OOMC.

Table 3.   Results of the system level measures.

| | Inheritance % | | Encapsulation % | | | | Size | |
	Attributes level	Method level	Attributes level	Method level	Coupling %	Polymorphism %	NOC	NOH
Project1	6.67	4.87	84.375	17.95	10.71	75	8	2
Project2	18.75	31.57	84.13	17.94	7.142	69.23	8	6
Project3	7.69	33.89	83.85	17.95	3.57	50	8	4

The ancestor and descendant classes for each of the classes are derived from the inheritance rela-tionship between the classes. For this the inheritance relationship identified through parsing is stored in a graph (data structure) representation (since C++ allows multiple inheritance).

4.3   Metric computation

The metrics given in the Table 2 are computed from the information stored in the text file for each of the classes. From them the class level and system level measures are computed and stored. The follow-ing Figure 2 shows the important modules and the information flow of the developed tool.

4.4   Results of OOMC

The measures computed by OOMC are presented in appropriate format in a text file for later analysis. For easy comparison of results, the metric values are represented visually using graph and bar charts.

5   EXPERIMENTATION AND RESULTS

The C++ projects developed by three different group of UG students for a given application problem have been taken for analysis. The resulted metric values are compared. Table 3 shows the summary of the system level measures obtained in the experiment. Due to the problem defined, number of classes defined in each project is found to be equal but the other metric values are not matching. This is due to the OO features used by the individual group. Commonly, in the three projects, attribute level encapsu-lation is more due to the declaration of attributes in private section and method level encapsulation is found to be less due to the declaration of most of the methods in public section.

579

6 CONCLUSION

The OO program complexity cannot be measured by a single metric. Hence, a two-dimensional measurement model is proposed. Using this model the OO features existing in the given program can be analyzed. As a future work, the inputted programs can be graded automatically by comparing the metric values with that of a model program. Thus, the proposed model and tool will be useful to assess the OO programs in real time laboratory courses at educational institutions.

REFERENCES

Abreau, F.B. & Melo, W. 1996. Evaluating the impact of Object-Oriented design on software quality, *Proceedings of the third international software metrics symposium*.

Benlarbi, S. & Melo, W. 1999. Polymorphism measures for early risk prediction, *Proceedings of the 21st International conference on software engineering*: 335–344: Los Angeles: USA.

Briand, L., Daly, J. & Wust, J. 1999. A unified framework for coupling measurement in Object-Oriented systems. *IEEE transactions on software engineering* **25**(1): 91–121.

Briand, L., WÜst, J., Daly, J.W. & Porter, D.V. 2000. Exploring the relationships between design measures and software quality. *Journal of systems software* **51**: 245–273.

Chidamber, S.R. & Kemerer, C.F. 1991. Towards a metric suite for Object-Oriented design. *Proceedinds of OOPSLA* **26**(11): 197–211: SIGPLAN Notices.

Chidamber, S.R. & Kemerer, C.F. 1994. A metric suite for Object-Oriented design. *IEEE transactions on software engineering* **20**(6): 476–493.

Halstead, M. 1977. Elements of software science: North Holland.

Henry, S. & Kafura, D. 1981. Software structure metrics based on information flow. *IEEE transactions on software engineering* SE-**7**(5): 510–518.

Kanmani, S., Sankaranarayanan, V. & Thambidurai, P. 2002. External system characteristics assessment using Object-Oriented inheritance metrics. *CSI Journal* **32**(1): 5–12.

Lorenz, M. & Kidd, J. 1994. Object-Oriented software metrics: A practical guide: Prentice Hall Inc.: New Jersey.

Li, W. & Henry, S. 1993. Object-Oriented metrics that predict maintainability. *Journal of systems and software* **23**(2): 111–122.

Xeons, M., Stavrinoudis, D., Zikouli, K. & Christodoulakis, D. 2000. Object-Oriented metrics – A survey. *Proceedings of FESMA 2000*: Madrid: Spain.

Computational Methods in Engineering and Science, Iu et al. (eds)
© 2003 Swets & Zeitlinger, Lisse, ISBN 90 5809 567 3

An object-oriented model for combined implementation of the finite element and meshless methods

P. Nanakorn & C. Somprasert
Civil Engineering Program, Sirindhorn International Institute of Technology,
Thammasat University, Thailand

ABSTRACT: This paper presents a design of an object-oriented model for combined implementation of the finite element and meshless methods. To this end, separate object-oriented models for the two analysis techniques are first considered and similarities and redundancies between the two are noted. Thereafter, the two models are merged into a new model that is able to handle both analysis techniques. The integration of the two original models enhances the efficiency and flexibility of the two original programs. It removes redundancies as well as promotes the reusability of the codes. Consequently, the integration results in, as a whole, a smaller program. In addition to allowing one of the two analysis techniques to be selected from within the proposed model, the model also provides possibilities of performing analysis methods that are hybrids of the two original analysis techniques.

1 INTRODUCTION

Among scientists and engineers, the finite element method has been, for more than two decades, considered as the most efficient and versatile computing tool for solving a wide variety of science and engineering problems. In the past, the implementation of the finite element method had been extensively carried out by using procedural programming languages. One particular procedural programming language worth being mentioned is FORTRAN, which had been the most favorite language for scientists and engineers even until recently. There are thousands of finite element analysis programs coded with FORTRAN. The procedural programming technique focuses on dividing computational tasks into many subroutines that can be reused. This technique helps improve the readability and maintainability of finite element analysis programs. However, as the finite element method is used to solve problems that are more complex, its data become difficult to manage. This renders the procedural programming technique inadequate and efficient data structuring is necessary. Although some procedural programming languages such as C allow, to some extent, structuring of data, they do not provide a direct way to define behavior of data structures. Around three decades ago, a new programming paradigm called the object-oriented programming technique was proposed. In this programming concept, software is organized as a collection of discrete and distinguishable objects that incorporate both data structures and behavior. The primary key of this approach is to consider tasks in software as communications between and operations within various objects. Each object encapsulates both its data and behavior. The structure of an object and its behavior are defined by using a template called "class." The concept greatly improves the maintainability, extendibility, and reusability of software. Since the technique allows problems with complex processes and data to be handled efficiently, it has naturally become a popular programming technique for implementing the finite element method (Forde et al. 1990, Mackie 1992, Donescu & Laursen 1996, Archer et al. 1999, Yu & Kumar 2001).

Mesh generation is one of the most important tasks that have to be done when the finite element method is used. For problems with difficult configuration, the meshing process in the finite element method can be very expensive. In addition, in several applications, remeshing and refining elements during computation are also required. Mesh generation in the finite element method becomes a bottleneck in the large scale analysis and is therefore considered as a shortcoming although it is ironically derived from the main idea of the method itself. In order to avoid difficulties of mesh generation, a number of new methods that need no mesh of elements for the interpolation of unknown fields have been proposed such as the Element-Free Galerkin Method (EFGM), the Partition of Unity Method, the Reproducing Kernel Particle Method (RKPM), and the Meshless Local Petrov-Galerkin (MLPG) Method (Belytschko et al. 1994, Babuska & Melenk 1997, Jun et al. 1998, Atluri et al. 1999). In this type of methods, which are collectively called the meshless methods, only nodes are required for the interpolation of unknown fields; thus, an input process that is much simpler than that of the finite element method can be achieved. However, the meshless methods have not been used generally in practical analysis because the methods are still appropriate to limited types of applications and require a larger computational resource in comparison with the finite element method.

The main difference between the finite element and meshless methods is the way unknown fields are interpolated; otherwise, there are a lot of similarities in terms of entities used in the finite element and meshless methods. Based on the concept of object-oriented programming, these similar entities can be designed as objects that are shared by both classes of methods. It is therefore advisable to design, based on the object-oriented method, a program that is capable of handling both analysis techniques. With the resulting program, it will be possible to allow the user to select appropriate analysis tools for different applications. In addition, it will also be possible to perform analysis methods that are hybrids of the two original classes of methods.

In this paper, a design of an object-oriented model that includes both the finite element and meshless methods is proposed. The design process commences with investigating two separate object-oriented models for the finite element and meshless methods. Here, similarities between the two models are observed. After that, the two models are merged into one new model. In the integration process, similarities and redundancies found in the two original models are removed by redesigning the existing classes and inventing necessary new classes. In addition, relationships between classes, including both the existing and new ones, are collectively designed in order that the most efficient model is obtained. Finally, improvements in various aspects after the integration of the two original models are discussed.

2 AN OBJECT-ORIENTED MODEL FOR THE FINITE ELEMENT METHOD

As mentioned earlier, various object-oriented models for the implementation of the finite element method have been proposed (Forde et al. 1990, Mackie 1992, Donescu & Laursen 1996, Archer et al. 1999, Yu & Kumar 2001). Most models consider entities in the finite element method that can be obviously identified, such as nodes, elements, materials, etc., and design corresponding objects accordingly. This approach of designing object-oriented models for the finite element method is natural, easy to understand and efficient. Other models may give their emphasis on different aspects of the implementation such as the finite element solver (Carey et al. 1994), sub-structuring (Mackie 2001), and calculation

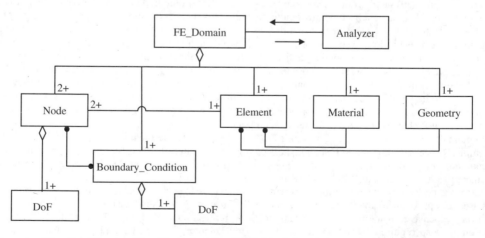

Figure 1. The class diagram of the object-oriented model for the finite element method.

control (Mackie 2002). Using the object-oriented programming technique to include a symbolic environment for finite element analysis has also been proposed (Zimmermann & Eyheramendy 1996). This symbolic environment enables finite element analysis to begin its calculation from the governing differential equations of the problem. Nevertheless, this added symbolic capability may not be much required for practical use. In this section, a simple but efficient object-oriented model for the finite element method will be shown. The model will be used as a basis for later discussion on a combined model for the finite element and meshless methods, being proposed in this study. Due to this fact, some necessary detailed information related to the model for the finite element method will be amply provided. Figure 1 shows the class diagram of the model for the finite element method. Note that, in this paper, Rumbaugh's notations are used in all object-oriented diagrams (Rumbaugh et al. 1991).

From the Figure, it can be seen that the key classes in the model include the *FE_Domain*, *Node*, *Element*, *Boundary_Condition*, *Material* and *Geometry* classes. The *FE_Domain* class defines objects that represent finite element domains. An *FE_Domain* object simply possesses sets of *Node*, *Element*, *Boundary_Condition*, *Material* and *Geometry* objects. While *Node* and *Element* objects represent nodes and elements, *Boundary_Condition*, *Material* and *Geometry* objects represent sets of boundary conditions, materials and geometrical properties. In C++, the declaration of the *FE_Domain* class can be presented as follows:

```
class FE_Domain {
  private:
    vector <Node *>                   pNode;       //Pointers to nodes
    vector <Element *>                pElement;    //Pointers to elements
    vector <Boundary_Condition *>     pBC;         //Pointers to BCs
    vector <Material *>               pMaterial;   //Pointers to materials
    vector <Geometry *>               pGeometry;   //Pointers to geometrical properties
    ...

  public:
    long AppendNode(const vector <double> &X);
    long RemoveNode(long nodeIndex);
    void MoveNode(long nodeIndex, const vector <double> &X);
    long AppendElement(const vector <long> &elemNodeList,
    Element::Class elementClass,
    Element::Type elementType);
    long RemoveElement(long elementIndex);
    ...
};
```

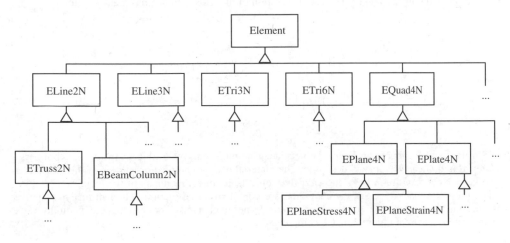

Figure 2   The hierarchical diagram of *Element* classes.

It can be seen from the above declaration that arrays of pointers to *Node, Element, Boundary_Condition, Material* and *Geometry* objects are used, not arrays of the objects themselves. This, however, does not mean that these member objects can exist after their *FE_Domain* object ceases to exist. In fact, these member objects are created within the *FE_Domain* object and are destroyed when the *FE_Domain* object is destroyed. This approach is employed based on the fact that nodes, elements, boundary conditions, materials and geometrical properties should not exist if there is no finite element domain. In other words, nodes, elements, boundary conditions, materials and geometrical properties are only parts of a finite element domain and they cannot exist by themselves. This kind of relationship between objects is called *aggregation*. The use of pointers increases the speed of the program when member objects are added or removed from an *FE_Domain* object during the preprocessing process. Also from the above declaration, it can be seen that the *FE_Domain* class may incorporate several methods that are used to manipulate member objects of an *FE_Domain* object, such as functions to add or remove *Node* objects.

A finite element node is defined by its coordinates and degrees of freedom. As a result, the coordinates and degrees of freedom must naturally be members of the *Node* class. In addition, a node must also possess information related to boundary conditions that are defined on its degrees of freedom. In C++, the declaration of the *Node* class may be presented as follows:

```
class Node {
    private:
        vector <double>               X;          //Coordinates
        vector <DoF>                  dof;        //Degrees of freedom
        vector <Element *>            pElement;   //Pointers to elements
        vector <Boundary_Condition *> pBC;        //Pointers to boundary conditions
        ...
    public:
        ...
};
```

In the declaration of the *Node* class, one of the member data is an array of pointers to *Element* objects. In a *Node* object, the *Element* objects pointed by the pointers are those that have the *Node* object as one of their nodes. Note that these *Element* objects are direct members of the *FE_Domain* object although they are accessed in the *FE_Domain* object by means of pointers. The *Node* object is not responsible for creating or destroying these *Element* objects. This kind of relationship is called *acquaintance*. In the *Node* class, there is also an array of pointers to *Boundary_Condition* objects. The *Boundary_Condition* objects pointed to by the pointers are those *Boundary_Condition* objects representing the boundary conditions that are applied to the *Node* object. Moreover, these *Boundary_Condition* objects are also direct members of the *FE_Domain* object. This implies that the relationship between the *Node* and *Boundary_Condition* objects is also of the acquaintance type.

Next, consider the *Element* class. The declaration of the class may be presented as follows:

```
class Element {
    private:
        vector <Node *>    pNode;      //Pointers to nodes
        Material *         pMaterial;  //Pointer to a material
        Geometry *         pGeometry;  //Pointer to a geometry set
        ...
    public:
        ...
};
```

Similar to *Node* objects, an *Element* object keeps pointers to its *Node* objects. In addition, pointers to *Material* and *Geometry* objects describing the material and geometrical properties of the element are also kept. Note that all *Material* and *Geometry* objects are direct members of the *FE_Domain* object. Since there are many types of elements that are commonly used in finite element analysis, derived classes must be used in order to introduce different types of elements. For example, the hierarchy of element classes shown in Figure 2 may be employed.

From Figure 2, it can be seen that elements are first categorized based on their geometry. For example, the *ELine2N* class represents line elements with two nodes while the *EQuad4N* class represents

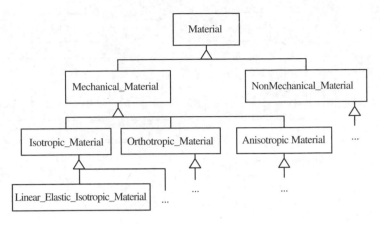

Figure 3. The hierarchical diagram of *Material* classes.

quadrilateral elements with four nodes. The classification based on geometry is used as the first level classification since, during the graphical input process, elements are always added via its geometry and detailed types of the added elements will have to be provided later. After the classification by geometry, other classification methods based on details that are more specific can be used. For example, the derived classes of the *ELine2N* class include the *ETruss2N* class for truss elements with two nodes and the *EBeamColumn2N* class for beam-column elements with two nodes.

The *Material* class is used to define *Material* objects used in finite element analysis. A *Material* object may be used by many different elements and, consequently, pointers to *Element* objects that employ the *Material* object are kept in the *Material* object. Similar to elements, there are many types of materials and derived classes are necessary. Figure 3 shows an example hierarchy of *Material* classes.

The last key class is the *Geometry* class. Objects of this class are used to store geometrical properties such as thickness, moments of inertia in different directions, etc. Similar to *Material* objects, in a *Geometry* object, pointers to *Element* objects that employ the *Geometry* object are stored as data.

3   AN OBJECT-ORIENTED MODEL FOR THE MESHLESS METHODS

In this section, an object-oriented model for the meshless methods will be considered. In the meshless methods, there is no element but nodes are still used. The integration to obtain the stiffness matrix equation in the meshless methods is naturally not performed on elements since there is no element. Rather, the domain is subdivided into many sub-domains called background cells, actually in a very similar way as the domain is subdivided into many elements in the finite element method. The integration is then performed on background cells, instead. A background cell is defined by using points in space in the same way as an element in the finite element method is defined by its nodes. These points, which are used to define background cells, are not used for the interpolation of unknown fields. For that purpose, nodes are used. The locations of nodes and points are completely independent. From this general information, the model shown in Figure 4 may be developed for the meshless methods.

Point objects in the meshless methods are, to a certain degree, similar to *Node* objects in the finite element method. A *Point* object possesses its coordinates and pointers to *Background_Cell* objects that employ the *Point* object. Oppositely, a *Background_Cell* object keeps pointers to its *Point* objects. *Point* objects do not possess *DoF* objects. *Node* objects possess *DoF* objects and have no relationship with *Background_Cell* objects. Another interesting point in this model is that *Material* and *Geometry* objects are associated with *Background_Cell* objects. This implies that, in addition to being used as sub-domains for integration, background cells are also used to define boundaries of materials and geometry sets. In the finite element method, these boundaries are usually defined by using boundaries of elements.

Although the object-oriented models for the finite element and meshless methods are not the same, similarities are evident. It is therefore advisable to merge the two models together in order to obtain one

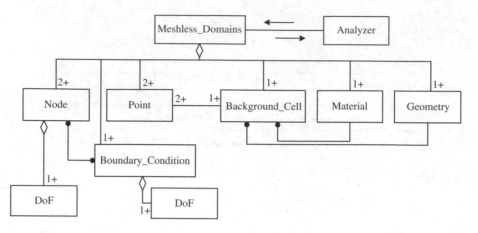

Figure 4.   The class diagram of the object-oriented model for the meshless methods.

model that can be used by both analysis techniques. This integration of the two models will especially promote the reusability of various components from the original models. As a whole, a smaller program can be expected. The merged model is proposed in the next section.

4   AN OBJECT-ORIENTED MODEL FOR COMBINED IMPLEMENTATION OF THE FINITE ELEMENT AND MESHLESS METHODS

From the object-oriented models for the finite element and meshless methods shown in Figures 1 and 4, the classes for both types of analysis techniques can be collectively considered. They can be categorized into two different types of class. The classes of the first type are those that represent the domain. These include the *FE_Domain* and *Meshless_Domain* classes. The classes of the second type are those whose objects are parts or components of the domain classes. These include the *DoF*, *Point*, *Node*, *Element*, *Background_Cell*, *Boundary_Condition*, *Material*, and *Geometry* classes. Most of these component classes are shared by both analysis techniques. Before merging the two original object-oriented models into one model, improvement on the organization of all available classes should be a priority. To begin with, relationships between objects of these classes must be carefully investigated. As objects of most component classes are used in both *FE_Domain* and *Meshless_Domain* classes, a new base class called *Domain* is introduced and the *FE_Domain* and *Meshless_Domain* classes can be derived from this new class. This is to allow objects of some component classes to be members of the *Domain* class and, as a result, they will automatically be members of both *FE_Domain* and *Meshless_Domain* classes. In addition, objects of all component classes always possess some common attributes. For example, if the application includes a graphical user interface, it is necessary to have graphical attributes for all components that can be drawn. Consequently, it is beneficial to create a base class called *Component* from which these component classes are derived. Moreover, a node can be thought of as a point with degrees of freedom. Therefore, the *Node* class can be derived from the *Point* class. If the *Node* class is derived from the *Point* class, the similarities between the *Element* and *Background_Cell* classes will be amplified. To utilize these similarities, a new base class called *Cell* from which the *Element* and *Background_Cell* classes are derived is introduced. The *Cell* class has pointers to *Point* objects as one of its members. Since the *Node* class is derived from the *Point* class, *Element* objects can utilize these pointers to keep its *Node* objects. Figure 5 shows the new hierarchy of the component classes.

Similar to the case of *Element* objects, there are many different types of *Background_Cell* objects. However, since background cells are used primarily to divide the whole domain into many sub-domains for integration, only classification based on geometry will suffice for background cells. In Figure 5, as examples, the *BCTri3P* and *BCQuad4P* classes represent two-dimensional background cells with three and four points, respectively. With the new hierarchy of the classes, the association between the classes in the proposed combined model for the finite element and meshless methods is shown in Figure 6.

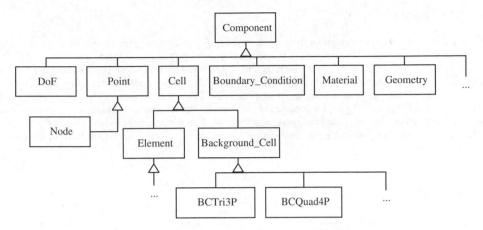

Figure 5. The new hierarchy of the component classes used in the combined model for the finite element and mesh-less methods.

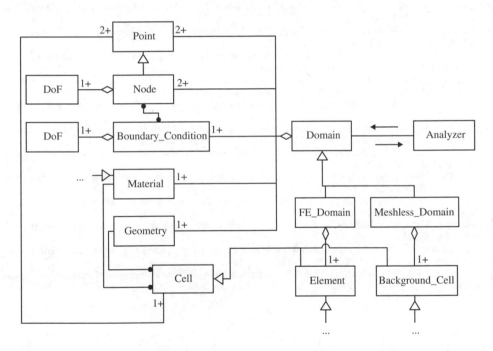

Figure 6. The combined object-oriented model for the finite element and meshless methods.

From the Figure, it can be seen that almost all component objects have become members of the *Domain* class with the exception of *DoF*, *Element*, and *Background_Cell* objects. *Element* objects are placed in the *FE_Domain* class while *Background_Cell* objects are placed in the *Meshless_Domain* class. In addition, the acquaintance-typed links between *Element* and *Background_Cell* objects and *Material* and *Geometry* objects are moved one level higher. In the proposed model, the *Cell* class is responsible for keeping these links. Information related to the links in the *Element* and *Background_Cell* classes is therefore removed. In addition, an acquaintance-typed relationship between *Cell* objects and *Point* object is also created. This link replaces the existing links between *Element* and *Node* objects and between *Background_Cell* and *Point* objects in the original models.

5 CONCLUSIONS

In this study, a new object-oriented model for combined implementation of the finite element and mesh-less methods is proposed. The new model is obtained by merging the two separate models of the finite element and meshless methods during which redundancies in the existing models are removed. To allow seamless integration of the two original models, some new classes are added as necessary and relationships between classes are also redesigned. The integration increases the reusability of the codes. As a result, the size of the resulting model becomes more compact than the combined size of the original two models. Furthermore, the proposed model is more flexible than the original models. One of the reasons is that it allows not only one of the two analysis techniques to be selected but also hybrids of the two to be implemented.

REFERENCES

Archer, G.C., Fenves, G. & Thewalt, C. 1999. A new object-oriented finite element analysis program architecture. *Computers and Structures* 70: 63–75.

Atluri, S.N., Kim, H.-G. & Cho, J.Y. 1999. A critical assessment of the truly meshless local Petrov-Galerkin (MLPG), and local boundary integral equation (LBIE) methods. *Computational Mechanics* 24: 348–372.

Babuska, I. & Melenk, J.M. 1997. The partition of unity method. *International Journal for Numerical Methods in Engineering* 40: 727–758.

Belytschko, T., Lu, Y.Y. & Gu, L. 1994. Element-free Galerkin methods. *International Journal for Numerical Methods in Engineering* 37: 229–256.

Carey, G., Schmidt, J., Singh, V. & Yelton, D. 1994. A prototype scalable, object-oriented finite element solver on multicomputers. *Journal of Parallel and Distributed Computing* 20: 357–379.

Donescu, P. & Laursen, T.A. 1996. A generalized object-oriented approach to solving ordinary and partial differential equations using finite elements. *Finite Elements in Analysis and Design* 22: 93–107.

Forde, B.W.R., Foschi, R.O. & Stiemer, S.F. 1990. Object-oriented finite element analysis. *Computers and Structures* 34(3): 355–374.

Jun, S., Liu, W.K. & Belytschko, T. 1998. Explicit reproducing kernel particle methods for large deformation problems. *International Journal for Numerical Methods in Engineering* 41: 137–166.

Mackie, R.I. 1992. Object-oriented programming of the finite element method. *International Journal for Numerical Methods in Engineering* 35: 425–436.

Mackie, R.I. 2001. Implementation of sub-structuring within an object-oriented framework. *Advances in Engineering Software* 32: 749–758.

Mackie, R.I. 2002. Using objects to handle calculation control in finite element modelling. *Computers and Structures* 80: 2001–2009.

Rumbaugh, J., Blaha, M., Premerlani, W., Eddy, F. & Lorensen W. 1991. *Object-oriented modeling and design*. New Jersey: Prentice Hall.

Yu, L. & Kumar, A.V. 2001. An object-oriented modular framework for implementing the finite element method. *Computers and Structures* 79: 919–928.

Zimmermann, Th. & Eyheramendy, D. 1996. Object-oriented finite elements I. Principles of symbolic derivations and automatic programming. *Computer Methods in Applied Mechanics and Engineering* 132: 259–276.

Computer-aided design, engineering and instruction

Computational Methods in Engineering and Science, Iu et al. (eds)
© 2003 Swets & Zeitlinger, Lisse, ISBN 90 5809 567 3

Construction planning tool based on virtual reality technology

P.G. Henriques & A.Z. Sampaio
Department of Civil Engineering and Architecture, IST/ICIST, Technical University of Lisbon,
Lisbon, Portugal

ABSTRACT: This paper describes a didactic application that is part of a research project whose main aim is to develop a computer-aided system which will assist design and construction processes. It is based on the visual simulation of construction activities. Geometric modelling and virtual reality techniques are used in the visualization of the design process and to define user-friendly interfaces in order to access construction information, which could prove useful to Civil Engineering professionals. As a first step, it was developed a prototype that serves as a didactic tool for Civil Engineering students of disciplines concerned with building construction. The construction of a double brick wall is the case studied. The wall is defined as a three-dimensional model formed with the several components needed to construct it. Using the wall's virtual model it is possible to show, in an interactive way, the sequence of the construction process and observe from any point of view the configurations in detail of the building components. This is then a didactic tool application in construction processes domain of great interest to Civil Engineering students.

1 INTRODUCTION

1.1 *Computer-aided design in construction*

Normally, academic and commercial applications of computer-aided design in construction provide a visual presentation of the final state of the project, that is, the three-dimensional (3D) representation of the building with an animated walk-through, allowing observation of both its interior and exterior. However, the tools of construction planning (scheduling, monitoring of progress and so on) work with the whole construction process, from its conception to its final phase. Not only are current computer tools and models unable to follow changes in the geometry of the building or structure during the construction process but also the representation of construction activities is not always compatible with, or even possible to be modelled by, these project aids. The visual simulation of the construction process needs to be able to produce changes to the geometry of the project dynamically.

1.2 *Research project*

The main aim of the research project: *Virtual reality in optimisation of construction project planning –* POCTI/1999/ECM/36300, ICIST/FCT (Henriques & Sampaio 1999) now in progress at the Technical University of Lisbon, within the program of activities of the Department of Civil Engineering and Architecture, is to develop a system for the planning and monitoring of construction activities by means of the visual simulation of its development (Henriques & Sampaio 2002). The innovative contribution lies in the application of 3D modelling techniques and virtual reality to the representation of information concerning construction, of practical use to civil construction professionals. As a first step it was developed a prototype that serves as a didactic tool for Civil Engineering students of disciplines concerned with building construction.

1.3 *Study case*

The study case is a common external wall composed with two brick panels. The wall's virtual model, developed along this work, allows the user to control the construction progression, in particular, the following actions:

- The interaction with the construction sequence by means of the production of 3D models of the building in parallel with the phases of construction;
- The accessing of qualitative and quantitative information on the status of the evolution of the construction;
- The visualization of any geometric aspect presented by the several components of the wall and the way they connect together to form the complete wall.

This communication is oriented to teaching construction techniques by means of virtual environments. The objective of this application is to show in which way new technologies afford fresh perspectives for the development of new tools in the training of construction process of edifices. It is then expected that the implementation of the prototype will be able to contribute to support teaching disciplines concerned with civil construction.

Using the developed virtual model, it permits students to study and to experiment alternative strategies for construction planning in the space provided by the virtual environment.

2 VIRTUAL REALITY TECHNOLOGY

2.1 Concept and historical background

Virtual reality can be described as a set of technologies, which, based on the use of computers, simulates an existing reality or a projected reality (Burdea & Coiffet 1993). This new tool allows computer-users to be placed in three-dimensional worlds, making it possible for them to interact with virtual objects at levels until now unknown in information technology: turning handles to open doors; switching lights on and off; driving a prototype car or moving objects in a house. To achieve this, elements of video, audio and three-dimensional modelling are integrated in order to generate reality, initially through specific peripheral devices (handles, helmets and gloves) and at present, through the Internet.

Its origin is attributed to flight simulators developed about fifty years ago by the American Army. The beginning of virtual reality is attributed to Ivan Sutherland, with the introduction in 1965 of the first three-dimensional immersion helmet, which was later divulged to the peripheral device industry with the designation, head mounted display. A precursor, Nicholas Negroponte (and collaborators), in the seventies, produced a virtual map of guided walks using a model of the city of Aspen, Colorado. In 1989, Jaron Lamier, an important driving force behind this new technology, designated it virtual reality (Vince 1998).

In the 90s, with the upsurge of the Internet, a specific programming language was defined, the virtual reality modelling language. It is a three-dimensional interactive language devised for the purpose of modelling and visually representing objects, situations and virtual worlds on the Internet, which makes use of mathematical co-ordinates to create objects in space (Burdea & Coiffet 1993). This constitutes a new means of communication that allows the construction of, and experimentation with, new computer-modelled realities. The virtual reality modelling language was the first standardised definition of three-dimensional space on the Internet.

2.2 Principal technological characteristics of virtual reality systems

Interaction and immersion can be considered the most important characteristics of virtual reality (Vince 1998): the immersion sensation is obtain by means of special physical devices, that allows the user to have the sensation of finding himself physically present in a world imagined and modelled by the system; the interactive characteristic is assumed because the virtual reality technology is not limited to passive visual representation of this world but enables the user to interact with it (touching or moving objects, for example). What is more, the virtual world responds to such actions in real time. In the developed application only the interactive propriety was explored.

Technically, the active participation of the user in a virtual environment, or, in other words, the sensation of immersion or presence in that environment, is achieved on the basis of two factors: the integration of information technology techniques (algorithms) used to obtain images of the highest degree of visual realism (ray-tracing, luminosity, application of textures etc.); the integration of a series of physical devices resulting from specific technologies like visual technology, sensorial technology (sensors of force and positioning) and mechanical technology (for transmitting movement such as a 3D mouse or gloves).

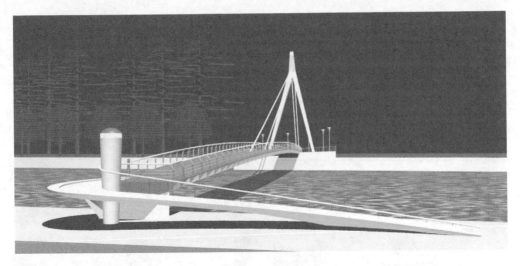

Figure 1.   A three-dimensional model of an overpass (a TRIEDE – Cons. Proj. Eng. Civil, SA courtesy).

One of the areas in which the incorporation of virtual reality technology as a means of geometric modelling and visual presentation of three-dimensional animated models is most often applied is Architecture (Fig. 1). The range of potential applications extends from buildings on a small scale to the development of town plans. However, virtual reality technology does not merely constitute a good interface but presents applications that provide the possibility of finding solutions to real problems in such diverse fields as Engineering, Medicine or Psychology.

3   DEVELOPING THE WALL VIRTUAL MODEL

3.1   Selecting the study case

As a study case in the building construction field a common part of a building was selected, an external wall with double brick panels. The developed virtual model allows the student to learn about the construction evolution concerning to an important part of a typical building.

The selected construction component focuses different aspect of the construction process: the structural part, the vertical panels and the opening elements. The 3D model of the wall was defined using the AutoCAD system (AutoDesk 2000), a computer-aided drawing system common in Civil Engineering offices. Next, the wall model was transposed to a virtual reality system based on a programming language oriented to objects, the EON system (EON Reality 2001).

3.2   Creating the wall 3D model

First, all building elements of the wall must be identified and defined as 3D models. Structural elements (demarking the brick panels), vertical panels of the wall and two standard opening elements, were modelled. In order to provide, later in a virtual space, the simulation of the geometric evolution of a wall in construction, the 3D model must be defined as a set of individual objects, each one representing a wall component.

3.2.1   Structural elements of the wall

Foundations, columns and beams, were considered as structural elements. The concrete blocks are defined as *box* graphic elements (available in the AutoCAD system) and the steel reinforcements as *cylinder* and *torus* graphic elements. Figure 2 shows some details of these components. In the image, it is possible to observe how to accommodate the steel reinforcements inside the structural elements. This is a real problem that is solved for each case in the work in loco. These elements were modelled taking account this kind of difficulty. So, it is an illustrative example.

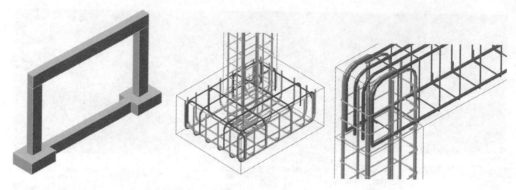

Figure 2.   Details of the structural elements of the wall 3D model.

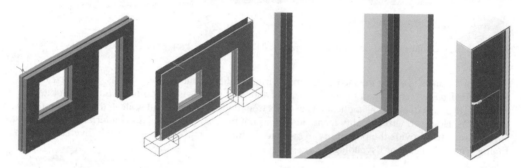

Figure 3.   Details of the vertical panels, window and door of the wall 3D model.

3.2.2   *Vertical panels of the wall*

Confined by the structural elements are two brick panels, a heating proof layer, two rendering coats and two painted surfaces. Initially, all these elements were modelled as boxes with different thickness as shown in Figure 3. The selection of thickness values for each panel is made according to the usual practice in similar real cases. Next, there were defined openings in the panels to place the window and the door elements (Fig. 3).

3.2.3   *Opening components of the wall*

Finally, the components of two common opening elements, a window and a door, were modelled (Fig. 3). The pieces of the window's and the door's frames were created as individual blocks. Each element was modelled taking in consideration the real configuration that such type of elements must present in real situations. By this, at the virtual animation of the wall construction, it is possible to observe each one separately and analyse conveniently the configuration details of those frames.

3.3   *Creating the virtual environment of the construction process*

One by one every part of each element considered as a building component of the wall was modelled. Figure 4 shows the complete wall model. Next, the 3D model was exported as a 3DStudio-drawing file (with the file extension.3sd) to the virtual reality system used, EON Studio (EON Reality 2001).

The virtual reality system should allow the manipulation of the elements of the wall model according to the planning prescribed for the carrying out of the construction. Supporting that, a range of nodes or function is available in the system to build up convenient virtual animations.

Figure 5 presents the work ambience of the EON system. In the left side of the image, is the nodes window (containing the virtual functions available in the system), in the central zone is the simulation tree (is the place where the drawing blocks hierarchy is defined and the actions are imposed to blocks)

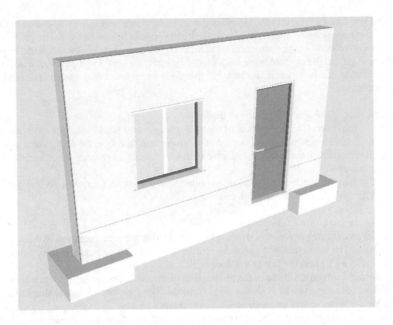

Figure 4. Projection of the developed 3D models of the wall.

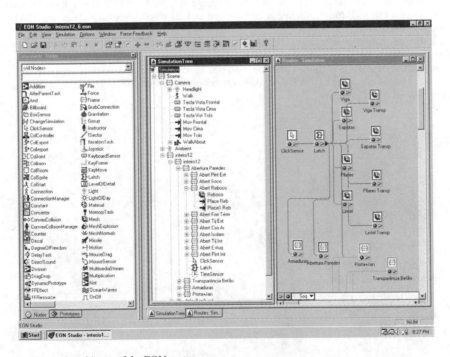

Figure 5. The work ambience of the EON system.

and, in the right side, the routes simulation window (the local where the nodes are linked). Essentially, the simulation is programmed using the last two windows.

When the wall model is inserted into the virtual system, the drawing blocks of the model are visualized in the central window. To programme an animated presentation the nodes or actions needed are picked from the nodes window and put into the simulation tree. Here, those nodes are associated to the blocks to be affected by the programmed animation. For instance, to impose a translation to the external rendering coat panel (named "reboco" in the simulation window in Fig 5), two *place* nodes are needed (one to define the place where to go and the other to bring back to it's original position). The *ClickSensor* and *Latch* nodes allow the initialisation of a programmed action. For that the user interacts with the virtual scenario pressing on a mouse button.

3.3.1 *Presentation of the vertical panels in explosion*

The exhibition of the several vertical panels of the wall in explosion is a kind of presentation with a great didactic interest. Figure 6 includes two steps of this animation, the opened and closed situations. The translation displacement value attributed to each panel was distinct from each other in order to obtain an adequate explosion presentation.

This type of presentation allows the student to understand the correct sequence of the panels in a wall and to observe the different thickness of the panels.

3.3.2 *Showing the evolution of construction phases*

The construction process was decomposed in 23 phases following the real execution of this kind of element in the work *in loco*. The first element to become visible, at the virtual scenario, is the steel reinforcement of the foundation (Fig. 7) and the last is the door pull.

The programmed animation simulates the progression of the wall edification. For each construction step the correspondent geometric model is shown (Fig. 7). In this way, the virtual model simulates the changes that really occur while the wall is in construction in a real work place.

For each new wall component becoming visible in a construction phase, the virtual model allows the user to pick the element and to manipulate the camera around it (Fig. 8). The user can then observe the element (displaced from the global model of the wall) from any point of view. Then all configuration details that the components of a real wall must present can be observed and analysed. This capacity is important in construction process training.

While the animation is in progress, a box text is presented, fixed at the upper right corner of the display (Fig. 8). It contains construction information about the step in exhibition. The text includes the number, the activity description and the material specification and quantification concerned to each phase. The visualization of this type of data following the virtual construction evolution is useful to students.

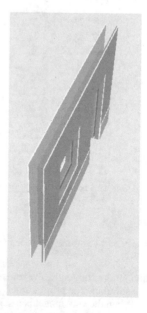

Figure 6.   Presentation in explosion of the vertical panels of the wall model.

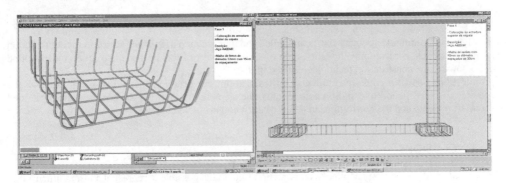

Figure 7.　Presentations of two steps of the virtual construction progress.

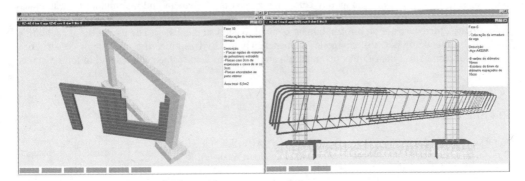

Figure 8.　Pictures presenting the box text with construction information and elements displaced from the global model of the wall.

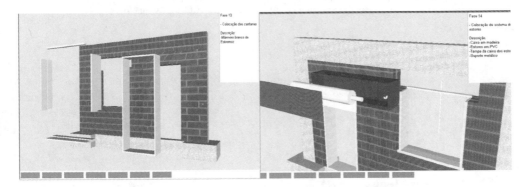

Figure 9.　Images presenting wall components in explosion.

The virtual animation presents, below the visualization area, a toolbar (Fig. 8). The set of small rectangles included in it shows the percentage, in relation to the wall fully constructed, up to the step visualized. To exhibit the next phase the user must click in any part of the model. To go back to an anterior step the user must click over the pretended rectangle in that progression toolbar.

Finally, the animation allows the user to visualize the peaces of wall elements in an exploded exhibition. The images included in Figure 9 show two elements presented in explosion. This type of presentation allows the student to know how the different parts connect each other to form wall components and can observe the configuration of those parts with detail. This capacity provided by the virtual model is also of great interest in construction domain instruction.

4 FUTURE PERSPECTIVES

Other type of building components can be modelled and manipulated in a virtual scenario for construction learning proposes. For example, the edification of several types of roofs or the construction underground walls showing the construction technologies used in it.

With those virtual models the students can learn about construction technologies and analyse the sequence of construction of building components, the steps required along the correspondent planned execution process and the configuration detail of each constructive element.

5 CONCLUSIONS

The virtual reality technology applied to construction field made possible to represent a three-dimensional space realistically. The visual simulation of the construction evolution of a common case was achieved. The user can interact with the virtual model of the wall and impose any sequence in the construction process, select from the wall model any component or parts of element and manipulate the camera as desire in order to observe conveniently any detail of the components configuration. While the animation is in process, the construction information associated to each step is listed. The use of these capacities, allowed by the developed virtual model, is beneficial to Civil Engineering student in construction process subjects. This type of didactic material can be used in face-to-face classes and in long distance learning.

REFERENCES

AutoDesk 2002. *AutoCAD – User manual, Release 2002*, AutoDesk, Inc.
Burdea, G. & Coiffet, P. 1993. *Virtual reality technology*, John Wesley & Sons.
EON Reality 2001. *Introduction to working in EON Studio*, EON Reality, Inc.
Henriques, P. & Sampaio, A.Z. 1999. *Project program: Virtual reality in optimisation of construction project planning*. POCTI/1999/ECM/36300, ICIST/FCT, Lisbon (Portugal).
Henriques, P. & Sampaio, A.Z. 2002. *Visual simulation in building construction planning*, 4th European Conference on Product and Process Modelling I, Portoroz (Slovenia), 9–11 September: 209–212.
Vince, J. 1998. *Virtual reality systems*, ACM SIGGRAPH Books series, Addison-Wesley.

Computational Methods in Engineering and Science, Iu et al. (eds)
© 2003 Swets & Zeitlinger, Lisse, ISBN 90 5809 567 3

Web-enabled quick response technology application in advanced product design

S.F. Wong
Department of Electromechanical Engineering, Faculty of Science and Technology,
University of Macau, Macao

W.W.C. Chung
Department of Industrial and Systems Engineering, The Hong Kong Polytechnic University, Hong Kong

ABSTRACT: Matching offers against customer needs is one the greatest challenges to electrical appliance firms to survive in the information age. Capturing the voices of customers for product design can enhance the value of a product. In the old economy, companies used Quality Function Deployment (QFD) to get the customer's voice. Much time would be spent on questionnaire surveys and paper work. In the new economy, the firms can innovate by leveraging IT and related technologies to acquire the relevant information and reuse it for product development. This paper describes a case to deploy QFD to increase the appeals of product design features to customers. A Web-Enabled Quick Response System (WEQRS) is proposed for to assist a firm intending to migrate to e-business. A prototype of a ventilating fan is used as an example to validate the concept of web-based QFD in operation.

1 INTRODUCTION

In information age, large companies have more resources such as IT technical personnel and financial resources. Moreover, they have better infrastructure to develop their e-business. So, bigger company is always ahead of Small and Medium Enterprise (SME) in terms of B2B development. This is base on the "Size matters" of e-business issue. So, SME how to survive in information age, this is quite important for them. The present-day strategic alliances of SME can help them to overcome this size matter of e-business issue. However, different SMEs are synthesized to develop their e-business. They will have the communication problems because different companies will have their data and knowledge format. How to consolidate their knowledge and data to common or standard format of knowledge this is necessary to involve knowledge management skill in SME to sharing information. Since SME cannot afford high cost investment in knowledge management, it is necessary to build up a low cost and simplified model to fit the case of SME.

Quick response (QR) is one of business strategy to maximize consumer satisfaction by implementing technologies (Ko 1997) (e.g. bar coding, electronic data interchange, and scanner). The information of product can be more easily transferred and communicated with the customer/business partner by these kinds of technologies. Moreover, Internet technology creates the global market share to the companies and customers. Customers/business partners can see and order the product anytime and anywhere. Meanwhile, it also is a good medium to connect the supplier chain for the collaborative companies. However, how to apply Internet technology and quick response system to benefit SME for surviving in information age? This is discovered sometimes when the end users/customers/business partners have their opinions about the products. The companies are hard to use the systematic method to get the customer's voice. Moreover, non-systematic customer voice, such as feedback their opinion of product design through e-mail, will involve the dirty data in the company database. They have to spend too much time filter the data and information coming from customers, and then catch the useful information and cleanse useless information. It will waste the manpower of the company and extend the product development time and level down the competitive advantages of product. In new economic age, how to apply Internet technology and knowledge management to enhance the communication speed with customers for the companies? Maybe creating web-enabled quick response system (WEQRS) can build up the bridge of company and customer or business partner to solve this kind of problems. Actually, WEQRS can apply in different area for business issues. However, the first step we consider to focus on product design area, because it will strongly affect the product life cycle. Shortening product life cycle can enhance the competitive advantages of SME. Therefore, they can survive in information age.

Sharing product information and electronic purchase reorder are most commonly mentioned in quick response technologies. Sometimes the customers/business partners can reorder the product and change the product colors and size through the companies' website. However, if the desired changes of product design will affect the product performance, intelligent elements have to be embedded in the website. Otherwise, general website of companies cannot satisfy customers' expectation.

In this research, we applied Active Server Page (ASP), Microsoft Access, Personal Web Server, and logical programming elements to build up the initial model of WEQRS design. The initial model of WEQRS was built up by the case study of ventilating fan. This virtual company was called to Company ABC. It was assumed to provide different kinds of ventilating fan, which can be ordered online, in their website. The Company ABC was to store the customers' order database for their business partners. Their business partners can check their order information on the website of Company ABC. If the business partner wants to order another new ventilating fan, they can directly design a new one on site by WEQRS of Company ABC. Moreover, they can also base on their checklist to modify their product from old design to new design by themselves. The modified parameters can be defined to two types that are performance change and product size change. The performance change will be considered to select different impellers or motors. A bigger power fan means the ventilating fan can match a higher flow rate. The WEQRS will be based on the product size criteria to search the suitable impeller and motor to meet the target use. The WEQRS will show the performance graphic result cause of human factor consideration for users' comprehension, so that the customers will more easily understand the analysis result and make decision. This system can also be used to enhance the performance for supply chain management by enabling the decision support for the stakeholders. Therefore, digitizing customer voice to feed the knowledge management system is considered one of critical success factor in manufacturing in the new economy for SME.

2 KNOWLEDGE SHARING AND EXCHANGE FOR ORGANIZATION

Nonaka and Takeuchi (1995) described four processes for the conversion of tacit and explicit knowledge (shown as Fig. 1), which they believe are crucial to creating value:

1. From tacit to tacit: the knowledge workers have to exchange or share their knowledge by face to face.
2. From tacit to explicit: through discussion and codifying to convert tacit knowledge into explicit knowledge means finding a way to express the inexpressible.
3. From explicit to explicit: combining different forms of explicit knowledge such as databases or EDI.

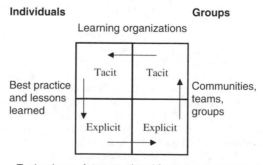

Figure 1. Exchanging and capturing tacit and explicit knowledge (source: Abell et al. 2001).

4. From explicit to tacit: interpret explicit knowledge using a person's frame of reference so that knowledge can be understood and then internalized or accepted by others.

Nowadays, the Internet and Video Conference technology are common to be used communication face to face. So, it can advance to share the tacit knowledge into tacit knowledge. However, this kind of knowledge sharing and exchange is very hard to re-use or systematically store in database. This is a disadvantage in this kind of knowledge. The business transaction is public to use explicit knowledge to explicit knowledge. However, most of this kind of knowledge only offers some standard product or material information. If we want to capture somebody experience for operations, decision making and so on, we have to capture the knowledge from tacit into explicit. So, most of consultant firms such as PricewaterhouseCoopers (PwC) will focus many of their KM activities on building explicit knowledge bases, because this process can enhance the quality of knowledge for organizations. If we can make high resolution of knowledge, we can more easily convert explicit knowledge to tacit knowledge. Maybe involving artificial intelligence elements in high resolution of knowledge will advance the result for converting it into tacit knowledge. So, the key factor of knowledge management can be focused on converting tacit knowledge into explicit knowledge. In this research, this is applied knowledge management technology by converting tacit knowledge into explicit knowledge in digitizing customer's voice.

3 THE WEQRS AND QFD

The concept of WEQRS (shown as Fig. 2) is based on Phase I (House of Quality) & II (Parts Deployment) of QFD application. The WEQRS interface can provide the engineering characteristics information of product design on the website to customer. The customer can articulate their needs through this system. The voice of customer can be digitized and analyzed to create value for the customer relationship management (CRM) of the company. The benefits of customers' voice capturing are: Minimizing the response time, Improving the response rate, Filtering the response information (Cleansing Dirty Data of Customers' Voice), Digitizing customers' needs, Enhancing the efficiency of customers' needs analysis, Easily creating CRM system, and Enhancing the precision level of forecasting market.

Based on the customer needs, the parts characteristics of a product have to be changed according to the engineering characteristics change. The parts characteristics change of product will have two cases. The first case is the changing part will not affect the performance of product such as changing color.

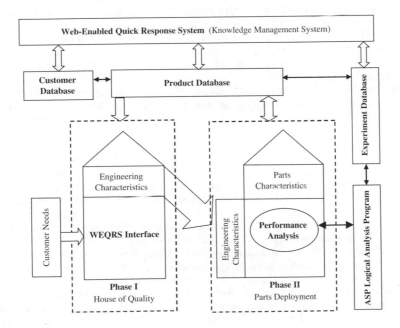

Figure 2. Architecture of WEQRS.

For this case, the system can directly provide the information of new product to the user and store it in customer information database. However, the second case is the changing part will be affected the performance of product such as changing material, component size or model. It is necessary to analyze the feasible report by ASP logical analysis program. The logical analysis program needs to call in some critical data from product database and experiment database, because it is necessary to simulate the analysis of product design engineer. So, the heart of WEQRS is the logical analysis program. Normally, each new product will be required to do the performance testing in research and development (R&D) department in the company. However, some predictable changes for product performance can be simulated. Meanwhile, the precision level of analysis result will depend on the quality of experiment database. The logical analysis program can quickly response to the customer requirement and automatically provide the feasible analysis to the users. So, it will help the customer to more easily make the decision in ordering the product. Moreover, the feasible result will be stored in customer information database. Their new order will be automatically updated. The customer information database will store the fresh information through this WEQRS system. The company can develop their customer relationship management (CRM) system through this updated customer information.

4 CASE STUDY OF VIRTUAL COMPANY OF VENTILATING FAN

The Company ABC is an e-company. They mainly provide different kinds of ventilating fan to sell on the website. Their engineering department needs to spend so many times to do feasibility analysis for different new product designs after they get the feedbacks of customers that desire to order different designs of ventilating fan. They discover this kind of problem will increase the workload of engineering department and reduce the speed of transaction. They worry that the customers may be lost in the future cause of this information overload problem. So, they tried to enhance the functions of their on-line order system.

The general e-commerce website it may offer the product redesign function, but it only focuses on the product size and color exchanging. However, this kind of system cannot satisfy the expectation of customers. If some performance analysis can be directly done by on-line system, it will increase the competitive advantages for the company and attract customers' motivation to voice their needs.

4.1 WEQRS interface

The primary model only designed for business partners with Company ABC. The database system of Company ABC had already stored the information about the order records of their business partners. When the business partners want to order the new product, they can login their account through the website of Company ABC (shown as Fig. 3). Then WEQRS will pass through ASP program and Microsoft Access to call in the data of their records (shown as Fig. 4). Their record will show the design code, motor code, impeller code, impeller diameter, shell code, login password, simple description, and company name. If the business partner wants to modify their products, they can directly select the target product base on the providing information. The user can access this website by the address:http://www.misr.mf. polyu.edu.hk/alfred/design.asp

4.2 Product modification

After the user selects their target product in record, then he can process the modification part. The user can modify their product by two parameters – Impeller Size and Motor Model, because these two parameters can also affect the product performance or size. If the user only expects to reduce the dimension of ventilating fan and keep the same flow rate, he can consider to change the impeller size first (shown as Fig. 5). Otherwise, he can select different motor models to change the product performance. The user can change or keep the product performance (flow rate) on the next step. After the user confirms the flow rate, the WEQRS can search the suitable impeller to match the performance as the customer requirement.

The logical analysis program (shown as Fig. 6) will analyze the different model of motor and impeller size to match the flow rate requirement such as the following program.

The customer can base on their requirement to select the suitable impeller model. After they confirm the impeller model, the system can analyze performance changing and show it both of the table format and graphic format (shown as Fig. 7). The user can see the ventilating fan will have different flow rate under different environment pressure. This kind of information can help the user to make the decision.

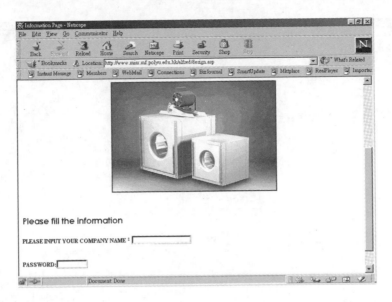

Figure 3.   WEQRS interface.

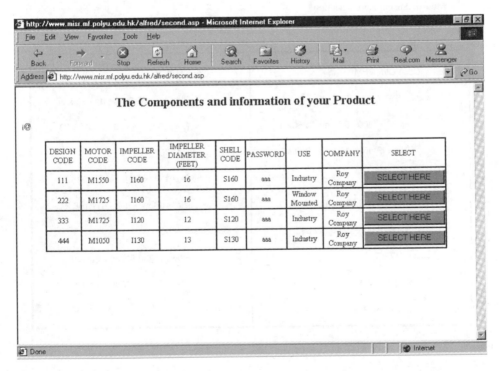

Figure 4.   Records of business partner.

4.3   Logical analysis of system

The critical point of system analysis is the flow rate of ventilating fan, because normally the flow rate is necessary to be confirmed by experiment testing to different size of impeller and motor model.

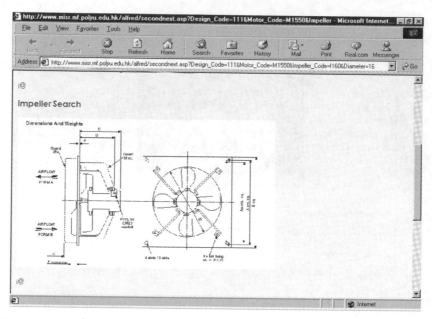

Figure 5. Modified parameters' selection.

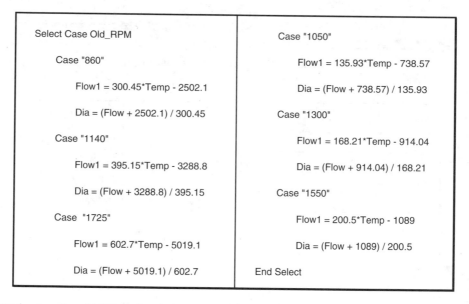

Figure 6. A portion of ASP logical analysis program of flow rate change.

Then experimental result can be drawn the performance graphic. However, the user may randomly change different kinds of impeller and motor on the website, how to analyze different performance result immediately without experiment? This can be considered to use interpolated method to solve this problem. WEQRS is used ASP to design interpolated program to analyze the experimental data of ventilating fan. So, the system can directly search the suitable impeller and motor model to fit the customer requirement.

604

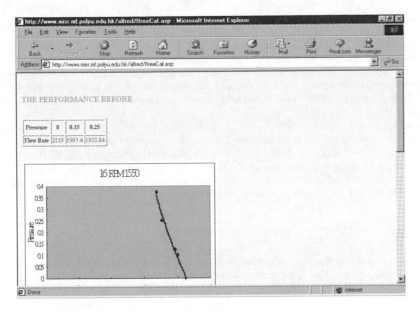

Figure 7.   Result of performance changing.

After the user can confirm the impeller size and motor model, the system can immediately search the suitable shell to fit the assembly of impeller and motor. When the user confirms the shell model, the system will create the new product record in the database for this business partner. The customer's voice can be acquired at that time. Company ABC can make their marketing strategic base on these voices.

5   LIMITATION OF INITIAL MODEL OF WEQRS

The initial model of WEQRS only provides the feasible product design information to customer. However, if the parts characteristics changing will affect the "Key Process Operations" or "Production Requirements", this is necessary to involve the elements of process planning and production planning. The full version of WEQRS will be expected to integrate Phase III (Process Planning) and IV (Production Planning) of QFD, because it can provide the detail feasible production report to both of the company and customer from WEQRS. Moreover, one of key factor of integrating operation process knowledge is necessary to involve and share the knowledge of workers for process planning. However, this is very hard to control this kind of knowledge, because normally sharing this kind of knowledge will meet the barriers as: lower education level of workers, language problems, computer operation problems of workers, and low motivation of workers by sharing knowledge. So, these kinds of factors will limit the development of WEQRS. The company has to be overcome these bottlenecks, if they want to develop full version WEQRS.

6   CONCLUSION

Integrating quick response technology and QFD to WEQRS can overcome the disadvantages of traditional quick response system, because the customer not only orders the standard product from companies' web-site. Moreover, they can involve their idea for the product through WEQRS. The system can analyze the customer needs and automatically provide feasible analysis to the customer by logical analysis program. So, the logical analysis program will be the core of this system. High quality experiment data will enhance the intelligence level of this logical analysis program. So, the satisfaction of customer will be increased because they can predict their new product performance through Internet and this system. Meanwhile, this system can increase the customers' motivation to share their information or needs to the company.

This system can achieve the small lot orders, electronic purchase reorder, enhancing electronic data interchange, and product planning with customer for the company. The customer relationship management system will directly update the customer needs from this system. This kind of information can help the decision maker planning the production strategy to market demand. Manpower will be saved for the company, because a part of engineering problems can be automatically solved by this web system.

The first tier's knowledge may be hard to capture from the workers. It will be affected the development of Phase III and IV of WEQRS. Enhancing workers' motivation of knowledge management may help to solve this barrier.

So, it is suggested to develop Phase I and II of WEQRS for extending competitive advantages of the firm special for small and medium enterprise, because this system is more economical and efficient. Moreover, this system can convert the customer needs to digital format to help the firm to make the market strategic. In this information age, this is trend to convert information from the free text style to the structure and standard data. Customer voice (tacit knowledge) is also necessary to digitizing to explicit knowledge for the knowledge management system.

ACKNOWLEDGEMENTS

The authors acknowledge the funding support by The Hong Kong Polytechnic University under project number G-W109, and Mr. K.S. Cheang to collect the data for the prototype.

REFERENCES

Abell, A. & Oxbrow, N. 2001. Competing with knowledge, *TFPL*.
Garstone, S. 1995. Electronic data interchange (EDI) in port operations, *Logistics Information Management* 8(2): 30–33.
Kalakota, R. & Robinson, M. 2000. e-Business 2.0 – Roadmap for Success: *Addison Wesley*.
Ko, E. & Kincade, D.H. 1997. The impact of quick response technologies on retail store attributes, *International Journal of Retail & Distribution Management* 25(2): 90–98.
Loshin, D. 2001. Enterprise Knowledge Management: The Data Quality Approach, *Academic Press*.
Romero, N.A., Pino, J.A. & Guerrero, L.A. 2000. Organizational Memories as Electronic Discussion By-products, *IEEE*: 11–18.
Schwartz, D.G. & Te'eni, D. 2000. Tying Knowledge to Action with kMail, *IEEE Intelligent Systems*: 33–39.
Stijn, E.V. & Wensley, A. 2001. Organizational memory and the completeness of process modeling in ERP systems – Some concerns, methods and directions for future research, *Business Process Management Journal* 7(3): 181–194.
Yang, P.C. & Wee, H.M. 2001. A quick response production strategy to market demand, *Production Planning & Control* 12(4): 326–334.

Computational Methods in Engineering and Science, Iu et al. (eds)
© 2003 Swets & Zeitlinger, Lisse, ISBN 90 5809 567 3

Design of coils with potential use in transcranial magnetic stimulation

A. Ferraz & L.V. Melo
Physics Dep., Instituto Superior Técnico, Lisboa, Portugal

D.M. Sousa
SMEEP, Instituto Superior Técnico, Lisboa, Portugal

ABSTRACT: Transcranial magnetic stimulation (TMS) is a new tool for neuroscience allowing for non-invasive stimulation of neurons. It has shown potential for treating several neuropsychiatric disorders such as hypokinetic movement, anxiety, schizophrenia and depression. It has also been used as a brain mapping tool. By inducing electric currents with a time-varying, strong magnetic field, TMS can modulate the neuronal circuits associated with these disorders. This modulation should be spatially-selective in order to minimize the stimulation of neighboring neuronal circuits. The improvement of this selectivity is then a key issue for this technology. We present an innovative coil configuration with two additional coils allowing for a significant improvement in spatial selectivity as compared to standard configurations. We simulated this coil configuration both by analytic computation of the Biot-Savart law. The results show one sharper, more intense spatial field peak resulting in improved field localization for this new coil configuration. The power losses, a very important parameter for actual implementation, were also computed using the analytical method and a finite element method. The results obtained from both methods are in agreement.

1 INTRODUCTION

Transcranial Magnetic Stimulation (TMS) is a tool for neuroscience which allows non-invasive stimulation of neurons (George 1996). It appeared in 1985 (Barker et al. 1985) as a brain mapping tool but it is nowadays considered as an established research tool in the cognitive neurosciences field, including the study of perception, attention, learning, plasticity, language and awareness (Walsh & Cowey 2000). TMS has also being used in the study and treatment of several neuropsychiatric disorders, such as movement disorders, epilepsy, both unipolar and bipolar disorders, anxiety disorders and schizophrenia (George et al. 1999, Topka et al. 1999, Fitzgerald 2002). Research aiming the use of TMS in other types of disorders has been reported for several times in the last few years (e.g. Barclay 2002).

The basic mechanism of this technique is Faraday's principle of electromagnetic induction (e.g. Jackson 1975). According to it, an oscillating electric current passing through a coil produces an oscillating magnetic field which in turn induces an electric current in an adjacent coil. Using the TMS method, the first coil is placed over somebody's scalp and his/her brain tissue reacts like the second coil. The intensity and duration of the current and the geometry of the primary coil are responsible for the amplitude and shape of the oscillating magnetic field produced and thus determines the intensity and location of the electric current induced in the brain. This is quite different from the electroconvulsive therapy (ECT), in which the application of an externally generated electric current is transmitted through the skull. In TMS, the magnetic pulses enter the brain painlessly and unimpeded – unlike the electric currents used in ECT – inducing an electric field responsible for the neuronal depolarisation in a localised volume under the skull. Reports on the therapeutic efficiency comparing ECT and TMS have been coming out (e.g. Janicak et al. 2002) and so far the latter has been referred as holding some important advantages over the former.

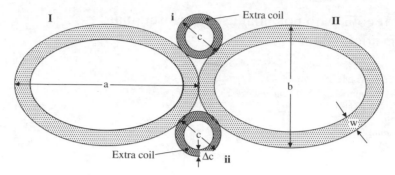

Figure 1. Generic geometry.

The coil design is limited by the combination of several physical parameters, such as weight and volume, the power it can dissipate, the spatial resolution and depth of the induced magnetic field (Ruohonen 1998) and the specific application that is required.

In Figure 1 we present the geometry used in this study.

In this work, we present the magnetic field obtained with different coil geometries and current densities. In a first approach we used a standard analytical method to evaluate the magnetic field for a simple geometry as a function of the current intensity.

To analyse the magnetic field generated by two pairs of coils with the geometry presented in Figure 1 and to complement the preliminary approach, the finite element method (FEM) was used (Reddy 1985). This method allows finding the solution of some problems usually associated with complex geometries, considering that the overall geometry was a collection of simple geometric elements. In our case, it is used to analyse the magnetic field generated by the proposed coils geometry using an appropriate software (Meeker 2002).

2   ANALYTICAL METHOD

The standard stimulating coils used in TMS are either circular or figure 8-shaped (Bohning 2000), which induces a magnetic field that is maximum under the intersection of the two windings (Ueno 1988). The peak magnetic field strengths are in the range 1–5 Tesla for currents' intensities between 5 and 10 kA and typical durations between 0.1 and 1 ms.

For the simplest geometry, a single circular coil, the magnetic field was evaluated using the referential depicted in Figure 2.

The magnetic field is given by (Wilson 1997):

$$B_r = \frac{\mu_0 I}{2\pi} \frac{z}{r} \frac{1}{\left[(e+r)^2 + z^2\right]^{1/2}} \left\{ -K(k) + \frac{e^2 + r^2 + z^2}{(e-r)^2 + z^2} E(k) \right\} \tag{1}$$

$$B_z = \frac{\mu_0 I}{2\pi} \frac{1}{\left[(e+r)^2 + z^2\right]^{1/2}} \left\{ K(k) + \frac{e^2 - r^2 - z^2}{(e-r)^2 + z^2} E(k) \right\} \tag{2}$$

where $K(k)$ and $E(k)$ are the elliptical integrals:

$$K(k) = \int_0^{\pi/2} \frac{d\theta}{\left(1 - k^2 \sin^2(\theta)\right)^{1/2}} \quad ; \quad E(k) = \int_0^{\pi/2} \left(1 - k^2 \sin^2(\theta)\right)^{1/2} d\theta \tag{3}$$

with $k^2 = 4er/[(e+r)^2 + z^2]$ and $2\theta = \pi - \phi$.

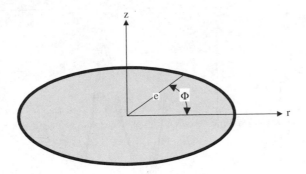

Figure 2.   Circular current loop referential.

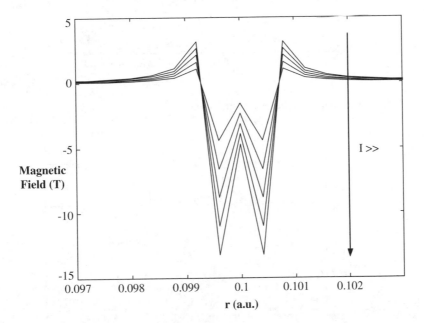

Magnetic Field (T)

Figure 3.   Magnetic field along r for the same current direction.

In Figures 3 and 4 we present the intensity of the magnetic field along r, without current in the extra (smaller) coils, for several intensities and different directions of the current in the larger coils. These results are compatible with previous results from other authors (e.g. Ruohonen 1998, George 2001).

In Figures 5–8 we show the influence of the extra coils on the intensity of the magnetic field along r for different directions of the current in the coils.

In Table 1 we present the resistance (R), self-inductance (L) and power losses (in the large coils and in the extra coils) results obtained for different parameters of the coils system. These parameters are very important for actual implementation.

3   FINITE ELEMENT METHOD

With the FEM the current in each coil is split into small, finite elements, which are used to compute the power losses. The main advantage of this method is that it can be easily adapted to real geometries (e.g. thick coils) allowing more realistic results. The results obtained for configurations similar to the previous ones are shown in Table 2.

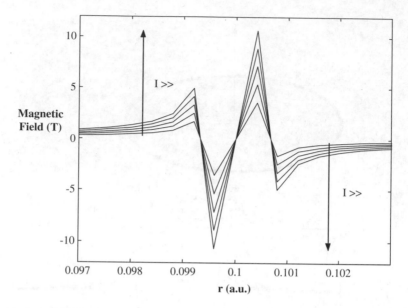

Figure 4. Magnetic field along r for different current directions.

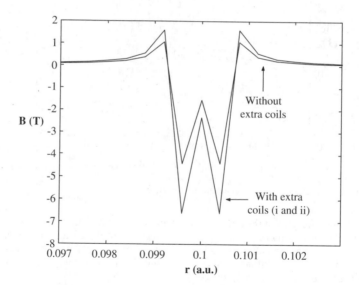

Figure 5. Magnetic field along r for the same current direction in all coils (clockwise).

4 DISCUSSION AND CONCLUSIONS

We propose an innovative design for a TMS coil configuration. This configuration adds a pair of smaller coils to the standard two-coil configuration. Analytic simulations show a better localized magnetic field allowing for a more accurate selection of the neural stimulation region. Our results show that consider-able improvements are obtained when the large coils are driven with currents in the same direction and

610

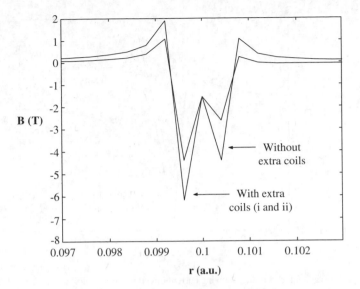

Figure 6.   Magnetic field along r for the following current directions: clockwise (cw) for coils I, II and i; counter-clockwise (ccw) for coil ii.

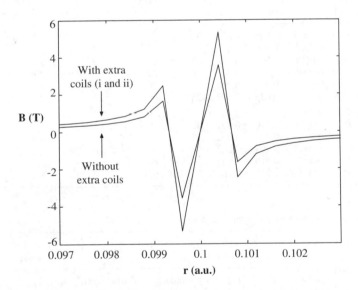

Figure 7.   Magnetic field along r for the following current directions: cw for coils I and i; ccw for coils II and ii.

the additional coils are driven with currents in opposite directions. The power losses in the coils could also be computed using the analytical method. The power losses were also calculated using a FEM. This method allows for realistic parameters (e.g. coil width) to be taken into account. The results obtained show good agreement with the previous ones.

Ongoing work concentrates on 3D field simulations using FEM and in optimising current conditions and control.

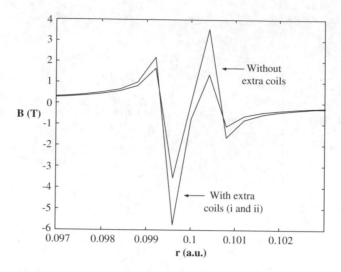

Figure 8. Magnetic field along r for the following current directions: cw for coil I; ccw for coils II, i and ii.

Table 1. Resistance (R), self-inductance (L) and power losses for different coil system parameters.

J (A/mm²)	a/b	a (m)	w (m)	c − Δc (m)	I and II loops	i and ii loops	R (mΩ)	L (μH)	I and II losses (kW)	i and ii losses (kW)
100	1	0.1	0.0028	0.08 − 0.002	36	50	49.9	14.4	2.0	1.6
100	0.5	0.1	0.0028	0.010 − 0.002	36	50	77.2	28.7	3.1	2.0
100	2	0.1	0.0028	0.012 − 0.002	36	50	39.9	7.2	1.6	2.4
100	1	0.1	0.0040	0.008 − 0.002	26	60	18.0	10.3	2.9	1.6
100	0.5	0.1	0.0040	0.010 − 0.002	26	60	27.7	20.5	4.4	2.0
100	2	0.1	0.0040	0.012 − 0.002	26	60	14.5	5.1	2.3	2.4
50	1	0.1	0.0028	0.08 − 0.002	70	50	99.8	28.7	1.0	0.8
50	0.5	0.1	0.0028	0.010 − 0.002	70	50	154.4	57.4	1.5	1.0
50	2	0.1	0.0028	0.012 − 0.002	70	50	79.7	14.4	0.8	1.2
50	1	0.1	0.0040	0.008 − 0.002	50	60	36.0	20.5	1.4	0.8
50	0.5	0.1	0.0040	0.010 − 0.002	50	60	55.3	41.1	2.2	1.0
50	2	0.1	0.0040	0.012 − 0.002	50	60	28.9	10.3	1.2	1.2

Table 2. Power losses for different coil system parameters.

J (A/mm²)	I and II loops	i and ii loops	a/b	a (m)	w (m)	c − Δc (m)	I and II losses (kW)	i and ii losses (kW)
100	36	50	1	0.1	0.0028	0.10 − 0.002	3.4	1.4
100	36	50	0.5	0.1	0.0028	0.10 − 0.002	4.4	1.4
100	36	50	2	0.1	0.0028	0.10 − 0.002	2.0	1.4
100	26	60	1	0.1	0.0040	0.12 − 0.002	5.0	1.8
100	26	60	0.5	0.1	0.0040	0.12 − 0.002	6.2	1.8
100	26	60	2	0.1	0.0040	0.12 − 0.002	2.6	1.8
50	36	50	1	0.05	0.0028	0.10 − 0.002	0.8	0.8
50	36	50	0.5	0.05	0.0028	0.10 − 0.002	1.0	0.8
50	36	50	2	0.05	0.0028	0.10 − 0.002	0.6	0.8
50	26	60	1	0.05	0.0040	0.12 − 0.002	1.2	1.0
50	26	60	0.5	0.05	0.0040	0.12 − 0.002	1.6	1.0
50	26	60	2	0.05	0.0040	0.12 − 0.002	0.6	1.0

REFERENCES

Barclay, L. 2002. Tinnitus may respond to transcranial magnetic stimulation. *Ann Neurol online (53)*.

Barker, A.T. et al. 1985. Non-invasive magnetic stimulation of human motor cortex. *Lancet* 1: 1106–1107.

Bohning, D.E. 2000. Introduction and Overview of TMS Physics. In George, M.S. & Belmaker, R.H. (eds.), *Transcranial Magnetic Stimulation in Neuropsychiatry*: 13–44. Washington DC: American Psychiatric Press, Inc.

Fitzgerald, P.B., Brown, T.L. & Daskalakis, Z.J. 2002. The application of transcranial magnetic stimulation in psychiatry and neurosciences research. *Acta Psychiatrica Scandinavica* 105(5): 324–340.

George, M.S. et al. 1996. Transcranial magnetic stimulation: a neuropsychiatric tool for the 21st century. *J Neuropsychiatry Clin Neurosci* 8: 373–382.

George, M.S. et al. 1999. Transcranial magnetic stimulation applications in neuropsychiatry. *Arc Gen Psychiatry* 56: 300–311.

Jackson, J.D. 1975. *Classical Electrodynamics* (2nd Edition). John Wiley & Sons.

Janicak, P.G. et al. 2002. Repetitive transcranial magnetic stimulation versus electroconvulsive therapy for major depression: preliminary results of a randomized trial. *Biol Psychiatry* 51: 659–667.

Reddy, J.N. 1985. *An Introduction to the Finite Element Method*. McGraw-Hill Book Co.

Ruohonen, J. 1998. *Transcranial Magnetic Stimulation: Modelling and New Techniques*, PhD Thesis, Espoo, Finland: University of Technology.

Topka, H. et al. 1999. Cerebellar-like terminal and postural tremor induced in normal man by transcranial magnetic stimulation. *Brain* 122: 1551–1562.

Ueno, S. et al. 1988. Localised stimulation of neural tissue in the brain by means of a paired configuration of time-varying magnetic fields. *J Appl Phys* 64: 5862–5864.

Wilson, M.N. 1997. *Superconducting Magnets*. Oxford: Oxford Science Publications.

Computational Methods in Engineering and Science, Iu et al. (eds)
© 2003 Swets & Zeitlinger, Lisse, ISBN 90 5809 567 3

Computer based techniques to design antennas in microwave diagnostics for fusion plasmas

M. Manso, J. Borreicho, L. Farinha, S. Hacquin, F. Silva & P. Varela
Centro de Fusão Nuclear, Associação EURATOM/IST, Lisboa, Portugal

D. Wagner
Max-Planck Institut für Plasmaphysik, Association EURATOM/IPP, Garching, Germany

ABSTRACT: Fusion research aims at large scale energy production with low environmental impact. A new machine (ITER) will be constructed in the frame of the collaboration between EU, Japan, RF, Canada, China and US. *Microwave reflectometry* is an essential diagnostic for ITER, one key issue being the design of antennas to launch and receive the probing microwaves. We develop a set of software tools to study those antennas. A mode matching FORTRAN code calculates the field distribution in the antenna aperture. The target fusion plasma is built and modeled by a MATLAB code and in order to solve the ray tracing differential equations we used the numerical Runge-Kutta method. A graphical user interface (GUI) was coded in MATLAB. A ray tracing approach computes the wave trajectory in the plasma and provides an estimation of the reflected energy. The developed software package proofed essential to design microwave antennas for reflectometry diagnostics.

1   INTRODUCTION

Fusion – the energy source of the sun – has the long range potential to serve as an abundant and clean source of energy. Fusion energy release involves reactions in a very hot gas – a plasma – in which two light atomic nuclei combine to form a heavier nucleus. Fusion fuels are deuterium and tritium, isotopes of hydrogen. Although tritium is not readily available it can be produced from lithium. Both deuterium and lithium are abundantly available to all nations for thousands of years. Plasmas that are hotter than the core of the sun have been produced and confined by strong magnetic fields in the laboratory. The challenge now is to make fusion energy practical.

With that purpose, a new machine – ITER-FEAT – has been designed and will be constructed in the frame of an international collaboration between EU, Japan, RF, Canada, China and US. Diagnostics are essential on ITER for machine operation, to guide the experimental program and to understand the physics issues. The requirements for diagnostics are much more demanding than in present devices and new capabilities must be demonstrated before being applied on ITER.

Microwave reflectometry (Hacquin et al. 2003) is an important diagnostic to measure the plasma electronic density profile and its fluctuations due to its robustness and high measuring capability. It is based on the radar principle and uses the reflection of microwaves with variable incident frequencies at different plasma cutoff layers. Present reflectometers are able to perform plasma density measurements with great accuracy but the access to the plasma core at high density is difficult due to refractive and scattering effects on the propagation path and at the reflecting region. To solve this problem one key issue is the design of the emitting/receiving antennas (Moresco et al. 1982) that launch the probing microwaves into the plasma and collect the reflected signals.

The propagation in the plasma is quite complex and the design of the antenna has to take into account the effect of the plasma. It is therefore necessary to simulate not only the antenna but also the propagation of the waves in the fusion plasma.

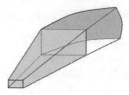

Figure 1.   The hog-horn antenna has two main parts: the taper and the ellipsoid reflecting mirror.

2   RAY TRACING CODE

Among the methods that provide approximate solutions for plasma wave propagation studies, ray tracing is a very attractive one, due to its relatively simple implementation and flexibility. It makes use of the W.K.B. approximation, which is based on the fact that the plasma refractive index varies slowly enough so that the propagation equation can always be solved locally. The slow variation of the refractive index at the scale of the wave length leads to considering the plasma as a set of layers with the same refractive indexes – stratified model. In the ray tracing theory the wave is therefore modelled as rays, similar to light rays. A ray is assumed to propagate along a straight line, bent only by refraction and reflection.

In this work we use ray tracing to study different antenna solutions for reflectometry diagnostics in fusion plasma machines. Using this method, the trajectory of a probing wave launched into the plasma can be estimated. Assuming the probing wavelength is small with respect to the dimensions of the system, the optical geometrics approximation can be used. This assumption fails near cut-offs and resonances and the ray tracing solution becomes invalid.

The ray tracing differential equations are shown in Equation 1 for a stationary regime and O-mode propagation (E//B). The developed ray tracing code uses the Runge-Kutta method. The solver was coded in C and a MATLAB graphical user interface (GUI) was developed.

$$\frac{dx}{d\tau} = 2\frac{k_x}{k_0^2} \quad \frac{dy}{d\tau} = 2\frac{k_y}{k_0^2} \quad \frac{dk_x}{d\tau} = -\frac{1}{n_c}\frac{dn}{dx} \quad \frac{dk_y}{d\tau} = -\frac{1}{n_c}\frac{dn}{dy} \tag{1}$$

The ray tracing method is also used to estimate the amount of energy returning to the antenna. In this work the same antenna is used both for emission and reception. Each ray is weighed at launch and upon returning to the antenna by a number defined by the radiation pattern of the antenna.

3   DETERMINATION OF THE ANTENNA RADIATION PATTERN

In addition to the operating frequency and some antenna parameters, such as the size of the antenna mouth, the ray tracing code also needs the antenna radiation pattern and the plasma model. A hog-horn antenna, shown schematically in Figure 1, is used in the simulation. To obtain the far field radiation pattern, which characterizes the antenna itself, three numerical codes were developed: (i) a mode matching code; (ii) a radiation pattern calculating code; and (iii) a radiation pattern formatting code.

The mode matching code is written in FORTRAN and takes into account all propagating and evanescent waveguide modes to analyze and design the antenna. This code accurately models step discontinuities in waveguides and continuous variations of the cross section by a staircase representation. The amplitude and phase distribution in the antenna aperture are calculated by a superposition of the resulting mode mixture. The field distribution is used to calculate the far field radiation pattern. The radiation pattern calculating code is written in FORTRAN and is used to determine the far field pattern, given the field distribution at the antenna mouth. Finally, the radiation pattern formatting code was also developed in MATLAB and is used to normalize the radiation pattern given by the radiation pattern calculating code. The normalization is achieved through spline interpolation and it is this formatted output file that is used as an input to the ray tracing code. Figure 2 shows a schematic of the codes interaction.

The fusion plasma is modelled after the ASDEX Upgrade plasma by a MATLAB code that receives as inputs the density profile and the magnetic flux elongation curves, and outputs the files with the density and gradient matrices. The ASDEX-Upgrade tokamak is installed in the Max-Planck Institut für Plasmaphysik, in Garching, Germany.

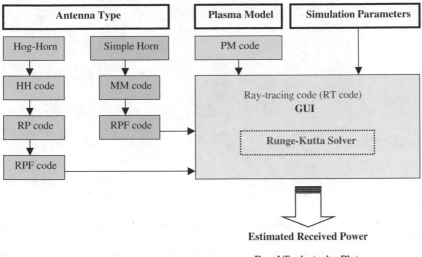

Figure 2. Schematic view of code interaction: RT code – ray-tracing code, PM code – plasma modeling code – RPF code – radiation pattern formatting code, HH code – hog-horn design code, MM code – mode matching code, RP code – radiation pattern calculation code.

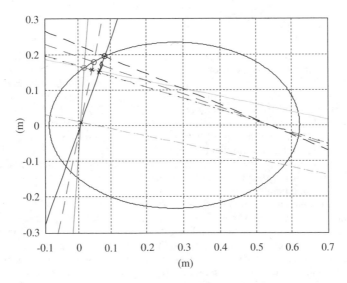

Figure 3. Geometrical construction of the hog-horn mirror. This figure is an output of the HH code.

4 ANTENNA DESIGN

A hog-horn antenna has two main parts: a pyramidal horn and an elliptical reflecting mirror. A hog-horn design code was developed in MATLAB. In this work two types of antennas were considered: (i) standard gain antennas, for which the gain and the maximum antenna length are specified as input data, and the code calculates the taper and respective mirror; (ii) pre-determined horn antennas, which are given as an input to the code for later mirror computation. The ellipsoid that defines the mirror of the antenna results from turning the generating ellipse around the x axis passing through its focal point. This procedure is illustrated in Figure 3. A known property of ellipsoids is that any ray going from one focal point towards

617

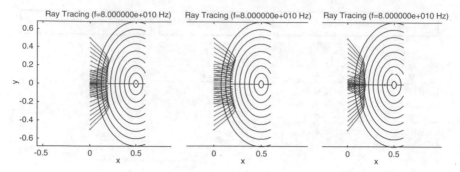

Figure 4. Influence of the focusing point location using an elliptic flux surface with focusing on (x_0, y_0), focusing after (x_0, y_0), and focusing before (x_0, y_0).

any point of the ellipsoid is reflected to the second focusing point and always travels the same distance. Keeping this geometrical rule in mind and making the feed of a pyramidal horn coincident with one of the surface focusing points, we guarantee that all rays launched from within the taper and reflected in the mirror will reach the other focusing point covering the same distance. Therefore, all rays will arrive in phase.

The taper of the hog-horn antenna can be found either using the mode matching code, which calculates the horn geometry, i.e. the aperture angles and the length, limiting the coupling of unwanted modes, or using the standard gain horn approach. This second method finds the pyramidal horn given its desired gain and dimensions (height and length) of the rectangular feed waveguide.

5 SIMULATION STUDIES

In order to study the problem of optimizing the antenna for typical ASDEX-Upgrade plasmas, several simulations were performed. Three hog-horn antennas with a standard gain taper were designed. For each antenna, the influence of the antenna focusing point and of vertical displacements of the plasma column on the amount of received energy were analysed. Several different probing frequencies covering the range 75–110 GHz (W band) were considered.

Simulations using different focusing points show that the received energy increases significantly when the focusing point is located after the cut-off layer corresponding to the probing frequency. This has to do with the way the launching rays illuminate the plasma: if the focusing point was located before the cut-off layer the rays would converge up to the focusing point and diverge afterwards, diminishing the number of rays returning to the antenna and thus, the returned energy. Figure 4 shows ray tracing results for three different focusing points.

Since a broadband solution is needed for the reflectometry diagnostic, after having carefully chosen the focusing point it was important to view the overall performance of the antennas in the whole frequency operating band. Results show that using highly directive antennas do not improve the level of received energy. In fact, the higher the antenna gain the lower the overall performance. Near 100 GHz the received signal drops to zero because no cut-off layer exists for frequencies higher than that limit. Since the rays are weighed by the radiation pattern of the antenna, certain characteristics of that diagram need to be verified. Thus, the antenna must have enough gain to concentrate the beam in the plasma probing area, but on the other hand it would be preferable if the main lobe was not too narrow.

The narrowing of the lobe and, therefore, the increase of the gain, means greater sensitivity to eventual perturbations and displacements in the plasma. For this reason, a compromise has to be reached between the needed directivity to probe in good conditions and the necessity to have a reasonably wide main lobe, along with the unwanted attenuated side lobes. This compromise had a positive effect on the ray tracing simulations.

Since the plasma is not still during probing it was vertically displaced in the simulations in order to evaluate its effect on the reflectometer signal. Figure 5 shows the received energy in function of the antenna gain for a standard horn. The wider the graph the better the antenna is, because it means that it can sustain greater displacements of the plasma column. Increasing the antenna gain revealed, once again, to worsen its performance.

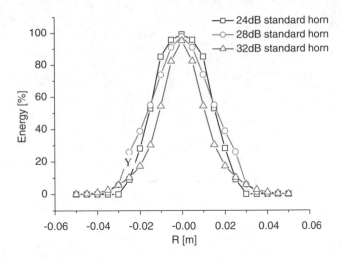

Figure 5. Ray-tracing received energy when vertically displacing by Y – the plasma from the mid-plane; using standard gain taper antennas shows that the lower gain antenna is a better choice.

6 CONCLUDING REMARKS

A complete numerical approach, including the propagation conditions in both antenna and plasma, has been developed to study the performance of antennas used in reflectometry diagnostics.

Numerical simulations lead to several important conclusions: (i) the hog-horn antenna is more efficient that non-focused standard horns; (ii) the antenna focusing point should be beyond the cut-off layer in order to maximize the received energy; (iii) the antenna gain cannot be too high or otherwise the system becomes too vulnerable to vertical displacements of the plasma column.

The developed software tools proved to be very helpful to design microwave antennas for reflectometry diagnostics of fusion plasmas and to estimate its performance in different plasma shapes. It is therefore a valuable tool to design the antennas for reflectometry diagnostics for the next step fusion device, ITER-FEAT.

ACKNOWLEDGMENTS

This work has been carried out in the framework of the Contract of Association between the European Atomic Energy Community and Instituto Superior Técnico. Financial support from Fundação para a Ciência e Tecnologia and Praxis XXI was also received.

REFERENCES

Balanis, C.A. 1982. Antenna Theory – Analysis and Design. Singapore: John Wiley & Sons.
Butcher, J.C. 1987. The Numerical Analysis of Ordinary Differential Equations – Runge Kutta and General Linear Methods. John Wiley & Sons.
Hacquin, S. et al. 2003. Optimization of the FM-CW reflectometry antenna for core density profile measurements on ASDEX-Upgrade. Accepted for publication in Review of Scientific Instruments.
Moresco et al. 1982. Focused aperture microwave antennas operating in the near field zone. International Journal of Infrared and Millimiter Waves 5(2): 279–292.
Press, W.H. et al. 1992. Numerical Recipes in C: The Art of Scientific Computing. Cambridge: Cambridge University Press.
Wagner, D. et al. 1999. International Journal of Infrared and Millimiter Waves 20: 567.

Computational Methods in Engineering and Science, Iu et al. (eds)
© *2003 Swets & Zeitlinger, Lisse, ISBN 90 5809 567 3*

Computer-aided design and evaluation of ground source heat pump systems for Macao

L.M. Tam & K.H. Lai
Department of Electromechanical Engineering, Faculty of Science and Technology,
University of Macau, Macao

ABSTRACT: Because of energy saving, ground-source heat pump systems installations have grown on a global basis (estimates range from 10 to 30% annually). A pseudo building with construction area equals to 10,000 ft^2 (500 × 200), a typical cooling load of 540,000 Btu/h and heating load of 160,000 Btu/h is used to design the ground-source heat pump system using the local soil conditions and to determine the borehole heat exchanger length. The total length of the borehole is one of the major indications of the effectiveness of the system. A simulation package called GLHEPRO using the numerical model recommended by Eskilson (1987), and a simple heat pump model is used for simulation purposes. The simulation results showed that the total borehole length is 10,125.75 ft. A 60 boreholes grid configuration with the depth for each borehole equal to 169 ft is selected. Since the design single borehole depth is very reasonable, system implementation can be achieved in engineering point of view.

1 INTRODUCTION

In recent decades, more and more sources of power are being developed such as solar energy, wind energy and hydraulic energy. But, there is another energy source – geothermal energy which is not originally innovative technology in this century. But, its application becomes more and more mature, especially in the field of HVAC. The use of geothermal resources can be broken down into three general categories: high temperature electric power production (>300°F), intermediate and low-temperature direct-use applications (<300°F) in power plant, and ground-source heat pump applications (generally <90°F). Ground-source heat pumps use the earth or groundwater as a heat source in winter and a heat sink in summer. Using resource temperatures of 4°C (40°F) to 38°C (100°F), the heat pump, a device which moves heat from one place to another, transfers heat from the soil to the house in winter and from the house to the soil in summer. Ground-source heat pumps were originally developed in the residential arena and are now being applied in the commercial sector. Many of the installation recommendations and design guides appropriate to residential design must be amended for large buildings. Kavanaugh and Rafferty (1997) and Caneta Research (1995) have a more complete overview of ground-source heat pumps. OSU (1988) and Kavanaugh (1991) provide a more detailed treatment of the design and installation of ground-source heat pumps, but the focus of these two documents is primarily residential and light commercial applications. Comprehensive coverage of commercial and institutional design and construction of ground-source heat pump systems is provided in CSA (1993). The term ground-source heat pump (GSHP) is applied to a variety of systems that use the ground, groundwater, or surface water as a heat source and sink. A GCHP which is a subset of GSHP refers to a system that consists of a reversible vapor compression cycle that is linked to a closed ground heat exchanger buried in soil (Fig. 1). Usually, the transfer of heat is done by burying a series of thermoplastic U tube piping network in the boreholes. Refrigerant is flowing inside the U tube and exchanges heat with the ground. In most of the cases, water or water-antifreeze solution is used as the refrigerant.

Vertical U tube configuration is used in this study. The advantages of vertical GCHP is (1) it requires relatively small plots of ground, (2) it is in contact with soil that varies very little in temperature and thermal properties, (3) it requires the smallest amount of pipe and pumping energy, and (4) it can yield

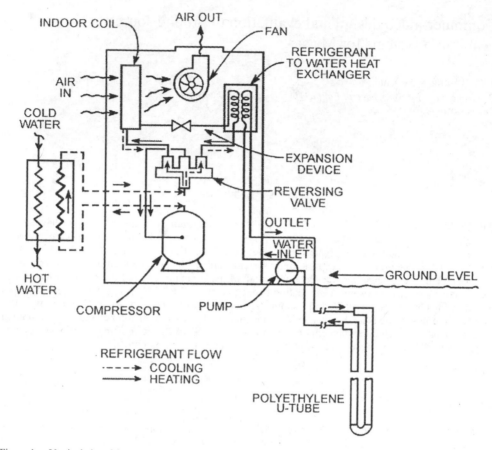

Figure 1.   Vertical closed-loop ground coupled heat pump system.

the most efficient GCHP system performance. Because of the above-mentioned characteristic, the GCHP is indeed a potential powerful tool for energy saving in HVAC system in Macau. Therefore, this study will focus on the investigation of the applicability of GCHP in Macau. A pseudo building with construction area equals to 10,000 ft^2 (500 × 200), a typical cooling load of 540,000 Btu/h is used to design the ground-source heat pump system and determine the borehole heat exchanger length.

2   METHOD OF APPROACH

Eskilson's model is used to determine the temperature at the borehole wall. His approach to the problem of determining the temperature distribution around a borehole is based on a hybrid model combining analytical and numerical solution techniques. A two-dimensional numerical calculation is made using transient finite-difference equations on a radial–axial coordinate system for a single borehole in homogeneous ground with constant initial and boundary conditions.

$$\frac{1}{a}\frac{\partial^2 T}{\partial r^2} = \frac{\partial^2 T}{\partial r^2} + \frac{1}{r}\frac{\partial T}{\partial r} + \frac{\partial^2 T}{\partial z^2} \tag{1}$$

To fix the boundary temperature, which is the ground temperature, measurements were made at three locations that are the wells of Instituto Salesiano, St. Antonio Church and St. Joseph hool (Calcada Igreja De S. Lazaro). The measurement duration of the first, second and third location is from April of 2001 to

April of 2002, April of 2001 to May of 2002 and April of 2001 to March of 2002. The average ground temperature is approximately equaled to 65°F. After solving the above equation, the borehole temperature, $T_{borehole}$ can be obtained. The fluid entering temperature to the heat pump can then be calculated from the total thermal resistance, which is the sum of the convective, pipe wall, and grout resistances.

$$T_f = T_{borehole} + Q_i R_{total} \qquad (2)$$

where $T_{borehole}$ = average borehole wall temperature in (°C); T_f = average fluid temperature in (°C); Q_i = current heat rejection pulse (W/m)s; and R_{total} = total thermal resistances m-s/kJ.

Several important parameters are necessary in the calculation of the borehole pipe length, which is index whether the system is particle or not. The shorter the pipe length, the easier implementation it would be. The monthly heating and cooling loads, the entering fluid temperature to the heat pump, the thermal properties of the ground, the geometric configuration of the ground loop heat exchanger, the borehole diameter, U-tube diameter, grout thermal properties, and the thermal properties of the working fluid, and the borehole configurations (in this case, a 6 × 10 rectangular configuration is used) are necessary to determine the length of the pipe. In the above inputs, the monthly peak heating and cooling loads are from assumption which is 540,000 Btu/h and 160,000 Btu/h respectively. The ground thermal physical properties are based on Ng (1999). The grout thermal physical properties are based on estimation by Kavanaugh and Rafferty (1997). The water entering temperature can be evaluated by the following equation.

$$T_{entering} = \frac{\dot{q}_{rejection,net}}{2\dot{m}C_p} + T_f \qquad (3)$$

where $\dot{q}_{rejection,\ net}$ = the net heat rejection rate (W); $\dot{m}$ = the mass flow rate of the working fluid (kg/s); C_p = specific heat of the working fluid (kJ/kg K); and $T_{entering}$ = the entering fluid temperature to the heat pump (°C).

3   RESULTS AND DISCUSSIONS

After simulation, the borehole length is 9469.56 ft. 15% safety factor is considered to eliminate the error from assumption of soil properties and manual calculation and the borehole length is 10,125.75 ft. Since 60 boreholes (6 × 10) grid configurations is applied in the design, each borehole depth is about 169 ft which can be easily achieved by existing technology in Macau. The schematic design diagram along with design parameters and potential HVAC equipment used is shown in Figure 2 and Figure 3.

As seen in the result, theoretically, it seems that the application of GCHP system should be practical in Macau since the length, depth and land requirements all fall in to a reasonable range. However, regarding to the implementation of this system, we should conduct a geotechnical survey at the site before the ground loop is designed. Sampling at shallow depths (10 to 40 ft, 3 to 12 m) should include sieve analysis and moisture content, which is typically performed with a hollow stem auger. This should be accompanied by deeper bores that identify the soil type (clay, sand, gravel, rock, etc.), static groundwater level, difficulty of drilling, and difficulty of inserting a vertical U-tube heat exchanger. Methods are currently being developed to bypass much of this activity by inserting a U-tube, imposing a thermal load in the field, and determining thermal properties with inverse methods.

4   CONCLUSIONS AND RECOMMENDATIONS

Numerical simulation in the design and calculations of GCHP system is conducted. The system is in vertical 6 × 10 grid configuration. Results with 15% safety factor recommended that the total borehole pipe length is approximately equal to 10,125.75 ft. Therefore, the length for the individual borehole length for the U tube is 169 ft which is considered to be reasonable and practical. Since the results are completely based on information from the literatures and the numerical simulation, the implementation

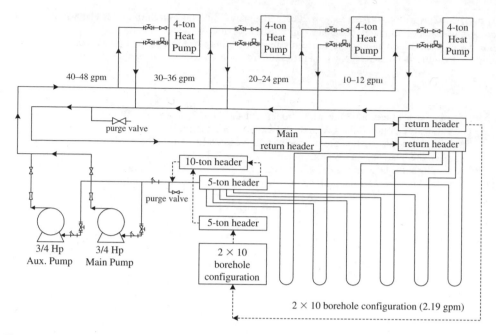

Figure 2.   Schematic diagram for the system.

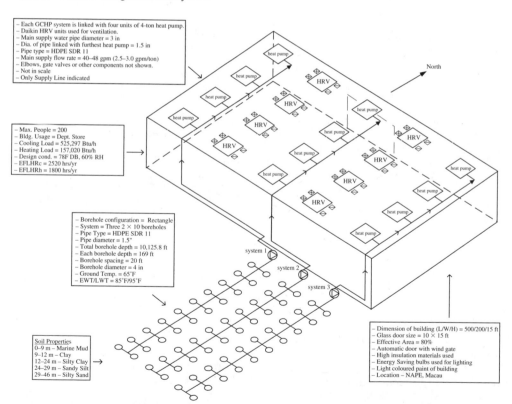

Figure 3.   Isometric diagram of piping system.

of actual system still need to have a more in-depth study. Moreover, other GCHP configurations, such as the horizontal configuration should also be tried.

REFERENCES

Caneta Research 1995. *Commercial/Institutional ground-source heat pump engineering manual*: ASHRAE.
CSA 1993. *Design and construction of earth energy heat pump systems for commercial and institutional buildings, Standard C447-93*: Canadian Standards Association.
Eskilson, P. 1987. *Thermal Analysis of Heat Extraction Boreholes*. Doctoral Thesis. Department of Mathematical Physics, University of Lund, Sweden.
Kavanaugh, S. P. and Rafferty, K. 1997. *Ground-source heat pumps – Design of geothermal systems for commercial and institutional buildings*. Atlanta: ASHRAE.
Kavanaugh, S. P. 1991. *Ground and water source heat pumps*. Oklahoma State University, Stillwater, OK, USA.
Ng, I. T. 1999. *Geotechnical Site Characterization Based on SPT*. Master Thesis. Civil Engineering Program, University of Macau.
OSU 1988. *Closed loop/ground-source heat pump systems installation guide*. Oklahoma State University, Stillwater, OK, USA.

Computational Methods in Engineering and Science, Iu et al. (eds)
© 2003 Swets & Zeitlinger, Lisse, ISBN 90 5809 567 3

Evaluation of interior thermal environment for light buildings

Z. Wang
Nanjing Normal University, Nanjing, China

J. Brau
INSA de Lyon, Villeurbanne, France

ABSTRACT: Certain fabrics are widely used in many kinds of buildings as construction materials. The evaluation of interior thermal and humidity environment for this kind of building is different from ordinary building. This project developed mathematical models for an exhibition hall constructed by fabrics and used software of TRNSYS to solve the models to obtain the solar radiant and transferred heat through wall, roof and floor with different transparences. We also analyzed the indoor humidity content when moisture-absorbing film was added to roof of the hall. Our research shows that for the fabric used in double layer, the surface of wall and roof is not too large, the heat resistance of floor is high enough, the heating and cooling energy is acceptable. The efficient of moisture-absorbing film is limited and ventilation is needed, especially in the case of high occupation density, which is often the case for exhibition building. The calculated results were compared with experimental results and good coincidence was showed.

1 INTRODUCTION

Certain fabrics are widely used as construction materials in many kinds of buildings, such as gymnastics, swimming halls and exhibition halls, etc. The fabric material is also used in many temporary and semi-temporary buildings or used as decorations of buildings. The walls of light buildings can be simple or double layer, with air in the middle. The interior thermal and humidity behavior of light buildings is different from ordinary buildings. This subject developed mathematical models for an exhibition hall constructed by fabric and solved the models to obtain the solar radiant and heat transferred through wall, roof and floor with different transparences.

2 CONSTRUCTION OF THE MODELS

2.1 *Entire models*

The entire model is constructed by several detailed models. In our case, there are wall, floor, interior space thermal balance, interior space humidity and meteorological model. The models of wall, floor and thermal balance are used for obtaining interior surface temperatures, heat flow by solar radiation, interior air temperature and mean radiant temperature. The humidity model is employed to evaluate the role of moisture-absorption behavior of roof materials. In meteorological model we input the meteorological data of Macon region of France as example.

2.2 *Wall model*

Wall model is divided into two kinds: simple and double layer material. In calculating procedure the orientation must be considered. The roof is treated as horizontal wall. Therefore, wall model can also be used to calculate heat transferred throw roof. The air movement within double layer wall is neglected.

Wall model must be able to calculate the solar-wall surface position. Meteorological data offer only the solar radiation to horizontal and vertical surface. Therefore the solar incline θ must be obtained:

$$\theta = \arccos(\cos \beta \sin X \cos(A - \alpha) + \sin \beta \cos X) \tag{1}$$

where β = solar altitude angle; X = inclination of wall normal line to zenith; α = wall surface azimuth; A = solar azimuth.

Directed solar radiation is calculated as follows:

$$f_{dir} = f_{dirh} \frac{\cos X}{\sin \beta} \tag{2}$$

where f_{dirh} = directed solar radiation intensity to horizontal surface.

Assuming that diffused solar radiation intensity for all the directions is equal to that for horizontal surface, we have:

$$f_{dif} = f_{difh} \tag{3}$$

The solar radiation heat flux throw the wall is:

$$f_{rs} = \tau(f_{dir} + f_{dif})S \tag{4}$$

where τ = transparence of wall; S = area of wall.

2.2.1 *Calculation of heat flux throw wall*

To analyze the heat transfer within semi-transparent double-layer materials is complicated. In our subject the internal temperature field is not interested, so we can treat it simply. The solar radiation absorbed by outside surface, inside surface of wall and the sum of both are calculated respectively as follows:

$$f_{rsea} = \alpha(f_{dir} + f_{dif})S \tag{5}$$

$$f_{rsia} = \alpha E_{int} S \tag{6}$$

$$f_{rsa} = f_{rsea} + f_{rsia} \tag{7}$$

where α = absorbency of wall.

$$f_{rle} = f_{rle1} + f_{rle2} = hr_c(t_c - t_{se})S + hr_e(t_{ext} - t_{se})S \tag{8}$$

where t_{se} = outside surface temperature of wall; t_{ext} = outside air temperature; hr_c = radiant heat transfer coefficient between sky and outside surface of wall; hr_e = radiant heat transfer coefficient between environment and outside surface of wall.

Radiant heat exchange between wall inside surface and other surfaces is:

$$f_{rli} = hr_i(t_{rm} - t_{si})S \tag{9}$$

where hr_i = radiant heat transfer coefficient of inside wall surface.

The convective and conductive heat transfer are calculated by ordinary formulas, therefore, not listed in this paper.

2.2.2 Heat balance equation

The inside and outside wall surface temperature can be calculated by use of heat balance equations. Based on the heat balance of outside wall surface following equation can be obtained:

$$f_{rle} + f_{rsea} + f_{ce} - f_{cd} = 0 \qquad (10)$$

where f_{ce}, f_{cd} = convective heat exchange of environment with outside wall surface and inside wall surface respectively.

Subject the meaning of all terms in formula (10), obtain:

$$\alpha(f_{dir} + f_{dif}) + hr_c(t_c - t_{se}) + hr_e(t_{ext} - t_{se}) + hc_e(t_{ext} - t_{se}) - \frac{\lambda}{e}(t_{se} - t_{si}) = 0 \qquad (11)$$

where hc_e = convective heat transfer coefficient between environment and outside surface. Based on the heat balance of inside wall surface we have:

$$f_{rsia} + f_{rli} + f_{ci} + f_{cd} = 0 \qquad (12)$$

where f_{ci} = convective heat transfer between inner air and inside surface.

Subjecting the meaning of terms in formula (12), obtain:

$$\alpha E_{int} + hr_i(t_{rm} - t_{si}) + hc_i(t_{int} - t_{si}) + \frac{\lambda}{e}(t_{se} - t_{si}) = 0 \qquad (13)$$

where hc_i = convective heat transfer coefficient between inner air and inside surface.

2.3 Floor model

2.3.1 Heat transferred from floor to earth

The heat transferred from floor with width of X to outside earth is:

$$df_{rg}(x) = \frac{1}{r(x)}(t_{sr} - t_{sg})dx \qquad (14)$$

where t_{sr} and t_{sg} = floor surface temperature and earth surface temperature respectively; $r(x)$ = heat transfer resistance, $r(x) = (\pi x / 2\lambda_g) + (\delta/\lambda_f)$; λ_g, λ_f = conductance of earth and floor respectively; δ = thickness of floor.

Heat transferred from unit length of floor to earth is:

$$f_{rg} = \int_0^x \frac{1}{r(x)}(t_{sr} - t_{sg})dx \qquad (15)$$

let $a = (\delta/\lambda_f)$, $b = (\pi/2\lambda_g)$, then $r(x) = a + bx$, from equation (15) can obtain:

$$f_{rg} = \log(\frac{a + bx}{a}) \frac{t_{sr} - t_{sg}}{b} = k(t_{sr} - t_{sg}) \qquad (16)$$

where k = floor–earth heat exchange coefficient, $k = \log[(a + bx)/a]/b$.

For a piece of floor with length L_1 and width L_2 the area can be calculated as follows:

$$f_{rg} = 2(L_1 + L_2)k(t_{sr} - t_{sg}) \qquad (17)$$

2.3.2 Floor model

The heat conduction between outside ground and earth is:

$$f_{cd1} = 2(L_1 + L_2)kC_{gp}(t_{sg} - t_g)$$ (18)

where C_{gp} = coefficient considering heat transfer behavior of earth with different densities, $0 < C_{gp} < 1$.
The heat conduction between floor surface and earth is:

$$f_{cd2} = 2(L_1 + L_2)k(1 - C_{gp})(t_g - t_{sf})$$ (19)

The rest of calculation is similar with wall model. We will not discuss again here.

2.4 Model of inner air humidity

When the film materials are used to construct buildings in witch the density of occupation is higher, like exhibition, etc., the verification for humidity contents must be taken.

2.4.1 Calculation of inside surface temperatures

The inside surface temperatures, include wall and roof surface, can be calculated by using related formulas in wall model. Outside meteorological data are input and inside surface temperatures in different outside air temperatures can be obtained.

2.4.2 Inner air humidity calculation

Inside surface temperature in winter can be taken as crisis of condensation: $t_{si} = t_{rl}$, the crisis of partial pressure of water vapor is:

$$p_{vl} = 10^5 \exp[\ln(140974) - \frac{3928.5}{231.667 + t_{rl}}]$$ (20)

The saturated air humidity can be calculated:

$$d_s = 622 \frac{p_{vl}}{p - p_{vl}}$$ (21)

Saturated partial pressure of water vapor of inside air

$$p_{vs} = 10^5 \exp[\ln(140974) - \frac{3928.5}{231.667 + t_{int}}]$$ (22)

Related humidity of inside air:

$$\varphi_i = \frac{p_{vl}}{p_{vs}} 100\%$$ (23)

In order to improve the situation of condensation of inside air, a layer of moisture-absorbing film can be added to film materials that construct the roof of the hall. If the humid absorbed is d_0, the critical equation of condensation is:

$$(d_s - d_e) + d_a - d_g = 0$$ (24)

when $[(d_s - d_e) + d_a - d_g] > 0$, no condensation, else condensation.

3 COMPUTATION METHOD AND VERIFICATION

We made use of software of TRNSYS (Transient System Analysis) and ESACAP to solve the mathematic models. TRNSYS is mainly used for thermal system analysis and ESACAP is for solving electric and thermal network problems. They both are famous for their accuracy, transparency and facility for using.
   An exhibition hall was simulated as an example. The input constants and values are as follows:

floor area: $50 \times 20 = 1000\,m^2$, floor thickness: 2 cm,
heat transfer calculation area of floor: $1.4 \times 1.4\,m^2$,
occupation: $500\ (1/2\,m^2)$,
thickness of calculated earth: 0.3 m,
wall height: 3, 4, 5.5 m respectively, wall conductivity: 1.0 W/m °C,
wall thickness: simple-layer: 0.003 m, double-layers: 0.1, 0.15, 0.2 m respectively.

The calculated time periods are:

- un hottest week in summer (July, within 4500–4700 h),for calculating the peak cooling load and space temperature in summer;
- un coolest week in winter (January, within 0–200 h), for calculating the peak heating load and space temperature in winter;
- un week in spring (May, within 3200–3400 h), for calculating the cooling and/or heating load and space temperature in the season.

In order to verify the calculated results un experimental chamber was constructed in CETHIL, INSA de Lyon. The calculated results were compared with experimental results and good coincidence was appeared. The results are very helpful for this kind of engineering equipment design and sizing. The results shows that for the double layer fabric construction, if the surface of wall and roof is not too large and the heat resistance of floor is high enough, the heating and cooling energy is acceptable. The efficiency of moisture-absorbing film is limited and ventilation is necessary, especially in the case of high occupation density, which is often the case in exhibition buildings.

REFERENCES

Roger, C. 1998. *Recueil des demarches et formules du Guide n° 2*. Edition PYC Livres.
Roger, C. 1998. *Calcul des charges de climatisation et conditinnement d'air*. Edition PYC Livres.
ASHRAE Handbook 1993. Fundamentals.
Duta, A. 1998. Etudes Thermiques et Aerauliques des Structures Legeres Duble Paroi avec Effect Parietodynamique. *These de doctorat*. INSA de Lyon: France.

Computational Methods in Engineering and Science, Iu et al. (eds)
© 2003 Swets & Zeitlinger, Lisse, ISBN 90 5809 567 3

HRBFS – a network simulator of fire in high-rise building

X.Y. Xie & A.Z. Ren
Institute of Engineering Disaster Prevention and Mitigation, Tsinghua University, China

X.Q. Zhou
Institute of Resource and Safety, China University of Mining Technology, Beijing, China

ABSTRACT: The proper decision-making of fire prevention and suppression, personnel evacuation in high-rise building fire depends on the exact forecast of smoke parameters' (airflow quantity, airflow direction, smoke spread area and variation of smoke concentration etc.) variation in building, especially these parameters' forecast in corridors at fire's early stage and the stage after smoke control measures are taken. The paper introduces the basic theory, simulation method, components, characters, framework and necessary data of HRBFS. The simulator is a powerful tool for design and verification of fire system, as well as decision-making of fire protection and construction fire accident analysis.

1 INTRODUCTION

Because of the danger of construction fire, many countries carried researches and achieved great success, especially in fire equipments, detection of ignition source, alarm apparatus, fire endurable materials, fire extinguishing chemical and fire code. However, the research and achievement lacked on law of ignition source's combustion behavior and smoke spread. Without relevant academic basis, the design and verification of fire system, decision-making of fire suppression and accident analysis is still in empirical stage. Because of the limit of ability, people are difficult to confirm the dynamic variation and spread law of smoke's parameters empirically, such as airflow quantity, airflow speed, concentration and temperature etc. So, the research on combustion behavior and smoke spread law in building is necessary.

Because of the inherent character, high-rise building fire will often cause great casualty and property loss if the fire can't be controlled effectively in fire's early stage. Over a long period of time, fire designers and fighters have achieved abundant experiences during high-rise building fire design, prevention and suppression with qualitative analysis method. The decision-making for rescue and evacuation are very difficult in complicated ventilation system.

With the wide application of new compositive materials, the major deadly cause has changed. Vast statistical information indicates that asphyxiation and poisoning has become the major deadly cause in high-rise building fire. According to American investigation, the person died from smoke in fire will reach to 80% in1994 (Lammiter et al. 1997). At the same time, two-thirds of those people died far from actual ignition source. To avoid more casualty and property loss, the smoke spread law and the effect of measures should be researched.

The proper decision-making of fire prevention and suppression and personnel evacuation in high-rise building fire depends on the exact forecast of smoke parameters' variation in building, especially the forecast of parameters in corridors at fire's early stage and after smoke control measures were taken. To instruct to take effective measures for fire prevention and suppression and personnel evacuation, the smoke flow law in building should be studied.

There are three models to simulate high-rise building fire: field model, zone model and network model. The network model considers the limited space as a unit, postulates that the state parameters (smoke temperature, concentration etc.) are even in a unit, postulates that the development of fire is reflected by change of parameters in a unit and each room can be reflected by an even parameter. The dynamic network model assumes the whole high-rise building as a ventilation network system, which is composed of rooms, halls, corridors, vertical shafts and pipelines. According to the research of ignition source's combustion behavior and smoke spread law, a mathematical model was set up and computer technology was used to simulate smoke spread law in building, such as dynamic distribution of airflow, temperature, concentration and pressure.

At present, more and more experts begin to research the network model. According to academic analysis and heat transfer model (Zhou et al. 2002), and at the base of dynamic network simulation technology, HRBFS (High-Rise Building Fire Simulator) was worked out (Xie 2002). HRBFS is developed from MFIRE (Zhou et al. 1984), a simulator of mine fire. Compared with other fire simulator, MFIRE has advantages in heat transfer and smoke tracking when airflow circulated or reversal happened (Zhou et al. 1998). Foreign experts applied MFIRE to simulate high-rise building fire and compared with other simulators. From the compared result, MFIRE has some advantages if it simulates high-rise building fire. HRBFS inherits the dynamic simulation technology of MFIRE, so it also has the same advantage compared with other simulator. To make the simulation close to realistic situation, HRBFS modified the combustion characteristic curve of ignition source, dynamic calculation of smoke temperature and the input and output of data. HRBFS can simulate the variation of airflow quantity, temperature, concentration of harmful gas, and pressure with time and space. It can provide fundamental data to simulate the effect of smoke controls system and determine the best evacuation route.

2 SIMULATION METHOD

Network model considers the construction as a network system that composed of channels with resistance. The physical character of intersections of channels is postulated to be uniform. The smoke flow in channels is postulated to be one dimension that is the parameters (air quantity, smoke concentration etc.), at the same cross-section keep uniform and proportion well. From happening to expand, high-rise building fire will reach to quasi-stable stage in short time, the parameters in the building distribute uniformly, so the precondition of one dimension simulation came into existence and the fire can be simulated with network model. The fundamental theories include: conservation law of mass, conservation law of energy, resistance rule and the calculation of smoke's temperature and concentration at the node.

Other models in fire should be taken into consideration, such as the parameters of ignition source, temperature distribution in room, heat transfer between smoke and wall. These models and the normal calculation model composed the network model under fire condition.

With computer numerical analysis method, the dynamic network simulation method [4] can calculate some parameters' value (airflow quantity, temperature of smoke, pressure and concentration of harmful gas in channels etc.) and their dynamic variation, as well as the time, place and influence of airflow reversal under fire condition. The theory and method of steady, quasi-steady, and non-steady are used in the system. According to the development of fire, the simulator includes four parts:

1. Simulation of smoke and injurant.
2. Steady simulation, which simulates airflow state under the normal situation.
3. Quasi-steady simulation, simulating airflow state in stable period after fire has happened for several hours.
4. Non-steady simulation, which simulates airflow state in fire development stage and weaken stage.

2.1 Simulation of smoke and injurant

The concentration of smoke and injurant can be calculated by the method of conservation law of mass. With a certain concentration, smoke and injurant that emit from ignition source flow along channels, and mix with smoke and injurant from other channels after reach to a node. With new concentration, the mixture flows out the node till out of construction.

To simulate the state of smoke and injurant, the airflow in channel is divided into some control volumes. With uniform concentration of smoke and injurant, the control volume flows with the speed of airflow. After reaches to the node, the control volume disappears and new control volume is formed. Because there are many nodes in network, the smoke flows into node with different concentration at different time, and the concentration of smoke and injurant in control volume at different node is different at the same time. Thus, the change of space and time augments the number of control volume to simulate the variation of smoke and injurant.

The core of the method is a three nested loop structure, the inner one calculates the state and position of control volume; the middle one chooses suitable time interval and the outer one outputs the result of every time interval. In the three loops, the inner one is the most important, which analyzes the move, disappearing and form of control volume. Control volume disappears and forms at the node when mix each other. So, the right order of control volume that reaches to the node should be known.

2.2 Steady simulation

At the early stage of fire, fire is small, the concentration and temperature of smoke is low, and smoke doesn't invade other area widely. At this stage, fire has little influence to construction's ventilation system, and the state of airflow in channel changes little, reversal airflow has not happened yet. So, the steady simulation model, a network calculation model that takes into consideration the pressure caused by thermal airflow, can be taken to simulate the parameters' state at this stage.

2.3 Quasi-steady simulation

When fire developed to a new stage, the power of mechanism system and airflow reach to a new balance state that is different from normal state. At this stage, the change of combustion state is little and the increasing rate of temperature in wall is slow. So, quasi-steady simulation technology can be taken to simulate the parameters' state.

Quasi-steady simulation includes two parts. The first part includes steady simulation and the distribution simulation of temperature and concentration of smoke. The second part includes the non-steady heat transfer calculation between smoke and wall, as well as the variation of temperature and concentration that initiated by buoyancy force, thermal stack force and airflow reversal.

Firstly, the network calculation of normal state was carried. Secondly, the quantity of heat and injurant that emitted from fire source was calculated. Thirdly, the variation of temperature can be calculated through airflow quantity calculation and heat transfer calculation between smoke and wall. Fourthly, the influence of pressure that caused by thermal airflow was analyzed in network calculation part. Repeat above calculations till the variation of pressure that caused by thermal airflow reaches to critical criteria. In other words, the ventilation system that influenced by thermal airflow reaches to a quasi-steady balance state, which with relative steady distribution of airflow quantity, pressure and temperature.

After fire occurred, the thermal draft will bring up variation of airflow quantity, temperature and concentration of injurant in channels at once. With the development of fire, the variation of airflow state in channels will become greater. Quasi-steady simulation only calculate the state parameters at the end time of a time period, and the dynamic change course at this time period was not taken into consideration, which is the main difference from non-steady simulation.

2.4 Non-steady simulation

When fire is at developing stage or weakening stage, the airflow quantity, temperature and concentration of smoke in channels changed acutely with time. The steady and quasi-steady simulation can't simulate the state variation of airflow at this stage. Because the state parameters of airflow change with time, so time interval method can be adopted to simulate parameters.

Time interval method divides simulation period into some time intervals. In every interval, the pressure that caused by thermal airflow and ventilator is assumed to balance with airflow resistance, and the state parameter in one location is constant.

The airflow in channel is divided into some control volumes. At a time interval, smoke concentration is const in a control volume. At the next time interval, the parameters of control volume is calculated by below method.

1. Confirm forward distance of control volume and calculate parameters of airflow. The calculation has below several steps:
   (a) According to airflow speed of last time interval and the simulation time interval, the forward distance of control volume, the temperature and concentration of smoke in control volume is calculated.
   (b) At the base of conservation law of mass and enthalpy balance law, the new value of parameters at the node is calculated.
   (c) With new value, the new control volume flows out node into another channel.

(d) Repeat steps from (a) to (c) till the control volume is convergent to critical criteria.
2. Calculate thermal pressure according to the distribution of temperature.
3. Calculate the distribution of airflow quantity in this time interval.
4. Repeat steps from (1) to (3), and till the control volume is convergent to critical criteria in this time interval.

In non-steady simulation, the speed and temperature of airflow change with time. So, the calculation of state parameter and forward distance of control volume becomes difficult. Because the forward distance is the function of time interval, airflow quantity, airflow temperature and parameters of channel, to calculate the forward distance of control volume, the average airflow quantity and temperature of smoke in channel in new time interval should be known. However, only the forward distance is calculated, the average airflow quantity and temperature of smoke can be calculated. So, two-nested loop construction is formed. In the inner one, rather exact average temperature of control volume can be calculated according to the defined airflow quantity and section parameters of channel. In the outer one, the airflow of last time interval is assumed to be the airflow quantity in this time interval, and the real value can be achieved after iteration calculation.

The non-steady simulation method not only can calculate the parameters in normal state, but also can calculate the variation of airflow quantity, temperature and smoke concentration in fire.

3 SYSTEM STRUCTURE

3.1 *Functions and structure*

The function of the system includes three parts: file operation, data input and fire simulation. The structure of the system is shown in Figure 1.

3.2 *File operation*

The type of dataset is ACCESS. There is an access file whose structure is defined in system. The main function has example opening, new example, save example as, channel's data input and node's data input. If there are channels' data and nodes' data in a text file, they can be input into the dataset.

3.3 *Data input*

Timetable method is adopted in inputting data. The method can simulate the condition changes with time, such as the increase of ignition source number, change of fire size, change of combustion characteristic, change of simulation time step and control measures. Thus the simulation is close to the real situation. The user-interface of data input and result output is simple and friendly. The user can set data in relevant window. Data should be checked before simulation, the input data includes the following parameters.

3.3.1 *System control parameters*
The parameters include number of channels, number of nodes, number of fans, simulation time period, default value of air temperature, default value of air density, simulation time step, precision criterion and alarm criterion.

3.3.2 *Parameters of channels*
The parameters of channel include name, serial number, start node and end node of channel, channel type, resistance, initial airflow quantity, frictional coefficient, geometry parameters, thermal exchange rate, thermal diffusion rate, initial temperature of wall and the personnel number in channel. In which, the pipeline of HAVC system is also looked as a part of channels.

3.3.3 *Parameters of nodes*
The parameters of node include as the height and initial temperature.

3.3.4 *Parameters of fans*
The parameters of fan include fan type, the channel's serial number in which fan locates, characteristic point, fit method of characteristic curve and dealing method of curve boundary. The fans of HAVC system and smoke control system can be defined here also.

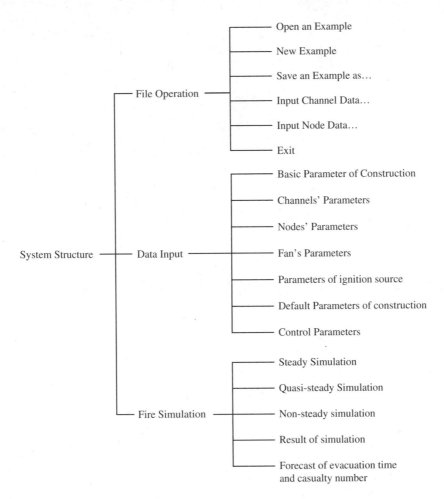

Figure 1. System structure.

3.3.5 *Parameters of ignition source*

According to development stage of fire, the combustion characteristic curve (heat–time curve and injurant–time curve) is set by conic curve. At the same time, the ignition source with fixed thermal quantity also can be set. The effect of design, running and proof of fire systems can be analyzed from simulation result of different fire size and location of ignition source. The influence of fire to ventilation system can be reviewed by set different fire size at a location.

3.3.6 *Default parameters*

The default parameters include the default value of construction, such as heat conduction coefficient, the length of channel, section area of channel, perimeter of channel and thickness of wall. The function is to reduce the quantity of data input. If most parameters are same to the default parameters, it is not to need input again in relevant window.

3.3.7 *Parameters' change*

After fire happened, we can set the change of parameters at a time, the parameters' change includes:

1. Change of resistance caused by some measures, such as open the fire door or fire damper.
2. Adding fans. To prevent the smoke into channel, the pressurize fans or smoke control fans should be used. Thus, the effect after fans are added can be simulated.

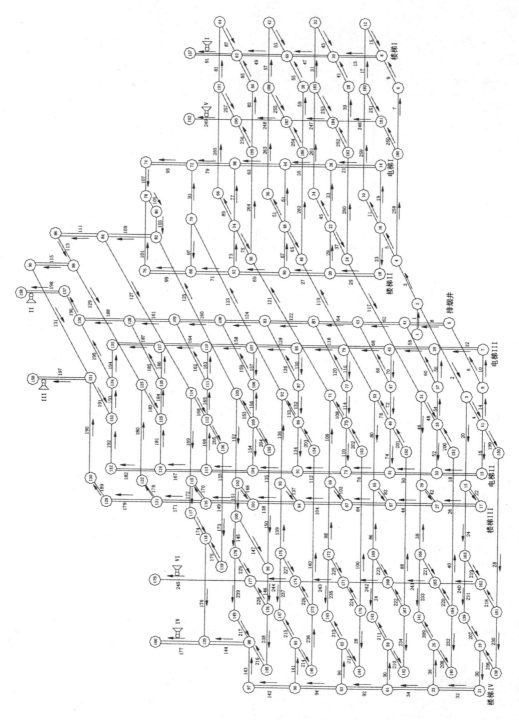

Figure 2.　The network figure of a compositive building.

3. Closing fans. At normal situation, the fans of HAVC system are running, However these fans should be closed in fire.
4. Change of result output time interval.
5. Change of simulation time step.

The system will save these essential data to dataset. The veracity of data is the key of simulation. The system has the function of data check, it provided some alarming information when running, and wrote error information to a file.

The user can change and delete data in relevant window. In window, the system tells user how to set the data and the format of data.

3.4 *Fire simulation*

This module is the core of system. In this module, high-rise building fire was simulated and the result is showed to user. The result includes:

1. The parameters' change in channels with time, such as airflow quantity, airflow speed, average temperature, injurant concentration and decrease of pressure. The change curve with time is drawn.
2. The location of smoke with time.
3. The reverse channel ant its parameters with time.
4. The best evacuation route.
5. The evacuation time and casualty number.

4 EXAMPLE

We simulated a supposed fire in a compositive building. The network is composed of 271 channels, 192 nodes and 6 fans (see Fig. 2). After simulation, the result can be shown as Figure 3, the parameters' curve with time in a channel is also been drawn in the system. Analyzing simulation result is an important job. For example, to simulate the influence of different fire size at a location and the effect of smoke control fans, the user should simulate more times and analyze the simulation result.

模拟结果显示

| 节点空气参数 | | 分支风流反向 | | | |
| 分支末端的温度和烟雾浓度及分支的压降 | | 分支中的烟雾性面数据 | | 热源数据 | |

选择通道 28   显示图形曲线   写入文件

时间	DELTA Q	风量	平均温度	末端温度	有害物	压降
30	-.022	0	18.37	18.25	0	0
60	.033	.1	18.37	18.25	0	0
120	.004	.1	18.37	18.25	0	0
480	2.848	2.9	21.76	18.25	0	2.36
510	-2.214	.7	76.8	18.25	0	.18
600	.076	.8	103.17	81.89	5.9564	.23
690	.048	.8	103.81	73.8	6.4123	.26
750	.239	1.1	93.95	84.05	7.2112	.41
780	.177	1.2	88.82	81.55	7.8566	55
990	-.479	.8	99.01	70.2	14.6717	.22
1140	.066	.8	118.47	81.78	17.7994	.27
1470	.597	1.4	201.44	151.86	18.6864	.93
1530	-.736	.7	215.87	150.23	19.3938	.24
1740	.322	1	289.97	231.22	20.6554	.57
1830	-.298	.7	294.98	211.73	22.2204	.3
1890	-.021	.7	297.07	231.45	21.4735	.28
1920	.615	1.3	314.46	263.61	22.0313	.98

读取模拟结果   关闭

Figure 3. The simulation result.

At the base of simulation result, the best evacuation route, which is the shortest evacuation route that took consideration of the concentration of smoke and personnel distribution. The evacuation personnel can evacuate the building by the route rapidly and safely. For example, at normal situation, the shortest length form node 142 to node 1 is 80.5 meters, and the shortest route is: 142→59→61→ 23→21→19→139→163→15→3→2→1. The fire is supposed to happen in channel 40, the channel 34 and channel 42 were polluted badly in short time. So, personnel would not choose the shortest route to evacuate out the building. According to a method which take consideration of the concentration of smoke and personnel distribution, the best evacuation route from node 142 to node 1 is: 142→59→143→ 169→55→57→67→94→93→92→70→68→52→50→20→18→16→4→2→1. Although, the best evacuation route that the personnel passed by is longer than the shortest route, the best evacuation route is safer and rapider to personnel.

5   CONCLUSION

The designer can evaluate the effect of smoke control system by HRBFS. The model can be used to validate the condition of fans opening of HAVC system and simulate the effect. The model can also be used to analyze the risk of smoke in construction and evaluate the scheduled evacuation scheme. At the base of smoke concentration and personnel density, the model can be used to define the best evacuation route, calculate the evacuation time, and provide fundamental data for the emergency evacuation and prevention design.

The model can simulate the parameters' variation. These parameters can provide reference data for design and service better for fire systems.

REFERENCES

Brian, Y. L. & Uri, V. 1997. *The Transport of High Concentrations of CarbonMonoxide to Locations Remote from the Burning Compartment*, Building and Fire Research Laboratory, 1997.4 .
U.S. Bureau of Mines 1994. *MFIRE User's Manual* 1994.3.
Xie, X. Doctor dissertation, The Technology of Network Computer Dynamic Simulation of High-Rise Building Fire, 2002.4.
Zhou, X. & Rudolf, E. G. 1991. Specialized FORTRAN Computer Programming and Analysis services to Upgrade Capability of MFIRE Program, Department of Mining Engineering Michigan Technological University, 1991.
Zhou, X. & Wu, B. 1997. The Application of Mine Fire Simulator MFIRE on Smoke Control During Building Fire, the FORUM for International Cooperation on Fire Research, 1997.10: 52–59.
Zhou, X., Xie, X. & Liu, G. 2002. Calculation of Smoke Temperature in Corridor during High-rise Building fire, Journal of China University of Mining & Technology, 2002.5, Vol.12 (3), 221–224.

Computational Methods in Engineering and Science, Iu et al. (eds)
© 2003 Swets & Zeitlinger, Lisse, ISBN 90 5809 567 3

BatchHeat – software for pinch analysis for batch processes

A.C. Pires
Instituto Superior de Engenharia de Lisboa, Sec. de Proc. Químicos e Reactores do DEQ,
Lisboa, Portugal

M.C. Fernandes, H.A. Matos & C.P. Nunes
Instituto Superior Técnico, Grupo de Integração de Processos do CPQUTL/DEQ,
Lisboa, Portugal

ABSTRACT: Pinch Analysis extensively used in continuous process was applied in this work to a discontinuous process. The intermittent nature of batch processes imposes a three-dimensional data treatment in terms of enthalpy, temperature and time, which brings up high complexity, making manual calculation very difficult and extensive. Therefore the main purpose of this work was to develop a software package, BatchHeat, incorporating Pinch Analysis and enabling the automatic application of this methodology to a discontinuous problem. The BatchHeat, may be used as a first tool during the course of a Process Integration study. The results obtained highlight the energy inefficiencies in the process and therefore enable to set the scope for possible heat recovery, through direct heat exchange or storage.

1   INTRODUCTION

Batch process interest emerged since the 80s but mainly in the areas of simulation, design, and process stage scheduling to minimize operation unit investment and energy costs. Process integration and energy integration are also very important to optimize this type of process, though the results obtained lead to smaller energy savings, when compared with the ones from application to continuous process (Linnhof et al. 1988). Generally, industrials are indifferent to batch process energy integration, because they are deeply concerned with the reduction of flexibility and process control. Besides that, generally in a batch process the energy costs are much lower than the costs related to the yield and the product quality.

Process Integration (PI) is widely used in continuous process (Gundersen et al. 1990). Among different methodologies, Pinch Analysis (PA), with a thermodynamic basis has been one of the most applied due to its easy implementation. The basic concepts of Pinch Analysis were therefore extended to batch process (Kemp et al. 1989). Different strategies can be applied depending on various scenarios: repeated batch when a cycle composed of various operations is repeated sequentially and single batch when the set of operations can not be considered to be repeated in time.

Literature review showed that there are several software packages using Pinch Analysis: Supertarget, and PinchExpress; Advent (Aspen Pinch); PRO_PI; HEXTRAN; PinchLeni and more recently HInt. To study batch processes, the available software concerning operation planning, simulation and resources, are for example: gBSS; BATCH; VirtECS Design; VirtECS Scheduler and Superbatch. BatchHeat is a computational application to apply Pinch Analysis to discontinuous processes. This program, which can also be applied to continuous processes, is an initial tool of a global energy analysis.

2   BatchHeat SOFTWARE

BatchHeat software presented in this work enables to determine the minimum energy consumption of continuous and batch processes, using different scenarios (Kemp et al. 1989).

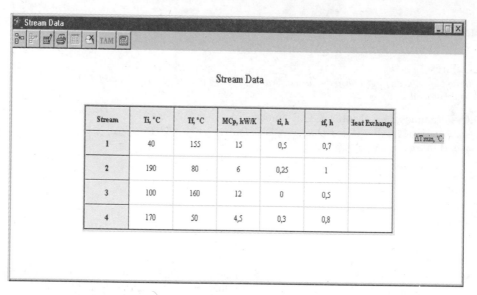

Figure 1. Process input data window.

The first step to start working with BatchHeat is to select the type of process: continuous or batch. The user should fill a table with characteristic variables for each stream:

Ti   – supply temperature – stream temperature before heat exchange
Tf   – target temperature – desired stream temperature after heat exchange
MCp   – heat capacity of the stream, constant in the temperature and time interval considered
ti   – starting time
tf   – finishing time
ΔT_{min} – minimum temperature difference.

In batch processes it is important to differentiate between external and internal heat exchange. In the latter the temperature is time dependent and the user should split that stream in several intervals in order to reflect real conditions.

Figure 1 presents the input data window, for an example problem involving a set of hot and cold streams and a ΔT_{min} value of 10°C. Hot streams are those requiring cooling and cold streams are those requiring heating.

The lower bound for energy target can be obtained for batch process using Time Average Model (TAM) presented by (Kemp 1990), that ignores the time and considers the process as whole. Therefore applying this model we assumed that heating and cooling needs are achieved in one cycle and consequently we have to consider an average enthalpy value for each stream. Results obtained using this method are 30 and 90 kWh respectively, for hot and cold utility minimum consumption during time cycle. These values correspond to the lower bound and could only be attained for a repeated batch scenario and an unlimited storage.

An alternative method to obtain minimum hot and cold utilities, uses heat cascade analysis, requiring splitting the process streams in different time intervals. This model allows following along the enthalpy variation, showing the temperatures and the time intervals, where it is necessary to use external heating and cooling.

BatchHeat generates a table, dividing the process at various time intervals in terms of shifted temperatures that are calculated based on hot or cold stream temperatures and according to ΔT_{min} value.

The program presents infeasible heat cascade and as a first step the energy, Q_{ij}, involved between two temperature levels and two time levels is given by

$$Q_{kj} = \left(T_k - T_{k-1}\right)\left(\sum_h MCp_h - \sum_c MCp_c\right)_{k,j} \left(t_j - t_{j-1}\right) \tag{1}$$

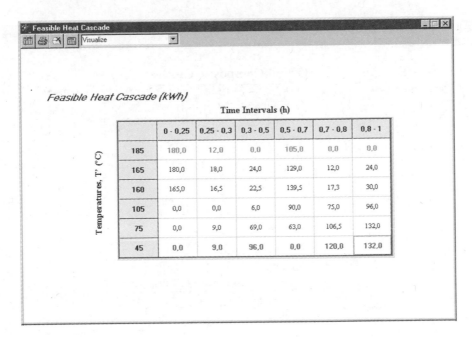

Figure 2.   Feasible heat cascade window.

where k and j are, respectively, the temperature and time level of each operation, k assumes values between zero and N, and j assumes values between zero and M. N and M are respectively the last temperature and the last time level. Note that h refers to hot streams and c refers to cold streams.

To obtain the infeasible cascade, the available energy at each temperature level and at each time interval, I_{kj}, is defined assuming a boundary condition:

$$I_{kj} = I_{k-1,j} + Q_{kj} \quad \text{and} \quad I_{0j} = 0 \quad \text{for} \quad j = [1, \dots, M] \tag{2}$$

The cumulative infeasible heat cascade using cumulative values for each temperature level:

$$IC_{kM} = \sum_j I_{kj} \tag{3}$$

leads to the same results as the infeasible cascade obtained using the Time Average Model.

Applying this model to the example data, we obtain minimum utility consumption as well as the pinch location. Feasible heat cascade, presented in Figure 2, is obtained by the infeasible one and is determined by:

$$F_{0j} = -(\min(I_{kj})) \quad \text{with} \quad 0 < k < N \tag{4}$$

This cascade shows pinch temperature point (temperature corresponding to a null enthalpy value), as well as the hot utility consumption (first line values) and cold utility needs (last line values). Therefore, from Figure 2 one can conclude that the minimum consumption of hot and cold utility is, respectively 297 and 357 kWh.

From the feasible heat cascade, it is possible to obtain the cumulative one, as done before:

$$FC_{kM} = \sum_j F_{kj} \tag{5}$$

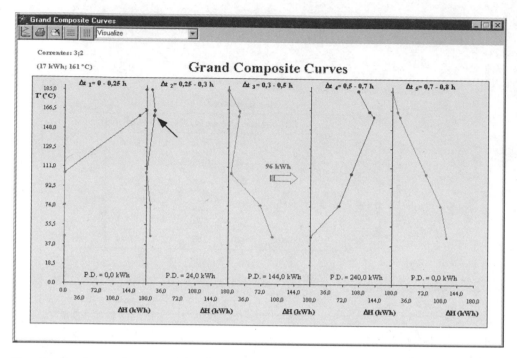

Figure 3. Grand composite curves window, for the first five time intervals (in figure, P.D. signifies direct heat exchange).

This cumulative feasible cascade allows the determination of the minimum storage value in a repeated batch scenario:

$$S_{RB} = min(FC_{kM})$$ (6)

For the example problem this value corresponds to 267 kWh.

From feasible heat cascades for each time interval, the program enables the calculation of the minimum heat consumption maximizing direct heat exchange and also the visualization of the Grand Composite Curves for each time interval. These curves are a graphic representation of temperature as a function of the enthalpy values, F_{kj}, obtained from feasible heat cascades.

To calculate the direct heat exchange, program identifies which streams allow heat exchange as well as the enthalpy involved in these exchanges, thus enabling the user to decide to carry them out or not.

Figure 3 exhibits the Grand Composite Curves obtained through BatchHeat for the example problem, thus allowing identifying direct heat exchanges as well as the storage potential in single or repeated batch.

From Grand Composite Curves, we can visualize that the direct heat exchange between streams two and four (S2 + S4) and streams one and three (S1 + S3) corresponds 408 kWh.

In this example, S3 stream, in interval 3, ΔT_3, can exchange 132 kWh with S2 stream, and 128 kWh with S4 stream that represents 40% of enthalpy of the cold stream, S3, and therefore justify the direct energy exchange.

Heat exchange between S1 and S2 stream is 132 kWh and between S1 and S4 stream is 108 kWh, i.e., 70% of total energy requirement. This direct heat exchange assumes a fixed operation scheduling, and therefore that these streams coexist in this time interval. Any delay in S2 or S4 streams would lead to a smaller heat reduction. The heat storage emerges as an alternative to direct heat exchange, leading to a more flexible process and thus allowing for little changes in the scheduling.

From feasible heat cascades for each time interval it is necessary to construct two separated cascades: one showing the available heat and another enthalpy needs. The available and required enthalpy, respectively, A_{kj}, and D_{kj}, at the temperature level k and at the time interval j are the result of the feasible heat cascade. Whenever there is a non-monotonous variation of the available or deficit energy, it means that there is a pocket allowing direct heat integration for ΔT greater than ΔT_{min}. For utility needs calculation these zones should be removed in each interval j.

The same procedure is repeated for all time intervals. Available energy at k temperature is given by the remaining cumulative enthalpy value for time intervals considered:

$$AC_{kM} = \sum_j A_{kj} \tag{7}$$

where AC – cumulative available heat above the pinch (e.g. kWh); and the heat deficit at k temperature is given by the sum of heat deficit:

$$DC_{kM} = \sum_j D_{kj} \tag{8}$$

where DC – cumulative heat deficit below the pinch (e.g. kWh).

Based on available and deficit heat cascades it is possible, using this software, to obtain the energy storage value for single batch, comparing the heat required in one time interval with the available cumulative energy in the previous time interval, attending to the temperature level.

$$\text{If } D_{k-1,j} < AC_{k-1,j-1} \Rightarrow Stor_{k-1,j-1} = D_{k-1,j}$$
$$DC_{k,j-1} = DC_{k-1,j-1} - Stor_{k-1,j-1} \tag{9}$$
$$Stor_j = \sum_k D_{k-1,j}$$

$$\text{If } D_{k-1,j} > AC_{k-1,j-1} \Rightarrow Stor_{k-1,j-1} = AC_{k-1,j}$$
$$AC_{k,j-1} = AC_{k-1,j-1} - Stor_{k-1,j-1} \tag{10}$$
$$Stor_j = \sum_k AC_{k-1,j}$$

where Stor – heat storage (e.g. kWh).

In single batch the maximum storage value corresponds for the example problem to 96 kWh (Fig. 3). In this scenario the hot utility will be therefore equal to 201 kWh. Industrial process work, sometimes, by repeated cycles. In this case it is possible to use the available energy at the end of one cycle in the next one. To achieve the minimum consumption of hot utility in repeated cycles the storage must be 267 kWh. This storage can be achieved through the heat recovery from S2 and S4 streams and afterwards supplied to the cold stream S1 in a repeated batch. The hot and cold utility consumption are in this case equal to the values obtained applying the TAM.

To better illustrate the global enthalpy behavior, the Batch Utility Curves are obtained using hot and cold utility cumulative values after direct heat exchange maximization, i.e., heat integration between coexistent streams in each time interval with ΔT values greater or equal than ΔT_{min}. Available energy and heat deficit at k temperature are given by the remaining cumulative enthalpy value for all time intervals. Therefore, cold and hot utility curves are, respectively, a graphic representation of temperature as a function of AC_{kM} and DC_{kM}.

Figure 4 shows Batch Utility Curves as a function of real temperatures, to turn the identification of storage areas easier. The area between hot and cold curve allows determination of maximum heat storage, 267 kWh in this example. The remaining heat must be supplied by utility, leading to 90 kWh of cold utility and 30 kWh of hot utility consumption. This value agrees with the energy target obtained from TAM.

Figure 5 presents hot and cold utilities for the different scenarios. Pinch Analysis applied to this process unable a 58% reduction of the hot utilities using only direct heat exchange. However, comparing the scenarios with and without storage, it is still possible to reduce the hot utility consumption in 32% and 90% respectively for single batch or repeated batch.

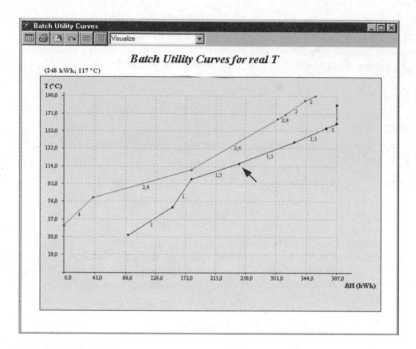

Figure 4. Batch utility curve window.

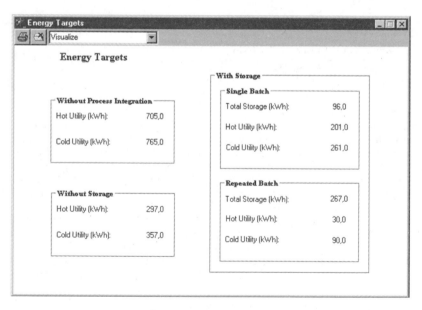

Figure 5. Energy targets summary table.

3 CONCLUSIONS

Pinch Analysis advantage is expressed by the simplicity and easiness of application to real continuous or batch process. This analysis allows to determine how far the process is from the minimum quantity of external utility and therefore to decide to invest on an energetic optimization.

Batch process having a three-dimensional dependency of enthalpy, temperature and time, turn the application of pinch analysis to this type of processes very complex. BatchHeat software using automatic calculation allows a global process view and enables to identify exceeding and deficit enthalpy areas and therefore to obtain Heat Exchangers network corresponding to the minimum heat consumption. It is also possible to study the influence of ΔT_{min} value on heat consumption, on heat storage potentiality in single and repeated batch and consequently on operation rescheduling.

BatchHeat is therefore an initial tool to global energetic analysis of all productive process, including continuous and discontinuous process, thus enabling to suggest process modifications to reduce heat consumption and gas effluents and utility units.

REFERENCES

Gundersen, T. and Naess, L. 1990. The Synthesis of Cost Optimal Heat Exchanger Networks. *Heat Recovery Systems & CHP*, 10, No. 4, 301–328.

Kemp, I.C. 1990. Application of the Time-Dependent Cascade Analysis in Process Integration. *Heat Recovery Systems & CHP*, 10, No. 4, 423–435.

Kemp, I.C. and Deakin, A.W. 1989. The Cascade Analysis for Energy and Process Integration of Batch Processes – Part 1. *Chem Eng Res Des*, 67: 495–509.

Kemp, I.C. and Deakin, A.W. 1989. The Cascade Analysis for Energy and Process Integration of Batch Processes – Part 2. *Chem Eng Res Des*, 67: 510–516.

Kemp, I.C. and Deakin, A.W. 1989. The Cascade Analysis for Energy and Process Integration of Batch Processes – Part 3. *Chem Eng Res Des*, 67: 517–525.

Linnhoff, B. and Smith, R. 1988. The Design of Separators in the Context of Overall Processes. *Chem Eng Res Des*, 66: 195–228.

Computational Methods in Engineering and Science, Iu et al. (eds)
© 2003 Swets & Zeitlinger, Lisse, ISBN 90 5809 567 3

EM Wave Simulation: a simulation and visualization package

W.K. Leung
Department of Electronic and Information Engineering, The Hong Kong Polytechnic University, Hong Kong

ABSTRACT: In this paper we describe our implementation of *EM Wave Simulation*, a windows-based electromagnetic wave simulation/visualization package. For the purpose of displaying electromagnetic wave images in real time, the package employs a 2½-dimensional finite-difference time-domain (FDTD) method, which assumes the computation domain is sandwiched between perfect electric conductors and is suitable for simulating simplified electromagnetic wave propagation in indoor environments. Frequency-domain information can be extracted with the built-in one- and two-dimensional Fourier transform tools. A user-friendly mouse-and-menu-driven interface allows user to set up simulations containing complex geometries in a straightforward manner. Simulated results, in both time and frequency domains, can be displayed in real-time on the computer screen, which makes the application ideal for interactive use, e.g. for undergraduate teaching. Some sample simulations are presented in the paper to illustrate its principal functionality.

1   INTRODUCTION

The rapid development of wireless communications over the past decades means that the study of electromagnetics (EM) is as important as ever to electronic engineers. Teaching EM to undergraduates, however, is no easy task. Many students often find the subject dull and abstract, loaded with mathematical operators that are difficult to understand. It is widely believed that the use of interactive educational software is highly effective in arousing the students' interests in the subject and assisting them in grasping the core concepts of EM (Iskander 1993). To this end, we have recently developed an interactive EM wave simulation/visualization package, which is being used for teaching EM to our undergraduate students.

2   IMPLEMENTATION DETAILS

Information technology has seen tremendous advances in recent years, and it is now possible to simulate complicated EM phenomena with a personal computer by using a suitable numerical technique. For this purpose, the finite-difference time-domain (FDTD) method is a very popular choice (Shlager & Schneider 1995). The FDTD method focuses on the time-domain Maxwell's curl equations:

$$\varepsilon \frac{\partial \mathbf{E}}{\partial t} = \nabla \times \mathbf{H} - \sigma \mathbf{E} - \mathbf{J} \tag{1}$$

$$\mu \frac{\partial \mathbf{H}}{\partial t} = -\nabla \times \mathbf{E} - \mathbf{M} \tag{2}$$

where **E** and **H** are the electric and magnetic field intensities, ϵ, μ and σ are the permittivity, permeability and conductivity of the medium (represented in matrix form for generality), and **J** and **M** are the externally imposed electric and magnetic current densities, respectively. The spatial and time derivatives are approximated by finite differences, and the electromagnetic fields are evolved iteratively according to a "leapfrogging" scheme:

$$E_x(t + \Delta t) = \frac{2\varepsilon - \sigma\Delta t}{2\varepsilon + \sigma\Delta t} E_x(t) + \frac{2\Delta t}{2\varepsilon + \sigma\Delta t} \left(\frac{\partial H_z\left(t + \frac{1}{2}\Delta t\right)}{\partial y} - \frac{\partial H_y\left(t + \frac{1}{2}\Delta t\right)}{\partial z} \right) \tag{3}$$

$$H_x\left(t + \frac{1}{2}\Delta t\right) = H_x\left(t - \frac{1}{2}\Delta t\right) - \frac{\Delta t}{\mu}\left(\frac{\partial E_z(t)}{\partial y} - \frac{\partial E_y(t)}{\partial z} \right). \tag{4}$$

The other update equations for E_y, E_z, H_y and H_z can be obtained by cyclic permutation of the x, y and z indices above. Following Yee's idea (Yee 1966), it is conventional to employ a staggered spatial grid that interleaves the electric and magnetic field components in each cell. This arrangement allows the simple finite-difference method to attain a second order accuracy.

We note that the time-stepping approach of FDTD makes the method ideal for interactive visualization, as the electromagnetic fields at each time step can be displayed directly in real-time during the simulation without any need of pre-computation and storage of many time frames. On the other hand, the FDTD method is often limited by its rather substantial demand of computer memory. As a rule of thumb, about 8–16 cells are needed per wavelength at the highest frequency of interest, and the total amount of computer memory required is proportional to the cube of the number of wavelengths in each direction of the computation domain. This explains why many previous FDTD visualization applications are not three-dimensional (Luebbers et al. 1990). Despite this limitation, the FDTD method is highly effective for electromagnetic simulations, particularly for those involving complicated geometries.

In order to reduce the memory requirement and to increase the simulation speed, our implementation employs a simplified FDTD method by focusing on a single standing mode in the vertical z-direction at each time. To facilitate this separation of the electromagnetic fields into independent modes, the ceiling and floor of the computation domain are taken to be perfect electric conductors (PEC), and all materials sandwiched between them must be homogeneous in the vertical direction, i.e. only obstacles extending across the whole height of the domain are modeled. Under these assumptions, it is easy to prove that the tangential electric field and perpendicular magnetic field vary as simple sinusoidal functions in the z-direction, vanishing at the two PEC surfaces, while the remaining electric and magnetic field components follow a co-sinusoidal function along the same direction. We therefore obtain the following governing equations for the six electric and magnetic field components:

$$E_x(x,y,z;t) = E_x(x,y;t)\sin(\frac{n\pi}{h}z), \quad H_x(x,y,z;t) = H_x(x,y;t)\cos(\frac{n\pi}{h}z), \tag{5}$$

$$E_y(x,y,z;t) = E_y(x,y;t)\sin(\frac{n\pi}{h}z), \quad H_y(x,y,z;t) = H_y(x,y;t)\cos(\frac{n\pi}{h}z), \tag{6}$$

$$E_z(x,y,z;t) = E_z(x,y;t)\cos(\frac{n\pi}{h}z), \quad H_z(x,y,z;t) = H_z(x,y;t)\sin(\frac{n\pi}{h}z), \tag{7}$$

where h is the height of the computation domain, and n is an integer counting the number of half-cycles between the ceiling and floor. This arrangement spares us from computing any finite differences along the z-direction, leaving us a practically 2-dimensional (2D) FDTD simulation within the x–y plane. Note however that the simulation still represents a 3-dimensional scenario, with the focus being on a single mode at a time. Realistic 3-dimensional problems can be solved by the superposition of multiple standing modes (Lauer 1995), although for teaching purposes it is usually not necessary. Nonetheless, the simplified scheme represents an improvement over the 2D FDTD method, and is sometimes referred to as 2½-dimensional (Lauer 1994, 1995).

There are six surfaces in a rectangular domain. With the ceiling and floor taken to be PEC, we are left with four surrounding sides at our disposal. Our application provides two possible choices of either reflective or absorbing boundary conditions (ABC). The former amounts to wrapping the whole domain with PEC and is suitable for simulating closed indoor environments, while the latter is appropriate for simulating open areas such as a concert hall (Mur 1981). Gaussian excitation of any electromagnetic field component can be set up conveniently via simple mouse clicks.

In FDTD, frequency-domain information can be extracted via discrete Fourier transforms (DFT). Our application provides built-in tools for both 1D and 2D DFT. In the 1D version, the application evaluates the full spectrum at a small number of selected cells, while in the 2D version, the spectral behavior at specified frequencies is calculated across the whole computation domain. Both 1D and 2D DFT results can be displayed directly inside the application.

3   SAMPLE SIMULATIONS

One of the most important features of wave propagation is interference. We can demonstrate the interference effect through a simplified Young's double slit simulation. First, we set up two identical EM sources, separated by a small distance d. The resultant interference pattern is then governed by:

$$\sin\theta = \frac{c}{2df} \tag{8}$$

where c is the velocity of light, f is the frequency of interest, d is the distance between the sources, and θ is the half-angle between the (first) destructive interference lines. As the interference pattern is frequency-dependent, it is best observed in the frequency domain with the 2D DFT tool. For instance, if we consider the red (4.3×10^{14} Hz) and blue (7.2×10^{14} Hz) light in the visible spectrum, a separation of 450 nm between the two light sources is appropriate, giving an angle of 50.8° (red light) and 27.6° (blue light), respectively. The simulated results are shown in Figure 1. The observed patterns are in very good agreement with the theoretical values predicted by Equation (8).

A natural extension to the previous example is the simulation of an antenna array. Antenna arrays are usually constructed to produce highly directional beams. We demonstrate the directivity by simulating a four-source $\lambda/2$-spaced antenna array with zero phase shift in Figure 2. It can be seen that two major lobes occurs along the direction perpendicular to the line of the array ($\theta = \pm 90°$), while almost no EM field is emitted towards the $\theta = 0°$ and 180° directions. In between, there are four minor side lobes indicated by the two dashed lines (Fig. 2). These observations are in excellent agreement with the theoretical prediction.

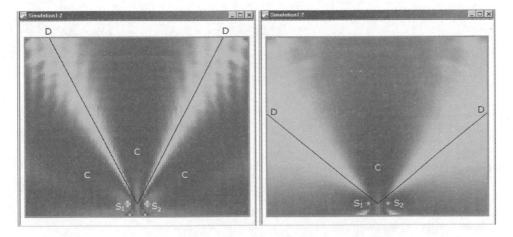

Figure 1.   Double-slit interference patterns of red (left) and blue (right) light. The solid lines (labeled with "D") indicate the theoretical positions of destructive interference, while S_1 and S_2 mark the locations of the two interfering sources and C marks the regions of constructive interference.

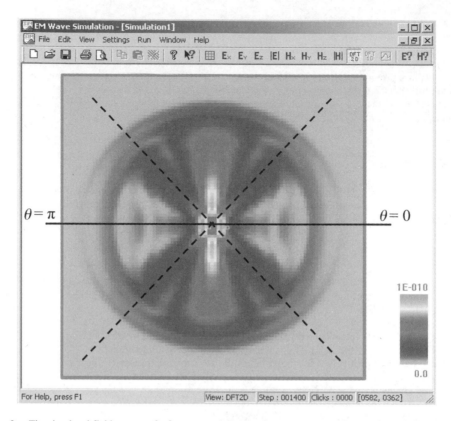

Figure 2.   The simulated field pattern of a four-source $\lambda/2$-spaced antenna array with zero phase shift.

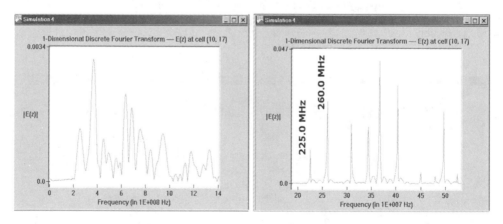

Figure 3.   1D DFT plots of an empty cavity resonator of 2 m × 1 m × 2 m after 32 ns (left) and 720 ns (right).

Finally, we consider an empty cavity resonator, which can demonstrate the 1D DFT tool of our application. The cavity is surrounded by PEC on all sides with a Gaussian excitation near the center. It can be shown that the cavity exhibits resonance at the following frequencies:

$$
f_{mnp} = \frac{c}{2} \sqrt{\left(\frac{m}{l}\right)^2 + \left(\frac{n}{w}\right)^2 + \left(\frac{p}{h}\right)^2}
\tag{9}
$$

652

where l, w, and h denote the length, width and height of the cavity, respectively, and m, n, and p are integers counting the number of half-cycles in the x-, y- and z-directions, respectively. In the 2½-D FDTD method, the vertical mode p is a parameter set by the user at the start of the simulation. The application will then simulate all the horizontal modes within the x–y plane. To separate the horizontal resonant modes, we apply the 1D DFT to an arbitrarily selected cell inside the cavity. As the simulation goes on, the non-resonant frequencies will fade out from the frequency spectrum. Screen captures displaying the resonance at various stages are shown in Figure 3. The two lowest resonant frequencies obtained from our simulation are 225.0 and 260.0 MHz respectively, which are in excellent agreement with the theoretical values of 225.00 and 259.81 MHz predicted by Equation (9).

4 CONCLUSIONS

In this paper, we report on our implementation of *EM Wave Simulation*, an interactive FDTD simulation and visualization application for electromagnetic waves (Leung et al. 2001). The application simplifies a 3D problem into a number of independent vertical modes, which are then simulated and visualized in real-time separately. Simulations are set up via simple mouse clicks, with built-in 1D and 2D DFT tools to extract frequency-domain information (if required). A few examples have been discussed to demonstrate the functionality of the application. We find that the application is useful for teaching electromagnetics to our undergraduate students in electronic and information engineering.

REFERENCES

Iskander, M.F. 1993. Computed-based electromagnetic education. *IEEE Trans. Microwave Theory Tech.* 41: 920–931.
Lauer, A., Bahr, A. & Wolff, I. 1994. FDTD simulations of indoor propagation. *44th IEEE VTC Conference Record* 2: 883–886.
Lauer, A., Wolff, I., Bahr, A., Pamp, J. & Kunisch, J. 1995. Multi-mode FDTD simulations of indoor propagation including antenna properties. *45th IEEE VTC Conference Record* 1: 454–458.
Leung, W.-K., Chen, Y. & Lee, Y.-S. 2001. EM Wave Simulation: An Animated Electromagnetic Wave Teaching Package. *Computer Applications in Engineering Education* 9(4): 208–219.
Luebbers, R.J., Kunz, K.S. & Chamberlin, K.A. 1990. An interactive demonstration of electromagnetic wave propagation using time-domain finite differences. *IEEE Trans. Educ.* 33: 60–68.
Mur, G. 1981. Absorbing boundary conditions for the finite-difference approximation of the time-domain electromagnetic-field equations. *IEEE Trans. Electromagn. Compat.* EMC-23(4): 377–382.
Shlager, K.L. & Schneider, J.B. 1995. A selective survey of the finite-difference time-domain literature. *IEEE Antennas and Propagation Magazine* 37(4): 39–56. (Also see the online FDTD database at http://www.fdtd.org/.)
Yee, K.S. 1966. Numerical solution of initial boundary value problems involving Maxwell's equations in isotropic media. *IEEE Trans. Antennas Propagat.* AP-14(4): 302–307.

Computational Methods in Engineering and Science, Iu et al. (eds)
© 2003 Swets & Zeitlinger, Lisse, ISBN 90 5809 567 3

Theory of digitized conjugate surface and solution to conjugate surface

L.Y. Xiao, D.X. Liao & C.Y. Yi
Huazhong University of Science and Technology, Wuhan, Hubei, China

ABSTRACT: In order to meet the needs of designing and processing digitized surfaces, the method to spreading digitized surface has been proposed. The key technique is to solve the problem of digitized conjugate surface. In our laboratory, the digitized conjugate surface was theoretically investigated, and the solution of conjugate surface based on digitized surface was also studied. The digitized conjugate surface theory was then proposed, and applied to build the model of solving conjugate surface based on digitized surface. A corresponding algorithm was developed. This study provides theoretical and technical bases for analyzing engagement of digitized surface, simulation and numerical processing technique.

1 INTRODUCTION

The development of modern technology is promoting the machine-building orientation towards synthesis, intelligence and digitalization. This leads to the investigations on the generating method of digitized surface. To solve the problem of generating digitized surface depends essentially on the theory of digitized conjugate surface.

Conjugate surface theory is a new basic technology that studies relationships and the mutual conversions between paired geometric graphs and paired motions under the condition of machinery processing and mechanical transmission. It involves mechanics, differential geometry, mechanism, and it can be used for the design of gear tooth face, cam contour, and cutting tools contour. It can be also applied to the motion analyses and synthesis of mechanism, processing simulation and so on. Recently, a number of works related to conjugate surface theory based on geometry analysis have been reported (Huang et al. 1995, Guo et al. 1999, Liu et al. 1999). The conjugation theory of analysis surface was then proposed. However, with the development of digitized design and process technique, the conjugate theory of analysis surface can not satisfy the need of modern design and processing technique. So it is urgent to study digitized conjugate surface theory based on discrete form, and to establish the theoretical system of digitized conjugate surface. In our laboratory, we tried to establish such a theoretical basis for conjugate generating process of digitized surface.

In this paper, we report the results of our studies on the theory of digitized conjugate surface and the solution to conjugate surface based on digitized surface. The system of digitized conjugate surface theory was developed. The model and algorithm of solving conjugate surface based on digitized surface were also proposed.

2 THEORY OF DIGITIZED CONJUGATE SURFACE

2.1 *Concept of conjugate surface*

For both a digitized surface and an analysis surface, conjugate relationship and conditions play the primordial roles, and are the important basis on conjugate theory study (Chen 1985).

Σ_1 and Σ_2 are two arbitrary conjugate surfaces, as shown in Fig. 1. Suppose Σ_1 is the mother surface, $S_1(o_1x_1y_1z_1)$ and $S_2(o_2x_2y_2z_2)$ are two coordinates. S_1 is fixedly connected with Σ_1, and S_2 is fixedly connected with Σ_2, $r_1^{(1)}$ is the position vector of the point on Σ_1 and $r_2^{(2)}$ is the position vector of the

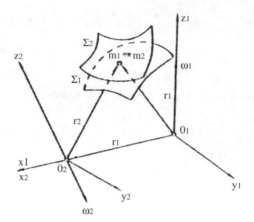

Figure 1.　The conjugate motion of two surfaces.

point on Σ_2. $n_1^{(1)}$ is unite normal vector of point $r_1^{(1)}$ on surface Σ_1 and $n_2^{(2)}$ is unite normal vector of point $r_2^{(2)}$ on surface Σ_2. For the terms mentioned above, the superscript represents the defined coordinate, while the subscript represents the surface it belongs to.

If the mother surface Σ_1 may be expressed by an equation with u and v as two parameters:

$$r_1^{(1)}=r_1^{(1)}(u,v) \tag{1}$$

The surface Σ_1 varies according to the law depending on θ parameters, and it can form a series of surfaces (surface cluster). The equation describing these surfaces is:

$$\{\Sigma_1\}: r_1^{(1)}=r_1^{(1)}(u, v,\theta) \tag{2}$$

u and v are the geometrical parameters of the mother surface, and θ is the variation parameter. When the mother surface doesn't change in shape, θ is the motion parameter. In this paper, θ is the motion parameter between surfaces.

There is a surface Σ_2, which has a common line L or a common point M with any surface Σ_1 of surface cluster $\{\Sigma_1\}$, (the line L is called as touch line, and the point M is called as touch point). In addition, at any point M of common line L, there are common tangential planes and common normal lines. Then the surface Σ_2 is the envelope of surface cluster $\{\Sigma_1\}$. Surface Σ_2 and surface Σ_1 are mutually conjugate surfaces. This touch phenomenon is called conjugate touching state or conjugate transmission.

These two surfaces, Σ_1 and Σ_2, are involved in conjugate touch motion only if they satisfy the following fundamental conditions:

1. The contact points M_1, M_2 (conjugate points) on the corresponding surface Σ_1, Σ_2 must be coincided (as shown in Fig. 1), that is:

$$r_2=r_1-r_0 \tag{3}$$

   For a practical problem the Equation 3 is equivalent to $r_2^{(2)} = r_2^{(2)} (u, v, \theta)$.
2. The two surfaces should tangent on contact point, that is, there is a common normal line, and they should contact on the outside of the region, that is

$$n_1=-n_2 \tag{4}$$

3. The relative speed of two surfaces v_{12} on the contact point should be located in common tangent plane, in order to ensure that the contact is continuous, and avoid the state of insertion or separation. That is

$$n \cdot v_{12}=0$$

   or

$$(r_{,u} \times r_{,u}) \cdot r_{,\theta}=0 \tag{5}$$

Generally, the Equation 5 is called conjugate condition (or enveloping condition).

By combining Equations 3 & 5, the solution gives us the conjugate surface Σ_2 of mother surface Σ_1. From conjugate condition (5) the relationship between motion parameters θ and mother surface geometry parameters (u, v) can be deduced: $\theta = \theta(u, v)$. Then substituting $\theta = \theta(u, v)$ in Equation 3, we obtain the conjugate surface Σ_2: $r_2^{(2)} = r_2^{(2)}(u, v)$. If the mother surface is digitized surface, the algorithm is quite complicated. But the obtained final conjugate surface is always a discrete digitized surface whether the surface is a digitized surface or an analysis surface.

2.2 The theory of digitized conjugate surface

The theory of conjugate surface based on analysis surface is undoubtedly an accurate and powerful tool, which can be used for the solution of conjugate surface and for analyzing conjugate contact. But there are critical defects in this theory. Firstly, its algebra and geometry conversion are complicated, and the computation is huge. Moreover the conversion can be only deduced manually, and it can't be done by computer directly. This largely limits the realization of computer simulation processing and dynamic optimization design. Secondly, for the digitized surface of non-analysis surface, the traditional conjugate surface theory and analysis method are not applicable. These considerations lead us to propose the theory of digitized conjugate surface, in the scope of solving different problems in modern digitized design, digitized process and various digitized inverse solving engineering.

In the digitized conjugate surface theory, we start from digitized discrete surface, and construct a cubic spline interpolating function through numerical analysis along two direction u and v of digitized surface. The function has continuous second derivative in whole space. That is $S(u) \in c^2_{[a,b]}$ and $S(v) \in c^2_{[a,b]}$. This converts the study of geometrical property for the points on the surface with double geometrical parameters to a problem of the intersection of two curved lines with single geometrical parameters. This conversion is based on Equations 3 & 5, and must respect to the conjugate relationship and condition in the process of motion on the surface. Therefore we can build a mathematical programming model for calculating the minimal value, and by means of the optimization algorithm, obtain the digitized surface Σ_2 conjugated with the digitized mother surface Σ_1.

The characteristic of this proposed digitized conjugate surface theory is that it gives up complicated derivation and conversion of traditional conjugate surface theory and only uses the frame relationship of conjugate conditions described in the second section. This new approach allows us to use the numerical methods and computer to solve various problems of conjugate surface theory. In addition, the dimension of the problem is reduced, and the algorithm is relatively simple. We can solve not only the problems of digitized mother surface, but also the problems of analysis mother surface. We realize the real theoretical analysis of digitized conjugate surface, it means the process from digitalization to analysis and solution of conjugate surface.

3 CALCULATION MODEL AND ALGORITHM OF DIGITIZED CONJUGATE SURFACE THEORY

3.1 Construction of cubic spline function with one-dimension on the digitized surface

If one digitized surface is known, we can construct cubic spline function with one-dimension along the direction of u and v respectively from node number. As an example, we construct cubic spline function along the direction of u.

Suppose we take, on the interval $[a, b]$, $n + 1$ nodes $(a = u_0 < u_1 < u_2 < \cdots < u_{n-1} < u_n = b)$ along the u direction. The function values of these points $r(u_i)$ are given as equal to r_i $(i = 0, 1, 2, ..., n)$.

Now we construct a function $S(u)$, and it satisfies the following conditions:

1 $S(u_i) = r_i$, $i = 0, 1, 2, ..., n$;
2 $S(u)$ is a third order polynomial on every small interval $[u_i, u_{i+1}]$;
3 $S(u)$ is second continuous differentiable on interval $[a, b]$ that is $S(u) \in c^2_{[a,b]}$, $S(u)$ is called cubic spline interpolating function of $r(u)$.

Suppose the cubic spline function $S(u)$ exists, and m_i represents the derivative value of $S(u)$ on the point u_i, that is $S'(u_i) = m_i - r_{,u}(i = 0, 1, 2, ..., n)$, the cubic spline function can be obtained on the small interval$[u_i, u_{i+1}]$, by using Hermite interpolating equation.

$$S(u)=\left(1+2\frac{u-u_i}{u_{i+1}-u_i}\right)\left(\frac{u-u_{i+1}}{u_i-u_{i+1}}\right)^2 r_i$$
$$+\left(1+2\frac{u-u_{i+1}}{u_i-u_{i+1}}\right)\left(\frac{u-u_i}{u_{i+1}-u_i}\right)^2 r_{i+1}$$
$$+\left(u-u_i\right)\left(\frac{u-u_{i+1}}{u_i-u_{i+1}}\right)^2 m_i$$
$$+\left(u-u_{i+1}\right)\left(\frac{u-u_i}{u_{i+1}-u_i}\right)^2 m_{i+1} \tag{6}$$

In Equation 6, the derivate value of m_i on the node u_i ($i = 0, 1, 2, ..., n$) are unknown. In order to get $m_0, m_1, m_2, ..., m_n$, by using second derivate continuity of function $S(u)$ on the node u_i, we can easily get a set of equations on every interior point, that is,

$$\left(1-\alpha_i\right)m_{i-1} + 2m_i + \alpha_i m_{i+1} = \beta_i, \left(i = 1,2,\cdots,n-1\right) \tag{7}$$

In Equation 7

$$\alpha_i = \frac{h_{i-1}}{h_{i-1}+h_i}$$
$$\beta_i = 3\left[\frac{1-\alpha_i}{h_{i-1}}\left(r_i - r_{i-1}\right)+\frac{\alpha_i}{h_i}\left(r_{i+1}-r_i\right)\right];$$
$$h_i = u_{i+1} - u_i$$

This is a problem of $(n - 1)$ linear equations with $(n + 1)$ unknown variable $m_0, m_1, ..., m_n$. If we complete two additional conditions or boundary conditions, we can obtain three pairs of angle equations. The matrix form is:

$$\begin{bmatrix} 2 & \alpha_0 & 0 & \cdots & & 0 \\ \left(1-\alpha_1\right) & 2 & \alpha_1 & \cdots & & \cdots \\ 0 & \left(1-\alpha_2\right) & 2 & \alpha_2 & & \cdots \\ \cdots & \cdots & \cdots & \cdots & & \cdots \\ 0 & \cdots & & \cdots & \left(1-\alpha_n\right) & 2 \end{bmatrix}\begin{Bmatrix} m_0 \\ m_1 \\ m_2 \\ \cdots \\ m_n \end{Bmatrix} = \begin{Bmatrix} \beta_0 \\ \beta_1 \\ \beta_2 \\ \cdots \\ \beta_n \end{Bmatrix} \tag{8}$$

we can get the unique solution $m_0, m_1, ..., m_n$ by using recursion relation. After substituting m_i ($i = 0, 1, ..., n$) into Equation 6, we get cubic spline function $S(u)$ on small interval $[u_i, u_{i+1}]$. The steps of calculating the cubic spline function may be summarized as below:

1 According to the given points (u_i, r_i) and corresponding boundary conditions, we calculate the set of Equations 8 and coefficients α_i and β_i under corresponding boundary conditions.
2 Solving three pairs angle equations on the given boundary conditions, we obtain the first derivate value $m_0, m_1, ..., m_n$ of every node.
3 After substituting $m_0, m_1, ..., m_n$ into Equation 6 of spline function, we get the final equation of cubic spline function.

3.2 Calculation model and algorithm for conjugate surface.

According to the digitized conjugate surface theory discussed above, we might construct the following model for solving the conjugate surface.

$$\min \; F(u) = \left(r_{,u}, r_{,v}, r_{,\theta}\right) \tag{9}$$

$$S.T \quad C_1(u) = \left(r_{,u}, r_{,v}, r_{,\theta}\right) \le \varepsilon \tag{10}$$

$$C_2(u) = r_2 + r_0 - r_1 = 0 \tag{11}$$

In these equations, ε is the given precision, the algorithm is as below:

1. According to node date on the given surface, we construct respectively the cubic spline functions $S_1^{(1)}(u)$ and $S_1^{(1)}(v)$ corresponding respectively to geometry parameters u and v. By using the coordinate transformation and motion relationship, we transform the functions into the coordinates $S_2(o_2 x_2 y_2 z_2)$ of conjugate surface Σ_2, that is $S_2^{(2)}(u, \theta)$, $S_2^{(2)}(v, \theta)$.
2. For a given θ_i with v_i, we perform a one-dimension search along direction u, then we get u_i which satisfy the Equations 9 & 10 (conjugate conditions). From u_i, we calculate the r_i of point M_i which meets the conjugate conditions. From the given θ, we can get the transform matrix A_θ. Finally, from Equation 11, we calculate the r_{2i} of the point on the conjugate surface, corresponding to r_{1i} on mother surface Σ_1.
3. Save the corresponding θ_i, v_i, u_i, r_{1i}, r_{2i} into date file.
4. Let $v_i = v_i + \Delta v$, continue the steps (2), (3) until the calculation is achieved on direction v.
5. Let $\theta_i = \theta_i + \Delta \theta$ repeat steps (2), (3), (4) until the calculation is completed on direction θ (or surpass the known surface).

From the above calculation, we get the conjugate digitized surface Σ_2 which conjugates with digitized surface Σ_1.

4 CONCLUSION

In this work, we proposed a theory of digitized conjugate surface and the solution to digitized conjugate surface. In general, when we try to solve conjugate surface based on digitized surface, we must construct digitized surface analysis equation. The proposed theory is, as the first time, out of this restriction. By means of the decomposition technique and reduced dimension, and using the optimization algorithm, we can largely reduce the complexity of solving conjugate surface. Furthermore, this theory will be applied as theoretical and technical basis, to further investigations of the digitized conjugate surface.

REFERENCES

Chen, C.H. 1985. Fundamentals of the Theory of Conjugate Surfaces. Beijing: Science Press.
Guo, W.Z. et al. 1999. Discretization and analysis principle of conjugating surfaces for adaptive synthesis. *Mechanical Engineering Academic Journal* 35(4):25–28.
Huang, K.Z. et al. 1995. The theory and application of decompose and recompose of solving conjugate surface. *Mechanical Engineering Academic Journal* 32(1):68–73.
Litvin, F.L. 1989. *Theory of Gearing*. Washington, DC: NASA.
Litvin, F.L. et al. 1991. Identification and minimization of deviations of real gear tooth surfaces. *ASME Journal of Mechanical Deign* 113(1):55–62.
Litvin, F.L. et al. 1993. Minimization of deviation of gear real tooth surfaces determined by coordinate measurement, *ASME Mechanical Design* 10(4):995–1001.
Litvin, F.L. et al. 1995. Computerized determination of curvature relations and contact ellipse for conjugate surface. *Comput. Methods Appl. Mech. Engrg.* 125(1–4):151–170.
Litivn, F.L. 1995. Applied theory of gearing state of the art. *Journal of Mechanical Design* (B):128–134.
Liu, J. et al. 1999. Theory of numerical simulation of conjugate surface. *Dalian University of Science and Technology Academic Journal* 39(2):259–267.
Shunmugam, M.S. et al. 1998. Establishing gear tooth surface geometry and normal deviation. *Mechanism and Machine Theory* 33(5):517–524.

Computational Methods in Engineering and Science, Iu et al. (eds)
© 2003 Swets & Zeitlinger, Lisse, ISBN 90 5809 567 3

Mathematical modeling and computer simulation for redundant manipulators

Y. Li
Department of Electromechanical Engineering, Faculty of Science and Technology,
University of Macau, Macao

ABSTRACT: This paper presents a methodology of analyzing the kinematics and dynamics characteristics of redundant manipulators. The mathematical models of a general planar redundant manipulator have been derived. Computer simulations have been carried out by MATLAB and SIMULINK software, which show the effectiveness of the algorithm obtained through this research.

1   INTRODUCTION

Since redundant manipulators have extra Degrees of Freedom (DOF) requested by finishing a task, they are extremely useful when the working space is narrow or impossible for a common robot or human to finish the mission. The application of redundant manipulator can be found in nuclear factory, space or deep sea exploration, environmental protection etc. (Li et al. 1999, 2000). Although the research work on the redundant manipulators has been carried out for many years, few papers can be found which focus on computer simulation with MATLAB and SIMULINK (Zlajpah 1998, 2000). By computer simulation, designer can not only understand the feasibility of mechanism during design period, but also can make an experiment which is difficult to perform in field. On the other hand, computer simulation can both save money and shorten the product developing period.

This paper is organized as follows. Kinematics of a general manipulator and measure criteria of redundancy are presented in Section 2. Dynamics analysis is presented in Section 3. Computer simulations with MATLAB and SIMULINK are given in Section 4, and conclusions are given in Section 5.

2   KINEMATICS OF A MANIPULATOR

2.1   *Forward kinematics modeling*

We assume that a general planar manipulator with n joint variables q, and working in m dimensional task spaces, the kinematics of the manipulator can be represented by following equations (Craig 1989):

$$x = P(q) \tag{1}$$

$$\dot{x} = J(q)\dot{q} \tag{2}$$

$$\ddot{x} = J(q)\ddot{q} + \dot{J}(q,\dot{q})\dot{q} \tag{3}$$

where J is the $m \times n$ Jacobian matrix; $\dot{J}(q,\dot{q})$ is its time derivative; $n > m$ in case of redundant manipulator.
    Let ϕ be an n-dimensional vector with components

$$\phi_i = \phi_{i-1} + q_i \tag{4}$$

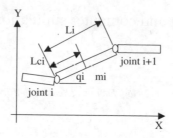

Figure 1. Two neighboring joints and links.

If $i = 1, ..., n$ and initial value $\phi_0 = 0$, and l_i denotes the length of the i-th link. For a planar manipulator, the following equations can be used to calculate the joint end-effector's positions

$$\begin{cases} x_i = x_{i+1} + l_i \cos(\phi_i) \\ y_i = y_{i+1} + l_i \sin(\phi_i) \end{cases} \tag{5}$$

where $i = n - 1, ..., 1$; initial values $x_n = l_n \cos(\phi_i)$; $y_n = l_n \sin(\phi_i)$. The pairs $[x_1, y_1]^T$ represent the position of the end-effector of the manipulator measured from joint i.

The Jacobian matrix for a planar manipulator is a $2 \times n$ matrix

$$J = \begin{bmatrix} \dfrac{\partial x_1}{\partial q_1} & \cdots & \dfrac{\partial x_1}{\partial q_n} \\ \dfrac{\partial y_1}{\partial q_1} & \cdots & \dfrac{\partial y_1}{\partial q_n} \end{bmatrix} \tag{6}$$

From equations (4) and (5), we can obtain

$$\begin{cases} \dfrac{\partial x_i}{\partial q_j} = -y_k \\ \dfrac{\partial y_i}{\partial q_j} = x_k \end{cases} \tag{7}$$

where $k = \max(i, j)$.

Therefore the components of J can be obtained from above equation

$$J = \begin{bmatrix} -y_1 & \cdots & -y_n \\ x_1 & \cdots & x_n \end{bmatrix} \tag{8}$$

The first derivative of J can be derived

$$\dot{J} = \sum_{k=1}^{n} \left(\frac{\partial J}{\partial q_k} \dot{q}_k \right) \tag{9}$$

Differentiating equation (7) with respect to q yields

$$\begin{cases} \dfrac{\partial^2 x_i}{\partial q_j \partial q_k} = -x_r \\ \dfrac{\partial^2 y_i}{\partial q_j \partial q_k} = -y_r \end{cases} \tag{10}$$

where $r = \max(i, j, k)$. Substituting equation (10) into (9) yields

$$
J = \frac{\dot{q}^T \begin{bmatrix} -x_1 & -x_2 & \cdots & -x_n \\ -x_2 & -x_2 & \cdots & -x_n \\ \cdots & \cdots & \cdots & \cdots \\ -x_n & -x_n & \cdots & -x_n \end{bmatrix}}{\dot{q}^T \begin{bmatrix} -y_1 & -y_2 & \cdots & -y_n \\ -y_2 & -y_2 & \cdots & -y_n \\ \cdots & \cdots & \cdots & \cdots \\ -y_n & -y_n & \cdots & -y_n \end{bmatrix}}
\tag{11}
$$

From above analysis, we can see that the only elements for a kinematical model of n–R planar manipulator to be calculated are x_i and y_i, other elements can be expressed in terms of x_i and y_i.

2.2 Performance measures

For a planar manipulator, the JJ^T is a 2×2 symmetric matrix

$$
JJ^T = \begin{bmatrix} a^{11} & a^{12} \\ a^{12} & a^{22} \end{bmatrix}
\tag{12}
$$

Hence the two singular values of JJ^T are

$$
\sigma_{1,2} = \frac{a^{11} + a^{22} \pm \sqrt{(a^{11} - a^{22})^2 + 4(a^{12})^2}}{2}
\tag{13}
$$

Supposed that components be $a^{ij} = a^{ij}_i$, which can be derived from the following equations

$$
a^{11}_i = a^{11}_{i+1} + y^2_i
\tag{14}
$$

$$
a^{22}_i = a^{22}_{i+1} + x^2_i
\tag{15}
$$

$$
a^{12}_i = a^{12}_{i+1} - x_i y_i
\tag{16}
$$

where the initial values are $a^{11}_{n+1} = a^{22}_{n+1} = a^{12}_{n+1} = 0$.
   Furthermore, differentiating a^{ij} yields

$$
\frac{\partial a^{11}}{\partial q_j} = 2(x_j \sum_{k=1}^{j-1} y_k - a^{12}_j)
\tag{17}
$$

$$
\frac{\partial a^{22}}{\partial q_j} = 2(-y_j \sum_{k=1}^{j-1} x_k + a^{12}_j)
\tag{18}
$$

$$
\frac{\partial a^{12}}{\partial q_j} = -x_j \sum_{k=1}^{j-1} x_k + y_j \sum_{k=1}^{j-1} y_k + a^{11}_j - a^{22}_j
\tag{19}
$$

2.3 Condition number

The condition number is defined as the ratio between the maximal and minimal singular value of J, that is

$$
\rho = \sqrt{\frac{\sigma_{max}}{\sigma_{min}}}
\tag{20}
$$

Substituting equation (13) into (20) yields

$$\rho = \sqrt{\frac{a^{11} + a^{22} + \sqrt{(a^{11} - a^{22})^2 + 4(a^{12})^2}}{a^{11} + a^{22} - \sqrt{(a^{11} - a^{22})^2 + 4(a^{12})^2}}} \tag{21}$$

The gradient of ρ is defined as

$$\nabla\rho = \frac{1}{2\rho} \frac{\sigma_2\nabla\sigma_1 - \sigma_1\nabla\sigma_2}{\sigma_2^2} \tag{22}$$

Equation (22) can be calculated by substituting (17)–(19) into it.

2.4 Manipulability

The manipulability measure is defined as

$$w = \sqrt{\sigma_1\sigma_2\cdots\sigma_m} = \sqrt{\det(JJ^T)} = \sqrt{a^{11}a^{22} - (a^{12})^2} \tag{23}$$

The gradient of w is defined as

$$\nabla w = \frac{a^{11}\nabla a^{22} + a^{22}\nabla a^{11} - 2a^{12}\nabla a^{12}}{2\sqrt{a^{11}a^{22} - (a^{12})^2}} \tag{24}$$

3 DYNAMICS OF A MANIPULATOR

The position of the mass center of the i-th link can be written as

$$\begin{bmatrix} x_{ci} \\ y_{ci} \end{bmatrix} = \begin{bmatrix} x_1 - x_i + l_{ci}\cos(\phi_i) \\ y_1 - y_i + l_{ci}\sin(\phi_i) \end{bmatrix} \tag{25}$$

Here $[x_1 - x_i, y_1 - y_i]^T$ denotes the position vector of joint i measured from the base of the manipulator. Differentiating above equation with respect to q yields

$$\begin{bmatrix} \dfrac{\partial x_{ci}}{\partial q_j} \\ \dfrac{\partial y_{ci}}{\partial q_j} \end{bmatrix} = \begin{bmatrix} -y_j + y_i - l_{ci}\sin(\phi_i) \\ x_j - x_i + l_{ci}\cos(\phi_i) \end{bmatrix} \quad \text{when} \ \ j \le i \tag{26}$$

$$\begin{bmatrix} \dfrac{\partial x_{ci}}{\partial q_j} \\ \dfrac{\partial y_{ci}}{\partial q_j} \end{bmatrix} = \begin{bmatrix} 0 \\ 0 \end{bmatrix} \quad \text{when} \ \ j > i \tag{27}$$

Classified by linear velocity and angular velocity, the Jacobian matrix can be written as

$$J = \begin{bmatrix} J_L \\ J_A \end{bmatrix} \tag{28}$$

The Jacobian matrices for the mass center of the i-th link can be derived

$$J_L^{(i)} = \begin{bmatrix} \partial x_{ci} / \partial q_1 & \cdots & \partial x_{ci} / \partial q_n \\ \partial y_{ci} / \partial q_1 & \cdots & \partial y_{ci} / \partial q_n \end{bmatrix} \tag{29}$$

while the derivatives of J_L with respect to q can be derived

$$\frac{\partial J_L^{(i)}}{\partial q_k} = \begin{bmatrix} \partial^2 x_{ci} / \partial q_1 \partial q_k & \cdots & \partial^2 x_{ci} / \partial q_n \partial q_k \\ \partial^2 y_{ci} / \partial q_1 \partial q_k & \cdots & \partial^2 y_{ci} / \partial q_n \partial q_k \end{bmatrix} \tag{30}$$

$$\begin{bmatrix} \dfrac{\partial^2 x_{ci}}{\partial q_j \partial q_k} \\ \dfrac{\partial^2 y_{ci}}{\partial q_j \partial q_k} \end{bmatrix} = \begin{bmatrix} -x_r + x_i - l_{ci}\cos(\phi_i) \\ -y_r + y_i - l_{ci}\sin(\phi_i) \end{bmatrix} \qquad \text{when } j \le i \tag{31}$$

$$\begin{bmatrix} \dfrac{\partial^2 x_{ci}}{\partial q_j \partial q_k} \\ \dfrac{\partial^2 y_{ci}}{\partial q_j \partial q_k} \end{bmatrix} = \begin{bmatrix} 0 \\ 0 \end{bmatrix} \qquad \text{when } j > i \tag{32}$$

where $r = \max(j, k)$.

The following symbols are introduced for simplification

$$\Theta^{(i)} = J_L^{(i)T} J_L^{(i)} \quad \text{and} \quad \Psi_k^{(i)} = \left(\frac{\partial J_L^{(i)}}{\partial q_k} \right)^T J_L^{(i)} + J_L^{(i)T} \left(\frac{\partial J_L^{(i)}}{\partial q_k} \right) \tag{33}$$

The Lagrange equation of the manipulator is expressed as

$$\frac{d}{dt} \frac{\partial T}{\partial \dot{q}} - \frac{\partial T}{\partial q} = \tau \tag{34}$$

where T is the total system energy; τ is the generalized force. The dynamics equation of a general manipulator can be written in matrix form as

$$\tau = H(q)\ddot{q} + h(q, \dot{q}) + B\dot{q} + g(q) \tag{35}$$

In particular, the matrix H is

$$H = \sum_{i=1}^{n} \left(m_i J_L^{(i)T} J_L^{(i)} + J_A^{(i)T} I_i J_A^{(i)} \right) \tag{36}$$

where m_i is the mass of the i-th link. Considering a planar revolute manipulator, the terms $J_A^{(i)T} I_i J_A^{(i)}$ can be simplified to

$$J_A^{(i)T} I_i J_A^{(i)} = I_i \begin{bmatrix} 1_{i \times i} & 0 \\ 0 & 0 \end{bmatrix}_{n \times n} \tag{37}$$

where I_i is the moment of inertia. 1 and 0 represent unity matrix and zero matrix respectively.

The Coriolis and centrifugal forces can be represented by

$$h_i = \sum_{j=1}^{n} \sum_{k=1}^{n} h_{ijk} \dot{q}_j \dot{q}_k \qquad \text{and} \qquad h_{ijk} = \frac{\partial H_{ij}}{\partial q_k} - \frac{1}{2} \frac{\partial H_{jk}}{\partial q_i} \tag{38}$$

The vector H can be expressed in a concise form

$$H = \sum_{i=1}^{n} m_i \left(\left(\sum_{j=1}^{n} \Psi_j^{(i)} \dot{q}_j \right) \dot{q} - \frac{1}{2} \begin{bmatrix} \dot{q}^T \Psi_1^{(i)} \\ \dot{q}^T \Psi_2^{(i)} \\ \cdots \\ \dot{q}^T \Psi_n^{(i)} \end{bmatrix} \dot{q} \right) \tag{39}$$

The gravity vectors g are given by

$$g_i = \sum_{j=1}^{n} m_j \vec{g}_{acc}^T J_{Li}^{(j)} \tag{40}$$

where $\vec{g}_{acc}^T$ is the acceleration of gravity. As for n-th revolute planar manipulator, the gravity force of the link i is equal to the gravity force of the link $i + 1$ contributed to the link i. Therefore

$$g_i = g_{i+1} + g_{cc} \left(m_i l_{ci} + \sum_{k=i+1}^{n} m_k l_i \right) \cos(\phi_i) \tag{41}$$

where $i = n - 1, \ldots, 1$; the initial value is $g_n = g_{acc} m_n l_{cn} \cos(\phi_n)$.

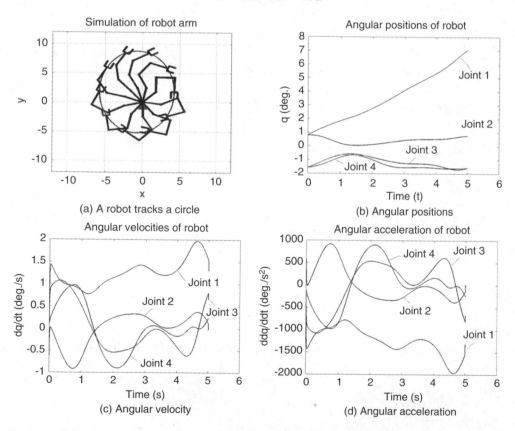

Figure 2. Kinematics analysis of a robot tracking a circle.

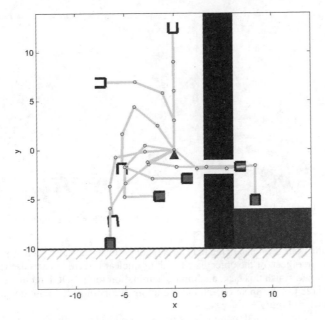

Figure 3. A robot passing through a hole on a wall to grasp an object.

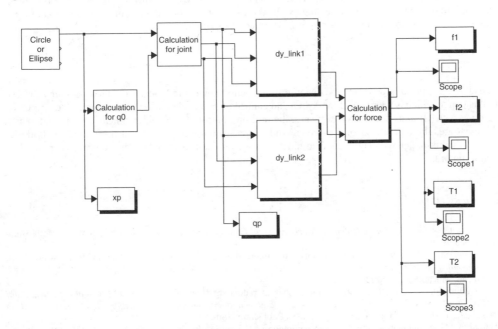

Figure 4. Control block diagram with SIMULINK.

4 COMPUTER SIMULATIONS

A four-link redundant planar robot who tracks a circle has been investigated. Using MATLAB language, the simulations are shown in Figure 2. The trajectory of this robot tracking a circle is shown in Figure 2(a), while the angular positions, velocities, and accelerations of each joint are shown in Figure 2(b), Figure 2(c), and Figure 2(d) respectively.

667

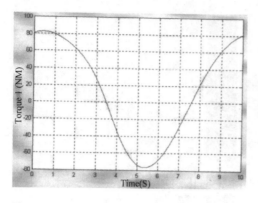

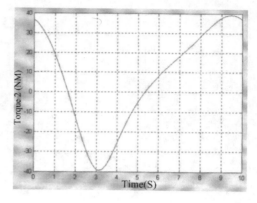

Figure 5.   Torque applied to joint 1.

Figure 6.   Torque applied to joint 2.

In some practical situations, a human being may be can not perform such task such as passing through a narrow hole to grasp an object, or the object is a bomb or nuclear infected material, then a redundant robot can replace human to finish this task, a computer simulation for a 4-DOF redundant manipulator inserting into a small hole to take up an object and retracting autonomously to put down the object on the ground can be shown in Figure 3.

The dynamics analysis for a 2-DOF manipulator is carried out. The manipulator tracks a circle in a plane, and the modeling is set up using MATLAB and SIMULINK as shown in Figure 4. In order to let the manipulator to track the circle accurately, the required control torque for joint 1 is shown in Figure 5, while the required control torque for joint 2 is shown in Figure 6.

5   CONCLUSIONS

The characteristics of kinematics and dynamics for a general planar redundant robot has been investigated in this paper, which is meaningful for design and control of a redundant robot. From computer simulation using MATLAB and SIMULINK software, the motion of a robot can be dynamically simulated, which demonstrates the correctness of theoretical study. The computer modeling and simulation method presented in this paper will lay a good foundation for practical construction and control of redundant manipulators.

REFERENCES

Craig, J.J. 1989. *Introduction to Robotics: Mechanics and Control, 2nd ed.* New York: Addison-Wesley Publishing Company.
Li, Y. and Sin, V.K. 1999. Redundant robots with applications to environmental protection. In Z. Wang (ed.), *Proceedings of First Macau Symposium on Environment and City Development, Macao, March 23–25, 1999.* Macao: Macau Foundation Press.
Li, Y. 2000. Study on the kinematic characteristics of redundant robot. In F. Liu & J. Sun (eds), *International Conference on Advanced Manufacturing Systems and Manufacturing Automation, Guangzhou, P.R. China, June 19–21, 2000.* Guangzhou: Guangdong People's Publishing House.
Zlajpah, L. 1998. Simulation of n–R planar manipulators, *Simulation Practice and Theory*, 6(3): 305–321.
Zlajpah, L. 2000. Integrated environment for modelling, simulation and control design for robotic manipulators. *3rd MATHMOD, IMACS Symp. on Mathematical Modelling*, Vienna, Austria.

Analysis of sheet forming for cold press die

Y.W. Ye & H. Liu
College of Electrical & Mechanical Engineering, Zhejiang University of Technology, Hangzhou, China

ABSTRACT: This paper discusses the mechanics specialties for press forming, and focuses on the use of computer visualization method to the simulation of metal sheet forming. Combining analysis with scientific calculation visualization simulates the process of press forming dynamically and correctly.

1   INTRODUCTION

Pressing process is a manufacturing method that obtained work piece with certain dimension, shape and capability through die forcing metal sheet to produce plastic deformation or make separation.

The process of sheet pressing forming is a very complex process of plasticity forming. Many factors directly or indirectly affect the result of shape. For ages the design and production of press die all depend on previous experience and need to repetitive debugging. After designed primarily by experience, traditional forming technical design needs continuous testing and mending die. So, this prolongs design cycle and increases die cost. With the press part becoming more and more complex, precision becoming more and more high, it is necessary finding a scientific sheet material forming analysis system to explore the forming status of press part and material flow action, which can act as the references of die design.

Because of the important signification in practical project, adopting finite element method to simulate the process of metal plasticity forming, have been valued generally by many scholars (for instance, Dong 2000). Finite element analysis of sheet forming process is an integrating technology, which comprises forming theory, material mechanics, computational mathematics, press process, computer graph and program design. It utilizes these theories and knowledge to carry out a series of numerical analysis and simulation for the process of sheet forming in computer in order to replace actual die test. It predicts problem, which is potential in the die design. It can provide scientific foundation for die designer and process worker in the design of die and process plan. Therefore it is a hot research topic that combining the numerical result of finite element numerical analysis with computing visualization simulation technique for pressing process.

2   KEY PROBLEMS

The character of sheet pressing forming is that the deformation of thickness direction is very small in contrast to other directions. The deformation mode of sheet forming mainly has the following kinds: bi-directional stretch, plane stress, stretches, depths extend, bending and counter-bending. But the procedure of sheet forming can be considered a standard static, so the velocity and acceleration can be ignored.

The sheet forming involves great deformation, great turn, material, nonlinear boundary, and dynamic friction. When the sheet is pressed, the stresses and strains in different parts of the sheet are different. On condition that the state is accordance with the region of plastic condition, the region can come into being plastic deformation, other regions don't occur plastic deformation. So the sheet could be separated into transferring region and ametabolic region. In addition, based on the state of force and deformation which meet a metabolic condition, the sheet can further separate into the region which have experienced deformation, the region that will be transferred and the region not involving to deformation in the whole course. For the ametabolic region bears force, it is called transferring force region.

As an integrating finite element numerical analysis for sheet forming, besides the problems of friction and touching it should be considered in those primary problems thereinafter: pivotal algorithm, subdued function, structure formula, finite element meshing, cracked rule, spring-back calculation and crinkly analysis.

To affirm the favoring process of computing and getting correct result, convergence and stability of the algorithm must be ensured. In fact there are two nonlinear finite element algorithms: explicit way (such as central difference way) and implicit way (such as Newark way). In general, implicit way in the static or standard static problem is adopted because it is absolute stable in the proper condition for parameters are chosen reasonably. But for dynamic touching problem, if implicit way is used its convergence can't be ensured because of continuous moving between touching force and limit condition. So explicit way of central difference array is used to resolve the problem of sheet forming currently.

For big deformation problem virtual work principle is adopted to use the stress and conjugated strain of Euler and two ways of Lagarange. Because the way of Euler takes accompanying coordinate system, so the best excellence rests with the superiority in meshing (it needn't re-meshing). But what the description of integral area of norm function in Euler is current structure. It should be the result, so it is iterative. The more important problem is the constitutive relationship is different because the Euler describes is different with the different substance point in time coordinate system. So the constitutive relationship can't be used in calculation formula. The way of Lagarange can overcome this shortcoming; moreover UL (Updated Lagaranian) is more useful and convenient than TL (Total Lagrangian), especially for tracking the changes of strain in the distortion process.

Finite element computing is a kind of numerical analyzing method. Its characteristic is producing immense number of digital information besides needing power process ability for computer. Only after carefully analyzing and understanding computing output information, the condition of the problem during the computing could be seen, and the object studied be comprehended. But this process is quite wasting time and troubling. For simple computing problem with low dimension correct conclusion could be obtained from computing results though workload is numerous. For complicated computing problem with high dimension, such as field distribution problem of three-dimensional continuum (stress distribution, pressure distribution, temperature distribution), it's "four dimension" in fact, its data is four-dimensional data. The results obtained from these four dimensional data are not easy to analyze and give out rational conclusion.

3 SCIENTIFIC VISUALIZATION COMPUTING

Visualization in scientific computing is a very useful technique for finite element analysis and simulation of engineering problem, which could be used in three ways: post processing, tracking and steering. Post processing divides computing and computing visualization as two steps, which couldn't intercross each other. Tracking demands displaying the computing results in time in order to let people know the current computing condition and decide whether to continue. Steering could intervene to computing in time, for example, refining mesh, modifying material, and changing step.

Because tracking and steering ways demand displaying the computing middle results in time and intervening to the process, more strict demand is put to the arithmetic for solving problem. It not only asks for correctly solving problem, but also asks for interactively outputting the computing results at the pointed time, and receives computing controlling parameters. These two ways need to study farther. Now visualization of finite element computing is mainly the post processing (Zhu 2001). Visualization technique for finite element analysis includes two parts: for scalar quantity field and vector field. Visualization method for scalar quantity field has isoline diagram, color nephogram, and slice up diagram. Visualization method for vector field mainly is arrowhead line expression method.

According to the above discussion the paper carries out finite element simulation for sheet pressing forming by use of Microsoft Visual C++ and OpenGL graphic technique. It establishes numerical simulation program, dynamically simulates the process of sheet pressing figuration, and gets good effects. Through the visualization simulation for pressing process the die design could be directed. The relative parameters for die design, such as minimum bending radius, position of possible crinkling, and plan for pressing steps, could be calculated.

The rigidity of the die is much greater than the sheet's in the sheet forming, the die is considered to be a rigid body. Therefore the die itself could be expressed by use of outline surface of its figure. For description of the sheet there are two models.

Discrete mesh model: The sheet middle surface is described by triangular or quadrilateral mesh. The description accuracy for formed sheet is depended on discrete accuracy. In general much discrete mesh

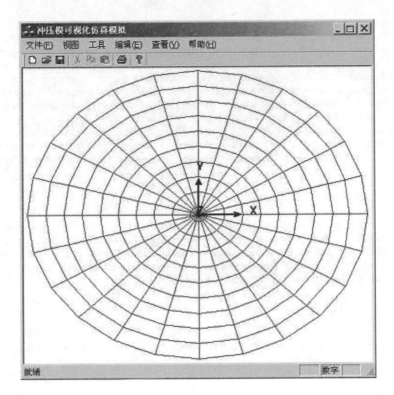

Figure 1.   Initial state of the sheet.

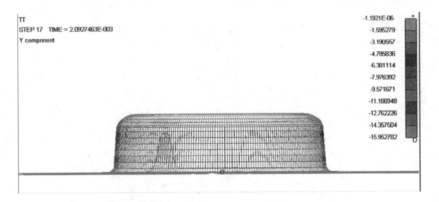

Figure 2.   The forming status of step 1.

is needed to the idea description of complex shape. But for the sake of efficiency it's better to reduce the number of discrete mesh in case of precision be guaranteed.

For model describing in computer it uses the data created from finite element method computing if using discrete mesh model to approximate the complex surface of the formed sheet, which is convenient to the engineering application. For example, it's comparatively simple and easy to deal with if scattering the surface to many linear patch in the contacting problem, which needs to judge the contacting state and compute the position of contacting points.

Parametric surface model: The sheet surface is expressed by use of Bezier, Coons, B Spline or BURBS surface. In the surface estimating parametric surface model is also used for it has obviously advantages, i.e., easy to express, design and modify, memory requirements small. So NURBS surface is used in the study.

The whole simulation analysis procedure is consisted following parts. (1) Program environment initialization. This realizes using VC++ programming based on OpenGL graphic database. Environment initialization contains: initializing pixel format, establishing OpenGL equipment environment, font realization for OpenGL and other settings. (2) Data reading model inputting the information of nodes and elements of the model. After reading the data file of the model and setting the model parameters, it analyzes the data to obtain the useful information for dir design, and display the obtained model. The visualization display process of the press die could be frame-by-frame or continuous (i.e., animation model).

Figures 1 and 2 are an example for die design. Figure 1 is the initial state of sheet, and Figure 2 is the forming status of step 1. Originally it has three dies (corresponding to three steps pressing). But the work piece often has problem of tearing up. After computing and analyzing, making sure the reason is that its bending radius is too small to forming, so four steps pressing is used, which solves the problem, and get the idea result.

4   CONCLUSIONS

The study of sheet pressing forming is very significant and difficult work. Introducing scientific computing visualization method is an efficient way to solve the problem of difficult understanding numerical data given by finite element analysis. Combining numerical computing and visualization to simulating dynamically and visually press process will be helpful not only to understand the course of pressing figuration, but also to guide the rational design of cold press die.

ACKNOWLEDGEMENT

Foundation item: Project supported by the Natural Science Foundation (599016) of Zhejiang Province.

REFERENCES

Dong, X.H., Zheng, Y., Lan, J. 2000. Some progress of computer simulation in metal plastic forming. *Met. Forming Process*, 18(1): 1–4.
Zhu, G.L., Liu, J., Wang, M.W. 2001. A dynamical visualization method for finite element post process. *Mech. Des. Study*, 17(3): 18–19.

Computational exploratory data analysis in cotton spinning

A.A. Cabeço Silva & M.E. Cabeço Silva
Departamento de Engenharia Têxtil, Universidade do Minho, Portugal

ABSTRACT: There are a vast number of useful analyses that can be solved by discriminant analysis. The main advantage these methods have is that they are generally easier to apply to blending cotton analysis than quantitative methods since they do not need any primary approximation data to build the model. They give simple "pass" or "fail" answers as to how well the samples match by comparing them to training sets of the desired quality samples. They learn to recognize the cotton blends based entirely on the cotton data itself without any other external information other than the analyst's logical grouping of the cotton into training sets. Discriminant analysis is the statistical technique that is most commonly used to solve complex problems. Its use is appropriate when we can classify data into two or more groups, and when we want to find one or more functions of quantitative measurements that can help discriminate among the known groups. The goal of this analysis is to present a method for predicting which group a new case is most likely to fall into, or to obtain a small number of useful predictor variables. In this paper, the intention is to classify cotton bales into well define groups or categories based on a training set of similar samples for grading cotton blends.

A new algorithm has been implemented using discriminant analysis and its advantages in quality design of cotton blends are verified.

1 INTRODUCTION

Computational exploratory data analysis methods include both simple basic statistics and more advanced, designated multivariate exploratory techniques designed to identify patterns in multivariate data sets.

The basic statistical exploratory methods include such techniques as examining distributions of variables (e.g., to identify highly skewed or non-normal, such as bi-modal patterns), reviewing large correlation matrices for coefficients that meet certain thresholds, or examining multi-way frequency tables (e.g., "slice by slice" systematically reviewing combinations of levels of control variables).

For many years the intersection of computing and data analysis contained menu-based statistics packages and not much else. Recently, statisticians have embraced computing, computer scientists are using statistical theories and methods, and researchers in all corners are inventing algorithms to find structure in vast online datasets. Data analysts now have access to tools for exploratory data analysis, decision tree induction, function finding, constructing, and visualization, and there are intelligent assistants to advice on matters of design and analysis. There are tools for traditional, relatively small samples and also for enormous datasets. In all, the scope for probing data in new and penetrating ways has never been so exciting.

In this work we will apply these techniques in cotton processing technology.

2 DISCRIMINANT ANALYSIS

Discriminant analysis is called a "general" discriminant analysis because it applies the methods of the general linear model to the discriminant function analysis problem. The discriminant function analysis problem is "recast" as a general multivariate linear model, where the dependent variables of interest are (dummy-) coded vectors that reflect the group membership of each case. The remainder of the analysis is then performed as described in the context of General Regression Models with a few additional features.

One advantage of applying the general linear model to the discriminant analysis problem is that you can specify complex models for the set of predictor variables. For example, we can specify for a set of continuous predictor variables, a polynomial regression model, response surface model, factorial regression, or mixture surface regression (without an intercept). Thus, we could analyze a constrained mixture experiment (where the predictor variable values must sum to a constant), where the dependent variable of interest is categorical in nature. In fact, discriminant analysis does not impose any particular restrictions on the type of predictor variable (categorical or continuous) that can be used, or the models that can be specified.

The value of discriminant analysis is the way it can be used as an exploratory tool – a variety of potentially useful variables can be submitted to several different analyses and the different models compared in order to determine those variables which are more or less important for group separation. However, the resulting model is not necessarily a good one: the discriminant function will include those variables which are good predictors of group membership but also some which are not, as well as some which may actually blur the distinction between groups. Consequently a step-wise analysis could be used, in which only those variables which contribute most to the discriminant function are included, one at a time. Even so, a good discriminant model is almost bound to be obtained for many different combinations of variables if all the available cases were used to generate the discriminant function. So in order to test the reliability of a function, each sample is randomly divided into two groups – the discriminant function is generated using one of these groups and tested on the other. In this way the value of the model could be readily assessed – any function which did not perform considerably better than chance in discriminating between groups in the second sample was rejected, regardless of its performance with the first sample.

There are several purposes for discriminant analysis:

- To investigate independent variable mean differences between groups formed by the dependent variable.
- To determine the most parsimonious way (the fewest dimensions) to distinguish between groups.
- To infer the meaning of the dimensions which distinguish groups based on discriminant loadings.
- To determine the percent of variance in the dependent variable explained by the independents.
- To determine the percent of variance in the dependent variable explained by the independents over and above the variance accounted for by control variables, using sequential discriminant analysis.
- To assess the relative importance of the independent variables in classifying the dependent variable.
- To discard variables which are little related to group distinctions.
- To classify cases into groups using a discriminant prediction equation.
- To test theory by observing whether cases are classified as predicted.

Discriminant analysis is appropriate for testing the hypothesis that the group means for two or more groups are equal. Each independent variable is multiplied by its corresponding weight, then the products are added together, which results in a single composite discriminant score for each individual in the analysis.

Averaging the scores derives a group centroid. If the analysis involves two groups there are two centroids; in three groups there are three centroids and so on. Comparing the centroids shows the distance of the groups along the dimension you are testing.

Applying and interpreting discriminant analysis is similar to regression analysis, where a linear combination of measurements for two or more independent variables describes or predicts the behavior of a single dependent variable. The most significant difference is that you use discriminant analysis for problems where the dependent variable is categorical versus regression where the dependent variable is metric.

Discriminant analysis involves three steps: derivation, validation, and interpretation.

The analysis assumes that the variables are drawn from populations that have multivariate normal distributions and that the variables have equal variances.

Computationally, discriminant function analysis is very similar to analysis of variance.

To summarize the discussion so far, the basic idea underlying discriminant function analysis is to determine whether groups differ with regard to the mean of a variable, and then to use that variable to predict group membership.

The discriminant function problem can be rephrased as a one-way analysis of variance (ANOVA) problem. Specifically, one can ask whether or not two or more groups are significantly different from each other with respect to the mean of a particular variable.

In the case of a single variable, the final significance test of whether or not a variable discriminates between groups is the F test. F is essentially computed as the ratio of the between-groups variance in the data over the pooled (average) within-group variance. If the between-group variance is significantly larger, then there must be significant differences between means.

Usually, one includes several variables in a study in order to see which one(s) contribute to the discrimination between groups. In that case, we have a matrix of total variances and covariances; likewise, we have a matrix of pooled within-group variances and covariances. We can compare those two matrices via multivariate F tests in order to determine whether or not there are any significant differences (with regard to all variables) between groups. This procedure is identical to multivariate analysis of variance.

When interpreting multiple discriminant functions, which arise from analyses with more than two groups and more than one variable, one would first test the different functions for statistical significance, and only consider the significant functions for further examination.

Next, we would look at the standardized b coefficients for each variable for each significant function. The larger the standardized b coefficient, the larger is the respective variable's unique contribution to the discrimination specified by the respective discriminant function. In order to derive substantive "meaningful" labels for the discriminant functions, one can also examine the factor structure matrix with the correlations between the variables and the discriminant functions.

Finally, we would look at the means for the significant discriminant functions in order to determine between which groups the respective functions seem to discriminate.

2.1 Linear discriminant model

For a set of p variables X_1, X_2, ..., X_p, the general model is:

$$Z_i = \sum_{j=1}^{p} a_{ij} X_j$$

where, the $X_j's$ are the original variables and the $a_{ij}'s$ are the discriminant function coefficients.

The principles to find the discriminant functions are:

The first discriminant function is that which maximizes the differences between groups compared to the differences within group which is equivalent to maximizing F in a one-way ANOVA.

$$F(Z) = \frac{MS_B(Z)}{MS_W(Z)}$$

$$Z_1 = \max\{F(Z)\}$$

The second discriminant function is that which maximizes the differences between groups compared to the differences within groups unaccounted for by Z_1 which is equivalent to maximizing F in a one-way ANOVA given the constraint that Z_1, Z_2 are uncorrelated.

$$F(Z) = \frac{MS_B(Z)}{MS_W(Z)},$$

$$Z_2 = \max\{F(Z) \mid r_{z_1, z_2} = 0\}$$

The total (T) SSCP matrix (based on p variables X_1, X_2, ..., X_p) in a sample of objects belonging to m groups G_1, G_2, ..., G_m with sizes n_1, n_2, ..., n_m can be partitioned into within-groups (W) and between-groups (B) SSCP matrices:

$$T = B + W$$

where

$$t_{rc} = \sum_{j=1}^{m} \sum_{i=1}^{n_j} \left(x_{ijr} - \bar{x}_r \right) \left(x_{ijc} - \bar{x}_c \right)$$

$$w_{rc} = \sum_{j=1}^{m} \sum_{i=1}^{n_j} \left(x_{ijr} - \bar{x}_{jr} \right) \left(x_{ijc} - \bar{x}_{jc} \right)$$

and

x_{ijk} Value of variable X_k for ith observation in group j,

$\bar{x}_{jk}$ Mean of variable X_k for group j,

$\bar{x}_k$ Overall mean of variable X_k, t_{rc}, w_{rc} element in row r and column c of total (T, t) and within (W, w) SSCP.

Analytic procedures to find discriminant functions:

- Calculate total (T), within (W) and between (W) SSCPs

  $T = B + W$

- Determine eigenvalues and eigenvectors of the product $W^{-1}B$.

  $$\lambda \left(B^{-1} W \right) = \left(\lambda_1, \lambda_2, ..., \lambda_p \right)$$

- λ_i is ratio of between to within SS's for the ith discriminant function Z,

  $$\lambda_i = \frac{SS_B \left(Z_i \right)}{SS_W \left(Z_i \right)}$$

and the elements of the corresponding eigenvectors are the discriminant function coefficients.

$$\xi_i \left(B^{-1} W \right) = \left(a_{i1}, a_{i2}, ..., a_{ip} \right)$$

3 MATERIALS AND METHODS

Usually in cotton spinning the blends are defined by grouping different cottons, leading in account the colour and length parameters, looking for always homogeneity of micronaire. The classification of these cottons was initially done using an algorithm developed with a spreadsheet.

However this tool analysis don't give us a good blend homogeneity from the point of view of regularity and constancy of properties related with its length and its uniformity, as well as of the resistance of the fibres. It is necessary to look for accurate method or algorithm that allows to predict membership of the different cotton bales in several exclusive groups leading in account the most important properties: micronaire, span length 2.5, strength, uniformity ratio, yellow degree, and reflectance.

The database used is composed by the parameters showed in the Table 1, representing 66 bales of different cottons, that has been evaluated by the HVI systems.

Table 1. Fibre properties.

Property/variable	Unit
Micronaire index (Mic)	$\mu g/''$
Upper half mean length (UHML)	mm
Span length 50 (SL 50)	mm
Span length 2.5 (SL 2.5)	mm
Uniformity index (UI)	–
Uniformity ratio (UR)	–
Tenacity (ST)	cN/tex
Elongation (EL)	%
Reflectance (Rd)	–
Yellow degree (+b)	–
Colour grade (CGrd)	–
Leaf (LF)	–

Using the conventional approach we defined 7 different blends, M1, M2, M3, M4, M5, M6 and M7 (the groups).

4   RESULTS AND DISCUSSION

In a previous work, concerning African cottons, we verified from the analysis of the discriminant function that the most significatives cotton properties, in relation to the discrimination of the groups, the bales, were: the micronaire, the span length 2.5, the uniformity ratio and the yellow degree. So, in the present study, we will not consider the influence of the strength and the reflectance.

The classification matrix shows the number of cases that were correctly and those that were misclassified (Table 2), the linear discriminant function for groups (Table 3), and the misclassified observations (Table 4) in the first phase.

Table 2.   Classification (phase 1).

Group	M1	M2	M3	M4	M5	M6	M7
M1	2	0	0	0	0	0	0
M2	1	3	0	3	0	1	2
M3	0	1	8	1	0	0	2
M4	0	0	4	4	0	0	4
M5	0	0	1	0	5	3	0
M6	0	0	0	1	0	8	3
M7	0	0	0	0	0	1	8
Total	3	4	13	9	5	13	19
N correct	2	3	8	4	5	8	8
Proportion	0.667	0.750	0.615	0.444	1.000	0.615	0.421

N = 66		N correct = 38		Prop. correct = 0.576

Table 3.   Linear discriminant function for group (phase 1).

	M1	M2	M3	M4	M5	M6	M7
Constant	−2653.2	−2749.3	−2799.3	−2768.0	−2923.2	−2939.5	−2949.8
Mic	17.4	19.7	19.4	19.4	24.7	25.0	22.1
SL 2.5	144.6	147.0	146.9	146.0	151.3	149.9	148.4
UR	9.0	8.9	10.2	9.9	9.2	10.1	9.6
+b	69.4	71.4	70.7	71.7	72.6	73.3	73.0

Table 4.   Misclassified observations (phase 1).

Observation	True group	Pred. group	Observation	True group	Pred. group
2	M4	M3	33	M7	M4
4	M3	M4	34	M7	M2
6	M3	M4	35	M7	M6
9	M3	M4	36	M7	M4
14	M3	M4	37	M7	M4
15	M4	M2	40	M7	M6
16	M4	M2	42	M7	M2
20	M1	M2	46	M7	M3
21	M4	M2	48	M7	M6
23	M2	M3	57	M6	M5
24	M3	M5	60	M6	M5
29	M4	M6	61	M6	M5
30	M7	M4	63	M6	M2
32	M7	M3	66	M6	M7

Table 5. Classification (phase 2).

Group	M1	M2	M3	M4	M5	M6	M7
M1	2	0	1	0	0	0	0
M2	0	10	0	0	1	0	0
M3	0	1	11	1	0	0	1
M4	0	0	0	10	0	1	0
M5	0	0	0	0	8	0	0
M6	0	0	0	0	0	10	1
M7	0	0	0	0	0	1	7
Total	2	11	12	11	9	12	9
N correct	2	10	11	10	8	10	7
Proportion	1.000	0.909	0.917	0.909	0.889	0.833	0.778

N = 66 N correct = 58 Prop. correct = 0.879

Table 6. Linear discriminant function for group (phase 2).

	M1	M2	M3	M4	M5	M6	M7
Constant	−4449.5	−4712.7	−4833.5	−4689.1	−5094.0	−5061.7	−4895.3
Mic	102.0	110.3	108.1	112.0	118.8	126.1	122.1
SL 2.5	241.9	250.2	250.7	242.9	258.5	251.2	250.0
UR	16.5	15.6	18.1	19.2	17.0	19.6	17.4
+b	88.9	92.6	91.5	92.8	95.1	97.0	96.6

Table 7. Misclassified observations (phase 2).

Observation	True group	Pred. group	Observation	True group	Pred. group
2	M3	M1	49	M5	M2
22	M4	M3	59	M6	M7
23	M2	M3	65	M7	M6
31	M7	M3	66	M6	M4

Table 8. Classification (phase 3).

Group	M1	M2	M3	M4	M5	M6	M7
M1	3	0	0	0	0	0	0
M2	0	10	0	0	0	0	0
M3	0	1	14	0	0	0	0
M4	0	0	0	11	0	0	0
M5	0	0	0	0	8	0	0
M6	0	0	0	0	0	11	0
M7	0	0	0	0	0	0	8
Total	3	11	14	11	8	11	8
N correct	3	10	14	11	8	11	8
Proportion	1.000	0.909	1.000	1.000	1.000	1.000	1.000

N = 66 N correct = 65 Prop. correct = 0.985

In this phase we can see that only 57.6% of the blends are correctly classified. We will correct this classification by introduction the new data information from the Table 4 (predicted groups).

Tables 5–7 present the results of the second phase of the discriminant analysis. We can verify now, that 87.9% of the observations (bales) are correctly classified; the observations misclassified are indicated in Table 7.

Table 9. Linear discriminant function for group (phase 3).

	M1	M2	M3	M4	M5	M6	M7
Constant	−6839.4	−7068.1	−7366.4	−7060.4	−7718.9	−7605.1	−7240.0
Mic	53.6	62.5	59.4	64.8	67.1	75.4	75.0
SL 2.5	379.0	387.4	392.5	378.5	403.0	392.7	386.1
UR	38.2	36.3	39.9	41.0	39.3	41.8	38.3
+b	88.4	92.2	91.2	93.1	94.7	96.9	97.0

Table 10. Misclassified observations (phase 3).

Observation	True group	Pred. group
21	M2	M3

Table 11. Summary of classification (final phase).

Group	M1	M2	M3	M4	M5	M6	M7
M1	3	0	0	0	0	0	0
M2	0	10	0	0	0	0	0
M3	0	0	15	0	0	0	0
M4	0	0	0	11	0	0	0
M5	0	0	0	0	8	0	0
M6	0	0	0	0	0	11	0
M7	0	0	0	0	0	0	8
Total	3	10	15	11	8	11	8
N correct	3	10	15	11	8	11	8
Proportion	1.000	1.000	1.000	1.000	1.000	1.000	1.000

N = 66   N correct = 66   Prop. correct = 1.000

Table 12. Linear discriminant function for group (final phase).

	M1	M2	M3	M4	M5	M6	M7
Constant	−7044.4	−7258.6	−7579.6	−7267.8	−7949.3	−7828.4	−7449.4
Mic	51.7	60.7	57.5	62.9	65.1	73.4	73.1
SL 2.5	391.6	399.7	405.6	391.3	416.4	405.9	398.9
UR	39.4	35.5	41.1	42.2	40.6	43.1	39.6
+b	88.3	92.1	91.0	92.9	94.5	96.7	96.9

Finally, only in a fourth step, we meet the goal of the correct classification (100%) (Table 11). Table 12 shows the final coefficients of the linear discriminant function for groups.

5 CONCLUSIONS

The variables with the largest (standardized) regression coefficients are the ones that contribute most to the prediction of group membership. Considering only the variables with the largest regression coefficients, we can have yet a good convergence of the discriminant algorithm, even if you apply a large number of samples (66 bales) and groups (7 blends), demanding only one more iteration.

Discriminant Analysis is a very useful tool for detecting the variables that allow the spinning engineer to discriminate between different (naturally occurring) groups, and for classifying cases into different groups with a better than chance accuracy, as we demonstrated in data analysis of the raw cotton blends of several origins.

REFERENCES

Brandt, S. 1989. *Statistical and Computation Methods in Data Analysis*, Northholland, Publishing Company, 5th edition.

Cabeço Silva, M.E. & Cabeço Silva, A.A. 1996. *Multivariate Analysis in Quality Design of Cotton Blends*, Beltwide Cotton Conferences, Vol. 2, p. 1467–1471, Nashville, USA, 1996, January.

Cabeço Silva, M.E., Marques, M.J.A. & Cabeço Silva, A.A. 1999. *Involving Statistical Factor Analysis for Disposable Surgical Clothing Properties Prediction*, EPMESC-Computational Methods in Engineering and Science, ELSEVIER SCIENCE Ltd, Edited by J. Bento, E. Arantes e Oliveira, E. Pereira, Vol. 2, p. 1085–1093, ISBN 008043570 X HC, Macau.

Cabeço Silva, M.E. 2000. *Análise Exploratória de Dados em Sistemas HVI de Metrologia Têxtil*, X Congreso Latino-Iberoamericano de Investigación de Operaciones y Sistemas – X CLAIO, México, D.F., Edição em CD, México.

Cabeço Silva, M.E. & Cabeço Silva, A.A. 2003. 'Computational Exploratory Data Analysis in Cotton Yarns', 2003 Beltwide Cotton Conferences, Cotton Textile Processing Research Conference, CD Proceedings, 8 p., Nashville, USA, January.

Hoaglin, D.C., Mosteller, F. & Tuckey, J.W. 1983. *Understanding Robust and Exploratory Data Analysis*, John Wiley & Sons, New York.

Huberty, C.J. 1994. *Applied Discriminant Analysis*, Wiley and Sons, New York.

Wool fabrics sensorial perception using factorial analysis

M.E. Cabeço Silva & A.C. Broega
Departamento de Engenharia Têxtil, Universidade do Minho, Portugal

ABSTRACT: Wool fabrics are one of the most traditional materials for garments in the world, and the most suitable fabrics for the cold winter in the Europe. Outwear clothes, particularly men's wear, are greatly influenced by our current way of life: our garments are much lighter than our fathers' and grandfathers'. We live in much better heated surroundings, and the fabric structure has become more open, less rigid, because we like to move in an easier way, we pay more attention to comfort and less to the formal aspects. As a consequence of this evolution, the global concept of quality to accomplish the exigencies of end users is dynamic, apart from the technical performance of the textile materials, it also takes care of parameters such as handle, thermophysiological comfort and appearances of new garments, during its life time, etc. In this work, 17 parameters of the physical/mechanical properties are to be measured, and based on multivariate statistical analysis we establish a fundamental understanding of the phenomenon of fabric handle and we maximize the KES-FB System information.

1   INTRODUCTION

The trend towards lighter and lighter fabrics has significantly influenced the recent evolution of the wool industry, since the wool products became suitable for all the seasons around. In the woollen sector, the weight (g/m) of traditional woven fabrics for jackets and suits has decreased in a probable irreversible way; in the worst sector, the existing technology had to be modified in many aspects particularly in spinning, finishing and apparel sectors, based on very innovative concepts of processes control.

The sensorial perception of touch or simply "fabric hand" has been considered as one of the most important performance attributes of textiles intended for use in garments. Methods for predicting fabrics handle and apparel manufacture performance from its physical, mechanical and dimensional properties have been investigated (Kawabata).

The goal of this paper is the objective evaluation of total hand value in wool fabrics using the multivariate statistical analysis, namely factor analysis to know the most important properties to define the handle of the fabrics.

Factor analysis is used to uncover the latent structure (dimensions) of a set of variables. It reduces attribute space from a larger number of variables to a smaller number of factors and as such is a "non-dependent" procedure (that is, it does not assume a dependent variable is specified). Factor analysis could be used for any of the following purposes:

- To reduce a large number of variables to a smaller number of factors for modeling purposes, where the large number of variables precludes modeling all the measures individually,
- To select a subset of variables from a larger set based on which original variables have the highest correlations with the principal component factors,
- To create a set of factors to be treated as uncorrelated variables as one approach to handling multicollinearity in such procedures as multiple regression,
- To validate a scale or index by demonstrating that its constituent items load on the same factor, and to drop proposed scale items which cross-load on more than one factor,
- To establish that multiple tests measure the same factor, thereby giving justification for administering fewer tests,

KES-FB1 – Tensile and shear

KES-FB2 – Pure bending

KES-FB3 – Compression

KES-FB4 – Surface

Figure 1. KES-FB System.

Table 1.   KES-FB System modulus.

Modulus	Properties	Parameters
KES-FB1	Tensile and shear	LT, WT, RT, G, 2HG, 2HG5
KES-FB2	Pure bending	B, 2HB
KES-FB3	Compression	LC, WC, RC, T
KES-FB4	Surface	MIU, MMD, SMD

- To identify clusters of cases and/or outliers,
- To determine groups by determining which sets of people cluster together.

For those reasons our treatment of factor analysis will be directed to its modern explication as a problem in statistical estimation.

For the factor model evaluation of such scores is neither so simple nor so uniquely defined, and it will consider two intuitively appealing methods for computing the values of the latent factors of a subject.

2   APPARATUS AND PROPERTIES

Fabric mechanical properties such as tensile, bending, shearing, compression and surface properties can be measured by the four modulus of KES-FB System (Fig. 1) under standard conditions. All these four apparatus are generators systems of big amount of data, parameters properties (Table 1) which needs to be analyzed, and this takes a lot of time and experts in this area, what is incomportable for the textile industry.

3   MATERIALS AND METHODS

For the objective evaluation of overall hand feeling in wool fabric, 18 kinds of fabrics were manufactured. We analyzed very light weighted wool fabrics, produced in three weave structure's, made from three different yarn count 2/52 Nm (76/2 tex), 2/60 Nm (62/2 tex), 2/80 Nm (50/2 tex), and finished in two different

Table 2. Fabrics characterization.

Cod	Weave	Count* (Nm)	Finishing	Spicks/cm Warp × weft		Tie factor	Weight/m² (g/m²)	Thickness 0,5 gf/cm² (mm)
31	Plain	2/80	Top dyeing	29.0	26.0	1	137	0.404
32	Plain	2/80	Piece dyeing	29.6	25.9	1	142	0.408
33	Twill2	2/80	Top dyeing	32.9	28.1	0.667	153	0.472
34	Twill2	2/80	Piece dyeing	33.1	27.3	0.667	153	0.484
35	Twill3	2/80	Top dyeing	35.3	31.0	0.624	165	0.559
36	Twill3	2/80	Piece dyeing	34.3	29.3	0.624	165	0.577
51	Plain	2/64	Top dyeing	27.5	23.7	1	152	0.412
52	Plain	2/64	Piece dyeing	27.2	23.3	1	155	0.440
53	Twill2	2/64	Top dyeing	30.1	25.6	0.667	167	0.500
54	Twill2	2/64	Piece dyeing	29.6	25.1	0.667	167	0.520
55	Twill3	2/64	Top dyeing	31.0	26.2	0.624	177	0.619
56	Twill3	2/64	Piece dyeing	31.1	26.4	0.624	178	0.612
71	Plain	2/52	Top dyeing	26.5	19.6	1	189	0.557
72	Plain	2/52	Piece dyeing	26.1	19.1	1	189	0.575
73	Twill2	2/52	Top dyeing	27.3	23.7	0.667	208	0.600
74	Twill2	2/52	Piece dyeing	26.9	23.1	0.667	208	0.649
75	Twill3	2/52	Top dyeing	29.4	25.5	0.624	224	0.744
76	Twill3	2/52	Piece dyeing	28.0	26	0.624	224	0.736

* The yarns are same at the warp and the weft.

Table 3. Properties.

Parameter	Properties	Unit
EM	Elongation (500 gf/cm)	%
LT	Linearity	–
WT	Tensile energy/unit area	gf·cm/cm²
RT	Resilience	%
B	Bending rigidity	gf·cm²/cm
2HB	Hysteresis	gf·cm/cm
G	Shear stiffness	gf/cm·grau
2HG	Hysteresis at $\Phi = 0.5°$	gf/cm
2HG5	Hysteresis at $\Phi = 5°$	gf/cm
LC	Linearity	–
WC	Compressional energy	gf·m/cm²
RC	Resilience	%
MIU	Coefficient of friction	–
MMD	Mean deviation of MIU	–
SMD	Geometrical roughness	μm
W	Weight per unit area	mg/cm²
T	Thickness at 0.5 gf/cm²	mm

finishing routines (top dyeing and piece dyeing). Table 2 shows the fabrics characterization where "cod" designs the code of the fabric, "weave" designs the manufacturing process.

Table 3 shows the different parameters analyzed and the different properties studied as well as the units of each parameter. In this work, we have different processes of production of weave structures and it is essential to know the most important properties in the handle of the weaves. So, we apply factor analysis to know these properties. The correlation matrix for all data was computed and the appropriateness of factor model was evaluated.

Kaiser–Meyer–Olkin index (KMO) measures the adequacy of the sampling and it predicts if data are likely to factor well, based on correlation and partial correlation. KMO can still be used, however, to assess which variables to drop from the model because they are too multicollinear. The value for the KMO is 0.82 in our case, so we can proceed with the factor analysis.

683

Table 4.   Factor matrix.

Variable	Factor 1	Factor 2	Factor 3
EM		0.56327	
LT		0.76389	
WT	0.74321		
RT	0.55321		
B	0.66577		
2HB	0.55324		
G		0.52396	
2HB		0.96335	
2HG5		0.66523	
LC			0.68324
WC			0.55384
RC	0.88532		
MIU		0.73256	
MMD	0.53352		
SMD	0.92335		
W			0.88925
T			0.76332

Table 5.   Communalities.

Variable	Communality	Eigenvalue	% Var
EM	0.85369	6.9368	63.9
LT	0.75392	2.3697	79.3
WT	0.96385	1.9673	87.3
RT	0.86324		
B	0.96358		
2HB	0.92926		
G	0.76325		
2HB	0.83697		
2HG5	0.88635		
LC	0.75369		
WC	0.93583		
RC	0.86283		
MIU	0.96824		
MMD	0.94369		
SMD	0.75361		
W	0.85369		
T	0.84637		

The principal components method has to be used to know haw many factors we can fix, so we run the analysis choosing the option of finding factors with eigenvalues more than 1. This analysis is some-what a trial run to see how many factors there could be. We use the most commonly rotation schema – the Laiser's Varimax rotation; hence we employ this in our analysis data to minimize the number of variables that have high loadings on a factor. The goal of the extraction step is to determine the number of factors we have retained. In this case, we obtain estimates of the initial factors from principal components analysis.

Table 4 shows the coefficients greater or equal to 0.5 used to express a standardized variable in terms of the factors.

Table 5 is labeled "final statistics", since it shows the communalities and factors statistics after the desired number of factors that has been extracted.

The total variance explained by each factor is listed in column labeled eigenvalue and the other column contains the percentage of the total variance attributable to each factor. This table shows that almost 87.3% of the total variance is attributable to the four factors. The other factors together, account for only 12.7% of the variance. Thus, a model with four factors may be adequate to represent the data. The factor matrix obtained in the extraction phase indicates the relationship between the factors and the individual variables;

Table 6. Rotated factor matrix.

Variable	Factor 1	Factor 2	Factor 3
EM			0.75426
LT			0.88426
WT			0.92157
RT			
B	0.86324		
2HB	0.75369		
G		0.95634	
2HB		0.88562	
2HG5		0.82457	
LC		0.75934	
WC		0.87536	
RC	0.63584		
MIU	0.91468		
MMD	0.92354		
SMD	0.96352		
W			0.82437
T		0.63814	

it is usually difficult to identify meaningful factors based on this matrix. Consequently, the difficulty of interpretation is an unrotated factor solution. So, the purpose of rotation is to achieve a simple structure to help us explain the factors. Thus, we use the varimax method rotation and Table 6 shows the factors greater or equal to 0.5 after rotation.

4 CONCLUSIONS

The factor analysis techniques have been applied to study the most important properties of the handle of the fabrics independently of the processing technologies. The analysis of results shows that the surface and bending properties are related with the factor 1 as well as the compression resilience. So, these properties are the most important. The parameter concerning the compression and shear properties and the thickness are more correlated with de factor 2. The factor 3 is related with the tensile parameters and the weight of the fabric. The application of the factors analysis allows concluding that the parameters most important for the definition of the handle of fabrics are the surface and bending properties. This work shows the importance of factor analysis application in the choice of the most appropriate properties to define the handle of the wool light fabrics.

REFERENCES

Araújo Marques, M.J., Cabeço Silva, M.E. & Cabeço Silva, A.A. 1998. Multivariate Analysis in the Optimisation and Modelling of the Quality Properties of Surgical Clothing. *2nd IMACS International Multiconference CESA 98, Computing Engineering in Systems Applications, Proceedings IEEE*, p. 254–258, Nabeul-Hammamet, Tunise, 1998, April.

Behera, B.K., Ishtiaque, S.M. & Chand, S. 1997. Comfort Properties of Woven from Ring, Rotor, and Friction-spun Yarns. *Journal of Textile Institute*, 88: 255–264.

Broega, A.C. 2001. Contribuição para a Quantificação do Toque e Conforto de Tecidos Super Finos de Lã, Universidade do Minho, Guimarães, Master Thesis.

Cabeço Silva, M.E. & Cabeço Silva, A.A. 1996. Multivariate Analysis in Quality Design of Cotton Blends. *1996 Cotton Beltwide Conferences Proceedings*, Vol. 2, p. 1467–1471, Nashville, USA, 1996, January.

Cabeço Silva, M.E., Marques, M.J.A. & Cabeço Silva, A.A. 1999. Involving Statistical Factor Analysis for Disposable Surgical Clothing Properties Prediction. *EPMESC–Computational Methods in Engineering and Science, ELSEVIER SCIENCE Ltd, Edited by J. Bento, E. Arantes e Oliveira, E. Pereira*, Vol. 2, p. 1085–1093, ISBN 00 80435 70 X HC.

Cabeço Silva, M.E. & Cabeço Silva, A.A. 2003. Exploratory Data Analysis in Cotton Quality Management. *2001 Beltwide Cotton Conferences Proceedings*, p. 1271–1273, Anheim, USA, 2001, January.

Cabeço Silva, A.A. & Cabeço Silva, M.E. 2003. Data Mining Approaches in Optimizing the Combed Yarns Processing. *2003 Beltwide Cotton Conferences, Cotton Textile Processing Research Conference*, CD Proceedings, 6 p., Nashville, USA, 2003, January.

Cabeço Silva, M.E. & Cabeço Silva, A.A. 2003. Computational Exploratory Data Analysis in Cotton Yarns. *2003 Beltwide Cotton Conferences, Cotton Textile Processing Research Conference*, CD Proceedings, 8 p., Nashville, USA, 2003, January.

Dunteman, George H. 1989. Principal components analysis. Thousand Oaks, CA: Sage Publications, Quantitative Applications in the Social Sciences Series, No. 69.

Fabrigar, L.R., Wegener, D.T., MacCallum, R.C. & Strahan, E.J. 1999. Evaluating the use of exploratory factor analysis in psychological research. *Psychological Methods*, 4: 272–299.

Fourt, L. & Hollies, N.R.S. 1970. Clothing: Comfort and Function. New York: Marcel Dekker, Inc., ISBN 0-8247-1214-5.

Kawabata, S., Postle, R. & Niwa, M. 1982. Objective Specification of Fabric Quality, Mechanical Properties and Performance. *Proc. of the Japan–Australia Joint Symposium of Textile Machinery Soc. of Japan*, Kyoto, Japan.

Computational Methods in Engineering and Science, Iu et al. (eds)
© *2003 Swets & Zeitlinger, Lisse, ISBN 90 5809 567 3*

Multimedia as a tool for distant learning of textile design

M. Neves & J. Neves
University of Minho, Textile Department, Portugal

ABSTRACT: Information and communication technologies progressed dramatically at the end of the XX century. The progress of the computation technology in conjunction with that of the telephone system (typing and transmission of optical fibre) has transformed computers into indispensable tools for both personal and professional communication. This communication revolution reflects the expectations of the younger generations who interact with these new technologies in an intuitive way. The advance in technology is also giving rise to changes in education, which is no longer restricted to a period of time in an individual's life, but is being faced as a continuous process.The project hereby presented tries to answer to the need for an interactive multimedia tool, to be used in the teaching of textile design in Portugal. The multimedia product created emerges as an alternative resource in long distance teaching or as an additional resource in loco teaching. The development and making of this product has been guided according to an instructive pedagogical perspective, always conjugating the content of the research with the primary objective of constructing a friendly interface.

1   INTRODUCTION

Our society (like any future society) is founded on work. There is no clue that may be used to foresee a decrease in the numbers of active individuals awaiting their own insertion in society through labour.

On the other hand the specialization and its diversification, allied to a need to be able to adapt to new forms of employment, demand of the worker a greater flexibility and capabilities to respond to different jobs. Thus, up to a few years ago, the worker could probably expect to keep the same job until retirement. Nowadays, his progress within the work market is progressively composed by a succession of jobs, of different levels of specialization.

Some decades ago it was expected that companies became training sites for the worker. And in fact, they fulfilled that role to some extent. Now the companies are starting to demand that schools prepare the individual in what the required specializations are concerned.

So, there is a need for an education oriented in terms of professional activities, before the active individual is inserted in the work market. And a work-oriented education implies the development of competences in the use of new working tools.

The real competing advantage of any individual, in what the work market is concerned, lies in his/her identification and problem-solving qualifications. It is up to the educational system to promote favorable conditions for the integration of the individual in the professional activity. It is within this principle that it is possible to answer to the functioning peculiarities of the Portuguese work market. For example, the search for human resources in the textile and clothing areas reveals a need for pre-defined competences that are strategical factors in the competitivity of the companies.

It can be concluded that the creativity abilities and the autonomy of an individual are determinant within the work market because they are determinant of the competitivity of a company in a market that is becoming less and less local and more trans-national.

The inclusion of modern technological resources in the educational areas of Communication and Computing allows for the reduction of the differences existing between our educational system and those of the more developed countries, thus turning it possible to improve the quality of the individual's training.

Therefore, the personal education, training and development should be continuous and simultaneously a search for new knowledge. For that to be possible, innovative and efficient proposals are needed. Proposals which being founded on the use of multimedia resources would contribute to the maximization of the teaching and learning possibilities by developing dynamic teaching materials.

2 AIMS

In the last twenty years Portugal has been steadily and regularly developed in the various economic areas. However, the growing market globalization demands from the Portuguese industry faster and more efficient answers.

Our industry is thus within a set of both opportunities and challenges and the competitive success of the textile companies depends, among other factors (such as production capabilities, product differentiation, client services) on the quality of its staff's training. The new computing and communication technologies of today allow access to interactive Multimedia resources, which in turn make teaching more efficient and motivating.

A growing democratization of knowledge is now evident, in which great emphasis is given to the continuous training of the individual, and to the acknowledgement of the specialized "know-how" as a bonus on the human resources of the textile company (or, for that matter, of any company).

Thus, the aims of this study are:

- to develop an interactive multimedia application for textile design.
- to develop that application accordingly to a pedagogical–didactical perspective.
- to use the new technologies in a practical applicability within the learning of textile design.
- to present an interactive document as an alternative/complementary resource in teaching textile design.
- to develop the application as a friendly interface.

In the search for an answer to these objectives, the use of CD-ROM Multimedia has proved to be the best solution. And determining that choice was:

- the possibility to integrate the different media (image, text, sound) so as to allow for a better understanding of the subject being presented.
- the low cost of editing a CD-ROM in contrast with that of editing an identical document in paper.
- the possibility of interaction, which besides allowing a self-imposed rhythm of the learning process, stimulates theme exploration and keeps motivation at higher levels.
- the easy availability of information, as personal computers become more and more present in our everyday lives.

The aim is therefore to demonstrate that Multimedia, as a pedagogical–didactical tool, brings about advantages to the learning of textile design. It allows relatively easy access to great quantities of information digitally stored in a CD-ROM disc.

3 PRACTICAL APPLICATION

The interactive Multimedia documents were designed considering these lines of conduct:

- sectioning of the document into information blocks in order to organize the wide variety of subjects in Textile Design into modules, and thus sequence its complexity.
- using different languages in the presentation of the concepts (through text, drawings, video, graphic representation – animated or otherwise – digitalized images of fabric and sound).
- facilitating the interaction and guidance throughout the study of the document.
- holding and maintaining the user's interest with a friendly interface.

It is this work's aim to demonstrate that the inclusion of new technologies in the teaching of Textile Design can not only improve the learning processes but also encourage potential pupils to study this subject.

Operationally wise a Multimedia degree was created as a backbone both for distant and personal teaching, which includes four curricula:

- Textile Design in Fabrics
- Textile Design in Knitting

- Textile Design in Printing
- Fashion Trends

These courses may function as autonomous units or integrated with one another. A coherent set of products – a book and a CD-ROM were developed for each of them.

4 PRACTICAL DEVELOPMENT OF THE FABRIC APPLICATION

The application Textile Design in Fabrics is presented. All the other applications were similarly developed.

4.1 *Preliminary study*

The preliminary study began with the setting of objectives and was developed throughout the following stages:

- bibliographic research on Textile Design
- selection and compilation of the information aforementioned
- bibliographic research on interactive Multimedia
- research on educational software
- research and experimentation of available authorship software
- selection of the technical help for the implementation of the authorship software in the application
- bibliographic research on the Multimedia applications' interface design

4.2 *Planning and development of the application*

4.2.1 *Requirements*

4.2.1.1 Necessary hardware
The hardware necessary for the project was: a Pentium III processor, 500 MHz, 64 Mb RAM, 8 Gb hard disk, CD-ROM player–recorder, sound board, microphone, video acquisition board and a 15″ monitor.

4.2.1.2 Software selection
The choice of software had to obey to certain requirements forcibly conditioned by the objectives set. Thus, the following components were selected:

1. Macromedia Flash 4
2. Sound Recorder Windows
3. Photoshop 5
4. Macromedia Director 8

At first Macromedia Flash was used to generate all the animations in vectorial drawing and the animation of the opening of the application. This program allows to create complex but easily executed animations in an accessible and friendly manner, because it contains little information for processing. The fact that this component is also completely compatible with Macromedia Director also influenced its choice (further on the reasons for this choice are explained). This program also allowed to synchronise the sound and the animations, by being possible to compress it in MP3 format, which originated on a compression of about 1/20.

The Sound Recorder was used to record the voices of the animations and videos. It was necessary to have them in digital format so that the authorship's software could later on be integrated.

The voice sampling was done at 22 kHz and recorded in WAVE format.

Photoshop plus a scanner were used in the digitalization of the drawing and the fabrics, as well as in the digital handling of those images.

Finally, all the elements were imported to Macromedia Director that integrated them and generated an executable file.

The choice of the Director was fundamental in the creation of the application and was due to a careful selection based on the following criteria:

- It allows for the publication in the Internet or an Intranet (e.g.: a school with a server and network posts) at a posterior date.
- It is a friendly user interface.

– It allows for the creation of scripts for automatic production processes.
– It is a modular program in which the time line is flexible and allows changes in order and content of the different components (sound, video, etc.) besides their reduction and enlargement.
– It is – as mentioned before – compatible with Flash.

4.2.1.3 Document's diagram

The CD-ROM is made up of several pages that are interconnected via a compounded struture, showing characteristics of linear, hierarchical and/or network structures.

The option for this structure (Fig. 1) was due to the facts that on the one hand, the subject of the document demands a logical sequence of concepts and on the other hand, one of the objectives was to give the user freedom of choice during the logical learning process.

4.2.2 *Conceptual design*

Graphic design creates a visual logic, a balance between the visual sensation and the text and image information. Without the visual impact of the form, colour and contrast, the pages are graphically boring and will not motivate the user to investigate their content.

Text pages that are too dense, lacking the visual aid offered by the carefully selected graphic design, layout and typography, become very difficult to read.

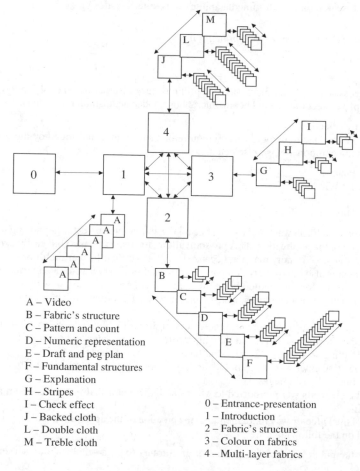

A – Video
B – Fabric's structure
C – Pattern and count
D – Numeric representation
E – Draft and peg plan
F – Fundamental structures
G – Explanation
H – Stripes
I – Check effect
J – Backed cloth
L – Double cloth
M – Treble cloth

0 – Entrance-presentation
1 – Introduction
2 – Fabric's structure
3 – Colour on fabrics
4 – Multi-layer fabrics

Figure 1.   Diagram.

To design a presentation image for a multimedia document is a personal experience in which it is inevitable for the designer to reach for past experiences – outside the sphere of computers – that may help him conceptualise the user's interface. It is common to use a theme or a metaphor to support the conceptual image of the issue at hand.

4.2.2.1 Metaphor

As the content of the application is about textile design, a metaphor of the textile yarn was used, as it is a primordial element of weaving. Besides being a dynamic visual element, it was also used to reinforce the idea of the weaving interlacing (taffeta, being the most basic structure of weaving, is always present in the left superior corner of the screen) and thus, lend further strength to the chosen metaphor.

It was likewise necessary to create navigation tools and interaction styles that would allow the users to interact with the information within the program in a way that was coherent with the chosen theme.

For the menu and navigation buttons, clothes buttons were chosen. They are presented in several forms and perspectives, so that the corresponding functions are differentiated: square for the four main chapters; round for the sub-chapters and triangular for the annotations.

In what animations were concerned, it was necessary to find a symbol, which would transmit the idea of movement. It was due to that that a spiral was chosen – because it evokes a cyclical continuity. The square is a stable and very anti-dynamic figure. It was thus attributed as icon for the static images (drawings and digitalized fabrics).

As the application is intended as a tool for learning, the care was taken to create a clean, simple design, which was at the same time entertaining and pleasant, without nevertheless distracting the user/student. Another reason for that simplicity was that there was a lot of written information that would visually weigh the screen down with the so-called "text stain".

4.2.2.2 Colours

The chromatic choice for the components of the application's layout was limited to four colours besides black.

A dominant colour was attributed to each «chapter» of the subject: blue for Fabric Structure; red for Colour on Fabrics and green for Multiple-layered fabrics. The aim behind this decision was to visually separate the different subjects. The three primary colours (RGH) of the additive mixture were used – (except for the colour given to Introduction, which was to be the identification colour of the product's packing).

As the foundation of any drawing is the line, the colours chosen were not used in their pure or saturated forms. It was therefore necessary to add a certain percentage of black to each colour so that the drawing was better defined, and also to minimize the impact saturated (and bright – on the screen) colours have on the eye.

4.2.2.3 Typography

Specific fonts, font sizes and font characteristics were used to represent certain kinds of information. For example, the capital letters of the chapter titles were drawn in such way as to give an identity to each page, chapter indexes and to the pages of the chapters themselves.

Specific fonts were attributed to specific locations so that users are helped when down-screening in search of the information they need. This is why sub-chapters' fonts are bigger and shadowed underneath.

The words in italics included in the text are active words that can be made to show their meanings within a small yellowish window, thus allowing the user not to have to access the glossary.

First only the general form of the words is perceptible, then the characters are together and thus the word becomes recognizable. Titles formed only by capital letters were avoided because their global format is perceived as rectangular, whereas the irregularities caused by the use of small characters helps in the immediate readability of the word.

4.2.2.4 Layout

Graphic design is the management of the visual information. That is done through the use of layout tools, typography and illustration, which guides the observer's eye throughout the page. They see the pages as masses of colour and form, with elements of contrast image/screen background. Later they start noticing specific information, first in drawings, graphics, photographs, etc., and only then do they see the text.

In the west, people read from left to right, from top to bottom and these are the fundamental visual axis that dominate most design decisions. The basis for most graphic design is conventional in press publications.

The basic structuring elements of the layout are lines, invisible but perceptible, which in a programmed way create interconnected spaces and sub spaces and which, at the same time, organize the page that will receive the various visual components (text, drawings, windows, buttons, etc.).

The coherence and harmony of a screen design is accomplished through balance, movement and rhythm among the integrated forms, the remaining empty spaces and the similarities between several aspects, which characterize those same elements.

In defining the layout, the location and style guidelines are established as well as those of the main titles, sub-titles, lines and navigation button, text and image. This process greatly helps – in terms of speed – the introduction of both text and images for each new page, as it is not necessary to stop to rethink the basic design for each of them.

It was also important to establish a layout and style grid in order to build up a rhythm both constant and unity-meaningful. Repetition was not intended as monotonous but as something that would give the document a coherent graphic identity.

This pattern of constant modular units, in which all the document pages share the same basic grid, maximizes its functionality and legibility.

The aim was to turn the document into something consistent and predictable, so that the user would feel comfortable while exploring it. Working with both constant layout and navigation allows the user to adapt much more quicker to the chosen design and to foresee the location of the information and of the navigation controls throughout the whole document.

5   FUTURE PERSPECTIVES

In the industrial age, the competitive foundations of a company were constituted by its capital and technology. These were presented as determining factors for the success of production. In the age of knowledge both capital and technology will be mere production tools and the competitive success of the companies will depend on how they are used by qualified individuals.

It will become necessary to give organizations the extra value of people's knowledge and creative ability and not merely their physical strength. This way, a smaller number of people, who think better and are helped by more flexible and more intelligent machines will be of greater value to a company than big quantities of non-thinking human resources. In the future it will be fundamental for the companies to possess knowledge – more than possessing natural resources, capital or laborers.

The 21st century will change the human resources' profile in the labor market, by profoundly modifying the relationship between trainee and trainer and that between teaching and the industrial world. In the knowledge society the information will not be concentrated in databases accessible to the masses. It will belong to the collective mind of the digital citizens who will have developed the capacity to exchange and combine information in order to create new knowledge. It will be a society where the creation of knowledge is done through more flexible and more varied electronic procedures, in the continuous and cumulative process that is the training of the individual.

The interface with the computers has also witnessed the beginning of a revolution. Up until now, the interaction with the computer was done via keyboard and mouse, and has been presented as being of practical access and – somewhat – satisfactory. In the future, it will be shown to be slow and inefficient.

The method to substitute both keyboard and mouse will be voice communication.

In fact, Portugal has already given its first steps in that area. This work for example, was entirely dictated to the computer by using Phillips' Free speech, which is, at the moment, the only software of voice recognition in the Portuguese market that allows work to be done in the Portuguese language. However, it is likely that new products will appear from the competition in a near future, because constant efforts are being done in the area of voice recognition research and of voice-transmitted analysis of information.

Not everything will be easy or bring benefits in the future. The digital age will bring negative issues such as the abuse of intellectual property, invasion of privacy, software piracy, theft of information and, at a more global scale, loss of jobs due to the automation of industries (factories will be able to work without interruptions and human interference will be minimal). This society will lead to a greater individuality and some people will not be able to stand the pressure, opting for apathy, while others will become well-prepared, competent workers.

Anyway, a more and more decentralized and global society can be glimpsed in the future. A society where there is a growing independence of the companies, where the circulation of information makes the system more permeable and allows it to transcend geographical borders.

Computers will be common and ordinary tools. They will be the transforming elements of the future society. These post-modern societies will, in turn, demand that its active individuals be prepared to live in continuously redefined realities.

REFERENCES

Araujo, M. de, e Castro & C.M.M. 1984. *Manual de Engenharia Têxtil*, Lisboa, Fundação Calouste Gulbenkian, volume II, cap. 8.

Araujo, M. 1995. *Engenharia e Design do Produto*, Lisboa, Universidade Aberta.

Argan, G.C. 1988. *Arte e Crítica de Arte*, Lisboa, Editorial Estampa.

Arnhei, R. 1986. *Arte e Percepção Visual: uma psicologia criadora*, 4.ª edição, Biblioteca Pioneira de Arte, Arquitectura e Urbanismo.

Barbrook, R. *The Digital Economy*, http://ma.hrc.wmin.ac.uk/ma.thcory.1.2.db.

Cadima, F.R. 1996. *História e Crítica da Comunicação*, Lisboa, Edições Século XXI.

Cameron, A. The Interactive Story. http://ma.hrc.wmin.ac.uk/ma.thcory.3.2.1db.

Carter, R. 1999. *Tipografia de Computador-3, Cor & Tipo* (Outubro), Lisboa.

Chevalier, J. & Gheerbrant, A. 1982. *Dicionário dos Símbolos – mitos, sonhos, costumes, gestos, formas, figuras, cores, número*, Edições Teorema.

Gianetti, C. (ed.). 1988. *Ars Telemática, Telecomunicação, Internet e Ciberspaço* (Janeiro), Relógio D'Água Editores.

Kerckhove, D. 1997. *A pele da Cultura* (Março), Relógio D'Água Editores.

Kahn, B. 1991. *Os computadores no ensino da ciência*, Lisboa, Nova Enciclopédia – Publicações Dom Quixote.

Kenedy, D.M. & McNaught, T.C. *Design elements for interactive multimedia*. http://www.asu.murdoch.edu.au/ajet/ajet13/wi97p1.html.

Mciver, G. *Introduction – in the Beginning* http://ma.hrc.wmin.ac.uk/ma.thcory.1.3.1.db.

Mciver, G. *Media for a Spectacular Society*, http://ma.hrc.wmin.ac.uk/ma.thcory.1.3.3.db.

McLuhan, M. 1969. *The Medium 15 the Message*, The Penguin Press.

Rita, S. & Roberta, S. 1998. *Creative News Leters and Annual Reports Designing Information*, Massachusetts, Rockport Publisher – Gloucester.

Computational Methods in Engineering and Science, Iu et al. (eds)
© *2003 Swets & Zeitlinger, Lisse, ISBN 90 5809 567 3*

Interactive animation and remote control in on-line courses

Y.H. Liu
South China University of Technology, Guangzhou, China

P. Lau
The Hong Kong Polytechnic University, Hong Kong

ABSTRACT: This paper introduces some applications of interactive animation and remote control technique in the on-line courses "Fundamental of Mechanical Manufacturing" and "Mechanical Engineering" which were jointly developed by Industrial Training Center, South China University of Technology (SCUT) and the Industrial Centre, the Hong Kong Polytechnic University (PolyU). Many interactive animations were created to show the working principle of complicated machines and manufacturing processes. Some virtual panorama visits were arranged for students by means of virtual panorama photos. More than 5 remote control experiments between SCUT and PolyU were carried out during 2001 and 2002.

1 DEVELOPING ON-LINE COURSES FOR INDUSTRIAL TRAINING

Science and Technique are advancing very quickly. Learning is a life-long process. Many people have to learn in their spare time. They can't go to traditional universities to study as formal students. Distant education on network is probably a good solution for them. In recent years, many on-line institutes have been set up in China. Many on-line courses were developed. Now there are more than 100 on-line courses in the On-Line Educational Institute of South China University of Technology (SCUT). Among them, "Fundamentals of Mechanical Manufacturing" and "Mechanical Engineering" were jointly developed by Industrial Training Center of SCUT and the Industrial Centre of the Hong Kong Polytechnic University (IC, PolyU).

2 DIFFICULTIES

The course "Fundamentals of Mechanical Manufacturing" is developed for students of the Department of Industry and Business Management. Another course "Mechanical Engineering" is developed for students of the Department of Scientific and Technological English. Both courses are linked to practical production process closely. Many complicated machines and technological operations are included in these courses. How to show these machines and operations lively on-line is a big problem during the development of these courses. Much time was spent to research in the ways to overcome difficulties mentioned above. Virtual reality and remote control were adopted with satisfactory result.

3 APPLICATION OF VIRTUAL REALITY IN OUR ON-LINE COURSES

Many interactive animations were created to show the principle of complicated machines and manufacturing processes. The animation "impact test" in Figure 1 is an example.

In this animation, two workpieces made of different materials are prepared. The yellow one is tougher and the green one is more brittle. When a student learns this session on-line, he/she can select either one of them using the mouse to show the result of the experiment on that workpiece (Fig. 2.). Then the experiment is repeated by selecting the other workpiece. The results are different (Fig. 3).

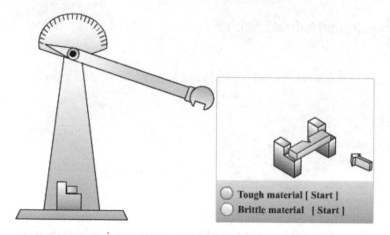

Figure 1.   Interactive animation "impact test".

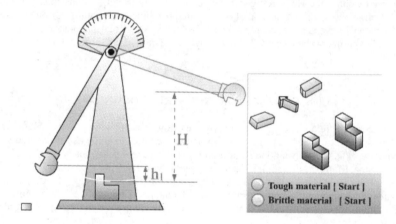

Figure 2.   Test with tough material.

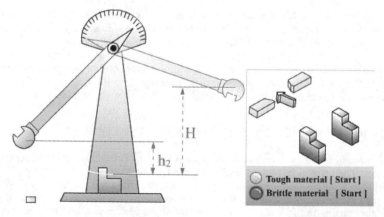

Figure 3.   Test with brittle material.

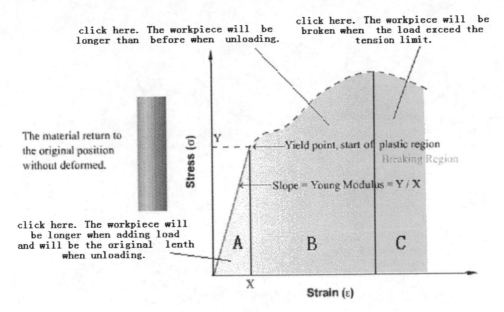

Figure 4. Interactive animation "tension test" when area A is selected.

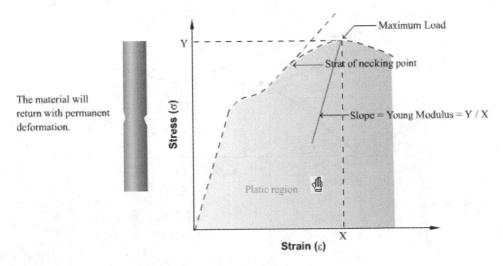

Figure 5. Interactive animation "tension test" when area B is selected.

Another example is the animation of "Analyze of the characteristic curve in tension test" (Fig. 4).

The third example is an animation on the relationship between microstructures of different carbon steel during cooling period. When clicking different kinds of steel, different change of microstructure is shown (see Figs 7–9).

Besides, some virtual panorama visits were arranged for students by means of virtual panorama photos. The photo of the CIMS training room in Figure 10 is an example.

When a student is watching the picture, he/she will have the feeling as if he/she were walking around an exhibition hall or visiting a lab. He/she can stop the movement of the image at anytime by click the moving view with the mouse to see it more clearly. Then he/she can move it again by moving the cursor of the mouse in opposite direction. He/she can also press the arrow key on the keyboard to control the size of the image. This technique can help students from a distant location to get some ideas of practical things.

697

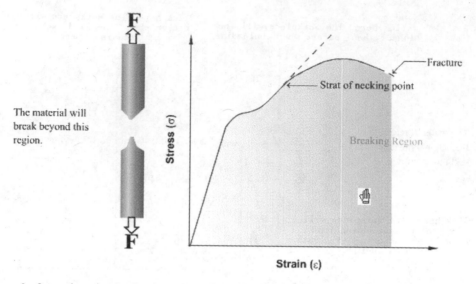

Figure 6. Interactive animation "tension test" when area C is selected.

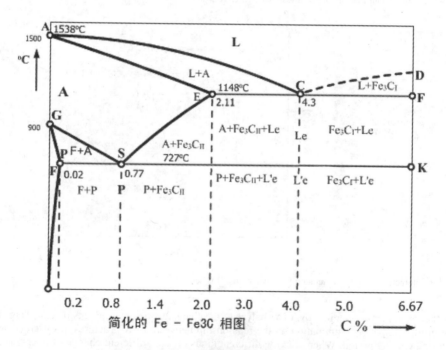

点击不同合金以显示相应区域的显微组织

合金	A	B	C	D	E	F
成份 %	0.77	<0.77	>0.77-2.11	4.3	>2.11- 4.3	>4.3

Figure 7. Interactive animation "relationship between microstructures and carbon quantity in carbon steels". (No carbon quantity is selected.)

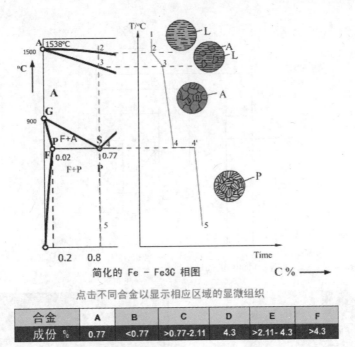

点击不同合金以显示相应区域的显微组织

合金	A	B	C	D	E	F
成份 %	0.77	<0.77	>0.77-2.11	4.3	>2.11- 4.3	>4.3

Figure 8. Interactive animation "relationship between microstructures and carbon quantity in carbon steels". (0.77% carbon quantity is selected.)

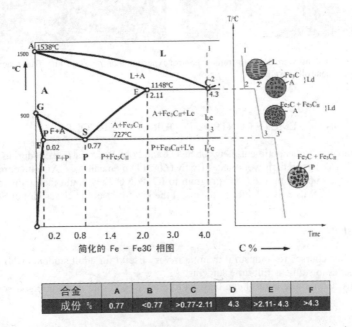

合金	A	B	C	D	E	F
成份 %	0.77	<0.77	>0.77-2.11	4.3	>2.11- 4.3	>4.3

Figure 9. Interactive animation "relationship between microstructures and carbon quantity in carbon steels". (4.3% carbon quantity is selected.)

Figure 10. The virtual photo of the CIMS training room in IC, PolyU.

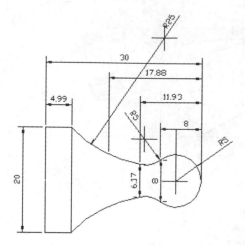

Figure 11.   Workpiece designed by students of SCUT.

Figure 12.   Several students were doing the remote control experiment one by one (left). The screens of the program and the CIMS being controlled (right).

4   APPLICATION OF REMOTE CONTROL IN OUR ON-LINE COURSES

More than 5 remote control experiments between SCUT and PolyU were carried out during 2001 and 2002. Students in SCUT designed their workpieces firstly (Fig. 11) by means of CAD and compiled the CNC program for it. After uploading the CNC program to IC, PolyU, they could then gain control and start the operation of the CIMS in IC, PolyU by means of the keyboard, at the classroom of SCUT (Fig. 12).

5   CONCLUSION

Developing on-line courses for industrial training is very useful for adult students who can't go to traditional universities to study as full time students.

Virtual reality and remote control can be used to play a good role in on-line courses for industrial training.

REFERENCES

Liu, Y.H., Lau, P. 2002. Application of video conference and remote control in experiments, *Proceeding of Seminar "research for modern education techniques"*, July, 2002.
Liu, Y.H., Lau, P. 2002. *Technique for developing on-line course*, Press of Higher Education, December, 2002.

Fluid mechanics and applications

Computational Methods in Engineering and Science, lu et al. (eds)
© 2003 Swets & Zeitlinger, Lisse, ISBN 90 5809 567 3

Optimization of aerodynamic shapes by means of inverse method

R.S. Quadros & A.L. de Bortoli
Federal University of Rio Grande do Sul, Graduate Program in Applied Mathematics – PPGMAp, Porto Alegre – RS, Brazil

ABSTRACT: The aim of this work is the attainment of an efficient technique for modeling aerodynamic bodies submitted to flows at high speeds. The inverse method, based on the Elastic Membrane Technique, is used to find a new geometry. Pressure coefficient comes from the Euler equations is solved by the finite volume with explicit Runge-Kutta five-stages scheme. Numerical tests are carried out for NACA 0012, 0009 and a new geometry for Mach 0.8 using the Euler equations and the results are found to compare favorably with experimental or numerical data available in the literature. Besides, it is worth pointing out that the technique build on earlier ones, because the coefficients needed to do the optimization shape calculations are adequately calculated.

1   INTRODUCTION

This work presents is the attainment of a technique for the optimized modeling of bodies, such as an airfoil, submitted to transonic flows. The technique is developed using Fourier series for a set of two linear equations with proper boundary conditions, based on the Elastic Membrane Technique (Baker 1999, Dulikravich & Baker 1999).

The design of optimal aerodynamic shapes can be classified in two categories: the direct and the inverse. The first intends to find the best global aerodynamic property. Otherwise, the inverse form requires a local property, such as the Cp (pressure coefficient), of the final configuration satisfying the development objective. The optimization algorithm implemented here modifies the form of an airfoil until finding the improved form for data coefficient (Cp). The advantage of the finite volume implementation is that the flux is locally verified, in agreement with the idea of the inverse project for the attainment of the geometric form of a body.

Numerical tests are carried out for the NACA 0012, 0009 and a new profile, based on NACA profiles for Mach 0.8 using the Euler equations. These results showed to compare well with experimental data found in the literature and this method is being extended to find new aerodynamic shapes in an efficient manner. Following, the numerical procedure turns clear the numerical implementation.

2   GOVERNING EQUATIONS AND SOLUTION PROCEDURE

The governing equations for non viscous flows are the Euler equations in which mass, momentum and energy are conserved. They can be written for unsteady two-dimensional compressible flows in differential form as

$$\frac{\partial \vec{W}}{\partial t} + \frac{\partial \vec{F}_1}{\partial x} + \frac{\partial \vec{F}_2}{\partial y} = 0 \tag{1}$$

where

$$\vec{W} = \left\{ \begin{array}{c} \rho \\ \rho u \\ \rho v \\ \rho E \end{array} \right\}, \qquad \overline{\overline{F}} = \left\{ \begin{array}{c} \rho \vec{q} \\ \rho u \vec{q} + p.\vec{i} \\ \rho v \vec{q} + p.\vec{j} \\ \rho H \vec{q} \end{array} \right\}, \qquad \overline{\overline{F}} = \vec{F_1}\,\vec{i} + \vec{F_2}\vec{j}$$

and ρ is the fluid density, $\vec{q}$ the velocity vector ($\vec{q} = u\,\vec{i} + v\vec{j}$) and p the pressure.

Since the integral form of conservation laws allow discontinuities, the approach is suitable for capturing shocks in the flow field (Kroll & Jain 1987). The total energy E and enthalpy H are given by

$$E = e + \frac{u^2 + v^2}{2}, \qquad H = E + \frac{p}{\rho}$$

where e is the internal energy. To close this system of equations the state relation for a perfect gas is employed

$$p = \rho R T = (\gamma - 1)\rho\left(E - \frac{u^2 + v^2}{2}\right) \tag{2}$$

where R is the gas constant and γ the specific heat ratio. Eq. (1) can be cast into the integral form (Kroll & Jain 1987)

$$\int_V \frac{\partial \vec{W}}{\partial t} dV + \int_S \left(\overline{\overline{F}}.\vec{n}\right) dS = 0 \tag{3}$$

One of the differences among the various finite volume formulations known in the literature is the arrangement of the control volume and update points for the flow variables (Kroll & Jain 1987). The most frequently used schemes are the cell-centered, cell-vertex and node-centered approach. Each of these schemes has its own advantages and disadvantages. The discretization used is based on the node-centered arrangement, as shown in Figure 1.

In the computational domain the cell vertices are identified by their indices (i, j). As Eq. 1 is valid for arbitrary control volume, it is also valid for $V_{i,j}$ that means

$$\frac{\partial \vec{W}_{i,j}}{\partial t} = -\frac{1}{V_{i,j}} \int_S \left(\overline{\overline{F}}.\vec{n}\right) dS \tag{4}$$

The finite volume discretization based on the central averaging is not dissipative. The numerical procedure does not converge to the steady state when the high frequency oscillations of error in the solution are not damped. To avoid these oscillations dissipative terms $\vec{D}_{i,j}$ are introduced.

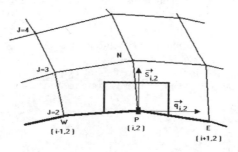

Figure 1. Node-centered arrangement.

The optimum amount of artificial viscosity is mainly determined by the smoothing properties of relaxation and is written as follows (Kroll & Jain 1987)

$$\frac{\partial \vec{W}_{i,j}}{\partial t} + \frac{1}{V_{i,j}} \left[\vec{Q}_{i,j} - \vec{D}_{i,j} \right] = 0 \tag{5}$$

This dissipation operator is a blend of second and fourth differences and is defined according to

$$\vec{D}_{i,j} = \vec{d}_{i,j+\frac{1}{2}} - \vec{d}_{i,j-\frac{1}{2}} + \vec{d}_{i+\frac{1}{2},j} - \vec{d}_{i-\frac{1}{2},j} \tag{6}$$

where dissipation coefficient is given by

$$\vec{d}_{i+\frac{1}{2},j} = \alpha_{i+\frac{1}{2},j} \left[\varepsilon^2_{i+\frac{1}{2},j} \delta_x \vec{W}_{i,j} - \varepsilon^4_{i+\frac{1}{2},j} \delta_{xxx} \vec{W}_{i-1,j} \right] \tag{7}$$

The difference operators of first and third order are δ_x and δ_{xxx}, respectively, and α is the scaling factor, which is written for the $\vec{i}$ direction in the following way

$$\alpha_{i+\frac{1}{2},j} = \frac{1}{2} \left(\lambda^{i^*}_{i,j} \right) + \left(\lambda^{j^*}_{i+1,j} \right) \tag{8}$$

where

$$\lambda^{i^*}_{i,j} = \lambda^i_{i,j} \phi^i_{i,j}$$

$$\phi^i_{i,j} = 1 + \left(\frac{\lambda^j_{i,j}}{\lambda^i_{i,j}} \right)^w$$

and w is a parameter used to scale the spectral radii.

The coefficients adapted to the local pressure gradients $\varepsilon^{(2)}$ and $\varepsilon^{(4)}$, needed to obtain the dissipation coefficient, are given by

$$\varepsilon^2_{i+\frac{1}{2},j} = k^2 \max(v_{max}), \qquad \varepsilon^4_{i+\frac{1}{2},j} = \max\left(0, K^{(4)} - \varepsilon^2_{i+\frac{1}{2},j} \right)$$

where

$$v_{max} = \left(v_{i+2,j}, v_{i+1,j}, v_{i,j}, v_{i-1,j} \right), \qquad v_{i,j} = \left| \frac{p_{i+1,j} - 2 p_{i,j} + p_{i-i,j}}{p_{i+1,j} + 2 p_{i,j} + p_{i-i,j}} \right|$$

and $0.5 \leq K^2 \leq 0.6$; $\frac{1}{128} \leq K^4 \leq \frac{1}{48}$.

The coefficient $\varepsilon^{(4)}$ provides the background dissipation in smooth parts of the flow and can be used to improve the capability of the scheme to damp high frequency modes. The spectral radius, used to control the amount of artificial dissipation, is defined for the $\vec{i}$ direction according to

$$\lambda^i = \frac{u(1 + M^2) + \sqrt{u^2(1 - M^2)^2 + \beta^2 c^2}}{2}$$

For a central difference scheme, zero artificial dissipation viscosity creates numerical difficulties. Therefore β^2 is chosen as $\beta^2 = \max(4 M^2, \psi)$, where $0.1 \leq \phi \leq 0.6$ (De Bortoli 2002).

705

2.1 Optimization

The optimization of aerodynamic geometries can be classified in two categories: the direct and the inverse (Rogalsky et al. 2000). The first intends to find the best global aerodynamic property. The inverse form requires a local property, pressure coefficient for example, of the final configuration satisfying the objectives of development.

For a given form of an airfoil, such as the NACA 0012, it is desired to find a new shape for a given coefficient. The advantage of the finite volume implementation here is that the flux is locally verified, in agreement with the idea of the inverse design for the attainment of the body geometry.

To the shape evolution, in the model used, one separates the upper and lower sides of the airfoil (Dulikravich 1995, Dulikravich & Baker 1999). On the upper side the Fourier series result

$$-\alpha_0 \Delta y^{top} + \alpha_1 \frac{d\Delta y^{top}}{ds} + \alpha_2 \frac{d^2 \Delta y^{top}}{d^2 s} = \Delta Cp^{top} \tag{9}$$

and on the lower airfoil contour

$$\alpha_0 \Delta y^{bot} + \alpha_1 \frac{d\Delta y^{bot}}{ds} + \alpha_2 \frac{d^2 \Delta y^{bot}}{d^2 s} = \Delta Cp^{bot} \tag{10}$$

where s is the airfoil contour coordinate, $\Delta Cp^{top,bot}$ the pressure coefficient difference between the desired and that at actual iteration and the $\alpha_i's$ are user constants, which control the rate of convergence of the shape evolution process.

These two ordinary differential equations, with constant coefficients, come from the well known forced mass-spring-damper system.

The ΔCp in equations (9) and (10) can be represented using the Fourier series expansion of the form

$$\Delta Cp^{top,bot} = a_0 + \sum_{n=1}^{n_{max}} [a_n \cos(N_n s) + b_n \sin(N_n s)] \tag{11}$$

where $N_n = \frac{2\pi}{L}$ and L is the total length of the airfoil contour.

The particular solution of either Eq. (9) and (10) can be represented in the general Fourier Series form as

$$\Delta y_p^{top,bot} = A_0 + \sum_{n=1}^{n_{max}} [A_n \cos(N_n s) + B_n \sin(N_n s)] \tag{12}$$

Since the Fourier coefficients of the particular solutions on the upper and lower airfoil contours are different, it can be expected that gaps will form at the leading and trailing edges of the airfoil. These gaps can be closed with appropriate homogeneous solutions of Equations (9) & (10). The upper contour homogeneous solution is

$$\Delta y_h^{top} = F^{top} e^{\lambda_1 s} + G^{top} e^{\lambda_2 s} \tag{13}$$

where

$$\lambda_{1,2} = \frac{\alpha_1 \pm \sqrt{\alpha_1^2 + 4\alpha_0 \alpha_2}}{2\alpha_2}$$

and F and G are as yet undetermined coefficients. Likewise, on the lower airfoil contour, the homogeneous solution is

$$\Delta y_h^{bot} = F^{bot} e^{-\lambda_1 s} + G^{bot} e^{-\lambda_2 s} \tag{14}$$

Thus, the overall displacement of the airfoil contour is given by the following equations:

$$\Delta y_h^{top} = F^{top} e^{\lambda_1 s} + G^{top} e^{\lambda_2 s} + \sum_{n=1}^{n_{max}} \left[A_n^{top} \cos(N_n s) + B_n^{top} \sin(N_n s) \right] \tag{15}$$

$$\Delta y_h^{bot} = F^{bot} e^{-\lambda_1 s} + G^{bot} e^{-\lambda_2 s} + \sum_{n=1}^{n_{max}} \left[A_n^{bot} \cos(N_n s) + B_n^{bot} \sin(N_n s) \right] \tag{16}$$

The four unknown constants $F^{bot,top}$ and $G^{bot,top}$ can now be found for the upper and lower airfoil contours such that the following four boundary conditions are met for:

- trailing edge displacement, $\Delta y^{bot}(0) = 0$
- trailing edge closure, $\Delta y^{bot}(0) = \Delta y^{top}(L)$
- leading edge closure, $\Delta y^{bot}(S_{le}) = \Delta y^{top}(S_{le})$ and
- smooth leading edge deformation, $(d/ds)\Delta y^{bot}(S_{le}) = (d/ds)\Delta y^{top}(S_{le})$

where Δy is the variation in the y-direction, L the total length and S_{le} is the value of the contour at the leading edge.

Using the previous equations simultaneously, for the unknown coefficients F and G, results in

$$\begin{bmatrix} F^{bot} \\ G^{bot} \\ F^{top} \\ G^{top} \end{bmatrix} \begin{bmatrix} 1 & 1 & 0 & 0 \\ 0 & 0 & e^{l\lambda_1} & e^{l\lambda_2} \\ e^{-S_{le}\lambda_1} & e^{-S_{le}\lambda_2} & -e^{S_{le}\lambda_1} & -e^{-S_{le}\lambda_2} \\ -\lambda_1 e^{-S_{le}\lambda_1} & -\lambda_2 e^{-S_{le}\lambda_2} & -\lambda_1 e^{S_{le}\lambda_1} & -\lambda_2 e^{S_{le}\lambda_2} \end{bmatrix} = \begin{Bmatrix} -\sum_{n=0}^{n_{max}} A_n^{bot} \\ -\sum_{n=0}^{n_{max}} A_n^{top} \\ \sum_{n=0}^{n_{max}} [\Delta A_n \cos(N_n S_{le}) + \Delta B_n \sin(N_n S_{le})] \\ \sum_{n=0}^{n_{max}} [-\Delta A_n \sin(N_n S_{le}) + \Delta B_n \cos(N_n S_{le})] N_n \end{Bmatrix}$$

The two-dimensional Fourier series shape evolution equation can be expanded to three dimensions, only expressing values to k-planes as in the two-dimensional case.

3 NUMERICAL RESULTS

In the following, numerical results for NACA 0012 and 0009 airfoils and a new form based on them are presented. The calculations have been carried out on a O-type grid which consists of 192×60 cells, as shown in Figure 2; it is refined in the η direction next the surface of the body (airfoil contour), where the flow variables will have the biggest gradients. The position of the other boundary is around thirty chord lengths away from the airfoil. For the pressure coefficient, calculated in nondimensional form, we have for each point

$$C_p = \frac{p - p_0}{\frac{1}{2}\rho V_0^2}$$

The pressure lines for NACA 0012 and new airfoil (without shock) is shown in Figure 3. One notices that our code can capture very well the shocks over both airfoils. The corresponding pressure coefficient over these geometries is shown in Figure 4a. One notices that the shock in the airfoil NACA 0012 occurs at approximately 50% of the chord-length.

To check the accuracy of the method, results for the NACA 0012 airfoil are compared to a solution obtained by the conforming mapping technique for incompressible flow and with experiments for

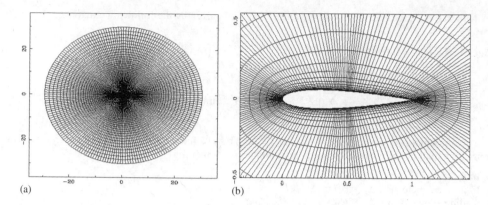

Figure 2. Structured O-type grid for NACA 0012 (a) and grid amplification (b), 192 × 60 cells.

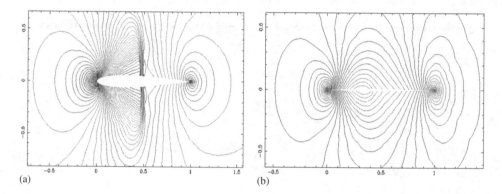

Figure 3. Pressure lines for NACA 0012 (a) and new profile (b), respectively; Mach = 0.8 and $\alpha = 0$.

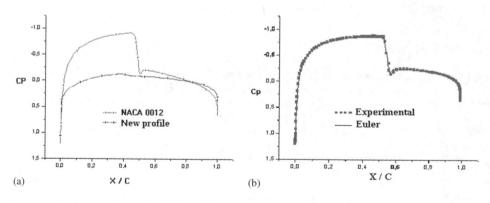

Figure 4. Pressure coefficient for NACA 0012 and new geometry (a) and numerical/experimental (Kroll & Jain 1987) Cp comparison for NACA 0012 (b), Mach = 0.8 and $\alpha = 0$.

compressible flows. Figure 4b shows the pressure coefficient computed for Mach = 0.8 and $\alpha = 0°$ (transonic flow). The overall accuracy is satisfactory; small deviations appear at shock position.

Figure 5 shows the geometry evolution. Starting with the NACA 0012 airfoil, after 25 iterations the NACA 0009 shape is obtained, using the inverse method for parameters $\alpha_0 = 1.2$, $\alpha_1 = 0.0$ and $\alpha_2 = 0.4$,

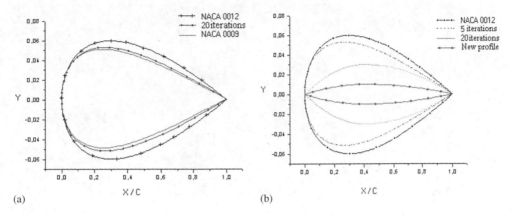

Figure 5.   Geometry evolution using the inverse method finding (a) NACA 0009 and (b) new profile.

which were obtained solving a system. Otherwise, starting the process with the NACA 0012 airfoil, after 25 iterations a new profile can be obtained.

4   CONCLUSIONS

The inverse shape design, based on the Elastic Membrane Technique, using Fourier series (Baker 1999), showed to be efficient finding of new aerodynamic geometries for given pressure coefficient differences. This method does not require modification of the flow analysis software and converges fast for transonic flows.

The method showed to be efficient finding not only the pressure coefficient for aerodynamic geometries, but also in the determination of a new geometry for a given pressure coefficient, even if the geometry is almost a flat plate at zero degrees angle of attack.

Such is done with around 25 iterations, when using the appropriate Fourier series coefficients obtained solving the corresponding system of equations. Besides, this method can be extended to three-dimensional situations, as wings (formed from a number of airfoils).

It is the author's opinion that the comparison between theoretical/experimental and numerical results is encouraging; therefore, the present technique is being extended for solving three-dimensional flows over aerodynamic geometries such as wings and complete airplanes. Besides, this method can also be extended to find appropriate geometry configurations in other areas besides the aeronautics, with the same efficiency.

REFERENCES

Baker, D.P. 1999. A Fourier Series Approach to the Elastic Membrane Inverse Shape Design Problem in Aerodynamics, M.Sc. Thesis, Dept. of Aerospace Eng., Pennsylvania State University.
De Bortoli, A.L. 2002. Multigrid Based Aerodynamical Simulations for the NACA 0012 Airfoil, *Applied Numerical Mathematics* 20, pp 337–349.
Dulikravich, G.S. 1995. Shape Inverse Design and Optimization for Three Dimensional Aerodynamics, *AIAA paper* 95-0695.
Dulikravich, G.S., Baker, D.P. 1999. Aerodynamic Shape Inverse Design Using a Fourier Series Method, *AIAA paper* 99-0185.
Kroll, N., Jain, R.K. 1987. Solution of Two-dimensional Euler Equations – Experience with a Finite Volume Code, Forschungsbericht, DFVLR-FB 87–41.
Rogalsky, T., Dersksen, R.W., Kocabiyik, S. 2000. Differential Evolution in Aerodynamic Optimization, University of Manitoba, Winnipeg, MB, Canada.

Computational Methods in Engineering and Science, Iu et al. (eds)
© 2003 Swets & Zeitlinger, Lisse, ISBN 90 5809 567 3

Parallel algorithms for compressible fluid flows in the subsurface

G. Alduncin & N. Vera-Guzmán

Institute of Geophysics, National Autonomous University of Mexico, Mexico City, Mexico

ABSTRACT: Parallel algorithms for the numerical simulation of compressible fluid flows in the sub-surface are presented. From a general velocity-pressure mixed model, dual macro-hybrid variational formulations are developed on the basis of nonoverlapping domain decompositions. Then, multidomain mixed finite element approximations are introduced as semidiscrete variational models, defined on independent subdomain meshes and compatible interface meshes, and numerical time integration schemes are applied for fully discrete versions. Specifically the implicit time marching scheme of Euler and the semi-implicit scheme of Peaceman-Rachford are considered, which can be implemented in parallel via resolvent or proximation characterizations, generalizing Uzawa and augmented Lagrangian numerical procedures.

1 INTRODUCTION

Domain decomposition methods have proved to be very efficient for generating parallel algorithms in the numerical treatment of partial differential equations governing physical processes. Such a methodology decomposes the global system at hand into subsystems of a simpler physical and geometrical local nature, communicating them through their interfaces by transmission conditions of continuity. Also, they permit constructing local spatial discretizations, with diverse interface models for synchronization. Here, we are interested in generating parallel algorithms for the numerical simulation of compressible fluid flows in the subsurface, based on nonoverlapping domain decompositions, multidomain mixed finite element approximations, and dual interface problems (Alduncin 1996, 1998, 2003).

From a general nonlinear velocity-pressure mixed model, whose structure resembles that of mono and multiphasic flow models in petroleum reservoir and environmental modelling, we formulate spatially decomposed compressible flow processes through nonoverlapping domain decompositions. Then, adopting the subdifferential methodology (Panagiotopoulos 1985, Alduncin 1989) for variational formulations, we are able to consider generalized external boundary conditions, as well as to incorporate internal interface continuity transmission constraints, via subdifferential equations relating the mass flow rate and pressure of the interacting subsystems. In this manner, upon dualization of the transmission conditions, we derive systematically the macro-hybrid mixed variational formulation of the nonlinear compressible flow problem, looking for solutions of mass flow rate and pressure in the interior of the subdomains, and of pressure on their internal boundaries. Further, taking advantage of the dual variational macro-hybrid structure of the problem, we can introduce in a natural form multidomain finite element spatial discretizations, defined on independent local meshes for the diverse subsystems, as well as on compatible interface meshes.

As numerical methods for the resolution of the semi-discrete macro-hybrid mixed finite element resulting problems, we apply specifically the implicit time marching scheme of Euler and the semi-implicit scheme of Peaceman-Rachford for numerical time integration. Then, applying resolvent methods for variational inequalities (Alduncin 1997), we derive proximation or projection characterizations, obtaining parallel algorithms of a macro-hybrid penalty-duality type. Importantly, these algorithms generalize Uzawa and augmented Lagrangian numerical procedures (Fortin & Glowinski 1982, Le Tallec 1994, Yang 1998). Also, they can be considered as mass-preconditioned iterative procedures for stationary flows, when the dynamical system evolves tending to a steady state (Alduncin 1996, 2003, Alduncin & Vera-Guzmán 2003).

Some works to be mentioned, closely related to this paper, are those of Douglas et al. (1994), Le Tallec & Sassi (1995), Arbogast & Yotov (1997), Wheeler et al. (1999) and Arbogast et al. (2000) where different implementations of multidomain finite element approximations are considered.

2 DECOMPOSED VELOCITY-PRESSURE MODELS

Let $\Omega \subset \Re^n$, $n \in \{1, 2, 3\}$, denote the spatial flow region in the subsurface, and $(0, T)$ the time interval, in which we are interested in simulating numerically compressible flow processes. As a representative Darcian flow model, we consider the following:

$$\left. \begin{array}{l} \boldsymbol{K}^{-1}(p)\boldsymbol{w} = -\boldsymbol{grad}p + \rho(p)\boldsymbol{g}, \\ c(p)\partial p/\partial t + \mathrm{div}\boldsymbol{w} = \widehat{q}, \end{array} \right\} \text{ in } \Omega \times (0, T). \tag{1}$$

Here, in the case of compressible monophasic flow, K, ρ and c stand for the pressure p-dependent mobility tensor, mass density and compressibility parameter, $\boldsymbol{g}$ is the gravity acceleration vector, and $\widehat{q}$ is the prescribed interior flow rate. In fact, the vector field w corresponds to the mass flux rate, $w = \rho u$, u being Darcy's velocity. An important aspect of this model is that its mathematical structure is that of more complex flow types; for example, miscible fluid flows (Wang et al. 2000), and compositional multiphase flows (Chen & Ewing 1997, Chen et al. 2000). In addition to the pressure initial condition of the system,

$$p(0) = p_0, \text{ in } \Omega, \tag{2}$$

boundary conditions are required to establish the physical interaction of the system with the exterior through its boundary $\partial\Omega$. In general, such conditions can be modelled via subdifferential equations of the form

$$p \in \partial\psi(\boldsymbol{w}\cdot\boldsymbol{n}), \text{ on } \partial\Omega \times (0, T), \tag{3}$$

where $\partial\psi$ denotes a monotone subdifferential operator expressing in a primal sense the relation between boundary mass flux and pressure (Panagiotopoulos 1985, Alduncin 1989). One of the advantages of this type of formulations is that by themselves they are variational formulations, which combined with the interior system (1) provide systematically variational mixed principles.

Next, for parallel computational purposes, we decompose the system by introducing nonoverlapping domain decompositions of the flow domain Ω,

$$\overline{\Omega} = \bigcup_{e=1}^{E} \overline{\Omega}_e, \quad \Omega_e \cap \Omega_f = \emptyset, \ 1 \le e < f \le E, \tag{4}$$

with connected and disjoint subdomains, and internal boundaries $\Gamma_e = \partial\Omega_e \cap \Omega$, $1 \le e \le E$, and interfaces $\Gamma_{ef} = \Gamma_e \cap \Gamma_f$, $1 \le e < f \le E$. In this manner, the initial and boundary value problem (1–3) is localized by the set of subproblems, for $e = 1, 2, \ldots, E$,

$$\left. \begin{array}{l} \boldsymbol{K}_e^{-1}(p_e)\boldsymbol{w}_e = -\boldsymbol{grad}p_e + \rho_e(p_e)\boldsymbol{g}, \\ c_e(p_e)\partial p_e/\partial t + \mathrm{div}\boldsymbol{w}_e = \widehat{q}_e, \end{array} \right\} \text{ in } \Omega_e \times (0, T),$$
$$p_e \in \partial\psi_e(\boldsymbol{w}_e\cdot\boldsymbol{n}_e), \qquad\qquad \text{ on } \partial\Omega_e / \Gamma_e \times (0, T),$$
$$p_e(0) = p_{0e}, \qquad\qquad\qquad \text{ in } \Omega_e, \tag{5}$$

communicated across their interfaces by the constraints of mass flux and pressure continuity,

$$\left. \begin{array}{l} \boldsymbol{w}_e\cdot\boldsymbol{n}_e = -\boldsymbol{w}_f\cdot\boldsymbol{n}_f, \\ p_e = p_f, \end{array} \right\} \text{ on } \Gamma_{ef} \times (0, T), \ 1 \le e < f \le E. \tag{6}$$

3 MACRO-HYBRID MIXED VARIATIONAL FORMULATION

In this section, we pass to formulate the decomposed problem variationally. That is, introducing an appropriate functional framework, the problem is posed in a weak sense, which allows its qualitative analysis as well as its systematic spatial discretization. The crucial aspect is how to express variationally transmission conditions (6); here, once again, a subdifferential approach turns out to be an appropriate one.

For $e = 1, 2, ..., E$, let $V(\Omega_e) = \{v \in L^2(\Omega_e): \text{div } v \in L^2(\Omega_e)\}$ and $Y(\Omega_e) = L^2(\Omega)$ be the mass flux rate and pressure field local spaces, and let $B(\Gamma_e) = L^2(\Gamma_e)$ be the common internal boundary space of mass flux and pressure traces. Then, transmission constraints (6) are variationally formulated in a dual sense by the subdifferential equation (Alduncin 1998), for $t \in (0, T)$,

$$\{w_e(t) \cdot n_e\} \in \partial I_{Q_D}(\{p_e(t)\}).\tag{7}$$

Here, I_{Q_D} denotes the indicator functional of the subspace of admissible internal boundary pressures, $Q_D \subset B(\Gamma) = \Pi_{e=1}^E B(\Gamma_e)$, defined by

$$Q_D = \{\{q_e\} \in B(\Gamma) : \ q_e = q_f \text{ on } \Gamma_{ef}, 1 \leq e < f \leq E\};\tag{8}$$

I_{Q_D} equals 0 on Q_D and $+\infty$ on $B(\Gamma)/Q_D$. It is important to observe that the conjugate functional of I_{Q_D} results to be precisely the indicator functional of the subspace of admissible internal boundary mass fluxes; i.e., $(I_{Q_D})^* = I_{Q_N}$, where $Q_N \subset B(\Gamma)$ is the primal subspace

$$Q_N = \{\{v_{ne}\} \in B(\Gamma) : \ v_{ne} = -v_{nf} \text{ on } \Gamma_{ef}, 1 \leq e < f \leq E\}.\tag{9}$$

In fact, subspaces Q_D and Q_N are orthogonal to each other.

Then, introducing the dual fields of internal boundary pressures

$$r_e = p_e, \text{ on } \Gamma_e \times (0, T), \ 1 \leq e \leq E,\tag{10}$$

the macro-hybrid variational formulation of the mixed compressible flow problem follows from the set of subproblems (5), synchronized variationally by (6) in a dual sense, upon the application of the subdifferential methodology (Panagiotopoulos 1985, Alduncin 1989): subdifferential equations are expressed as variational inequalities, then integrated in their domains, and boundary conditions incorporated to the primal equation via Green's formula.

(MHM) Find $(w_e, p_e) : (0, T) \rightarrow V(\Omega_e) \times Y(\Omega_e)$, for $e = 1, 2, ..., E$:
$\int_{\Omega_e} K_e^{-1}(p_e) w_e \cdot (v - w_e) \, d\Omega + \int_{\partial\Omega_e/\Gamma_e} \psi_e(v \cdot n_e) \, d\partial\Omega$
$\quad \geq \int_{\partial\Omega_e/\Gamma_e} \psi_e(w_e \cdot n_e) \, d\partial\Omega + \int_{\Omega_e} p_e (\text{div} v - \text{div} w_e) \, d\Omega$
$\qquad + \int_{\Omega_e} \rho_e(p_e) g \cdot (v - w_e) \, d\Omega - \int_{\Gamma_e} r_e(v \cdot n_e - w_e \cdot n_e) \, d\Gamma, \quad \forall v \in V(\Omega_e),$
$\int_{\Omega_e} c_e(p_e) \partial p_e / \partial t \, q \, d\Omega + \int_{\Omega_e} \text{div} w_e \, q \, d\Omega = \int_{\Omega_e} \hat{q}_e \, q \, d\Omega, \qquad \forall q \in Y(\Omega_e),$
$p_e(0) = p_{0e};$
and $\{r_e\} : (0, T) \rightarrow Q_N$ satisfying the synchronizing condition
$0 \geq \sum_{e=1}^E \int_{\Gamma_e} w_e \cdot n_e(s_e - r_e) \, d\Gamma, \qquad \forall\{s_e\} \in Q_N.$

We note that the synchronizing variational inequality of the above problem is in fact equivalent to a variational equality since its set of variations, Q_N is a subspace. Moreover, it can be further expressed in terms of interface integrals, instead of internal boundary integrals, in the form $\Sigma_{1 \leq e < f \leq E} \int_{\Gamma_{ef}} (w_e \cdot n_e + w_f \cdot n_f) s_{ef} \, d\Gamma = 0, \forall\{s_{ef}\} \in \Pi_{1 \leq e < f \leq E} L^2(\Gamma_{ef})$, showing the weak implementation of the interface mass flow equilibrium. However, our presentation, natural in the context of subdifferential formulations, is the appropriate one for the derivation of proximal interpretations and Uzawa type approximations as we shall see below.

4   NONCONFORMING MACRO-HYBRID FINITE ELEMENT APPROXIMATIONS

Next, taking advantage of the macro-hybrid structure of variational problem (**MHM**), we introduce, in a natural fashion, locally conforming finite element approximations on the basis of independent subdomain meshes and compatible internal boundary meshes. Thus, in general, corresponding local finite element meshes do not need to match at the interfaces, nor their traces to coincide with the internal boundary meshes.

Hence, for $e = 1, 2, ..., E$, let

$$\boldsymbol{w}_{h_e} = \sum_{i=1}^{n_{h_e}} \alpha_{e,i} \boldsymbol{\phi}_{e,i} \in \boldsymbol{V}_{h_e} = [\boldsymbol{\phi}_{e,1}, \boldsymbol{\phi}_{e,2}, ..., \boldsymbol{\phi}_{e,n_{h_e}}] \subset \boldsymbol{V}(\Omega_e),$$

$$p_{h_e} = \sum_{k=1}^{m_{h_e}} \lambda_{e,k} \zeta_{e,k} \in \boldsymbol{Y}_{h_e} = [\zeta_{e,1}, \zeta_{e,2}, ..., \zeta_{e,m_{h_e}}] \subset Y(\Omega_e),$$

(11)

be mesh-independent locally conforming mixed finite element approximations (Brezzi & Fortin 1991, Roberts & Thomas 1991) of the mass flux rate and pressure fields, $\{w_e\}$ and $\{p_e\}$, and let

$$r_{h_e} = \sum_{s=1}^{d_{h_e}} \pi_{e,s} \xi_{e,s} \in \boldsymbol{B}_{h_e} = [\xi_{e,1}, \xi_{e,2}, ..., \xi_{e,d_{h_e}}] \subset B_u(\Gamma_e)$$

(12)

be locally conforming hybrid finite element approximations of the internal boundary pressures $\{r_e\}$. In this manner, the flow problem is spatially discretized, and we now look for the coefficient vectors $\{\alpha_e\}$, $\{\lambda_e\}$ and $\{\pi_e\}$ of the mass flux rate, pressure and internal boundary pressure, respectively, as solutions of the following macro-hybrid mixed semi-discrete problem.

(**MHM**$_h$) Find $(\alpha_e, \lambda_e) : (0, T) \to \Re^{n_{h_e}} \times \Re^{m_{h_e}}$, for $e = 1, 2, ..., E$:
$$A_e(\lambda_e)\alpha_e \cdot (\beta - \alpha_e) + \Psi_e(\beta)$$
$$\geq \Psi_e(\alpha_e) - \{L_e^T \lambda_e + T_e^T \pi_e - f_e(\lambda_e)\} \cdot (\beta - \alpha_e), \quad \forall \beta \in \Re^{n_{h_e}},$$
$$\{M_e^*(\lambda_e)d\lambda_e/dt - L_e\alpha_e\} \cdot \nu = q_e \cdot \nu, \qquad\qquad \forall \nu \in \Re^{m_{h_e}},$$
$$\lambda_e(0) = \lambda_{0e};$$
and $\{\pi_e\} : (0, T) \to \boldsymbol{Q}_{N_h}$ satisfying the synchronizing condition
$$0 \geq \sum_{e=1}^{E} T_e \alpha_e \cdot (\mu_e - \pi_e), \qquad\qquad \forall \{\mu_e\} \in \boldsymbol{Q}_{N_h}.$$

Here, the matrices, vectors and functionals are defined, for $i, j = 1, 2, ..., n_{h_e}$, $k, l = 1, 2, ..., m_{h_e}$, $s = 1$, $2, ..., d_{h_e}$, by $A_{e,ij}(\lambda_e) = \int_{\Omega_e} K_e^{-1}(\sum_{m=1}^{m_{h_e}} \lambda_{e,m}\zeta_{e,m})\phi_{e,j} \cdot \phi_{e,i} \, d\Omega$, $\Psi_e(\beta) = \int_{\varpi\Omega_e/\Gamma_e} \psi_e(\sum_{i=1}^{n_{h_e}} \beta_i\phi_{e,i} \cdot n_e) \, d\partial\Omega$, $M^*_{e,kl}(\lambda_e) = \int_{\Omega_e} c_e(\sum_{m=1}^{m_{h_e}} \lambda_{e,m}\zeta_{e,m}) \zeta_{e,l}\zeta_{e,k} = -\, d\Omega$, $L_{e,ki} = -\int_{\Omega_e} \zeta_{e,k} \, \text{div}\phi_{e,i} \, d\Omega$, $T_{e,si} = \int_{\Gamma_e} \xi_{e,s} \, \phi_{e,i} \cdot n_e$ $d\Gamma_e$, $f_{e,i}(\lambda_e) = \int_{\Omega_e} \rho_e(\sum_{m=1}^{m_{h_e}} \lambda_{e,m}\zeta_{e,m}) \, g \cdot \phi_{e,i} \, d\Omega$, $q_{e,l} = \int_{\Omega_e} \hat{q}_e \zeta_{e,l} \, d\Omega$. Also, assuming that hybrid approximations (12) coincide at the interfaces, the discrete version of the primal transmission admissibility subspace (9) can be defined by the natural one,

$$\boldsymbol{Q}_{N_h} = \{\{\mu_e\} \in \prod_{e=1}^{E} \Re^{d_{h_e}} : \sum_{s=1}^{d_{h_e}} \mu_{e,s}\xi_{e,s} = -\sum_{s=1}^{d_{h_f}} \mu_{f,s}\xi_{f,s} \text{ on } \Gamma_{ef}, \ 1 \leq e < f \leq E\}.$$

(13)

For the well posedness or stability of semi-discrete problem (**MHM**$_h$), we assume that corresponding mixed and hybrid compatibility LBB conditions are fulfilled (Brezzi & Fortin 1991, Roberts & Thomas 1991); i.e, the discrete divergence operators $[L_e] \in \mathfrak{L}$ ($\Pi_{e=1}^{E} \Re^{n_{h_e}}$, $\Pi_{e=1}^{E} \Re^{m_{h_e}}$) and the discrete Neumann internal boundary trace operators $[T_e] \in \mathfrak{L}$ ($\Pi_{e=1}^{E} \Re^{n_{h_e}}$, $\Pi_{e=1}^{E} \Re^{d_{h_e}}$) are surjective.

5   TIME MARCHING SCHEMES

In order to complete the discretization of the compressible flow model problem, we now proceed to integrate numerically in time. The schemes to be considered are of implicit and semi-implicit type, and have in common the property of being implementable in terms of proximation or projection operators. This algorithmic approach generalizes Uzawa and augmented Lagrangian methods to macro-hybrid formulations (Alduncin 1998, 2003).

Let us first express problem ($\mathbf{MHM}_h$) in its subdifferential form:

($\mathbf{MHM}_h$) Find $(\alpha_e, \lambda_e) : (0, T) \to \Re^{nh_e} \times \Re^{mh_e}$, for $e = 1, 2, ..., E$:
$$-L_e^T \lambda_e - T_e^T \pi_e \in A_e(\lambda_e)\alpha_e + \partial\Psi_e(\alpha_e) - f_e(\lambda_e),$$
$$L_e \alpha_e = M_e^*(\lambda_e)d\lambda_e/dt - q_e,$$
$$\lambda_e(0) = \lambda_{e0};$$
and $\{\pi_e\} : (0, T) \to Q_{N_h}$ satisfying the synchronizing condition
$$\{T_e \alpha_e\} \in \partial I_{\mathbf{Q}_{N_h}}(\{\pi_e\}).$$

As an implicit time marching scheme, we consider (Alduncin 1998), for $m \geq 0$, and $r > 0$ denoting the time step,

The implicit Euler scheme

$$-L_e^T \lambda_e^{m+1} - T_e^T \pi_e^{m+1} \in A^e(\lambda_e^{m+1})\alpha_e^{m+1} + \partial\Psi_e(\alpha_e^{m+1}) - f_e(\lambda_e^{m+1}),$$
$$L_e \alpha_e^{m+1} = M_e^*(\lambda_e^{m+1})(\lambda_e^{m+1} - \lambda_e^m)/r - q_e^{m+1}, \quad e = 1, 2, ..., E;$$
$$\{\pi_e^m\} + r\{T_e \alpha_e^{m+1}\} \in \{\pi_e^{m+1}\} + r\partial I_{\mathbf{Q}_{N_h}}(\{\pi_e^{m+1}\}).$$

On the other hand, as a semi-implicit scheme for problem ($\mathbf{MHM}_h$), we consider (Alduncin 1998, 2003)

The Peaceman–Rachford scheme

$$-L_e^T \lambda_e^m - T_e^T \pi_e^m \in A_e(\lambda_e^m)\alpha_e^m + \partial\Psi_e(\alpha_e^m) - f_e(\lambda_e^m),$$
$$L_e \alpha_e^m = M_e^*(\lambda_e^{m+1/2})(\lambda_e^{m+1/2} - \lambda_e^m)/(r/2) - q_e^{m+1/2}, \quad e = 1, 2, ..., E;$$
$$\{\pi_e^m\} + r/2\{T_e \alpha_e^m\} \in \{\pi_e^{m+1/2}\} + r/2 \, \partial I_{\mathbf{Q}_{N_h}}(\{\pi_e^{m+1/2}\});$$
$$-L_e^T \lambda_e^{m+1} - T_e^T \pi_e^{m+1} \in A_e(\lambda_e^{m+1})\alpha_e^{m+1} + \partial\Psi_e(\alpha_e^{m+1}) - f_e(\lambda_e^{m+1}),$$
$$L_e \alpha_e^{m+1} = M_e^*(\lambda_e^{m+1})(\lambda_e^{m+1} - \lambda_e^{m+1/2})/(r/2) - q_e^{m+1/2}, \quad e = 1, 2, ..., E;$$
$$\{\pi_e^{m+1/2}\} + r/2 \, \{T^e \alpha_e^{m+1}\} \in \{\pi_e^{m+1}\} + r/2 \, \partial I_{\mathbf{Q}_{N_h}}(\{\pi_e^{m+1/2}\}).$$

Notice that the synchronizing condition of the flow problem has been implemented sequentially in the time marching schemes as an Uzawa or proximal-point approximation, which can be interpreted as a pseudo-numerical time integration. Indeed, for example, for the implicit Euler scheme we have the interpretation $\{T^e \alpha_e^{m+1}\} \in (\{\pi_e^{m+1}\} - \{\pi_e^m\})/r + \partial I_{\mathbf{Q}_{N_h}}(\{\pi_e^{m+1}\})$. As we shall see in the next section, this is an appropriate strategy for proximal-point interpretations.

6 PARALLEL PENALTY-DUALITY ALGORITHMS

We finally proceed to implement the time marching schemes in terms of proximation or projection characterizations (Alduncin 1998, 2003). Towards this end, we introduce the resolvent operator of the synchronizing subdifferential $\partial I_{Q_{Nh}}$, $J_{\partial I_{Q_{N_h}}}^r \equiv (I + r\partial I_{Q_{Nh}})^{-1}$, $r > 0$, which is a single valued firm contraction (Gabay 1982). Then, considering the implicit Euler scheme, the synchronizing condition has the resolvent interpretation, $\{T_e^{m+1}\} = J_{\partial I_{Q_{N_h}}}^r (\{\pi_e^m + rT^e \alpha_e^{m+1}\})$. Furthermore, introducing as an approximation of (8), the discrete dual transmission admissibility subspace of (13), in the sense of the conjugate $(I_{Q_{Nh}})^* = I_{Q_{Dh}}$, or the orthogonal subspace to Q_{Nh},

$$Q_{D_h} = \{\{\mu_e\} \in \prod_{s=1}^{E} \Re^{d_{he}} : \sum_{s=1}^{d_{he}} \mu_{e,s}\xi_{e,s} = \sum_{s=1}^{d_{hf}} \mu_{f,s}\xi_{f,s} \text{ on } \Gamma_{ef}, \, 1 \leq e < f \leq E\}, \tag{14}$$

the resolvent operator has the projection characterization (Alduncin 1997) $J^r_{\partial I_{Q_{N_h}}} = I - Proj_{Q_{Dh}}$, where

$Proj_{Q_{Dh}}(\nu) = \mathbf{arg}\,(\inf_{\mu \in Q_{Dh}} 1/2\,\|\mu - \nu\|^2)$, $\nu \in \Pi_{e=1}^E \,\Re^{dh_e}$. Therefore, the implicit Euler scheme has the following algorithmic interpretation (Alduncin 1997, 1998, 2003).

Algorithm 1

Given $\{\lambda_e^0\} \in \Pi_{e=1}^E \Re^{m_{h_e}}$ and $\{\pi_e^0\} \in Q_{Nh}$,
known $\{\lambda_e^m\}$ and $\{\pi_e^m\}$, $m \geq 0$,
calculate in parallel α_e^{m+1} and λ_e^{m+1}, $e = 1, 2, ..., E$, such that
$$-L_e^T(\lambda_e^m + rM_e^{-*}(\lambda_e^{m+1})q_e^{m+1}) - T_e^T(\pi_e^m - Proj_{Q_{Dh}}(\{\pi_f^m + rT_f\alpha_f^m\})_e)$$
$$\in (A_e(\lambda_e^{m+1}) + rL_e^T M_e^{-*}(\lambda_e^{m+1})L_e + rT_e^T T_e)\alpha_e^{m+1} + \partial\Psi_e(\alpha_e^{m+1}) - f_e(\lambda_e^{m+1}),$$
$$\lambda_e^{m+1} = \lambda_e^m + rM_e^{-*}(\lambda_e^{m+1})(L_e\alpha_e^{m+1} + q_e^{m+1});$$
and $\{\pi_e^{m+1}\} \in Q_{Nh}$ satisfying the synchronizing condition
$$\{\pi_e^{m+1}\} = \{\pi_e^m + rT_e\alpha_e^{m+1}\} - Proj_{Q_{Dh}}(\{\pi_e^m + rT_e\alpha_e^{m+1}\}).$$

On the other hand, introducing the intermediate family of vectors $\{\kappa_e^{m+1}\} \in \Pi_{e=1}^E \Re^{dh_e}$ such that

$$\{\kappa_e^{m+1}\} \in \partial I_{Q_{N_h}}(\{\pi_e^{m+1/2}\}), \tag{15}$$

the Peaceman-Rachford scheme has the projection implementation (Alduncin 1997, 1998, 2003)

Algorithm 2

Given $\{\alpha_e^0\} \in \Pi_{e=1}^E \mathcal{D}(\Psi_e^*), \{\lambda_e^0\} \in \Pi_{e=1}^E \Re^{m_{h_e}}$ and $\{\pi_e^0\} \in Q_{Nh}$,
known $\{\alpha_e^m\}, \{\lambda_e^m\}$ and $\{\pi_e^m\}$, $m \geq 0$,
calculate $\{\kappa_e^{m+1}\}$ satisfying the (primal) synchronizing condition
$$\{\kappa_e^{m+1}\} = Proj_{Q_{Dh}}(\{2/r\,\pi_e^m + T_e\alpha_e^m\});$$
and, in parallel, $\lambda_e^{m+1/2}, \pi_e^{m+1/2}, \alpha_e^{m+1}, \lambda_e^{m+1}, \pi_e^{m+1}$, $e = 1, 2, ..., E$, such that
$$\lambda_e^{m+1/2} = \lambda_e^m + r/2\,M_e^{-*}(\lambda_e^{m+1/2})(L_e\alpha_e^m + q_e^{m+1}),$$
$$\pi_e^{m+1/2} = \pi_e^m + r/2\,(T_e\alpha_e^m - \kappa_e^{m+1}),$$
$$-L_e^T\left(\lambda_e^{m+1/2} + r/2\,M_e^{-*}(\lambda_e^{m+1})q_e^{m+1}\right) - T_e^T\left(\pi_e^{m+1/2} - r/2\,\kappa_e^{m+1}\right)$$
$$\in \left(A_e(\lambda_e^{m+1}) + r/2\,L_e^T M_e^{-*}(\lambda_e^{m+1})L_e + r/2\,T_e^T T_e\right)\alpha_e^{m+1} + \partial\Psi_e(\alpha_e^{m+1})$$
$$-f_e(\lambda_e^{m+1}),$$
$$\lambda_e^{m+1} = \lambda_e^{m+1/2} + r/2\,M_e^{-*}(\lambda_e^{m+1})(L_e\alpha_e^{m+1} + q_e^{m+1}),$$
$$\pi_e^{m+1} = \pi_e^{m+1/2} + r/2\,(T_e\alpha_e^{m+1} - \kappa_e^{m+1}).$$

Therefore, we have derived fully parallel algorithms for nonlinear compressible flow models, that generalize Uzawa and augmented Lagrangian procedures. For another approach in the context of linear compressible models, we refer to the work of Gaiffe et al. 2002 (See also Lions & Pironneau 2000.).

REFERENCES

Alduncin, G. 1989. Subdifferential and variational formulations of boundary value problems. *Comput. Methods Appl. Mech. Engrg.* 72: 173–186.
Alduncin, G. 1996. Numerical resolvent methods for constrained problems in mechanics. *Approx. Theory & its Appl.* 12(4): 1–25.
Alduncin, G. 1997. On Gabay's algorithms for mixed variational inequalities. *Appl. Math. Optim.* 35: 21–44.
Alduncin, G. 1998. Numerical resolvent methods for macro-hybrid mixed variational inequalities. *Numer. Functional Anal. Optimiz.* 19: 667–696.
Alduncin, G. 2003. Parallel proximal-point algorithms for constrained problems in mechanics. In L.T. Yang & M. Paprzycki (eds.), *Practical Applications of Parallel Computing*. Hauppauge: Nova Science.

Alduncin, G. & Vera-Guzmán, N. 2003. Parallel proximal-point algorithms for mixed finite element models of flow in the subsurface. *Int. J. Numer. Meth. Engng.*

Arbogast, T., Cowsar, L.C., Wheeler, M.F. & Yotov, I. 2000. Mixed finite element methods on non-matching multiblock grids. *SIAM J. Numer. Anal.* 37: 1295–1315.

Arbogast, T. & Yotov, I. 1997. A non-mortar mixed finite element method for elliptic problems on non-matching multiblock grids. *Comput. Methods Appl. Mech. Engrg.* 149: 255–265.

Brezzi, F. & Fortin, M. 1991. *Mixed and Hybrid Finite Element Methods.* New York: Springer-Verlag.

Chen, Z. & Ewing, R.E. 1997. From single-phase to compositional flow: Applicability of mixed finite elements. *Transport Porous Media* 27: 225–242.

Chen, Z., Qin, G. & Ewing, R.E. 2000. Analysis of a compositional model for fluid flow in porous media. *SIAM J. Appl. Math.* 60: 747–777.

Douglas, J.Jr., Pereira, P. & Yeh, L.-M. 1994. Domain decomposition for immiscible displacement in single porosity systems. In M. Křižek, P. Neittaanmäki & R. Stemberg (eds.), *Finite Element Methods, Fifty Years of the Courant Element*: 191–199. New York: Marcel Dekker.

Fortin, M. & Glowinski, R. (eds.) 1982. *Méthodes de Lagrangien Augmenté: Applications à la Résolution Numérique de Problèmes aux Limites.* Paris: Dunod-Bordas.

Gabay, D. 1982. Application de la méthode des multiplicateurs aux inéquations variationnelles. In M. Fortin & R. Glowinski (eds), *Méthodes de Lagrangien augmenté*: 279–307. Paris: Dunod-Bordas.

Gaiffe, S., Glowinski, R. & Masson, R. 2002. Domain decomposition and splitting methods for Mortar mixed finite element approximations to parabolic equations. *Numer. Math.* 93: 53–75.

Le Tallec, P. 1994. Domain decomposition methods in computational mechanics. *Comput. Mech. Advances* 1: 121–220.

Le Tallec, P. & Sassi, T. 1995. Domain decomposition with nonmatching grids: Augmented Lagrangian approach. *Math. Comp.* 64: 1367–1396.

Lions, J.-L. & Pironneau, O. 2000. Non-overlapping domain decomposition for evolution operators. *C.R. Acad. Sci. Paris, Série I* 330: 943–950.

Panagiotopoulos, P.D. 1985. *Inequality Problems in Mechanics and Applications.* Boston: Birkhäuser.

Roberts, J.E. & Thomas, J.-M. 1991. Mixed and hybrid methods. In P.G. Ciarlet & J.L. Lions (eds), *Handbook of Numerical Analysis, vol. II*: 523–639. Amsterdam: North-Holland.

Wang, H., Liang, D., Ewing, R.E., Lyons, S.L. & Qin, G. 2000. An ELLAM-MFEM solution technique for compressible fluid flows in porous media with point sources and sinks. *J. Comput. Phys.* 159: 344–376.

Wheeler, M.F., Arbogast, T., Bryant, S., Eaton, J., Lu, Q., Pezynska, M. & Yotov, I. 1999. A parallel multiblock/multidomain approach for reservoir simulation. In *Reservoir Simulation Symposium* SPE 51884: 51–61. Houston: Society of Petroleum Engineers.

Yang, D. 1998. Simulation of miscible displacement in porous media by a modified Uzawa's algorithm combined with a characteristic method. *Comput. Methods Appl. Mech. Engrg.* 162: 359–368.

Computational Methods in Engineering and Science, Iu et al. (eds)
© 2003 Swets & Zeitlinger, Lisse, ISBN 90 5809 567 3

A micro mechanics model for cavitation

J.M. Sánchez & D. Sá López
Escuela Técnica Superior de Ingenieros Navales, Universidad Politécnica de Madrid, Madrid, Spain

ABSTRACT: Our objective in this paper is to model the evolution, convection, growth, collapse and nucleation, of a dilute suspension of a large number of spherical small bubbles dispersed in an incompressible Newtonian liquid continuum. We introduce the time-dependent local number density of bubbles in order to describe the time evolution of the micro bubble suspension. The micro mechanics evolution law expresses a relationship between an objective time rate of the number density function and the time rate of nucleation of bubbles. The mathematical model is numerically solved using a Bubnov–Galerkin spectral method that seeks the approximation to the solution function using a Chebyshev expansion in the bubble volume growth variable. The proposed model is tested with an application to a uniformly dispersed population of expanding spherical bubbles as a consequence of an imposed reduction of pressure. The corresponding population evolutions have been graphically displayed showing a behaviour that agrees with the expected.

1 INTRODUCTION

To model the interaction between viscous fluid and bubble dynamics a compressible fluid continuum with microstructure whose constituents are spherical bubbles of microscopic dimension that are allowed to expand or contract independently, is considered. A bubbly fluid, comprising a dilute suspension of a large number of spherical micron-sized gas bubbles dispersed in an incompressible Newtonian liquid continuum of density ρ_L and viscosity μ_L.

The compressibility of gas bubbles within the effective bubbly fluid continuum causes the mixture of liquid and gas to behave as compressible Newtonian fluid with density $\rho(x, t) = \rho_L(1 - \wp(x, t))$ and viscosity $\mu(x, t) = \mu_L(1 - \wp(x, t)) + \mu_G\wp(x, t)$ where μ_G is the gas viscosity, $\wp(x, t) = V_G(x, t)/V(t)$ is the local void fraction with $V_G(x, t)$ the local gas phase volume and $V(t)$ the volume of bubbly liquid.

According to our assumptions $\wp(x, t) \ll 1$. We also disregard the hydrodynamic interactions between neighbouring bubbles.

The system liquid-gas micro bubbles, occupies an open and bounded domain $B \subset R^3$ in which the 3D-bubble suspension will be dragged along the flow of a regular motion $t \to \phi_t$ with spatial velocity v_t.

Our point of view considers a micro space F attached to each point $X \in B$ in which the internal parameters needed to describe the bubble suspension vary.

In the general case, the extra microscopic variables needed to describe the bubble pattern would be the volume of the bubble, the three material co-ordinates of the bubble centre defining the position of each bubble in the cluster, the gas component of each bubble and any other thermo-mechanical property carried by the bubble over the evolutionary process.

In the present model, the volume is the only microscopic property that will be considered consequently, the radius r of the bubble is the only internal parameter and $F = R_+$.

In order to describe the time evolution of the micro bubble suspension, we postulate the existence of a time-dependent number density function representing the time evolution of the number of bubbles with radius r per unit volume, per unit length, around a fixed particle $X \in B$.

The formulation of the micro mechanical model consist of the definition of the time dependent vector field, defining the micro scale motion of the bubble, and the establishment of the objective relationship governing the time rate of the number density function describing the internal structure of our continuum.

The growth velocity of the bubble $W_t(r)$ is given by the Rayleigh-Plesset equation (Plesset & Prosperetti 1977), defining the time evolution of the radius r of a single spherical bubble in an unlimited viscous incompressible liquid continuum.

The micro-mechanics evolution law is a conservation law stated as an objective time rate obtained Lie-dragging along the flow of the time dependent vector field W_t, the number density function. This time rate should equal zero in the no nucleation hypothesis or the estimated time rate of nucleated bubbles when nucleation is considered.

The model incorporates nucleation, but the lack of experimental information about the time rate of nucleation of bubbles makes it impossible the testing of the model with reasonable data.

The numerical treatment of the mathematical model, an initial value problem defined by the micro-mechanics evolution equation and the initial number density distribution, uses a Bubnov–Galerkin spectral method (Canuto et al. 1988) that seeks the approximation to the solution function using a Chebyshev expansion in the bubbles volume growth variable.

The time dependence on the evolution of the bubble population equation comes through the stresses in the present model and thereby, it is coupled to the Navier-Stokes and continuity equations of the continuum.

Our numerical approximation treats the micro mechanics evolution model as uncouple at each time step to the conservation laws by changing the coefficients of the evolution equation in due manner at the next time step. The result is an updated version of the number density of the micro bubble distribution defining the state of our continuum at time t, which is used to update the local void fraction, the density and the viscosity of the effective continuum, which will feed back the conservation laws program.

Although it is clear how the model should be used at the micro mechanic scale in hydrodynamic cavitating flows, it has been tested so far, in a bubbly fluid continuum with a uniformly dispersed population of micro bubbles in which the history of pressures and the initial number density distribution function have been imposed.

2 THE MODEL

2.1 Description of the internal structure of the continuum

To take into account the behaviour of the internal structure of the bubbly fluid continuum, we modify the containing space R^3 by "adding" the micro space R_+ to allow the extra microscopic degree of freedom necessary to define the size of the bubble.

The micro bubbles in the suspension vary in size and are random in space distribution, but we assume the existence of a time-dependent number density distribution function h_t defined at time t on $\phi_t(B) \times R_+$, representing the time evolution of the average density of bubbles of radius r per unit volume, per unit length, in the neighbourhood of a fixed particle $X \in B$.

The number density function h_t is the unique component of a four differential form in the 4D manifold $\phi_t(B) \times R_+$ and if X is taken fixed

$$r \rightarrow h_t(\phi_t(X), r) \tag{1}$$

is a one-differential form β_t in R_+. At each time t we assume there is a maximum radius of bubble $R_{max}(t)$.

The number of bubbles at the point $x = \phi_t(X)$ of all size at time t is

$$h(x,t) = \int_0^{R_{max}} h_t(x,r)dr \tag{2}$$

and the total number of bubbles in B in that instant is

$$N(t) = \int_0^{R_{max}} dr \left(\int_{B_t} h_t(x,r)dv_x \right) \tag{3}$$

720

The local void fraction per unit volume of bubbly liquid is then given by

$$\wp(x,t) = \frac{4\pi}{3} \int\limits_{0}^{R_{max}} h_t(x,r) \, r^3 \, dr \qquad (4)$$

The function $\Gamma_t(x, r)$ will represent for a fixed particle x, the per unit macro volume, rate at which bubbles of size r are nucleated. A property that we assume is known experimentally.

2.2 Lie derivative of the number density distribution function

A state of our physical system at time t is a couple $(x, r) \in B_t \times R_+$, defined by the spatial point x and the radius r of the bubble. Let $\varphi_{t,s}$ be the evolution operator that maps a state at time s to what the state would be at time t after time t-s has elapsed

$$\varphi_{t,s} = (\phi_{t,s}, \psi_{t,s}) : B_s \times R_+ \to B_t \times R_+.$$

$\varphi_{t,s}$ is made up with the flow $\phi_{t,s}$ of v_t and the flow $\psi_{t,s}$ of W_t, the time dependent vector field on R_+ defining the growth velocity of the bubble.

Assuming no nucleation of new micro bubbles, we can state the conservation law,

$$\frac{d}{dt} \int\limits_{\psi_{t,s}(P)} \beta_t = 0 \qquad (5)$$

where P is any nice open set (any open interval) of R_+.

This time rate will be equal in general to the estimated rate of nucleated bubbles $\Gamma(t, r)$.

Using the generalised transport theorem (Marsden & Hughes 1983)

$$\frac{d}{dt} \int\limits_{\psi_{t,s}(P)} \beta_t = \int\limits_{\psi_{t,s}(P)} L_{W_t} \beta_t \qquad (6)$$

we have

$$\int\limits_{\psi_{t,s}(P)} L_{W_t} \beta_t = 0 \qquad (7)$$

for any P, condition that we will express in local differential form as

$$L_{W_t} \beta_t = 0 \qquad (8)$$

or $L_{W_t} \beta_t = \Gamma$ in the case of non-zero nucleation.

The non-autonomous Lie derivative of W_t and β_t is defined by (Sánchez et al. 1997)

$$L_{W_t} \beta_t = \frac{\partial \beta_t}{\partial t} + \left((W_t | \text{grad}_r \, h_t) + h_t \, \text{div}_r W_t \right) dr = \frac{\partial \beta_t}{\partial t} + \left(W_t \frac{\partial h_t}{\partial r} + h_t \frac{\partial W_t}{\partial r} \right) dr \qquad (9)$$

Using the Hodge star operator (Abraham et al. 1988) we get

$$\left(L_{w_t} \beta_t \right)^* = \frac{\partial h_t}{\partial t} + W_t \frac{\partial h_t}{\partial r} + h_t \frac{\partial W_t}{\partial r} \qquad (10)$$

2.3 The mathematical model

The conservation law defines in the general case a non-homogeneous linear convection PDE in the density function

$$\frac{\partial h}{\partial t}(t,r) + W_t(r) \frac{\partial h}{\partial r}(t,r) + h(t,r) \frac{\partial W_t}{\partial r}(r) = \Gamma(t,r) \qquad (t,r) \in [0,T] \times R_+ \qquad (11)$$

with convection velocity $W_t(r)$.
Equation (11) together with the initial condition

$$h(0,r) = h_o(r)$$ (12)

defining the initial bubble density distribution, a data experimentally provided at some chosen starting moment, define the initial value problem that model our physical problem.

2.4 The growth law

The growth law is based on the behaviour of a single bubble in an infinite domain of liquid whose temperature and pressure far from the bubble are T_∞ and $p_\infty(t)$ respectively. The evolution law of the radius of the bubble is given by the Rayleigh-Plesset equation for bubble dynamics (Brennen 1995). It is assumed that the bubbles contain some quantity of gas. Depending on the characteristics of the process under study the growth law is chosen. In this paper the inertially controlled dynamic behaviour of the bubble in the absence of significant thermal effects with a polytropic behaviour of the gas content is considered. Under these circumstances the temperature in the continuum liquid and bubbles, is assumed uniform equals to T_∞. For simplicity, only the inviscid case is considered at the micro scale.

2.5 The nucleation law

In water at normal temperatures homogeneous nucleation theory becomes virtually irrelevant, so that the only type of nucleation to be considered in the process under study is heterogeneous nucleation that can occur on the micro bubbles of the effective continuum freely suspended within the liquid. This observation excludes sub-micron sized solid particles of the free stream nuclei population, limiting the character of the nuclei that will be uniform in composition simplifying greatly the process.

The model incorporates this type of nucleation, but the lack of experimental information about the rate of nucleation of bubbles makes it impossible the testing of the model with reasonable data.

3 THE NUMERICAL APPROXIMATION

Our numerical approach uses a Chebyshev–Galerkin weighted residual method (Canuto et al. 1988). We assume that at time t the radii of the micro bubbles in the continuum, belong to the interval $(0, R_{max}(t))$ where $R_{max}(t)$ is defined according to the evolutionary process under study.

We consider the time-dependent positive co ordinates transformation

$$\xi \to r = \frac{R_{max}(t)}{2}(1+\xi)$$ (13)

mapping $(-1, 1)$ onto $(0, R_{max}(t)) \subset R_+$, with inverse $r \to \xi = 2r/R_{max}(t)$.
We accept the usual abuse of notation.

$$h(t,r) = h\left(t, \frac{R_{max}(t)}{2}(1+\xi)\right) = h(t,\xi)$$ (14)

With this, Equation (11) expressed in the new co ordinates will be

$$\frac{\partial h}{\partial t} + \frac{2}{R_{max}(t)}\left[\left\{W_t(\xi) - \frac{dR_{max}(t)}{dt}\frac{1+\xi}{2}\right\}\frac{\partial h}{\partial \xi} + h\frac{\partial W_t(\xi)}{\partial \xi}\right] = \Gamma(t,\xi)$$ (15)

The approximation to the solution $h(t, \xi)$ is sought in the complex space P_N of algebraic polynomials in the real variable ξ of degree up to N.

$$F_N(t,\xi) = \sum_{n=0}^{N} c_n(t)T_n(\xi)$$ (16)

A basis of P_N is described by the $N + 1$ elements $(\xi \rightarrow T_n(\xi))$ with $n = 0, 1, ..., N$, where $T_n(\xi)$ represents the Chebyshev polynomials.

The $N + 1$ expansion coefficients $(t \rightarrow c_n(t))$ are determined from the condition expressing the orthogonality of the residual and each of the corresponding basis functions.

Taking into consideration the orthogonal properties of the trial and test functions and Equation (15), we get the system of ODE.

$$\varepsilon_p \frac{\pi}{2} \dot{c}_p(t) + \frac{2}{R_{\max}(t)} \left\{ \sum_{q=0}^{N-1} \left(\frac{2}{\varepsilon_q} \sum_{\substack{i=q+1 \\ i+q \text{ odd}}}^{N} i c_i(t) \right) \Phi_{p,q}(t) + \sum_{n=0}^{N} c_n(t) \Psi_{n,p}(t) \right\} = \varepsilon_p \frac{\pi}{2} \gamma_p(t) \tag{17}$$

$p = 0, 1, ..., N$; and $\varepsilon_o = 2$ and $\varepsilon_q = 1$ if $q \geqslant 1$, where

$$\Phi_{p,q}(t) = I_{p,q}(t) - \varepsilon_p \frac{dR_{\max}(t)}{dt} \frac{\pi}{8} \left(\delta_{p,q-1} + 2\delta_{p,q} + \delta_{p,q+1} \right) \tag{18}$$

with

$$I_{p,q}(t) = \int_{-1}^{1} W(t,\xi) T_q(\xi) T_p(\xi) w(\xi) d\xi \quad ; \quad \Psi_{n,p}(t) = \int_{-1}^{1} \frac{\partial W(t,\xi)}{\partial \xi} T_n(\xi) T_p(\xi) w(\xi) d\xi \tag{19}$$

$w(\xi) = (1 - \xi^2)^{-\frac{1}{2}}$ is the Chebyshev weight, and the $\gamma_p(t)$ are the expansion coefficients in P_N of the experimentally provided nucleation law $\Gamma(t, \xi)$.

The initial conditions for the SODE are also experimentally provided. Again in P_N

$$h_o(\xi) = \sum_{p=0}^{N} c_p^o T_p(\xi) \tag{20}$$

and the expansion coefficients c_p^0 will define the initial conditions $c_p(0)$ that complete the initial value problem that we must solve to define approximately the micro bubble density.

4   NUMERICAL APPLICATION

In the application the dilute suspension of spherical gas bubbles is uniformly dispersed in an incompressible viscous Newtonian fluid. Expansion of these compressible bubbles is achieved by reducing the pressure on the system as a whole thereby giving rise to a micro scale motion of the otherwise quiescent interstitial liquid. The behaviour of the gas in the bubble is isothermal and the no nucleation hypothesis is assumed.

The history of pressures $p_\infty(t)$ is linearly decreasing with time and $p_\infty(t) < p_V(\forall t > 0)$ for explosive growth to occur.

The $p_\infty(t)$ will be the local pressure of the liquid surrounding the bubble.

At time $t = 0$ we know from the initial bubble size distribution $h_0(r)$, the maximum radius $R_{\max}^0$ of all the bubbles. At time t we define the critical radius $R_c(t)$ as the radius that makes $p_\infty(t)$ the Blake threshold pressure for the bubbles of a certain mass of gas.

The critical radius $R_c(t)$ is given by

$$R_c(t) = \frac{4S}{3(p_V - p_\infty(t))} \tag{21}$$

Regardless the content of gas in the bubbles of the population, all of the nuclei whose size r is greater than $R_c(t)$ will become unstable, grow explosively and cavitate, whereas those nuclei with radii smaller than that critical, will react passively and will not become visible to the eye.

We define, where $R_{max}(t) = R_{max}^0 + \int W_\infty(t)dt$, where $W_\infty(t)$ is the asymptotic growth rate for $r \gg R_c$ estimated by

$$W_\infty(t) \approx \sqrt{\frac{2}{3} \frac{(p_V - p_\infty(t))}{\rho_L}} \tag{22}$$

and Equation (15) will be now

$$\frac{\partial h}{\partial t} + \frac{2}{R_{max}(t)}\left[\left\{W_t(\xi) - W_\infty(t)\frac{1+\xi}{2}\right\}\frac{\partial h}{\partial \xi} + h\frac{\partial W_t(\xi)}{\partial \xi}\right] = \Gamma(t,\xi) \tag{23}$$

The growth law is given by

$$W_t(r) = \begin{cases} 0 & \text{if} \quad r \leq R_c \\ \sqrt{f(r)} & \text{if} \quad r \geq R_c \end{cases} \tag{24}$$

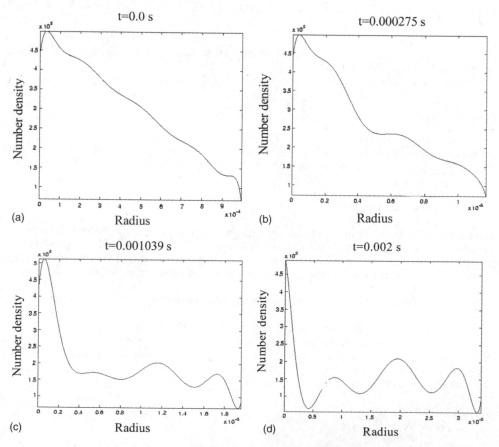

Figures 1a–d. Evolution of the number density distribution function under explosive growth. Initial size distribution and number density distributions corresponding to three instants of the process. (Number of bubbles/m^4 vs. radius m).

with

$$f(r) = \frac{2(p_V - p_\infty(t))}{3\rho_L}\left(1 - \frac{R_c^3}{r^3}\right) + \frac{4SR_c^2}{3\rho_L r^3}\ln\left(\frac{r}{R_c}\right) - \frac{2S}{\rho_L r}\left(1 - \frac{R_c^2}{r^2}\right) \tag{25}$$

This equation shows that for each t the velocity is smaller than the asymptotic growth rate $W_\infty(t)$ and that no bubble in the population can have a radius greater than $R_{max}(t)$ at time t.

The lower the pressure level $p_\infty(t)$, the greater the tension $p_V - p_\infty(t)$, the smaller the instantaneous critical size $R_c(t)$ and the larger the number of nuclei that are activated.

The evolution of the number distribution resembling the typical nuclei number distributions from water tunnels is shown in Figures 1a–d below. For an initial tension of 2800 Pa, the critical radius is 0.000259 m. All the bubbles with radii greater than this value grow. At all time, the part of the number density curve to the left of the instantaneous critical radius, remain constant. There, the number of bubbles decreases because they grow and no other bubbles with smaller radii occupy their place until the tension is increased.

The increasing tension brings micro bubbles of smaller size into the explosive growth process regardless their gas content as it is shown in the corresponding evolution curves.

The evolution of the local void fraction and the density of the effective continuum fields, that couple the macro and micro scales, fully agree with the expected.

Table 1. Numerical values of the different variables and functions involved in explosive growth.

t [s]	$p_\infty(t)$ [Pa]	$R_c(t)$ [m]	$R_{max}(t)$ [m]	$\wp(t)$	$\rho(t)$ [kg/m^3]
0	2800	0.000259	0.001	0.000007	999.993092
0.000006	2794.6	0.000256	0.001003	0.000007	999.993025
0.000031	2772.1	0.000241	0.001016	0.000007	999.992729
0.00014	2674.284297	0.000194	0.00108	0.000009	999.991061
0.000275	2552.62429	0.000155	0.001176	0.000012	999.987921
0.000478	2370.105157	0.00012	0.001349	0.00002	999.98003
0.000733	2140.016149	0.000093	0.001608	0.000039	999.961196
0.001039	1864.466564	0.000074	0.00197	0.000085	999.914884
0.001453	1492.01515	0.000057	0.002537	0.00023	999.770311
0.002	1000	0.000044	0.003406	0.000734	999.265608

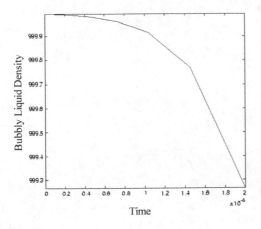

Figure 2. Evolution of the continuum density under ($\rho(t)$[kg/m^3] vs. time r).

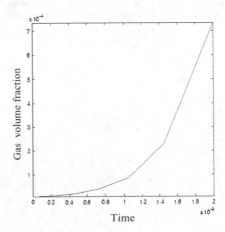

Figure 3. Evolution of the local void fraction under explosive growth ($\wp(t)$ vs. time s).

5 CONCLUSIONS

Assuming that the mixture incompressible viscous liquid/non-interacting spherical gas bubbles is a compressible viscous fluid continuum with microstructure described by a number density function, a micro mechanics equation that takes into account the evolution of the internal structure of our continuum has been stated.

The numerical treatment of the mathematical model defined by the micro-mechanics evolution equation and the initial number density function, using a spectral method is included.

The model has been tested in a bubbly fluid continuum with a uniformly dispersed population of micro bubbles in the no nucleation hypothesis. In the test, the history of pressures and the initial number density distribution have been imposed.

The corresponding evolution curves show a behaviour that agrees with the expected. The present model is a necessary step in the numerical modelling of cavitating flows where it should be used to model the microscopic aspects of cloud cavitation in multiphase complex flows (See Kubota et al. 1992).

REFERENCES

Abraham, R., Marsden, J.E. & Ratiu,T. 1988. *Manifolds Tensor Analysis and Applications*. Second edition, Applied Mathematical Sciences 75, New York: Springer-Verlag.

Brennen, C.E. 1995. *Cavitation and Bubble Dynamics*. New York, Oxford: Oxford University Press.

Canuto, C., Hussaini, M.Y., Quarteroni, A. & Zang, T.A. 1988. *Spectral Methods in Fluid Dynamics*. Berlin. Heidelberg: Springer-Verlag.

d'Agostino, L. & Brennen, C. 1989. Linearized dynamics of spherical bubble clouds, *J. Fluid Mech.* 199: 155–176.

Edwards, D.A., Brenner, H. & Wasan, D.T. 1991. *Interfacial transport processes and Rheology*. Butterworth-Heinemann.

Kubota, A., Kato, H. & Yamaguchi, H. 1992. A new modelling of cavitating flows-a numerical study of unsteady cavitation on a hydrofoil section. *J. Fluids Eng.*, 111: 204–210.

Kodama, Y., Take, N., Tamiya, S., & Kato, H. 1981. The effect of nuclei on the inception of bubble and sheet cavitation on axisymmetric bodies. *J. of Fluids Eng.* 103: 557–563.

Marsden, J.E. & Hughes, T.J.R. 1983. *Mathematical Foundations of Elasticity*. Prentice-Hall, Inc.

Plesset, M.S. & Prosperetti, A. 1977. Bubble dynamics and cavitation. *Ann. Rev. Fluid Mech.* 9: 145–185.

Sánchez, J.M. 2000. Computational micro fracture model for brittle cracking. *In Proc. of the 16th IMACS World Congress*, electronic version, Code 415 Computational Physics, Chemistry and Biology, Continuum Mechanics. Lausanne.

Sánchez, J.M. & Vega Miguel, J.L. 2002. Computational micro mechanics model for the convection of a cracks population in a brittle material. *Int. J. of Sol. & Str.* 39: 797–817.

Sánchez, J.M, Vega Miguel, J.L. & Garbayo, E. 1997. General evolution laws of a micro cracks population in a brittle material. In *Proc. of the 15th IMACS World Congress*, Volume 5: 513–518. Berlin.

van Wijngaarden, L. 1964. On the collective collapse of a large number of gas bubbles in water. In *Proc. of the 11th Int. congress of Applied Mechanics*: 854–861. Springer.

Computational Methods in Engineering and Science, Iu et al. (eds)
© 2003 Swets & Zeitlinger, Lisse, ISBN 90 5809 567 3

Simulation of mixing and chemical reacting flows over a flat wall

A.L. de Bortoli
UFRGS – Department of Pure and Applied Mathematics, Porto Alegre – RS, Brazil

ABSTRACT: The aim of this work is the numerical solution of laminar flows over a flat wall perturbed by mixing and temperature increase due to a chemical reaction. The model approximates the overall, single step, binary, irreversible reaction between two species, resulting a third, which is taken to be a process of first order with respect to each of the reactants, being the specific reaction rate controlled by temperature-dependent Arrhenius kinetics. The simulations are performed using the finite differences explicit Runge-Kutta three-stages scheme for second order time and space approximations. Consistent results, for reactant and product concentration fields, as well as for temperature of reaction, are obtained, showing that the model is able to follow nonlinear behavior of the mixing and reaction progress, for Schmidt and Prandtl numbers of order 1, Zel'dovich 1, Damköhler 300 and heat release parameter 1, which are values for gaseous hydrocarbon chemistry. Besides, it is worth pointing out that these results build on earlier ones, which were obtained from thin-layer model approximations.

1 INTRODUCTION

Combustion is a difficult multidisciplinary topic which occurs, for example, in chambers of rockets or engines, where the temperature and initial concentration of the substances and combustion products changes progressively with time. As the analytical solutions exist only for simple idealized flow situations and DNS (direct numerical simulation) is not easily applicable to high Reynolds number flows, the conventional numerical simulation of the combustion equations remains as an alternative. Therefore, simplifications of the general case usually take the form of a restricted problem definition and/or a simpler treatment of the reaction kinetics.

Turbulent combustion flows include many complicated features as swirl, recirculation and combustion. To have stable combustion it must be rapid and all chemical time scales of the process must be rapid. They vary from the largest scales, which depends on the geometry, to the smallest scales, that depends on the process of energy dissipation (Persson 2001). Usually, three different scales are considered in turbulent flow fields: the integral scale, the Taylor and the Kolmogorov microscales.

In revision, when using the Favre-averaged equations, which are indicated mainly to incompressible flows (Persson 2001), the integral length and time scales can be written as $l = c_d u'^3/\varepsilon$, $t_T = l/u'$, where c_d is a constant of value 0.37, ε the turbulence dissipation rate and u' the velocity fluctuation. The Kolmogorov's length and time scales can be expressed as $\eta = (v^3/\varepsilon)^{1/4}$, $t_\eta = (v/\varepsilon)^{1/2}$, respectively. The Taylor length and time scales are $\lambda = (vu'^2/\varepsilon)^{1/2}$, $t_F = l_F/u_L$, where l_F is the flame thickness (of order 1 mm), v is the kinematic viscosity and u_L is the laminar burning velocity, which can be assumed to follow the expression from Metgalchi & Keck (1980) for gasoline, as example: $u_L = [26.32 - 84.72 (\phi - 1.13)^2] (T/298)^a (p/10^5)^b$, $a = 2.18 - 0.8(\phi - 1)$, $b = 0.22(\phi - 1) - 0.16$, where ϕ is the equivalence ratio, T the local gas temperature and p the pressure inside the kernel of combustion region. Turbulent burning velocity can be written as $u_T = u_L + u'$ and, when isotropic turbulence is assumed, the following expression results $u_T = u_L + [1 + (u'/u_L)^2]^{1/2}$. For laminar premixed flamelet (when chemical time scales are much smaller than turbulence time scales) continuity can be approximated as $\rho u_T = const.$

Therefore, the turbulent burning velocity can be written as $u_T = u_b/(1 + H_e)$, where u_b is the burned gas velocity and H_e is the heat release parameter: $H_e = T_b/(T_u - 1)$. It is easy to show that the range of length scales increases rapidly with the increase of Reynolds number because $l/\eta \propto R_e^{3/4}$ and three dimensional flows will require a number of grid points following the relation $N \propto (l/\eta)^3 \propto R_e^{9/4}$. Therefore, fine grids are required to solve all length scales of interest.

Mass, momentum, energy and chemical species for reacting flows complete the set of governing equations; here the diffusion coefficient (viscosity) is considered to be temperature dependent.

The solution of this set of equations gives us information about the chemical reaction and its coupling with fluid dynamics.

There are many methods which can be employed to the solution of reacting flow equations, such as the finite differences and the finite element. Each of these methods has its own advantages and disadvantages, which are not discussed in this work; both methods are being successfully used by well-known researches all over the world.

This work develops a numerical technique for chemical reacting flows over a flat plate. The model is based on the finite differences explicit Runge-Kutta three-stage scheme for second order time and space accuracy. Many computations were performed and special cases are chosen to focus the effect of mixture/reaction layer increase for given Damköhler, Zel'dovich, Prandtl, Reynolds and Schmidt numbers.

Obtained results are consistent with many analytical and experimental data or numerical results found in the literature using other techniques.

2 GOVERNING EQUATIONS AND SOLUTION PROCEDURE

The governing partial differential equations for mass, momentum, energy and chemical species for two-dimensional laminar reacting flows are following indicated. The model is based on a one-step, finite rate Arrhenius kinetics (Chang et al. 1991)

$$v_f[F] + v_o[O] \Rightarrow v_p[P] + heat \tag{1}$$

Usually, for combustion of fuels such as methane, hydrogen, gasoline (here as an octane-air mixture) etc., nitrogen is clearly not participating in the reaction. For gasoline, then, one can write the reaction as (Fan et al. 1999)

$$C_8H_{18} + 12.5O_2 + 47N_2 \Rightarrow 8CO_2 + 9H_2O + 47N_2 \tag{2}$$

The works from Lawrence Livermore National Laboratory (2002) indicates, for example, that the complete reaction mechanism for iso-octane oxidation includes 3600 elementary reactions among 860 chemical species, which can be classified (arranged) in around 25 major classes of elementary reactions (Curran et al. 2002).Therefore, the necessity of simplifications is obvious.

As the reaction heat release is supposed known and the temperature changes are significant, it is sufficient to predict the overall structure behavior of the flame. Therefore, the governing equations are written in indicial form as (compressible case) (Wang et al. 1999):

Continuity

$$\frac{\partial \rho}{\partial t} + \frac{\partial (\rho u_j)}{\partial x_j} = 0 \tag{3}$$

Momentum

$$\frac{\partial (\rho u_i)}{\partial t} + \frac{\partial (\rho u_i u_j)}{\partial x_j} = -\frac{\partial p}{\partial x_i} + \frac{1}{R_e} \frac{\partial \tau_{ij}}{\partial x_j} + \frac{1}{F_r} \rho F_i \tag{4}$$

Energy

$$\frac{\partial (\rho T)}{\partial t} + \frac{\partial (\rho u_i T)}{\partial x_i} = \frac{1}{R_e P_r} \frac{\partial}{\partial x_i} \left(\mu \frac{\partial T}{\partial x_i} \right) + v_p H_e D_a \rho C_f \rho C_o e^{-Z/T} \tag{5}$$

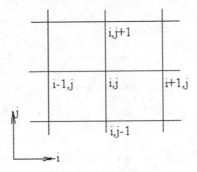

Figure 1. Finite differences conventional node-centered grid.

Chemical species: sub-index k corresponds to f for fuel, o for oxidizer and p for product

$$\frac{\partial(\rho C_k)}{\partial t} + \frac{\partial(\rho u_i C_k)}{\partial x_i} = \frac{1}{R_e S_c} \frac{\partial}{\partial x_i}\left(\mu \frac{\partial C_k}{\partial x_i}\right) \pm v_k H_e D_a \rho C_f \rho C_o e^{-Z/T} \tag{6}$$

where μ is the viscosity, D_a the Damköhler (the ratio of the turbulent time to the flame time, $D_a = t_T/t_F$, Z the Zel'dovich, P_r the Prandtl number S_c the Schmidt number, H_e the heat release parameter and $\tau_{i,j} = \mu(du_i/dx_j + du_j/dx_i)$ the stress tensor. Viscosity is temperature dependent and given by $\mu \approx T^{3/4}$ in nondimensional form. A fixed Prandtl number gives a relationship for the temperature dependence on the viscosity. Fixed Lewis (S_c/P_r) number provides temperature dependencies to the mass diffusivities of each of the species. The unknown quantities are the mixture density ρ, the fluid velocities u_i, the temperature T, the chemical species C_f, C_o and C_p and the pressure p. Pressure can be obtained using the state relation

$$p = \rho R T \tag{7}$$

When the central finite differences scheme is employed, for a point (i,j) (see Fig. 1 for details) in the computational domain, it results the following second order space derivative approximations for a general variable Ψ

$$\frac{\partial \psi}{\partial x} \approx \frac{\Psi_{i+1,j} - \Psi_{i-1,j}}{2\Delta x} \tag{8}$$

$$\frac{\partial^2 \psi}{\partial x^2} \approx \frac{\Psi_{i+1,j} - 2\Psi_{i,j} + \Psi_{i-1,j}}{\Delta x^2} \tag{9}$$

and similarly for other derivatives. After approximation, results the following form for energy equation, as example

$$\frac{(\rho T)_{i,j}^{t+\Delta t} - (\rho T)_{i,j}^{t}}{\Delta t} + \frac{(\rho u T)_{i+1,j}^{t} - (\rho u T)_{i-1,j}^{t}}{2\Delta x} + \frac{(\rho v T)_{i,j+1}^{t} - (\rho v T)_{i,j-1}^{t}}{2\Delta y} =$$

$$\frac{E_{i+1,j}^{t} - E_{i-1,j}^{t}}{2\Delta x} + \frac{F_{i+1,j}^{t} - F_{i-1,j}^{t}}{2\Delta y} + v_p H_e D_a\left[\rho C_f \rho C_o e^{-Z/T}\right]_{i,j}^{t} \tag{10}$$

with

$$E_{i,j} = \left(\frac{\mu}{R_e P_r}\right)_{i,j} \frac{T_{i+1,j} - T_{i-1,j}}{2\Delta x} \tag{11}$$

and

$$F_{i,j} = \left(\frac{\mu}{R_e P_r}\right)_{i,j} \frac{T_{i,j+1} - T_{i,j-1}}{2\Delta y} \tag{12}$$

The other equations are approximated in a similar manner; they are solved after application of boundary conditions.

2.1 Boundary conditions

The proper implementation of boundary conditions is always important when solving flow problems. Although velocities are equal to zero at solid boundaries, density, pressure, temperature and chemical specie values must be known or estimated. Although extrapolations (of the same order of domain discretization) are an alternative when we don't know the real value of a variable, usually not all variables are extrapolated because of code stability restrictions. Therefore, the boundary conditions can be summarized as follows for:

North boundary: $\partial u/\partial y = \partial v/\partial y = \partial T/\partial y = \partial p/\partial y = \partial(\rho C_i)/\partial y = \partial\rho/\partial y = 0$
South boundary: similar to north except that $u = v = 0$ and for $x \geqslant L/2$, $C_f = 1$.
West boundary: $u = p = T = 1$; $v = 0$; $\partial(\rho C_i)/\partial x = \partial\rho/\partial x = 0$
East boundary: $\partial u/\partial x = \partial v/\partial x = \partial T/\partial x = \partial p/\partial x = \partial(\rho C_i)/\partial x = \partial\rho/\partial x = 0$

As one see (Fig. 2), for velocities and concentration at south boundary as well as for velocity, pressure and temperature at west boundary the conditions are of Dirichlet type; otherwise are of Neumann type.

2.2 Time-stepping scheme

To obtain numerical solutions of high accuracy, the Runge-Kutta method is chosen. This method is characterized by its low operation count and low storage requirements. More than two stages are employed to extend the stability region. Therefore, the following multistage scheme, which requires few computational storage, is employed

$$\vec{W}_{i,j}^{(0)} = \vec{W}_{i,j}^{(n)}$$
$$\vec{W}_{i,j}^{(r)} = \vec{W}_{i,j}^{(0)} - \alpha_r \Delta t \vec{R}_{i,j}^{(r-1)} \tag{13}$$
$$\vec{W}_{i,j}^{(n+1)} = \vec{W}_{i,j}^{(r)}$$

where $\vec{R}_{i,j}$ corresponds to the residual, with $r = 0,1,2$, $(n = 3)$ and the following second order time coefficients $\alpha_1 = 1/2$, $\alpha_2 = 1/2$ and $\alpha_3 = 1$. $\vec{W}$ is the vector of flux variables given by $\vec{W} = [\rho, \rho u, \rho T, \rho C_f, \rho C_o, \rho C_p]^T$ and the pressure p comes from the state relation.

When using explicit methods, time-step restrictions are natural; however small time-steps are needed to well represent the mixture/reaction time scales. For the compressible case, the CFL condition results

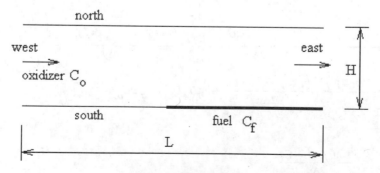

Figure 2. Boundary conditions for mixing and chemical reacting flows over a flat wall.

for the wave equation

$$\Delta t \le CFL \frac{V}{\lambda^i + \lambda^j} \tag{14}$$

where CFL is the Courant–Friedrich–Lewy number, V is the cell area (or volume for three-dimensional situations) and λ^i the spectral ratio of the Jacobian matrix related to the i direction.

3 NUMERICAL RESULTS

The numerical calculations, using finite difference schemes, do not always permit a view of the detailed physical pattern of the flow field. Internal dissipation, shock smearing, mesh resolution, etc, impose certain restrictions. However, the reaction front position, product formation and temperature and pressure increase can be calculated using such method.

Therefore, following numerical results for mixing flows including chemical reaction over a flat wall are presented. This situation is chosen because of its resemblance to the flat wall boundary layer flow. Figure 3 shows the product C_p after 0.01, 0.05 and 0.1 seconds for Prandtl number of 1 and Schmidt number of 1, Damköhler 300, Zel'dovich 1, H_e 1 (heat release parameter) and Reynolds 100 (laminar flow). As one see, the diffusion effect appears due to the mixture/reaction starting at plate surface.

The reactants F (fuel) and O (oxidizer) are placed in contact, as shown in Figure 2. Figure 3 presents the formation of P (a mixture of CO_2 and H_2O) after 0.01, 0.05 and 0.1 seconds; reaction zone increase is similar to that of common boundary layer type flows.

Figure 4 shows the temperature increase and product formation. Obviously, a corresponding oxidizer to fuel quantity is required, as by equation (2) indicated. Such behavior would remain the same when increasing H_e (heat release) or Damköhler numbers, however its variation would be greater.

4 CONCLUSIONS

A computational model is presented for coupling fluid dynamics, chemistry and heat transfer over a flat wall. Computations are performed to obtain a physical understanding of the mixture/reaction macroscopic behavior, for special nondimensional numbers.

Numerical tests showed a flow movement which grows with the increase of heat release parameter. When this parameter decrease, the chemical reaction occurs at lower temperatures and the flow velocity also decrease. There are indications, for still higher parameter values, of the vortices formation and a potential source of disturbance due to the motion and development of these vortices.

Obtained results are qualitatively consistent with many experimental observations and measurements for flat walls boundary layer type flows; they indicate the boundary influence due to the mixing-reaction effect. The reaction diffusion occurs in the neighborhood of the surface and remains close to it for small times. This understanding is essential if real progress is to be made for more complex problems involving turbulence, whose applications in the area of combustion are obvious.

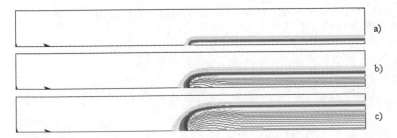

Figure 3. C_p for flow over a flat plate including chemical reaction for $D_a = 300$, $Z = 1$, $H_e = 1$, $G_r = R_m^2$, $R_n = 100$ (a) $t = 0.01$ s, (b) $t = 0.05$ s and (c) $t = 0.1$ s.

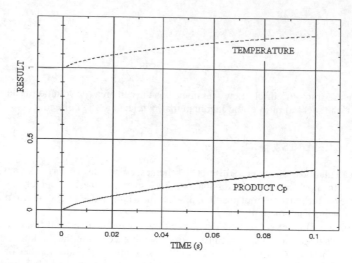

Figure 4. Temperature increase and product C_p formation for flat wall mixing flow including chemical reaction for $D_a = 300$, $Z = 1$, $H_e = 1$, $G_r = R_e^2$ and $R_e = 100$.

It is the author opinion that the finite difference with a proper discretization is useful to obtain a suitable (fast, accurate, simple and cheap) method for the solution of complex flows without introducing many complications to the conventional code structure.

REFERENCES

Boersma, B.J. 1998. Direct Simulation of a Jet Diffusion Flame, Annual Research Briefs, Center for Turbulence Research, pp 47–56.

Chang, C.H.H., Dahm, W.J.A. & Trygvason, G. 1991. Lagrangian Model Simulations of Molecular Mixing Including Finite Rate Chemical Reactions, in a Temporally Developing Shear Layer, *Phys. Fluids* A 3(5), pp 1300–1311.

Curran, H.J., Gaffuri, P., Pitz, W.J. & Westbrook, C.K. 2002. A Comprehensive Modeling Study of iso-Octane Oxidation, *Combustion and Flame* 129, pp 253–280.

De Bortoli, A.L., Thompson, M. & Calderón, A.U.Z. 2002. Numerical Simulation for Rotating Internal Weakly Viscoelastic Flows in Rectangular Ducts, *Int. J. Numer. Meth. Fluids* 39, pp 485–496.

Persson, M. 2001. Predictive Tools for Turbulent Reacting Flows, MSc Dissertation, Lulea University of Technology, 2001.

Fan, L., Li, G., Han, Z. & Reitz, R.D. 1999. Modeling Fuel Preparation and Stratified Combustion in a Gasoline Direct Injection Engine, *SAE* 1999-01-0175, Detroit, Michigan.

Lawrence Livermore National Laboratory: www.llnl.gov (2003).

Metgalchi, M. & Keck, J.C. 1980. Laminar Burning Velocity of Propane-Air Mixtures at High Temperature and Pressure, *Combust. flame* 38, 1980, 143–154.

Wang, X., Suzuki, T., Ochiai, Y. & Oda, T. 1999. Numerical Studies of Reacting Flows over Flat Walls with Fuel Injection, *JSME*, Series B, Vol. 42, Nr. 1.

Computational Methods in Engineering and Science, Iu et al. (eds)
© 2003 Swets & Zeitlinger, Lisse, ISBN 90 5809 567 3

Numerical simulation of Moffatt Vortices in a V-shaped notch

C. Lin & G. Yang

Nanjing Institute of Technology, Nanjing, China
Inner Mongolia University for Nationalities, Tongliao, China

ABSTRACT: A numerical study for low-Reynolds-number incompressible viscous fluid flow near a V-shaped notch by boundary element method is presented. The numerically simulated velocity field shows that rolling a cylinder over the V-shaped notch can form Moffatt vortices. Comparison of the simulated velocity fields with Taneda's experimental results shows good agreement. The numerical results indicate that the fluid flow in the wedge-shaped region consists of an infinite sequence of vortices of rapidly diminishing strength. It is also found that the maximum n is 4 for $y_{n+1} = y_n/2$, $Y_{n+1} = Y_n/2$ of a V-shaped notch with $\theta = 28.5°$ and $h = 5$ cm, because the strength of fourth vortex is 10^{-4}. In particular, the method is very effective in researching and solving the problem of fluid flow near a V-shaped notch with variable θ and h, as well as in finding the condition for the appearance Moffatt vortices.

1 INTRODUCTION

Low-Reynolds-number viscous fluid flow is an important subject due to its general application in diverse areas such as micro vascular fluid mechanics, cellular biophysics, suspension rheology, colloids, aerosols, and polymers (Happel 1983). As there are few exact analytical solutions for the problems, most studies are focused on computational methods. There are Finite Element Method (Reacher 1976), Finite Difference Method (Walfang 1980) and Boundary Element Method (Wu 1984, Weinbaum & Ganatos 1990, Lin & Han 1991). One of the most interesting phenomena in this kind of flows is the existence of vortex. Taneda (1979) presented an experimental method to visualize low-Reynolds-number flow around bodies or near sharp corner. He provides photographs of the fluid flows passing bodies placed near a plane wall, flows passing a fence, a step, wall cavity, and near a sharp corner. We have performed studies with Boundary Element Method (BEM) for linear and non-linear problems of incompressible viscous fluid flows. We investigated numerically the flow passing cylinder arcs (Lin 1992), the flow passing rectangular cavity (Lin & Yang 1994), the flow passing fence (Lin 1994), and the flow passing two cylinders (Lin & Yang 2002) with the BEM (Wu 1984, Lin & Han 1991). In this paper, we will present the boundary element method first. Secondly, we will describe the discrete forms of the boundary element method. Then we will apply this method to investigate numerically the velocity fields of the flow passing a V-shaped notch. Finally, comparison of the numerical results with Taneda's experimental results is carried out.

2 BOUNDARY ELEMENT METHOD

The boundary element method used for studying the problem of the low-Reynolds-number viscous fluid flow passing a V-shaped notch is described in the following sections. The idea of the method is based on a fundamental solution of Stokes equation as Green's function and to convert the Navier-Stokes equations to nonlinear boundary integral equations. For the low-Reynolds number problems, the nonlinear boundary integral equations are reduced to solving linear boundary integral equations. With the boundary values, the velocity fields in the flow domain can be simulated numerically.

2.1 Mathematical statement of the problem

Let the origin of the Cartesian coordinates be in the corner of the v-shape notch with the x-axis pointing in the direction of the velocity at infinity. Set Ω as the flow domain and Γ as its boundary, there are

$$\Gamma = \Gamma_f \cup \Gamma_\infty \tag{1}$$

and

$$\Gamma_f \cap \Gamma_\infty = \Phi_0 \tag{2}$$

where Γ_f is a fixed boundary and Γ_∞ is an outer boundary of fiction (Lin 1991).

The velocity field can be expressed as $\{v_x, v_y\}$, so there is $\{1, 0\}$ in the boundary Γ_∞ while the boundary condition on the Γ_f will satisfy the no-slip boundary condition.

And there is

$$\Gamma_f = \Gamma_c \cup \Gamma_v \tag{3}$$

which is the surface of the cylinder Γ_c and the surface of the v-shape notch Γ_v.

2.2 Boundary integral equation

For Ω is the flow domain in Euclidean space R^2 with boundary Γ, the low Reynolds number incompressible viscous motion satisfies the following non-dimensional equations

$$\begin{cases} (T_{ij}(\vec{V})),_j = 0 \\ (v_i),_i = 0 \end{cases} \qquad (\text{in } \Omega) \tag{4}$$

and the boundary conditions are

$$\begin{cases} (v_i) \big|_{\Gamma_f} = v_{fi} \\ (v_i) \big|_{\Gamma_\infty} = v_{\infty i} \end{cases} \qquad (\text{on } \Gamma) \tag{5}$$

where v_i is the known component of the boundary velocity, T_{ij} is the shear-stress tensor which can be determined by the velocity field $\vec{V}$ as follows

$$T_{ij}(\vec{V}) = -\delta_{ij} p + \frac{1}{R_e}(v_{i,j} + v_{j,i}) \tag{6}$$

where δ_{ij} is the Kronecker delta, R_e is Reynolds number and p is pressure.

Suppose the fundamental singular solutions of (4) and (5) are $\vec{W}^k$ and q^k which correlate respectively with the velocity vector $\vec{V}$ and the pressure p, they satisfy

$$\begin{cases} (T_{ij}(\vec{W}^k)),_j = \delta(\vec{X} - \vec{X}_0) \cdot \vec{e}^k \\ (\vec{W}_i^k),_i = 0 \end{cases} \tag{7}$$

where $\delta(\vec{X} - \vec{X}_0)$ is the Dirac Delta function, $\vec{e}^k$ is the unit vector along the k-axis. The fundamental solutions are then given by

$$w_j^k(\vec{X} - \vec{X}_0) = -\frac{R_e}{4\pi}[\delta_{jk} \ln \frac{1}{|\vec{X} - \vec{X}_0|} + \frac{(x_j - x_{j0})(x_k - x_{k0})}{|\vec{X} - \vec{X}_0|^2}] \tag{8}$$

$$q^k(\vec{X} - \vec{X}_0) = -\frac{1}{2\pi} \frac{(x_k - x_{k0})}{\left|(\vec{X} - \vec{X}_0)\right|^2} \quad k = 1, 2 \tag{9}$$

Using Green's formula, the boundary integral equation can be expressed as

$$cv_k(\vec{V}) = \oint_\Gamma n_i(\vec{X}_0) T'_{ij}(\vec{W}^k(\vec{X} - \vec{X}_0))_0 v_j(\vec{X}_0) d\Gamma_0$$
$$-\oint_\Gamma n_i(\vec{X}_0) T_{ij}(\vec{V}(\vec{X}_0))_0 w_j^k(\vec{X} - \vec{X}_0) d\Gamma_0 \tag{10}$$

$$cp(\vec{X}) = \oint_\Gamma q^k(\vec{X} - \vec{X}_0) T_{kj}(\vec{V}(\vec{X}))_0 n_j(\vec{X}_0) d\Gamma_0$$
$$-\frac{2}{R_e} \oint_\Gamma \frac{\partial q^k(\vec{X} - \vec{X}_0)}{\partial x_{j0}} v_k(\vec{X}_0) n_j(\vec{X}_0) d\Gamma_0 \tag{11}$$

where

$$T'_{ij}(\vec{W}^k(\vec{X} - \vec{X}_0))_0 = \delta_{ij} q^k(\vec{X} - \vec{X}_0) + \frac{1}{R_e}[w_i^k(\vec{X} - \vec{X}_0)_{,j} + w_j^k(\vec{X} - \vec{X}_0)_{,i}] \tag{12}$$

and the coefficient is given as $c = 1$ for $\vec{X} \in \Omega$ or $c = 1/2$ for $\vec{X} \in \Gamma$ and $d\Gamma_0$ indicates that the integration over Γ is with respect to the point $\vec{X}_0$.

Let us set

$$T_{kj}(\vec{V}_0(\vec{X}_0))n_j(\vec{X}_0) = t_k(\vec{X}_0) \tag{13}$$

which is the component of the local surface-stress force along the k-axis (for $k = 1$ is x-axis, $k = 2$ is y-axis).

3 NUMERICAL INTEGRAL TECHNIQUES

3.1 *Discretization and algebraic equations*

Equations (10) and (11) can be solved numerically by applying the standard method. We transform the integral equations into a linear system of algebraic equations. From equation (10) we set

$$F^k = \oint_\Gamma n_i T_{ij}(\vec{V}(\vec{X}_0))_0 w_j^k(\vec{X} - \vec{X}_0) d\Gamma_0 \tag{14}$$

$$U^k = \oint_\Gamma n_i T'_{ij}(\vec{W}^k(\vec{X} - \vec{X}_0))_0 (v_j(\vec{X}_0) d\Gamma_0 \tag{15}$$

Hence (10) reduces to

$$F^k = U^k - \frac{v_k}{2} \qquad (\ for\ \vec{X} \in \Gamma\ ). \tag{16}$$

For the numerical implementation of the boundary integral equations, it is necessary to divide the boundary Γ into N boundary elements Γ_β $(\beta = 1, 2, ..., N)$. All of the boundary elements are small relative to Γ_β and over which the components of $t_{j\beta}$ and $v_{j\beta}$ may, for the purposes of the integral equation, be considered constant and equal to their values at the center of the element, thereby yielding the numerical approximations to (14) and (15) for the αth boundary element,

$$F_\alpha^k = \sum_{\beta=1}^N T_{j\beta} \int_{\Gamma_\beta} w_j^k(\vec{X}_\alpha - \vec{X}_0) d\Gamma_0 \tag{17}$$

$$U_\alpha^k = \sum_{\beta=1}^{N} n_{i\beta} v_{j\beta} \int_{\Gamma_\beta} T_{ij}'(\vec{W}^k(\vec{X}_\alpha - \vec{X}_0)) d\Gamma_0 \quad \alpha = 1,2,3,\cdots\cdots N \tag{18}$$

Now, the integral representation in the formulas have been determined explicitly by (8) and (12). They are expressed in the form

$$G_{k\alpha j\beta} = \int_{\Gamma_\beta} w_j^k(\vec{X}_\alpha - \vec{X}_0) d\Gamma_0$$

$$= -\frac{R_e}{4\pi} \int_{\Gamma_\beta} [\delta_{kj} \ln \frac{1}{(\vec{X}_\alpha - \vec{X}_0)} + \frac{(x_{i\alpha} - x_{i0})(x_{j\alpha} - x_{j0})(x_{k\alpha} - x_{k0})}{|\vec{X}_\alpha - \vec{X}_0|^2}] d\Gamma_0 \tag{19}$$

$$L_{k\alpha ij\beta} = \int_{\Gamma_\beta} T_{ij}'(\vec{W}^k(\vec{X}_\alpha - \vec{X}_0)) d\Gamma_0$$

$$= -\frac{1}{\pi} \int_{\Gamma_\beta} \frac{(x_{i\alpha} - x_{i0})(x_{j\alpha} - x_{j0})(x_{k\alpha} - x_{k0})}{|\vec{X}_\alpha - \vec{X}_0|^4} d\Gamma_0 \tag{20}$$

Thus equation (16) can be rewritten as

$$\sum_{\beta=1}^{N} \sum_{j=1}^{2} G_{k\alpha j\beta} T_{j\beta} = F_\alpha^k \tag{21}$$

This is a matrix equation, which can be written as $GT = F$. Then equations (10) and (11) can be rewritten as

$$c_\alpha v_{k\alpha} = \sum_{\beta=1}^{N} \sum_{i=1}^{2} \sum_{j=1}^{2} L_{kij\alpha\beta} n_{i\beta} v_{j\beta} - \sum_{\beta=1}^{n} \sum_{j=1}^{2} G_{kj\alpha\beta} t_{j\beta} \tag{22}$$

$$c_\alpha p_\alpha = \sum_{\beta=1}^{n} \sum_{i=1}^{2} \sum_{j=1}^{2} S_{\alpha ij\beta} n_{i\beta} v_{j\beta} + \sum_{\beta=1}^{n} \sum_{j=1}^{2} R_{\alpha j\beta} t_{j\beta} \tag{23}$$

where

$$S_{\alpha ij\beta} = \frac{2}{R_e} \int_{\Gamma_\beta} \frac{\partial q^j(\vec{X}_\alpha - \vec{X}_0)}{\partial x_{i0}} d\Gamma_0$$

$$= \frac{1}{R_e\pi} \int_{\Gamma_\beta} [\delta_{ij} \frac{1}{|\vec{X}_\alpha - \vec{X}_0|^2} - \frac{2(x_{i\alpha} - x_{i0})(x_{j\alpha} - x_{j0})}{|\vec{X}_\alpha - \vec{X}_0|^4}] d\Gamma_0 \tag{24}$$

$$R_{\alpha j\beta} = \int_{\Gamma_\beta} q^j(\vec{X}_\alpha - \vec{X}_0) d\Gamma_0 = -\frac{1}{2\pi} \int_{\Gamma_\beta} \frac{x_{j\alpha} - x_{j0}}{|\vec{X}_\alpha - \vec{X}_0|^2} d\Gamma_0 \tag{25}$$

In the course of the numerical computation, the surface-stress force on the boundary Γ can be solved from the matrix equation (21) by applying LU method decomposition. Then taking $c = 1$ and using the boundary values, the velocity field and the pressure field in the flow domain can be determined from $v^k = U^k - F^k$, for $\vec{X} \in \Omega$ and equation (23).

3.2 Raising computational accuracy

In order to raise computational accuracy, we transform the integral equations into the formulas of analytic solutions. When $i = j$, there are singular integrals which have been treated with various ways. In this paper, using coordinate transformation method we calculated the singular integrals and obtained the integral formulas of the analytic solutions. When $i \neq j$, the integral equations can be computed approximately with Gauss' integral method.

4 NUMERICAL RESULTS

The present numerical computational method is applied to calculate the problems of the fluid flow passing a v-shape notch with $\theta = 28.5°$ and $h = 5$ cm. The numerical solutions are then compared with Taneda's (1979) experimental results.

4.1 Distributions of the surface–stress force

In the course of numerical simulation, the surface–stress forces on the V-shaped notch and on the cylinder are calculated accurately first. Then the velocity fields are simulated numerically.

4.2 Velocity fields

To examine the effectiveness of the present numerical method, simulation of velocity fields of a flow passing a V-shaped notch is generated and compared with Taneda's (1979) experimental results.

The numerical simulation of velocity field in Figure 1 shows that Moffatt vortices can be formed by rolling a cylinder over the V-shaped notch. Figure 2 shows the streamlines of the flow. We can see that the largest vortex produced by the rotating cylinder induces a smaller one next to it. The direction of rotation of the small vortex is opposite to that of the large vortex, and the distance between the center of the small vortex and the corner point is 0.5 times the distance between the center of the large vortex and the corner point. We also found that the distance between the dividing streamline of two vortices and the corner point is 0.5 times the distance between the corner point and the dividing streamline, which is between the large vortex and the cylinder. By comparison, the numerical simulation of the velocity fields is in good agreement with Taneda's (1979) experiment. The numerical results indicate that the fluid flow in the wedge-shaped region consists of an infinite sequence of vortices of rapidly diminishing strength. We also found that

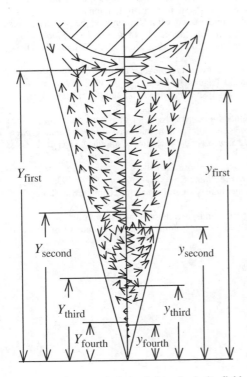

Figure 1. The numerical simulation of velocity field.

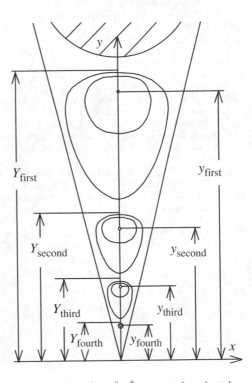

Figure 2. Streamlines for flow past v-shaped notch.

$y_{n+1} = y_n/2$, $Y_{n+1} = Y_n/2$, where y_n is the center coordinate of the vortex, Y_n is the coordinate of dividing streamline between two vortices, and for a V-shaped notch with $\theta = 28.5°$ the maximum n is 4 (i.e. $n_{max} = 4$) because the strength of the fourth vortex is 10^{-4}. In particular, the method will be researched and used to solve numerically the problems of the fluid flow near a V-shaped notch with variable θ and h, and that can lead to the identification of the condition for appearance of Moffatt vortices.

5 CONCLUSIONS

A boundary element method is described in the present study. Numerical investigations of it show that the method is very effective in researching and solving V-shaped notch with variable θ and h, so that the method can be used to find the condition for the appearance of Moffatt vortices. It is also important to note that the present method not only is more suitable for application in problems with arbitrary boundary shape but it can also be treated numerically for problems of flow passing multiple bodies. In addition, the method has major advantages in computational convenience and superior accuracy hence leading to a significant decrease in numerical computational work.

ACKNOWLEDGMENTS

This work was supported in part by the Nanjing Institute of Technology Foundation for Natural Sciences and the Inner Mongolia Foundation for Natural Sciences.

REFERENCES

Happel, J. & Bernner, H. 1983. *Low Reynolds Number Hydrodynamics*, The Hague: Nijhoff Publishers.
Lin Changsheng & Han Qingshu. 1991. Boundary element analysis of low-Reynolds-number viscous fluid flow, In Y.K.Cheung & J.H.W Lee (eds), *Computational Mechanics*, 1481–1486. Rotterdam: Balkema.
Lin Changsheng. 1992. Numerical simulation of low-Reynolds-number viscous fluid flow around the cylindrical arcs, In G. Cheng & J. Lin (eds), *EPMESCIV; International conference on education, practice and promotion of computational methods in engineer using small computer*, Dalian, 30 July to 2 August 1992, Dalian: The Dalian University of Technology Press.
Lin Changsheng & Yang Guizhi. 1994. Numerical analysis of the flow past rectangular cavity, In Du Qingshua (ed), *Application in Mechanics and Engineering*, Taiyuan: Shangxi Technology Press.
Lin Changsheng. 1994. Numerical analysis of Stokes flow passing over a fence on a flat plate with BEM, *Shanghai Journal of Mechanics*, 15(1): 67–73.
Lin Changsheng & Yang Guizhi. 2002. A boundary element method for numerical solution of viscous fluid flow past two cylinders, In Yao Zhenhan & Aliabadi M.H (eds), *Boundary Element Techniques, Proceedings of the third international conference on boundary element techniques, 10–12 September 2002*. Beijing: Tsinghua University Press & Springer.
Reacher, P.J. 1976. *Computational Fluid Dynamics*, New Mexico: Hermosa Publishers.
Taneda S. 1979. Visualization of separating Stokes flows, *J. of the Physical Society of Japan*, 46(6): 1935–1942.
Weinbaum, S. & Ganatos, P. 1990. Numerical multipole and boundary integral equation techniques in Stokes flow, *Annual Review of Fluid Mechanics*, 22: 275–316.
Wolfang, K. 1980. *Computational Fluid Dynamics*, Washington: Hemisphere Publishing Corporation.
Wu, Jianghang. 1984. A Green's Function Method for the Numerical Solution of Steady-State Incompressible Viscous Fluid Flow, *Acta Mechanica Sinica*, 16(5): 425–433.

Computational Methods in Engineering and Science, Iu et al. (eds)
© *2003 Swets & Zeitlinger, Lisse, ISBN 90 5809 567 3*

Numerical study for multi-solitary-wave interaction of regularized long-wave equation

C. Lin & G. Yang
Nanjing Institute of Technology, Nanjing, China
Inner Mongolia University for Nationalities, Tongliao, China

ABSTRACT: The numerical scheme of the cubic spline difference approximations for the regularized long-wave (RLW) equation, $u_t + u_x + uu_x + u_{xxt} = 0$, is described with second-order and fourth-order accuracies for the time and spacing variables respectively. The interactions of head-on and catch up collisions of four solitary wave solutions of the RLW equation are examined numerically. The numerical studies show that the interaction collision of the multi-solitons exhibit true solitons behavior, being stable on collision with Korteweg-de Vries (KdV) solitary waves and the oscillatory tail might be by numerical error. The oscillatory tail is appearance but that is within the numerical error of <2.3‰. Since this error is less than the error of the single solitary waves test program, the RLW solitary waves are considered to behave as true solitons to within the numerical error of the calculation.

1 INTRODUCTION

Nonlinear physics has always provided a fertile area for interaction between experimental results, numerical studies and analytic work. The discovery of the elastic effect of a collision of solitary waves followed from the numerical and analytic study of solitary wave solitons of the appropriate equations. For example, the Korteweg-de Vries (KdV)(1895) equation

$$u_t + u_x + u_{xx} + u_{xxx} = 0 \tag{1}$$

was originally derived as a model for the unidirectional propagation of nonlinear waves on the water surface. It has attracted a great deal of research interest because it describes in fact a wide class of nonlinear wave processes taking into account of weak-dispersion and it also possess some unusual mathematical properties. In this context, $u = u(x, t)$ represents the wave amplitude and x and t are the space and time coordinates, respectively, with x pointing in the direction of wave propagation. The equation admits a family of solitary wave solutions having the property that the result of the nonlinear interaction of a pair of unequal solitary waves leaves the waves unaltered, except for a phase shift. This so-called solitons property was first observed in numerical studies made by Zabusky & Kruskal (1965) and the proof of it was one of the triumphs of the inverse-scatterical method for solving partial-differential equations (Miura 1976). For water waves, an alternative model was proposed by Peregrine (1964) and by Benjamin et al. (1972), it is the so-called regularized long-wave (RLW) equation

$$u_t + u_x + uu_x - u_{xxt} = 0 \tag{2}$$

Equation (2) has solitary-wave solutions, similar to those for (1), of the form

$$u(x,t) = 3c \, \mathrm{sech}^2 (k\,x - \omega\,t + \delta) \tag{3}$$

where

$$k = \sqrt{c/(1+c)}/2 \quad , \quad \omega = \sqrt{c(1+c)}/2 \qquad (4)$$

and c and δ are arbitrary constants. It is an exact solution of the regularized long-wave equation. This exact analytic solution is used as a test solution for the numerical schemes. Eilbeck & McGuire (1975, 1977) derived a three level finite difference scheme which is stable for all practical values of k, and in fact works best if $k = h$ (Eilbeck & McGuire 1975, 1977). This means that a considerably greater time step can be taken in the numerical study of equation (2) as compared to the KdV equation (1). The scheme can be made conservative by a space-averaged nonlinear term as in the following leapfrog finite scheme

$$\omega_{i-1}^{m+1} - (2+h^2)\omega_i^{m+1} + \omega_{i+1}^{m+1}$$
$$= \omega_{i-1}^{m-1} - (2+h^2)\omega_i^{m-1} + \omega_{i+1}^{m-1} + \tau h(1+\omega_i^m)(\omega_{i+1}^m + \omega_{i-1}^m) \qquad (5)$$

The numerical study by Eilbeck & McGuire (1975, 1977) showed that after a collision of two-solitary waves the pulses reappeared with amplitudes within 3% of their original amplitude. On the basis of this evidence, the waves exhibit true soliton behavior of solutions of the RLW equation. But Abdulloev, Bogolubsky & Makhankov (1976) and Bona (1980) pointed out that when very large solitary waves (~ 10) collide in the RLW equation, a very small oscillating tail ($\sim 10^{-3}$) appeared. They believed that this higher order oscillating tail could not be adequately explained as numerical error. Unfortunately these studies were not accurate enough to pick up the elastic radiation occurring after the two-solitary wave collision in the model. We have performed studies numerically with the cubic spline difference method for two-solitary wave collision (Lin & Yang 2001) and three solitary wave collisions (Lin & Yang 2002). In this paper, a more striking demonstration of these effects has been achieved. Numerical scheme of the cubic spline difference approximations with the second-order and fourth-order accuracies is developed to study the interaction of head-on collision and catch up collision of multi-solitary wave solutions of the RLW equation with the goals of quantifying the elastic radiation and the small oscillatory tail predicted by lower order schemes so that the cause of the oscillatory tail could be identified.

This paper is set out as follows: In section 2 we give a summary of the numerical scheme, and discuss the numerical accuracy of the method. Details of the four-soliton interaction are presented in section 3 with four sample cases. Finally, in section 4 we summarize the results and report on the progress of investigations in the nature of the analytic form of the multi-soliton solution of the regularized long wave equation.

2 NUMERICAL SCHEME

The conservation form of the regularized long-wave equation (2) is

$$u_t + (u + u^2/2)_x = \delta u_{xxt} \qquad (6)$$

By putting

$$f = u + u^2/2 \quad , \qquad (7)$$

the equation (6) becomes form as following

$$u_t + f_x - \delta u_{xxt} = 0 \qquad (8)$$

The following cubic spline difference approximations are used for the first and second derivatives

$$\left.\frac{\partial u}{\partial x}\right|_i = D_i \quad , \quad \left.\frac{\partial^2 u}{\partial x^2}\right|_i = S_i \qquad (9)$$

According to the theory of cubic spline difference approximations, they can be implemented conveniently into a numerical scheme as follows

$$D_{i+1} + 4D_i + D_{i-1} = \frac{3}{h}(F_{i+1} - F_{i-1}) \tag{10}$$

$$S_{i+1} + 4S_i + S_{i-1} = \frac{6}{h^2}(F_{i+1} - 2F_i + F_{i-1}) \tag{11}$$

where F_i is a discrete function. According to the cubic spline difference method, scheme (9) is of second-order accuracy and scheme (11) is of fourth-order accuracy.

Equation (8) is discretized in time t by using equation (10) for $h = \Delta t = \tau$

$$\frac{3}{\tau}(u^{n+1} - u^{n-1}) + (f_x^{n+1} + 4f_x^n - f_x^{n-1}) - \frac{3\delta}{\tau}(u_{xx}^{n+1} - u_{xx}^{n-1}) = 0 \tag{12}$$

The above equation is discretized in space x by using the (10) and (11) for $\Delta x = h$, and

$$\because \quad f_x^{n+1} = \frac{3}{h}(f_{i+1}^{n+1} - f_{i-1}^{n+1}) \tag{13}$$

$$f_x^n = \frac{3}{h}(f_{i+1}^n - f_{i-1}^n) \tag{14}$$

$$f_x^{n-1} = \frac{3}{h}(f_{i+1}^{n-1} - f_{i-1}^{n-1}) \tag{15}$$

$$u_{xx}^{n+1} = \frac{6}{h^2}(u_{i+1}^{n+1} - 2u_i^{n+1} + u_{i-1}^{n+1}) \tag{16}$$

$$u_{xx}^{n-1} = \frac{6}{h^2}(u_{i+1}^{n-1} - 2u_i^{n-1} + u_{i-1}^{n-1}) \tag{17}$$

$$\therefore \quad \begin{aligned} &\frac{3}{\tau}[(u_{i+1}^{n+1} + 4u_i^{n+1} + u_{i-1}^{n+1}) - (u_{i+1}^{n-1} + 4u_i^{n-1} + u_{i-1}^{n-1})] \\ &+ \frac{3}{h}[(f_{i+1}^{n+1} - f_{i-1}^{n+1}) + 4(f_{i+1}^n - f_{i-1}^n) - (f_{i+1}^{n-1} - f_{i-1}^{n-1})] \\ &- \frac{18\delta}{\tau h^2}[(u_{i+1}^{n+1} - 2u_i^{n+1} + u_{i-1}^{n+1}) - (u_{i+1}^{n-1} - 2u_i^{n-1} + u_{i-1}^{n-1})] = 0 \end{aligned} \tag{18}$$

Note that

$$f_{i+1}^{n+1} = u_{i+1}^{n+1}(1 + u_{i+1}^{n+1}/2) \tag{19}$$

$$f_{i-1}^{n+1} = u_{i-1}^{n+1}(1 + u_{i-1}^{n+1}/2) \tag{20}$$

$$f_{i+1}^n = u_{i+1}^n (1 + u_{i+1}^n / 2) \tag{21}$$

$$f_{i-1}^n = u_{i-1}^n (1 + u_{i-1}^n / 2) \tag{22}$$

$$u_{i+1}^{n-1} = u_{i+1}^{n-1} (1 + u_{i+1}^{n-1} / 2) \tag{23}$$

$$u_{i-1}^{n-1} = u_{i-1}^{n-1} (1 + u_{i-1}^{n-1} / 2) \tag{24}$$

Then scheme (18) may be written as the following

$$\omega_{i-1}^{n+1} u_{i-1}^{n+1} + \omega_i^{n+1} u_i^{n+1} + \omega_{i+1}^{n+1} u_{i+1}^{n+1}$$
$$= \omega_{i-1}^n u_{i-1}^n + \omega_{i+1}^n u_{i+1}^n + \omega_{i-1}^{n-1} u_{i-1}^{n-1} + \omega_i^{n-1} u_i^{n-1} + \omega_{i+1}^{n-1} u_{i+1}^{n-1} \tag{25}$$

where

$$\omega_{i-1}^{n+1} = 1 + \frac{6\delta}{h^2} - \frac{\tau}{h} (1 + u_{i-1}^{n+1} / 2) \tag{26}$$

$$\omega_i^{n+1} = 4 + \frac{12\delta}{h^2} , \tag{27}$$

$$\omega_{i+1}^{n+1} = 1 - \frac{6\delta}{h^2} + \frac{\tau}{h} (1 + u_{i+1}^{n+1} / 2) \tag{28}$$

$$\omega_{i+1}^n = -\frac{4\tau}{h} (1 + u_{i+1}^n / 2) \tag{29}$$

$$\omega_{i-1}^n = -(1 + u_{i-1}^n / 2) \tag{30}$$

$$\omega_{i+1}^{n-1} = 1 - \frac{6\delta}{h^2} + \frac{\tau}{h} (1 + u_{i+1}^{n-1} / 2) \tag{31}$$

$$\omega_i^{n-1} = 4 + \frac{12\delta}{h^2} \tag{32}$$

$$\omega_{i-1}^{n-1} = 1 - \frac{6\delta}{h^2} - \frac{\tau}{h} (1 + u_{i-1}^{n-1} / 2) \tag{33}$$

3 HEAD-ON COLLISIONS OF FOUR SOLITONS

Using the cubic spline difference scheme described in section 2, the exact analytic solution of the RLW was used as a test solution for the numerical schemes, with the initial conditions given by

$$u_0(x) = \sum_{i=1}^4 3(c_i - 1) \sec h^2 [\sqrt{\frac{c_i - 1}{4\delta c_i}} (x - d_{i0})] \tag{34}$$

742

where, δ, c_i and d_{i0} are arbitrary constants. In testing the reemergence of the RLW solitary wave after a head-on collision with another solitary wave, we are interested in the accuracy of the numerical values of wave amplitudes given by the program. Tests (Lin & Yang 2001) show that the order of magnitude of the numerical errors will be at the very most 2.3‰. The catch-up and head-on collision of four solitary wave solutions of the RLW equation have been examined numerically for four different cases. In the first case, the catch-up collision of four solitary wave solutions (noted as 1, 2, 3, 4) happens at the same time. In the second case, the evolution of collision is that 1 catches up with 2 and then with 3 and then with 4. The actions follows these are 2 catching up with 3 and then with 4. Finally, it is 3 catching up with 4. In the third case, there are head-on and catch up collisions. It is 4 and 3 have head-on collision first, then 4 and 2 have head-on collision followed by 4 and 1 colliding each other head-on. Then the catch-up collisions happen between 2 and 3, 1 and 3, 1 and 2. In the fourth case, the interactions are 3 head on 2, 3 head on 1, 4 head on 2, 4 head on 1, and finally soliton 1 catches up with soliton 2, and soliton 4 catches up with soliton 3.

3.1 Catch-up collision at the same time

In this case the arbitrary constants c_i and d_{i0} are given as $c_1 = 8.0$, $c_2 = 6.0$, $c_3 = 4.0$, $c_4 = 2.0$, $d_{10} = 4.0$, $d_{20} = 12.0$, $d_{30} = 20.0$, $d_{40} = 28.0$, so the solitary wave amplitudes are $A_1 = 21.0$, $A_2 = 15.0$, $A_3 = 9.0$, $A_4 = 3.0$ and δ is 0.1. A graphical expression of the four solitons catch-up collision is plotted for difference time steps in Figure 1. Please note that the catch-up collision of the four solitary wave solutions (as 1, 2, 3, 4) happens at the same time.

3.2 Catch up collision one by one

A graph of RLW four-soliton catch-up collision one by one with $\delta = 0.1$ is shown in Figure 2. In this case, the arbitrary constants c_i and d_{i0} are given as $c_1 = 10.0$, $c_2 = 6.0$, $c_3 = 4.0$, $c_4 = 2.0$, $d_{10} = 4.0$, $d_{20} = 8.0$, $d_{30} = 16.0$, $d_{40} = 28.0$, so the solitary wave amplitudes are $A_1 = 27.0$, $A_2 = 15.0$, $A_3 = 9.0$, and $A_4 = 3.0$. The evolution of the catch-up collisions is plotted for difference time steps in Figure 2. In the course of the interaction, catch up collision of soliton 1 and soliton 2 is the first to happen, then soliton 1 catches up and collide with soliton 3 and after that is with soliton 4. The actions follows these are 2 catching up with 3 and then with 4. Finally, it is 3 catching up with 4.

3.3 Head-on and catch up collision one by one

Letting the arbitrary constants c_i and d_{i0} be $c_1 = 8.0$, $c_2 = 6.0$, $c_3 = 4.0$, $c_4 = -2.0$, $d_{10} = 4.0$, $d_{20} = 12.0$, $d_{30} = 20.0$, $d_{40} = 26.0$, the solitary wave amplitudes are $A_1 = 21.0$, $A_2 = 15.0$, $A_3 = 9.0$, $A_4 = 3.0$, and $\delta = 0.1$.

The evolution of head-on and catch up collision of solitons for this case is plotted for difference time steps in Figure 3. In the course of the interaction, it is 4 and 3 have head-on collision first, then 4 and 2

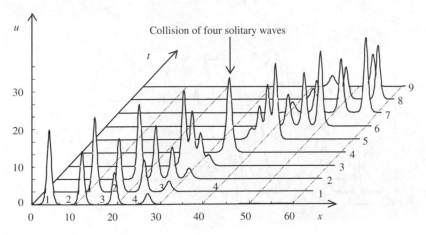

Figure 1.   Evolution of four solitons catches up one by one for the first case.

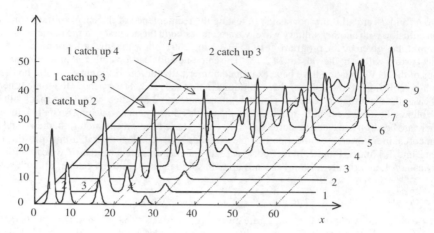

Figure 2. Evolution of collision four solitons catches up one by one for the second case.

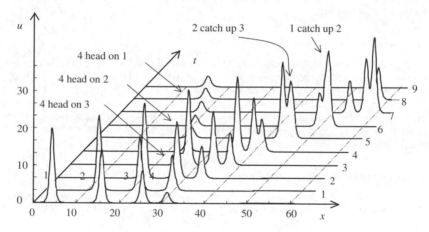

Figure 3. Evolution of four solitons head-on and catch up collision for the third case.

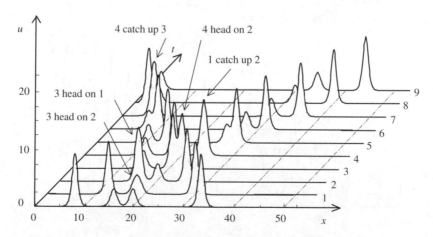

Figure 4. Evolution of four solitons head-on and catch up collision for the fourth case.

have head-on collision followed by 4 and 1 colliding each other head-on. Then catch-up collisions happen between 2 and 3, 1 and 3, 1 and 2.

3.4 Head-on and catch up collision

Letting the arbitrary constants c_i and d_{i0} be $c_1 = 4.0, c_2 = 2.0, c_3 = -2.0, c_4 = -4.0, d_{10} = 8.0, d_{20} = 16.0,$ $d_{30} = 20.0, d_{40} = 34.0$, the solitary wave amplitudes are $A_1 = 9.0, A_2 = 3.0, A_3 = 3.0, A_4 = 9.0$, and $\delta = 0.1$.

Figure 4 shows the collisions happen at difference time steps. There are head-on collisions of 3 and 2, 3 and 1, 4 and 2, 4 and 1 first. Then catch-up collisions between 1 and 2, 4 and 3 follow.

The solution was then allowed to develop in time. The amplitudes and phases of each solitary wave were calculated periodically by interaction from the grid point values. It was found that the solitary waves passed through a strongly nonlinear interaction region and reappeared with their original amplitudes. The oscillatory tail is appearance but that is within the numerical error of $<2.3‰$ (Lin & Yang 2001, 2002). Since this error is less than the error for the single solitary waves test program, the RLW solitary waves were considered to behave as true solitons to within the numerical error of the calculation.

4 CONCLUSIONS

Our purpose here is to present a numerical method describing a particular wave interaction and to show the details evolution of the collisions. The results of our own numerical computations broadly speaking, confirm those results of Eilbeck & McGuire (1975, 1977). That is the catch up and head-on collisions of four-soliton waves are elastic. The RLW solitary waves were considered to behave as true solitons to within the numerical error of the calculation.

ACKNOWLEDGMENTS

This work was supported in part by the Nanjing Institute of Technology Foundation for Natural Sciences and the Inner Mongolia Foundation for Natural Sciences.

REFERENCES

Abdulloev, Kh.O. Bogolubsky, I.L. & Makhankov, V. G. 1976. One more example of inelastic soliton interaction; *Phys. Lett.*, 56: 427–428.

Benjamin, T.B. Bona, J.L. & Mahony, J.J. 1972. Model equations for long waves in nonlinear depressives systems; *Phil. Trans. Roy. Soc.*, 272: 47–78.

Bona, J.L. Pritchard, W.G. & Scott, L.R. 1980. Solitary-wave interaction; *Phys. Fluids*, 23, 438–441.

Eilbeck, J.C. & McGuire, G.R. 1975. Numerical study of the regularized long-wave equation I: Numerical methods; *J. Comput. Phys.*, 19: 43–57.

Eilbeck, J.C. & McGuire, G.R. 1977. Numerical study of the regularized long-wave equation II: Interaction of solitary waves; *J. Comput. Phys.*, 23: 63–73.

Korteweg, D.J. & de Vries, G. 1895. On the change of form of long waves advancing in a rectangular canal, and on a new type of long stationary waves; *Phil. Mag.*, 39: 422–443.

Lin Changsheng & Yang Guizhi. 2001. On numerical calculation of two solitary waves solutions RLW equation; *Journal of Inner Mongolia University for Nationalities*, 16(4): 356–360.

Lin Changsheng & Yang Guizhi. 2002. Head-on collision of three solitary wave solutions of regularized long-wave equation; In Lin S. Mao R. Shen H. Sun G. and Sun Y. (eds.). *EPMESC VIII; The 8th International Conference on Enhancement and Promotion of Computational Methods in Engineering Science*, Shanghai, 15–25 July 2001. Shanghai: Shanghai Sanlian Publishers.

Miura, R.M. 1976. The Korteweg-de Vries equation: a survey of results, *SIAM Rev.*, 18: 412–459.

Peregrine, D.H. 1966. Calculation of the development of an undular bore, *J. Fluid Mech.*, 25: 321–330.

Zabusky, N.J. & Kruskal, M.D. 1965. Interaction of solitons in a collision less plasma and the recurrence of initial states, *Phys. Rev. Lett.*, 15: 240–243.

Computational Methods in Engineering and Science, Iu et al. (eds)
© *2003 Swets & Zeitlinger, Lisse, ISBN 90 5809 567 3*

Numerical–experimental interaction in hydrodynamics: an integrated approach for the optimal management of hydraulic structures and hydrographic basins

M. Pirotton, A. Lejeune, P. Archambeau & S. Erpicum
Laboratories of Fluid Mechanics, Applied Hydrodynamics and Hydraulic Constructions,
University of Liege, Belgium

B. Dewals
Laboratories of Fluid Mechanics, Applied Hydrodynamics and Hydraulic Constructions,
University of Liege, Belgium
Research Fellow of the National Fund for Scientific Research, Belgium

ABSTRACT: The present paper presents studies and tools devoted to the reduction of uncertainties in the scientific knowledge of various unsteady flows, including solid transport effects, in the perspective of applications such as restoration, rehabilitation and enhancement of riverbeds, rivers and waterways. This global approach will be tested and fitted in a dynamic and integrated experimental model of reference flows in the laboratory. The relevance of using of physical experimental scale models will be highlighted for the identification of constitutive laws used in numerical modelling.

Practical applications prove that the integration of powerful hydrodynamic software for water management is realistic and will lead to compute transient flooding, sedimentation and erosion events in natural compound channels, as well as to design and simulate the behaviour of hydraulic regulation works or locks. It will also help in working out new projects or reconstructions of hydraulic infrastructures, as well as ecological rehabilitations. Even if the sole quantitative plan will be taken into account, this optimisation project will also lead to highlight favourable solutions, from a qualitative point of view.

1 INTRODUCTION

Practitioners in hydro-engineering are highly interested in the development of powerful tools for the prediction of a large scope of hydraulic behaviours. Scale models have been considered for a long time as an efficient approach for the design of hydraulic structures. Provided the right similitude law is used, they give access to reliable and uncontested results. Such experimental tests on scale models remain attractive for highly complex situations while hydrodynamic models fail to predict with confidence complex effects such as turbulence.

However, physical modeling suffers from several intrinsic defaults: high cost for building and measurement equipments, low geometrical flexibility once the model has been constructed, etc. In order to circumvent these problems, physical experimentations are more and more advantageously completed by numerical simulations. In this way, the quasi-three-dimensional flow solver WOLF is currently used at the University of Liege in interaction with physical models.

This approach leads to a suitable complementary work both at the stages of designing the scale model and during its exploitation. Typical applications include various aspects of hydrographic basins sustainable management and hydraulic structures design in the scope of the global change. The aim of this paper is to illustrate the benefits gained with the combination of both physical and numerical modeling.

2 FREE-SURFACE FLOW SOFTWARE WOLF 2D

WOLF 2D is an efficient analysis and optimization tool, which has been completely developed for several years in the Service of Applied Hydrodynamics and Hydraulic Constructions (HACH) at the University of Liege (http://www.ulg.ac.be/hach). WOLF 2D is part of WOLF free surface flows computation package, which includes in the same development environment the resolutions of the 1D and 2D depth-integrated Navier-Stokes equations as well as a physically based hydrological model, along with powerful optimization capabilities based on Genetic Algorithms (WOLF AG). See Figure 1 for the general organization of WOLF computation units. Each code handles structured or unstructured grids, dealing with natural topography and mobile bed simultaneously, for any unsteady situation with mixed regimes (including moving hydraulic jumps).

WOLF 2D software solves the 2D shallow-water equations on multi-block grids, dealing with natural topography and mobile bed. WOLF 2D is part of WOLF free surface flows computation package, which has been completely developed at the University of Liege. A mass balance for bed load sediments is coupled to the hydrodynamic model (see Dewals et al. 2002a or Dewals et al. 2002b). Side slope stability analysis are systematically performed to take gravity induced solid discharges into account.

The same finite volume technique in each computer code solves the equations, formulated in a conservative form to ensure exact mass and momentum balance, even across moving hydraulic jumps. An original splitting of the convective terms has been specifically developed for the model, in order to handle properly transient discontinuities. The computation core has reached now a high degree of reliability. Its stability, robustness and accuracy have been widely highlighted. Indeed the validation of the model has been performed continuously by comparisons with analytical solutions, field and laboratory measurements available in the literature or collected in the Hydraulic Laboratories in Liege.

2.1 *Mathematical model*

The governing equations for hydrodynamic free surface flows are the depth-integrated Navier-Stokes equations for an incompressible fluid:

$$\frac{\partial h}{\partial t} + \frac{\partial hu_j}{\partial x_j} = 0 \tag{1}$$

$$\frac{\partial u_i}{\partial t} + u_i \frac{\partial u_j}{\partial x_j} + g \frac{\partial H}{\partial x_i} \cos\theta_i + g\, n^2 \frac{\sqrt{u_j^2}}{h^{4/3}} u_i = g \sin\theta_i + \frac{\partial}{\partial x_j}\left(v_t \frac{\partial u_i}{\partial x_j}\right) + G_i \tag{2}$$

where the Einstein notation has been used (sum over repeated subscripts). This system expresses the mass balance and the momentum balance along both space directions. The following symbols have

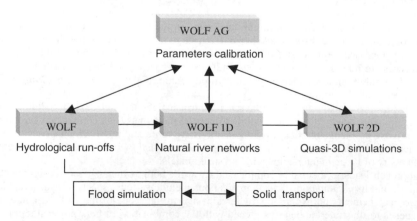

Figure 1.   General organisation of WOLF computation units.

been used: g (gravity acceleration), h (water height), n (Manning roughness coefficient), u_i (velocity components), t (time), x_i (space coordinates), θ_i (bed slope), G_i (external forces), ρ (water density) and ν_t (turbulent viscosity).

Equation (1) expresses the mass conservation, and equation (2) the momentum conservation along the x- and y-axis.

The total kinematic viscosity is given by the following expression:

$$\nu_t = \nu + \gamma \sqrt{2\left(\frac{\partial u_j}{\partial x_j}\right)^2 + \left(\frac{\partial u_1}{\partial x_2} + \frac{\partial u_2}{\partial x_1}\right)^2} \ . \tag{3}$$

It must be outlined that no restrictive assumption is required for the bottom slope (see Pirotton 1997 or Mouzelard 2002). In order to simulate flows on very steep topographies (such as spillways or torrents for example), local references are defined with the x- and y-axis following locally the mean bottom slope (see Figure 2). The effect of the local bed curvature is taken into account thanks to a curvilinear system of coordinates. This technique ensures the water depth to be orthogonal to the local main flow direction.

According to the well-known Exner equation, the mass balance for sediments can be stated as

$$(1-p)\frac{\partial z_b}{\partial t} + \frac{\partial q_{bx}}{\partial x} + \frac{\partial q_{by}}{\partial y} = 0 \tag{4}$$

where z_b stands for the bed level, p is the porosity and q_{bx}, q_{by} represent the solid discharges in both horizontal directions. The bed load discharge is evaluated for instance using the Meyer-Peter and Müller formula, known to be widely reliable:

$$q_b = 8\sqrt{(s-1)g\,d^3}\left[\xi\frac{R_h\,J}{(s-1)d} - 0.047\right]^{\frac{3}{2}} \tag{5}$$

s represents the relative density of the sediment particles, d is the mean grain diameter, J stands for the energy slope and R_h for the hydraulic radius. Several other solid discharge laws are available within the computation program and the user is free to choose any of them.

2.2 Space discretisation

The finite volume method is used to achieve the space discretization. This approach is recognized to be especially adequate for highly advective flow conditions because it allows an easy implementation of upwind schemes. Moreover the numerical treatment by finite volume ensures exact mass and momentum

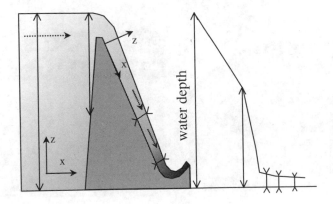

Figure 2. Axis inclination.

conservation even across moving discontinuities. These two properties are of first interest for computing dam-break induced flows.

A remaining challenge lies in the design of a stable and efficient flux evaluation scheme. The stability requirements imply that whatever the flow regime no numerical oscillation may be generated, especially in the vicinity of discontinuous solutions (e.g. hydraulic jumps or sediment bores). Roe's approximate Riemann solver as well as an original Flux Vector Splitting technique are implemented in WOLF's numerical model. Both methods showed their ability to simulate sharp transitions without excessive smearing. Our original scheme is based on an upstream evaluation of purely advective terms (Dewals et al. 2002a, Mouzelard 2002), hence it takes into account the basic physical meaning of these transport terms in the mathematical model. The scheme, first developed for pure hydrodynamics, has been extended to the fully coupled system including the Exner equation. One key advantage is that the flux evaluation in the coupled model remains in accordance with the numerical scheme developed so far for the clear-water hydrodynamics. The numerical scheme has also proven its ability to reproduce automatically the phenomena of dunes and antidunes propagation observed in the nature (depending on the flow regime). The scheme is Froude independent, so that the critical transition doesn't call for any particular treatment.

2.3 Time discretisation

As we are mostly interested in transient flows, an accurate and non-dissipative temporal scheme has to be chosen. The explicit Runge-Kutta schemes are used in WOLF both for their relatively low computation cost in the case of transient phenomena and for the high order of precision they allow to reach.

Implicit time integration is now available within WOLF software. This technique constitutes a brilliant strategy for computing highly accurate steady state solutions as it dramatically cuts the computation time for such a case. However due to the large algebraic systems to be solved at each time step implicit methods remain inappropriate for very unsteady flow conditions, where small time steps have to be used to assure a sufficient time accuracy and to track reliably each wave movement.

3 THE NAM THEUN LARGE DAM SCALE MODEL

The Nam Theun II dam is a part of a set of three dams located on the Nam Theun river in Laos. A gated spillway is integrated in the main structure, and is followed downstream by a stilling basin. The Laboratories of the University of Liege have undertaken the study of the hydraulic scale model of the spillway and the stilling basin of the Nam Theun II dam.

The topography of the upstream basin shows that the dam is located near an important meander, which should *a priori* be included in the geometry of the scale model to ensure a correct stream orientation at the approach of the spillway.

However, the need of high precision measurements on the spillway induced the physical model to be scaled at 1 to 60. Therefore, the geometry of the scale model was restricted to the last part of the

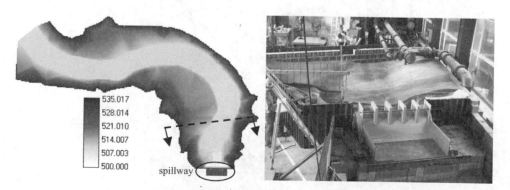

Figure 3. Topography (m) of the upstream basin and general view of the scale model.

upstream basin (see the dotted line on Figure 3), the spillway with its stilling basin, and the first part of the downstream river (see Figure 3).

In order to respect the real upstream flow orientation and repartition, some numerical simulations were decided to be performed to delineate the upstream scale model boundaries.

3.1 Upstream part

The building of a scale model including a large upstream area was impossible due to the scale used and the space limitation. A first series of numerical simulations covering the whole reservoir flow (Figure 3) of the dam was performed in order to compute the flow field near the spillway without influence of the upstream boundary conditions.

The maximum discharge was numerically introduced by an upstream filling basin. No downstream boundary condition was needed (supercritical flow), thanks to the use of properly inclined axes, as seen in paragraph 2.1.

The results computed show that the main stream leaves the river path, and flows in the floodplain. This causes a highly asymmetric flow near the spillway. This latter is indeed found to be more effective through its right side (considered in the stream direction), rather than through the left one as it could be expected by the sole topography observation (see Figure 3).

3.2 Geometry of the scale model

A second set of simulations was performed based on the geometry selected for the scale model. The results strongly differ from those obtained on the global geometry, and confirm the poor analogy of the discharge distribution at the upstream part of the scaled geometry, see Figure 4. It is obvious that the main flow follows the river bed path, in absence of the upstream bend.

To improve the behavior of this reduced geometry in comparison with the whole reservoir and to ensure correct upstream flow conditions for the scale model, several modifications were decided for the upstream part of the reservoir.

The first modification concerns the topography. In concordance with the scale model, the computation simulates an upstream filling basin. This means that there is no upstream discharge or speed boundary condition, but only a solid wall. The discharge is introduced in the geometry by a bottom negative infiltration process, which avoids any arbitrarily imposition of velocity directions, and ensures a perfect fitting with the scale model conditions.

As the filling procedure depends on the topography, it was decided to modify this latter in order to get the same discharge distribution as in the global, previous simulation. The river bed of the filling basin was thus fundamentally modified (Figure 4).

The second modification concerns the adjunction of deflectors to the upstream basin. The deflectors are shown on the Figure 5. Several simulations were performed to determine their number and to estimate their optimal length and orientation in the scope of minimizing the difference with the large scale simulation. The bed modification ensures a good upstream discharge distribution, while the deflectors forces the flow into the right orientation.

The different measurements on the scale model enable a comparison between physical and numerical simulations. Figure 5 proves the good agreement obtained in the estimation of the height/discharge curve related to the spillway.

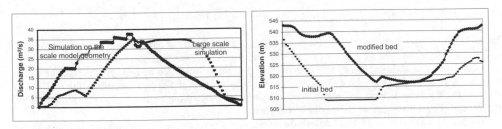

Figure 4. Discharge repartition along the transversal section corresponding at the upstream limit of the scale model geometry (left) and bed elevation of the scale model stilling basin (right).

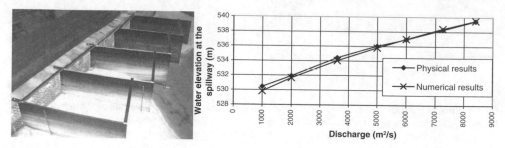

Figure 5.   Deflectors of the scale model (left) and discharge rating curve of the spillway (right).

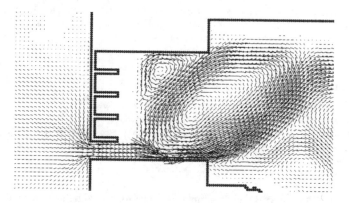

Figure 6.   Velocity field in the case of a single gate opened.

3.3   *Stilling basin and river restitution*

The dam spillway is completed by five large gates to regulate accurately the discharge and the upstream level. However, the operations with asymmetric gate opening focused the attention because of the erosion risk in the downstream natural river. A complete numerical and physical hydrodynamic study of the stilling basin design was thus performed to induce and control dissipation for a wide range of discharges and gate opening configurations.

Conclusions obtained from physical tests and numerical simulations showed that some opening configurations should be avoided to keep erosion process at an acceptable level. For example, Figure 6 shows the velocity field in the situation of a low discharge flowing through the right gate only. Such situation produces an extreme erosion downstream of the stilling basin and must be rejected.

4   THE KOL DAM SERIES OF NUMERICAL AND SCALE MODELS

The Kol Dam project includes the building of a very large embankment dam, with a crest 500 m long at the level 648 m, as well as a complex of desilting structures. The exploitation level will vary between 642 and 636 m. The spillway is made up of 6 bays 17.1 m wide, separated by 6 m wide piers. The opening of the bays is regulated by sector gates. The spillway has a length of 420 m, a variable width of 108.5 m upstream to 70 m downstream, and is ended by a flip bucket. An approach channel 230 m wide is located upstream of the spillway and of the desilting structure.

This structure is made of 14 identical desilting chambers placed side by side, perpendicularly to the spillway axis, upstream of it. These huge chambers are 16.42 m wide, 12 m high and 180 m long. Their bottom is made of pyramidal shaped hoppers.

Figure 7.   Global scale model (left) and detailed study of the scouring intensity induced by the impinging jet downstream of the flip-bucket (right).

4.1   *General strategy*

Any hydraulic project of such a high level of complexity involves both critical hydrodynamical considerations and solid transport aspects. Moreover those two types of interacting features include intricate processes such as fully three-dimensional turbulent flows (water intakes), possibly laden with suspended load (desilting structures, upstream reservoir), and even aerated flows (spillway, flip-bucket). It is thus clear that no numerical model is currently available to perform a reliable analysis of the design as a whole.

On the other hand, a design methodology based exclusively on physical modeling may not be seen as the most attractive approach. Indeed, the spatial extension of the dam with its upstream reservoir, as well as the crucial need to reproduce experimentally essentially unspoiled physical phenomena forces the scale of the physical model to remain large enough. Besides the faithful scaling of complex physical processes (such as sediment load capacity, turbulence spectrum or air entrainment), a preserved accuracy of the various measurements to be performed also forces the scale model to remain particularly large. Fulfilling those two requirements would force the modeler to undertake the building of an extraordinary big and costly scale model, especially at the stage of the design, when the final shape and geometry of the structure are not yet known.

As a consequence, the global project has been studied at the University of Liege thanks to a whole set of combined numerical and physical models. Both shades of modeling are carried out in continuingly interacting and complementary perspective. For each part of the whole structure to be designed, the modeling method bringing most advantages for the specific case has been selected: numerical, physical or a combination of both.

4.2   *Hydrodynamic processes*

Physical scale models have been used to perform a complete analysis of the following items:

– water intakes of the diversion channels, flow through them and downstream restitution
– water intake for the hydroelectric power plant
– approach conditions of the spillway.

Indeed those flow fields are mainly driven by turbulence as well as three-dimensional free surface flow conditions and are thus hardly impossible to handle reliably numerically at a reasonable cost. Figure 7 shows the global model and a more detailed view of the spillway.

Numerical modelling has been used simultaneously for instance to optimise the design of the water inflow into the upstream part of the global model, in such a way that effects related to the upstream reservoir topography, although not reproduced experimentally, remain properly reproduced. Figure 8 shows three-dimensional views of the Digital Elevation Model used in the numerical simulations.

4.3   *Solid transport features*

Solid transport aspects of the project involve the design of the desilting chambers and their hopers as stated above. The trap efficiency of those structures is mainly influenced by the turbulence spectrum of the

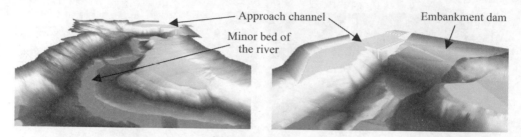

Figure 8.   3D view of the most downstream part of the reservoir topography (left), as well as details of the dam topography and the evacuation structures (right).

Figure 9.   Bottom topography in the downstream part of the reservoir initially, after 12 years and after 24 years of silting.

sediment-laden flow crossing the chambers. Once more no available numerical model can perform such a simulation at an acceptable level of accuracy and cost. Two large scale models have thus been exploited.

On the other hand, the long-term sedimentation time in the whole reservoir can only be estimated thanks to a numerical approach because of the huge space- and time-scales of the process. This has been performed with WOLF in a quasi-steady approach (Figure 9). The *transient* aggradation and bed level evolution have been computed on the basis of successive steady-state hydrodynamic simulations.

Finally, the rapidly transient flow with highly erosive power during a flushing operation have been analyzed with the fully-coupled and unsteady numerical tools available in WOLF. The mains goals of this numerical study consists in predicting the effect of a given flushing scenario in terms of changes in bathymetry in the downstream area of the reservoir of Kol. The efficiency of the flushing operation has been be evaluated with respect to the recovered storage capacity, its extension in space and the location of the channel generated by the flushing.

5   CONCLUSION

An integrated approach for global studies on scale and numerical models was described. It demonstrates that the increase of the computers performance, and the efficiency of the actual free-surface flow solvers lead to useful numerical information for optimal scale model design and studies.

In this way, the software package WOLF developed by the Service of Applied Hydrodynamics and Hydraulic Constructions at the University of Liege is described as an efficient tool in the scope of the study and design of hydraulic structures or, more generally, of river flow management. The detailed examples of two large dams in Asia showed that the numerical simulations have allowed to reach precious time savings by circumscribing the scale model in the minimum part where a physical model is required. The numerical approach also brought valuable guidelines for the analysis of aspects of the project impossible to handle in a laboratory experiment.

REFERENCES

Dewals, B., Archambeau, P., Erpicum, S., Mouzelard, T. & Pirotton, M. 2002a. Dam-Break Hazard Mitigation with Geomorphic Flow Computation, *Using Wolf 2D Hydrodynamic Software. International Conference on Risk Analysis*, Sintra, Portugal, WIT Press.

Dewals, B., Archambeau, P., Erpicum, S., Mouzelard, T. & Pirotton, M. 2002b. An Intergated Approach for Modelling Gradual Dam Failures and Downstream Wave Propagation. *First IMPACT Workshop*, Wallingford (UK), HR Wallingford.

Mouzelard, T. 2002. Contribution à la Modélisation des Écoulements Quasi Tridimensionnels Instation-naires à Surface Libre. *PhD thesis*, University of Liege.

Pirotton, M. 1997. Une Approche Globale pour Modéliser la Genèse et la Propagation des Crues Naturelles ou Accidentelles, *Classe des Sciences*, Académie Royale de Belgique, 182 pages.

Rogiest, P. 1997. An implicite finite volume scheme for the computation of unsteady compressible flows on multi-block structured grids. *PhD thesis*, University of Liege.

Roe, P.L. 1981. Approximate Riemann solvers, parameter vectors and difference schemes, *Journal of Computational Physics*.

Venkatakrishnan, V. 1993. On the accuracy of limiters and convergence to steady state solutions, *AIAA paper*.

A methodology for delineating wellhead protection areas

J.P. Lobo Ferreira
Laboratório Nacional de Engenharia Civil, Lisbon, Portugal

B. Krijgsman
Free University Amsterdam, Amsterdam, Holland

ABSTRACT: In this paper the mathematical concepts used for the development of a new methodology for wellhead protection and zoning are presented. The methodology is based on analytical solutions. The wellhead protection limits are a function not only of the local geology, of the parameters and the regional hydraulic characteristics of the aquifer but also of the extraction rates (or of the productivity of the aquifer). This new mathematical tool, that we consider to be robust and user friendly, aims to facilitate the work of those that have to apply national legislation regarding groundwater protection, e.g. the Portuguese DL n° 382/99 of September 22, 1999. The paper is based in the work developed by Krijgsman and Lobo-Ferreira (2001). The application of the methodology is exemplified for a case study developed in the coastal aquifer of the Bardez, Goa State, India.

1   INTRODUCTION

Groundwater resources constitute an important origin of water, effective or potential, and are therefore important to be preserved. The quality of groundwater resources is susceptible to be affected by social-economic activities, especially in the case of using and occupying the soil, such as with urban areas, infrastructures, agriculture etc.

The contamination of the groundwater resources is, in general, persistent, so that the recuperation of the water quality is a slow and difficult process (cf. Leitão 1997). The protection of the groundwater resources is therefore a strategic objective of major importance. It is important to develop an equilibrated and durable use of these resources.

One preventive instrument to assure the protection of the groundwater resources used for abstraction is setting up protection zones around wells extracting groundwater for public supply. The limits of these areas are a function of the geology, hydraulic characteristics of the concerned aquifer and amount of extracted water. In these defined areas around wells, restrictions should be set up concerning public use and transformation of soil, in order to protect the quality of the groundwater resources beneath it.

Protection zones around wells are generally defined as time of travel zones, either to allow for the attenuation of concentrations of contaminants in the aquifer or to provide a monitoring zone. If a contamination is detected in a monitoring zone, it could be dealt with before it enters the well. The objective of this project (cf. Krijgsman & Lobo-Ferreira 2001) was to develop a methodology to delineate the dimensions of such protection zones, in this special case the one corresponding with a travel time of 50 days. This in order to follow the Portuguese Decreto-Lei n° 382/99 of September 22, 1999, which states that groundwater extraction wells should be protected against pollutants, giving for this three zones, one of which corresponds with a traveltime of 50 days.

Based on experimental studies developed e.g. in German labs and according to several authors the limit of 50 days is chosen since this is generally accepted as a limit within which virus and pathogenic bacteria are naturally eliminated in groundwater, in particular *E. coli*. A 99.9% elimination in groundwater of E. coli is reached after a time ranging from 10 to 100 days, a function of the soil type, of incubation temperature and of soil moisture. The limit of 50 days was selected for the development of this paper methodology because, in general, it is accepted that this is the travel that allows the elimination of virus and pathogenic bacteria in groundwater, in particular E. coli.

For reaching this goal, hydrogeological studies have to be carried out, in order to determine the perimeters of protection zones. It is however often not possible to conduct detailed studies of individual wellfields, since this involves normally considerable time and costs. Instead, it would be easier, faster and cheaper if a more general methodology was available, with which one could quickly and without much effort give ranges of the perimeters of the required protection zones.

A semi-confined aquifer is by definition separated from superficial strata by a semi-impermeable layer, which implies that considerable time may pass before a pollutant can enter the aquifer after entering the soil.

A confined aquifer is by definition secluded from superficial strata by an impermeable layer, such as clay, by which it is fully protected from pollutants entering the soil above it.

The output of the use of this methodology should be a map, made by using Geographic Information Systems (GIS), on which one could see in colors the required dimension of a protection zone, without having to study the hydrogeological setting of an area into details.

In order to reach this goal, a general relationship between hydrological parameters and the dimensions of the required 50 day zone was set up, which was validated using a numerical computer program to simulate groundwater flow, Visual Modflow v. 2.7.2.

Once this relationship was validated, it was applied on a case study area in Goa, India.

2 METHODOLOGY FOR DEFINING PERIMETERS OF PROTECTION ZONES

2.1 Analytical solution

In the handbook 'Ground Water and Wellhead Protection' of EPA (1994) the following equation can be found:

$$t_x = \frac{n}{Ki}\left[r_x - \left(\frac{Q}{2\pi Kbi}\right)\ln\left\{1+\left(\frac{2\pi Kbi}{Q}\right)\cdot r_x\right\}\right] \tag{1}$$

In this equation: t_x = time of travel (days); n = effective porosity; K = hydraulic conductivity (m/d); r_x = distance over which groundwater travels in t_x before entering a pumping well (m), being negative $(-)$ if downgradient and positive $(+)$ if upgradient; Q = discharge of pumping well (m^3/d); b = aquifer thickness (m); i = hydraulic gradient before pumping.

Equation (1) can be used to calculate traveltime from a point x to a well, in case of a sloping hydraulic gradient, for both up- and downgradient points.

To calculate the distance as function of t, this equation should be written for r as function of t, n, K, Q, b and i.

2.2 Krijgsman and Lobo-Ferreira (2001) equations

To solve Equation (1) Krijgsman and Lobo-Ferreira (2001) developed a new methodology that may be consulted in http://www.dha.lnec.pt/nas/english/projects/BK_LF_ICT2001.pdf. Equations (2), (3) and (4) have been deducted for the evaluation of the upgradient and downgradient distances and also the distance perpendicular to the direction of flow:

- For the upgradient protection distance equation:

$$r = (0.00002x^5 - 0.0009x^4 + 0.015x^3 + 0.37x^2 + x)/F \tag{2}$$

with $x = \sqrt{\dfrac{2Ft}{A}}$

and $F = 2\pi Kbi / Q$

- For the downgradient protection distance equation:

$$r = (0.042x^3 + 0.37x^2 + 1.04x)/F \tag{3}$$

Table 1. Minimum values for the establishment of the minimum required radii.

Aquifer system type	1st protection zone/ immediate zone	2nd protection zone/ near zone	3rd protection zone/ far zone
Type 1	$r = 20$ m	r is the highest value of 40 m and r_1 (t = 50 days)	r is the highest value of 350 m and r_1 (t = 3500 days)
Type 2	$r = 40$ m	r is the highest value of 60 m and r_2 (t = 50 days)	r is the highest value of 500 m and r_2 (t = 3500 days)
Type 3	$r = 30$ m	r is the highest value of 50 m and r_3 (t = 50 days)	r is the highest value of 400 m and r_3 (t = 3500 days)
Type 4	$r = 60$ m	r is the highest value of 280 m and r_4 (t = 50 days)	r is the highest value of 2400 m and r_4 (t = 3500 days)
Type 5	$r = 60$ m	r is the highest value of 140 m and r_5 (t = 50 days)	r is the highest value of 1200 m and r_5 (t = 3500 days)
Type 6	$r = 40$ m	r is the highest value of 60 m and r_6 (t = 50 days)	r is the highest value of 500 m and r_6 (t = 3500 days)

Type 1 – confined aquifer system with lithological support formed by porous formations
Type 2 – unconfined aquifer system with lithological support formed by porous formations
Type 3 – semi confined aquifer system with lithological support formed by porous formations
Type 4 – aquifer system with carbonated lithological formations
Type 5 – fissured aquifer system with metamorphic and igneous formations
Type 6 – aquifer system with igneous and metamorphic formations having a small degree of fissures and weathering

- For the protection distance perpendicular to the direction of flow equation:

$$r = 4\sqrt{\frac{Q}{n \cdot b}} \tag{4}$$

2.3 Limitations

The method can not be applied, in reliable conditions, in all cases. It depends on the availability of data and on the local characteristics of the aquifer systems.

First of all, the area under analysis must be that of an unconfined aquifer. For confined aquifers the confining strata significantly increase the time required for the pollutant to penetrate the aquifer. The travel time thought the confining strata probably would exceed 50 days. In these cases, protection zones around the well should have a minimum value, e.g. those of the already mentioned Portuguese Decreto-Lei nº 382/99 of September 1999, shown in Table 1.

A second set of requirements is related with the need of having reliable input data regarding the following parameters:

- hydraulic conductivity (K)
- water levels, from which a gradient (i) is derived
- aquifer thickness (b)
- effective porosity (n)

A problem can be what to use for the value of b. The law defines this as the saturated thickness of the aquifer in the well. It should be noted that total screen length should not be used as aquifer thickness:

- First, obtained K-values from pumping tests are an average of the total aquifer thickness, including less permeable layers within the aquifer.
- Second, if for example only the top of the aquifer is screened, than still the total thickness of the aquifer will contribute to well-inflow as a result of non-horizontal flow near the screened part of the well.

With large-diameter wells, vertical flow is especially important, since an important constituent of total inflow will be bottom-inflow.

Data of extraction rates of wells (Q) are not necessarily required of all wells, since an output map can be made assuming average extraction rates. Extraction rates are never constant per well throughout a year and can vary considerably between wells even close to each other. Depending on the area the methodology is applied depending on the density of data. Different approaches can be made:

a) If a study is made on a single well field, the output map should be based on the period (season or 50 day period) with the maximum extraction rate over a season in the well field.
b) If a study is done on a greater area, one should have some data of extraction rates to know in what range the extractions are. In that case several maps can be made, using on one map for example the average and on another the maximum value of Q.
c) If plenty of data of extraction rates are available, a map can be made of the whole area showing the distribution of extraction rate and using this map in the calculation of the distribution of the needed protection area. The limitation on this is that when using the map to define protection areas for a new to be drilled well, this well should have an extraction rate corresponding with the extraction rate on the input map of Q-distribution.

3 THE METHODOLOGY APPLIED ON A CASE STUDY AREA IN GOA, INDIA

3.1 *Introduction*

The developed methodology of determining the perimeters of protection zones needed to be applied on a case study area. For this, an area had to be chosen with enough reliable data available. The methodology is developed for use on unconfined aquifers, since these are the most directly vulnerable for pollutants entering from the surface.

First consideration was the Palmela County on the Setúbal Peninsula, near Lisbon, Portugal. This area consists roughly of a superficial (unconfined) aquifer and a deep confined aquifer, separated by an impermeable layer. There are numerous wells with hydrogeological data available, however all of them relate to the confined aquifer; no data are available of the superficial aquifer.

Other areas considered within Portugal were Torres Vedras and Ribeiras do Oeste. These areas displayed the same problem as with the Palmela County.

Another area on which a research is being set up, is Goa, India, in cooperation with the local university of Goa. LNEC received detailed data of an area of 8 by 15 km, consisting of a superficial aquifer. Data are available of all the needed parameters: waterlevels, aquifer thickness, screenlength, extraction rates and porosity. Data are available of a maximum of 59 wells in this area; not all wells have data of each parameter.

Goa State, which has a land area of 3702 km^2, has a tropical climate with three seasons: a wet monsoon period from June to September, providing a precipitation of 2500 to 4300 mm, a winter season from October to January and a summer season from February to May. The population density of Goa is about 316/km^2 (Census 1991).

The case study area, which consists of a coastal area of 120 km^2, is situated in the northwestern corner of Goa in the district Bardez, having a population density of 717/km^2. The lithology consists of superficial laterites and sands, which are used as aquifers, underlain by precambriam metamorphic and crystalline rocks.

The studied area is rural, apart from the coastline, where many tourist resorts are located. The water demand is estimated on 150 liters per day in rural areas, while in tourist resorts this is about 500 liters (cf. Chachadi & Raikar 2000). To supply this demand, many large diameter wells are dug in the unconfined lateritic and sandy aquifers. A small part of the wells is in lithologies of (weathered) metagraywackes and phyllites. The well density is approximately 25 per km^2. The wells are normally shallow, not more than 15 m, with a diameter of up to 8 meters.
Available data:

- Waterlevels (taken on the same date), needed for deriving a hydraulic gradient (i): data are available on 57 wells for all seasons.
- Saturated aquifer thickness (b): data are available on 53 wells.
- Hydraulic conductivity (K): data are available on 6 wells, ranging between 1.4 and 31 m/day.
- Extraction rate (Q): Data are available for two types of aquifers:
  - For lateritic aquifers: Q varies between 86 and 216 m^3/d, from which an average of 151 m^3/d is taken.

– For sandy aquifers: Q varies between 155 and 259 m³/d, from which an average of 207 m³/d is taken.
• Effective porosity (n): data are available for 2 types of aquifers:
  – For lateritic aquifers: n varies between 0.20 and 0.30, from which an average of 0.25 is taken.
  – For sandy aquifers: n varies between 0.15 and 0.35, from which an average of 0.25 is taken.

3.2 Application of the methodology

3.2.1 Input data

For application of the methodology for delineation of wellhead protection areas, the program *Surfer, version 6.01* of *Golden Software* is used.

As input, a data file is needed with information of coordinates of wells, together with data of i, b, K, Q and n. From these data, continuous grids are extrapolated, all with the same dimensions to be able to make calculations with several grids.

• An hydraulic gradient, i, (rise over run) is derived from an extrapolated grid of water levels.
• For b the saturated aquifer thickness is used. Using the screen length would be inadequate, since a large part of the extracted water in the wells originates from bottom inflow.
• For hydraulic conductivity, K, just six values are known, showing no clear correlation with lithology. From these six values a continuous grid is extrapolated.
• For extraction rate, Q, only two average values are known for the two lithologies. All wells with a lateritic lithology (37 wells) are given a value of 151 m³/d, all wells in a sandy lithology (15 wells) are given the value of 259 m³/d.
• For effective porosity, n, a value of 0.25 is taken as constant over the whole area. No grid is made since n has in this case a constant value.

Wells with data are well distributed throughout the whole area, except the southwestern corner of the area, which is the Indian Ocean. Input data is shown in Figures 1–4, and the wells are identified in Figures 1, 2 and 4 by the white dots. In Figure 3 the white dots represent the wells used for hydraulic conductivity assessment. Up to now there is however no topographical information available linked to the concerning area, because of this information being classified. Therefore, no physical boundaries of

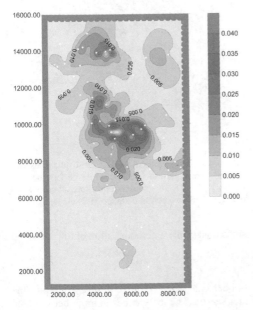
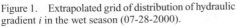

Figure 1. Extrapolated grid of distribution of hydraulic gradient i in the wet season (07-28-2000).

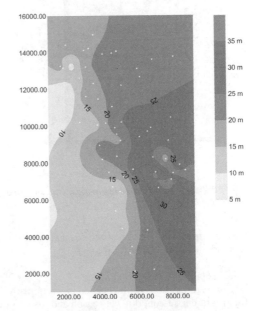

Figure 2. Extrapolated grid of distribution of saturated aquifer thickness, b, for the wet season (07-28-2000).

761

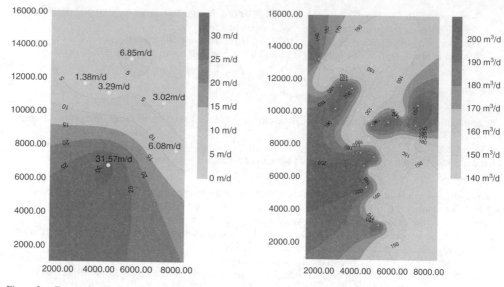

Figure 3.   Extrapolated grid of distribution of hydraulic conductivity, K.

Figure 4.   Extrapolated grid of distribution of extraction rate or aquifer productivity Q.

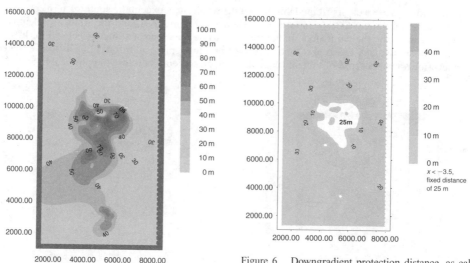

Figure 5.   Upgradient protection distance, as calculated with equation (1).

Figure 6.   Downgradient protection distance, as calculated with equation (2). The white part is the area where $x < -3.5$ and consequently has given a fixed downgradient protection distance of 25 m.

the area are known. In this case this is no real obstacle, since this is only a demonstration of the use of the method on a non-hypothetical area.

3.2.2   Output of the methodology

The three dimensions of the needed protection area have been calculated for three different seasons: the (dry) summer season, the wet season and the (dry) winter season. The differences in input between the seasons are the saturated thickness and the hydraulic gradient, both depending on the varying water-levels. This could cause a difference in calculation of the protection area, depending on the season the data are used of. Figures 5–7 show the results obtained for the summer season.

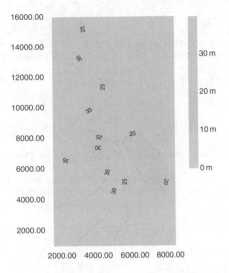

Figure 7. Protection distance perpendicular to direction of flow, as calculated with equation (3).

3.3 Conclusions

The results obtained for the summer season, the wet season and the winter season, do not show significant differences per season in the dimensions of the needed protection areas.

The water levels can have a variation throughout the year of up to 6 m, but normally this is not more than 2–3 m. The hydraulic gradient derived from the water levels does not change much throughout the year. The maximum value is about 0.046 in the summer; in the wet season the maximum gradient is just slightly lower with a value of 0.043. Apparently the watertable rises or drops quite uniformly over the area with the change of season.

Due to the varying waterlevels, the saturated thickness will also vary throughout the year, but the effect on the needed protection area does not seem to be more than a few meters.

Whenever applying the methodology it should of course always be tried to estimate the maximum 50-day distance that is possible to occur in a certain time span. Theoretically, if calculating the upgradient protection distance one should therefore use, if available, the data of a season or year that create the highest hydraulic gradient, have the highest extraction rates or smallest aquifer thickness. The opposite is the case with the downgradient protection distance concerning the hydraulic gradient, thereby making it all more complex which data to use for which calculation.

In this case however, it proves not to result in considerable differences when using data of different seasons, which does not mean that this is always the case. After applying the methodology on more areas it will be possible to say more about this.

3.3 Discussion

As grid interpolation method in Surfer®, the default method kriging has been used. A different method could well give different results, as well as changing the options within the kriging method.

No physical boundaries are concerned. It is known that the southwestern part of the case study area is the Indian Ocean. The exact location is not known; therefore it is treated as being land where data are missing and interpolated. In this case that does not matter much, since the area is only used for demonstrating the methodology.

Extrapolation of the grid of K-values is based on just six values. It would be a more logical approach to use a lithological map, from which a constant value of hydraulic conductivity per lithological unit is given. In this case there was no lithological map available.

A gradient is derived from data of levels in the wells, while the watertable in between the wells is probably higher. In other words, the gradient is derived from the maximum drawdown values, which are in the wells.

ACKNOWLEDGMENT

We acknowledge the support of the International Cooperation for Development Programme of the European Union that partially financed this project under research project "COASTIN – Measuring, Monitoring and Managing Sustainability: The Coastal Dimension", (4th Framework Programme, INCO-DEV, Contract No IC 18-CT98-0296 (Dec. 1998–Nov. 2002).

REFERENCES

Chachadi, A.G. and Raikar, P.S. 2000. *Fresh water demands for 2021 AD in Goa*. Department of Earth Science, Goa University, Goa.
Krijgsman, B. and Lobo-Ferreira, J.P. 2001. A Methodology for Delineating Wellhead Protection Areas. Lisboa, Laboratório Nacional de Engenharia Civil, Informação Científica de Hidráulica INCH 7, Nov. 2001.
Leitão, T. 1997. *Metodologias para a reabilitação de aquiferos poluídos*, Lisboa, Laboratório Nacional de Engenharia Civil, Teses e Programas de Investigação Nº 11.
US Environmental Protection Agency. 1994. *Ground Water and Wellhead Protection*, EPA/625/R-94/001.

Computational Methods in Engineering and Science, Iu et al. (eds)
© 2003 Swets & Zeitlinger, Lisse, ISBN 90 5809 567 3

Combining stochastic simulations and inverse modelling for delineation of groundwater well capture zones

C. Rentier
National Fund for Scientific Research of Belgium, Hydrogeology, Department of GEOMAC, University of Liège, Belgium

M. Roubens
MATH-(logics), Department of Mathematics, University of Liège, Belgium

A. Dassargues
Hydrogeology, Department of GEOMAC, University of Liège, also at Hydrogeology & Engineering Geology, Department of Geology–Geography, Katholieke Universiteit Leuven, Belgium

ABSTRACT: In hydrogeology, protection zones of a spring or a pumping well are often delimited by isochrones that are computed using calibrated groundwater flow and transport models. In heterogeneous formations, all direct and indirect data, respectively called hard and soft data, must be used in an optimal way. Approaches involving in situ pumping and tracer tests, combined with geophysical and/or other geological observations, are developed. In a deterministic framework, the calibrated model is considered as the best representation of the reality at the current investigation stage, but result uncertainty remains unquantified. Using stochastic methods, a range of equally likely isochrones can be produced allowing to quantify the influence of our knowledge of the aquifer parameters on protection zone uncertainty. Furthermore, integration of soft data in a conditioned stochastic generation process, possibly associated with an inverse modelling procedure, can reduce the resulting uncertainty. A stochastic methodology for protection zone delineation integrating hydraulic conductivity measurements (hard data), head observations and electrical resistivity data (soft data) is proposed.

1 INTRODUCTION

Time-related capture zones delimited by isochrones (contour lines of equal travel time to the well) are only parts of the well capture zone. In many regions, protection zones, corresponding to particular isochrones, are foreseen in local regulations providing a time-related protection. For example, in Walloon Region of Belgium, as in other regions, a protection zone corresponding to the 50-day isochrone must be delimited which brings essentially an effective protection against bacterial pollution. Ensuring a delay between an eventual injection of pollutant and its arrival at the pumping well, it allows to decide in each case on a priority intervention scheme. In heterogeneous formations, numerical computational methods are used to obtain a delimitation (Kinzelbach et al. 1992, Dassargues 1994, Derouane & Dassargues 1998). If many and various data are available in terms of geological and hydrogeological information in the studied domain, a very detailed geological interpretation is possible with a possibly reasonable but unquantified error. Measured parameters can be extrapolated consistently based on geological interpretation to condition model calibration. In a deterministic frame, even if the model is calibrated accurately on many data, these computed protection zones cannot be known exactly due to the limited knowledge of the aquifer parameters.

To assess the uncertainty in the delineation of time-related capture zones due to imperfect knowledge of hydraulic conductivity, different stochastic methodologies using a Monte Carlo simulation approach have been developed (Bair et al. 1991, Varljen & Shafer 1991, Vassolo et al. 1998, Evers & Lerner 1998,

Levy & Ludy 2000, Kunstmann & Kinzelbach 2000). The influence of the variance and correlation length of hydraulic conductivity on the location and extent of the resulting stochastic capture zone uncertainty have been studied (van Leeuwen et al. 1998, Guadagnini & Franzetti 1999). Further developments integrating conditioning procedures on hydraulic conductivity values (Varljen & Shafer, 1991, van Leeuwen et al. 2000), on head observations (Gomez-Hernandez et al. 1997, Vassolo et al. 1998, Feyen et al. 2001) and on additional soft data (Kupersberger & Bloschl 1995, Mc Kenna & Poeter 1995, Anderson 1997, Nunes & Ribeiro 2000, Rentier et al. 2002) allow to decrease prior uncertainty of hydraulic conductivity and therefore to reduce the uncertainty of the well protection zone.

2 STOCHASTIC METHODOLOGY

Even if the question of whether a stochastic approach which treats aquifer heterogeneity as a random space function is applicable to real aquifers under field conditions has not been definitively answered (Anderson 1997), more and more use is made of stochastic description of aquifer heterogeneity. In porous aquifers, most of the solute spreading is governed by the hydraulic conductivity (K) spatial variability, which is generally considered as the uncertain parameter. In practice, due to the few available hydraulic conductivity measurements (hard data), it can be useful to integrate soft data, like piezometric heads or geophysical data, in the conditional stochastic generation of hydraulic conductivity fields. This conditioning allows to reduce the variance of the distribution and consequently decrease the uncertainty of the results.

In order to characterize the uncertainty of well capture zones due to aquifer heterogeneity and to show how additional soft data might be used to constrain this uncertainty, a stochastic approach integrating different sorts of data is presented. For the purpose of the demonstration, a synthetic but realistic case was designed involving three data sets: hydraulic conductivity measurements, head observations and electrical resistivity data.

2.1 Set-up of a synthetic aquifer

A hypothetical groundwater flow domain was constructed using geological and hydrogeological conditions similar to actual alluvial sites. The domain was chosen large enough to avoid boundary effects. Two layers were distinguished: an upper fine sand and clay layer and a lower coarse sand and gravel layer. A "true" hydraulic conductivity field representing the "reality" was created: a uniform hydraulic conductivity value of $10^{-5}\,\mathrm{ms}^{-1}$ was applied to the upper layer whereas for the lower layer a non-conditional simulation was generated using the Turning Band algorithm (Mantoglou & Wilson 1982) with an isotropic, exponential correlation structure identical to the one found for the alluvial sediments of the Meuse River valley downwards to Liège (Belgium). From a dense grid of more than 10000 cells, 15 hydraulic conductivity values were selected, representing pumping test results in virtual piezometers and providing the hard data set. A regular pattern of sampling locations was rejected for the purpose of keeping consistent with a probable real-world situation.

A single pumping well (pumping rate of $60\,\mathrm{m}^3\mathrm{h}^{-1}$) located in the heterogeneous sand and gravel unconfined aquifer was used for the simulation. Pure deterministic groundwater flow conditions were then computed, providing the synthetic "measured" heads at the 15 virtual piezometers (first set of soft data).

The resistivity data set (second set of soft data) was created based on the observed correlation ($r = 0.9$) existing between electrical resistivity (ρ) and hydraulic conductivity (K) in the alluvial sediments of the Meuse River. Considering $N(0,1)$ as a random draw within a standard normal distribution and $\mathring{\sigma}$, the standard deviation of the regression residual, 300 resistivity values, distributed on 12 tomographic profiles were generated by the following equation:

$$\ln \rho = 6.836 + 0.345 \ln K + \mathring{\sigma}\, N(0,1) \tag{1}$$

Considering advective transport time to the well, the "true" 20-day isochrone line (associated with the concerned pumping rate) resulting from the "true" hydraulic conductivity field was also computed and used as reference for further comparisons.

2.2 Monte Carlo stochastic conditional simulations

The stochastic methodology developed by Varljen & Shafer (1991) is applied in order to determine and quantify the uncertainty in the location and extent of the 20-day capture zone due to imperfect knowledge

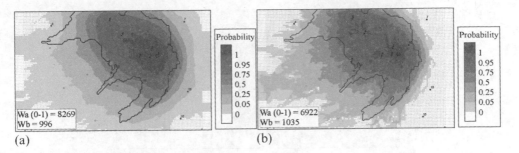

(a)   (b)

Figure 1.   (a) Comparison of the CaPD determined by stochastic realizations conditioned on 15 hydraulic conductivity values, and the reference 20-day isochrone (black line). (b) Comparison of the CaPD determined by stochastic realizations conditioned on 15 hydraulic conductivity values and inverse modelling using 15 piezometric heads and the reference 20-day isochrone (black line).

of the hydraulic conductivity field. According to a Monte-Carlo analysis, four hundred stochastic simulations of equally likely hydraulic conductivity fields for the lower layer were generated using the Turning Bands method. Each of them were subsequently conditioned on hydraulic conductivity measurements by a kriging technique. Groundwater flow and a particle tracking process were computed for each realization. The ensemble of obtained capture zones was then treated statistically to infer the capture zone probability distribution (CaPD). This CaPD gives the spatial distribution of the probability that a conservative tracer particle released at a particular location is captured by the well within a specified time span (van Leewen 2000), in this particular case 20 days. Figure 1a compares the CaPD with the "real" isochrone and show how the stochastic approach reveals the uncertainty about the location of the protection zone. In order to quantify and evaluate the performance of additional conditioning, two measures among several others have been selected (van Leeuwen et al. 2000). Wa is a measure of uncertainty based on the extent of the uncertainty zone for which the probability P of capture is $0 < P < 1$. Wb compares the location of the reference isochrone with the location of isoline $\Gamma(0.5)$ for which 50% probability of capture is obtained. Units of both measures are expressed in number of 5 m × 5 m cells.

2.3   Additional conditioning by head measurements (first set of soft data)

This additional conditioning was performed by calibrating the groundwater flow on head measurements for each stochastic conditional simulation generated previously by inverse modelling (PEST code, Doherty 1994). Resolution of the inverse problem required to carry out a parameterisation reducing the number of adjustable parameters. Therefore a zonation was applied that consist, based on four specified threshold values (S_i), in dividing the hydraulic conductivity variation interval in five classes (C_i) of uniform value (KC_i), representing the adjustable parameters. The threshold values were defined by determining the best hydraulic conductivity data combination that minimize the variability within each class, by minimizing the following equation:

$$f = \sum_{i=1}^{N_c} \sum_{j=1}^{N_{d_i}} \left(\ln K_{ij} - \overline{\ln K_i} \right)^2 \quad \text{with} \quad \overline{\ln K_i} = \frac{1}{N_{d_i}} \sum_{j=1}^{N_{d_i}} \ln K_{ij} \quad i = 1, N_c \tag{2}$$

where N_c is the number of classes, N_{d_i} is the number of data in each class (varying from one combination to an other) and $\Sigma N_{d_i} = N_D$, the total number of K data.

The inverse procedure was applied to optimize the uniform hydraulic conductivity values for each parameter. However, for some realizations, these optimized hydraulic conductivity values did not respect the relative order $KC_i < KC_{(i+1)}$ defined by the thresholding. Considering these realizations as geologically erroneous, they were rejected. For each remaining realization, the 20-day capture zone was determined and consequently the CaPD for the ensemble of possible capture zones was estimated. Results show how the use of head conditioning by inverse modelling combined with the use of a selection criterion reduce the uncertainty of the probability distribution (decrease in Wa).

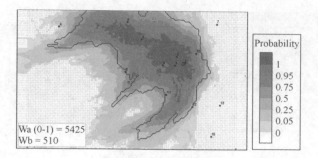

Figure 2. Comparison of the CaPD determined by realizations conditioned on 15 hydraulic conductivity values, 15 piezometric heads and 300 resistivity values and the reference 20-day isochrone (black line).

2.4 *Additional conditioning by geoelectrical resistivity data (second set of soft data)*

The geophysical data set, as additional useful information, was directly integrated in the generation process by conditioning each stochastic simulation on both hydraulic conductivity measurements and resistivity values by a cokriging technique. Four hundred stochastic "co-conditional" simulations were then generated based on a linear model of coregionalisation that was adjusted on experimental simple and cross-covariances. Following the same procedure as described in the previous paragraph (parameterization and inverse modeling), an advective 20-day isochrone was computed for each remaining realization and the resulting CaPD was calculated (Fig. 2). The use of additional available data decrease the prior uncertainty of the parameters and, in consequence, reduce the uncertainty of the well CaPD (decrease in Wa). It can also be observed that the $\Gamma(0.5)$ isoline tends to approach the reference isochrone (decrease in Wb). Comparing results of Figure 2 to those of Figure 1, a clear improvement is observed in the description of the medium heterogeneity leading to more realistic isochrones with regards to the reference one.

3 CONCLUSION

Advances in the delimitation of protection zones are made by the use of stochastic methodologies. Moreover introduction of additional available data decreases the prior uncertainty of the parameters and, in consequence, reduces the uncertainty of the well capture zone probability distribution (CaPD). Since geophysical data and head observations are easier to collect on the field then hydraulic conductivity measurements, they are generally more abundant. The methodology presented can be used in real applications to quantify the uncertainty in the location and extent of well capture zones when little or no information is available about the hydraulic properties, through the conditioning on geophysical data and/or head observations.

The different stochastic approaches listed previously considers purely advective transport (particle tracking procedure). They do not take the concept of macrodispersivity into account. A topic for further research will be the assimilation of information from solute concentrations (i.e. tracer tests) to further reduce uncertainty in capture zone delineation. Further applied research is needed to design optimal strategies of soft data acquisition like, for example, selection of locations for conditioning measurements.

ACKNOWLEDGEMENTS

We thank the National Fund for Scientific Research of Belgium for the PhD research grant given to C. Rentier, a part of this work has also been funded through the EU-DAUFIN project (EVK1-1999-00153) and the EU-MANPORIVERS project (ICA4-CT-2001-10039).

REFERENCES

Anderson, M. P. 1997. Characterization of geological heterogeneity. In: *Subsurface flow and transport* (Proc. of the Kovac's Colloquium), 23–43.

Bair, E. S., Safreed, C. M. & Stasny, E. A. 1991. A Monte Carlo-based approach for determining traveltime-related capture zones of wells using convex hulls as confidence regions. *Ground Water* 29(6): 849–855.

Dassargues, A. 1994. Applied methodology to delineate protection zones around pumping wells. *Journal of Environmental Hydrology* 2(2): 3–10.

de Marsily, G. 1984. Spatial variability of properties in porous media: a stochastic approach. In Bear J. & Corapcioglu M. (eds), *Fundamentals of Transport Phenomena in Porous Media*: 721–769. NATO-ASI Series E(82).

Derouane, J. & Dassargues, A. 1998. Delineation of groundwater protection zones based on tracer tests and transport modelling in alluvial sediments. *Environmental Geology* 36(1–2): 27–36.

Doherty, J., Brebber, L. & Whyte, P. 1994. *PEST – Model-independent parameter estimation. User's manual.* Watermark Computing, Corinda. Australia. 122 p.

Evers, S. & Lerner, D. N. 1998. How uncertain is our estimate of a wellhead protection zone? *Ground Water* 36(1): 49–57.

Feyen, L., Beven, K. J., De Smedt, F. & Freer, J. 2001. Stochastic capture zone delineation within the generalized likelihood uncertainty estimation methodology: Conditioning on head observations. *Water Resources Research* 37(3): 625–638.

Gomez-Hernandez, J. J., Sahuquillo, A. & Capilla, J. E. 1997. Stochastic simulation of transmissivity fields conditional to both transmissivity and piezometric data, I, Theory. *Journal of Hydrology* 203: 162–174.

Guadagnini, A. & Franzetti, S. 1999. Time-related capture zones for contaminants in randomly heterogeneous formations. *Ground Water* 37(2): 253–260.

Kinzelbach, W., Marburger, M. & Chiang, W.-H. 1992. Determination of groundwater catchment areas in two and three spatial dimensions. *Journal of Hydrology* 134: 221–246.

Kunstmann, H. & Kinzelbach, K. 2000. Computation of stochastic wellhead protection zones by combining the first-order second-moment method and Kolmogorov backward equation analysis. *Journal of Hydrology* 237: 127–146.

Kupfersberger, H. & Blöschl, G. 1995. Estimating aquifer transmissivities – on the value of auxiliary data. *Journal of Hydrology* 165: 85–99.

Levy, J. & Ludy, E. E. 2000. Uncertainty quantification for delineation of wellhead protection areas using the Gauss-Hermite quadrature approach. *Ground Water* 38(1): 63–75.

Mantoglou, A. & Wilson, J. L. 1982. The turning bands method for simulation of random fields using line generation by a spectral method. *Water Resources Research* 18(5): 1379–1394.

Mc Kenna, S. A. & Poeter, E. P. 1995. Field example of data fusion in site characterization. *Water Resources Research* 31(12): 3229–3240.

Nunes, L. M. & Ribeiro, L. 2000. Permeability field estimation by conditional simulation of geophysical data. *Calibration and Reliability in Groundwater Modelling* (Proc. of the ModelCARE 99 Conf.), IAHS Publ. 265: 117–123.

Rentier, C., Bouyère, S. & Dassargues, A. 2002. Integrating geophysical and tracer test data for accurate solute transport modelling in heterogeneous porous media. In: *Groundwater Quality: Natural and Enhanced Restoration of Groundwater Pollution* (Proc. GQ'2001 in Sheffield). IAHS Publ. n°275:3–10.

van Leeuwen, M., te Stroet, C. B. M., Butler, A. P. & Tompkins, J. A. 1998. Stochastic determination of well capture zones. *Water Resources Research* 34(9): 2215–2223.

van Leeuwen, M., Butler, A. P., te Stroet, C. B. M. & Tompkins, J. A. 2000. Stochastic determination of well capture zones conditioned on regular grids of transmissivity measurements. *Water Resources Research* 36(4): 949–957.

van Leeuwen, M. 2000. *Stochastic determination of well capture zones conditioned on transmissivity data.* PhD thesis, Department of Civil and Environmental Engineering, University of London, 154 p.

Varljen, M. D. & Shafer, J. M. 1991. Assessment of uncertainty in time-related capture zones using conditional simulations of hydraulic conductivity. *Ground Water* 29(5): 737–748.

Vassolo, S., Kinzelbach, W. & Schäfer, W. 1998. Determination of a well head protection zone by inverse stochastic modelling. *Journal of Hydrology* 206: 268–280.

Computational Methods in Engineering and Science, Iu et al. (eds)
© 2003 Swets & Zeitlinger, Lisse, ISBN 90 5809 567 3

The diffusion model for long-term coastal profile evolution: numerical validating

A.V. Avdeev & E.V. Goryunov
Institute of Computational Mathematics and Mathematical Geophysics of SB RAS, Novosibirsk, Russia

M.M. Lavrentiev Jr.
Sobolev Institute of Mathematics of SB RAS, Novosibirsk, Russia

R. Spigler
Dipartimento di Matematica, Universit' a di "RomaTre", Italy

ABSTRACT: An algorithm for simultaneous determination of *two* coefficients in an inverse problem for equations of parabolic type is studied. The model under investigation was proposed by De Vriend et al. to describe the long-term coastal profile evolution. The iterative inversion procedure is based on minimization of an appropriate cost functional. Results of numerical tests are presented to illustrate the performance of the algorithm. A file of real *measured* data on the coastal profile evolution was processed numerically to validate the aforementioned diffusion-type model.

1 INTRODUCTION

In the paper of De Vriend et al. (1993), a model of diffusion type was proposed to describe the evolution of the coastal profile. The diffusion coefficient in the governing equation (its physical dimension is length squared divided by time) corresponds to the time scale of shoreline change following a disturbance (wave action).

The aforementioned diffusion model can be described by the following basic equation:

$$\frac{\partial(\delta X)}{\partial t} = D^2(z)\frac{\partial^2(\delta X)}{\partial z^2} + g(t,z,\delta X \frac{\partial(\delta X)}{\partial z}) \tag{1}$$

Here $\delta X(z, t)$ represents the change of the cross-shore position (i.e., change in the depth at the distance z from the shore line) of the coastal profile and $D(z)$ is the diffusion coefficient.

Following De Vriend et al. (1993), we give some explanations for special cases of the term $g(t, z, \delta X, \partial(\delta X)/\partial z)$. If $g = S(z, t)$ (an external source function), it is possible to introduce the effects of random forcing, along-shore transport gradients, and human interference such as nourishment and sand mining (De Vriend et al. 1993).

The linear choice $g = B(z)\partial(\delta X)/\partial z$ or $g = B(z)\delta X$ is also interesting in view of applications. In these models, the coefficient $B(z)$ represents the speed of along-shore sand wave movement. We assume that $g = B(z)\delta X$ in (1). So, we shall consider the following.

Inverse problem. Given the function $\delta X_0(t)$ (the change of the cross-shore position at the point $z = 0$), find the coefficients $D(z)$ and $B(z)$ such that the solution $\delta X(z, t)$ to the problem

$$\frac{\partial(\delta X)}{\partial t} = D^2(z)\frac{\partial^2(\delta X)}{\partial z^2} + B(z)\delta X, \tag{2}$$

$$\delta X\big|_{t=0} = 0, \tag{3}$$

$$\left.\frac{\partial(\delta X)}{\partial z}\right|_{z=0} = \varphi_0(t), \quad \delta X\big|_{z=H} = 0 \tag{4}$$

satisfies the equation (surface measurements)

$$\delta X\big|_{z=0} = \delta X_0(t). \tag{5}$$

Note that the parameter H can be thought of as an estimate of the depth of closure, i.e., the location where the diffusive and transport phenomena virtually end.

2 NUMERICAL RECONSTRUCTION OF TWO COEFFICIENTS OF THE EQUATION

2.1 *Description of the algorithm*

As the first step to solve the aforementioned inverse problem numerically, we apply the Fourier transform. Thus, the original dynamic problem is replaced by a Helmholtz equation with a complex-valued coefficient. The point is that recovering the space-dependent coefficients $D(z)$ and $B(z)$ in the frequency domain, we do not need to go back to the time domain. Here we have used the term *frequency domain* for retaining a formal analogy with inverse problems for the wave equation, in which case the frequency domain concerning Fourier images of solutions has a clear physical meaning (see, e.g., Aki & Richards 1980, Avdeev et al. 1999).

For numerical processing we apply to the problem (2)–(4) the formal Fourier transform under assumption that the coefficients are smooth: $D(z)$, $B(z) \in C^2(0, H)$ and $\omega \in [\omega_1, \omega_2]$.

So, we consider the problem

$$\frac{d^2V}{dz^2} + \frac{B(z) - i\omega}{D^2(z)} V = 0, \tag{6}$$

$$\left.\frac{dV}{dz}\right|_{z=0} = F(\omega), \quad V\big|_{z=H} = 0, \tag{7}$$

where $V(z,\omega) = \int_0^\infty e^{-i\omega t} \delta X dt$ and $F(\omega) = \int_0^\infty e^{-i\omega t} \varphi_0(t) dt$.

The inverse problem we are interested in consists in reconstructing both functions $D(z)$ and $B(z)$ from the additional information

$$V_0(\omega) = V(0,\omega), \quad \omega_1 \le \omega \le \omega_2, \tag{8}$$

where $[\omega_1, \omega_2]$ is the interval of available frequencies.

We solve the inverse problem (6)–(8) by minimizing the cost functional

$$\Phi[D,B] = \int_{\omega_1}^{\omega_2} \left|V_0(\omega) - K[D,B](\omega)\right|^2 d\omega + \beta \sup_z \left|D - D^{\text{est}}\right| + \gamma \sup_z \left|B - B^{\text{est}}\right|, \tag{9}$$

where the operator $K[D, B](\omega)$ maps the current "test" values $D(z)$ and $B(z)$ into the trace of the solution of the boundary value problem (6), (7) at $z = 0$. Here β and γ are some weighted regularization parameters, $0 \le \beta \le 1$, $0 \le \gamma \le 1$, and D^{est} and B^{est} are the estimated values of D and B, respectively, that can be obtained from physical measurements.

In the modification of the numerical algorithm which is used here, we operate with the gradient of the "uniform" cost functional (9), i.e., the functional with $\beta = 0$ and $\gamma = 0$. In this case, after rather technical calculations (omitted here since, e.g., a similar approach can be found in McGillivray & Oldenbourgh 1990), we obtain the following formulas for the gradients of the cost functional with respect to D and B:

$$\left(\nabla_D \Phi[D,B]\right)(z) = -4\text{Re} \int_{\omega_1}^{\omega_2} \frac{B(z) - i\omega}{D^3(z)} \overline{F(\omega)} \cdot \left[V_0(\omega) - F(\omega) G(z,0;\omega)\right] \times$$
$$\times \overline{G}(z,z;\omega) \overline{G}(z,0;\omega) d\omega \tag{10}$$

$$\left(\nabla_{B}\Phi\left[D,B\right]\right)(z)=2\mathrm{Re}\int_{\omega_{1}}^{\omega_{2}}D^{-2}\left(z\right)\overline{F}\left(\omega\right)\cdot\left[V_{0}\left(\omega\right)-F\left(\omega\right)G\left(z,0;\omega\right)\right]\times$$
$$\times\overline{G}\left(z,z;\omega\right)\overline{G}\left(z,0;\omega\right)d\omega \tag{11}$$

Here $G(z, \zeta; \omega)$ is the Green function of the problem (6), (7) and the bar denotes the complex conjugation.

Here we do not consider a rather nontrivial theoretical question of uniqueness as well as existence of the *global* minimum point of the cost functional (9). Instead, we refer the reader to the paper Alekseev et al., 1993 in which similar questions are discussed for a more simple statement of the inverse problem.

To minimize the cost functional (9), the conjugate direction method was used (see Karmanov 1986).

2.2 *Numerical experiments for synthetic data*

Computer codes were prepared in C++; and we used Mathematica 3.0 for verification and visualization. Only resources of a personal computer are needed. On each iteration, the function $V(z, \omega)$ was computed by the so-called semi-analytical method described in Alekseev et al., 1993.

We approximated the coefficients $D(z)$ and $B(z)$ by piecewise constant functions, i.e., 6 layers with equal width 200 meters. According to the results of auxiliary numerical tests, the regularization parameters β and γ were chosen equal to 0.3 and 0.2, respectively. The results of reconstruction (obtained after 98 iterations) are shown in Figure 1.

In order to test the robustness of the algorithm, the inversion data were artificially corrupted by adding a randomly distributed white noise. A normally distributed random noise with average fluctuations equal to 5% of the data amplitudes was added to the inversion data. The result of simultaneous identification of *two* coefficients is shown in Figure 2. The initial approximations for $D(z)$ and $B(z)$ were the same as in the previous test, see Figure 1.

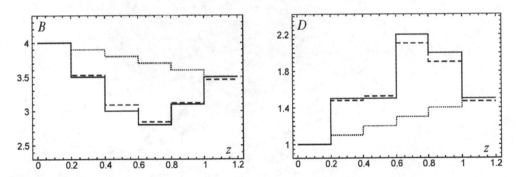

Figure 1. Simultaneous reconstruction of both coefficients $D(z)$ and $B(z)$. True functions are shown by solid lines, the initial guess by dotted lines, and the result of reconstruction by dashed lines.

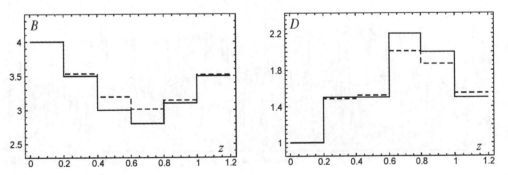

Figure 2. Results of simultaneous reconstruction of two coefficients $D(z)$ and $B(z)$ from corrupted data.

To conclude this section we would like to stress that the proposed version of the numerical inversion algorithm has demonstrated quite reasonable performance and accuracy. Thus, a basis for real data processing has been established.

3   VALIDATING THE DIFFUSION MODEL: NUMERICAL EXPERIMENTS FOR REAL DATA

Real data representing the long-term evolution of the cross-shore position of coastal profiles were collected over a period of 10 years from 1981 to 1991 at Duck, North Carolina, and are presented in Figure 3. Real data $\delta X_r(z, t)$ (the subscript "r" stands for "real") for the cross-shore position consist of a 100×250 array. This means 100 observation points in the spatial variable z, each being an average over 80 meters, and 250 observation times, each being an average over 15 days. In addition, the source function $f(t)$ was measured at the same observation times. This function represents the average height of waves over the observation periods and is also shown in Figure 3.

It was supposed that the coastal profile evolution is described by the model equations (2)–(5), where, in equation (2), the function $f(t)$ is added to the right-hand side.

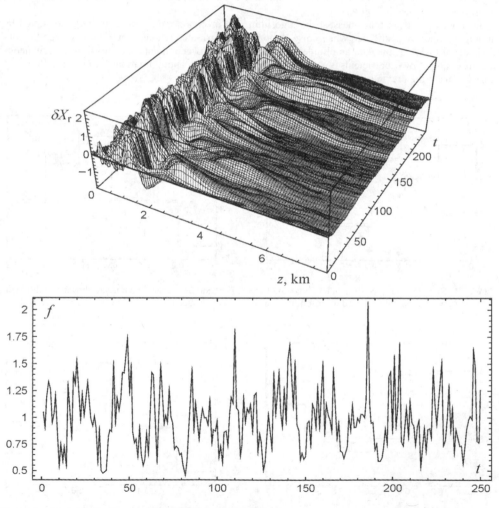

Figure 3.   Real data for the cross-shore position and the source term.

First, the data were numerically recalculated in terms of the Fourier Transform $\mathcal{F}$ in order to obtain the data $V_r(z, \omega) = \mathcal{F}[\delta X_r(z, t)]$ and $\Psi(\omega) = \mathcal{F}[f(t)]$. More precisely, to solve the problem we use the Fourier-type representation

$$V(z,\omega) = \int_0^T \delta X(z,t)^{-i\omega t} dt, \tag{12}$$

where T is the whole period of observation; with the inverse transform given by the formula

$$\delta X(z,t) = \frac{1}{T} \sum_n V(z,\omega_n) e^{i\omega_n t}, \tag{13}$$

where $\omega_n = n\pi/T$, $n = 1, 2, \dots, N$.

Then the cost functional was computed by the formula

$$\Phi = \int_{\omega_1}^{\omega_N} \int_0^H \left| V_r - V_{\text{synth}} \right|^2 dz d\omega, \tag{14}$$

where the data V_{synth} ("synthetic") were obtained as the solution to the direct problem

$$\frac{d^2 V}{dz^2} = \frac{i\omega - B}{D^2} V + \frac{\psi(\omega)}{D^2}, \tag{15}$$

$$\left. \frac{dV}{dz} \right|_{z=0} = F(\omega), \quad V\big|_{z=H} = h(\omega). \tag{16}$$

The boundary data were obtained as follows:

$$F(\omega) = \frac{V_r(\Delta z, \omega) - V_r(0, \omega)}{\Delta z}, \quad h(\omega) = V_r(H, \omega) \quad (\Delta z = 80\,\text{m}).$$

First, the cost functional was studied in the case where both coefficients of the equation are *constants*: and $D(z) \equiv$ const and $B(z) \equiv$ const. Then a contour plot of the cost functional on the plane (B, D) was calculated (Fig. 4). Each point in Figure 4 corresponds to the value of the functional (14) for a pair of constants (B, D). The following step-sizes for numerical computations were chosen: $\Delta B = 10$, $\Delta D = 0.02$. The intervals for B and D were taken $(0, 800)$ and $(0, 0.5)$, respectively.

We can see that the cost functional has a rather complicated structure even in the simplified case of the governing equation (2) with *constant* coefficients D and B.

Usual gradient methods provide very poor convergence in such cases and it is practically impossible to find the global minimum.

Therefore the global minimum of the functional (14) was found numerically by exhaustive search over a rough mesh. We denote the corresponding values of the parameters of the problem by $B_{\min}$ and $D_{\min}$, $B_{\min} = 400$ and $D_{\min} = 0.25$. Next, the solution $V_{\min}(z, \omega)$ to the direct problem (15), (16) (with constants $B_{\min}$ and $D_{\min}$ substituted for D and B, respectively) was computed.

Finally, the time-dependent profile, $\delta X_{\min 1}(z, t)$ was determined after inverse Fourier transformation, see (13).

A 3D plot of the profile $\delta X_{\min 1}(z, t)$ is shown in Figure 5. For comparison with the measured data, we calculated the time integral of the relative error:

$$\delta(z) = \int_0^T \left| \frac{\delta X_r - \delta X_{\min 1}}{\delta X_r} \right| dt. \tag{17}$$

775

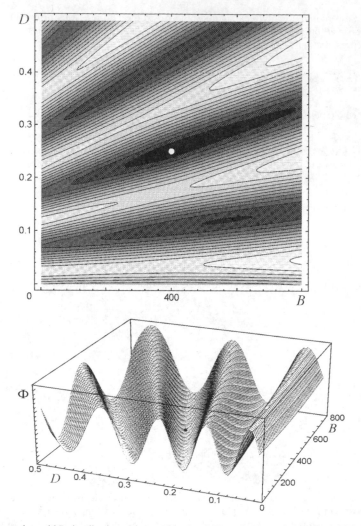

Figure 4. Contour plot and 3D visualization of the cost functional. The point shows the position of the global minimum.

The average relative error is 15.8%. We can see that the rough assumption that the parameters of the problem are just constants, $B = B_{min}$ and $D = D_{min}$, turns out to be in a rather good qualitative agreement with the measured data, see Figure 3. It is important that in our case the relative error $\delta(z,\ t) = |(\delta X_r - \delta X_{min1})/\delta X_r|$ does not increase with the growth of time.

Then, using the procedure of minimization of the cost functional by the conjugate direction method we reconstructed the coefficients $D(z)$ and $B(z)$ under the assumption that they are piecewise constant functions (see Fig. 6). The range of the spatial variable was divided into 9 layers of equal width. The values $B = B_{min}$ and $D = D_{min}$, which were obtained in the previous numerical test, are shown by dashed line.

The corresponding 3D plot of the profile $\delta X_{min2}(z,\ t)$ is shown in Figure 7. Comparison with the measured data was made as above, see (17). The average relative error is 5.4%. We can see that piecewise constant coefficients provide a better agreement with the measured data, see Figure 3, 5, and 7.

So, the results obtained can validate the diffusion model which was proposed by De Vriend and Capobianco in De Vriend et al. 1993 for description of the long-term coastal profile evolution.

This research was supported in part by the Russian Foundation for Basic Research under grants 01-05-64704, 00-07-90343, and 00-15-99092; the GNFM of the Italian INdAM; UNESCO under contracts UVO–ROSTE. © A.V. Avdeev, E.V. Goryunov, M.M. Lavrentiev (Jr.), R. Spigler, 2003.

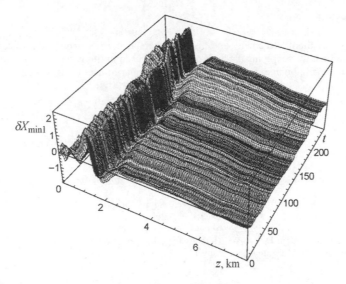

Figure 5.   Results of numerical reconstruction of the cross-shore position for constant values of D and B.

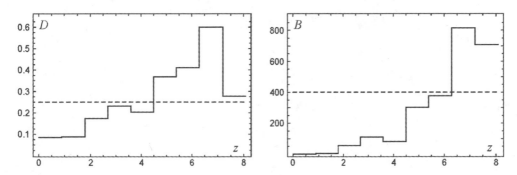

Figure 6.   Results of reconstruction of $D(z)$ and $B(z)$ for 9 layers of equal width.

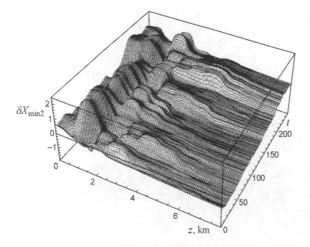

Figure 7.   Results of numerical reconstruction of the cross-shore position for piecewise constant $D(z)$ and $B(z)$ (Figure 6).

ACKNOWLEDGEMENT

We are grateful to William Birkemeier and his coworkers at the US Army Field Research Facility in Duck, North Carolina, for supplying the high-quality data used in this research. We also thank Michele Capobianco (Tecnomare, Venice) for his interest and a number of useful suggestions.

REFERENCES

Aki, K. & Richards, P.G. 1980. *Quantitative Seismology: Theory and Methods*. San Francisco: Freeman.

Alekseev, A.S., Avdeev, A.V., Fatianov, A.G. & Cheverda V.A. 1993. Wave processes in vertically in homogeneous media: a new strategy for velocity inversion. *Inverse Prob.* 9(3): 367–390.

Avdeev, A.V. & Lavrentiev, M.M. (Jr.), Priimenko V.I. 1999. *Inverse Problems and Some Applications*. Novosibirsk: ICM&MG.

De Vriend, H.J., Capobianco, M. & Chesher, T. et al. 1993. Approaches to long-term modeling of coastal morphology: a review. *Coastal Eng.* 21(1/3 [special issue]): 225–269.

Karmanov, V.G. 1986. *Mathematical Programming*. M.: Nauka, (in Russian).

McGillivray P.R. & Oldenbourgh D.W. 1990. Methods for calculating Frechét derivatives and sensitivities for the nonlinear inverse problems: a comparative study. *Geophys. Prosp.* 38(5): 499–524.

Computational Methods in Engineering and Science, Iu et al. (eds)
© 2003 Swets & Zeitlinger, Lisse, ISBN 90 5809 567 3

Characterisation of the sediment dynamics of Hac-Sá beach, Macao

F.S.B.F. Oliveira, C. Vicente & M. Clímaco
Laboratório Nacional de Engenharia Civil, Lisboa, Portugal

ABSTRACT: The recent evolution of the physics of Hac-Sá beach system is characterised based on
the analysis, processment and interrelation of different types of field data and mathematical modelling
of the hydrodynamic and sediment dynamics processes. The methodology established was primarily the
analysis and quantification of the natural agents that intervene in the beach dynamics, followed by the evaluation of the major beach dynamics processes and their interaction mechanisms and finally, the assessment of the recent evolution of the system.

During the last two decades, the significant morphological changes occurred in the coastal area of
Hac-Sá beach and the land reclamation works at the north and south headlands of Hac-Sá bay introduced changes in the nearshore wave propagation, that together with the progressive occupation of the
beach backshore zone with infrastructures and other constructions affected the shoreline equilibrium.

1 INTRODUCTION

The objective of the present study is to characterise the sediment dynamics of Hac-Sá beach system, in
Macau, and its recent evolution. The study is part of an extended study (Oliveira 2002, Oliveira et al.
2002) aiming the rehabilitation of Hac-Sá beach, requested by Laboratório de Engenharia Civil de
Macau (LECM) to Laboratório Nacional de Engenharia Civil (LNEC), Portugal.

Hac-Sá beach, confined between the two headlands of Hac-Sá bay, in Figure 1, is located in the SE
coast of Coloane island (part of Macau, a Special Administrative Region in the south coast of People's
Republic of China). It is a narrow pocket beach with 1.2 km long and a NNE-SSW general orientation.
The northern side of Hac-Sá headland and the southern side of Ká-Hó headland have in the recent
decades advanced over the sea through reclamation works. The slopes of the enlarged banks are protected against wave action by rock revetments. The beach backshore zone is occupied with constructions and infrastructures, particularly advanced in the southern stretch of the beach.

This paper is organized in four main chapters. In the present chapter the objective of the study is
pointed out and the area of study is presented. Chapter 2 describes the main aspects of the methodology
applied. Chapter 3, the main chapter, characterises the meteorological and hydrodynamic agents, and
the physical processes determinant for the sediment dynamics of the beach: section 3.1 characterises
wind and typhoons; section 3.2 characterises tides, storm surges and currents; section 3.3 characterises
the wave climate and its recent evolution in the nearshore of Hac-Sá beach; section 3.4 characterises the
morphology and sediments of the beach and nearshore and their recent morphological evolution; section 3.5 characterises the longshore and cross-shore sediment transport processes and the shoreline evolution. Finally in chapter 4, conclusive remarks are presented.

2 STUDY METHODOLOGY

The main aspects of the methodology applied were: the analysis of the natural agents that intervene in
the Hac-Sá beach dynamics; the identification and evaluation of the major beach dynamics processes
and their interaction mechanisms; and the characterisation of the recent evolution of the beach system.
The study was developed based on the analysis of: meteorological, wave and sea level observations;

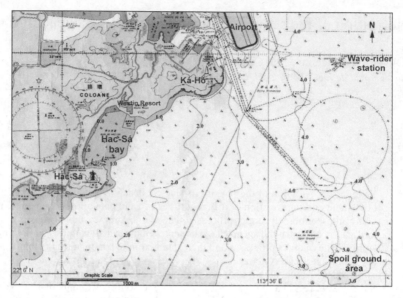

Figure 1.   Location of the area of study (based on chart "Macau, China. Portos de Macau, Taipa e Coloane.", 2001, by Capitania dos Portos).

bathymetric, topographic and sedimentological surveys; and aerial and land photographs. The data were processed, analysed and interrelated either based on analytical methodologies or mathematical models of beach dynamics. Specific coastal engineering mathematical models of wave propagation, short-term beach profile evolution, and medium- and long-term shoreline evolution were applied. Other types of numerical models not exclusively applied in coastal engineering studies were also used, like the case of a digital terrain model applied to study the morphological evolution of the coastal region of Hac-Sá beach.

3   COASTAL DYNAMICS

3.1   Wind and typhoons

The most frequent wind conditions and patterns recorded at Ká-Hó meteorological station are described in Instutito Hidrográfico & Capitania dos Portos de Macau (1998) and in Autoridade de Aviação Civil de Macau (1997). The predominant wind directions are NNE and N with almost 25% of occurrence and mean speed intensity of about 26 and 21.5 km/h respectively, followed by the S to SSW sector, based on data from the 1986–94 period. The prevailing seasonal wind is: from N during the monsoon months from November to February; from SW from March to May; from S to SW in June and July; from SW in August and September; and changing from SW to N in the transitional month of October. The maximum wind speed intensity variation was from 44 to 110 km/h taking typhoons into account, and from 34 to 75 km/h excluding typhoons, based on data from the 1952–81 period.

Typhoons and tropical storms can produce strong wind, intense rainfall, very low atmospheric pressure and storm surges. In the region of Macau they are more frequent from June to October and their usual effect in the coast is an incidence of locally generated waves and a rising of the sea level above the astronomic prediction of tidal levels. This sea level rise is the result of atmospheric pressure decrease and accumulation of water near the coast due to wind stress. The average sea level rise, resulting from atmospheric pressure decrease associated to typhoons is 20 cm, and the maximum value can reach 50 cm. The maximum value of wind set-up can reach 2 m.

3.2   Tides, storm surges and currents

Tides in the area of Macau are semi-diurnal, with strong diurnal inequalities. However in some occasions of the year, one of the tides is so small that there seems to exist only a tide per day. The mean and

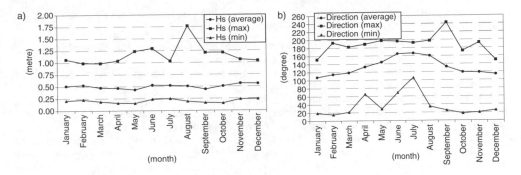

Figure 2.   Annual variation of: (a) Hs; and (b) Wave direction. Based on data from the 1996–99 period.

maximum tidal range are 1.13 and 2.84 m respectively (Instutito Hidrográfico & Capitania dos Portos de Macau 1998).

The highest sea levels in Macau occur during the passage of typhoons and result from the association of spring tides with storm surges. Maximum annual levels recorded during the 1925–89 period vary between 3.03 and 4.74 m above CD, with a mean value of 3.41 m. Statistic calculations gave values of about 3.9 and 4.3 m for return periods of 10 and 50 years.

Patterns of ebb and flood tidal currents resulting from mathematical modelling (Portela 1998) show that the current intensity decays near the coast and is very small inside Hac-Sá bay.

3.3   Wave regime and nearshore propagation

Being the dominant forcing agent of the sediment dynamics process, the wave climate in the coastal region of Hac-Sá bay was estimated. The wave regime at the wave-rider station was treated and characterised statistically. Such allowed establishing a representative annual wave regime, which was transferred through numerical modelling to the nearshore. The numerical results of propagation of the representative wave regime in front of the beach were used to evaluate the longshore beach exposure to the incoming wave energy and as input to the longshore and cross-shore sediment transport numerical models applied.

Wave data (recorded at the wave-rider station located opposite Ká-Hó) of four complete years, from the 1996–99 period, for which simultaneous information on the wave parameters height, period and direction is known, were statistically treated to characterise the wave regime at the wave-rider station position. These results not only gave information on the inter-annual variation of the wave regime, but also allowed to estimate monthly and annual variations of each parameter, and consequently, to identify seasonal variation tendencies of the wave regime. Two more energetic seasons associated to the larger wave periods were identified along the year, one from October to February and the other from June to August (Fig. 2a). Two directional wave seasons were also identified (Fig. 2b): from April to September the incoming waves have a higher obliquity relatively to the general beach alignment, generated by a SSE incident direction, than during the rest of the year. The lowest directional dispersion occurs in April, June and July.

A representative regime at the wave-rider station position was established based on the analysis of the distribution of the frequency of occurrence of the wave characterisation parameters, in which main conclusions are:

- The directional wave sector with greater incidence, about 36%, is N105–135°, followed by the sector N165–195° with about 19.6%, and the sector N135–165° with about 14.2% of occurrence.
- The significant wave height (Hs) distribution showed that about 38% of the incoming waves have an Hs between 0.25 and 0.50 m. This Hs class is followed by the class [0.50–0.75], with about 27.8% of occurrence. Waves with Hs lower than 0.25 m occurred 5.4% of the period of analysis. Waves with Hs within the class [0.75–1.00] have an occurrence of 6.5%, and finally waves with Hs higher than 1.00 m occurred only 1.4% of the time, and mostly from the directional sector N105–135°.
- The zero upcrossing period (Tz) distribution showed that the largest percentage of waves, about 39.3%, has a Tz between 3 and 4 s. The second highest class of occurrence, about 23.2%, correspond

to waves with Tz between 4 and 5 s. The longest waves, with Tz higher than 5 s, occurred 9.8% of the time and waves with Tz within the class [2–3] have 4% of occurrence.

In order to assess the recent wave climate evolution in the nearshore area of Hac-Sá beach, two cases of propagation of the representative wave regime towards the coast have been studied: one correspondent to the situation in 1985 (prior to the construction of Macau International Airport) and the other correspondent to the situation in 2002. The methodology established to investigate the sea wave transformations and resultant wave climate in the nearshore area for both cases, was based on a deterministic wave propagation numerical model (Oliveira 1997), which simulates simultaneously the physical processes of refraction, diffraction, reflection and energy dissipation due to breaking. The results allow the quantification of the parameters that characterise the wave regime in any location of the area of study, the evaluation of areas of energy concentration, the identification of protected or shadow areas due to effects of diffraction and the analysis of the wave action variation along the longshore extension of Hac-Sá beach. In general, the alongshore exposure of Hac-Sá beach to wave energy is rarely uniform. The northern part of the beach tends to be protected by the northern headland (Ká-Hó) against incoming wave energy from the directional sectors N45–75° and N75–105°, due to the occurrence of diffraction, which effect is the generation of a shadow zone (Fig. 3), which extension decreases as the incident wave angle increases. Also due to the occurrence of diffraction, the southern part of the beach tends to be protected by the southern headland (Hac-Sá) against incoming wave energy from the directional sectors N135–165° and N165–195°.

The comparison of the results for the two cases, 1985 and 2002, allows concluding on the effect of the recent morphological (depth and sea bottom geometry) and land contour changes on the wave climate in the nearshore area. One of the major impacts of the wave propagation changes on the beach was the significant increase of the gradient of the incident wave energy in the longshore direction due to the presence of a recent irregular shoal resultant from disposal of dredged material in the nearshore region in front of the beach, which generates a complex wave transformation pattern with several convergence zones, i.e. peaks of energy concentration, between this new morphological feature and the shoreline.

3.4 *Morphological evolution*

The morphological evolution of Hac-Sá beach adjacent area in the last three decades was characterised based on comparisons between hydrographic data available from several chart surveys from the 1966–2001 period. The data were processed with a digital terrain model and the volumetric results obtained from the comparisons allowed to conclude on the rates of erosion and accretion. The morphological evolution of the coastal area in the vicinity Hac-Sá beach is characterised by a general deposition, with an average accretion rate of 2–3 cm.year^{-1}. Since 1985, this accretion rate observed showed a tendency to increase offshore and decrease nearshore Hac-Sá bay. Most likely, this evolution tendency is a consequence of the changes on the hydrodynamic conditions introduced by the construction of Macau

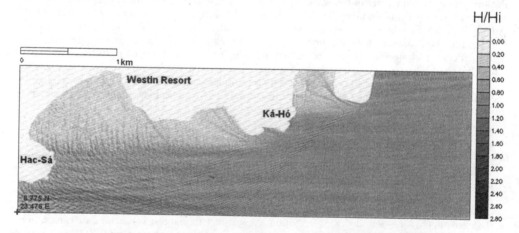

Figure 3. Numerical results of wave propagation. Incident wave: Tz = 3s; Direction = N60°.

International Airport. An evidence of its effect is the increase of deposition observed in the surround-ings of Ká-Hó cape, which can be interpreted based on the sheltering effects of the airport, favouring sediment deposition. Another new morphological feature is the existence of a distinct accumulation zone of disposal of dredged material, the spoil ground area, in front of Hac-Sá beach, already mentioned in section 3.3.

The beach morphological evolution was analysed based on the shoreline positions at MSL and the backshore occupation of the beach. These data were obtained from aerial photographs of Hac-Sá beach, shoreline surveys and charts showing the occupation of the backshore area for the 1966–2002 period. The main evolution aspects are: a significant retreat of the shoreline at the southern stretch; remarkable land reclamation at the north and south headlands; and a progressive occupation of the backshore zone with infrastructures and other constructions resulting in the narrowing of the beach.

3.5 Beach dynamics

The Hac-Sá beach sediment motion was characterized for the situations of 1985 and 2002. The main differences between the two situations concern to the following aspects: shoreline shape and alignment changing due to the erosion process occurred in October 2000; modification of backshore limits due to progressive occupation of beach periphery; and variation of beach exposure to wave action due to the changes on the wave propagation pattern resultant from the nearshore bottom changes and land contour changes.

The numerical results obtained from the propagation of the representative annual wave regime were used as input to the sediment transport numerical models. Numerical simulations of short-, medium- and long-term sediment transport were performed to: interpret long term longitudinal equilibrium configurations of the beach; evaluate the cross-shore distribution of the longshore transport along the year in both directions; characterized seasonal patterns of morphological variations; evaluate the influence of the previously referred morphological and wave propagation changes on the recent erosion events; and analyze the impact of maritime storm events on the beach profile development.

3.5.1 Longshore transport

Hac-Sá beach longshore transport, confined between the north and south headlands, produces a constant displacement of the granular sediments in both longitudinal directions, according to the varying incident wave directions. Along the beach, the longshore transport is liable to a constant spatial and time variation: each incident wave reaches the beach with height and direction that vary alongshore, thus, producing different local transport (spatial variation); each beach position is reached by a succession of waves with varying characteristics (time variation).

Two numerical models (Vicente 1991, DHI 2000a), based on different mathematical approaches, were applied for the past equilibrium situation of the beach (1985) to simulate the annual and seasonal longshore sand transport and its cross-shore distribution for a profile at the central part of the beach, and to identify the cross-shore extension of the area of predominant longshore sediment motion. Since no field data is available for validation of the results, the major agreement observed from the comparison of the results from both numerical models granted their reliability. The shoreline evolution for this past equilibrium was also simulated (Fig. 4). The main aspects of the results are:

- The shoreline presents a long-term stability with a net transport almost null, as expected, taking into account the good confinement of the beach by the headlands;
- There are distinct shoreline oscillations along the year as result of the wave climate conditions, which present a seasonal variation of wave directions. During the period from April to September there is a net transport directed northwards that produces enlargement of this part of the beach, and beach width reduction at the southern part. From October to March occurs an opposite sand mass displacement, with beach enlargement at southern part and decreasing of the beach width at north (Fig. 4);
- The average annual cumulative transport is approximately $120 \times 10^3 \, \text{m}^3$ with about $60 \times 10^3 \, \text{m}^3$ directed to north and about $60 \times 10^3 \, \text{m}^3$ directed to south (Fig. 5);
- The longshore sediment transport occurs within the surf zone and is not significant 1 m below CD.

The long-term evolution of the shoreline was also simulated for the beach conditions in 2002 (Fig. 6). When comparing these results of the new equilibrium configuration with the results of the past equilibrium configuration, it can be observed a retreat of the shoreline positions in the southern stretch of the beach, counterbalanced by shoreline advance at the northern stretch.

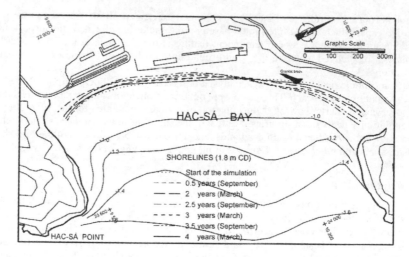

Figure 4.   Numerical results of shoreline evolution (1985).

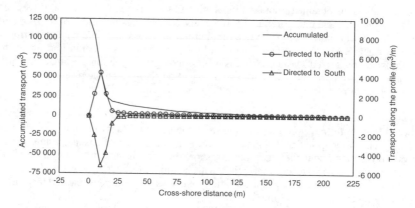

Figure 5.   Average annual longshore transport for a beach profile.

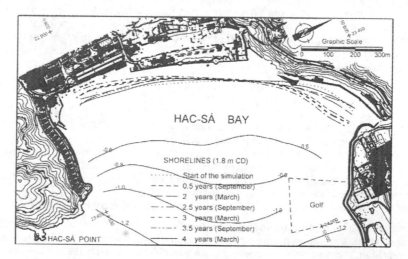

Figure 6.   Numerical results of shoreline evolution (2000).

3.5.2 Cross-shore transport

The cross-shore evolution of Hac-Sá beach is characterized by medium-term cyclic morphological changes due to a seasonal wave regime, and short-term morphological changes due to the occurrence of maritime storm events. The periodicity of the characteristics of the wave regime generates a sand shift from the offshore to the onshore of the beach and vice versa. The result is the occurrence of beach profiles with different morphological characteristics: during the most energetic seasons, the sand shifts from the berm to form one or a series of bars parallel to the shoreline in the offshore region; during the less energetic seasons, the reverse takes place, the sand shifts back onshore and the berm grows.

The short-term morphological changes in Hac-Sá beach are the result of a fast interaction between the hydrodynamics associated to a storm event (more energetic waves and sea level rise in the surf zone) and the beach morphology, during a short period of time. The result is a sudden increase of the offshore sediment transport due to the increase of the seaward gravity driven current (the undertow) and consequent development or reinforcement of the offshore bar. In order to assess Hac-Sá beach cross-shore behaviour, a numerical model (DHI 2000b) was applied to simulate the beach profile response to a maritime storm event. Two cross-shore profiles were used to represent the northern and southern sectors of the beach. The evolution of each profile was simulated for two situations, 1985 and 2002. In 2002, at MSL, the profile in the northern sector shows a beach retreat that is about 63% of the retreat occurred in 1985, and the profile in the southern sector, which in 1985 was not significantly affected, shows now a retreat about 2.2 times higher than the one observed in the northern sector of the beach. Such results show the increase, from 1985 to 2002, of the retreat of the southern sector relatively to the retreat of the northern sector in the presence of a storm event. These results illustrate the vulnerability of the southern sector of the beach, where the presence of advanced infrastructures (vertical seawall interrupted by beach access stairs) can limit the natural expansion of the beach profile in the presence of a more severe maritime storm event, which can cause a larger beach retreat. In fact, the presence of a highly reflective structure like a vertical seawall (absence of dissipative conditions) can even worsen the erosive process by increasing the offshore sediment transport due to the increase of the energy locally involved.

4 CONCLUSIONS

The sediment dynamics of Hac-Sá beach was characterized through the analysis and quantification of the natural forcing agents, followed by the evaluation of the major beach dynamics processes and their interaction mechanisms and finally, the assessment of the recent evolution of the system.

A detailed sea wave study was performed to obtain the wave climate in the beach nearshore region, since wave action is the determinant forcing factor in the shaping of shoreline and beach profile. The statistic characterisation of the wave regime at the wave-rider station was performed: energetic and directional seasons of the annual wave regime were identified; and a representative annual wave regime was established. The propagation of the representative wave regime towards the coast was simulated through numerical modelling. The evolution analysis of the wave climate from 1985 (prior to the construction of Macau International Airport) to 2000, showed significant changes due to the morphological and land contour changes occurred. The morphological changes are the accretion processes occurred in the nearshore region, with an average rate of $2-3 \, cm \cdot year^{-1}$, and the development of a distinct accumulation zone of disposal of dredged material, the spoil ground area, in front of Hac-Sá beach. The land contour changes are the significant shoreline retreat at the southern stretch of the beach and the remarkable land advance at the north and south headlands. These changes introduced changes in the wave propagation that affected the shoreline equilibrium.

The results of the numerical simulations of the sediment dynamics showed an annual accumulated longshore transport of sand of about $60 \times 10^3 \, m^3$ in each direction. Significant shoreline oscillations along the year occur in response to wave action, which presents an expressive seasonal variation of wave directions. During the period from April to September the net transport is directed northwards and produces enlargement of the north part of the beach and narrowing of the southern stretch. From October to March, an opposite sand mass displacement occurs, with enlargement of the beach width at south and narrowing at the northern stretch. The predominant longshore transport occurs within the surf zone. Sediment displacement is not significant 1 m below CD.

Short-term morphological evolutions in Hac-Sá beach were numerically simulated. The retreat of the cross-shore beach profile, as response to a relatively frequent maritime storm event (not of the most severe type), can easily reach about 10 m, even without the occurrence of exceptional sea levels.

During the last two decades, the significant morphological changes occurred in the coastal area of Hac-Sá beach and the land reclamation works at the north and south headlands of Hac-Sá bay introduced changes in the nearshore wave propagation, that together with the progressive occupation of the beach backshore zone with infrastructures and other constructions affected the shoreline equilibrium.

ACKNOWLEDGMENTS

The authors thank to LECM for authorizing the publication of the present study.

REFERENCES

Autoridade de Aviação Civil de Macau 1997. *Macau International Airport Technical Handbook*. LECM, Macau.

DHI 2000a. *LITDRIFT. Longshore Current and Littoral Drift – LITDRIFT user guide*, Danish Hydraulic Institute, Denmark.

DHI Software 2000b. *LITPROF. Profile development. – LITPROF user guide*, DHI, Denmark.

Instituto Hidrográfico and Capitania dos Portos de Macau 1998. *Meio Hídrico de Macau*. 2ª edição, Lisboa, Portugal.

Oliveira, F. S. B. F. 1997. *Numerical modelling of irregular wave propagation in the nearshore region*, Ph.D. Dissertation, Imperial College of Science Technology and Medicine, University of London, UK.

Oliveira, F. S. B. F. 2002. *Rehabilitation of Hac-Sá beach (Macau). Phase 1. Volume 1: Sea wave study on the east coast of Coloane*. Report 282/02 – NET, Laboratório Nacional de Engenharia Civil, Lisboa, Portugal.

Oliveira, F. S. B. F., Clímaco, M. & Freire, P. M. S. 2002. *Rehabilitation of Hac-Sá beach (Macau). Phase 1. Volume 2: Characterisation of the sediment dynamics and the erosion causes*. Report 286/02 – NET, Laboratório Nacional de Engenharia Civil, Lisboa, Portugal.

Portela, L. I. 1998. *Improvement of the Area Surrounding Macao in the Pearl River Estuary: Study of Three Configurations*. Rel.263/98-NET, LNEC, Lisboa.

Vicente, C. 1991. *Aperfeiçoamento de métodos de modelação matemática e física aplicáveis a problemas de dinâmica costeira*. Programa de investigação para Investigador Coordenador. Departamento de Hidráulica, Laboratório Nacional de Engenharia Civil, Lisboa, Portugal.

Computational Methods in Engineering and Science, Iu et al. (eds)
© 2003 Swets & Zeitlinger, Lisse, ISBN 90 5809 567 3

Numerical experiment of thermal comfort in a bedroom equipped with air-conditioner

V.K. Sin, H.I. Sun, L.M. Tam & K.I. Wong
Department of Electromechanical Engineering, Faculty of Science and Technology, University of Macau, Macao

ABSTRACT: More than 6 months of a year, the weather of Macao is hot and humid. This uncomfortable weather induces the importance of operating air-conditioner in bedrooms. However, improper location of the air-conditioner will affect the performance of cooling effect, the thermal comfort of occupants, and the indoor air quality (IAQ). This paper is to simulate the air temperatures, velocities, and air diffusion performance index (ADPI) of a typical bedroom equipped with air-conditioner in Macao by using the computational fluid dynamics (CFD) software FLOVENT 3.2. Four simulations with each of the air-conditioner installed at four different locations are being studied. Analyses and comparison of results lead to the identification of proper location of the air-conditioner installation with best thermal comfort.

1 INTRODUCTION

Increasing concern has been paid on the indoor air quality (IAQ) all over the world due to its serious effect on human being. However, systematic research for IAQ has never being conducted in Macao and guideline in IAQ has not yet been established. Most of the residential and commercial buildings need to be operating with air-conditioners for more than half of a year because of the hot and humid weather of Macao. Therefore, IAQ is in fact an important issue to be investigated in Macao. Air temperature and velocity inside a building operated with air conditioner are the two important variables to be studied as they directly affect the thermal comfort of the occupants. Choice of location for installing the air-conditioner is crucial as it can affect the air temperature and velocity distribution which in turn affect the thermal comfort of the occupants and IAQ of the building.

Another important parameter to measure the thermal comfort for occupants is air diffusion performance index (ADPI). As stated in Anonymous b (2001), ADPI is a parameter that measures the uniformity of the space in terms of the proportion of the volume with velocity lower than 0.35 m/s and draft temperature between $-1.7°C$ and $+1.1°C$ from the mean temperature. For example, an ADPI of 0.8 means that 80% of the locations in the simulated area have velocity lower than 0.35 m/s and draft temperature between $-1.7°C$ and $+1.1°C$ from the mean temperature. The most comfortable condition of the room is obtained when the ADPI is 1.0 where 100% of the locations in the simulated area meet the comfort criteria as stated.

The objective of the paper is to investigate the choice of location for air-conditioner installation in order to accomplish the best thermal comfort for occupants in a bedroom. Distributions of air temperature, velocity, as well as ADPI are simulated with the commercial CFD code FLOVENT 3.2. Results in the paper can be used as reference for room occupants to make decision on where to install their air-conditioner.

As a first attempt to the analysis, the following assumptions are made. Steady state environment is assumed which means that the indoor and outdoor air temperatures are assumed to be unchanged with time and the supply temperature of the air-conditioner is also assumed to be fixed. Thermal radiation through the windows and that between the objects inside the room are not concerned. The generation of carbon dioxide from the occupants is not considered since air velocity and temperature are the main parameters interested in this simulation

2 MATHEMATICAL MODEL AND SIMULATION OF TURBULENCE-FLOVENT

In FLOVENT, Reynolds-averaged Navier-Stokes equations are solved with finite volume technique and turbulence effects upon the mean flow are modeled through the eddy viscosity concept. Steady state analysis is to be used and the governing equations are

$$\frac{\partial U_i}{\partial x_i} = 0 \tag{1}$$

$$\rho U_i \frac{\partial U_j}{\partial x_i} = \frac{\partial}{\partial x_i}\left(\mu\left(\frac{\partial U_i}{\partial x_j} + \frac{\partial U_j}{\partial x_i}\right) - \overline{\rho u_j u_i}\right) - \frac{\partial p}{\partial x_j} + g_j\left(\rho_r - \rho\right) \tag{2}$$

$$\rho C_p U_i \frac{\partial T}{\partial x_i} = \frac{\partial}{\partial x_i}\left(\lambda \frac{\partial T}{\partial x_i} - \rho C_p \overline{u_i T'}\right) + S \tag{3}$$

where ρ is density of air; ρ_r is reference density of air; μ is viscosity of air; U_i is mean velocity; u_i is fluctuating velocity; T is mean temperature; T' is fluctuating velocity; C_p is specific heat at constant pressure of air; λ is thermal conductivity; S is heat generation; g_j is gravitational acceleration. $-\rho\overline{u_j u_i}$ is the Reynolds stress which is defined as:

$$-\rho\overline{u_j u_i} = \mu_t\left(\frac{\partial U_i}{\partial x_j} + \frac{\partial U_j}{\partial x_i}\right) - \frac{2}{3}\rho k \delta_{ij} \tag{4}$$

where μ_t is turbulent viscosity; δ_{ij} is Kronecker delta. $-\rho C_p \overline{u_i T'}$ is the Reynolds flux which is defined as:

$$-\rho C_p \overline{u_i T'} = \lambda_t \frac{\partial T}{\partial x_i} \tag{5}$$

Turbulent viscosity, μ_t, defined from dimensional analysis as:

$$\mu_t = C_\mu \rho \frac{k^2}{\varepsilon} \tag{6}$$

where $C_\mu = 0.09$; k is turbulent kinetic energy; ε is dissipation rate of k.
Turbulent conductivity, λ_t, is related to turbulent viscosity by the following equation:

$$\lambda_t = \frac{c_p \mu_t}{\sigma_t} \tag{7}$$

where σ_t is the turbulent Prandtl number (equal to 0.9).
Transport equations for k and ε are:

$$\frac{\partial(\rho U_i k)}{\partial x_i} = \frac{\partial}{\partial x_i}\left(\left(\mu + \frac{\mu_t}{\sigma_k}\right)\frac{\partial k}{\partial x_i}\right) + P + G - \rho\varepsilon \tag{8}$$

$$\frac{\partial(\rho U_i \varepsilon)}{\partial x_i} = \frac{\partial}{\partial x_i}\left(\left(\mu + \frac{\mu_t}{\sigma_\varepsilon}\right)\frac{\partial \varepsilon}{\partial x_i}\right) + C_1 \frac{\varepsilon}{k}\left(P + C_3 G\right) - C_2 \rho \frac{\varepsilon^2}{k} \tag{9}$$

where $C_1 = 1.44$; $C_2 = 1.92$; $C_3 = 1.0$; $\sigma_k = 1.0$; $\sigma_\varepsilon = 1.217$; P is the shear production defined as:

$$P = \mu_{eff} \frac{\partial U_i}{\partial x_j}\left(\frac{\partial U_i}{\partial x_j} + \frac{\partial U_j}{\partial x_i}\right) \tag{10}$$

G is the production of turbulence kinetic energy due to buoyancy, and is given by:

$$G = \frac{\mu_{eff}}{\sigma_t} \beta g_i \frac{\partial T}{\partial x_i} \tag{11}$$

where $\mu_{eff} = \mu + \mu_t$, is the effective viscosity; β is coefficient of volumetric expansion.

The draft temperature, T_d, is calculated using the following formula as stated in Anonymous b (2001):

$$T_d = T_a - T_m - 7.66 \times (Speed - 0.15) \tag{12}$$

where T_a is air temperature (i.e. fluid temperature); T_m is mean air temperature in °C over all empty regions of all mean flow regions; and $Speed$ is the local air-stream centerline speed which is in m/s.

3 SIMULATION OF BEDROOM WITH AIR-CONDITIONER

Simulations of installing the air-conditioner at four different locations in a typical furnishing bedroom in Macao as shown in Figure 1 and Figure 2 are being performed. The overall dimension of the bedroom is 3.03 m × 3.33 m × 2.5 m with one external wall, three internal walls, door, floor and ceiling. Part of the external wall is installed with two external windows made of single-glazed glass. The overall thermal transmission coefficient (U) of the external wall is 2.35 W/m^{2}K and that for the window is 5.9 W/m^{2}K. It is assumed that all internal walls, ceiling, floor are thermally insulated. Simple furniture which includes a bed with a bed table and a bed lamp beside it, a wardrobe, a TV set and a dressing table are put inside the bedroom. The room is equipped with a split-type air-conditioner which supplies air at a temperature of 16°C, and at a flow rate of 200 m^3/hr (7.9 ACH). Air change per hour (ACH) is the number of

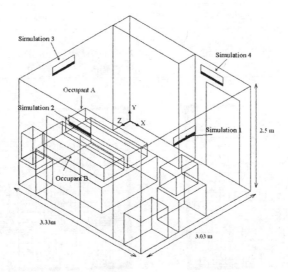

Figure 1. 3-D layout of the simulated bedroom which shows the 4 different locations of the air-conditioner for simulations.

changes of total room volume of air per hour. 10% of the supply air is fresh air while the rest of it is the return air. Two occupants who totally generate metabolic heat of 140 W as stated by Stocker and Jones (1982) are sleeping on the bed where occupant A is close to the wardrobe and occupant B is close to the windows. During summer in Macao, the outdoor air temperature is set as 33°C and the initial temperature of the simulated bedroom is set as 30°C. Thermal radiation from the outdoor through the windows is not considered in this simulation as Venetian blinds are assumed to be used. Only convective and conductive heat transfer from the occupants and the air-conditioner are being concerned. Any heat released from the other furniture and walls are not taken into account.

For simulation 1, the air-conditioner is located opposite to the bed and above the TV set. For simulation 2, the air-conditioner is located beside the bed and is above the windows. For simulation 3, the air-conditioner is located behind the bed and is above the heads of the occupants. For simulation 4, the air-conditioner is located next to the wardrobe and is above the door. The exact location of the air-conditioner and the number of grids being used in each of these four simulations are given in Table 1. It should be emphasized that non-uniform grids as shown in Figure 3 are used where the grids around the occupants and the air-conditioner is fine enough to obtain the accurate numerical results.

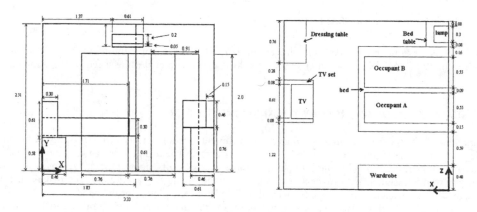

Figure 2.   Front view (left) and top view (right) of simulation 2.

Table 1.   Summary of location for air-conditioner and number of grids being used for the 4 simulations.

	Simulation 1	Simulation 2	Simulation 3	Simulation 4
No. of grids used:	57,387	51,408	52,640	52,632
Location of air-conditioner (x,y,z):	(3.33, 2.13, 1.52)	(1.37, 2.13, 3.03)	(0, 2.13, 1.52)	(1.98, 2.13, 0)

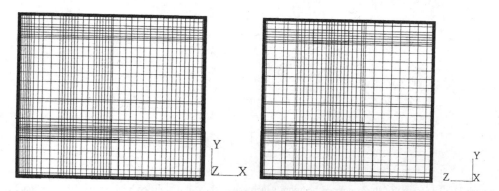

Figure 3.   Non-uniform grids (grid number = 57,387) of z-view (left) and x-view (right) for simulation 1.

4 RESULT AND ANALYSIS

Two vertical z-planes are constructed to show the contours of temperature and velocity. The first one called plane A is set at z = 1.6 m where occupant A is lying and the other called plane B is set at z = 2.1 m where occupant B is lying. When the occupants sleep, they usually cover their bodies with blanket except exposing their heads to the air. This makes the head of the occupants to be the most sensitive to the changes of air temperature and velocity and so the area near the head of the occupants are the most important part in the analysis.

In simulation 1, the temperature and velocity contours of planes A and B are shown in Figure 4 and Figure 5. The temperature ranges from 21.9°C to 24.6°C for the room and that around the head of occupant A is 23.1°C and around the head of occupant B is 23.8°C. These give a moderate and comfortable temperature for the occupants. The velocity around the head of occupant A is 0.16 m/s while for occupant B is 0.23 m/s. The ADPI of the bedroom for this simulation is 0.84. Simulation with increasing number of grids up to 67,914 has been performed and the results show no significant difference with that obtained in 57,387 grids.

In simulation 2, the temperature and velocity contours of planes A and B are shown in Figure 6 and Figure 7. The temperature ranges from 21.7°C to 25.8° for the room and that around the head of occupant A is 24.6°C and around the head of occupant B is 25.8°. These give a moderate and comfortable temperature for the occupants. The velocity around the head of occupant A is 0.08 m/s while for occupant B is 0.05 m/s. The ADPI of the bedroom for this simulation is 0.82.

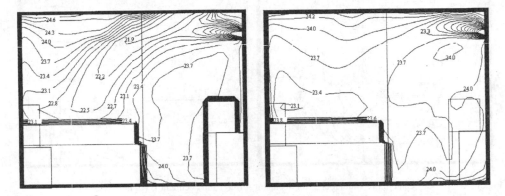

Figure 4.   Temperature contours of simulation 1 for occupant A (left) and occupant B (right).

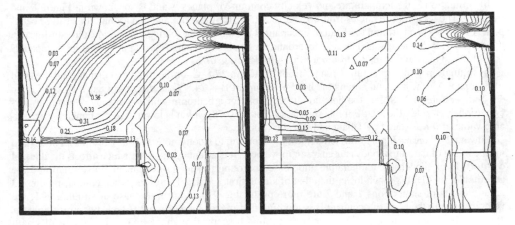

Figure 5.   Velocity contours of simulation 1 for occupant A (left) and occupant B (right).

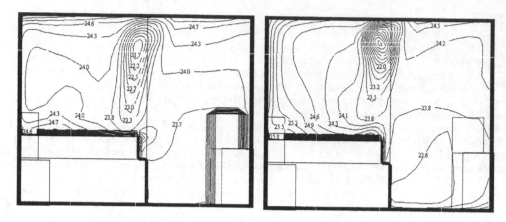

Figure 6. Temperature contours of simulation 2 for occupant A (left) and occupant B (right).

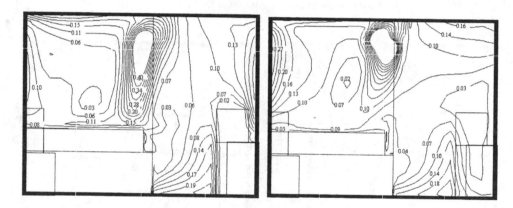

Figure 7. Velocity contours of simulation 2 for occupant A (left) and occupant B (right).

In simulation 3, the temperature and velocity contours of planes A and B are shown in Figure 8 and Figure 9. The temperature ranges from 19.7°C to 25.5°C for the room and that around the head of occupant A is 23.9°C and around the head of occupant B is 24.5°C. These give a moderate and comfortable temperature for the occupants. The velocity around the head of occupant A is 0.11 m/s while for occupant B is 0.11 m/s. The ADPI of the bedroom for this simulation is 0.76.

In simulation 4, the temperature ranges from 21.9°C to 27.8°C for the room and that around the head of occupant A is 27.8°C and around the head of occupant B is 24.8°C as shown in Figure 10 and Figure 11. Temperature as 27.8°C is too hot and unacceptable for a bedroom which is already operated by air-conditioner. Besides, the ADPI of this simulated room is 0.78, which is lower than the generally acceptable value of 0.8 as stated by Tibaut and Wiesler (2002).

Lower ADPI means higher draft temperature. According to Anonymous a (2001), draft is an undesired local cooling of the human body caused by air movement, it is an annoying factor which affects the thermal comfort. Even though the temperature and velocity around the head of occupants in simulation 3 are acceptable, its ADPI is much lower than that of simulation 1 and 2. Therefore, when comparing these 3 cases, results of simulations 1 and 2 are more preferable. However, in the case of simulation 1, the velocity around the head of occupant B is too high that makes the occupant B feels uncomfortable for exposing his head at this air velocity during sleeping. It seems that simulation 2 gives the results with best thermal comfort.

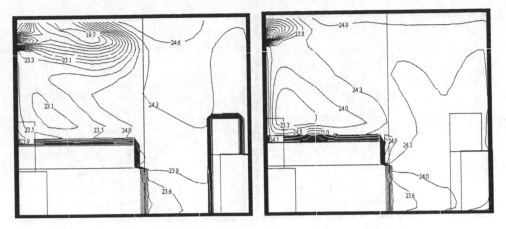

Figure 8.   Temperature contours of simulation 3 for occupant A (left) and occupant B (right).

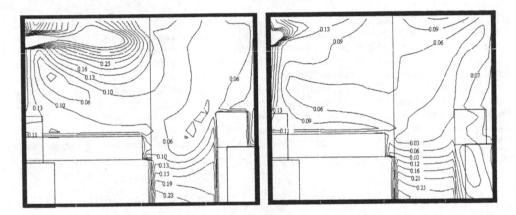

Figure 9.   Velocity contours of simulation 3 for occupant A (left) and occupant B (right).

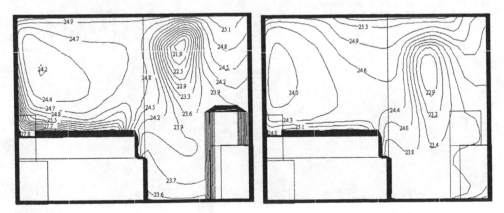

Figure 10.   Temperature contours of simulation 4 for occupant A (left) and occupant B (right).

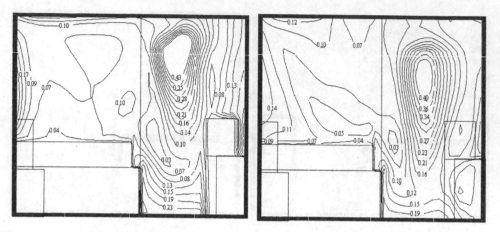

Figure 11.   Velocity contours of simulation 4 for occupant A (left) and occupant B (right).

5   CONCLUSIONS

Four different locations for installing the air-conditioner in a bedroom in Macao are studied with the aid of CFD software FLOVENT 3.2 to obtain the best thermal comfort for the occupants. Among these four simulations, the fourth one that the air-conditioner is placed above the door is the worst case due to the high temperature produced around the occupant's face. Among the other three cases which can obtain moderate temperature, the third simulation that the air-conditioner is placed behind the bed produces draft temperature which affects the thermal comfort of the occupants during sleeping, so choice of location of air-conditioner corresponding to simulation 3 is not a good one when comparing with the other two. In the simulation 1 where the air-conditioner is placed opposite to the bed, the velocity around the head of occupant B is too high that makes him feel uncomfortable when sleeping at this environment. The best situation in terms of thermal comfort is identified to be simulation 2 in which air-conditioner is located beside the bed and above the windows.

REFERENCES

Anonymous a 2001. *2001ASHRAE handbook fundamentals*. Atlanta: American Society of Heating, Refrigerating and Air-conditioning Engineers, Inc.
Anonymous b 2001. *Flovent reference manual, FLOVENT/OL/MM/0801/1/0*. Surrey: Flomerics Limited.
P. Tibaut / B. Wiesler 2002. Thermal comfort assessments of indoor environment by means of CFD. *8th International Conference Air Distribution in Rooms*: 97.
Wilbert F. Stocker / Jerold W. Jones 1982. *Refrigeration & Air Conditioning*, Singapore: Mcgraw Hill Book Company.

Computational Methods in Engineering and Science, Iu et al. (eds)
© *2003 Swets & Zeitlinger, Lisse, ISBN 90 5809 567 3*

Pilot study on transboundary pollution analysis of Huai River

A.Q. Zhang, S.K. Han & L.S. Wang
Department of Environmental Science, School of the Environment, Nanjing University, Nanjing, China

H.S. Wu
Jiangsu Provincial Environmental Science Institute, Environmental Protection Department of Jiangsu Province, Nanjing, China

ABSTRACT: Monitoring results of some monitoring sites on Huai River in Jiangsu section were employed to make pilot analysis on the transboundary pollution state of the river and the policy influence on it. Data exploration showed that the intensive administrative intervention in pollution control action during 1996–1997 period in Huai River Basin significantly reduced the transboundary pollution problem of Huai River at the Jiangsu-Anhui boundary, while the major pollutant indices of Huai River in Xuyi section changed as well. Moreover, the predomination of the index of NH_3–N implied more improvements should be made in Xuyi to meet the strict requirements of Ecological Agriculture Demonstration.

1 INTRODUCTION

Huai River ranks among seven largest river in China, yet the deficit of water resource in Huai River Basin is critical. Now known as one of the dirtiest rivers in China, the water body was plagued by grave pollution problems as well. Even some drinking water sources along it are choked by nutriment and micro-pollution of toxic organic compounds. In response to the severe pollution situation, many regulations were carried into execution to alleviate problems. *Interim Ordinance of Water Pollution Prevention and Control for Huai River Basin* that was promulgated by the State Council on August, 1995 was a typical example, and the water quality in 1998 along the mainstream of Huai River has shown some improvement consequently. Nevertheless, the water quality of the River is still bad. Moreover, its large drainage area, covering four provinces, Henan, Shandong, Anhui and Jiangsu, makes the situation more complicated.

As the last inland region in Huai River Basin before the river arrives at its estuary, Jiangsu Province, the economically most developed province in China, acts as final recipient of all water pollutants discharged from human activities in the upper reaches of the river. The classical bone of contention between Jiangsu and other Huai River Basin-related provinces has always laid on the allocation of total amount control indices for water pollutant discharge, and such inter-provincial fight against pollution is a laborious process, which often turns out a failure without scientific data analysis. Since the main stream of Huai River enters Jiangsu from Anhui Province at Daliuxiang and extends 80 km in Xuyi County to the Hongze Lake, one major fishery base in South China, the conflict between Jiangsu and Anhui Province seems fiercer.

In this paper, reported results of four monitoring sites located on this 80 km river section, Daliuxiang (Point 1), Dashishan (Point 2), Huai River Bridge (Point 3), Hongguang Chemical Plant (Point 4) were collected to make pilot analysis on the transboundary pollution state of the river and the policy influence on it. Furthermore, evaluation of present pollution status of Xuyi County was also made on the basis of such analysis.

2 DATA EXPLORATION

2.1 *Monitoring area*

All four monitoring points stand in Xuyi County, which is located at the lower riches of Huai River, lying in the west part of Jiangsu Province. Besides, it is adjacent to Tianchang County in Anhui Province

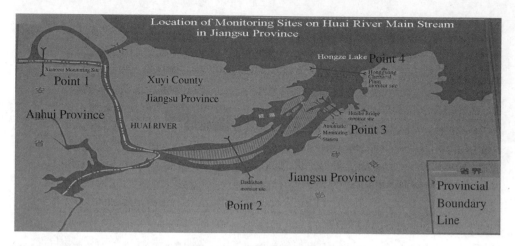

Figure 1. Location of monitoring sites on the trunk stream of Huai River basin.

at the east and borders upon Laian County and Mingguang City in Anhui Province at the southwest. 11 types of pollution data including permanganate index (PI), BOD_5, ammonia (NH_3–N), nitrite (NO_2–N), volatile phenols (VOP), total CN content (TCN), total arsenic content (TAs), total content of sexivalence-chromium (TCr), total lead content (TPb), total cadmium content (TCd), and Oil were recorded. The locations of the monitoring sites are illustrated in Figure 1. Point 1 lies on the boundary line of Jiangsu and Anhui Province, while point 4 locates in the lake inlet of Trunk stream of Huai River. In addition, from the Figure 1 it also can be seen that there is transboundary pollutant input between point 1 and point 2.

2.2 Data exploration methods

Some statistical methods such as simple sequential analysis, linear display methods (principal component analysis), and hierarchical clustering were adopted to extract intrinsic information from the experimental data normalized to zero mean and unit variance in order to avoid misclassifications arising from the different order of magnitude of both numerical value and variance. Since the methods of classification used here are non-parametric, they make no assumptions about the underlying statistical distribution of the data and therefore no evaluation of normal (Gaussian) distribution of the data is necessary. As for PCs, the characteristic roots (eigenvalues) of the PCs are a measure of their associated variances, and the sum of eigenvalues coincides with the total number of variables. Correlation of PCs and original variables is given by loadings, and individual transformed observations are called scores.

Cluster analysis is an unsupervised pattern recognition technique that uncovers intrinsic structure or underlying behaviour of a data set to classify the objects of the system into categories or clusters based on their similarity without making a priori assumptions about the data. Part of popularity of hierarchical cluster analysis cannot only be attributed to its relative computational simplicity but also to its ability of visually representing the result as a so-called dendrogram. All the data processing was accomplished by using SPSS 9.0 for Windows (SPSS INC. 1998). The detailed mathematic principles will not be discussed here since there are innumerable books and scientific papers focusing on it. Concisely, hierarchical clustering starts with M particles or clusters of which the most similar ones are joined successively into new clusters. The Squared Euclidean distance was chosen to measure the similarity between different particles, while Ward's strategy was followed to unit clusters since it possesses relatively small space distortion and yields the most meaningful clusters. Finally, a maximum internal homogeneity in the separated groups was achieved in this way (Andersen et al. 1998, Fletcher 1980, Goldfarb 1970, Manual 1985, Massart & Kanfman 1983). As cluster analysis allows the grouping of river water samples on the basis of their similarities in chemical composition, the dendrogram of cluster analysis may present illustrative profiles of transboundary pollution contribution from Anhui to Jiangsu section. The whole method was applied to normalized data.

Table 1. Descriptive statistics of point 1.

Pollution index	Minimum	Maximum	Mean	Std. deviation
PI	4.00	8.30	5.4933	1.7091
BOD_5	1.000	2.200	1.65000	0.54681
NH_3-N	0.010	0.650	0.12250	0.25853
NO_2-N	0.029	0.189	$8.6500E-02$	$5.3743E-02$
VOP	0.001	0.006	$3.1667E-03$	$2.3166E-03$
TCN	0.000	0.001	$1.6667E-04$	$4.0825E-04$
TAs	0.011	0.022	$1.5667E-02$	$4.9261E-03$
TCr	0.000	0.002	$3.3333E-04$	$8.1650E-04$
TPb	0.000	0.001	$1.6667E-04$	$4.0825E-04$
TCd	0.0000	0.0001	$1.66667E-05$	$4.08248E-05$
Oil	0.100	0.640	0.24833	0.20961

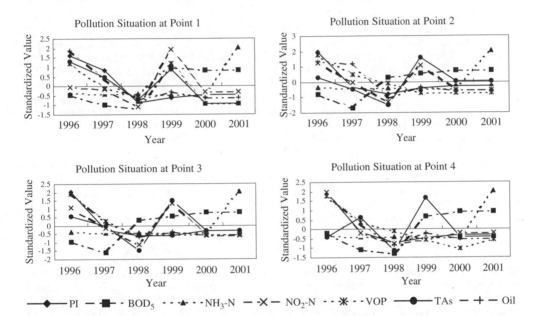

Figure 2. Sequence charts for four monitoring points.

3 RESULT AND DISCUSSION

3.1 Selection of sensitive pollution indices

Since the quality of monitoring data plays a very important role in statistical analysis, the very first step in data processing was to select sensitive pollution indices that well-describe the pollution status of case area through logistic quality screening process. Normally, large variations were highly expected for a sensitive index. Therefore, basic descriptive statistics of monitoring data were performed to eliminate the insensitive indices. Table 1 lists the result of descriptive statistics of data for Point 1, and obviously qualified indices passing such screening test were PI, BOD_5, NH_3-N, NO_2-N, VOP, TAs, TCN, TPb, and Oil. Considering the measuring methods of TCN, TCr, TCd and TPb proved to be modified in 2001, such data types were not taken into account in the following research.

3.2 Influence of intensive administrative intervention

Simple sequential analysis was employed to analyse the influence of intensive administrative intervention on pollution status, and Figure 2 showed the temporal variation profile of different pollution indices

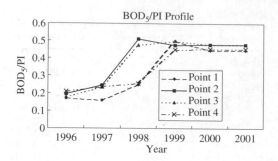

Figure 3.   The ration of BOD_5 to PI.

at each sampling site. Of all the PI data investigated in this study, the whole temporal variation profile presented a lowest value appeared in 1998 which was about 50 per cent less than in 1996, while the PI level showed no obvious variation between 1999 and 2001. Furthermore, the same kind of temporal profiles can be found in analyzing data of VOP, TAs, and Oil, and the flood in 1999 led to a relatively high pollution value of TAs, and Oil. As for NO_2–N, the lowest pollution data also occurred in 1998 with a significant increase in 1999, but the pollution levels in 2000 and 2001 kept balance with those in 1996 and 1997. Obviously, it can be seen from Figure 2 that intensive administrative intervention in pollution control action during 1996–1997 period in Huai River basin significantly reduced the pollution input, while the relatively serious pollution in 1999 may result from the flood disaster. However, the monitoring data of BOD_5 and NH_3–N took on a quite different temporal profile.

Unlike the other indices, ammonia level in the studied watercourse decreased gradually from 1996 to 2000, and in 2001 NH_3–N jumped sharply by more than 20 times, which indicate a possible new agriculture-related pollution input. Considering the variation profile of ratio of BOD_5 to PI shown in Figure 3, there is a significant promotion in the biodegradability of the water since the level of BOD_5 after 1998 showed an obvious increase when being compared with that before 1998. Both indices implied the impact of administrative intervention before 1998 and industrial restructuring after 1998. Before 1998, untreated industrial discharges, especially the small-size papermaking factories, contribute mainly to the severe pollution of Huai River. The implementation of ZERO ACTION successfully closed many small-scale plants of heavy pollution and also spurred other plants to carry out necessary technical innovation and renovation to reduce pollutant discharge, and that is why most pollution indices reached their minimums in 1998. Moreover, all levels of local governments along Huai River turned to develop no-pollution or light pollution industries to meet the strict requirement of water quality since 1998, and Xuyi County even set Ecological Agriculture Demonstration Area as its development target. As a result, the major pollutant indices of Huai River in Xuyi section changed. Permanganate index is no longer the major pollution index for it after 1998, while the index of NH_3–N becomes one of the predominant indices in river pollution since 1998. Generally, the high level of NH_3–N is suspected to originate from overland runoff of riverine agricultural fields where the use of inorganic fertilizers is rather frequent. Besides, ammonia may also originate from decomposition of nitrogen-containing organic compounds such as proteins and urea occurring in municipal wastewater discharges. Thus, the domination of NH_3–N in present water samples implies that discharge of municipal wastewater and misapplication of fertilizer and pesticides has constituted the highest proportion in the present pollution sources of Huai River, which is totally different from former pollution pattern.

However, the abnormal high concentration of ammonia also warned us that more improvements should be made in Xuyi to prevent possible eutrophication in water body, and the obtained data is far from enough to trace the source of ammonia such as organic or inorganic precursors. Since the high variability of such environmental data results from not only a couple of possible anthropogenic sources, such as different emitters and dischargers but also geogenic, hydrological, and meteorological influences, serious lack of sufficient supporting data makes time series analysis impossible. Further work should be done on it when necessary data are available.

Besides, the pollution situations of Point 1 and 4, a site on Jiangsu-Anhui boundary and a farmost sampling site to the boundary showed different rules although the pollution profiles of four sites appeared in quite similar models. The value of PI in 1999 was the best example. Further analysis on spacial variation will be made by use of clustering analysis.

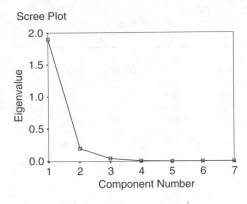

Figure 4.   Scree plot of principal component analysis.   Figure 5.   Component plot of first 3 PCs.

3.3   *Profile of transboundary pollution contribution to Huai River in Jiangsu section*

On the basis of such small database, hierarchical clustering technique was tried to find the clue which may shed more light on the transboundary pollution problem on Jiangsu-Anhui boundary. By applying the Bartlett's sphericity test, a value of 87.597 for the Bartlett chi-square statistic was found, confirming that variables are not orthogonal but correlated at the 99% significance level. In addition, the correlation matrix of the 7 analyzed indices was calculated, while the correlation coefficients and corresponding significant level were listed in Figure 2. Note that data of all sampling points were used to calculate the correlation matrix. Therefore, the correlation coefficients should be interpreted with caution as they are affected simultaneously by spatial and temporal variations. Nevertheless, some clear intrinsic relationships can be readily inferred. For example, high and positive correlation can be observed between BOD_5 and Ammonia index (at the 95% significance level), while BOD_5 is negatively correlated with volatile phenols (at the 95% significance level). Thus it is necessary to extract principal components to explain the n data variability.

Principal component analysis (PCA) can mathematically transform the original data with no assumptions about the form of the covariance matrix. Moreover, such analysis allows a grouping of different pollution indices based on their pollution similarities. Since the Scree plot in Figure 4 showed a sharp change of slope after the third eigenvalue, the first three PCs, which can explain 99% of the variance or information contained in the original data set, were retained in order to comprehend the underlying data structure. Note that only PC1 had eigenvalue greater than unity and explained 88.4% of the variance. In addition, loadings of the three retained PCs are presented in Figure 5. PC1 is highly contributed by PI, NO_2–N, VOP, TAs, and Oil, which were correlated at the 95% significance level and related to anthropogenic pollution resulting from heavy-pollution industries (see Table 2). BOD_5 and ammonia have a negative participation in PC1. PC2 explains 9.2% of the variance and highly participated by BOD_5 and ammonia. Finally, PC3 (2.1% of the variance) is also positively contributed by ammonia. The result of PCA indicated that the organic pollutants might be the predominant pollutants affecting the water quality of Huai River at Xuyi Section. Since normally a rotation of principal components can achieve a simpler and more meaningful representation of the underlying factors, various rotation methods were tried in this study. Figure 6 showed different component plots for different rotation methods, separately. Clearly, it can be seen that direct-oblimin method gave a simplest loading plot, in which PI, BOD_5 and Ammonia constituted high portion of PC1, PC2 and PC3, individually.

Unlike PCA that used only three PCs for display purposes, cluster analysis used all the information contained in the original data set. Hierarchical agglomerative clustering by the Ward's method was selected for sample classification. Figure 7 illustrated the dendrogram of collected data, and two well-differentiated clusters can be found in it. Furthermore, both clusters consisted of two or more subgroups with river water quality decreasing from top to bottom.

The main cluster near the bottom (labeled as N5) only consisted of annual pollution data in 1996 at 4 sampling sites, and it represented the most serious water contamination situation characterized by high positive scores on PC1 and extremely low scores on PC2 and PC3 as shown in Figure 8. Within the group, there were two subgroups. One subgroup included point 2 and point 4 of relative high flow and

Table 2. Correlation matrix of seven studies indices.

	PI	BOD$_5$	NH$_3$–N	NO$_2$–N	VOP	TAs	Oil
Correlation coefficient							
PI	1.000	−.332	−.139	0.468	0.764	0.375	0.885
BOD$_5$	−.332	1.000	0.372	0.053	−.451	.042	−.449
NH$_3$–N	−.139	.372	1.000	−.189	−.303	−.193	−.239
NO$_2$–N	.468	.053	−.189	1.000	.595	.520	.469
VOP	.764	−.451	−.303	.595	1.000	.352	.830
TAs	.375	.042	−.193	.520	.352	1.000	.460
Oil	.885	−.449	−.239	.469	.830	.460	1.000
Significant level (1-tailed)							
PI		.057	.259	.010	.000	.035	.000
BOD$_5$	.057		.037	.403	.014	.422	.014
NH$_3$–N	.259	.037		.189	.075	.184	.130
NO$_2$–N	.010	.403	.189		.001	.005	.010
VOP	.000	.014	.075	.001		.046	.000
TAs	.035	.422	.184	.005	.046		.012
Oil	.000	.014	.130	.010	.000	.012	

Component Plot in Rotated Space (Varimax)

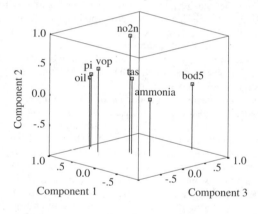

Component Plot in Rotated Space (Direct Oblimin)

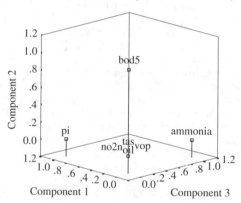

Component Plot in Rotated Space (Quartimax)

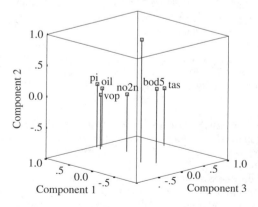

Component Plot in Rotated Space (Equamax)

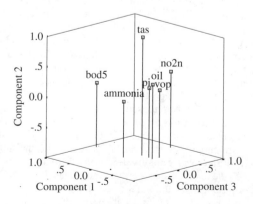

Figure 6. Component plot in rotated space.

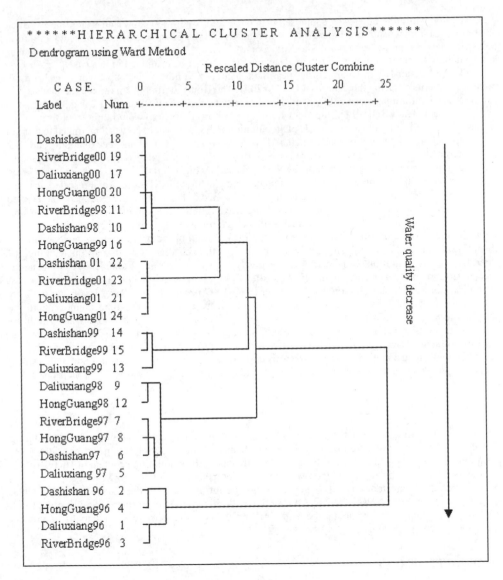

Figure 7. Hierarchical clustering dendrograms for 6-year monitoring data.

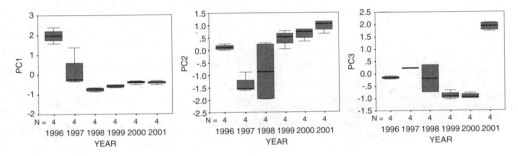

Figure 8. Boxplots of factor scores.

water quality, while the other was composed of point 1 and 3 with low flow and water quality. Such clustering result may result from the dilution of branches and the possible existence of heavy-pollution industries between point 2 and point 3 in Xuyi County. Since the flow data at 4 river cross sections is not available, the contribution of transboundary pollution cannot be evaluated in this case.

Unlike the cluster N5, the main cluster near the top contained more subgroups, which means more pollution patterns covering the data from 1997 to 2001. The first subgroup from the top (labeled as N1) comprised the pollution states in 2000 at all sampling points, those in 1998 at point 2 and 3, and that in 1999 at point 4. High BOD_5/PI value (shown in Fig. 3) and relatively low ammonia level formed the feature of N1, indicating the lowest levels of all pollution indices. Cluster N1 was then linked to a small subgroup including all pollution states in 2000 (labeled as N2) at a rescaled distance of about 7, and the pollution profiles in N2 presented high level of both BOD_5/PI ratio and NH_3–N ammonia. Subsequently, the third subgroup from the bottom (label as N3), specified by high BOD_5/PI ratio, low ammonia level, and high nitrite concentrations, was grouped into one big cluster with N1 and N2 at a rescaled distance of about 10. There were three pollution states in 1999 contained in N3. Finally, the remained one (labeled as N4) joined with the three clusters at a rescaled distance of about 12, and it included all pollution situations in 1997 and two in 1998 with a characteristic of medium BOD_5/PI ratio and ammonia level.

Furthermore, some useful information about spacial variation of the data can be extracted from the clustering dendrogram by analyzing the pollution profiles in each year. Note that all the data investigated here is concentration. Considering the relatively high flow at Point 2 and 4, the homogeneous spacial variation pattern can be found in the year after 1997 (except that in 1999). That may indicate that the policy intervention in pollution control during 1996–1997 period in Huai River Basin significantly reduced the transboundary pollution problem of Huai River at the Jiangsu-Anhui boundary and the pollution input of Xuyi County to Huai River showed a increasing trend. As an Ecological Agriculture Demonstration area in Jiangsu Province, more improvements should be made in Xuyi to control the level of pollutant discharge to meet the strict requirements. In addition, it is necessary to emphase it again that the conclusion made here is just a working hypothesis since flow data is missing in this study. Further exploration on transboundary pollution should be done when the new information is collected.

4   CONCLUSION

Data exploration using some statistical methods such as sequential analysis and hierarchical clustering showed that the intensive administrative intervention in pollution control action during 1996–1997 period in Huai River basin significantly reduced the transboundary pollution problem of Huai River at the Jiangsu-Anhui boundary, while the major pollutant indices of Huai River in Xuyi section changed as well. Permanganate index is no longer the major pollution index in the Huai River after 1998. Amazingly, the index of NH_3–N becomes one of the predominant indices in river pollution since 1998, implying more improvements should be made in Xuyi to meet the strict requirements of Ecological Agriculture Demonstration.

ACKNOWLEDGEMENT

The work was funded by 863 High-Tech Research and Development Program of P.R. China (Grant No. 2002AA649130) and EU project (EC Contract ICA4-CT-2001-10039).

REFERENCES

Andersen, per K., Borgan, O., Gill, R.O. & Keiding, N. 1998. Statistical Models Based on Counting Processes. New York: Springer.
Fletcher, R. 1980. Practical Methods of Optimization, Vol. 1. New York: Wiley.
Goldfarb, D. 1970. A family of variable-metric methods derived by variational means. *Math. Comp.* 24, 23–26.
Manual, S. 1985. MedChem. Manual. Release 3. Claremont, CA: Pomona College.
Massart, D.L. & Kanfman, L. 1983. The interpretation of analytical chemical data by the use of cluster analysis. New York: Wiley.

Physics and material science

Computational Methods in Engineering and Science, Iu et al. (eds)
© 2003 Swets & Zeitlinger, Lisse, ISBN 90 5809 567 3

Perturbed periodic waves for a system of coupled nonlinear Schrödinger equations (CNLS)

S.C. Tsang & K.W. Chow

Department of Mechanical Engineering, The University of Hong Kong, Hong Kong

ABSTRACT: Studies of solitons interactions of CNLS are of both theoretical and practical importance. In applications to long distance communications by optical fibers, any overlapping or merging of
adjacent solitons should be prevented. Introduction of a phase difference between adjacent solitons can
be used as means to control the problem. In this paper, the effects of an initial phase difference on the
long time evolution of perturbed periodic waves will be investigated. In elliptical polarization, a phase
difference would cause the long-time dynamics to break away from a periodic pattern. On the other
hand, in a linearly polarized mode, a phase difference does not have an important influence on the long-
time evolution.

1   INTRODUCTION

The dynamics of slowly varying wavepackets is important in several branches of physics, for examples,
fluid dynamics and nonlinear optics. The nonlinear Schrödinger equation arises as the governing model
equation if second order dispersion and cubic nonlinearity constitute the leading order balance (Ablowitz &
Segur 1981). When the linear theory admits two or more modes, the coupled nonlinear Schrödinger
equations (CNLS) would be the relevant model.

In hydrodynamics, the propagation of two wavepackets along a direction where the group velocity
projections overlap will lead to CNLS, if the amplitude expansion is performed up to the third order in
amplitude (Dhar & Das 1991). Similarly, in optics, the propagation of short pulses in birefringent fibers
will also lead to a form of CNLS (Menyuk 1988).

CNLS equations can only be solved analytically in special cases, e.g. the Manakov model
(Radhakrishnan et al. 1997), where the self phase and cross phase modulations coefficients are equal.
For other cases, numerical methods should be applied.

Two important criteria can be used to choose an appropriate numerical method (Greig & Morris
1976). The first criterion is that the method must not have damping properties as time increases, and the
second criterion is that the method must capture the positions of the wave fronts. Many numerical methods were proved to be appropriate for solving CNLS. Taha & Ablowitz (1984) compared the Hopscotch
method with other finite difference methods for solving CNLS. They found that the algorithm of the
Hopscotch method is relatively simpler than others finite difference methods in this case. The algorithm
is totally explicit, but it can still attain unconditional stability. Due to this advantage, the Hopscotch
method is applied to solve CNLS in our present work.

The present work is motivated by the studies of solitons interactions of CNLS. In application to long
distance communications by optical fibers, successive solitons should maintain their separation. When
adjacent solitons overlap or merge together, the situation is undesirable and may lead to signal degradation.

This problem can be controlled by introducing a phase difference between perturbations of adjacent
solitons. The effects of an initial phase difference on the long time evolution of perturbed periodic
waves of CNLS will be studied. However, for brevity, we would only focus our attention on the elliptical and linear polarization modes.

The accomplishments of the paper will be explained in the following manner. The background
theory will be introduced in Section 2. The numerical results will be presented in Section 3 and the

discussion of the results will be given in Section 4. In the following tasks, we have used x, t in accordance with some earlier theoretical works (Ablowitz & Segur 1981, Taha & Boyd 2000, 2001). The roles of x and t are actually reversed in optical application.

2 BACKGROUND THEORY

The main goal of the present work is to consider the CNLS in the bright soliton regime

$$i\frac{\partial A}{\partial t} + \frac{\partial^2 A}{\partial x^2} + \left(|A|^2 + \sigma|B|^2\right)A = 0 \tag{1}$$

$$i\frac{\partial B}{\partial t} + \frac{\partial^2 B}{\partial x^2} + \left(\sigma|A|^2 + |B|^2\right)B = 0. \tag{2}$$

A, B are slowly varying envelopes. Coefficients of the second order dispersion and self phase modulation terms are set to be one by suitable scalings (Tan & Boyd 2000, 2001). σ is a wave-wave interaction coefficient, and describes the cross phase modulation of the wave packets. The value of σ will vary depending on the nature of the underlying physical system, for examples, polarization in nonlinear optics.

Hopscotch method is applied to solve Equations 1 and 2. Both explicit and implicit schemes would be involved for each equation.

We shall discretize the x-domain into N equal parts of size h, or $x = nh$, $n = 0, 1, 2, \ldots$. This x-domain used in our study must be a multiple of the period of the periodic initial condition. The time step of the t-domain is k, or $t = mk$, $m = 0, 1, 2, \ldots$. x and t will be represented in the subscript and superscript respectively. Discretization is accomplished by a first order Euler scheme in t, and a second order central difference in x.

The explicit schemes are:

$$i\left(\frac{A_n^{m+1} - A_n^m}{k}\right) + \left(\frac{A_{n+1}^m - 2A_n^m + A_{n-1}^m}{h^2}\right)$$
$$+ \frac{1}{2}\left[A_{n-1}^m\left(\left|A_{n-1}^m\right|^2 + \sigma\left|B_{n-1}^m\right|^2\right) + A_{n+1}^m\left(\left|A_{n+1}^m\right|^2 + \sigma\left|B_{n+1}^m\right|^2\right)\right] = 0, \tag{3}$$

$$i\left(\frac{B_n^{m+1} - B_n^m}{k}\right) + \left(\frac{B_{n+1}^m - 2B_n^m + B_{n-1}^m}{h^2}\right)$$
$$+ \frac{1}{2}\left[B_{n-1}^m\left(\sigma\left|A_{n-1}^m\right|^2 + \left|B_{n-1}^m\right|^2\right) + B_{n+1}^m\left(\sigma\left|A_{n+1}^m\right|^2 + \left|B_{n+1}^m\right|^2\right)\right] = 0. \tag{4}$$

The implicit schemes are:

$$i\left(\frac{A_n^{m+1} - A_n^m}{k}\right) + \left(\frac{A_{n+1}^{m+1} - 2A_n^{m+1} + A_{n-1}^{m+1}}{h^2}\right)$$
$$+ \frac{1}{2}\left[A_{n-1}^{m+1}\left(\left|A_{n-1}^{m+1}\right|^2 + \sigma\left|B_{n-1}^{m+1}\right|^2\right) + A_{n+1}^{m+1}\left(\left|A_{n+1}^{m+1}\right|^2 + \sigma\left|B_{n+1}^{m+1}\right|^2\right)\right] = 0, \tag{5}$$

$$i\left(\frac{B_n^{m+1} - B_n^m}{k}\right) + \left(\frac{B_{n+1}^{m+1} - 2B_n^{m+1} + B_{n-1}^{m+1}}{h^2}\right)$$
$$+ \frac{1}{2}\left[B_{n-1}^{m+1}\left(\sigma\left|A_{n-1}^{m+1}\right|^2 + \left|B_{n-1}^{m+1}\right|^2\right) + B_{n+1}^{m+1}\left(\sigma\left|A_{n+1}^{m+1}\right|^2 + \left|B_{n+1}^{m+1}\right|^2\right)\right] = 0. \tag{6}$$

To initiate the Hopscotch method, we replace m by $m - 1$ in Equations 5 and 6. Together with Equations 3 and 4 at time level m, the explicit schemes become

$$A_n^{m+1} = 2A_n^m - A_n^{m-1} \; ,$$
(7)

and

$$B_n^{m+1} = 2B_n^m - B_n^{m-1} \; .$$
(8)

The procedure for determining solutions at every spatial locations at each subsequent time step will be described below:

The appropriate scheme to be used at each spatial location is determined by the sum of the time step m and the spatial location n. For $(m + n)$ equal to an odd number, the explicit scheme should be used. For $(m + n)$ even the implicit scheme would be utilized. The explicit schemes are applied first in each time step, so that the values at every other spatial location become known. The implicit schemes can then be applied. As can be seen from Equations 5 and 6, unknowns to be found by the implicit scheme depend on the values at the adjacent spatial locations. As the values at these adjacent spatial locations are already determined by the explicit scheme, unknowns can be found by directly substituting these values into the implicit scheme. As the implicit schemes only involve direct substitution, they are said to be "explicit".

3 NUMERICAL RESULTS

We solve Equations 1 and 2 numerically with these initial conditions:

$$A = a_0\left(1 - \varepsilon \cos lx\right)$$
(9)

$$B = b_0\left[1 - \varepsilon \cos l(x + \theta)\right]$$
(10)

where a_0 and b_0 are the initial amplitudes of the two unperturbed periodic waves. $\varepsilon \ll 1$ is a small parameter which represents the strength of the perturbation. l is the wave number of the perturbation, and θ represents the initial phase difference between the two perturbations.

The goal is to study the effects of phase difference on the evolution of periodic waves of CNLS. The Hopscotch method will be used, and the attention is confined to several special cases of σ of interest in optical applications.

The first two cases represent elliptical polarization, corresponding to $\sigma = 1$ and the third and fourth cases represent linear polarization, corresponding to $\sigma = 2/3$. Table 1 summarizes the parameters used in the numerical experiments of the four cases.

4 DISCUSSIONS AND CONCLUSIONS

The stability criteria of the plane wave solution in linear and elliptical polarization modes can be established (Tan & Boyd 2001). For linear polarization, the plane wave solution is linearly stable only if the

Table 1. Summary of the parameters used in the numerical experiments.

Case	a_0	b_0	ε	l	θ
1	0.5	0.5	0.1	0.5	0
2	0.5	0.5	0.1	0.5	$\pi/2$
3	0.5	0.5	0.1	0.5	0
4	0.5	0.5	0,1	0 5	$\pi/2$

807

perturbation wave number is above the critical value

$$l_c = \left(a_0^2 + b_0^2 + \sqrt{\left(a_0^2 + b_0^2 \right)^2 - \frac{20}{9} a_0^2 b_0^2} \right)^{1/2} ;$$

otherwise the plane wave is unstable.

For Case 1, l_c is 0.91. From Table 1, we choose $l = 0.5$, which implies that the plane wave is unstable. In Figure 1a, the plane view of the long-time evolution of $|A|$ is shown, and the darker the color in the Figure, the larger the amplitude of $|A|$ it represents. It is found that $|A|$ oscillates between the nearly-uniform state and the one-hump state in the t-direction. In the spatial direction, there is only one single peak within the spatial period. In this case, the spatial period is $2\pi/l = 2\pi/0.5 = 4\pi \approx 12.57$. It is expected that the peak is located at the centre of the period, i.e. $x_c = 4\pi/2 \approx 6.29$. Figure 1b is basically the same as Figure 1a, except it represents the amplitude of B.

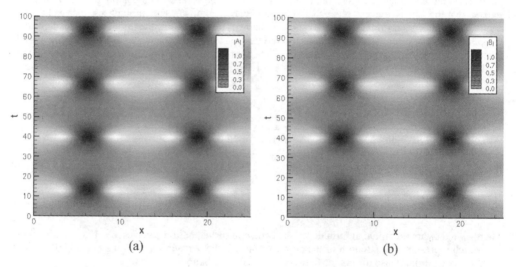

Figure 1. Plane view of the long-time evolution of the destabilized wave solution for the linearly polarized mode.

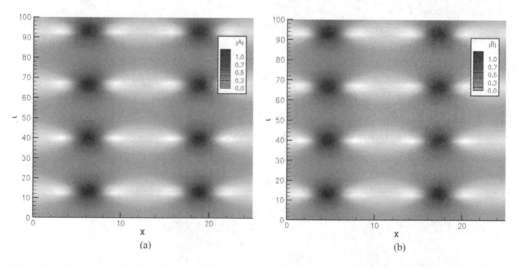

Figure 2. Plane view of the long-time evolution of the destabilized wave solution for the linearly polarized mode.

In Case 2, the initial conditions used are the same as those in Case 1, except that there is a phase angle between the perturbations of A and B. l_c in Cases 1 and 2 are the same, so it is expected that the oscillation pattern in t-direction is basically the same in these two cases (Figs 1, 2). However, from Figure 2b, it can be observed that the location of the peak of $|B|$ is at $x_{SP} \approx 4.71$, which is $\pi/2$ shifted from the centre of the spatial period (i.e. $x_c \approx 6.29$). This shift is actually introduced by the phase shift of the perturbation in the initial condition.

Cases 3 and 4 represent elliptic polarization mode. In this polarization mode

$$l_c = \sqrt{a_0^2 + b_0^2}\,.$$

For Case 3, as in Cases 1 and 2, there is also one single peak within the spatial period. $|A|$ and $|B|$ both oscillate periodically along the t-direction between the nearly-uniform state and the one-hump state (Fig. 3). However, the oscillation period is shorter when compared with Case 1, i.e. the linear polarization mode.

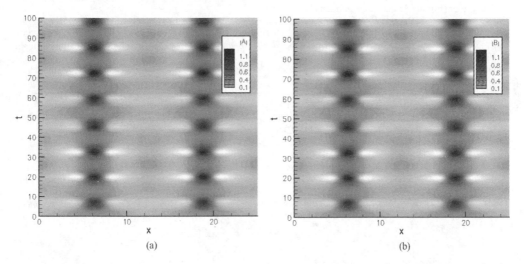

Figure 3. Plane view of the long-time evolution of the destabilized wave solution for the elliptically polarized mode.

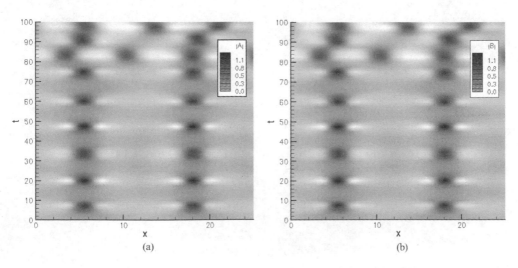

Figure 4. Plane view of the long-time evolution of the destabilized wave solution for the elliptically polarized mode.

For Case 4, the oscillations in t-direction remain the same as in Case 3 in the first few cycles after introducing a phase angle. However, this periodic structure is destroyed as time goes on. Moreover, the location of the peak is not at the centre of the spatial period. Both the peaks of $|A|$ and $|B|$ are located at $x_m = 5.5$, which is the mean position of x_{SP} and x_c (Fig. 4).

From these four cases, it can be concluded that the influence of an initial phase difference between the perturbations on the long-time evolution depends on the parameters of the underlying physical system. In the case of linearly polarized mode, a phase difference does not have an important influence on the long time evolution. However, in the case of elliptical polarization, a phase difference would cause the long-time dynamics to break away from a periodic pattern.

REFERENCES

Ablowitz, M.J. & Segur, H. 1981. *Solitons and the inverse scattering transform*. Philadelphia: SIAM.

Chu, P.L. & Desem, C. 1992. *Soliton-soliton interactions*, in Optical solitons: theory and experiments, J.R. Taylor (Editor). England: Cambridge University Press.

Dhar, A.K. & Das, K.P. 1991. *Fourth order nonlinear evolution equation for two Stokes wave trains in deep water*. Physics of Fluids 3: 3021–3026.

Greig, I.S. & Morris, J.L. 1976. *A Hopscotch method for the Korteweg-de Vries equation*. Journal of Computational Physics 20: 64–80.

Ismail, M.S. & Taha, T.R. 2001. *Numerical simulation of coupled nonlinear Schrödinger equations*. Mathematics and Computers in Simulation 56: 547–562.

Menyuk, C.R. 1988. *Stability of solitons in birefringent optical fibers II. Arbitrary amplitudes*. Journal of the Optical Society of America B, Optical Physics 5: 392–402.

Radhakrishnan, R., Lakshmanan, M. & Hietarinta, J. 1997. *Inelastic collision and switching of coupled bright solitons in optical fibers*. Physical Review E 56: 2213–2216.

Taha, T.R. & Ablowitz, M.J. 1984. *Analytical and numerical aspects of certain nonlinear evolution equations II, nonlinear Schrödinger equation*. Journal of Computational Physics 55: 203–230.

Tan, B. & Boyd, J.P. 2000. *Coupled-mode envelope solitary waves in a pair of cubic Schrödinger equations with cross modulation: Analytical solution and collisions with application to Rossby waves*. Chaos, Solitons and Fractals 11: 1113–1129.

Tan, B. & Boyd, J.P. 2001. *Stability and long time evolution of the periodic solutions to the two coupled nonlinear Schrödinger equation*. Chaos, Solitons and Fractals 12: 721–734.

Computational Methods in Engineering and Science, Iu et al. (eds)
© 2003 Swets & Zeitlinger, Lisse, ISBN 90 5809 567 3

The calculation of the optical constants of indium-tin-oxide films

L.J. Meng
Departamento de Física, InstitutoSuperior de Engenharia do porto, Portugal

F. Placido
Thin Film Centre, University of Paisley, Paisley, Scotland

V. Teixeira
Departamento de Física, Universidade do Minho, Portugal

M.P. dos Santos
Departamento de Física, Universidade de Évora, Portugal

ABSTRACT: ITO thin films were deposited on the glass substrates by microwave-enhanced d c reactive magnetron sputtering technique at different oxygen partial pressures ($3.8 - 11.7 \times 10^{-4}$ mbar). The optical properties were studied by measuring the transmittance and the ellipsometric spectra. The optical constants of the films in the telecommunication wavelengths ($1500-1600$ nm) were obtained by fitting the transmittance combined with spectroscopic ellipsometry measurement.

1 INTRODUCTION

Indium-tin-oxide (ITO) thin film is one of the widely used transparent conducting oxides because of its good conductivity and high transmittance in the visible and near IR regions. Many applications of the ITO films have been found in solar cells, flat panel displays and anti-static coatings on monitor tubes. In the telecommunication area, there are many applications, such as modulators and some switching applications, which need conducting and transparent films for 1500–1600 nm wavelengths. Although ITO films are conducting, normally they are highly absorbing in these wavelengths. Therefore, in order to make ITO films suitable for this area, the deposition technique and the deposition parameters must be carefully selected.

Many techniques, such as chemical vapor deposition, evaporation, sputtering, and laser ablation and so on, have been used for depositing the ITO films. Among these techniques, reactive magnetron sputtering plays a very important role. By this technique, the relatively cheap metallic target, comparing to the ceramic oxide target, can be used and the deposition parameters can be easily controlled and repeated. However, the conventional reactive magnetron sputtering has a disadvantage, the poisoning or oxidation of the target which will result in an abrupt decrease of the deposition rate. Microwave-enhanced reactive magnetron sputtering system can avoid this problem as the sputtering zone and reactive zone are separated. In this work, the ITO films have been prepared using the microwave-enhanced dc reactive magnetron sputtering system and the optical properties of the films have been studied.

2 EXPERIMENTAL DETAILS

ITO films were deposited on the glass substrates by a commercial microwave-enhanced dc reactive magnetron sputtering system. The system is equipped with four sputtering sources and controlled by the computer. The box in the top of the chamber is the microwave chamber for ionizing the oxygen and argon gases. In this work, only one dc source in the left side was used. The chamber was pumped with

Table 1. Deposition conditions.

	H1H13154	H1K06123	H1L18113
Oxygen partial pressure ($\times 10^{-4}$ mbar)	3.8	5.9	11.7
Total pressure ($\times 10^{-3}$ mbar)	2.8	4.1	5.0
Oxygen flow (sccm)	36	47	62
Ar flow (sccm)	80	115	130
Sputtering current (A)	4	4	4
Sputtering voltage (V)	411	387	350
Sputtering power (kW)	1.64	1.54	1.39
Sputtering time (Sec.)	600	900	890

turbomolecular pump backed by a rotary pump. The target is In-Sn (90:10) alloy of 99.99% purity (120 × 372 mm^2). The distance between target and substrate was about 122 mm. The substrate holder is rotated with the speed of 1 cycle/s. The cathode was cooled by running water.

The vacuum chamber was evacuated down to pressure 1×10^{-5} mbar prior to the deposition. Then the oxygen reactive gas and the argon gas were introduced into the chamber and the microwave chamber is operated for cleaning the substrate and the holder. This process will last about 5 min. After that the dc power will be applied to the cathode and film deposition process starts. The constant current mode has been used for all the deposition. Three ITO films with different oxygen partial pressure have been prepared. The deposition conditions are listed in Table 1. The film transmittance was measured by Hitachi 3501 spectrophotometer. The ellipsometric parameters Psi and Delta were measured using a Jobin-Yvon UVISEL variable angle ellipsometer. The ellipsometric Psi and Delta data were acquired at 55° of incidence angle over the spectral range 400–800 nm in steps of 1 nm. The fitting has been done using our own software and a Levenberg-Marquardt least squares routine.

3 RESULTS AND DISCUSSION

Figure 1 shows the transmittance of the ITO films prepared at different oxygen partial pressures. It can be seen that the transmittance in the infra-red region is lower for the film prepared at low oxygen partial pressure compared to the film prepared at high oxygen partial pressure. It has been well know that this decrease in the transmittance is because of the absorption of the free electron (Hamberg & Granqvist 1986). The optical band gaps have been calculated from α^2 vs hυ plots by suggesting a direct allowed transition for ITO films prepared at different oxygen partial pressures. The optical band gaps calculated by this method have been given in Table 2. All the films give the same value of 3.84 eV. This value is higher than the other reported values (Meng & Santos 1998, George & Menon 2000). The film thickness may be the reason for this high optical band gap value. The ITO films studied in this work are quite thin (about 100 nm), and normally, the thinner the film is, the higher the optical band gap is (Meng & Santos 1994).

In order to get the dispersion curves of the refractive index n and the extinction coefficient k, the transmittance spectra and the ellipsometric spectra have been fitted together using the Lorentz oscillator model (classical model). The dielectric function for classical model can be described as follows:

$$\varepsilon(\omega) = \varepsilon_\infty + \frac{(\varepsilon_s - \varepsilon_\infty)\omega_t^2}{\omega_t^2 - \omega^2 + i\Gamma_0\omega} + \frac{\omega_p^2}{-\omega^2 + i\Gamma_D\omega} + \sum_{j=1}^{2} \frac{f_j\omega_{0j}^2}{\omega_{0j}^2 - \omega^2 + i\Gamma_{0j}\omega} \tag{1}$$

The third term is the Drude term which is related with the free electron gas. From Figure 1 it can be seen that all the films behave like a good dielectric at the wavelength below the 2000 nm, so the Drude term can be ignored. We fit the transmittance and the ellipsometric spectra using the classical model with two oscillators instead of the three oscillators (Voronov & Placido 2002) – one at about 4 eV to model the direct band gap and the other at about 0 eV to model the absorption in the infra-red region. Therefore, the dielectric function for our fitting is as follows:

$$\varepsilon(\omega) = \varepsilon_\infty + \frac{(\varepsilon_s - \varepsilon_\infty)\omega_t^2}{\omega_t^2 - \omega^2 + i\Gamma_0\omega} + \frac{f_1\omega_{01}^2}{\omega_{01}^2 - \omega^2 + i\Gamma_{01}\omega} \tag{2}$$

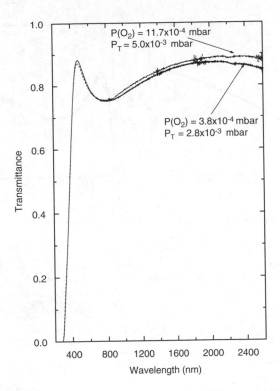

Figure 1. Specular transmittance spectra of two ITO films prepared at different oxygen partial pressures.

Table 2. Some properties of the ITO films.

	H1H13154	H1K06123	H1L18113
Thickness (nm)	106	104	103
Deposition rate (nm/s)	0.18	0.12	0.12
Sheet resistance Rs ($\times 10^3 \Omega/\square$)	0.209	0.969	8.104
Resistivity ρ ($\times 10^{-5} \Omega$-m)	2.2	10.1	83.5
Transmittance T at $\lambda = 1550$ nm	85%	86%	86%
Figure of merit Φ (T/Rs) at $\lambda = 1550$ nm ($\times 10^{-3} \Omega^{-1}$)	4.1	0.9	0.1
n at $\lambda = 1550$ nm	1.79	1.73	1.84
k at $\lambda = 1550$ nm	0.03	0.03	0.01
Optical band gap (eV)	3.84	3.84	3.84

where ε_s is static dielectric constant, ε_∞ is high-frequency dielectric constant, ω_t is the characteristic frequency of the main oscillator, Γ_0 and Γ_{01} are the damping factors, f_1 is the strength of the second oscillator and ω_{01} is its characteristic frequency.

Figure 2 gives the fitting results for the sample prepared at the oxygen pressure of 5.9×10^{-4} mbar (H1K06123) and the fitting parameters are given in Table 3. It can be seen from the Figure 2 that the transmittance generated by the two oscillators classical model is a little bit lower than the real transmittance. That is the limitation of the model. The Lorentzian functions have long tails which add together through the visible spectral range and results in a low transmittance. This limitation can be avoided by fitting the spectra starting a little bit away from the optical band gap.

The spectra of the two samples prepared at low oxygen partial pressures can be fitted very well without considering the inhomogeneity. However, for sample prepared at high oxygen partial pressure, the

813

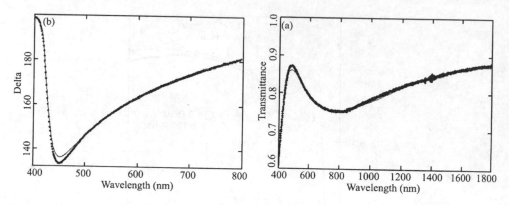

Figure 2. Measured and fitted delta and transmittance for sample prepared at the oxygen partial pressure of 5.9×10^{-4} mbar (H1K06123).

Table 3. Fitting parameters.

	ε_∞	ε_s	ω_t (eV)	Γ_0 (eV)	f_1	ω_{01} (eV)	Γ_{01} (eV)	d_T (nm)	d_{Delta} (nm)	Inhomogeneity
H1H13154	4.27	4.43	3.34	0.30	8.60	0.28	0.05	106	105	0
H1K06123	4.28	4.42	3.31	0.35	4.36	0.39	0.04	104	104	0
H1L18113	3.02	4.03	4.04	0.22	4.44	0.26	0	104	102	−0.07

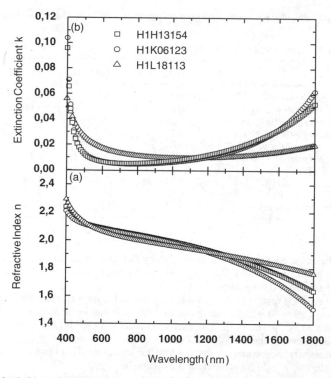

Figure 3. Refractive index n (a) and extinction coefficient k (b) with wavelength for ITO films prepared at different oxygen partial pressures.

inhomogeneity is about 7%. As it can be seen from the X-ray diffraction spectra that the samples prepared at low oxygen partial pressure show an amorphous structure and the film prepared at high oxygen partial pressure show a polycrystalline structure. It means the structure of the films changes from the amorphous into polycrystalline structure as the oxygen partial pressure is increased. However, for the sample prepared at high oxygen partial pressure, it may contain both amorphous and polycrystalline structure and then an inhomogeneity.

From Table 3 it can be seen that the film thickness obtained from the transmittance fitting and the delta fitting has small difference. This difference may result from the surface overlayer of the films as the ellipsometric spectra are sensitive to it. The dispersions of the refractive index n and the extinction coefficient k of the films prepared at different oxygen partial pressures are shown in Figure 3. It can be seen that the sample prepared at high oxygen partial pressure shows a different behavior comparing to the samples prepared at low oxygen partial pressure. That can be related with the variation of the structure of the films. The refractive index and the extinction coefficient at the wavelength 1550 nm have been listed in Table 2. The figure of merit of the films at this wavelength has also been given in the table. It can be seen that the film prepared at low oxygen partial pressure gives a high figure of merit.

4 CONCLUSIONS

ITO thin films have been deposited on the glass substrates at different oxygen partial pressures using microwave enhanced dc reactive magnetron sputtering technique. ITO film with electrical resistivity of $2.2 \times 10^{-5} \Omega$-m and the transmittance 85% at the wavelength 1550 nm which is interesting for telecommunications have been obtained. All the films give the high transmittance in the visible and near infra-red region. Two oscillators classical model has been used for fitting the transmittance and the ellipsometric spectra. The film prepared at the oxygen partial pressure 1.2×10^{-3} mbar shows an inhomogeneity of 7% and the different behaviors on the dispersion of the refractive index and the extinction coefficient.

REFERENCES

George, J. & Menon, C.S. 2000. Electrical and optical properties of electron beam evaporated ITO thin films. *Surface and Coatings Technology* 132: 45–48.
Hamberg, I. & Granqvist, C.G. 1986. Evaporated Sn-doped In_2O_3 films: Basic optical properties and applications to energy-efficient windows. *J. Appl. Phys.* 60(11): R123–R159.
Meng, L.J. & Santos, M.P. 1994. The influence of oxygen partial pressure and total pressure (O_2 + Ar) on the properties of tin oxide films prepared by dc sputtering. *Vacuum* 45(12): 1191–1195.
Meng, L.J. & Santos, M.P. 1998. Properties of indium-tin-oxide films prepared by rf reactive magnetron sputtering at different substrate temperature. *Thin Solid Films* 322: 56–62.
Voronov, A. & Placido, F. Transparent, conductive ITO for telecommunication wavelengths. *45th Annual Technical Conference Proceedings of Society of Vacuum Coaters.*

Computational Methods in Engineering and Science, Iu et al. (eds)
© 2003 Swets & Zeitlinger, Lisse, ISBN 90 5809 567 3

Monte-Carlo modeling of structure and properties of interphase layers of polymer matrix composites

Yu.G. Yanovsky & A.V. Teplukhin

Institute of Applied Mechanics of Russian Academy of Sciences, Moscow, Russia

ABSTRACT: New approach and software for modeling and for studying the configuration of very large molecular systems in frame of Monte-Carlo method have been developed. The software allows: i) to model systems which include some chain-type molecules placed in immobile rigid nano-particles, ii) to investigate these systems at different temperatures and densities, iii) to calculate the different energetically characteristics, iv) to realize visual analysis of some most important molecular conformations. Some configurational systems containing the fragments of graphite surfaces and the molecules of $n-C_{40}H_{82}$ and $n-C_{50}H_{102}$ at different temperatures (300 and 900 K respectively) and rates of deformation are discussed.

1 INTRODUCTION

As it is known a practice of the making and the using of polymeric composites stimulates the solution of general problem of mechanics of heterogeneous continua viz. determination of quantitative relation between microstructure and mechanical properties of composite materials. As elaboration of the methods of description of mechanical properties of composites it is necessary to involve into the calculation procedure more detailed characteristics of microstructure, for example, morphology and interfacial properties, spatial structure of filler particles etc. (Yanovsky & Obraztsov 1998).

The necessity of calculation of the interphase layer's characteristics and microstructure has been discussing in the literature fairly long. In connection with evolution of the molecular physics methods and appearance of the new generation computational techniques the direct quantitative numerical investigations of the interphase layer's structure and properties for the real polymeric media became a reality.

According to problem, we are going to consider and to solve, the objects of our calculation had to become the aggregates of composite structures, which consist of hundred thousand atoms. But computer experiments, which will operate with such kind of systems, will require very large computational capacity and a process of calculations seems to be many weeks long. Thus, problem of development of effective methods and algorithms of modeling of such operands appears as very actual.

2 MODELING OF HEAT MOTION OF POLYMERIC MACROMOLECULES BY MONTE-CARLO APPROXIMATION

2.1 *Some algorithmic aspects*

One of very perspective approach, which allows to substantially enhance an efficiency of calculation is based on the molecular models with rigid bonds and valent angles. In frame of these models an atomic heat motion takes place by means of rotation around the chemical bonds, which one can characterize low torsion potentials. In spite of seeming simplicity, realization of this method by means of molecular dynamic approximation requires a solution of certain systems of differential equations of motion jointly with huge number of non-linear algebraic equations in order to take into consideration the rigid bonds (Lemak et al. 1998).

The use of Monte-Carlo method is to study a behavior of macromolecules with arbitrary configuration does not eliminate the problem of solution of system of algebraic equations but substantially decreases an amount of those. At the same time Monte-Carlo method is particularly efficient in case of modeling of chain-type polymers, for instance polyethylene, polybutadiene etc. Calculation of coordinates of atoms for novel and new conformations, for such types of molecules lead us to batches of elementary turnings and translations.

To develop the method of direct computer experiments for studying of the mechanical properties of polymer composites filled with rigid nano-particles new approach and software for modeling and for studying the configuration of very large molecular systems in frame of Monte-Carlo method (Allen & Tildesley 1987) have been proposed. The software allows: (i) to model systems which include some chain-type molecules placed in immobile rigid nano-particles, (ii) to investigate these systems at different temperatures and densities, (iii) to calculate the different energetically characteristics, (iv) to realize visual analysis some most important molecular conformations.

The basis of the method is an original algorithm of space decomposition, which allows to make a classical Metropolis sampling (Metropolis et al. 1953) for numerous molecules (a number of molecules depend on a number of accessible processors). According to present-day classification for parallel algorithm of modeling of molecular systems (Heffelfinger 2000) above algorithm belong to program using the space decomposition. Energy intermolecular interaction is calculated by method of atom-atomic potential functions and electrostatical forces are also taken into account (Poltev et al. 1984, Poltev & Shulyupina 1986). Software was prepared by FORTRAN 77 together with standard MPI routines. Computational experiments for model molecular systems (from 10^4 to 6×10^5 atoms, see Fig.1) were made on super computer system MVC 1000 M (Moscow, Computational Center of Russian Academy of Sciences).

Potential energy of rotation around bonds C–C was calculated according to standard expression: $E = U_0/2(1 + \cos(3\varphi))$, where $U_0 = 2.45$ Kcal/mol. Coordinates of the graphite atoms and the molecules of n-alkanes were estimated using some crystallographic information about structure of those (Bokii 1971).

2.2 Some particulars of calculation

As it was mentioned energetical and structural characteristics the systems under consideration are calculated according to statistical notional sample of molecular configurations of elementary cell by Monte-Carlo method using Metropolis sampling (2). The algorithmic realization goes to classical schema (1). The temperature, the size of cell and the number of molecules during process of modeling do not alter. An initial molecular collocation is arbitrary but limited by size of cell.

Practically a process of modeling goes according to next procedure. At given temperature T an arbitrary alteration of location the bunch of atoms (elementary test) from energy E_1 to E_2 is defined. This test is positive if $E_2 \leqslant E_1$, or, otherwise, if $\exp[(E_1 - E_2)/KT] > \xi$ is arbitrary number which uniformly allocated in space $(0, 1)$ and K is Boltzman's constant. If the elementary test do not adopt we once again embrace into sample above bunch of atoms.

Elementary test consists of displacement the bunch of atoms on $\delta(1 - 2\alpha_i)$ A distance and turn around an arbitrary selected axis which passes trough center mass of these at the angle $\gamma(1 - 2\beta)$ radian of those atoms. Here α_i and β are arbitrary selected numbers evenly distributed into interval $(0, 1)$ and i- x- y- and z- are coordinates of atoms of given molecules. The magnitudes of δ and γ must chosen such a way that only half of new configurations during investigations were rejected. If, as a result of modeling, a group of atoms deserts the cell, that according to periodical boundary conditions they have to return from the opposite side. A complex of configurations we have gained according to multifold iterations (by infinite number of tests) form a canonic Gibbs ensemble (in this case NVT-ensemble) where likelihood of option of configuration at given energy E is proportional to Boltzman's factor $\exp(-E/KT)$ and transitions between configurations acquire the properties of Mark's chains.

2.3 Samplings and discussion

Figures 1 and 2 show us some configurational systems containing the fragments of graphite surface (two sheets by 1008 atoms of C) and the molecules of n–$C_{40}H_{82}$, which were indicated during modeling at different temperatures (300 and 900 K respectively).

Using the original program software, we have developed, average amount of monomeric links of n–$C_{40}H_{82}$ molecules in stratum of 0.25 Å thickness versus distance from graphite surface at different temperature were evaluated (see Fig.3).

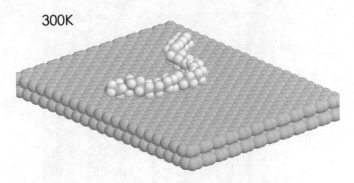

300K

Figure 1. Fragment of graphite surface and molecule of n–$C_{40}H_{82}$ at temperature 300 K.

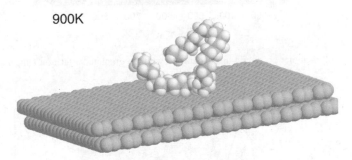

900K

Figure 2. Fragment of graphite surface and molecule of n–$C_{40}H_{82}$ at temperature 900 K.

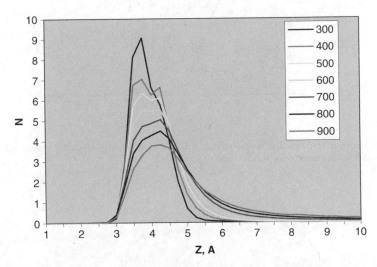

Figure 3. Average amount of monomeric links of n–$C_{40}H_{82}$ molecules versus distance from graphite surface at different temperature. Thickness of section 0.25 Å.

Format of a sample for each examined temperature was $\sim 10^8$ molecular configurations. Distance was evaluated from flatness passing through centers of carbon atoms of first graphite sheet.

As it can be seen from the Figure 3 the dependencies are looking as the bimodal curves. Area to the left from 4 Å corresponds to direct Van-der-Vaals contacts of the monomers of n-alkanes and the graphite atoms, but area to the right occupies "molten" or "stcamy" monomer links.

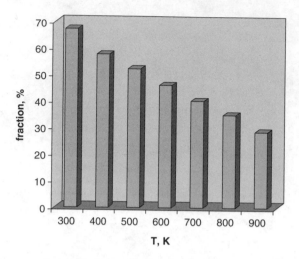

Figure 4. Fraction of monomer polymeric links contacted with the graphite surface on temperature. (According to Fig. 3).

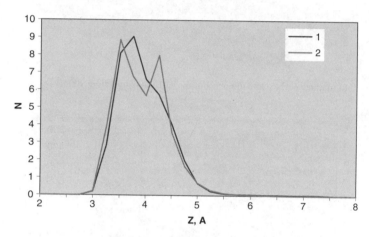

Figure 5. Dependence of number of monomers on distance to graphite surface for system with one or two n–$C_{40}H_{82}$ molecules. Temperature 300 K. Thickness of section 0.25 Å.

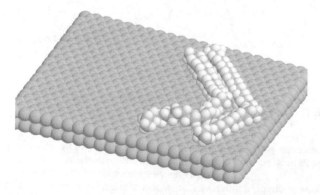

Figure 6. Fragment of graphite surface and two molecules of n–$C_{40}H_{82}$ at temperature 300 K. Monte-Carlo simulation.

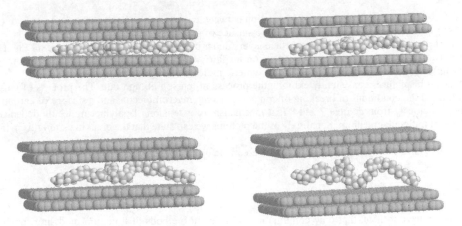

Figure 7. Conformation of fragment of $C_{50}H_{102}$ molecule located between to graphite surfaces under process of tension by high rate. Temperature 300 K.

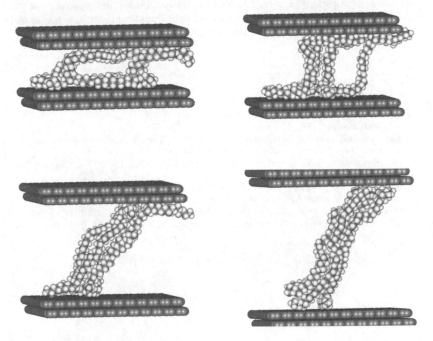

Figure 8. Conformation of fragment of four $C_{50}H_{102}$ molecule located between to graphite surfaces under process of tension by high rate. Temperature 300 K.

Field under appropriate curves in range from 0 to 4 Å let us to evaluate a portion of monomer polymeric links, which contact with the surface of rigid nano-particles at different temperatures. The appropriate dependencies depict on Figure 4.

In order to get more accurate appraisals above mentioned characteristics we need, of course, to take into account an interaction between the whole amount of molecular monomers of model composite system. Because the real composite materials store very huge amount of polymeric macromolecules we can not consider the conformational properties of those in frame of Gaussian coil model.

Figure 5 shows the dependencies of some number of monomers in stratum 0.25 Å on distance to graphite surface for system with one or two n–$C_{40}H_{82}$ molecules at temperature 300 K.

As it can be seen from the Figure 6 interaction between two molecules leads to the decreasing of content of monomer links, which contact with the surface of graphite.

Alteration of structure of model interface layers under process of extension we can see on Figures 7 and 8. Both pictures show the effects of tension for fairly simple model of rubber composite, viz. two particles of carbon's filler and one or four polymeric molecules ($C_{50}H_{102}$).

The set of pictures consistently exhibits the process of tension at high rate. The process of tension was limited the conditions of breaking of contacts between macromolecule and a surface of carbon.

It is obviously from Figures 7 and 8 that mechanism of extension of polymer molecule depends on density of molecular medium. Seemingly of the pictures we can state that the breaking strengths in these experiments are different.

We also have to note that all calculations were made by computer claster (PIII/800).

3 CONCLUSIONS

Finally we have to note, that only combination of different methods of imitation modeling for direct numerical simulations of conformational and micromechanical properties of a macromolecular stratum near the solid surfaces will be able to take the more realistic and truthful information about the peculiarities of behavior within the interphase layers of polymer composites.

In case of future investigations for more complex and realistic models of composite materials we are going to develop a new special and based on the parallel algorithms the computer software and to go on our calculations by means of multiprocessor supercomputers.

REFERENCES

Allen, M.P. & Tildesley, D.J. 1987. *Computer simulation of liquids*, Oxford University Press, N.Y.
Bokii, G.B. 1971. *Crystallochemistry*, Moscow, Nauka.
Heffelfinger, G.S. 2000. *Comput. Phys. Commun.* v.128: pp. 219–237.
Lemak, A.S., Balabaev, N.K., Karnet, Yu.N. Yanovsky, Yu.G. 1998. The effect of a solid wall on polymer chain behavior under shear flow. *J. Chem. Physics.* v.108, N2: pp. 797–805.
Metropolis, N.A., Rosenbluth, A.W., Rosenbluth, M.N., Teller, A.H. Teller, E. 1953 *J. Chem. Phys.* v.21: pp. 1087–1092.
Poltev, V.I., Grokhlina, T.I., Malenkov, G.G. 1984. *J. Biomol. Struct. Dyn.* v.2: pp. 413–429.
Poltev, V.I. & Shulyupina, N.V. 1986, *J. Biomol. Struct. Dyn.* v.3: pp. 739–765.
Yanovsky, Yu.G. & Obraztsov, I.F. 1998. Computational modeling of structure and mechanical properties of polymer composites. *Proc. 4th World Congr. on Computational Mechanics – New Trends and Application* S.Idelson, et al. (eds), Barcelona, Part 3, Section 5, 13.

Computational Methods in Engineering and Science, Iu et al. (eds)
© 2003 Swets & Zeitlinger, Lisse, ISBN 90 5809 567 3

Quantum Monte Carlo calculations of silicon self-interstitial defects

W.K. Leung
Department of Electronic and Information Engineering, The Hong Kong Polytechnic University, Hong Kong

R.J. Needs
TCM Group, Cavendish Laboratory, University of Cambridge, Cambridge, England, UK

ABSTRACT: We apply the fixed-node diffusion quantum Monte Carlo method to the study of self-interstitial defects in silicon. The best estimate of the formation energies of the most stable configurations is about 4.9 eV, which is significantly higher than the values computed by density functional theory methods. The quantum Monte Carlo calculations suggest a value of approximately 5 eV for the formation plus migration energies of the silicon self-interstitial defects. This value is consistent with recent experimental data.

1 BACKGROUND

Silicon is at the heart of modern electronics industry. One of the major problems in the manufacture of submicron devices is the diffusion of dopant atoms during thermal processing, which limits how small the devices can be made. It might seem that diffusion in silicon is a simple problem because the classical diffusion equation is well-known:

$$\frac{\partial C(\mathbf{r},t)}{\partial t} - D\nabla^2 C(\mathbf{r},t) = 0 , \tag{1}$$

where $C(\mathbf{r}, t)$ is the concentration of the diffusing atoms, $\mathbf{r}$ is spatial position, t is time, and D is the diffusion constant, whose value has been measured for almost all possible impurities. However, these values of D describe only macroscopic diffusion under equilibrium conditions, i.e. intrinsic defects are at their thermal concentrations, whereas in a real production environment, several processes would damage the crystal structure of the silicon substrates. For example, ion implantation is known to create excess point defects. Diffusion of dopant atoms in silicon is critically influenced by intrinsic defects such as self-interstitials and vacancies, e.g. phosphorus and boron are interstitial-mediated diffusers, while antimony diffusion is vacancy-mediated. To understand these effects requires detailed knowledge of diffusion on the microscopic scale in non-equilibrium situations. It is therefore very important to improve our understanding of intrinsic point defects in silicon. Measurements of the self-diffusion constant of silicon at high temperatures using radioactive tracers have established an Arrhenius relation:

$$D_{SD}(T) = D_{SD}(0) \exp(-Q/k_\mathrm{B}T) \tag{2}$$

where Q is the activation energy (in the range of 4.1–5.1 eV), T is the absolute temperature, k_B is Boltzmann's constant, and $D_{SD}(0)$ is a pre-exponential factor $\sim 10^3 \, \mathrm{cm}^2\,\mathrm{s}^{-1}$ that is two orders of magnitude larger than typical values in metals and semiconductors (Frank 1984). In recent experiments, the activation energy of the interstitial self-diffusive mechanism is given as 4.68–4.95 eV (Gösele et al. 1996, Bracht et al. 1998, Ural et al. 1999).

The picture emerging from theoretical studies, however, is far from clear. For instance, in two recent density functional theory (DFT) calculations with the local density approximation (LDA), the lowest-energy self-interstitials in silicon are predicted to have a formation energy of about 2.2 eV in one study and 4.2 eV in the other (Clark & Ackland 1997, Lee et al. 1998). Despite the discrepancies, it is widely believed that the split-[110], hexagonal and tetrahedral defects are low in energy. The objective of this study is to apply the highly accurate fixed-node diffusion quantum Monte Carlo method to determine the formation energies of these neutral silicon self-interstitials, and compare their values with the DFT values obtained from LDA or PW91-generalized gradient approximation (Perdew 1991). We have also studied the newly-proposed "caged" structure (Clark & Ackland 1997) and Pandey's concerted-exchange (CE) self-diffusion mechanism (Pandey 1986). Both of them were predicted to have a very low formation or activation energy.

2   FIXED-NODE DIFFUSION MONTE CARLO (DMC) METHOD

We have applied the standard fixed-node DMC method to calculate the formation energies of the defect structures above. DMC is a stochastic method for solving the imaginary-time N-body Schrödinger equation:

$$-\frac{\partial \Psi(\mathbf{R},t)}{\partial t} = (\hat{H} - E_{\text{Ref}})\Psi(\mathbf{R},t) \tag{3}$$

where Ψ is the wave function, $\hat{H}$ is the Hamiltonian, $\mathbf{R}$ denotes a $3N$-dimensional vector of electronic positions, t is a real variable measuring the progress in imaginary time, and E_{Ref} is an energy offset. This equation has the property that any initial wave function will decay towards the ground state wave function (provided that they are not orthogonal), and it is well established that DMC can give an excellent description of electron correlation in the ground state. The time evolution of Equation (3) can be simulated using a Monte Carlo technique, which represents the wave function by the number density of a set of $3N$-dimensional vectors of electronic positions $\{\mathbf{R}_i\}$ – usually called *walkers* in DMC. These walkers move in the $3N$-dimensional space via diffusive jumps arising from the kinetic energy term and are also evolved via a removal/duplication process arising from the potential energy term (Foulkes et al. 2001). The application of this simple algorithm to many-electron problems is, however, far from straightforward. First, a fermionic wave function has both positive and negative regions, but negative values cannot be represented by the number density of a set of walkers $\{\mathbf{R}_i\}$, which is always positive. Second, the removal/duplication rate diverges whenever two electrons or an electron and a nucleus coincide, leading to extremely poor statistical behaviour. These problems can be circumvented by simulating the mixed distribution $\Psi\Phi_G$ instead of Ψ, where Φ_G is a pre-determined guiding wave function, and restricting the wave function Ψ to have exactly the same nodal surface as Φ_G. This arrangement means that Ψ cannot evolve into the exact ground state wave function and is commonly known as the *fixed-node approximation*. In this study we have used a guiding wave function of the Slater-Jastrow form:

$$\Phi_G(\mathbf{R}) = D^{\uparrow}D^{\downarrow}\exp[J(\mathbf{R})] \tag{4}$$

where $D^{\uparrow}$ and $D^{\downarrow}$ are Slater determinants of up-spin and down-spin DFT-LDA orbitals (calculated at the L-point of the Brillouin zone with a basis set cutoff of 18 Ry), and $\exp[J(\mathbf{R})]$ is the Jastrow factor (which provides electron correlation). The overall guiding wave function contains 32 parameters for the perfect crystals, and up to 80 parameters for the defect structures. These parameters are then determined by minimizing the variance of the local energy.

The DMC calculations were performed on a 64-node Hitachi SR2201 supercomputer, and the entire study took about 4 months of computational time. More technical details of the DMC calculations can be found in other references (Foulkes et al. 2001, Leung et al. 1999).

3   STRUCTURES OF THE SELF-INTERSTITIAL DEFECTS IN SILICON

Silicon naturally occurs in the very stable diamond lattice structure. Within this structure, all silicon atoms are tetrahedrally fourfold-coordinated with a bond length of 2.35 Å. To model the defect structures, two *fcc* supercell sizes have been used: 16(+1) and 54(+1) atoms. The structures are almost identical using the different system sizes. All the structures have been fully relaxed under DFT-LDA (Needs 1999), and we have checked the validity of the LDA structures using PW91-GGA and found that they are almost identical, suggesting that the LDA structures are accurate. This arrangement also facilitates a direct comparison of the DMC, DFT-LDA and DFT-GGA methods. The formation energy of each defect is then calculated as the difference in energy between the defect and perfect structures:

$$E_F = E_{Defect}^{N+1} - \frac{N+1}{N} E_{Perfect}^{N} \tag{5}$$

where E^{N+1} and E^N are the DMC energies of the defect and perfect simulations cells, respectively. Finite-size effects are corrected using the results of our DFT-LDA calculations.

3.1 The hexagonal interstitial

The hexagonal interstitial earns its name from the hexagonal shape formed by the six nearest neighbours surrounding the interstitial atom (Fig. 1). The six neighbouring atoms, however, are not lying on the same plane. The bond length between the interstitial atom and its six nearest neighbours is 2.36 Å, and the relaxation of the six nearest neighbours is rather substantial.

3.2 The split-[110] interstitial

The split-[110] interstitial defect is formed by placing a pair of atoms about a single vacant lattice site along the [110] direction. The two interstitial atoms have four nearest neighbours, two of which become fivefold coordinated while the others remain fourfold coordinated (Fig. 2). The bond between the two interstitial atoms is 2.44 Å. They are joined by bonds of length 2.47 Å to the fivefold-coordinated neighbours and 2.33 Å to the fourfold-coordinated neighbours.

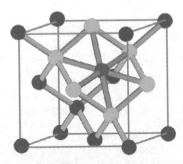

Hexagonal:	0.000	0.000	0.000
Neighbour 1:	0.684	-0.684	-2.153
Neighbour 2:	2.153	0.684	-0.684
Neighbour 3:	0.684	2.153	0.684
Neighbour 4:	-0.684	0.684	2.153
Neighbour 5:	-2.153	-0.684	0.684
Neighbour 6:	-0.684	-2.153	-0.684

Figure 1. The hexagonal defect structure (left) and the atomic coordinates of the six nearest neighbours relative to the interstitial atom (right). All lengths are given in Å.

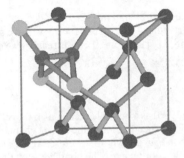

Interstitial 1:	0.862	0.862	-0.787
Interstitial 2:	-0.862	-0.862	-0.787
Neighbour 1:	1.449	-1.449	-1.433
Neighbour 2:	-1.449	1.449	-1.433
Neighbour 3:	1.413	1.413	1.406
Neighbour 4:	-1.413	-1.413	1.406

Figure 2. The split-[110] defect structure (left) and the atomic coordinates of the four nearest neighbours relative to the centre of the interstitial pair (right).

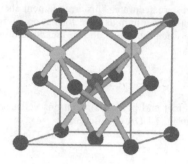

Tetrahedral:	0.000	0.000	0.000
Neighbour 1:	1.410	-1.410	-1.410
Neighbour 2:	1.410	1.410	1.410
Neighbour 3:	-1.410	-1.410	1.410
Neighbour 4:	-1.410	1.410	-1.410

Figure 3. The tetrahedral defect structure (left) and the atomic coordinates of the four nearest neighbours relative to the interstitial atom (right).

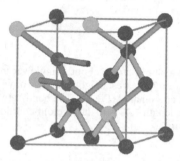

Exchange 1:	0.000	0.759	-0.759
Exchange 2:	0.000	-0.759	0.759
Neighbour 1:	-1.980	0.831	-2.069
Neighbour 2:	1.980	-0.831	2.069
Neighbour 3:	0.592	-1.907	-1.907
Neighbour 4:	-0.592	1.907	1.907
Neighbour 5:	1.980	2.069	-0.831
Neighbour 6:	-1.980	-2.069	0.831

Figure 4. The saddle-point structure of Pandey's concerted-exchange mechanism (left) and the atomic coordinates of the six nearest neighbours relative to the exchanging pair (right).

3.3   The tetrahedral interstitial

The tetrahedral interstitial defect is formed by placing an interstitial atom at the centre of the conventional unit cell of the diamond structure. The interstitial atom is joined by four relatively weak bonds of length 2.44 Å to four surrounding atoms in a tetrahedral shape (Fig. 3). The four neighbouring atoms then become fivefold coordinated. The relaxation in this structure is somewhat smaller than that in the hexagonal structure. Recent DFT-LDA studies suggest that this structure is unstable towards the hexagonal structure. This is important because it means that the tetrahedral structure is actually a saddle point of the diffusion path connecting adjacent hexagonal sites via an intermediate tetrahedral site.

3.4   The "caged" interstitial

The "caged" interstitial is a new defect proposed by Clark & Ackland in 1997. This defect has very low symmetry and seems to relax towards the split-[110] structure. The exact structure we have studied is a partially relaxed structure and is quite similar to the split-[110] structure. For this reason, we do not provide a separate illustration of the "caged" interstitial here.

3.5   The concerted-exchange mechanism

The concerted exchange (CE) was proposed by Pandey (1986) as a possible mechanism for self-diffusion in silicon without the mediation of an intrinsic defect. The crucial point of the CE mechanism is that at no stage of the process are there more than two broken bonds in the whole structure. At the saddle point, the exchanging pair and two near neighbours are threefold coordinated. The bond between the exchanging pair is very strong and is only 2.15 Å long (cf. 2.35 Å in bulk silicon). All other atoms remain fourfold coordinated.

Table 1. LDA, GGA and DMC formation energies in eV of the silicon self-interstitials and saddle point of the concerted exchange mechanism. Numbers in parentheses denote statistical errors in the final digits.

Defect structure	LDA	PW91-GGA	DMC (16 + 1 atoms)	DMC (54 + 1 atoms)
Hexagonal	3.31	3.80	4.70 (24)	4.82 (28)
Split-[110]	3.31	3.84	4.96 (24)	4.96 (28)
"Caged"	3.34	3.85	5.26 (24)	5.17 (28)
Tetrahedral	3.43	4.07	5.50 (24)	5.40 (28)
Concerted exchange	4.58	4.93	5.85 (24)	5.78 (28)

4 DEFECT FORMATION ENERGIES

The results of our DMC and DFT calculations for the defect formation energies are shown in Table 1. First, we note that the DMC energies for the two supercell sizes are similar, indicating that the residual finite-size effects are small, probably in the range of 0.1–0.2 eV. Second, the DMC formation energies are about 1 eV larger than the GGA results and 1.5 eV larger than the LDA results. According to our DMC and GGA calculations, the hexagonal interstitial has the lowest formation energy, followed closely by the split-[110] interstitial. The "caged" structure is also low in energy, whereas the tetrahedral interstitial has a significantly higher energy. The saddle point of the CE mechanism has an energy that is far too high to explain self-diffusion in silicon. Our DMC results therefore support that the interstitial mechanism is more important than the CE mechanism for silicon self-diffusion.

Because mapping the energy barriers to self-diffusion is computationally prohibitive using DMC, we can only estimate the activation energy as follows. We note that the tetrahedral interstitial is a saddle point of the possible diffusion path between adjacent hexagonal sites. The DMC formation energy of the tetrahedral interstitial (5.4 eV) is therefore a rigorous upper bound to the self-diffusion activation energy. The true activation energy is expected to be significantly lower. Based on the DFT-LDA energy landscape, we arrive at the best estimate of ~5 eV for the hexagonal–hexagonal diffusion path. This value is in excellent agreement with the recent experimental value of 4.95 eV (Bracht et al. 1998).

5 CONCLUSIONS

We have demonstrated the importance of a proper treatment of electron correlation when calculating the defect formation energies in silicon. The LDA and GGA do not provide a satisfactory explanation of the recent experimental data as they predict too small an activation energy. The larger defect formation energies in DMC indicate a possible resolution of this problem, and this may be an important step in improving our understanding of self-diffusion in silicon.

REFERENCES

Bracht, H., Haller, E.E. & Clark-Phelps, R. 1998. *Phys. Rev. Lett.* 81(2): 393–396.
Clark, S.J. & Ackland, G.J. 1997. *Phys. Rev. B* 56(1): 47–50.
Foulkes, W.M.C., Mitas, L., Needs, R.J. & Rajagopal, G. 2001. *Rev. Mod. Phys.* 73(1): 33–83.
Frank, W., Gösele, U., Mehrer, H. & Seeger, A. 1984. In G.E. Murch & A.S. Nowick (eds), *Diffusion in Crystalline Solids*: 63–142. New York: Academic Press.
Gösele, U., Plössl, A. & Tan, T.Y. 1996. In G.R. Srinivasan, C.S. Murthy & S.T. Dunham (eds), *Process Physics and Modeling in Semiconductor Technology*: 309. New Jersey: Electrochemical Society.
Lee, W.-C., Lee, S.-G. & Chang, K.J. 1998. *J. Phys. Condens. Matter* 10(1998): 995–1002.
Leung, W.-K., Needs, R.J., Rajagopal, G., Itoh, S. & Ihara, S. 1999. *Phys. Rev. Lett.* 83(12): 2351–2354.
Needs, R.J. 1999. *J. Phys. Condens. Matter* 11(50): 10437–10450.
Pandey, K.C. 1986. *Phys. Rev. Lett.* 57(18): 2287–2290.
Perdew, J.P. 1991. In P. Ziesche & H. Eschrig (eds), *Electronic Structure of Solids '91*: 11. Berlin: Akademie Verlag.
Ural, A., Griffin, P.B. & Plummer, J.D. 1999. *Phys. Rev. Lett.* 83(17): 3454–3457.

Computational Methods in Engineering and Science, Iu et al. (eds)
© *2003 Swets & Zeitlinger, Lisse, ISBN 90 5809 567 3*

Quantum Monte Carlo calculations of the pair correlation functions in jellium

W.K. Leung
Department of Electronic and Information Engineering, The Hong Kong Polytechnic University, Hong Kong

R.J. Needs
TCM Group, Cavendish Laboratory, University of Cambridge, Cambridge, England, UK

ABSTRACT: We apply the variational and fixed-node diffusion quantum Monte Carlo methods to calculate the pair correlation functions and correlation energy of the unpolarized and fully polarized (non-relativistic) homogeneous electron gas. We fit our PCF data to simple polynomials and compare them with the widely-used Perdew-Wang model. We also compare the correlation energies with the classic results by Ceperley and Alder, as well as the more recent ones by Ortiz and coworkers.

1 INTRODUCTION

The homogeneous electron gas (HEG), or jellium, is the simplest but yet an extremely important many-electron system. The correlation energy of HEG is the key ingredient of the local density and local spin density approximations (LDA/LSDA) (Perdew & Zunger 1987, Perdew & Wang 1992) for use in density functional theory (Hohenberg & Kohn 1964, Kohn & Sham 1965), which is currently the most popular computational tool for studying condensed matter systems. The success of LDA and LSDA is partly due to its simplicity, but their inaccuracies have led people to search for more sophisticated methods. Most "beyond-LDA" schemes, such as the generalized gradient approximation (Perdew et al. 1986, Perdew 1991) and weighted density approximation (Alonso & Girifalco 1978, Gunnarsson et al. 1979), require input data from the HEG, in particular the pair correlation functions (PCF). The objective of this study is to apply the variational and fixed-node diffusion quantum Monte Carlo methods (QMC) to obtain accurate results for the correlation energy and PCF of the unpolarized and fully polarized HEG (Foulkes et al. 2001).

2 DETAILS OF CALCULATIONS

The HEG can be characterized by two parameters alone, namely the density $n = n_\uparrow + n_\downarrow$ and spin polarization $\xi = (n_\uparrow + n_\downarrow)/n$, where $n_\uparrow$ and $n_\downarrow$ are the up-spin and down-spin electron densities, respectively. In condensed matter physics it is customary to express the density n by a density parameter $r_s = (3/4\pi n)^{1/3}$. The spin-resolved PCF, $g_{\alpha\beta}(r)$, is defined as the probability of finding a pair of electrons with spins α and β at a separation r, normalized to unity by the uncorrelated Hartree distribution. Within QMC, the spin-resolved PCF is most conveniently defined as:

$$g_{\alpha\beta}(r) = \lim_{\Delta r \to 0} \left\langle \frac{V \cdot \Delta N_{\alpha\beta}}{4\pi r^2 \Delta r \cdot N_\alpha (N_\beta - \delta_{\alpha\beta})} \right\rangle , \tag{1}$$

where V is the volume of the simulation cell, N_α and N_β are the total numbers of electrons with spins α and β in the entire system, $\delta_{\alpha\beta}$ is the delta function, $\Delta N_{\alpha\beta}$ is the number of electron pairs with the prescribed spins α and β separated by a distance between $r \pm \Delta r$, and the whole average is to be taken over the ground state distribution. By definition, $g_{\uparrow\downarrow}(r) = g_{\downarrow\uparrow}(r)$, and for a partially polarized HEG, there are altogether three PCF, namely $g_{\uparrow\uparrow}(r)$, $g_{\downarrow\downarrow}(r)$ and $g_{\uparrow\downarrow}(r)$. They are reduced to two distinct functions in the unpolarized HEG [because $g_{\uparrow\uparrow}(r) = g_{\downarrow\downarrow}(r)$] and only one in the fully polarized HEG [because only $g_{\uparrow\uparrow}(r)$ exists].

The definition of $g_{\alpha\beta}(r)$ in Equation (1) lends itself a direct method of evaluation in QMC simulations, whereby the inter-electron distances are worked out in all the simulated configurations, sorted according to the relative spins and then collected in a histogram with a bin size Δr. In this study we have used a bin size of $L_{WS}/200$, where L_{WS} is the radius of the largest sphere that can be inscribed inside the Wigner-Seitz face-centred cubic simulation cell.

In this study, two system sizes have been studied: 178 electrons for the unpolarized HEG, and 181 electrons for the fully polarized HEG. The supercell technique is used to mimic the infinite nature of the HEG. We have used a trial/guiding wave function of the Slater-Jastrow type with a localized, spherically symmetric two-body Jastrow function, whose parameters are determined by minimizing the variance of the local energy. We have also used the truncated Coulomb potential (Foulkes et al. 2001), which eliminates the need for expensive Ewald summations and minimizes the Coulomb finite-size effects. Residual finite-size effects are found to be very small, as the PCF computed from different system sizes are almost identical. This is to be expected as the widest exchange-correlation (XC) hole extends to about $3.5r_s$ only, while our simulation cell covers more than $5r_s$.

To the best of our knowledge, this QMC study represents the most accurate calculations of the PCF to date. Currently, the most widely-used model of the PCF is devised by Perdew & Wang (1992), who fitted their models to the spin-averaged QMC data of Ceperley (1978) and Ceperley & Alder (1980). More recently, Gori-Giorgi et al. (2000) proposed a new model based on the QMC data of Ortiz et al. (1994, 1999). However, the Perdew-Wang model is essentially a spin-averaged formulation, while the model due to Gori-Giorgi et al. is available for the unpolarized systems only. Our new QMC model, on the other hand, is available for both $\xi = 0.0$ and $\xi = 1.0$, and it is fitted to simple polynomials, which make the model much simpler to use.

3   RESULTS

We have studied the HEG at six different densities between $r_s = 0.1$ and $r_s = 10.0$ and two polarizations. The QMC results for the PCF of the unpolarized and fully polarized HEG at $r_s = 5.0$ and 10.0 are shown in Figures 1–2, compared with the Hartree-Fock (HF) and Perdew-Wang (PW) models. A few interesting points can be noted from the figures. First, contrary to Ortiz & Ballone (1994), we observe a relatively minor difference between the variational and diffusion QMC results. This is probably attributable to the higher quality of the trial wave function we have used.

Second, it can be seen that the PW model is only in fair agreement with the QMC models. This is particularly true at low densities ($r_s \geqslant 5$) and the PW model verges on a breakdown at $r_s \sim 10$. This can be explained by the fact that, it is more difficult to represent the short-ranged part of $g_{\uparrow\downarrow}(r)$ at very low densities because their values are very close to zero (but we cannot let it become negative at any time because a probability can never be negative). In general, the PW model predicts too flat a bulge at $r \sim 1.7r_s$ for parallel spins and too high a bulge at $r \sim 1.5r_s$ for antiparallel spins. These discrepancies are sometimes explained, quite correctly, in terms of the crudity of the PW formulation of spin resolution, as their formulation is based on the *spin-averaged* QMC data of Ceperley (1978) and Ceperley & Alder (1980). However, it should be noted that the PW model does not fit the QMC data very well even for the fully polarized HEG, which contains only one type of spins and therefore requires no spin resolution at all. In light of this, we believe that apart from the inaccurate spin resolution scheme, the limited flexibility of the PW functional form for $g_{\alpha\beta}(r)$ is also a major reason for the observed discrepancies.

We have also calculated the correlation energies of the HEG for the density range $r_s = 0.1$–10. Their values are important ingredient of LDA and LSDA, and they are given in Table 1, where we also show the QMC results of Ortiz et al. (1994, 1999) and Ceperley et al. (1978, 1980). We note that our results are in very good agreement with the Ortiz values ($\sim 3\%$ discrepancy), while the widely-used Ceperley results seem to overestimate the magnitude of the correlation energy by as much as 5–10% at the intermediate densities $r_s \sim 2.0$, especially for $\xi = 1$.

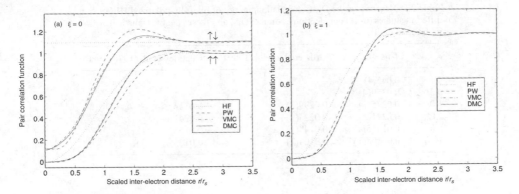

Figure 1. Pair correlation functions of the HEG at $r_s = 10.0$, and (a) $\xi = 0$ and (b) $\xi = 1$. The antiparallel-spin pair correlation functions have been raised by 0.1 in (a) to avoid clutter. HF denotes the Hartree-Fock pair correlation functions. PW is from Perdew & Wang 1992. VMC and DMC are variational and fixed-node diffusion QMC results from this study, respectively.

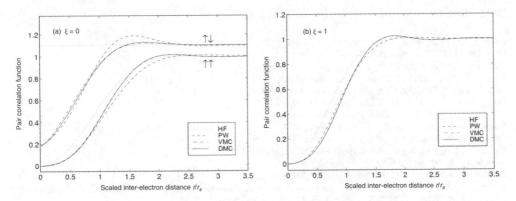

Figure 2. Pair correlation functions of the HEG at $r_s = 5.0$, and (a) $\xi = 0$ and (b) $\xi = 1$. The antiparallel-spin pair correlation functions have been raised by 0.1 in (a) to avoid clutter. HF denotes the Hartree-Fock pair correlation functions. PW is from Perdew & Wang 1992. VMC and DMC are variational and fixed-node diffusion QMC results from this study, respectively.

Table 1. Negative of the DMC correlation energy of the unpolarized HEG from different studies (in Ha per electron). Numbers in parentheses denote statistical errors in the final digits.

r_s	This study	Ortiz et al. (1994, 1999)	Ceperley et al. (1978, 1980)
0.1	0.22 (1)	–	–
0.3	0.0843 (1)	–	–
1.0	0.0580 (1)	0.0563 (2)	0.0595
2.0	0.0446 (1)	0.0438 (2)	0.0451
5.0	0.0283 (1)	0.0283 (1)	0.0283
10.0	0.0187 (1)	0.0183 (2)	0.0186

4 CONCLUSIONS

We have successfully applied the variational and fixed-node diffusion Monte Carlo methods to calculate the PCF and correlation energy of the unpolarized and fully polarized HEG. We find that the PW

Table 2.   Negative of the DMC correlation energy of the fully polarized HEG from different studies (in Ha per electron). Numbers in parentheses denote statistical errors in the final digits.

r_s	This study	Ortiz et al. (1994, 1999)	Ceperley et al. (1978, 1980)
0.1	0.045(4)	–	–
0.3	0.0359(4)	–	–
1.0	0.0275(1)	0.0299(2)	0.0316
2.0	0.0221(1)	0.0223(2)	0.0241
5.0	0.01455(1)	0.0148(1)	0.0155
10.0	0.01008(1)	0.0104(2)	0.0105

model is in poor agreement with the QMC data at low densities ($r_s \geqslant 5$). We also find that the correlation energies given by Ceperley and Alder seem to be too negative in the important density range $r_s \sim 1.0$–2.0, particularly for the fully polarized systems.

REFERENCES

Alonso, J.A. & Girifalco, L.A. 1978. *Phys. Rev. B* 17: 3735.
Ceperley, D.M. 1978. *Phys. Rev. B* 18: 3126.
Ceperley, D.M. & Alder, B.J. 1980. *Phys. Rev. Lett.* 45: 566.
Foulkes, W.M.C., Mitas, L., Needs, R.J. & Rajagopal, G. 2001. *Rev. Mod. Phys.* 73(1): 33–83.
Gior-Giorgi, P., Sacchetti, F. & Bachelet, G.B. 2000. *Phys. Rev. B* 61: 7353.
Gunnarsson, O., Jonson, M. & Lundqvist, B. 1979. *Phys. Rev. B* 20: 765.
Hohenberg, P. & Kohn, W. 1964. *Phys. Rev.* 136: B864.
Kohn, W. & Sham, L.J. 1965. *Phys. Rev.* 140: A1133.
Ortiz, G. & Ballone, P. 1994. *Phys. Rev. B* 50: 1391.
Ortiz, G., Harris, M. & Ballone, P. 1999. *Phys. Rev. Lett.* 82: 5317.
Perdew, J.P. & Zunger, A. 1981. *Phys. Rev. B* 23: 5048.
Perdew, J.P. & Wang, Y. 1986. *Phys. Rev. B* 33: 8800.
Perdew, J.P. 1991. In P. Ziesche & H. Eschrig (eds), *Electronic Structure of Solids '91*: 11. Berlin: Akademie Verlag.
Perdew, J.P. & Wang, Y. 1992. *Phys. Rev. B* 45: 13244.
Perdew, J.P. & Wang, Y. 1992. *Phys. Rev. B* 46: 12947.

Computational Methods in Engineering and Science, Iu et al. (eds)
© 2003 Swets & Zeitlinger, Lisse, ISBN 90 5809 567 3

Quantum Monte Carlo study of the relativistic spin-polarized jellium

W.K. Leung
Department of Electronic and Information Engineering, The Hong Kong Polytechnic University, Hong Kong

R.J. Needs
TCM Group, Cavendish Laboratory, University of Cambridge, Cambridge, England, UK

ABSTRACT: We apply the variational and fixed-node diffusion quantum Monte Carlo methods to calculate the ground state energy of the unpolarized ($\xi = 0$) and fully polarized ($\xi = 1$) homogeneous electron gas at zero temperature within the density range $r_s = 0.1$–10.0, incorporating relativistic effects via first-order perturbation theory. The relativistic correction terms are compared with the Hartree-Fock expectation values. It is found that electron correlation is significant at intermediate densities $r_s \sim 1.0$–5.0. This demonstrates the importance of a proper treatment of electron correlation in RDFT calculations.

1   INTRODUCTION

Relativistic effects are important at high velocity or, equivalently, in heavy elements. They can be calculated in a number of ways, e.g. the Dirac-Fock method, and the relativistic density functional theory (RDFT) (Rajagopal & Callaway 1973). Applying RDFT with a local spin density approximation (LSDA) requires input data from the homogeneous electron gas (HEG), or jellium (MacDonald 1978, 1983, Rajagopal 1978, Xu et al. 1983). Previous studies of the relativistic HEG have used methods that are equivalent to the summation of "ring diagrams", which are accurate for high densities but lack a proper treatment of electron correlation. In this paper we have chosen a different approach – we incorporate relativistic effects in our quantum Monte Carlo (QMC) calculations via first-order perturbation theory, including terms of the order $1/c^2$, where c is the speed of light. This approach provides a proper treatment of both correlation and relativistic effects for densities up to $r_s \sim 0.1$, and has previously been applied to produce highly accurate results for molecules (Vrbik et al. 1988), atoms (Kenny et al. 1995) and unpolarized HEG (Kenny et al. 1996). The objectives of this study are to obtain more accurate energies of the unpolarized HEG ($\xi = 0$) and to extend the study to spin-polarized systems ($\xi = 1$) too. Note that the density parameter r_s and spin polarization ξ are defined as:

$$r_s = \left(\frac{3}{4\pi n}\right)^{1/3} \quad \text{and} \quad \xi = \frac{n_\uparrow - n_\downarrow}{n}, \tag{1}$$

where $n_\uparrow$ and $n_\downarrow$ are the up- and down-spin electron densities, and $n = n_\uparrow + n_\downarrow$ is the total electron density, respectively.

2   RELATIVISTIC HAMILTONIAN

An effective relativistic Hamiltonian for many-electron systems subject to a static electric field has been derived by Bethe and Salpeter from the Breit equation (Bethe & Salpeter 1957) and by Itoh from quantum electrodynamics (Itoh 1965). The Hamiltonian, however, is not bounded from below and therefore can only be used as a perturbation operator. The full Hamiltonian is fairly complex and contains a large number of terms. Fortunately, many of those terms vanish in the HEG for symmetry reasons. In the unpolarized HEG, only the following relativistic correction terms survive (written in atomic units):

$$\Delta H_{rel} = -\sum_i \frac{1}{8c^2}\vec{\nabla}_i^4 + \sum_{i<j}\frac{\pi}{c^2}\delta(\mathbf{r}_i - \mathbf{r}_j) - \sum_{i<j}\frac{1}{2c^2}\vec{\nabla}_i \cdot \left[\frac{(\mathbf{r}_i - \mathbf{r}_j)(\mathbf{r}_i - \mathbf{r}_j)}{r_{ij}^3} + \frac{1}{r_{ij}}\right]\cdot\vec{\nabla}_j$$

where $\mathbf{r}_i$ denotes the position vector of the ith electron, r_{ij} is the distance between the ith and jth electrons, and $c = 137.036$ is the speed of light in atomic units. The first term of the Hamiltonian is the mass–velocity correction, the second term is the combined electron–electron Darwin and spin–spin interaction corrections (we call it the "contact" term here), and the final term is the retardation correction. The last two terms are collectively known as the Breit interaction. In the fully spin-polarized HEG, the contact term vanishes because of the Pauli exclusion principle, and we are left with only two terms to evaluate.

Before we discuss the QMC results, it is useful to derive the expectation value of ΔH_{rel} with respect to the Hartree–Fock (HF) wave function of the HEG, which is just a Slater determinant of simple plane waves $\exp(i\mathbf{G}\cdot\mathbf{r})$, where $\mathbf{G}$'s are wave vectors up to the Fermi level in the reciprocal space. For the mass–velocity correction, which accounts for the variation of mass with velocity, the HF expectation value (in Hartrees per electron) is given by:

$$\Delta E_1^{HF} = -\frac{1}{84\pi r_s^4 c^2}\left[\left(\frac{9\pi(1+\xi)}{4}\right)^{7/3} + \left(\frac{9\pi(1-\xi)}{4}\right)^{7/3}\right]$$

$$= \begin{cases} -\dfrac{3}{56 r_s^4 c^2}\left(\dfrac{9\pi}{4}\right)^{4/3} & \text{for } \xi = 0 \\[3mm] -\dfrac{3}{56 r_s^4 c^2}\left(\dfrac{9\pi}{2}\right)^{4/3} & \text{for } \xi = 1 \end{cases} \tag{2}$$

Similarly, for the contact term, the HF expectation value is given by:

$$\Delta E_2^{HF} = -\frac{3(1-\xi^2)}{16 r_s^3 c^2}$$

$$= \begin{cases} -\dfrac{3}{16 r_s^3 c^2} & \text{for } \xi = 0 \\[3mm] 0 & \text{for } \xi = 1 \end{cases} \tag{3}$$

Finally, for the retardation term, the HF expectation value is given by:

$$\Delta E_3^{HF} = \frac{9(1+\xi^2)}{16 r_s^3 c^2}$$

$$= \begin{cases} \dfrac{9}{16 r_s^3 c^2} & \text{for } \xi = 0 \\[3mm] \dfrac{9}{8 r_s^3 c^2} & \text{for } \xi = 1 \end{cases} \tag{4}$$

We have also computed the HF expectation value of the same finite HEG systems as in the QMC calculations. The difference between these values and the infinite-system values given in (2)–(4) is useful for correcting finite size effects in the QMC results.

3   QUANTUM MONTE CARLO CALCULATIONS

The expectation values of ΔH_{rel} are calculated using both the variational (VMC) and fixed-node diffusion quantum Monte Carlo (DMC) methods with a face-centered cubic supercell (Kenny et al. 1996, Foulkes et al. 2001). In the DMC calculations, we have applied the mixed estimator ΔE_{DMC} as well as the extrapolated estimator $\Delta E_{ext} = 2\Delta E_{DMC} - \Delta E_{VMC}$, which is accurate to second order. We have used a trial/guiding wave function of the Slater-Jastrow type with a total of 20 variational parameters determined by minimizing the variance of the local energy. The Slater determinants are constructed with simple plane waves $\exp(i\mathbf{G} \cdot \mathbf{r})$, where $\mathbf{G}$'s are reciprocal lattice vectors of the supercell. In order to mimic the translational and rotational invariance of the HEG as closely as possible, we always include complete shells of $\mathbf{G}$'s that are related under the symmetry of the cell. This restricts the number of electrons of either spin to a well-defined sequence determined by the symmetry of the cell. In this study, we have simulated 178 and 338 electrons for the unpolarized HEG, and 181 and 307 electrons for the fully polarized HEG.

The exact formulas for computing the different terms of ΔH_{rel} within QMC have been given elsewhere (Vrbik et al. 1988, Kenny et al. 1995, 1996). For the mass–velocity term, it can be evaluated as:

$$\Delta E_1^{QMC} = -\frac{1}{8c^2} \int \Psi \sum_i \bar{\nabla}_i^4 \Psi \, d\mathbf{R}$$

$$= -\frac{1}{2c^2} \int \sum_i \left(-\frac{1}{2} \frac{\bar{\nabla}_i^2 \Psi^2}{\Psi} \right)^2 \Psi^2 d\mathbf{R} \tag{5}$$

which means that the mass–velocity term can simply be computed as the sum of squares of the single-particle kinetic energies (multiplied by the constant $-1/2c^2$) over the Monte Carlo simulated distribution. For the contact term, we collect statistics about the inter-electron distances in the Monte Carlo simulations and construct the pair-correlation functions. The QMC contact term is then given by scaling the HF results with the values of the antiparallel-spin pair-correlation function at zero distance, $g_{\uparrow\downarrow}(0)$. (The QMC pair-correlation function of the HEG is reported in a separate publication.) Finally, for the retardation term, Vrbik et al. (1988) gave the following formula:

$$\Delta E_3^{QMC} = -\frac{1}{2c^2} \int \Psi \sum_{i<j} \bar{\nabla}_i \cdot \left[\frac{(\mathbf{r}_i - \mathbf{r}_j)(\mathbf{r}_i - \mathbf{r}_j)}{r_{ij}^3} + \frac{1}{r_{ij}} \right] \cdot \bar{\nabla}_j \Psi \, d\mathbf{R}$$

$$= -\frac{1}{c^2} \int \sum_{i<j} \left[\frac{(\mathbf{r}_{ij} \cdot \mathbf{F}_i)(\mathbf{r}_{ij} \cdot \mathbf{F}_j)}{r_{ij}^3} + \frac{\mathbf{F}_i \cdot \mathbf{F}_j}{r_{ij}} \right] \Psi^2 d\mathbf{R} \tag{6}$$

where $\mathbf{F}_i$ is an operator associated with the calculation of kinetic energy in QMC (i.e. we compute $\mathbf{F}_i$ even if we do not intend to calculate the retardation correction – therefore, this formulation incurs very little extra cost):

$$\mathbf{F}_i = -\frac{1}{\sqrt{2}} \vec{\nabla}_i \ln \Psi . \tag{7}$$

We have included six densities in this study: $r_s = 0.1, 0.3, 1.0, 2.0, 5.0,$ and 10.0. Finite-size effects are found to be small because we have used the model periodic Coulomb potential instead of the Ewald summation (Foulkes et al. 2001). Residual finite-size effects are corrected using the Hartree-Fock (HF) results obtained from the finite system (system size n) and the infinite system (system size ∞), i.e.

$$\Delta E^{QMC}(\infty) \approx \Delta E^{QMC}(n) - \Delta E^{HF}(n) + \Delta E^{HF}(\infty) . \tag{8}$$

The corrected results are then averaged between the two system sizes simulated in order to obtain a more reliable value.

4   RESULTS

The final QMC values of the relativistic corrections are given in Tables 1 & 2. As expected, the mass–velocity term dominates at high density, while at low density the retardation term is the most significant correction. More importantly, electron correlation strongly modifies all the relativistic terms. At

Table 1. Extrapolated estimator QMC results of the relativistic corrections for the unpolarized HEG (in Ha per electron). Numbers in parentheses denote statistical errors in the final digits.

r_s	Mass–velocity	Contact	Retardation
0.1	$-3.90(2) \times 10^{-1}$	$0.94(3) \times 10^{-2}$	$3.01(4) \times 10^{-2}$
0.3	$-4.87(1) \times 10^{-3}$	$0.30(1) \times 10^{-3}$	$1.13(2) \times 10^{-3}$
1.0	$-4.45(1) \times 10^{-5}$	$0.54(3) \times 10^{-5}$	$3.13(4) \times 10^{-5}$
2.0	$-3.42(2) \times 10^{-6}$	$0.42(3) \times 10^{-6}$	$4.42(4) \times 10^{-6}$
5.0	$-1.43(1) \times 10^{-7}$	$0.07(2) \times 10^{-7}$	$3.84(8) \times 10^{-7}$
10.0	$-1.48(1) \times 10^{-8}$	$0.01(1) \times 10^{-8}$	$6.22(5) \times 10^{-8}$

Table 2. Extrapolated estimator QMC results of the relativistic corrections for the fully polarized HEG (in Ha per electron). Numbers in parentheses denote statistical errors in the final digits.

r_s	Mass–velocity	Contact	Retardation
0.1	$-9.753(2) \times 10^{-1}$	0.00	$5.89(9) \times 10^{-2}$
0.3	$-1.209(2) \times 10^{-2}$	0.00	$2.18(4) \times 10^{-3}$
1.0	$-1.008(1) \times 10^{-4}$	0.00	$5.8(1) \times 10^{-5}$
2.0	$-6.725(8) \times 10^{-6}$	0.00	$7.3(2) \times 10^{-6}$
5.0	$-2.158(6) \times 10^{-7}$	0.00	$5.4(1) \times 10^{-7}$
10.0	$-1.841(1) \times 10^{-8}$	0.00	$7.3(2) \times 10^{-8}$

$r_s = 10$ and $\xi = 0$, the QMC results are about 3 times the HF values for the mass–velocity term and 2 times the HF values for the retardation term. In general, electron correlation effects are more pronounced in the unpolarized HEG and at low density. At high density, the QMC results are well approximated by the HF values.

An immediate application of our QMC results is to construct an LSDA to the exchange-correlation (XC) energy functional for use in RDFT calculations. We define a relativistic correction to the XC energy ΔE_{XC} as the expectation value of ΔH_{rel} minus those terms that are generated by the Dirac-Kohn-Sham equations themselves. These self-generated terms include the mass–velocity correction of the non-interacting HEG and the "direct" part of the Breit interaction. As the mass–velocity term is relatively similar between the interacting and non-interacting HEG, we note that the contribution to the relativistic XC energy from the mass–velocity term is actually *smaller* than that from the Breit interaction at all the densities studied here. Note also that the relativistic XC energy contains some purely relativistic effects which are not commonly perceived as electron correlation effects, e.g. retardation correction. The nomenclature, however, remains unchanged in order to emphasize the similarity between the relativistic and non-relativistic DFT treatments.

For convenience we have fitted the QMC XC energies divided by the HF exchange-only results to a simple polynomial expression $S(r_s, \xi)$ as follows:

$$\Delta E_{XC}^{QMC}(r_s, \xi) = \Delta E_{XC}^{HF}(r_s, \xi) \cdot S(r_s, \xi) \tag{9}$$

where

$$\Delta E_{xc}^{HF} = \frac{9 + 3\xi^2}{8c^2 r_s^3} \quad \text{and} \quad S(r_s, \xi) = \begin{cases} \sum_{i=0}^{4} c_i r_s^i & \text{for } r_s < 5 \\ a + b r_s & \text{for } r_s \geq 5 \end{cases} . \tag{10}$$

The parameters a and b are fixed by a linear interpolation between $r_s = 5$ and 10, while c_i are determined by a least-squares fitting process. Their values at $\xi = 0$ and $\xi = 1$ are given in Table 3.

Table 3. Values of the parameters a, b and c_i in (10), fitted to the extrapolated QMC results.

Fitted parameters	Unpolarized ($\xi = 0$)	Fully polarized ($\xi = 1$)
c_0	1.0000	1.0000
c_1	−0.26630	−0.10556
c_2	0.14345	0.043811
c_3	−0.027741	−0.0057928
c_4	0.0019089	0.00023468
a	0.77	0.913
b	0.042	0.0154

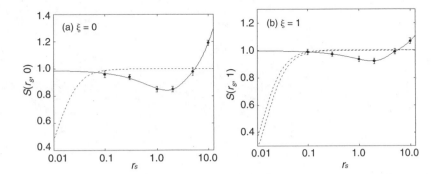

Figure 1. The relativistic XC energy divided by the HF expectation value at $\xi = 0$ and $\xi = 1$ (Eqs 9–10). Solid lines and error bars: extrapolated QMC results. Dashed line in (a): Rajagopal 1978, and MacDonald & Vosko 1979. Upper dashed line in (b): MacDonald 1983. Lower dashed line in (b): Xu et al. 1984.

5 CONCLUSIONS

The fitted QMC scale factor $S(r_s, \xi)$ is shown in Figure 1 for $\xi = 0$ and $\xi = 1$, compared with the exchange-only "ring-diagrams" expressions given by MacDonald et al. (1979, 1983), Rajagopal (1978) and Xu et al. (1984), which include higher-order relativistic corrections but lack in electron correlation. It can be seen that the two curves are very similar in shape. This suggests it is reasonable to use a simple quadratic interpolation for intermediate polarizations:

$$c_i(r_s, \xi) = c_i(r_s, 0) + \xi^2 [c_i(r_s, 1) - c_i(r_s, 0)]. \tag{11}$$

Also, we note that at low densities ($r_s > 0.1$), the "ring-diagrams" expressions are almost identical to the HF theoretical values $S = 1$ (not shown), while the QMC results differ from the HF values by as much as 20% in the important medium density range $r_s \sim 1.0$–5.0. This demonstrates the importance of a proper treatment of electron correlation in RDFT calculations.

REFERENCES

Bethe, H.A. & Salpeter, E.E. 1957. *Quantum Mechanics of 1- and 2-Electron Atoms*. Berlin: Springer.
Foulkes, W.M.C., Mitas, L., Needs, R.J. & Rajagopal, G. 2001. *Rev. Mod. Phys.* 73(1): 33–83.
Kenny, S.D., Rajagopal, G. & Needs, R.J. 1995. *Phys. Rev. A* 51: 1898.
Kenny, S.D., Rajagopal, G., Needs, R.J., Leung, W.-K., Godfrey, M.J., Williamson, A.J. & Foulkes, W.M.C. 1996. *Phys. Rev. Lett.* 77: 1099.
MacDonald, A.H. & Vosko, S.H. 1979. *J. Phys. C* 12: 2977.
MacDonald, A.H. 1983. *J. Phys. C* 16: 3869,

Rajagopal, A.K. & Callaway, J. 1973. *Phys. Rev. A* 7: 1648.
Rajagopal, A.K. 1978. *J. Phys. C* 11: L943.
Vrbik, J., DePasquale, M.F. & Rothstein, S.M. 1988. *J. Chem. Phys.* 88: 3784.
Xu, B.X., Rajagopal, A.K. & Ramana, M.V. 1984. *J. Phys. C* 17: 1339.

Computational Methods in Engineering and Science, Iu et al. (eds)
© 2003 Swets & Zeitlinger, Lisse, ISBN 90 5809 567 3

Molecular structure and elastic modulus of an oriented crystalline polymer

U. Gafurov

Institute of Nuclear Physics, Tashkent, Uzbekistan

ABSTRACT: It is suggested a modeling method on the base of measurable structural and deformation parameters to estimate the interrelation between the interconnecting chains over amorphous segments conformation structure distribution and the elastic modulus in tension experimental value for a loaded oriented linear amorphous – crystalline polymer. Macromolecule chain pulling out value of polymer crystallite was defined using Frenkel-Kontorowa dislocation model.

Having defined the sizes and the distribution of super-molecular elements (e.g. of crystalline and amorphous regions) and of interconnecting amorphous sections over lengths and conformation structure, having measured the deformation curve and having calculated with help of the curve the elasticity modulus one could find the local load on the chain sections). On experimental value of elastic modulus can be to elevated the content of the interconnecting amorphous molecular chains in straightening conformation states.

1 INTRODUCTION

Macroscopic properties of amorphous – crystalline polymers are complexly and the fundamentals of stress and strain distribution in the heterogeneous solid are still of essential interest (Braham 1979, Kausch 1989, Pantelides 1992, Ward 1993, Reneker 1994, Strobl 1999, Peacock 2000). The structural features lead to the irregular wide distribution of local loads, especially for the flexible-chain polymers since the value of the elasticity modulus of the macromolecular sections with coiled rotary isomers in the polymers is low.

In work (Kennedy et al. 1994) presented force – length relations of linear polyethylene for different molecular mass samples. The ultimate properties are found to depend on the weight-average molecular weight indicating the importance of the non-crystalline regions. The role of individual chains on polymer deformation and chain stretching was also studied (Kausch 1994) as well as the role of tie molecules on yielding deformation of isotactic polypropylene (Koh-He Nitta & Takayanagi 1999).

The purpose of the present study is to suggest the molecular model of the stretching and local loads distribution over the amorphous section of molecular chains in a loaded an oriented linear amorphously-crystalline polymer, to estimate connection of polymer elastic modulus experimental values and the amorphous sections molecular (conformation) structure.

2 ELASTIC MODULUS, LOCAL LOAD DISTRIBUTION AND MOLECULAR STRUCTURE OF AMORPHOUS SECTIONS

2.1 Structure model

Let us consider two-phase model of an oriented flexible-chain crystalline polymer type of linear polyethylene microfibrils of which consist of interchanging crystalline and amorphous regions. The longitudinal length is the sum of the lengths of amorphous and crystalline regions

$$L = L_{cryst} + L_{amorph}.$$

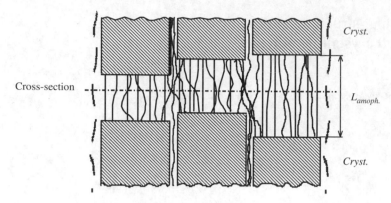

Figure 1.   Scheme of a fibrillar fragment of a loaded highly-oriented amorphously-crystalline polymer.

A cross-section pass through amorphous regions of polymer sample. In this cross-section can realize high local loads and takes place more local deformation and considerable develops of creep and fracture molecular processes.

After some initial stress the deformation of a flexible-chain crystalline polymer is rotary isomeric and the macromolecule segments of the polymer are primarily in straightened conformation state (Gafurov 2001). To this portion of the deformation curve corresponds the low value of the elasticity modulus.

If the conformation straightening of some part of strained passing chains has taken place, further stretching of the polymer leads to the considerable increasing of the elasticity modulus. Since the value of deformation in tension modulus of macromolecules being in *trans*-conformation state is high even when the concentration of such a type of the interconnecting sections is relatively low, then they give remarkable contribution in the elasticity modulus of the sample.

3   INTERRELATION OF ELASTIC MODULUS AND MOLECULAR STRUCTURE OF INTERCONNECTING SEGMENTS

Let us suppose that the relative middle number of interconnecting sections being in straightened *trans*-conformation state is n_i and the relative number of the sections including coiled isomers is n_j. The lengths and the elasticity modulus of the two groups of passing chains are E_i and E_j.

We shall believe that in some section of the sample the amorphous regions and longitudinal periods have the same length and the elastic deformation of the highly oriented polymer is primarily defined by stretching of the amorphous sections being inside fibril. Actually, for highly oriented flexible-chain crystalline polymers the assumption is valid; it has been shown by X-ray analysis (Kuksenko 1976).

Since straightened regions can be considerably loaded then one should take into account the stretching of the chains in the crystalline regions. So the deformation is the sum of the amorphous section lengthening and of the value of the pulling out of polymer crystallites of the strained chain in *trans*-conformation structures out of both sides of the amorphous section, e.g.

$$\Delta L_{amorph} = \Delta L_i + 2x_i$$

Let us believe that in some cross-section of a polymer the longitudinal period of is lengthened on due to external load. Then for average load on the sample in Hookean approximation we can write

$$\sigma = E\frac{\Delta L}{L} = n_i E_i \frac{\Delta L_i}{L_i} + n_j E_j \frac{\Delta L_j}{L_j} = n_i E_i \frac{\Delta L - 2x_i}{L_i} + n_j E_j \frac{\Delta L_j}{L_j} \qquad (1)$$

where ΔL_i and ΔL_j – lengthening of the straightened interconnecting amorphous sections and the sections with coiled rotary isomers correspondingly, L_i and L_j – correspondingly lengths of inter-connecting amorphous sections in straighten and with coiled rotary isomers, and E is elastic modulus experimental value in tension of the big period or as we consider, of the sample.

In continual approximation with the use of Frenkel-Kontorova's dislocation model and semi-infinitive polymer crystallite it has been found before (Guschina & Chevichelov 1974, Kaush 1989, Kaush 1994) following expressions for the pulling out value of this chain out of crystalline region is defined from

$$x_i = \frac{a}{\pi} \arcsin \frac{f_i}{\sqrt{2kU_0}} \tag{2}$$

where a – is the period of the crystal lattice, f_i – local load on straightened interconnecting amorphous sections, U_0 – is the height of intermolecular interaction barrier, x_i – is the displacement of the molecular group from the crystalline lattice site. The value of the elasticity constant k has been calculated with the use of the elasticity constant values for the valence angle and covalence bound taken from literature.

Using for the approximations

$$\arcsin \frac{f_i}{\sqrt{2kU_0}} \approx \frac{f_i}{\sqrt{2kU_0}},$$

we get

$$\Delta L = f_i \left[\frac{L_i}{E_i} + \frac{2a}{\pi \sqrt{2kU_o}} \right] \tag{3}$$

One can find relationship between molecular structural parameters for the passing chains and elasticity modulus of the polymer sample. Substituting (2, 3) in (1) for E we get

$$E = E_i n_i \frac{L}{L_i} \left(1 - \frac{2a}{\pi} \frac{1}{\frac{L_i}{E_i} \sqrt{2kU_0} + \frac{2a}{\pi}} \right) + n_j E_j \frac{L}{L_j} \tag{4}$$

For the value calculation it is necessary to know the distribution of the interconnecting amorphous molecular chains over lengths and conformation structure. The main contribution to the second term of the sum gives the interconnecting macromolecular sections with the lowest number of coiled rotary isomers in their structure. Let us estimate the value for some values n_i and n_j for the above presented parameters: $L = 2 \cdot 10^{-6}$ cm, L_i, $L_j = 5 \cdot 10^{-7}$ cm, $E_t = 200$ GPa, $E_j = 4$ GPa, $k = 3.2 \cdot 10^5$ dyn/cm, $U_0 = 2$ kDJ/mole and $a = 2.5 \cdot 10^{-8}$ cm. The some result of estimations are given in Table 1.

As one could see, the interconnecting sections with coiled isomers give considerable contribution to the value of the modulus only for a negligible number of straightened passing chains. When the latter's concentration is just several percent the contribution of the chains to the value of becomes comparatively significant and the most loaded will be straightened interconnecting chains. Notice that for the values of the elasticity modulus the concentration of completely straightened polymer chains might be only several percent on which considerable load is localized.

Local loads distribution and overstrain coefficients:

From (3) for the local load we have

$$f_i = \frac{\Delta L}{\frac{L_i}{E_1} + \frac{2a}{\pi \sqrt{2kU_o}}} = \eta_i \sigma \tag{5}$$

where η is the coefficient of overstrain for i-th chain. From Hookean approximation

$$\sigma = E \frac{\Delta L}{L},$$

we get

Table 1. The values of elastic modulus for some values content in interconnecting amorphous molecular chains.

n_i	n_j	E, GPa
0.02	0.2	13.2
0.02	0.4	16.4
0.05	0.2	28.2
0.05	0.4	31.4
0.1	0.2	53.2
0.1	0.4	56.4

Table 2. The values of overstrain coefficients in dependence on different lengths of amorphous interconnecting sections and experimental values of elastic modulus.

L_a/E	$E = 10\,\text{GPa}$	$E = 20\,\text{GPa}$	$E = 30\,\text{GPa}$
L_a	η	η	η
20	57	28	19
49	44	22	15
50	40	25	13
70	33	16	11
100	27	13	9

$$\eta_i = \frac{L}{E\left(\dfrac{2a}{\pi\sqrt{2kU_o}} + \dfrac{L_i}{E_t}\right)} \tag{6}$$

According to (6) the less the value of the modulus the higher the value of the overstressed coefficient and the higher value of the coefficient for the straightened passing chains in amorphous sections of shorter length. In Table 2 we list using the numerical values of the parameters placed above.

4 CONCLUSION

Having defined the sizes and the distribution of super-molecular elements (e.g. of crystalline and amorphous regions) and of interconnecting amorphous sections over lengths and conformation structure, having measured the deformation curve and having calculated with help of the curve the elasticity modulus one could find the local load on the chain sections correspondingly overstrain coefficients.

On experimental value of elastic modulus can be to elevated the content of the interconnecting amorphous molecular chains in straightening conformation states.

REFERENCES

Barham, P.J. & Keller, A.J. 1979. *Polym. Sci., Polym. Lett. Edn.* 17, 591.
Gafurov, U. 2001. *Macromolecular Symposia*, 176, 119.
Guschina, N.A., Kosobukin, V.A. & Chevichelov, A.D. 1974. *Physics state solidy* 16, 3000.
Kausch, H.H. 1989. *Polymer Fracture, Polymer Properties and Applications*, Springer Verlag, Berlin.
Kausch, H.H. & Plummer, J.G. 1994. *Polymer*, 35, 18, 3848.
Kennedy, M.A., Peacock, H.J. & Mandelkern, L. 1994. *Macromolecules*, 27, 5297.
Koh-He Nitta, M. & Takayanagi M. 1999. *J. Polymer Sci.* (B) Polymer Phys, 37, 357.
Kuksenko, V.S. et al. 1976. *Visokomolekulyarnie ssoedineniya*, 6.
Pantelides, S.T. 1972. *Phys. Today*, September, 67.
Peacock, A.J. 2000. *Handbook of Polyethylene: Structures, Properties, and Applications*, Marcel Dekker, Texas.
Reneker, D.H. 1994. *Polymer*, 35, 8, 1909.
Strobl G.R. 1999. *The Physics of Polymers*, 2nd ed., Springer, Berlin.
Ward, I.M., Hadley, D.W. 1993. *An Introduction to the Mechanical Properties of Solid Polymers*, Wiley & Sons, New York.

Computational Methods in Engineering and Science, Iu et al. (eds)
© 2003 Swets & Zeitlinger, Lisse, ISBN 90 5809 567 3

Low velocity impact on laminates reinforced with Polyethylene and Aramidic fibers

M.A.G. Silva & C. Cismaşiu
Centro de Investigação em Estruturas e Construção – UNIC, Faculdade de Ciências e Tecnologia,
Universidade Nova de Lisboa, Quinta da Torre, Portugal.

C.G. Chiorean
Faculty of Civil Engineering, Technical University of Cluj-Napoca, Cluj-Napoca, Romania

ABSTRACT: The present study reports low velocity impact tests on composite laminate plates reinforced either with Kevlar 29 or Dyneema. The tests are produced using a Rosand Precision Impact tester. The experimental results obtained for Kevlar 29 are simulated numerically. The deflection history and the peak of the impact force are compared with experimental data and used to calibrate the numerical model.

1 INTRODUCTION

The intrinsic analytical difficulty of treatment of impact problems on composite laminates of polymeric matrix partially caused by the large number of parameters involved advises experimental testing to allow adequate certification of computational models that estimate the delamination, dynamic contact deformation, depth and area of damage, crashworthiness and correlated topics. Recent advances toward understanding damage mechanisms and mechanics of laminated composites (Abrate 1998, Choi *et al.* 1992, Choi *et al.* 1991a,b, Fukuda *et al.* 1996), coupled with the development of advanced anisotropic material models (Clegg *et al.* 1999, Hayhurst *et al.* 1999, Hiermaier *et al.* 1999) offer the possibility of avoiding many experimental tests by using impact simulation. However, the numerical results should be used with precaution and must always be validated by experimental tests.

The present study reports low velocity impact, normal to the surface of laminated plates reinforced either with Kevlar 29 or with Dyneema, i.e. aramidic or tough polyethylene fibres. The paper concentrates essentially on macroscopic phenomenology and factors like stacking sequence, curing conditions, relative mass plate-impactor, stiffness or shape of striker are not examined. Calibration of the numerical model is described for the case of the Kevlar 29 plates.

All the simulations presented in the paper have been carried out by using the hydrocode AUTODYN (Autodyn, version 4.2), specially designed for non-linear transient dynamic events such as ballistic impact, penetration and blast problems. The software is based on explicit finite difference, finite volume and finite element techniques which use both grid based and gridless numerical methods. A new material model, specifically designed for the shock response of anisotropic material (Hayhurst *et al.* 1999), has been implemented.

2 EXPERIMENTAL TESTS

Impact tests were performed using a Rosand Precision Impact Tester (Software manual, version 1.3), with a hemi-spherical headed striker indenting laminated plates at the centre of a circle of 100 mm diameter rigidly held by a steel ring. The experimental tests were conducted at INEGI Porto and the results reported in (Silva 1999). Two sets of tests were considered, one on Kevlar 29 plates and the other on Dyneema plates.

Table 1. Characteristic values for impact on Kevlar 29 and Dyneema plates.

Material	Dyneema			Kevlar 29		
Nominal energy at impact (J)	1	10	20	1	10	20
Maximum deflection (mm)	1.98	6.20	8.92	1.95	5.68	7.15
Deflection peak time (ms)	4.26	3.46	2.66	–	4.02	3.54
Maximum impact force (kN)	0.8	2.0	2.70	0.7	3.4	5.5
Force at peak time (ms)	–	–	–	–	3.10	2.68

The first set considered low speed impact on Kevlar 29 laminates. The plates were made from prepreg fabric impregnated with vinylester resin, at a curing temperature of 125°C, with eleven plies originating a thickness of 1.8 mm.

The second set of tests corresponds to low speed impact on plates reinforced with high tenacity polyethylene fibres, Dyneema, processed with eight plies, and a thickness of 1.7 mm and real density of 150 g/m^2. The plates were pressed with a thermoplastic stamilex film with a melting temperature of 120°C. Averaged results indicated Young's modulus E = 3540 MPa, failure stress σ_u = 289 MPa and strain at failure ε_u = 5.6%. Thermoplastic matrices respond nonlinearly, cause higher damping and spread damage into larger regions than corresponding thermosetting matrices and are associated with significant damage on the tension face, even at low energies, and the strain on that face is believed to control damage initiation, but this property was less evident in tests, perhaps due to the ductility of the Dyneema matrix and its high toughness.

The results obtained for impact tests on Kevlar 29 and Dyneema plates with different impact energy are shown in Table 1.

Transmitted force is considerably higher for Kevlar 29 as energy level increases. The deflection peak time values appear with some delay with respect to Dyneema, material that exhibits higher damping.

3 NUMERICAL MODEL

The experimental tests obtained with Kevlar 29 are simulated numerically. In order to reduce the size of the problem, the numerical model contains only the head of the striker, modelled as a 5 mm radius steel sphere. Its density was adjusted as to ensure the same impacting mass as the one in the experimental tests. The target is modeled as a circular plate of 100 mm diameter and 1.8 mm thickness, firmly clamped on the edges. The relatively small diameter of the plate and the low velocity of the strikes make the boundary conditions extremely important and therefore must not be neglected.

The resulting model, see Figure 1, with two planes of symmetry, was obtained using the general purpose mesh generation program, TrueGrid (TrueGrid, version 2.1.0). The numerical analysis was performed taking advantage of axial symmetry. Both striker and target were modelled using the Lagrange processor, hexahedron brick elements, with 1.4 mm uniform cells size in the impact area.

In the Lagrangian mesh, the time step is automatically setup as to ensure that a disturbance does not propagate across a cell in a single time step.

As the solid Lagrange elements in AUTODYN have only one integration point at the centre of element they are sensitive to the hourglassing problem. These modes produce rigid body motion and the mesh starts self-straining, destroying the solution. In order to prevent this phenomenon, hourglass coefficient had to be increased to 0.15, the minimum value which prevents the generation of spurious modes.

4 MATERIAL PROPERTIES AND MODEL CALIBRATION

The most important characteristics and phenomena governing the behaviour of composite materials under ballistic impact are: material anisotropy, shock response, coupling of volumetric and deviatoric behaviour, anisotropic strength degradation, material compaction, phase changes. In the case of anisotropic materials, there is a strong coupling between the equation of state and the constitutive relations, as volumetric strain leads to deviatoric stress and similarly, deviatoric strain leads to spherical stress. An advanced material model (Hayhurst et al. 1999, Hiermaier et al. 1999), specially designed to simulate the shock response of anisotropic materials, has recently been implemented as mentioned above, and

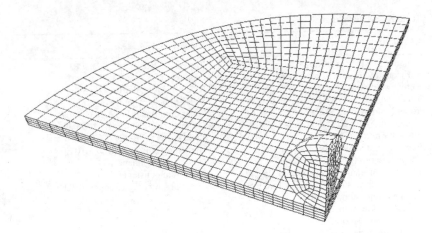

Figure 1. Numerical model.

couples the non-linear constitutive relations with the equation of state. The coupling is based on the methodology proposed by Anderson *et.al* (Anderson *et al.* 1994). The model can additionally include compaction and orthotropic brittle failure criteria to detect directional failure such as delamination.

Composite materials of polymeric matrix subject to impact exhibit complex behaviour. Experimentally, the dominant tensile material failure modes were identified as extensive delamination, due to matrix cracking and/or matrix-fibre debonding, in-plane fibre failure and punching shear failure caused by a combination of delamination and fibre failure leading to bulk failure. In the numerical model the composite material is considered to be homogeneous. Kevlar fibres and matrix resin are not separately modelled and the main phenomena of relevance are accounted for in an macro-mechanical model.

Delamination is assumed to result from excessive through-thickness tensile stresses or strains and/or from excessive shear stresses or strains in the matrix material. In the incremental constitutive relation

$$
\begin{Bmatrix} \Delta\sigma_{11} \\ \Delta\sigma_{22} \\ \Delta\sigma_{33} \\ \Delta\sigma_{23} \\ \Delta\sigma_{31} \\ \Delta\sigma_{12} \end{Bmatrix} = \begin{bmatrix} C_{11} & C_{12} & C_{13} & 0 & 0 & 0 \\ C_{21} & C_{22} & C_{23} & 0 & 0 & 0 \\ C_{31} & C_{32} & C_{33} & 0 & 0 & 0 \\ 0 & 0 & 0 & \alpha C_{44} & 0 & 0 \\ 0 & 0 & 0 & 0 & \alpha C_{55} & 0 \\ 0 & 0 & 0 & 0 & 0 & \alpha C_{66} \end{bmatrix} \begin{Bmatrix} \Delta\varepsilon_{11} \\ \Delta\varepsilon_{22} \\ \Delta\varepsilon_{33} \\ \Delta\varepsilon_{23} \\ \Delta\varepsilon_{31} \\ \Delta\varepsilon_{12} \end{Bmatrix}
\tag{1}
$$

the stress $\Delta\sigma_{11}$ normal to the laminate and the corresponding orthotropic stiffness coefficients C_{ij} are instantaneously set to zero, whenever the failure is initiated in either of those two modes, $j = 1$ in equation(2):

$$
\Delta\sigma_{jj} = 0 \text{ and } C_{ij} = C_{ji} = 0 \text{ for } i=1,3
\tag{2}
$$

Delamination may also result from reduction in shear stiffness of the composite, via parameter α in equation (1). In-plane fibre failure is assumed to result from excessive stresses and/or strains in the 22 or 33 directions, $j = 2$ or $j = 3$ in equation (2). The combined effect of failure in all three material directions is represented changing the material stiffness and strength to isotropic characterization, with no stress deviators or material tensile stresses. A fractional residual shear stiffness is maintained through the parameter α, whose value is obtained by experimental tests.

Composite material cell failure initiation criterion is assumed to be based on a combination of material stress and strain failure. Subsequent to failure initiation, the cell stiffness and strength properties are modified in agreement with the failure initiation modes.

The volumetric response of the material is defined through the solid equation of state. The polynomial sub-equation of state used in the numerical simulation, allows non-linear shock effects to be coupled with the orthotropic material stiffness.

Table 2. Material data.

Kevlar 29	4340 Steel
Equation of state: orthotropic	Reference density (g/cm^3): 7.83
Sub-equation of state: Polynomial	Bulk modulus (kPa) 1.59 E + 07
Reference density (g/cm^3): 1.40	Reference temperature (K) 300
Young's modulus 11 (KPa): 2.392 E + 05	Specific heat capacity (J/kgK) 477
Young's modulus 22 (KPa): 6.311 E + 06	Shear modulus (kPa) 8.18 E + 07
Young's modulus 22 (KPa): 6.311 E + 06	Melting temperature (K) 1793
Poisson's ratio 12: 0.115	Yield stress (kPa) 7.92 E + 05
Poisson's ratio 23: 0.216	Hardening constant (kPa) 5.10 E + 05
Shear modulus (kPa): 1.54 E + 06	Hardening exponent 0.34
Tensile failure stress 11 (KPa) 5.00 E + 04	Strain rate constant 0.014
Maximum shear stress 12 (KPa) 1.00 E + 05	Thermal softening exponent 1.03
Tensile failure strain 11: 0.01	
Tensile failure strain 22: 0.08	
Tensile failure strain 33: 0.08	
Post failure response: Orthotropic	
Failure: material stress-strain	
Residual shear stiff. Frac.: 0.20	

All the reported tests were performed on Kevlar 29 composite target. The 4340 steel was represented using the Johnson-Cook strength model, which include strain and strain rate hardening and thermal softening effects. Material data for Kevlar 29 target and 4340 steel impactor are shown in Table 2.

The values characterizing the orthotropic strength of the target were obtained in experimental tests carried out at Ernst-Mach-Institut in Germany. Quasi-static tensile tests were used to provide data on in-plane stiffness and failure strains. The through thickness stiffness was obtain in quasi-static compression tests. However, due to the fact that the sample thickness was less than 2 mm and because of instantaneous through thickness delamination, it was not possible to determine the Poisson's ratio, ν_{12}.

The value in the Table was derived iteratively through numerically low speed impact calibration. The calibration is based on the observation that the impact performance of target is dominated by the in-plane stiffness coefficients C_{22} and C_{33} in equation 1. It can be shown (Tsai 1988) that these coefficients depend on the in-plane and through thickness Young's moduli E_{22} and E_{11} and on Poisson's ratio ν_{12} and ν_{23}:

$$C_{22} = C_{33} = \frac{(1 - \nu_{12}\nu_{21})}{(1 + \nu_{23}) \cdot (1 - \nu_{23} - 2\nu_{21}\nu_{12})} E_{22}; \quad \nu_{21} = \frac{E_{22}}{E_{11}}\nu_{12} \tag{3}$$

On the other hand, the positiveness of the stiffness $\mathbf{C}$ and the compliance $\mathbf{S} = \mathbf{C}^{-1}$ tensors in anisotropic materials is imposed by thermodynamic principles based on the fact that the elastic potential should remain always a positive quantity. The positive definiteness of these two tensors for the transversely isotropic materials implies that the following system of relations must hold (Jones 1975):

$$|\nu_{12}| = |\nu_{13}| < \left(\frac{E_{11}}{E_{22}}\right)^{1/2}, \quad |\nu_{23}| < 1 \tag{4}$$

and

$$\nu_{12}^2 \nu_{23} < \left(1 - \nu_{23}^2\right)\frac{E_{11}}{2E_{22}} - \nu_{12}^2 \tag{5}$$

It can readily be derived from these relations in conjunction with the material data depicted in Table 2, that the extreme positive limit of ν_{12} Poisson's ratio is $\nu_{12}^{max} = 0.1219$.

The graph in Figure 2 illustrates the high sensitivity of the stiffness coefficient C_{22} for Poisson's ratio between 0.1 and $\nu_{12}^{max} = 0.1219$. Therefore, the value of ν_{12} from Table 2 was obtained iteratively, equating the numerical response of the target with the experimental results.

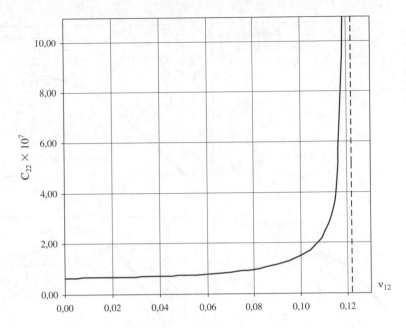

Figure 2.   Variation of stiffness coefficient C_{22} with Poisson's ratio ν_{12}.

5   COMPARISON OF RESULTS

The numerical estimates for the deflection history under the punch and the maximum impact force computed according to Hertz contact law (Hertz 1982), were in good agreement with the experimental data for all energy levels, as shown in Figure 3. The numerical results predict a larger peak at an earlier stage than experimental tests. This trend is more evident for higher nominal impact energy. The increase of the peak value is clearly recognizable as is the earlier unloading with increase of the nominal energy at impact. The maximum deflection, δ_{max}, and the corresponding peak time, t^{δ}_p, are shown in Table 3, for each energy level.

The maximum impact force, F_{max}, obtained for each level of nominal energy, and the corresponding peak time, t^F_p, are compared in Table 4, with the experimental values reported in (Silva 1999).

6   CONCLUSIONS

The experimental results show that laminates with lower bending stiffness allow higher *radiation damping*, reducing the impact forces, fact evidenced by Dyneema versus Kevlar. This effect is strengthened for laminates of lower mass which also exhibit lower contact force. Forces transmitted by impactor reach maximum values earlier than the corresponding maximum deflection.

Energy dissipation in Dyneema plates is achieved through *plastic* deformation, whereas Kevlar plates deform more locally, facing delamination for higher forces.

Simulations of low speed impact on composite laminate plates reinforced with Kevlar 29 were performed using the finite difference numerical code AUTODYN-3D, based on an advanced mode for orthotropic materials (Hayhurst *et al.* 1999). Its main draw is the ability to use a non-linear equation of state in conjunction with an orthotropic stiffness matrix which allows an accurate modelling of the response of composite materials under impact conditions.

The deflection history and the peak of the impact force are compared with experimental data, for four single strikes with energy levels of 2,5, 10 and 20J, respectively. The estimates for the displacements were in good agreement with the experimental data for all energy levels. The maximum relative error in the maximum displacement under the head of the striker, 11%, took place for the lowest energy level, 2J.

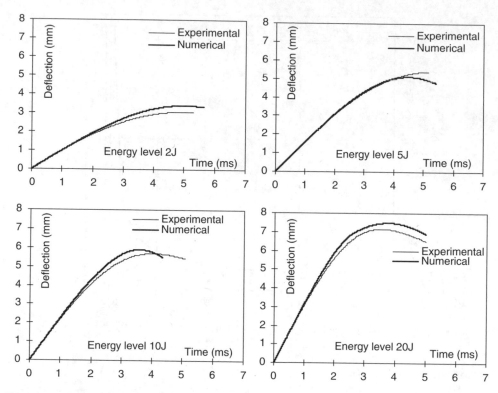

Figure 3. Deflection of the head of the striker.

Table 3. Maximum deflection and corresponding peak time.

	2		5		10		20	
Energy (J)	Exp	Num	Exp	Num	Exp	Num	Exp	Num
δ_{max} (mm)	3.05	3.41	5.40	5.20	5.68	5.86	7.15	7.40
t^δ_p (ms)	4.96	5.01	5.14	4.74	4.02	3.85	3.54	3.68

Table 4. Maximum impact force and corresponding peak time.

	2		5		10		20	
Energy (J)	Exp	Num	Exp	Num	Exp	Num	Exp	Num
F_{max}(kN)	1.10	1.25	1.70	1.78	3.50	3.40	5.50	5.60
t^F_p (ms)	4.12	3.80	3.84	3.92	3.10	2.70	2.68	2.40

The accuracy of the estimates increase with the energy level, as shown by a relative error of 2% for 20J. The peak of the impact force was computed according to the Hertzian contact model.

Compared with the experimental value, the relative error was 18% and 1.8% for the lowest and highest energy level, respectively. An explanation of these results could be the higher sensitivity of the numerical solutions to boundary conditions and hourglass effects in the case of low energy levels.

Future work is envisaged considering the numerical simulation of impact in Dyneema plates.

ACKNOWLEDGEMENT

This work is part of the research developed at DEC, Faculdade de Ciências e Tecnologia, supported by contract 43228/EME/2001 with Fundação para a Ciência e Tecnologia. Cooperation with colleagues from INEGI, Porto and Comd. F. Neto from Navy School of Lisbon is gratefully acknowledged.

REFERENCES

Abrate, S. 1998. *Impact on Composite Structures*. Cambridge University Press.
Anderson, C.E., Cox, P.A., Johnson, G.R., Maudlin, P.J. 1994. A constitutive formulation for anisotropic materials suitable for wave propagation computer program. *Computational Mechanics*, 15:201–223.
Century Dynamics, Inc. *AUTODYN*. Interactive Non-Linear Dynamic Analysis Software, version 4.2, 1997.
Choi, H.Y., Chang, F.K. 1992. A model for predicting damage in graphite/epoxy laminated composites resulting from low-velocity point impact. *Journal of Composite Materials*, 26(14):2134–2169.
Choi, H.Y., Downs, R.J., Chang F.K. 1991a. A new approach toward understanding damage mechanisms and mechanics of laminated composites due to low-velocity imapct: Part I – Experiments. *Journal of Composite Materials*, 25:992–1011.
Choi, H.Y., Downs, R.J., Chang F.K. 1991b. A new approach toward understanding damage mechanisms and mechanics of laminated composites due to low-velocity imapct: Part II – Analysis. *Journal of Composite Materials*, 25:1012–1038.
Clegg, R., Hayhurst, C.J., Leahy J., Deutekom, M. 1999. Application of coupled anisotropic material model to high velocity impact response of composite textile armour. In *18th Int. Symposium and Exhibition on Ballistics*, San Antonio, Texas USA, November 15–19.
Fukuda, H., Katoh, F., Yasuda, J. 1996. Low velocity impact damage of carbon fiber reinforced thermoplastics. In *Progress in Durability Analysis of Composite Systems*, pages 139–145, Bakelma.
Hayhurst, C.J., Hiermaier, S.J., Clegg, R.A., Riedel, W., Lambert, M. 1999. Development of material models for nextel and kevlar-epoxy for high pressures and strain rates. In *Hypervelocity Impact Symposium*, Huntsville, AL, Nov. 16–19.
Hertz, O. 1982. Ueber die beruhrung fester elastischer korper. *Journal für die Reine und Angewandte Mathematik*, (92):156–171.
Hiermaier, S.J., Riedel, W., Clegg, R.A., Hayhurst, C.J. 1999. Advanced material models for hypervelocity impact simulations. Technical report, ESA/ESTEC Contract No. 12400/97/NL/PA(SC).
Jones, R.M. 1975. *Mechanics of composite materials*. Washington: Scripta Book Co.
Rosand Precision Impact Tester. Software Manual. Version 1.3.
Silva, M.A.G. 1999. Low speed impact on polyethilene and aramidic FRP laminates. In *Proceedings of Conference on Integrity, Reliability and Failure*, Porto, 19–22 July.
Tsai, S.W. 1988. *Composite Design,* Think Composites, 4th edition.
XYZ Scientific Applications, Inc. *TrueGrid*, version 2.1.0, 2000.

Computational Methods in Engineering and Science, lu et al. (eds)
© 2003 Swets & Zeitlinger, Lisse, ISBN 90 5809 567 3

Numerical study of an electret filter composed of split-type fibers

Y.H. Cao & C.S. Cheung
Department of Mechanical Engineering, The Hong Kong Polytechnic University, Hong Kong

Z.D. Yan
Zhejiang University, Hangzhou, Zhejiang Province, China

ABSTRACT: In this paper, an electret filter is modeled as a staggered, parallel array of rectangular fibers. Simulation based on the model is conducted by numerically solving the flow field through the model filter, the electric field around the electret fiber and particle trajectories. All the major particle capturing mechanisms are considered. Simulation results are validated against published experiment-based empirical expressions. Simulations are conducted for two typical fiber arrangements to evaluate the effect of orientation of the fiber on the collection efficiency.

1 INTRODUCTION

An electret filter contains fibers that have a quasi-permanent electric charge. In comparison with deep bed filter, an electret filter has higher collection efficiency and lower flow resistance because of attracting forces imposed by charged fibers on the particles flowing nearby. This paper is focused on modeling the collection process of a split-type electret fiber. A brief review of the characteristics and collection mechanisms of an electret filter can be found in Brown (1993).

Numerical models of electret filter can be found in Brown (1979), Lathrache et al. (1986), and Oh et al. (2001). They modeled the electret fiber as line dipole charged cylinder or cylinder with uniform charge distribution. Numerical models for non-circular fibers were also developed, which include those of Emi et al. (1987) and de Haan et al. (1987). Because of the lack of models for simulating viscous flow around rectangular electret fibers, Fardi and Liu (1992) developed numerical solution of flows through a staggered array of rectangular fibers. Despite of such progresses in the modeling of electret filer, there is still a demand to develop a numerical model which simulates the flow field, and the particle deposition due to both mechanical and electrostatic effects, for rectangular fibers.

In this paper, the flow field around a rectangular fiber is simulated with the numerical model of Chen et al (2002). The particle trajectory and deposition are then simulated by numerically integrating the particle equation of motion, taking into account the effect of Coulombic force, polarization force, Brownian diffusion, interception and inertial impaction.

2 SIMULATING THE FLOW FIELD

Fardi & Liu (1992) modeled an electret filter composed of rectangular fibers as a staggered array of fiber blocks arranged perpendicular to the main flow direction. Following their concept, Chen et al. (2002) numerically simulated aerosol collection in filters with staggered parallel rectangular fibers. They solved a two-dimensional flow field through the filter with the control volume method and with periodic boundary conditions. In the present study the flow field is simulated following the same approach of Chen et al. (2002). The geometry of the staggered array of fibers is schematically shown in Figure 1, in which the space between fibers is determined by the packing density of the filter, α. In addition, two orientations of the electret fibers will be considered, which are designated as the vertical-fiber array and

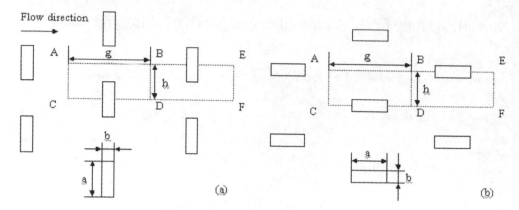

Figure 1.　Geometry of staggered array of rectangular fibers: (a) vertical-fiber array; (b) horizontal-fiber array.

the horizontal-fiber array, as shown in Figure 1a and Figure 1b. A vertical-fiber array has fibers with their broad side perpendicular to the main flow direction whereas a horizontal-fiber array has fibers with their broad side parallel to the main flow direction. For the flow field, the Navier-Stokes equations governing the flow pattern can be written as:

$$\frac{\partial u}{\partial x} + \frac{\partial v}{\partial y} = 0 \tag{1}$$

$$\rho(u\frac{\partial u}{\partial x} + v\frac{\partial u}{\partial y}) = -\frac{\partial p}{\partial x} + \mu(\frac{\partial^2 u}{\partial x^2} + \frac{\partial^2 u}{\partial y^2}) \tag{2}$$

$$\rho(u\frac{\partial v}{\partial x} + v\frac{\partial v}{\partial y}) = -\frac{\partial p}{\partial y} + \mu(\frac{\partial^2 v}{\partial x^2} + \frac{\partial^2 v}{\partial y^2}) \tag{3}$$

The equations above assume steady, incompressible, Newtonian and laminar two-dimension flow with constant fluid physical properties. The flow is assumed to be periodic within the domain along the flow direction. Furthermore, because of the geometry and physical symmetry of the problem, computation can be simplified by considering only the cell ABCD. Harmonic means method (Ould-Amer et al. 1998) is used to handle the abrupt changes in physical properties at fluid-fiber interface.

The boundary conditions are given as follows:

$$\frac{\partial u}{\partial y} = 0, v = 0 \tag{4}$$

along the upper and lower boundary AB and CD;

$$u = 0, v = 0 \tag{5}$$

along the surface of the fiber;

$$u(x,y) = u(x+g,h-y), v(x,y) = -v(x+g,h-y) \tag{6}$$

at the inflow entrance; and

$$u(x,y) = u(x-g,h-y), v(x,y) = -v(x-g,h-y) \tag{7}$$

at the outlet.

In these equations, g and h are the width and height of the computational domain. Furthermore, the pressure p is expressed in the following form:

$$p(x,y) = P(x,y) - \beta x \tag{8}$$

where β is a numerical constant, representing the average pressure gradient along the main flow direction and P is a periodical variable:

$$P(x,y) = P(x+g, h-y) \tag{9}$$

With the governing equations (1)–(3) and the boundary conditions given in equations (4)–(9), the velocity and pressure terms can be solved with a finite volume approach combined with the SIMPLE algorithm, with details given in Chen et al. (2002).

3   CALCULATION OF SINGLE FIBER COLLECTION EFFICIENCY

For a particle-laden flow in which the particle concentration is low, the trajectory of a particle can be determined by solving the following equations of motion of the particle as it travels through the fluid under the influence of various forces:

$$\frac{du_p}{dt} = \frac{18\mu}{\rho_p D_p^2 C_c}\left(u - u_p\right) + f_{Cx} + f_{Px} + n_x(t) \tag{10a}$$

$$\frac{dv_p}{dt} = \frac{18\mu}{\rho_p D_p^2 C_c}\left(v - v_p\right) + f_{Cy} + f_{Py} + n_y(t) \tag{10b}$$

The first terms on the right-hand side of equations (10a) and (10b) are the acceleration components due to the drag force and C_c is the Cunningham slip factor. The components $f_{Cx}, f_{Cy}, f_{Px}, f_{Py}$, are the acceleration components due to Coulombic force and polarization force. The terms $n_x(t)$, $n_y(t)$ are the Brownian force components per unit particle mass. In a known flow field, equations (10a) and (10b) can be numerically integrated with the fourth-order Runge-Kutta algorithm to obtain the trajectory of a particle. In computing the flow filed, a domain of half of a fiber block is considered due to symmetry. However, the trajectories of particles may not have symmetry and hence particle trajectories are calculated in a domain enclosing the entire fiber. The single fiber efficiency, η_{simu}, is then calculated from the total number of particles traced and the number of particles captured, as follows, when a large number of particles are traced:

$$\eta_{simu} = \frac{\text{captured particle number}}{\text{total particle number}} \cdot \frac{2h}{\text{fiber height}} \tag{11}$$

3.1   *Coulombic force and polarization force*

The modeling and simulation of the electrostatic field around an electret fiber can be found in Emi et al. (1987) and Baumgartner et al. (1993). The interaction between a particle and an electret fiber is shown in Figure 2, in which the fiber surface charge is considered as an arrangement of many parallel line charges of infinite extent along the fiber axis. For the vertical-fiber array, there are two typical arrangements (Figs 2a, b); while for the horizontal-fiber array, there is only one arrangement (Fig. 2c). The electrostatic interaction between fibers is assumed to be negligible for packing density lower than 0.1 (Rao & Faghri 1990).

The electric field strength $\mathbf{E}$ at the position of the particle shown in Figure 2 is:

$$\mathbf{E} = \sum_{i=1}^{n} \mathbf{E}_i^+ + \mathbf{E}_i^- = \frac{1}{2\pi\epsilon_0}\sum_{i=1}^{n}\frac{\lambda^+ \cdot \mathbf{r}_i^+}{\left(r_i^+\right)^2} + \frac{\lambda^- \cdot \mathbf{r}_i^-}{\left(r_i^-\right)^2} \tag{12}$$

where $\mathbf{r}_i^+$, $\mathbf{r}_j^-$ are displacement vectors; n is the line number of charge pair; λ^+, λ^- are line charge densities, which are related to the surface effective charge density σ_d as follows:

$$\lambda^+ = -\lambda^- = \frac{b\sigma_d}{n} \tag{13}$$

Thus the Coulombic force $\mathbf{F}_C$ for a singly charged particle and the polarization force $\mathbf{F}_P$ for a neutral particle can each be expressed as:

$$\mathbf{F}_C = e\mathbf{E} \tag{14}$$

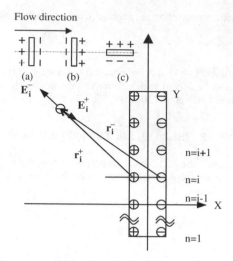

Figure 2. Electrostatic field and electrostatic forces simulation model (Three typical fiber charge configurations used in the simulation are shown on the upper left corner: (a) positively charged side facing inflow; (b) negatively charged side facing inflow; (c) positively charged side facing up.)

$$F_P = \frac{\pi}{4}\left(\frac{\varepsilon_p - 1}{\varepsilon_p + 2}\right)\varepsilon_0 D_p^3 \text{ grad } E^2 \tag{15}$$

where e is an elementary charge of -1.6×10^{-19}C; ε_p is the dielectric constant of aerosol particle; ε_0 is the space permittivity.

3.2 Brownian diffusion

Brownian diffusion is a major particle collection mechanism for submicron particles. The force can be modeled as a Gaussian random process (Abuzeid et al. 1991). This approach was adopted by Chen et al. (2002) in which the force for unit mass, n_i (n_x or n_y), is given by

$$n_i(t_j) = Z_i \sqrt{\frac{2\pi S_0}{\Delta t}} \tag{16}$$

In equation (15), Z_i is a random number with zero mean value and unit variance and S_0, is the spectral intensity of a white noise process.

$$S_0 = \frac{216\mu k T}{\pi^2 D_p^5 \rho_p^2 C_c} \tag{17}$$

This model has an obvious advantage in that the Brownian force is treated like the other fluid or external forces, and hence can be directly coupled in the computational scheme.

4 RESULTS AND DISCUSSION

Kanaoka et al. (1987) conducted experiments on electret filters and generated from their experimental data two semi-empirical expressions, one for the single fiber collection efficiency of neutral particles (equation 18) and the other for charged particles (equation 19), referring to the vertical-fiber array arrangement:.

$$\eta_{InD} = 1.07 Pe^{-2/3} + 0.06 K_{In}^{2/5} \tag{18}$$

$$\eta_{CD} = 1.07 Pe^{-2/3} + 0.06 K_{In}^{2/5} + 0.067 K_C^{3/4} - 0.017 K_{In}^{1/2} K_C^{1/2} \tag{19}$$

where Pe is the Peclet number, while $K_{In} = (\varepsilon_p - 1)/(\varepsilon_p + 2) \cdot C_c Q^2 d_p^2/3\pi^2 \varepsilon_0 \mu a^3 U_\infty$ and $K_C = C_c n_p e Q/ 3\pi^2 \varepsilon_0 \mu d_p a U_\infty$ are dimensionless parameters indicating the polarization force effect and the Coulombic force effect respectively. The simulation parameters shown in Table 1 are based on those used by Kanaoka (1987). Simulated results for the vertical-fiber array are shown in Figures 3a and 3b, while those for the horizontal-fiber array are shown in Figures 4a and 4b.

Figure 3a shows the variation of the simulated efficiency, η_{simu}, along with results of η_{InD} calculated from the equation (18), with K_{In}, for neutral particles. The value of σ_d is unknown and is selected to be

Table 1. Summary of model filter properties and simulation conditions.

Fiber size a × b [mm × mm]	Packing density α[−]	Filtration velocity U_∞ [m/s]	Particle diameter D_p [μm] and charge number [−]	
			0 Charge	1 Charge
38 × 10	0.031	0.1, 0.2, 0.3, 0.5	0.01~0.2	0.03~0.2

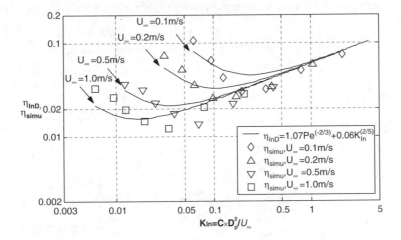

Figure 3a. Single fiber efficiency of neutral particle, vertical-fiber array; $C = (\varepsilon_p - 1)/(\varepsilon_p + 2) \cdot C_c Q^2/3\pi^2 \varepsilon_0 \mu a^3$.

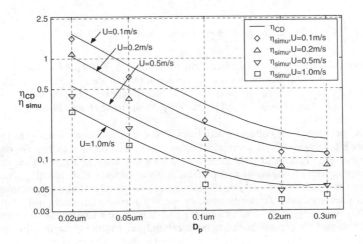

Figure 3b Single fiber efficiency of singly charged particle, vertical-fiber array.

855

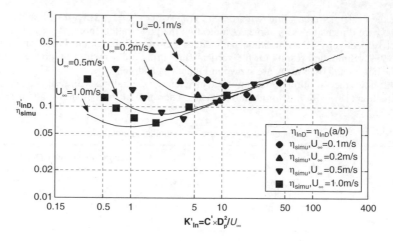

Figure 4(a). Single fiber efficiency of neutral particle, horizontal-fiber array; in which $C' = (\varepsilon_p - 1)/(\varepsilon_p + 2) \cdot C_c Q^2/3\pi^2 \varepsilon_0 \mu b^3$.

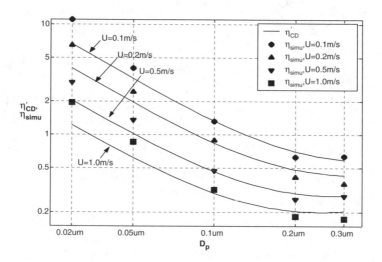

Figure 4(b). Single fiber efficiency of singly charged particle, horizontal-fiber array.

$200\,\mu As/m^2$ (corresponding to a total surface charge density, $\sigma_{t,d} = 460\,\mu As/m^2$) so that the two set of results match well in the straight-line portion. The order of magnitude of the chosen value of σ_d or $\sigma_{t,d}$ is comparable to those obtained or used by other investigators: $775\,\mu As/m^2$ in Brown (1979), $200–500\,\mu As/m^2$ in Lathrache and Fissan (1986), $500\,\mu As/m^2$ in de Haan (1987), $258–957\,\mu As/m^2$ in Baumgartner et al. (1993).

In Figure 3a, the single fiber collection efficiency first decreases almost linearly as the polarization force parameter K_{In} decreases until a minimum value is reached and then increases non-linearly with further decrease in K_{In}. The linear portion corresponds to the range of K_{In} in which the polarization force is dominating, while in the non-linear part, the diffusional force is dominating. For neutral particles, larger particles are polarized to higher level and therefore are subjected to larger polarization force, while small particles have lower Pe number and hence more intensive Brownian diffusion effect. Lower filtration velocity results in longer resident time of a particle around a fiber and lower Pe number, which thus enhances both particle deposition due to polarization force and the Brownian diffusion effect.

Figure 3b shows corresponding results for singly charged particles. The same value of $\sigma_d = 200\,\mu As/m^2$ is used in the simulation. The variation of the simulated results, as well as the results calculated from

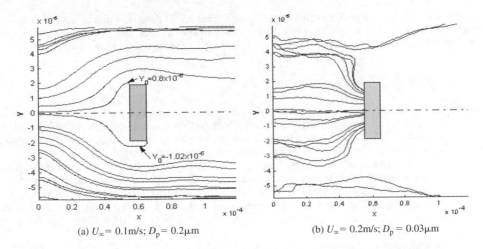

(a) $U_\infty = 0.1$m/s; $D_p = 0.2\mu$m

(b) $U_\infty = 0.2$m/s; $D_p = 0.03\mu$m

Figure 5(a)–(b). Trajectories of particle: (a) for neutral particles; (b) for singly charged particle, fiber is positively charged on the inflow side.

the equation (19), with particle diameter, is shown. For charged particles, there is a almost linear decrease in efficiency with particle diameter with particle diameters less than 0.1 μm; and a subsequent non-linear decrease with particle diameter. The former indicate the region in which the Coulombic force is dominating while the latter indicate the region in which diffusion force is dominating. In regions in which the Coulombic force is dominating, the single fiber collection effect can be higher than 1 indication that strength of the Coulombic force. Both of the Coulombic force and the diffusion force decrease with increase in particle size and filtration velocity.

As mentioned previously, a vertical fiber can have two charge configurations (Figs 2a, b) that should be considered, one with the positive charges facing the inflow and the other with negative charges facing the inflow. It is assumed that there is equal opportunity for each of them to exist and hence the simulated results presented for the vertical-fiber array and charged particles are the averaged values obtained for the two configurations.

The comparisons shown in Figure 3a and Figure 3b serve to validate our simulated results against the experimental results of Kanaoka et al. (1987). It can be concluded that the numerical model is able to predict the collection efficiency of an electret fiber, for both charged and neutral particles, in regions where diffusional and electrostatic effects are dominating.

The corresponding results for the horizontal-fiber array are shown in Figure 4a and Figure 4b. In the horizontal-fiber array, the single fiber efficiency are related to those for the vertical-fiber array as follows:

$$\eta'_{InD} = (1.07Pe^{-2/3} + 0.06K_{In}^{2/5})(a/b) \tag{20}$$

$$\eta'_{CD} = (1.07Pe^{-2/3} + 0.06K_{In}^{2/5} + 0.067K_C^{3/4} - 0.017K_{In}^{1/2}K_C^{1/2})(a/b) \tag{21}$$

Again, there is good matching between the simulated results and the calculated based on equation (20) for neutral particles or equation (21) for singly charged particles, with the same $\sigma_d = 200$ μAs/m^2 used throughout the simulation. The simulated results for the horizontal fibers' array are similar to those for the vertical fibers' array. Comparison of Figure 3a against 4a, and Figure 3b against Figure 4b, shows that the horizontal-fiber array produces much higher single fiber collection efficiency, indicating that the orientation of a fiber does affect the particle collection process.

One of the advantages of numerical simulation is that the particle trajectories can be traced and visualized. Figure 5(a) shows the trajectories of neutral particles under different filtration velocities and sizes of particle, with the fiber in the vertical position. It can be observed that the particle trajectories turn sharply towards the corner of the fiber where the inhomogeneity of the electric field is largest (Baumgartner 1993). These sharp turns represent the influence of the polarization force effect that enhances the chance of a particle to be captured and hence, the particle is attracted towards its front side. The trajectories of singly charged particles are shown in Figure 5(b) for vertically positioned fiber.

The fiber has positive charges on the front side facing the incoming fluid flow and hence the particle is attracted towards its front side. Coulombic force effect is much stronger than polarization force effect as it is clear that particles divert much more from the streamline at region near the fiber.

5 CONCLUSION

The particle collection processes in an electret filter, composed of rectangular split-type fibers arranged in a staggered-parallel array, is numerically simulated. Both the flow field and the electric field around the fiber are determined. All the major collection mechanisms have been coupled in the equation of particle motion. The single fiber collection efficiency is obtained by tracing the particle trajectories.

Two typical orientations of fiber, namely the vertical-fiber array and the horizontal-fiber array, are considered, for comparing the efficiency of each arrangement. The simulation results for neutral particles and singly charged particles based on the vertical-fiber array are first validated against results calculated from the semi-empirical expressions developed by Kanaoka, et al. (1987). Simulation based on the horizontal-fiber array is then carried out to evaluate the effect of orientation of the fibers. Finally typical particle trajectories are presented.

The following results are observed from the present study:

1. The horizontal-fiber array has better single fiber collection efficiency than the vertical-fiber array when diffusional effect and electrostatic effects are taken into consideration.
2. In the case of neutral particles, diffusional effect dominates when the particle velocity is low and when the particle diameter is small. Polarization force effect dominates when the particle velocity is low but the particle diameter is large.
3. In the case of charged particles, the Coulombic force can be so significant that the single fiber collection efficiency can have a value higher than one. It decreases with increase in particle velocity or particle diameter.

REFERENCES

Abuzeid, S., Busnaina, A.A. & Ahmadi, G. 1991. Wall deposition of aerosol particles in a turbulent channel flow. *J. Aerosol Sci.* 22(1): 43–62.
Amano, R.S., Herlee, A.B., Smith, R.J. & Niss, T.G. 1987. Turbulent heat transfer in a corrugated-wall channel with and without fins. *J. Heat Transfer* 109: 62–67.
Baumgartner, H., Piesch, C. & Umhauer, H. 1993. High-Speed Cinematographic Recording and Numerical Simulation of Particle Depositing on Electret Fibers. *J. Aerosol Sci.* 24: 945–962.
Brown, R.C. 1979. Electric effects in dust filters. *Proc. 2nd World Filtration Congress*, London, 1979, pp. 291–301.
Chen, S., Cheung, C.S., Chan, C.K. & Zhu, C. 2002. Numerical simulation of aerosol collection in filters with staggered parallel rectangular fibers. *Computational Mechanics.* 28: 152–161.
de Haan, P.H., Wapenaar, K.E.D. & Van Turnhout, J. 1987. Simulation of electret filters made of non-cylindrical fibers. *Proc. Filtech Conf.,* Utrecht, 1987, pp. 366–375.
Fardi, B. & Liu, B.Y.H. 1992. Flow field and pressure drop of filters with rectangular fibers. *Aerosol Sci. Technol.* 17: 36–44
Kanaoka, C., Emi, H., Otani, Y. & Ishiguro, T. 1987. Effect of Charging State of Particles on Electret Filtration. *Aerosol Sci. Technol.* 7: 1–13.
Lathrache, R. Fissan, H. and Neumann, S. 1986. Deposition of submicron particles on electrically charged fibers. *J. Aerosol. Sci.* 17: 446–449.
Ould-Amer, Y., Chikh, S., Bouhadef, K. & Lauriat, C. 1998. Forced convection cooling enhancement by use of porous materials. *Int. J. Heat Fluid Flow* 19: 251–258.
Rao, N. & Faghri, M. 1990. Computer modeling of electrical enhancement in fibrous filters. *Aerosol Sci. Technol.* 13 (2): 127–134.
Zhu, C., Lin, C.H. & Cheung, C.S. 2000. Inertial impaction dominated fibrous filtration with rectangular or cylindrical fibers. *Powder Technol.* 112: 149–162.

Student paper competition

Computational Methods in Engineering and Science, Iu et al. (eds)
© 2003 Swets & Zeitlinger, Lisse, ISBN 90 5809 567 3

Inverse problem of liquid stream with free surface

I.E. Platonova & Yu.L. Menshikov
Dnepropetrovsk National University, Dnepropetrovsk, Ukraine

ABSTRACT: In this work the inverse problem of restoring the channel bottom shape according to experimental measuring of the level of free surface is examined. The problem is investigated in two-dimensional statement for the case of stationary flow of incompressible liquid along the channel of the finite depth. This problem is reduced to the integral Fredholm equation of the first kind with the kernel which is determined through the improper integral. In order to receive the solution it is proposed to use the known Tikhonov method of regularization with the choice of special mathematical model.

1 INTRODUCTION

Designing of effective means of protection from wave erosion, problems of detection of bodies moving under water, designing of vehicles on underwater wings and others are reduced to the solution of inverse problems of a liquid flow with a free surface (Lonyangapuo, Elliott, Ingham & Wen 1998, 1999, Iljinskij 1999). The algorithms of solution of such problems can only be used in the prospective purposes. In the given work the inverse problem of restoring channel bottom shape according to experimental measuring the level of free surface of liquid is studied (Menshikov 2001).

2 PROBLEM STATEMENT

Let's consider the flat stationary motion of an incompressible ideal liquid along the channel (Fig. 1).

It is known that velocity potential $\Phi(x, y, t)$ satisfies the Laplace equation (Loitcyansky 1978):

$$\frac{\partial^2 \Phi}{\partial x^2} + \frac{\partial^2 \Phi}{\partial y^2} = 0 . \tag{1}$$

In this case the boundary and initial conditions take the forms (Sretenskij 1936):

$$\left[\frac{\partial \Phi}{\partial y} \right]_{y=-h} = 0 , \tag{2}$$

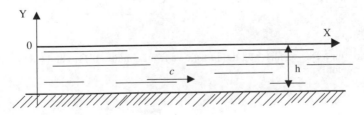

Figure 1. Motion of liquid along the channel.

$$\left[\frac{\partial^2 \Phi}{\partial t^2} + g\frac{\partial \Phi}{\partial y}\right]_{y=0} = 0, \tag{3}$$

$$\frac{1}{g}\left[\frac{\partial \Phi}{\partial t}\right]_{t=0,y=0} = f(x), \tag{4}$$

$$[\Phi]_{t=0,y=0} = F(x). \tag{5}$$

The rise u(x, t) of a point of moving liquid surface above undisturbed level is determined as:

$$u(x,t) = \frac{1}{g}\left[\frac{\partial \Phi}{\partial t}\right]_{y=0}. \tag{6}$$

Let's assume that the open surface remains completely smooth if along the channel of finite depth h the flow of a liquid goes with c speed. Some perturbation changes this law of motion, causing additional speeds of potential character. Let us denote two components of the velocity of unperturbed motion as u and v:

$$u = c - \frac{\partial \Phi}{\partial x}, \quad v = -\frac{\partial \Phi}{\partial y}, \tag{7}$$

where Φ = potential of velocity of unperturbed flow.

The Bernoulli equation has a form (Loitcyansky 1978):

$$\frac{p}{\rho} = N - gy - \frac{1}{2}\left[\left(c - \frac{\partial \Phi}{\partial x}\right)^2 + \left(\frac{\partial \Phi}{\partial y}\right)^2\right], \tag{8}$$

where p = pressure; ρ = density; g = acceleration of gravitation; N = const.

If to believe that the speeds are small in comparison with c we shall have:

$$\frac{p}{\rho} = N - gy + c\frac{\partial \Phi}{\partial x}. \tag{9}$$

Let's write down (9) for an open surface at stationary motion:

$$\eta(x) = \frac{c}{g}\left[\frac{\partial \Phi}{\partial x}\right]_{y=0}. \tag{10}$$

Derivative to a form of a wave along an X axis should coincide with a vector of speed on a surface of a liquid:

$$\frac{\partial \eta}{\partial x} = \frac{-\dfrac{\partial \Phi}{\partial x}}{c - \dfrac{\partial \Phi}{\partial x}} = -\frac{1}{c}\left[\frac{\partial \Phi}{\partial y}\right]_{y=0}. \tag{11}$$

The condition for a free surface of a liquid follows from conditions (10) and (11) (Sretenskij 1936) (by stationary motion):

$$\frac{c^2}{g}\left[\frac{\partial^2 \Phi}{\partial x^2}\right]_{y=0} + \left[\frac{\partial \Phi}{\partial y}\right]_{y=0} = 0. \tag{12}$$

862

Suppose that the bottom of channel has a form:

$$y = -h + \varepsilon \cdot \cos \alpha x, \tag{13}$$

where ε = small positive value.

The boundary condition on the bottom of channel is of the following form:

$$\left[\frac{\partial \Phi}{\partial y} \right]_{y=-h} = 0. \tag{14}$$

The condition on a free surface remains as (12). The wave is defined by expression (10). From boundary conditions (12), (14) for the form of bottom as (13) we receive potential of speeds for the channel of finite depth:

$$\Phi(x,y) = \frac{\varepsilon c^3}{2\alpha \operatorname{ch} \alpha h (c^2 - c'^2)} \left[\left(\frac{g}{c^2} + \alpha \right) e^{\alpha y} + \left(\frac{g}{c^2} - \alpha \right) e^{-\alpha y} \right] \sin \alpha x, \tag{15}$$

where $c'^2 = \alpha^{-1} g \operatorname{th} \alpha h$ is a square of speed of distribution of progressive waves on a surface of the channel of finite depth h.

The equation of a wave has a form:

$$\eta(x) = \frac{c}{g} \left[\frac{\partial \Phi}{\partial y} \right]_{y=0} = \frac{c^2 \varepsilon}{\operatorname{ch} \alpha h (c^2 - c'^2)} \cos \alpha x. \tag{16}$$

Suppose now that the bottom of channel has profile

$$y_1 = -h + \delta \cdot \sin \alpha x = -h + z_1(x), \tag{17}$$

where δ = small positive value. From boundary conditions (12), (14) for the profile of bottom as (17) we finally obtain:

$$\Phi(x,y) = \frac{\delta c^3}{2\alpha \operatorname{ch} \alpha h (c^2 - c'^2)} \left[\left(\frac{g}{c^2} + \alpha \right) e^{\alpha y} + \left(\frac{g}{c^2} - \alpha \right) e^{-\alpha y} \right] \cos \alpha x. \tag{18}$$

The equation of a wave has a form:

$$\eta_1(x) = \frac{c}{g} \left[\frac{\partial \Phi}{\partial y} \right]_{y=0} = \frac{c^2 \delta}{\operatorname{ch} \alpha h (c^2 - c'^2)} \sin \alpha x. \tag{19}$$

Depending on sizes c^2 and c'^2 the mutual arrangement of heights and hollows of surface are being changed. If $c^2 > c'^2$, the mutual arrangement coincides; if $c^2 < c'^2$, then there will be a inverse picture (height there corresponds a hollow).

Suppose that $f(x)$ is the continuous periodic function with the period T. Then it can be represented as a trigonometrical series (Zorich 1984):

$$f(x) = \frac{1}{c} a_0 + \sum_{k=1}^{\infty} \left[a_k \cos \frac{2\pi}{T} kx + b_k \sin \frac{2\pi}{T} kx \right], \tag{20}$$

where a_k, b_k = Fourier coefficients.

On sense of a problem the function $f(x)$ should have zero average value on the period 1, i.e. $a_0 = 0$.

The wave of a kind (16) or (19) on a free surface corresponds to every component ($a_k \cos 2\pi k/T$ + $b_k \sin 2\pi k/T$) of function $f(x)$:

$$\eta(x) = c^2 \sum_{k=1}^{\infty} \frac{1}{\mathrm{ch}\dfrac{2\pi}{T}\,kh\,(c^2 - c_k^2)} \left[a_k \cos\frac{2\pi}{T}kx + b_k \sin\frac{2\pi}{T}kx \right], c_k^2 = \frac{gT}{2\pi k}\,\mathrm{th}\frac{2\pi k}{T}\,h. \tag{21}$$

From expression (21) it is easy to see that the free surface will represent in this case periodic continuous function of the period T.

Suppose that the free surface of a liquid is measured by experimental way and it is written down as:

$$\eta(x) = \sum_{k=1}^{\infty} \lambda_k \cos\frac{2\pi k}{T}x + \sum_{k=1}^{\infty} \sigma_k \sin\frac{2\pi k}{T}x, \tag{22}$$

where

$$\lambda_k = \frac{2}{T}\int_0^T \eta(\tau)\cos\frac{2\pi}{T}k\tau\,d\tau, \; \sigma_k = \frac{2}{T}\int_0^T \eta(\tau)\sin\frac{2\pi}{T}k\tau\,d\tau. \tag{23}$$

If we assume that corresponding component in decomposition (22) is assigned to every component in decomposition (20) functions $f(x)$ then we have:

$$a_k = \frac{\lambda_k}{c^2}\,\mathrm{ch}\frac{2\pi}{T}kh(c^2 - c_k^2), \; b_k = \frac{\sigma_k}{c^2}\,\mathrm{ch}\frac{2\pi}{T}kh(c^2 - c_k^2),$$

$$f(x) = \frac{1}{c^2}\sum_{k=1}^{\infty}\left[\lambda_k \,\mathrm{ch}\frac{2\pi}{T}kh(c^2 - c_k^2)\cos\frac{2\pi}{T}kx + \sigma_k \,\mathrm{ch}\frac{2\pi}{T}kh(c^2 - c_k^2)\sin\frac{2\pi}{T}kx \right]. \tag{24}$$

Expression (21) can be written as:

$$\eta(x) = \frac{2c^2}{T}\sum_{k=1}^{\infty}\frac{1}{\mathrm{ch}\dfrac{2\pi}{T}kh(c^2 - c_k^2)}\int_0^T f(\tau)\cos\frac{2\pi}{T}k(\tau - x)\,d\tau =$$

$$= \int_0^T f(\tau)\frac{2c^2}{T}\sum_{k=1}^{\infty}\frac{1}{\mathrm{ch}\,\alpha_k h(c^2 - c_k^2)}\cos\alpha_k(\tau - x)\,d\tau = \int_0^T f(\tau)K(x - \tau)\,d\tau = A_\Pi f, \tag{25}$$

where A_Π = integral operator.

Let's present expression (24) in the similar form:

$$f(x) = \frac{2}{Tc^2}\sum_{k=1}^{\infty}\mathrm{ch}\,\alpha_k h(c^2 - c_k^2)\int_0^T \eta(\tau)\cos\alpha_k(\tau - x)\,d\tau = C_\Pi \eta, \tag{26}$$

where C_Π = integral operator.

Obviously, that $C_\Pi = A_\Pi^{-1}$, as $f(x) = C_\Pi \eta = C_\Pi A_\Pi f = Ef = f$, where E = unit operator.

Let $f(x)$ be any absolutely integrability function at $(-\infty, \infty)$. Then it is possible to present $f(x)$ as:

$$f(x) = \int_0^{\infty}[a(\xi)\cos\xi x + b(\xi)\sin\xi x]\,d\xi. \tag{27}$$

864

The form of a free surface of a liquid from component $a(\xi)\cos\xi x$ has a kind:

$$\eta_0(x,\alpha) = \frac{c^2 a(\alpha)}{ch\alpha h(c^2 - c'^2)}\cos\alpha x,$$

and from component $b(\xi)\sin\xi x$ is the following

$$\eta_1(x,\alpha) = \frac{c^2 b(\alpha)}{ch\alpha h(c^2 - c'^2)}\sin\alpha x.$$

From two components the surface has a form:

$$\eta(x) = \int_0^\infty \eta_0(x,\alpha)\,d\alpha + \int_0^\infty \eta_1(x,\alpha)\,d\alpha =$$

$$= \int_0^\infty [a(\alpha)\cos\alpha x + b(\alpha)\sin\alpha x]\frac{c^2}{(c^2 - c'^2)ch\alpha h}\,d\alpha, \qquad (28)$$

where

$$a(\alpha) = \frac{2}{\pi}\int_0^\infty f(\xi)\cos\alpha\xi\,d\xi, \quad b(\alpha) = \frac{2}{\pi}\int_0^\infty f(\xi)\sin\alpha\xi\,d\xi,$$

$$\eta(x) = \frac{2c^2}{\pi}\int_0^\infty \frac{1}{(c^2 - c'^2)ch\alpha h}\int_0^\infty f(\xi)\cos\alpha(x-\xi)\,d\xi\,d\alpha = \int_0^\infty f(\xi)K(x-\xi)\,d\xi = Af, \qquad (29)$$

where

$$K(x-\xi) = \frac{2c^2}{\pi}\int_0^\infty \frac{1}{(c^2 - c'^2)ch\alpha h}\cos\alpha(x-\xi)\,d\alpha, \quad c'^2 = \frac{g}{\alpha}th\alpha h,$$

A = integral operator.

Let's present $\eta(x)$ as Fourier integral:

$$\eta(x) = \int_0^\infty [\lambda(\alpha)\cos\alpha x + \sigma(\alpha)\sin\alpha x]\,d\alpha, \qquad (30)$$

$$\eta(x) = \int_0^L f(\xi)K(x-\xi)\,d\xi = \tilde{A}(L)f, \qquad (31)$$

where $\tilde{A}(L)$ = integral operator which depends from parameter L.

Besides at the approached solution of the equation (29) we shall suppose that the spectrum of function f(x) is limited to a constant N:

$$\alpha \le N. \qquad (32)$$

The inequality (32) can be proved from the other point of view too. The level of surface $\eta(x)$ is being measured experimentally by some sensor in inverse problem for flow with free surface. For any real device the strip of perceived frequencies is limited from above. Therefore in any case (even if f(x) has a unlimited spectrum) we receive function $\eta(x)$ with the limited spectrum.

From here we have:

$$\tilde{\eta}(x) = \int_0^L \tilde{f}(\xi) K_N(x - \xi) d\xi = \tilde{A}(L,N) \tilde{f}, \tag{33}$$

where

$$K_N(x - \xi) = \frac{2c^2}{\pi} \int_0^N \frac{1}{(c^2 - c'^2) \mathrm{ch}\alpha h} \cos\alpha(x - \xi) d\alpha .$$

For function $\tilde{f}(x)$ we receive expression:

$$\tilde{f}(x) = \int_0^L \tilde{\eta}(\xi) K_{1,N}(x - \xi) d\xi = \tilde{C}(L,N)\tilde{\eta}, \tag{34}$$

where

$$K_{1,N}(x - \xi) = \frac{2}{\pi c^2} \int_0^N (c^2 - c'^2) \mathrm{ch}\alpha h \cos\alpha(x - \xi) d\alpha , \tilde{C}(L,N) = \text{integral operatorn}$$

which depends from L and N.
By analogy with previous it is possible to consider that $\tilde{C}(L, N) = \tilde{A}^{-1}(L,N)$, as

$$\tilde{f}(x) = \tilde{C}(L, N)\tilde{\eta}(x) = \tilde{C}(L, N)\tilde{A}(L,N)\tilde{f}(x) = E\tilde{f}(x) = \tilde{f}(x). \tag{35}$$

Let's use designations $u_\delta(x) = \tilde{\eta}(x)$, $z(x) = \tilde{f}(x)$, with the purpose of transition to traditional designations in the theory of the incorrectly formulated problems. Then the equation (33) will have a form:

$$A_{\bar{p}} z = u_\delta, \tag{36}$$

where $u_\delta \in U = L_2[c,d]$, $z \in Z = C[a,b]$, $A_{\bar{p}}: Z \to U$ – integral operator in (33) which depends on vector-parameters of mathematical model of process $\bar{p} = (h, g, c, N, L)*$ (* – mark of transposition). The operator $A_{\bar{p}}$ will be completely continuous under these conditions (Tikhonov & Arsenin 1979). Then the problem of the solution of the equation (36) will be unstable to small changes of the initial data (Tikhonov & Arsenin 1979).

3 METHOD OF SOLUTION

We assume now that the right-hand part of the equation (36) (function u_δ) and operator $A_{\bar{p}}$ are given approximately with some fixed error δ and h:

$$\|u_\delta - u_T\|_U \le \delta, \quad \|A_T - A_{\bar{p}}\|_{Z \to U} \le h,$$

where u_T = exact right-hand part; A_T = exact operator in the equation (36). The error of the operator $A_{\bar{p}}$ thus is defined in this case by error of vector-parameters of mathematical model $\bar{p} = (h, g, c, N, L)*$.
The Tikhonov method of regularization for an equation with the approximately given operator is used for obtaining of steady solution (Tikhonov & Arsenin 1979). As stabilizing functional was chosen the following:

$$\Omega[z] = \int_a^b [\dot{z}^2(s) + z^2(s)] ds . \tag{37}$$

The solution obtained by method of regularization gives the solution of an extreme problem:

$$\Omega[\tilde{z}] = \inf_{z \in Q_{\delta,h} \cap Z_1} \Omega[z],$$ (38)

where, $Q_{\delta,h} = \{z: z \in Z, A_{\bar{p}} \in K_A, \|A_{\bar{p}} z - u_\delta\|_U \leq \delta + h\|z\|_Z\}$, $Z_1 =$ the set everywhere dense in Z. The preliminary computations have shown that the accuracy of the approximated solution is very low and that solution coincides practically always with the trivial solution (Menshikov 2001).

Generally vector-parameters of mathematical model has the kind: $\bar{p} = (p_1, p_2, \ldots, p_n)^*$.

Let's believe that every component of a vector $\bar{p}$ can accept values in the given interval:

$$\check{p}_i \leq p_i \leq \hat{p}_i, i = 1, n,$$ (39)

where $\check{p}_i, \hat{p}_i$ – are known. Then $\bar{p} \in \bar{D} \subset R^n$ ($\bar{D}$ – closed parallelepiped in R^n, n – dimension of a vector $\bar{p}$). To each vector $\bar{p}$ corresponds the certain operator $A_{\bar{p}}$ in the equation (36). The set of such completely continuous operators $A_{\bar{p}}$ will be designated as a class $K_A = \{A_{\bar{p}}\}$. The absolutely exact operator in the equation (36) does not belong to a class K_A as its structure is unknown and also it can not be written down in the finite form (i.e. by means of finite number of parameters). Let's assume however that at the fixed level of idealization of the description of real process there is some operator A_T which has structure similar to the structure of the operators $A_{\bar{p}}$ and for which the equality $A_T z_T = u_T$ is carried out. Let's believe therefore that the parameters appropriate to the operator A_T also satisfy to inequalities (39). This operator can be treated as the exact operator (but not the absolutely exact operator). In all further reasonings the operator A_T ($A_T \in K_A$) will be understood in such sense.

Exact solution z_T of the equation (36) will be understood as function from Z which satisfies to a condition $A_T z_T = u_T$.

For reduction of an error of approximate regularized solution of the equation (36) the method of special minimal (maximal) mathematical model is offered (in abbreviated form SMinMM (SMaxMM)) (Menshikov 1991). Suppose that among vectors $\bar{p} \in \bar{D}$ it is possible to define some vector $\bar{p}_0 \in \bar{D}$ that satisfies the inequality:

$$\Omega\left[A_{\bar{p}}^{-1} u\right] \geq \Omega\left[A_{\bar{p}_0}^{-1} u\right]$$

for any allowable function $u \in U$ and any vector $\bar{p} \in \bar{D}$ ($A_{\bar{p}}^{-1}$ is a inverse operator to $A_{\bar{p}}$). The operator $A_{\bar{p}} \in K_A$ which corresponds to a vector $\bar{p}_0 \in \bar{D}$ will be named the special minimal operator. Mathematical model with vector $\bar{p}_0 \in \bar{D}$ will be named CMinMM.

It is possible to define CMaxMM in the same way.

The opportunity of use of special mathematical model in calculations will depend on properties of function:

$$\Omega[N] = \Omega\left[\tilde{f}(x)\right] = \int_0^T \left\{ \left[\int_0^L \tilde{\eta}(\xi) K_{1,N}(x-\xi)d\xi\right]^2 + \left[\int_0^L \tilde{\eta}(\xi)\frac{\partial}{\partial x}\left[K_{1,N}(x-\xi)\right]d\xi\right]^2 \right\} dx \ .$$

It may be proved that derivative of this function with respect to N has positive value at any values of other parameters of MM of researched process. Hence CMinMM will correspond to the minimal value of N, and CMaxMM – maximal value of N.

For check of offered algorithm of the solution of integral equation the test calculation was executed.

The solution of problem about stream which flow around a flat profile under a free surface of ideal incompressible liquid was chosen as test (Jasko 1995). The parameters of mathematical model in this case have the following values: $H = 10\,m$, $h_1 = 1.5\,m$, $g = 9.82\,m/s^2$, $c \approx 70\,m/s$ (h is the depth of profile immersion). The equation (36) will be considered in dimensionless variables choosing as scale the depth of channel H. Then $h_1 = 0.15$, $g = 1\,s^{-2}$, $c = 7\,s^{-1}$, $Fr = v^2/gL = 100$, $[0, L] = [0; 0.5]$, where $Fr =$ Froude number.

Table 1. The level of free surface of liquid.

X	0	1	2	3	4	5	6	7	8	9	10	11	12	13	14	15	16
U_δ	1.7	2.0	2.5	2.6	3.2	3.9	4.0	4.0	3.9	3.2	2.9	2.4	1.9	1.4	1.0	1.0	1.0

Table 2. The form of restored bottom of channel.

X	0	1	2	3	4	5	6	7	8	9	10	11	12	13	14	15	16
$\tilde{Z}$ x0.01	0	0	0	0	0	5	7	8	5	3	2	0	0	0	0	0	0

The numerical values of function $u_\delta(x)$ are submitted in Table 1.
The error of function $u_\delta(x)$ in relation to exact function $u_T(x)$ is determined by size δ:

$$\left\| u_T - u_\delta \right\|_U \leq \delta = 0{,}7 \cdot 10^{-5}.$$

The error of the operator $A_{\bar{p}}$ in the equation (36) in relation to the exact operator A_T was determined by accuracy of approximation of kernel $K(x - \xi)$ and by accuracy of the chosen mathematical model of process: $\left\| A_T - A_{\bar{p}} \right\|_{C \to L_2} \leq h = 0.1$.
The results of calculations (function $\tilde{z}(x)$) are presented in Table 2.

4 CONCLUSION

The results of calculations are shown that at rather small error of measurement of a level of a free surface of a liquid the restitution of bottom of the channel is possible which creates perturbations of a free surface similar to perturbations from a wing. The use of a method of special mathematical model has allowed to increase the accuracy of the approximated solution in comparison with traditional algorithm. The construction of special mathematical model in general remains unsolved problem.

REFERENCES

Iljinskij, N.B. 1999. Inverse boundary problems of the mechanics of a liquid and gas. *Modern problems of Mechanics.: Abst. of Conf. MSU, Moscow, 22–26 November 1999*: 100–101.
Jasko, N.N. 1995. The numerical solution of a nonlinear problem on movement flat profile of a structure under a free surface of an ideal incompressible liquid. *Herald of Science Akademy: Mechanics of Liquid and Gas* 4: 100–107. Moscow.
Loitcyansky, L.G. 1978. *Mechanics of Liquid and Gas*. Moscow: Science.
Lonyangapuo, J.K., Elliott, L., Ingham, D.B. & Wen, X. 1998. Identification of the shape of the bottom surface from a given free surface profile. *Boundary Element Research in Europe, Computational Mechanics Publications*: 81–90.
Lonyangapuo, J.K., Elliott, L., Ingham, D.B. & Wen, X. 1999. Retrieval of the shape of the bottom surface of a channel when the free surface profile is given. *Engineering Analysis with Boundary Elements* 23(5–6): 457–470.
Menshikov, Y.L. 1991. Choice of optimal mathematical model in problems of recognition of external impacts. *The differential equations and their applications in physics*: 25–33. Dnepropetrovsk: Dnepropetrovsk University.
Menshikov, Y.L. 2001. Inverse problem for stream of liquid with free surface. *Proc. of GAMM, 2–7 April 2001*. Zurich: Switzerland.
Sretenskij, L.N. 1936. *The theory of wave motions of a liquid*. Moscow.
Tikhonov, A.N. & Arsenin, V.J. 1979. *Methods of solution of incorrectly formulated problems*. Moscow: Science.
Zorich, V.A. 1984. *Mathematical analysis*. Moscow, p.II.

Nonlinear analysis of spatial member structures

G. Nie & Z. Zhong
Key Laboratory of Solid Mechanics of MEC, Department of Engineering Mechanics and Technology,
Tongji University, Shanghai, China

ABSTRACT: A special nonlinear program *NASMS* based on the Chinese building codes is presented in this paper for the analysis of spatial member structures constructed by thin-walled members or solid members. A precise FEM model considering the warping deformation and the effect of the secondary shear stress is used in the linear elastic module of *NASMS*. In the geometrically nonlinear module of *NASMS*, the effects of axial force, shearing force, biaxial bending moment and bimoment are involved in the geometrical stiffness matrix of element and the tracing methods of nonlinear equivalent path supplied by the program are the modified Newton-Raphson method and arc-length method. The elasto-plastic calculating module of *NASMS* is composed by elasto-plastic hinges method and elasto-plastic zone method. Mathematical programming is also provided to implement the elasto-plastic analysis of structures. The analysis of ultimate bearing capacity of spatial member structures can be accomplished by this program.

1   INTRODUCTION

Spatial member structures consisting of beams and columns are the most widely used structures. Widespread computer usage in engineering offices allows us to now take into account the nonlinear behavior of structures and to gain a better idea of the actual safety factor against failure. Moreover in some cases, such as the stability analysis of single layer dome, second-order computations are compulsorily prescribed by construction codes. With the development of computer-aided design, a growing demand exists for general-purpose programs of analysis of spatial member structures, and such programs should be flexible enough to treat various spatial structures and give a fine response of the structure to various limit states (ultimate, bucking, serviceability). However up to now, there are few such programs. Almost all of the large standard FEM programs available are not designed according to the Chinese Codes. Furthermore, these programs are not very professionally developed and are not convenient to use. Therefore, based on the Chinese building codes, a special program *NASMS* has been developed in this paper. *NASMS* is composed of linear elastic calculating module, geometrical nonlinear module and elasto-plastic calculating module and it also provides with the selection of the element calculating model and calculating method in every module. Based on a precise FEM model by considering the warping deformation and the effect of the secondary shear stress, the analysis of ultimate bearing capacity of spatial member structures constructed by thin-walled members accomplished by this program has a perfect accuracy.

2   LINEAR ELASTIC CALCULATING MODULE

Since the linear elastic program module of *NASMS* forms the base of other analysis module, the accuracy and preciseness of this module are particularly important. The default element calculating model in linear elastic module of program *NASMS* is a two-node thin-walled beam model with fourteen degrees of freedom (Hu 1988, Nie 2001, 2002). The elastic stiffness matrix of the element is shown in Equation (1).

$$\mathbf{k_e} =$$

$$
\begin{bmatrix}
k_{11} & & & & & & & & & & & & & \\
0 & k_{22} & & & & & & & & & & & & \\
0 & k_{32} & k_{33} & & & & & & & & & & & \\
0 & k_{42} & k_{43} & k_{44} & & & & & & & & & & \\
k_{51} & k_{52} & k_{53} & k_{54} & k_{55} & & & & & & & & & \\
k_{61} & k_{62} & k_{63} & k_{64} & k_{65} & k_{66} & & & & & & & & \\
k_{71} & k_{72} & k_{73} & k_{74} & k_{75} & k_{76} & k_{77} & & & & & & & \\
k_{81} & 0 & 0 & 0 & k_{85} & k_{86} & k_{87} & k_{88} & & & & & & \\
0 & k_{92} & k_{93} & k_{94} & k_{95} & k_{96} & k_{97} & k_{98} & k_{99} & & & & & \\
0 & k_{02} & k_{03} & k_{04} & k_{05} & k_{06} & k_{07} & k_{08} & k_{09} & k_{00} & & & & \\
0 & k_{a2} & k_{a3} & k_{a4} & k_{a5} & k_{a6} & k_{a7} & k_{a8} & k_{a9} & k_{a0} & k_{aa} & & & \\
k_{b1} & k_{b2} & k_{b3} & k_{b4} & k_{b5} & k_{b6} & k_{b7} & k_{b8} & k_{b9} & k_{b0} & k_{ba} & k_{bb} & & \\
k_{c1} & k_{c2} & k_{c3} & k_{c4} & k_{c5} & k_{c6} & k_{c7} & k_{c8} & k_{c9} & k_{c0} & k_{ca} & k_{cb} & k_{cc} & \\
k_{d1} & k_{d2} & k_{d3} & k_{d4} & k_{d5} & k_{d6} & k_{d7} & k_{d8} & k_{d9} & k_{d0} & k_{da} & k_{db} & k_{dc} & k_{dd}
\end{bmatrix}
\tag{1}
$$

From Equation (1), it can be seen that most of the non-diagonal elements in the element stiffness matrix of thin-walled beam is different from zero. The non-zero non-diagonal elements express the coupling effect of deformations. The effect of bending deformations and warping deformation on axial deformation of beam, the effect of transverse shearing deformation and the effect of warping deformation and quadratic shearing stress can be considered in the default element model. This default calculating model can be used in arbitrary cross-section including open cross-section, closed cross-section and combined section of the thin-walled members. The explicit expression of element in the element stiffness matrix of thin-walled beam is omitted.

When the members of structure are not thin-walled beam or the degrees of freedom of either end of member is released, this default model can be modified by using the static force cohesion method to deal with all kinds of different joint connection of structures and various cross-section of members.

When the inner force of members of structures has been attained, the stress of cross-section is calculated according to Chinese building codes.

3 GEOMETRICAL NONLINEAR MODULE

The selection of element calculating model and the selection of tracing method of nonlinear equivalent path are provided in geometrical nonlinear module of program *NASMS*.

For thin-walled members, the effect of axial force, shearing force, biaxial bending moment and bimoment shall be taken into consideration in the geometrical stiffness matrix of nonlinear element model (Nie 2002). The geometrical stiffness matrix of element is shown in Equation (2).

$$
K_G =
\begin{bmatrix}
k_{11} & & & & & & & & & & & & & \\
k_{21} & k_{22} & & & & & & & & & & & & \\
k_{31} & 0 & k_{33} & & & & & & & & & & & \\
k_{41} & k_{42} & k_{43} & k_{44} & & & & & & & & & & \\
k_{51} & 0 & k_{53} & k_{54} & k_{55} & & & & & & & & & \\
k_{61} & k_{62} & 0 & k_{64} & 0 & k_{66} & & & & & & & & \\
k_{71} & k_{72} & k_{73} & k_{74} & k_{75} & k_{76} & k_{77} & & & & & & & \\
k_{81} & k_{82} & k_{83} & k_{84} & k_{85} & k_{86} & k_{87} & k_{88} & & & & & & \\
k_{91} & k_{92} & 0 & k_{94} & 0 & k_{96} & k_{97} & k_{98} & k_{99} & & & & & \\
k_{1001} & 0 & k_{1003} & k_{1004} & k_{1005} & 0 & k_{1007} & k_{1008} & 0 & k_{1010} & & & & \\
k_{1101} & k_{1102} & k_{1103} & k_{1104} & k_{1105} & k_{1106} & k_{1107} & k_{1108} & k_{1109} & k_{1110} & k_{1111} & & & \\
k_{1201} & 0 & k_{1203} & k_{1204} & k_{1205} & 0 & k_{1207} & k_{1208} & 0 & k_{1210} & k_{1211} & k_{1212} & & \\
k_{1301} & k_{1302} & 0 & k_{1304} & 0 & k_{1306} & k_{1307} & k_{1308} & k_{1309} & k_{1310} & k_{1311} & 0 & k_{1313} & \\
k_{1401} & k_{1402} & k_{1403} & k_{1404} & k_{1405} & k_{1406} & k_{1407} & k_{1408} & k_{1409} & k_{1410} & k_{1411} & k_{1412} & k_{1413} & k_{1414}
\end{bmatrix}
\tag{2}
$$

870

where

$$k_{11} = \frac{N}{l}, \quad k_{21} = -\frac{M_{zi}}{l^2} + \frac{M_{zj}}{l^2} - \frac{Q_y}{l}, \quad k_{22} = \frac{6N}{5l} + \frac{12N}{\lambda_z^2 l},$$

$$k_{31} = -\frac{M_{yi}}{l^2} + \frac{M_{yj}}{l^2} - \frac{Q_z}{l}, \quad k_{33} = \frac{6N}{5l} + \frac{12N}{\lambda_y^2 l}, \quad k_{41} = -\frac{B_i - B_j}{l^2},$$

$$k_{44} = \frac{12N}{\lambda_\omega^2 l} + \frac{6Nl}{5\lambda_y^2} + \frac{6Nl}{5\lambda_z^2}, \quad k_{55} = \frac{2Nl}{15} + \frac{4Nl}{\lambda_y^2}, \quad k_{66} = \frac{2Nl}{15} + \frac{4Nl}{\lambda_z^2},$$

$$k_{71} = -\frac{B_i}{l}, \quad k_{77} = \frac{4Nl}{\lambda_\omega^2} + \frac{2Nl^3}{15\lambda_y^2} + \frac{2Nl^3}{15\lambda_z^2}, \quad k_{84} = \frac{B_i}{l^2} - \frac{B_j}{l^2}, \tag{3}$$

$$k_{88} = \frac{N}{l}; \quad k_{99} = \frac{6N}{5l} + \frac{12N}{\lambda_z^2 l}; \quad k_{1010} = \frac{6N}{5l} + \frac{12N}{\lambda_y^2 l},$$

$$k_{1101} = \frac{B_i - B_j}{l^2}, \quad k_{1108} = -\frac{B_i - B_j}{l^2}, \quad k_{1212} = \frac{2Nl}{15} + \frac{4Nl}{\lambda_y^2},$$

$$k_{1313} = \frac{2Nl}{15} + \frac{4Nl}{\lambda_z^2}, \quad k_{1401} = \frac{B_j}{l}, \quad k_{1414} = \frac{4Nl}{\lambda_\omega^2} + \frac{2Nl^3}{15\lambda_y^2} + \frac{2Nl^3}{15\lambda_z^2}$$

In the equations above, N represents the axial force; M_{yi}, M_{yj}, M_{zi}, M_{zj} represent the bending moments; M_x represents torsion moment; B_i, B_j represent bimoments; Q_y, Q_z represent shear forces; λ_y, λ_z, λ_ω represent ratio of slenderness of member. The process of derivation of geometrical stiffness matrix of element is omitted.

The tracing methods of nonlinear equivalent path supplied by program *NASMS* are modified Newton-Raphson method and spherical arc-length method (Bruce et al. 1987, Carrera 1994, Li 2000, Zeng 2001). And the efficiency and veracity of arc-length method is better. The aim of the structural design according to the special project can be achieved with the only necessity to choose some appropriate parameters in *NASMS*.

4 ELASTO-PLASTIC CALCULATING MODULE

In general, elasto-plastic analysis of spatial member structures may involve the formation of (a) elasto-plastic zone and (b) elasto-plastic hinges. To study the former, each member is subdivided into finite elements, and the cross-section of each finite element is further subdivided into many blocks. The material nonlinearity is considered by the stress–strain relationship of the blocks in a section. As a result the elasto-plastic zone solution is known as the "exact solution". Because of its highly intensive computation, the elasto-plastic zone solution is not often used in daily engineering design. In the latter, only one element per member is used to model the actual strength and stiffness of members. Therefore the elasto-plastic hinges method has clear advantage over the elasto-plastic zone method in computation. Both methods can be selected according to the actual need of users.

4.1 *Elasto-plastic hinges analysis*

Three elasto-plastic hinges per member are likely formed due to member forces and three hinges may occur simultaneously or successively. Considering the geometric nonlinearity, several kinds of calculating modules are provided in the elasto-plastic hinges module of *NASMS*, namely the first-order elasto-plastic hinges analysis, the second-order elasto-plastic hinges analysis, the first-order elasto-plastic holonomic analysis and the second-order elasto-plastic holonomic analysis. The geometric stiffness matrix shown in Equation (2) is used in the second-order analysis. The unloading and reload cycles are performed in holonomic analysis.

4.2 Elasto-plastic zone analysis

In elasto-plastic zone (Espion 1986, Izzuddin 2000) module, analysis starts with the use of only one elastic element per member. Subsequently, automatic refinement (or subdivision) of elastic elements into expensive elasto-plastic elements is performed only when and where necessary. By using the method of finite-cut of element, the gradual developments of yielding both across the section and along the axis of the member can be taken into consideration. During the elasto-plastic analysis of structures, the Young's modulus of element stiffness matrix should be replaced by elasto-plastic constitutive matrix calculated according to the stress state of infinitesimal and the simplex stress–strain relationship of material. For the ideal elasto-plastic material, the constitutive matrix after the infinitesimal has entered the plastic stage is

$$
\mathbf{D}_{ep} = \begin{bmatrix} D_{11} & D_{12} & D_{13} \\ D_{21} & D_{22} & D_{23} \\ D_{31} & D_{32} & D_{33} \end{bmatrix} = \frac{E}{3(\sigma_{xx}^2 + 3(\tau_{xz}^2 + \tau_{xy}^2))} \begin{bmatrix} 9(\tau_{xz}^2 + \tau_{xy}^2) & -3\sigma_{xx}\tau_{xz} & -3\sigma_{xx}\tau_{xy} \\ -3\sigma_{xx}\tau_{xz} & \sigma_{xx}^2 + 3\tau_{xy}^2 & -3\tau_{xz}\tau_{xy} \\ -3\sigma_{xx}\tau_{xy} & -3\tau_{xz}\tau_{xy} & \sigma_{xx}^2 + 3\tau_{xz}^2 \end{bmatrix}
\tag{4}
$$

From Equation (4), it can be seen that the effect of the normal stress and shear stress of infinitesimal on the change of material performance is considered.

The stiffness matrix of the member can be obtained by integrating the stiffness of infinitesimal. The explicit expression of stiffness matrix of the element in cannot be obtained and only the integrated expression can be given. The integrated expression of partial main elements in stiffness matrix is shown as follows:

$$
k_{22} = \int_v \frac{36D_{11}(l-2x)^2 y^2 + l^2 \phi_y(-12D_{13}(l-2x)y + D_{33}l^2\phi_y)}{l^6(1+\phi_y)^2} dv = \frac{12EI_z}{l^3(1+\phi_y)}
\tag{5}
$$

$$
k_{44} = \int_v \frac{1}{(l^3 + 12l\lambda)^2}(36(D_{11}\omega^2(l-2x)^2 + (lx - x^2 + 2\lambda)(-2D_{12}\omega(l-2x)y
$$
$$
+ D_{22}y^2(lx - x^2 + 2\lambda) + z(2D_{13}\omega(l-2x) - (2D_{23}y - D_{33}z)(lx - x^2 + 2\lambda)))))dv
$$
$$
= \frac{6(GJl^4 + 20GJl^2\lambda + 120GI_p\lambda^2 + 120GJ\lambda^2 - 120GJ_b\lambda^2 + 10EI_\omega l^2)}{5l(l^2 + 12\lambda)^2}
\tag{6}
$$

$$
k_{77} = \int_v \frac{1}{(l^3 + 12l\lambda)^2}(4D_{11}\omega^2(2l^2 - 3lx + 6\lambda)^2 + (l^3 - 4l^2x - 12x\lambda + 3l(x^2 + 2\lambda))
$$
$$
(4D_{12}\omega y(2l^2 - 3lx + 6\lambda) + D_{22}y^2(l^3 - 4l^2x - 12x\lambda + 3l(x^2 + 2\lambda)) + z(-4D_{13}\omega
$$
$$
(2l^2 - 3lx + 6\lambda) - (2D_{23}y - D_{33}z)(l^3 - 4l^2x + 3lx^2 + 6l\lambda - 12x\lambda))))dv
$$
$$
= \frac{2GJl^6 + 15GJl^4\lambda + 270GI_p l^2\lambda^2 + 90GJl^2\lambda^2 - 270GJ_b l^2\lambda^2}{15l(l^2 + 12\lambda)^2} +
$$
$$
\frac{30EI_\omega l^4 + 180EI_\omega l^2\lambda + 1080EI_\omega l^2}{15l(l^2 + 12\lambda)^2}
\tag{7}
$$

Gaussian numerical integration method is used to obtain the integrated elements in stiffness matrix in *NASMS*. In order to exactly consider the plastic zone formation, much more Gaussian integrating point should be employed. This module can determine the ultimate bearing capacity of spatial member structures.

4.3 *Mathematical programming module*

Mathematical programming (Tin-Loi 1996) is also provided to implement the elasto-plastic analysis of spatial member structures in program *NASMS*. The Mathematical programming procedures devised have the capability to search the whole solution set of possibly existing equilibrium paths and to provide an identification of plastic yielding (activation) or unloading (deactivation).

One can compare the preciseness and effectiveness of the iterative increment methods and the mathematical programming.

5 EXAMPLES

Figure 1 shows a single layer dome with thirteen nodes and eighteen elements. The rise of dome is 6.1 m. The load-displacement curve at vertex of the single dome is shown in Figure 2 when different calculating modules are introduced.

From Figure 2, it can be seen that the geometrically nonlinear solution calculated with *NASMS* is the same as that conducted by many researchers (Li 1998, Zeng 2001). This indicates the accuracy and effectiveness of *NASMS* provided in this paper.

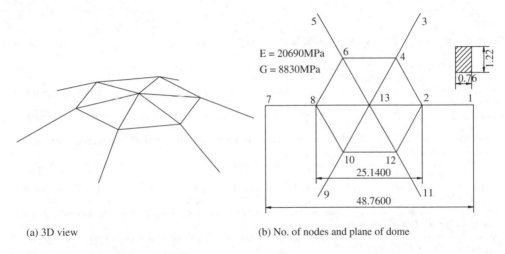

(a) 3D view   (b) No. of nodes and plane of dome

Figure 1.   Single layer dome (unit: m).

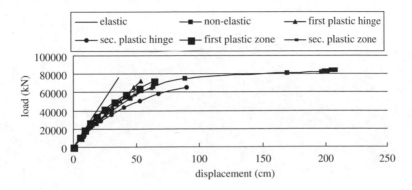

Figure 2.   Load-displacement curve at vertex of structure.

873

6 CONCLUSIONS

In general, the nonlinear calculating program *NASMS* studied in this paper meets the need of the design of spatial member structures in line with the Chinese building codes. A precise FEM model considering the restrained torsion is used in the linear elastic analysis. The effect of axial force, shearing force, biaxial bending moment and bimoment was involved in the geometrical stiffness matrix of element and the tracing method of nonlinear equivalent path supplied by the program is the modified Newton-Raphson method and arc-length method. The elasto-plastic hinges analysis and elasto-plastic zone analysis can be achieved by *NASMS* and Mathematical programming is also provided to implement the elasto-plastic analysis of structures. *NASMS* can give adequate insight into the structural behavior. Numerical Example proves the feasibility of *NASMS*.

ACKNOWLEDGEMENTS

This work was supported by the Teaching and Research Award Fund for Outstanding Young Teachers in High Education Institutions of MOE, PRC, and the National Natural Science Foundation of China (Project no. 10125209).

REFERENCES

Bruce, W.R & Siegfried, F.S. 1987. Improved arc length orthogonality methods for nonlinear finite element analysis. *Computers & Structures* 27(5): 625–630.

Carrera, E. 1994. A study of arc-length-type methods and their operation failures illustrated by a simple method. *Computers & Structures* 50(2): 217–229.

Espion, B. 1986. Nonlinear analysis of framed structures with a plasticity minded beam element. *Computers & Structures* 22(5): 831–839.

Hu, Y.R. & Chen, B.Z. 1988. A new torsional stiffness matrix of thin-walled bar element. *Computational Structural Mechanics and Application* 5(3): 19–28.

Izzuddin, B.A. & Smith, D.L. 2000. Efficient nonlinear analysis of elasto-plastic 3D R/C frames using adaptive techniques. *Computers & Structures* 78: 549–573.

Li, Y.Q. 1998. *Stability research on large-span arch-supported reticulated shell structures.* Shanghai: Tongji University.

Nie, G.J. & Qian, R.J. 2001. An improved element stiffness matrix of prismatic thin-walled beam. *The proceedings of IASS 2001 Sympsium*: 150–151.

Nie, G.J. 2002. *Research on the nonlinear analytical model and analytical methods of spatial steel member structures.* Shanghai: Tongji University.

Tin-Loi, F. & Misa, J.S. 1996. Simplified geometrically nonlinear elastoplastic analysis of semirigid frames. *Mech. Struct. & Mach.* 24(1): 1–20.

Tin-Loi, F. & Misa, J.S. 1996. Large displacement elastoplastic analysis of semirigid steel frames, *Int. J. for Numerical Methods in Engineering* 39: 741–762.

Zeng, Y.Z. 2001. *On the Research of Stability Behaviors of Single-layer Spherical Domes.* Shanghai: Tongji University.

Computational Methods in Engineering and Science, Iu et al. (eds)
© 2003 Swets & Zeitlinger, Lisse, ISBN 90 5809 567 3

FEM analysis of the spatial reticulated structure constructed by rectangular pipes

G. Nie & Z. Zhong
Key Laboratory of Solid Mechanics of MEC, Department of Engineering Mechanics and Technology
Tongji University, Shanghai, China

R. Qian
Department of Structural Engineering, Tongji University, Shanghai, China

ABSTRACT: During the deformation of spatial reticulated structure, the rectangular-pipe member may generate the warping deformation in addition to axial tension or compression, shear, bending and torsion. At the same time the warping shearing strain will affect the warping deformation again. As a result it is necessary to consider warping deformation of section in the spacial reticulated structure constructed by rectangular pipes. A new FEM model for rectangular-pipe members is derived and the inter-related structural calculating program is worked out in this paper. The new FEM model has been testified by the numerical example. The comparison of the space beam model with the thin-walled beam model also shows the effect of warping on spatial reticulated structure constructed by rectangular pipes. The FEM model presented in this paper can be used for the design of the spatial structure constructed by rectangular pipes.

1 INTRODUCTION

Recently the spatial reticulated structures constructed by rectangular pipes have been widely used in various construction engineering. The main reason is that the rectangular pipe is of unique superiority. From the point of view of a structural engineer, the rectangular-pipe section is of a relatively large inertia moment. Thus its ability to resist compression, bending and torsion is superior to circular-pipe section. When the rectangular pipes are used in spatial reticulated structures, the steel consuming of the whole structure can be reduced. Moreover it is convenient to manufacture the spatial structures constructed by rectangular pipes and to install them on site. Therefore, it is necessary to give adequate insight into the behavior of this kind of structure (Liu et al. 2000, Yin et al. 2000).

2 DEFORMATION OF RECTANGULAR PIPE

There are four basic deformations for solid member, namely, axial tension or compression, shearing, bending and torsion. Since the warping deformation of solid section is very small, the performance analysis is based on the hypothesis of planar sections after bending. But this hypothesis is not always true for thin-walled member. Generally the warping deformation of thin-walled member under load is quite remarkable, and the stress generated by warping should be considered when the bearing capacity of member is calculated (Chan et al. 1987, Suresh 1997, Jonsson 1999). The warping deformation as well as the plane deformation of section may result in the yield or shock of member. So the section stress of the rectangular pipe is more complicated. In order to accurately analyze rectangular member, all the possible deformations shown in Figure 1 should be considered.

From Figure 1, it is seen that restrained torsion and warping deformation are the characteristics of rectangular section. The normal stress and shearing stress will be generated by warping in the member.

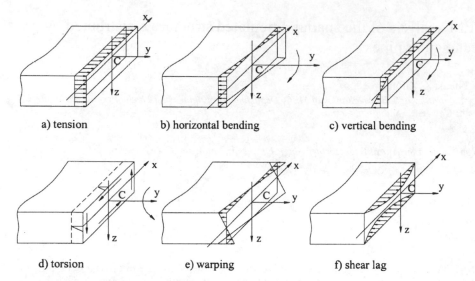

| a) tension | b) horizontal bending | c) vertical bending |
| d) torsion | e) warping | f) shear lag |

Figure 1. Deformations of a rectangular pipe.

At the same time the additional warping shearing strain will affect the warping deformation again (Bao 1991, Li 1990, Ni 2000). The distribution of sectional normal stress of thin-walled members is the same as the sectorial area and the distribution of sectional shearing stress is the same as the sectorial moment.

3   FEM MODEL OF MEMBER

Based on the above analysis of deformation of a rectangular pipe, it is necessary to consider warping deformation of cross-section in the spacial reticulated structure constructed by rectangular pipes. For this purpose this paper presents a new FEM model applicable for thin-walled member. The new model is based on the space beam model (Qian 1990, Zeng 2001) of six degrees of freedom at each end, each node of an element has an additional degree of freedom for the warping deformation of cross-section. Thus in the new model, each node of the thin-walled element has seven degrees of freedom namely u, v, w, θ_x, θ_y, θ_z, θ_ω, and accordingly there are seven nodal forces for each node namely P_x, P_y, P_z, M_x, M_y, M_z, B, where θ_ω is a warping function mainly considering the effect of the secondary shear stress (Hu 1988, Nie 2001, 2002). Except that the pattern of the torsion deformation uses the cubic polynomial, the patterns of other deformations adopt the same function as the space beam model. The displacement patterns of each element can be written as follows:

$$u = c_1 + c_2 x \tag{1}$$

$$v = c_3 + c_4 x + c_5 x^2 + c_6 x^3 \tag{2}$$

$$w = c_7 + c_8 x + c_9 x^2 + c_{10} x^3 \tag{3}$$

$$\theta_x = c_{11} + c_{12} x + c_{13} x^2 + c_{14} x^3 \tag{4}$$

$$\theta_y = c_8 + 2c_9 x + 3c_{10} x^2 + l^2 \phi_z \cdot c_{10} / 2 \tag{5}$$

$$\theta_z = c_4 + 2c_5 x + 3c_6 x^2 + l^2 \phi_y \cdot c_6 / 2 \tag{6}$$

$$\theta_\omega = c_{12} + 2c_{13} x + c_{14}(6\lambda + 3x^2) \tag{7}$$

The last terms on the right-hand sides of Equations (5)–(6) represent the effect of the shearing deformation. The last term of Equation (7) represents the effect of secondary shear stress. According to the Equations (1)–(7), the shape functions of each element can be obtained:

$$\mathbf{a} = \begin{bmatrix} a_{11} & 0 & 0 & 0 & 0 & 0 & 0 & a_{18} & 0 & 0 & 0 & 0 & 0 & 0 \\ 0 & a_{22} & 0 & 0 & 0 & a_{26} & 0 & 0 & a_{29} & 0 & 0 & 0 & a_{213} & 0 \\ 0 & 0 & a_{33} & 0 & a_{35} & 0 & 0 & 0 & 0 & a_{310} & 0 & a_{312} & 0 & 0 \\ 0 & 0 & 0 & a_{44} & 0 & 0 & a_{47} & 0 & 0 & 0 & a_{411} & 0 & 0 & a_{414} \\ 0 & 0 & a_{53} & 0 & a_{55} & 0 & 0 & 0 & 0 & a_{510} & 0 & a_{512} & 0 & 0 \\ 0 & a_{62} & 0 & 0 & 0 & a_{66} & 0 & 0 & a_{69} & 0 & 0 & 0 & a_{613} & 0 \\ 0 & 0 & 0 & a_{74} & 0 & 0 & a_{77} & 0 & 0 & 0 & a_{711} & 0 & 0 & a_{714} \end{bmatrix} \qquad (8)$$

where

$$a_{22} = \frac{(l-x)(lx - 2x^2 + l^2(1+\phi_y))}{l^3(1+\phi_y)}, \quad a_{33} = \frac{(l-x)(lx - 2x^2 + l^2(1+\phi_z))}{l^3(1+\phi_z)},$$

$$a_{44} = \frac{(l-x)(l^2 + lx - 2x^2 + 12\lambda)}{l(l^2 + 12\lambda)}, \quad a_{77} = \frac{(l-x)(l^2 - 3lx + 12\lambda)}{l(l^2 + 12\lambda)}, \text{ etc.} \qquad (9)$$

The strain of a thin-walled element subject to the external force consists of normal strain and shearing strain. The expression of strain can be written as:

$$\epsilon = \begin{bmatrix} \varepsilon_L \\ \gamma_{xz} \\ \gamma_{xy} \\ \gamma_\rho \end{bmatrix} = \begin{bmatrix} \dfrac{du_x}{dx} - y\dfrac{d\theta_z}{dx} - z\dfrac{d\theta_y}{dx} - \omega\dfrac{d\theta_\omega}{dx} \\ \dfrac{dw}{dx} - \theta_y \\ \dfrac{dv}{dx} - \theta_z \\ (\dfrac{\varphi}{t} - \rho)\theta_\omega \end{bmatrix} \qquad (10)$$

The normal strain, the shearing strain caused by transverse load, and the shearing strain caused by torsion and warping are included in Equation (10). Furthermore the normal strain consists of axial strain, bending strain and warping strain. According to the principle of virtual work, 14×14 element stiffness matrix can be attained. The expressions of element in stiffness matrix related to torsion and warping are listed as follows:

$$\mathbf{k_e} = \begin{bmatrix} k_{11} & 0 & 0 & 0 & k_{15} & k_{16} & k_{17} & k_{18} & 0 & 0 & 0 & k_{1b} & k_{1c} & k_{1d} \\ 0 & k_{22} & k_{23} & k_{24} & k_{25} & k_{26} & k_{27} & k_{28} & k_{29} & k_{20} & k_{2a} & k_{2b} & k_{2c} & k_{2d} \\ 0 & k_{32} & k_{33} & k_{34} & k_{35} & k_{36} & k_{37} & k_{38} & k_{39} & k_{30} & k_{3a} & k_{3b} & k_{3c} & k_{3d} \\ 0 & k_{42} & k_{43} & k_{44} & k_{45} & k_{46} & k_{47} & k_{48} & k_{49} & k_{40} & k_{4a} & k_{4b} & k_{4c} & k_{4d} \\ k_{51} & k_{52} & k_{53} & k_{54} & k_{55} & k_{56} & k_{57} & k_{58} & k_{59} & k_{50} & k_{5a} & k_{5b} & k_{5c} & k_{5d} \\ k_{61} & k_{62} & k_{63} & k_{64} & k_{65} & k_{66} & k_{67} & k_{68} & k_{69} & k_{60} & k_{6a} & k_{6b} & k_{6c} & k_{6d} \\ k_{71} & k_{72} & k_{73} & k_{74} & k_{75} & k_{76} & k_{77} & k_{78} & k_{79} & k_{70} & k_{7a} & k_{7b} & k_{7c} & k_{7d} \\ k_{81} & 0 & 0 & 0 & k_{85} & k_{86} & k_{87} & k_{88} & 0 & 0 & 0 & k_{8b} & k_{8c} & k_{8d} \\ 0 & k_{92} & k_{93} & k_{94} & k_{95} & k_{96} & k_{97} & k_{98} & k_{99} & k_{90} & k_{9a} & k_{9b} & k_{9c} & k_{9d} \\ 0 & k_{02} & k_{03} & k_{04} & k_{05} & k_{06} & k_{07} & k_{08} & k_{09} & k_{00} & k_{0a} & k_{0b} & k_{0c} & k_{0d} \\ 0 & k_{a2} & k_{a3} & k_{a4} & k_{a5} & k_{a6} & k_{a7} & k_{a8} & k_{a9} & k_{a0} & k_{aa} & k_{ab} & k_{ac} & k_{ad} \\ k_{b1} & k_{b2} & k_{b3} & k_{b4} & k_{b5} & k_{b6} & k_{b7} & k_{b8} & k_{b9} & k_{b0} & k_{ba} & k_{bb} & k_{bc} & k_{bd} \\ k_{c1} & k_{c2} & k_{c3} & k_{c4} & k_{c5} & k_{c6} & k_{c7} & k_{c8} & k_{c9} & k_{c0} & k_{ca} & k_{cb} & k_{cc} & k_{cd} \\ k_{d1} & k_{d2} & k_{d3} & k_{d4} & k_{d5} & k_{d6} & k_{d7} & k_{d8} & k_{d9} & k_{d0} & k_{da} & k_{db} & k_{dc} & k_{dd} \end{bmatrix} \qquad (11)$$

where

$$k_{71} = k_{17} = -k_{d1} = -k_{1d} = k_{d8} = k_{8d} = -k_{87} = -k_{78} = -\frac{ES_\omega}{l},$$

$$k_{44} = \frac{6(GJl^4 + 20GJl^2\lambda + 120GI_p\lambda^2 + 120GJ\lambda^2 - 120GJ_b\lambda^2 + 10EI_\omega l^2)}{5l(l^2 + 12\lambda)^2},$$

$$k_{47} = k_{74} = k_{d4} = k_{4d} = k_{da} = k_{ad} = \frac{(GJl^4 + 720GI_p\lambda^2 - 720GJ_b\lambda^2 + 60EI_\omega l^2)}{10(l^2 + 12\lambda)^2},$$

$$k_{77} = \frac{2GJl^6 + 15GJl^4\lambda + 270GI_pl^2\lambda^2 + 90GJl^2\lambda^2 - 270GJ_bl^2\lambda^2 + 30EI_\omega l^4 + 180EI_\omega l^2\lambda + 1080EI_\omega l^2}{15l(l^2 + 12\lambda)^2} \tag{12}$$

where I_y, I_z, I_p, J_s, J_b, I_ω and S_ω are the moment of inertia about the y-axis, the moment of inertia about the z-axis, polar moment of inertia, torsion constant, *Bredt* torsion constant, warping constant and sectorial moment of section, respectively.

The possible deformations of rectangular pipe are considered in the stiffness matrix above. While using it, the warping deflection and bimoment of member can be calculated. At the same time more insight into the behavior of the spatial structure constructed by rectangular pipes can be provided. In addition, this new model can be used in designing arbitrary cross-sections including open cross-section, closed cross-section and combined section of the thin-walled members.

4 PROGRAM OF SPATIAL STRUCTURE

All the possible deformations of a member in spatial structure, such as axial tension or compression, bending, shear, torsion and warping can be considered in the new model above. Whereas some of members of spatial structure may only undergo bending and shear without warping deformation. In order to predict the behavior of structure composed by the members of different kinds of deformation patterns, the static force cohesion method is adopted in the program to form the stiffness matrix of the element of different degrees of freedom at the nodes.

5 EXAMPLES

5.1 *Verification of the new model*

Using the program established in this paper, a cantilever constructed by a rectangular pipe shown in Figure 2 is calculated to testify the new model. The cantilever is divided into sixteen elements with equal length. Comparison of the results obtained with the new model with the exact solution obtained with restraint torsion differential equation is shown in Figure 3 and Figure 4. It is shown that the new model has perfect accuracy. Hence, it can be used in the analysis of spatial structure constructed by rectangular pipes.

5.2 *Use of the model*

A single-layered reticulated shell constructed by the rectangular pipes shown in Figure 5 is calculated with the new model. The span of the shell is 5.0 m and the rise of the shell is 1.0 m. The dead load acting on the shell is 12 kN/m². The height of the rectangular cross-section is 30 mm, the width is 20 mm, and the thickness of web and flange is 2 mm. Figure 6 shows the warping stresses and the compressive stresses superposed by the axial compression and bending of the members. Simultaneously the comparison of the results from the FEM model presented in this paper with those from the space beam model is made. From Figure 6 it can be seen that the normal stress from warping cannot be ignored in the structural design especially for the braced member. The new FEM model given in this paper can

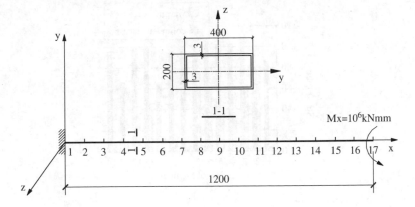

Figure 2.   Cantilever (Unit: mm).

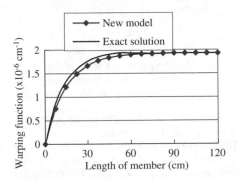

Figure 3.   Warping function of cantilever.

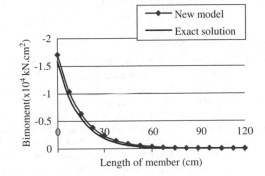

Figure 4.   Bimoment of cantilever.

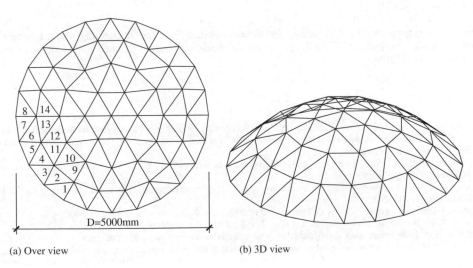

(a) Over view   (b) 3D view

Figure 5   Single reticulated shell constructed by rectangular pipes.

879

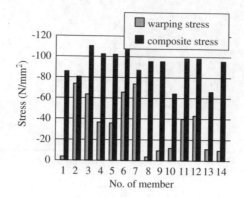

Figure 6.   Stresses of elements 1~14.

be used to calculate the sectional warping stress of rectangular pipes and to reflect the characteristics of deformation of thin-walled member when being used in the analysis of spatial reticulated structure constructed by rectangular pipes.

6   CONCLUSIONS

(a) The element stiffness matrix of thin-walled beam presented in this paper can analyze the torsion deformation and warping deformation in thin-walled structures. It can further be used to analyze the thin-walled members with arbitrary cross-sections including open cross-sections, closed cross-sections and combined sections.

(b) In thin-walled structures, warping of sections should be taken into account in the analysis of strength, stiffness and stability of the structures, especially for the braced members.

(c) The numerical example verifies the element model. The comparison between the space beam model and the thin-walled beam model indicates the effect of warping in spatial reticulated structures constructed by rectangular pipes. The new model presented in this paper can be used to calculate the warping deformation and warping stress, which provides adequate insight in the structural behavior. Therefore, it can be used in the design of the spatial structures constructed by rectangular pipes.

ACKNOWLEDGEMENTS

This work was supported by the Teaching and Research Award Fund for Outstanding Young Teachers in High Education Institutions of MOE, PRC, and the National Natural Science Foundation of China (Project no. 10125209).

REFERENCES

Bao, S.H. & Zhou, J. 1991. *Thin-walled Bar Structural Mechanics*. Beijing: China Architecture & Building Press.
Chan, S.L. & Kitipornchai, S. 1987. Geometric Nonlinear Analysis of Asymmetric Thin-walled Beam-columns. *Engrg Struct.* 9(4): 243–254.
Hu, Y.R. & Chen, B.Z. 1988. A new torsional stiffness matrix of thin-walled bar element. *Computational Structural Mechanics and Application* 5(3): 19–28.
Jonsson, J. 1999. Distortional Warping Functions and Shear Distributions in Thin-walled Beams. *Thin-Walled Structures* 33: 245–268.
Jonsson, J. 1999. Distortional Theory of Thin-walled Beams. *Thin-Walled Structures* 33: 269–303.
Li, K.X. 1990. *Warping of elastic thin-walled bars*. Beijing: China Architecture & Building Press.
Liu, W.B. et al. 2000. Impression of reticulated structures used in walls and application of square or rectangular pipe in spatial reticulated structures. *Symposiums on conference of the ninth spatial structures*: 637–644.
Ni, Y.Z. & Qian, Y.Q. 2000. *Analysis of elastic thin-walled beams & bridges*. Beijing: People's Communications Publishing House.

Nie, G.J. & Qian, R.J. 2001. An improved element stiffness matrix of prismatic thin-walled beam. *The proceedings of IASS 2001 Sympsium*: 150–151.

Nie, G.J. 2002. *Research on the nonlinear analytical model and analytical methods of spatial steel member structures*. Shanghai: Tongji University.

Park, S.W., Fujii, D. & Fujitani, Y. 1997. A Finite Element Analysis of Discontinuous Thin-Walled Beams Considering Nonuniform Shear Warping Deformation. *Comput. and Struct.* 65(1): 17–27.

Qian, R.J. 2000. *Analytical Theory and Calculating Method of Space Grid Structures*. Beijing: China Architecture & Building Press.

Suresh, R. & Malhotra, S.K. 1997. Some Studies on Static Analysis of Composite Thin-walled Box Beam. *Comput. & Struct.* 62(4): 625–634.

Yin, D.Y. et al. 2000. Discuss on application of square or rectangular pipes in spatial reticulated structures. *Symposiums on conference of the ninth spatial structures*: 587–593.

Zeng, Y.Z. 2001. *On the Research of Stability Behaviors of Single-layer Spherical Domes*. Shanghai: Tongji University.

A three steps method for structural damage diagnosis based on FEM and vibration data

J. Xie & D.J. Han
Department of Civil Engineering, South China University of Technology, Guangzhou, China

ABSTRACT: Structural damage diagnosis is one of core content of the structure health monitoring. A new three steps structural damage diagnosis procedure, damage detection, damage element select & search and iterative quantity at element level, is developed on the base of partial works of previous authors. The procedure is illustrated and testified by an example of a 3-span continuous RC beam with single/ multi-damage locations, considered modeling uncertainty (structural parameter random variation) and measure errors (modal random noise).The result shows that the procedure is effective and robust, suited for practical application.

On the background of new infrastructures constructing continuously and lots of existing structures deteriorating, structure health monitoring has become a research focus in civil engineer communities at home and aboard. Some initial explorations for practical structures have been done, such as the comprehensive investigation, both experimental and theoretical study for the Interstate 40 and Alamosa Canyon bridge in the states (Farrar, C.R. et al 2000), as well as the experimental research project for 17 damage levels of Z-24 bridge in Swiss (Brincker, R. et al 2001) and so on. Vibration-based damage diagnosis (VBDD) method has become a main method in structure health monitoring research, because this method is relatively easy to realize automatic and quick diagnosis. It is noticed that a civil structure has a feature that damage is possible to be detected at component level. An important branch of VBDD Method based on finite element model (FEM) and measured dynamic behavior is developed. This approach is theoretically rigorous and has great potential for damage detection and damage localization as well as damage quantification. However, although various techniques for VBDD have been proposed, there is no a single technique is completely effective and reliable because of the complexity and variety of structural damage. There is no exception for the above mentioned method. Therefore, to develop a procedure which combines several techniques and takes the advantage of each single technique is reasonable. In this paper a three steps method based on FEM and the measured dynamic behavior is proposed for damage diagnosis for a civil structure. It is noticed that a civil structure has a feature with damage diagnosis at element level, i.e., damage detection and damage location & search are all at element level. The procedure is illustrated and testified by an example of a 3-span continuous RC beam with single/multi-damage locations. In this example, the FEM modeling uncertainty as well as the errors caused by random noise in frequency and mode measurement are considered. The results show that the assumed damage can be detected satisfactorily. Also, the computing time is relatively short as considering only seldom elements are damaged. The procedure is effective and robust. Therefore it is suited for damage diagnosis in practical application for structure health monitoring.

1 THEORETICAL BACKGROUND

1.1 *Mode shape expansion*

In general, the number of measured degrees of freedom (DOF) is much less than that of in FEM model. For the sake of convenience in using a FEM dependent method, it is necessary to make the measured number of DOF be equal to that in FEM. There are three approaches to realize this requirement. First, the number of measured mode shapes is expanded to that of the FEM. Second, the number of DOF for FEM is condensed to that of measured ones. Third, both number of the measured mode shapes are expanded and that of the FEM is condensed to a proper number simultaneously. Since model condensation destroys connectivity of structural matrixes (stiffness and mass), which is adverse for damage location. However, by using mode shape expansion technique, a relatively higher resolution for indicating damage location can be gained. Therefore, a mode shape expansion method (Alvin, K.F.1996) is used in this paper to obtain unmeasured component of the mode shape $\{\phi\}$

$$\{\phi_e^d\} = -(([D_e]^T[D_e])^{-1}[D_e]^T[D_m]\{\phi_m^d\}$$ (1)

$$[D] = [K] - \lambda^d[M] = [D_m \quad D_e]$$ (2)

where $\{\phi_m\}$ is measured part of the mode shape. $\{\phi_e\}$ is the unmeasured portion. Superscript d refers to damage case. λ is equal to the square of the measured circular frequency. It is worth to notice that an undamaged stiffness is used in Equation (2) as damaged stiffness is often unknown.

1.2 Three steps method for structural damage diagnosis

FEM dependent methods for damage diagnosis, such as sensitivity based model updating method etc., are considered at element level. If it is used for a large structure with many elements, a calculation cost will be considerable. In fact damage always occurs at the seldom places of a structure, which means if a coarse estimation for potential damage locations is provided before inversion calculation, undoubtedly, computation efficiency will be improved massively. Therefore, the first step of the three steps procedure is to obtain prior information of the damage. In this paper mode residual force is applied for this purpose. The mode residual force for mode i is defined as

$$\{E\}_i = [Z^d]_i\{\phi^d\}_i$$ (3)

where $[Z^d]_i = -(\omega_i^d)^2[M^u] + [K^u]$ and $[M^d] = [M^u] - [\Delta M]$, $[K^d] = [K^u] - [\Delta K]$. Superscript u refers to undamaged case

It is noticed that only one mode is needed to recognize single/multi-damage location. Of course, in order to weaken the effects of various noises and to improve the reliability of damage localization more modes are generally used. Since different mode has different sensitivity to the same damage, a weight index is proposed to consider the effect of multi modes.

$$E^*_j = \sum_{i=1}^{m} \frac{|E'_{ij}|}{\sum_{k=1}^{m}|E'_{kj}|} E'_{ij} = \sum_{i=1}^{m} W_{ij} E'_{ij}$$ (4)

where W_{ij} is weight factor for the jth DOF of the ith mode. m is the number of measured modes. E_i is the result of normalizing vector E_i by using its maximum component, in this way the effect of order difference of the norm magnitude of row vectors of matrix $[Z]$ is eliminated.

Theoretically, mode residual force for undamaged elements obtained from Equation (3) should be equal to zero, however, because of the measurement noise, mode shape expansion error and uncertainties of finite element model. The mode residual forces are usually not exactly equal to zero. Therefore, it is necessary to take a search & selection procedure to determine the most likely damage part of the structure. This is the second step of the proposed method. Before selecting damaged elements homologous DOF must be found, which is due to calculation for modal residual force implemented at DOF. Fahat, C. & Hemez, F.M. (1993) proposed an error threshold as:

$$T = ER([s \times (\alpha / \alpha^*) \times n])$$ (5)

where n is total number of DOF of FEM. Error function ER (in this paper ER refers to E^*) is arranged from large to small. α is the sum of all $ER \cdot \alpha^*$ is the product of the maximum ER(namely $ER(1)$) and n. s is a controller factor for computation cost. [] means to get integer. Then all the DOF whose corresponding ER is greater than or equal to threshold T is selected. After that, a search & selection strategy is established to obtain damaged element in potentiality. In this paper the possible element is found by the following approach. If an element whose whole nodes are related to the above selected DOF then it is considered as a possible damage element. If no element satisfies with above requirement, then the elements that have one or more node related to damaged DOF are chosen. In this way some disturbance of noises can be reduced. The third step of the procedure is to find damage location and damage quantity according to the prior information provided by the second step. The third step can also be divided into two parts, the first is inversion calculation and the second is to get damage quantity through an iterative algorithm. For $[K^d] = [K^u] - [\Delta K]$ and using modal orthogonality conditions as

$$[\phi^d]_{m \times n}^T [K^d]_{n \times n} [\phi^d]_{n \times m} = [\phi^d]_{m \times n}^T [M^u]_{n \times n} [\phi^d]_{n \times m} [\Lambda_t^d]_{m \times m} = [\phi^d]^T [K^u + \Delta K][\phi^d] \qquad (6)$$

It is assumed that the lose of stiffness is caused only by damaged elements, thus

$$[\Delta K]_{n \times n} = \sum_{i=1}^{w} \beta_i [K_i^u]_{n \times n} \qquad (7)$$

where w is the total number of the possible damage elements determined by element search in the second step. $[K_i^u]$ is the stiffness matrix of the ith undamaged element expanded to full n rank. β_i is damage degree of the ith element. Substitute Equation (7) into Equation (6) and rearrange by β, then

$$[S]_{g \times w} \{\beta\}_{w \times 1} = \{\gamma\}_{w \times 1} \qquad (8)$$

where the ith column of matrix $[S]$ consists of the components of upper triangular part of matrix $[\phi]^T$ $[K_i^u] [\phi]$. There are $g = m(m+1)/2$ independent equations in Equation (8) because of the symmetry of the stiffness and mass matrixes. In general Equation (8) is overdetermined in the condition of selecting damaged elements. In order to get a unique inversion result optimal algorithm is adopted, Thus Equation (8) is transformed into a least square problem with constraints as follows:

$$\min \frac{1}{2} \|[S]\{\beta\} - \{\gamma\}\|_2^2 \quad s.t. \quad -1 \le \beta \le 0 \qquad (9)$$

The advantages of proposed inversion algorithm lie in that the searching procedure is limited to a few suspected elements. The magnitude of the damage degree can be obtained without any iteration. Thus, the computing time is greatly reduced. However, because of incompletion of measured structural mode with noises, the obtained damage degree is not exact. Therefore an iteration algorithm is needed to adjust the inversion results. Of course if it is just to detect damage rather than quantify damage. The iteration includes 3 steps. First, the result of the Equation (9) is used to update the undamaged stiffness matrix. Second, the updated matrix is introduced into Equation (1) to expand mode shape again, and third, Equations (3)–(9) are used to get a new inversion result to updating undamaged stiffness matrix. Repeat the above steps, until convergence is reached. Finally, positions and magnitudes of damages can be obtained.

2 EXAMPLE OF A 3-SPAN CONTINUOUS RC BEAM AND SIMULATION OF UNCERTAINTIES

A 3-span continuous RC beam is showed in Figure 1. The beam section is $0.3 \text{ m} \times 0.6 \text{ m}$. Density and elastic modulus of the beam are 2500 kg/m^3 and $3 \times 10^{10} \text{N/m}^2$ respectively. The FEM of the beam is established using 2-D elastic beam element and element length is 0.5 m, which results 48 elements in all. There are two damage positions, as illustrated in Figure 1. Damage is described by reducing elastic

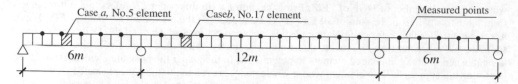

Figure 1. Damage location and arrangement of measured points of the continuous beam.

Table 1. Natural frequencies of every damage cases (Hz).

Mode rank	Undamaged case	Damage case a	Damage case b	Damage case c $(a + b)$
1	10.2245	10.1920	10.2192	10.1862
2	26.1800	25.8551	25.8811	25.5523
3	33.1342	32.5002	33.0443	32.4457
4	40.8984	40.4817	40.7767	40.3306
5	69.1330	68.9287	67.8203	67.6133
6	104.725	104.033	104.066	103.4512
7	118.227	117.332	118.135	117.1737
8	132.547	132.175	131.290	130.9045

modulus for 40%. There are 21 measured DOF that are expanded to 94 DOF of the whole FEM. The first 8 bending modes are used in diagnosis procedure and natural frequencies of the modes in different damage cases are showed in Table 1.

In this example, the influence of the FEM uncertainty and measurement errors upon the damage diagnosis is studied. The objective is to examine the damage prediction capacity of the proposed method. To this end, it is set that the elastic modulus P with 5% variation, namely

$$p^* = p(1 + 5\% \times rand(-1,1)) \tag{10}$$

in which $P*$ is the elastic modulus used in FEM. $rand(-1,1)$ refers to random number distribute uniformly between $[-1,1]$.

Parameters $e1$ and $e2$ are introduced to describe the random noises of the frequency and mode shape respectively.

$$\hat{f}_j^d = f_j^d + f_j^d \cdot e1 \cdot rand(-1,1) \qquad \hat{\phi}_{ij}^d = \phi_{ij}^d + \phi_{ij}^d \cdot e2 \cdot rand(-1,1) \tag{11}$$

in which $e1$ and $e2$ are nominal percentage levels for noise controls. Subscript j is the jth mode and i is the ith component of the mode shape.

3 RESULTS AND DISCUSSION

Some typical calculation results are listed in Table 2–4 and Figure 2–3. Nominal noise percentage $e1$, $e2$ are taken as 1% and 5% respectively, which is reasonable according to the fact that frequencies can be measured more accurate than mode shape. Comprehension errors means that simultaneously to consider 5% random variance of elastic modulus and 1% frequencies measure random error as well as 5% mode shape measure random error. It is noticed that only those damage element corresponding to its damage case is listed. In fact after damaged elements are selected out the degree of damage of other elements almost approach to zero such as that showed in Figure 2 and Figure 3. From the tables it can be seen that the proposed three steps procedure is sensible and effective. When there is no noise, the results converge to the correct damage location and to its damage quantity whatever the conditions for damage in different span, single or multi-damage location etc. When there is some noises, a few iteration such

886

Table 2. Verified results of damage case a (Element No. 5).

	1		2		5	
Iteration number	Selected elements	Damage degree β	Selected elements	Damage degree β	Selected elements	Damage degree β
No noise	3,4,5,6	−0.5142	5	−0.4073	5	−0.4000
Modeling uncertainty	1,2,3,4 5,6,7	−0.5099	5	−0.4029	5	−0.3953
Measure errors	3,4,5,6 7,8,14 15,47,48	−0.5974	5,28,29 30	−0.4619	5,22	−0.4505
Comprehension errors	3,4,5,6,7 24,25,26 27,28,29, 30,31	−0.5326	5,22,23 24	−0.4136	5,22,23 24,25	−0.4056

Table 3. Verified results of damage case b (Element No.17).

	1		2		5	
Iteration number	Selected elements	Damage degree β	Selected elements	Damage degree β	Selected elements	Damage degree β
No noise	14,15,16 17,18	−0.4991	17	−0.4049	17	−0.4000
Modeling uncertainty	13,14,15 16,17,18 26	−0.4804	17	−0.4237	17	−0.4220
Measure errors	16,17,18 19, 26	−0.5826	17,45,46	−0.4562	17	−0.4445
Comprehension errors	12,13,14 15,16,17 18,19	−0.5763	17,23,28 45	−0.4645	17,23,28 45	−0.4530

Table 4. Verified results of damage case c (Element No. 17 & No. 5).

	1		2		5	
Iteration number	Selected elements	Damage degree β	Selected elements	Damage degree β	Selected elements	Damage degree β
No noise	3,4,5,6,14, 15,16,17,18	−0.5156 −0.4992	5,17	−0.4077 −0.4053	5,17	−0.4000 −0.4000
Modeling uncertainty	1,2,3,4,5,6,14 15,16,17,18	−0.5039 −0.4487	5,17	−0.3846 −0.3813	5,17	−0.3813 −0.3798
Measure errors	3,4,5,6,7,14 15,16,17,18 28,29,30,31 34,35	−0.5711 −0.5597	5,6,7 17,29 30,31	−0.4608 −0.4672	5,17,29 30,31	−0.4325 −0.4566
Comprehension errors	3,4,5,13,14 15,16,17,18	−0.6567 −0.5105	5,16,17 23,28,45 46	−0.4994 −0.4716	5,16,17 28,45,	−0.4794 −0.4814

as 5 times or so is needed to identify the damage quantity. However, there are much more elements selected for potential damage locations than actual damaged element at last, but the identified damage degree approximate to zero or relatively smaller magnitude except actual damaged elements such as showed in Figure 3 and Table 4.

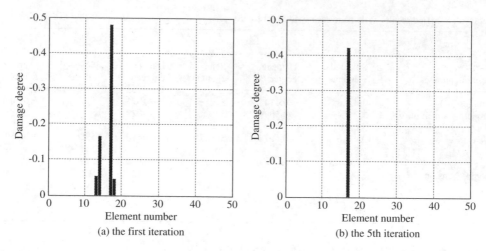

(a) the first iteration   (b) the 5th iteration

Figure 2.   Convergence of damage case *b* existing modeling uncertainty.

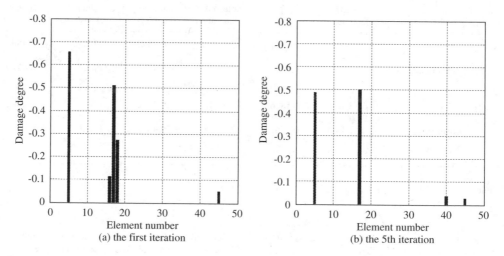

(a) the first iteration   (b) the 5th iteration

Figure 3.   Convergence of damage case *c* existing comprehension errors.

Verification about the performance of the procedure under the errors and uncertainties shows that the damages can be located correctly and the damage quantity can approach to a right degree. Herein three random disturbances including 5% random variance of elastic modulus and 1% random error of frequencies measure as well as 5% random error of mode shape measure are considered by single or combination. Relatively, measure errors have a larger effect to damage diagnosis than random variation of the structural parameter.

In fact the first iteration is the most important step as the exact damage location is determined, the effect of sequent iteration is just to eliminate the irrelative undamaged elements and to reach the actual damage stiffness by elaborately adjusting structural stiffness matrix (see Fig. 2). Selecting mode and mode number as well as reducing error caused by mode shape expansion can help to realize a correct iteration in the first step. In this study, cases of damage degree varying from 10% to 90%,of single/multi-damage location, of various random levels for measurement errors and model uncertainties have been considered. The results show that for most cases the proposed three steps procedure can get a relatively satisfied result. It should be pointed out that when there is noise existing the procedure may not be successful always. In such cases the results should be calibrated by damage detection probability under certain noise level. And in the future if more robust statistical techniques are combined to deal with errors and uncertainties, especially in estimating potential damage elements, the proposed method would perform better.

4 CONCLUSION

Damage cases are variable for different structure style. Therefore, there are some limitations and defects for any structural health monitoring/damage detection methods. There is no a single method which is completely effective and reliable. To develop a procedure which combine several methods and take the advantage of each single method is a reasonable way. In this paper a three steps method including damage detection, damage element select & search and damage quantity at element level is proposed. Simulation verification of a 3-span continuous RC beam shows the procedure owns relatively good accuracy to detect damage and robustness under the errors at the same time inversion computation cost is relatively low because of searching for potential damages. The procedure is suited for damage diagnosis in practical application for structure health monitoring.

REFERENCES

Alvin, K.F.1996. Finite Element Model Update via Bayesian Estimation and Minimization of Dynamic Residuals. Proc.of 14th International Modal Analysis Conference, Dearborn, Michigan.
Brincker, R., Andersen, P. & Cantieni, R. 2001. Identification and level 1 damage detection of the Z24 Highway Bridge by frequency domain decomposition. Experimental Techniques, 25(6): 51~57.
Farhat, C. & Hemez, F.M.1993. Updating finite element dynamic models using an element-by-element sensitivity methodology. AIAA Journal, 31(9): 1702~1711.
Farrar, C.R., Cornwell, P.J., Doebling, S.W. & Prime, M.B. 2000. Structural health monitoring studies of the Alamosa Canyon and I-40 Bridge. Los Alamos National Laboratory Report, LA-13635-MS.

Computational Methods in Engineering and Science, Iu et al. (eds)
© 2003 Swets & Zeitlinger, Lisse, ISBN 90 5809 567 3

The post-buckling behavior of open cross-section thin-walled columns in the context of the Generalized Beam Theory (GBT)

P. Simão & L. Simões da Silva
Civil Engineering Department, University of Coimbra, Coimbra, Portugal

ABSTRACT: GBT is a very powerful and elegant theory to analyze thin walled prismatic members. It combines the classical Vlasov theory and the folded plate theory, and its most important concepts are: (i) the member global behavior is obtained through a linear combination of pre-established deformation modes, and (ii) the diagonalization of the system of equilibrium equations is obtained through an orthogonalization procedure. In this paper, the extension of GBT into the post-buckling domain is presented for open cross-section columns, using an energy formulation and a Rayleigh-Ritz approach, making resource to a set of orthonormal polynomials. An illustrative example is presented, exploring to some extent the interaction of the deformation modes.

1 INTRODUCTION

Depending on the member's length, thin-walled open section columns exhibit several types of buckling phenomena, like local, distortional or global buckling. Some of these phenomena involve the deformation of the cross section in its own plane, where some plates experience significant transverse bending stresses and displacements. If the folding lines have relative displacements, the buckling mode is said to be distortional. With the exception of sophisticated numerical methods like finite element and finite strip methods, the common way to deal with distortional buckling is to consider simplified models of cross section parts (Lau & Hancock 1987). Consequently, to the present, the limited attempts at the study of the behavior of thin-walled members in the post-buckling range were either experimental (Young & Rasmussen 1998) or numerical (Kwon & Hancock 1993), post-buckling analysis being heavy and cumbersome, with analytical sensitivity to the relevant phenomena being lost.

In the 60's, combining the traditional thin walled members theory with folded plate theory, Schardt created GBT, which became a very powerful and elegant tool to study thin walled sections enabling cross section distortion. Its most important concepts are (i) the member global behavior is obtained through a linear combination of pre-established deformation modes, each multiplied by its respective amplitude modal function ^kV (these functions being the only unknowns of the problem, one-dimensional and defined along the member's length), and (ii) the diagonalization of the system of equilibrium equations is obtained through an orthogonalization procedure. Due to space limitations, GBT fundamentals must be found elsewhere (Shardt 1999). Concerning the geometrically non-linear behavior of thin walled members, a GBT energy formulation was developed for open or closed sections (Simão & Simões da Silva 2002), containing the most relevant non-linear coefficients and thus enabling post-buckling analysis.

This paper presents an application of GBT's energy formulation to the analysis of the post-buckling behavior of open section columns, applying the traditional stability procedures (Thompson & Hunt 1973 and 1984) and the Rayleigh-Ritz method. Instead of trigonometric functions, orthonormal polynomials are used as coordinate functions, and an algorithm to generate these polynomials is also presented. Finally, to illustrate the above concepts, a channel section is analyzed in the post-buckling range, and some conclusions are presented.

2 STABILITY ANALYSIS OF COMPRESSED COLUMNS

2.1 General energy formulation

Taking into account the usual GBT assumptions (Schardt 1989), the total potential energy (TPE) for open cross section members with an axial load P at the end $x = L$, accounting only for the membrane longitudinal stresses, is defined as (Simão & Simões da Silva 2002):

$$A = \sum_{i=1}^{n_{MD}} \sum_{k=1}^{n_{MD}} \frac{1}{2} \int_0^L \left({}^{ik}C^M + {}^{ik}C^B \right) {}^iV'' \, {}^kV'' dx + \sum_{i=1}^{n_{MD}} \sum_{k=1}^{n_{MD}} \frac{1}{2} \int_0^L {}^{ik}D_1 \, {}^iV' \, {}^kV' dx +$$

$$+ \frac{1}{2} \sum_{i=1}^{n_{MD}} \sum_{k=1}^{n_{MD}} \int_0^L {}^{ik}D_2 \, {}^iV \, {}^kV'' dx + \frac{1}{2} \sum_{i=1}^{n_{MD}} \sum_{k=1}^{n_{MD}} \int_0^L {}^{ik}D_{2T} \, {}^iV'' \, {}^kV \, dx + \frac{1}{2} \sum_{i=1}^{n_{MD}} \sum_{k=1}^{n_{MD}} \int_0^L {}^{ik}B \, {}^iV \, {}^kV \, dx +$$

$$+ \frac{1}{4} \sum_{i=1}^{n_{MD}} \sum_{k=1}^{n_{MD}} \sum_{l=1}^{n_{MD}} \int_L {}^{ikl}\kappa_{\sigma 2} \, {}^iV'' \, {}^kV' \, {}^lV' dx + \frac{1}{4} \sum_{i=1}^{n_{MD}} \sum_{j=1}^{n_{MD}} \sum_{k=1}^{n_{MD}} \int_L {}^{ijk}\kappa_{\sigma 3} \, {}^iV' \, {}^jV' \, {}^kV'' dx +$$

$$+ \frac{1}{8} \sum_{i=1}^{n_{MD}} \sum_{j=1}^{n_{MD}} \sum_{k=1}^{n_{MD}} \sum_{l=1}^{n_{MD}} \int_L {}^{ijkl}\kappa_{\sigma 4} \, {}^iV' \, {}^jV' \, {}^kV' \, {}^lV' dx \quad - P \times {}^iV' \Big|_{x=x_1} \tag{1}$$

where

$$^{ik}C^B = \int_s \frac{E t^3}{12(1-\mu^2)} {}^if \, {}^kf \, ds \qquad {}^{ik}D_1 = \int_s \frac{G t^3}{3} {}^i\dot{f} \, {}^k\dot{f} \, ds$$

$$^{ik}D_2 = \int_s \frac{E t^3 \mu}{12(1-\mu^2)} {}^i\ddot{f} \, {}^kf \, ds \qquad {}^{ik}D_{2T} = \int_s \frac{E t^3 \mu}{12(1-\mu^2)} {}^if \, {}^k\ddot{f} \, ds \tag{2, a–e}$$

$$^{ik}B = \int_s \frac{E t^3}{12(1-\mu^2)} {}^i\ddot{f} \, {}^k\ddot{f} \, ds$$

are the generalized cross section properties due to bending effects, and

$$^{ik}C^M = \int_s E t \, {}^iu \, {}^ku \, ds \quad (\text{1}^{\text{st}} \text{ order term}) \qquad {}^{ikl}\kappa_{\sigma 2} = \int_s E t \, {}^iu \left({}^kf_s \, {}^lf_s + {}^kf \, {}^lf \right) ds$$

$$^{ijk}\kappa_{\sigma 3} = \int_s E t \, {}^ku \left({}^if_s \, {}^jf_s + {}^if \, {}^jf \right) ds \qquad {}^{ijkl}\kappa_{\sigma 4} = \int_s E t \left({}^if_s \, {}^jf_s + {}^if \, {}^jf \right) \cdot \left({}^kf_s \, {}^lf_s + {}^kf \, {}^lf \right) ds \tag{3, a–d}$$

are the generalized cross section properties related to the membrane effects, these terms already containing the geometrical non-linear effects up to 3rd order, thus enabling the post-buckling analysis.

2.2 Application of the Rayleigh-Ritz method – the use of a set of orthonormal polynomials

In order to perform the system discretization, the Rayleigh-Ritz method will be used, approximating each amplitude modal function ^kV by a set of coordinate functions $^k\phi_i$:

$$^kV = {}^ka_1 \, {}^k\varphi_1 + {}^ka_2 \, {}^k\varphi_2 + {}^ka_3 \, {}^k\varphi_3 + \ldots \tag{4}$$

coefficients ^ka_i being the unknowns of the problem. For most practical applications, the boundary conditions may be taken mode by mode and can be written in the form:

$$\sum_{i=0}^{n_{parts}} {}^j\bar{a}_i \, {}^kV^{(i)} \Big|_{x=\bar{x}} = 0 \, , j=1,\ldots,{}^kn_{BC} \tag{5}$$

where $^{k}n_{BC}$ denotes the number of independent boundary conditions for mode k, n_{parts} is the number of cross-sectional parts in each boundary condition and $^{j}\bar{a}_i$ are known coefficients. These boundary conditions can be either kinematic or static and, in the context of the Rayleigh-Ritz method, the coordinate functions must respect the kinematic boundary conditions. If, in addition, they also respect the static boundary conditions, usually (but not always) the convergence will be improved. So, the efficiency of this method depends on the correct choice of the coordinate functions (Richards 1977) and choosing orthonormal functions $^{k}\phi_i$ over the member's length, as shown in eqs. (6) (Courant & Hilbert 1953):

$$\int_L {}^{k}\varphi_i \, {}^{k}\varphi_j \, dx = \begin{cases} c \text{ if } i = j \\ 0 \text{ if } i \neq j \end{cases}, \tag{6}$$

(c being a real non zero constant usually equal to 1), will accelerate the convergence of the method. In eqs. (6), the first condition is a normalization rule, whereas the second constitutes an orthogonality condition. Grouping together the boundary conditions for each mode of deformation, a homogeneous system is obtained, which has a trivial solution of no numerical interest. However, if one adds eqs. (6), the resulting system is no longer homogeneous and can yield a set of coordinate functions for each mode of deformation k. In order to exemplify this method, it is applied to determine the coordinate functions to be used in the post-buckling analysis of a thin-walled compressed column. Applying this procedure to the 1st mode of deformation, for a simply supported column – axial displacement restricted at $x = 0$ and free at $x = L$ – the boundary conditions are:

i) kinematic conditions: $\left. {}^{1}V \right|_{x=0} = 0$ and $\left. {}^{1}V' \right|_{x=0} = 0$; $\tag{7}$

ii) static conditions: $\left. {}^{1}V'' \right|_{x=L} = 0$ and $\left. {}^{1}V''' \right|_{x=L} = 0$. $\tag{8}$

giving:

$${}^{1}\varphi_1 = \frac{x^2 \sqrt{5}}{L^2} \tag{9}$$

For modes 2 and higher and allowing warping at the end sections – end plate with negligible bending inertia at both ends – the boundary conditions are:

i) kinematic conditions: $\left. {}^{k}V \right|_{x=0} = 0$ and $\left. {}^{k}V \right|_{x=L} = 0$; $\tag{10}$

ii) static conditions: $\left. {}^{k}V'' \right|_{x=0} = 0$ and $\left. {}^{k}V'' \right|_{x=L} = 0$. $\tag{11}$

In this case the first three orthonormal polynomials are:

$${}^{k}\varphi_1 = \sqrt{\frac{70}{31}} \left(\frac{3}{L} x - \frac{6}{L^3} x^3 + \frac{3}{L^4} x^4 \right)$$

$${}^{k}\varphi_2 = \sqrt{\frac{462}{5}} \times \left(-\frac{1}{L} x + \frac{6}{L^5} x^5 \right) + \sqrt{2310} \times \left(\frac{2}{L^3} x^3 - \frac{3}{L^4} x^4 \right) \tag{12, a–c}$$

$${}^{k}\varphi_3 = \sqrt{\frac{2730}{7781}} \times \left(\frac{27}{L} x - \frac{736}{L^3} x^3 + \frac{2073}{L^4} x^4 \right) + \sqrt{\frac{84630}{251}} \times \left(-\frac{66}{L^5} x^5 + \frac{22}{L^6} x^6 \right).$$

After the discretization process, the TPE becomes a function of a set of n_c coordinates ^{i}a and a global numbering for all the coordinates can be used. It must be stressed that the use of a larger number of coordinate functions per mode increases the method's convergence, but for the present purpose three functions per mode are sufficient.

2.3 Buckling and post-buckling analysis

The n_c equilibrium conditions for the analyzed member are determined by (Thompson & Hunt 1973):

$$\frac{\partial A}{\partial\,^i a} = 0\,, \quad i = 1,\dots,n_c\,. \tag{13}$$

For a simply supported column under axial compression, equations (13) yield the following fundamental path (denoted by subscript FP):

$$^1 a_{1,FP} = \,^1 a_{FP} = P\frac{L^3\,\sqrt{5}}{10\,^1 C}\,, \tag{14}$$

all other coordinates being zero. Now, a sliding coordinate transformation for all coordinates is applied in the form:

$$^i a = \,^i a_{FP} + \,^i q\,, \tag{15}$$

the trivial solution being given by:

$$^i q = 0\,, \quad i = 1,\dots,n_c\,. \tag{16}$$

After this transformation, the TPE becomes a function of all the $^i q$ coordinates and will be designated by W. The critical loads are easily evaluated by setting to zero the determinant of the Hessian matrix of the TPE,

$$\det\left(H_{FP}\right) = \left|\frac{\partial^2 W}{\partial\,^i q\, \partial\,^j q}\bigg|_{FP}\right| = 0\,. \tag{17}$$

For open sections, expression (17) consists on a linear generalized eigenvalue problem, where the eigenvalues are the requested critical loads, and their correspondent eigenvectors represent the buckling modes. For each buckling mode, its non-zero coefficients indicate the active coordinates and, if properly normalized, each eigenvector coefficient can be regarded as the participation of its correspondent coordinate, whereas the matrix containing all eigenvectors enables a coordinates transformation that diagonalizes the Hessian matrix (Hangai & Kawamata 1972). After finding the active coordinates and the critical loads, the post-buckling behavior is determined by searching alternative equilibrium paths in the neighborhood of the critical point. These paths can be easily found by searching non-trivial solutions of the non-linear equilibrium system, making resource to the appropriate numerical techniques (Crisfield 1980).

3 APPLICATION TO A LIPPED-CHANNEL COLD-FORMED MEMBER

3.1 Introduction and bifurcational analysis

In order to enable a better grasp on the procedures described above, the orthonormal polynomials of equations (9) and (12) and the numerical technique for post-buckling analysis will be used to study the lipped-channel cold-formed column shown in figure 1. The corresponding modes of deformation are presented in figure 2. The critical behavior is presented in figure 3 for member lengths between 50 and 3500 mm, for three different cases. All cases use just one coordinate function for mode 1 given by formula (9) but they use different approximations for the remaining modes: case 1 uses just the first coordinate function of expression (12), case 2 takes the first two functions and case 3 takes all three functions, and will be used in the post-buckling analysis.

The lowest eigenvalue was used here to describe the member critical behavior. In the present problem and due to the shape of H_{FP}, it must be stressed that coordinate 1 (related to axial elongation mode) is passive for all values of the load factor P and thus can be ignored in the eigenvalue analysis. Figure 3

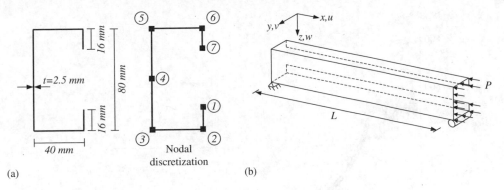

(a)

(b)

Figure 1. a) Cross section properties and nodal discretization of the analyzed member; b) the simply supported column.

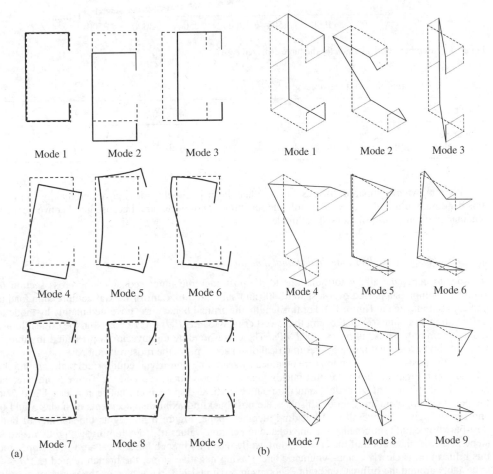

(a)

(b)

Figure 2. a) modal cross section displacements shape; b) modal warping displacements shape.

illustrates the decrease of critical load with the number of approximating polynomials, this decrease being more relevant when local or distortional buckling are dominant – significant decrease occurs for the smaller lengths (lengths up to about 250 mm) when local plate buckling of the web rules the bifurcational behavior. For example, the use of just one half-wave curve stiffens the member up to 46% for

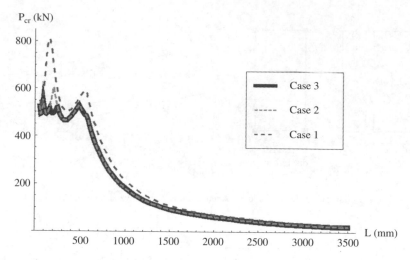

Figure 3. Member's critical loads for the three analyzed cases.

Table 1. Coefficients of the 1st eigenvector.

Mode n	3			5			7			9		
Coordinate	^5a	^6a	^7a	^{11}a	^{12}a	^{13}a	^{17}a	^{18}a	^{19}a	^{23}a	^{24}a	^{25}a
Coeff. value	0.96	0	0	98.13	0	0.40	-0.49	0	0	-0.02	0	0

$L = 160$ mm, when compared to cases 2 or 3. When the critical behavior is governed mainly by minor axis bending, the use of additional coordinate functions is not necessary, because good convergence is obtained with only one polynomial per mode.

3.2 Post buckling results in the distortional domain

In order to apply the above formulations to the post-buckling range, consider the cross section of figure 1a) simply supported over a span of 400 mm submitted to a uniform axial compression load at one end, as indicated in figure 1b). For this length, the critical behavior is governed mainly by mode of deformation 5 (symmetric distortion), the lowest critical load is $P_{cr} = 481.05$ kN and the critical longitudinal membrane stress is $\sigma_{cr} = 1.002$ MPa. The corresponding eigenvector is presented in table 1, where it can be seen that mode 5 – symmetrical distortion – plays the most relevant role.

Next, the post-buckling solutions are obtained by searching non-trivial equilibrium paths around the critical point. Figure 4 shows the post-buckling paths for coordinates ^1a and ^{11}a, figure 5 presents the horizontal displacement of node 4 and the relative vertical displacement between nodes 1 and 7 in the post-buckling domain, figure 6 displays the post-buckling axial displacements of nodes 5 and 6, and figure 7 gives an overall post-buckling member shape. Figure 6 highlights the stability of both post-buckling equilibrium paths, the one related to negative values of ^{11}a exhibiting a slightly negative slope in the neighborhood of the critical point, while the other branch has always a positive slope, and this bifurcation is clearly a non-symmetric one. Having negative slope, the branch related to $^{11}a < 0$ is unstable around the bifurcation point (Thompson & Hunt 1973). However, as the values of coordinate ^{11}a move away from 0, this path gains a positive slope and the equilibrium becomes stable.

Looking at figure 8, where the longitudinal membrane stresses along two illustrative cross sections are presented for the critical point and for two points of both post-buckling branches, it can be concluded that, mainly due to the modal interaction between modes 5 and 3, the path relative to $^{11}a < 0$ is associated with a stress increase in the flanges near the lips, while the other branch exhibits a stress increase at the web, this configuration being more stable than the first.

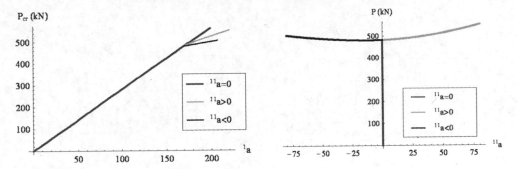

Figure 4. Post-buckling equilibrium paths for coordinates ^{I}a and ^{II}a.

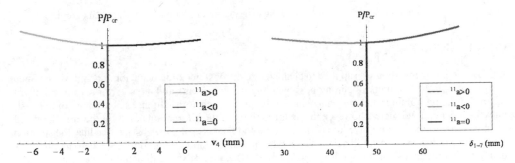

Figure 5. Post-buckling displacements (v_4: horizontal displacement of node 4; δ_{1-7}: relative vertical displacement between nodes 1 and 7).

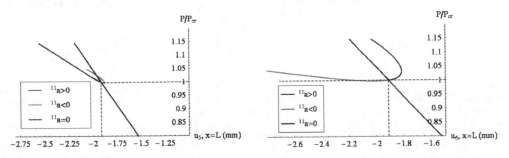

Figure 6. Post-buckling axial displacements for nodes 5 and 6.

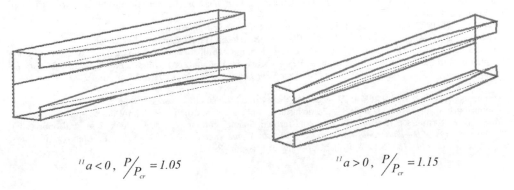

$^{II}a < 0$, $P/P_{cr} = 1.05$ $^{II}a > 0$, $P/P_{cr} = 1.15$

Figure 7. Post-buckling member's configuration.

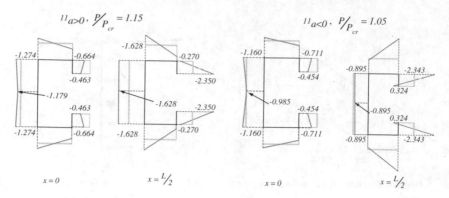

Figure 8. Longitudinal stresses for the critical state (light grey) and for post-buckling states (dark grey).

CONCLUSIONS

A GBT energy formulation was applied to characterize the post buckling behavior of a thin walled open section compressed column, obtaining post-buckling displacements and stresses for several points of the member, highlighting some relevant behavioral aspects, such as the slightly stable nature of the post-buckling behavior. It is noted, however, that in the vicinity of the bifurcation point, for the branch related to an increase of the compressive longitudinal stresses in the parts of the webs near the lips, the solution is unstable, followed by restabilization. The Rayleigh-Ritz method was used to discretize the problem, making resource to a set of orthonormal polynomials. Because of space limitations, the advantages of these orthonormal polynomials, in comparison to the commonly used trigonometric functions, were not fully explored. Nevertheless it is worth saying that these polynomials can model a wide range of boundary conditions, favor easy integrations and, for the common applications in practice, a small number of polynomials per mode leads to very precise results.

REFERENCES

Courant R. & Hilbert D. 1953. Methods of mathematical physics (Vol I). New York: Interscience Publishers.
Crisfield M. A. 1982. Numerical analysis of structures. In J. Rhodes & A. C. Walker (eds), Developments in thin-walled structures – 1: 235–284. London: Applied Science Publishers.
Hangai Y. & Kawamata S. 1972. Perturbation method in the analysis of geometrically nonlinear and stability problems. In J. T. Oden, R. W. Clough & Y. Yamamoto (eds), Advances in Computational Methods in Structural Mechanics and Design: 473–489. Huntsville: UAH Press.
Kwon Y. B. & Hancock G. J. 1993. Post-buckling analysis of thin-walled channel sections undergoing local and distortional buckling. Computers & Structures 49(3): 507–516.
Lau S. C. & Hancock G. J. 1987. Distortional buckling formulas for channel columns. Journal of Structural Engineering (ASCE) 133(5): 1063–1078.
Richards T. H. 1977. Energy methods in stress analysis. Chichester: Ellis Horwood.
Schardt R. 1989. Verallgemeinerte Technische Biegetheorie. Berlin Heidelberg: Springer Verlag.
Simão P. & Simões da Silva L. 2002. Comparative analysis of the stability of open and closed thin-walled cross section members in the framework of Generalized Beam Theory. In A. Lamas & L. Simões da Silva (eds), Eurosteel – Proceedings of the 3rd European Conference on Steel Structures, Coimbra 19–20 September 2002. Guimarães: CMM.
Thompson J. M. T. & Hunt G. W. 1973. A general theory of elastic stability. London: John Wiley & Sons
Thompson J. M. T. & Hunt G. W. 1984. Elastic instability phenomena. Chichester: John Wiley & Sons.
Young B. & Rasmussen K. 1998. Tests of cold-formed channel columns. In W. W. Yu & R. A. LaBoube (eds), Proceedings of the Fourteenth International Specialty Conference on Cold-Formed Steel Structures, St. Louis, Missouri, October 15–16 1998. Rolla: University of Missouri-Rolla.

Computational Methods in Engineering and Science, Iu et al. (eds)
© *2003 Swets & Zeitlinger, Lisse, ISBN 90 5809 567 3*

On the weighted residual method for rectangular flat plates

K.M. Lo, D.W. Xue & K.H. Lo
Department of Electromechanical Engineering, University of Macau, Macao

ABSTRACT: This paper uses the least-square and the Galerkin schemes of the weighted residual method to solve rectangular flat plates. This paper points out that if the assumed solution does not satisfy the boundary conditions, then the results may be erroneous. In addition, this paper shows that using more terms in the assumed solution might not give better results of the maximum deflection and this is opposite to the common belief. Suggestions for choosing the collocation points in the collocation scheme are also made in this paper.

1 INTRODUCTION

The Weighted Residual Method (WRM) is very useful for obtaining approximate solutions to complex problems. China is one of the countries that is at the forefront of this branch of computational mechanics.

The governing equation for flat plates is $\nabla^2\nabla^2 w = q/D$, where w is the deflection, q is the external load, and D is the flexural rigidity. To avoid solving the cumbersome high order partial differential equation, the WRM can be applied. Firstly, an approximate solution $W = \Sigma C_i W_i$ is assumed, where C_i and W_i are the *i-th* undetermined coefficient and function, respectively. Secondly, W is substituted into the governing equation, thereby obtaining the residual $R = \nabla^2\nabla^2 W - q/D$. Finally, the residual R is multiplied with a certain weight function and then minimised over the domain Ω in certain ways, i.e., $\int WRd\Omega = 0$ and the integration is calculated over Ω. Different residual-minimisation schemes give rise to different weight functions. Typical residual-minimisation schemes are the collocation ($W: \delta$ function), the least-square ($W: \partial R/\partial C_i$), and the Galerkin schemes ($W: W_i$).

The Ritz method in solid mechanics involves the assumption of solutions. The authors pointed out that when solving beams by using the Ritz method, both the displacement boundary conditions (displacement and angle of rotation) and the force boundary conditions (shear and bending moment) could be important, though theoretically the method requires the displacement boundary conditions only. As the WRM also involves the assumption of solutions, the effects of boundary conditions are bound to be strong.

In the literature, the criteria for choosing the approximate solution W in the WRM have not been talked on too much. This paper focuses on two aspects: 1. Boundary conditions, and 2. The number of terms in the assumed solution W. This paper points out that if W violates the boundary conditions, then the results can be completely wrong. In addition, it is commonly believed that the more terms W contains, the better the results of the maximum deflection would be. This paper shows that this is not always true. For the collocation method, suggestions for choosing the collocation points are made.

2 EFFECTS OF BOUNDARY CONDITIONS AND NUMBER OF TERMS IN W

The effects of boundary conditions and number of terms in W shall be investigated in this section.

Effects of boundary conditions

Boundary conditions typically can be categorised into two types. Lets use the simply supported side y = 0 as an illustration. The displacement boundary conditions and force boundary conditions are listed in the following order(deflection, angle of rotation, shear and bending moment).

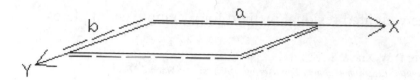

Figure 1. A simply supported rectangular plate.

$$w = 0, \frac{\partial w}{\partial y} \neq 0, \frac{\partial}{\partial y} \nabla^2 w + (1-\mu) \frac{\partial}{\partial x} (\frac{\partial^2 w}{\partial x \partial y}) \neq 0, \frac{\partial^2 w}{\partial y^2} = 0$$

Boundary conditions for other types of supports can be obtained in a similar fashion.

The simply supported plate in Figure 1. under a uniform distributed load was solved by using the least square scheme with the following assumed solutions:

$$W_1 = C \cdot Sin(\frac{\pi x}{a}) Sin(\frac{\pi y}{b})$$

$$W_2 = C[x(x-a)]^2 [y(y-b)]^2$$

W_1 satisfies all the boundary conditions, but W_2 violates the angle-of-rotation and bending moment boundary conditions. For instance, at the side of y = 0:

$$\partial W_2 / \partial y = 0 \text{ and } \partial^2 W_2 / \partial y^2 \neq 0$$

The maximum deflections obtained from the two assumed solutions are shown in Table 1. It can be seen that W_1 gives satisfactory results when the ratio of (b/a) is not big, whereas the results given by W_2 are unacceptable. From this example, it can be seen that the WRM, like other approximate methods, e.g., the Ritz method, depends strongly on whether the assumed solutions satisfy the boundary conditions or not. Similar conclusions can be drawn for plates of other support conditions.

2.1 Effects of the number of terms

In this section, the maximum deflections given by the different weighted residual schemes will be compared. Theoretically speaking, the approximate solution W and the external load q can be expanded into infinite series, then W can be obtained by comparing coefficients. In practice, only a finite number of terms in the infinite series are used. It is commonly believed that the more terms the truncated series contains, the better the maximum deflections will be. However, the results follow show that this is not always true.

Firstly, the plate in Figure 1 was solved again by using the least-square and the Galerkin schemes with following assumed solutions

$$W_3 = C_{11} Sin(\frac{\pi x}{a}) Sin(\frac{\pi y}{b}) + C_{13} Sin(\frac{\pi x}{a}) Sin(\frac{3\pi y}{b}) + C_{31} Sin(\frac{3\pi x}{a}) Sin(\frac{\pi y}{b}) + C_{33} Sin(\frac{3\pi x}{a}) Sin(\frac{3\pi y}{b})$$

$$W_4 = C_{11} Sin(\frac{\pi x}{a}) Sin(\frac{\pi y}{b}) + C_{33} Sin(\frac{3\pi x}{a}) Sin(\frac{3\pi y}{b})$$

$$W_5 = C_{11} Sin(\frac{\pi x}{a}) Sin(\frac{\pi y}{b}) + C_{33} Sin(\frac{3\pi x}{a}) Sin(\frac{3\pi y}{b}) + C_{55} Sin(\frac{5\pi x}{a}) Sin(\frac{5\pi y}{b}) + C_{77} Sin(\frac{7\pi x}{a}) Sin(\frac{7\pi y}{b})$$

The maximum deflections obtained from W_3, W_4, W_5 are shown in Table 2. From Table 1 and Table 2, it can be seen that in terms of maximum deflection, the ranking is: $W_3 > W_1 > W_4 > W_5$.

900

Table 1. Maximum deflections: $w = \alpha(qa^4/D)$ obtained from W_1 and W_2 by using the least-square scheme for a simply supported plate under a uniform distributed load.

b/a	Theoretical solution (α)	$W_1(\alpha)$	Error (%)	$W_2(\alpha)$	Error (%)
1	0.00406	0.00416	2.46	0.00110	−72.91
1.1	0.00485	0.00499	2.89	0.00133	−72.58
1.2	0.00564	0.00580	2.84	0.00156	−72.34
1.3	0.00638	0.00657	2.98	0.00178	−72.10
1.4	0.00705	0.00730	3.55	0.00199	−71.77
1.5	0.00772	0.00798	3.37	0.00219	−71.63
1.6	0.00830	0.00861	3.73	0.00237	−71.45
1.7	0.00883	0.00919	4.08	0.00252	−71.46
1.8	0.00931	0.00972	4.40	0.00266	−71.43
1.9	0.00974	0.01020	4.72	0.00277	−71.56
2	0.01013	0.01065	5.13	0.00287	−71.67
3	0.01223	0.01348	10.22	0.00329	−73.10
4	0.01282	0.01474	14.98	0.00338	−73.63
5	0.01297	0.01539	18.66	0.00340	−73.79
inf	0.01302	0.01664	27.80	0.00342	−73.73

Table 2. Maximum deflections: $w = \alpha(qa^4/D)$ obtained from W_3, W_4 and W_5 by using the least-square and the Galerkin schemes for a simply supported plate under a uniform distributed load (Both schemes give the same results for the same assumed solution).

b/a	Theoretical solution (α)	$W_3(\alpha)$	Error (%)	$W_4(\alpha)$	Error (%)	$W_5(\alpha)$	Error (%)
1	0.00406	0.00406	0.00	0.00417	2.71	0.00417	2.71
1.1	0.00485	0.00486	0.21	0.00500	3.09	0.00500	3.09
1.2	0.00564	0.00564	0.00	0.00580	2.84	0.00580	2.84
1.3	0.00638	0.00638	0.00	0.00658	3.13	0.00658	3.13
1.4	0.00705	0.00707	0.28	0.00731	3.69	0.00731	3.69
1.5	0.00772	0.00770	−0.26	0.00799	3.50	0.00799	3.50
1.6	0.00830	0.00828	−0.24	0.00852	2.65	0.00862	3.86
1.7	0.00883	0.00881	−0.23	0.00920	4.19	0.00920	4.19
1.8	0.00931	0.00928	−0.32	0.00973	4.51	0.00973	4.51
1.9	0.00974	0.00970	−0.41	0.01022	4.93	0.01022	4.93
2	0.01013	0.01008	−0.49	0.01067	5.33	0.01067	5.33
3	0.01223	0.01205	−1.47	0.01350	10.38	0.01350	10.38
4	0.01282	0.01242	−3.12	0.01476	15.13	0.01476	15.13
5	0.01297	0.01234	−4.86	0.01541	18.81	0.01541	18.81
inf	0.01302	0.01105	−15.13	0.01667	28.03	0.01667	28.03

In terms of maximum deflection, if only W_1 and W_3 are compared, then it can be seen that the latter is better. So, in this case, using more terms has given better results. However, it is obvious that W_1 gives better results of the maximum deflections than W_4 and W_5 which have more terms. Hence, in this case, using more terms has not given better results of the maximum deflections. As a result, it is wrong to say that the more terms one uses, the better the maximum deflections one gets.

If W_3 and W_5 are compared (both have got the same number of terms), it can be seen that the former is better. This can be explained as follows: both W_3 and W_5 are part of the Navier solution, which is an infinite series. W_3 contains the "low-order" terms of the Navier series, whereas terms of W_5 are of higher order than that of W_3. As it has been shown that using the first few low-order terms in the Navier series can give satisfactory results, it implies that these low-order terms are of more significance in the Navier solution. So, it is natural for W_3 to be better than W_5 in this case. However, this conclusion might not be true in other cases. This is because it has not been rigorously proved that terms in the front of an infinite series are of more significance than the terms in the rear. Also, without a theoretical solution as the reference, it is almost impossible to tell which terms are of more significance in an infinite series.

901

Table 3. Maximum deflections: $w = \alpha(qa^4/D)$ obtained from W_6 and W_7 by using the Galerkin scheme for a clamped plate under a uniform distributed load.

b/a	Theoretical solution (α)	$W_6(\alpha)$	Error (%)	$W_7(\alpha)$	Error (%)
1	0.00126	0.00128	1.59	0.00122	−3.17
1.1	0.00150	0.00153	2.00	0.00145	−3.33
1.2	0.00172	0.00176	2.33	0.00166	−3.49
1.3	0.00191	0.00196	2.62	0.00184	−3.66
1.4	0.00207	0.00214	3.38	0.00199	−3.86
1.5	0.00220	0.00229	4.09	0.00211	−4.09
1.6	0.00230	0.00242	5.22	0.00220	−4.35
1.7	0.00238	0.00253	6.30	0.00228	−4.20
1.8	0.00245	0.00263	7.35	0.00234	−4.49
1.9	0.00249	0.00271	8.84	0.00238	−4.42
2	0.00254	0.00278	9.45	0.00241	−5.12
inf	0.00260	0.00342	31.54	0.00205	−21.15

This conclusion on the number of terms and maximum deflections is also true for other support conditions. Suppose that the plate in Figure 1 is clamped on all sides and is acted on by a uniform distributed load. The problem was solved by using the Galerkin scheme with the following assumed solutions:

$$W_6 = C_{22}[1 - Cos(\frac{2\pi x}{a})][1 - Cos(\frac{2\pi y}{b})]$$

$$W_7 = C_{22}[1 - Cos(\frac{2\pi x}{a})][1 - Cos(\frac{2\pi y}{b})] + C_{24}[1 - Cos(\frac{2\pi x}{a})][1 - Cos(\frac{4\pi y}{b})] +$$

$$C_{42}[1 - Cos(\frac{4\pi x}{a})][1 - Cos(\frac{2\pi y}{b})]$$

Both expressions violate the shear condition, but the authors showed in another paper that the shear condition usually did not have a strong effect. Results of the maximum deflections obtained from these assumed solutions are shown in Table 3.

Even though W_7 has more terms than W_6, W_{7n} gives worse maximum deflections for (b/a) <1.5. However, for bigger values of (b/a), W_7 would give better maximum deflections. Again, this example shows that using more terms does not guarantee better results for the maximum deflections.

3 SUGGESTIONS FOR CHOOSING COLLOCATION POINTS IN THE COLLOCATION SCHEME

The simply supported plate under a uniform distributed load in Figure 1 was solved again with $\overline{W}_1$ by using the collocation scheme. The collocation points were chosen as follows: at the centre, at (0.5a, 0.25b), at (0.5a, 0.75b), at (0.25a, 0.5b), and at (0.75a, 0.5b).

In the literature, when the collocation scheme is used, the centre is almost always chosen. But usually, no explanations are given on why the centre should be used. In terms of maximum deflection, it was found in this example that if only one collocation point is used, then the four locations listed above would give exactly the same results, and they are better than the results given by the centre (the errors in Table 4 are still relatively big, but this is attributable to the fact that only one collocation point was used). This inferior results given by the centre can be explained as follows: in terms of boundary conditions, external load, the assumed solution, and geometry, the four locations (0.5a, 0.25b), (0.5a, 0.75b), (0.25a, 0.5b), and (0.75a, 0.5b) are "identical". Hence, choosing any one of them as the collocation point is equivalent to forcing the residual R to zero at ALL of the four locations. However, if only the centre is used, then the collocation scheme only forces the residual to zero at the centre, this is why the results given by the centre are worse than those given by the other four locations.

Table 4. Maximum deflections: $w = \alpha(qa^4/D)$ obtained from W_1 by using the collocation scheme at different locations for a simply supported plate under a uniform distributed load.

b/a	Theoretical solution (α)	W_1 (calculation done at centre) (α)	Error (%)	W_1 (calculation done at (0.5a, 0.25b) or (0.5a, 0.75b) or (0.25a, 0.5b) or (0.75a, 0.5b)) (α)	Error (%)
1	0.00406	0.00257	−36.70	0.00363	−10.59
1.1	0.00485	0.00308	−36.49	0.00435	−10.31
1.2	0.00564	0.00358	−36.52	0.00506	−10.28
1.3	0.00638	0.00405	−36.52	0.00573	−10.19
1.4	0.00705	0.00450	−36.17	0.00637	−9.65
1.5	0.00772	0.00492	−36.27	0.00696	−9.84
1.6	0.00830	0.00531	−36.02	0.00751	−9.52
1.7	0.00883	0.00567	−35.79	0.00801	−9.29
1.8	0.00931	0.00599	−35.66	0.00848	−8.92
1.9	0.00974	0.00630	−35.32	0.00890	−8.62
2	0.01013	0.00657	−35.14	0.00929	−8.29
3	0.01223	0.00832	−31.97	0.01176	−3.84
4	0.01282	0.00909	−29.10	0.01286	0.31
5	0.01297	0.00949	−26.83	0.01342	3.47
inf	0.01302	0.01027	−21.12	0.01452	11.52

So, in the collocation scheme, it is better to choose locations that exhibit the highest symmetry as the collocation points. In the above example, the centre is "singly symmetric", whereas any one of the other four locations is "four-fold symmetric" and better results will be obtained if these four-fold symmetric locations are used.

4 CONCLUSIONS

In the weighted residual method, the assumed solutions must satisfy as many of the boundary conditions as possible in order to get satisfactory results.

In terms of the maximum deflection, more terms might not give better results.

In the collocation scheme, it is better to choose locations that exhibit the highest symmetry as the collocation points.

REFERENCES

Xu, C. 1987. *The Weighted Residual Method in Solid Mechanics*, Tongji University (In Chinese).
Timoshenko, S.P., Woinowsky-Krieger, S. 1959. *Theory of Plates and Shells*[M], McGraw Hill College Div; 2nd edition.
Lo, K.H., Lo, K.M. On the Boundary Conditions of the Ritz Method, *Mechanics in Engineering* (Accepted for Publication) (In Chinese).

Computational Methods in Engineering and Science, Iu et al. (eds)
© 2003 Swets & Zeitlinger, Lisse, ISBN 90 5809 567 3

A unified failure criterion for concrete under multi-axial loads

P.E.C. Seow & S. Swaddiwudhipong
National University of Singapore, Singapore

ABSTRACT: A survey of existing literature shows that researchers have used different sets of failure criteria to predict the stress at failure for concrete under biaxial, triaxial and axi-symmetrical loads. Furthermore, attempts at developing a failure surface for concrete have generally been for normal strength concrete only. Therefore a new and unified 5-parameter failure surface is developed and presented in this paper. It is suitable for use with normal strength, high strength and steel fiber-reinforced concrete ranging from 20 MPa to 165 MPa. A method to modify the failure surface for plain concrete to take into account the effect of fibers in steel fiber-reinforced concrete is also presented. Comparison with experimental results shows that the proposed failure surface is suitable for the above mentioned concrete types under any load combination.

1 INTRODUCTION

Advances in concrete technology have resulted in the production of High Strength Concrete (HSC), as well as Steel Fiber-Reinforced Concrete (SFRC), ranging in strength from 40 MPa to 165 MPa, and possibly even greater. While the usage of high-strength concrete increases a structure's load-bearing capacity, the introduction of steel fibers increases the tensile strength and ductility of plain concrete. In addition, concrete-steel composites are increasingly used in the construction industry. By confining concrete with steel tubes, a structure's ductility and ability to withstand cyclic or earthquake loads is increased. As some concrete structures such as dams or bridge piers may be subjected to complex states of stress, it is imperative to predict the state of stress at failure accurately to enable a realistic Finite Element Analysis and optimum design of such structures. A survey of existing literature shows that researchers have used different sets of failure criteria to predict the stress at failure for concrete under biaxial, axi-symmetrical and triaxial loads. Kupfer & Gerstle (1973) proposed a series of curves for concrete under biaxial loads, while others have used 2-parameter equations for concrete under axi-symmetrical loads (Xie et al. 1995, Setunge et al. 1993). These equations can only be used for a limited number of load cases, and not for concrete under triaxial loads, where the applied load is different in all three directions. Therefore some researchers adopt a 5-parameter failure surface in the 3-dimensional Haigh-Westergaard stress space proposed by Willam & Warnke (1975) for predicting the state of stress in concrete under triaxial loads (Chern et al. 1992).

As research in HSC or FRC under multi-axial loads is relatively new, attempts at developing a failure surface for concrete have generally focused on normal strength concrete (NC). Therefore, the failure surface proposed by Willam & Warnke (1974) for NC is commonly adopted by researchers for use with HSC as well as FRC. However, it has been observed that this failure surface cannot be used to predict the stress at failure when concrete is subjected to high triaxial compressive loads (Guo 1997). In addition, the Willam and Warnke surface is unable to take into account the effects of steel fibers in FRC. The surface must also first be defined using experimental points of concrete under multi-axial loads before it can be used for analysis, thus researchers without experimental data for concrete under multi-axial loads will have difficulty in defining a suitable failure surface corresponding to the type of concrete being analysed. In view of the above mentioned difficulties, a new and unified failure surface is developed and presented in this paper for concrete under any combination of multi-axial loads. A method for modifying the failure surface for plain concrete to make it suitable for predicting the ultimate stress in FRC is also proposed.

This new surface is suitable for the analysis of normal strength, high strength and steel fiber-reinforced concrete ranging from 20 MPa to 165 MPa.

2 PROPOSED FAILURE SURFACE

2.1 Plain concrete

A proposed 5-parameter failure surface is depicted in Figure 1. It satisfies all the requirements of being smooth, convex, and having the ratio of the tensile to compressive meridians close to fi near the apex, and approach 1 asymptotically as the octahedral stress tends to infinity. The proposed failure surface is defined by a tensile meridian, ρ_t, where the angle of similarity, $\theta = 0°$, and a compressive meridian, ρ_c where $\theta = 60°$. An elliptical curve, $\rho(\xi, \theta)$ is used to interpolate and determine the stress at failure for any state of stress with angles of similarity between $0°$ and $60°$ and lying between ρ_t and ρ_c. Due to symmetry of the surface, these three equations are sufficient to define the entire failure surface:

$$\frac{\xi}{f_c} = a_2\left(\frac{k\rho_t}{f_c}\right)^2 + a_1\left(\frac{k\rho_t}{f_c}\right) + a_0 \qquad , \quad k \leq 1 \tag{1}$$

$$\frac{\xi}{f_c} = b_2\left(\frac{\rho_c}{f_c}\right)^2 + b_1\left(\frac{\rho_c}{f_c}\right) + b_0 \tag{2}$$

$$\rho(\xi, \theta) = \frac{2\rho_c(\rho_c^2 - \rho_t^2)\cos\theta + \rho_c(2\rho_t - \rho_c)[4(\rho_c^2 - \rho_t^2)\cos^2\theta + 5\rho_t^2 - 4\rho_t\rho_c]^{1/2}}{4(\rho_c^2 - \rho_t^2)\cos^2\theta + (\rho_c - 2\rho_t)^2} \tag{3}$$

where

$$\cos\theta = \left[\frac{3(\sigma_3 - \sigma_m)}{\sqrt{6}\sqrt{\sigma_1^2 + \sigma_2^2 + \sigma_3^2 - 3\sigma_m^2}}\right] \qquad \text{For } \sigma_3 \geq \sigma_2 \geq \sigma_1 \tag{4}$$

σ_i ($i = 1$ to 3) is the principal normal stress in the ith direction, σ_m is the mean stress, $\xi = \sqrt{3}\,\sigma_{oct}$, $\rho = \sqrt{3}\,\tau_{oct}$, ($\sigma_{oct}$ and τ_{oct} are the octahedral normal and shear stress respectively), f_c is the uniaxial compressive strength of concrete and $a_2, a_1, a_0, b_2, b_1, b_0$ and k are constants to be determined. The coefficients, a_0 and b_0, are the points of intersection of the tensile and compressive meridians with the hydrostatic axis. At this points, concrete is subjected to equal, triaxial tension (f_{ttt}), which is taken to be equal to f_t, the uniaxial tensile strength of concrete (Guo 1997). This results in a closer fit to the rest of the experimental data, thus a_0 and b_0 are determined by:

$$a_0 = b_0 = \frac{\sqrt{3}f_{ttt}}{f_c} = 0.1732 \tag{5}$$

Regression analysis was carried out on 296 experimental data points of cubes and cylinders under multiaxial stresses for both HSC and NC failing on the tensile and compressive meridian. Based on the experimental results shown in Figure 2, as reported by Cadappa et al. (2001) & (1999), Liu & Foster (2000), Taliercio et al.(1999), Guo (1997), Taliercio & Gobbi (1997), Imran & Pantazopoulou (1996), Xie et al. (1995), Setunge et al. (1993), Chern et al. (1992), Lahlou et al. (1992), Jiang et al.(1991), Kotsovos (1979), Hobbs (1970), and Mills & Zimmerman (1970), the following values of the remaining coefficients in Equations 1 and 2 are determined to be:

$a_2 = -0.1597$, $\qquad\qquad a_1 = -1.455$

$\qquad\qquad\qquad\qquad\qquad\qquad\qquad\qquad\qquad\qquad\qquad\qquad$ (6)

$b_2 = -0.1746$, $\qquad\qquad b_1 = -0.778$

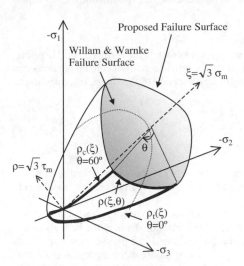

Figure 1. Proposed unified failure surface.

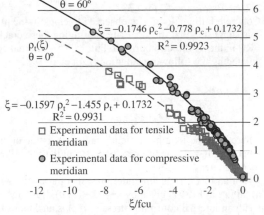

Figure 2. Regression of tensile and compressive meridian to experimental results.

The value of k is taken to be equal to 1 for plain concrete, while the method of determining k for fiber-reinforced concrete is presented in the next sub-section.

2.2 Fiber-reinforced concrete

It has been suggested that the addition of steel fibers provide an additional confining effect to the concrete as the fibers delay the onset of tensile cracks by restraining the propagation of micro-cracks in concrete (Yin et al. 1992, Traina & Mansour 1991). The benefits of adding fiber to concrete are more obvious when concrete fails in tension than in compression (Chern et al. 1992), thus the addition of fibers would influence the tensile meridian, ρ_t, more than the compression meridian, ρ_c. A constant, k, is therefore introduced in Equation (1) to account for the increase in the observed stress at failure along ρ_t, due to the confining effect of different amount and types of fibers, and to rotate the tensile meridian accordingly. As the failure surface is determined by interpolating between ρ_t and ρ_c, an increase in ρ_t will also result in a change in the shape of the failure surface, thus accounting for the effect of fiber in FRCs subjected to other load ratios.

2.2.1 Determining the value of k

The constant, k, in Equation 1 is determined by fitting the tensile meridian of FRC, $\rho_{t(FRC)}$, through a known control point where concrete is subjected to biaxial compression with a load ratio of 1:1. The observed state of stress at failure in FRC at this point is $(\sigma_1, \sigma_2, \sigma_3) = (-f_{cc}, -f_{cc}, 0)$, where f_{cc} is the biaxial compressive strength of FRC. If the experimental value of f_{cc} is known for a type of FRC, then $\rho_{t(FRC)}$ and $\xi_{(FRC)}$ may be calculated for this particular state of stress of $(-f_{cc}, -f_{cc}, 0)$. By substituting the values of $\rho_{t(FRC)}$ and $\xi_{(FRC)}$ into Equation 1, k may be determined as follows:

$$k = \frac{-a_1 - \sqrt{a_1^2 - 4a_2\left[a_0 + \frac{2}{\sqrt{3}}\left(\frac{f_{cc}}{f_c}\right)\right]}}{a_2\sqrt{\frac{8}{3}\left(\frac{f_{cc}}{f_c}\right)}} \tag{7}$$

where the coefficients a_2, a_1, a_0, are given in Equations 5 and 6. Due to constraints in testing equipment, it is understandable that the experimental value of f_{cc} may not be readily available to some researchers or engineers. Therefore, a method to predict f_{cc} for FRC in the absence of experimental values is presented in the sub-section that follows.

2.2.2 Predicting f_{cc} for FRC

Concrete failing along ρ_t is observed to crack in the plane perpendicular to the σ_3 direction. Therefore, it is suggested that the steel fibers bridging these cracks actually provide internal confining stresses in FRC in the σ_3 direction prior to failure. This internal confining effect due to the fibers, σ_{tu}, may be calculated by the following equation (Lim et al. 1987):

$$\sigma_{tu} = \frac{\eta_l \eta_o V_f l_f \tau_u}{2r'} \tag{8}$$

where η_l is the length efficiency factor, η_o is the orientation factor, V_f is the volume fraction, l_f is the fiber length, τ_u is the ultimate bond strength of the fiber, and r' is the ratio of fiber cross-section area to its perimeter.

For an FRC specimen under biaxial compression with a load ratio of 1:1, the observed state of stress at failure in FRC is given by $(-f_{cc}, -f_{cc}, 0)$. It is suggested that this state of stress in the FRC specimen (with an internal confining stress of σ_{tu}) is analogous to the state of stress in a plain concrete specimen subjected to an external confining pressure of σ_{tu}. The state of stress at failure in the analogous plain concrete specimen is therefore given by $(-f_{cc}, -f_{cc}, -\sigma_{tu})$. Thus, determining the biaxial compressive strength of FRC is analogous to determining the value of stress at failure in the σ_1 and σ_2 directions in plain concrete when an external confining pressure of $-\sigma_{tu}$ is applied. The values of $\rho_{t(Plain)}$ and $\xi_{(Plain)}$ when the state of stress is $(-f_{cc}, -f_{cc}, -\sigma_{tu})$ may thus be expressed in Equations 7 and 8 as follows:

$$\rho_{t(Plain)} = \sqrt{\frac{2}{3}} (f_{cc} - \sigma_{tu}) \tag{9}$$

$$\xi_{(Plain)} = \frac{-2f_{cc} - \sigma_{tu}}{\sqrt{3}} \tag{10}$$

By substituting Equations 9 and 10 into Equation 1 when $k = 1$, the analytical value of f_{cc} may be determined through an iterative process. This analytical value of f_{cc} can then be used to evaluate k for FRC in Equation 5.

3 EXPERIMENTAL VERIFICATION

3.1 Plain concrete

To verify the accuracy of the elliptical interpolation in Equation 3 for the stress at failure lying between $\theta = 0°$ and $60°$, 80 other triaxial test results of Guo (1997) and Mills & Zimmerman (1970), which were not used in the determination of ρ_t and ρ_c, were compared with the proposed failure surface. The comparison of experimental and predicted values of ρ along the deviatoric plane in Figure 3, shows that the proposed model provides a close and conservative estimate of the experimental values.

Figure 4 depicts a biaxial failure curve which was degenerated from the proposed failure surface. Comparison with another 54 experimental data points of NC and HSC plates under biaxial loads by Hussein & Marzouk, (2000), and Kupfer & Gerstle (1973) shows that the proposed failure surface can also reasonably predict the stress at failure for concrete under biaxial loads.

36 experimental points of NC and HSC confined by steel tubes by Hong et al. (2001), O'Shea & Bridge (2000), and Lahlou et al. (1992) were selected and plotted against the proposed failure surface in Figure 5. These experiments were used as care was duly taken to ensure that only the concrete was loaded during the testing and the steel tube contribution was only for confinement purposes. The range of unconfined uniaxial cylinder strength of the specimens was from 38 MPa to 113 MPa. Apart from a small scatter due to slight experimental differences, it can be seen that the proposed failure surface is able to predict the ultimate stress of confined concrete under axi-symmetrical loading relatively well.

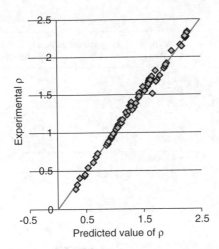

Figure 3. Comparison of experimental values of ρ.

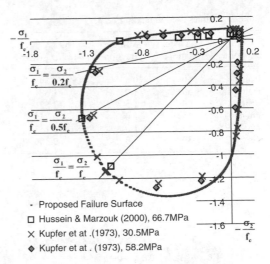

Figure 4. Biaxial curve for normal and high strength concrete.

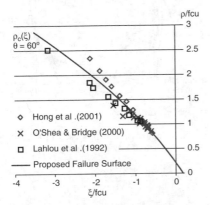

Figure 5. Comparison of confined concrete test results with ρ_c.

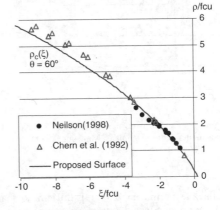

Figure 6. Comparison of FRC results with ρ_c.

Table 1. Details of experiments on FRC under multi-axial loads.

Researcher	Loading condition	Specimen type	Fiber type	Aspect ratio, l/d	Volume fraction, $Vf\,(\%)$
Chern et al. (1992)	Triaxial	Cylinder	Hooked	44	1.0 & 2.0
Neilsen (1998)	Triaxial	Cylinder	Straight	30 & 40	6.0
Traina & Mansour (1991)	Biaxial	Cube	Hooked	60	0.5, 1.0 & 1.5
Yin et al. (1989)	Biaxial	Plate	Straight	45 & 59	1.0 & 2.0

3.2 *Fiber-reinforced concrete*

A total of 79 experimental data points from 4 different sources were used in the verification of the proposed failure surface for FRC. A summary of the salient variables in each series of experiments is given in Table 1.

The experimental results obtained for FRC under triaxial loads failing along the compression meridian, ρ_c, are plotted and compared with the proposed ρ_c for plain concrete in Figure 6. Based on the limited

909

Table 2. Values used in rotating ρ_t.

Experiment	Aspect ratio, l/d	Volume fraction, Vf (%)	σ_{tu} (MPa)	k	Analytical f_{cc} (MPa)	Experiment f_{cc} (MPa)	Analytical/ experiment
Traina 0.5	60	0.5	0.84	0.978	54.2	60.0	0.90
Traina 1.0	60	1.0	1.68	0.961	59.3	62.8	0.94
Traina 1.5	60	1.5	2.52	0.951	74.9	74.3	1.01
Yin 45–1	45	1.0	0.756	0.979	49.1	57.5	0.85
Yin 59–1	59	1.0	1.153	0.972	56.3	61.4	0.92
Yin 59-2	59	2.0	2.306	0.947	60.4	59.9	1.01

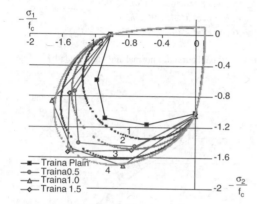

Figure 7. Comparison of proposed FRC curve with experimental data of Traina & Mansour (1991).

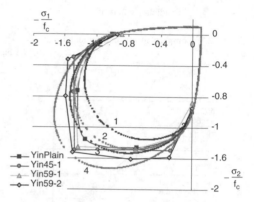

Figure 8. Comparison of proposed FRC curve with experimental data of Yin et al. (1989).

available experimental data, it is difficult to quantify the effects of adding different types and quantity of fiber on ρ_c. Since the proposed compression meridian is able to fit the experimental data relatively well, it is postulated that the same equation for ρ_c may be used for both plain concrete as well as FRC in this study.

The values of σ_{tu}, analytical f_{cc} and k used in rotating ρ_t for various mixes of FRC under biaxial loads are presented in Table 2.

A comparison with the experimental values of ρ_t shows that most of the predicted biaxial compressive strength of FRC are conservative. It can also be seen that the analytical f_{cc}, and consequently, ρ_t, is sensitive to changes in both the volume fraction as well as the aspect ratio of fiber used. As σ_{tu} increases, so does ρ_t. This is reflective of the observed experimental trend.

As the biaxial failure curves are obtained from the intersection of the 3-dimensional failure surface with the σ_1–σ_2 plane, a small increase in ρ_t results in a visible elongation of the biaxial curve in the direction where the load ratio of σ_1:σ_2 is 1:1. Figures 7 and 8 show a comparison of the experimental biaxial curves obtained by Traina & Mansour (1991) and Yin et al. (1989), respectively, with the analytical biaxial curves which were obtained using the values in Table 2. The analytical curves labeled 1 to 4 in Figures 7 and 8 correspond to the increasing values of σ_{tu} in Table 2, with curve 1 corresponding to plain concrete and curve 4, to the experiment "Traina 1.5" in Figure 7 and "Yin 59-2" in Figure 8. It can be seen that the proposed biaxial curves for FRC are able to elongate as σ_{tu} increases, and they are able to capture the trend observed in the experimental curves reasonably well. This shows that the proposed failure surface for plain concrete can be modified using the method described in this paper to account for the effects of adding fiber to the concrete.

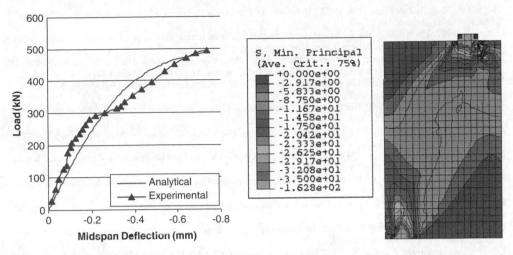

Figure 9.   Load–deflection curve.

Figure 10.   Contour plot showing stress distribution in beam.

3.3   *Finite element analysis*

The proposed failure surface was also implemented in a constitutive model for concrete proposed by Tho et al. (submitted for review), and experiments of cubes under triaxial loads, as well deep beams by Kong et al. (1970), were modeled using finite element analysis. The load–deflection curve for a deep beam is shown in Figure 9, while the corresponding contour plot showing the stress distribution in the deep beam is presented in Figure 10. The analytical results are able to approximate the experimental behaviour reasonably well in terms of the deformation under load and ultimate stress at failure. More results of the finite element analysis on other concrete specimens will be presented during the conference.

4   CONCLUSION

A new unified 5-parameter failure surface has been proposed for use with normal strength, high strength and fiber-reinforced concrete, with strengths ranging from 20 MPa to 165 MPa. The associated coefficients are established through regression analysis of 296 experimental data points for NC and HSC and verified against 170 additional experimental data. A method for modifying the failure surface to account for the effect of fiber has been proposed. The resulting failure surface for FRC has been compared with 79 more data points obtained from tests on FRC under multi-axial loads, and the proposed failure surface is able to reflect the trend observed experimentally. This failure criterion has also been successfully implemented into a constitutive model for concrete and used in the finite element analysis of concrete under multi-axial loads. A comparison of analytical and experimental results show that the proposed unified failure surface is applicable to confined and unconfined NC, HSC and FRC subjected to biaxial, triaxial or axi-symmetrical loads. The use of this failure surface in finite element analysis will enable engineers to model and predict the behaviour of concrete under complex states of stress, thus aiding them in the analysis and design of such structures.

REFERENCES

Cadappa, D.C., Sanjayan, J.G. & Setunge, S. 2001. Complete Triaxial Stress–Strain Curves of High-Strength Concrete. Journal of Materials in Civil Engineering, ASCE 13: 209–215.

Cadappa, D.C., Setunge, S. & Sanjayan, J.G. 1999. Stress Versus Strain Relationship of High-Strength Concrete under High Lateral Confinement. Cement and Concrete Research 29: 1977–1982.

Chern, J.C., Yang, H.J. & Chen, H.W. 1992. "Behaviour of Steel Fiber Reinforced Concrete in Multiaxial Loading", ACI Materials Journal 89: 32–40.

Guo, Z.H. 1997. The Strength and Deformation of Concrete – Experimental Results and Constitutive Relationship. Tsinghua University Press, Beijing: China (In Chinese).

Hobbs, D.W. 1970. Strength and Deformation Properties of Plain Concrete Subjected to Combined Stress-Part 1: Strength Results Obtained on One Concrete. Cement and Concrete Association Technical Report 42.451: 1–12.

Hong, M., Kiousis, P.D., Ehsani, M.R. & Saadatmanesh, H. 2001. "Confinement Effects on High-Strength Concrete", ACI Structural Journal 98: 548–553.

Hussein, A. & Marzouk, H. 2000. Behavior of High-Strength Concrete under Biaxial Stresses. ACI Materials Journal 97: 27–36.

Imran, I. & Pantazopoulou, S.J. 1996. Experimental Study of Plain Concrete under Triaxial Stress. ACI Materials Journal 93: 589–601.

Jiang, L., Huang, D. & Xie, N. 1991. Behaviour of Concrete under Triaxial Compressive–Compressive–Tensile Stresses. ACI Materials Journal 88: 181–185.

Kong, F.K., Robins, P.J. & Cole, D.F. 1970. Web reinforcement effects on deep beams. American Concrete Institute, Proceedings 67(73): 1010–1017.

Kupfer, H.B. & Gerstle, K.H. 1973. Behaviour of Concrete under Biaxial Stresses. Journal of the Engineering Mechanics Division, ASCE 99: 853–866.

Lahlou, K., Aictin, P.C. & Chaallal, O. 1993. Behaviour of High-strength Concrete under Confined Stresses. Cement and Concrete Composites 14: 185–193.

Lim, T.Y., Paramasivam, P. & Lee, S.L. 1987. Analytical Model for Tensile Behaviour of Steel–Fiber Concrete. ACI Materials Journal 84: 524–536.

Liu, J. & Foster, S.J. 2000. A 3-Dimensional Finite Element Model for Confined Concrete Structures. Computers & Structures 77: 441–451.

Mills, L.L. & Zimmerman, R.M. 1970. Compressive Strength of Plain Concrete under Multiaxial Loading Conditions. ACI Journal 67: 802–807.

Nielsen, C.V. 1998. Triaxial Behaviour of High-Strength Concrete and Motar. ACI Materials Journal 95:144–151.

O'Shea, M.D. & Bridge, R.Q. 2000. Design of Circular Thin-Walled Concrete Filled Steel Tubes. Journal of Structural Engineering, ASCE 126: 1295–1303.

Setunge, S., Attard, M.M. & Darvall, P.L. 1993. Ultimate Strength of Confined Very High-Strength Concretes. ACI Structural Journal 90: 632–641.

Taliercio, A.L.F., Berra, M. & Pandolfi, A. 1999. Effect of High-Intensity Sustained Triaxial Stresses on the Mechanical Properties of Plain Concrete. Magazine of Concrete Research 51: 437–447.

Taliercio, A.L.F & Gobbi, E. 1997. Effect of Elevated Triaxial Cyclic and Constant Loads on the Mechanical Properties of Plain Concrete. Magazine of Concrete Research 49: 353–365.

Tho, K.K, Seow, P.E.C. & Swaddiwudhipong, S. (in prep.), Numerical Method for Analysis of Concrete Under Multi-Axial Loads. Magazine of Concrete Research.

Traina, L.A. & Mansour, S.A. 1991. Biaxial Strength and Deformational Behaviour of Plain and Steel Fiber Concrete. ACI Materials Journal 88:354–362.

William, K.J. & Warnke, E.P. 1974. Constitutive Model for the Triaxial Behaviour of Concrete. Proceedings of the IABSE Seminar on Concrete Structures Subjected to Triaxial Stresses 19. Bergamo: Italy.

Xie, J., Elwi, A.E. & MacGregor, J.G. 1995. Mechanical Properties of Three High-Strength Concretes Containing Silica Fume. ACI Materials Journal 92: 135–145.

Yin, W.S., Su, E.C.M., Mansur, M.A. & Hsu, T.T.C. 1989. Biaxial Tests of Plain and Fiber Concrete. ACI Materials Journal 86: 236–243.

Computational Methods in Engineering and Science, Iu et al. (eds)
© *2003 Swets & Zeitlinger, Lisse, ISBN 90 5809 567 3*

Axial capacity prediction for driven piles at Macao using artificial neural network

W.F. Che
Civil Engineering Laboratory of Macau, Macao

T.M.H. Lok & S.C. Tam
Faculty of Science and Technology, University of Macau, Macao

H. Novais-Ferreira
National Laboratory of Civil Engineering, Lisbon, Portugal

ABSTRACT: A back-propagation neural network model is proposed for estimation of pile bearing capacity from dynamic stress wave data together with the properties of the driven pile. The bearing capacity predicted by TNOWAVE was employed as the desired output in training. Three network models were used to predict the total, shaft and toe resistance of the pile, respectively. The study showed that the neural network model predicted total bearing capacity reasonably well compared with TNOWAVE solution, and the prediction for shaft resistance and toe resistance were also in acceptable range. The neural network provides a method for prediction of pile bearing capacity, which is reliable but less dependent on experienced personnel than the signal matching method.

1 INTRODUCTION

Driven piles are the common foundation systems for high-rise buildings and crucial infrastructures in Macao. Prediction of axial pile capacity is the most important task in the execution control of pile foundations. There are many methods for determination of pile capacity using static and dynamic load tests. Due to the high cost of static load tests, dynamic load tests are employed more frequently for the estimation of pile capacity. Results from dynamic load test can be analyzed using pile driving formulas, wave equation methods, and signal matching with pile driving analyzer, in the order of increasing complexity. However, study showed that these methods generally require some empirical parameters such as the hammer efficiency or damping factor to be provided by trained and experienced personnel. Therefore, the degree of uncertainty is still very high for the application of dynamic load tests to the prediction of axial pile capacity.

This study focused on the application of the artificial neural networks to the analysis of dynamic pile tests and the prediction of axial capacity of driven piles at Macao. Chan et al. (1995), Chow et al. (1995a, 1995b), Goh (1996), and Teh et al. (1997) had demonstrated that the bearing capacity of piles could be predicted by a trained neural network with measured stress-wave data as the input parameters. In the present study, a training set is selected from the database of 81 PHC driven piles available from more than 50 construction projects at Macao. The measured stress wave data, the pile head displacement and the properties of the pile were used as the input data while the bearing capacity predicted by TNOWAVE (TNO 1998) was the target value.

2 ARTIFICIAL NEURAL NETWORK AND BACK-PROPAGATION ALGORITHM

Artificial neural networks (ANN) are computer models, the objective of which is to mimic the knowledge acquisition skill of human brain (Haykin 1994). A feedforward neural network consists of an input

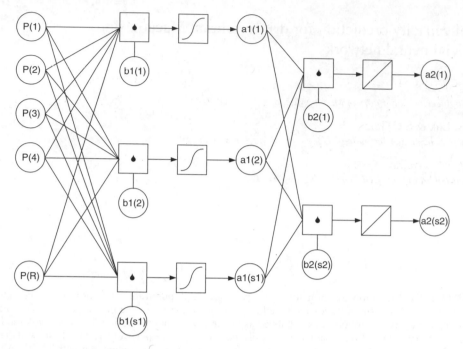

Figure 1.   Neural network structure.

layer, an output layer and one or more hidden layers as shown in Figure 1. Each layer comprises of several processing units, which is connected to all the neurons in next layer. The neurons in each layer interact with neurons in other layers through weighted connections.

Before a neural network can be used to handle a certain problem, it must be trained by representative input and output patterns. At the beginning of the training, the connecting weights are assigned with default or random values. The input data are propagated through the network by multiplying the values of neurons in the input layer by the connecting weights. These summed products, the net inputs of neurons, are calculated using a transfer function to determine the output of each neuron. Normally, the S-shaped sigmoid differentiable function and linear function could be selected as transfer function.

The network output value is compared with the target output of the training data set. The square of the relative error represents the errors. In the back-propagation algorithms, the connecting weights of the neural network are then modified in response to these errors using a gradient descent strategy by minimizing the overall mean error of all the output units. Training is carried out iteratively until the mean square error over the entire training pattern reaches a pre-defined acceptable level.

A separate set of data is used to validate the trained neural network. Once the trained neural network is tested to be acceptable, no more adjustment of the connecting weights is necessary.

The neural network used in this study was developed on the Matlab platform with the neural network toolbox installed (Demuth & Beale 1998). The present work is focused on using back propagation neural networks. For higher training performance, the Levenberg-Marquardt optimization algorithm (Martin et al. 1994) would be employed, and the corresponding Matlab Neural Network toolbox function is available for implementation.

3   MODEL FORMULATION OF THE PROBLEM

Assuming that a unique correlation exists between the measured stress wave data and the soil resistance during a hammer impact, a general functional can be proposed to relate the two sets of parameters as follows:

$$R_s = \Gamma\left[F_{\max}, F_{\min}, Z \cdot v_{\max}, Z \cdot v_{\min}, Disp_{\max}, d, T, L\right] \tag{1}$$

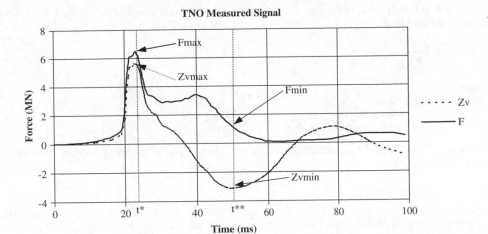

Figure 2. Typical measured force and velocity history.

where:

Γ: A general functional relates the pile driving data to the soil resistance, which is approximated by a neural network in this study.

R_s: Soil resistance (shaft, toe, or total);

F_{max}:   Measured force at time t*;

F_{min}:   Measured force at time t**;

$Z \cdot v_{max}$:   Measured maximum velocity multiplied by the pile impedance at time equal t*;

$Z \cdot v_{min}$:   Measured minimum velocity multiplied by the pile impedance at time equal to t**;

t*:   Time at maximum velocity; (see Fig. 2)

t**:   Time at minimum velocity within [t*,t* + 2L/c]; (see Fig. 2)

Z: pile impedance (Z = EA/c);

A: Cross section Area of the pile;

E: Young's modulus of pile;

c: Stress wave velocity;

$Disp_{max}$: Measured maximum displacement;

d: Pile diameter;

T: Pile wall thickness;

L: Length of pile;

In this study, the amount of input parameters selected for the ANN model is comparable to those required by the CASE method. In fact, the input parameters for the ANN were chosen with the CASE method (Rausche et al.) as a guideline.

4 DETAILS OF ARTIFICIAL NEURAL NETWORK

4.1 Network structure

A feedforward neural network with one hidden layer was selected in this study for the estimation of axial pile capacity. The number of neurons in the input, hidden and output layer were R = 8, S1 = 25 and S2 = 1, respectively, as shown in Figure 1. The eight process units of the input layer were the input parameters as indicated in equation (1). The output layer had only one unit corresponding to the capacity (shaft, toe or total capacity), of which target values were taken from the TNOWAVE calculations.

4.2 Normalization of input and output data

As a conventional practice, normalization of the data to values to lie within a certain range is generally performed to facilitate training. In this study, the measured force and velocity data were normalized into

915

the interval [0.1, 0.9] to avoid the slow rate of learning near the end points of the range. Hence, for a variable with maximum value of V_{amax}, each value, V_a, was normalized to the value, $\bar{V}_a$, using:

$$\bar{V}_a = 0.1 + 0.8 \frac{V_a}{V_{a\,max}}$$

(2)

where:

V_a: Original value
V_{amax}:   Corresponding maximum value
$\bar{V}_a$:   Corresponding normalized value

All values of F_{max}, F_{min}, $Z \cdot v_{max}$ and $Z \cdot v_{min}$ will be considered as an independent variable and normalized by equation (2).

4.3   Database for the neural network

A database of 101 dynamic pile tests with signal matching using TNOWAVE was used in this study. The tests were performed on PHC piles located in more than 50 different sites in various soil formations in Macao. All the tests were carried out on precast circular PHC piles. The piles in the database were divided into two groups. Eighty-one of the dynamic pile tests were randomly selected to form the training set. The remaining twenty tests were used as the testing set to validate the network.

4.4   Determination of the number of neurons

Hornik et al. (1989) postulated that a feedforward neural network with a single hidden layer of sufficient number of neurons is capable of approximating any continuous function. However, there was not any guideline existed for determining the appropriate size of the hidden layer. In this study, the size of the hidden layer was determined through a process of trial and error. The mean square error of the test set was adopted for determining the adequate number of neurons in the hidden layer. Various numbers of neurons in the hidden layer, ranging from 20 to 30, had been tested in the study. It was found that 25 neurons in the hidden layer were adequate to achieve the best performance.

4.5   Training of the neural network

Three similar neural networks, ANN_M01, ANN_M02, and ANN_M03, of the structure described previously were set up to predict the total capacity, the shaft resistance, and the toe resistance, respectively. The corresponding calculated values of total capacity, shaft resistance and toe resistance from TNOWAVE were used as the target values. For the 81 piles in the training set, all three network models were able to learn all the patterns presented with an Mean Square Error (MSE) (Demuth and Beale, 1998) of less than 1e-20.

4.6   Validating of the neural network

To verify that the trained network can generalize the association between stress-wave data and pile capacity, the 20 piles in the testing set TS1 were used. The relative Root Mean Square percentage error (EN) (Chan et al. 1995) was calculated for the predictions of each neural network using the following expression:

$$EN = \sqrt{\frac{\sum_{n=1}^{N} \left(\frac{R_N - R_D}{R_D} \right)^2}{N}}$$

(3)

where R_N and R_D represent the network output and the desired output, respectively, N is the number of examples used.

Figure 3 presents the comparison of total pile capacity of the piles in the testing set TS1 predicted by trained ANN_M01 and TNOWAVE. It is found that good and reasonable predictions were obtained using ANN_M01. The Root Mean Square percentage error (EN) is only about 0.068 for the total bearing capacity of the piles in the testing set TS1. The maximum error is about 13% with most of them underestimate.

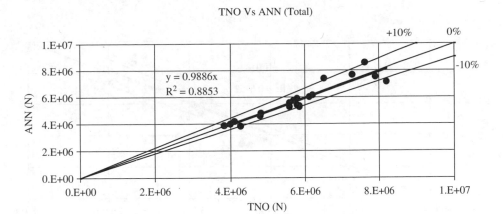

Figure 3. Comparison of pile total capacity predicted by ANN_M01 and TNOWAVE.

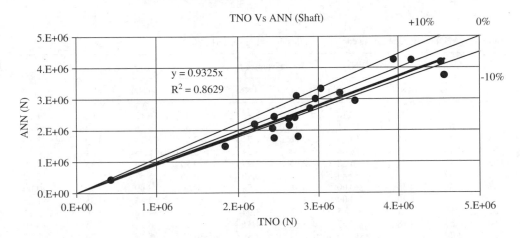

Figure 4. Comparison of pile shaft resistance predicted by ANN_M02 and TNOWAVE.

Figure 4 presents the comparison of total shaft resistance of the piles in the testing set TS1 predicted by trained ANN_M02 and TNOWAVE. It is found that reasonable predictions were obtained using ANN_02. The Root Mean Square percentage error (EN) is 0.14 for the total shaft resistance prediction of the piles in the testing set TS1 and the maximum error is about 34%.

Figure 5 presents the comparison of toe resistance of the piles in the testing set TS1 predicted by trained ANN_M03 and TNOWAVE. It is found that reasonable predictions were obtained using ANN_M03. The Root Mean Square percentage error (EN) is 0.18 for the toe resistance prediction of the piles in the testing set TS1 and the maximum error is about 36%.

5   DISCUSSION

The EN defined previously, the coefficient of determination R^2, and the equation of linear regression were calculated for the predictions of the neural network models for the total capacity, shaft resistance, and toe resistance. The results are summarized in the Table 2.

As shown by the values of R^2 in Table 2, the proposed neural networks provided good predictions for total bearing capacity and shaft resistance of driven piles. However, the predictions for toe resistance were not as good as the others and greater scatter was found. Similar performance for prediction of toe

917

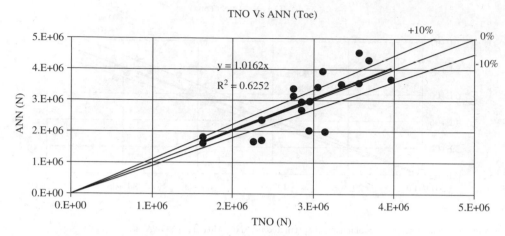

Figure 5.   Comparison of pile toe resistance predicted by ANN_M03 and TNOWAVE.

Table 1.   ANN predicted bearing capacity of piles in testing set TS1.

Pile No.	Shaft resistance ($\times 10^3$ kN)			Toe resistance ($\times 10^3$ kN)			Total capacity ($\times 10^3$ kN)		
	TNO	ANN	Error (%)	TNO	ANN	Error (%)	TNO	ANN	Error (%)
T01	3.27	3.20	2.14	2.85	2.95	−3.40	6.12	6.02	1.69
T02	2.63	2.16	17.75	1.63	1.61	1.06	4.26	3.84	9.83
T03	2.42	2.06	14.79	2.36	1.71	27.35	4.78	4.60	3.83
T04	2.44	1.75	28.08	3.34	3.51	−5.23	5.78	5.41	6.34
T05	2.89	2.71	6.32	2.95	2.97	−0.80	5.84	5.35	8.38
T06	2.62	2.37	9.44	2.95	2.96	−0.46	5.57	5.29	5.02
T07	4.30	4.30	−0.00	3.56	3.56	−0.00	3.99	3.99	0.00
T08	2.20	2.20	−0.09	1.63	1.81	−11.24	3.83	3.88	−1.25
T09	4.17	4.24	−1.75	3.11	3.94	−26.71	7.28	7.66	−5.21
T10	4.53	4.18	7.67	3.68	4.31	−17.02	8.21	7.13	13.10
T11	2.74	1.80	34.19	2.94	2.03	30.99	5.68	5.74	−1.08
T12	3.03	3.34	−10.07	2.75	3.37	−22.70	5.78	5.90	−2.01
T13	2.70	2.41	10.69	3.14	1.99	36.57	5.84	5.28	9.56
T14	2.96	3.01	−1.73	3.56	4.55	−27.67	6.52	7.39	−13.38
T15	3.94	4.27	−8.25	3.96	3.68	7.12	7.90	7.52	4.84
T16	4.57	3.76	17.80	3.05	3.43	−12.35	7.62	8.57	−12.51
T17	1.84	1.49	19.02	2.26	1.66	26.33	4.10	4.19	−2.25
T18	2.44	2.44	−0.00	2.36	2.36	−0.00	4.80	4.80	0.00
T19	3.46	2.95	14.86	2.75	3.13	−13.98	6.21	6.15	0.99
T20	2.72	3.10	−14.13	2.85	2.68	5.99	5.57	5.57	−0.09

Table 2.   Comparison of the prediction results of neural networks.

	Shaft	Toe	Total
EN	0.140	0.180	0.068
Linear regression	Y = 0.9325x	Y = 1.0162x	Y = 0.9886x
R^2	0.8629	0.6252	0.8853
Conclusion	Good	Acceptable	Good

resistance was also found by other researchers (Chow et al. 1995 & Teh et al. 1996). This is also expectable for this study since most of the driven piles at Macao are predominated friction piles. As the toe capacity is only a small fraction of the total capacity, a tiny error in the absolute value of the toe capacity implies a large relative error.

6 CONCLUSION

The study showed that the proposed Artificial Neural Network Model provided good predictions of pile bearing capacity using some specified points from the measured force and velocity history from the dynamic pile tests together with the properties of the PHC pile. The predictions were reasonably good for predicting the total capacity and shaft resistance of pile. However, the predictions for the toe resistance were less accurate than the other two but still in the acceptable range.

One must be noted that the Artificial Neural Network Model was trained using the TNOWAVE predicted bearing capacity. As such, the network cannot predict the true static capacity of pile, but only the TNOWAVE predicted capacity. And the accuracy of the network prediction is only as good as the TNOWAVE solution.

The significant advantage of the proposed neural network model is its prediction of pile bearing capacity is reliable but less dependent on experienced personnel than the signal matching methods of pile driving analyzers.

REFERENCES

Chan, W. T., Chow, Y. K., & Liu, L. F. 1995, Neural network: an alternative to pile driving formulas, *Computer and Geotechnics*, 17, pp. 135–156.

Chow, Y. K., Chan, W. T., Liu, L. F. & Lee, S. L. 1995a. Prediction of pile capacity from stress-wave measurements: a neural network approach, *International Journal for numerical and analytical methods in geomechanics* Vol. 19, pp. 107–126.

Chow, Y. K., Liu, L. F. & Chan, W. T. 1995b. Applications of neural networks in stress wave problems in piles, *Numerical Models in Geomechanics, Balkema, Rotterdam*, pp. 429–434.

Demuth, H. & Beale, M. 1998. *Neural Network Toolbox User's Guide – Version 3.0*, The Mathworks, Inc.

Goh A. T. C. 1996. Pile driving records reanalyzed using neural networks, *Journal of Geotechnical Engineering*, Vol. 122, No. 6, June 1996, pp. 492–495.

Haykin, S. 1994. *Neural Networks: A comprehensive foundation*, Macmillan College Publishing Company, Inc.

Hornik, K., Stinchcomebe, M. & White, H. 1989. Multilayer feedforward networks are universal approximators, *Neural Networks*, 2(3), pp. 356–366.

Martin, T. Hagan & Mohammad, B. Menhaj 1994. Training feedforward networks with the Marquardt algorithm, *IEEE Transactions on neural networks*, Vol. 5, No. 6, November, 1994, pp. 989–993.

Rausche, F., Goble, G. G. & Likins, G. E. *Dynamic Determination of Pile Capacity*.

Rumelhart, D. E., Hinton G. E. & William R. J. 1986. Learning internal representation by error propagation, *Parallel distributed processing*, Vol. 1, pp. 318–362, MIT press, Cambridge, Mass.

The, C. I., Wong, K. S., Goh, A. T. C.& Jaritngam, S. 1997. Prediction of pile capacity using neural networks, *Journal of Computing in Civil Engineering*, Vol. 11, No. 2, April 1997, pp. 129–138.

TNO Building and Construction Research. 1998. *Foundation Pile Diagnostic System, Pile Driving Analysis/ Dynamic Load Testing*.

Computational Methods in Engineering and Science, Iu et al. (eds)
© 2003 Swets & Zeitlinger, Lisse, ISBN 90 5809 567 3

Modeling of settlement of embankment on soft marine clay with vertical drains

W.N. Sam
Civil Engineering Laboratory of Macau, Macao

T.M.H. Lok
Department of Civil and Environmental Engineering, Faculty of Science and Technology, University of Macau, Macao

H. Novais-Ferreira
National Laboratory of Civil Engineering, Lisbon, Portugal

ABSTRACT: Landfills of sand over marine deposits are widely used in Macau. Hence, prediction of settlement of soft marine clay due to the construction of landfill becomes valuable. In this study, observed settlements were compared with calculated settlements for a newly constructed embankment on soft marine clay. Finite difference computer software, FLAC, was adopted for numerical simulation. Modified Cam-clay model was chosen as the constitutive model for the marine clay. To facilitate the numerical simulation of soft marine clay with installation of vertical drains to promote rapid consolidation, effects of vertical drains were simulated using equivalent permeability for the soft marine clay. Using the equivalent permeability, the soft marine clay without drains achieved similar average degree of consolidation as those predicted with drains.

1 INTRODUCTION

Land reclamation by hydraulic sand fill over marine deposits represents an important percentage of land in Macau. For those new reclaimed area, settlement is an important aspect to be taken into consideration. Normally the sand fill itself does not have much settlement after construction. Therefore, the most crucial part of settlement is due to the consolidation of the existing soft soil layers underneath the sand fill.

The consolidation of soft soil layers will usually take place over a long period of time. They will influence the infrastructure over the landfill. As a result, roads, drainage pipes and other structures without pile foundations will be damaged if they were constructed before the completion of the consolidation of the underneath soft soil layers. Also the consolidation process in the soft soil layers will cause negative friction on piles.

One common goal for projects involving land reclamation is to start subsequent construction over the landfill at the earliest. For this reason, vertical drain is a popular technique to increase the consolidation rate of soft soil layers. In particular, synthetic vertical drain, made of polyethylene core and covered with non-woven textile filter, had been commonly accepted for soft ground improvement till now.

The two important issues about design and installation of vertical drains are to estimate the magnitude and the rate of settlement for the soft soil layers. The magnitude of settlement will affect the final level of the construction, while the rate of settlement will affect the time of subsequent construction. In the past, predictions were usually made with one-dimensional consolidation theory, which, however, did not represent the actual geometry of the project. Nowadays, there are a number of computer programs, which can simulate not only one-dimensional but also two-dimensional, or even three-dimensional problems.

There are a number of previous studies on the problem of consolidation of soft clay with vertical drains. Novais-Ferreira & Shen (1997) and Tang (2001) studied the consolidation of clayey soils with

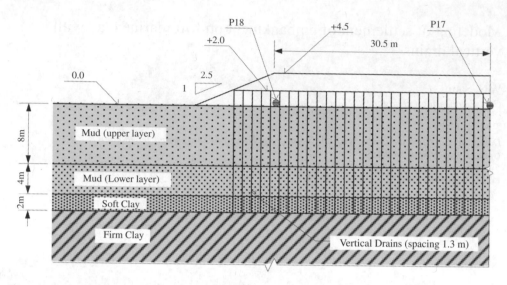

Figure 1.   Typical section with vertical drain layout.

vertical drains at the Macau International Airport. Indraratna & Redana (1997) presented a methodology to include the smear effect of vertical drains in a two-dimensional plane strain finite element model employing Modified Cam-clay theory. Lin et al. (2000) presented an interface element to simulate vertical drain with well resistance by imposing a specified cross sectional area and permeability to the interface elements.

In this study, a simplified method was proposed to predict the settlements of an embankment over soft soil layers with vertical drains.

2   FIELD INVESTIGATION

2.1   General

The project was located in a large area called COTAI between Coloane and Taipa of Macau. This area was planned to be a new urban area of Macau for future infrastructures. Since large area of reclaimed land was needed, landfill was the major work for this project. In this study, the newly constructed "VU3.3 Road" in COTAI was chosen for the purpose of case study. (G.D.I., 2000)

The total length of this road is 1.212 km, connecting the existing causeways between Taipa and Coloane. Half width of this road is 30.5 m at the crest and about 42 m at the base.

Vertical drains of 1.3 m spacing in hexagonal pattern were used as the ground improvement technique during construction of this road. The typical section with vertical drain layout is shown in Figure 1.

2.2   Geological profile

Site investigation in COTAI area was performed by Civil Engineering Laboratory of Macau (LECM 1994). The investigation included 20 boreholes totally with undisturbed samples for laboratory testing for engineering properties. However, the boreholes were not made especially for this road and were spread over the entire COTAI area. During construction of this road, some additional boreholes were made specifically for this project. However, only SPT (Standard Penetration Test) was made for those boreholes. From the above information, the subsoil properties were estimated and summarized in Table 1.

In the typical profile, the soft marine mud was divided into two layers. The upper layer was from 0 to 8 m, and the lower layer was from 8 m to 12 m of depth. A layer of soft clay, overlying another deposit of firm clay, was found below the marine mud from 12 m to 14 m. In this study, the soil below the soft clay layer was ignored. It was not only because the vertical drains were driven only 16 m into the bottom

Table 1. Summary of soil parameters.

Soil layer	ρ_d kg/m^3	λ	κ	ν_λ	C_h m^2/s	C_v m^2/s	k_h m/s	k_v m/s	n	ν
Mud (upper)	920	0.291	0.0387	3.26	7.89e-8	6.35e-8	5.60e-10	5.58e-10	0.638	0.3
Mud (lower)	1260	0.182	0.0329	3.05	1.10e-7	7.64e-8	4.38e-10	3.53e-10	0.609	0.3
Soft clay	1177	0.170	0.0416	2.88	1.12e-7	8.11e-8	6.68e-10	3.51e-10	0.564	0.3

ρ_d = Dry density
κ = Slope of elastic swelling line
C_h = Horizontal coefficient of consolidation
k_h = Horizontal coefficient of permeability
n = Porosity

λ = Slope of normal consolidation line
ν_λ = Specific volume at $p' = 1$
C_v = Vertical coefficient of consolidation
k_v = Vertical coefficient of permeability
ν = Poisson's ratio

of the soft clay layer (including 2 m of sand fill), but also the settlements of the soil below the soft clay layer were little. It was judged that the magnitude of the settlements was relatively smaller, and the rate of consolidation was much slower for the soil layers without vertical drains. This was verified from the readings of the Sondex settlement system, which showed that the settlements underneath the soft clay layer were insignificant. The soil profile was illustrated in Figure 1.

2.3 Measurements of settlements during construction

During the construction of this road, 19 settlement plates (P1–P19) had been installed at the site. The settlement plates were installed after a layer of sand fill was placed. The readings for the elevations of the settlement plates as well as the levels of sand fill were taken weekly till the end of the construction. In this study, two settlement plates, P17 and P18, were chosen for comparison with numerical modeling. Settlement plate P17 was in the middle of the embankment, and P18 was at the edge of the embankment, as shown in Figure 1.

3 NUMERICAL MODELING

3.1 Computer code

The two-dimensional explicit finite difference program, FLAC (Fast Lagrangian Analysis of Continua and Itasca 1998), was adopted for numerical modeling in this study. FLAC was developed by Itasca Consulting Group Inc. originally for geotechnical and mining engineering. It contains a built-in programming language, FISH, which can be used for creating functions to extend the usefulness of FLAC. In summary, the latest version offers now a wide range of capabilities to solve complex problems in continuum mechanics.

The default mode of simulation in FLAC is static mechanical analysis. However, groundwater flow analysis can be performed independently, or coupled to the analysis as used in this study.

3.2 Constitutive model

The Modified Cam-clay model was used by many researchers successfully for modeling of soft soil under consolidation. Since it considered the plastic volumetric deformation of soil during consolidation, it was adopted as the constitutive model for the marine mud in this study.

The Cam-clay model was originally proposed by Roscoe et al. (1958). It was widely used since it gives a simple and unified framework to model the behavior of clays. However, some deficiency was found and modifications were made to the Cam-clay model. The Modified Cam-clay model was introduced as an enhancement (Roscoe & Burland 1968). The detail description about implementation of this model into a computer code can be found in Britto & Gunn (1987).

3.3 Equivalent permeability method

Since vertical drains are usually installed in a very dense pattern, it will be a tremendous task to model every vertical drain in a numerical model of an embankment. Instead of simulating the actual vertical drain pattern in the site, equivalent permeability method was proposed in this study.

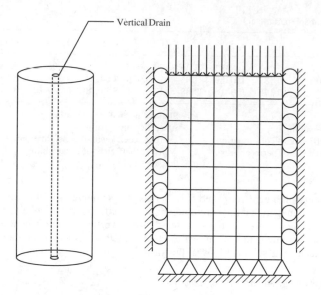

Figure 2. Cylindrical unit and mesh for the cylindrical unit.

In the theoretical method, the rate of consolidation is controlled by k_v (vertical coefficient of permeability) and k_h (horizontal coefficient of permeability). In the actual site situation, horizontal permeability is more important since the major drainage will be horizontal drainage from the soft soil to the vertical drains.

In the Equivalent Permeability Method, the permeability of the subsoil layers without vertical drains is multiplied by a factor called the Vertical Drain Factor (F_{VD}). This factor is multiplied to both vertical permeability and horizontal permeability. However, the vertical permeability here is more important since there is no significant horizontal drainage. By choosing the Vertical Drain Factor carefully, the consolidation rate of the subsoil layers without vertical drains would be similar to the one with vertical drains installed. Using this equivalent permeability method, the details of the vertical drains need not be simulated in the numerical modeling. The mesh can be reduced to a much simpler one, with which the time for solving the problem will be reduced significantly.

The equivalent permeability is expressed by:

$$k_{eq} = F_{VD}k \tag{1}$$

where k_{eq} = equivalent coefficient of permeability; k = original coefficient of permeability of the soil; and F_{VD} = vertical drain factor.

4 VERIFICATION OF THE EQUIVALENT PERMEABILITY METHOD

A simple verification was performed for the equivalent permeability method. Assume that a surcharge of 30 kPa was applied at the top of a cylindrical unit with a vertical drain installed at the center as shown in Figure 2. The soil properties of this cylindrical unit were the same as the upper mud layer in this site (Table 1). The cylindrical unit was assumed to be 8 m in depth and the water level was at the top level. The effective diameter of the cylinder is estimated based on the 1.3 m spacing of the vertical drains in hexagonal pattern. The dimensions of the vertical drain are 0.1 m in width and 0.0035 m in thickness. The smear effect and well resistance were ignored. The degree of consolidation by horizontal drainage could be calculated by the following theoretical solution (Hansbo 1981):

$$U_h(t) = 1 - e^{-\frac{8T_h}{F(n)}} \tag{2}$$

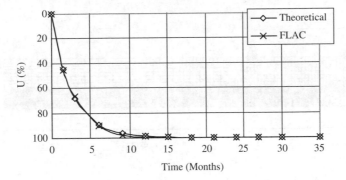

Figure 3. Degree of consolidation versus time for theoretical solution and numerical model using equivalent permeability method (with $F_{VD} = 54$).

where

$$T_h = \frac{C_h t}{d_e^2}; \quad F(n) = \frac{n^2}{n^2 - 1} \ln(n) - \frac{3n^2 - 1}{4n^2}; \quad n = \frac{d_e}{d_w}$$

and T_h = horizontal time factor; C_h = horizontal coefficient of consolidation; d_e = effective diameter of unit cell; d_w = diameter of drain well; U_h = horizontal degree of consolidation.

Considering both horizontal and vertical drainage, the total degree of consolidation would be (Carillo 1942):

$$U = 1 - (1 - U_h)(1 - U_v) \tag{3}$$

where U = total degree of consolidation; U_h = degree of consolidation by horizontal drainage calculated by equation (2); and U_v = degree of consolidation by vertical drainage calculated by one-dimensional consolidation theory.

The numerical model using the equivalent permeability method is also shown in Figure 2. The vertical drain was not included in the model, but, instead, the permeability for the soil was substituted by the equivalent permeability. Parametric study using different value of F_{VD} was performed to find out the best match between the theoretical solution and numerical modeling with FLAC. The best match was obtained when $F_{VD} = 54$. The degree of consolidation with vertical drains for both the theoretical solution and the numerical modeling with FLAC using the equivalent permeability method is shown in Figure 3. Very close approximations to the theoretical solution were obtained with the proposed equivalent permeability method.

5 CASE STUDY

5.1 Numerical modeling for the embankment

The vertical drains were driven from the surface of 1st phase sand filling (about 2 m above Mean Sea Level) to a depth of 16 m. Only the soil layers with the vertical drains installed were included in the model since the consolidation for the underneath soil layer without vertical drains would be very slow and negligible as explained previously in Section 2.

For analyzing the behavior of the embankment on the soft marine clay, only half of the section was considered because of symmetry. As shown in Figure 4, the grid was 75 by 14 elements. The size of the 50 by 14 elements under the embankment was 1 m by 1 m. The size of the 25 by 14 elements away from the embankment was 2 m by 1 m. Boundary conditions were such that the horizontal movements were

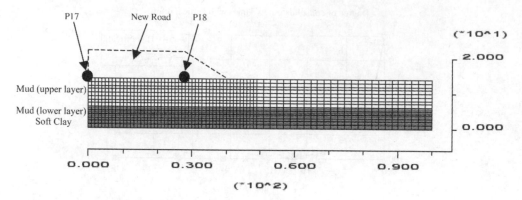

Figure 4.   Mesh for analyzing in FLAC.

Table 2.   List of results for observed and calculated settlements.

	P17			P18				P17			P18		
Days	G.L.	Obs.	Cal.	G.L.	Obs.	Cal.	Days	G.L.	Obs.	Cal.	G.L.	Obs.	Cal.
0	0.896	0.000	0.000	0.817	0.000	0.000	207	3.430	−1.451	−1.423	3.400	−1.630	−1.492
20	0.896	−0.242	−0.302	0.817	−0.355	−0.369	222	3.617	−1.464	−1.455	3.439	−1.638	−1.513
40	1.792	−0.593	−0.529	1.633	−0.621	−0.630	235	3.190	−1.502	−1.483	3.300	−1.684	−1.526
55	1.792	−0.698	−0.690	1.633	−0.698	−0.729	249	3.510	−1.515	−1.487	4.540	−1.737	−1.610
69	1.792	−0.726	−0.778	1.633	−0.717	−0.792	264	3.550	−1.525	−1.508	4.600	−1.750	−1.684
82	1.850	−0.778	−0.827	1.850	−0.751	−0.849	279		−1.535	−1.529		−1.760	−1.730
96	3.500	−0.906	−0.866	2.950	−0.900	−0.890	291		−1.543	−1.544		−1.767	−1.757
110	2.780	−1.106	−1.087	3.380	−1.087	−1.052	306		−1.556	−1.560		−1.775	−1.784
124	3.700	−1.174	−1.107	3.300	−1.300	−1.241	320		−1.562	−1.573		−1.781	−1.803
138	3.245	−1.212	−1.197	3.605	−1.373	−1.314	334		−1.566	−1.586		−1.786	−1.819
152	3.260	−1.246	−1.244	3.420	−1.446	−1.362	348		−1.570	−1.596		−1.790	−1.832
166	3.200	−1.282	−1.283	3.400	−1.497	−1.399	361		−1.577	−1.606		−1.798	−1.844
180	3.510	−1.346	−1.343	3.500	−1.547	−1.442	396		−1.584	−1.618		−1.807	−1.861
193	3.530	−1.384	−1.385	3.470	−1.578	−1.467							

* G.L. = Ground Level (M.S.L.); Obs. = Observed Settlements (m); Cal. = Calculated Settlements (m)

fixed at both vertical sides, and the vertical and horizontal movements were fixed at the bottom of the model. The water level was at the ground surface.

The filling process was modeled as pressure acting on the ground surface. The magnitude of the pressure was calculated as the product of the density and the thickness of the sand fill. The density of the hydraulic sand fill was $17\,kN/m^3$, based on field measurements. The thickness of the sand fill was estimated based on the levels of the sand fill recorded in field as shown in Table 2.

The displacements at the locations of settlement plates from the analysis were of major interests. Settlements at the corresponding time intervals of the settlement plate readings were obtained from FLAC so that comparison between the predictions and the observations could be made conveniently. However, the results from the numerical analysis using FLAC were not limited to those time intervals only.

5.2   Comparing observed and calculated settlements

The observed and calculated settlements at settlement plates P17 and P18 were selected for comparison. Based on the study of a cylindrical unit, a vertical drain factor of 54 was used to model the effect of vertical drains. Since the soil properties listed in Table 1 were obtained based on laboratory testing of samples at a distance from the site, they may not be the most representative values. A range of λ values for

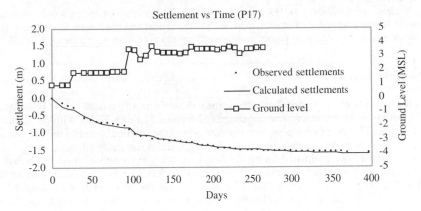

Figure 5. Observed and calculated settlement curves of P17,
($\lambda = 0.32$, $F_{VD} = 54$ and $k = 1.45 \times 10^{-9}$ m/s for upper mud layer)

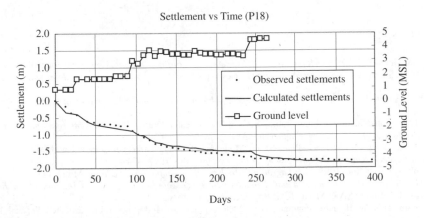

Figure 6. Observed and calculated settlement curves of P18.
($\lambda = 0.32$, $F_{VD} = 54$ and $k = 1.45 \times 10^{-9}$ m/s for upper mud layer)

the upper part of mud (0 to −8 MSL) and a range of permeability values for all the layers were used in the parametric study to determine the optimal values of these parameters.

In this construction, the last layer of sand fill finished at 264 days. By comparing the sum of square errors of the calculated and observed settlements between 0 day and 264 days, it was found that the best match could be achieved with $\lambda = 0.32$ and permeability = 1.45×10^{-9} m/s for the upper mud layer.

A comparison between the observed and calculated settlements using the optimal values of λ and permeability was listed in Table 2. The differences between the observed and calculated settlements at the end of monitoring (396 days) were 0.034 m (2.1%) for P17 and 0.054 m (2.9%) for P18. The settlements after 264 days were not used for the back analysis but were used for prediction using the values of λ and permeability obtained for the settlements before 264 days. Therefore, the comparison for 396 days represents the predictability of the model based on the parametric study for the information before 264 days. However, this can be further validated when observations over a longer period become available in the future.

In Figure 5 and Figure 6, the observed and calculated settlement curves are presented. It could be seen that the settlement curves using the optimal values of λ and permeability were very close to the observations. Closer examinations showed that the prediction for P17 was better than P18. This was probably because the settlement at P17 was at the middle of the embankment with only vertical movements, and P18 was at the edge of embankment with both vertical and horizontal movements. It is commonly known that horizontal movements are more difficult to predict in such cases.

6 CONCLUSIONS

- Based on the study of a cylindrical unit, the vertical drain factor (F_{VD}) used for the best match curve was 54, which means the equivalent permeability becomes 54 times higher with the installation of vertical drains. The use of equivalent permeability allows the detailed layout of the vertical drains to be ignored. The mesh was much simpler, and the computation was more efficient.
- The soil parameters were adjusted during the parametric study since only estimated values of the soil parameters at this site could be obtained. These soil parameters were the same during the analyses for P17 and P18. The optimal value of λ and permeability for the best match were about 1.1 and 2.6 times the values estimated from the laboratory test, respectively. This may reflect the discrepancy between the field and lab values but also deserve further investigation in the future.
- The predicted settlement curves obtained using FLAC gave a good match to the actual measurements. The differences between the observed and the calculated settlements at the end of monitoring (396 days) are 2.1% and 2.9% for P17 and P18, respectively.

REFERENCES

Britto, A.M. & Gunn, M.J. 1987. *Critical State Soil Mechanics via Finite Elements*. New York: John Wiley & Sons.
Carillo, N. 1942. Simple two- and three-dimensional cases in the theory of consolidation soils. *Jour. Math. Phys.* Vol. 21: 1–5.
G.D.I. 2000. *Project design for Vu3.3 road between NU4 roundabout and NT2 roundabout.* Macau: Profabril Asiaconsult Lda.
Hansbo, S. 1981. Consolidation of Fine-Grained Soils by Prefabricated Drains. *Proc. 10th International Conference on Soil Mechanics and Foundation Engineering*, Vol. 3. Stockholm: 677–682.
Indraratna, B. & Redana, I.W. 1997. Plane-Strain Modeling of Smear Effects Associated with Vertical Drains. *Journal of Geotechnical and Geoenvironmental Engineering*, May 1997: 474–478.
Itasca Consulting Group, Inc. 1998. *Fast Lagrangian Analysis of Continua: User's Guide.* Minnesota.
LECM 1994. *Reclamation between Taipa and Coloane Geotechnical Site Investigation and Analysis – Final Report.* Report No. 495, Macau SAR: Macau Civil Engineering Laboratory.
Lin, D.G., Kim, H.K. & Balasubramaniam, A.S. 2000. Numerical Modeling of Prefabricated Vertical Drain. *Geotechnical Engineering Journal*, Vol. 31, No. 2, August 2000, SAGS: 109–125.
Novais-Ferreira, H. & Shen, Q. 1997. Ground Monitoring of the Artificial Island during Consolidation. Macau International Airport Technical Handbook, Part IV, Chapter 2: 443–456.
Roscoe, K.H. & Burland, J.B. 1968. On the Generalized Stress–Strain Behaviour of "Wet" Clay. *Engineering Plasticity*. Cambridge University Press: 535–609.
Roscoe, K.H., Schofield, A.N. & Wroth, C.P. 1958. On the yielding of soils. *Geotechnique*, Vol. 8: 22–53.
Tang, U.M. 2001. *Consolidation of Clayey Soils by Prefabricated Vertical Drains*. Master Thesis. Macau: University of Macau.

Computational Methods in Engineering and Science, Iu et al. (eds)
© *2003 Swets & Zeitlinger, Lisse, ISBN 90 5809 567 3*

Parametric study on constitutive modeling for predictions of diaphragm wall behavior

C.H. Ng & T.M.H. Lok
Department of Civil and Environmental Engineering, Faculty of Science and Technology,
University of Macau, Macao

ABSTRACT: The usage of diaphragm wall has been increasing dramatically in highly populated cities to facilitate excavations in congested area. For the safety of the excavation and protection of the surrounding structures, reliable prediction of ground movements is very crucial. In order to predict ground movements realistically, the stress-strain-strength characteristics of soil must be represented in a reasonable manner. In this study, three well-known soil constitutive models, namely, the Mohr-Coulomb, the Duncan Hyperbolic, and the Modified Cam Clay model, were adopted, and the finite element program, CRISP, was employed to investigate the behavior of a diaphragm wall for an excavation project in Macau using these three constitutive models. Parametric studies on the excavation depth and system stiffness were performed for a case history of diaphragm wall project.

1 INTRODUCTION

Rapid development of the densely populated city of Macau has led to the great demand of living space. Consequently, diaphragm walls become the primary structural elements for supporting deep excavations and creating underground space in proximity of existing buildings. For the safety of the excavation and the efficiency of the construction process, reliable prediction of ground movements is very crucial, which, in turn, depends very much on the soil constitutive models used in the analysis. In this study, a case history in Macau was used to illustrate the behavior of diaphragm wall as well as the related ground movements. Firstly, three well-known soil constitutive models of various degree of complexity were chosen for the numerical modeling of a diaphragm wall for an excavation project in Macau. Secondly, influences of different design parameters on the behavior of diaphragm wall were examined.

2 DESCRIPTION OF THE EXCAVATION PROJECT

The diaphragm wall for the excavation project for the market of S. Domingos in the city of Macau was investigated. The construction site was located at old urban area of Macau where there are historical buildings of masonry nearby. In order to provide a safe excavation, diaphragm wall was adopted as the earth retaining system. The geometry of the excavation was 10.5 m in depth and 19 m by 84 m in area. During the construction process, 3 levels of struts were installed within four stages of excavation. The struts were spaced at 6 m intervals horizontally. The initial ground water table was located at 1.5 m below the ground surface. Before each stage of excavation, dewatering in the excavation was performed to maintain the water table 0.5 m below the excavation surface. The cross section of the excavation is shown in Figure 1a. A previous study of the diaphragm wall was conducted using the Mohr-Coulomb model for the soil (Subrahmanyam & Leong, 1999, Leong, 2001). The material parameters for the three constitutive models used in this study were derived from those used by Leong (2001) and are summarized in Tables 1, 2, and 3, respectively. The overall construction process is described in Table 4, while the material properties for the structural members are summarized in Table 5.

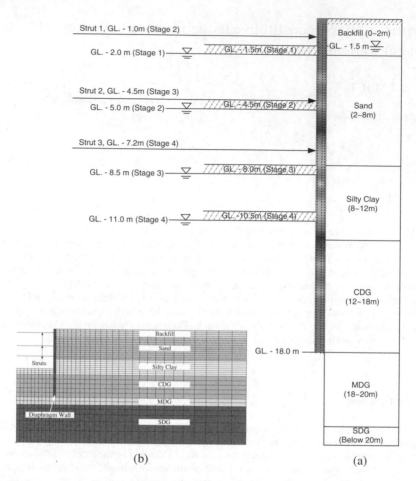

Figure 1.   (a) Cross section of the excavation; (b) Finite element mesh for the excavation.

Table 1.   Material properties (Mohr-Coulomb model).

Soil layer	1 Backfill	2 Sand	3 Silty clay	4 CDG	5 MDG	6 SDG
E (MPa)	40.9	54	28.8	77.1	72000	72000
v	0.36	0.35	0.44	0.286	0.20	0.20
C′ (kPa)	0	0	40	4	4000	5000
$\phi'(°)$	23	26	3	35	45	47
γ' (kN/m³)	10.99	11.43	10.19	12.30	15.13	15.44

3   FINITE ELEMENT PROGRAM

Numerical analyses in this study were performed using the finite element program CRISP (Britto and Gunn, 1987), which is a geotechnical finite element program incorporating critical state soil mechanics theory. The latest version, which was used in this study, called SAGE CRISP, was released by CRISP Consortium (Woods and Rahim, 1992). The program is able to perform drained, undrained and fully coupled consolidation analysis of static problems. A number of soil constitutive models are available in the program. In particular, the Mohr-Coulomb, Duncan Hyperbolic (Duncan et al. 1980), and Modified Cam Clay (Roscoe and Burland, 1968) models were used in this study. However, since the Duncan

Table 2. Material properties (Duncan Hyperbolic model).

Soil layer	1 Blackfill	2 Sand	3 Silty clay	4 CDG
K	330	340	260	1200
K_{ur}	820	830	650	2400
K_b	150	150	140	150
C' (kPa)	0	0	40	4
ϕ' (°)	23	26	3	35
R_f	0.7	0.7	0.7	0.7
m	0.25	0.25	0.15	0.20
n	0.35	0.35	0.20	0.40

Table 3. Material properties (Modified Cam Clay).

Soil layer	1 Backfill	2 Sand	3 Silty clay	4 CDG
K	0.0005	0.0005	0.002	0.003
λ	0.01	0.01	0.02	0.009
ecs	0.65	0.58	1.32	0.89
M	0.898	1.027	1.278	1.418
G (MPa)	15	20	10	30

Table 4. General construction process.

Stage #	Construction process
1	Excavation (1.5 m) and dewatering (2.0 m)
2	Installation of strut 1 (1.0 m) → Excavation (4.5 m) and dewatering (5.0 m)
3	Installation of strut 2 (4.5 m) → Excavation (8.0 m) and dewatering (8.5 m)
4	Installation of strut 3 (7.2 m) → Excavation (10.5 m) and dewatering (11.0 m)

Table 5. Material properties for structural members.

	E (GPa)	ν	A (m^2)	I (m^4)
Strut	210	–	222×10^{-4}	5.12×10^{-4}
Wall	20	0.15	0.6 m thick	

Hyperbolic model included in the software was based on an older formulation, the model was re-implemented into CRISP by the authors to conform to that by Duncan et al. (1980).

4 ANALYTICAL MODEL

In this study, two-dimensional plane-strain finite element models were created using the finite element code CRISP as shown in Figure 1b. The soil and wall were discretized into eight-node quadrilateral elements having two degrees of freedom at each node. Struts were modeled by two-node beam elements having three degrees of freedom at each node. The joints between the strut and wall were modeled by pinned connection while the interaction between the wall and the soil was modeled by interface elements, of which the behavior followed Mohr-Coulomb failure criteria. Structural elements and the two layers of underlying rock were assumed to be linear elastic. The loading on the diaphragm wall consists of the soil pressure and the pore water pressure. Pore water pressure was assumed to be hydrostatic and directly prescribed as a distributed load on the wall, while the earth pressure was calculated using the

931

Table 6. Brief comparison of three different soil constitutive models.

Mohr-Coulomb (M-C)	Duncan Hyperbolic (DH)	Modified Cam Clay (MCC)
Main characteristics		
A linear elastic-perfectly-plastic model; For frictional materials that ruptures at a critical combination of normal stress and shearing stress on a plane.	An incremental nonlinear stress-dependent models with failure criteria based on Mohr-Coulomb; 3 important characteristics of soil are modeled: non-linearity, stress-dependence and inelasticity;	A popular elasto-plastic model based on critical state soil mechanics; Virgin compression and recompression lines are linear in the ln p′–v space.
Advantages		
Only c and ϕ are required to describe the plastic behavior; It can be applied to three-dimensional stress space.	Soil parameters can be obtained directly from standard triaxial test; Simple and obvious enhancement to the Mohr-Coulomb model.	It provides a unified framework between the mean effective stress, the deviator stress, and the specific volume.
Limitations and drawbacks		
Only linear elastic behavior is modeled before failure; Intermediate principal stress is not taken into account; A constant rate of negative volumetric strain would be predicted if associated flow rule is applied.	Only volumetric shrinkage can be predicted; Potential numerical instability may occur when shear failure is approached; Ad hoc solutions must be applied when the model is implemented in three-dimensional stress space.	Only linear elastic behavior is modeled before yielding; Unreasonable values of v can be predicted due to log-linear compression lines.

effective unit weight of the soil. The loading from the surrounding buildings was modeled by an equivalent surcharge with intensity equal to 30 kPa acting on the ground surface of the retaining side.

5 SOIL CONSTITUTIVE MODELS

Three different constitutive models were used for modeling the behavior of the soil in three independent sets of analyses, namely, the Mohr-Coulomb (M-C), the Duncan Hyperbolic (DH), and the Modified Cam Clay (MCC) model. While the two lower layers of soil were modeled as linear elastic, all the other layers were modeled using a single constitutive model with the corresponding modeling parameters listed in Tables 1, 2, and 3. A brief comparison of three different soil constitutive models is presented in Table 6.

6 PREDICTED GROUND MOVEMENTS

6.1 Prediction of lateral wall displacements

Figure 2a shows the lateral wall displacements calculated using the M-C, DH, and MCC models. Only the information for the final stage is presented here, but information for other stages can be found in Ng (2003). In general, the predictions using those three models are reasonable comparing to the observations.

In Figure 2a, the predictions calculated using the M-C model are very satisfactory in terms of the magnitude and the shape of wall displacements. Lateral wall displacement predicted by DH model also provided similar predictions as the M-C model did. However, the shape of the displacement curve shifted slightly upper which is probably caused by the increasing stiffness with depth as predicted by

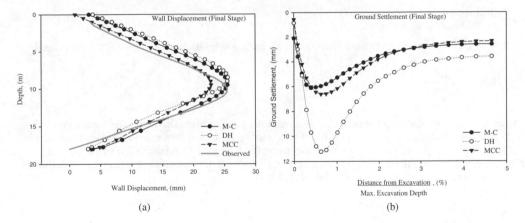

Figure 2.   Predictions of (a) Horizontal wall displacement; (b) Ground settlements.

the DH model. Furthermore, it was found by Ng (2003) that DH model provided better predictions for the earlier stages because soil nonlinear behavior was considered before failure.

The predictions using MCC model were less satisfactory than those by the M-C and DH models. This was probably due to the inaccuracy in the critical state parameters derived from the limited information available. In addition, it was observed by Ng (2003) that MCC model generally under-predicted the displacements for earlier stages because of its inherent linear elastic behavior prior yielding.

6.2   Prediction of ground settlement

Figure 2b shows the ground settlements predicted using the M-C, DH, and MCC models. In general, the shapes of the predicted settlement profiles using those models were reasonable. For example, the distance from excavation for maximum ground settlement was about half of the depth of excavation, which is consistent with the settlement envelope proposed by Clough et al. (1989). However, the magnitude of the predicted settlements varied with the model used.

Comparing to the observed maximum settlements of around 15–25 mm recorded by extensometers at different cross sections, the magnitude of settlements predicted using the M-C and MCC models was too small. On the other hand, the corresponding maximum settlement predicted using the DH model was closer to the observations. The under-prediction for the ground settlement was probably due to the pre-yielding characteristics of the M-C and MCC models. Since linear elastic moduli were used in those two models, the bulk modulus before the yielding of soil would have been overestimated, which resulted in smaller settlements. In addition, for the M-C model, a constant rate of negative volumetric strain (dilation) would be predicted as associated flow rule was employed. This would lead to even smaller ground settlements.

In general, the predicted settlements were smaller than the observations even with the DH model. As indicated by Mana and Clough (1981), this is probably due to the effects of consolidation and traffic loading behind the wall, which was not modeled in the analyses.

7   PARAMETRIC STUDY USING THE PROPOSED ANALYTICAL MODEL

7.1   Parameters investigated

The objective of this study is to examine the influences of the design parameters of deep excavation on the behavior of diaphragm walls for the case history described earlier. The parameters investigated in this study were the depth of excavation, the stiffness of the strut, and the stiffness of the wall. To provide a unified framework for comparison, the effects of design parameters were presented in the form of dimensionless parameters:

– Factor of safety about base heaving (FS,$_{heave}$)

– Stiffness of the bracing system (K_s)
– Stiffness of the retaining wall (K_w)

For the factor of safety about base heaving, the FS proposed by Bjerrum and Eide (1956) was used:

$$FS_{,heave} = \frac{C_{ub} N_{cb}}{\gamma_m H + q} \tag{1}$$

where C_{ub} is the undrained shear strength around the base of excavation, q is the surcharge loading, H is the depth of excavation, γ_m is the mean bulk unit weight of soil above the excavation level, and N_{cb} is the bearing capacity factor as a function of width (B), length (L), and depth (D) of the excavation.

The stiffness of the bracing system and the retaining wall are dimensionless parameters defined as follows (Mana and Clough 1981, Clough et al. 1989):

$$K_s = \frac{S}{\gamma_w h_{avg}} \tag{2}$$

$$K_w = \frac{EI}{\gamma_w h_{avg}^4} \tag{3}$$

where EI and S are the flexural stiffness of the retaining wall and the stiffness of the strut per horizontal unit length, respectively, γ_w is the unit weight of water, and h_{avg} is the average vertical spacing of struts.

7.2 Effects of depth of excavation

To investigate the effects of the depth of excavation, various depths ranging from $0.43\,H_{,origin}$ to $1.48\,H_{,origin}$ (where $H_{,origin} = 10.5\,\text{m}$) were adopted in the simulations. According to equation (1), the value of $FS_{,heave}$ was then calculated using the corresponding value of H. The influence of the $FS_{,heave}$, on the behavior of the diaphragm wall is shown in Figure 3. In general, both the lateral wall displacements and settlements increase with decreasing $FS_{,heave}$.

A dramatic change of the behavior of the diaphragm wall was observed when the $FS_{,heave}$ was at about 1.5. For $FS_{,heave}$ larger than 1.5, both the lateral displacements and settlements decrease slowly with increasing $FS_{,heave}$. In fact, the magnitude of those quantities increases by less than 1% when the $FS_{,heave}$ decreases from 3.4 to 1.5. On the other hand, both the lateral displacements and settlements increase rapidly for $FS_{,heave}$ smaller than 1.5. This was consistent with the implication of failure by the limit equilibrium analysis of base heaving.

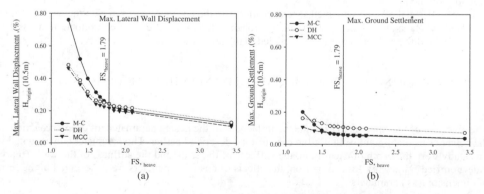

Figure 3.   Effects of depth of excavation: (a) Max. lateral wall displacement; (b) Max. ground settlement.

Carefully examination both Figures 3a and 3b, one interesting phenomenon was also discovered. When the $FS_{,heave}$ was less than 1.5, the increases in both lateral wall displacements and ground settlements were more dramatic by the M-C model than by the other two models. This may be due to the sudden change of the stiffness inherent in the linear-elastic perfectly plastic formulation.

7.3 Effects of stiffness of bracing

To investigate the effects of the stiffness of bracing, various values of the bracing stiffness ranging from $0.3 K_{s,origin}$ to $5.7 K_{s,origin}$ (where $K_{s,origin} \approx 2700$) were adopted in the simulations. As show in Figure 4, the wall displacements and the ground settlements increase with decreasing values of K_s.

The trend of the curves predicted using the three different models was also similar. There were just slight decreases in both the maximum lateral wall displacements (about 0.07% of the depth of excavation) and the maximum ground settlements (about 0.05% of depth of the excavation) when the stiffness of the bracing increases by about 5.7 times.

7.4 Effects of wall stiffness

The effects of the system stiffness of retaining system on the behavior of diaphragm are shown in Figure 5. Various values of K_w ranging from $0.5 K_{w,origin}$ to $10 K_{w,origin}$ (where $K_{w,origin} \approx 400$) were used in the simulations. As shown in Figure 5, when the stiffness of the retaining system increases, both the maximum lateral wall displacements and the ground settlements decrease almost log-linearly.

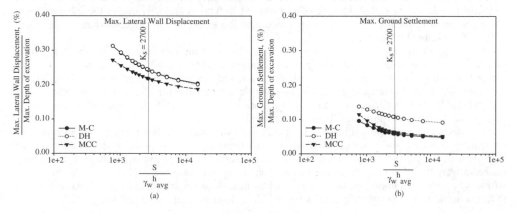

Figure 4. Effects of stiffness of bracing: (a) Max. lateral wall displacement; (b) Max. ground settlement.

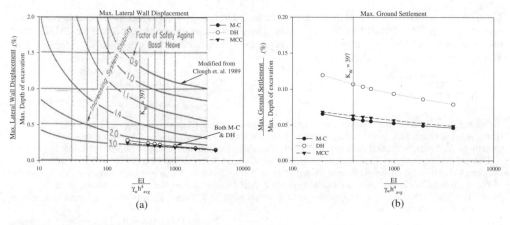

Figure 5. Effects of stiffness of wall: (a) Max. lateral wall displacement; (b) Max. ground settlement.

935

Also, the trend of the curves predicted using the three different models was similar. In general, the effects of the stiffness of the wall were very similar to those of the stiffness of the bracing. There were just slight decreases in both the maximum lateral wall displacements (about 0.15% of the depth of excavation) and the maximum ground settlements (about 0.03% of depth of the excavation) when the wall stiffness increases by ten times.

For comparison, the curves obtained by Clough et al. (1989) were superimposed on Figure 5. As shown in Figure 5, the small influence of the wall stiffness on the behavior of the diaphragm was probably due to the high value of FS for base heaving. As shown by Clough et al. (1989), when the factor of safety against base heave was greater than 2.0, the wall stiffness would have insignificant effects on the lateral wall displacements. However, it should be noticed that Terzaghi's method for factor of safety was used by Clough et al. (1989), which usually overestimated the value of FS.

8 CONCLUSION

A parametric study with various basic design parameters for deep excavation project was performed using three different soil constitutive models. In general, the predictions compared more favorable to the observations if the constitutive model such as the Duncan Hyperbolic model takes into account the nonlinear behavior of soil before yielding. The magnitude of the lateral wall displacements toward the end of excavation depends very much on the ultimate shear strength of soil. Therefore, a model with reasonable description of the shear strength such as Duncan Hyperbolic and Mohr-Coulomb model would provide reasonable predictions. The Modified Cam Clay model, though famous of its unified framework, was less favorable than the other two models. This is probably due to the inherent linear elastic behavior of the Modified Cam Clay model and the inaccuracy in the critical state parameters derived from limited information available.

The effects of the depth of excavation and the stiffness of the retaining wall on the behavior of diaphragm were examined using the dimensionless parameters $FS_{,heave}$ and K, respectively. It was found that the wall displacements and ground settlements generally increase with decreasing $FS_{,heave}$ and K. However, the increase in those quantities became very dramatic when the $FS_{,heave}$ dropped below a critical value. On the other hand, the variation of those quantities with K remained almost log-linear throughout the range of values of K being investigated.

The effects of the system stiffness on the behavior of diaphragm wall were relatively insignificant comparing to those of the factor of safety against base heaving. This was probably due to the relatively high value of $FS_{,heave}$ in the original condition of the excavation. In other words, the diaphragm wall was conservatively designed in terms of the factor of safety against base heaving. As a result, the system stiffness becomes less important in a system such as this. However, it should be noted that the system stiffness would be more important when a lower $FS_{,heave}$ is associated with a soft soil profile.

ACKNOWLEDGMENTS

The authors would like to express their sincere thanks to Dr. L.M.N. Lamas and Mr. Ao, Peng Kong, former and current President of LECM; Mr. Leong, Hong Sai, Geotechnical Engineer of LECM, for their immense helps rendered to carry out this investigation. Financial assistance from the University of Macau is also greatly acknowledged.

REFERENCES

Bjerrum, L. and Eide, O. 1956. Stability of strutted excavations in clay, *Geotechnique*, 6(1), 32–47.
Britto, A.M. & Gunn, M.J. 1987. *Critical State Soil Mechanics Via Finite Elements*, John Wiley & Sons.
Clough, G.W., Smith, E.M., and Sweeney, B.P. 1989. Movement Control of Excavation Support Systems by Iterative Design, *Proceedings, ASCE, Foundation Engineering: Current Principles and Practices*, Vol. 2, pp.869–884.
Duncan, J.M., Byrne, P., Wong, K.S., and Mabry, P. 1980. *Strength, Stress-Strain and Bulk Modulus Parameters for Finite Element Analyses of Stresses and Movements in Soil Masses*, Report No. UCB/GT/80-01, UCB.
Leong, H.S. 2001. *Prediction of Deformation of Diaphragm Wall*, M.S. Thesis, Faculty of Science and Technology, University of Macau.
Ng, C.H. 2003. *Comparative Study of Constitutive Models on Predictions of Diaphragm Wall Behavior.* M.S. Thesis, Faculty of Science and Technology, University of Macau.

Mana, A.I. and Clough, G.W. 1981. Prediction of Movements for Braced Cuts in Clay. *J. Geot. Eng., ASCE*, 109(GT6), June, 759–777.

Roscoe, K.H., & Burland, J.B. 1968. *On the generalized stress-strain behaviour of 'wet' clay*, Engineering plasticity, Cambridge University Press, pp.535–609.

Subrahmanyam, M.S. and Leong, H.S. 1999. Prediction of Deformation of Diaphragm wall, *The 7th International Conference on Enhancement and Promotion of Computational Methods in Engineering and Science, EPMESC VII*, Volume II, PP, 1415–1422.

Woods, R. & Rahim, A. 1992. *SAGE CRISP Technical Reference Manual*, The CRISP Consortium.

Computational Methods in Engineering and Science, Iu et al. (eds)
© 2003 Swets & Zeitlinger, Lisse, ISBN 90 5809 567 3

Modeling of hydrogen diffusion and γ-hydride precipitation process at a blunt notch in zirconium

X.Q. Ma, S.Q. Shi & C.H. Woo
Department of Mechanical Engineering, The Hong Kong Polytechnic University, Hong Kong

L.Q. Chen
Department of Materials Science and Engineering, The Pennsylvania State University, PA, USA

ABSTRACT: Stress distribution near a stress concentrator such as a notch will directly affect the hydrogen distribution and the hydride precipitation morphology around the notch in zirconium. In this report, two-dimensional diffusion of interstitial hydrogen atoms and γ-hydride precipitation around a blunt notch in a zirconium specimen were simulated using the phase-field method. The stress distribution around the notch in the specimen was calculated by finite element method. The elastic energy for both hydrogen diffusion and hydride formation were described by Khachaturyan's elastic theory. For the hydrogen distribution around the notch, Cahn-Hilliard diffusion equation was solved by explicit finite difference method. For the hydride formation, both Cahn-Hilliard equation and time-dependent Ginsberg-Landau equations were solved simultaneously by spectrum method. The result shows that hydrogen atoms diffuse to the high tensile hydrostatic stress region near the notch tip and hydrides precipitate in this region at a higher density in the direction perpendicular to the notch surface.

1 INTRODUCTION

Zirconium and its alloys are structure material in nuclear industry. They often work in the environment that contain hydrogen or deuterium. When the H plus D concentration in the material exceeds the terminal solid solubility (TSS), the hydrides will form. Since the brittleness of the hydrides, the materials are susceptible to a crack initiation and propagation process called delayed hydride cracking (DHC). The DHC process involves three steps: diffusion of H or D along a stress gradient to elevated-tensile-stress regions in the material such as at the crack tips or notch tips loaded in tension; formation and growth of hydrides at the crack tip; development of the crack through the hydride. The whole process can repeat itself. This process may threaten the safety of the nuclear reactor. It is believed that critical conditions for fracture initiation at the hydride are controlled by the morphology and microstructure of hydride precipitate at the flaw tip (Shi et al. 1994a,b). Because of the difficulty and high cost of in-situ observation of dynamic process of H diffusion and hydride growth, theoretical modeling is an important means to investigate the DHC process. However, due to the complexity of the precipitation pattern and difficulty in analysis, theories about hydride precipitation in the past can not predict a realistic morphology. Simplified assumptions were often imposed in these theories (Dutton & Puls 1975, Eadie et al. 1993, Shi et al. 1994c), for example, the assumption that only one hydride exists at the flaw tip region. In the past few years, we have applied the elastic theory by Khachaturyan (1983) to study the DHC process. This theory considers not only the elastic energy caused by the misfit strain of a hydride, but also the elastic interaction between hydrides, as well as the interaction between hydrides and an external load. Therefore, it can be used to describe an elastic system that has arbitrary precipitation pattern. It has been demonstrated that the phase-field kinetic model that incorporates Khachaturyan's elastic theory is very powerful for describing both the diffusion process and morphology/microstructural evolution in solid–solid phase transformation. The methodology has been successfully applied to many different systems (Chen & Wang 1996), including the γ-hydride precipitation in single and bi-crystalline zirconium with or without an uniformly applied external load (Ma et al. 2002a,b).

Most materials used in engineering applications are polycrystalline materials under non-uniformly applied load. When the average grain size is small as compared to the average hydride size, the matrix may be treated as a continuum with randomly distributed crystal orientation from one location to another. In effect, the anisotropic properties of each grain may be ignored. It is demonstrated in the experiments that in this situation, the orientation of the hydride precipitate is determined by the applied stress (Ells 1970). The objective of this work is to develop a framework to simulate hydrogen diffusion and γ-hydride precipitation in a continuum media under non-uniformly applied stress near a notch tip by phase field method.

2   SYSTEM DESCRIPTION AND CALCULATION OF APPLIED STRESS

To investigate the hydrogen diffusion and hydride precipitation near a notch tip, we construct a system as follows to mimic the cantilever condition: the sample is a zirconium plate with uniformly distributed dilute hydrogen. The upper and bottom surfaces of the specimen have a small curvature to mimic a part of a pressure tube. The dimension of the specimen is shown in Figure 1. It is 38 mm in length and 5 mm in thickness. A notch with a flank angle of 45° and a depth of 1.0 mm is cut at the center of upper edge. The root radius of notch tip is 0.3 mm. The yield strength of the specimen was assumed to be about 850 MPa at 250°C, a value for fully irradiated Zr-2.5 Nb alloy (Puls 1990). A clamp fixes one end of the specimen and a constant weight of 50 N is applied at the other end to generate a stress concentration at the area around the notch. The commercial finite element software, NASTRAN, is used to calculate stress and strain distributions. The average mesh size in the area around the notch is 0.03 mm. Figure 2 is the hydrostatic stress distribution around the notch. The resulting maximum normal stress σ_{xx} near the notch tip is about 670 MPa, which is still within the elastic regime. The gradient of the stress field is the driving force for hydrogen diffusion and stress distribution around the notch will affect the orientation of the precipitate hydride. The stresses obtained from NASTRAN are then mapped in a uniform

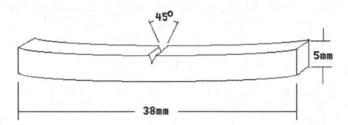

Figure 1.   Dimensions of the specimen.

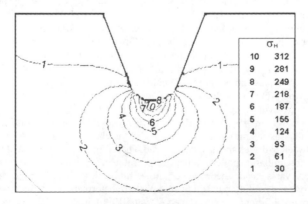

Figure 2.   Distribution of hydrostatic stress (σ_H) around the notch.

grid by interpolation and used as the applied stresses in the calculation of hydrogen diffusion and hydride precipitation in the phase-field model.

3 MODELING OF DIFFUSION PROCESS

In order to simulate the hydrogen diffusion process in a zirconium specimen with a notch, the compositional field $c(r,t)$, atomic fraction of hydrogen atoms, is adopted as a phase-field variable. We assume the initial composition is constant, i.e. $c(r,0) = c_o$. The governing equation of $c(r,t)$ for the diffusion process is Cahn-Hilliard equation (Allen et al. 1979):

$$\frac{\partial c}{\partial t} = \nabla \cdot (M \nabla \mu) \tag{1}$$

where M is the mobility and μ is the chemical potential of hydrogen respectively. The mobility of hydrogen is defined as $M = Dc(r,t)/RT$, where D is the diffusivity of hydrogen in zirconium, R is the gas constant and T is the absolute temperature. If D and T are not a function of location, then equation (1) can be further expended to

$$\frac{\partial c}{\partial t} = \frac{D}{RT}[c\nabla^2 \mu + \nabla c \cdot \nabla \mu] \tag{2}$$

The chemical potential of hydrogen is defined by

$$u = \frac{\partial G}{\partial n_H} = \frac{\partial G}{n \partial c} \tag{3}$$

where G is the Gibbs free energy, n_H represents the mole number of hydrogen atoms, and n is the mole number of all atoms, including zirconium atoms. The Gibbs free energy for hydrogen in solid solution is defined by

$$G = E + W - TS \tag{4}$$

where E is internal energy that includes mainly two parts: a chemical part (E_c) related to bonding properties between hydrogen and metal matrix, and another part (E_e) related to elastic distortion energy of the lattice due to hydrogen interstitials without externally applied stress. W is the work done by external forces and S is the entropy of the system, which is the sum of the entropy of vibration (S_v) and entropy of mixing (S_m). Therefore, (4) becomes

$$G = E_c + E_e + W - T(S_v + S_m) \tag{5}$$

Using lower cases of the letters to represent the quantities for one mole of hydrogen, and assuming dilute solution, one can write Gibbs free energy as

$$G = n_H (e_c + e_e + w - Ts_v) - TS_m \tag{6}$$

where S_m needs a special treatment because it involves mixing of hydrogen with zirconium atoms. The result is,

$$S_m = -nR \{c(r,t)ln[c(r,t)] + [1-c(r,t)]ln[1-c(r,t)]\} \tag{7}$$

From (3), the chemical potential can be found as

$$u = \mu_o + e_e + w + RT \ln \frac{c(r,t)}{1-c(r,t)} \tag{8}$$

where μ_o is the combination of e_c and $-Ts_v$, which is not a function of r, based on the assumptions of uniform temperature and dilute solution. Therefore, μ_o will not contribute to the solution of Cahn-Hilliard equation. The terms e_e and w are related to the elastic energy of hydrogen interstitials under externally applied stress. From Khachaturyan's theory, it can be calculated as:

$$\mu = \mu_0 - V_{mol}C_{ijkl}\varepsilon_{ij}^c\varepsilon_{kl}^a + V_{mol}C_{ijkl}\varepsilon_{ij}^c\varepsilon_{kl}^c\delta c(\mathbf{r})$$

$$+ [\int\frac{d^3g}{(2\pi)^3}n_i\sigma_{ij}^c(g)\Omega_{jk}(n)(\sigma_{kl}^a(g))^*n_l]_r - [\int\frac{d^3g}{(2\pi)^3}n_i\sigma_{ij}^c(g)\Omega_{jk}(n)(\sigma_{kl}^c(g))^*n_l]_r \qquad (9)$$

$$+ RT\{\ln[c(r,t)] - \ln[1-c(r,t)]\}$$

where g and $n = g/g$ are a reciprocal vector and its unit vector in reciprocal space respectively, ε_{ij}^c is the eigenstrain of hydrogen atom, $\sigma_{ij}^c = C_{ijkl}\varepsilon_{kl}^c$, $\sigma_{ij}^a(g)$ is the applied stress in reciprocal space, V_{mol} is the mole volume of zirconium, and $[\ldots]_r$ represents the backward Fourier transformation of quantities in $[\ldots]$. For an elastically isotropic system,

$$\Omega_{ij}(\mathbf{n}) = \frac{\delta_{ij}}{G} - \frac{n_in_j}{2G(1-v)} \qquad (10)$$

where G is the shear modulus and v is the Poisson's ratio. According to (9), the chemical potential of hydrogen is a function of both hydrogen concentration and stress.

The Cahn-Hilliard equation was solved by explicit Euler finite difference method. The system is divided into a 512×512 uniform Cartesian grid with the distance between the nearest grid point Δx or Δy along x or y direction respectively, $\Delta x = \Delta y = h$. In the finite difference method, Cahn-Hilliard equation (2) takes the form of

$$c_{i,j}^{n+1} = c_{i,j}^n + \frac{D\Delta t}{RT}\{\frac{1}{\Delta x^2}c_{i,j}^n \cdot [(\mu_{i+1,j} + \mu_{i-1,j} + \mu_{i,j+1} + \mu_{i,j-1} - 4\mu_{i,j})]$$

$$+ (\frac{1}{4\Delta x^2})[(c_{i+1,j}^n - c_{i-1,j}^n)(\mu_{i+1,j} - \mu_{i-1,j}) + (c_{i,j+1}^n - c_{i,j-1}^n)(\mu_{i,j+1} - \mu_{i,j-1})]\} \qquad (11)$$

The flux of hydrogen at any point in the matrix is equal to

$$J = -M\nabla\mu \qquad (12)$$

The boundary condition at the notch surface is that the hydrogen flux across the boundary is zero. By numerically solving the Cahn-Hilliard equation, we can obtain the temporal and spatial evolution of hydrogen concentration.

Parameters used in the calculations are given in Table 1. Hydrogen diffusion was calculated by solving (11). Figure 3 is the equilibrium hydrogen distribution after 80 hours of diffusion at 250 °C. The

Table 1. Parameters used in the simulation of hydrogen diffusion.

Parameters	Value	References
Initial hydrogen concentration, c_0	0.6 at %	
Diffusivity of H in Zr, D_H	$5.468 \times 10^{-7}\exp(-45293/RT)$ m^{2}s^{-1}	(Shi 1994c)
Mole volume of Zr, V_{mol}	1.4×10^{-5} m^3/mole	(Singman 1985)
Young's modulus of Zr, E	81.5 GPa at T = 250°C	(Puls 1990)
Poisson's ratio, v	0.33 at T = 250°C	(Puls 1990)
strains ε^c caused by a hydrogen	$\varepsilon_{11}^c = \varepsilon_{22}^c = 0.0329$	(MacEwen1985)
atom* in Zr	$\varepsilon_{33}^c = 0.0542$ (ε_{33}^c in c direction of hcp lattice)	

*Values are for the hydrogen isotope, deuterium.

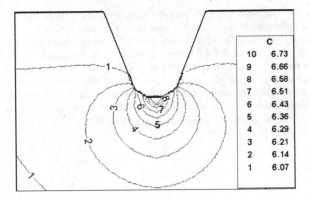

Figure 3. Hydrogen concentration ($c \times 10^{-3}$) around the notch after 80 hours of diffusion.

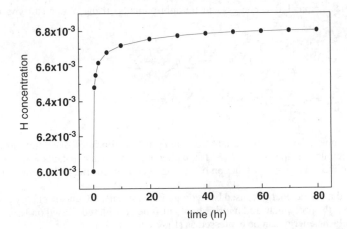

Figure 4. Change of the peak hydrogen concentration versus diffusion time.

shape of the hydrogen distribution is similar to that of hydrostatic stress. As being expected, hydrogen atoms diffuse from lower tensile hydrostatic region to higher tensile hydrostatic region, and the peak concentration of hydrogen is reached at the position corresponding to the maximum tensile hydrostatic stress. Figure 4 is the peak concentration versus time. We can see that the diffusion rate is fast in the first 10 hours, while slows down in the later stage and reach equilibrium at about 60 hours.

According to (9), the chemical potential of hydrogen is a function of both hydrogen concentration and applied stress. Interstitial hydrogen atoms cause the distortion of the lattice of the matrix, thus increase the chemical potential of the system. An applied tensile stress can ease the lattice distortion thereby lower the chemical potential. Since the chemical potential of the system must reduce to the minimum in a spontaneous process, the diffusion process for hydrogen from the regions with lower tensile hydrostatic stress to higher tensile hydrostatic stress regions will continue until the chemical potential everywhere in the system reach the same value.

4 MODELING OF HYDRIDE PRECIPITATION

In a homogeneous continuum system, the anisotropic properties of each grain are neglect, the orientation of the hydride is determined by the stress field. To simulate this situation, a multi-variant system, which can represent the structural change of the precipitates growing along any direction in the specimen, is constructed. We use η_p, the long-range order (lro) parameter to denote the structure of pth variant. The more lro the system have, the more closer to mimic the isotropic system. In our simulation, 12 lro were adopted. The angular difference between the two successive variants is 15°.

According to thermodynamics, the equilibrium state of a multiphase system corresponds to minimum free energy. The driving force for the temporal evolution of a multiphase microstructure consists of the following four major contributions: the reduction in chemical free energy; the change in the total interfacial energy between different phases; the relaxation of the strain energy caused by the lattice mismatch between the matrix and precipitates; and the reduction in the interaction energy between the eigenstrains and external load. Consider all these contributions to the total free energy in terms of the field variables, $c(r, t)$ and $\eta_p(r, t)$, the total chemical free energy of a multi-phase system may be expressed as (Cahn & Hilliard 1958):

$$F_c = \int_v \left[f(c, \eta_p(r)) + \sum_{p=1}^{v} \frac{\alpha_p}{2} (\nabla \eta_p(r))^2 + \frac{\beta}{2} (\nabla c)^2 \right] d^3r \tag{13}$$

where α and β are gradient energy coefficients. The integration is performed over the entire system. The last two terms are related to the interfacial energy between different phases. The first term, $f(c, \eta_p)$, the local specific chemical free energy, can be approximated using a Landau-type of free energy polynomial

$$f(c, \eta_p) = \frac{A_1}{2}(c - c_1)^2 + \frac{A_2}{2}(c - c_2)\sum_{p}^{v}\eta_p^2 - \frac{A_3}{4}\sum_{p}^{v}\eta_p^4 + \frac{A_4}{6}\sum_{p}^{v}\eta_p^6$$

$$+ A_5\sum_{q \neq p}^{v}\eta_p^2\eta_q^2 + A_6\sum_{p \neq q, p \neq r}^{v}\eta_p^4(\eta_q^2 + \eta_r^2) + A_7\sum_{p \neq q \neq r}^{v}\eta_p^2\eta_q^2\eta_r^2 \tag{14}$$

where v is the number of orientation variants, c_1 and c_2 are equilibrium concentrations of hydrogen in matrix and in precipitate, respectively; $A_1 \sim A_7$ are phenomenological constants which are chosen to fit the local specific free energies as a function of hydrogen content for the zirconium matrix and for the hydride.

In addition to the elastic energy caused by hydrogen interstitials, the elastic energy caused by lattice mismatch between the precipitate and matrix should also be considered. Based on Khachaturyan's theory, this part of elastic energy can be expressed as (Li & Chen 1997):

$$E_{el} = \frac{V}{2}C_{ijkl}\bar{\varepsilon}_{ij}\bar{\varepsilon}_{kl} - VC_{ijkl}\bar{\varepsilon}_{ij}\sum_{p}^{v}\varepsilon_{kl}^0(p)\overline{\eta_p^2(r)} + \frac{V}{2}C_{ijkl}\sum_{p}^{v}\sum_{q}^{v}\varepsilon_{ij}^0(p)\varepsilon_{kl}^0(q)\overline{\eta_p^2(r)\eta_q^2(r)}$$

$$- \frac{1}{2}\sum_{p}^{v}\sum_{q}^{v}\int\frac{d^3g}{(2\pi)^3}B_{pq}(n)\{\eta_p^2(r)\}_g^*\{\eta_q^2(r)\}_g \tag{15}$$

where $\overline{(\dots)}$ represent the volume average of $(\dots)$, V is the total volume of the system, C_{ijkl} is the elastic constant tensor, $\varepsilon^0(p)$ is the stress-free transformation strain for the pth variant when $\eta_p(r) = 1$. Here we can assume $\varepsilon_{ij}^0(r) = \sum_p \varepsilon_{ij}^0(p)\eta_p^2(r)$, which denotes the local stress-free transformation strain. $B_{pq}(n) = n_i\sigma_{ij}(p)\Omega_{jk}(n)\bar{\sigma}_{kl}(q)n_l$, $\{\eta_q^2(r)\}_g$ is the Fourier transform of $\eta_q^2(r)$ and $\{\eta_q^2(r)\}_g^*$ is the complex conjugate of $\{\eta_q^2(r)\}_g$.

In the phase-field model, the temporal evolution of the microstructure during phase transformation is determined by simultaneously solving the time dependent Ginzburg-Landau equations (Cahn 1961, Wang et al. 1993) for $\eta_p(r, t)$ and the Cahn-Hilliard diffusion equation (Allen et al. 1979) for $c(r, t)$,

$$\frac{\partial \eta_p(r,t)}{\partial t} = -L_p\frac{\delta F}{\delta \eta_p(r,t)} + \varsigma_p(r,t) \tag{16}$$

$$\frac{\partial c(r,t)}{\partial t} = \nabla \cdot M\nabla\frac{\delta F}{\delta c(r,t)} + \xi(r,t) \tag{17}$$

944

Table 2. Parameters used for the simulation of the hydride precipitation.

Parameters	Value	References
Initial hydrogen concentration, c_0	0.1	
Dynamic constant, L, M	$0.4\,J^{-1}sec^{-1}$	
strains ε^0 caused by a hydride misfit in Zr	$\varepsilon^0_{11} = 0.00551, \varepsilon^0_{22} = 0.00564$	(Carpenter 1973)
	$\varepsilon^0_{33} = 0.0570\ (\varepsilon^0_{33}$ in c direction of hcp lattice)	
$A_1 \sim A_7$	$A_1 = 18.5; A_2 = -8.5; A_3 = 11.5$	
	$A_4 = 4.5; A_5 = A_6 = A_7 = 1$	

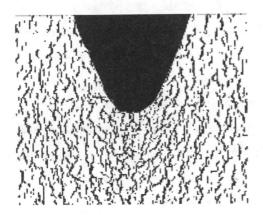

Figure 5. Phase-field simulation of γ-hydride precipitation.

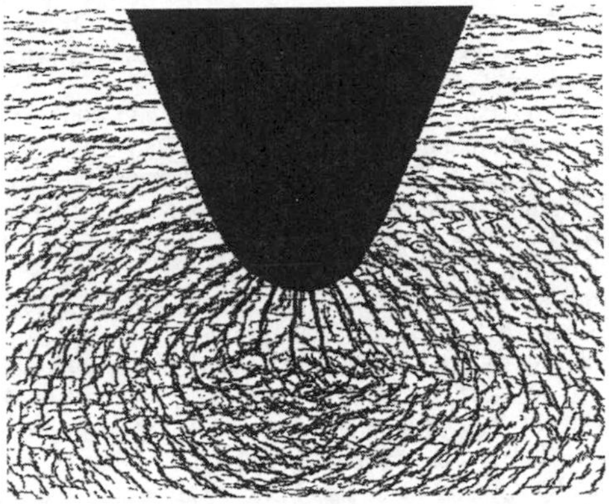

Figure 6. Optical microscopy of hydride precipitation in zirconium.

where L and M are kinetic coefficients characterizing structural relaxation and diffusional mobility. To simplify the calculation, we approximate these coefficients as constants. F is the total free energy of the system; $\zeta_p(r, t)$ and $\xi(r, t)$ are Langivin random noise terms that are related to fluctuations in the long-range order parameters and hydrogen concentration, respectively. The noise terms satisfy Gaussian distribution and meet the requirement of the fluctuation–dissipation theorem. Parameters used in the simulation of hydride precipitation were shown in Table 2. Since the above equations are non-linear with respect to field variables, we solve them numerically using a semi-implicit Fourier Spectral method (Chen & Shen 1998). The simulation is conducted in a 512×512 uniform grid, and one of the results is shown in Figure 5.

It can be seen that density of the hydrides is higher at the notch tip area and hydrides precipitate in the direction perpendicular to the notch surface. Figure 6 shows a real example of hydride precipitation patterns developed after many temperature cycles. Even though our simulation was done at a constant temperature and at a different hydrogen concentration, the precipitation pattern obtained by the simulation resembles the real case.

ACKNOWLEDGEMENTS

This work was supported by grants from Hong Kong Polytechnic University (G-V851) for Ma, from the Research Grants Council of Hong Kong (B-Q471) for Shi and Woo, and from the U.S. National Science Foundation (DMR 96-33719) for Chen.

REFERENCES

Allen S.M and Cahn J.W. 1979. A microscopic theory for antiphase boundary motion and its application to antiphase domain coarsening. Acta Metal. 27. 1085–1095.

Cahn J.W. and Hilliard J.E. 1958. Free energy of a nonuniform system. I. Interfacial free energy. *J. Chem. Phys.* 28: 258–267

Cahn J.W. 1961. On spinodal decomposition. *Acta Metall.* 9: 795–801

Carpenter G.J. 1973. Dilatational misfit of zirconium hydrides precipitated in zirconium. *J. Nucl.Mater.* 48: 264–266

Chen L.Q. and Wang Y.Z. 1996. Mathematically modeling mesoscale microstructural evolution. *J.O.M.* 48: 11–18

Chen L.Q. and Shen J. 1998. Application of seni-implicite Fourier-spectral method to phase-field equations. *Comp. Phys. Comm.* 108: 147–158

Dutton R. and Puls M.P. 1975. A theoretical Model for hydrogen induced sub-critical crack growth, In: A.W. Thompson and I.B. Bernstein (eds) *Effect of Hydrogen on Behavior of Materials.* New York: The Metallurgical Society of AIME. 516–524

Eadie R.L. Metzger D.R. and Lèger M. 1993. The thermal ratchetting of hydrogen in zirconium niobium – an illustration using finite-element modeling. *Scr. Metall.* 29: 335–340

Ells C.E. 1970. The stress orientation of hydride in zirconium alloys, *J. Nucl. Mater.* 35: 306–315

Khachaturyan A.G. 1983. Theory of Structural Transformations in solids, John Wiley & Sons. Inc., New York.

Li D.Y. and Chen L.Q. 1997. Selective variant growth of coherent $Ti_{11}Ni_{14}$ precipitate in a TiNi alloy under applied stresses. *Acta Metal.* 45: 471–480

Ma X.Q., Shi S.Q., Woo C.H. and Chen L.Q. 2002a. Effect of applied load on nucleation and growth of γ-hydrides in zirconium. *Comp. Mater. Sci.* 23: 283–290

Ma X.Q., Shi S.Q., Woo C.H. and Chen L.Q. 2002b. Simulation of γ-hydride in bi-crystalline zirconium under uniformly applied load. *Mat. Sci. Eng A-Struct.* 334: 6–10

MacEwen S.R, Coleman C.E, Ells C.E. and Faber J. 1985. Dilation of hcp zirconium by interstitial deuterium. *Acta Metall.* 33: 753–757

Puls M.P. 1990. Effect of crack tip stress states and hydride-matrix interaction stress on delayed hydride cracking. *Metall. Trans.* A . 21: 2905–2917

Shi S.Q. and Puls M.P. 1994a. Criteria of fracture initiation at the hydrides in zirconium alloys. I. Sharp crack tip *J. Nucl. Mater.* 208: 232–242.

Shi S.Q., Puls M.P. and Sagat S. 1994b. Criteria of fracture initiation at the hydrides in zirconium alloys. II. Shallow notch. *J. Nucl. Mater.* 208: 243–251.

Shi S.Q., Liao M. and Puls M.P. 1994c. Modeling of time dependent hydride growth at the crack tips in zirconium alloys. *Model. Simul. Mater. Sci.* 2: 1065–1078

Singman C.N. 1984. Atomic volume and allotropy of the elements. *J. Chem. Edu.* 61: 137–142

Wang Y., Chen L.Q.and Khachaturyan A.G. 1993. Shape evolution of a coherent tetragonal precipitate in partially stabilizer cubic ZrO_2: A computer simulation. *J. Am. Ceram. Soc.* 76: 3029–3033

Computational Methods in Engineering and Science, Iu et al. (eds)
© *2003 Swets & Zeitlinger, Lisse, ISBN 90 5809 567 3*

The influence of long-range action and boundary effect on vibration of the atom-string with finite atoms

X. Zhang
College of Aerospace Engineering, Nanjing University of Aeronautics and Astronautics, Nanjing, China

ABSTRACT: Two models of the atom-strings with finite atoms are established to research the influence of long-range action and boundary effect on their vibration. Series of programs are made to calculate the models at different conditions. The results are satisfying and interesting, and offer reference for the larger scale computation in the future.

1 INTRODUCTION

Nano-structured materials have excited considerable interest in the community of materials research in the last few years due to their scientific and commercial values. In order to reveal the physical properties of the nano-structured materials, many methods have been developed, among which molecular modeling probably is the most popular. However, even though the molecular modeling method was advanced many years ago and is an excellent tool, some problems still remain unsolved. For instance, how to precisely calculate the long-range action and boundary effect in the molecular modeling is still unclear.

This paper discusses the influence of boundary effect and long-range action on the vibration of system which consists of finite nano-particles. In this paper, firstly Lennard–Jones potential is introduced to describe the action between atoms. Then, two models are established to depict the conditions of boundary effect and long-range action. Finally, modeling results are shown as well as supposed explanation being given.

2 MODELS FOR COMPUTATION

In this section two kinds of models with finite atoms are introduced to depict the forms of action among atoms. By using the models, the problem is simplified to a series of non-linear equations which can be easily processed by computer.

2.1 *Selection of inter-atomic potential*

In molecular modeling it is important to choose an appropriate description of the forces between individual atoms prior to the actual calculations. Many methods have been developed to describe the action of atoms, among of which Lennard–Jones potential is widely used. In this paper, Lennard–Jones potential is supposed to correctly depict the actions between atoms:

$$u(r_{ij}) = 4\varepsilon \left[\left(\frac{\sigma}{r_{ij}} \right)^{12} - \left(\frac{\sigma}{r_{ij}} \right)^{6} \right] \tag{1}$$

where ε is a minimal value of potential and in this case, the distance between atoms equals to $2^{1/6}\sigma$. In equation (1) ε and σ are units of energy and length, respectively.

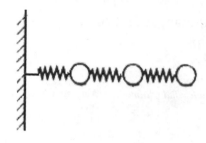

Figure 1-1.

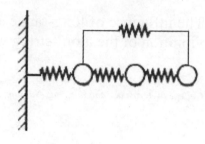

Figure 1-2.

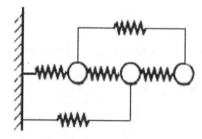

Figure 1-3.

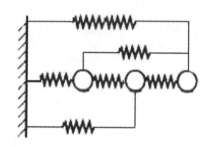

Figure 1-4.

Thereafter, the force that atom i acting on atom j is:

$$F(r_{ij}) = 48\left(\frac{\varepsilon}{\sigma^2}\right)(x_i - x_j)\left[\left(\frac{\sigma}{r_{ij}}\right)^{14} - \frac{1}{2}\left(\frac{\sigma}{r_{ij}}\right)^{8}\right] \tag{2}$$

However, the expression above is not suitable for computation. In order to get the appropriate forms, new units should be introduced. Let $(m\sigma^2/48\varepsilon)^{\frac{1}{2}}$ and σ represent units of time and length respectively, here m is the mass of an atom. Then the deductive form of equation (2) can be given as:

$$\ddot{x}_{ij} = (x_i - x_j)\left[(x_i - x_j)^{-14} - \frac{1}{2}(x_i - x_j)^{-8}\right] \tag{3}$$

2.2 Model I: fixed boundary

In these models boundary effect is the major subject that we concern. For convenience, only one-dimensional case is considered, so concrete models are as shown in Figures 1-1 to 1-4.

Figure 1-1 shows the case that only forces between immediate atoms are considered. In Figures 1-2, 3 and 4, the long-range action and boundary effect are gradually introduced.

Besides these 3-atom models, the method used here may be also dealt with the models with four or five atoms. The major concerns of those computations are to find whether the long-range action and boundary effect obviously affect the vibration of systems.

2.3 Model II: period boundary

This model is similar to model I except that period boundary is introduced instead of the fixed one. Once the period boundary is used, the boundary effect disappears automatically and the long-range action becomes the important factor influencing the vibration of atom-string.

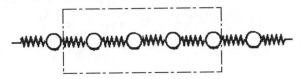

Figure 2-1.   4-atom model.

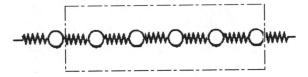

Figure 2-2.   5-atom model.

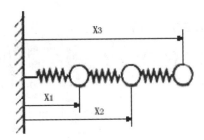

Figure 3-1

To introduce period boundary, a representative atom-sting element (RAE) is used in this study. Thus, systems are composed of several RAEs, which are of the same vibration state. The 4-atom and 5-atom models are shown in Figures 2-1 to 2-2.

In computation the long-range action across three atoms and four atoms are calculated gradually in each RAE so as to find the influence of long-range action under the period boundary condition.

3   MODELING METHOD AND RESULTS

3.1   *Modeling method*

By using Lennard–Jones potential, the vibration equations of different models are established. These equations can be solved by Runge–Kutta method. When calculating, the most important procedure is to set appropriate initial displacement and velocity, since any improper initial data will cause divergent results. Generally, the initial data may be given according to the displacement and velocity of atoms at the positions of equilibrium. Consequently, the first step is to determine the position of equilibrium of atoms by Newton–Raphson method. For example, if we select the coordinates as shown in Figure 3-1, then by using Newton–Rapson method, equilibrium positions of atoms are obtained, which can be written as:

$$X_1^0 = 1.12246 \quad X_2^0 = 2.24492 \quad X_3^0 = 3.36427 \tag{4}$$

3.2   *Results of model I*

As mentioned above, different cases in model I are considered so as to study the influence of long-range action and boundary effect. In order to correctly show the results, identical initial data should be set to start the program. The results based on Model I are shown in Figures 4-1 to 4-4 in which abscissa indicates time and ordinate displacement.

949

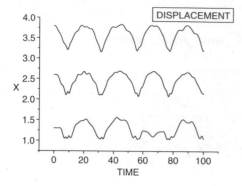

Figure 4-1. Result of Figure 1-1.

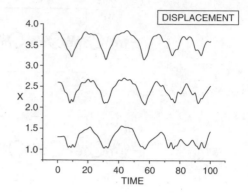

Figure 4-2. Result of Figure 1-2.

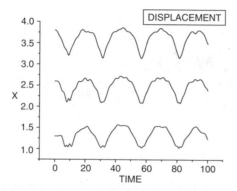

Figure 4-3. Result of Figure 1-3.

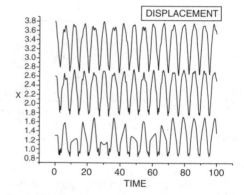

Figure 4-4. Result of Figure 1-4.

As can be seen from figures above, the "frequency" of the system with 3 atoms is obviously affected by the boundary effect. This influence mainly comes from the attractive force between the boundary and the farthest atom. On the contrary, the long-range force between the atoms affects slightly on the vibration of system.

In 3-atom models the modeling, in some degree, did not entirely reveal the relation between boundary effect and long-range action and the vibration conditions of the systems. Therefore, 4-atom systems and 5-atom systems are also calculated in this study, parts of the results obtained are shown in Figures 5-1 to 5-3, which are the results of 4-atom models; and Figures 6-1 and 6-2, which are the results of 5-atom models.

The state depicted in Figure 5-1 does not consider the boundary effect and long-range action. The states described in Figures 5-2 and 5-3 consider the long-range action and both the long-range action and the boundary effect, respectively. The state described in Figure 6-1 does not consider boundary effect and long-range action. The state depicted in Figure 6-2 considers both the boundary effect and the long-range action.

It is interesting that the influence of boundary effect seems to disappear compared with 3-atom system. The results shows, as the number of the atoms increases, boundary effect may be ignored.

3.3   Results of model II

In the modeling of Model I the long-range action is not obvious compared to the boundary effect, but it does not mean that long-range action can be neglected. By using period boundary, the long-range action can be introduced solely and the influence of it may be apparent. For the 4-atom systems, identical initial data are set to start the modeling programs, so the same as the 5-atom systems. The results based on Model II are shown in the following figures.

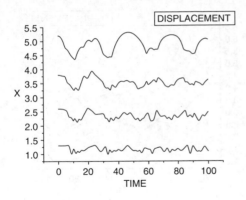

Figure 5-1.

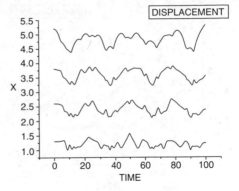

Figure 5-2.

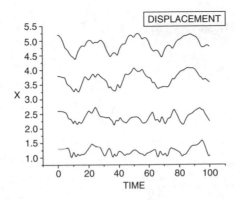

Figure 5-3.

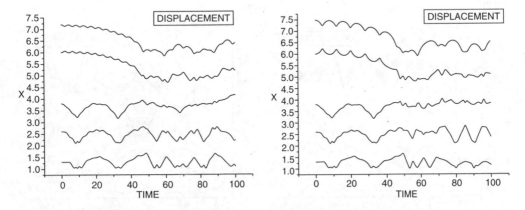

Figure 6-1.

Figure 6-2.

Figures 7-1 to 7-3 are the results of the model shown in Figure 2-1, which is in the state without the long-range action. The state depicted in Figure 7-2 only considers long-range action across 3 atoms; the state described in Figure 7-3 considers the long-range action across both 3 and 4 atoms. Figures 8-1 to 8-3 are the results of the model shown in Figure 2-2, which is also in the state without the long-range

951

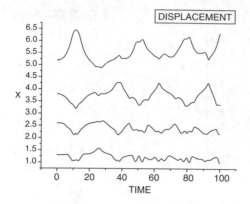

Figure 7-1.

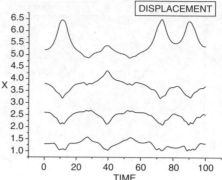

Figure 7-2.

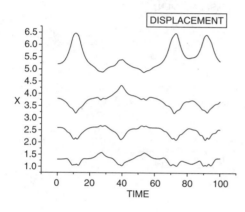

Figure 7-3.

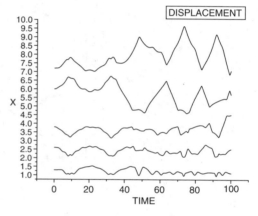

Figure 8-1.

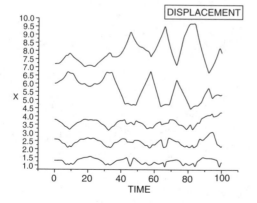

Figure 8-2.

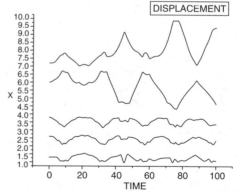

Figure 8-3.

action. The state depicted in Figure 8-2 only considers long-range action across 3 atoms; the state described in Figure 8-3 considers the long-range action across both 3 and 4 atoms.

As can be seen from the figures above, long-range action affects the vibration of the systems without boundary effect. It means that when calculating the vibration properties of systems with finite atoms, long-range actions should be considered. This agrees well with the results of many molecular modeling using the period boundary.

4 CONCLUSIONS AND DISCUSSION

In this paper models of different boundaries are established to research long-range action and boundary effect. Computational results show that boundary effect can be neglected in the systems with more than 4 atoms, because as the number of atoms increases, the influence of which the boundary acting on the vibration of systems decrease drastically and forces among atoms contribute most. Long-range action, on the other hand, should not be omitted because the vibration of the system, though is not much obvious, varies as the number of atoms in the system increases. So, for system that contains more than 4 atoms, boundary effect can be ignored, while long-range action should be considered especially as the number of the particles increase. Another important aspect should be mentioned is that all the conclusions concern with the system with finite atoms.

However, the study of this paper is only an initial research on the inter-action in the systems with finite atoms, for both the models and the description of force field among atoms have been simplified. For example, the vibration of the systems is actually not in one dimension and the arrangement of atoms is also not confined along a line. Besides that, not all the force fields can be described appropriately by Lennard–Jones potential. All these problems wait for further research.

ACKNOWLEDGMENT

The author gratefully acknowledges helpful discussions with Dr. Z. Huang.

REFERENCES

He, G.Y. Zhou, Y.B. & Zhang, G.F. 1993. *FORTRAN'77 Handbook*, Beijing: Sciencepress
Heermann, D.W. 1995. *Computer Simulation Methods in Theoretical Physics*. New York: Springer-Verlag
Leach, A.R. 1996. Molecular Modeling: Principles and Applications. London: Addison Wesley Longman limited

A 2D FDTD full-wave code for simulating the diagnostic of fusion plasmas with microwave reflectometry

F. da Silva, M. Manso & P. Varela
Associação EURATOM/IST Centro de Fusão Nuclear, Instituto Superior Técnico, Lisboa, Portugal

S. Heuraux
*Laboratoire de Physique des Milieux Ionisés et Applications, Université Henri Poincaré,
Vandœuvre Cedex, France*

ABSTRACT: A two-dimensional finite-differences time-domain full-wave code has been developed
to simulate microwave reflectometry diagnostic in fusion plasmas, aiming at the following objectives:
understanding physical effects, improving profile measurements, assessing signal processing techniques
and devising new applications for reflectometry. The bases of the code as well as its computational
implementation are described. An example is presented where the code is used to simulate density profile
measurements in the presence of turbulence and to assess the behavior of the burst-mode data analysis
tool, used in ASDEX Upgrade tokamak, to improve the accuracy of profile measurements distorted by
plasma turbulence. It is shown that the code validates the burst mode analysis and establishes the domain
where this data processing is applicable.

1 INTRODUCTION

Reflectometry is a useful tool to reach plasma density profile, coherent plasma events and some information on turbulence in natural and fusion plasma as well. Reflectometry is based on the radar principle
and it can use several techniques, namely Frequency Modulated Continuous Wave (FM-CW) broadband
reflectometry. In the case of the O-mode reflectometry, where the electric field is parallel to the external
magnetic field, the reflection point is directly given by the equality of the plasma frequency ($f_{pe} = 8.98$
$n_e^{1/2}$, n_e in m^{-3}) and the probing frequency f. In general for a given polarization of the probing wave the
reflection point is determined by the nullity of the real part of the propagation index. This region is
called cut-off layer. As a FM-CW broadband reflectometer is able to give group delay $\tau_g(f) = 1/(2\pi)$
$\partial\varphi/\partial f$ for all the probing frequency where $\varphi(f)$ is the phase difference between the reflected wave and
the reference wave. Then it is possible by Abel's inversion to reconstruct the plasma density profile
assuming monotonic profile where the distance to the cut-off position $d(f)$ is given as follows:

$$d(f) = \frac{c}{2\pi^2} \int_0^f \frac{\partial\varphi}{\partial f} \left(f^2 - \upsilon^2\right)^{-1/2} d\upsilon, \tag{1}$$

where c is the speed of light in a vacuum and f the probing frequency.

Fusion plasmas in next step tokamaks such as ITER will be extremely complex media with harsh electromagnetic machine environment. In this severe condition the control of the plasma position will be
made by reflectometers. This medium is subject to turbulence and local density mode, which impose an
earmark on the phase of the reflected wave. While these perturbations carry important information either
for the study of turbulent phenomena or to the understanding of plasma modes, from the point of view
of density profile evaluation they represent a nuisance. The understanding of these phenomena and how
to obviate them is of prime importance to obtain accurate profile evaluation and it is one of the ground
reasons for the use of simulation codes. With these, we can simulate several plasma scenarios; check their
influence on the reflected signals and the impact on profile reconstruction. The possibility and limitations

of the data analysis tools with under a controlled environment have been explored here for the burst mode method (Varela 2003). The implementation of 2D full have code obeys to the following objectives: understanding physical effects, improving profile measurements, assessing signal processing techniques and devising new applications for reflectometry.

This code has been already used in several studies, such as preliminary studies of the impact of profile movements (Silva 2001), destructive interference, and the effects of magnetic modes with turbulence have on the reflectometry signal (Silva 2003 and reference therein). In the present work we evaluate the behavior of data analysis tools used at ASDEX Upgrade: *best path analysis* (Varela 1999) and *burst mode analysis* (Varela 2001). We also do a preliminary study on the recovery of density profiles in the presence of turbulence using these signal processing techniques.

2   DESCRIPTION OF THE CODE

2.1   *Reflectometer modeling and base equations*

The reflectometer modeled is one of ASDEX homodyne broadband reflectometers. Within the simulating box, the end part of the active arm waveguide and the antenna are modeled by imposing the electric field to a null value $E_z = 0$. The oscillator corresponds to the method of injecting the signal in the simulation grid and the end part of the active arm, the reference arm, the mixer and low-pass filter are done in the post-FDTD simulation as part of the signal processing analysis of the results. To simulate the experiment a single horn antenna both for emission and reception of the signal is used. This antenna is a 2D H-plane horn with a half power beam width of $\approx 30^\square$ (on these simulations). The signal is excited in the waveguide as a TE_{10} mode with a method for unidirectional signal injection (transparent injection) based on the bootstrapping method (Schneider 1998) is used allowing to extract the return signal alone. The input signal in the simulation has an evolving instantaneous frequency $f(t) = f_c + f_\Delta x(t)$, with $f_\Delta < f_c$, where f_Δ is the frequency deviation, f_c is the carrier frequency and $x(t)$ is the modulating signal, in our case a ramp. The return signal is directly pick-up in the waveguide (since we are using unidirectional injection) and multiplied (mixed) with the reference signal. This results on a signal with the double of the probing frequencies range (which will be filtered-out) and a low frequency component carrying the phase difference, the reflectometry signal

$$s(t) = S(t)\cos\left[\varphi(t)\right]. \tag{2}$$

This signal will be analyzed using signal processing techniques, presented later, to obtain the relevant information, in particular the phase derivative $\partial\varphi/\partial f$ needed for density profile reconstruction.

The code being developed is a two-dimensional full-wave Maxwell code. It solves the Maxwell curl equations together with the plasma current equation, for ordinary mode propagation in a cold plasma, using a Finite-Differences Time-Domain (FDTD) method. The propagation is considered on the $x-y$ plane with no gradients on the z axis. The plasma static magnetic filed is assumed in the z direction, restricting the current flow to this same direction. Evolutions of the density n_e occur on a slow time scale. The effects of plasma have been included in the response of the current density to the electric field. This summarizes a transversal magnetic propagation (TM), together with the equation for the plasma current.

$$\frac{\partial B_x}{\partial t} = -\frac{\partial E_z}{\partial y}; \quad \frac{\partial B_y}{\partial t} = \frac{\partial E_z}{\partial x}; \quad \frac{\partial E_z}{\partial t} = \frac{1}{\mu_0\varepsilon_0}\left(\frac{\partial B_y}{\partial x} - \frac{\partial B_x}{\partial y}\right) - \frac{1}{\varepsilon_0}J_z;$$

$$\frac{\partial J_z}{\partial t} = \frac{e^2}{m_e}n_eE_z. \tag{3}$$

using a Yee scheme (Yee 1966), we arrive at the FDTD expressions used in the code. For the electric field:

$$E_z^{n+1/2}(i,j) = E_z^{n-1/2}(i,j) + \frac{\Delta t}{\mu_0\varepsilon_0\Delta y}\left[B_y^n\left(i+\frac{1}{2},j\right) - B_y^n\left(i-\frac{1}{2},j\right)\right] - $$

$$-\left[B_x^n\left(i,j+\frac{1}{2}\right) - B_x^n\left(i,j-\frac{1}{2}\right)\right], \tag{4}$$

and half time step latter the magnetic field and the current density are updated:

956

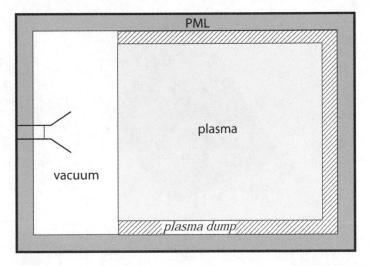

Figure 1.   Schema of the simulation set-up. The box is surrounded with PML boundary conditions. One antenna is used for both emission (using an unidirectional transparent source) and reception. The plasma is damped on the borders to match the vacuum propagation conditions suited to the PML technique.

$$B_x^{n+1}\left(i, j+\frac{1}{2}\right) = B_x^n\left(i, j+\frac{1}{2}\right) - \frac{\Delta t}{\Delta y}\left[E_z^{n+1/2}\left(i, j+1\right) - E_z^{n+1/2}\left(i, j-1\right)\right]$$

$$B_y^{n+1}\left(i+\frac{1}{2}, j\right) = B_y^n\left(i+\frac{1}{2}, j\right) + \frac{\Delta t}{\Delta x}\left[E_z^{n+1/2}\left(i+1, j\right) - E_z^{n+1/2}\left(i-1, j\right)\right]$$

$$J_z^{n+1}\left(i, j\right) = J_z^{n+1}\left(i, j\right) + \frac{\Delta t e^2}{m_e} n_e(i, j) E_z^{n+1/2}. \tag{5}$$

To ensure the stability of the scheme, the Courant-Friedrichs-Lewy (CFL) conditions $\delta x, \delta y \geqslant 2c\delta t$, must be met, where δx, δy and δt are the spatial and temporal increments, respectively. In this work $\delta x = \delta y = \lambda_M/20$ and $\delta t = T_M/40$, where λ_M and T_M being the wavelength and the period correspondent to the highest probing frequency f_M. The choice of the number of points per wavelength results from a compromise. The higher the spatial sampling the better the phase accuracy and the longer the code runs. The increase is twofold: (i) In every iteration more points in the grid have to be calculated; (ii) Due to the CFL conditions the time sampling must be concomitantly augmented resulting in an increase of iteration to run the same time interval. The values presented here represent a good compromise between phase accuracy and speed of execution. The simulations in this work were done on a grid of 1023×723, corresponding to $38.4\,\text{cm} \times 27.1\,\text{cm}$ during $120{,}000$ iterations corresponding to 75 ns.

On the borders of the grid, a perfectly matched layer (PML) (Bérenger 1994) is being used as a boundary condition. These conditions outperform the traditional one-way equation methods such as proposed by (Higdon 1986) and are becoming the state of the art boundary conditions. The plasma is *damped* close to the box borders to match the vacuum propagation conditions suited to the PML technique.

This code can also describe all kinds of reflectometers as the amplitude modulation reflectometer, pulse reflectometer, Doppler reflectometer and correlation reflectometer.

2.2   *Modeling of the plasma*

The coupling between the electromagnetic (EM) wave and the plasma is taken into account by the current density, which depends on plasma density and the plasma velocity induced by the EM wave. The density is usually a function of space and time since the code allows time evolution of the density profile due to modification of the base plasma with time, local plasma movements, plasma coherent modes and turbulence.

Along one radial cut of the isodensity curves the plasma has generic shape given by $n_e(r) = n_0 \{1 - [(R - r)/R]^n\}^m$. Density modes $\delta n_{e_{MOD}}$ and turbulence $\delta n_{e_{TRB}}$ are added to the unperturbed background plasma $n_e = n_{e0} + \delta n_{e_{MOD}} + \delta n_{e_{TRB}}$ and all of these parts can be a function of time.

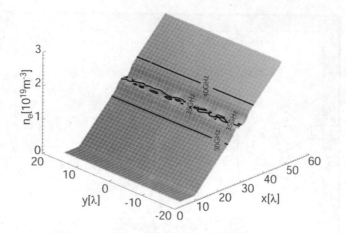

Figure 2. Model of a plasma density profile n_{e_0} with a $q = 2$-type island at the o-point (the plateau) $\delta n_{e_{MOD}}$ with turbulence $\delta n_{e_{TRB}}$ at the plateau (deforming the cut-off positions).

In Figure 2, the plasma is modeled along curves of isodensity with elliptical shape allowing the description of curved plasmas. We depict an example of a typical plasma setup. Added to the base plasma n_e, there is a mode $\delta n_{e_{MOD}}$, in this case a plateau due to the o-point of a $q = 2$ island, and a the plateau turbulence $\delta n_{e_{TRB}}$ is present as can be seen by the distorted cut-off positions for $f = 35\,\mathrm{GHz}$.

One way to model the turbulence is to compute the sum of modes with random phase according to a given wavenumber spectrum (Heuraux 2003). The density perturbation an each point obeys to:

$$\delta n_{e_{TRB}}(x, y) = \sum_{i=i_{min}}^{i_{max}} \sum_{j=j_{min}}^{j_{max}} A(i, j) \cos\left[k_x(i)x + k_y(j)y + \varphi(i, j) \right].$$

(6)

The amplitude $A(i, j)$ is chosen in agreement with experimental data (Devynck 1993) and may be modified during the simulation to accommodate several plasma scenarios. Keeping the same amplitude spectrum and varying $\delta n_{e_{TRB}}$ with time allows the setting of several turbulence *snapshots* with the same spectral scenario.

Here we have assumed that the samples of the density fluctuation are independent from each other. But in experiments, one has to know if it is justified using this simplification. However, it is also possible to simulate the situation where the density fluctuations are correlated in time over the measurement duration. The same limitation arrives when the plasma rotates at high velocity and we look at a localized density perturbation. That obeys to the following criterion on the validity domain: object velocity time measurement duration is much less than typical length of the object.

2.3 *Modeling of the signal processing tool*

The burst-mode data processing method (Varela 2001) was first applied to the analysis of broadband reflectometry data obtained in the ASDEX Upgrade tokamak to reduce the profile distortion due to plasma turbulence. The method takes advantage of the ultra-fast sweeping capability of the diagnostic to analyze data in bursts of frequency sweeps. A typical setup is to use sets of eight closely spaced ($10\,\mu s$) ultra-fast ($25\,\mu s$) sweeps to obtain an *average* profile with a temporal resolution of $270\,\mu s$. The method uses all the information in the individual spectrograms of the reflectometry signals, enhancing the features common to the sweeps in the burst and strongly reducing the spurious perturbations that might be present in some sweeps due to the effects of plasma turbulence. The burst-mode group delay is then obtained from the combined spectrogram using the best-path algorithm (Varela 1999).

Figure 3 (Varela 2002) shows how burst-mode analysis can avoid perturbations presented during plasma measurement and how it recovers the *underlying* profile and illustrates the difference between burst mode and a *normal* average. The burst mode technique applies statistics only to the coherent part in the measurements with the advantage of not needing *a priori* knowledge of this coherence. Average methods on the other hand will include also non-coherent parts of the signals.

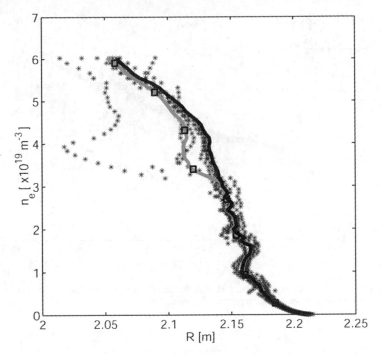

Figure 3. Density profiles obtained with the burst mode (continuous curve), average (gray with squares) and single sweep (asterisk) analysis. The phase derivative perturbations present in some sweeps cause large deviations in both the average and single-sweep profiles. The burst profile, however is not affected.

3   SIMULATIONS OF PROFILE RECONSTRUCTION AND COHERENT MODE TOOLS IN THE CASES OF A FLUCTUATING TOKAMAK PLASMA

The aim of this part is to show that the simulation of the diagnostic tools is a very good way to determine the validity domain of the tools used to evaluate the phase derivative for profile reconstruction, the limitations introduced by the density fluctuations, strong turbulence in this case.

3.1   Broadband sweep simulation and phase derivative evaluation

Simulations were made on a $51\lambda_{40\,\mathrm{GHz}} \times 36\lambda_{40\,\mathrm{GHz}}$ grid, probing a linear plasma whose maximum value is $2.5 \times 10^{19}\,\mathrm{m}^{-3}$ ($f_{\mathrm{pe}} = 45.0\,\mathrm{GHz}$) at $x = 33\lambda_{40\,\mathrm{GHz}}$. The normalized magnitude k-spectrum is constant ($S_{amp} = 1.0$) for small $k = (k_x^2 + k_y^2)^{1/2}$ satisfying $|k_x|,|k_y| \leqslant 400\,\mathrm{rad\,m}^{-1}$ from where it decays proportionally to k^{-3}.

The frequency is swept probing a density range $[1-2] \times 10^{19}\mathrm{m}^{-3}$, corresponding to [30–40 GHz] frequency band. Since experimental sweeping times are in the order of 20 μs, this density fluctuation can be considered frozen. Sweeps are made in groups of 8, each group at a different fluctuation level. We define the fluctuation level in percentage of the ratio between the root mean square (RMS) value of the fluctuation matrix $\delta n_{e_{TRB}}$ and value of the density at the cut-off position for a frequency of 35 GHz (middle-band) $n_{e-35\,\mathrm{GHz}}$. Each of the 8-burst runs (frequency sweeps with different fluctuations snapshot), the reflectometry signals [see Equation (2)] is processed using the methods described in 2.3.

Figure 4 shows the burst mode phase derivatives $\partial\phi/\partial f$ obtained for several levels of turbulence $n_{e-RMS}/n_{e-35\,\mathrm{GHz}}$, namely 0.5%, 1.0%, 1.5%, 2.0%, 2.5%, 5.0% and 10.0%, together with the theoretical Wentzel-Kramers-Brillouin (WKB) solution for the unperturbed plasma. Up to 2.5% the burst mode technique soundly recovers the base profile *under the turbulence*. For 5.0% and 10.0% the obtained phase derivatives detach form the WKB solution. In order to evaluate the error bars and to know the behavior of the burst mode phase derivative, the root mean square error (RMSE).

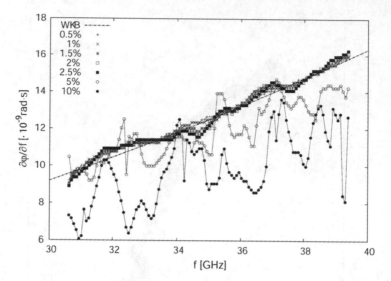

Figure 4. Phase derivatives for several levels of turbulence. The continuous line is the theoretical Wentzel-Kramers-Brillouin (WKB) solution for the unperturbed plasma.

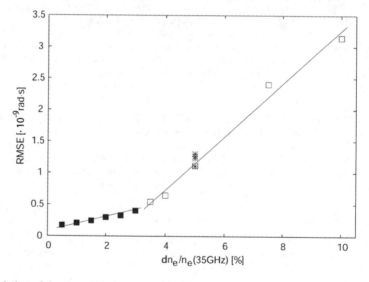

Figure 5. Evolution of the error with the level of turbulence. Around ~3.5% there is a change in the type of response of the system which goes from a linear to a nonlinear behavior.

$$\text{RMSE} = \sqrt{\frac{1}{N-1} \sum_{i=0}^{N-1} \left[\left(\frac{\partial \varphi_{\text{BRST}}}{\partial f} \right)_i - \left(\frac{\partial \varphi_{\text{WKB}}}{\partial f} \right)_i \right]^2} \tag{7}$$

of the burst mode phase derivative $\partial \phi_{\text{BRST}}/\partial f$ when referenced to the $\partial \phi_{\text{WKB}}/\partial f$ were computed and presented on Figure 5, for several levels of turbulence. Around $\delta n_{e-RMS}/n_{e-35\,\text{GHz}} \square 3.5\%$ there is a change in the evolution of the error, corresponding to a change in the type of response of the system (which goes from linear to non linear), rather than a failure of the analysis method. For a turbulence of 5% the graphic shows also RMSE points obtained for four more burst calculated for independent sets of turbulence matrixes. The statistical dispersion seems to be of the same order of magnitude as the fluctuations observed on the *main set* of samples. It indicates that the obtained error represent a characteristic of the system rather than a statistical fortuity.

4 DISCUSSION

As shown here, the presented simulations have enabled to evaluate the capability of a reflectometry data processing method in plasma physics using relevant features of the experimental set-up, namely the antenna arrangement. Especially, in this case, the simulations have been very helpful because they have exhibited a change in the plasma response when the amplitude of the density fluctuation increases. However, we have to keep in minds the physics of the problem, and what you want to study, because there are always some other physical or technical criteria to be satisfied. For example, one criterion for the burst-mode analysis presented here is that the data acquisition time has to be lower than the evolution time of the studied physical parameter (density profile here). However the simulations can help us to find a set of corrections to make an extension of the validity domain for the burst-mode data analysis and to justify the experimental choice of 6–8 single sweeps used in the burst mode for typical ASDEX plasma scenarios. The developed numerical tools are useful to prepare and study some new idea on plasma diagnostics using reflectometry as it has been done to extract the radial wave number spectrum of the density fluctuation (Heuraux 2003) or to identify the physical effect associated to the lost of the reflectometer signal (Silva 2003) or with similar tools to validate theoretical works (Gusakov 2002). Simulation can also provide information on the MHD events, which can occur in the plasma. The determination of MHD event position is particularly relevant for ITER to provide information about magnetic equilibrium and on the earmarks of the MHD events on the reflectometer signal. The FDTD code can also be used to simulate several kinds of reflectometry techniques such as frequency modulated reflectometry, amplitude modulation reflectometry, pulse reflectometry, correlation refelcometry, Doppler reflectometry and also the propagation of electromagnetic wave in fluctuating plasma for basic studies needing a full-wave description.

ACKNOWLEDGEMENTS

This work has been carried out within the framework of the Contract of Association between the European Atomic Energy Community and Instituto Superior Técnico. Financial Support from Fundação para a Ciência e Tecnologia and Praxix XXI was also received.

REFERENCES

Bérenger, J.P. 1994. A Perfectly Matched Layer for the Absorption of Electromagnetic Waves. *Journal of Computational Physics.* **114**: 185–200.

Devynck, P. et al. 1993. Localized measurements of turbulence in the TORE SUPRA tokamak. *Plasma Phys. Control. Fusion* **35**(1): 63–76.

Gusakov, E.Z. et al. 2002. Small angle scattering and spatial resolution of fluctuation reflectometry (comparison of the two-dimensional analytical theory with numerical calculations). *Plasma Phys. Cont. Fusion* **44**, 1565.

Heuraux, S. et al. 2003. Radial wave number spectrum of density fluctuations deduced from reflectomety phase signals. *Rev. Sci. Instrum.* **74**(3): 1501–1505.

Higdon, R.L. 1986. Absorbing Boundary Conditions for Difference Approximations to the Multi-Dimensional Wave Equation. *Math. Comput.* **47**: 437.

Schneider, J.B. et al. 1998. Implementation of Transparent Sources in FDTD Simulations. *IEEE Tran. Tran. Antennas and Propagation* **46**(8): 1159–1168.

Silva, F. et al. 2001. Simulation of reflectometry density changes using a 2D full-wave code. *Rev. Sci. Instrum.* **72**(1): 311–314.

Silva, F. et al. 2003. Simulations of amplitude and phase variations induced by magnetic islands with turbulence on reflectometry signals. *Rev. Sci. Instrum,* **74**(3): 1497–1500.

Varela, P. et al. 1999. Automatic evaluation of plasma density profiles from microwave reflectometry on ASDEX upgrade based on the time-frequncy distribution of the reflected signals. *Rev. Sci. Instrum.* **70**(1): 1060–1063.

Varela, P. et al. 2001. Assessment of density profile automatic evaluation from broadband reflectometry data. *Rev. Sci. Instrum.* **71**(1): 315–318.

Varela, P. 2002. *Automatic time-frequency analysis for plasma density profile evaluation from microwave reflectometry.* Lisbon: Universidade Técnica de Lisboa-Instituto Superior Técnico.

Yee, K.S. 1966. Numerical solution of initial boundary value problems involving Maxwell's equations in isotropic media. *IEEE Tran. Antennas and Propagation.* **14**: 302–307.

Effective improvement of Internet-based intelligent medical consultative system

B.N. Li, M.C. Dong & M.I. Vai
Department of Electrical & Electronics Engineering, Faculty of Science and Technology,
University of Macau, Macao

ABSTRACT: In this paper, the traditional rule-based medical Expert System is improved in two ways. The first is to diagnose with weighted symptoms so that the system can realize the inference under uncertainty. The second is to auto-customize rule base in Client, which can speed up the rule matching procedure comparatively. Both of them were applied in our prototyping system and acquired the expected experimental results.

1 INTRODUCTION

The rule-based Expert System belongs to the first generation Expert System (ES), which owns some unique advantages such as explicit inference, easy construction and so on. But the shortcomings are also obvious, for example, low ability of self-learning, low speed of rule matching in mass rule base. Therefore, many new concepts and methods were proposed to improve it, e.g. Neural-networks, Genetic algorithm, Data mining, etc. Among them, Neural-networks are helpful in improving the fault tolerance and the reasoning speed of Expert System, Genetic algorithm and Data mining are mainly used to extract the useful knowledge as rules from the patient's medical data automatically.

The objective of our research project, "Internet-based Intelligent Home Healthcare System", is to set up a reliable, cost-effective Expert System to support home body-checking and healthcare decision-making. It is combined with the specifications of Health Level Seven (HL7) and applied within the communities so that the unnecessary hospital-visits and hospitalizations could be reduced, more efficient and better quality healthcare could be delivered through Internet. It is intended to be a front-line intelligent, computerized clinical service to the patient and acts as a supporting or somewhat a substituting role for the clinician. New techniques, which realize the diagnosis in incomplete symptoms as well as auto-update the customized rule base in the Client terminal, are introduced in this paper.

2 IMPROVEMENT I: TREATING THE UNCERTAINTY WITH WEIGHTED RULES

Well-structured and precise inferring procedure is one of important advantages of rule-based Expert System, but in another side, it means lacking of flexibility in uncertain case. In practical medical diagnosis, it is unreasonable to always acquire all of the parameters needed for inference in once time. It is may be the first reason why there are still so little Expert Systems applied in our life, especially in medical domain. Of course, we can increase the number of the rules in knowledge base to have more careful inference, but it will result in another problem, namely, "knowledge explosion". How to deal with these problems boosts the development of many theories, e.g. Artificial Neural-networks, Naïve Bayesion Algorithm. The concept of inference under uncertainty in traditional rule-based ES includes the uncertain messages and the uncertain rules. The traditional rule-based ES just matches the whole antecedent condition, active if it's matched and negative if not. It is difficult to deal with the problem of uncertainty. Neural-networks are strongly tolerable for it stores the knowledge in the huge network and infers the result with weights. Here we extract the "weighting" idea from Neural-networks and propose a new method to improve the traditional rule-based ES: treating the uncertainty with weighted symptoms.

2.1 Principle

As to medical rule-based ES, it's feasible and reasonable to assume each symptom as a meta-term of the rule's antecedent. Therefore, a rule from the rule base of our system can be described as following:

$$IF \; (Symptom_1 \; weight_1, Symptom_2 \; weight_2, \cdots, Symptom_n \; weight_n) \tag{1}$$

$$THEN \; Disease_k : CF$$

where "$Symptom_i$" consists of the physiological parameter and its rank after discretization, "$weight_i$" is the corresponding importance-degree of "$Symptom_i$" in this rule, "$Disease_k$" is the subsequent result and "CF" is the confidence factor of this rule. Compared with the traditional predication rule, the rule here owns extra terms "$weight_i$" in its antecedent items. Once a message comes including the user's symptoms, our system will infer the result in such a way: firstly, it searches how many meta-terms are included in this message; secondly, it accumulates the weight if a meta-term is matched with the one of the rule's antecedent; at last, the final confidence factor is calculated by the following formula:

$$CR = (weight_{K_1} + weight_{K2} + \cdots + weight_{Km}) * CF \tag{2}$$

where "CR" is the confidence of the result, "$weight_{ki}(i = 1, 2, \cdots, m)$" is the weight of the ith meta-term matched.

2.2 Further discussion

2.2.1 How to deal with the threshold?

It's impossible to avoid the problem of threshold in terms of dealing with uncertainty. Neural-networks can adjust its thresholds automatically supervised by the training examples. As we know, one of the necessary jobs to operate the rule base of traditional rule-based Expert System is to keep the consistency of all rules. It is the foundation that the determined result can be inferred by rule-based ES. But due to now we need to treat the cases with incomplete symptoms, it is possible that there appear several inferring results in once diagnosing procedure. Thus we try to give out all reasonable, possible results so that the patients could assess and learn more about their health condition. All possible results are displayed to users in the descending sort of their CR. But it doesn't mean as more as better. On the contrary, sometimes it may confuse the users. Thus we need to set a predefined threshold recommended by medical specialists and now it is well equipped in our prototyping system. As a result, the system will abandon those possible results whose CR are less than the threshold. In the case when the number of final results exceeds 5, we just give out the first 5 possible results in our prototyping system.

2.2.2 How to obtain the weights?

Our developed prototyping system is still lack of self-learning function till now. But the ability of self-learning is an important feature of Intelligent System. As to Neural-networks, it keeps adjusting its coefficients based on the feedback till it has learned all the given training examples. Thus we need also add this self-learning function to make our system be able to adjust the "$weight$" and "CF" of the rules automatically. In fact, our system will install the module of self-learning in Server side and make good use of the calculating capability of Server. In the Server side of our prototyping system, the rules are generated automatically with the help of Data mining or Genetic algorithm. As to the weight of every meta-term in the rule, we adopt the statistic method. For example, if there are 25 kinds of physiological parameters collected, each parameter has 6 ranks ("Dangerous_High", "Fairish_High", High, Low, "Fairish_Low", "Dangerous_Low") and 10 kinds of cardiovascular diseases totally, then the space of rules is $10 * 6^{25}$. After Data mining, 500 classical rules are generated automatically in Server illustrated in the following format:

$$IF \quad (Symptom_{1,1} \quad Symptom_{1,2} \quad \cdots \quad Symptom_{1,k_1})$$
$$THEN \quad Disease_{n1} \quad CF_1$$
$$IF \quad (Symptom_{2,1} \quad Symptom_{2,2} \quad \cdots \quad Symptom_{2,k_2})$$
$$THEN \quad Disease_{n2} \quad CF_2$$
$$\cdots$$
$$IF \quad (Symptom_{500,1} \quad Symptom_{500,2} \quad \cdots \quad Symptom_{500,k_{500}})$$
$$THEN \quad Disease_{n500} \quad CF_{500} \tag{3}$$

where $n_1, n_2, \cdots, n_{500} \in \{1, \cdots, 10\}$ and $Symptom_{i,j}$ denotes the j-th symptom of the i-th rule. Assuming there were p *Symptoms* relative to $Disease_n$ and q rules including $Symptom_m$ according to $Disease_n$, we could adopt the method of statistics to find out the contribution of $Symptom_m$ ($m \in \{1, \cdots, p\}$) to $Disease_n$ ($n \in \{1, \cdots, 10\}$). The $weight_m$ may be worked out by accumulating the *CF* of the corresponding rules, namely,

$$weight_m = CF_1 + CF_2 + \cdots + CF_q \tag{4}$$

Of course, it is not enough because if $Symptom_m$ is a trivial factor but its amount is prior, $weight_m$ will be fairly big. In fact, it's not reasonable in terms of most medical cases. Therefore, the $weight_m$ needs to be generalized as following:

$$weight_m = \frac{CF_1 + CF_2 + \cdots + CF_q}{q} \tag{5}$$

$$weight_m = \frac{weight_m}{weight_1 + weight_2 + \cdots + weight_p} \tag{6}$$

If there is a new case encountered, the rules in the Server's rule base are updated accordingly. Then the updated rules will be transferred to the Client's rule base, which is discussed in the following section.

3 IMPROVEMENT II: AUTO-UPDATING THE CUSTOMIZED RULE BASE IN CLIENT SIDE

One of the problems of traditional ES is "knowledge explosion" mentioned previously. If there are hundred and thousands of rules in the rule base, the rule-matching speed turns to be a big problem. Much effort has been involved in it and some new methods or ideas have been developed, e.g. "blackboard controlling mechanism", "meta-rule", etc. In fact, most of them are based on the hierarchical structure. Therefore it is just the problem of how to schedule and operate the tree and sub-trees. As to medical Expert System, there are some unique features. Within a certain period, the symptoms for one person's often-occurred disease are not varied frequently and also the body status degrades gradually. Thus, in terms of a particular person, it is not necessary to search all rules in the mass rule base. The theories of "blackboard controlling mechanism" and "meta-rule" are based on this idea. These methods are effective because they avoid to search and match the rules in whole rule base sequentially. But all of them are lack of the self-learning capability. As to our prototyping system, there are some new features introduced by fully using the network.

3.1 Principle

Similar to the mechanism of "blackboard controlling mechanism" and "meta-rule", the rule base in Client side consists of the typical rules selected from the rule base in Server side and rules deduced from the user's Electronic Healthcare Records (HER). Therefore, it can reduce the inference time because the amount of rules has been refined markedly. As shown in Figure 1, the rule base in Server is set up including the rules generated from the medical historical data by the means of Data mining. At the same time, the "*weight*" and the "*CF*" are also generated and updated in this way. Once there are some new symptoms that cannot be treated by the Client, the Client will report the new symptoms to the Server and ask for its help. Then the Server diagnoses them in the mass rule base. If the results are obtained, the results and relative rules will be sent back to the Client. The results are displayed to user and the rule base in Client is updated with the rules received from the Server. This improvement owns two advantages at least. The first is to relieve the burden of the network and Server because they just response while there are demands from Clients. The second is to depress the requirements of the Client's processor and storage space, which makes it even possible to transplant the system from the Client to the portable apparatus. All in all, this technique tries to reduce the cost while keeping a higher performance of the whole system.

3.2 Further discussion

3.2.1 Evaluating the performance

The traditional evaluating principle is that to separate the examples into two sections: One is for training and another is for evaluating. After training, the Artificial Intelligent (AI) system is tested with the

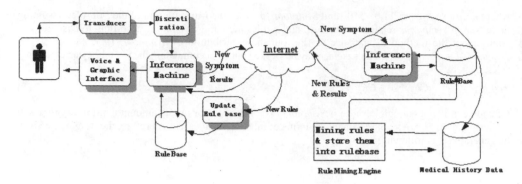

Figure 1.   The framework of customized rule-based medical Expert System.

evaluating set. If the accuracy ratio is less than an acceptable threshold, it indicates that the previous self-learning procedure fails and we need to train the system again. We also evaluate the performance of the customized rule base in Client based on this idea. As mentioned previously, the rules in the Client's rule base consist of the typical rules and the rules customized based on user's "Electronic Healthcare Record (EHR)". When a new user is registered into our system, we must assess the performance of its rule base before distributing it, as well as assess its performance while updating the rule base in Client. Since we take the rule base in Server as the normative, comprehensive reference, we just need to diagnose the new user's case by means of the rule base in Client and the rule base in Server separately. Then the two sets of diagnosing results are compared. We can define the result similarity and result dissimilarity as:

$$S_{Simi} = \frac{N_{same}}{N_{Server}} \times 100\% \tag{7}$$

$$S_{Diss} = \frac{N_{diss}}{N_{Server}} \times 100\% \tag{8}$$

where S_{Simi} stands for the result similarity and S_{Diss} stands for the result difference; N_{same} ($0 \le N_{same} \le _{Server} \le 5$) is the amount of the same diagnosing results between Client and Server; N_{diss} ($0 \le N_{diss} \le N_{Server} \le 5$, $0 \le (N_{same} + N_{diss}) \le N_{server} \le 5$) is the amount difference of the results in Client with the results in Server and N_{Server} is the amount of diagnosing results from Server. Here we set the acceptable threshold as 60%. For example, if 3 diagnosing results are worked out in Server, the Client must find out 2 diagnosing results compliant to the results from Server.

3.2.2   Increasing the intelligence in the Internet-based circumstance

The traditional concept of intelligence in rule-based ES includes modifying the searching strategies, adjusting the rules' position, adding and deleting rules from the rule base. For example, "meta-rule" can be used to adjust some rules if they are used scarcely or frequently. "Blackboard controlling mechanism" accelerates the rule searching and matching procedure by exchanging messages among multi-processing. But since our prototyping system is based on Internet, the concept of increasing intelligence is broadened comparatively. It is required not only to own the capabilities mentioned previously, but also be further enhanced to deal with "multi-occurrence", "concurrent" and "priority" issues etc. The Server may communicate with many Clients simultaneously so that the case should be handled according to their urgency-grades automatically. The whole system is based on the idea of "Information sharing" so that it could customize the rule base on Client side once-for-ever learning to improve their intelligence.

4   THE PRELIMINARY EXPERIMENTAL RESULTS

The developed system so far could assess users' cardiovascular system with their hemodynamic parameters. The commercial transducer in Client side would measure and collect the user's 30 physiological parameters as shown in following Table 1.

Table 1. The physiological parameters used for Data mining.

Acronym	Meaning	Acronym	Meaning	Acronym	Meaning
HR	Heart Rate	DP	Difference of Pressure	ABS	Area of Body Surface
MSP	Mean Systolic Pressure	MDP	Mean Diastolic Pressure	MAP	Mean Artery Pressure
CAP	Coronary Artery Pressure	PAWP	Pulmonary Artery Wedge Pressure	PAP	Pulmonary Artery Pressure
ABK	Aorta Blood leave K-value	AFEK	Artery Flexible Expanding K-value	AC	Artery Compliance
TPR	Total Peripheral Resistance Cardiac	STPR	Standard Total Peripheral Resistance	SV	Stroke Volume
CO	blood Output per Minute	SI	Stroke Index	CI	Cardiac Index
LVPE	Left Ventricle Pump Effect	LVPEI	Left Ventricle Pump Effect Index	LVR	Left Ventricle Resistance
EWK	Energy Workings K-value of Left Ventricle	HOV	Heart Oxygen Volume of Consumption	HOI	Heart Oxygen consumption Index
V	Blood Viscosity	OV	Original blood Viscosity	BV	Total Blood Volume
ALK	Microcirculation Blood Half Renew K-value	ALT	Microcirculation Blood Half Renew Time	TM	Microcirculation Blood Mean Stay Time

Then the collected data set will be sent to our system to do Data mining and generate the rules. Actually, the continuous analogue data sent by transducer will be discreted into 7 ranks: "Dangerous_High", "Fairish_High", "High", "Normal", "Low", "Fairish_Low", "Dangerous_Low". Then we save these data into Database and mark the "Normal data" as "Ignore". Afterwards, we use "ODBCHook (Trial Edition)" to prepare the data file for "Scc 5.0 (Trial Edition)" to do Data mining. Finally, the system calculates the corresponding weights of meta-terms, such as "SV", "TPR", "V" etc. in rule base. With these data, we set up the Server's rule base and customized the Clients' rule base. The preliminary test shows the satisfactory results. Next, we plan to integrate the module of Neural-networks with our system and compare it with the system integrated with Data mining.

REFERENCES

C.D. Cooke, et al. 2000. Validating Expert System rule confidences using Data mining of Myocardial perfusion SPECT Database. *Computers in Cardiology* 27: 785–788.
M. Gou, et al. 2002. A framework for dynamic evidence based machine using Data mining. *Proceedings of the 15th IEEE Symposium on Computer-Based Medical Systems (CBMS 2002).*
M.V. Fidelis, et al. 2000. Discovering comprehensible classification rules with a Genetic Algorithm. IEEE Computer Society.
Wenwei, Chen. 1998. *Intelligent decision-making technology.* Beijing: Publishing House of Electronics Industry.
Y. Liu. 1998. The Design and Implementation of a Virtual Medical Centre for Patient Home Care. *Proceedings of the 20th Annual International Conference of the IEEE Engineering in Medicine and Biology Society*, Vol. 20, No. 3.
Z. Z. Shi. 2002. *Knowledge Discovery.* Beijing: Tsinghua University Press.

Computational Methods in Engineering and Science, Iu et al. (eds)
© 2003 Swets & Zeitlinger, Lisse, ISBN 90 5809 567 3

Variability of the solution of the redundant muscle force sharing in biomechanics due to the choice of physiological criteria

M.P.T. Silva & J.A.C. Ambrósio

Instituto de Engenharia Mecânica – Instituto Superior Técnico, Lisboa, Portugal

ABSTRACT: The muscle system redundancy gives the human body the ability to recruit different muscles to produce one same movement or posture. Due to their redundant nature, the muscle forces are the solution of an optimization problem. Sequential quadratic optimization tools are used here to calculate muscle forces, internal forces and joint moments-of-force. The solutions minimize specified cost functions representing a physiological criteria adopted by the central nervous system. In this work the differences in the results obtained using different cost functions in the analysis of a gait cycle are analyzed. The muscle forces in the locomotion apparatus, net joints moments of force of the upper body and joint reaction forces are calculated using an inverse dynamic analysis formulation. The biomechanical model is described using a general-purpose multibody formulation with fully Cartesian coordinates and a Hill-type muscle model, with the muscle forces calculated using the maximum isometric contraction forces, activation patterns, muscle lengths and change of length rates. The lower extremities muscle apparatus is described using 43 muscles per leg, each one having origin and insertion points in the skeletal structure. Curved muscles are also included to represent complex paths that wrap around joints, bones or other muscles.

1 INTRODUCTION

Multibody-based methodologies are currently being used to simulate and analyze a wide variety of mechanical systems. In biomechanical applications, in particular, these methodologies present several advantages since they provide a way of calculate the internal forces developed by a specific biomechanical model during the execution of a specified task, without using intrusive force measuring devices. In particular when associated with proper optimization tools, these methodologies can be used to solve the "redundant problem in biomechanics" (Yamaguchi 2001), calculating the muscle forces produced by a specified muscle apparatus.

It is the purpose of this work to present a multibody based methodology that together with the use of optimization tools, allows for the calculation of the redundant muscle forces, generated in a particular muscle apparatus of the human body. The proposed methodology uses a multibody formulation with fully Cartesian coordinates where rigid bodies and kinematic joints are modeled using the Cartesian coordinates of a set of anatomical points and unit vectors located at the joints and extremities of the subject under analysis (Jalón & Bayo 1994). Using this general-purpose methodology, a whole-body-response biomechanical model is constructed.

Two different types of actuators are used to drive the biomechanical model through a previously acquired motion: joint actuators and muscle actuators. The use of muscle actuators to drive the degrees-of-freedom of the joints crossed by the muscles forming the muscle apparatus under analysis generates a mechanical system with a redundant nature.

Optimization tools are used to resolve this indeterminate problem. These tools consider that muscle forces are generated according to the minimization of some performance criteria and that any optimal solution obtained for the muscle forces must satisfy the equations of motion of the biomechanical system. These performance criteria are analytical expressions that simulate the decisions taken by the central nervous system when executing of the prescribed task. Several different physiological criteria are used in the present work in order to identify their influence on the solution of the redundant problem.

A muscle model is associated to each muscle actuator in order to simulate the muscle activation-contraction dynamics (Kaplan 2000, Zajac 1989). In the present work, a Hill type muscle model is applied, being the force produced by the muscle contractile element calculated as a function of the muscle activation, maximum isometric peak force, muscle length and muscle shortening rate (Yamaguchi 2001).

2 MULTIBODY FORMULATION AND EQUATIONS OF MOTION

A multibody formulation, using natural or fully Cartesian coordinates, is applied to the study and analysis of the human body movement. With this formulation, rigid bodies are constructed using the Cartesian coordinates of a set of points and unit vectors. A vector of generalized coordinates is constructed with the Cartesian coordinates of the points and vectors used in the definition of the mechanical system:

$$\mathbf{q} = \left\{ q_1 \ q_2 \ q_3 \cdots q_n \right\}^T \tag{1}$$

where $n = 3(np + nv)$ is the total number of natural coordinates and np and nv are, respectively, the total number of points and unit vectors of the model.

The set of generalized coordinates, presented in Equation (1), is not sufficient to the definition of the mechanical system. To completely define its kinematic structure and topology, some kinematic constraint equations must be supplied. There are several types of constraint equations that are used in formulations using fully Cartesian coordinates: rigid body constraints, joint constraints and driving constraints. Rigid body constraints, however, are the most common type since the number of natural coordinates that defines a rigid body is always larger then the number of its degrees-of-freedom (Jalón & Bayo 1994). This means that not all the coordinates are independent and that kinematic constraint equations need to be added to express these dependences. Rigid body constraints, express physical properties of rigid bodies that are supported by the scalar product equation, given by:

$$\Phi^{(SP,1)}(\mathbf{q}, t) = \mathbf{v}^T \mathbf{u} - L_v L_u \cos\left(\langle \mathbf{v}, \mathbf{u} \rangle (t) \right) = 0 \tag{2}$$

where $\mathbf{v}$ and $\mathbf{u}$ are two generic vectors used in the definition of rigid bodies, L_v and L_u are the respective norms and $<\mathbf{v},\mathbf{u}>(t)$ is the angle between them. Equation (2) is also applied to the definition of kinematic joints and driving constraints. In the case of driving constraints, used to prescribe the motion of the system, this angle is a function of time.

There are other types of kinematic constraints, such as the linear combination constraint or the cross product constraint that, are also used when modelling with natural coordinates (Jalón & Bayo 1994). All the constraint equations associated with the mechanical system are assembled in a single vector and written as:

$$\Phi(\mathbf{q}, t) = \mathbf{0} \tag{3}$$

The equations of motion of a general multibody system, with relative motion between rigid bodies constrained by kinematic joints and acted upon by external applied forces are given by:

$$\mathbf{M}\ddot{\mathbf{q}} + \Phi_\mathbf{q}^T \lambda = \mathbf{g} \tag{4}$$

where $\mathbf{M}$ is the global mass matrix of the system, $\Phi_\mathbf{q}$ is the Jacobian matrix of the constraints, $\ddot{\mathbf{q}}$ is the vector of generalized accelerations, $\mathbf{g}$ is the generalized force vector and λ is the vector of Lagrange multipliers (Jalón & Bayo 1994, Nikravesh 1988). When performing inverse dynamic analyses of mechanical systems, the Lagrange multipliers vector is the unknown present in Equation (4). In the specific case of biomechanical applications, this vector is associated with the reaction forces, muscle forces and net moments of force at the model joints. All other quantities are calculated from kinematic data and force measuring devices.

3 THE BIOMECHANICAL MODEL

In this work, a three-dimensional, whole-body response biomechanical model of the human body is used (Ambrósio et al. 1999, Silva et al. 1997). The model, presented in Figure 1, is described using the

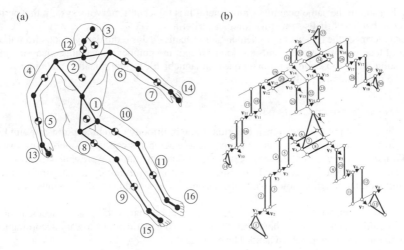

Figure 1.　(a) Sixteen anatomical segments;　　　　(b) Kinematic structure of rigid bodies and kinematic joints.

Figure 2.　Joint actuator associated with the knee.

general multibody formulation. It has a kinematic structure made of 33 rigid bodies, interconnected by revolute and universal joints, adding to the 16 anatomical segments.

A collection of physical characteristics, including the mass, moments of inertia or segment lengths among others, is associated to each anatomical segment of the biomechanical model. These properties are obtained from the literature and scaled (Ambrósio et al. 1999).

In order to perform the inverse dynamic analysis, the motion of the subject needs to be acquired. The trajectories of a set of points, located at the joints and extremities of the biomechanical model (Silva & Ambrósio 2002a) are obtained using three-dimensional motion reconstruction techniques (Addel-Aziz & Karara 1971). The trajectories are filtered, to reduce the noise levels introduced during the motion reconstruction procedure (Winter 1990), and corrected to be consistent with the kinematic structure of the biomechanical model. Velocity and acceleration curves are calculated, for each point, using the time differentiation of the cubic spline trajectory curves. The externally applied forces over the biomechanical model are the feet ground reaction forces, which are collected using three force plates.

In order to drive the biomechanical model throughout the inverse dynamic analysis, joint actuators such as the one represented in Figure 2 for the knee joint are specified. Joint actuator equations are kinematic constraints, of scalar product type as presented in Equation (2), in which the angle between the two vectors is a function of time and describes, for each time step, the motion of the kinematic joint associated with that degree-of-freedom. The number of non-redundant constraint equations is now equal to the number of natural coordinates describing the model, therefore, the inverse dynamics problem is determined.

4　REDUNDANT MUSCLE FORCES AND OPTIMIZATION TECHNIQUES

In complex biomechanical systems almost every joint is crossed by several muscles or muscle groups. Different muscle activation patterns generate forces that produce the same net moments-of-force at the

971

joints, which result in the same posture or movement. It is the central nervous system that activates the set of muscles that best fulfills some physiological criteria.

Historically termed "the redundant problem in biomechanics", this redundancy results from the fact that the number of load-transmitting elements at a joint exceeds the number of available equations and consequently the solution for those forces is not unique (Yamaguchi 2001). Optimization techniques are applied to resolve the indeterminate problem.

4.1 Muscle actuators in multibody systems

Muscles are introduced in the equations of motion of the multibody system as point-to-point kinematic driver actuators. Depending on their complexity, muscle actuators can be defined using two or more points. In Figure 3, two muscles of the lower extremity muscle apparatus are presented to illustrate the two types of muscle definition: the *semimembranosus* and the *tensor fasciae latea*. The first is defined only with an origin and an insertion point, while the second uses two additional via points due to its more complex path.

To each muscle actuator is associated a constraint equation that specifies the muscle action during the analysis period, which constrain the distance between two generic points to vary according to a specified length history, calculated before hand. The constraint is

$$\Phi^{(MA,1)}(\mathbf{q},t) = \left(\mathbf{r}_m - \mathbf{r}_n\right)^T \left(\mathbf{r}_m - \mathbf{r}_n\right) - L_{nm}^2(t) = 0 \tag{5}$$

where $\mathbf{r}_m$ and $\mathbf{r}_n$ are respectively the global position vectors of the origin and insertion points and $L_{nm}(t)$ is the muscle total length, calculated for each time step of the analysis.

A Lagrange multiplier is associated to each muscle actuator. The physical meaning of this multiplier is of a force per unit of length. The muscle driver actuator equations are introduced in the Jacobian matrix of the constraints together with the remaining kinematic constraint equations. An integrated solution of the problem is obtained for muscle and reaction forces.

4.2 Dynamics of muscle tissue

The dynamics of muscle tissue can be divided into activation dynamics and muscle contraction dynamics (Zajac 1989), as schematically indicated in Figure 4. Activation dynamics generates a muscle tissue state

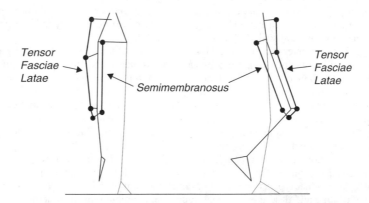

Figure 3. Muscle actuators defined with two or more points.

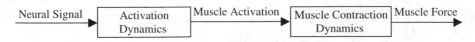

Figure 4. Dynamics of muscle tissue.

that transforms the neural excitation produced by the central nervous system, into activation of the contractile apparatus. Activation dynamics, although not implemented in this work, describes the time lag between neural signal and the corresponding muscle activation (Kaplan 2000). To simulate the muscle contraction dynamics a Hill-type muscle model is applied (Zajac 1989).

The model, depicted in Figure 5, is composed of an active Hill *contractive element* (CE) and a *passive element* (PE). Both elements contribute to the total muscle force $F^m(t)$. A *series elastic element* (SEE), associated with cross-bridge stiffness, is not included in the model because the motion targeted for the applications does not involve short-tendon actuators (Zajac 1989).

In a Hill-type muscle model, the contractile properties of the muscle are controlled by its current length $L^m(t)$, rate of length change $\dot{L}^m(t)$ and activation $a^m(t)$. The force produced is

$$F^m_{CE}(a^m(t), L^m(t), \dot{L}^m(t)) = \frac{F^m_L(L^m(t)) F^m_L(\dot{L}^m(t))}{F^m_0} a^m(t) \qquad (6)$$

where F^m_0 is the maximum isometric force and $F^m_L(L^m(t))$ and $F^m_L(\dot{L}^m(t))$ are functions that represent the muscle force-length and force-velocity dependency, respectively (Zajac 1989).

The passive element is independent of activation and it only starts to produce force when the muscle is stretched beyond its resting length L^m_0. The force produced by the passive element is only function of the muscle length, being its value always determined.

A muscle apparatus of thirty-five muscle actuators is used to simulate the right lower extremity intermuscular coordination. The physiological information regarding muscle definition is obtained from literature (Yamaguchi 2001) and compiled in a muscle database. The whole muscle apparatus is presented in Figure 6.

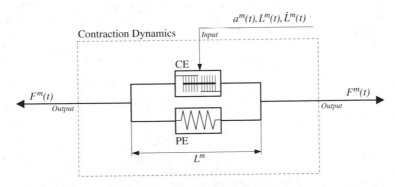

Figure 5.   Contraction dynamics using a Hill-type muscle model.

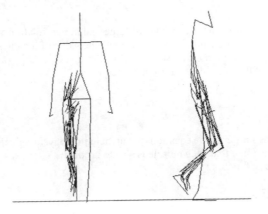

Figure 6.   Lower extremity muscle apparatus.

4.3 Static optimization: cost functions and constraints

Indeterminate systems present an infinite set of possible solutions, being the aim of optimization techniques to find, from all the possible solutions, the one that minimize a prescribed objective function, subjected to a certain number of restrictions or constraints. Mathematically an optimization problem can be stated as follows:

$$\text{minimize} \quad F_0(u_i) \quad \text{subject to:} \begin{cases} f_j(u_i) = 0 & j = 1,...,n_{ec} \\ f_j(u_i) \geq 0 & j = (n_{ec}+1),...,n_{tc} \\ u_i^{lower} \leq u_i \leq u_i^{upper} & i = 1,...,n_{sv} \end{cases} \tag{7}$$

where u_i are the state variables bounded by u_i^{lower} and u_i^{upper}, $F_0(u_i)$ is the objective or cost function to minimize and $f_j(u_i)$ are constraint equations that restrain the state variables. In Equation (7), n_{st} represents the total number of state variables and n_{tc} the total number of constraint equations in which n_{ec} are of equality type.

The minimization of cost functions simulates the physiological criteria adopted by the central nervous system when deciding which muscles to recruit as well as the level of activation that produce the adequate motion or posture for the task being undertaken. Several cost functions have been used by researchers in the study of the redundant problem in biomechanics (Crowninshield & Brand 1981, Yamaguchi 2001). The selection of the most appropriate criterion to use in the optimization process resides upon several important aspects such as the type of motion under analysis, the objectives to achieve or the presence of any type of pathology. In normal gait, the central nervous system is probably more interested in maximizing the comfort, i.e., minimizing the muscle fatigue. Two possible criteria could be the minimization of the total muscle stress or the minimization of the total muscle force, which is generally accepted to be closely related with the minimization of the muscular fatigue (Crowninshield & Brand 1981).

Each human task is thoroughly controlled by the central nervous system using a particular criterion or a set of criteria. A cost function should be able to reflect the inherent physical activity or pathology as well as to include relevant physiological characteristics and functional properties such as the maximum isometric force or electromyographic activity (Tsirakos *et al.* 1997). Some of most commonly used (non-linear) cost functions are (Tsirakos *et al.* 1997):

(*i*) Sum of the square of the individual muscle forces:

$$F_0 = \sum_{m=1}^{n_{ma}} \left(F_{CE}^m \right)^2 \tag{8}$$

This cost function is considered to fulfill the objective of energy minimization. This cost-function does not include physiological or functional capabilities.

(*ii*) Sum of the cube of the average individual muscle stress:

$$F_0 = \sum_{m=1}^{n_{ma}} \left(\sigma_{CE}^m \right)^3 \tag{9}$$

This cost function is based on a quantitative force-endurance relationship and on experimental results. It includes physiological information, namely the value of the physiological cross sectional area of each muscle and it predicts muscle co-activation.

(*iii*) Sum of the square of the normalized muscular forces:

$$F_0 = \sum_{m=1}^{n_{ma}} \left(\frac{F_{CE}^m}{F_0^m} \right)^2 \tag{10}$$

This cost function is similar to the equation (8) but including physiological information, namely the maximum isometric force that each muscle is able to produce.

(*iv*) "Soft saturation" cost function:

$$F_0 = \sum_{m=1}^{n_{ma}} \sqrt{1 - \left(\frac{F_{CE}^m}{F_0^m} \right)^2} \tag{11}$$

974

This cost function, that includes physiological information, produces a more realistic synergistic function of the muscle for activation and co-activation of muscles.

Cost functions can also be sum of the instantaneous muscle power or the sum of the square of the total reaction forces at the joints. In the present work the principle of minimization of the sum of the square of the muscle forces and the principle of minimization of the sum of the cube of the individual muscle stresses (Crowninshield & Brand 1981) are used in applications involving human locomotion. Substituting Equation (6) in Equations (8) and (9), the following mathematical expressions for the referred cost functions are obtained:

$$F_0 = \sum_{m=1}^{n_{ma}} (F_{CE}^m)^2 = \sum_{m=1}^{n_{ma}} (\frac{F_L^m F_L^m}{F_0^m} a^m)^2 \quad ; \quad F_0 = \sum_{m=1}^{n_{ma}} (\sigma_{CE}^m)^3 = \sum_{m=1}^{n_{ma}} (\bar{\sigma} \frac{F_L^m F_L^m}{F_0^{m^2}} a^m)^3 \tag{12}$$

where n_{ma} are the number of muscle actuators and $\bar{\sigma}$ is the specific muscle strength with a constant value of $31.39\,N/cm^2$ (Yamaguchi 2001). Using these cost functions, the state variables associated with muscle actuators represent muscle activations that are bounded to assume values between 0 and 1. No bounds are specified for all other state variables. The optimal solution must fulfill the equations of motion of the system, given by Equation (4). These equations are supplied to the optimizer as constraint equations for the state variables.

5   APPLICATION CASE

The methodologies and the biomechanical model described before are applied to the gait analysis of a subject performing a normal cadence stride period. The subject under analysis is a 25-year-old male with a height of 1.70 m and a total body mass of 70 kg. The subject is wearing running shoes. During the stride period, the subject has to walk over three force plates that measure the ground reaction forces for both feet (Ambrósio et al. 1999b). A total number of 66 frames are recorded using four cameras with a sampling frequency of 60 Hz.

The net moments of force developed at the joints of the biomechanical model are calculated. In particular, the net moments of force for the joint actuators of the right ankle and knee joints are presented in Figure 7. These results are within the expected values reported in the literature for a normal cadence stride (Winter 1990).

The muscle forces developed in the right leg during the stride period are calculated using the procedures described before. The results obtained with two optimization packages (Vanderplaats 1999), are presented in Figure 8 for the muscle *Gluteus Minimus* and for the *Gluteus*. The forces present almost the same behavior although different force levels can be identified. Note that the results shown in Figure 8 are two possible solutions that fulfill the equations of motion of the system obtained by different optimization packages.

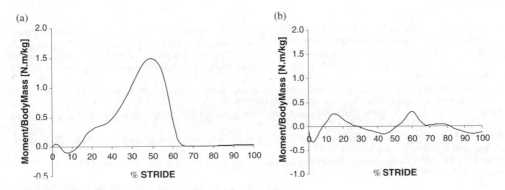

Figure 7.   Net moment-of-force (scaled by the body mass) for the right ankle and knee joints.

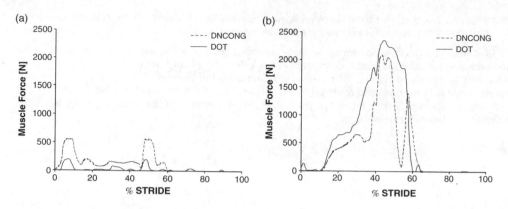

Figure 8.   Muscle Forces: (a) *Gluteus Minimus*, (b) *Soleus*.

6   CONCLUSIONS

In this work, a methodology was presented that allows for the calculation of the net moments-of-force and reactions at the joints, and also for the calculation of the muscle forces developed in a specific muscle apparatus of a subject describing a prescribed motion. The subject is simulated using a whole-body-response biomechanical model constructed using a multibody formulation with natural or fully Cartesian coordinates.

It was showed that the use of muscle actuators usually introduces indeterminacy in the equations of motion of the biomechanical system. When this occurs, optimization techniques allow for the selection, among an infinite set of possible solutions, of the one that minimizes a specific cost-function associated, from the physiological point of view, with the criteria used during the trial by the central nervous system of the subject.

ACKNOWLEDGEMENTS

The work developed in this article was supported by Fundação para a Ciência e Tecnologia through the project PRAXIS/P/EME 14040/98, entitled *Human Locomotion Biomechanics Using Advanced Mathematical Models and Optimization Procedures* and PhD scholarship GGPXXI/BD/2851/96. The authors want to gratefully acknowledge the valuable inputs and discussions had with Dr. Matthew Kaplan.

REFERENCES

Addel-Aziz, Y. and Karara, H. 1971. Direct linear transformation from comparator coordinates into object space coordinates in close-range photogrammetry, *Symposium on Close-range Photogrammetry*, Falls Church, Virginia, 1–18.

Ambrósio, J.A.C., Silva, M.P.T. and Abrantes, J. 1999. Inverse Dynamic Analysis of Human Gait Using Consistent Data, *IV International Symposium on Computer Methods in Biomechanics and Biomedical Engineering*, Lisbon, Portugal, 275–282.

Crowninshield, R.D. and Brand, R.A. 1981. Physiologically Based Criterion of Muscle Force Prediction in Locomotion, *Journal of Biomechanics*, 14(11): 793–801.

Jalón, J.G.D. and Bayo, E. 1994. Kinematic and dynamic simulation of multibody systems: the real-time challenge, Springer-Verlag, New York; Hong Kong.

Kaplan, M.L. 2000. "Efficient Optimal Control of Large-Scale Biomechanical Systems", Ph.D. Dissertation, Stanford University, Stanford.

Nikravesh, P.E. 1988. *Computer-aided analysis of mechanical systems*, Prentice-Hall, Englewood Cliffs, New Jersey.

Numerics, V. 1995. "IMSL FORTRAN Numerical Libraries – Version 5.0", Microsoft Corp.

Silva, M.P.T., Ambrósio, J.A.C. and Pereira, M.S. 1997. Biomechanical Model with Joint Resistance for Impact Simulation, *Multibody System Dynamics*, 1(1): 65–84.

Silva, M.P.T. and Ambrósio, J.A.C. 2002a. Kinematic Data Consistency in the Inverse Dynamic Analysis of Biomechanical Systems, *Multibody System Dynamics*, 8(2): 219–239.

Silva, M.P.T. and Ambrósio, J.A.C. 2002b. A Multibody Based Methodology for the Solution of the Redundant Nature of the Muscle Forces Using Static Optimization, *V Congreso de Métodos Numéricos en Ingeniería*, Madrid, Spain.

Tsirakos, D., Baltzopoulos, V. and Bartlett, R. 1997. Inverse Optimization: Functional and Physiological Considerations Related to the Force-Sharing Problem, *Critical Reviews in Biomedical Engineering*, 25(4&5): 371–407.

Vanderplaats 1999. "DOT – Design Optimization Tools – USERS MANUAL – Version 5.0."

Winter, D.A. 1990. Biomechanics and motor control of human movement, Wiley, New York.

Yamaguchi, G.T. 2001. *Dynamic Modeling of Musculoskeletal Motion*, Kluwer Academic Publishers, Boston, Massachussetts.

Zajac, F.E. 1989. Muscle and tendon: properties, models, scaling, and application to biomechanics and motor control, *Critical Reviews in Biomedical Engineering*, 17(4): 359–411.

Computational Methods in Engineering and Science, Iu et al. (eds)
© 2003 Swets & Zeitlinger, Lisse, ISBN 90 5809 567 3

Design and implementation of machine translation system for web pages

C.W. Tang, F. Wong & Y.P. Li
Faculty of Science and Technology, University of Macau, Macao

ABSTRACT: No machine translation is perfect or almost perfect, but it can give us a draft translation or basic idea of original document. Currently, most of machine translations focus on documents with English as the source language. In this paper, we introduce a WebPages Translation System from Portuguese to Chinese. The system is based on the *Local Application Approach*, where the translation process is carried out in local machine.

1 INTRODUCTION

There are a lot of languages in the world; communication between two different languages requires translation. Generally speaking, human translation is the best solution for high translation quality, however, it is also the most inconvenient and expensive no matter in time or money. It is the reason why researchers develop machine translation systems. Since English is the most common language, almost all machine translation systems focus between English and other native language. Nowadays, MT systems have been extended from individual application to the translation of Internet webpage.

There are various web page translation systems; each of them has their own translation mechanism. According to the system architecture, web page translation systems are mainly developed with alternative approaches: *Middle Ware Approach* and *Application Based Approach*. The architectures of these systems are illustrated in Figure 1. A typical representative of Middle Ware Approach in commercial product is the IBM WebSphere Translation Server (IBM) and Yoden's system (Yoden 1996) is another in the research domain. This kind of software is setting between the client browser and Internet. When translating the active pages in client browser, the system requires the page's URL to be resent for translation, or the pages

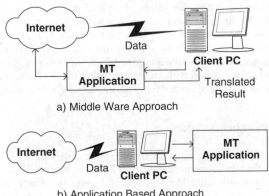

a) Middle Ware Approach

b) Application Based Approach

Figure 1. Different webPage machine translation system architectures.

are automatically translated by middle ware proxy server after user has requested for translation. This approach has its advantage in retrieving the full web content according to the specified URL links provided. But the weakness of this approach is that, data transmission between the client side and the server is necessary, hence it causes the time propagation to deliver the translation content. On another hand, our system is belonging to second approach of Application Based manner. Compare with the first approach, our system has the following features: (1) the translation of web content is processed based on a local translation engine, (2) no data transmission over the Internet is required, and (3) the translation of the browsing web content can be reflected immediately. Furthermore, for dynamic web page, where the content of the web page is generated dynamically, system with Middle Ware approach cannot translate such pages. They can only process the web with static information. But with our system, this limitation can be resolved. Instead of replying on the page's URL, we adopt the methodology to directly communicate with web browser.

Unlike common translation application, web page translation system focus on not only translation quality, but also the source of page content and its format in displaying. Therefore, it needs to deal with the HTML Syntax. The HTML Tags in a page are used to format the layout; it also provides information to identify the sentence boundary. Therefore, the first step in web page translation process is to extract the real content from the page, and also record down all the format tags from the original page. This information is used to restore the original page format to the translation in target language. For MT system between alphabetic and non-alphabetic languages like Portuguese and Chinese, we also need to consider the length of text between these two languages. Therefore, a good methodology to handle HTML text is very important. In next session, we will discuss the detail in how the HTML content is handled in web page MT system.

The rest of the paper is organized as follows: an introduction system structure for retrieving the page content is provided in next section. The processing of HTML text and frame embedded web page are discussed in sections 3 and 4. Section 5 demonstrates the translation to a web page based on our developed system, and a conclusion is drawn to end this paper.

2   WEB TRANSLATION SYSTEM

The web translation system (WT) mainly consists of two modules: the front-end module that extracts the textual information from the browsing page, and the backend translation engine that does the translation for the input text. For the translation module, we adopt the translation engine of PCT Assistant (Wong et al. 2002), which is a machine aids human translation system based on the methodology of example-based approach (EBMT). Since it is a kind of Corpus-based Machine Translation (CBMT), it is there are mainly three key issues. Building the Corpus-base sentence. Retrieving the best matches example for input sentence. Produce translation sentence (Li et al. 1999). And it allows the new translated examples to be stored into the case base during the system lifetime or in the training phase. This allows the MT system to gradually accumulate the new knowledge and hence to improve its quality in translation. In contrast to Yoden's translation system (Yoden et al. 1996), the translation knowledge are invariant once the system is developed, and the self-learning ability is not provided by the system.

Besides translation engine, getting the source content of pages is another core component of web page translation system. As mentioned, our system is based on local application approach. The system employs the Component Object Model (COM) as back end module to communicate with the web browser. With this COM technology, system can obtain the source of HTML text directly from browser; no matter the current pages are static or dynamic. Therefore, with this design, a real time machine translation to the web page can be realized without any limitation to the various types of web pages. Even, offline content can also be translated since data retransmission of a dedicated page is not required.

3   HTML SYNTAX PREPROCESSING

HTML analysis is a vital part in WT system. A simple way to obtain the textual content from the hypertext format is to extract the text between the pairs of HTML elements (<body>...</body>). That is to simply ignore all the HTML elements from the file of a web page. However, this method will ignore also the connection of a sentence when part of the words are being separated by format tags, thus it is impossible to retrieve a completed sentence. Table 1 illustrates the problem, which is a sample of HTML code fragment and its translation result produced by DrEye (Dreye 2002).

Table 1. HTML format of source text and its translation generated by DrEye system.

Source HTML code (in English)	DrEye translation result (in Chinese)
<BODY> <B>When</B> I came in, he is eating. </BODY>	<BODY> <B><B>何時</B>我進來，他是吃。 </BODY>

Table 2. List of inner tags.

Name	Description	Name	Description
A	Anchor	ABBR	Abbreviated form
ACRONYM		B	Bold text style
BIG	Large text style	CITE	Citation
CODE	Computer code fragment	DEL	Deleted text
DFN	Instance definition	DIV	Generic language/style container
EM	Emphasis	I	Italic text style
INS	Inserted text	KBD	Text to be entered by the user
SAMP	Sample program output, scripts, etc.	SMALL	Small text style
SPAN	Generic language/style container	STRONG	Strong emphasis
SUB	Subscript	SUP	Superscript
TT	Teletype or monospaced text style	U	Underlined text style
VAR	Instance of a variable or program argument		

Therefore, we found that the translation of hypertext will depend on how the format of it is analysed. In order to correctly obtain the original text from a page, we use a syntax analyzer for preprocessing the HTML Syntax. According to W3C HTML 4.01 Specification (HTML 4.01), there are 91 different HTML elements. Some of them may appear within a sentence to modify the appearances of special characters or words, such as "bold text style <B>". We classify this kind of elements as "*inner tag*". The rest are used as structural syntax to modify a whole sentence or paragraph of text. Thus, this kind of elements seldom appears within a sentence and we classify it as "*outer tag*". Here, we summarize the set of inner tags of HTML 4.01 elements in Table 2.

In order to preserve the web page contents and its original layout, during the translation process, all outer and inner tags information are maintained and applied to the target text, so that the content structures and appearances of the original page can be retained and be consistent with the translated page. Unfortunately, Chinese and Portuguese are not the languages from the same cognate; the translation between the source and the target text cannot be completely aligned. Some of the inner tags that modify the partial content of a sentence in the source cannot re-locate the same content in the target sentence. Which, as a result, cannot always be set to modify the same content in the target sentence after translated.

In our system, if this situation occurs, we will extend the inner tag to modify the whole sentence instead of looking for the fragment of translation to modify. Here, a demonstration is presented to describe the problem. Followings are the source sentence of Portuguese (a) in HTML format, and its translation in Chinese (b).

<BODY><B>Senhor António**</B>**, vamos falar do assunto da cortiça.**</BODY>** (a)

<BODY><B>安東尼奧先生，讓我們談談軟木事情吧。 **</BODY>** (b)

In these sentences, elements, *<BODY>* and *</BODY>* are classified as out tags, while the *<B>* and *</B>* are inner tags. In sentence analyzing process, the system will try to extract the sentence based on the outer tags as boundary information. Therefore, in the first pass, the sentence together with inner tags, "*<B>Senhor António</B>, vamos falar do assunto da cortiça*", is retrieved. As the extracted sentence is

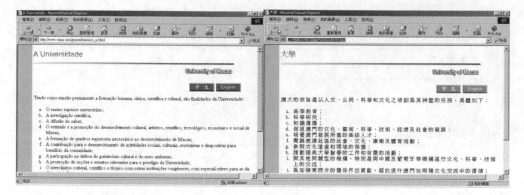

Figure 2.   Simple translation demo and result.

embedded with inner tags that modify the partial words of it, the second pass of the analyzing is to remove the inner tags and generates the sentence in plaintext. Then the sentence is fed into the translation engine for translation. As described, although we maintain the list of outer and inner tags that analyzed from the source of page content, and the data of their modifications. To the Chinese translation, there is no any algorithm to align the fragments from original Portuguese. We simply extend the modification of inner tags *<B>* and *</B>* to the whole sentence. Moreover, based on described outer–inner tag analysis method, the web page that consists of table, bullets, and etc. can be handled properly, as shown in (Fig. 2).

Image and picture are very common in webpage construction, since webpage author would like to use picture in his page to present information through this sophisticate format, and on the other hand, to attract people's attention. Figure 2 shows a very simple one that uses image as control buttons for navigating and linking to other pages. According to syntax of hypertext, image file location can be expressed in relative or absolute URL. If image file location is set as absolute, then to our processing, it is much easier for us to keep track of the content resources of a webpage. When re-constructing a new page with target translation, all the image and picture files can be re-directed according to the absolute URLs provided in the source file. Otherwise, it needs a way to finding out the base URLs for relative links where the image files are located. In resolving this situation, it needs a parser to find out all URLs for image files, and back up these base URLs. After the translation, the target text together with other layout objects like table, bullet items, pictures, controls, and etc. will be generated into a new page, and presented to user. Figure 2 illustrates the source and target pages, in Portuguese and Chinese respectively, preserving all the layout and appearances of contents in the page.

4   FRAME HANDLING

Frame is a multi-view environment in web browsing. Where the content of a page is divided into several sections in different browsing windows. For example, Figure 3 shows a web page that consists of three frames for different purposes, one for banner, one for navigation and another uses for displaying the main content. The source code for frame embedded page, as Figure 3, is shown in Table 3.

When translating frame embedded web pages, it makes system design more complex, no matter in client web browser and translation system. In multiple frame environments like Table 3, the main page only contains <FRAMESET> and sub frame information. Since main page does not contain sentence, the translation result will remain the same as original. In our system, we keep track of frame information and trace "<FRAME>" link in order to overcome this common situation.

When translating frame page, system will search each frame link in main page. Since user already load those pages in browser. We used the frame link as our input source and send request to browser. Now browser return pages content as normal translation. Because of frame property, all of translation result will be stored in temporary storage instep of show in browser directly. For regenerating final result, system will modify the frame information in main page so as to display suitable translation. Table 4 shows Pseudo code and (Figure 4) shows a frame embedded page and its translation.

982

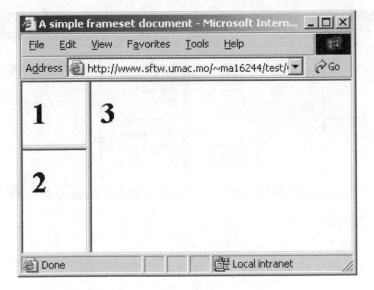

Figure 3.   A sample frame output.

Table 3.   A sample frame web page.

```
<HTML>
<HEAD>
<TITLE>A simple frameset document</TITLE>
</HEAD>
<FRAMESET cols = "20%, 80%">
    <FRAMESET rows = "100, 200">
        <FRAME src = "frame1.htm">
        <FRAME src = "frame2.htm">
    </FRAMESET>
    <FRAME src = "frame3.htm">
</FRAMESET>
</HTML>
```

Table 4.   Pseudo code for frame handle.

```
Address = Trace all frame link
For all Add in Address
    Contents = Get content of Add
    Pre-process_HTML(contents)
    Result = Translate
    Stack->Add(result)
End for
ModifyMainFrame(stack)
```

5   PAGE TRANSLATION

Inside translation system, system engine is a kernel part of all. Review that our translation engine is a part of PCT Assistant (Wong 2002), it is based on Example-Based Machine Translation (KBMT) Mechanism. The main idea of this system is an interactive based, self-improved, and machine aided translation system. In PCT Assistant, there is two training method. First one is aligning bilingual corpus or pre-translated documents. Second one is running the MS Word embed translating program.

983

Figure 4. Frame handling and translation.

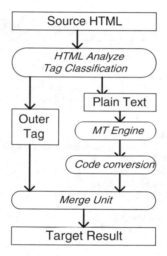

Figure 5. Page translation data flow.

Word embed program is an easy way to use and learn new translation for normal user. When user translates a sentence, our engine first search from knowledge database. If pervious already translated, engine will directly return translated result, otherwise, return the most similar sentence to user. User can modify the translated result for correction. Since modification are made and verified by user, system can treat it as new knowledge and learn from it. Repeat this training step, system become mature. After the training phase, PCT Assistant has enough knowledge and can be run as full automatic translation.

In our Web Translation System, we employ PCT Assistant as backend engine. During translation process, Web Translation System extract sentence from HTML coding pages (see session 2.2). After extraction, sentence become plain text and can be used for PCT Assistant. PCT Assistant returns its translation. If our translation engine cannot translate, it will return original Portuguese sentence. Because the Codepages of Portuguese and Chinese are overlap, we have to convert these Portuguese sentences to suitable encoding. In our system, we convert them into Unicode as our solution. Since Outer Tag and Inner Tag are saved at the beginning stage, our regenerate module can combine these as output result (Fig. 5).

5.1 Discussion

Since translation knowledge comes from pervious training in PCT Assistant, knowledge may not be fully covering all sentences on Internet. As a result, sometime we will get back an original sentence mix

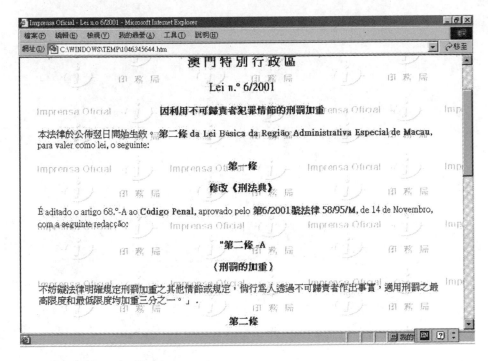

Figure 6.   Translation result with unknown sentence.

with translated Chinese sentence. In Figure 6, it is one of the Macau Governmental documents. It shows that when our system cannot find a target translation, the source sentence is not changed.

6   CONCLUSION

In this paper, we proposed a web page machine translation system in application based approach for Portuguese to Chinese translation. In the above discussion, we pointed out that translation accuracy is directly affected by HTML handling technique and the way we solve. During system testing, we found that some original structures of web page cannot be restored. It is because not all web-page designers follow the W3C specification and we haven't considered their situations. We believe that if the designer improves web pages following the standard, the translation result will be much better.

REFERENCES

DrEye 2002. http://www.dreye.com/tw_2002/product/pc/dreye2002.htm
HTML 4.01 Specification. W3C Recommendation, 24-Dec-1999. http://www.w3.org/TR/html4
IBM WebSphere Translation Server. http://www-3.ibm.com/software/pervasive/products/voice/translation_server.shtml
Li, Y.P., Pun, C.M. & Wu, F. 1999. Portuguese–Chinese Machine Translation in Macao. MT Summit VII.
Wong, F., Dong, M.C. & Li, Y.P. 2002. *Portuguese to Chinese Machine Translation System.* Proceedings of Symposium on Technological Innovation in Macau 2002. Page(s): 134.
Wu, F. & Li, Y.-P. 1999. *A Portuguese–Chinese On-Line Domain-Based Machine Translation System.* EPMESC VII, Page(s): 1075–1084.
Yoden, N., Harada, K. & Yumura, T. 1996. *A Personal Machine Translation Software for WWW browser.* Consumer Electronics, 1996. Digest of Technical Papers, International Conference on, 5–7 Jun-1996. Page(s): 234.

The conflict resolution at connected railway junctions

C.W. Tsang & T.K. Ho
Department of Electrical Engineering, The Hong Kong Polytechnic University, Hong Kong

ABSTRACT: Traffic conflicts at railway junctions occur when trains arrive at the junctions at the same time. Conflicts result in train delays, leading to a loss of punctuality, thus hampering the quality of service. A sensible right-of-way assignment will reduce the total weighted delay of all trains, but the number of possible assignments is vast, especially at connected railway junctions. Genetic algorithm provides a quick means to locate a near-optimal assignment within a limited time, but crossover (a genetic operator) often produces right-of-way assignments that violate the constraints imposed by the definitions of the chromosomes. This paper proposes two scanning methods to locate the violations and modify them before they are admitted to the next generation. Simulations are carried out to study the performances of the scanning methods.

1 INTRODUCTION

1.1 *Importance of punctuality in railways*

Railways are major means for passenger and freight transportation in most cities. In order to survive in a competitive market, one of the prime concerns for railway industries is to improve the quality of service. Among all the aspects of services, punctuality plays an important role in satisfying the demands of consumers. The consequence of a loss of punctuality may be costly for people in this modern era, especially in the commercial sector. Railway industries therefore strive to improve punctuality of their train services so that they may attract more consumers with more reliable services.

1.2 *Conflict resolution at railway junctions*

A train may deviate from its schedule due to excessive station dwell time or tractive equipment faults, resulting in delay of services. Generally, the delay of a leading train may propagate to other trains in the network through the signaling and interlocking system, especially if the trains are running under a tight schedule. This delay may even be magnified in some areas, such as converging junctions, because trains on the incoming routes are in conflicts in using the junction. Trains are likely to be required to slow down or even stop completely to give way to the delayed train. Therefore, it is necessary to reduce the total delay of trains by assigning them to pass the junction in a suitable sequence (a right-of-way assignment).

First-come-first-serve (FCFS) is a common right-of-way assignment due to its simplicity, but it does not guarantee minimal delay. Theoretically, there exists one assignment such that the total delay of trains is minimal. However, this assignment may not be overwhelmingly superior to other less-than-optimal alternatives. In fact, in addition to the optimality of the solution, the assignment decision must be sufficiently quick to allow the implementation of the decision in real-time traffic control.

It can be shown that the number of feasible right-of-way assignments not only grows exponentially with the number of trains approaching the junction, but also the number of junctions connected together. This results in a large solution space for a multi-junction conflict resolution problem. For example, with 15 trains arriving at a connected areas of two junctions, the possible number of assignments reaches 0.75 millions. Although deterministic searches guarantee the location of the optimal solution (Carter & Price 2001), they require high computation demand. On the other hand, heuristic searches can provide a near-optimal solution in a relatively short duration (Carter & Price 2001) and thus they offer useful means for this application.

1.3 *Objectives*

This paper describes the application of the heuristic method, Genetic Algorithm (Haupt & Haupt 1998) to tackle the conflict resolution at two connected junctions. Genetic algorithm (GA) is based on evolutionary principle and has been successfully applied in numerous engineering areas. Crossover, the traditional genetic operator, poses difficulties in this application as it often produces right-of-way assignments that are outside the solution space because of the definitions of the chromosomes. Two scanning methods have been developed so that crossover can be applied directly and its advantages can be retained. The performance of the scanning methods has been investigated and the results reveal interesting characteristics in deciding the solution for the right-of-way assignment problem.

2 BACKGROUND

2.1 *Railway junction and right-of-way assignments*

A railway junction only allows two incoming routes and one outgoing route. Two connected-junctions, A and B, therefore consist of three incoming routes with route labels 0, 1 and 2 respectively as illustrated in Figure 1.

When trains on different routes approach the junction, they must be arranged to pass the junction orderly. The leading train on an incoming route must be ahead of other trains on the same route before and after crossing the junction. In other words, it is assumed that no sidings are available at the vicinity of the junctions so that overtaking of trains is forbidden.

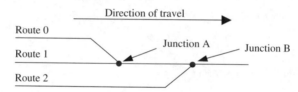

Figure 1. Track layout of 2 connected junctions.

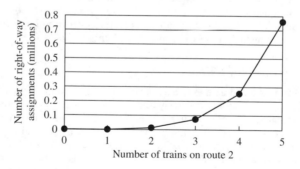

Figure 2. Relationship between number of trains involved and size of solution space.

Table 1. Size of solution space of single- and double-junctions.

Number of junctions	Number of trains				
	1	5	10	15	20
1	1	10	252	6435	184,756
2	1	30	4200	756,756	133,024,320
Ratio*	1:1	1:3	1:17	1:118	1:720

*Ratio of size of solution space between single-junction and double-junction.

Assume n is the total number of trains, and n_0, n_1 and n_2 are the numbers of trains on routes 0, 1 and 2 respectively, a right-of-way assignment S is defined as a sequence of route labels, which represents the order of trains leaving this two-junction area (i.e. moving out of junction B).

$$S = \{u_i\} : u_i \in \{0, 1, 2\} \quad \text{for} \quad 1 \leq i \leq n : n = n_0 + n_1 + n_2 \tag{1}$$

If $N(a)$ is defined as the number of occurrence of a, all feasible S's must satisfy the constraints:

$$N(u_i : u_i = 0) = n_0, \quad N(u_i : u_i = 1) = n_1 \quad \text{and} \quad N(u_i : u_i = 2) = n_2 \tag{2}$$

The right-of-way assignments at the two local junctions, S_A and S_B, can be obtained from S easily. When all elements "2" are eliminated from S, the remaining sequence is S_A. On the other hand, S_B is obtained when all elements "0" and "1" in S are regarded as the same.

2.2 Size of the solution space

The size of the solution space (total number of assignments) depends on the number of trains on the routes and the number of connected junctions involved. This can be computed by Equations 3 and 4 defined below.

If $F_1(x, y)$ is the number of possible right-of-way assignments for a single-junction with x and y trains on routes 0 and 1 respectively, it is defined by the following iterative equation:

$$F_1(x, y) = F_1(x - 1, y) + F_1(x, y - 1) \qquad \text{for} \quad x, y \geq 1 \tag{3}$$

where $F_1(1, 0) = F_1(0, 1) = 1$ and $F_1(0, 0) = 0$.

For a double-junction, if the number of trains on route 2 is z, the number of possible right-of-way assignments $F_2(x, y, z)$ is given by:

$$F_2(x, y, z) = F_1(x, y) F_1(x + y, z) \qquad \text{for} \quad x, y, z \geq 1 \tag{4}$$

where $F_2(x, 0, 0) = F_2(0, y, 0) = F_2(0, 0, z) = 1$.

Figure 2 and Table 1 show how the size of the solution space changes with respect to the number of trains and the number of junctions involved. From Figure 2, the number of right-of-way assignments increases exponentially with the number of trains on route 2, while there are five trains on each of route 0 and route 1. Table 1 further compares the increase in size of solution space between single-junction conflicts and double-junction conflicts.

Exhaustive search for the optimal solution in such a huge solution space is time-consuming and indeed infeasible. Even with the more intelligent deterministic optimization method (Ho et al. 1997), the computation demand is still considerable, which hinders real-time control. In practice, the optimality may be traded off by shorter computation time (Ho & Yeung 2001), and heuristic searches provide alternatives for this optimization problem.

2.3 Genetic algorithm

Genetic algorithm is one of the most popular heuristic searching techniques. Its concept is based on biological evolution processes.

With GA, a set of variables is encoded in a set of genes, which is embedded in a chromosome. Each chromosome is rated according to their fitness, which is measured by the objective function. Evolution of chromosomes takes place over generations, in which new chromosomes are reproduced by genetic operators. While chromosomes with better fitness values will be selected in each generation, the remaining ones are discarded. The intention of the algorithm is then to produce fitter chromosomes in successive generations so that the chromosome containing the optimal solution might eventually be obtained (Davis 1991).

Genetic operator is one of the important factors on the performance of GA. Crossover and mutation are the two most commonly used operators. During crossover, segments of two parent chromosomes are interchanged to produce two children chromosomes. Under this operation, the parental qualities are integrated and inherited in their children, so that the solution space near the parents might be explored. However, as the existing pool of population may not contain chromosomes near the optimal solution, mutation may be used to randomly change parts of the chromosomes so that the algorithm can explore new regions of the solution space, providing escapes from converging to a local optimum.

3 MODELING IN GENETIC ALGORITHM

3.1 Chromosomes

The total delay of train is governed only by the right-of-way assignment. The chromosome therefore contains only one gene in the form of S.

3.2 Objective function

The objective is to minimize the cost of the chromosomes. The cost of a right-of-way assignment C, is defined as the sum of the weighted delay of all trains subjected to S. The objective function is thus defined as follows:

$$\text{Minimize} \quad C = \sum_{i=1}^{n} w_i(\hat{t}_i - \tilde{t}_i) \quad \text{for} \quad 0 \le w_i \le 1 \tag{5}$$

where $\hat{t}_i$ is the traveling time of train i suffered from delay; $\tilde{t}_i$ is the traveling time of train i not suffered from delay and w_i is the weighting of train i representing the relative priority of trains.

3.3 Crossover

Two-point crossover is used. However, crossover of two parent chromosomes S_a and S_b often results in violations of the necessary constraints defined in Equation 2 at the children chromosomes S_c and S_d.

For example if S_a and S_b are {2 1 0 0 1 2} and {0 0 2 2 1 1} respectively, and crossover occurs between the second and fourth elements inclusively, the children S_c and S_d will be {2 0 2 2 1 2} and {0 1 0 0 1 1} respectively.

S_c implies there are two too many trains on route 2 and one short on both routes 0 and 1. Conversely, S_d indicates a deficit of two trains on route 2 and one train more on both routes 0 and 1. Both of them are therefore non-feasible solutions. In order to proceed with the algorithm, the children chromosomes must be discarded unless they are modified by one of the scanning methods defined in the next section.

3.4 Scanning methods

Two scanning methods have been developed to locate the violations in the children chromosomes and convert them into feasible solutions. The algorithms are described as follows:

3.4.1 Forward scanning

Step 1: Obtain c_r and d_r, the numbers of r contained in two parent chromosomes S_c and S_d respectively, for $r = 0, 1, 2$.

Step 2: If $\forall r \begin{Bmatrix} p_r = c_r - n_r \\ q_r = d_r - n_r \end{Bmatrix} = 0$, stop, else goto Step 3.

Step 3: Start from u_1, scan the elements in S_c and S_d in turn to locate the first encountered u_c and u_d respectively which satisfies:

$$N(u_i : u_i = r) = n_r + 1 \qquad \forall r, \ 1 \le i \le c, d$$

Step 4: Interchange u_c and u_d, goto Step 1.

In words, S_c and S_d are scanned from the first element and the first invalid elements in the parents are located and then exchanged. The first few elements of the chromosomes are therefore protected from modifications, but the remaining elements are likely to be modified.

3.4.2 Rotation scanning

In this method, S_c and S_d are scanned from the first element of the exchanged segments. If the ends of the chromosomes are reached, the scan is continued from the first elements of the chromosomes. Step 3 is thus replaced by:

Step 3: Start from u_k (where the first crossover point occurs between u_{k-1} and u_k), scan the elements in S_c and S_d in turn to locate the first encountered u_c and u_d respectively, satisfying:

$$N(u_i : u_i = r) = n_r + 1 \qquad \forall r, \ k \le i \le n \ \text{ then } \ 1 \le i \le k-1$$

In contrast to forward scanning, rotation scanning allows modifications of the first few elements of chromosomes. In fact, the locations of modification are related to the position of the exchanged segments, which is randomly generated.

3.5 Seeding

Seeding is a technique that can be adopted by GA to improve convergence. A solution, which is predicted to be an optimal or near-optimal solution, is included in the first generation. If this is the case, convergence may be improved. Otherwise, there will be a possibility that the algorithm chases a local optimum and requires a few generations to break away from it, slowing down the rate of convergence.

From the previous study (Ho et al. 1997), FCFS is likely to be a near-optimal solution for dealing with single converging junction, especially for traffic involving single type of train. FCFS is therefore a sensible seed to be adopted in the algorithm.

4 SIMULATION RESULTS

4.1 Setup

The above model of genetic algorithm has been implemented by computer simulation. Consistent with the objective of adopting heuristic optimization in this application, an event-based train simulator (Ho et al. 1997) has been developed specifically for simulating train movements over two connected railway junctions. It has the advantage of simulating train movements without exact details along the track, while still taking into account of the constraints imposed by the signaling and interlocking systems. As a result, the computation demand is reduced while the train movements still resemble closely to practical situations.

The traffic condition simulated has been based on a total of six trains, two on each route. The resulting size of solution space is 90, which is not too small to demonstrate the capability of the heuristic search but is also not too large for performing an exhaustive search in order to locate the optimal solution to compare with the effectiveness of the heuristic method.

Table 2 shows the simulation parameters used for three simulation cases. These cases have been setup to investigate:

- the relative merits of the two scanning methods;
- the performance of FCFS as a seed under uniform traffic condition; and
- the performance of FCFS as a seed under mixed traffic condition.

In each case, the simulation has been repeated over 100 trials and the mean costs of the best chromosomes evolved at each generation are computed. The intention is to average out the effect caused by the randomly chosen initial population, which have a significant effect on the rate of convergence.

4.2 Forward scanning and rotation scanning

Figure 3a illustrates the comparison of the performance of the two scanning methods. Both methods enable convergence toward the optimal solution, but rotation scanning is found to be more effective.

Rotation scanning is superior to forward scanning because it provides more radical modifications on the chromosomes, thus offering more exit routes from the local optima. In forward scanning, the first few elements will be preserved, meaning that the order of the first few trains passing the junctions is unlikely to be changed. As a result, if the parent population does not contain "fit" portion of this gene (which is usually the case), this genetic deficiency is more likely to be passed onto the children population, resulting in difficulties in converging to the optimal solution. In other words, the algorithm is trapped at local optima.

On the other hand, rotation method is able to explore new regions of solution space without restrictions on the location of modification. Thus, this method has a higher tolerance on a poor parent population.

4.3 Seeding with FCFS under normal traffic

Figure 3b shows how seeding with FCFS improves the costs of initial population when compared to Figure 3a, and shows how seeding improves the performance of forward scanning.

The initial costs are reduced by almost 10 s in both scanning methods, so the algorithm can obtain a solution close to the optimal solution. Inspection of Table 2 reveals that FCFS is indeed a near-optimal solution and thus is a good seed for the algorithm.

Table 2. Simulation parameters.

Case	Pool size	Number of generations	FCFS*	Train weighting**	Optimal solution	
					$\{S\}$	(C)
1	16	5	No	Equal	{1 2 0 0 2 1}	(350.9)
2	16	5	Yes	Equal	{1 2 0 0 2 1}	(350.9)
3	16	5	Yes	Unequal	{2 0 2 0 1 1}	(302.6)

* $S = \{1\ 2\ 0\ 1\ 2\ 0\}$, $C = 352.9$.
** Equal: $w_i = 1, \forall i$
Unequal: $w_i = 1, \forall i2$

$w_i = 0.5, i = 2$.

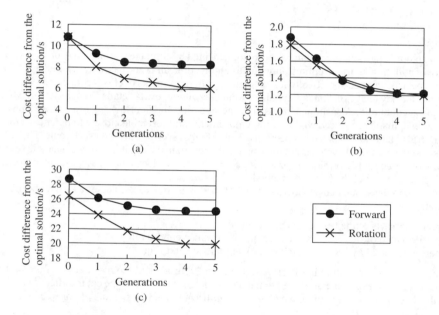

Figure 3. Simulation results. (a) Relative performance from optimal cost with crossover only; (b) relative performance from optimal cost with crossover and FCFS under uniform traffic; (c) relative performance from optimal cost with crossover and FCFS under mixed traffic.

With the help of such seed, forward scanning performs almost exactly as rotation scanning. The parent population now contains an individual with a "fit" portion {1 2 0 ...}. Modifications at the beginning of the chromosomes become unnecessary.

4.4 Seeding with FCFS under mixed traffic

Under this case study, the first arriving train on route 1 is deliberately assigned a weighting factor of 0.5, representing a train with low priority. In reality, this train may represent a freight train, competing the right-of-way with other higher priority passenger trains. Figure 3c shows the result of the simulation. Rotation scanning again outperforms forward scanning, even incorporated with FCFS.

From Table 2, FCFS is no longer a good seed as compared to the optimal solution. The freight train, albeit its earlier arrival, is allowed to suffer from a larger delay, in order to allow through passage of passenger trains on the other routes. As a result, seeding with FCFS does not improve the performance of forward scanning.

5 CONCLUSIONS

We have presented the application of genetic algorithm to generate a near-optimal solution for conflict resolution over two connected railway junctions. Junction conflicts usually involve large solution space and a quick decision is necessary to reduce the delay of trains. Although GA is capable of producing a fast near-optimal solution, crossover often results in violation of constraints on the chromosomes. Two scanning methods have been proposed to modify and correct the chromosomes before they are admitted to the next generation. Rotation scanning outperforms forward scanning because it has the ability to provide escapes from local optima.

The scanning methods have been used to demonstrate the feasibility of FCFS for connected railway junction. It has been shown that FCFS is indeed an effective solution under uniform traffic. However, under mixed traffic condition, it fails to provide a good approximation to the optimal solution and seeding with FCFS does not necessary assist convergence in the algorithm.

Although this study focuses on the use of crossover in genetic algorithm, other genetic operators such as mutation are certainly potential candidates for improving the convergence of the algorithm. Further works therefore involves the feasibility of involving and cooperating other genetic operators.

REFERENCES

Carter, M.W. & Price, C.C. 2001. *Operations Research – A Practical Introduction*. Boca Raton: CCD Press.
Davis, L. 1991. *Handbook of Genetic Algorithms*. New York: Van Nostrand Reinhold.
Haupt, R.L. & Haupt, S.E. 1998. *Practical Genetic Algorithms*. New York: Wiley.
Ho, T.K., Norton, J.P. & Goodman, C.J. 1997. Optimal traffic control at railway junctions. *IEE Proc. – Electr. Power Appl.* 144(2): 140–148.
Ho, T.K. & Yeung, T.H. 2001. Railway junction traffic control by heuristic methods. *IEE Proc. – Electr. Power Appl.* 148(1): 77–84.

Computational Methods in Engineering and Science, Iu et al. (eds)
© *2003 Swets & Zeitlinger, Lisse, ISBN 90 5809 567 3*

Regulation of train service by coasting control in metro railway system

K.K. Wong & T.K. Ho
Department of Electrical Engineering, The Hong Kong Polytechnic University, Hong Kong

ABSTRACT: To maximize the capacity of the rail line and provide a reliable service for passengers throughout the day, regulation of train service to maintain steady service headway is essential. In most current metro systems, train usually starts coasting at a fixed distance from the departed station to achieve service regulation. However, this approach is only effective with respect to a nominal operational condition of train schedule but not necessarily the current service demand. Moreover, it is not simply to identify the necessary starting point for coasting under the run time constraints of current service conditions since train movement is attributed by a large number of factors, most of which are non-linear and inter-dependent. This paper presents an application of classical measures to search for the appropriate coasting point to meet a specified inter-station run time and they can be integrated in the on-board Automatic Train Operation (ATO) system and have the potential for on-line implementation in making a set of coasting command decisions.

1 INTRODUCTION

Railway is currently the major mass transportation means in most of the countries around the world. With the increasing population and expanding commercial, industrial and social activities, safe, comfortable and reliable railway service is the most desirable. Indeed, any delays or interruptions on railway service may bring a city to a standstill, which may also carry a significant economic loss. As the metro systems are usually electrified, one of the major expenses on operational cost in railway systems is the electricity bill. As a result, it is more attractive from railway operators' viewpoint even if a few percentages of electricity saving can be achieved at a reasonable expense of travelling time.

Under a typical flat-out inter-station run, a train accelerates from a station to maximum speed and maintains the train speed as much as possible until it is necessary to brake to a halt for the next station. Usually, the running time is the shortest and the energy consumption is the highest as the train is travelling very close to the maximum permissible speed throughout the trip. However, the traction motors are allowed to turn off once the train accelerates above a certain speed if coasting is allowed. With the application of coasting, the momentum of the train carries it through and the brake is still needed to bring the train to a stop at the next station. Inter-station run-time is longer but the energy saving can be achieved because the train spends less time on motoring.

A flat-out inter-station run is still necessary during rush-hours and recovery of train service from disturbance. However, certain measures can be introduced to reduce the energy consumption when time is not an important issue. From the viewpoint of adjusting headway, a longer travelling time is more preferable than a longer station waiting time at off-peak hours because of the energy loss from the air conditioning system when the train-doors have to be kept open at stations, and it accounts for a substantial proportion of the electric bill. Coasting control (Chang & Sim 1997, Wong 2001) is one of the topical areas to attain a trade-off between energy consumption and travelling time and has been commonly used. For current practices, coasting can only be started at a specific point between stations, which is predetermined to suit the nominal operational condition and hence not flexible for dynamic service regulation.

In metro systems, the stations are usually a few kilometers apart. It is not unusual that there is not adequate room to accommodate multiple coasting points. This paper presents the application of the classical searching methods to look for the single coasting point to meet a specified run time in an inter-station run with the aid of a single-train simulator, which takes into account all factors attributed to train movement. The software has been developed in Visual Basic and the details will be discussed in the Section 3.

In practice, there are about 30 seconds or less to derive the location of the coasting point for the next inter-station run when a train stops at a station. A fast solution is important for real-time control or supervision of the operation. Thus, it is also the objective of this study to explore the possible ways to improve the trade-off between computation time and the quality of the solution.

In general, a number of parameters are involved in this coasting point identification problem, such as track geometry and traction equipment. Theoretically, any point between the two stations is a possible coasting point. The solution set may contain all points between stations. Even if a certain distance res-olution is imposed so that the solution set is finite, the solution space is still too large for any searching method to attain the necessary coasting point in reasonable time. Moreover, it is difficult to link track geometry, traction equipment to every possible coasting point with the consideration of the correspond-ing run-time and energy consumption analytically, because of non-linearities in traction equipment characteristics and interactions among train through power and signalling systems. Usually, the run-time increases monotonically and the energy consumption monotonically decreases as the coasting point moves away from the starting station. Direct searching techniques (de Cuadra et al. 1996) and heuristic approaches have the potential to obtain the possible coasting point without using the mathe-matical model. Heuristic methods, e.g., Genetic Algorithm (Haupt & Haupt 1998), requires initial pop-ulation for evolution and this population also accounts for the total number of iterations in the searching process. However, initial population is not essential with the direct searching techniques. As a result, a fast and reliable solution is possible obtained with the direct search techniques. The coasting point iden-tification by classical methods, Golden, Fibonacci and Gradient (Gottfried & Weisman 1973, Reklaitis et al. 1983) has been tested under various traffic conditions and the results are encouraging.

2 METHODOLOGY

In this section, three classical searching methods are introduced to identify the coasting point for real time train scheduling control. Two bi-section methods, Golden Section and Fibonacci search, are high-lighted, and the idea of how to fix the necessary coasting point with the gradient method is also presented. In order to determine the quality or fitness of the chosen solution, the objective function is essential and it is defined as follows.

$$F = \left| \frac{T_g - T_D}{T_D} \right| \tag{1}$$

T_D is the desired run-time (sec) and T_g is the run-time achieved by the updated solution (sec). Since the run time may be either above or below the desired values in a particular run, the absolute sign is in place to nullify the polarity effect. F is a non-negative quantity and a smaller value implies a better solution. Other definitions for F are equally valid if other consideration is taken into account.

2.1 Golden section search

With the application of this algorithm on coasting control to regulate the train schedule, the fitness of two initial coasting points are determined in advance. These two values will then be useful for further search of new coasting point. The basic idea of the Golden section search is that the solution space is divided into two unequal parts, the ratio of the larger of the two segments to the total length of the interval should be the same as the ratio of the smaller to the larger segment.

Assume the solution space consists of a length Z which composes of two segments z_1 and z_2 as shown in Figure 1, the Golden section implies that

$$\frac{z_1}{z} = \frac{z_2}{z_1} \tag{2}$$

$$z = z_1 + z_2 \tag{3}$$

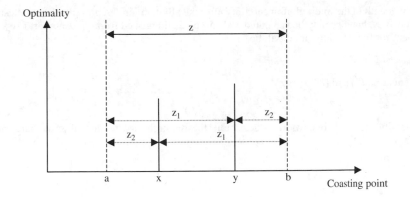

Figure 1.   Golden section search deviation.

Equation (2) then gives

$$z_1^{\,2} = z \times z_2 \qquad (4)$$

Substituting for z from equation (3) into (4) and dividing by z_2^2 becomes

$$\left(\frac{z_1}{z_2}\right)^2 + \left(\frac{z_1}{z_2}\right) = 1 \qquad (5)$$

This quadratic can be solved for the ratio z_1/z_2. The positive root will be

$$\left(\frac{z_1}{z_2}\right) = 0.618033989 \qquad (6)$$

If the two coasting points are placed with this fractional spacing from either end on the solution space as shown in Figure 1, the solution space will then be reduced to a length of 0.618 times the previous of uncertainty of interval. It is obvious that one of the two evaluations of coasting point is available for the next step by the virtue of the golden ratio. Thus only one additional "golden-spaced" evaluation is required to reduce the solution space by another 0.618 fraction. The process is repeated again and the uncertainly of interval is further reduced by the golden ratio until the obtained coasting point satisfies the expected train operational requirement in an inter-station run. To achieve the maximum reduction in the subsequent solution space, the evaluations should be placed symmetrically about the centerline of the solution space. When this is done, a new evaluation provides an additional reduction in the solution space.

2.2   Fibonacci search

The concept of Fibonacci search is very similar to the Golden section search. The main difference is the reduction ratio on the solution space in each iteration is fixed at 0.618 with Golden search, whilst the reduction ratio varies with the previous uncertainty of interval in Fibonacci search. The arrangement of the search points within the new search interval is shown in Figure 2. The search point is placed symmetrically within any interval and it always has sub-interval of the same length, regardless of the quality of the search point.

From Figure 2, the interval of uncertainty L_{n-1} is $(y - x_{n-2})$ and the new search interval can be defined as,

$$L_n = \frac{L_{n-1} + \varepsilon}{2} \qquad (7)$$

L_n is the length of the interval of uncertainty after the nth iteration. represents the smallest distance by which two evaluations may be separated and still be distinguished from one another. The symmetry requirement for the search interval will be

$$L_{n-2} = L_{n-1} + L_n \qquad (8)$$

From equation (7) and (8),

$$L_{n-2} = 3L_n - \varepsilon \qquad (9)$$

It is possible to work backward to determine the required size for any intermediate interval of uncertainty.

$$L_{n-3} = 5L_n - 2\varepsilon$$

$$L_{n-4} = 8L_n - 3\varepsilon$$

Then, it can be generalized as

$$L_{n-k} = F_{k+1}L_n - F_{k-1}\varepsilon \qquad (10)$$

The coefficients F_{k+1} and F_{k-1} can be obtained by

$$F_{k+1} = F_k + F_{k-1} \qquad k = 1,2,3,\ldots, \qquad (11)$$

and $F_0 = F_1 = 1$

Thereby, Fibonacci search provides a better reduction ratio on the solution space in each iteration if the maximum number of iterations is predetermined in advance.

2.3 Gradient based search

The Gradient method uses the derivative as illustrated in equation (12) to locate the necessary coasting point. Similarly, the fitness of two initial coasting points are predetermined and the search direction of the updated coasting point depends on the polarity and the magnitude of the gradient,

$$Gradient = \frac{\Delta \ Run \ Time}{\Delta \ Coasting \ Location} \qquad (12)$$

Then the step length can be calculated by,

$$Step \ length = \left\{ Gradient^{-1} \times \left[Run \ time_{Flat-out} - Run \ time_{Expected} \right] \right\} \qquad (13)$$

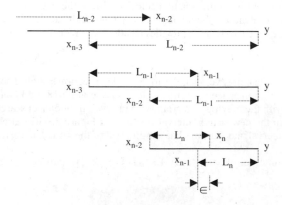

Figure 2.   Sequence of uncertainty intervals in a Fibonacci search.

And the new coasting point can be obtained by the following equation:

New coasting point = Old coasting point + Step length (14)

In general, the step length becomes larger when the run-time of the current coasting location is far away from the expected one, or vice versa. Therefore, a smaller number of iterations are likely to be required in a particular iteration run with Gradient method. However, the drawback of this algorithm is the step length and search direction cannot be defined in the searching process if there is no change on the run-time between the current and previous guess of the coasting points (i.e. slope is not available) and the searching process will be terminated.

3 SOFTWARE IMPLEMENTATION

The software has been developed in Visual Basic and it is mainly divided into two components. The first component is the single-train simulator, which is used to calculate the train performance in a particular run. The other is the coasting point identification module and its function is to fix the necessary coasting location under the constraints of the current traffic conditions.

3.1 *Single-train simulator*

The single-train simulator is mainly divided into five stages and the principal loop in the program is the incrementing time. At the beginning of each update period, it is assumed that the position and speed of the train are known. An initial setting of all data of train required by the simulation is defined at this stage. Based on the current train position and speed as well as the track-based data, the train mode (motoring, coasting and braking) is determined. Once the train mode is established, the position and speed of the train in the next time interval can then be update with simple application of Newton's Second Law. This process requires an effective representation of track gradient and curvature, motor efficiency and train loading within the simulator. Finally, the calculated speed and position of train will then be used as the initial values for the next time update. The structure of the single train simulator is shown in Figure 3.

3.2 *Coasting point identification model*

Once the train performance with the "flat-out" operation is initially obtained from the single train simulator, the coasting-point identification model starts. The ultimate purpose of this model is to identify

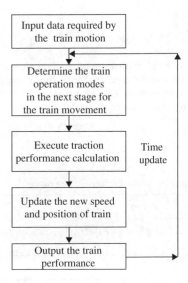

Figure 3.   Structure of single train simulator.

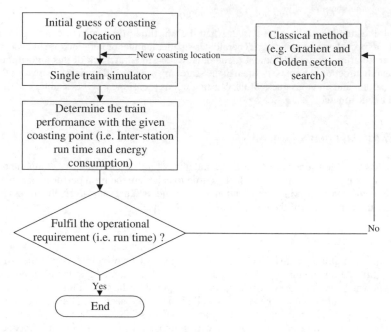

Figure 4.   Structure of coasting control of train movement.

the necessary coasting point under the specified operational requirements. Generally, the run-time and the energy consumption of the train can be extended and reduced respectively when an early single coasting point is so required. A new coasting point will be produced by the classical searching method if the train output performance does not satisfy the expected requirements (i.e. run time and energy consumption performance). The same process repeats until either the new coasting point satisfies the expected requirements or the maximum number of program loop by the user is reached. The structure of the model is illustrated in Figure 4.

4   CASE STUDIES

This section presents the results in the identification of a single coasting location in a particular inter-station run with the three proposed searching methods and thus compares the quality of the searched outcomes for the purpose of real-time control. Two case studies are discussed and certain reduction on the run time is required for train regulation in an inter-station run in each case.

4.1   Short inter-station distance

4.1.1   Experimental setup

Table 1.   Simulation setup for train regulation in short inter-station run.

| Inter-station distance (m) | Flat out run | | Expected inter-station run (s) | Time extension for coasting* (%) |
	Run time (s)	Energy consumption (MJ)		
1100	81.8	192.52	88	7

* The flat-out run-time is extended for coasting operation.

4.1.2 Results

Table 2.   Average number of iterations and the obtained train performance.

Searching method	Number of iterations	Energy consumption (MJ)	Normalised* computation unit	Coasting point from the starting station (m)
Golden	11	131.8	2.9	200
Fibonacci	8	131.8	2.19	200
Gradient**	3–6	131.8	1–1.78	198–201

* 1 unit = 3.7 s computation time; ** 10 tests have been carried out with Gradient method.
A cost value F of 0.001 is required with each searching method.

4.2   Long inter-station distance

4.2.1   Experiment setup

Table 3.   Simulation setup for train regulation in long inter-station run.

Inter-station distance (m)	Flat out run		Expected inter-station run (s)	Time extension for coasting* (%)
	Run time (s)	Energy consumption (MJ)		
9025	310.8	1378.87	348	12

* The flat-out run-time is extended for coasting operation.

4.2.2   Results

Table 4.   Average number of iterations and the obtained train performance.

Searching method	Number of iterations	Energy consumption (MJ)	Normalised* computation unit	Coasting point from the starting station (m)
Golden	5	651.4	1.7	2683
Fibonacci	5	651.4	1.69	2685
Gradient**	2–7	647.2–664.2	1–1.7	2637–2746

* 1 unit = 12.9 s computation time; ** 10 tests have been carried out with Gradient method.
A cost value F of 0.001 is required with each searching method.

4.3   Discussions

Simulation result shows that the classical method can provide the necessary coasting point with a reasonable time for real time train scheduling control. The attained coasting locations from each method are quite close to each other. Besides, the results also reveal that the number of iterations and computation time are smaller with the Gradient method when comparing with the Golden and Fibonacci method. However, one of the main drawbacks with the Gradient method is that the slope is not available in a search if there is no change on the run-time between the current and previous guess of the coasting points, and then the updated coasting location can be deduced.

In addition, the Golden and Fibonacci methods find the solution within 11 iterations in both cases and the initial interval is reduced to a smaller interval in which it contains the solution. Although Golden and Fibonacci methods are not the fastest means to obtain the solution, they are not likely limited by the track layout characteristics in the searching process since the solution space is reduced to a certain value in each iteration. Therefore, the Golden and Fibonacci method is more robust and reliable to locate the coasting point in a particular run when compared with the Gradient method. Moreover, the average number of iterations for a long inter-station run is smaller than that in the case of a short one since the rate of reduction on the solution space for a long inter-station run in each iteration is larger.

5 CONCLUSIONS

An application of classical methods on the single coasting point searching in a metro railway system has been presented. With the aid of the train simulator, the searching methods can provide a fast solution for real-time train scheduling control. In practice, they can be integrated in the on-board Automatic Train Operation (ATO) system and the coasting control command for the next inter-station run can be obtained when a train stops at a station. In addition, dynamic coasting control is more flexible and efficient in the regulation of train schedule as it adapts to the current train service demand and its additional advantage is the energy saving can be achieved.

REFERENCES

Chang, C.S. & Sim, S.S. 1997. Optimising train movements through coast control using genetic algorithms. *IEE Proceedings – Electrical Power Applications*, 144(2): 65–73.
de Cuadra, F., Fernandez, A., de Juan, J. & Herrero, M.A. 1996. Energy-saving automatic optimization of train speed commands using direct search techniques. Computer in Railway V, vol. 1: 337–346.
Gottfried, B.S. & Weisman, J. 1973. *Introduction to Optimization Theory*. New Jersey: Prentice Hall, Inc.
Haupt, R.L. & Haupt, S.E. 1998. *Practical Genetic Algorithm*. New York: Wiley.
Reklaitis, G.V., Ravindran, A. & Ragsdell, K.M. 1983. *Engineering Optimization Methods and Applications*. New York: John Wiley & Sons, Inc.
Wong, K.K. 2001. *Optimisation of Run Time and Energy Consumption of Train Movement*. MSc dissertation, Hong Kong Polytechnic University.

Author index